Mass Transfer with Chemical Reaction in Multiphase Systems
Volume I: Two-Phase Systems

NATO ASI Series

Advanced Science Institutes Series

A series presenting the results of activities sponsored by the NATO Science Committee, which aims at the dissemination of advanced scientific and technological knowledge, with a view to strengthening links between scientific communities

The series is published by an international board of publishers in conjunction with NATO Scientific Affairs Division

A	Life Sciences	Plenum Publishing Corporation
B	Physics	London and New York
C	Mathematical and Physical Sciences	D. Reidel Publishing Company Dordrecht and Boston
D	Behavioural and Social Sciences	Martinus Nijhoff Publishers The Hague/Boston/Lancaster
E	Applied Sciences	
F	Computer and Systems Sciences	Springer Verlag Berlin/Heidelberg/New York
G	Ecological Sciences	

Series E: Applied Sciences – No. 72

Mass Transfer with Chemical Reaction in Multiphase Systems

Volume I: Two-Phase Systems

edited by

Erdoğan Alper, B.Sc., Ph.D. (Cantab)

Professor of Chemical Engineering
University of Ankara, Besevler, Ankara, Turkey
Anadolu University, Eskişehir, Turkey

1983 **Martinus Nijhoff Publishers**
The Hague / Boston / Lancaster
Published in cooperation with NATO Scientific Affairs Division

Proceedings of the NATO Advanced Study Institute on
Mass Transfer with Chemical Reaction in Multiphase Systems,
Çeşme - İzmir, Turkey, August 10 - 21, 1981

Library of Congress Cataloging in Publication Data

NATO Advanced Study Institute on Mass Transfer with Chemical Reaction in Multiphase Systems (1981 : Çeşme, Turkey)
Mass transfer with chemical reaction in multiphase systems.

(NATO ASI series. Series E, Applied sciences ; no. 72-73)
"Published in cooperation with NATO Scientific Affairs Division."
"Proceedings of the NATO Advanced Study Institute on Mass Transfer with Chemical Reaction in Multiphase Systems, Çeşme--Izmir, Turkey, August 10-21, 1981"--T.p. verso.
Includes bibliographical references.
Contents: v. 1. Two-phase systems -- v. 2. Three-phase systems.
1. Mass transfer--Congresses. 2. Chemical reactions--Congresses. I. Alper, Erdoğan. II. North Atlantic Treaty Organization. Scientific Affairs Division. III. Title. IV. Series: NATO advanced science institutes series. Series E, Applied sciences ; no. 72-73.
TP156.M3N38 1981 660.2'8423 83-13285
ISBN 978-90-247-2874-9 (set)
ISBN 978-90-247-2872-5 (v. 1)
ISBN 978-90-247-2873-2 (v. 2)

ISBN 978-90-247-2872-5 (this volume)
ISBN 978-90-247-2689-9 (series)
ISBN 978-90-247-2874-9 (set)

Distributors for the United States and Canada: Kluwer Boston, Inc., 190 Old Derby Street, Hingham, MA 02043, USA

Distributors for all other countries: Kluwer Academic Publishers Group, Distribution Center, P.O. Box 322, 3300 AH Dordrecht, The Netherlands

NATO ADVANCED STUDY INSTITUTE
on
"MASS TRANSFER WITH CHEMICAL REACTION IN MULTIPHASE SYSTEMS"

DIRECTOR

E. Alper Department of Chemical Engineering,
Faculty of Sciences,
Ankara University, Beşevler, Ankara, Turkey.

SCIENTIFIC ADVISOR

W.-D. Deckwer Institut für Technische Chemie,
Universitaet Hannover,
D -3000 Hannover 1, F.R.Germany.

HONORARY SCIENTIFIC ADVISORS

P.V. Danckwerts Chemical Engineering Department,
University of Cambridge,
Pembroke Street, Cambridge, England.

M.M. Sharma Department of Chemical Technology,
University of Bombay,
Matunga Road, Bombay, India.

LECTURERS

E. Alper	Department of Chemical Engineering, Fâculty of Sciences, Ankara University,Beşevler,Ankara,Turkey.
G.Astarita	Istituto di principi di Ingegneria Chimica, Piazzale Tecchio, 80125 Napoli.Italia.
J.C.Charpentier	CNRS,Laboratoire des Sciences du Genie Chimique, 1, rue grandville, 54042 Nancy Cedex,France.
W.-D.Deckwer	Institut für Technische Chemie, Universitaet Hannover. D -3000 Hannover 1, F.R.Germany.
A.Germain	Universitè de Liège, Facultè des Sciences Appliquèes, Chimie Industrielle, Le Rue A.Stevart,2 B -4000 Liège,Belgique.
S.Hartland	Technisch-Chemisches Laboratorium ETH - Zentrum CH -8092 Zurich, Switzerland.
H.Hofmann	Institut für Technische Chemie, Egerlandstrasse 3, d -8250 Erlangen, F.R.Germany.
G.A.L'Homme	Universitè de Liège, Laboratoire de Gènie Chimique, Institut de Chimie-Mètallurgie, 2, rue A.Stevart, B -4000 Liège,Belgique.
R.Mann	UMIST,The University of Manchester, PO Box 88,Manchester M 60 1QD England.
H.Sawistowski	Imperial College of Science and Technology, Department of Chemical Engineering, London SW7 2BY, England.
K. Schügerl	Institut für Technische Chemie Univesitaet Hannover, D -3000 Hannover,F.R.Germany.
Y.T. Shah	University of Pittsburgh, School of Engineering, Chemical and Petroleum Engineering, 1249 Benedum Hall,Pittsburgh, P.A. 15261, U.S.A.

PREFACE

The phenomenon of "mass transfer with chemical reaction" akes place whenever one phase is brought into contact with one r more other phases not in chemical equilibrium with it. This henomenon has industrial, biological and physiological importance. n chemical process engineering, it is encountered in both separ- tion processes and reaction engineering. In some cases, a hemical reaction may deliberately be employed for speeding up the ate of mass transfer and/or for increasing the capacity of the olvent; in other cases the multiphase reaction system is a part f the process with the specific aim of product formation. inally, in some cases, for instance "distillation with chemical eaction", both objectives are involved. Although the subject is learly a chemical engineering undertaking, it requires often a ood understanding of other subjects, such as chemistry and fluid echanics etc., leading to publications in diversified areas. On he other hand, the subject has always been a major field and one f the most fruitful for chemical engineers.

It is for these reasons that the editor decided to organise NATO Advanced Study Institute covering all aspects, with the ltimate aim of an overview of the landscape to identify features hat provide orientation. After many discussions with Professors .-D. Deckwer, P.V. Danckwerts, C. Hanson and M.M. Sharma, it was ecided to limit the ASI to (1) gas-liquid, (2) liquid-liquid, and 3) gas-liquid-solid systems. Thus, the only really important rea left out was fluid-solid systems, part of which was however ealt with in another NATO Advanced Study Institute on "Analysis f Fluid-Solid Catalytic Systems" under the directorship of Prof. .F. Froment. The originally planned date for the Institute had o be postponed for one year in order to prevent a clash with nother NATO Advanced Study Institute.

This two-volume book consists entirely of the proceedings f the NATO Advanced Study Institute, which was held in Cesme, zmir, Turkey during August 10-22, 1981. It includes review ectures of the eminent scientists as presented during the nstitute. Although every attempt was made by the director/ ditor, it was not altogether possible to realise absolute unifor- ity as these reviews were written in a relatively short time by uthors who did not have the chance of coming together prior to he meeting. During the Institute, some short original contri- utions were also presented by interested participants on areas losely related to the invited reviews.

Due to the large amount of material, these Proceedings are ivided into two volumes. The first volume includes the general ntroductory reviews involving the mathematical lay-out, physico-chemical data, reaction kinetics and transport data, gas-liquid and liquid-liquid systems, and biochemical systems. The second volume is devoted entirely to the three-phase system and its application to coal technology and Fischer-Tropsch synthesis.

Special features of this Institute reflected fully in these Proceedings, are the treatments of biological reactions, facilitated transport, reactive distillation, solvent extraction of metals and some related aspects of coal utilisation.

Here, I would very much like to compliment and thank all lecturers not only for their very clear oral and written contributions but also for wholeheartedly supporting the Institute. I feel particularly obliged to make a special acknowledgement to Prof. W. -D. Deckwer, who was involved from the very beginning to the very end, to Professors P.V. Danckwerts and M.M. Sharma who contributed immensely to the scientific organisation, and to Prof. M.M. Sharma who was also very kind in providing material prior to publication.

I would also like to thank all participants for their contributions to the Advanced Study Institute. Indeed, it was their active participation which brought a real sense of satisfaction to the director/editor.

I express, of course above all, my deepest gratitude to the Scientific Affairs Division of NATO and their officers, in particular Dr. M. di Lullo and Mr. M. Sudarskis, who not only almost entirely supported the Institute financially, but also helped a local objective of mine, i.e. promoting scientific affairs of Turkish chemical engineers. I gratefully acknowledge also the financial contributions of the Turkish Scientific and Technological Research Council and the Ankara Office of the British Council.

I would also like to thank my assistants and co-workers at various universities in Turkey for doing many tedious chores, and to thank Mrs. Bilge Göksidan for the drawings.

Last, but by no means least, my efforts in organising this ASI could not have succeeded without the patience and the understanding of my wife, Ayse, and our daughter, Gizem, who have on too many occasions been neglected during the last two years; for their co-operation and inspiration I am particularly grateful.

Ankara, Turkey

ERDOGAN ALPER

TABLE OF CONTENTS: Volume I

LECTURERS VI
PREFACE VII

E. ALPER
Introduction to Mass Transfer with Chemical Reaction Operations (with Heavy Emphasis on Gas-Liquid Systems) 1

G. ASTARITA
General Mathematical Layout of Multiphase Systems 17

G. ASTARITA
Chemical Desorption 37

B.I. MORSI and J.C. CHARPENTIER
Review of Obtaining and Estimation Methods of Physico-Chemical and Related Data: Part 1 - Solubilities and Diffusivities of Gases in Liquids 53

B.I. MORSI and J.C. CHARPENTIER
Review of Obtaining and Estimation Methods of Physico-Chemical and Related Data: Part 2 - Gas-Liquid Mass Transfer Parameters. Measurement and Some Data in Several Types of Reactors 101

R. MANN
Heat and Mass Transfer in Exothermic Gas Absorption 189

R. MANN
Absorption with Complex Reaction in Gas-Liquid Reactors 223

E. ALPER
Process Design Aspects of Gas Absorbers 291

J. ANDRIEU and J.M. SMITH
Gas-Liquid Rate Constant Measurements by Chromatography 341

R. SICK, P. WEILAND and U. ONKEN
Determination of Gas-Liquid Mass Transfer by Oxidation of Hydrazine 351

P.M.M. BLAUWHOFF, G.J.B. ASSINK and W.P.M. VAN SWAAIJ
Simultaneous Mass Transfer of Two Gases with Complex Reversible Reactions: An Example Being the Simultaneous Absorption of H_2S and CO_2 into Aqueous Solutions 357

J.H. MELDON
Facilitated Gas Transport in Liquids 369

J.H. MELDON and J.E. ROBERTS
Theory of Membrane CO_2 Transport with Equilibrium Reaction 381

H. SAWISTOWSKI
Distillation with Chemical Reaction 391

K. SCHÜGERL
Introduction to Biochemical Systems. Formal Treatment of Biochemical Reactions and Characterisation of Fermentation Systems 415

W.-D. DECKWER
Physical Transport Phenomena in Biological Tower Reactors 459

K. SCHÜGERL
Biochemical Reactions and Oxygen Transfer into Different Fermentation Broths and Reactors 497

K. SCHÜGERL
Process Design Aspects and Comparison of Different Bioreactors 525

A. LÜBBERT
Turbulence Measurements in Bubble Columns 553

A. SCHUMPE, K. NGUYEN-TIEN and W.-D. DECKWER
Gas/Liquid Mass Transfer Parameters (ε_G, $k_L a$, a) for Non-Newtonian Two-Phase Flow in a Bubble Column 565

E. ALPER
Introduction to Liquid-Liquid Extraction with Chemical Reaction 577

H. SAWISTOWSKI
Physical Aspects of Liquid-Liquid Extraction 613

S. HARTLAND and L. STEINER
Experience with Liquid/Liquid Test Systems in Extraction 637

H. SAWISTOWSKI
Some Aspects of Metal Extraction 667

Participants 677

INTRODUCTION TO MASS TRANSFER WITH CHEMICAL REACTION OPERATIONS (WITH EMPHASIS ON GAS-LIQUID SYSTEMS)

Erdogan Alper

Department of Chemical Engineering
Ankara University, Besevler, Ankara, Turkey

1. INTRODUCTION

Chemical processes which incorporate diffusion usually involve chemical reactions. Often diffusion and reaction occur in the same region, and the two rate phenomena are coupled so closely that they have to be treated simultaneously. "Mass transfer with chemical reaction" is, indeed, an example of a topic which does not fall entirely within the province of either a chemist or the conventional engineer, because it requires simultaneous consideration of molecular diffusion, fluid mechanics and chemical reaction kinetics, thus becoming a typical and classical chemical engineering topic.

Simultaneous mass transfer and chemical reaction of a soluble solute in two or three-phase systems has considerable importance not only in the chemical and process industries but also in biological and physiological processes (1). Among others, process metallurgy and enviromental sciences provide also many challenging problems. In process engineering, it is encountered both in separation processes and in chemical reaction engineering. In the former, a reactant is introduced deliberately to react with the transferring solute for speeding up the rate of mass transfer and increasing the capacity of the solution. Gas scrubbing, for instance, is typical of such application. On the other hand, there are large numbers of examples where these heterogenous reactions are a part of the process to obtain a desired product; here mass transfer has merely an effect on supply or removal of reactants and products from the reaction zone. However this distinction is not always clear-cut and there are also processes where the aim is simultaneous attainment of both objectives, i.e. product formation and separation. Destillation with chemical reaction is typical of such processes.

When a chemical reaction is employed for separation purposes, it involves usually reactions which in the terminology of chemists are referred to as extraordinarily fast reactions; for instance, a first order reaction having a half-life of the order of 10^{-4} - 10^{-5} s is typical of many gas-liquid reactions.

2. CLASSIFICATION OF MULTIPHASE SYSTEMS

"Mass transfer with chemical reaction in multiphase systems" covers, indeed, a large area. Table 1 shows a general classification of the systems encountered. From the possible two-phase systems, solid-solid reactions, liquid-solid (reactive or catalytic) and gas-solid (reactive or catalytic) reactions are not discussed here. The first one was reviewed by Tamhankar and Doraiswamy (2) and gas-solid (reactive) systems, such as, coal gasification, calcination of limestone, reduction of ores, etc. have been treated in some detail in recent reviews (3-5). The industrially important fluid-solid catalytic processes were the topic of a previous Advanced Study Institute (6) and have been also discussed authoritatively elsewhere (5,7). Concerning solid (reactive)-liquid two-phase systems, only some interesting examples are presented in Table 2 (1).

The gamut of the problems in the remaining gas-liquid, liquid-liquid and gas-liquid-solid systems may be considered as in Table 3, and they have many aspects which can be examined in a satisfactorily coherent manner.

Table 1. A GENERAL CLASSIFICATION OF PRACTICAL MULTIPHASE SYSTEMS

1. Two-phase systems :
   - Solid-solid reactions
   - Fluid-solid systems : 1.Gas-reactive solid, 2.Gas-catalytic solid, 3.Liquid-catalytic solid,4.Liquid -sparingly soluble solid (solid dissolution accompanied by chemical reaction 5.Liquid-insoluble reactive solid
   - Gas-liquid systems(including vapour-liquid)
   - Liquid-liquid systems

2. Three-phase systems :
   - Fluid-fluid-fluid systems (e.g. desorption of a volatile compound in liquid systems)
   - Fluid-fluid-solid systems : 1.Gas-liquid-sparingly soluble reactive solid, 2.Gas-liquid-insoluble reactive solid, 3.Gas-liquid-catalytic solid

TABLE 2. EXAMPLES OF LIQUID-SOLID REACTIONS

1. Sparingly soluble solid :
   - Alkaline hydrolysis of solid esters
   - Reaction between benzyl chloride and dry sodium salts(e.q. acetate) to manufacture benzyl esters
   - Reaction between terephtalic acid and ethylene oxide (with or without solvent) in the presence of homogeneous catalysts
   - Reaction between cyanuric chloride and amines to manufacture reactive dyes
   - Dyeing with reactive dyes

2. Insoluble solids :
   - Recation between RCl and sodium cyanate
   - Leaching of various minerals by acidic or basic solutions
   - Cementation reactions in hydrometallurgy
   - Reduction of vat dyes by basic solution of sodium dithionite
   - Production of acetylene from calcium carbide
   - Production of certain organometallic compounds, e.g. Grignard reagent

TABLE 3. MASS TRANSFER WITH CHEMICAL REACTION (SEVERAL COMMON ASPECTS)

- Mass transfer accompanied by an irreversible reaction of general order
- Mass transfer accompanied by a reverbisle reaction of general order
- Mass transfer in a medium containing more than one reactant
- Consecutive reactions : role of mass transfer
- Simultaneous mass transfer of more than one solutes accompanied by chemical reaction
- Simultaneous absorption/desorption with reaction
- Distillation with reaction
- "Facilitated" transport of dissolved solutes
- Liquid-liquid reactions
- Mass transfer with reaction in a slurry containing sparingly soluble particles
- Absorption with catalytic reaction(Slurry reactors and trickle beds)

3. PRACTICAL EXAMPLES AND APPLICATIONS

3.1. Gas-liquid systems

Examples of processes of industrial importance, where gas absorption is accompanied by chemical reaction are numerous. The object may be either the removal of one component from a mixture of gases, as in the case of CO_2 removal from synthesis gas or the preparation of products like H_2SO_4, HNO_3, adipic acid etc. The processes include absorption of gases, such as CO, CO_2, CS_2, H_2S, O_2, O_3, NO, NO_2, PH_3, Cl_2, Br_2, $COCl_2$, HCl, HBr, SO_2, SO_3 and olefins. A thorough listing of such processes may be found, among others, in (9). Some additional examples of industrial interest are given in Table 4. There are also some important examples in biochemical systems and these gas-liquid reaction (in some cases, they are, however, distincly a three phase system) deserve a detailed treatment.

In most cases of gas scrubbing processes, it is necessary to have also a regenerative process, that is , the spent solution from the absorption column should be regenerated and used again. The topic has been reviewed authoritatively recently by Shah and Sharma (10), and some important examples of desorption preceeded by a chemical reaction is given in Table 5. It is interesting to note that many gas-liquid (indeed, liquid-liquid) reactions produce a volatile product resulting in simultaneous absorption/desorption with chemical reaction. Although they are not representative of the situation prevailing in the desorption unit of a gas treating plant, where a volatile component is stripped from the liquid phase without any gaseous component being transferred from the gas to the liquid, there are still many industrially important processes. For instance substitution chlorination of organic compounds produces hydrogen chloride which must simultaneously desorb back in the gas phase to prevent supersaturation of the liquid phase. Another industrially important process involves "supported liquid phase catalyst", where the reactants have to be transferred from a bulk gas to a liquid reaction phase while the products are released back into the gas phase. Here the catalyst is in the form of a melt on a solid support and it finds applications in alkylation, carbonylation, hydroformylation and oxidation of inorganic and organic compounds. The subject matter was recently reviewed delicately by Villadsen and Livberg (11,12). Other examples of these interesting systems are shown in Table 6.

ABLE 4. ADDITIONAL EXAMPLES OF GAS-LIQUID REACTIONS

- Precipitation of Cu,Ni, and Co from their sulphate solutions by hydrogen
- Precipitation of UO_2, V_2O_3, MoO_3 and WO_3 from solutions of their respective salts by hydrogen
- Precipitation of Ag,Cu,$Ni(CO)_4$, $Fe(CO)_4$ and $Co(CO)_4$ by CO from solutions of their complex salts
- Air oxidation of black liquor containing Na_2S
- Oxidation of aluminium trialkyls
- Oxidation of trialkyl boranes
- Liquid phase oxidation of petrochemicals
- Liquid phase oxidation of benzoic acid and substituted benzoic acids (molten) to make phenols
- Reaction between NH_3 and caprolactone for manufacture of caprolactam
- Absorption of CO in organic medium containing cuprous aluminium complexes; in methanol to produce acetic acid or metyl formate ; in dimethylamine to make dimethyl formamide

---

ABLE 5. SELECTED EXAMPLES OF INDUSTRIALLY IMPORTANT
ESORPTION WITH CHEMICAL REACTION PROCESSES

- Desorption of CO_2 and H_2S from carbonated potash and alkanolamine solutions
- Desorption of Cl_2 from brine
- Desorption of NO from $FeSO_4$-NO complex
- Desorption of CO from ammoniacal cuprous chloride complex or organometallic complexes
- Desorption of HCl in the manufacture of fatty alcohol sulphates by reaction between fatty alcohols and chlorosulphonic acid

---

ABLE 6. SELECTED EXAMPLES OF INDUSTRIAL IMPORTANCE
OR SIMULTANEOUS ABSORPTION/DESORPTION WITH REACTION

- Absorption of SO_2, HCl, etc. in aqueous solutions of carbonate (desorption of CO_2)
- Chlorination of aqueous HCN (desorption of CNCl)
- Absorption of NO_2 and N_2O_4 in water (desorption of NO)
- Ozonolysis of unsaturated fatty acids and esters (desorption of O_2)
- Conversion of HCl to Cl_2 by Kcl chlor process
- Absorption of O_2 in carbon monoxide complex of haemoglobin (desorption of CO)
- Reaction between HCl and CH_3OH in the presence of $ZnCl_2$ as catalyst (desorption of CH_3Cl)
- Absorption of C_2H_2 in aqueous solutions of CuCl NH_4Cl and HCl (desorption of vinyl acetylene

Distillation with reaction, where the normal process is coupled with a liquid phase reaction, is also interesting and esterifications of certain alcohols with acids are typical industrial applications. These include, among others the homogeneously catalyzed butyl acetate process and the production of the plasticizer di-octyl-phthalate from phthalic anhydride and 2-ethyl-hexanol. However, the subject which involves both product formation and separation aspects has not usually been treated in the literature relating specifically to "mass transfer with reaction".

Depending on the relative rates of mass transfer and the chemical reaction (as well as many process-specific conditions), the appropriate equipment for gas-liquid systems may conform to many different geometries. The conventional types include bubble columns, spray columns, multistage contactors (sieve or bubble plate, mechanically agitated etc.) and packed towers whose design aspects will be discusses in the two published volumes of this ASI.

3.2. Liquid-liquid systems

Although such reactions are common, particularly in the organic process industry, they have received limited attention until very recently (1,13). One particular reason for this negligence may be that a good number of liquid-liquid reactors operate discontinuously and are of a very modest size, thus they have less need of carefully engineered designs.

In liquid-liquid systems, a chemical reaction is encountered for three distinct purposes. Firstly, the reaction may be a part of the process, such as nitration and sulphonation of aromatic substances, alkylation, hydrolysis of esters, oximation of cyclohexanone, extraction of metals and pyrometallurgical operations involving melts and molten slag. Secondly, a chemical reaction is deliberately introduced for separation purposes (e.g. removal of dissolved acidic solutes from a variety of hydrocarbons). Finally, the yield and the rate of formation of many single phase reactions are affected and often can be favourably increased by the deliberately controlled addition to the reaction system of an immiscible extractive phase, whose major purpose is to extract the product from the reactive phase. Such operations are sometimes referred to as "extractive reactions and have been discussed previously in some detail (14-17).

Recently many novel applications, such as phase transfer catalysis, were also reported in this field and further details on liquid-liquid reactions are given by Alper (18).

3.3. Gas-liquid-solid systems

Reactions involving gas, liquid and solid are very important in the process industries. In some cases, the solid may be reactive (e.g. thermal coal liquefaction, absorption of SO_2 into limestone etc.), but mostly it functions as catalyst so that gases like hydrogen, water, ammonia, or oxygen are involved. The processes can be classified on the basis of these gases as hydrogenation, hydration, amination, or oxygenation. Out of these processes, hydrogenation is by far the most important process.

The subject matter is further discusses by L'Homme (19) and other contributors in the second volume of this ASI.

4. EFFECTS OF CHEMICAL REACTIONS AND MODELING

In multiphase systems, chemical reaction affects the mass transfer rate in two distinct ways (20). At low reaction rates, it only serves to change the bulk concentration of the transferring solute, thus increasing the driving force. On the other hand, for reasonably fast reactions, the concentration gradient near the interface is affected leading to an "enhancement" of mass transfer rate. The effect of the latter, i.e. the enhancement factor, can be predicted very satisfactorily using the well known models such as the film or surface renewal models of Higbie and Danckwerts (9) for fluid-fluid systems. Indeed, they all lead numerically to almost the same predictions except for some extraordinarily unusual cases (9).

Table 7 shows rate expressions for various regimes for fluid-fluid systems; it is seen that the irreversible chemical reaction can dramatically alter the functional dependence of the specific rate of mass transfer of solute, R, on the physico-chemical properties and hydrodynamical factors. Indeed, under certain conditions, R can be independent of the interfacial concentration of solute, A* or the bulk concentration of reactive species, B^o despite the intrinsic kinetic showing first order dependencies on the reactant concentrations (see Table 7). These intriguing situations, which are similar to the so-called "diffusion falsification" regime of fluid-porous catalytic solid systems (5), can be successfully handled by the "theory of mass transfer with chemical reaction". Indeed, they can be deployed to obtain kinetics of exceedingly fast reactions in simple apparatuses, which in the normal investigations in homogeneous systems would have required sophisticated and expensive equipment. Further, it is possible, under certain conditions, to obtain values of rate constants without knowing the solubility and diffusivity. In addition, simple experiments yield diffusivity and solubility of reactive species which would otherwise have been - indeed, if possible - extremely difficult.

Simple theories have also been extremely useful in differentiating the "essentials" of experimental models from unnecessary details. The procedures have resulted in a sense of satisfaction, as in some cases of great industrial importance, the scale-up has been successfully accomplished from lab-scale models which do not bear any resemblance to the large scale unit (21-23).

Table 7. RATE EXPRESSIONS FOR AN IRREVERSIBLE (m,n)th ORDER REACTION FOR FLUID-FLUID SYSTEMS (9)

$$A + zB \longrightarrow \text{Products} \qquad r_A = k_{mn}\, A^m\, B^n$$

Regime	Conditions	Rate expression
Very slow	$k_L a \gg v k_{mn} (A^*)^{m-1} (B^o)^n$	$R = (\frac{v}{a}) k_{mn} (A^*)^m (B^o)^n$
Slow	$k_L a \ll v\, k_{mn} (A^*)^{m-1} (B^o)^n$ $M \ll 1$	$R = k_L\, A^*$
Very fast	$3 < M \ll \frac{B^o}{zA^*}\sqrt{\frac{D_B}{D_A}}$	$R = A^* \sqrt{\frac{2}{m+1} D_A (A^*)^{m-1} (B^o}$
Instantaneous	$M \gg \frac{B^o}{zA^*}\sqrt{\frac{D_B}{D_A}} \gg 3$	$R = \frac{k_L\, B^o}{z}\sqrt{\frac{D_B}{D_A}}$

For many gas-liquid-solid systems where the particles are large, reactants have to overcome various diffusional resistances to arrive at the site of reaction so that various steps occur consecutively leading to a series of resistances (6). Here, the effect of reaction on gas-liquid mass transfer rate, can only be through reducing the bulk concentration of the dissolved gas, therefore increasing the overall driving force. However, in the case of finely powdered catalyst, the situation is better modelled by a gas-pseudo-homogeneous liquid (i.e. suspension of solid) phase, hence, leading to considerable "enhancement" of absorption rate (24).

Finally, it may be pointed out that various aspects of mass transfer modelling of multiphase systems are discussed in recent reviews (25,26).

5. EXAMPLES OF RECENT PROGRESS AND SCOPE FOR FURTHER WORK

The subject matter is indeed an area where academicians have contributed immensely, and in many cases ingeniously. Most of the current highlights are of course discussed throughout this book and it is not at all intended to cover the whole area by the following examples.

It has been, for instance, possible to obtain rate constants, diffusivities and solubilities from measured "mass transfer with chemical reaction" rates in a simple model equipment (9). Nevertheless, there are still controversies on the physicochemical properties and kinetics of some common systems, such as reaction of CO_2 with ethanolamines (27,28), and catalyzed oxidation of sodium sulphite (29). Indeed, the kinetics of CO_2-diethanolamine reaction still admit room for speculation (28) and certain aspects of CO_2 mass transfer in carbonate solutions have yet to be settled (27,30). On the other hand, many complex schemes, mostly theoretical, have been analyzed successfully so that selectivities etc. can be predicted.

Chromatography has also been introduced in obtaining physicochemical data for some gas-liquid systems (31,33).

Chemical methods have moreover been successfully used to measure interfacial area and true mass transfer coefficients for various equipments under different conditions (34,35). However, in many cases, there are still very few data to predict even the most essential parameters (36).

Recently, many novel aspects of desorption with chemical reaction (or chemical desorption) as distinct from absorption have been studied and the basic points of difference emphasized. In the interesting area of simultaneous absorption and reaction in reacting systems the possibility of supersaturation in the intermediate vicinity of the interface was pointed out and properly analyzed. An industrially fully exploited case of simultaneous absorption and desorption into molten catalysts, i.e. a supported liquid phase catalyst, was discussed in another Advanced Study Institute with a heavy emphasis upon a further development of the theoretical basis (12).

Although the interactions between chemical and physical rate processes causing the occurrence of steady-state multiplicity in a variety of single phase chemical reactors have been well known for some time, these ideas have been only recently applied to gas-liquid reactions (37,38). Steady-state multiplicity and stability problems in stirred tanks, bubble columns, trickle beds etc. were illustrated and exotic behaviour, such as sustained oscillations, were demonstrated (37).

"Facilitated" transfer of solutes in liquid membranes has been also an active field of research. This process, which has been investigated in physiological systems, may be used for separation purposes; indeed, there has been considerable interest in CO_2 selective liquid membranes, but so far no commercial application have been reported. "Facilitated" transport in a liquid membrane has also been applied in extraction of metals and progress in this area was the topic of a recent seminar (39).

Biochemical systems have also been examined in some detail (40). It may however be concluded in aerobic fermentation, that the reaction is not fast enough to "enhance" the mass transfer (41). On the other hand, for some enzymatic gas-liquid reactions absorption enhancement is observed and they behave, in general, like other ordinary chemical reactions (42).

Three phase systems have been the main focus of activities in chemical reaction engineering, and the many novel aspects of them are too numerous to cover here, hence only a few examples will be referenced. In the case of gas-liquid-sparingly soluble solid, it has been demonstrated that particles substantially smaller than the diffusion film thickness of film model can enhance the specific rates of mass transfer if the reaction is sufficiently fast (45). Work in this area has been persistently pursued by Sada and coworkers (46,47). Recently Alper et al. (24) has pointed out and demonstrated that in catalytic slurry reactors similar enhancement can be observed if the catalyst particles are sufficiently small. There is however some dispute on the order of magnitude of the enhancement (48,49). Another aspect is complex reactions and in the case of slurry reactors the product distribution may well depend on the degree of diffusional resistance (50). Dynamic methods have been ingeniously employed to obtain physicochemical parameters in slurry reactors (51).

The topic of trickle beds involves also many novel investigations, some of which can be found in reference (52).

There has been a considerable amount of work in three-phase systems associated both with direct (53) and indirect (54) coal liquefaction and the chemical cleaning of coal (55). It is however evident now that the topic, in particular coal liquefaction, provides little by way of challenge for the mass transfer specialist.

The technique of "triphase catalysis", where liquid-liquid reactions are catalyzed by "phase transfer catalysts" chemically fixed on inert polymer supports, is an extremely interesting example of three-phase systems. They may be potentially of great interest technically since the catalysts can be recovered or used continously (56).

Finally, "mass transfer with chemical reaction" deserves still considerable attention and Table 8 lists some areas of interests.

7. CONCLUSIONS

This brief introduction to the subject matter reveals many problems not only of interest in academia, but also of considerable practical relevance covering the entire spectrum of conventional chemical industry, physiology, biotechnology, process metallurgy and environmental sciences. Indeed, whilst academicians can see its great importance and ability to contribute usefully, industry should appreciate the potential merit of these investigations. Hence, the worthwhile aim of converting laboratory data into commercial plants with reasonable confidence, is not at all an optimistic or unrealistic endeavour.

NOTATION

a	interfacial area per unit volume of reactor
A^*	interfacial concentration of A in the reactive phase
B^o	bulk concentration of reactive species B
D_A, D_B	diffusivities of dissolved A and B
k_L	physical mass transfer coefficient
k_{mn}	reaction rate constant
M	$\dfrac{\sqrt{\frac{2}{m+1} D_A k_{mn} (A^*)^{m-1} (B^o)^n}}{k_L}$
m	reaction order in A
n	reaction order in B
v	volume of reactive phase per unit volume of reactor
z	stoichiometric coefficient

TABLE 8. SELECTED TOPICS FOR FURTHER STUDIES

- Reactions at very low temperatures
- Reactions involving supercritical solutes
- Use of laboratory models for design purposes
- Photochemical gas-liquid reactions
- Membrane reactors and "facilitated" transport
- Desorption with reaction under conditions of flashing,into growing bubbles etc.
- Simultaneous absorption/desorption with reaction (including "supported liquid phase catalysis")
- Steady-state multiplicity and stability
- Physicochemical data from new techniques(e.g.chromatography)
- Interfacial phenomena coupled with reaction
- Fast reactions with phase transfer catalysis
- Fast reactions with micellar catalysis
- Fast reactions in ion-exchangeresins including triphase catalysis
- Mass transfer with electrochemical reaction
- Solid-liquid reactions:Dyeing with reactive dyes
- Selectivity of slurry reactors
- Four-phase (two liquids) reactors
- Multiphase reactors involving non-newtonian materials
- Fast reactions in slurry reactors containing fine particles as catalyst
- Improved design methods(consideration of effect of backmixing etc.)

REFERENCES

1. Doraiswamy,L.K. and M.M.Sharma. Heterogeneous Reactions: Analysis, Examples and Reactor Design (Wiley Interscience, New York, 1981)

2. Tamhankar,S.S. and L.K.Doraiswamy. "Analysis of solid-solid reactions. A review". AIChEJl 25 (1979) 561-582.

3. Schumpe,A. and Y.Serpemen. "Stoffübertragung mit chemischen Reaktionen". Fortschritte Verfahrenstechnik. 18 (1981) 75-115.

4. Kunii,D. "Chemical reaction engineering and research and development of gas-solid systems". Chem.Engng.Sci. 35 (1980) 1887-1911.

5. Froment,G.F. and K.B.Bischoff. Chemical Reactor Analysis and Design (John Wiley, New York), 1979).

6. Froment,G.F. (Director) Analysis of Fluid-Solid Catalytic Systems (NATO ASI, Rijkuniversiteit, Gent, Belgium, 1974)

7. Satterfield,N. Heterogeneous Catalysis in Practice (McGraw Hill Co., New York, 1980).

8. Sharma,M.M. "Absorption with reaction". (Plenary Lecture, CHISA, Prague, 1975.

9. Danckwerts,P.V. Gas-Liquid Reactions(Mc Graw Hill Co., New York, 1970).

10. Shah,Y.T. and Sharma,M.M. "Desorption with or without chemical reaction". Trans.Instn.Chem.Engrs. 54 (1976) 1-41.

11. Villadsen,J. and H.Livberg. "Supported liquid phase catalysis" (Proceedings of NATO ASI on "Multiphase Reactors", Portugal 1980).

12. Villadsen,J. and H.Livberg. "Supported liquid phase catalysis". Cat.Rev.Sci.Eng. 17 (1978) 203.

13. Laddha,G.S. and T.E.Degaleesan. Transport Phenomena in Liquid Extraction (Tata Mc Graw Hill Co. New Dehli, 1978).

14. Trambouze,P. "Calcul des reacteurs pour al misc en ocuvre de réactions on deux phases liquides interviennent." (Chemical Reaction Engineering 2nd Symposium, Supplement to Chem.Engng.Sci. 14 (1961) 161-170).

15. Piret,E.L., W.M.Penney and P.Trambouze. "Extractive Reaction: Batch or continous flow chemical reaction systems. Dilute case" AIChEJl 6 (1960) 394-402.

16. Trambouze,P. and E.L.Piret "Continous stirred tank reactors". AIChEJl 5 (1959) 384-389.

17. Schonemann,K. "Der derzeitige Stand bei der Vorausberechnung der Verweilzeitverteilung in technischen Reaktoren" (Proceedings of 2nd Symposium on Chemical Reaction Engineering, Supplement to Chem.Engng.Sci. 14 (1961) 193-203.

18. Alper,E. "Introduction to liquid-liquid systems". (Proceedings of NATO ASI on "Mass transfer with chemical reaction in multiphase systems", Turkey, 1981).

19. L'Homme,G.A. "Introduction to gas-liquid-solid systems" (Proceedings of NATO ASI on "Mass transfer with chemical reaction in multiphase systems", Turkey, 1981)

20. Alper,E. "Comments on "Gas-liquid reactions: Formulation as initial value problems." Chem.Engng.Sci. 34 (1979) 1076-1078.
21. Danckwerts,P.V. and E.Alper. "Design of gas absorbers: Part III Laboratory "point" model of a packed column absorber." Trans.Instn.Chem.Engrs. 53 (1975) 34-40.
22. Alper,E. and P.V.Danckwerts. "Laboratory scale model of a complete packed column absorber". Chem.Engng.Sci. 31 (1976) 599-607.
23. Alper,E. "Auslegung von Füllkörperkolonnen als chemische Absorber durch Simulation in Labormodellen." Chemie-Ing.Tech. 51 (1979) 1136-1138.
24. Alper,E., B.Wichtendahl and W.-D.Deckwer. "Gas absorption mechanism in catalytic slurry reactors". Chem.Engng.Sci. 35 (1980) 217-223.
25. Thoenes,D. "Current problems in the modeling of chemical reactors." Chem.Engng.Sci. 35 (1980) 1840-1853.
26. Van Landegehm,H. "Multiphase Reactors: Mass transfer and modeling". Chem.Engng.Sci. 35 (1980) 1912-1942.
27. Laddha,S.S. and P.V.Danckwerts. "Reaction of CO_2 with ethanol amines: Kinetics from gas absorption". Chem.Engng.Sci. 36 (1981) 479-482.
28. Danckwerts,P.V. "The reaction of CO_2 with ethanolamines." Chem.Engng.Sci. 34 (1979) 443-446.
29. Linek,V. and V:Vacek. "Chemical engineering use of catalyzed sulfite oxidation kinetics for the determination of mass transfer characteristics of gas-liquid contactors". Chem.Engng.Sci. 36 (1981) 1747-1768.
30. Alper,E. "Kinetics of absorption of CO_2 into buffer solutions containing carbonic anhydrase". Entropie 17 No: 101 (1981) 40-47.
31. Plöcker,U. H.Kaupp and J.M.Prausnitz "Calculation of high pressure vapor-liquid equilibrium from a corresponding states: Correlation with emphasis on asymetric mixtures." Ind.Engng.Chem. Proc.Des.Dev. 17 (1978) 324-331.
32. Clever,H.L. and R.Battino. Techniques of Chemistry Series vol. 8, part 1. (M.R.J.Dack (ed.), Wiley, New York, 1975), p. 379
33. Andrieu,J. and J.M.Smith. "Gas-liquid reactions in chromatographic columns". Chem.Engng.Jl. 20 (1980) 211-218.
34. Sharma,M.M. and P.V.Danckwerts. "Chemical methods of measuring interfacial area and mass transfer coefficients in two-phase systems." Brit.Chem.Eng. 15 (1970) 522-528.
35. Alper,E. "Measurement of effective interfacial area in a packed-column absorber by chemical methods". Trans.Instn.Chem. Engrs. 57 (1979) 64-68.
36. Alper,E. "Absorption with chemical reaction: Design methods and effect of viscosity for packed columns". I.Chem.Engrs. (London) Jubilee sympsium (1982)
37. Luss,D. "Steady-state multiplicity and stability". (Proceedings of NATO ASI on "Multiphase Reactors", Portugal, 1980).

38. Sharma,S., L.A.Hofmann and D.Luss. "Steady-state multiplicity of adiabatic gas-liquid reactors: II. The two consecutive reaction case." AIChEJl. 22 (1976) 324-331.
39. Seminar on "Liquid membrane applications in waste water treatment and metals recovery." UMIST, England (1980).
40. Schügerl,K. "Introduction to biochemical systems." (Proceedings of NATO ASI on "Mass transfer with chemical reaction in multiphase systems." Turkey, 1981)
41. Alper,E., Y.Serpemen and W.-D.Deckwer. "Gas absorption with simultaneous biochemical reaction" (Proceedings of VI IFS, Canada, Pergamon Press 1981), p. 311-316.
42. Alper,E., M.Lohse and W.-D.Deckwer. "On the mechanism of enzyme catalyzed gas-liquid reactions: Absorption of CO_2 into buffer solutions containing carbonic anhydrase." Chem.Engng.Sci. 35 (1980) 2147-2156.
43. Hanson,C. Recent Advances in Liquid-Liquid Extraction (Pergamon Press, London, 1971)
44. Sharma,M.M. "Extraction with Reaction", chapter 2a in Handbook of Extraction (To be published, Wiley and Sons, New York).
45. Bailes,P.J., Hanson,C. and M.A.Hughes. "Liquid-liquid extraction: Metals." Chem.Engng.83. No: 18 (1976) 86-94.
46. Ramachandran,P. and M.M.Sharma. "Absorption with reaction in a slurry containing sparingly soluble fine particles." Chem.Engng. Sci. 24 (1969) 1631.
47. Sada,E., H.Kumazawa and M.A.Butt. "Simultaneous absorption with reaction in a slurry containing fine particles." Chem.Engng. Sci. 32 (1977) 1493-1503.
48. Alper,E. and W.-D.Deckwer. "Comments on "Gas absorption with catalytic reaction." Chem.Engng.Sci. 36 (1981) 1097-1099.
49. Pal,S.K., M.M.Sharma and V.A.Juvekar. "Fast reactions in slurry reactors". Paper submitted to Chem.Engng.Sci.
50. Acres,G.J.K., A.J.Bird and P.J.Davidson. "Recent developments in platinium metal catalyst systems". Chem.Engr. (London) No: 283 (1974) 145-158.
51. Komiyama,H. and J.M.Smith. "Sulfur dioxide oxidation in slurries of activated carbon. Part II. Mass transfer studies." AIChEJl 21 (1975) 664-676.
52. L'Homme,G.A. (Ed.) Chemical Engineering of Gas-Liquid-Solid Catalyst Reactions (Proceedings of an International symposium CEBEDOC, Liege, 1979).
53. Shah,Y.T., Singh,P.C. and A.Calimli. "Direct coal liquefaction" (Proceedings of NATO ASI on "Mass transfer with chemical reaction in multiphase systems," Turkey, 1981).
54. Deckwer,W.-D. "Coal liquefaction via indirect routes." (Proceedings of NATO ASI on "Mass transfer with chemical reaction in multiphase systems", Turkey, 1981).
55. Shah,Y.T. and R.S.Albal. "Chemical cleaning of coal." (Proceedings of NATO ASI on "Mass transfer with chemical reaction in multiphase systems", Turkey, 1981).
56. Regen,S.L."Triphase catalysis". J.Am.Chem.Soc.97(1975)5956-5957

GENERAL MATHEMATICAL LAYOUT OF MULTIPHASE SYSTEMS

G. ASTARITA

Istituto di Principi di Ingegneria Chimica, University of Naples, I-80125

This article presents a general analysis of the rate of mass transfer within an agitated liquid phase in contact with a gas phase, with particular attention to the case where diffusion and chemical reaction are occurring simultaneously. At the end of the lecture, some elementary concepts are discussed for the case where the liquid phase contains a suspended solid.

It should be borne in mind that the gas-side resistance to mass transfer, though occasionally negligible, is never zero, and that therefore the overall transfer rate is in general influenced by it. Attention is here restricted to the liquid phase, and the value of the interface partial pressure of the transferring components is regarded as a known quantity. The equations which are derived should be coupled with the equation for mass transfer in the gas phase in order to calculate the overall transfer rate.

The aim is to introduce basic concepts and to establish the general mathematical background. Therefore attention is restricted here to those conditions of simultaneous diffusion and chemical reaction which can be regarded as limiting or asymptotic cases. Significant simplifications may arise, for example, when the chemical reaction is extremely fast or extremely slow, as compared to diffusion phenomena.

Consider the case where a chemical solvent is used in gas treating, so that the liquid phase contains one or more components

$B_j (j = 1, 2, \ldots N)$ which may react chemically with the component A which is being either absorbed or desorbed. Usually the component B_j are non-volatile, so that all the chemical reactions take place in the liquid phase.

The occurrence of the chemical reactions has two distinct effects on the overall behavior of the system. The first one is, for the case of absorption, as follows. When component A is absorbed into the liquid phase, it is consumed by the chemical reactions and therefore its concentration in the bulk of the liquid, a_o, is kept low. This in turn implies that the driving force for additional absorption remains higher than it would be if no chemical reaction were taking place. Conversely, in desorption the chemical reaction continuously produces the component to be desorbed, thus providing a high concentration of it in the liquid and a large driving force for the desorption.

The second effect is more subtle. At a given level of driving force, the actual rate of mass transfer may be very significantly larger when chemical reactions are taking place than it would be in the absence of chemical reactions. The rate enhancement may be very large, up to two orders of magnitude or even more. Indeed, the enhancement may be so large as to actually reduce the mass transfer resistance in the liquid phase to the point at which it is negligible as compared to the resistance in the gas phase.

The concept of rate enhancement introduced above is formalized as follows. In the absence of chemical reactions, the mass transfer rate in the liquid phase is given by

$$N = k_L^{\circ} (a_i - a_o) \tag{1}$$

where N is the mass transfer flux and a_i is the interface concentration of A in the liquid phase. The actual rate in the presence of chemical reactions may be larger than the value given by Eq.1; a "chemical" mass transfer coefficient, k_L, can be defined as follows :

$$N = k_L (a_i - a_o) \tag{2}$$

The rate enhancement factor, I, is defined as the ratio of the actual rate and the rate which would be observed under the same driving force in the absence of chemical reactions :

$$I = \frac{N}{k_L^{\circ} (a_i - a_o)} = \frac{k_L}{k_L^{\circ}} \tag{3}$$

It is important to realize that the value of the mass transfer coefficient k_L^o is determined by the fluid mechanics of the liquid phase in the neighborhood of the gas-liquid interface, and in fact that its value cannot in general be predicted from first principles due to the complexity of the fluid mechanics problem involved. Since predicting the value of k_L would involve an even more complex problem, there is in practice no hope of calculating k_L from first principles.

Fortunately, however, although both k_L^o and k_L are strongly influenced by the details of the fluid mechanics involved, their ratio I turns out to be almost independent of it. Therefore, the theory of coupled mass transfer and chemical reaction can be developed on the basis of very crude models of the fluid mechanics involved. The film theory model will be used throughout this lecture.

The rate enhancement factor is in general a complicated function of the composition of the gas and liquid phases. However, simplified equations emerge from the consideration of limiting conditions. The basic concepts underlying these simplifications are discussed below.

The intrinsic rate of a chemical reaction is measurable by means of a yardstick which is called the reaction time, t_r. A precise definition of t_r will be given later; at this stage, it is sufficient to say that t_r is a measure of the time required by the chemical reaction in order to change by a significant amount the concentration of the limiting reactant.

Diffusion phenomena can also be characterized by a time scale, the diffusion time, t_D. The exact definition of t_D will be given later. The diffusion time is a measure of the time available for molecular diffusion phenomena to take place before mixing of the liquid phase makes the concentration uniform. Therefore, t_D decreases as the mixing or turbulence of the liquid phase is increased.

Having established the time scales t_D and t_r, their dimensionless ratio ϕ is the measure of the relative rates of reaction and diffusion :

$$\phi = t_D/t_r \qquad (4)$$

When $\phi \ll 1$, the reaction is too slow to have any significant influence on diffusion phenomena, and essentially no rate enhance ment will take place. This is the situation referred to in the following as the "slow reaction regime". Of the two effects of chemical reactions discussed above, only the first one takes place in the slow reaction regime.

Conversely, if $\phi >> 1$, the reaction is fast enough to result in a significant rate enhancement. However, there is an upper bound to the possible rate enhancement, which is discussed in qualitative terms in the following.

The assumption that fugacity is continuous across an interface, holds also for chemically-reactive systems. Therefore, the value of a_i, the interface concentration of physically dissolved transferring component is related to the interface partial pressure, p_i, by the condition that physical equilibrium prevails at the interface :

$$a_i = p_i/H \tag{5}$$

where H is Henry's law constant.

However, the liquid at the interface is in general not in chemical equilibrium with the gas phase; the value of α, the total concentration of the transferring component (in both physically dissolved and chemically combined forms) at the interface is not the value α_i^* corresponding to equilibrium with p_i :

$$\alpha_i \neq \alpha_i^* \tag{6}$$

However, in the limit of infinitely fast reactions (mathematically, when $\phi \to \infty$), chemical equilibrium is established instantaneously, and therefore α_i will indeed be equal to α_i^* . This is the condition referred to in the following as the "instantaneous reaction regime", and it represents the upper bound to the rate enhancement, which is attained when all resistance to mass transfer due to chemical kinetics has been eliminated.

In the instantaneous reaction regime, the transferring component can diffuse in both its physically dissolved and its chemically combined form, with no kinetic resistance to the transformation from one form to the other. Therefore, if diffusion in either form is equally easy (i.e., if the diffusivities of all species are equal), the rate of mass transfer will be governed by a driving force measured in terms of the total concentration of the transferring component, rather than in terms of the concentration of its physically dissolved form :

$$N = k_L^\circ \left(\alpha_i^* - \alpha_o \right) \tag{7}$$

The quantity α_o is the value of α in the bulk of the liquid. Eq.7 represents an upper bound for the rate of mass transfer. The corresponding value of the rate enhancement factor, I_∞ (i.e., the value of I when $\phi \to \infty$) is calculated from Eq's 3 and 7 :

$$I_\infty = \frac{\alpha_i - \alpha_o}{a_i - a_o} \tag{8}$$

Values of I calculated from Eq.8 can be very large indeed, and are typically of the order 10^2 to 10^4.

We have qualitatively established so far that, when $\phi \ll 1$, the rate enhancement factor is unity, while when $\theta \to \infty$ its value is very large and is given by Eq.8. Clearly, an intermediate region exists where $\phi \gg 1$, and therefore I is appreciably larger than unity, yet the reaction is not so fast as to be instantaneous, and therefore I is appreciably less than I_∞. This intermediate region will be referred to in the following as the "fast reaction regime".

While the issue of chemical kinetics can be avoided in the slow reaction regime (since the rate of reaction is so slow that its actual value needs not be known), and in the instantaneous reaction regime (since the rate of reaction is so fast that, again, its value needs not be known), it cannot be avoided in the case of the fast reaction regime. However, considerable simplifications arise also in this limiting case, and the following simple equation is obtained for the enhancement factor I :

$$I = \sqrt{\phi} \tag{9}$$

Once the basic concepts discussed above have been introduced, a more detailed analysis needs to face the issue of modeling of the phenomena under investigation. Mass transfer in an agitated liquid phase is obviously influenced by the hydrodynamics of the liquid phase near the gas-liquid interface. At the interface itself, the tangential stress must be continuous; in other words, the drag that the gas exerts on the liquid must be equal in value and opposite in sign to the drag that the liquid exerts on the gas. Since the viscosity of the gas, is very much less than that of the liquid, it follows that, sufficiently near the interface, the velocity in the liquid phase is constant, while that in the gas phase is a linear function of the distance from the interface. Correspon dingly, hydrodynamic models for mass transfer analysis are different for the two phases; in particular those for the liquid phase lead to the very important concept of the diffusion time, t_D.

The simplest hydrodynamic model proposed in the literature is the film theory model. This assumes the existence, near the gas-liquid interface, of a stagnant film of thickness δ , through which mass transfer can only take place by molecular diffusion. The rest of the liquid phase is assumed to be perfectly well mixed. Therefore, the concentration at a depth α from the interface is equal to the bulk-liquid concentration for every species.

In the absence of any chemical reaction, the concentration profile of the volatile component in the film is linear, and the mass transfer flux is given by :

$$N = D \frac{a_i - a_o}{\delta} \tag{11}$$

or, equivalently,

$$k_L^o = \frac{D}{\delta} \tag{12}$$

Eq.12 does not have predictive value, since the value of the film thickness δ (into which the whole ignorance about the true fluid mechanics has been lumped) is not known. However, if a problem of mass transfer with chemical reaction is analyzed on the basis of the film theory model, the value of I will usually turn out to depend on δ , and Eq.12 can then be used to express I as a function of k_L^o. The value of the latter will of course need to be estimated on the basis of available correlations for classical mass transfer.

The film theory has an important drawback. Although, the value of δ is not known, one should regard it as uniquely determined by the hydrodynamics of the liquid phase. On the basis, Eq.12 would predict k_L^o to be proportional to the diffusivity D. Empirical mass transfer coefficient correlations available in the literature for a liquid in contact with a gas consistently indicate that in fact k_L^o is proportional to the square root of D. Therefore, analyses based on the film theory model are not expected to predict correctly the influence of diffusivity values on the enhancement factor I. Therefore, one is lead to a more complex model of the fluid mechanics involved, the penetration theory model. This model leads, in its several variations, to the correct prediction of the proportionality between k_L° and $\sqrt{D}$. By a purely dimensional argument, the proportionality constant must have the dimensions of the square root of a frequency, and therefore the following equation can be used as a definition of the diffusion time t_D :

$$k_L^\circ = \sqrt{D/t_D} \tag{13}$$

Comparison with Eq.12 shows the relationship between the diffusion time t_D and the empirical parameter δ to be :

$$t_D = \frac{\delta^2}{D} \tag{14}$$

The concept of diffusion time, as will be seen in the following, is very useful in establishing conditions of asymptotic behavior of phenomena of coupled mass transfer and chemical reaction. These will be discussed in the next sections. Physically, the value of t_D may be regarded as a yardstick of the time scale of mixing of the liquid phase: as the intensity of mixing increases, the diffusion time (and the parameter δ) decreases.

Values of t_D in industrial mass-transfer units may be estimated either directly, by actual inspection of the hydrodynamic conditions of the liquid phase, or, more usually, from Eq.13 and empirical correlations for the mass transfer coefficient k_L°. When this is done, the values of t_D turn out to be in a comparatively narrow range :

$$4 \times 10^{-3} < t_D < 4 \times 10^{-2} \text{ sec} \tag{15}$$

for all units where the liquid phase is reasonably well mixed.

We now turn attention to the analysis of coupled mass transfer and chemical reaction, and in particular we try to establish conditions of asymptotic behavior. For the sake of simplicity, we begin by considering the case where only one chemical reaction may take place in the liquid; let r be the rate at which the reaction takes place, i.e., the number of moles of component A which are consumed per unit time and per unit volume. Notice that, by definition, r is positive in absorption and negative in desorption. In general, the rate r will be given by some kinetic equation of the following general form :

$$r = r\,(a, b_1, b_2, \ldots\, b_N) \tag{16}$$

where b_j is the concentration of component B_j (recall that B_j's are non-volatile components present in the liquid phase).

Eq.16 implies that the rate of reaction is not constant throughout the liquid, since its value depends on the local composition of the liquid phase. In the presence of the chemical reaction, the diffusion equation for component A becomes, for the film-theory model,

$$D \frac{d^2 a}{dx^2} = r \tag{17}$$

While in the absence of chemical reactions the concentration profile has a zero curvature, Eq.17 shows that the curvature will be non-zero when reactions take place. In particular, the curvature will be positive in absorption and negative in desorption, but in both cases the effect of the chemical reaction is to make the concentration gradient at the interface steeper that it would be in the absence of chemical reactions. Hence, one expects an enhancement factor larger than unity.

However, there are conditions where the enhancement effect is negligible, namely, when the curvature of the concentration profile is very small, say :

$$\text{Curvature} << \frac{\text{average gradient}}{\text{film thickness}} \tag{18}$$

Equation 17 shows that the curvature is r/D, i.e., it is not constant throughout the film. However, an average value of the reaction rate can be defined as follows :

$$r_{avg} = \frac{1}{a_i - a_o} \int_{a_o}^{a_i} r \, da \tag{19}$$

which will be negative in desorption and positive in absorption. The average curvature is r_{avg}/D, while the average gradient is $(a_i - a_o)/\delta$; therefore, Eq.19 reduces to :

$$\frac{a_i - a_o}{r_{avg}} >> \frac{\delta^2}{D} = t_D \tag{20}$$

When condition 20 is satisfied, no significant enhancement is expected; this condition will be referred to as the slow-reaction regime, where the enhancement factor is expected to be unity :

$$I = 1 \tag{21}$$

The left hand side of Eq.20 can be interpreted as a "reaction time", t_r. Indeed, it represents the time required by the reaction to change the concentration over the interval $|a_i - a_o|$, should the reaction take place at the average rate over that interval. Therefore, we define a reaction time as follows :

$$t_r = \frac{a_i - a_o}{2\, r_{avg}} = \frac{(a_i - a_o)^2}{2 \int_{a_o}^{a_i} r\, da} \tag{22}$$

The factor 2 is introduced so that, for a simple first-order reaction, t_r is simply the inverse of the kinetic constant. Condition 20 implies that the reaction requires much more time than is available during the diffusion phenomenon-hence, the reaction does not appreciably influence the diffusion, and no enhancement is observed.

If condition 20 is not satisfied, the term r cannot be neglected in Eq.17. Yet that term depends not only a, but also on the concentration b_j of the non-volatile components: hence in principle Eq.17 is coupled with the diffusion equations of all other liquid phase components. The problem formulated in such a general form is very difficult to solve.

However, a major simplification arises from the following considerations. The concentrations b_j in the film will in general be different from b_{jo}, and the difference is likely to be greatest at the interface itself. However, suppose that the following condition is satisfied for all the b_j's:

$$|b_{ji} - b_{jo}| << b_{jo} \tag{23}$$

Condition 23 implies that the concentration b_j can, at all points in the liquid phase, be approximated with their bulk-liquid values. Therefore, the term r in Eq.17 could be approximated with :

$$r = r(a, b_{1o}, b_{2o}, \ldots\ldots b_{No}) = r_o(a) \tag{24}$$

so that the diffusion equation for a would be uncoupled from those for the b_j's.

The question which arises is whether condition 23 is likely to be fulfilled. One may notice that the analogous condition for a is certainly not fulfilled, since if it were the liquid phase composition would be uniform everywhere, and no mass transfer would take place. However, the chemical capacity of the liquid by far exceeds the physical solubility of component A, and therefore the bulk-liquid concentrations b_{jo} are likely to be much larger than either a_i or a_o; therefore, condition 23 is not an unrealistic one. Of course, the requirements to be satisfied in order that condition 23 holds remain to be determined, and will be discussed later.

Since we are now considering the case where condition 20 is not satisfied, we may go one step further and assume that :

$$t_r << t_D \tag{25}$$

The asymptotic behavior observable when both 23 and 25 hold will be referred to as the "fast-reaction regime". The diffusion equation reduces to :

$$D \frac{d^2a}{dx^2} = r_o(a) \tag{26}$$

and its solution leads, after some algebra, to Eq.9.

One may recall that the values of t_D of practical interest are in a rather narrow range, see Eq.15. In contrast with this, reaction times are very different for different reactions; t_r may well be, for some fast reactions, of order 10^{-4} sec, or even less. Correspondingly, Eq.9 predicts enhancement factors of order 10 or more.

Eq.9 indicates that, in the fast reaction regime, the mass transfer rate increases with increasing rate of the chemical reactions, though less than linearly. However, this is true only as long as the condition in Eq.23 is fulfilled. As the rate of reaction becomes very large, the concentrations of the non-volatile components near the interface become appreciably different from their bulk-liquid values, and the Eq.24 ceases to be valid.

It is comparatively easy, however, to envisage the asymptotic behavior of the process considered in the limit of extremely large rates of chemical reactions. As the kinetic constants keep increasing, a situation is reached where chemical equilibrium will prevail everywhere in the liquid phase, rather than only in the bulk of the liquid. Further increase of the kinetic constants will then not have any effect, and the rate of mass transfer is therefore expected to become independent of the chemical kinetics, i.e., I is expected to reach some asymptotic upper limit I_∞ which is independent of the reaction time t_r. This situation is called the instantaneous reaction regime.

The calculation of I_∞ offers, in the general case, some rather subtle mathematical difficulties, as can be understood by considering the essentially singular character of the limit $t_r \to 0$, and the non-linearity induced by the requirement of chemical equilibrium. In this lecture, we focus attention on a rather simple special case, the analysis of which, however, reveals the essential features of the instantaneous reaction regime.

Let the main reaction which A undergoes in the liquid phase be :

$$A + \sum_j \nu_j B_j = 0 \tag{27}$$

where the ν_j's are stoichiometric coefficients, positive for the reactants in absorption and negative for the products in absorption. The equilibrium condition for reaction 27 is :

$$K = \frac{\prod_j (b_j^{-\nu_j})}{a} \tag{28}$$

In the bulk of the liquid, the concentrations b_{jo} are determined by the local values of the molarity m_o and the chemical saturation y_o; Eq.28 then determines the bulk-liquid value of a_o. In the instantaneous reaction regime, Eq.28 will hold at all points in the liquid phase, and not only in the bulk.

The diffusion equation for the non-volatile species B_j is :

$$D_j \frac{d^2 b_j}{dy^2} = \nu_j r \tag{29}$$

where r is the reaction rate. Notice that, although chemical equilibrium prevails everywhere in the liquid, r is not zero, since the reaction takes place at whatever rate is required to maintain chemical equilibrium in the presence of diffusion.

At every point in the liquid, the molarity m is given by a linear combination of the concentrations :

$$m = \sum_j \lambda_j b_j \qquad (30)$$

The coefficient λ_j and ν_j are not independent of each other, since the progressing of reaction 27 by itself (i.e., in the absence of diffusion phenomena) does not change the molarity; hence :

$$\sum_j \lambda_j \nu_j = 0 \qquad (31)$$

If Eq.30 is differentiated twice with respect to x, and Eq's. 29 are substituted into the result, one obtains :

$$\frac{d^2 m}{dx^2} = r \sum_j \frac{\lambda_j \nu_j}{D_j} \qquad (32)$$

Notice that, in general, the right hand side of Eq.32 is not zero; this implies that the molarity is not constant in the liquid phase. This induces a very significant difficulty, since the solution of the equilibrium equations requires m to be known.

A considerable simplification, however, arises if all the D_j are equal. Since molecular diffusivities in ordinary liquids are not very different from each other, in this section we restrict attention to the case where :

$$D_1 = D_2 \; ..... = D' \qquad (33)$$

so that Eq.32 implies that :

$$\frac{d^2 m}{dx^2} = 0 \qquad (34)$$

Since the boundary conditions on 34 are :

$$x = \delta \,, \quad m = m_o \qquad (35)$$

$$x = 0 \,, \quad \frac{dm}{dx} = 0 \qquad (36)$$

the solution is :

$$m = m_o \qquad (37)$$

i.e., if the diffusivities of all the non-volatile species are equal, the molarity is a constant throughout the liquid phase. (The same conclusion is reached if one considers the penetration theory differential equations).

The total contenent of A, α , is also related linearly to the b_j's :

$$\alpha = a + \sum_j \mu_j b_j \tag{38}$$

Again, the coefficient μ_j and ν_j are not independent of each other, since the progressing of reaction 27 does not by itself change the value of α . Hence :

$$1 + \sum_j \nu_j \mu_j = 0 \tag{39}$$

The diffusion equation for a is :

$$D \frac{d^2 a}{dx^2} = r \tag{40}$$

Differentiating Eq.38 twice with respect to x, and substituting Eq's 29 and 40, one obtains :

$$\frac{d^2 \alpha}{dx^2} = r \left[\frac{1}{D} + \frac{1}{D'} \sum_j \nu_j \mu_j \right] \tag{41}$$

We now introduce an additional simplification, namely, the assumption that :

$$D = D' \tag{42}$$

If Eq.42 holds, Eq's. 39 and 41 imply that :

$$\frac{d^2 \alpha}{dx^2} - 0 \tag{43}$$

Notice that, as said before, should the chemical reaction take place in the absence of diffusion phenomena, it would not change the value of α , though of course the values of a and b_j would change.

When the chemical reaction takes place in the presence of diffusion phenomena, however, the value of α could change. In fact, suppose e.g., that D' > D; before the reaction takes place, component A must diffuse in the physically dissolved form, i.e., with diffusivity D; after some reaction has taken place, some of component A will be in a chemically combined form, and can therefore diffuse more easily. If, however, Eq.42 holds true, diffusion of A in any form is equally easy, and therefore the diffusion equation for α becomes independent of the kinetic term r.

Since the diffusion equation for α has the same form as that for a in the absence of chemical reaction, the rate of mass transfer in the presence of the chemical reaction will be equal to the rate without chemical reaction which would be observed under a driving force $|\alpha_i - \alpha_o|$, i.e. :

$$I = \frac{\alpha_i - \alpha_o}{a_i - a_o} \tag{44}$$

Subject to the condition that Eq. 42 holds true, the result in Eq. 44 is of general validity; however, it cannot in general be used for predictive purposes, since the value of α_i is in general not known. However, in the instantaneous reaction limit, α_i can be calculated from the requirement that equilibrium should prevail at the interface, i.e., one obtains Eq.8. Furthermore, since the molarity at the interface is the same as in the bulk, the value of α_i^* can simply be read off an equilibrium curve determined at the bulk-liquid value of the molarity.

The difference $\alpha_i^* - \alpha_o$ represents the total amount of component A which would need to be absorbed (or desorbed if $\alpha_i^* < \alpha_o$) in order to bring the liquid to chemical equilibrium with the gas. In this sense, $\alpha_i^* - \alpha_o$ can be regarded as the "chemical driving force", as compared to the physical driving force $a_i - a_o$. Turning back to the definition of I in Eq. 3 one sees that the enhancement of mass transfer rate in the instantaneous reaction regime is equal to the capacity enhancement.

It is useful to introduce the following definition of ξ_j :

$$b_j = b_{jo} - \nu_j \xi_j \tag{45}$$

When this is substituted into the diffusion equations and boundary conditions for b_j, the result is obtained that both are the same for all values of the index j, so that, necessarily :

$$\xi_1 = \xi_2 = \ldots\ldots = \xi \tag{46}$$

where ξ can be interpreted as an extent of reaction with respect to the bulk-liquid composition. It is important to realize that the result in Eq.46 is based only on the assumption that Eq.33 holds true: i.e., if the diffusivities of the non-volatile species are all equal, their concentrations are everywhere related to each other by Eq's. 45 and 46.

It takes some tedious but straight forward algebra to obtain, on the basis of Eq.46, the following result: condition 23 is satisfied provided that :

$$\sqrt{\phi} \ll I_\infty \tag{47}$$

i.e., there is no regime intermediate between the fast and the instantaneous reaction one.

In summary three asymptotic behaviours have been discussed: slow, fast and instantaneous reaction. These are best summarized by considering the parameter ϕ, defined by Eq.4, which increases with increasing rate of the chemical reactions.

As long as ϕ is significantly smaller than unity, no enhancement takes place, i.e. $I = 1$. When ϕ exceeds unity by a significant amount, the enhancement factor is given by Eq.9, i.e. $I=\sqrt{\phi}$ However, an upper bound to the value of I is reached in the instantaneous reaction regime; i.e. when $\phi \to \infty$, I approaches some asymptotic value I_∞ to be calculated from the solution of the appropriate instantaneous reaction problem. The value of I_∞ depends only on the liquid-phase composition and the value of a_i, but is independent of the fluid mechanics of the liquid phase and of the parameter ϕ.

Given a value of a_i, and a composition of the liquid phase, e.g., given the conditions at a particular point of an industrial unit, the value of I_∞ is fixed, and I may be regarded as a unique function of ϕ. The asymptotic behaviour of this function has been discussed in this lecture. Of course, the actual behavior of the function is more complex, since the two cusps at $\phi = 1$ and $\phi = I_\infty^2$ are in fact smoothed out; these transition regions have been discussed at length in the literature. However, particularly if $I_\infty \gg 1$, the three asymptotic equations $I = 1$, $I = \sqrt{\phi}$, and $I = I_\infty$ will be quite acceptable approximations over rather wide ranges of values of ϕ.

We now turn attention to the problem of experimental investigation of diffusion-reaction problems. The main purpose of gathering mass transfer rate data in the laboratory is to understand the chemical mechanism and the corresponding kinetics well enough to allow the development of a reliable model of the rate enhancement phenomenon. From such a model, the enhancement factor I can be calculated as a function of temperature and composition of both gas and liquid phases. This function should form the basis for design procedures.

When a set of rate data is obtained, the first step in the analysis is to ascertain which regime of mass transfer applies to the data. The quantity which is measured is the total mass transfer rate $\bar{V}$; the chemical mass transfer coefficient k_L can be extracted from the data provided the driving force and the interface area are known. The key to the identification of the regime is the dependency of $\bar{V}$ on the operating variables. The latter are: the degree of mixing as represented by k_L°; the interface area A; the liquid volume V; temperature; and the physical driving force $a_i - a_o$

The dependency of $\bar{V}$ on the operating variables can be calculated on the basis of the Equations given above; the results are summarized in Table 1. The entries in the Table are sufficiently different from one regime to another to permit identification of the regime of the available set of data (of course, data may fall in a transition region).

Once the first step of the analysis has been completed and the regime has been identified the data can be scrutinized to gain insight into the chemical mechanisms involved. The kind of information which can be obtained is briefly reviewed in the following paragraphs.

First, consider the case where the data indicate fast reaction regime behavior. A straight forward calculation yields :

$$\bar{V} = v \sqrt{Dk_o}\, A \frac{p_i - p^*}{H} \tag{48}$$

where v is the molar volume of the gas and p^* is the equilibrium partial pressure corresponding to the bulk-liquid composition. The value of the equivalent kinetic constant k_o can therefore be extracted from the data, provided the VLE behavior is known (so that p^* is known), as well as the value of H. The value of k_o depends on the liquid-phase composition; the assumption is here made that the function $r(\cdot)$ in Eq.24 is linear in a, so that :

$$r(a, b_{1o}, b_{2o}, \ldots\ldots b_{no}) = k_o\, a \tag{49}$$

If a thermodynamic model is available, the values of the bulk-liquid concentrations $b_{1o}.....b_{No}$ can be calculated as functions of the molarity m and the degree of saturation y. Therefore, the experimentally determined dependency of k_o on m and y should yield the form of the g(·) function, from which the chemical mechanism and the true kinetic constant can be obtained.

Next consider the case where the data fall in the instantaneous reaction regime. The equation for $\bar{V}$ becomes in this case :

$$\bar{V} = v I_{\infty} k_L^{\circ} A \frac{p_i - p^*}{H} \tag{50}$$

from which values of I_{∞} can be calculated and compared with predictions of models such as those discussed before.

It is often useful to make the tentative assumption that the diffusivities of all solutes are equal to each other. If that is the case, Equation 8 can be used for I_{∞} and Equation 50 reduces to :

$$\bar{V} = v k_L^{\circ} A (\alpha_i^* - \alpha_o) \tag{51}$$

The value of α_i^* can be simply read off a VLE curve as the value of α corresponding to the interface partial pressure of the volatile component, at the bulk-liquid molarity. Therefore, a quantitative check of the validity of Equation 51 can be performed.

Often in actual fact the diffusivities are not all equal and therefore the measured value of $\bar{V}$ will not be equal to the right hand side of Equation 51. However, while $\bar{V}$ itself will depend strongly on the composition of both gas and liquid phases, its ratio to the right hand side of Eq.51 will be almost constant and close to unity: values between 0.7 and 1.3 are typical. This ratio can be taken as an empirically determined correction factor for non-equal diffusivities.

Finally, consider the case where the data appear to fall in the transition region between fast and instantaneous reaction behavior. In this case, it is useful to perform experiments with a stirred cell, so that different values of k°_L (and hence of ϕ) can be obtained by changing the stirrer speed. This allows a region of ϕ values at a given liquid phase composition to be scanned. The experimentally determined I versus ϕ curve can then be compared with a theoretical model of the transition region.

By appropriate choice of the laboratory units to be used, it is in general possible to obtain a reliable model for the prediction of I values for any given system.

Before concluding this lecture, a few elementary concepts concerning 3-phase systems will be discussed. Consider a liquid phase containing a suspension of (usually sparingly soluble) solid particles, in contact with a gas phase containing a component A which may be absorbed in the liquid phase and react chemically with the dissolved solid B (or possibly with the solid itself), forming a product C (which is also in general sparingly soluble). The example of greatest industrial interest is flue gas desulfurization, where A is SO_2, B is lime of limestone, and C is calcium sulphite. The sequence of reactions may be written as :

$$B(s) \rightleftarrows B(l) \tag{52}$$

$$A+B(l) \rightarrow C(l) \tag{53}$$

$$C(l) \rightleftarrows C(s) \tag{54}$$

where 52 and 54 may usually be regarded as being at equilibrium, while 53 may or may not. The following reaction is also possible:

$$A + B(s) \rightarrow C(l) \tag{55}$$

First consider the case where the reaction consuming A is 53. In this case, all the possible mechanisms discussed in the earlier part are possible. However, two major differences with the 2-phase case should be considered. First, if the absorbing medium is a liquid, its capacity is related to the total concentration of reactive solute in the feed liquid, and therefore b_o is necessarily high (unless possibly at the rich end of the absorber). In contrast with this, in the case of a slurry the capacity is related to the content of reactive solid, and b_o may have a comparatively low value (determined by the equilibrium of reaction 52), thus resulting in low values of I_∞ . Rate enhancement in slurried liquids may be low even if reaction 53 is instantaneous.

The second point is that, as long as both B(s) and C(s) are present, the concentrations of B(l) and C(l) are fixed by the equilibrium of reactions 52 and 54, so that absorption of A does not result in a change of composition of the liquid phase. This in turn implies that the relative flow pattern of gas and slurried liquids is irrelevant, and indeed in industrial practice cocurrent, and well-mixed slurry flow patterns are used as commonly as the more traditional countercurrent flow pattern.

Let us now turn attention to the case where A is consumed by reaction 55 rather than 53. The question which arises here is whether the reaction at the liquid-solid interface may or may not influence the mass transfer rate of A from the gas-liquid interface to the bulk of the liquid.

In order to discuss this problem, it is useful to refer back to the discussion at the beginning of this lecture. Any chemical reaction taking place in the liquid phase can influence the mass transfer rate only if it takes place appreciably in the gas-liquid interface region, say at $x < \delta$. Values of δ of practical interest can be calculated from Eq's. 12 and 15; taking for D a typical value of 10^{-5} cm^2/sec, one obtains :

$$2 \times 10^{-4} < \delta < 4 \times 10^{-4} \text{ cm} \tag{56}$$

A reaction taking place on the liquid-solid interface could therefore result in appreciable rate enhancement only if a significant number of solid particles are present at distances from the gas-liquid interface less than 10^{-4} cm. This in turn would require particle diameters no more than 10^{-5} cm, an unrealistically low value. It therefore appears that, whenever the reaction takes place at the liquid-solid interface, no significant rate enhancement will be observed for the gas-liquid mass transfer process: the latter will essentially proceed in the slow-reaction regime.

In conclusion, one may say that mass transfer from the gas-liquid interface to the bulk of the liquid in slurry reactors is unlikely to exhibit any major enhancement effect unless the solubility of the solid is appreciably larger than the physical solubility of the gas. Of course, the value of a_o in a slurry reactor is determined by the kinetics of both solid dissolution and chemical reaction; discussion of these effects is beyond the scope of this lecture.

Acknowledgment

This lecture is based in large part on material due to appear in Chapters 3, 4 and 13 of a book which will be published by J. Wiley: "Gas Treating with Chemical Solvents", by G.Astarita, D.W. Savage, A.L.Bisio. I am indebted to my coauthors of this book for many of the ideas developed in this lecture. Those ideas have of course originated from a very wide spectrum of published literature; the are no bibliographical references to this article, and the reader is referred to the book quoted above for any such references.

Table 1

Dependency of $\bar{V}$ on Operating Variables
(Rate of Reaction First Order With Respect to Transferring Component)

REGIMES / OPERATING VARIABLES	SLOW		FAST	INSTANTANEOUS	
	Kinetic	Diffusional		Large Driving Force	Small Driving Force
k_L°	i[(a)]	k_L°	i	k_L°	k_L°
A	i	A	A	A	A
V	V	i	i	i	i
$a_i - a_o$	$a_i - a_o$	$a_i - a_o$	$a_i - a_o$	almost i	$a_i - a_o$
Apparent Activation Energy for k_L	Energy of Activation of Reaction	Very Small	1/2 Energy of Activation of Reaction	Very Small	Heat of Reaction
	$k_L \simeq k_L^\circ$	$k_L = k_L^\circ$	$k_L >> k_L^\circ$	$k_L >> k_L^\circ$	$k_L >> k_L^\circ$

(a) Independent.

CHEMICAL DESORPTION

G. ASTARITA

Istituto di Principi di Ingegneria Chimica, University of Naples, I-80125 Naples

In the industrial practice of gas treating, removal of one or more components of a gas stream by means of a chemically reactive liquid is almost always accompanied by a process which regenerates the rich liquid solution by stripping out the absorbed gas. The only important exception is flue gas desulfurization, where, often the product of the reaction (usually calcium sulfite and sulfate) is simply disposed of as waste;however, also for flue gas desulfurization processes where a regeneration step is included, such as the Wellman-Lord process, are in common industrial use.

Regeneration of a chemically reactive solvent is a process of chemical desorption: the chemical reaction which has taken place in the absorption step takes place in the reverse direction in the regeneration step, and the absorbed gaseous components are desorbed back to a stripping gas phase, which is usually steam. The inversion of the process of absorption is accomplished by means of a difference in pressure (lower in the regeneration step) and temperature (higher in the refeneration step), as well as by the stripping action of steam.

Chemical desorption is very important from the industrial viewpoint. The efficiency of desorption determines how lean a solution can be fed to the top of the absorption unit; this in turn determines the lowest residual partial pressure attainable in the treated gas stream. Should the latter be too high for meeting the specifications on the treated gas, a costly final purification unit will be required.

In spite of its industrial importance, chemical desorption has been studied by far less than chemical absorption. This is probably due to the fact that experiments of chemical desorption are much more difficult to perform than those of chemical absorption.

As soon as serious attention is given to the analysis of chemical desorption, it is immediately apparent that classical concepts of chemical absorption theory cannot be carried over directly to chemical desorption. The most striking example is the concept of irreversibility, which is the asymptotic behavior most commonly analyzed in the literature on chemical absorption. Applying the same ideas to chemical desorption leads immediately to paradoxes. Furthermore, the very idea of irreversibility appears to require some rethinking, for should the chemical reactions taking place in the absorber unit really be irreversible, the desorption step would be impossible to perform.

Consider a gas-liquid system where the reaction taking place in absorption is reaction 27 (reference is made here to the previuos lecture on the mathematical layout. Equations are numbered consecutively with that lecture). The equilibrium constant K, see Eq.28, is in actual practice always very large (values of the order of 10^{-3} for equimolar reactions are typical) which is not surprising since K is a measure of the ratio of the chemical capacity of the reactive liquid to the physical capacity of the non-reactive solvent: should such a ratio not be large, there would be no reason to use a chemically reactive liquid rather than simply the non-reactive solvent.

Desorption is generally performed at a higher temperature than absorption, and since reaction 27 is generally exothermic the value of K in the desorption unit will be smaller. However, it will still in general be a comparatively large number, say of the order of $10 \div 10^2$ for equimolar reactions. Since in the desorption unit the reaction takes place in the reverse direction, it takes place against a chemical thermodynamics barrier of an equilibríum constant of the order of $10^{-1} \div 10^{-2}$ for equimolar reactions. Desorption is still possible because, although a_o will in geleral be quite a bit smaller than the b_{jo}'s, the equilibrium vapor pressure p^* corresponding to a_o may well be appreciably large (again, the same symbols are used as in the preceeding lecture on the mathematical layout).

An analysis of chemical desorption has recently been published (Chem.Eng.Sci., 35, 649, (1980)), which is based on a number of simplifying assumptions: the film theory model is assumed, the diffusivities of all species are taken to be equal to each other, and in the solution of the differential equations an approximation which is second order with respect to distance from the gas-liquid interface is used; this approximation was introduced as early as 1948 by Van Krevelen and Hoftizer. However, the assumptions listed above are not at all drastic, and two crucial elements are kept in the analysis: reversibility of the chemical reactions and arbitrary chemical mechanisms and stoichiometry.The result is a methodology for developing, for any given chemical mechanism, a highly non-linear, implicit, but algebraic equation for the calculation of the rate enhancement factor as a function of temperature, bulk-liquid composition, interface gas partial pressure and physical mass transfer coefficient. The method of solution is easily generalized to the case of unequal diffusivities and corrections for differences between the film theory and the penetration theory models can be calculated.

The theoretical analysis of chemical desorption leads to the following general conclusions :

i. Chemical absorption theory can be applied to desorption up to and including the fast reaction regime, provided the chemical reaction which takes place during absorption has a forward rate which is linear in the transferring component's concentration.

ii. In the fast-to-instantaneous transition region, and in the instantaneous reaction regime, the methodology of analysis is the same for chemical absorption and desorption, but the actual results of the analysis are different because a different range of parameters in involved.

iii. The most useful measure of the driving force is not the one most commonly adopted. Traditionally, the driving force is taken as the difference between the interface and bulk-liquid concentrations of the transferring component, $a_i - a_o$. However, it is preferable for the analysis to consider the parameter which is defined as the ratio of these two concentration :

$$\Psi = a_i / a_o \tag{57}$$

The parameter Ψ is of course also expressible as the ratio of the interface partial pressure of the transferring compo nent, p_i, and the equilibrium vapor pressure p^* of the same component corresponding to the bulk-liquid composition :

$$\psi = p_i/p^* \tag{58}$$

The value of p_i can be calculated by solving the mass transfer equation in the gas phase and the value of p^* can be calculated from vapor-liquid equilibrium data. It is important to realize that, in order to calculate ψ, one does not need to know the value of Henry's law constant in the reactive medium.

iv. Of course, chemical absorption occurs in the range $1 < \psi < \infty$, and chemical desorption in the range $0 < \psi < 1$. The "irreversible" asymptote in absorption emerges in the limit $\psi >> 1$; an "irreversible" asymptote emerges also for desorption in the limit $\psi << 1$. Both conditions ($\psi >> 1$ and $\psi << 1$) may occur independently of what the value of the equilibrium constant may be provided it is finite, which of course it always is in practice.

The statements above require some clarification. In the instantaneous reaction regime, Eq.28 must hold both in the bulk of the liquid and at the interface. Furthermore, if all diffusivities are equal, Eq.45 holds true. Straightforward algebra then leads to the following equation :

$$\psi = \prod_j (1 - \nu_j \xi_i / b_{jo})^{-\nu_j} \tag{59}$$

where ξ_i is the value of ξ at the interface. Furthermore, the following equation is easily derived :

$$I_\infty = 1 + \frac{\xi_i}{a_i - a_o} \tag{60}$$

Now consider the case where $\psi \to \infty$. Since none of the b_{jo}'s can be zero, the solution of equations 59-60 is :

$$I_\infty = 1 + \frac{b_{1o}}{\nu_i a_i} \tag{61}$$

where B_1 is that component among those having a positive ν_j which has the lowest value of b_{1o}/ν_1. Eq.61 is the classical result of the so called "irreversible" theory of absorption.

Conversely, consider the case where $\psi \to 0$. Again none of the b_{jo}'s can be zero, and hence the solution of Eq.s 59-60 is :

$$I = 1 + \frac{b_{2o}}{|\nu_2| a_o} \qquad (62)$$

where B_2 is that component among those having a negative ν_j which has the lowest value of $b_{jo}/|\nu_j|$. It is important to observe that the "irreversible" limit in desorption, $\psi \to 0$, exists in spite of the fact that the equilibrium constant for the desorption reaction, 1/K, is in all practical cases a very small number.

The conclusions reached so far lead to a very interesting speculation. The conditions for "irreversibility" in absorption and in desorption, $\psi \to \infty$ and $\psi \to 0$, correspond to operating conditions for away from a pinch (region of very low driving force). Under such conditions, the mass transfer rate is large and the bulk of the gas treating process takes place over a comparatively small height of packing or number of trays. However, in industrial operation one wants to approach a pinch condition at several points, namely:

i. At the absorber lean end, so as to reduce the concentration of the impurity in the exit gas stream to as low a value as thermodynamically compatible with the composition of the liquid feed.

ii. At the absorber rich end, so that for a given amount of impurities removed, the liquid flowrate can be as small as thermodynamically compatible with the composition of the feed gas.

iii. At the desorber lean end, so as to recycle to the absorber as lean a liquid solution as is feasible.

As pinch points are approached, the mass transfer rate becomes very small and large packing heights or numbers of trays are required. In other words, most of the total packed height or number of trays is in fact used to effect mass transfer near pinches; therefore, from the viewpoint of design of the height of an industrial unit, understanding of the mass transfer rate near pinches is crucial. In terms of the parameter ψ , a near-pinch condition may be defined as :

$$\psi = 1 + \varepsilon, \qquad |\varepsilon| \ll 1 \qquad (63)$$

with $\varepsilon > 0$ corresponding to absorption and $\varepsilon < 0$ to desorption.

Practically, all the experimental mass transfer rate data available in the literature are for unpinched conditions; the reason being of course that experiments near pinches are very hard to carry out. Furthermore, a large fraction of the theoretical analysis in the literature refers to the "irreversible" limit, which is, of course, as far away from a pinch as possible. The following question arises : how much of the available information is useful for predicting mass transfer rates near a pinch, i.e. under those conditions which are the most relevant for industrial design ?

It turns out that the problem is trivial up to and including the fast reaction regime provided the reversibility of the chemical reactions is taken into account in the description of the kinetics, i.e., provided both the forward and the reverse rates are considered. The problem is not trivial in the fast-to-instantaneous transition and in the instantaneous reaction regime. These are the regimes under which most industrial treating towers operate.

The detailed analysis of this problem for the instantaneous reaction regime has been published recently (Chem.Eng.Sci., 35, 1755, (1980)). The resulting equations are, in the general case, highly non-linear because they imply the solution of one or more polynomial equations describing chemical equilibrium. However, three asymptotic cases emerge naturally and are easy to solve. The first two ones are the "irreversible" limits $\psi \to \infty$ and $\psi \to 0$. The third one is the "pinched" limit, i.e., the limiting case where Eq.63 holds true.

Although the value of I_∞ at $\psi = 1$ is the largest one of those corresponding to any other value of ψ, the value one would calculate from the "irreversible" limit equations in the pinched region is far in excess of the true value. In fact analysis of a specific industrial unit (H_2S absorption/desorption in aqueous amine) shows that, in the pinched region, the "irreversible" limit equations would lead to the conclusion that mass transfer is entirely gas-phase controlled, while in fact the pinched limit equations show that liquid-side resistance to mass transfer is still significant.

There are important qualitative differences between the behavior in the "irreversible" limit and in the pinched limit. In the "irreversible" limit, the enhancement factor is independent of temperature, and the mass transfer rate is almost independent of the driving force, a_i-a_o. Conversely, in the pinched limit the enhancement factor is a strongly decreasing function of temperature and the mass transfer rate is proportional to the driving force, a_i-a_o.

It is interesting to observe that in the pinched limit the more complex penetration theory equations can be solved explicitly. The solution one obtains coincides almost exactly with the film-theory solution, provided the square root of any diffusivity ratio is substituted in the latter for the ratio itself.

An important point to be discussed with regard to chemical desorption is the question of the required steam rate. The steam injected into the regeneration unit serves two purposes: it provides the sensible and latent heat required for the desorption operation, and it represents the diluent gas needed to keep the partial pressure of acid gas in the gas phase low enough to allow stripping to take place. Consequently, the required steam rate may be dictated either by the heat balance, or by stripping operation. The minimum steam rate needs to be calculated for both requirements, and the actual minimum is the larger one of the two.

First, consider the stripping requirement. Let Y be the mole fraction in the gas phase. In a Y- α plane, the equilibrium line is strongly curved upwards. The operating line also has a positive curvature, since steam condenses along the regeneration tower to supply sensible heating as well as the heat of desorption. However, the curvature of the operating line, particularly at low α values, is much less than that of the equilibrium line, and the minimum steam rate generally corresponds to an internal pinch point at which the equilibrium and operating lines are tangent to each other.

The pinch point is likely to occur at a very low value of Y, and therefore, the analysis can be based on the simplified forms that the equilibrium equation takes in the limit $Y \rightarrow 0$:

$$p^* = K_o(T)\, m^q F_o(y) \qquad (64)$$

Furthermore, since the pinch point occurs at a low value of Y, the temperature at the bottom of the regeneration unit, T_1, may be used in Eq.64; in other words, the problem of non-isothermal behavior is not an important one as far as the determination of the minimum steam rate is concerned.

The equation of the operating line is, if S is the molar steam rate :

$$p = \pi Y = \frac{L\, m\, \pi}{S\, \alpha} (y - y_B) \tag{65}$$

(with α = my being certainly a very good approximation near the lean end). Therefore, the condition that the equilibrium and operating lines are tangent to each other can be expressed by the following two equations :

$$y_p - y_B = \frac{S_m K_o(T_1) m^{q-1}}{L} \tag{66}$$

$$1 = \frac{S_m K_o(T_1) m^{\nu-1}}{L} F'_o(y_p) \tag{67}$$

Eqs.66 and 67 require the equilibrium and operating lines to have the same value and the same tangent at the pinch point $y = y_p$. The function $F'_o(y)$ is the first derivative of the function $F_o(y)$.

Elimination of y_p between Eqs.66 and 67 allows the calculation of the minimum steam rate S_m. This will, of course, depend on the value of y_B, the fractional saturation of the lean solution The maximum value of the latter can in turn be calculated from the requirement that a pinch occurs at the top of the absorber, i.e. by setting the left hand side of Eq.64 equal to the partial pressure in the exit gas from the absorber, and solving for y.

The equations simplify very considerably whenever $F_o(y)$ is expressible as :

$$F_o(y) = y^2 \tag{68}$$

which is very likely to be the case. When Eq.68 applies, the solution to Eqs.66-67 is

$$y_p = 2y_B \tag{69}$$

$$\frac{S_m}{L} = \frac{\pi}{4y_B K_o(T_1) m^{q-1}} \tag{70}$$

Let the allowable partial pressure in the clean gas be p_T. Then the pinch at the lean end of the absorber condition ($p^*=p_T$ in Eq.64) yields :

$$y_B = \sqrt{\frac{p_T}{K_o(T_2)m^q}} \tag{71}$$

where T_2 is the temperature at the top of the absorber. Elimination of y_B between Eqs.70 and 71 yields :

$$\frac{S_m}{L} = \frac{m^{1-q/2}}{4\sqrt{p_T K_o(T_2)}} \exp\{-\frac{Q_o}{R}(\frac{1}{T_2}-\frac{1}{T_1})\} \tag{72}$$

where use has been made of the identity :

$$\frac{K_o(T_2)}{K_o(T_1)} = \exp\{-\frac{Q_o}{R}(\frac{1}{T_2}-\frac{1}{T1})\} \tag{73}$$

with Q_o the heat of solution at $y \to 0$.

Eq.72 is very important and should be discussed in some detail. First of all, it shows that strong absorbents,characterized by a very low value of $K_o(T_2)$, require a large steam rate for stripping. Notice, however, that S_m increases only proportionally to the square root of $1/K_o(T_2)$. The equation also reflects the beneficial effect of a large temperature swing between regenerator bottom and absorber top which can be achieved by deeper solvent cooling or by the use of solvents with high boiling point.

Attention is now turned to the heat balance limitation. The minimum steam rate required for the heat balance is :

$$\frac{S_m}{L} = \frac{\rho_L L C_L(T_1-T_3)+ m \bar{Q} \Delta y}{Q_s} + (\frac{S}{L})_R \tag{74}$$

where T_3 is the temperature of the liquid leaving the rich-lean heat exchanger, $\bar{Q}$ is the average heat of absorption over the absorber, Δy is the change of fractional saturation over the absorber, Q_s is the latent heat of steam, and $(S/L)_R$ is the steam rate required at the regenerator top. Eq.74 states that the heat supplied to the reboiler needs to: (i) supply sensible heat to raise the liquid temperature from T_3 to T_1; (ii) supply the heat of desorption of acid gases; and (iii) supply a residual steam rate at the regenerator top such that the water vapor pressure over the solution entering the regenerator is matched by the partial pressure of steam in the vapor upflow.

The overhead gas product from the regeneration unit of an industrial gas treating process contains a large amount of the gaseous components removed from the gas stream to be treated, and the practical problem of its disposal needs to be considered. The most important example is the one where the components to be removed are carbon dioxide and hydrogen sulfide.

When a gas stream contains significant amounts of both CO_2 and H_2S the traditional gas treating processes using an alkaline solvent remove most of both acid gases from the gas stream. The gas exiting from the regenerator unit can be vented to the atmosphere only if the H_2S content is practically zero; in all other cases it must be sent to a Claus plant where H_2S is transformed into sulfur.

In the last few years processes which can selectively absorb H_2S while leaving a large fraction of CO_2 in the treated gas have become increasingly important. The reasons for the industrial interest in selective H_2S removal are listed below.

i) When the gas stream to be treated is a low-heat content fuel gas, H_2S needs to be removed, but removal of CO_2 would only marginally increase the heat content of the gas. Since the energy requirements of a gas treating process are roughly proportional to the amount of acid gases to be stripped out in the regeneration step, removal of CO_2 will result in a significant increase of the energy requirement.

ii) When the H_2S/CO_2 ratio is the gas stream is too low, the acid gas exiting from the regeneration step in a conventional process will have an H_2S content low enough to make operation of the Claus plant difficult if not impossible. In this case selective removal of H_2S may be utilized in two modes: either as a first step of the basic gas treating process, yielding a high-H_2S content stripped gas to be fed to the Claus plant, and followed by a CO_2-removal step; or alternately, a conventional process stripping both gases simultaneously followed by selective removal of H_2S from the acid gas stream.

Selectivity for H_2S has been demonstrated experimentally and explained theoretically since the late sixties for some special chemical solvents which are not in common industrial use. Experimental techniques are rather difficult and few sets of reliable data are available. Theoretical work has been fairly abundant but limited to rather simplified cases.

Analysis of selective H_2S removal needs to be developed along the following lines :

i. Thermodynamic modeling.

ii. Theoretical analysis for the case where both gases are absorbed in the instantaneous reaction regime.

iii. Experimental work and theoretical analysis on a specific system in industrial use.

A thermodynamic selectivity for H_2S can be defined as :

$$S_T = \frac{p^*_{CO_2} / p^*_{H_2S}}{\alpha_{CO_2} / \alpha_{H_2S}} \qquad (75)$$

where p^* are the partial pressures at equilibrium and α are the total loads in the liquid phase.

In the absence of chemical reactions, the selecticity S_T is simply the ratio of Henry's law constants for CO_2 and H_2S which is approximately independent of temperature and has for water the value 3.05. Other physical solvents exhibiting higher selectivity for H_2S do exist, but none is known with thermodynamic selectivity much in excess of 10.

In the case of aqueous chemical solvents, chemical equilibrium of the following reaction is always established :

$$HCO_3^- + H_2S \rightleftarrows H_2O + CO_2 + HS^- \qquad (76)$$

and therefore the selectivity can be expressed as :

$$S_T = \frac{H_{CO_2}}{H_{H_2S}} \; K \frac{[\,HCO_3^-\,]}{[\,HS^-\,]} \; \frac{\alpha_{H_2S}}{\alpha_{CO_2}} \qquad (77)$$

where K is the equilibrium constant of reaction 76, i.e., the ratio of the first dissociation constant of H_2S to that of H_2CO_3.

The latter is a marginally stronger acid and therefore $K < 1$; reaction 76 is mildly endothermic, and therefore K increases with temperature.

If the concentrations of physically dissolved acid gases are neglected with respect to the amounts which are chemically combined, $[HS^-]$ can be identified with α_{H_2S}. In fact, S^{--} is formed only in extremely strong alkali which cannot be regenerated, and there are no other forms of chemically combined H_2S in the alkaline solutions in use. In contrast with this, $[HCO_3^-]$ is in general less than α_{CO_2}, since chemically combined CO_2 can also exist in the form of the CO_3^{--} ion or in the case of amines, in the form of carbamate. One therefore obtains :

$$S_T \leq \frac{H_{CO_2}}{H_{H_2S}} K \tag{78}$$

with the equal sign applying only to those alkaline solvents in which all the chemically combined CO_2 is in the form of the bicarbonate ion. This leads to the selection of tertiary amines, such as methyldiethanolamine (MDEA) which is in industrial use.

Even the highest possible thermodynamic selectivity, i.e., the right hand side of Equation 78, is rather low, in fact lower than that of water. It follows that processes which are highly selective for H_2S must rely on kinetic selectivity, and therefore their design must be based on a thorough understanding of the rates of coupled mass transfer and chemical reaction. An overdesigned packed height will result in approaching the unfavorable conditions of thermodynamic selectivity.

The thermodynamic limitations discussed above turn out to be reflected also in kinetic limitations when both gases are absorbed in the instantaneous reaction regime. The analysis of simultaneous mass transfer of two volatile components which both undergo instantaneous chemical reaction leads to the interesting result that one of the two gases may actually be desorbed in spite of the fact that its partial pressure in the gas phase is larger than the equilibrium vapor pressure corresponding to the bulk-liquid composition. The physical mechanism responsible for this result is briefly discussed in the following.

When the chemical reactions are instantaneous both physical and chemical equilibrium are established across the gas-liquid interface; therefore, the liquid at the interface has the composition corresponding to equilibrium with the gas-phase composition at the interface. Now suppose that both interface partial pressure p_i and p_i', are larger than the equilibrium vapor pressures p^* and $p^{*\prime}$, corresponding to the bulk-liquid composition, say :

$$p_i > H a_o \qquad (79)$$

$$p_i' > H' a_o' \qquad (80)$$

It may however be that the equilibrium values of a and a' corresponding to p_i and p_i' are not both larger than a_o and a_o'. The equilibrium equations for the two reactions are sufficiently coupled to make this possible. When conditions are such that at the interface say a' is less than a_o', a driving force for desorption of the primed component may develop and that component may desorb in spite of the fact that Equation 80 holds true.

Several years ago, Kohl anf Blohm presented gas-phase concentration profiles measured in an industrial absorber where H_2S and CO_2 were absorbed simultaneously in a monoethanolamine/ethylene glycol/water solution. These data clearly show that in the lower part (rich end) of the absorber H_2S is actually desorbed from the liquid phase (the concentration of H_2S in the gas phase initially increases as the gas moves up the lower part of the absorber). Analysis of these data shows that in the bottom part of the absorber, Equations 79 and 80 hold true, but the equilibrium conditions are such that the "reverse mass transfer" phenomenon may take place; furthermore, in the bottom part of the absorber conditions are indeed such that the reactions may be regarded as instantaneous.

However, the same absorber exhibits at the lean end an actual selectivity for H_2S which by far exceeds the thermodynamic selectivity; in fact by several orders of magnitude. The reason for this is that at the absorber's lean end conditions, the reactions which H_2S undergoes may be regarded as instantaneous, but those which CO_2 undergoes are not. By itself, CO_2 could be absorbed in the fast reaction regime. The selectivity observed at the lean end is a kinetic one and therefore it cannot be predicted by what is essentially an equilibrium theory, i.e. the theory of mass transfer accompanied by instantaneous reactions.

The reaction which H_2S undergoes in any aqueous alkaline solvent is simply a proton-transfer reaction with the base in solution which can be written as :

$$H_2S + B \rightleftarrows BH^+ + HS^- \tag{81}$$

Reaction 81 can always be regarded as instantaneous. In contrast with this, CO_2 undergoes reactions which only under special conditions may be regarded as instantaneous. If the solvent considered is a thermodynamically-selective one (i.e., one where S_T equals the right hand side of Equation 78), the chemically combined form of CO_2 is the bicarbonate ion and therefore the reaction for CO_2 can be written as :

$$CO_2 + H_2O + B \rightleftarrows BH^+ + HCO_3^- \tag{82}$$

Reaction 82 is in general slow enough as to result in the mass transfer of CO_2 taking place at most in the fast reaction regime (the kinetics of Reaction 82 will be discussed in some more detail below). This means that the occurrence of Reaction 82 does not appreciably influence the concentration distribution of the non-volatile components B and BH^+ near the interface. Consequently, that distribution in governed only by the occurrence of Reaction 81.

The argument above shows that the rate of H_2S absorption can be calculated from instantaneous reaction theory as if CO_2 were not simultaneously absorbed; the converse is certainly not true since the concentration distributions governed by H_2S absorption will in fact strongly influence the CO_2 mass transfer rate: the chemical driving force for Reaction 82 is influenced by the occurrence of Reaction 81. Data on simultaneous absorption of H_2S and CO_2 into aqueous MDEA solutions which support this conclusion quantitatively The measured H_2S mass transfer rate, which changes sign during the experiment, agrees within experimental accuracy with the value calculated according to the procedure outlined above.

In order to predict the actual kinetic selectivity for H_2S, however, one would need to calculate both the H_2S and the CO_2 mass transfer rate; the latter can only be calculated if the kinetics of Reaction 82 are well understood. A possible mechanism for Reaction 82 is as follows :

$$H_2O + B \rightleftarrows BH^+ + OH^- \tag{83}$$

$$CO_2 + OH^- \rightleftarrows HCO_3^- \tag{84}$$

with Reaction 83 being an instantaneous proton-transfer one and Reaction 84 being the slow step. The kinetics of Reaction 83 have been established firmly.

If the mechanism 83-84 is the actual one, the rate of Reaction 82 in proportional to the OH^- concentration. The latter, in turn, is simply determined at any given liquid composition by the alkalinity of B, say by the pK of the base. The equilibrium of Reaction 81 is also entirely determined by the pK of the base. It follows from this argument that two thermodynamically selective alkalis having the same pK should exhibit the same kinetic selectivity.

However, there is evidence that Reaction 82 proceeds in aqueous solutions of tertiary amines, faster than can be accounted for on the basis of the mechanism 83-84. The kinetics of Reaction 82 in aqueous solutions of tertiary amines are not yet well understood and its understanding is preliminary to any attempt to predicting kinetic selectivity.

REVIEW OF OBTAINING AND ESTIMATION METHODS OF PHYSICO-CHEMICAL AND RELATED DATA : PART 1 - SOLUBILITIES AND DIFFUSIVITIES OF GASES IN LIQUIDS

B.I. MORSI and J.C. CHARPENTIER

Laboratoire des Sciences du Génie Chimique - Centre National de la Recherche Scientifique - ENSIC, 1, rue Grandville - 54042 NANCY Cedex - France.

Gas-liquid reactions and absorptions are widely used in five main fields of the chemical, petrochemical, agro-food, pharmaceutical and energetical industries that are

- the liquid phase processes : oxidation, hydrogenation, sulfonation, nitration, halogenation, hydro-halogenation, alkylation, sulfation, polycondensation, polymerization...
- the gas scrubbing : CO_2, H_2S, CO, SO_2, NO, NO_2, N_xO_y, HF, SiF_4, Cl_2, P_2O_5, Phosgene, Hydrocarbons... (very often to fight against the air pollution)
- the manufacturing of pure products : H_2SO_4, HNO_3, $BaCO_3$, $BaCl_2$, adipic acid, nitrates, phosphates...
- biology : aerobic fermentation, oxidation of sludges, manufacturing of proteins from hydrocarbons, biological waste treatment...
- multi- and two-phase flow in petroleum and nuclear engineering.

The heart of these processes is an absorber or a reactor, of particular configuration best suited to the chemical absorption or reaction being carried out. Indeed gas-liquid contactors present different shapes such as tubular spray columns, wetted falling film columns, packed columns, bubble columns, plate columns, mechanically agitated reactors, jet and venturi reactors, gas lift loop reactors... These different configurations are mainly due to the competition between the various phenomena involved with the *chemical thermodynamics and the physical and chemical kinetics* that intervene simultaneously and that determine the *choice of the equipment* which will have to work in the *optimal energetic and hydrodynamic* conditions.

Solubilities and diffusivities of gas are practically always required for design of gas-liquid process and obtaining solubility and diffusivity data for the gas-liquid system under consideration may be a chalenging problem so wide is the range of solutes and solvent the chemical engineer or researcher may encounter. Moreover the choice of a suitable gas-liquid contactor is also a question of matching these data, those concerning the reaction kinetics and the physical kinetics characteristics of the proposed reactor, i.e., specific gas-liquid interfacial area, heat and mass transfer coefficients and gas or liquid holdup. Some considerations on solubility and diffusivity will be proposed in part 1 of this review and on gas-liquid mass transfer in part 2.

1 CHEMICAL ENGINEERING APPLICATION OF SOLUBILITY DATA

1.1 Introduction

The solubility of gases in liquids is both practically important in the most diverse fields of technology and science and theoretically interesting.

Knowledge of the solubility of gases in liquids is of *practical* importance in the reliable process design, in the scale-up purposes, in the determination of mass transfer effects, in the verification of mass transport limitations and conditions, in various industrial processes, particularly petroleum industries, geological investigations, waste water treatment and aerobic fermentation, in the study of artificial atmospheres for divers and astronauts, in the interaction of gases with our environment (as in the biological oxygen demand in natural waters), in the study of oxygen transport by inert organic liquids, in processes for saline water demineralization, in the study of various biological fluids and tissues. Both dilute solutions resulting from the small solubility of many gases in liquids and the available large variety of sizes, shapes and polarities of gas molecules to act as "probes" have made the solubilities of gases in liquids an excellent *theoretical* tool to investigate liquid and solution structure and properties. (Wilhelm and Battino, 1973 [1], Chappelow and Prausnitz, 1974 [15], Cysewski and Prausnitz, 1976 [11], Osburn, 1970 [2], Beutier et al., 1978 [18], Cukor and Prausnitz, 1971 [27], Lohse and Deckwer, 1981 [30], Charpentier, 1981 [32], De Ligny et al., 1976 [3], Clever and Battino, 1975 [33], Rivas and Prausnitz, 1971 [28], De Ligny and Van der Veen, 1972 [21], Schumpe et al., 1978 [5]).

A satisfactory theory for solutions of gases in liquids is much harder to design than for solution of liquids or solids in liquids, the letter components differing from one another far less in significant properties such as molar volumes and intermolecular attractive forces.

1.2 Basic Thermodynamic Considerations

Thermodynamic equilibrium between the liquid and gas are remarkably reviewed in many papers [13, 20, 23, 24, 25, 26, 31, 33]. Consider a gas phase (subscript g) and a homogeneous liquid phase (subscript ℓ) in thermodynamic equilibrium. For any component i, the chemical potential in both phases must be equal :

$$\mu_i^g = \mu_i^\ell \tag{1}$$

Connection with the observable variables total pressure P_t, mole fraction composition y_i of gas and x_i of liquid, respectively, is usually established with the aid of the two auxiliary functions : the gas-phase fugacity coefficient ϕ_i and the liquid phase activity coefficient γ_i

$$\mu_i^g(P_t,T,y_i) = \mu_i^{*g}(T) + RT\ \mathrm{Lin}\ P_t + RT\ \mathrm{Lin}\ y_i + RT\ \mathrm{Lin}\ \phi_i$$

$$= \mu_i^{*g}(T) + RT\ \mathrm{Lin}\ f_i \tag{2}$$

$$\mu_i^\ell(P_t,T,x_i) = \mu_i^{o\ell}(P_t,T) + RT\ \mathrm{Lin}\ x_i + RT\ \mathrm{Lin}\ \gamma_i$$

$$= \mu_i^{o\ell}(P_t,T) + RT\ \mathrm{Lin}\ a_i \tag{3}$$

where f_i is the gas phase fugacity = $(P_t y_i \phi_i)$

a_i is the liquid phase activity = $(x_i \gamma_i)$

$\mu_i^{*g}(T)$ is the chemical potential (depends only on the temperature)

$\mu_i^{o\ell}(P_t,T)$ is the chemical potential (depends on both temperature and pressure).

For the *gas phase*, this is generally the ideal gas state at the same temperature and pressure, P_t = 1 atm. For the *liquid phase* of solutions of gases (where the pure component "gas" is often supercritical), most frequently the so-called unsymmetric convention for normalization of activity coefficient is adopted. Thus for a binary system :

for solvent $(i = 1)$: $\gamma_1 \to 1$ as $x_1 \to 1$

for solute $(i = 2)$: $\gamma_2 \to 1$ as $x_2 \to 0$

The standard state potential of the subcritical component (solvent) is the potential of the pure liquid at system temperature and pressure. On the other hand the activity coefficient of the solute is taken as approaching unity at infinite dilution. Hence $\mu_2^{o\ell}$ can be interpreted as the chemical potential of pure solute in a hypothetical liquid state corresponding to extrapolation from infinite dilution (which serves as reference state) to $x_2 = 1$ along a line where $\gamma_2 = 1$, that is, along the Henry's law line. In physical terms, it might be regarded as a hypothetical state in which the mole fraction of solute is unity (pure solute), but some thermodynamic properties are those of the solute 2 in the reference state of infinite dilution in solvent 1 (e.g., partial molar heat capacity). Since from the context it should always be clear whether the superscript circle denotes "standard state" or "pure substance", no further distinction is introduced.

Substituting from equations (2) and (3) into equation (1), one gets after rearrangement

$$\frac{\phi_2 y_2 P_t}{\gamma_2 x_2} = \frac{f_2}{\gamma_2 x_2} = H_{2,1}(P_t,T) \tag{4}$$

with
$$H_{2,1}(P_t,T) = e^{\left[\frac{\Delta\mu_2^{\circ}(P_t,T)}{RT}\right]} \tag{5}$$

where $\Delta\mu_2^{\circ}(P_t,T) = \mu_2^{o\ell} - \mu_2^{*g}$ is the standard change of the partial molar Gibbs energy upon solution.

$H_{2,1}(P_t,T)$ is called the Henry's law constant of substance 2 in solvent 1.

Its value depends strongly on the nature of the solvent. Clearly, this well-defined an experimentally accessible quantity, may be evaluated (at saturation pressure of the solvent, p_{1S}) by extrapolating to $x_2 = 0$ a plot of (f_2/x_2) vs x_2 :

$$\lim_{x_2 \to 0} \left(\frac{f_2}{x_2}\right) = H_{2,1} \tag{6}$$

In general, the effect of pressure on Henry's law constant, as on other properties of condensed phases, is rather small. From its definition it follows that $(\partial \text{Lin} H_{2,1}/\partial P)_T = \overline{V}_2^{o\ell}$, and hence

$$\mathrm{Lin}\ H_{2,1}(P_t,T) = \mathrm{Lin}\ H_{2,1}(p_{1S},T) + \int_{p_{1S}}^{P} \frac{\overline{V}_2^{o\ell}}{RT}\, dp$$

$$\simeq \mathrm{Lin}\ H_{2,1}(p_{1S},T) + \frac{\overline{V}_2^{o\ell}(P-p_{1S})}{RT} \qquad (7)$$

where $\overline{V}_2^{o\ell}$ being the partial molar volume of component 2 at infinite dilution which may be assumed, as a first approximation to be independent of pressure.

1.3 Some Useful Approximations

In *low pressure*, it is frequently possible to adopt various approximations in evaluating the functions of equations (5) and (6) without seriously reducing numerical accuracy :

1.3.a Liquid phase. Empirically it is well established that, for a sparingly soluble gas the solubility is proportional to its vapour phase fugacity (partial pressure)

$$f_2 = K\ x_2 \qquad (8)$$

provided the gas pressure is not too large. By comparison with equation (4), the significance of the proportionality constant K is immediately revealed, as

$$K = \gamma_2\ H_{2,1} \qquad (9)$$

At a given temperature and pressure, $H_{2,1}$ is independent of composition.

Thus the constancy of K requires constancy of γ_2, which is in fact, the essential feature of Henry's law. Since the activity coefficient has been normalized to 1 for $x_2 \to 0$, equation (8) is tantamount to stating that for the particular system, the plot of fugacity vs x_2 may be replaced by its tangent at infinite dilution [see equation (6)].

1.3.b Gas phase.

b.1 Estimation of fugacity coefficient ϕ_2. For many purposes one may rely upon a frequently used approximation commonly known as the Lewis fugacity rule, which should be useful up to pressure of order of 5-10 % of the critical. The rule assumes that at constant T and P_t the fugacity is independent of composition, i.e.,

$$\phi_2 = \phi_2^o \text{ at same T and } P_t \qquad (10)$$

where ϕ_2^o denotes the fugacity coefficient of pure gaseous solute.

Generally, it appears that for many gases, at temperatures below the normal boiling temperature of water and pressures of the order of a few atmospheres, the fugacity correction is rather small and often negligible.

b.2 Estimating gas-phase mole fraction y_2. With the usual simplifying assumptions, the mole fraction of solvent in the gaseous phase at low pressures may be calculated from the vapour pressure of the liquid via Raoult's low, i.e.,

$$y_2 = 1-y_1 \simeq \left[P_t-(1-x_2)p_{1S}/P_t\right] \tag{11}$$

For the solvent, water, x_2 is in general negligibly small and hence,

$$y_2 \simeq (P_t-p_{1S})/P_t \tag{12}$$

Assuming an ideal gas-phase and replacing the fugacity by the partial pressure p_2 of solute may often yield satisfactory results, provided p_2 is small, the solubility of gas in the liquid sufficiently low, and the temperature well below the critical temperature of the solvent. In general, $p_2 \leq 1$ atm and $x_2 \leq 10^{-3}$. Under these circumstances, the solution may be regarded as being effectively infinitely dilute. These approximations lead to the most familiar and simplest form of Henry's law,

$$\frac{p_2}{x_2} = H_{2,1} \tag{13}$$

1.4 Methods of Expressing Gas Solubility

The solubilities of gases in liquids have been expressed in terms of many different units, depending on the particular application (1, 2, 31, 32, 33). The principal ways are :

1.4.1 Bunsen's coefficient, α. Is defined as the volume of gas dissolved, reduced to standard conditions (0 °C and 1 atm) that is absorbed by a unit of solvent at the temperature of the measurement under a gas partial pressure of 1 atm, that is,

$$\alpha = \frac{V_2}{V_1}\left(\frac{273.15}{T}\right) \tag{14}$$

where V_1 is the volume of gas absorbed at (273.15 K and P = 1 atm

V_2 is the volume of gas absorbed at (T K and the total pressure of the measurement)
T is the temperature of the measurement in K.

In the literature, this coefficient is sometimes called the absorption coefficient or the coefficient of absorption.

1.4.2 Kuenen's coefficient, S. Is defined as the volume of gas (in cm^3) at a partial pressure of (1 atm) reduced to 273.15 K and 1 atm, dissolved by a quantity of solution containing 1 gm of solvent.

1.4.3 The Ostwald's coefficient, L. Is defined as the actual uncorrected volume of gas dissolved to the volume of the absorbing solvent, both measured at the same temperature, that is

$$L = V_2/V_1 \tag{15}$$

It is helpful to specify the temperature and total pressure of the measurements when reporting this coefficient.

1.4.4 The absorption coefficient, β. Is defined as the volume of gas absorbed (reduced to 273.15 K and 1 atm via the ideal-gas equation of state) per unit volume of liquid if the total pressure is always kept at 1 atm. Since β and α are very similar.

1.4.5 The mole fraction, x_2. Is defined as the ratio of the number of moles of gas dissolved to the total number of moles of both the dissolved gas and the absorbing solvent :

$$x_2 = \frac{n_2}{n_1+n_2} \tag{16}$$

where n_1 is the number of moles of the solvent,
n_2 is the number of moles of the dissolved gas.

Both the partial pressure of the gas and the temperature of measurement must be specified to fix the mole fraction x_2.

At any partial pressure of gas, p_g, the mole fraction solubility may be calculated from Ostwald's coefficient L as,

$$x_2 = \frac{1}{\left(1 + \frac{RT}{L p_g V_1^o}\right)} \tag{17}$$

where R is the gas constant

V_1^o is the molar volume of the solvent at T, ($V_1^o = V_1/n_1 = M_1/\rho_1$)

M_1 is the molecular weight of the solvent

ρ_1 is the density of the solvent at T, K.

1.4.6 The Henry's law constant, K_H, K_2 and K_c. When a gas is in equilibrium with a solution under the gas partial pressure p_g, one can drive an expression for Henry's law constants, which are always concentration dependent. Henry's law actually is strictly applicable only in the extrapolation to infinite dilution, that is

$$\lim_{x_\ell \to 0} \frac{p_g}{x_\ell} = \lim_{x_\ell \to 0} \frac{f_g}{x_\ell} = K_H \tag{18}$$

where x_ℓ is the mole fraction of the gas in liquid phase
f_g is the gas phase fugacity

At about atmospheric pressure, for most gases, equating f_g with p_g involves negligible errors in gas solubility estimations. For dilute solutions, up to mole fraction solubility limit of x_2 (p_g = 1 atm) < 0.01, Henry's law is applicable, and one can write

$$K_H = \frac{p_g}{x_\ell} = \frac{p_g}{x_2} \tag{19}$$

when p_g = 1 atm, equation (19) becomes :

$$K_H = \frac{1}{x_2} \tag{20}$$

Two other ways to represent Henry's law constants, *first* :

$$K_2 = \frac{p_g}{c_\ell} \tag{21}$$

where c_ℓ is the concentration of the gas in the liquid phase ; and *second* :

$$K_c = \frac{c_g}{c_\ell} \tag{22}$$

where c_g is the concentration of the gas in the gaseous phase.

It is to emphasize that, the method of estimating Henry's ̣aw constants, and the gas partial pressure or concentration units ̣ust be specified. Corrections should be made for the possible ̣on-ideality of the gas phase or the non-applicability of Henry's ̣aw, particularly, for cases of high pressure and elevated tempe-̣ature measurements.

.4.7 The weight solubility, c_w. Is defined as the number of mo-̣es of gas with partial pressure of 1 atm per gram of solvent, as

$$c_w = \frac{n_g(p_g=1)}{w_1} \tag{23}$$

̣here w_1 is the weight in grams of the solvent ($w_1 = \rho_1 V_1$).

.4.8 Other gas solubility units. Depending on the various appli-̣ations, units such as volume fraction, concentration, molarity, ̣olality and weight fraction have been used to express gas solubi-̣ity in liquids. For example,

$$c_A^* = \frac{p}{He} \tag{24}$$

̣here p is the partial pressure of the solute gas in atmospheres (atm)

He is the Henry's law constant (atm.lit/mol)

c_A^* is the solubility of the solute gas at a given temperatu-re (mol/lit).

̣quation (24) is used by [42, 43, 44].

.5 Factors Affecting the Precision of Gas Solubility Measurements

The factors affecting the precision and accuracy of gas solu-̣ility in liquids are numerous. A precision of ± 1 % in gas solu-̣ility measurements appears to be adequate for most practical and ̣heoretical applications. These factors are :

.5.1 Purity of both solvent and gas. The purity of the solvent ̣s of relatively minor importance, 99 mol % purity being more ̣han adequate, since impurities tend to be of the same molecular ̣ature (size, shape, polarity, ...) as the solvent. This minimize ̣he effect of differences in solubility between the solvent and ̣he impurities. The same criteria essentially hold for gas purity, ̣9 mol % being generally adequate.

.5.2 Pressure measurement and control. The measurement of pres-̣ure is one of the most accurate parts of the gas solubility mea-̣urements, since the pressure can be readily determined via manometers and barometers with good precision to ± 0.01 atm. Pressure control is critical in some procedures and its importance depends actually on the apparatus and technique used.

1.5.3 Temperature measurement and control. To understand the effect of temperature on gas solubility measurements, there are three factors to consider :

a. the temperature effect on the solvent vapour pressure,
b. the change in the equilibrium partial pressure of the dissolved gas with temperature at an approximately constant concentration,
c. the temperature-pressure level of the experiment.

These factors obviously depend on the particular system under investigation and on the type of apparatus used. The measurements of the temperature are readily done with a good precision. The temperature control to ± 0.1 °C (fairly easy to achieve), provides a more than adequate margin of error.

1.5.4 Attainment of equilibrium (i.e., saturation) and incomplete degassing of the solvent. The wide divergence of reported values of gas solubility for some systems are very probably due to failure to attain equilibrium. The attainment of equilibrium is of prime importance. It is well to remember that in any approach to saturation, the equilibrium condition to be reached asymptotically with time. In flow systems, the attainment of equilibrium can be checked by determining the solubility over a range of flowrates. For non-flow systems, the vigor and duration of stirring or shaking can be varied, and adjustment of the pressure above and below the saturation pressure may be tried. The testing of an apparatus should include sufficient varying of the operating parameters to test for the attainment of equilibrium. Moreover, adequate degassing of the solvent is necessary for almost all gas solubility measurements. A solvent has been degassed when sufficient gas has been removed such that the outgassing of any residual gas will have no effect on the measurement. Several criteria have been used for checking on the completeness of degassing. Among other procedures of degassing, the "reasoning-by-analogy", where the reproducibility of the literature data is the sole control for adequate degassing ; the boiling away of 10 to 20 % of the solvent under vaccum ; the technique of spraying a solvent through a fine nozzle into an evacuated chamber ; and the ultrasonic shaking bath.

1.6 Experimental Methods

1.6.a Physical methods. These methods may be divided into two broad classifications.

1 *Saturation methods* : wherein a previously degassed solvent is saturated with a gas under conditions in which the necessary pressures, volumes and temperatures may be determined,

2 *Extraction methods* : wherein the dissolved gas in a previously saturated solution is removed under conditions in which the pressure, volume and temperature may be determined.

Equilibrium saturation conditions have been attained for the gas and liquid phases by shaking a mixture of the two ; and it can be verified by the way of test methods explained in (1.5.4).

Determination of the amount of dissolved gas has been carried out by various physical and chemical methods in different apparatuses :

a.1 *Low-pressure gas solubility apparatus* : the working pressure is ≃ 1 atm. Determination of the gas solubility in liquids at low pressures is often carried out by *volumetric measurements* of both the gas and solvent volumes under a reference pressure and constant temperature. The measurement of the volumes may be fulfilled by Burets system, Markham and Kobe apparatus (8, 46), Truebore tubing, modified Markham and Kobe apparatus (34), Burets and microgasometer, Douglas microgasometric solubility apparatus (35), Microburets, Morrison and Billett apparatus (33), Burets and pressure readings, Dymond and Hildebrand (36), Syringes and gas chromatographic analysis, by using a calibration gas (42, 43, 44).

a.2 *High-pressure gas solubility apparatus* : determination of the solubility of gases in liquid at high pressure has become of increasing importance.

The problems of adequate mixing of the gas and liquid phases to ensure saturation, pressure and temperature control and sampling and measurement of the gas dissolved at high pressure present greater difficulties than in apparatuses that operate at atmospheric pressure. These problems were solved in Smith and Gardiner apparatus (37), by a modern autoclave design and connections of stainless-steel, a magnetically driven bladed turbine stirrer, a modern temperature and pressure measurements and control, and a meniscus volume correction for the liquid in the buret measurement. In this apparatus, a volumetric method was applied to measure the gas and liquid volumes in buret system at atmospheric pressure.

a.3 *Pressure drop method* : it is an apparatus in which the gas absorbed by a given volume of solvent is measured by determining the pressure change in the gas reservoir of known volume. Apparatuses have been designed for use in all pressure ranges. This method has been used to determine the solubility of hydrocarbon in water between 0.5 and 1.5 atm (38), of hydrogen, ethylene, ethane and propane in toluene at pressure up to 12 atm (39) and of methane in n-decane at pressure up to 68 atm (40).

a.4 *Gas chromatography* : it is applied to the determination of gas solubility in liquids in two ways :

I - It is used to determine the amount of gas dissolved in a liquid by passing the carrier gas successively through a known volume of gas-saturated liquid to purge the liquid of the dissolved gas, through a preabsorbing column or columns to dry and sometimes concentrate the gas, and finally through the chromatograph to measure quantitatively the amount of solute gas removed from the gas-saturated solvent. This approach was used to determine the solubility of 65 gaseous and liquid hydrocarbons in water at room temperature (41), solubility of N_2O and CO_2 in cyclohexylamine and toluene (43, 44), carbon dioxide in monoethanolamine (42, 44), carbon dioxide in diethanolamine and ethyleneglycol (44) and carbon dioxide in ethanol, ethyleneglycol and mixtures of the two (42).

II - The theory of gas-liquid chromatography has been applied to the behaviour of low concentration of various vapours as the solute gas, in the column-supported liquid as the solvent, to obtain the vapour activity coefficient in the solvent at infinite dilution by elution chromatography and at higher concentrations by frontal analysis (23, 33, 45).

a.5 *Mass spectrometry* : it is applied for gas solubility determinations in solvent. The procedure is to outgas a gas-saturated solvent sample, trap the gas, and then analyze the gas by mass spectrometry. It has been used to determine the solubility of methane, oxygen and nitrogen in water, and various gases in blood (33).

1.6.b Chemical methods. Standard chemical analytical methods can often be used to determine the concentration of a solute gas that has either acid, base, or redox properties. Acidic gases include the hydrogen halides and carbon dioxide, basic gases include ammonia and methylamine. Iodimetric methods have been used with sulfur dioxide, hydrogen sulfide, ozone and chlorine. Both chlorine and hydrogen cyanide have been precipitated as silver salts. Phosgene has been determined by absorption in silver nitrate solution followed by back-titration of the acid librated (46). For oxygen, the procedure requires a good technique to control the pH and iodide-ion concentration, the dissolved oxygen oxidize freshly precipitated manganous hydroxide to manganic hydroxide at high pH

$$Mn^{2+} + 2\ OH^- = Mn\ (OH)_2$$

$$2\ Mn\ (OH)_2 + \frac{1}{2}\ O_2 + H_2O = 2\ Mn\ (OH)_3$$

The solution is made acidic, under which condition, the manganic ion reacts with excess iodide ion. The resulting iodine (I_3^-) is titrated with thiosulfate :

$$2\ Mn(OH)_3 + 6H^+ + 3I^- = 2Mn^{2+} + I_3^- + 6H_2O$$

$$I_3^- + 2S_2O_3^{2-} = 3I^- + S_4O_6^{2-}$$

This procedure is called Winkler method and it is often used to determine dissolved oxygen in natural waters (33).

It is to note that there are many versions of commercially and miscellaneous apparatuses available to measure the gas solubility in liquids.

.7 Solubilities of gases in electrolyte solutions

Consider a non polar solute 2 (such as oxygen) dissolved in a polar liquid solvent 1 (e.g., water). The chemical potential of the solute, according to equation (2) is

$$\mu_i^2(P_t,T,y_i) = \mu_i^{*2}(T) + RT\ Lin\ f_1$$

If a salt is now added to the solution, the fugacity f_i will be changed. This change may be an increase (*salting-out*) or a decrease (*salting-in*), and is often a large effect. If the liquid phase is in contact with another phase (gas, liquid or solid) there will be a transfer of component i between phases until the chemical potential is again equal in all phases (4, 6).

Such salt effects are of practical importance in stagemise separation processes and in pollution abatement. Certain salts increase the solubility by more than an order of magnitude (saling-in), and also change the solvent selectivity for various solutes ; others decrease the solubility (salting-out) (4, 6, 7, 10) artial molal properties of the dissolved gas are also profoundly ffected by the addition of salt. Thermodynamic properties of gas-electrolyte solution are also an important consideration in he design and operation of fuel cells, where mass transfer of reactants to reaction sites controls the power output ; thus the solubility of oxygen in the electrolyte contained in a typical fuel cell used for space applications is about 100 times less than the solubility in pure water. Tiepel and Gubbins (4) applied a method based on perturbation theory for mixtures to predict the thermodynamic properties of gases dissolved in electrolyte solutions. The theory was compared with their experimental data for the dependence of the solute activity coefficient on concentration, temperature and pressure. The theory was also compared with previous theories for salt effects and found to be superior. The calculations were best for salting-out systems. The qualitative feature of salting-in was predicted by the theory, but quantitative predictions were not satisfactory for such systems, this was attributed to approximations made in evaluating the perturbation terms. The theory pointed out that, salting-out occured when the molecules and ions were not very large, provided that chemical association between ions and solute did not occur. Salting-in occured when the ions were large or when association forces occured ; under such conditions the perturbation terms were large and the theory gave poorer results.

Schumpe et al., (5) presented a model to calculate more reliable oxygen solubilities in single and mixed electrolyte solutions. The model was established from literature data and their own experiments with NH_4Cl, $CaCl_2$, K_2SO_4, $KHSO_4$, $MnSO_4$, Na_2HPO_4 and several mixtures of these salts in temperature range of 10 to 40 °C. The salting-out effect of most electrolytes was described by a relation originally proposed by Sechenov, 1892 (49), this relation is given by

$$\text{Log}\left(\frac{c_o}{c}\right) = K\, c_{e\ell} \tag{25}$$

where K is the Sechenov's constant (lit/mol)
c_o is the oxygen solubility in water (mol/lit)
c is the concentration of dissolved oxygen in salt solution (mol/lit)
$c_{e\ell}$ is the concentration of the salt in water (mol/lit).

It is obvious from this previous equation that the solubility values of some salts could be represented by a straight line relationship. According to the model of Van Krevelen and Hoftijzer (50), the ionic strength I is introduced as a better measure of electrolyte activity and the salting-out constant is considered to be the result of contributions from the various gas species (h_G), and ions (h_+, h_-) present :

$$\text{Log}\,(c_o/c) = hI \tag{26}$$

with $h = h_G + h_+ + h_-$ (27)

Values of h_G, h_+ and h_- are given by Danckwerts, 1970 (51) and recently by Charpentier, 1981 (32) for various gases and ions.

For the case of mixed electrolyte solutions, Danckwerts estimated the solubility by an expression of the form :

$$\mathrm{Log}(c_o/c) = \sum_{i=1}^{n} h_i I_i \qquad (28)$$

where I_i is the ionic strength attributable to electrolyte species i, and h_i is a constant derived from equation (27) and is characteristic for that electrolyte. Thus the partial ionic strengths of each salt are multiplied with the corresponding h values ($= h_G+h_++h_-$) and summed.

This previous equation is, however, difficult to understand because the contributions of each ionic species have to be multiplied by the ionic strength of the salt. This results in different solubility predictions for mixed electrolyte solutions.

For dilute solutions one should at least expect that the salting-out effect of each ion should be independent from the other ions present in the solution. Hence solubilities of mixed electrolyte solutions will be better represented by the following modification of the Van Krevelen-Hoftijzer's model :

$$\mathrm{Log}\ (c_o/c) = \sum_{i=1}^{n} h_i I_i + h_G \sum I_i \qquad (29)$$

where h_i is the specific constant of ionic species i and I_i is the contribution of the ionic strength of that ion

$$I_i = \frac{1}{2} c_i z_i^2 \qquad (30)$$

where c_i is the concentration of ions i
z_i is the valency of i.

Schumpe et al. (5) considered the effect of one gas (O_2) on the constants of the ionic species and simplified equation (29) to

$$\mathrm{Log}(c_o/c) = \sum_{i=1}^{n} H_i I_i \qquad (31)$$

where H_i values are now the ion specific constants for salting-out oxygen.

For comparison with Sechenov constants determined experimentally for solutions of one salt only, equation (29) is written

$$\mathrm{Log}(c_o/c) = c_{e\ell} \frac{1}{2} \sum_{i=1}^{n} H_i N_i z_i^2 \tag{32}$$

where N_i is the number of ions of type i in the electrolyte.

The values of H_i are developed from multiple regression analysis. It is required that the function

$$\delta = \sum_{i=1}^{m} \left(K_{i\,expt} - \frac{1}{2} \sum_{j=1}^{n} H_i N_{ij} \; z_i^2 \right)^2 \tag{33}$$

becomes a minimum.

m presents the entire number of variable measured data (K_{expt}. values)

n presents the number of ionic species considered in the evaluation.

By differentiating δ with respect to the unknown H_i values and setting the derivations to zero, the set of simultaneous normal equations results from which the H_i are obtained by Gaussian elimination.

According to the results obtained by Schumpe et al. (5) the anions always have positive H_i values, while the H_i values for cations are negative. These authors concluded that equation (31) together with their ionic specific constants for salting-out oxygen at 25 °C (data are tabulated in their article), could be applied to estimate oxygen solubilities in mixed electrolyte solutions in the temperature range of 10 to 40 °C. Moreover, Bidner and Santiago (9) studied the solubility of non-electrolyte liquids in aqueous solutions of electrolytes.

1.8 Some Literature Works on the Solubility of Gases in Aqueous, Non-polar and Polar Solvents

1.8.a Aqueous solvents. Solubility of oxygen in aqueous sucrose solution was studied by Hikita and Azuma (8) at temperature range of 15 to 45 °C and at atmospheric pressure. Solubilities of nitrous oxide and ethylene were measured in aqueous solutions of diethanolamine, triethanolamine and ethylenediamine at 25 °C and 1 atm by Sada et al. (10). Their obtained data for the solubility of non-reacting gas (N_2O) in aqueous amine solutions were useful to evaluate the solubility of reacting gas (CO_2) into the amine solution under consideration by the following relationship

$$\text{Log}\left(\frac{\alpha}{\alpha_w}\right)_{CO_2} = \text{Log}\left(\frac{\alpha}{\alpha_w}\right)_{N_2O} \tag{34}$$

where α, α_w : Bunsen absorption coefficient defined by equation (14) for the solute in solvent, in water.

That, if the reacting gas has almost the same interaction parameters as the non-reacting one.

Gotoh (19) determined theoretically the solubilities of non-reacting gases in liquids from the free theory.

Alvarez (44) recently, measured the solubility of CO_2 in aqueous monoethanolamine and diethanolamine at 20 °C and at pressure range of 1 to 2.5 atm.

1.8.b Non-polar solvents. Mole fraction solubilities of ethylene at atmospheric pressure and temperature ranging from -9 to 70 °C were reported in heptane, dodecane, carbon tetrachloride, carbon sulfide and chlorobenzene by Sahgal and Hayduk (12). They observed a relation between ethylene solubilities in non-polar solvents and those of methane, ethane and propane, along with the corresponding energy of vaporization at the normal boiling point of those gases.

Cysewski and Prausnitz (11) proposed a semi-empirical correlation for estimating gas solubilities over a wide range of temperature. Their correlation provided reasonable estimates of solubility for a variety of gases (CH_4, C_2H_6, C_3H_8, H_2S, N_2O, SO_2) in typical non-polar solvents (carbon tetrachloride, n-propane, n-butane, n-pentane).

1.8.c Polar solvents. Sahgal and Hayduk (12) measured the solubilities of ethylene at atmospheric pressure and temperature range of -9 to 70 °C in isopropanol, butanol and ethyleneglycol. Hydrogen bonding (H-bonding) factors were used to relate the solubilities in one hydrogen bonding solvent to those in other hydrogen bonding solvents. The semi-empirical correlation proposed by Cysewski and Prausnitz (11) was also used to predict the solubility of gases in polar solvents and in water too.

Rivas and Prausnitz (13) measured the solubilities of ethane, carbon dioxide and hydrogen sulfide in propylene carbonate, N-methyl-2-pyrrolidone and tetramethylene sulfone ; and in mixtures of these polar solvents with monoethanolamine and diglycolamine, in temperature range of -10 to 100 °C. They pointed out the economic advantages of the mixed solvent when compared with the single one ; the mixed solvent can be regenerated more easily, and requires less steam or a smaller number of trays in the stripping columns.

Hayduk and Laudie (14) studied the effect of hydrogen bonding on gas solubilities in polar solvents. They reported that, the H-bonding factors which were based on ideal gas solubilities and solubilities in water, appeared to be closely to H-bonding factors in the simple alcohols ; and similarly, H-bonding factors in solvents containing a carbonyl group or dihydrox group were related to those in acetone. The relation between the various H-bonding could be used to estimate solubilities in these and other associated solvents.

Alvarez (44) measured the solubility of CO_2 in ethanol, ethyleneglycol at 20 °C ; Bigeard (42) recently, measured the solubility of CO_2 in mixtures of ethanol and ethyleneglycol at 20 °C and 1-2.5 atm.

1.9 Solubility of Gases in Liquids at elevated Temperature

For industrial applications, it is interesting to study the effect of temperature on gas solubility in liquids. Some literature investigations are given.

Cukor and Prausnitz (16, 27) developed two apparatuses for rapid and accurate measurement of the solubility of gases in liquids at pressures in the vicinity of 1 atm over the temperature range 25-200 °C. Equilibrium compositions were determined from the total gas pressure and from a material balance. Their apparatuses with a careful operation, yielded solubilities of about 1 % accuracy. They reported values of solubility expressed in mole fraction (x_2) of methane in n-hexadecane at 1 atm partial pressure and some experimental values of Henry's constant for methane, ethane and hydrogen in n-hexadecane, bicyclohexyl and diphenylmethane for temperature range 25-200 °C.

Chappelow and Prausnitz (15) measured the low-pressure solubilities of methane, ethane, propane, n-butane, iso-butane and hydrogen in n-hexadecane, n-eicosane, squalane, bicyclohexyl, octamethylcyclotetraxiloxane, diphenylmethane and 1-methylnaphthalene over the temperature range 25 to 200 °C. They used an apparatus as that used by Cukor and Prausnitz (27).

Maloney and Prausnitz (29) used high-pressure, gas-liquid chromatography to measure Henry's constants and infinite-dilution partial molar volumes of ethylene in liquid polyethylene.

Their data yielded Henry's constants from 130 to 300 °C to 600 atm. These results are useful for the design of separation equipment in the high-pressure polyethylene process.

Hayduk and Laudie (14) observed that all gases solubilities in a given solvent have a common value as the solvent critical temperature is approached. By extrapolating bilogarithmic curves of gas solubilities vs temperature to the solvent critical temperature, they determined reference solubilities in a number of polar and non-polar solvents. Beutier and Renon (18) confirmed the previous observations of Hayduk and Laudie (14) and they derived from thermodynamic considerations an exact value of the reference solubility.

Rivas and Prausnitz (28) designed an apparatus for measuring the solubilities of gases in pure or mixed solvents at pressures below 1 atm and at temperatures ranging from -30 to 200 °C. They obtained accurate values of the solubilities of ethane, carbon-dioxide and hydrogen sulfide in propylene carbonate.

Preston et al. (17) studied the effect of temperature on Henry's constant in simple mixtures. They showed that for nitrogen-ethane system, the Henry's constant went through a maximum at about temperature of 190 K.

1.10 Calculation of Gas Solubility from Literature Data

The solubility of gases in liquids is highly temperature dependent that is why the mole fraction solubility x_2 was generally correlated by plotting -RT Lin x_2 vs T. From a least-squares analysis, many fitting equations were proposed. Wilhelm and Battino (1) used the following relationship

$$-RT \text{ Lin } x_2 = A_o + A_1 T \qquad (35)$$

where A_o and A_1 are constants for some systems.

And for some systems where the experimental data were particularly precise over a range of temperatures, the following quadratic expression was used

$$-RT \text{ Lin } x_2 = A'_o + A'_1 T + A'_2 T^2 \qquad (36)$$

where A'_o, A'_1 and A'_2 are constants for some systems and their values are tabulated in ref. (1).

The temperature dependence of the solubility was also accounted for by fitting to an expression of the form

$$-RT\ \mathrm{Lin}\ x_2 = A + \frac{B}{T} + C\ \mathrm{Lin}\ T + D\ T \qquad (37)$$

where T is the temperature in K,
A, B, C and D are constants for some systems.

In this previous relationship, proposed by Glew (47) the inclusion of the fourth term depended on the overall precision and the number of points.

Benson and Krause (48) recommended an equation of the form

$$-\mathrm{Lin}x_2 = a_o + \frac{a_1}{T} + \frac{a_2}{T^2} + \ldots \frac{a_n}{T^n} \qquad (38)$$

as providing the best fit with the least number of constants for their high-precision data ; (a_o, a_1, ... a_n are constants).

However, Wilhelm et al. (31) noted that the advantage of equation (37) over polynomial fits with an equal number of coefficient is that, it correctly correlates solubility and temperature with a significantly smaller standard deviation. The differentiation of equation (37) yields the different thermodynamic functions, as

$$\Delta\overline{H}_2^o = -B + CT + DT^2 \qquad (39)$$

$$\Delta\overline{S}_2^o = A + C + C\ \mathrm{Lin}\ T + 2\ D\ T \qquad (40)$$

$$\Delta\overline{G}_2^o = \Delta\overline{H}_2^o - T\Delta\overline{S}_2^o \qquad (41)$$

where $\Delta\overline{H}_2^o$ is the standard enthalpy change or enthalpy of solution
$\Delta\overline{S}_2^o$ is the standard entropy change or entropy of solution
$\Delta\overline{G}_2^o$ is the Gibbs' energy change or energy of solution

For not too large temperature intervals where $\Delta\overline{H}_2^o$ can be regarded as being constant, the mole fraction solubility $x_2(T)$ at temperature (T) may be calculated from

$$\mathrm{Lin}\ \frac{x_2(T)}{x_2(T_o)} = \frac{\Delta\overline{H}_2^o}{R}\left[\frac{1}{T_o} - \frac{1}{T}\right] \qquad (42)$$

with reasonable accuracy. $x_2(T_o)$, T_o and $\Delta\overline{H}_2^o$ must be known.

Recently, Lohse and Deckwer (30) measured the solubility of chlorine in various aromatics (benzene, toluene, ethylbenzene, o-, m- and p-xylene and 2-, 3- and 4-chlorotoluenes) for the temperature range of 15-75 °C. They plotted the solubility expressed in (mol/lit) vs (1/T) and straight lines were found. Their solubility data were represented by an equation of the same type as equation (42). Moreover, many correlations and theories for gas solubility in liquids have been proposed. In almost, all correlations, some function of solubility is plotted vs a parameter characteristic of the solute or of the solvent. Three commonly used plots are, Log x_2 vs (ε/K), Log x_2 vs (α') and -RT Log x_2 vs ($\Delta\overline{E}_b^2$) where (ε/K) holds for Lennard-Jones force parameter, (α') holds for the polarizability and ($\Delta\overline{E}_b^2$) holds for the energy of vaporization of the gas at its normal boiling point [1, 33]. Some theories were applied too with some success to predict the solubility of gases in liquids [1].

Osburn [2] proposed the following relationship

$$\text{Log } L = -0.025\ (\sigma_L)^4 - 0.025 \left(\frac{(P)\rho_L}{M_L}\right)^4 \qquad (43)$$

where L is Ostwald's coefficient

σ_L is the surface tension of the pure liquid at 20 °C

(P) is the parachor ; (it can be calculated from the molecular structure by adding the contributions of the atoms and the bonds - tabulated values of (P) are shown in ref. [2])

M_L is the molecular weight of the pure liquid

ρ_L is the density of the pure liquid at 20 °C.

This previous equation holds for the determination of the solubility of O_2, CO, and N_2 in normal pure liquids. The term "normal" was used here to describe a liquid in which a gas dissolves by physical solution only, with no complexing or other interaction between the molecules.

.11 Some Experimental Values of The Solubilities of Gases in Liquids

We present here some experimental data on gas solubilities in liquids obtained in our laboratory. The liquid previously degassed is saturated by the unreacted gas, in a thermostated autoclave, provided with a mechanically bladed stirrer under a solute partial pressure p. After saturation attainment, a sample of the saturated liquid is taken via a syringe of high precision and injected into a gas-chromatograph in order to extract the solute dissolved in a known volume of the liquid sample V_l. By the way of calibration gas of known solute mole fraction, the number of moles of the solute dissolved in V_1 may be determined at given temperature and pressure. Alvarez (44) measured the solubility of CO_2 in aqueous and organic (viscous and non viscous) liquids at 20 °C and a pressure of 1-2.0 atm. Bigeard (42) studied the solubility of CO_2 in ethanol, ethyleneglycol and mixtures of these liquids at the experimental conditions previously used by Alvarez (44). These workers defined Henry's law constant as the ratio of the solute gas partial pressure p in gaseous phase to the concentration of the dissolved gas in the liquide phase, c_A^* at the same temperature, as

$$He = \frac{p}{c_A^*}$$

Their results are schematically given in Fig. (1) ; one may notice on one hand that the solubility of CO_2, c_A^* in ethanol is the highest, while in viscous liquid, ethyleneglycol, it is the lowest and on the other hand the solubility of CO_2 increases with decrease in the solution viscosity. Moreover Belhaj (43) investigated the solubilities of CO_2 and N_2O in isopropanol, toluene and toluene + 10 % by volume of isopropanol and defined the solubility as the dissolved gas mole fraction, x_2. According to his experimental results obtained at 20 °C and, shown in Fig. (2), one may conclude that the addition of 10 % by volume of isopropanol to the toluene does not change the solubility mole fraction x_2.

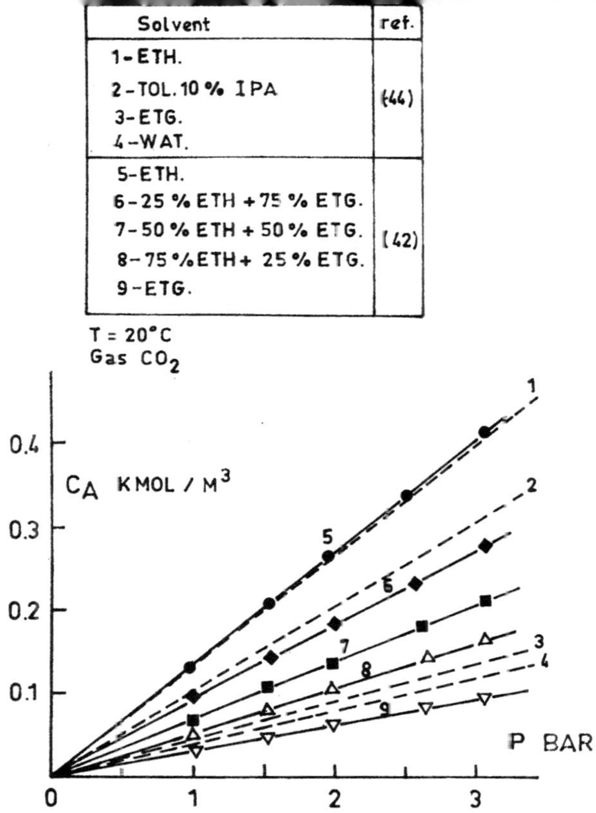

Fig. 1 Solubility of CO_2 in aqueous and organic solvent at 20°C

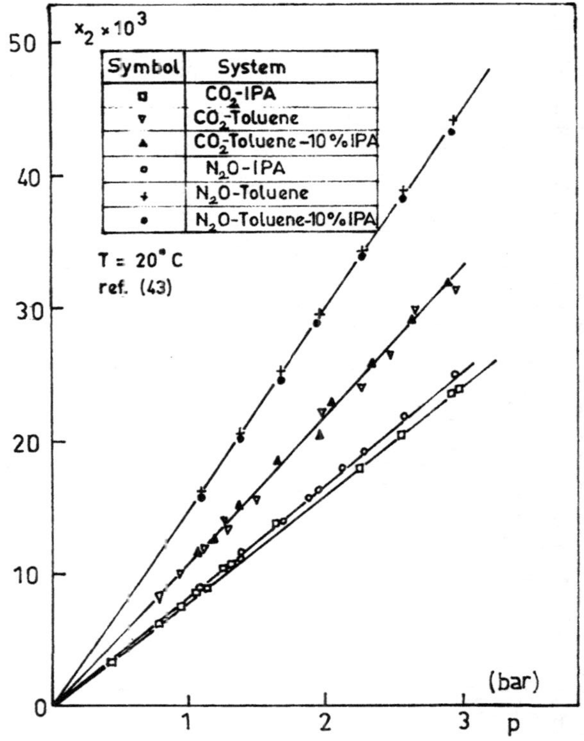

Figure. 2 Solubilities of CO_2 and N_2O in pure or mixture of solvents

LITERATURE CITED

(1) Wilhelm, E. and R. Battino, Chem. Reviews, 73, 1 (1973).
(2) Osburn, J.O., Federation Proceedings, 29, 1704 (1970).
(3) De Ligny, C.L., Van der Veen, N.G. and J.C. Van Houwelingen, Ind. Eng. Chem. Fundam., 15, 336 (1976).
(4) Tiepel, E.W. and K.E. Gubbins, Ind. Eng. Chem. Fundam., 12, 18 (1973).
(5) Schumpe, A., Alder, I. and W.D. Deckwer, Biotechnology and Bioengineering, XX, 145 (1978).
(6) Sada, E. and S. Kito, Ind. Eng. Chem. Fundam., 14, 143 (1975).
(7) Meissner, H.P. and C.L. Kusik, Ind. Eng. Chem. Process. Des. Dev., 18, 391 (1979).
(8) Hikita, H., Asai, S. and Y. Azuma, Cand. J. Chem. Eng., 56, 371 (1978).
(9) Bidner, M.S. and M. De Santiago, Chem. Eng. Sci., 26, 1484 (1971).
(10) Sada, E., Kumazawa, H. and M.A. Butt, J. Chem. Eng. Data, 22, 277 (1977).
(11) Cysewski, G.R. and J.M. Prausnitz, Ind. Eng. Chem. Fundam., 15, 304 (1976).
(12) Sahgal, A., La, M.M. and W. Hayduk, Cand. J. Chem. Eng., 56, 354 (1978).
(13) Rivas, O.R. and J.M. Prausnitz, AIChE J., 25, 975 (1979).
(14) Hayduk, W. and H. Laudie, AIChE J., 19, 1233 (1973).
(15) Chappelow, C.C. and J.M. Prausnitz, AIChE J., 20, 1097 (1974).
(16) Cukor, P.M. and J.M. Prausnitz, J. Phys. Chem., 76, 398 (1972).
(17) Preston, G.T., Funk, E.W. and J.M. Prausnitz, Physics and Chemistry of Liquids, 2, 193 (1971).
(18) Beutier, D. and H. Renon, AIChE J., 24, 1122 (1978).
(19) Gotoh, K., Ind. Eng. Chem. Fundam., 15, 269 (1976).
(20) Monfort, J.P. and J.L. Perez, Chem. Eng. J., 16, 205 (1978).
(21) De Ligny, C.L. and N.G. Van Der Veen, Chem. Eng. Sci., 27, 391 (1972).
(22) Leroi, J.C. and J.C. Masson, Ind. Eng. Chem. Process. Des. Dev., 16, 139 (1977).
(23) Plöcker, U., Knapp, H. and J.M. Prausnitz, Ind. Eng. Process. Des. Dev., 17, 324 (1978).
(24) Anderson, T.F. and J.M. Prausnitz, Ind. Eng. Process. Des. Dev., 19, 1 (1980).
(25) Anderson, T.F. and J.M. Prausnitz, Ind. Eng. Process. Des. Dev., 19, 9 (1980).
(26) Prausnitz, J.M., Trans. I. Chem. E., 59, 3 (1981).
(27) Cukor, P.M. and J.M. Prausnitz, Ind. Eng. Chem. Fundam., 10, 638 (1971).
(28) Rivas, O.R. and J.M. Prausnitz, Ind. Eng. Chem. Fundam., 18, 289 (1979).
(29) Maloney, D.P. and J.M. Prausnitz, Ind. Eng. Chem. Process. Des. Dev., 15, 216 (1976).

(30) Lohse, M. and W.D. Deckwer, J. Chem. Eng. Data, 26, 159 (1981).
(31) Wilhelm, E., Battino, R. and R.J. Wilcock, Chem. Reviews, 77, 219 (1977).
(32) Charpentier, J.C., Adv. Chem. Eng., 11, 1 (1981).
(33) Clever, H.L. and R. Battino, Techniques of chemistry Series, Vol. 8, part 1, M.R.J. Dack, Ed., Wiley, New-York, N.Y., p. 379 (1975).
(34) Clever, H.L. and C.J. Holland, J. Chem. Eng. Data, 13, 411 (1968).
(35) Douglas, E., J. Phys. Chem. 68, 169 (1964).
(36) Dymond, J. and J.H. Hildebrand, Ind. Eng. Chem. Fundam., 6, 130 (1967).
(37) Gardiner, G.E. and N.O. Smith, J. Phys. Chem., 76, 1195 (1972).
(38) Kresheck, G.C., Schneider, H. and H.A. Sharaga, J. Phys. Chem., 69, 1316 (1965).
(39) Waters, J.A., Mortimer, G.A. and H.E. Clement, J. Chem. Eng. Data, 15, 174 (1970).
(40) Kounce, C.E. and B.B. Benson, J. Phys. Chem., 67, 933 (1963).
(41) Mc Auliffe, C.A., J. Phys. Chem., 70, 1267 (1966).
(42) Bigeard, P., Microthèse, INPL-ENSIC - Nancy, France (1981).
(43) Belhaj, M.S., Thèse, INPL - ENSIC, Nancy, France (1980).
(44) Alvarez, F.C., Thèse, INPL - ENSIC, Nancy, France (1980).
(45) Tewari, Y.B., Martire, D.E. and J.P. Sheridan, J. Phys. Chem., 74, 2345 (1970).
(46) Markham, A.E. and K.A. Kobe, Chem. Rev., 28, 519 (1941).
(47) Glew, D.N., J. Phys. Chem., 66, 605 (1962).
(48) Benson, B.B. and D.J. Krause, J. Chem. Phys., 64, 489 (1976).
(49) Sechenov, M., Ann. Chem. Phys., 25, 226 (1892).
(50) Van Krevelen, D.W. and P.J. Hoftijzer, Chem. et Ind. Numero spécial du XXIe Congrès International de Chimie Industrielle Bruxelles, p. 168 (1948).
(51) Danckwerts, P.V., Gas-Liquid Reactions, Mc Graw Hill, New York (1970).

2 CHEMICAL ENGINEERING APPLICATION OF DIFFUSIVITY DATA

2.1 Introduction

Diffusivities in liquids are of major importance to chemical engineers in all processes involving mass transfer problems, through the unit operation of gas-liquid absorption, liquid-liquid extraction, dissolution, crystallization, distillation, condensation, mixing, chemical reaction and isotopes separation (2, 4, 5, 15, 14). The design of industrial process plant is often based on limited experimental diffusion data. Therefore the knowledge of diffusion coefficients is of great practical value as well as important for the understanding of diffusion mechanisms.

Diffusion in liquids is more difficult to assess than either solids or gases, because of additional complexities introduced by way of liquid structure, molecular notions and intermolecular forces into any representative model. It is interesting to emphasize that, the first recorded observation of diffusion was in 1815 by Parrott, the first measurements were made by Graham in 1850 and the first theoretical contribution by Fick in 1855. Nevertheless, there may be no truly reliable methods yet for predicting diffusivity values from other known physical data and although a vast amount of experimental work over the last century has gone into determining diffusivity, many of the values cited in the literature, particularly those measured before 1940, are of low accuracy. Because of this, the total of the accurately known experimental data is very meagre. In the case of gas-liquid diffusion, the scant data available are invariably expressed by empirical correlations since generally applicable equations for diffusion have not yet been fully developed. The main handicap in predicting diffusivity values is the lack of a satisfactory comprehensive, adequate liquid state theory.

The problems in experimental work are many, the main ones are due to turbulence, vibration, convection and the difficulties of determining concentration within the system without disturbing the experiment. These difficulties were all enhanced by the length of time required to carry out a determination in the earliest methods. The lack of accurate data also prevented development of a liquid state theory, since it was impossible to test any suggested improvements. The final difficulty, which was not realized for a long time, is that diffusivity values are very concentration-dependent and that, depending on the measuring technique employed ; differential (instantaneous) or integral (average) values will be obtained. The differential diffusivity is the one required in all theoretical derivations and is basically the diffusivity between two solutions differing by an infinitely small amount. The integral diffusivity is an average value for concentration range. Of course, all experimentally determined coeffiients must be integral values. However if the concentration difference can be made sufficiently small, the diffusivity value calculated can be taken as the differential one at the mean concentration range where the relationship between the two can be expressed as

$$D_{int} = \frac{1}{C_2-C_1} \int_{C_1}^{C_2} D_{dif}\, dC \qquad (1)$$

Since 1955, many diffusivity experimental data and techniques have been published both for non-electrolytic and electrolytic, organic and inorganic solutions ; Davis et al., 1967 (1) published some values of diffusivity coefficient of CO_2 in organic liquids obtained by means of short wetted-wall column. Leffer and Cullinan, 1970 (13) studied the variation of liquid diffusion coefficients with composition for both binary and dilute ternary systems. Dim and Ponter, 1971 (22) measured the diffusivity of CO_2 in polymeric solutions. Dim et al., 1971 (3) studied the diffusion coefficient of CO_2 in primary alcohols and methyl cellulose ether solutions by means of laminar jet technique at 25 °C. Simon and Ponter, 1975 (9) determined the diffusion coefficient of CO_2 in aqueous ethanol-water mixtures at 25 °C and atmospheric pressure. Using a short wetted-wall column, they showed experimentally that diffusivity-viscosity relationships fail when generally applied especially for their highly non-ideal systems. Sada et al., 1975 (21) studied the diffusivities of O_2, N_2 and H_2 in binary alcohol-water mixtures by a bubble solution method at 20 and 30°C. Sovova and Prochazka, 1976 (12) proposed a method for measuring diffusivity of gases in liquids by measuring the rate of absorption of a gas in a laminar film of liquid flowing radially over the surface of a horizontal disc. The absorption cell worked continuously with recirculation of liquid and the profile of velocities in the liquid film was investigated by photochromic technique and was found to be parabolic. This method yields directly the value of the diffusion coefficient of the gas in the liquid. These authors proposed values of diffusion coefficient of CO_2 into water and toluene. Yasumishi and Hoshida, 1979 (6) made some measurements of diffusivity of CO_2 in aqueous solution of 15 electrolytes containing NaCl and Na_2SO_4 at 25 °C (measurements were also taken at 15 and 35 °C in case of 4 electrolytes). Mazerei et al.,1980 (19) studied experimentally diffusivity of sparingly soluble gases CO_2, H_2 and He in H_2O at 25 °C and Hikita et al., 1980 (25) measured the diffusivities of mono- di and tri-ethanolamines in aqueous solution at 25 °C and atmospheric pressure as a function of ethanolamine concentration (up to 3.5 mol/lit) by means of diaphragm cell technique. Many interesting reviews (2, 14, 23, 45, 48) have also appeared.

Now we will try to explain what diffusion is and for both non electrolytic and electrolytic solution, what methods of estimating diffusivity are.

2.2 Terminology and Definition

Diffusion is the mass transport of species relative to environment on a molecular scale. The driving force for diffusion can be :

1. a concentration gradient (ordinary diffusion)
2. a temperature gradient (thermal diffusion)
3. a pressure gradient (pressure diffusion)
4. unequal external forces (forced diffusion).

Generally, thermal and pressure diffusion are extremely small forced diffusion occuring predominantly as a result of concentration gradient. For any defined system, generally, diffusion coefficient is a function of pressure, temperature and composition ; also a viscosity dependence has been usually assumed.

Fick (5) observed a linear relationship between the flux of a species and its concentration gradient and defined the diffusion coefficient (being the proportionality constant). One of the most frequent formulation of Fick's law is

$$J_i = -D_{im} \nabla C_i \qquad (2)$$

Describing the diffusion of species i through a medium m, ∇C_i being the molar concentration gradient and the reference frame for the flux J being the molar average velocity of the bulk. Many equations have been derived from Fick's first law of diffusion depending on the choice of units for the flux and composition gradient and on the reference frame for the flux. The experimental values reported are generally consistent with Fick's definition according to equation (2). It has been established from irreversible thermodynamics that the deriving force of diffusion is the gradient of chemical potential, instead of the concentration gradient.

As a consequence, the diffusion coefficient D should be corrected by a thermodynamic factor α, giving an activity corrected diffusivity D' as,

$$D' = D/\alpha \qquad (3)$$

α was shown to be

$$\alpha = \left(\frac{\partial \text{ Lin } a_i}{\partial \text{ Lin } x_i}\right)_{p,\theta} = 1 + \left(\frac{\partial \text{ Lin } f_i}{\partial \text{ Lin } c_i}\right)_{p,\theta} \qquad (4)$$

where a_i : molar fraction activity coefficient
f_i : concentration activity coefficient
x_i : c_i/c_t
c_t : total concentration

Equations (2), (3) and (4) combine to give

$$J_i = - D' \frac{c_i}{a_i} \nabla a_i \qquad (5)$$

The activity-corrected diffusion coefficient D' is much less often used than the Fick's diffusivity D. For ideal liquid mixtures and for small concentrations of diffusing species, $a_i = c_i$ making D and D' identical.

2.3 Estimation Methods of the Diffusivity

In the estimation of diffusivities for applications to chemical engineering problems, various approaches, kinetic theory, absolute theory, hydrodynamic theory, statistical-mechanical theory and both empirical and semi-empirical correlation had been employed for the calculation of binary diffusion coefficient. It is essential to appreciate that different theories are necessary for non-electrolytes and electrolytes solutions, therefore, different estimation methods are required for each case. A fact that all methods had been overlooked and limitations of each one were recognized too.

2.3.1 Non-electrolytes. a- Dilute solutions. Taking into account the hydrodynamic theory of diffusion, Nernst (27) considered a relationship between pressure in gases and osmotic pressure in liquids and derived,

$$J_i = \frac{-u_i}{N.F} \nabla P_i \qquad (I.2)$$

when (u_i/F) is the "mobility" or the steady-state velocity of the diffusing particle under influence of a unit force (F/u_i is the friction coefficient).
∇P_i is the osmotic pressure gradient
N is Avogadro's number.

For non-electrolytic solution, Einstein (2) assumed

$$p_i = C_i \; R \; T \qquad (I.2)$$

Equations (I.1) and (I.2) combine to give

$$J_i = \frac{-R}{N} \cdot \frac{u_i}{F} T \nabla C_i = kT \frac{u_i}{F} \nabla C_i \qquad (I.3)$$

using equation (2), $D_i = kT(u_i/F)$ (I.4)

where $k = (-R/N)$ = Boltzmann's constant.

Equation (I.4) is also known as the Nernst-Einstein's equation and is the starting point of the hydrodynamic theory.

Basset (2) derived an equation for the drag force acting on a rigid sphere A moving in "creeping flow" through medium B, from hydrodynamics

$$F = 6\pi\mu_B u_A R_A \left(\frac{2\mu_B + R_A \beta_{AB}}{3\mu_B + R_A \beta_{AB}}\right) \qquad (I.5)$$

where β_{AB} is called the coefficient of sliding friction, and R_A is the diameter of the rigid sphere.
Two limiting cases are of interest :

Case 1 - $\beta_{AB} = 0$ when the fluid slips over the surface of the sphere, and equation (I.5) becomes

$$F = 4\pi\mu_B u_A R_A \qquad (I.5.1)$$

Reducing equation (I.4) to

$$D_{AB} = \frac{kT}{4\pi\mu_B R_A} \qquad (I.6)$$

Equation (I.6) should be expected to give a better correlation of data for systems where molecular sizes are comparable, however large deviation have also been noted (18).

Case 2 - $\beta_{AB} = \infty$ when there is no tendency for the fluid to slip over the surface of the sphere and equation (I.5) reduces to Stokes' law

$$F = 6\pi\mu_B u_A R_A \qquad (I.5.2.)$$

Equations (I.5.2) and (I.4) combine to give

$$D_{AB} = \frac{kT}{6\pi\mu_B R_A} \qquad (I.7)$$

which is the well-known "Stokes-Einstein"'s equation.

This equation should be applicable to describe diffusion of spherical molecules in solvent B of considerably smaller molecules. One must note that, the range of the validity of this equation is therefore limited and its accuracy rapidly decreases as the diffusing particle size decreases.

Although the hydrodynamical theory is based on an over-simplification of liquid structure,equation (I.4) is often used as a starting point to describe diffusion processes and to predict diffusivities.

Arnold, 1930 (28) derived an expression for gaseous diffusivity based on the classical kinetic theory for gases and applied this to the liquid state. He proposed three assumptions relative to the collision rate :

a. all collisions are binary
b. the collision rate is unaffected by the volume occupied by the molecules
c. the intermolecular attraction do not come into play.

He recognized also that in the liquid state none of them can be considered valid and introduced a factor 0 to take account of this failure and suggested

$$D_{AB} = \frac{g\sqrt{1/M_A+1/M_B}}{0.S^2} \qquad (I.8.a)$$

where g : a constant of proportionality to allow for the error in the basic assumptions

$0 = A_A.A_B.\mu_B^{1/2}$

$S = V_A^{1/3}+V_B^{1/3}$ (to be calculated by the Kopp-Le Bas' method).

A_A, A_B : association factor for diffusing solute and for solvent respectively.

g was found to be 0.01 at 20 °C, $A_A = A_B = 1$, for *non-associating systems* reducing equation (I.8.a) to

$$D_{AB} = \frac{0.01\sqrt{1/M_A+1/M_B}}{\mu_B^{1/2}\left[V_A^{1/3}+V_B^{1/3}\right]^2} \qquad (I.8)$$

For *associated systems*, the association factors have been determined by Arnold, 1930 (28) but no way has been shown in the literature to predict them. Thus only for ideal dilute non associating systems equation (I.8.a) could be used to predict diffusivity. This equation was used by Davis et al., 1967 (1) with satisfactory agreement with their experiments for diffusion of CO_2 in organic liquids (alcohols).

The other and better known is the absolute rate approach, based on the hole theory of liquid state. Eyring, 1936 (29) assumed that during a chemical reaction there is one step which can be identified with an "activated state" and that the process is unimolecular. The reaction rate constant k' can be expressed as

$$k' = \tau \frac{kT}{h} e^{-\Delta E/RT} \qquad (I.9.1)$$

where τ is the transmission coefficient which is the probability that a chemical reaction takes place after the system has reached the activated state and is usually assumed to be unity.

h = Planck's constant

ΔE = energy of activation of diffusion.

The first application of reaction rate theory to transport phenomena was given by Eyring, 1936 (29). He assumed the liquid to have a lattice configuration and considered both diffusion and viscosity as activated rate controlled processes, taking place by molecular jumps from one position to another. When the viscosity is independent of the applied forces, i.e. Newtonian flow, Eyring derived the following relationships :

$$\mu = \left(\frac{F}{F^*}\right)_\mu \cdot \frac{\lambda_1}{\lambda_2\lambda_3\lambda^2} \frac{kT}{k'} \qquad (I.9.2)$$

and $$D = \left(\frac{F^*}{F}\right)_D \cdot \lambda^2 \cdot k' \qquad (I.9.3)$$

Eliminating k' between equations (I.9.1) and (I.9.2) and between equations (I.9.1) and (I.9.3) results respectively in

$$\mu = \left(\frac{F}{F^*}\right)_\mu \frac{\lambda_1}{\lambda_2\lambda_3\lambda^2} \frac{h}{\tau} e^{\Delta E_\mu/RT} \tag{I.10.1}$$

and $$D = \left(\frac{F^*}{F}\right)_D \lambda^2 \frac{kT\tau}{h} e^{-\Delta E_D/RT} \tag{I.10.2}$$

which combine to give

$$D = \frac{kT}{\mu} \left(\frac{F^*}{F}\right)_D \left(\frac{F}{F^*}\right)_\mu \frac{\lambda_1}{\lambda_2\lambda_3} e^{(\Delta E_\mu - \Delta E_D)/RT} \tag{I.11}$$

Eyring and Co-workers, 1941 (30) made the following assumptions :

1. The processes of diffusion and viscous flow are identical then, $\left(\frac{F^*}{F}\right)_D \cdot \left(\frac{F}{F^*}\right)_\mu = 1$ and $\Delta E_\mu = \Delta E_D$, so equation (I.11) becomes

$$D = \frac{kT}{\mu} \frac{\lambda_1}{\lambda_2\lambda_3} \tag{I.12}$$

where λ_1, λ_2, λ_3 are intermolecular distances in liquid lattice for each of three directions.

2. The intermolecular distances can be related to the molar volume by

$$\frac{\lambda_1}{\lambda_2\lambda_3} = \left(\frac{N}{V}\right)^{1/3}$$

thus $$D_{AB} = \frac{kT}{\mu_B} \left(\frac{N}{V_B}\right)^{1/3} \tag{I.13}$$

where V = molecular volume of solvent = μ_B/ρ_B

An investigation of existing diffusivity data by Akgerman and Gainer, 1972 (31) showed that equation (I.13) predicts diffusion coefficients too high by a factor 6 when compared to equations (I.6) and (I.7). It is interesting to note that most equations developed to predict diffusion coefficient contain the variable parameters, temperature and viscosity. As the viscosity of liquid is highly dependent upon temperature a test of these equations for changes of only the temperature parameter is not possible.

In the literature only two relations could be found relating D to T without using parameter dependent on T and one relation relating μ to T.

The form of Arrhenius' equation for viscosity is widely accepted and used

$$\mu = A_1 \; e^{B_1/T} \qquad (I.14)$$

and to describe the temperature dependence of diffusivity the same Arrhenius' form was assumed by many workers [18, 25, 8, 23, 48]

$$D = A_2 \; e^{-\Delta E/RT} \qquad (I.15)$$

however, all work derived from absolute rate theory based on

$$D = A_3 \; T \; e^{-\Delta E/RT} \qquad (I.16)$$

Thus, over limited temperature ranges, plot of Lin D vs 1/T do, in fact, give quite good straight line following equation (I.15) and a positive departure from straight line at higher temperature should indicate the validity of equation (I.16).

It was from basic equation, unsuitable as they stood for predicting diffusivity data, that the *empirical correlations* commonly used in chemical engineering were derived.

The first attempt to obtain a general correlation was that of Wilke, 1949 [39], he showed the temperature dependence on group $(D\mu/T)$, the so-called diffusion factor for a defined system, where the diffusing component is dilute.

Wilke and Chang, 1955 [32] studied the diffusivity of both iodine and toluene in alkanes and included in their analysis other systems from the literature. They investigated the influence of solvent properties, such as, viscosity, molar volume, molecular weight and heat of vaporization and found a linear relationship between Log $(D\mu_B/T)$ and Log μ_B with a slope of (0.5). They examined also the influence of solute properties, by collecting diffusion data for a variety of solutes in the solvents, water, methanol, ethanol, hexane, toluene and carbon tetrachloride and they observed a linear relationship between Log $(D\mu_B/T)$ and log V_A, the slope being (-0.6). They proposed the following equation

$$D_{AB} = 7.4\times10^{-8}\left(\frac{T\,M_B^{0.5}}{\mu_{AB}V_A^{0.6}}\right) \qquad (I.17)$$

For unassociated solvents (no hydrogen bonding), they proposed the following general correlation, in which the degree of association of the solvent α_L was taken into account

$$D_{AB} = 7.4\times10^{-8}\,\frac{T(\alpha_B M_B)^{0.5}}{\mu_{AB}\,V_A^{0.6}} \qquad (I.18)$$

where V_A = molar volume of solute at its normal boiling point (cm^3/mol)

M_B = molecular weight of solvent

μ_{AB} = viscosity of solution in cP

α_B = association factor of solvent

α_B = 1.0 for unassociated solvents ; α_B = 1.9 for methanol

α_B = 2.6 for water ; α_B = 1.5 for ethanol

$(\alpha_B M_B)$ represents the effective molecular weight.

Although these authors did not indicate a method for estimating α_B, other than from diffusion data, equation (I.18) is one of the most widely used and referred to in the literature.

Reddy and Doraiswamy, 1967 (24) modified equation (I.18) by replacing the association factor in this equation for the square root of the solvent molar volume, they proposed

$$D_{AB} = A_o\,\frac{T\,M_B^{1/2}}{\mu_B V_A^{1/3} V_B^{1/3}} \qquad (I.19)$$

The constant A_o is this previous equation has been found to depend on the molecular volumes of solvent and solute, thus

1. when $\frac{V_B}{V_A} \le 1.5$, $A_o = 10\times10^{-8}$

2. when $\frac{V_B}{V_A} > 1.5$, $A_o = 8.5\times10^{-8}$

Equation (I.19) has been tested for 96 systems (aqueous and organic) and found to hold satisfactorily. The average error are in case 1 of 13.5 % for 76 systems and 18 % in case 2 for 20 systems. However this correlation does not hold satisfactorily for highly viscous liquids.

Scheibel, 1954 (17), by using the hydrodynamic theory confirmed the constant behavior of ($D\mu_B/T$) factor for any system and examined the diffusion in water, methanol and benzene, neglecting any association and proposed

$$D_{AB} = 8.2\times10^{-8}\ \frac{T}{\mu_{AB}V_A^{1/3}}\left[1 + \left(\frac{3\ V_B}{V_A}\right)^{2/3}\right] \tag{I.20}$$

where μ_{AB} is the viscosity of solution in cP.

This equation appeared to respect the behavior of large molecules in dilute solution diffusing through a solvent of small molecules and appeared to correct adequately for the effect of solvent volume down to a solute volume equal to about twice the solvent volume, although in some solvents the range may extend down to solute volumes equal to solvent volumes.

- In case of water, equation (I.20) holds to $V_A = V_B$
- In case of methanol, equation (I.20) holds to $V_A = 1.5V_B$
- In case of benzene, equation (I.20) holds to $V_A = 2V_B$
- In case of miscellaneous solvent, equation (I.20) holds to $V_A = 2.5V_B$.

Shrier, 1967 (16) recommended equations (I.18) and (I.20) to estimate liquid diffusivities in dilute solutions.

Lusis and Ratcliff, 1971 (33) proposed the following empirical correlation

$$D_{AB} = \frac{8.52\times10^{-8}}{\mu_B\ V_B^{1/3}}\left[1.4\left(\frac{V_B}{V_A}\right)^{1/3} + \frac{V_B}{V_A}\right] \tag{I.21}$$

which is not recommended for diffusion of solutes in water ; however a good prediction was noted for alkane-alkane systems.

In equations (I.20) and (I.21) no association factors are necessary.

Sovova, 1976 (10) proposed a correlation for predicting the diffusivity coefficient of gases in water and in two groups of organic liquids. The general form of the correlation is

$$D_{AB} = a_o \, f \, \mu_L^{b_o} / V_A^{0.6} \qquad (I.22)$$

where a_o = constant = 14.8×10^{-5}
V_A^o = molar volume of the solute gas, cm^3/mol
μ_L = viscosity of the liquid in cP
f = correlation factor
b_o = correlation exponent.

The correlation factor and exponent varies as a function of solvent nature, thus

f = 1 and b_o = −1.15 for water

f = 1.8 and b_o = −1.15 for 1st organic group, which contains among others aromatic hydrocarbons and their derivatives - or more specifically : benzene, toluene, chlorobenzene, nitrobenzene, aniline, acetone, tetrachloromethan, 4-methyl, 2-pentanone and methanol.

f = 2.28 and b_o = −0.50 for 2nd organic group, which contains aliphatic hydrocarbons and alcohols - or more specifically : n-paraffins C_6-C_{16}, alcohols of n-paraffins C_2-C_8 cyclohexane and cyclohexene.

Othmer and Thaker, 1953 (34) proposed the following empirical relationship

$$D_{AB} = \frac{1.4 \; 10^{-4}}{\mu_B V_A^{0.6}} \; \mu_w^{(1.1 \Delta H_B . \Delta H_w)} \qquad (I.23.1)$$

w refers to water and the constants are based on empirical data for diffusion in water as a solvent equation(I.23.1) becomes

$$D_{AB} = \frac{1.4 \; 10^{-4}}{V_A^{0.6} \mu_w^{2.1}} \qquad (I.23.2)$$

As for equation (I.18), Hayduk and Laudie, 1974 (35) tested the reliability of this equation for different aqueous solutions and they concluded that a revised equation of the following form should relate diffusion data best

$$D_{AB} = \frac{13.26\ 10^{-5}}{V_A^{0.589}}\ \mu_B^{-1.4} \tag{I.24}$$

where μ_B is the viscosity of solvent in cP.

Both equations (I.23.1) and (I.24) do not take into account any association of the solute with the solvent, therefore large deviation can be expected.

Hayduk and Cheng, 1971 (36) proposed

$$D_{AB} = A_o \mu^{B_o} \tag{I.25}$$

For diffusion at infinite solution under isothermal conditions, the constants A_o and B_o depend only on the solute properties. Some cited values in the literature,

B_o = -0.44 for diffusion of CO_2 in water, Hayduk and Cheng, 1971 (36)

B_o = -0.45 for diffusion of CO_2 in organic liquids, Davis et al. (1)

B_o = -0.47 and $A_o = 1.41\times10^{-10}$ for diffusion of CO_2 in organic liquids, Mc Manamey and Woollen (9)

B_o = -0.92 and $A_o = 4.25\times10^{-9}$ for diffusion of CO_2 in organic solutions (ethanol, ethyleneglycol), Alvarez (26).

King et al., 1965 (37) proposed

$$D_{AB} = 4.4\times10^{-8}\ \frac{T}{\mu_B}\left(\frac{V_B}{V_A}\right)^{1/6}\left(\frac{\Delta H_{vB}}{\Delta H_{vA}}\right)^{1/2} \tag{I.26}$$

where ΔH_v = heat of vaporization at normal boiling point, cal/gm mol.

This equation is not suitable for cases where the solvent is viscous (the recommended limit of $D_{AB}\mu_B/T$ is 1.5×10^7 cP cm^2/sK), it is not particularly accurate for diffusion coefficient in aqueous systems.

Nakanishi, 1978 (20) proposed a general correlation to predict D_{AB} with improved accuracy for non electrolytes in dilute solutions at 25 °C as

$$D_{AB} = \frac{2.9726}{\mu_B(I_1 Q_1 V_A)^{1/3}}\ \frac{0.716 A_B S_2 V_B}{\mu_B I_1 S_1 Q_1 V_A} \times 10^{-5} \tag{I.27}$$

where A_B association parameter for associated solvents
I_1 interaction parameter for polar solutes
S_1, S_2 shape factors for paraffins
Q_1 quantum correction factor for light gases.

In order to apply this equation, it is necessary to assign the values of 5 empirical parameters as well as those for molar volumes V_A and V_B and the viscosity of solvent at 25 °C. μ_B and V_B at 25 °C may be found in the tabulation of an appropriate handbook. If the solution in question is also liquid, V_A may be calculated from their density data at 25 °C. The author recommended the use of molar volume of solute at its normal boiling temperature $(V_A)_{nbt}$ to get V_A at 25 °C as

$$(V_A)_{25°C} = \alpha_o (V_A)_{nbt}$$

where α_o = 0.894 for solids
α_o = 1.065 for gases.

the values of other parameters I_1, Q_1, S_1 and S_2 are listed in author's paper (1978).

Experimental tests carried out indicated that the average deviation for 149 data points in 18 solvents was only ± 9.1 %.

b- Concentrated solutions. For ideal concentrated solutions, Powell et al., 1941 (38) and Wilke, 1949 (39) had used the following relationship

$$\left(\frac{D_{AB}\mu_{AB}}{T}\right)_{conc.} = \frac{D_{AB}\mu_B}{T} X_B + \frac{D_{BA}\mu_A}{T} X_A \qquad (b.1)$$

where X_A, X_B = mole fraction of solute, and solvent respectively
and $X_A + X_B$ = 1.

Equation (b.1) expresses a linear variation of the quantity $(D_{AB}\mu_{AB})$ with composition at a given temperature and provides at least a crude approximation to the dependence of diffusivity on concentration in the measurements carried out by Garner and Marchant, 1961 (40) for associated compounds in water.

The deviation from ideality has to be allowed for by means of activity coefficients, the most usable form of the correlation is

$$\left(\frac{D_{AB}\mu_{AB}}{T}\right)_{conc.} = \left(\frac{D_{AB}\mu_B}{T} X_B + \frac{D_{BA}\mu_A}{T} X_A\right) \left(1 + \frac{d\ \mathrm{Lin}\gamma_A}{d\ \mathrm{Lin}X_A}\right) \qquad (b.2)$$

where γ_A is the solute activity coefficient
X_A is the mole fraction of solute.

Equation (b.2) was tested for nonideal solutions by Wilke, 1949 (39). However this correlation is fairly unreliable and is not applicable for the solution of solids in water. For example, the diffusivity data of aqueous sucrose at 1 and 25 °C in water below 60 gm/lit obtained by Gosting and Morris, 1949 (41) could be represented by the following relationship

$$(D_{AB})_{conc} = \frac{D_{AB}\mu_B}{\mu_{AB}} \left(1 + \frac{d(\mathrm{Lin}\gamma_A)}{d(\mathrm{Lin}X_A)}\right) \qquad (b.3)$$

where μ_B, μ_{AB} : viscosity of solvent and solution respectively.

This previous equation has been already proposed by James et al., 1939 (42) and correlated closely their experimental results.

For non associated solutions, both ideal and non-ideal, also associated solution if the degree of association is constant, Vignes, 1966 (7) had proposed an empirical equation for two-component system, that is completely miscible and free from association as

$$(D_{AB})_{conc} = (D_{AB})^{X_B} (D_{BA})^{X_A} \left(1 + \frac{d(\mathrm{Lin}\gamma_A)}{d(\mathrm{Lin}X_A)}\right) \qquad (b.4)$$

This equation was modified by Leffler and Cullinan, 1970 (13) to give improved correlation by incorporating the viscosity of the solution and of the pure components resulting in the following relationship

$$(D_{AB}\mu_{AB})_{conc} = (D_{AB}\mu_B)^{X_B} (D_{BA}\mu_A)^{X_A} \left(1 + \frac{d(\mathrm{Lin}\gamma_A)}{d(\mathrm{Lin}X_A)}\right) \qquad (b.5)$$

Equations (b.4) and (b.5) are particularly poor for binaries of n-alkanes. All previous equations (b.1), (b.2), (b.3), (b.4) and (b.5) while satisfactory for binary ideal concentrated solutions, are not reliable for non-ideal systems or those in which molecular association is significant. It is recommended when considering a new system to check, whether it has an experimental measured diffusivity and in the absence of experimental values equations (b.4) and (b.5) appear to be the preferred relationships for estimating the effect of concentration on diffusivity.

N.B. In the case of miscible liquids, the term $d(\mathrm{Lin}\gamma_A)/d(\mathrm{Lin}X_A)$ may be evaluated from vapor-liquid equilibrium data. Thus for ideal vapors

$$1 + \frac{d(\text{Lin}\gamma_A)}{d(\text{Lin}X_A)} = \frac{d(\text{Lin}P_A)}{d(\text{Lin}X_A)}$$

where P_A is the partial vapour pressure of the solute in solution.

2.3.2 Electrolytic solutions. Diffusion in electrolyte solutions is more complicated by dissociation of molecules into ions (cations and anions) and the resulting effects of the charge producing forces between these ions, the solution and ion clusters. In electrolytics, the small sizes of ions allow molecules to diffuse more rapidly than the undissociated molecules. Both cations and anions, despite differences between the sizes, diffuse at the same rate, so that the electrical neutrality of a given solution is preserved.

The attempts to derive empirical relationships for predicting diffusivity in electrolytes have been little satisfactory, whilst, the proposals for non electrolytes have been succesful to a certain extent, especially for dilute solutions.

a- Dilute electrolyte solutions. The theory of diffusion of salts at low concentration is well developed. For dilute solutions of a single salt (strong electrolyte) at infinite dilution, the molecular diffusion coefficient may be calculated from an equation obtained by Nernst-Haskell, 1888 (27) on the assumption of complete dissociation,

$$D_{AB} = \frac{RT}{F^2}\left(\frac{\lambda_+^\circ \lambda_-^\circ}{\lambda_+^\circ+\lambda_-^\circ}\right)\left(\frac{Z_+ + Z_-}{Z_+ Z_-}\right) \qquad \text{(II.a.1)}$$

where D_{AB} = diffusivity of molecule

λ_+°, λ_-° = cationic and anionic conductance at infinite dilution (zero concentration) mho/equivalent

Z_+, Z_- = absolute value of cation, anion valence

T = absolute temperature

R = gas constant, Joule/gm mole K = 8.314

F = Faraday = 96 500 Coulombs/gm equivalent

The values λ_+° and λ_-° in solvent (water) can be obtained for many ionic species at 25 °C and at temperature other than 25 °C may be estimated with aid of the following relationship

$$\lambda^\circ(\theta) = \lambda^\circ(25^\circ\text{C})+a(\theta-25) + b(\theta-25)^2+c(\theta-25)^3 \qquad \text{(II.a.2)}$$

where θ in °C.
Values of a, b and c for some of the more common ions are tabulated in Perry, 1963, (p. 14-24).

An approximation correction factor may be introduced as

$$D_{AB}(\theta) = \left(\frac{T}{334\ \mu_w}\right) D_{AB}(25°C) \tag{II.a.3}$$

Diffusivities of *weak electrolytes* in water were measured by Bidstrup and Geankoplis, 1963 (43). The experiments were for concentration up to 0.1 N in the carboxylic acid series (formic, acetic, propionic, butyric, valeric and caproic). Their resulting correlation, which was shown to be equally applicable to the corresponding α-amino carboxylic acids, is simply the Wilke and Chang's equation (I.18) but with the constant 7.4 replaced for 6.6, i.e.

$$D_{AB} = 6.6\times10^{-8}\ \frac{T(\alpha_B M_B)^{0.5}}{\mu_{AB}\ V_A^{0.6}} \tag{II.a.4}$$

b- Concentrate electrolytic solutions. In electrolytic solution, as the salt concentration becomes finite and increases, the diffusion coefficient decreases rapidly and then usually rises, often becoming greater than the diffusion coefficient at infinite dilution D_{AB} at high normalities. The initial decrease at low concentrations is proportional to the square root of the concentration but deviations from this trend are usually significant above 0.1N.

No reliable method has yet been proposed to relate D_{AB} to concentration. However diffusivities of electrolytes at higher concentration may, therefore, be estimated from semi-empirical equation proposed by Gordon, 1937 (44)

$$(D_{AB})_{conc} = D_{AB}\left(\frac{\mu_B}{\mu_{AB}}\right)\left(\frac{1}{\rho_B \overline{V}_B}\right)\left(1 + m\,\frac{\partial \mathrm{Lin}\gamma_\pm}{\partial\ m}\right) \tag{II.b.1}$$

with $m\,\frac{\partial \mathrm{Lin}\gamma_\pm}{\partial m} = \frac{m}{\gamma_\pm}\,\frac{\partial \gamma_\pm}{m}$, $\quad \rho_B = \frac{1000\rho_{AB}}{(1000 + mM_A)M_B}$ and $m = \frac{1000 w_A}{M_A(1-w_A)}$

where D_{AB} = diffusion coefficient at infinite dilute solution (estimated from equation II.a.1)
m = molalitly of solute (gm mol/kg solvent)
μ_B = viscosity of solvent (water) cP
μ_{AB} = viscosity of solution, cP
$\gamma_\pm$ = mean ionic activity coefficient of solute based on molality
ρ_{AB} = density of solution, gm/cm^3

ρ_B = molal density of solvent (gm mole of water/cm^3 of solution)
$\overline{V}_B$ = partial molal volume of solvent in solution, cm^3/gm mole
M_A = molecular weight of solute
w_A = mass percent of the solute

Equation (II.b.1) has been applied to systems at concentrations up to 2N, Reid and Sherwood, 1966 (45). In many cases the product ($\rho_B\overline{V}_B$) is close to unity, as is the viscosity ratio (μ_B/μ_{AB}), so that equation (II.b.1) provides an activity correction to the diffusion coefficient at infinite dilution, hence

$$(D_{AB})_{conc} \simeq D_{AB}\left(1 + m\frac{\partial \mathrm{Lin}\gamma\pm}{\partial m}\right) \qquad \text{(II.b.2)}$$

Harned and Owen, 1950 (46) provided tabulation of $\gamma\pm$ as a function of m for several aqueous solutions and also, a method for estimating partial molal volumes $\overline{V}_B$ is described by Lewis and Randall, 1923 (47).

Equation (II.b.1) then is reasonably reliable and not difficult to apply. To get the values of $(D_{AB})_{conc}$ at temperature θ and if the values of $\gamma\pm$, λ_+° and λ_-° are not available at this temperature, calculate D_{AB} at 25 °C by means of equations (II.a.1) and (II.b.1) and multiply this by a factor $(T/298)\left|\mu_{AB}(25)/\mu_{AB}(\theta)\right|$ to obtain,

$$[D_{AB}(\theta)]_{conc} = D_{AB}(25°C)\,\frac{T}{298}\,\frac{\mu_{AB}(25°C)}{\mu_{AB}(\theta)}$$

The ratio of the solution viscosity at 25°C to that at θ may be assumed to be the same as the corresponding ratio for water.

Yasunishi and Hoshida, 1979 (6) proposed, for the diffusivity of CO_2 in aqueous solutions of electrolytes at 25°C, the following relationship

$$-\frac{d}{dC_s}\left(\frac{D_{AB}}{D_{Aw}}\right) = 0.02 + 0.483\left(\frac{\phi_w - \phi_1}{\phi_w}\right) \qquad \text{(II.b.3)}$$

where D_{AB} is the diffusivity of CO_2 in liquid phase
D_{Aw} is the diffusivity of CO_2 in water
ϕ_w is the fluidity of water $(1/\mu_w)$, cm s/gm
ϕ_1 is the fluidity of electrolyte at C_s = 1 mol/lit
C_s is the concentration of electrolyte, mol/lit

These authors added also that their diffusivity values at 25°C could also be correlated with equation of the form

$$D_{AB}\mu^{n} = \text{constant} \qquad \text{(II.b.4)}$$

where $n = 1$ for all monovalent electrolytes
$n = 0.5$ for electrolytes containing one or two divalent ions.

Hikita et al., 1980 (25) suggested the use of Scheibel's equation (I.20) to correlate their diffusivity values of mono-, di- and tri-ethanol amines in aqueous solution at 25°C and proposed

$$(D_{AB})_{conc} = D_{AB}\left(\frac{\mu_{AB}}{\mu_{W}}\right)^{-2/3} \qquad \text{(II.b.5)}$$

where D_{AB} is the diffusivity of ethanolamines at infinite dilution
μ_{AB}, μ_{W} are the viscosities of aqueous ethanolamines solution and of water.

2.4 Some Experimental Values of the Diffusivity of Gases in Liquids

Alvarez (26) measured by physical absorption (without reaction) in a wetted wall falling film absorber, the diffusivity of non reacting gas N_2O in different aqueous and organic liquids and solutions at 20 °C. These systems are given in Table (I).

Gas	Liquid
N_2O	water
	ethanol
	ethyleneglycol
	toluene + 10% IPA
	MEA + WAT.
	MEA + ETH.
	MEA + ETG.
	DEA + WAT.
	DEA + ETH.
	DEA + ETG.
	CHA + tol. + 10% IPA
	CHA + ETG.

Table I : Gas-liquid systems used for the determination of the gas diffusivity at 20 °C by physical absorption

This author assumed that N_2O (non reacting solute) and CO_2 (reacting solute) have almost the same molecular properties and consequently they have the same diffusivity values in a given solution at the same temperature. Figure (3) represents some values of the diffusivity of CO_2 in amine solutions.

Finally we would present some useful tested formulae in Table (II), to predict with allowable accuracy the diffusivity of gases in different electrolyte and non electrolyte solutions.

	Non electrolytes		Electrolyte
	Aqueous	Organic	
dilute	Wilke and Chang Scheibel Hayduk and Laudie Sovova	Wilke and Chang Scheibel Reddy and Doraiswamy Laudie and Ratcliff King, Hsuch and Mao Sovova	Nernst and Haskell
concen-trated	Vignes		Gordon

Table II : Recommended correlations for diffusivities.

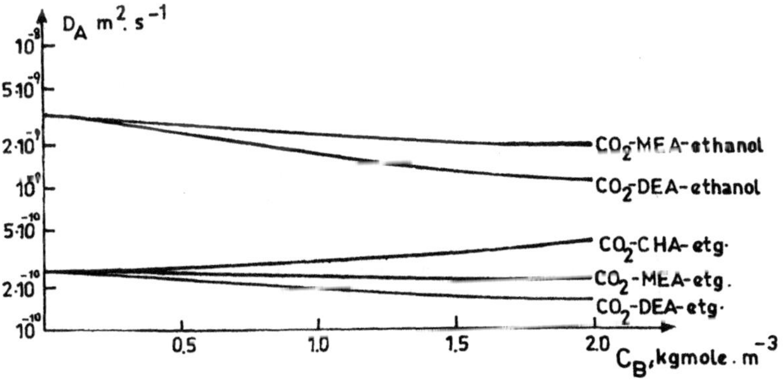

Fig.3.Diffusion coefficients of CO_2 in MEA-ethanol,DEA-ethanol, MEA-ethylenglycol,DEA-ethyleneglycol and CHA-ethylenglycol solutions.

LITERATURE CITED

(1) Davis, G.A., A.B. Ponter and K. Craine, Cand. J. Chem. Eng., 45, 372 (1967).

(2) Simon, J. and A.B. Ponter, Cand. J. Chem. Eng., 53, 541 (1975).

(3) Dim, A., G.R., Gardner, A.B. Ponter and T. Wood, J. Chem. Eng. (Japan), 4, n° 1, 92 (1971).

(4) Ratcliff, G.A. and J.G. Holdcroff, Trans. Inst. Chem. Engrs., 41, 513 (1963).

(5) Nienow, A.W., Brit. Chem. Eng., 10, n° 12, 827 (1965).

(6) Yasunishi, A. and F. Hoshida, Int. Chem. Eng., 19, n° 3, 498 (1979).

(7) Vignes, A., Ind. Eng. Chem. Fund., 5, n° 2, 189, (1966).

(8) Nijsing, R.A.T.O., R.H. Hendriksz and H. Kramers, Chem. Eng. Sci., 10, 88 (1959).

(9) Simon, J. and A.B. Ponter, J. Chem. Eng. (Japan), 8, n° 5, 347 (1975).

(10) Sovova, H., Collection Czechoslov. Chem. Commun., 41, 3715 (1976).

(11) Frederick, C.T. and O.C. Sandall, Chem. Eng. Commun., 3, 147 (1979).

(12) Sovova, H. and J. Prochazka, Chem. Eng. Sci., 31, 1091 (1976).

(13) Leffler, J. and H.T. Cullinan, Ind. Eng. Chem. Fund., 9, n° 1, 84, (1970).

(14) Charpentier, J.C., Adv. Chem. Eng., 11, 1 (1981).

(15) Kamal, M.R., and L.N. Canjar, Chem. Eng. Prog., 62, n° 1, 82 (1966).

(16) Shrier, A.L., Chem. Eng. Sci., 22, 1391 (1967).

(17) Scheibel, E.G., Ind. Eng. Chem., 46, n° 9, 2007, (1954).

(18) Wise, D.L. and G. Houghton, Chem. Eng. Sci., 21, 999 (1966).

(19) Mazarei, A.F. and O.C. Sandall, AIChE J., 26, n° 1, 154 (1980).

(20) Nakanishi, K., Ind. Eng. Chem. Fund., 17, n° 4, 253, (1978).

(21) Sada, E., S. Kito, T. Oda and Y. Ito, Chem. Eng. J., 10, 155, (1975).

(22) Dim, A. and A.B. Ponter, Chem. Eng. Sci., 26, 1301, (1971).

(23) Skelland, A.H.P., "Diffusional mass transfer", A. Wiley Interscience Publication, New-York, (1974), chapter 3.

(24) Reddy, K.A. and L.K. Doraiswamy, Ind. Eng. Chem. Fundam., 6, 77, (1967).

(25) Hikita, H., H. Ishikawa, K. Uku, T. Murakami, J. Chem. Eng. Data, 25, 324, (1980).

(26) Alvarez, F., Thèse, INPL, Nancy, (1980).

(27) Nernst, W., Phys. Chem., 2, 613, (1888).

(28) Arnold, J.H., Ind. Eng. Chem., 22, 1091, (1930).

(29) Eyring, H., J. Chem. Phys., 4, 283 (1936).

(30) Eyring, H., J. Glasston, S. and K.J. Laidler, "Theory of rate processes", Mc Graw Hill Book Co. Inc., New-York, (1941).

(31) Akgerman, A. and J.L. Gainer, J. Chem. Eng. Data, 17, 372, (1972).
(32) Wilke, C.R. and P. Chang, AIChE J., 1, 264, (1955).
(33) Lusis, M.A. and G.A. Ratcliff, AIChE J., 17, 1492, (1971).
(34) Othmer, D.F. and M.S. Thakar, Ind. Eng. Chem., 45, 589, (1953).
(35) Hayduk, W. and H. Laudie, AIChE J., 20, 611, (1974).
(36) Hayduk, W. and S.C. Cheng, Chem. Eng. Sci., 26, 635, (1971).
(37) King, C.J., L. Hsueh and K.W. Mao, J. Chem. Eng. Data, 10, 348, (1965).
(38) Powell, R.E., W.E. Roseveare and H. Eyring, J. App. Phys., 12, 669, (1941).
(39) Wilke, C.R., Chem. Eng. Prog., 45, 218, (1949).
(40) Garner, F.H. and P.J.M. Marchant, Trans. Inst. Chem. Eng. London, 39, 397, (1961).
(41) Gosting, L.J. and M.S. Morris, J. Am. Chem. Soc., 71, 1998, (1949).
(42) James, J., J. Hollingshead and A.R. Gordon, J. Chem. Phys., 7, 89, (1939).
(43) Bidstrup, D.G. and C.J. Geankoplis, J. Chem. Eng. Data, 8, 170, (1963).
(44) Gordon, A.R., J. Chem. Phys., 5, 522, (1937).
(45) Reid, R.C. and T.K. Sherwood, "The properties of gases and liquids", 2nd ed., Mc Graw Hill, New-York, (1966), chapter 11.
(46) Harned, H.S. and B.B. Owen, "The physical chemistry of electrolytic solutions", ACS Monogr., 95, (1950).
(47) Lewis, G.N. and M. Randall, "Thermodynamics and the free energy of chemical substances", Mc Graw Hill, New-York, p. 39, (1923).
(48) Reid, R.C., J.M. Prausnitz and T.K. Sherwood, "The properties of gases and liquids", 3rd ed., Mc Graw Hill, New-York, (1977), chapter 11.

GENERAL CONCLUSION

Solubility and diffusivity of gases in liquids have continued to be an area of active interest both from the practical and theoretical standpoints. Basic thermodynamic considerations on solution and diffusion of gases in liquids were briefly discussed. Solubility of gases in non-polar, polar, aqueous and electrolyte solvents and diffusivity of gases in dilute and concentrated, electrolyte and non-electrolyte solutions were investigated. Effect of temperature, pressure and salts on both solubility and diffusivity was encountered too.

However, we would like to emphasize three points :

first, the need for more adequate experimental work,
second, in the absence of experimental data, Table (II) summarizes some tested relationships, available for the prediction of the diffusivity of the gases in different liquids. However, in our knowledge, there is no general explicite correlation for direct prediction of the gas solubility in certain type of liquids, except some graphical representations and for electrolyties, equations (31) and (43) may be applied to a certain extent.
third investigating these physico-chemical data with different reaction phenomenon and under industrial operating conditions, and more experimental work.

REVIEW OF OBTAINING AND ESTIMATION METHODS OF PHYSICO-CHEMICAL AND RELATED DATA : PART 2 - GAS-LIQUID MASS TRANSFER PARAMETERS. MEASUREMENT AND SOME DATA IN SEVERAL TYPES OF REACTORS

by J.C. Charpentier and B.I. Morsi

Laboratoire des Sciences du Génie Chimique - Centre National de la Recherche Scientifique - ENSIC, 1, rue Grandville - 54042 NANCY Cedex - France.

The gas holdup, the interfacial area and the mass transfer coefficient are the main variables determining the mass transfer rates in gas-liquid contacting device. The methods used to measure these parameters can be classified into two categories : local measurements with physical techniques such as light scattering and reflection techniques, photographical and electrical and electrochemical techniques, and global measurements with chemical techniques. Each method has its advantages and its drawbacks. Let us describe and comment them and give some data with a major emphasis on packed columns and trickle-bed reactors, mechanically agitated reactors, bubble columns, spray towers, jet reactors and plate columns.

1. PHYSICAL TECHNIQUES

These methods are mainly devoted to the measurement of gas holdup, bubble size and specific surface area in gas-liquid dispersions usually encountered in bubble columns, plate columns, mechanically agitated tank and spray towers. Any two of these interfacial parameters are sufficient to define all three since they are interrelated as,

$$a' = \frac{6\alpha}{d_{SM}} \qquad \text{and} \qquad d_{SM} = \frac{\sum_i n_i d_b^3}{\sum_i n_i d_b^2} \qquad (1)$$

where α is the gas holdup and d_{SM} is the volume-surface mean diameter of Sauter mean diameter. d_b is the diameter of a single bubble or drop and n_i is the number of bubbles or drops of diameter d_b.

1.1 Gas Holdup

The gas holdup α is determined by directly measuring the height of aerated liquid Z_a and that of the clear liquid without aeration, Z. The average gas holdup is then evaluated from the relation,

$$\alpha = \frac{Z_a - Z}{Z_a} \tag{2}$$

This method, most often used for plate columns, vertical bubble column and for mechanically agitated tank (1) is not time consuming but is not very accurate (15-20 % accuracy) especially when waves or foams are occuring on the top of the dispersion.

An alternate and more accurate manometric technique has been used by Reith et al. (2) and Burgess and Calderbank (3) where the gas holdup in the dispersion is computed from measurements of the clear liquid height in the dispersion at successive manometer tappings on the side of the froth container. Linek and Mayrhoferova (4) used an electrical technique to measure the dispersion height. The method is based on measuring the surface elevation at certain selected points by means of an electrically conductive tip. The height is determined by the vertical position of the tip at which the sum of contact times equals one half of the measurement period. The accuracy of the measured value of the total surface elevation is claimed by the authors to be $\pm$ 0.2 mm, and the gas holdup is then calculated from the total surface elevation and the cross section of the reactor.

The gamma-ray transmission technique previously employed by Vermeulen et al. (5) has been used to determine the point gas holdup in mechanically agitated tank by Calderbank, in plate columns and in packed columns (1). The principles of the application of gamma-ray absorption to holdup measurements depend on the use of the relationship,

$$\text{Lin} \frac{I_o}{I} = \lambda \rho s \tag{3}$$

I_o/I is the intensity ratio between incident beam of radiation (I_o) and transmitted beam (I), λ is the mass absorption coefficient, values of which for most atoms have been published (6), ρ is the density to be measured (simply related to gas holdup) and s is the thickness of the absorbing medium.

Thus Calderbank (7) used a cesium 137 source in conjunction with a scintillation counter and scaler. Traverses of the reactor were made, and readings taken of the gamma-radiations transmitted through the empty reactor, the reactor filled with liquid and the reactor containing the dispersion. The point gas holdup was then calculated by,

$$\alpha = \frac{\log t_2/t_1}{\log t_o/t_1} \qquad (4)$$

where t_o, t_1 and t_2 are the times for a fixed number of counts respectivelly for the reactor empty, fulled with liquid and containing the dispersion.

1.2 Bubble Size

The Sauter mean particle size of dispersion, Eqn. (1) is evaluated directly by a statistical analysis of photomicrographs or high speed flash photographs when the dispersion is dynamically maintained. Photographs are taken through the wall of the transparent reactor or in the interior of the reactor with the aid of an intrascope.

To avoid any wall effect or perturbation effect that may occur with these methods, a sampling apparatus for the photographic technique has been proposed by Kawecki et al. (8). Bubbles are extracted from the tank containing the dispersion by means of a tube connected to a small square-section column through which a continuous flow of liquid and bubbles rises. The flowrate is choosen high enough so that the small differences in the free rise velocity of the bubbles do not affect the mean residence time of the bubbles in the column. The bubbles in the column are then photographed and their diameters are determined from enlargements of the negatives projected on a screen. An alternative sampling method has been used by Todtenhauft (9) where bubbles extracted from the dispersion are photographed through a calibrated capillary tube.

It should be noted that the photographic technique for measurement of bubble size most often gives local values, does not take into account large bubbles and is applicable only in the case of small gas holdup, that is usually a low gas velocity.

.3 Interfacial Area

The local interfacial contact area is determined directly by the light transmission and reflection techniques.

In the light transmission technique, a parallel beam of light is passed through the dispersion and a photocell is placed at a large distance from it. Light scattered by the bubbles passes outside the photocell and is lost while that part of the incident parallel beam which passes through the dispersion without meeting any obstacle is recorded by the photocell placed at the extremity of an internally blackened tube. Calderbank (7) showed that for scattering bubbles, which are large in comparison with the wavelength of light, the scattering cross-section is equal to its projected area. Furthermore, the total interfacial area per unit volume of the dispersion equal four times the projected area per unit volume, thus giving the following equation,

$$\mathrm{Lin}\,\frac{I_o}{I} = \frac{a'L}{4} = \mathrm{Lin}\,\frac{t}{t_o} \qquad (5)$$

where L is the optical path length. By connecting the photocell to a light quantity meter and electric timer, it is possible to measure the time for a given quantity of light to be received by the photocell when the light passes through the liquid (t_o) and $a'L < 25$ when multiple scattering is negligible and for bubble diameter larger than 50 μ. Landau et al. (10) extended this technique to conditions when the light source and photocell are placed outside the column and multiple scattering is taken into account. An interesting empirical correlation based on anisotropic scattering in all six mutually perpendicular directions is given,

$$\frac{a'L}{4} = \frac{\mathrm{Lin}(I_o/I)}{I-\Phi}$$

with $\Phi = 1 - 6.59\left[\frac{1-\exp(-0.233\tau)}{\tau}\right]$

which is valid for a'L up to 100. This represents a fourfold increase of the range of the applicability of the light attenuation technique which is not time consuming. The range of interfacial area is up to 8 cm^{-1} with fractions of light transmitted less than 0.02 and values of a' are approximating by 5 % those obtained simultaneously by the photographic method.

With optically dense dispersion such as those occuring in plate columns, the transmission method may fail because of intense multiple scattering. In such cases, a reflectivity probe may be used (11) where the optical reflectivity or backward scattered light from dispersion is measured. The specific interfacial area is calculated by,

$$\left[(R_\infty/R)-1\right] = (46.5/a') \qquad (6)$$

where R is the intensity of light reflected back by the dispersion and R_∞ is the intensity of light reflected back when a' is infinite (as obtained by extrapolation of a plot of 1/R against 1/a' to 1/a' = 0). It was found that for a given dispersion R_∞ is a function of the refractive index ratio.

1.4 Comments on Physical Methods. Need of Chemical Methods as a Complementary and not an Alternative Tool

The specific contact surface area necessitated for mass transfer in a gas-liquid dispersion or more generally in any type of gas-liquid reactor is defined as the interfacial contact area of all the bubbles or drops or phase element such as films or rivulets within a volume element divided by the volume of that element.

It is necessary to distinguish between the overall specific contact area S for the whole reactor with volume V_R and the local specific contact area S_i for a small volume element ΔV_i. In practise ΔV_i is directly analysed by the physical methods. The main difficulty in the determination of the overall specific contact area from the local specific contact area is that S_i depends strongly on the position of ΔV_i in the reactor which is a consequence of a variation in the local gas holdup and/or the local Sauter mean diameter as seen in Eqn. (1). So there is a need for a direct determination of the overall interfacial contact area over the whole reactor which is possible with the use of the chemical technique. But it can already be stated that this technique that allows for overall values cannot be used without certain restrictions that arise from the results observed with the physical methods. For example the chemical method can hardly be used with certain restrictions for fast-coalescing systems since the presence of a chemical compound will probably reduce the coalescence rates and in fast-coalescing systems, as it is observed with the physical methods, the value of the specific contact area may depend strongly on the position in the reactor, which reduces the meaning of an average value obtained with the chemical methods. In fact both physical and chemical techniques should be simultaneously necessary to qualify the phenomenae that occur in gas-liquid reactor. While the chemical methods offer overall values of interfacial area that are practical and immediatly available for design a complementary information is however desirable to follow the variations of the local interfacial parameters such as α, d_{SM} within the reactor, which is very important for the knowledge of the trends to follow in the scaling-up design. But this complementary information that is only possible with the physical methods, should be obtained from local simultaneous measurements of two of the three interfacial parameters and not only from one single local measurement of d_{SM} or α. In order to gather these complementary informations simultaneously, the electroresistivity probe technique, proposed by Burgess and Calderbank (3) for the measurement of bubble properties in bubble dispersion is very promising.

A three dimensional resistivity probe with five channels was designed in order to sence the bubble local interface approache angle as well as measure bubble size and velocities in sieve tray froths. Moreover the probe system accepts only those bubbles whose central axes are coincident with the vertical probe axis, this discrimination function being achieved with the aid of an on-line computer, receiving signals from the five channels communicating with the probe array. Gas holdup, gas-liquid specific interfacial area and even gas and liquid-side mass transfer efficiencies have been calculated directly from the local measured distributions of bubbles size and velocity. The derived values of the dispersion parameter for air-water system have been found in excellent agreement with independently observed and previously published data. This promising method has revealed an interesting result : the magnitude of the interfacial areas reported compare very favourably with those computed from the chemical technique but are lower than those measured using photography. Whereas photography through the container wall appears to truncate data above equivalent diameters of about 15 mm, the probe used by Burgess and Calderbank only deletes data below a diameter of 4 mm. So the obvious difference between the interfacial values measured by chemical technique and photographic technique may be due to the presence of the small number fraction of large bubbles that dominate the interfacial area parameters of an assembly of small and large bubbles such as that met in sieve tray, and to their apparent inadvertent omission by photography owing to observational effects, as confirmed later by Kurten and Zehner in mechanically agitated reactor (12).

This is a good example of the limitations of the physical methods when they provide on single information within a local volume (d_{SM}) while the other information α, is obtained for the global volume of the reactor.

So before to present now the chemical methods that are time consuming, it is important to insist and to keep in mind that a perfect knowledge of the mass transfer phenomenae that occur inside a gas-liquid reactor necessitates in fact simultaneous local physical and global chemical techniques whereas the chemical methods offer ex-abrupto global values without any indication on the in situ spatial variations. Ignoring it may be misleading for reactors that present important inhomogeneity (13).

2. CHEMICAL TECHNIQUES

The chemical techniques to determine gas-liquid interfacial area and mass transfer coefficient have been intensively developed for the 15 last years. The principles of these techniques are the measurement of the absorption rates where an absorbed gas undergoes chemical reaction with precisely known kinetics. A gas A is absorbed into a liquid and there undergoes a reaction with a dissolved reactant B,

$$A + zB \xrightarrow{k} \text{Products}$$

By suitably choosing the solubility, the concentration of the reactant and the rate of reaction, either the mass transfer coefficients, or the interfacial area or both groups of parameters can be deduced from the overall rate of absorption (14). Generally but not always, a steady flow of each phase through the reactor is assumed. Indeed the competition between the phsyical and chemical kinetics at the level of mass transfer between gas and liquid (the mass transfer reaction regime where the reaction belongs) may allow for the choice of the type of gas-liquid contactor (1). This is clearly shown in Fig. 1 that represents schematically the concentration profiles for A and B on each side of the interface.

2.1 Determination of $k_L a$

If the resistance to transfer of component A is entirely in the liquid phase ($k_L a \ll He\ k_G a$) the global rate of absorption per unit volume of a gas-liquid contactor, in the absence of chemical reaction is,

$$\Phi = \psi a = k_L a(C_A^* - C_{Ao}) = \Phi_t / V_R \qquad (7)$$

where C_A^* and C_{Ao} are the concentrations at the interface and in the bulk of liquid respectively (mol/L^3)
Φ_t is the global rate of absorption (mol/T)
V_R is the volume of the gas-liquid contactor (L^3)
ψ is the specific rate of absorption = $\Phi/a = \Phi_t/(aV_R)$, (mol/(L^2T)

The value of $k_L a$ can in certain circumstances be determined by purely physical experiments in the reactor. For instance, $k_L a$ may be evaluated from the observed total rate of absorption in the case of a piston-like countercurrent flow of the two-phases where C_A^* is a known function of C_{Ao} or when the gas and the liquid are well stirred so that C_A^* and C_{Ao} are the same at all points. However there are two possible difficulties about the determination of $k_L a$ by the physical absorption method. One is that the flow

Regime	Interface concentration profiles	Rate equations	
Instantaneous	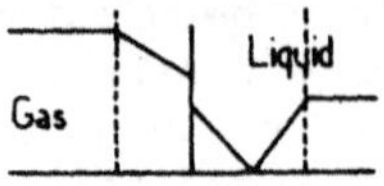	$\psi = \dfrac{\frac{p}{He} + \frac{D_B}{D_A}\frac{C_{Bo}}{Z}}{\frac{1}{k_L} + \frac{1}{He\,k_G}}$	A
Instantaneous and surface		$= k_G\,p$	B
Rapid		$\psi = \dfrac{p}{\frac{1}{k_G} + \frac{He}{E\,k_L}}$	C
Rapid pseudo 1st or mth order	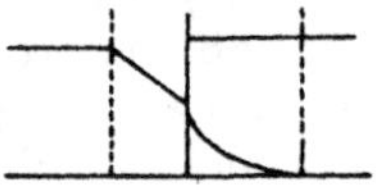	$\psi = \dfrac{p}{\frac{1}{k_G} + \frac{He}{\sqrt{D_A\,k_2\,C_{Bo}}}}$	D
Intermediate	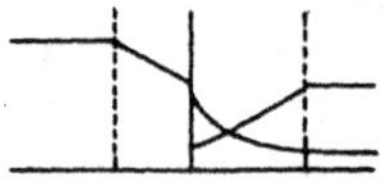	No exact general expression developed	E
Intermediate	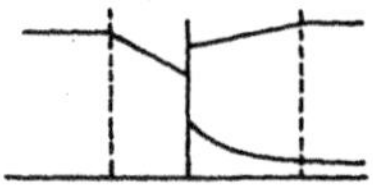	No exact general expression developed	F
Slow diffusional process	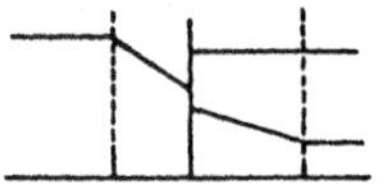	$\psi = \dfrac{p}{\frac{1}{k_G} + \frac{He}{k_L} + \frac{He\,a}{\beta\,k_2\,C_{Bo}}}$	G
Very slow chemical process in the bulk of the liquid	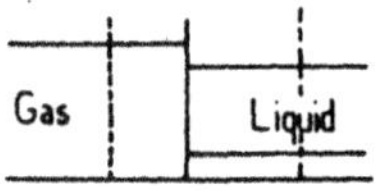	$R_A = k_2\,C_A^{*}\,C_{Bo}\,\beta\,\varepsilon$	H

Fig. 1 - Interface concentration profiles for the eight distinct kinetic regimes for mass transfer with reaction.

patterns and residence time distribution of the phases may be undetermined (C_{Ao} is not a known function of C_A^*) and hence the value of k_La cannot be deduced from Φ. The other is that in efficient contacting devices the gas and liquid may approach equilibrium quite closely. As the determination of k_La depends on the difference between the actual and the equilibrium extent and this necessitates an extremely accurate measurement of the flowrates, this method may become impracticable.

2.2 Determination of k_La with an irreversible slow reaction

The problem raised by the saturation of the liquid can be avoided by using a solution which reacts with the dissolved gas in the slow reaction regime i.e. the reaction is too slow to affect the rate of absorption directly, but, on the other hand, the reaction is fast enough to reduce the bulk concentration of dissolved gas effectively to zero. If the considered reaction is irreversible and second-order (first order with respect to both components A and B) this leads to,

$$\Phi = \Psi a = k_2 C_A^* C_{Bo} \varepsilon\beta \tag{8}$$

The conditions to be satisfied are,

$$k_L a << \varepsilon\beta k_2 C_{Bo} \tag{9a}$$

and $\dfrac{a\, D_A}{\varepsilon\beta\, k_L} << 1$ and $$\frac{D_A k_2 C_{Bo}}{k_L^2} = Ha^2 << 1 \tag{10a}$$

Thus the rate of absorption Φ is the same at all points in the reactor and the residence time distribution of the liquid is irrelevant.

It is possible similarly to make use of reaction which is m,nth-order, the local rate being thus $k_{mn}C_A^m C_B^n$. The condition for $C_{Ao} = 0$ to be zero is,

$$k_L a << \varepsilon\beta k_{mn} C_A^{*m-1} C_{Bo}^n \tag{9b}$$

and for no reaction in the film

$$Ha^2 = \frac{2}{m+1} \frac{D_A k_{mn} C_A^{*m-1} C_{Bo}^n}{k_L^2} << 1 \tag{10b}$$

It is not always easy to satisfy both conditions (9a) or (9b) and (10a) or (10b) simultaneously ; thus, if $k_{mn}C_{Bo}^{n}$ is made large enough to satisfy condition (9b) (i.e. $C_{Ao} = 0$) it may become too large for condition (10b) to be satisfied (no reaction in the film). When condition (10b) is satisfied but condition (9b) is not, Eqn. (9) may be written,

$$\Phi = k_R a C_A^*$$

with $$\frac{1}{k_R a} = \frac{1}{k_L a} + \frac{1}{k_{mn} C_{Ao}^{m-1} C_{Bo}^{n} \beta\varepsilon}$$

If $k_{mn}C_{Bo}^{n}$ is varied, keeping $k_L a$ constant, a plot of $1/k_R a$ against $1/k_{mn} C_{Bo}^{n}$ will be a straight line of slope $1/C_{Ao}^{m-1}\beta\varepsilon$ with intercept $1/k_L a$. This offers a method of determining $k_L a$ when it is not possible to satisfy condition (9b).

Some suitable chemical systems for the determination of $k_L a$ in the slow reaction regime are presented in Table I [1].

Solute gas A	Reactant B	Catalyst in the absorbent
CO_2	$K_2CO_3 + HKCO_3$	NaClO
O_2 diluted with air	CuCl	—
O_2 diluted with air	Na_2SO_3	$CuSO_4$
O_2 diluted with air	Na_2SO_3	$CoSO_4$
O_2	Glucose	Glucose oxidase

Table I - Chemical systems used to determine $k_L a$ in the slow reaction regime.

2.3 Determination of $k_L a$ with an Irreversible Instantaneous Chemical Reaction

There is another and quite different method whereby a chemically reacting system can be used to determine $k_L a$. This consists in the use of an irreversible instantaneous reaction where the rate of absorption is

$$\Phi = k_L a C_A^* E_i = k_L a C_A^* \left(1 + \frac{D_B}{zD_A} \frac{C_{Bo}}{C_A^*}\right) \tag{11}$$

The condition to be satisfied is $Ha > 10E_i$. If in addition $C_{Bo} << C_A^*$, then the rate of absorption is,

$$\Phi = k_L a \frac{C_{Bo}}{z} \frac{D_B}{D_A}$$

The rate of absorption is independent of the concentration of component A in the gas phase. It is also independent of the residence time distribution in the gas phase. In practise it is found that the use of Eqn. (10) is available when $E_i > 4$. Some suitable chemical systems for the determination of k_La in the instantaneous regime are presented in Table II (1).

Solute gas A	Reactant B
NH_3	H_2SO_4
SO_2, Cl_2, HCl	NaOH
H_2S, HCl, CO_2	Amines
O_2 diluted with air	$Na_2S_2O_4$

Table II - Chemical systems used to determine k_La in the instantaneous reaction regime.

2.4 Determination of a

When the reaction between the components A and B in the liquid phase is mth-order in A and nth-order in B, under certain circumstances the concentration of component B is the same everywhere as it is in the bulk of the solution ($k_{mn}C_{Bo}^n$ is constant). The reaction is said rapid pseudo-mth order in A (Fig. 1, D). The conditions for this situation are $3 < Ha << E_i$ and the rate of absorption is expressed by,

$$\Phi = Ek_L aC_A^* = Hak_L aC_A^* = a \left(\frac{2}{m+1} k_{mn} D_A C_A^{*m+1} C_{Bo}^n\right)^{1/2} \quad (12)$$

Thus, the rate of absorption is independent of k_L, that is, of the hydrodynamic conditions. Thus, provided that the average specific rate of absorption,

$$\Psi = \left(\frac{2}{m+1} k_{mn} D_A C_A^{*m+1} C_{Bo}^n\right)^{1/2} \quad (13)$$

is known and C_A^* and C_{Bo} have effectively the same value in all parts of the system, the specific interfacial area is determined directly from the measurement of the total rate of absorption

It is not always necessary to know the kinetics of the reaction in order to determine φ. Indeed the values of φ may be measured by absorbing the component A into the same solution in some laboratory apparatus with a known interfacial area, such as laminar jet, wetted wall column, stirred cell... The agitation speed or the flowrate should be varied to confirm that φ is really independent of k_L and β. In reactors when the residence time distribution of the gas phase is unknown, it is sometimes possible to choose conditions such that there is practically no change in the partial pressure of the gas by using a high gas flowrate and keeping the value of φ low. Moreover if component A is diluted by a carrier-gas, it must be confirmed that the gas-side resistance is negligible. Some suitable aqueous, organic and viscous chemical systems for the determination of a in the pseudo-mth regime are presented in Table III (1).

2.5 Determination of k_Ga

If the resistance to transfer of component A is entirely in the gas phase ($Hek_Ga << k_La$), the rate of absorption, per unit volume of a gas-liquid system is, in the absence of a chemical reaction,

$$\Phi = k_G a(p-p_i) \tag{14}$$

where p and p_i are the pressure respectively in the bulk of gas and at the interface.

The value of k_Ga can in certain circumstances be determined by purely physical experiments in the reactor, for instance in the case of piston-like countercurrent gas-liquid flow or when both phases are well-stirred. However, the rate of absorption depends on the residence time distribution in both phases that may be undetermined and in addition there is normally an appreciable resistance on the liquid side that must be taken into account. Thus the liquid side resistance can be eliminated and the rate of absorption can be made independent of the liquid side residence time distribution by using a solution which reacts instantaneously and irreversibly with the dissolved gas so that there is no back-pressure. Therefore,

$$\Phi = k_G a(p-p_i) = k_L a C_A^* \left(1 + \frac{D_B}{zD_A}\frac{C_{Bo}}{C_A^*}\right) = k_L a C_A^* E_i \tag{15}$$

The condition to satisfy is Ha > $10E_i$ and in this case, the dissolved gas A reacts instantaneously with the component B in a plane beneath the interface where the concentration of both components is zero (Fig. 1, A). If the reaction plane is now at the

Solute gas A	Reactant B	Catalyst
CO_2 diluted with air	$Na_2CO_3 + HNaCO_3$, $K_2CO_3 + HKCO_3$	$As(OH)_2O^-$, ClO^-
	Aqueous or organic solutions of amine	—
	LiOH–NaOH, KOH–$Ba(OH)_2$	—
	Na_2S	—
COS diluted with air	Aqueous solutions of amine	—
O_2 in air	$Na_2S_2O_3$	—
O_2 diluted with air	Na_2SO_3	$CoSO_4$, $CuSO_4$
Isobutylene in C_4 fraction or air	H_2SO_4	—
CO_2 diluted with air	Cyclohexylamine in toluene or xylene containing 10% isopropanolamine	—
	Cyclohexylamine in cyclohexanol	—
	Monoethanolamine in aqueous di- or polyethylene glycol	—
CO_2 and O_2 diluted with air	Aqueous cuprous amine complex solution	
O_2 diluted with air or N_2	Propionaldehyde	Manganese propionate
O_2 in air	C_{10} trialkylaluminum dissolved in organic solvents	—
Cl_2	*p*-Cresol dissolved in dichlorobenze	—
H_2	Edible oil	Ziegler–Natta
Desorption of isoamylene into N_2	*n*-Heptane, toluene	—

Table III - Chemical systems used for the determination of a in rapid pseudo m-th order regime.

interface (surface reaction, $p_i = C_A^* = 0$), it may be written,

$$p_i = HeC_A^* = \frac{He\left(k_G ap - k_L a \frac{D_B}{zD_A} C_{Bo}\right)}{Hek_G + k_L} = 0 \tag{16}$$

where $\Phi = k_G ap = k_L a \frac{D_B}{zD_A} C_{Bo}$.

This is the situation where the dissolved component B reaches the interface by diffusion through the liquid as fast as gaseous component A reaches it by diffusion through the gas and the transfer process is controlled by diffusion in both phases. Moreover if at all points in the reactor the following condition is satisfied,

$$k_G ap - k_L a \frac{D_B}{zD_A} C_{Bo} < 0 \tag{17}$$

the global rate of absorption per unit volume of the reactor remains equal to $k_G ap$ but the interfacial concentration of component B is greater than zero (Fig. 1, B). Thus provided Eqn. (16) the absorption process is entirely controlled by the transport of component A across the gas-film and,

$$\Phi = k_G ap$$

This forms the basis of methods for measuring $k_G a$. Indeed if the entrance and exit low partial pressure of the soluble gas p_e and p_s are measured for the reactor with a length Z, $k_G a$ is calculated by

$$k_G a = \frac{G_m}{PZ} \text{ Lin } \frac{p_e}{p_s} \tag{18}$$

where (G_m) is the superficial molar mass flowrate of the insoluble gas (in gmole/cm^2s) and P is the total pressure (in atm).

It is important to note that in all cases, the calculation of $k_G a$ should be based on analysis of the gas stream as a small error in the analysis of the liquid stream can lead to large errors in the calculated value of $k_G a$. Some suitable chemical systems for the determination of $k_G a$ in the instantaneous regime are presented in Table IV (1).

2.6 Simultaneous Measurement of Mass Transfer Coefficients and Interfacial Area

In certain circumstances, it may occur mass transfer resistance in both phases which necessitates the knowledge of both $k_G a$ and $k_L a$. Moreover, for the purpose of calculating the effect of a chemical reaction on the rate of absorption of a gas, it is generally necessary to know the parameters k_L and a separately. In using not only the two-film model but also the surface-renewal

Solute gas A	Insoluble gas diluent	Reactant B
SO_2 or Cl_2	Air, Freon 12, Freon 22, Freon 114	NaOH
NH_3	Air, Freon 12, Freon 22, Freon 114	H_2SO_4
Triethylamine	Air, Freon 12, Freon 22, Freon 114	H_2SO_4
I_2	Air	NaOH
SO_2	Air	Na_2SO_3
Propylene, CO	Air	Cuprous amine complex solution

Table IV - Chemical systems used to determine k_Ga in instantaneous and surface regimes.

model, the separate mass transfer parameters can be determined by the chemical methods with pseudo-1st order regime or instantaneous regime.

Thus the condition to satisfy the pseudo-1st order in component A, in using the Danckwerts' model (14),

$$\sqrt{1+Ha^2} < \left(\sqrt{\frac{D_A}{D_B}} + \sqrt{\frac{D_B}{D_A}}\,\frac{C_{Bo}}{zC_A^*}\right) = E_i \sqrt{\frac{D_A}{D_B}} \tag{19}$$

whence, if there exist a mass transfer resistance in both phases,

$$\Phi = p\left[\frac{1}{k_Ga} + \frac{He}{k_La\sqrt{1+Ha^2}}\right]^{-1} \tag{20}$$

Two important cases are now presented whether k_L and a or k_G and a are necessitated.

— Simultaneous determination of k_L and a. When the gas phase resistance is negligible, the rate of absorption is given by Eqn. (20) with $k_Ga = 0$, that is,

$$\Phi = \frac{p}{He}\,k_La\sqrt{1+Ha^2} = aC_A^*\sqrt{D_Ak_2C_{Bo}+k_L^2} \tag{21}$$

If the rate of absorption is measured with different values of k_2C_{Bo} and the hydrodynamic conditions remain constant, a plot of Φ^2 against k_2C_{Bo} gives a straight line with slope $D_Aa^2C_A^{*2}$ and intercept $(k_L a C_A^*)^2$. This is the so-called Danckwerts' plot. If C_A^* and D_A are known both $k_L a$ and a can be determined.

Note that a similar result is obtained using the double-film model to interpret an irreversible (m,n)-th order when the condition Ha << E_i can be satisfied but the condition Ha > 3 cannot be satisfied (1).

In using the Danckwerts' plot, it is necessary to ensure that the physical properties of the system do not alter as $k_2 C_{Bo}$ is changed. For this reason it is more convenient to make use of a catalytic reaction and to change k_2 by additing small amounts of catalyst. If the catalyst is sufficiently powerful, the reaction rate can be varied over a wide range without substantially altering the concentration of the solution. The rate of absorption is thus,

$$\Phi = aC_A^* \sqrt{D_A k_c C_{Cat} + k_L^2} \tag{22}$$

The concentration C_{Cat} of the catalyst can be varied and a and k_L determined plotting Φ^2 against C_{Cat}. Some suitable chemical systems for the determination of k_L and a with the use of Danckwerts' plot are presented in Table V (1).

Solute gas A	Reactant B	Catalyst
CO_2 diluted with air	$HNaCO_3 + Na_2CO_3$	Arsenite
CO_2 diluted with air	$HNaCO_3 + Na_2CO_3$	Hypochlorite
O_2 in air	CuCl	—
O_2 in air	Na_2SO_3	$CoSO_4$

Table V - Chemical systems for determination of k_L and a, with use of Danckwerts plot.

It is interesting to know that the hydrodynamic conditions (i.e. k_L) may be influenced by the chemical reaction, causing a change in a and k_L with a change in the reaction rate. *Thus, complementary informations of those given by the Danckwerts' plot can be offered when the rates of simultaneous chemisorption of one gas (giving a with a pseudo-mth order reaction) and of physical absorption (or desorption) of another (giving k_L or $k_L a$) are determined experimentally* (15-19).

— Simultaneous determination of k_G and a. If it is not possible to keep the gas-side resistance negligible, it is still possible to determine k_G and a by use of a fast irreversible pseudo-mth order reaction (Ha >> 1 and $\sqrt{1+Ha^2} \approx Ha$). Thus the rate of absorption expressed by Eqn. (20) becomes,

$$\Phi = p\left(\frac{1}{k_G a} + \frac{He}{k_L a Ha}\right)^{-1} \qquad (23)$$

whence, $$\frac{p}{\Phi} = \frac{1}{a}\left(\frac{1}{k_G} + \frac{He}{k_L Ha}\right) = \frac{1}{a}\left[\frac{1}{k_G} + \frac{He}{\left(\frac{2}{m+1} D_A C_A^{m-1} C_{Bo}^n\right)^{0.5}}\right] \qquad (24)$$

Thus, if $k_2 C_{Bo}$ or $k_{mn} C_{Bo}^n$ is varied a plot of $\frac{p}{\Phi}$ against $He/\sqrt{D_A k_A C_{Bo}}$ or against

$$\frac{He}{\left(\frac{2}{m+1} D_A C_A^{*m-1} C_{Bo}^n\right)^{0.5}}$$

will give a straight line of intercept ($1/k_G a$) and of slope ($1/a$), so that a and k_G can be calculated simultaneously. Convenient systems for this method are the absorption of dilute CO_2 into aqueous solutions of NaOH or amines.

2.7 Laboratory Equipment to Determine the Physico-Chemical Parameters Necessitated by the Chemical Methods

It has just been seen that the chemical determination of the interfacial mass transfer parameters necessitates the knowledge of

- kinetics of the choosen chemical gas-liquid system
- diffusivities of the soluble gas or gases and the reactant or reactants in the liquid solution
- solubility of the gas or the gases
- practically a mathematic grouping of these parameters

$$C_A^* \sqrt{D_A} \qquad \text{or} \qquad \left(\frac{2}{m+1} k_{mn} C_{Bo}^n C_A^{*m+1}\right)^{1/2} \qquad \text{or...}$$

These physico-chemical parameters are determined in laboratory equipments in which the contact time between the gas and the liquid and the interfacial area are carefully controlled. The principles of the technique, the laboratory equipments used and the application to the oxidation of aqueous sodium sulphite solution will be now presented.

Table VI - Factors affecting the overall rate of absorption-reaction

Variable	Symbol	A	B	C	D	E	F	G	H
Concentration of reactant B in the bulk of the liquid	C_{B0}	+	−	+	+	?	?	+	+
Partial pressure of component A in the bulk of the gas	p	+	+	+	+	?	?	+	+
Interfacial area	a	+	+	+	+	+	+	+	−
Liquid holdup	β	−	−	−	−	+	+	+	+
Liquid-side mass-transfer coefficient	k_L	+	−	+	−	?	?	+	−
Gas-side mass-transfer coefficient	k_G	+	+	+	+	?	?	+	−
Rate constant for second-order chemical reaction	k_2	−	−	+	+	?	?	+	+

(Kinetic regime: columns A–H)

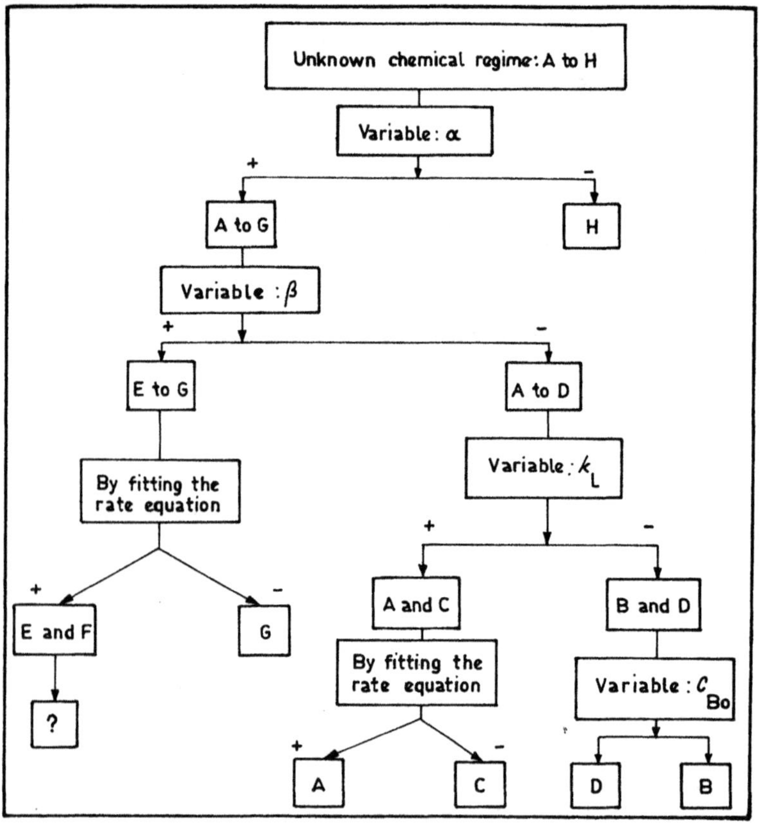

Table VII - Organigram for identification of the kinetic regimes

2.7.1 Determination of the kinetic regime of the chemical gas-liquid system. If again we consider the irreversible second order reaction

$$A + zB \xrightarrow{k_2} \text{Products}$$

eight kinetic regimes are identified in terms of the two-film theory, as seen in Fig. (1). The knowledge of the variables whose variations led to the identification of one of these regimes is due to the noticeable distinct difference in the form of the rate equations. Thus, as explained by Levenspiel and Godfrey [20], Table VI [1] shows which factors affect the rate in each regime. For a given regime, a "+" sign indicates that a change in that particular variable affects the rate, a "-" sign indicates that it does not and a "?" sign means a probable effect, but that the defining rate equations are not available.

Moreover the schematic organigram of Table VII [1] shows how to identify the kinetic regime for a given gas-liquid reaction, from a series of systematically planned experiments where the variables are varied independently and leads to the determination of the kinetic regime of the chemical gas-liquid system used [20, 21]. Very often the corresponding rate is a function of the physico-chemical parameters that must therefore be determined.

Determination of the physico-chemical parameters. The physico-chemical parameters are determined in the laboratory equipment presented in Fig. 2 by dynamic methods, in which a jet or a film of liquid moves continuously through the gas, to which it is exposed for a known length of time θ (this contact-time being from 10^{-4} s to a few seconds). Moreover in these equipments, the interfacial area A_m is known and the movement of the liquid is carefully controlled so that it can be considered as quiescent during its passage through the gas. Thus, during the contact-time θ each element of liquid absorbs the same amount $Q(\theta)$ of gas per unit area as through it were stagnant and infinitely deep (as in the Higbie's model). In each laboratory equipment, the different values of the contact-time θ are generally obtained in varying the liquid volumetric flowrate Q_L and the length of the liquid film or jet h that is the geometrical parameters. Indeed the contact-time is defined as

$$\theta = \frac{h}{u_s} \qquad (25)$$

where u_s is the velocity at the surface that is a function of Q_L and the geometry of the equipment (diameter of the jet or film, angle of the cone...). Expressions of θ for different laboratory equipments are presented in Table VIII.

TYPE	LAMINAR JET	CYLINDRICAL WETTED WALL	CONIC WETTED WALL	SPHERICAL WETTED WALL	STRING OF DISKS	ROTATING DRUM	STIRRED VESSEL	STIRRED VESSEL
SCHEME	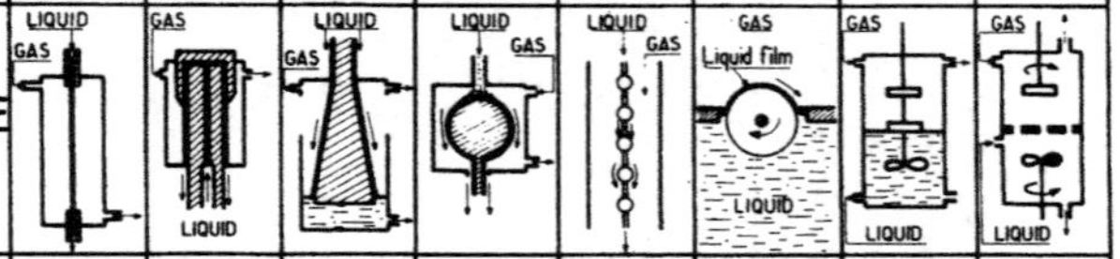							
k_L cm/sec	0.016 -0.16	$3.6 \cdot 10^{-3}$ -0.016	$5 \cdot 10^{-3}$ -0.011	$5 \cdot 10^{-3}$ -0.016	$3.6 \cdot 10^{-3}$ -0.016	0.016 -0.36	$1.6 \cdot 10^{-3}$ -0.02	$2 \cdot 10^{-3}$ -0.02
CONTACT TIMES	10^{-3}–10^{-1} sec	10^{-1}–2 sec	0.2–1 sec	0.1–1 sec	10^{-1}–2 sec	2.10^{-4}–10 sec	0.06–10 sec	0.08–10 sec
INTERFACIAL AREA	0.3–10 cm^2 high precision	10–100 cm^2 high precision	80 cm^2 high precision	10–40 cm^2 high precision	30-360 cm^2 moderate precision	diameter 10cm length 12cm high precision	80 cm^2 good precision	diameter 10cm length 15cm 2-30% open

Fig. 2 - Principal types of laboratory equipment.

Table VIII - Characteristic parameters of laboratory equipment [(a)]

Equipment	θ	$\gamma(\theta)$
Laminar jet	$\frac{\pi d^2 h}{4Q_L}$	$4(Q_L h)^{1/2}$
Cylindrical wetted wall	$\frac{2h}{3}\left(\frac{3\mu_L}{\rho_L g}\right)^{1/3}\left(\frac{\pi d}{Q_L}\right)^{2/3}$	$d(6h)^{1/2}\left(\frac{Q_L}{d}\right)^{1/3}\left(\frac{\pi\rho_L g}{3\mu_L}\right)^{1/6}$
Conic wetted wall	$\frac{4\pi R^2}{5\sin\alpha}\left(\frac{3\mu_L}{2\pi\rho g R\cos\alpha}\right)^{1/3}\frac{1}{Q_L^{2/3}}\left(1+\frac{\sin\alpha}{R}l\right)^{5/3}-1$	$\frac{2(\pi l^2\sin\alpha+2\pi R l)}{\sqrt{\pi}}\frac{Q_L^{1/3}}{\sqrt{\delta}}$ $\delta = \frac{4\pi R^2}{5\sin\alpha}\left(\frac{3\mu_L}{2\pi\rho g R\cos\alpha}\right)^{1/3}\left[\left(1+\frac{\sin\alpha}{R}l\right)^{5/3}-1\right]$
Spherical wetted wall	$\left(\frac{8}{4.5}\right)^2\pi R_s^{5/6}\left(\frac{2\pi\rho_L g}{3\mu_L}\right)^{-1/6}Q_L^{-1/3}$	$4.5\left(\frac{2\pi\rho_L g}{3\mu_L}\right)^{1/6}Q_L^{1/3}R_s^{7/6}$
Rotating drum	$\frac{Z_r}{u}$	$4Q_L^{1/2}\left(\frac{W_r Z_r}{2\pi\delta_r}\right)^{1/2}$

[a] α, angle of the conic wetted wall; l, generating line of the cone; R, radius of the basis of the cone; R_s, radius of the spherical wetted wall; W_r, belt width of the rotating drum; Z_r, exposed belt length; δ_r, film thickness; θ, contact time.

If $Q(\theta)$ is the amount of gas absorbed by unit interfacial area during the contact-time θ, the average rate of absorption during this time is $Q(\theta)/\theta$. Since the total area exposed in the laboratory equipment is A_m, the measured rate of absorption $\Phi(\theta)$ into the film is related to $Q(\theta)$ by,

$$\frac{\Phi(\theta)}{A_m} = \frac{Q(\theta)}{\theta} = \Psi \tag{26}$$

The absorption rate $\Phi(\theta)$ (in gmole/s or cm^3/s of component A) is measured experimentally, and $Q(\theta)/\theta$ is calculated for different kinetic regimes from the Higbie's theory [14]. The contact-time θ is calculated from Eqn. (25) and can be altered by altering Q_L and the geometrical parameters of the considered laboratory equipment. Thus, in carrying out experiments with the same chemical systems and with the same kinetic regime as those used to determine the mass transfer parameters in the industrial gas-liquid reactor, $\Phi(\theta)$ can be determined as a function of θ in the laboratory equipment and the variations of $Q(\theta) = [\theta\Phi(\theta)/A_m]$ for the different values of θ allow for the determination of the necessary physico-chemical parameters. This is done as follows whether the absorption is accompanied by a chemical reaction or not.

For the case of purily physical absorption into liquid initially gas-free ($C_{Ao} = 0$), from the Higbie's theory, the amont of component A absorbed per unit area during the time θ is [14]

$$Q(\theta) = 2C_A^* \sqrt{\frac{D_A\theta}{\pi}} \tag{27}$$

It follows from Eqn. (26) that,

$$\Phi(\theta) = \frac{Q(\theta)}{\theta} A_m = 2A_m C_A^* \sqrt{\frac{D_A}{\pi\theta}} = \gamma(\theta) C_A^* \sqrt{D_A} \tag{28}$$

where $\gamma(\theta)$ is a characteristic parameter of each laboratory apparatus. A plot of measured $\Phi(\theta)$ against $\gamma(\theta)$ at various liquid flow-rates Q_L and film or jet lengths h should give a straight line through the origin of slope $C_A^*\sqrt{D_A}$. If the solubility of gas is known or determined separately, its diffusivity may be determined. Values of $\gamma(\theta)$ for several equipments are presented in Table VIII.

For the case of absorption accompanied by a chemical reaction, from the Higbie's theory, $Q(\theta)$ is deduced from the solution of the equation,

$$D_A \frac{\partial^2 C_A}{\partial x^2} - r(x,\theta) = \frac{\partial C_A}{\partial \theta} \tag{29}$$

where $r(x,\theta)$ is the rate per unit volume of liquid at which the reaction is destroying the solute gas at time θ and at distance x from the interface. Analytical or numerical solutions of the diffusion-reaction equations are available for a number of kinetic regimes. For an irreversible first or pseudo-first order reaction, with the Hatta's number called Ha',

$$Ha' = \sqrt{\frac{\pi}{4} k_2 C_{Bo} \theta} << \sqrt{\frac{D_A}{D_B}} + \frac{C_{Bo}}{zC_A^*} \sqrt{\frac{D_B}{D_A}} = E_i \qquad (30)$$

Danckwerts (14) has proposed the following approximative solutions,

$$Q(\theta) = C_A^* \sqrt{D_A k_2 C_{Bo}} \left(\theta + \frac{1}{2k_2 C_{Bo}}\right)$$

to within 3 % when $k_2 C_{Bo} \theta > 2$ and,

$$Q(\theta) = 2C_A^* \sqrt{\frac{D_A \theta}{\pi}} \left(1 + \frac{k_2 C_{Bo} \theta}{3}\right)$$

to within 5 % when $k_2 C_{Bo} \theta < 1/2$. Thus, the experimental determination of the amount $\Phi(\theta)$ of the gas absorbed in function of the time θ leads to the following results :

- for long contact-time ($\theta > 2/k_2 C_{Bo}$) a plot of $Q(\theta) = |\theta\Phi(\theta)/A$ against θ will give a straight line of slope $C_A^* \sqrt{D_A k_2 C_{Bo}}$ and of intercept $(C_A^*/2)\sqrt{D_A/k_2 C_{Bo}}$ and the ratio slope to intercept will give $k_2 C_{Bo}$.

- for short contact-time $[\theta < (1/2)k_2 C_{Bo}]$ a plot of $Q(\theta)/\sqrt{\theta}$ against θ will give a straight line of slope $(2/3)(k_2 C_{Bo} C_A^*)\sqrt{(D_A/\pi)}$ and of intercept $2C_A^* \sqrt{D_A/\pi}$.

In principle, therefore, both k2 and $C_A^* \sqrt{D_A}$ can be estimated from each set of experiments but in practice it is found that the measurements a long contact-time give more accurate values of $C_A^* \sqrt{D_A k_2}$ while those at short contact-time give more accurate values of $C_A^* \sqrt{D_A}$ (14).

If, in addition of Eqn. (30), the condition,

$$Ha' > 3 \quad (\text{or } k_2 C_{Bo} \theta > 12)$$

is satisfied, then, within 5 %,

$$Q(\theta) = \theta C_A^* \sqrt{D_A k_2 C_{Bo}} \qquad (31a)$$

This is the case of the fast pseudo-first condition where the rate of absorption $Q(\theta)/\theta$ is the same at all points on the surface and therefore independent of the hydrodynamics.

Similar considerations apply to fast pseudo-mth order reaction of component A, in which case [14]

$$Q(\theta) = \theta \sqrt{\frac{2}{m+1} D_A k_{mn} C_A^{*m+1} C_{Bo}^n} \qquad (31b)$$

Under these circumstances the measurements of $Q(\theta)$ from experimental $\Phi(\theta)$ in Eqn. (26) do not lead to separate values of $C_A^*\sqrt{D_A}$ and k_2 or of $C_A^{*(m+1)/2}\sqrt{D_A}$ and k_{mn}. Nevertheless they lead to the value of the grouped parameters

$$C_A^*\sqrt{D_A k_2 C_{Bo}} \qquad \text{or} \qquad \sqrt{\frac{2}{m+1} D_A k_{mn} C_A^{*m+1} C_{Bo}^n}$$

that are those necessitated for the determination of the interfacial area in the gas-liquid reactor (Eqn. 11) without a detailed knowledge of each physico-chemical parameter.

At last for an irreversible instantaneous reaction, that is when Ha' >> E_i the solution of the diffusion-reaction equations governing this case have been given by Danckwerts [14]. This leads to

$$Q(\theta) = 2E_i C_A^* \sqrt{\frac{D_A\theta}{\pi}} = 2C_A^* \sqrt{\frac{D_A\theta}{\pi}} \left[\sqrt{\frac{D_A}{D_B}} + \frac{C_{Bo}}{zC_A^*}\sqrt{\frac{D_B}{D_A}}\right] \qquad (32)$$

In such case, in addition to the quantities C_A^*, D_A and C_{Bo} (which enter into the pseudo-first order case), the diffusivity D_B of the reactant B is also involved. It is possible to infer the value of one of these quantities from the measurements of $Q(\theta)$ from experimental $\Phi(\theta)$, if the values of the others are known and can be estimated. For example, if C_{Bo} and z are known and if C_{Bo} is sufficiently small for C_A^* and D_A to have substantially the same value as in pure solvent, then the diffusivity D_B can be determined. An interesting possibility is the reduction of C_A^* to a value much less than C_{Bo}, in which circumstances (D_B/D_A) (C_{Bo}/zC_A^*) may be much more than one (in practise very often $C_{Bo}/zC_A^* >> 1$) and thus the quantity of component A absorbed per unit area of the laboratory equipment in time θ is,

$$Q(\theta) = 2C_A^* \sqrt{\frac{D_A\theta}{\pi}} \sqrt{\frac{D_A}{D_B}} \left(1 + \frac{D_B}{D_A}\frac{C_{Bo}}{zC_A^*}\right) = \frac{2C_{Bo}}{z}\sqrt{\frac{D_B\theta}{\pi}} \qquad (33)$$

Thus the diffusivity D_B can be determined without the uncertainties involved in the estimation of C_A^* and D_A.

2.7.2 Limits of the chemical technique when applied to gas-liquid systems inhibiting bubble coalescence in such equipment as bubble columns, mechanically agitated tanks and plate columns. Interfacial parameters in these reactors (especially stirred tanks) are generally determined in assuming the liquid phase perfectly mixed, which is realistic, and the gas phase either in piston flow or perfectly mixed. When the gas is dispersed into clean liquids such as water, the coalescence and the redispersion due to bubble interaction either inside the dispersion or with the cavities behind the stirrer blades assure a perfect mixing of the gas phase. On the contrary for liquids involving the presence of dissolved electrolytes or surface active agents that diminish or totally inhibit the coalescence, neither the perfectly nor the piston flow model may represent conveniently the behaviour of the gas, even though the bubbles may be considered independent in a first approximation.

To define the limits of the chemical technique to measure interfacial parameters in such inhibiting coalescence or foaming dispersions, Midoux et al. [22] propose a flow model for shrinking and not shrinking bubbles that seems more realistic than the piston flow. Then criteria are deduced to insure, when the real behaviour of the gas phase is ignored, the independence within 10 % relatively to this behaviour of the mass transfer parameter data experimentally measured in a laboratory-scale well stirred tank by the chemical technique.

Assumptions for not shrinking bubbles are :

(1) The bubbles originated by the distributor do not loose their identity up to the moment they leave the reactor. The distribution function $f(\delta_o)$ of the size of the bubbles $\delta_o = d_b/d_{SM}$ is represented by the Bayens' distribution :

$$f(\delta_o) = K\delta_o^2 \exp(-k^2\delta_o^2) \tag{34}$$

with $k = 8/3\sqrt{\pi}$ and $K = 4k^3/\sqrt{\pi}$

(2) Each bubble is perfectly mixed except for instantaneous chemical regime.

(3) Mass transfer is located in the liquid phase and the true liquid-side mass transfer coefficient only depends on the square root of the bubble diameter.

(4) For each bubble, the liquid phase composition is invariable, i.e. the liquid evolution time is sufficiently slow to be neglec-

ted when compared to the residence time of a bubble inside the dispersion. Moreover the residence time of each bubble reduced by the gas residence time in dispersion, depends on the magnitude of its diameters as,

$$\theta_o = \frac{t_{\delta_o}}{\tau_G} = \frac{\lambda}{\delta_o^v} \qquad \text{with} \qquad \lambda = \frac{\int_o^\infty \delta_o^{v+3} f(\delta_o) d\delta_o}{\int_o^\infty \delta_o^3 f(\delta_o) d\delta_o}$$

The probability for the bubble to leave or to stay inside the dispersion depends on the competition between the drag force and the buoyancy force acting on it, this normally leads to $1 \leq v \leq 2$ (23) with the limiting values $v \to 0$ corresponding to a piston flow of identical bubbles and $v \to \infty$ corresponding to a short circuit behaviour.

The previous assumptions together with the classical relationships of the different specific absorption rates, depending on the chemical reaction regime used to determine interfacial parameters, lead to the theoretical mean reduced values of the gaseous reactant in the gas exit stream and to absorption efficiency E_A for a gas-liquid dispersion, where a_o is the geometrical specific interfacial area and $\overline{k_L}$ is the mean true liquid-side mass transfer coefficient defined for a bubble of diameter d_{SM}.

Thus, with this value of, E_A, apparent values a_{app} and $(k_La)_{app}$ may be calculated in assuming that the gas phase is perfectly micromixed and then compared with a_o and $\overline{k_L}a_o$. Then the curve representing the variation of the ratio between the apparent values and the geometrical values, with the characteristic depletion time of a bubble reduced by the gas residence time, $\overline{\theta}_o = d_{SM}\gamma_o/6\overline{\Psi}_o\tau_G$ will approximate the unity if certain conditions are fulfilled (γ_o is here the inlet solute gas concentration in gas-bulk and $\overline{\Psi}_o$ is the corresponding absorption rate for a bubble of diameter d_{SM}). More precisely the estimated mass transfer interfacial parameters will approximate the true value within $\pm$ 10% if certain conditions for $\overline{\theta}_o$ and E_A are fulfilled. These conditions are a minimum value of the depletion time to exceed or a maximum value of the efficiency to do not exceed. These extrema are presented on Table IX where it is seen that the limiting values depend on the order m with respect to the gaseous reactant and on the reduced characteristic diffusion time $\overline{\theta}_D = d_{SM}^2/4D_G\tau_G$.

Though these results may be considered only like sufficient conditions, they state precisely the limits of the chemical technique. So in the slow chemical regime (used to determine k_La) it is easy to satisfy the condition $\overline{\theta}_o > 3.3$. Let us consider, for example, the sodium sulfite system in the presence of Co^{++} ions as a catalyst with the currently used experimental conditions $C_{Bo} = 0.2$ kgmole/m^3, $\overline{\Psi}_o = 8.27\times10^{-9}$ kgmole/m^{2}s, $\gamma_o = 8.32\times10^{-3}$ kgmole/m^3, $(Co^{++}) = 10^{-6}$ kgmole/m^3, and $d_{SM}=10^{-3}$ m. This gives $\overline{\tau}_o = 168$ s and then $\tau_G<51$s to satisfy $\overline{\theta}_o = \overline{\tau}_o/\tau_G > 3.3$. Such a gas residence time has been used in practice in industrial size equipment.

Table IX - Limiting values of $\overline{\theta}_o$ and E_A for the various regimes used to measure the mass transfer parameters

Regime No	Regime	$\overline{\phi_0}$
1	Physical absorption or slow chemical	$\overline{k}_L \frac{\gamma_0}{He}$
2	Intermediary pseudo 1-nth-order chemical	$(\overline{k}_L^2 + D_L k_{1n}(C_B^0)^n)^{1/2} \frac{\gamma_0}{He}$
3	Rapid pseudo m-n^{th}-order chemical	$\left(\frac{2}{m+1} D_L k_{mn}(C_B^0)^n\right)^{1/2} \left(\frac{\gamma_0}{He}\right)^{\frac{m+1}{2}}$
4	Instantaneous chemical	$\overline{k}_L \frac{D_{BL}}{D_L} \frac{C_B^0}{z}$
5	Instantaneous chemical at the interface	$k_G p$

Regime No	Parameter to determine	Order m	Minimum value of $\overline{\theta}_0$	Maximum value of E_A
1	$k_L a$		3.3	0.25
2	a and $k_L a$	1	4.0	0.20
3	a	0	1.35	0.48
		½	1.80	0.41
		1	2.60	0.30
		2	3.70	0.21
		2	4.80	0.16
4	$k_L a$		3.3	0.27
5	$k_G a$		$\overline{\theta}_D > 20$	0.55

On the contrary, the same gas-liquid reaction with the condi-ions of rapid pseudo second-order regime in oxygen (to determine), that is, C_{Bo} = 0.8 kgmole/m^3, (Co^{++}) = 3.56×10^{-4} kgmole/m^3, $_o$ = 1.56×10^{-7} kgmole/m^{2}s, cannot be used easily. Indeed, it is alculated from the condition $\overline{\theta}_o$ > 3.7 proposed in Table IX, that $_o$ = 8.9 s and τ_G < 2.4 s. Such a small gas residence time limits he applicability of this chemical method to the small scale nits. This may explain why interfacial areas have been measured n tanks the volume of which are not bigger than 1 m^3, with gas uperficial velocity as high as 0.047 m^3/m^{2}s (24). Complementary nformation for the case of shrinking bubbles may be found in ef. (22).

Moreover for the cases where the solute gas concentration hanges considerably from inlet to outlet, a representative dri-ing force must be defined in order to evaluate accurately the ass transfer parameters. Hassan and Robinson (25) have described model of gas bubble coaslescence-redispersion interactions that roposes that the effective driving force for the mass transfer s the most significantly affected by coalescences between bub-les of minimum solute gas partial pressure, i.e. those having xit gas composition, and bubbles of near maximum solute composi-ion (closest to inlet gas composition) which are about to under-o their first coalescence subsequent to having been freshly spar-ed into the dispersion. This model leads to evaluation of a mass ransfer effective coalescence frequency F (F = 0 for bubble burs-ing at the free liquid surface, F = 100 for fairly well mixed) nd to an integral average mass transfer driving force correction actor $f = \overline{p}/p_s$ when $\overline{p}$ is the gas residence time average partial ressure and p_s is the exit stream partial pressure. This factor s used to correct the apparent driving force based on exit gas omposition for the imperfect mixing of the dispersed gas phase esulting from the restricted coalescence and redispersion inter-ction that is characteristic of electrolyte solutions in which he chemical technique is applied. Using the equation of the spe-ific absorption rate to determine interfacial areas, the correc-ion factor is applied to the experimentally-measurable exit gas omposition in order to determine the relevant residence time ave-age driving force. Thus conditions for the efficiency E_A and the oalescence frequency F are researched so that the gas dispersion an be considered as perfectly mixed (f = 1). It is obtained that or all practical purposes the gas dispersion can be considered to e perfectly mixed (f = 1 ± 0.1) as long as E_A < 0.8 and F > 40 re satisfied simultaneously. Both these criteria generally will e met in the case of sparingly-soluble gas absorption without hemical reaction in turbulent liquids such as water which do not ppreciably hinder bubble coalescence. However, for E_A > 0.8 as an often be the case in chemically-reactive systems and for < 20 as has been found to be the case in coalescence-inhibiting liquids such as aqueous electrolyte solutions, f is significantly greater than 1.0 particularly for $E_A > 0.90$ (typical of the absorption of CO_2 by aqueous solutions of strong alkalis in mechanically agitated tanks with a relatively high specific interfacial area). In such latter cases, the specific area cannot be computed accurately from specific absorption rate equations merely using $p_i = HeC_A^* = p_s$; rather one must substitute $p_i = \bar{p} = fp_s$ in such cases which will yield a lower value of the specific area than that calculated using p_s alone. Finally application of the correction factor to the chemical technique leads to an improved estimation of the mass transfer effective specific interfacial area in mechanically agitated tanks compared to merely assuming that the gas phase is perfectly mixed where it may be overestimated by a factor 1.4 to 3.5.

In conclusion, before using for scaling-up the literature reported data for aqueous ionic liquids, it is suggested to verify if the complementary conditions proposed by the previous authors (22, 25) *are verified.*

2.7.3 Application of the chemical technique to the case of an organic liquid phase. The gas-liquid reaction systems that are used to determine the interfacial area and the mass transfer coefficients in gas-liquid reactors are almost exclusively constituted of reactants in an aqueous solution. In fact, many industrial gas-liquid reactors work with organic liquids, and it would be desirable to have some organic reactant systems in order to be able to measure the mass transfer parameters or at least to estimate the difference in their values when applying the data obtained with non-organic systems to reactors operating with organic liquids. There are a few investigations on such organic reactions on their application to gas-liquid reactors (26-32). We want to present here some succinct considerations on the reaction most often suggested these last years (26, 28, 29) : the reaction between carbon dioxyde and cyclohexylamine (CHA) in a toluene (TOL) plus 10 % isopropanol (IPA) solution. The addition of small amount of IPA does not alter the physical properties of the solution appreciably and enables the carbonated product to be kept in solution (29). The apparent rate of reaction is expressed by

$$r = k_{mn} C^m_{CO_2} C^n_{CHA}$$

with an overall stoichiometry

$$CO_2 + 2R_2NH \rightarrow R_2NCOO^- + R_2NH_2^+$$

The results of the chemical thermodynamic and kinetic study by Alvarez (30) will be briefly commented here as an application of the previous considerations on the laboratory apparatuses. All the experimental work was carried out in a laboratory cylindrical wetted-wall which was also used to determine the diffusion coefficient of CO_2. The height h was modified between 6 and 14 cm and the cylinder diameter d is 1.56 cm. The liquid flowrate Q_L was varied in the range 1-4.6 cm^3s^{-1}. Except for the solubility and diffusivity, all the measurements were performed at 20 °C. The experimental conditions are

$$0.01 < p = p_{CO_2} < 0.16 \text{ Atm}$$

$$0.09 < C_{Bo} = C_{CHA} < 2.7 \text{ kgmoles m}^{-3}$$

The use of the contact-time analysis together with the value of the contact-time θ presented in Table VIII gives,

$$\theta = \frac{h}{u_s} = \frac{2h}{3}\left(\frac{3\mu_L}{\rho_L g}\right)^{1/3}\left(\frac{\pi d}{Q_L}\right)^{2/3} \tag{35}$$

Then the knowledge of the interfacial area $A = \pi dh$ and the measurement of the rate of physical absorption with initial free gas-liquid Φ leads to the following expression (including Eqn. 34)

$$\Phi_t = \psi A = 2\, C_A^* \sqrt{\frac{D_A}{\pi\theta}} = Kh^{0.5}Q_L^{0.33}C_A^*\sqrt{D_A} \tag{36}$$

that allows for the determination of $C_A^*\sqrt{D_A}$ and then for the CO_2 solubility when the diffusivity is known. Some data are presented in Figs. (3a, b, c). Moreover it has been observed that the solubility is practically independent of the amine concentration C_{Bo} ($C_A^* = 0.104$ Kmole/(m^3 atm) at 20 °C) while the diffusivity could be represented by $D_A = 4.1\sqrt{1-1.2X}$ where X is the molar fraction of amine (28).

Now, when the gas phase resistance is considered, the specific rate of absorption for a rapid m,nth order reaction ($3 < Ha << E_i$) may be expressed as,

$$\frac{p}{\psi} = \frac{1}{k_G} + \frac{He}{\left[\frac{2}{m+1} D_A k_{mn} (C_A^*)^{m-1} C_{Bo}^n\right]^{0.5}} \tag{37}$$

Though this equation is not explicite ($p_i = HeC_A^*$), if the conditions for a rapid reaction are verified, the representation of (p/ψ) versus $(D_A C_{Bo}^n)^{-1/2}$ will give, for the right values of m and n, a straight line. The slope and intercept of this line lead to the calculation of the kinetic constant k_{mn} and the resistance in gaseous phase ($1/k_G$) respectively. Figure 4 presents some experimental data to verify the order of some chemical reactions, in aqueous and organic solvents, with respect to the solute gas (CO_2). It is obvious that the order of these chemical reactions with respect to the solute is always 1. Figures (5a, b, c) show experimental representation of the average value of the left handside of eqn. (37), $(\overline{p/\psi})$ against $(D_A C_{Bo}^n)^{-1/2}$ for different reactants in both aqueous and organic, non viscous and viscous solvents. According to these figures it is clear that, the resistance in gaseous phase to the chemical reaction is negligible. So such reactions may be used to determine the gas-liquid interfacial area in reactors working with organic viscous and non viscous liquids except if the equipment is very efficient.

Indeed in well agitated reactors or in too long packed columns, the efficiency may be very high and the conditions of a rapid chemical might not be respected everywhere in the contactor. In fact the solubility and the specific rate are usually much higher with purely organic systems than with purely aqueous or aqueous organic systems. This had led Alvarez (30) to study these systems taking into consideration the foaming effect or the viscous effect encountered with organic liquids and which are not taken into consideration when the mass transfer parameters are measured with aqueous ionic systems such as sulfite oxidation and CO_2-aqueous NaOH. These systems are the rapid pseudo m,nth order reaction between CO_2 and monoethanolamine (MEA), or diethanolamine (DEA) or cyclohexylamine (CHA) contained in different organic solvent solutions. These are presented in Table X.

Each time the carbon dioxide partial pressure $p_{CO_2} = p$ has been varied between 10^{-2} and 0.2 atm. except for the system CO_2-DEA in polyethyleneglycol where it has been varied up to 1 atm. The chemical systems giving the smallest specific rate are thus recommended and seem more appropriate than the chemical system CO_2-CHA in toluene for very efficient gas-liquid reactor (see fig. 6).

SYSTEM	$Kmole\ m^{-2}\ s^{-1}$	C_{Bo} $Kmole\ m^{-3}$	μ_L cp	foaming effect
CO_2-MEA in water	$2.43\ pC_{Bo}^{0.5}D_A^{0.5}$	0.20-2.00	$1+0.32C_{Bo}$	-
CO_2-DEA in water	$1.10\ pC_{Bo}D_A^{0.5}$	0.20-0.82	$1+0.54C_{Bo}$	+
CO_2-CHA in toluene	$5.80\ pC_{Bo}D_A^{0.5}$	0.20-2.00	$0.61+0.08C_{Bo}$	-
CO_2-MEA in ethanol	$6.30\ pC_{Bo}D_A^{0.5}$	0.20-1.00	$1.54e^{0.18C_{Bo}}$	-
CO_2-DEA in ethanol	$1.40\ pC_{Bo}D_A^{0.5}$	0.30-2.20	$1.54e^{0.438C_{Bo}}$	+
CO_2-MEA in etg.	$4.10\ pC_{Bo}D_A^{0.5}$	0.20-1.10	$21+2.084C_{Bo}$	-
CO_2-DEA in etg.	$0.80\ pC_{Bo}D_A^{0.5}$	0.25-2.10	$21e^{0.258C_{Bo}}$	-
CO_2-CHA in etg.	$3.86\ pC_{Bo}^{0.5}D_A^{0.5}$	0.25-2.00	$21e^{0.098C_{Bo}}$	-
CO_2-NaOH in water	$2.84\ pC_{Bo}^{0.5}D_A^{0.5}$	0.20	1.0	-
O_2-Na_2SO_3 in water	$7.42\times10^{-2}P^{1.5}D_A^{0.5}$	0.40-0.80	1.5	+

etg. = ethyleneglycol - p in atm

Table X

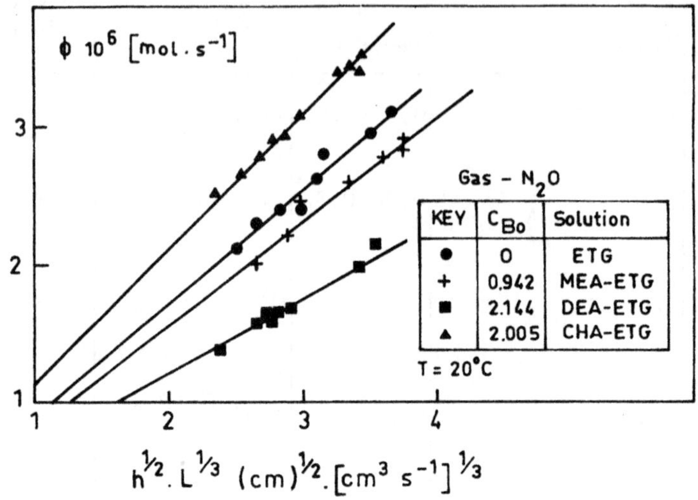

Fig. 3a - Determination of $C_A^* \sqrt{D_A}$ by physical absorption in cylindrical wetted wall falling film absorber

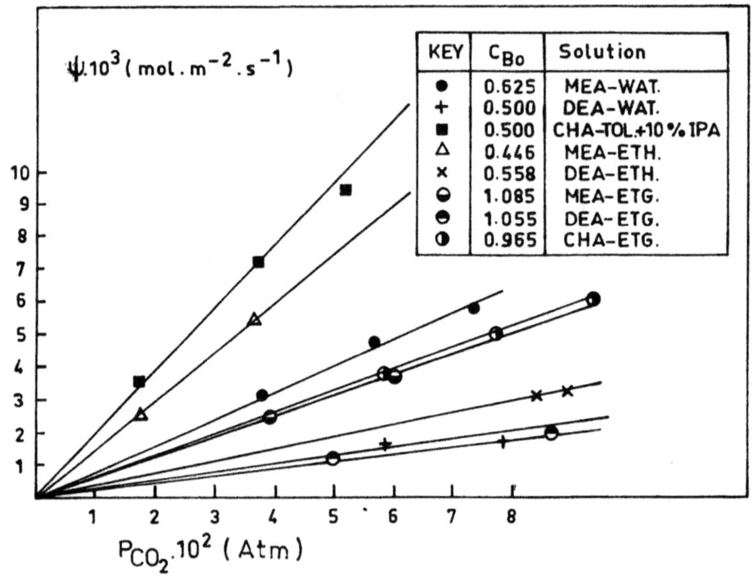

Fig. 3b - Determination of $C_A^* \sqrt{D_A}$ by physical absorption in cylindrical wetted wall falling film absorber

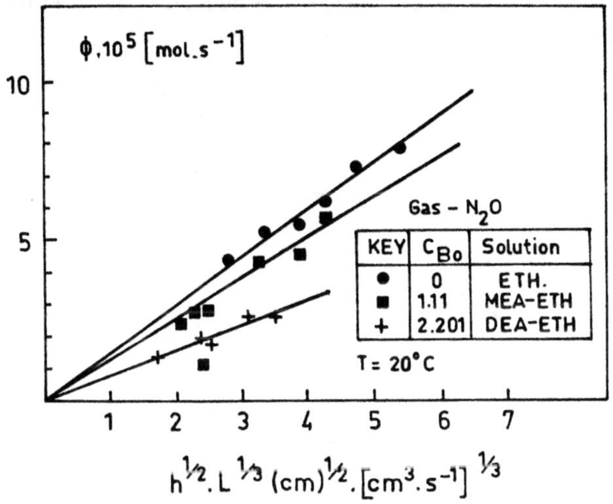

Fig. 3c - Determination of $C_A^* \sqrt{D_A}$ by physical absorption in cylindrical wetted wall falling film absorber

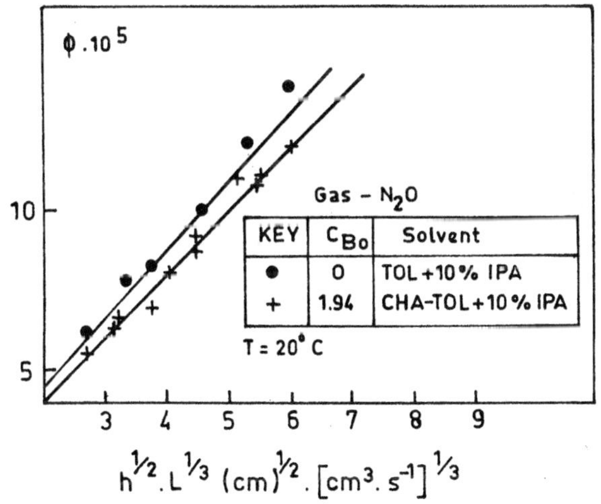

Fig. 4 - Verification of the order of the chemical reaction with respect to the solute gas.

Fig. 5a - Gas absorption with chemical reaction - determination of k_{mn} and $(1/k_G)$

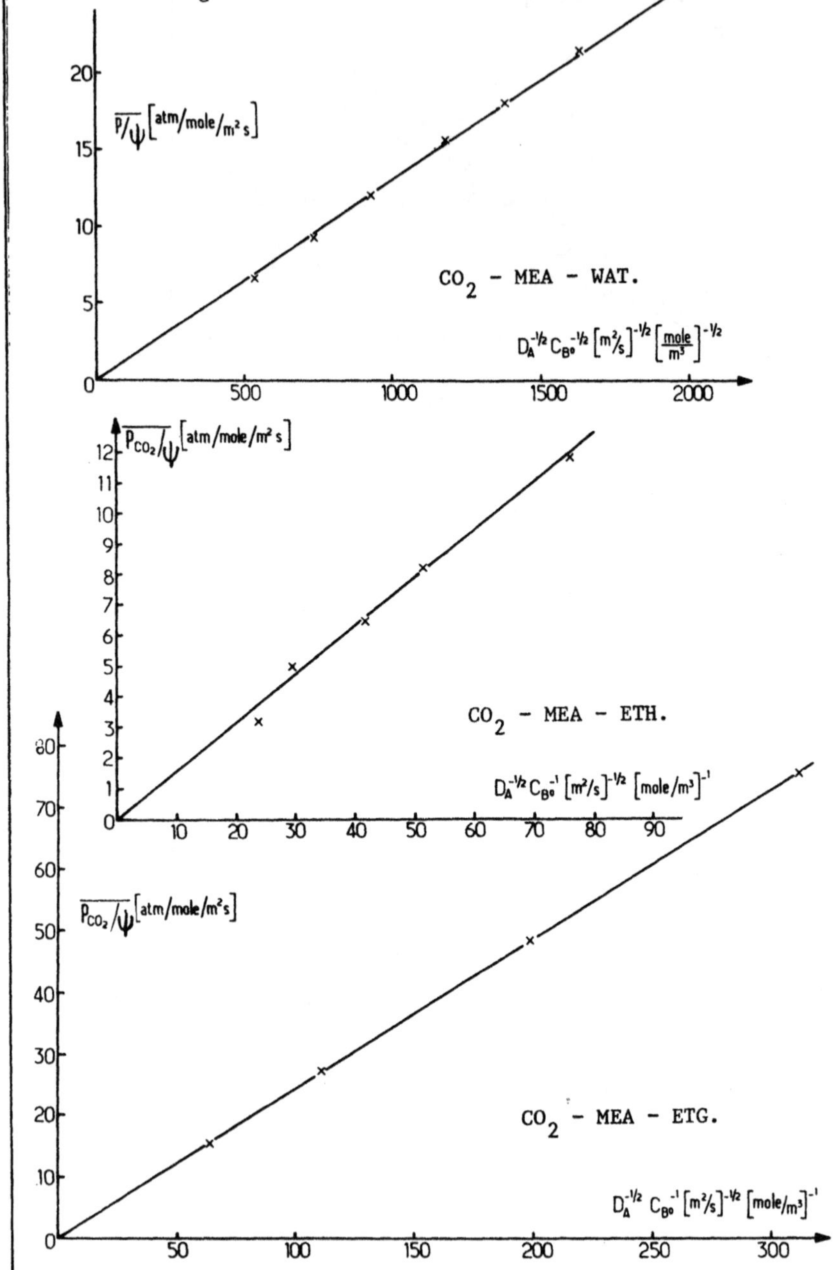

Fig. 5b - Gas absorption with chemical reaction - determination of k_{mn} and $(1/k_G)$

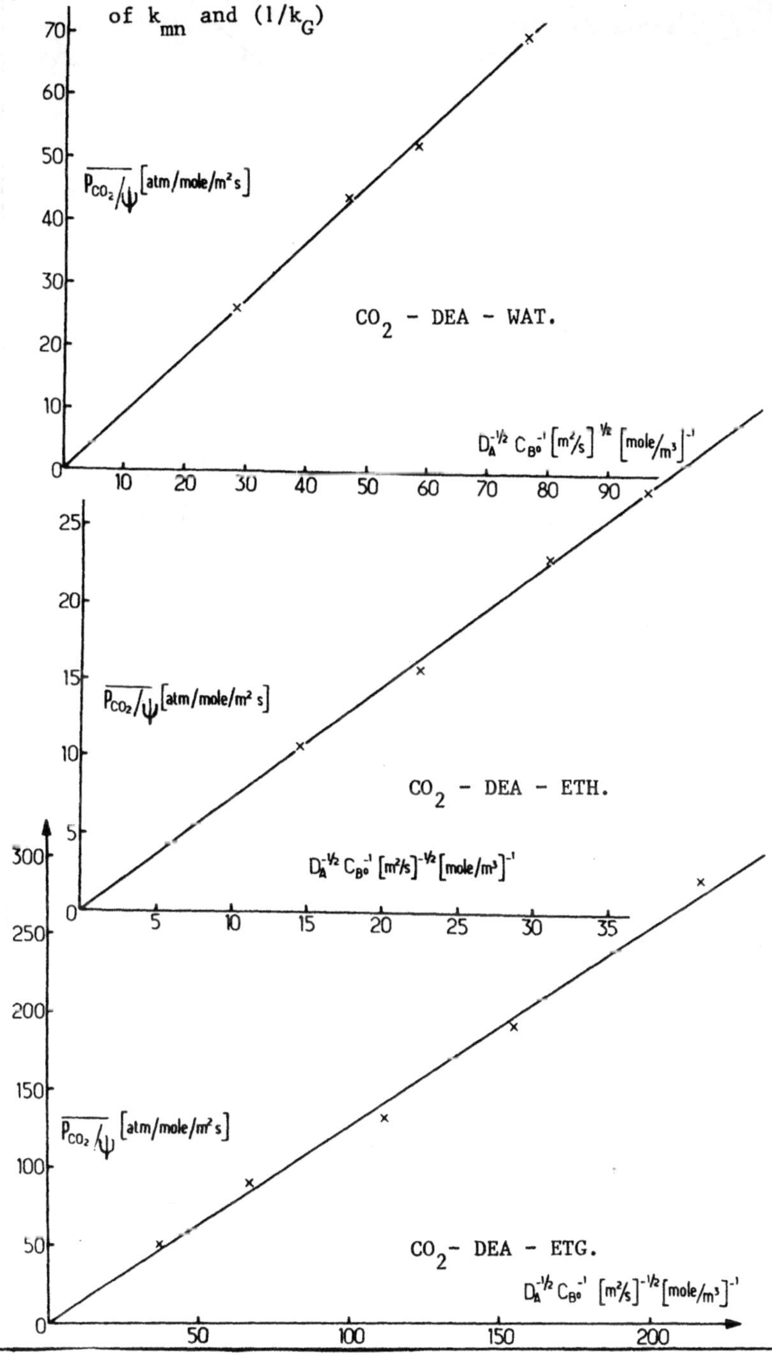

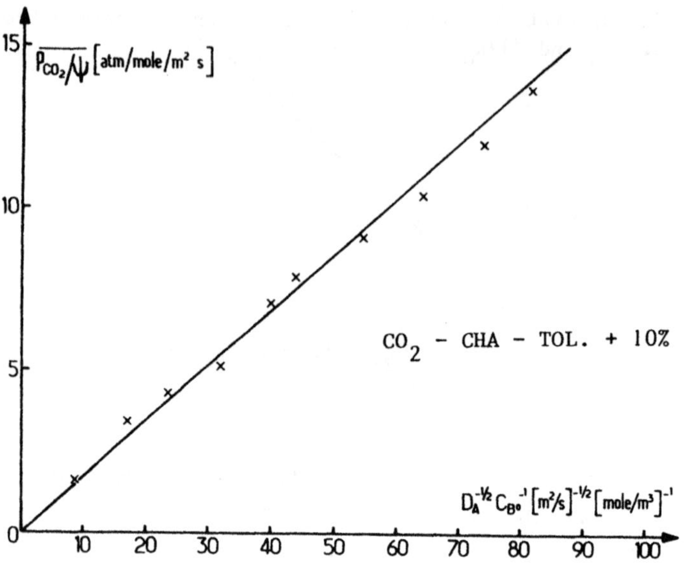

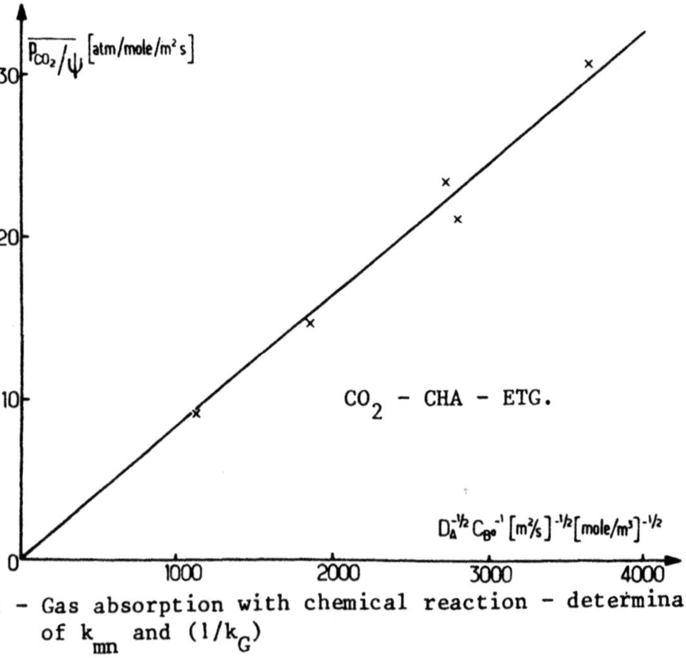

Fig. 5c - Gas absorption with chemical reaction - determination of k_{mn} and $(1/k_G)$

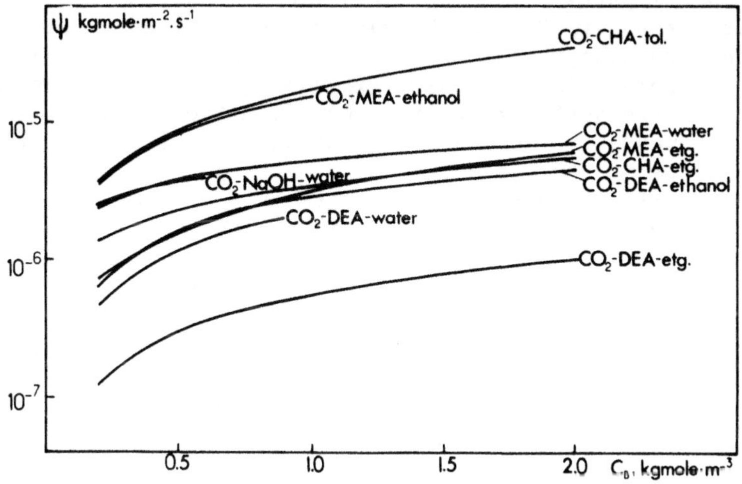

Fig. 6. Specific rates of absorption for eight amine chemical systems and CO_2-NaOH system.

3 MASS TRANSFER DATA WITH CHEMICAL AND PURILY PHYSICAL PROCESSES

In order to characterize the gas-liquid mass transfer performances of a reactor, k_La, k_Ga and a are determined by the above methods. But the question may arise whether the values of these parameters so obtained are valid only for similar mass transfer conditions and how they can be extrapolated to other operating conditions (for example from chemical absorption to physical absorption or vaporization). This means in other words, is the value of k_L or a, corresponding to the hydrodynamic conditions of chemical absorption, the same as the value of k_L or a corresponding to the hydrodynamic conditions of physical absorption ?

In fact in the case of a packed column some zones of liquid in the packing are almost motionless and they probably will be saturated by the absorbing gas during a physical absorption and hence almost ineffective to mass transfer. When the absorbing capacity of the liquid is increased by a chemical reactant or in evaporation operations, these zones will be still effective and the measurement of the amount of gas absorbed will give the feeling that k_L and a have greater values. Also in a mechanically agitated reactor the effective interfacial area in the case of a reacting system may be not equal to the effective interfacial area in the case of a physical absorption or desorption where the absorbing capacity of the liquid is not increased. Morevoer in the case of the absorption system with a fast chemical reaction the mass transfer coefficient is independent of the hydrodynamics and equal at every point in the vessel and all the interfacial distributed in all parts of the agitated vessel contribute equally to the mass transfer. But in the case of physical desorption or absorption, the mass transfer coefficient cannot be assumed to be equal at all points in the vessel (it can have quite different values around the agitator and far away from the agitator or in the region towards the top surface) and the interfacial region in different parts of the reactor do not contribute equally to the mass transfer (13). The same considerations apply for any gas-liquid reactor that offer hydrodynamic inhomogeneity.

For the above reasons it can be seen for example why the assumption of the same value of the interfacial area in physical and chemical absorption can lead to some incertainty especially if the mass transfer coefficient is obtained from the ratio k_La/a where k_La is measured by physical absorption or desorption and a from chemical absorption in two different series of experiments. The effective interfacial area in the case of the fast reaction system where the absorbing capacity is increased by a chemical reactant is probably much larger than the effective interfacial area for physical absorption or desorption and in fact the semi-stagnant liquid zones in a packing or in the bulk of an agitated tank can be more and less effective to mass transfer depending on the ratio between the absorbing capacity and the rate of absorption, as pointed out by Joosten and Danckwerts (33). These authors introduced a parameter γ

$$\gamma = \frac{1 + \dfrac{C_{Bo}}{zC_A^*}}{E}$$

that is the ratio between the increase of liquid absorption capacity and the increase of mass transfer due to chemical reaction. The experimental results showed that for a physical absorption ($\gamma = 1$) and absorption with instantaneous chemical reaction ($\gamma \simeq 1$) the effective areas are the same while for $\gamma >> 1$ the interfacial area increased. The different effective areas found by Joosten and Danckwerts (33) or by Laurent (21) using equal packing and similar absorbing solutions, can be explained by the different γ values in the different cases.

However the different values of k_La which depend on the type of mass transfer process (vaporization, chemical and physical absorption) are not only due to variations of liquid areas involved in mass transfer operations, but they are also due to variations of local mass transfer coefficients inherent to these zones (12, 13, 18) or variation of the mass transfer coefficient by the existence of interfacial turbulence caused by chemical absorption nhancing the physical absorption (34).

So the technique of conducting simultaneous absorption with ast pseudo-mth order reaction and physical absorption or desorp-ion concurrently is certainly a promising attempt to understand he whole complex problem of transfer in gas-liquid reactor (15, 7, 19). This leads to the simultaneous measurement of k_La and a hence k_L. But it may let some doubt on the value of k_L that, for xample, can be changed by the presence of the simultaneous chemi-al reaction. As explained by Prasher (18), it will be still more romising to lead these simultaneous experimental measurements in sing a chemical reaction in a regime where both hydrodynamics nd reaction have comparable effects (a Danckwerts plot where k_L nd a are determined by the chemical reaction simultaneously with $_La$ determined by the physical absorption or desorption).

In conclusion it is important to know that the effective in-erfacial area and the liquid mass transfer coefficient determined y chemical methods using fast reactions are probably larger than he effective interfacial area for physical desorption or absorp-ion. A correction relative to the interfacial area may be reali-ed with the parameter γ previously defined but it must be kept n mind that also the values of k_L may vary simultaneously and hould also necessitate a correction.

So finally, a perfect knowledge of gas-liquid mass transfer ata theoretically necessitates simultaneous measurements of mass ransfer parameters by two ways : local physical measurement of , k_L and k_La and global chemical or physical measurement of a, $_L$ and k_La to follow the variation in situ of the mass transfer arameter. It seems important to remark that the techniques exist ow and that the next years will provide fruitful data on that opic including nuclear engineering (35).

4 MASS TRANSFER COEFFICIENTS AND INTERFACIAL AREAS IN ABSORBERS AND REACTORS - SCALE-UP

The choice and the design of a suitable reactor for gas-liquid reaction or absorption is very often a question of matching the chemical thermodynamics and the reaction kinetics with the capabilities of the proposed reactor. Specific interfacial area a, liquid holdup β or gas holdup α and mass transfer coefficients k_La and k_Ga are the most significant characteristics of a reactor. Some published values of the mass transfer parameters will be presented now. Our objective here is to help to answer the following questions :

For a proposed type of gas-liquid contactor compatible with the properties and flowrates of the phases and with the reaction type, what are the likely values of the specific interfacial area and the gas and liquid mass transfer coefficients by which the contact performance can be predicted ? And what is the expected accuracy of these values ? *Table XI gives typical values of these parameters in typical contactors shown in Fig. 7 for fluids with properties not very different from those of air and water (especially liquid viscosity under 5 cP where the liquid is non-foaming).*

Recent data concerning more viscous and organic liquids will be also presented. Because this review is especially concerned with the chemical technique of determining mass transfer parameters, experimental data obtained by this technique will be given in subsequent figures and tables.

4.1 Packed Columns

Packed columns are used conventionally to obtain a low pressure drop or low liquid holdup when there is practically no heat to remove or supply or when the gas or the liquid is corrosive. They are not used when solids are present in the feed or are formed in the reaction. Although packed columns or reactors can be operated cocurrently, their operation is usually countercurrent. In particuler, countercurrent use is preferred when a higher concentration driving force is needed, that is, for distillation or for most physical absorption. However, when irreversible reaction occurs between dissolved gases and the absorbent, the mean concentration driving force is the same for both modes of operation. In this case the capacity of cocurrent columns is not limited by flooding, and at any given flowrates of gas and liquid the pressure drop in a cocurrent column is less. Also, in three-phase reactors, with packing serving as a catalyst, it is advantageous to use cocurrent operation.

Table XI - Mass transfer coefficients and effective interfacial areas in gas-liquid reactors

Type of reactor	β (% gas–liquid volume)	k_G (gm moles/cm^2 sec atm) $\times 10^4$	k_L (cm/sec) $\times 10^2$	a (cm^2/cm^3 reactor)	k_La (sec^{-1} $\times 10^2$)
Packed columns					
Countercurrent	2–25	0.03–2	0.4–2	0.1–3.5	0.04–7
Cocurrent	2–95	0.1–3	0.4–6	0.1–17	0.04–102
Plate columns					
Bubble cap	10–95	0.5–2	1–5	1–4	1–20
Sieve plates	10–95	0.5–6	1–20	1–2	1–40
Bubble columns	60–98	0.5–2	1–4	0.5–6	0.5–24
Packed bubble columns	60–98	0.5–2	1–4	0.5–3	0.5–12
Tube reactors					
Horizontal and coiled	5–95	0.5–4	1–10	0.5–7	0.5–70
Vertical	5–95	0.5–8	2–5	1–20	2–100
Spray columns	2–20	0.5–2	0.7–1.5	0.1–1	0.07–1.5
Mechanically agitated bubble reactors	20–95	—	0.3–4	1–20	0.3–80
Submerged and plunging jet	94–99	—	0.15–0.5	0.2–1.2	0.03–0.6
Hydrocyclone	70–93	—	10–30	0.2–0.5	2–15
Ejector reactor	—	—	—	1–20	—
Venturi	5–30	2–10	5–10	1.6–25	8–25

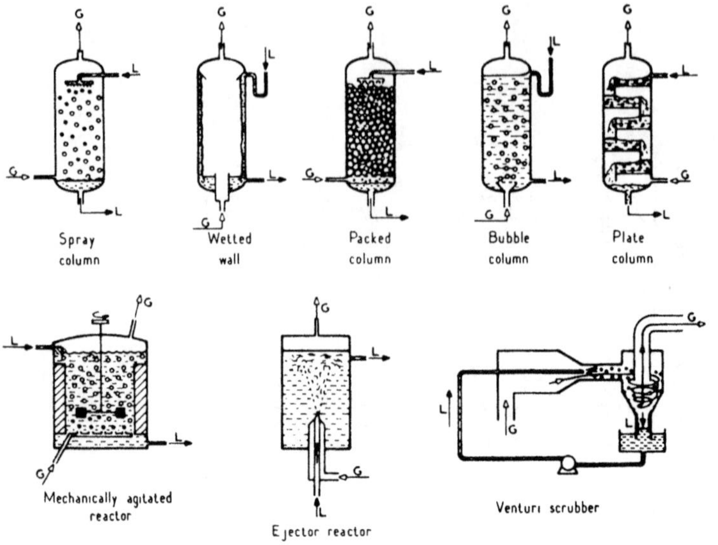

Fig. 7 - Principal types of industrial equipment.

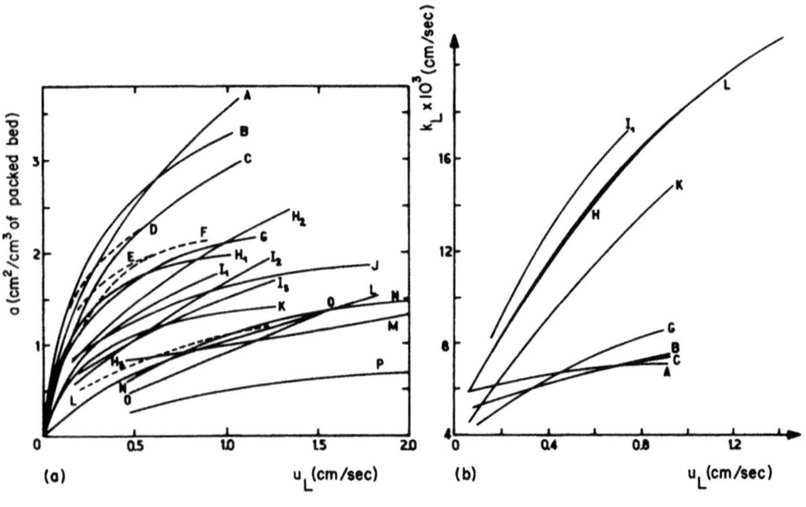

Fig. 8 - Interfacial areas (a) and true liquid-side mass transfer coefficients (b) in countercurrent packed columns.

	Packing		Temperature °C	Column diameter D	Number of particules by unit of packed volume	Packing surface area (cm^{-1})	Chemical systems
	Type	Nominal diameter (inch)					
A	Ceramic Intalox saddles	$\frac{1}{2}$	25°C	4 in	640 000	4.7	Absorption CO_2
B	Ceramic Pall rings	$\frac{1}{2}$	25°C	4 in	360 000	4.2	Absorption CO_2
C	Steel Pall rings	$\frac{5}{8}$	25°C	6 in	220 000	3.5	Absorption CO_2
D	Ceramic Raschig rings	$\frac{3}{8}$	30°C	10-20 cm	1 070 000		CO_2-NaOH
E	Ceramic Raschig rings	$\frac{3}{8}$	30°C	4.37 cm	980 000		9 different systems
F	Ceramic Intalox saddles	$\frac{1}{2}$	30°C	10 cm	385 000		CO_2-NaOH
G	Ceramic Raschig rings	$\frac{1}{2}$	25°C	4 in	370 000	3.8	CO_2-$HNaCO_3$-Na_2CO_3-H_2AsO_2
H_1	Ceramic Pall rings	1	25°C	9 in	49 000	2.2	Absorption CO_2
H_2	Inox Steel Pall rings	1	25°C	20 cm	49 000	2.0	CO_2-NaOH Air-Dithionite
H_3	Polypropylene Pall rings	1	25°C	20	51 000	2,0	CO_2-NaOH
I_1	Ceramic Intalox saddles	1	25°C	9 in	84 000	2.5	Absorption CO_2
I_2	Ceramic Intalox saddles	1	25°C	20 cm	75 300	2.5	CO_2-NaOH, CO_2-DEA Air-Dithionite
I_3	Polypropylene Intalox saddles	1	25°C	20 cm	53 500	2.0	CO_2-NaOH
J	Ceramic Raschig rings	1	25°C	12 in		2.03	O_2-Na_2SO_3-Co^{++}
K	Ceramic Raschig rings	1	25°C	9 in	48 000	1.8	Absorption CO_2
	Ceramic Raschig rings	1	25°C	20 cm	50 600	1.9	Air+Dithionite
	PVC Raschig rings	1	25°C	20 cm	51 400	1.9	Air+Dithionite
L	Ceramic Raschig rings	$1\frac{1}{2}$	20°C	18 in	14 000	1.3	CO_2-NaOH-CO_2-$HKCO_3$-K_2CO_3-ClO^-
M	Ceramic Intalox saddles	$1\frac{1}{2}$	11 à 34°C	0.5 m	21 000	1.6	O_2-Na_2SO_3-Co^{++}
N	Ceramic Raschig rings	$1\frac{1}{2}$	11 à 34°C	0.5 m	13 000	1.3	O_2-Na_2SO_3-Co^{++}
O	Polypropylene Pall rings	$1\frac{1}{2}$	11 à 34°C	0.5 m	15 000	1.3	O_2-Na_2SO_3-Co^{++}
P	Polypropylene Intalox saddles	2	11 à 34°C	0.5 m	6 200	1.1	O_2-Na_2SO_3-Co^{++}

Table of Fig. 8

4.1.1 Countercurrent packed columns. A great number of published values of interfacial area per unit packed volume a and true liquid-side mass transfer coefficient k_L have been compiled in ref. (1) for different packings in columns operating in the trickle flow regime before the loading state (fig. 8). The interfacial area depends on the type and size of the packing and is independent of the column height when reported per unit packed volume and is independent of the gas superficial velocity u_G when the column operates below the loading conditions. For any type of packing, a decreases when the size of the packing increases. For a given value of packing diameter, pall rings give the highest interfacial area. Moreover for a given packing shape, plastic materials often offer the smallest interfacial area. In large-scale columns, it is usually recommended to use 2 - in packings as being the most economical.

When the results of Fig. 8 do not apply, a suitable equation may be used. The most convenient one is probably the relationship of Onda et al. (37)

$$\frac{a}{a_c} = 1-\exp\left|-1.45\left(\frac{\sigma_c}{\sigma}\right)^{0.75}\left(\frac{L}{a_c\mu_L}\right)^{0.1}\left(\frac{L^2 a_c}{\rho_L^2 g}\right)^{-0.05}\left(\frac{L^2}{\rho_L \sigma a_c}\right)^{0.2}\right|$$

where a is here the wetted area per unit of packed volume, a_c is the total dry area of the packing per unit packed volume, L is the mass superficial liquid flowrate ($0.25 < L < 12$ kg/m^2.s). ρ_L and μ_L are the density and the viscosity of the liquid ($0.8 < \rho_L < 1.9$ g/cm^3, $0.5 < \mu_L < 13$ cP) ; σ and σ_c are the surface tension of the liquid and the critical surface tension of liquid for a particular packing material ($0.3 < \sigma/\sigma_c < 1.3$). The above equation correlates results for packing diameter comprised between 1 and 3.8 cm within a maximum uncertainty of $\pm$ 20 %, except in the case of pall rings where it is conservative. The construction of the Pall rings disperses part of the liquid as small droplets not taken into account and this effect may double the values of a. For such metal 1 and 1.5 m rings, an intensive work published by Mahajani and Sharma (39) has led to correlate the effective interfacial area in trickle flow as a function of the fluid superficial velocity $a = C u_L^{\alpha} u_G^{\beta}$ (CGS units) with C, α and β varying respectively in the range 0.85-1.17 ; 0.38-0.46 and 0.03-0.11 depending on the shape and the material of the packing.

The true liquid side mass transfer coefficient k_L all lie between 9.4×10^{-2} and 2×10^{-2} cm/s and are usually assumed independent of u_G. A representative relationship for k_La (in sec^{-1}) has been proposed by Mohunta et al. (42) ($\pm$ 20 %)

$$k_La = 25\times10^{-4}\left(\frac{g\rho_L}{a_c\mu_L}\right)^{0.66}\left(\frac{g^2\rho_L}{\mu_L}\right)^{1/9}\left(\frac{\mu_L L^3 a_c^3}{g^2\rho_L^4}\right)^{0.25}\left(\frac{\mu_L}{\rho_L D_L}\right)^{-0.5}$$

within the range $0.1 < L < 42$ kg/m^{2}s, $0.015 < G < 1.22$ kg/m^{2}s, $0.7 < \mu_L < 1.5$ cP, $140 < (Sc_L = \mu_L/\rho_L D_L) < 1030$, $6 < d < 50$ mm in column diameters comprised between 6 and 50 cm. For viscous liquid up to 26 cP (Sc_L up to 220.000) Mangers and Ponter (40) extended the plot proposed by Sherwood and Holloway (41) between $k_La/D\ Sc^{0.5}$ and L/μ_L. For these liquids, when the liquid flowrate is increased, two hydrodynamical behaviours are observed depending the packing is only partially wetted or completely wetted. The transition rate is given by

$$\left(\frac{L}{\mu}\right)_{tr} = 4.0\left[1.12(1-\cos\theta)^{0.6}\left(\frac{\rho_L\sigma_L^3}{\mu_L^4 g}\right)^{0.2}\right]^{0.87} = 4(M.W.R)^{0.87}$$

M.W.R is the minimum wetting rate and θ is the contact angle of the system.

Thus two correlations are applicable, proposed for water and water-glycerol mixtures should be applied for viscous liquids :

- for partially wetted packing,

$$\frac{k_La}{D_L} = 0.0039\left(\frac{L}{\mu_L}\right)^{\alpha} Sc_L^{0.5}Ga^{0.27}Ka^{0.33}(M.W.R)^{-1.67}$$

with $\alpha = 0.484(M.W.R)^{0.108}$, $Ga = \rho_L^2 gd^3/\mu_L^2$ and $Ka = \rho_L\sigma_L^3/\mu_L^4 g$

- where complete wetting is achieved, a further increase in liquid flowrate increases the liquid film thickness on the packing and leads to enhanced transfer rate due to interfacial friction and then

$$\frac{k_La}{D} = 2.03\left(\frac{L}{\mu_L}\right)^{1.44} Sc_L^{0.5}Ga^{-0.18}$$

It should be noted that these correlations are applicable for high surface tensions ($\sigma_L > 67$ dyn/cm), especially nonfoaming liquids.

Moreover most of the reported mass transfer results are confined to aqueous systems. As explained before, Sridharan and Sharma (43) developped new chemical systems that have led to the measurement of a and k_L in organic solvents. For example, values of interfacial areas obtained in a reaction of CO_2 with cyclohexylamine in xylene in 10 cm diameter column packed with 0.5 in ceramic Raschig rings are higher than those obtained in comparable working conditions with aqueous systems (43). This is due to a lower surface tension leading to a better wetting of the particles

Finally the true gas-side mass transfer coefficients k_G values may be correlated within a range of $\pm$ 30 % by

$$\frac{k_G P}{G} = \frac{C}{M} (a_c d)^{-1.7} \left(\frac{Gd}{\mu_G}\right)^{-0.3} Sc_G^{-0.5}$$

where P is the total pressure in atm, M is the gas molecular weight in gm/mole and Sc_G is the Schmidt's number for the gas phase, C = 2.3 for d < 1.5 cm and C = 5.23 for d > 1.5 cm.

4.1.2 Packed bubble columns. As seen later, bubble columns are frequently operated because of their low cost, simplicity of operation, high heat transfer rates and the ease with which the liquid residence time can be varied. However they involve the disadvantage of severe gas back-mixing and bubble coalescence phenomena which can be reduced substantially by packing the column. Packed bubble columns show 15-100 % improvement in effective interfacial area a_d and mass transfer coefficient $(k_L a)_d$ based on void volume, over those for empty bubble columns under otherwise similar conditions (44, 45, 46). However, when mass transfer results are based on total column volume (that is the actual design volume), the improvement over an empty bubble column is less, because of the substantial part of the column volume occupied by solid. To avoid this, while maintaining the advantages of the packed bubble columns, it is desirable to use packing with high porosities ε such as screen packings (47) and gauze packings (48).

Chen and Vallabh (47) have obtained data on gas holdup and liquid side mass transfer coefficients from 68 to 144 mm i.d. columns using cylindrical (0.5 in and 1 inch) screen packings (ε = 0.97). Sahay and Sharma (45) have reported a detailed study for 100 to 380 mm i.d. columns with packings of different sizes and shape (ceramic, plastic, metal). Recently Sawant et al. (48) reported results concerning wire gauze packings (diameter equal to the inner diameter of the column - ε = 0.95) in 100 and 200 i.d. columns. All these data may be regrouped in Fig. 9 (48). It may be stated that in the range of gas and liquid superficial velocities covered ($1 < u_G < 30$ cm/s ; $0.1 < u_L < 0.5$ cm/s) the *data obtained in a small column diameter with the specified system and the packing can be used for a reasonable scale-up on a basis that both interfacial area and liquid side mass transfer coefficients (reported on total column volume) vary as* $u_G^{0.5}$ independently of the liquid velocity, the type of gas distributor and the ratio H'/D between the height of dispersion and the diameter of the column. Moreover the performance characteristics of wire gauze packings are comparable with those of Pall rings and superior to other conventional packings. Note that most of the results reported in Fig. 9 concern experimental values in the presence of antifoaming agents (45, 48) which are found to be about 50 % lower than those obtained in its absence (47).

4.1.3 Cocurrent packed columns - Trickle-bed reactors. Cocurrent gas-liquid flow in packed beds, packing being either catalytic or inert, is advantageously employed in the petroleum and chemical industries. Successful modeling of mass transfer in packed-bed reactors requires careful study of the three-phase hydrodynamics - fluid flow patterns, pressure drops, and liquid holdup.

Because cocurrent flow is not bounded by the phenomenon of flooding, it offers a greater range of hydrodynamic patterns, which must be specified before considering the mass transfer behavior (49-53).

a. *Hydrodynamics*. In the case of cocurrent downflow, the packed-bed reactor works in two main regimes : first, the *trickle-flow regime* or gas continuous flow regime in which for initially a zero gas rate the liquid phase trickles over the packing in a network of films, rivulets, and drops adjacent to a stagnant continuous gas phase ; and second, the *single-phase liquid regime* in which for initially a zero gas rate the liquid phase fills the packing voids. When the liquid mass flowrate L is kept constant and the mass gas flowrate G is started and increased, the following flow patterns are observed.

The initial trickle flow of liquid ($L < 20$ kg/m^2 sec) through beds of rings, spheres, beads and pellets, as the gas flow increases, gives way to an alternate gas-rich or liquid-rich flow downward through the column (pulse flow) and finally to a turbulent state which appears to involve a continuous gas phase with part of the liquid suspended as a mist and with the other part covering the packing as a film (spray flow or blurring flow) or, for high liquid flowrates to a liquid continuous regime where the gas flows in the form of small bubbles (dispersed bubble flow regime). For small values of L ($L < 2-5$ kg/m^{2}s), there is not enough liquid to wet the whole packing surface and pulse flow cause by the liquid obstructing the gas flow is not encountered. When the liquid phase *foams* during gas flow, two new regimes arise, foaming flow and foaming-pulsing flow, which appear between trickling flow and pulsing flow.

. Effect of superficial gas velocity on effective interfacial area. air–sodium dithionite.

Symbol	*Column dia.* (mm)	*H'/D*	*Packing*
○	100	1.5 - 13	Wire gauze
×	200 (single tube sparger)	3	Wire gauze
■	200 (two tube sparger)	3	Wire gauze
□	200	4.2	1.5 in. ceramic Intalox saddles
▽	200	4.0	1 in. stainless steel Pall rings
●	200	4.4	1 in. ceramic Intalox saddles
⊕	200	4.1	Empty

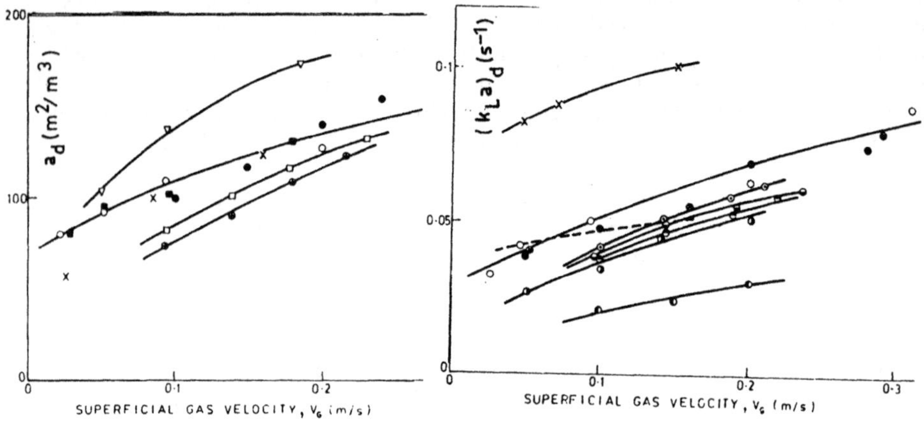

Effect of superficial gas velocity on liquid side mass transfer coefficient.

	Column dia. (mm)	*H'/D*	*Packing*	*System*
○	100	1.5 - 13	Wire gauze	Lean CO_2–Na_2CO_3 + $NaHCO_3$
●	200 (single tube sparger)	3	Wire gauze	Lean CO_2–Na_2CO_3 + $NaHCO_3$
⊗	200 (two tube sparger)	3	Wire gauze	Lean CO_2–Na_2CO_3 + $NaHCO_3$
◐	100	4.8	3/8 in. ceramic Raschig rings	Lean CO_2–Na_2CO_3 + $NaHCO_3$
⊙	200	3.2	5/8 in. stainless steel Pall rings	Lean CO_2–Na_2CO_3 + $NaHCO_3$
◑	200	4.3	1 in. stainless steel Pall rings	Lean CO_2–Na_2CO_3 + $NaHCO_3$
◒	200	4.3	1 in. ceramic Raschig rings	Air–CuCl + HCl
◓	200	4.3	1.5 in. ceramic Intalox saddles	Air–CuCl + HCl
×	69.85	17.45	0.5 in. × 0.5 in. cylindrical screen packings	Air + CO_2–water

Fig. 9 - Packed bubble columns (48)

Starting from a liquid-full system at constant liquid flow ($L > 20\ kg/m^2s$ downward, or at any L upward), the following patterns are observed as the gas flowrate increases : first bubble flow in which bubbles appear unbroken in the continuous liquid phase ; then distorded bubble flow in which the bubbles begin to coalesce and to surround several packing elements, then pulse flow ; and finally spray flow.

In industrial practice, applications of the trickle bed reactor are found in the gas continuous as well as in the pulsing flow regime. However there is a considerable lack of information concerning the hydrodynamic behaviour of the pulses. Quite recently an intensive work has been published by Block (56) which shows clearly for the air-water system that the transition to pulsing flow occurs at the same real liquid velocity u_{LR} corresponding to a Froude number $Fr = 0.09 = u_{LR}^2/gd_p$ with Raschig rings and ceramic cylinders. During the pulsing flow, the liquid periodically blocks the gas channel between the packing particles and this such created liquid plug is busequently blown away by the gas flow and moves downward at a relatively high speed (1 m/s). On its course the pulse takes in continuously fresh liquid at the front and leaves liquid behind it at its back. So a pulse can more and less be described as a wave moving downward *increasing heat and mass transfer*. The pulse frequency is linearly controlled by the difference between the real liquid velocity and the real liquid velocity at pulsing onset occuring at the bottom of the column. The holdup at the front of a pulse is generally 60 % higher than the holdup between pulses and this independently of gas and liquid flowrates. Finally Block has also shown that the pulse characteristics are not very dependent on the column diameter.

Flow patterns and transition from one pattern to another as the flowrates changes, have been described by several authors. For example, this behaviour is summarized in Fig. 10 for cylindrical 1×5 mm and spherical 2.4 and 3 mm spherical alumine catalyst with air and different foaming and nonfoaming hydrocarbons and viscous organic liquids (32). This diagram was formerly proposed for different gas and liquids (51, 54). It covers the fluid physico-chemical range $0.77 < \rho_L < 1.2\ g/cm^3$; $0.3 < \mu_L < 67$ cP ; $19 < \sigma_L < 75$ dyn/cm and $0.15 < \rho_G < 2\ kg/m^3$. Complementary diagrams with fundamental basis or phenomenological approach and also for other shapes and sizes of packings are proposed by Talmor (55) and Hofmann et al. (53, 57). They have been recently extensively reviewed and tested by Perez Sosa (58).

These different flow patterns provide several different gas-liquid geometric configurations. Hence the mass transfer properties in downward cocurrent flow are related to gas and liquid energy dissipation rates and depend on pressure drop and total liquid holdup.

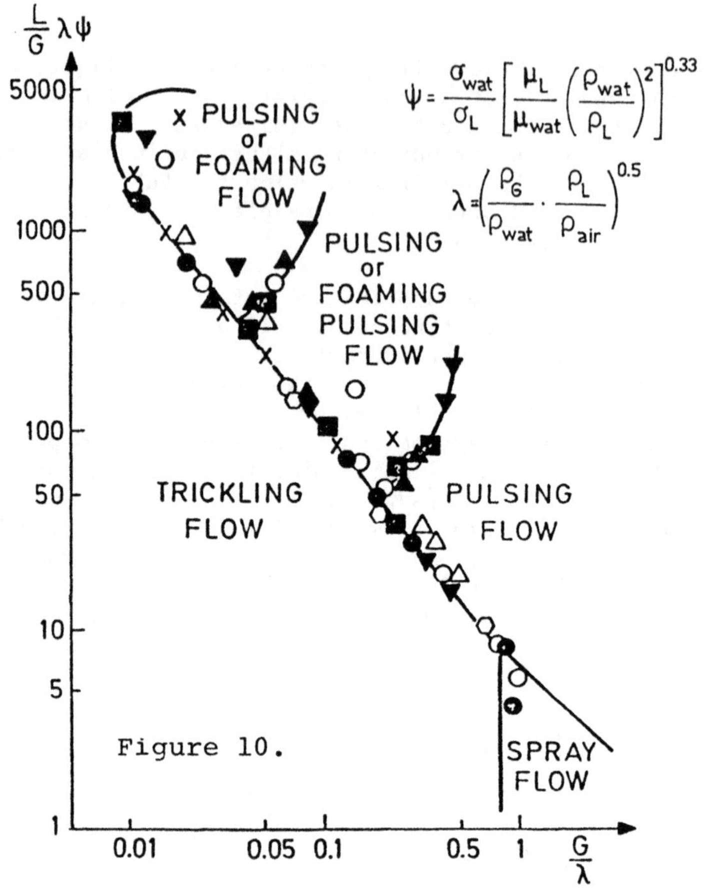

Figure 10.

LIQUID	CATALYST PACKING	β_c	σ_L dynes/cm	ρ_L g/cm^3	μ_L cp	A	a	λ	Ψ	FOAMING (*)	KEY
Methanol	Spherical catalyst d = 2.4 mm	0.071	25.08	0.805	0.698	0.237	0.408	0.897	2.970	+	■
Cyclohexane		0.078	25.55	0.769	0.904	0.275	0.402	0.877	3.274	-	○
Kerosene		0.079	25.80	0.810	1.346	0.290	0.366	0.900	3.571	+	●
Desulfurized gas-oil		0.105	29.00	0.840	5.793	0.424	0.242	0.917	5.022	+	▼
Ethyleneglycol		0.2220	48.96	1.118	17.218	0.560	0.138	1.057	3.051	-	✱ or X
Polyethyleneglycol		0.2223	39.36	1.146	66.258	0.616	0.103	1.071	6.737	-	⬡
Kerosene	Spherical catalyst d = 3mm	0.077	25.3	0.80	0.99	0.26	0.53	0.88	3.3	+	□
Cyclohexane		0.077	25	0.78	0.93	0.23	0.46	0.88	3.3	-	▽
Desulfurized gas-oil	Cylindrical catalyst 3 d = 1x5 mm	0.13	29.50	0.85	5.75	0.445	0.29	0.922	4.86	+	▲
Cyclohexane		0.097	25.55	0.771	0.904	0.311	0.50	0.877	3.274	-	△

(*) (+) foaming liquid; (-) nonfoaming liquid

There are several correlations to predict these parameters, the most famous being proposed by Larkins et al. (59) for cocurrent bubble packed reactors (± 20 %)

$$\log \frac{\delta_{LG}}{\delta_L+\delta_G} = \frac{0.416}{(\log X)^2+0.666}, \qquad \delta_{LG} = \left(\frac{\Delta H}{Z}\right)_{LG} + \frac{\beta\rho_L+(1-\beta)\rho_G}{\rho_m}$$

$$\log \beta = -0.774 + 0.525(\log X) - 0.109(\log X)^2$$

$$0.05 < X = \sqrt{\frac{\delta_L}{\delta_G}} < 30$$

δ_L are δ_G are the friction pressure loss that would exist if the liquid and the gas were assumed to flow alone separately in single phase flow with the same rates as those in two phase flow. For design, values for δ_L and δ_G may be obtained by an Ergun's type relationship as a first approximation,

$$\frac{f}{2} = \frac{\Delta H}{Z}\,\frac{\varepsilon^3}{\rho u^2 a_g(1-\varepsilon)} = h_K\,\frac{\mu a_g(1-\varepsilon)}{\rho u} + h_B$$

with different values of h_K and h_B for different packings, but a quick and easy experimental measurement of h_K and h_B for the packing to be used, will lead to more accurate values. Note that Larkins and al's relationship is not valuable when liquid is trickling but several other relationships have been proposed including this regime. For air-water flows through beds of spheres, Sato et al. (60) proposed a similar expression which is symetric about X = 1.2 instead of X = 1 :

$$\log \frac{\delta_{LG}}{\delta_L+\delta_G} = \frac{0.70}{(\log X/1.2)^2+1.00} \quad \text{and} \quad \beta = 0.42\left|a_g(1-\varepsilon)\right|^{0.33} X^{0.22}$$

$$0.1 < X < 20$$

An alternate relationship has also been proposed by the same authors,

$$\left(\frac{\delta_{LG}}{\delta_L}\right)^{0.5} = 1.30 + 1.85X^{-0.85} \quad \text{for} \quad 0.1 < X < 20$$

Turpin and Huntington (61) also proposed the empirical correlations

$$\ln f_{LG} = 7.96 - 1.34 \ln Z + 0.0021(\ln Z)^2 + 0.0078(\ln Z)^3$$

$$\beta = -0.017 + 0.132 \left(\frac{L}{G}\right)^{0.24}$$

with $f_{LG} = \dfrac{d_e \delta_{LG}}{2\rho_G u_G^2}$, $\quad d_e = \dfrac{2}{3}\dfrac{\varepsilon d}{(1-\varepsilon)}$, $\quad 0.2 < Z = \dfrac{Re_G^{1.167}}{Re_L^{0.767}} < 500$

$$1 < \left(\frac{L}{G}\right)^{0.24} < 6$$

In the previous equations, u is the superficial velocity, ε the intergranular porosity of the packing, a_g the specific area of the packing, μ and ρ the viscosity and density of the considered phase, ρ_m is the density of the manometric fluid.

Instead of using the Larkins et al's two phase parameter δ_{LG} proportional to the resultant of the friction forces, Charpentier et al. (54, 62) have suggested for the high gas-liquid interaction systems to use the two phase parameters ξ_{LG}, ξ_L and ξ_G proportional to the frictional power,

$$\xi_{LG} = \frac{1}{\varepsilon}\left(\frac{L}{\rho_L} + \frac{G}{\rho_G}\right)\left(\frac{\Delta H}{Z}\right)_{LG} + \frac{L+G}{\varepsilon\rho_m}$$

with the limiting cases whether the gas is flowing alone in single phase flow through the packing or the liquid phase is flowing alone as a trickling flow over the packing. Thus a representation was adopted between these parameters with ψ_L or β versus χ',

$$\psi_L = \sqrt{\frac{\xi_{LG}}{\xi_L}}, \qquad \chi' = \sqrt{\frac{\xi_L}{\xi_G}}$$

L and G are the liquid and gas superficial mass flowrates. Finally for the case of foaming and nonfoaming hydrocarbons and viscous liquids together with the packing diameter (d_p = 1-3 mm) of the catalyst encountered in the trickle-bed reactors, the following correlations are suggested to determine the pressure drop and the liquid holdup :

- for nonfoaming and non viscous liquids (cyclohexane, petroleum ether, gazoline...) or for foaming liquids in trickle flow, (within $\pm$ 20 %)

$$\Phi_L = \left(\frac{\delta_{LG}}{\delta_L}\right)^{0.5} = 1 + \frac{1}{X} + \frac{1.14}{X^{0.54}} \qquad \beta = \frac{0.66X^{0.81}}{1+0.66X^{0.81}}$$

with $0.1 < X < 80$

- for foaming hydrocarbons (gas oils, kerosene...), except at trickle flow (within ± 20 %)

$$\psi_L = \left(\frac{\xi_{LG}}{\xi_L}\right)^{0.5} = 1 + \frac{1}{X'} + \frac{6.55}{X'^{0.43}} \qquad \beta = \frac{0.92X'^{0.30}}{1+0.92X'^{0.30}}$$

for $0.05 < X' < 100$.

- for viscous organic liquids (ethylene and polyethyleneglycol..) within ± 20 %,

$$\psi_L = 1 + \frac{1}{X'} + \frac{7.11}{X'^{0.55}} \qquad \beta = \frac{4.83X'^{0.58}}{1+4.83X'^{0.58}}$$

for $0.05 < X' < 100$.

For trickle-bed reactors and also for plate columns and bubble reactors, it is seen for both ionic and organic liquids that very often the correlations to use to determine either the hydrodynamic parameters or the mass transfer parameters depend on the foaming or the coalescing ability of the dispersion. The "a priori" knowledge of this ability which is not related to the surface tension σ_L may be qualitatively obtained by an equipment in which two bubbles coalescates (62). Indeed it is well known that the foaminess of a liquid in the presence of a gas phase is caused by the decreased coalescence of gas bubbles trapped in the liquid. Therefore the study of the coalescence of gas bubbles trapped in liquids may be carried out as a qualitate mean to determine the foaming ability of these liquids. In studying bubble coalescence, the objective is to contact pairs of gas bubbles and determine the percentage of the coalescing pairs. The schematic diagram of such an equipment used for organic liquids is shown on Fig. 11. The experimental procedure consists in switching on the piston injection drive and photographing, with a moving camera, 200-250 contacted pairs of nitrogen bubbles generated and injected into a thermostated rectangular tank filled with the organic liquids. The injection tubes were side by side and had identical orifice diameters and the rate of the gas injection was in the range 0.07-17 cm^3/mn. The degree of coalescence was determined by counting on the projected film, the number of coalescing pairs. This was reported as "% coalescence" by dividing the number of coalescing pairs by the total number of pairs injected. The reproductibility of the system was very good (within 4 %). Figure 11.a shows the results of experiments carried out with the tank filled with pure cyclohexane in which small amounts of desulfurized gas-oil were added. The gas injection rate was 8 cm^3/mn. It has been observed that for pure cyclohexane or for mixtures of cyclohexane and small weight percentages of desulfurized gas-oil there is 100 % coalescence and such liquids are not foaming. Then between 6 and 14 % of added desulfurized gas-oil, the coalescence rate is decreasing abruptly from 100 % to 0 % while the surface tension and the viscosity are very slightly varying (σ_L = 25.55 - 26 dyn/cm and μ_L = 0.950 - 1.105 cP).

The pressure drop and liquid holdup have been measured simultaneously with the previous mixtures in a reactor packed with the 2.4 mm spherical catalyst 2 for L = 5 kg/m^2s. Figure 11.b shows that for a constant gas flowrate the pressure drop continuously increases with the percentage of added gas-oil. The foams appear between 5 % and 8 % of added gas-oil which corresponds to the start of the decrease of % coalescence in Fig. 11.a. Note that the observed flow regimes of the data of Fig. 11.b are the trickling and pulsing flows for the nonfoaming mixtures (when the % coalescence is smaller than 6 %) while the trickling, foaming, foaming-pulsing and pulsing flows are observed for the foaming liquids (especially when the % coalescence rate was between 8 % and 14 %, that is for nearly quite the same values of μ_L and σ_L as those of the nonfoaming mixtures).

It is important to note that the data usually obtained with air and water are not very representative especially for foaming fluids. Thus it is suggested to use complementary specific correlations depending the aqueous or the non aqueous nature of the liquid - Gianetto et al. (52). *Moreover it should be also emphasized that most of the data concern small diameter column (D < 10 cm) with a good initial distribution. For reactors up to as much as 3 m in diameter, it may be that some segregation in the flow of the phases leads to less important gas-liquid interaction and therefore to smaller pressure loss especially for the nonfoaming liquids* (36).

b. *Gas-liquid mass transfer correlations.* Liquid phase mass transfer coefficient k_La is affected both by the gas rate and the liquid rate. The following correlation is recommended by Reiss (64) and Satterfield (63)

$$k_L a = 0.0173\, E_L^{0.5} \left(\frac{D_A}{2.4\times10^{-9}}\right)^{0.5}$$

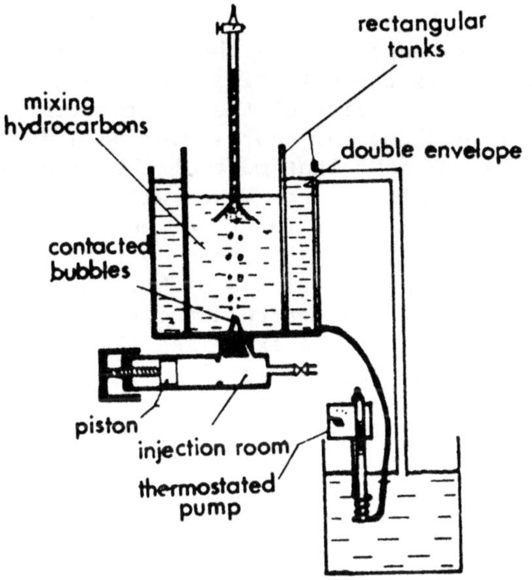

Schematic diagram of the coalescence equipment.

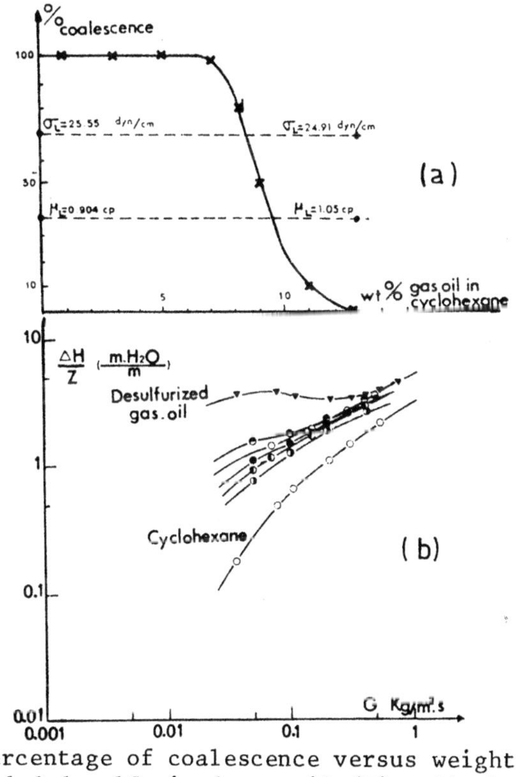

ig. 11 - Percentage of coalescence versus weight percentage of added desulfurized gas-oil (Fig. 11.a). Pressure loss versus weight percentage of added desulfurized gas-oil (fig. 11.b) where E_L is an energy dissipation term (in W/m^3) for liquid flow evaluated as $E_L = |(\Delta P/Z)_{LG} u_L|$; u_L is the superficial liquid velocity (in m/s) and D_A is the diffusivity of the gas in the liquid (in m^2/s). If the viscosity of the liquid differs much from that of water, a complementary correction should be applied. However it is important to note that most of the results tested to obtain Reiss' relationship are concerned with pulse flow and spray flow (E_L > 60-100 W/m^3). For lower gas and liquid rates corresponding to a trickling flow of liquid over the packing, for which reported values of k_La vary from 0.01-0.1 s^{-1}, reported experimental values of k_La are smaller than those predicted by the above equation. A comparison of the experimental data published by the different authors has led Charpentier to recommend, as a first approximation, for ionic liquids

$$k_La = 0.0011\ E_L \left(\frac{D_A}{2.4\times 10^{-9}}\right)$$

for $5 < E_L < 100\ Wm^{-3}$ with eventualy a complementary correction for viscosity. For smaller values of u_L and u_G, the pressure loss is equal to a few cm H_2O/m of packing in trickle beds and the use of an energy dissipation term is irrelevant in trickle flow. In this case, mean value of 0.008 s^{-1} will be a good approximation for packing with d_p > 2 mm. *Finally, it is important to note that at high flowrate of gas and liquid, values of k_La may exceed 1 sec^{-1}, which can hardly be attained in other types of gas-liquid contactors such as a bubble column or an agitated vessel. On the contrary, when the liquid is trickling over the packing, k_La values are of the same order of magnitude as those obtained in countercurrent under the same working conditions.*

Gas phase mass transfer coefficient k_Ga is also affected both by the gas rate and the liquid rate. Most extensive results have been proposed by Reiss (64) for 12.5, 25 and 76 mm polyethylene Raschig rings and 25 mm intalox saddles, by Gianetto et al. (65) for 6 mm spheres, Berl saddles and glass and ceramic Raschig rings and by Shende and Sharma (66) for 25 mm ceramic - intalox saddles and propylene Pall rings and 16 mm stainless steel Pall rings. In these different studies, the values of k_Ga vary between 2 and 70 s^{-1} for liquid rates ranging from 0.5 to 30 cm/s and gas rates from 40-450 cm/s but there exists some discrepancy in the way of representating the results. So for packings not studied by the previous worker, use of the Reiss' relationship will be a first approximation for any flow pattern except when liquid is trickling

$$k_Ga = 2.0 + 0.1\ E_G^{0.66}$$

here E_G is an energy dissipation term (in W/m^3) for gas flow valuated as $E_G = |(\Delta P/Z)_{LG} u_G|$. *As for the trickling flow regime, he $k_G a$ values for concurrent flow are higher than those for coun- ercurrent flow, when comparable, so the use of the correlations aluable for countercurrent will give conservative results.*

Effective interfacial area a increase with an increase in the as as well as the liquid superficial velocity, unlike the counter- urrent operation where the effective interfacial area is indepen- ent of the superficial gas velocity up to the loading point. The alues of effective interfacial area offered by various packings iffer substantially. For example, the extensive work of Shende nd Sharma [66] leads to the conclusion that polypropylene Pall ings show better performance than polypropylene Intalox saddles nd that the highest values of interfacial area are offered by tainless steel Pall rings for each size packing. In fact inter- acial areas are much larger than those obtained with countercur- ent flows and very much greater than the geometric packing area $_c = a_g(1-\varepsilon)$ itself for high liquid and gas flowrates (it may ex- eed 100 % for spray flow pattern). Following Gianetto et al. [65] t is likely that for pulse and spray flow, where a great frac- ion of the liquid phase is in the shape of droplets, there xists a relationship between a and the energy supplied per feed luid volume unit and per column unit, that is the pressure gra- ient if gravity is ignored. The correlation proposed by the au- hors is,

$$\frac{a}{a_g(1-\varepsilon)} = 0.25 \left| \left(\frac{\Delta P}{Z}\right)_{LG} \frac{\varepsilon}{a_g(1-\varepsilon)} \right|^{0.5}$$

Unfortunately most other authors did not present experimen- al pressure loss data simultaneously with interfacial area data. o in using flow pattern diagrams and in calculating pressure oss with energetical relationships, a comparison of the above quation obtained for 6 mm packings with experimental results of ther packings has been proposed by Charpentier [51]. It is con- luded that if for pulse flow and spray flow, this equation is a ood approximation, whatever the shape and packing size, however, uch equation may fail for a trickling flow of liquid. In this ase, a conservative representation *for aqueous liquids such as ulfite or sodium hydroxide solution* is given by,

$$\frac{a}{a_g(1-\varepsilon)} = 0.05 \left| \left(\frac{\Delta P}{Z}\right)_{LG} \frac{\varepsilon}{a_g(1-\varepsilon)} \right|^{1.2}$$

$$\text{for } \left(\frac{\Delta P}{Z}\right)_{LG} \frac{\varepsilon}{a_g(1-\varepsilon)} < 12\ N/m^2$$

This equation is proposed mainly for spheres and pellets packings with the porosity $\varepsilon < 0.50$.

More recently, Morsi et al. (68, 69) compared the interfacial area results obtained in trickle flow with organic and aqueous systems for spherical glass spheres (d_p = 1.16 and 2.4 mm), spherical aluminous catalyst (d_p = 2.4 mm) and glass Raschig rings (d_p = 6.48 mm).

It has appeared that for very small packings, the ionic solutions may give interfacial area 5 to 10 times higher than for non foaming organic liquids. So except for small d_p, all the other data are better represented by the correlation (within $\pm$ 30 %)

$$a = 7.75\times10^{5} \frac{\varepsilon}{a_g(1-\varepsilon)} \delta_{LG}$$

$$\text{for} \quad 3\times10^{-5} < \frac{\varepsilon}{a_g(1-\varepsilon)} \delta_{LG} < 7\times10^{-4}$$

When the holdup and pressure drop are correlated with the correlations ψ_L and β versus X, proposed previously.

These seem the first published correlation for organic systems with aluminous catalytic packings and confirm the non always applicability of the mass transfer parameters determined with aqueous chemical systems to the case of organic liquids in gas-liquid reactors.

Complementary informations for carbon pellets and granules may be found in the work by Mahajani and Sharma (67) where a and $k_L a$ are correlated in function of the phase velocities.

4.2 MECHANICALLY AGITATED BUBBLE REACTORS

Mechanically agitated bubble contactors are very effective gas-liquid reactors when viscous liquids or slurries, or very low gas flowrates or when large volume of liquid are necessitated for carrying out the chemical process. They are also used for the ease with which the intensity of agitation and the residence time of the liquid can be varied and the heat can be removed. Their principal disadvantage is that both the liquid and the gas phase are almost completely backmixed. A variety of agitators are employed but the most common types are the turbine with straight or inclined blades and the flow turbine. The gas may be introduced through a perforated tube or through a porous or perforated plate. The surface aerators will not be considered here.

a. *Hydrodynamics*

The three main parameters are the flooding of the stirrer (the stirrer speed N must be sufficient to assure a good dispersion of the gas inside the whole reactor (70)), the agitation power P_g and the gas holdup $\alpha = 1-\beta$.

If the correlations are well established to determine the agitation power in absence of a gas phase, P_o, for different stirrers (71), the situation is more complex for the case of aerated tanks. Most authors propose to represent the decrease of the agitation power in presence of a gas phase at the same stirring velocity, by correlations between P_g/P_o and the aerated number $N_A = Q_G/Nd_A^3$. Q_G is here the volumetric gas flowrate and d_A is the agitator diameter. However Dickey (72) has reviewed the literature data and experimented air-water systems in a tank of diameter D = 1.52 m with three turbines agitators, d_A = 0.38, 0.51 and 0.61 m for $0.02 < N_A < 0.4$ which corresponds to industrial applications ($N_A > 0.06$ for large diameter tanks). The data, plotted in Fig. 12 are not very encouraging. It may just be concluded that for $N_A > 0.1$, a good approximation is $0.3 < P_g/P_o < 0.6$ with a majority of data $P_g/P_o = 0.4$. For smaller diameter tanks (D < 0.3 m) and for liquids having physico-chemical properties different of water, Hassan and Robinson (73) and Luong and Valesky (74) introduce a Weber's number

$$\frac{P_g}{P_o} = CN_A^m We^n, \qquad We = N^2 d_A^3 \rho_L/\sigma_L$$

with C = 0.497 ; m = -0.38 ; n = -0.18 for newtonian liquids and C = 0.514 ; m = -0.38 ; n = -0.194 for non newtonian liquids (with $N_A < 0.06$). A purely empirical but more general correlation is the Michell and Miller plot represented by

$$P_g = \text{Const}\left(\frac{P_o^2 N d_A^3}{Q_G}\right)^m = \text{Const } M^m \qquad \text{for } 1 < M < 10^9$$

the constant and m depend on the tank-stirrer system used. For example Midoux and Charpentier (71) have shown that for coalescing liquids, m = 0.45 and moreover Const = $0.34\sqrt{n_p}$ if the agitator is of turbine type with n_p blades. It has also been shown that this correlation could be used for scaling-up to 800 times in volume, for non newtonian liquids and for light suspensions as long as the agitation regime is turbulent ($R_A > 2000$) (71, 76).

Besides for ionic liquids inhibiting coalescence or foaming or for a gas distribution by-passing the stirrer (a perforated plate distributor). it is observed that $m \simeq 0.33$ (71).

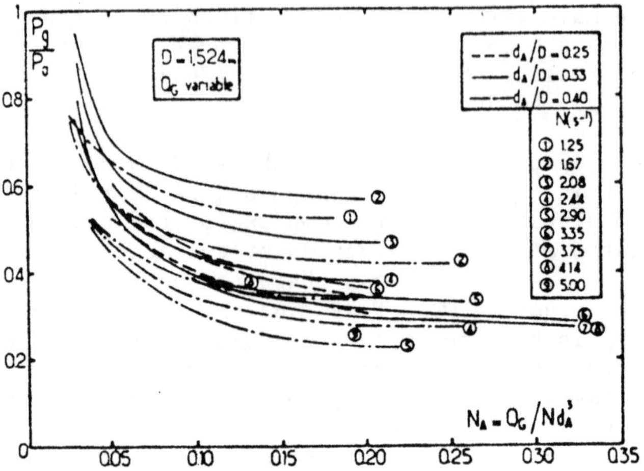

Fig.12- P_g/P_0 against N_A; variables are N and Q_G (according to DICKEY)

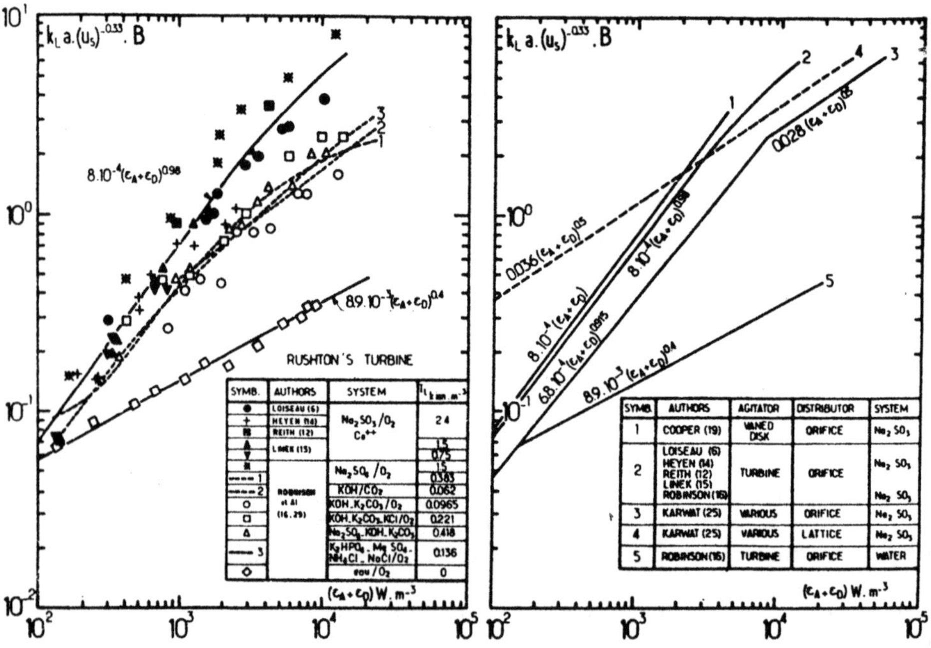

Fig. 13 - Liquid-side mass transfer coefficient versus mechanical power.

The variation of the gas holdup with the working parameters is also very complex and depend on the coalescence or foam ability of the gas-liquid dispersion. *For pure coalescing liquids*, Calderbank (77) has proposed,

$$\alpha = \left(\frac{u_s}{u_t}\alpha\right)^{0.5} + 2.16\times10^{-4}\left(\frac{\varepsilon_A\rho_L^{0.2}}{\sigma_L^{0.6}}\right)\left(\frac{u_s}{u_t}\right)^{0.5}$$

where ε_A is the agitation power per unit liquid volume, u_t is the terminal velocity of a single bubble and u_s the gas superficial velocity. This correlation concern tanks diameter up to 100 l. Sridhar and Potter (78) have shown recently that this correlation could be applied for experiments under pressure (up to 10 atm) with a gas density correction and using the total energy ε_T given to the dispersion by the agitator, the inlet gas kinetic energy and the expansion of the sparged gas. So the second term of the previous equation has to be multiplied by $(\varepsilon_T/\varepsilon_A)(\rho_g/\rho_A)^{0.16}$, ρ_A being the density of air at the operating conditions.

For larger diameter tanks with one or several stirrers and for non foaming liquids with a large range of physico-chemical properties different of air-water, the literature data can be regrouped as

$$\alpha \sim \varepsilon_T^m u_s^n$$

with m and n depending on the agitator geometry.

For ionic and inhibiting coalescence system the results may be regrouped as

$$\alpha \text{ or } \frac{\alpha}{1-\alpha} = 0.0051\ \varepsilon_T^{0.57}\ u_s^{0.24}$$

This correlation furnishes α or 1-α depending that the dispersion height or the clear liquid height is equal to the tank diameter (71). The ionic strenght does not seem to have a great influence and actually only the surface tension gradients in the liquid film located between two bubbles govern the behaviour of the dispersion.

Moreover the gas holdup is very much influenced by additives such as coalescence inhibitor or micronic particles that decrease the sizes of the bubbles and increases α.

b. *Mass transfer*

Most of the data concern the liquid-side mass transfer coefficient k_La and the interfacial area a reported either to the dispersion (or tank) volume or to the liquid volume.

For k_La values, the data may be regrouped as $k_La \sim \varepsilon_T^m u_s^n$ or $k_La \sim \varepsilon_T^p u_s^q$, the values of exponents m, n, p and q depending on the coalescing ability of the dispersion and the geometry of the stirrer. So Van Riet (79) has proposed for the literature data concerning non viscous liquids two correlations : one *for pure water* (within 40 % and for tank diameter D up to 2.6 m^3) :

$$k_La = 0.026\ \varepsilon_A^{0.4} u_s^{0.5}$$

the second for *ionic solutions inhibiting coalescence* (within 40 % and for tank diameter up to 4.4 m^3)

$$k_La = 0.002\ \varepsilon_A^{0.7} u_s^{0.2}$$

The data concern $0.5 < H/D < 1.5$; $0.5 < \varepsilon_A < 10$ kw/m^3. They are independent of the agitator geometry, its position in the tank and concern a well dispersed solution. The agitation power and k_La are reported to the clear liquid volume. H is the tank height.

Takin into consideration the shape of the distributor, the gas-liquid physico-chemical properties, the fomaing effects and high gas flowrates, Midoux and Charpentier (71) have proposed comparable correlations in which the specific agitation energy ε_A was replaced by the specific total energy ε_T and a coefficient B was introduced,

$$B = \left(\frac{\rho_L D_A^2}{\mu_L}\right)^{1/3}_{O_2-H_2O} \cdot \left(\frac{\rho_L D_A^2}{\mu_L}\right)^{-1/3}$$

to precise the influence of the nature of the dispersion on k_L. ε_D is the energy per unit liquid volume due to the expansion of the sparged gas.

It must be said that it is very difficult to predict with a good accuracy the k_La values for foaming systems which are much higher than the values obtained with water or non foaming liquids. The behaviour is comprised between the two extreme curves of fig. 13a. The difficulty is also important when solids are present. For particles diameter 50-200 μ, it has been observed that k_La is markly decreased when the solid volumetric concentration leads o an apparent viscosity which is more than four times the viscosity of the liquid alone - Joosten et al. (80). Note that the presence of polymers additifs leads also to a considerable decrease of k_La (81).

The interfacial areas are correlated by relation involving either the specific agitation energy (77, 82, 90) or the total energy (71) or the minimum efficient stirring speed (83-88).

Midoux and Charpentier have shown that there are large discrepancies between the values of a obtained with the different correlations even with the classical system air-water. However the variations of the interfacial area with the stirring velocities are the same (Fig. 14). *So the correlations will have to be used for determining the tendencies with accuracies.* They should be employed with great care for determining a priori the values of a.

Together with the gas holdup, Calderbank proposed,

$$a_d = a(1-\alpha) = 1.44 \left(\frac{\varepsilon_A^{0.4}\ \rho_L^{0.2}}{\sigma_L^{0.6}} \right) \left(\frac{u_s}{u_t} \right)^{0.5}$$

with $a_d \leq 120\ m^{-1}$ and $\alpha \leq 0.08$ and $u_s = 0.265\ m\ s^{-1}$

or large range of surface tension σ_L and density ρ_L including water, alcohol, CCl_4, ethylacetate, nitrobenzene and toluene. This correlation was modified by Sridhar and Potter (78) for reactors working under pressure up to 10 atm. The data concern small volume reactors (less than 100 litres). *More generally, many literature results can be regrouped as $a \sim \varepsilon_T^m u_s^n$ with $m = 0.32 \pm 0.08$, $n < 0.5$ for coalescing fluids and $m = 0.8$-1 and $n = 0$ for foaming and inhibiting coalescence fluids.* For example Fig. 15 presents some data concerning ionic solutions in different size tanks (92).

Finally for both a and k_La, it would be advisable to carry out experiments with a new gas-liquid system in a small volume reactor (less than 100 ℓ) and the scaling-up must be led at the same superficial velocity with a and $k_La \sim \varepsilon_T^{0.8-1}$ for foaming and ionic dispersions and a and $k_La \sim \varepsilon_T^{0.4}$ for pure and non coalescing organic liquids.

.3 BUBBLE COLUMNS

Bubble columns where a gas is dispersed through a deep pool of liquid are commonly used in industry as absorbers, strippers or reactors when a large liquid holdup, large liquid residence

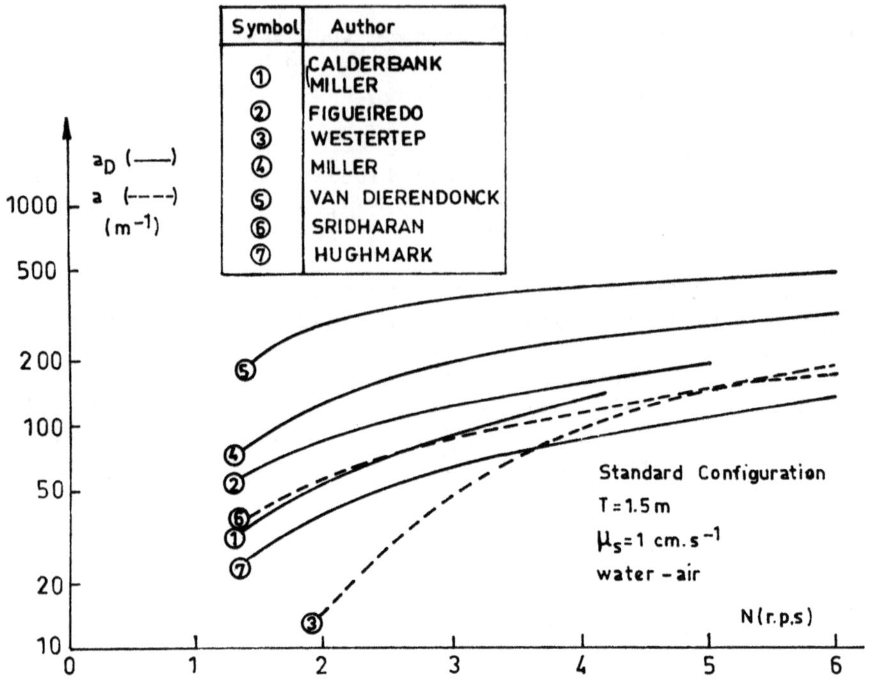

Figure 14

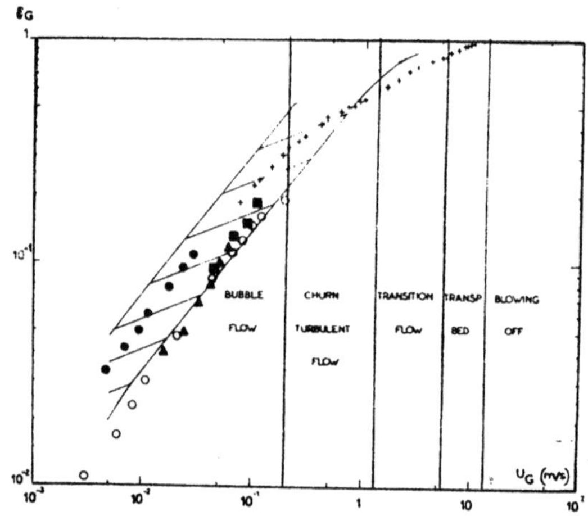

Figure 16. Gas holdup as a function of the superficial gas velocity : shaded area ,zone of the literature data(Table 1). Present data for the air-water combination:● ,D_T=0.02 m; + ,D_T=0.075 m; o , D_T=0.25 m; ▲,D_T=0.48 m; ■ , column with draught tube.

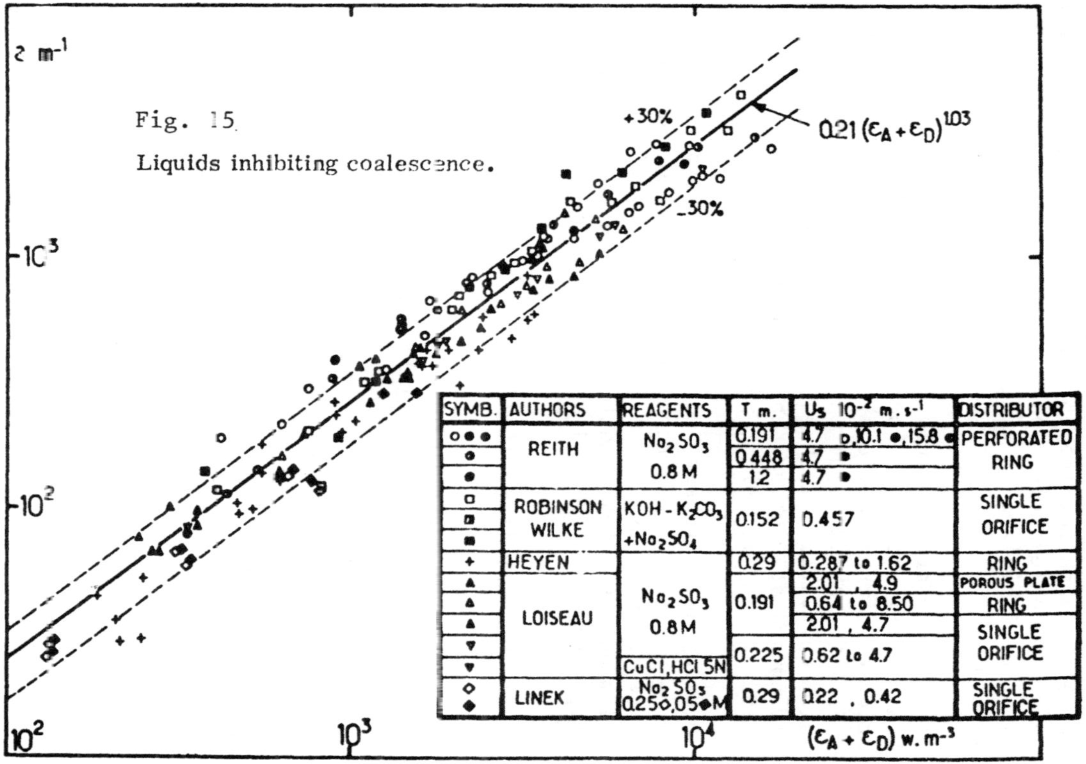

Fig. 15

Liquids inhibiting coalescence.

time, or large heat transfer is needed. They may be operated either countercurrently, cocurrently or semibatch. Other advantages of bubble columns are the absence of moving parts, minimum maintenance, small floor space, ability to handle solids, relatively low cost, large interfacial area, and large mass transfer coefficient.

The principal disadvantages of bubble columns are a large extent of liquid-phase back-mixing, a high pressure drop of the gas due to the high static head of liquid, and a decrease in the specific interfacial area for length/diameter ratios greater than 12-15 because of coalescence. Coalescence may be minimized by inserting fixed or fluidized packings, grids, or perforated plates or by pulsation.

Bubble columns are commonly used in industry where they operate with a superficial gas velocity corresponding to the flow configuration of either bubble flow (u_G < 0.2-0.3 m/s) or churn turbulent flow (u_G < 1.35 m/s) when working countercurrently but they may also operate with other flow configurations when working cocurrently upward at high gas throughputs up to 14 m/s as reviewed by Botton et al. (96, 97) (Fig. 16).

The hydrodynamic property most frequently measured is the gas holdup $\alpha = \varepsilon_G$ which is the percentage of the volume of the two-phase mixture occupied by gas. A correlation covering a wide range of column dimensions, flow conditions, and system properties has been developed by Hughmark (98) and modified by Mashelkar (93) :

$$\alpha = \left[u_G/(30+2u_G)\right] \left[(1/\rho_L)(72/\sigma_L)\right]^{0.33}$$

This equation is valid for $0.6 < \rho_L < 1.3$ gm/cm^3, $0.9 < \mu_L < 150$ cP and $25 < \sigma_L < 76$ dyn/cm, for bubble diameters greater than 0.1 cm and for column diameters up to 1.5 m. However, electrolytes or foaming systems may give holdup values 30 % higher than those of nonelectrolyte systems predicted by the previous equation, as shown by Akita and Yoshida (99) and by the extensive review of Hikita et al. (100). These last authors define a correction factor f by which the fractional gas holdup is increased by the presence of electrolytein water with

$$f = 10^{0.04I} \qquad \text{for} \quad 0 < I < 1 \text{ g.ion/litre}$$

and $$f = 1.1 \qquad \text{for} \quad I > 1$$

Gas holdup is not significantly affected as diameter D_T of the column is further increased above 0.15 m (96, 101, 102) neither by the gas sparger except at very low gas flowrate.

For mixtures of liquids, the situation is more complex (103). Although the superficial liquid velocity u_L does not have a large effect, it can easily be taken into account. For example, in the case of cocurrent flow, α is corrected to the true holdup α' by the relation $u_G/\alpha = u_G/\alpha' - u_L/(1-\alpha')$ or by the relations given by Nicklin et al. (104) where the terminal velocity is taken into account. An alternate, and general correlation, that has been tested in industrial equipment for the oxidation of toluene, cyclohexane, and alipharic acids under pressure, has been proposed by Van Dierendonck et al. (84)

$$\alpha = 1.2(\mu_L u_G/\sigma_L)^{3/4}(\rho_L \sigma_L^3/g\mu_L^4)^{1/8}$$

for $\alpha < 0.45$; $3 < u_G < 40$ cm/sec ; $0 < u_L < 2$ cm/sec ;
$0.8 < \rho_L < 1.3$ gm/cm^3 ; $0.5 < \mu_L < 5$ cP ;
$20 < \sigma_L < 75$ dyn/cm ; and $D_T > 15$ cm

A systematic study of mass transfer in bubble columns by Mashelkar and Sharma (93, 94, 95) is summarized in Fig. 17. Increasing the superficial gas velocity increases the gas holdup α, the volumetric mass transfer coefficients, and the interfacial area per unit volume of dispersion, but not the true mass transfer coefficients. Sharma and Mashelkar (105) found good agreement between their experimental values of k_G and the values from Geddes' stagnant sphere model equation (106)

$$k_G = -\frac{d_B}{6t_c} \ln \left| \frac{6}{\pi^2} \sum_1^{\infty} \frac{1}{n^2} \exp\left(- \frac{D_G n^2 \pi^2 t_c}{(d_B/2)^2} \right) \right|$$

in using the first term of that series ($t_c = Z/v_B$ where v_B is the velocity of the bubble rising through a height Z). The interfacial area (107, 108, 109) is increased by an increase in viscosity or in temperature, the presence of solids, a decrease in surface tension, or the presence of electrolytes. However, the physical properties of the gas appear to have no effect.

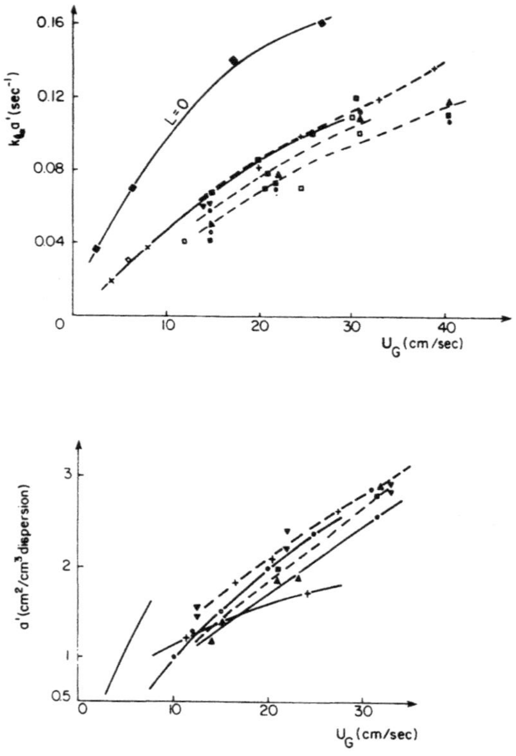

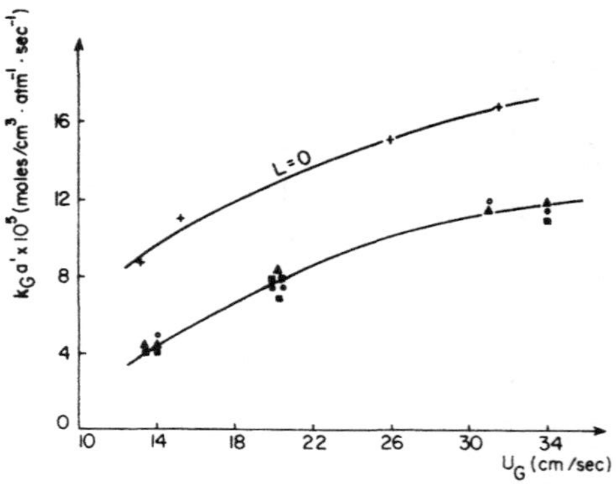

Symbol	Column diameter (cm)	Type of run	L (cm · sec^{-1})	Chemical system	References
+	6.6	semibatch	0	15 different chemical systems	M8 M9 S23
○		counter-current	0.2		
△			0.3		
□			0.5		
▽		co-current	0.2		
▼			0.3, 0.5		
●	10		0.12		
▲			0.24		
■			0.48		
●					
×	14	cocurrent	0.6 - 3	CO_2-Air NaOH	V12
——	15.3	cter. current	0.09-0.32	Air. Na_2SO_3-$CoSO_4$	C8
◆	7.7 15 30 60	semibatch	0	Air. Na_2SO_3.$CuSO_4$	Y3 M8

Fig. 17 - Variation in mass transfer data per unit volume of dispersion with superficial gas velocity in bubble columns.

Indeed, as for hydrodynamics, mass transfer depends strongly n the physico-chemical properties of the gas-liquid system and any correlations have been proposed to predict the interfacial reas a' and liquid mass transfer coefficient k_La', reported to he unit volume of dispersion. They have been recently reviewed y Botton et al. (97) and Hikita et al. (111). It seems that for he scale-up prevision in bubble flow regime (u_G < 0.3 m/s), small cale experiments with the system of interest will allow scale-up n the basis of equal superficial velocity of the gas. So the data n Fig. 17, or those found in the many literature references, or f specific experiments can be used noting that α, k_Ga', k_La' and ' vary approximatively as $u_G^{0.75}$. For other flow regimes and for raught tubes, see Fig. 18.

Complementary and extensive informations on the behaviour of ubble column bioreactors with the influence of gas distributor ype and composition of liquid have been published by Schügerl et l. (110). For most of the data corresponding to u_G < 0.08 m/s nd for all the systems investigated (demineralized water, culture salt solution, methanol and ethanol salt solutions with and ithout salt additives, n-propanol and n-butanol solutions and 10% a_2SO_4 solution), it is observed that the mean relative gas holdup nd the specific interfacial area increase with increasing coalesence hinderance if nozzle and porous plates aerators are employed. ith perforated plates the coalescence rate only slightly influenes α and a'. In coalescence promoting media the aerator type does ot have any influence on these parameters. In coalescence hindeing media, aerators with the highest local energy dissipation ate, e.e., nozzles give the smallest bubbles, the highest α and ' and perforated plates the largest bubbles and the smallest α nd a'. Supplementary informations on such bubble bioreactors will e found in the texts published by Deckwer and Schügerl in this ATO Summer School Program.

.4 SPRAY TOWERS

Spray towers are of particular interest because of their abiity to handle corrosive and solid-laden fluids when only one or most two theoretical stages of contact are required or when the as pressure drop must be kept to a minimum. They are also used or exothermic chemical reactions or for the absorption of highly oluble gases, where large volumes of liquid must flow through the olumn to avoid an excessive temperature rise. The disadvantages spray towers lie in the power consumption needed for pressuring he absorbing liquid through spray nozzles, in the height necessa- to achieve one theoretical stage of absorption, and in the need or installing mist eliminators.

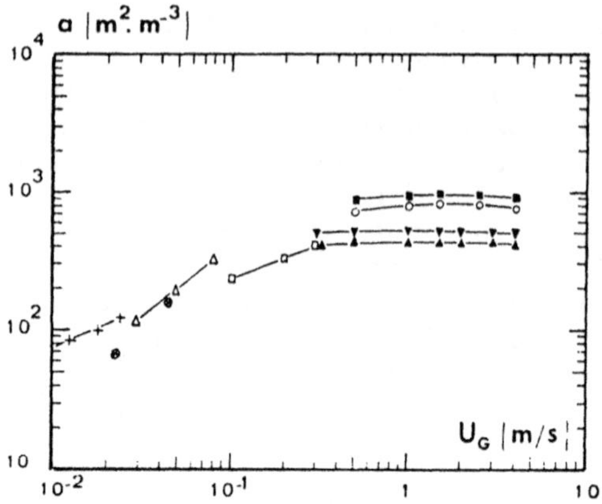

Fig.18a. Specific interfacial area as function of the superficial gas velocity. +, D_T=0.02 m; ⊗, column with draught tube (D_T=0.19 m; D_I=0.13 m); Δ, column with draught tube (D_T=0.48 m, D_I=0.14 m); □, square cross section, 0.1x0.1 m. Circular cross section, D_T=0.075 m: ▲, U_L=5x 10^{-3} m s^{-1}; ▼, U_L=25x10^{-3} ms^{-1}; o, U_L=12x10^{-2} ms^{-1}; ■, U_L=18x10^{-2} ms^{-1}.

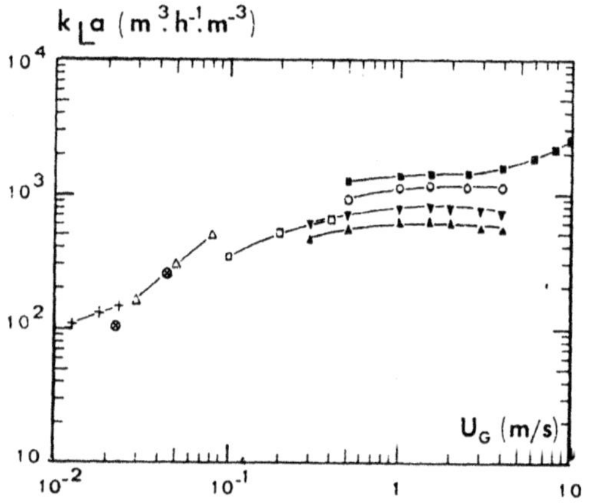

Fig.18b. k_La as a function of the superficial gas velocity.

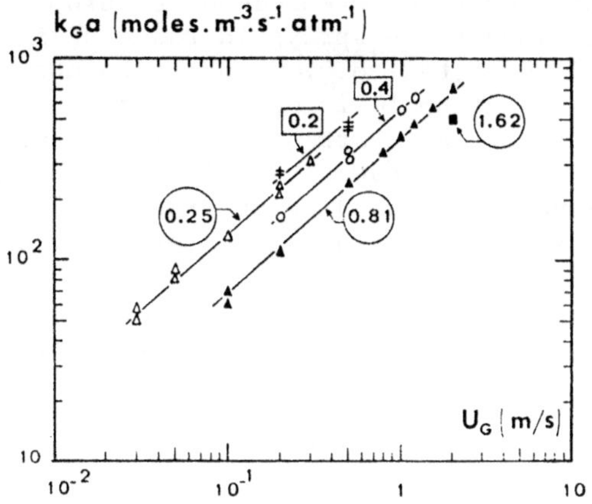

Fig.18c. k_Ga as function of the superficial gas velocity. ⓧ, circular column in which the emulsion height is x ; [y], rectangular column in which the emulsion height is y.

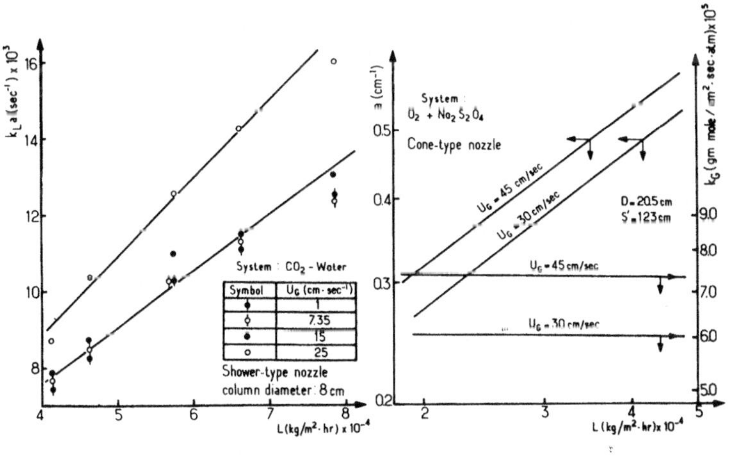

Fig.19. Influence of gas and liquid flow rates on k_La and a in spray towers (112).

The large number of variables, variety in methods of operation and differences in mechanical details such as spray nozzle design have led to scattered mass transfer data. Mass transfer correlations for single drops or even clouds of drops do not apply because normal production of the sprays introduces effects that cannot be interpreted quantitatively. In order to obtain good contact and avoid bypassing of upflowing gas, the spray must cover the entire tower cross section. Part of the liquid spray impinges on the twoer wall and trickles down as a film, which reduces the mass transfer efficiency that would be predicted from drop-type contact. Also, as no spray nozzle produces perfectly uniform drops, drop coalescence occurs as the drops fall through the tower. Mehta and Sharma (112) are the only ones to report comprehensive research into the effect of commercial nozzles, column height, gas and liquid flowrates, and physical properties of the liquid on spray column performance, with diameters up to 0.4 m. These studies provide the following conclusions useful for design.

The values of k_La per unit volume of column are practically independent of the gas velocity up to a critical value which depends on nozzle type, column diameter, and physical properties (112, 113) ; then k_La increases with increasing gas velocity. Typical results are given in Fig. 19 for a shower nozzle. Moreover k_Ga, k_G and a all increase as the gas velocity increases independently of the type of nozzle, the column size, the liquid flowrate and probably the physical properties, as shown in Fig. 20. The interfacial area and true liquid side coefficient increase with the liquid flowrate L, owing to increased surface area, higher drop velocity and increased circulation or turbulence within the drops. The interfacial area varies exponentially with the liquid flowrate the exponent increasing with decreases in the nozzle orifice diameter, that is, decreases in droplet size. In contrast, k_G is not influenced by the liquid rate.

The values of interfacial area and of overall mass transfer coefficient increase with decreasing distance S' between the spray nozzle and gas inlet, whatever the nozzle type, column dimensions and flowrates. Indeed the spray provides a large interfacial area in the vicinity of the nozzle, where there is intensive circulation. Then a decreases quickly away from the nozzle, as a result of both coalescence of droplets and collection of liquid on the column walls. k_Ga and a are approximately proportional to $(S')^{-0.4}$ for absorption and desorption processes, which shows that k_G is practically independent of the column height. Moreover Mehta and Sharma indicate that a is unaffected by ionic strength and viscosity but may decrease about 20 % when solids are generated by the reaction of gas with the liquid. Thus the following correlations may be used for design (112)

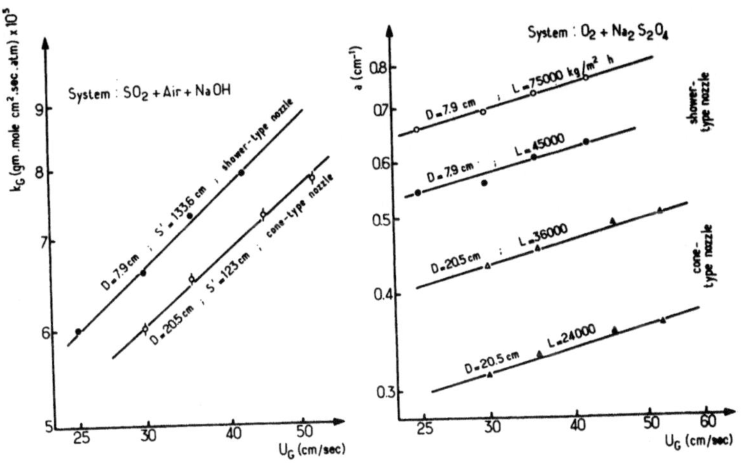

Fig.20. Influence of gas velocity on k_G and a in spray towers (112).

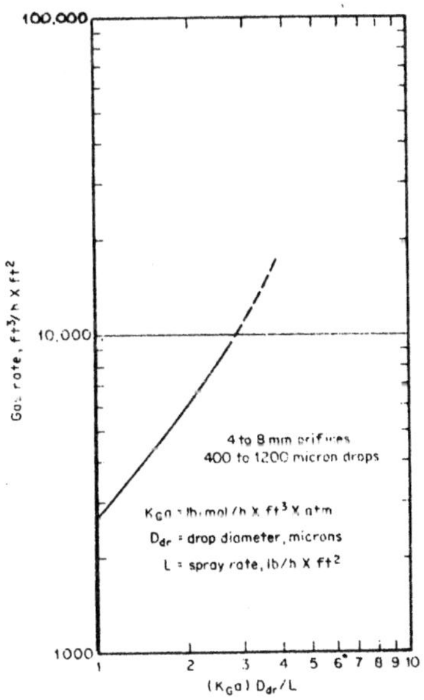

Fig.21.Emprical correlation of solid cone nozzle spray tower performance tests.

$$a = \alpha' L^{\beta'} u_G^{0.3} (S')^{-0.4}$$

$$k_G = m u_G^{0.5}$$

where a is in cm^2/cm^3, k_G is in gm moles/cm^2 sec atm, L is in kg/m^2 hr, u_G is in cm/sec, and S' is in cm. Constants α', β' and m are listed in the following Table for nozzle type, orifice dia-

D(cm)	S'(m)	Type of nozzle	Orifice diameter (mm)	$\alpha'(\times 10^4)$	β'	$m(\times 10^5)$
8	1.3	Shower	69 holes of 1.2 mm	246	0.38	1.02
21	1.2	Solid cone	5.5	2.12	0.81	0.95
21	1.2	Solid cone	4.4	0.51	0.93	2.2
39	2.8	Solid cone	5.5	8.97	0.62	2.2
39	2.8	Solid cone	4.4	4.9	0.70	2.2
39	2.8	Solid cone	8.4	42.8	0.47	2.2

Values of α', β' and m for the spray tower mass transfer equation (112).

meter, and other variables investigated by Mehta and Sharma. For other specific cases, the books of Ramm (114) and Kohl and Riesenfeld (115) will be of help. Scale-up often utilizes a general correlation in which the number of transfer units $k_G aH/G$ is proportional to $(P_L/P_G^{0.1})$ where P_L and P_G are the power introduced in the liquid and in the gas, H is the height, and G is the superficial molar gas flowrate.

Complementarily, industrial experience with solid cone-nozzles for a maximum height of 1.33 m has led Zenz to conservatively suggest the empirical correlation shown in Fig. 21 (116).

4.5 JET REACTORS

During the last few years, interest has increased in reactors in which high mass transfer is realized by means of liquid jet or injection devices providing gas or liquid entrainment : we are concerned here only with jet reactors and venturi-jet reactors.

The ejector reactor conceptually is a bubble reactor into which a liquid jet entraining gas is injected (117) (Fig. 22). The ejector provides high liquid velocities (over 20 m/sec) which entrain gas by suction through a mixer device placed inside the reactor where there is intense gas-liquid contact. Also, gas and liquid are circulated inside the reactor by a pumping effect in

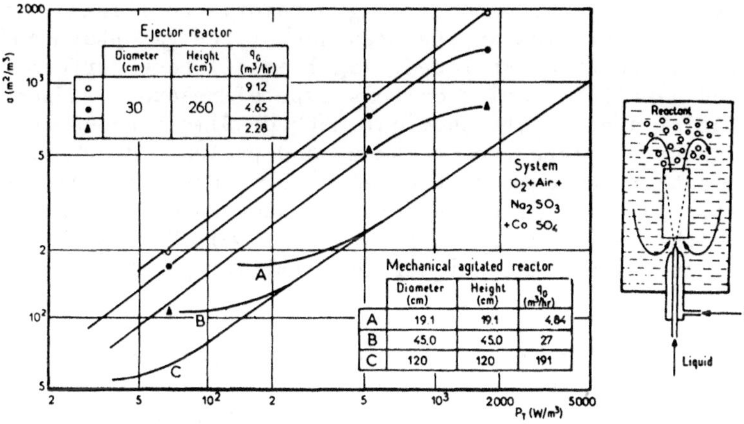

Fig.22. Interfacial area in ejector reactor:comparison with mechanically agitated reactor

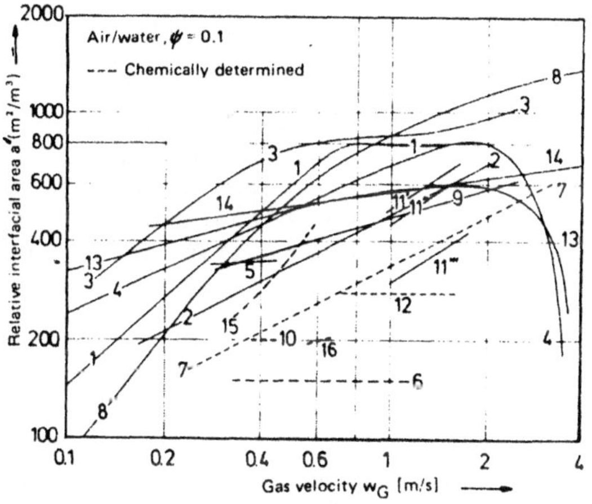

Fig.23. Relative interfacial area in the two-phase layer on trays.Comparison of data from the literature for air/ water system. (Numbers refer to various authors (121).

such equipment, and the ejector and mixer devices act jointly as the reactor agitator and gas bubble generator. The liquid leaves the equipment by overflow in the presence of the gas phase and the two phases are then separated. Nagel et al. have performed complete and detailed studies on this type of reactor, measuring the interfacial area by the chemical method. Simultaneous measurement of phase flowrates, entrainment and pressure drop has enabled these authors to compare the values of interfacial area with those of stirred tanks [118] in terms of same specific input power relative to the volume of the reactor (Fig. 22). Values obtained with the ejector are always higher for similar operating conditions.

In venturi equipment, the liquid is injected through a nozzle into a high-velocity gas stream. The liquid is then atomized by the formation and subsequent shattering of attenuated, twisted filaments and thin cuplike films which provide a high degree of turbulence and large interfacial areas for heat and mass transfer. In the breakup, nearly spherical droplets are formed which also provide a large surface area per unit volume of liquid, but then the degree of turbulence is decreased. Venturi scrubbers are often used for simultaneous removal of gaseous and particulate pollutants. Their advantages are high volumetric flowrate ; simplicity, compactness, and absence of moving parts, all leading to a low first cost of equipment ; and the ability to handle slurry absorbents. Disadvantages are the short gas-liquid contact time, high pressure drop on the gas side, the need for a phase separator after the scrubber, and the limitation to applications with large volumetric gas-liquid ratios (at small ratios, efficient atomization of the liquid does not occur).

Two principal modes of operation commonly employed are liquid injection into the throat of the venturi and liquid introduction so as to wet the entire convergent section. The second mode of operation gives a lower pressure drop under similar operating conditions [119]. Many studies have been reported on mass transfer in venturi scrubbers in terms of efficiencies for specific equipment and specific problems (absorption of SO_2, NO_2, NH_3 ; desorption of CO_2, O_2, and so on). So only qualitative relations of a, k_La and k_Ga to the flow parameters can be proposed.

The gas side mass transfer coefficients k_Ga and k_G increase with liquid feed rate or with gas velocity at each given position in the venturi scrubber and decrease at constant liquid rate and gas velocity with increasing distance from the point of liquid injection [119]. The values of k_La generally increase with increasing liquid flowrate or gas velocity (often referred to as the velocity at the throat). However k_La will sometimes exhibit a maximum when the gas velocity increases ; the explanation is that, at higher gas velocities, an increase in turbulence in the throat of the venturi results in the formation of droplets smaller than the thin filaments first formed at lower gas velocities. Internal circulation is reduced in these smaller droplets, and there is also a reduction in the size of the zone of intense turbulence. These two phenomena lead to a maximum for the values of k_L. The values of the effective interfacial area a increase with both gas and liquid flowrates.

Virkar and Sharma (119) are the only investigators to have reported systematic measurements of a, k_La and k_Ga by the chemical method. Their work utilized laboratory-scale vertical venturis, operating with liquid injection either at the throat or ahead of the convergent section. For example, for wetted-approach operation, in the range of gas velocities from 50 to 90 m/sec and liquid flowrates from 7 to 30 cm^3/sec, the following equations were proposed and are given here for illustration :

$$a = 2(\Delta P/\Delta h) + 0.65$$

and $$k_G a \times 10^{+10} = 0.25 \times 10^{-2}\ \Delta P\sqrt{r} + 1.56$$

where ΔP and Δh are the pressure drops across the venturi and across the convergent section in N/m^2, r is the liquid/gas ratio in liters/m^3, a is in cm^2/cm^3 and k_Ga is in kg moles/m N sec. The reactor volume is taken to be the volume of the divergent cone plus the volume of the spray zone in the separator. The venturi had a circular throat of 16 mm diameter and 9 mm length, and the angles of the convergent and divergent sections were 35° and 8°, respectively. The entrance and exit sections were each 5 cm in diameter. These equations are specific to one type of equipment of laboratory scale and are of doubtful validity for other types and sizes.

That is the reason why recently Laurent et al. (120) *simulated* a such type of gas-liquid reactor, *a pilot-scale liquid motivated, venturi-jet scrubber* 33 mm throat diameter and 700 mm length with three ejector nozzles of 3, 4 and 5 mm diameter by *a laboratory-scale laminar jet* of 0.5 mm diameter and 1.5 cm height. In both equipment, the hydrodynamic conditions at the level of the interface were fixed the same (i.e., same k_G or same k_L) for the flowrates of each phases which were in the ratios of approximatively 250 and 34000 respectively for the values of L and G). This technique of simulation will be presented by Alper in the present NATO Summer School Program. It has necessitated previously the chemical determination of a, k_La and k_Ga by the chemical technique in both apparatuses and it allows for the prediction of the new performances of the venturi reactor when the and/or the liquid are changed just in carrying out experiments with this new gas-liquid system in the laminar-jet.

4.6 PLATE COLUMNS

Plate columns are used for operations requiring a large number of transfer units, high pressure, high gas flowrates and low liquid flowrates, when it is necessary to supply or to remove heat, when solids are present in the liquid (or gas), and when the diameter is greater than 70 cm. They have the ability to handle large variations in gas and liquid flowrates. Figure 23 compares the data of various authors for the relative interfacial area referred to two-phase dispersion volume as reported in the extensive review by Stichlmair and Mersmann (121). The references on this figure are the references of the review (121). The comparison concerns only the air-water system, since data for other systems are scare. The broken lines concern the chemical technique and it is seen that the data are clearly smaller than the others which shows once again that the data found in literature should be used with a great care when applied to one another system or equipment. Mass transfer data will be presented here for the most common plate designs, bubble cap plates and sieve plates.

4.6.1 Bubble-Cap Plates

For systematic study of several gas-liquid chemical reactions using a laboratory model bubble-cap column, Sharma et al. (95) have shown that the presence of electrolytes, size of caps, type of slots, ionic strength, liquid viscosity and presence of solids do not affect the mass transfer rates. These rates chiefly depend on the gas and liquid flowrates (95, 122). The influence of the superficial gas flowrate on k_L and k_G is indicated in Fig. 24. Interfacial area a" per unit area of plate (or per unit floor area) for plate diameters varying between 0.15 and 1.20 m have been grouped in Fig. 25 (123) which with the following correlations (95) can be used to scale up bubble cap plates up to 2 or 3 m in diameter

$$k_G a' = 6.2 u_G^{0.75} S^{-0.67} D_G^{0.5}, \qquad k_L a' = 7 u_G^{0.75} S^{-0.67} D_L^{0.5}$$

$$k_G = 11.5 u_G^{0.25} S^{-0.5} D_G^{0.5}, \qquad a'' = 0.54 u_G^{0.5} S^{0.83}$$

$$k_L = 13 u_G^{0.25} S^{-0.5} D_L^{0.5}$$

$k_G a'$ and $k_L a'$ are in sec^{-1}, k_G is in cm/sec, and a" is in cm^2/cm^2; S is the submergence in cm, defined as the height of the bubble

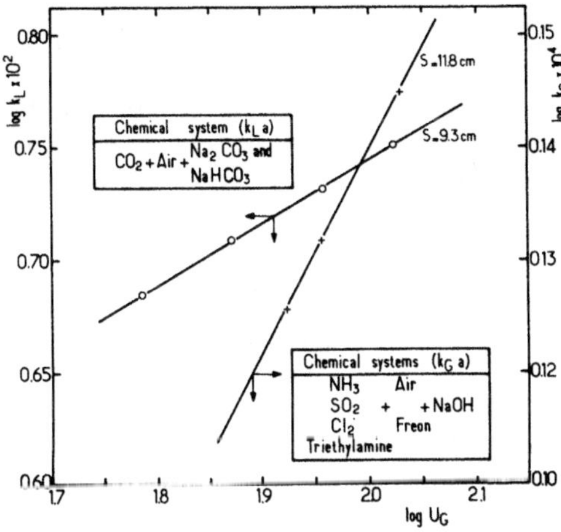

Fig.24. Effect of superficial gas velocity on k_L and k_G in 22.5 cm model bubble cap plate column (95).

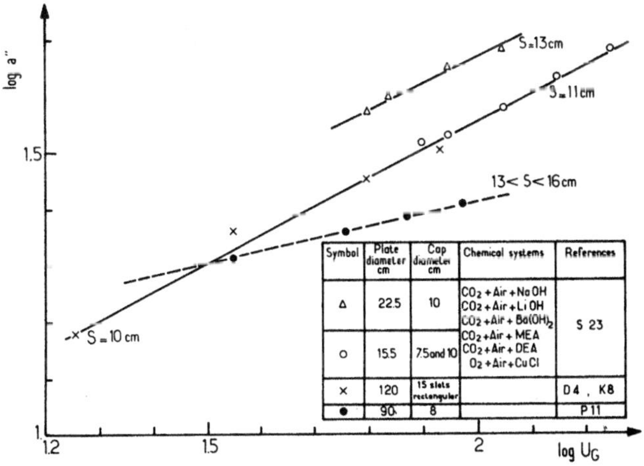

Fig.25. Influence of u_G on the interfacial area per unit floor area in a bubble-cap plate column (123).

travel and measured from halfway up the slot height to the top of the dispersion ; and a' is the interfacial area per unit volume of dispersion in cm^2/cm^3 (a" = a'×S).

4.6.2 Sieve Plates

On a perforated plate the liquid side mass transfer coefficient k_La and gas side mass transfer coefficient k_Ga, based on the column volume, vary linearly with the dispersion height. The true liquid- and gas-side mass transfer coefficients k_L and k_G first increase with the dispersion height and then go through a maximum and decrease slightly (123). Sharma and Gupta (124) attribute this to different behavior of the density of dispersion and the average bubble size with increase in gas flowrate, which leads to a phase inversion point. These authors correlate their experimental data for 10 cm i.d. perforated plates without downcomers by the following expressions

$$k_Ga = 2.6\times10^{-4}\ F^{-1.75}L^{0.6}u_G^{1.2}\ ;\ k_La = 4.2\times10^{-4}\ F^{-2.2}L^{0.6}u_G^{1.2}$$

k_Ga is in moles/sec atm cm^3, k_La is in sec^{-1}, u_G is in m/sec, L is in kg/m^2hr, and F is the percentage free area of the plate (14-30 %).

We note that k_Ga and k_La are nearly independent of the free area of the plate if the gas and liquid velocities are based on perforated area rather than column cross-sectional area. Moreover the values of the volumetric gas and liquid-side mass transfer coefficients based on dispersion volume, k_Ga' and k_La', are each practically constant whatever the gas and liquid mass flowrate. Perforated plates, especially those with a high free area, can handle relatively higher liquid and gas flowrates and provide higher values of overall mass transfer coefficients than at corresponding flow rates through bubble-cap plate columns. However, discrepancies exist between different reported values of k_L because of varying ionic strength and the presence of solids and antifoaming agents. Therefore the formulas proposed by different authors should be carefully studied before use, as already said.

The data on interfacial area are more homogeneous. Interfacial area a' based on dispersion volume is in the range of 2-2.5 cm^2/cm^3 for dispersion heights of 8-16 cm ; it is not influenced much by the liquid flowrate, especially for sieve plates without downcomers. Values of a' increase with increasing gas velocity, percentage free area, ionic strength, and liquid viscosity (124). Laurent and Charpentier (123) have regrouped the literature data for plates with and without downcomers, and for turbogrid plates, in Fig. 26 which applies for a mean value of the dispersion height of 12 cm.

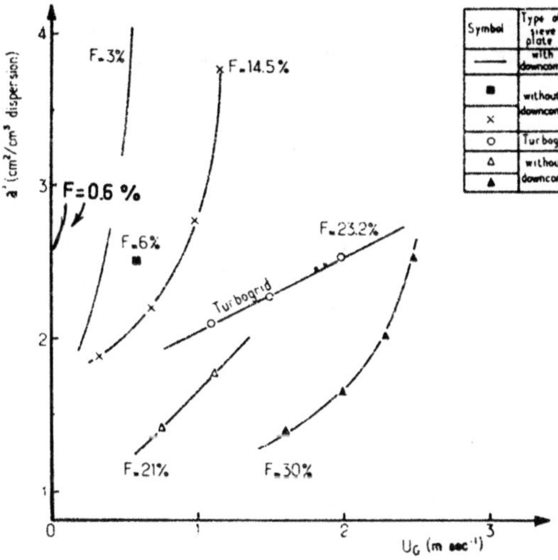

Symbol	Type of sieve plate	Plate diameter cm	Porosity F %	Chemical systems	References
———	with downcomer	22.8	3	CO_2 - Air - NaOH	B9
■	without downcomer	5x15 / 30x30	6 / 5.5		P3, R16
×		10	14.5		S20
○	Turbogrid		23.2		
△	without downcomer		21		
▲			30		

Fig.26. Variation in interfacial area per unit volume of dispersion with u_G for sieve plates (123).

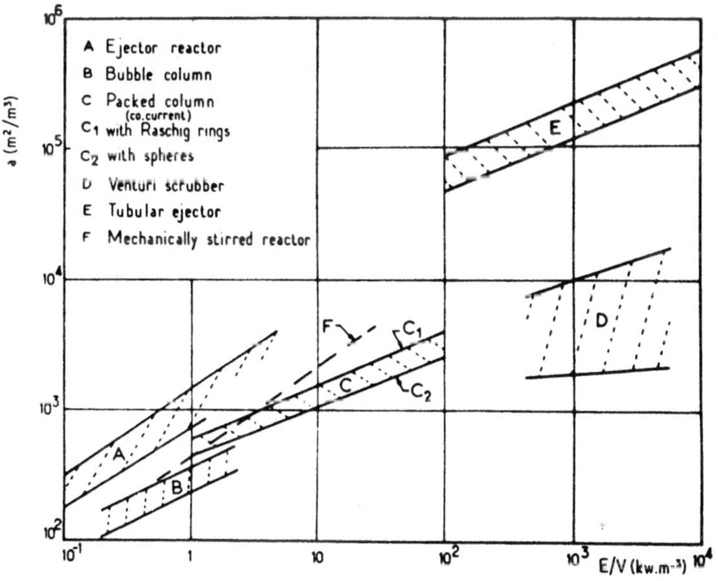

Fig.27. Interfacial area in several types of reactors (1).

In conclusion, reliable values of interfacial area per unit floor area are given by the equation

$$a'' = 30G^{0.5}\rho_G^{-0.25}$$

where a" is in m^2/m^2, G is in kg/m^2sec, and ρ_G is in kg/m^3.

CONCLUSION : SOME SCALE-UP RECOMMANDATIONS

In this section, data for the mass transfer parameters most often measured by the chemical method, that is, the integral values of a, k_La and k_Ga (and their dependence on fluid and equipment parameters), have been presented. In practice, their mode of use depends upon which of the following problems confronts the design engineer :

1. *The equipment has already been in operation,* but the problem is to change the gas-liquid system in the existing reactor for economic reasons. Although the procedure is time-consuming, it seems reasonable to predict the performance from experiments carried out in a small-scale laboratory apparatus with the same k_L, k_G and a/β as the existing industrial equipment (1). In most practical cases, these laboratory experiments simultaneously will provide the needed knowledge of the reaction kinetics.

2. *The type of gas-liquid reactor has been chosen* (packed column, spray column, plate column, mechanically agitated tank, and so on), and the problem is to size the equipment, using published mass transfer data. In such case, the diagrams and correlations given above for contactors with diameters smaller than 0.4 m can be used with a fair degree of confidence to scale-up tubular packed, spray, and plate columns to 2 or 3 m in diameter. For mechanically agitated or bubble reactors, small-scale experiments are recommended with the given gas-liquid system in the laboratory apparatus (D = 5-20 cm) similar in shape, agitation and contact time to the chosen reactor type ; and then, to scale-up the system, the same specific interfacial area is ensured by a constant total power input per unit liquid volume.

In all the cases, it will be assumed that for the actual gas-liquid system, k_G and k_L vary as the power 0.5 of the solute gas diffusivity.

3. *A suitable type of reactor has to be chosen prior to sizing.* This is an economic problem, with competition between the value of the interfacial area and the energy expense required to create it. In such cases, Fig. 27 in Nagel et al. (125) and Fig. 28 pre-

sented in the same plot by Schügerl et al. for complementary equipment provides a comparison between the extents of interfacial area provided by the major types of equipment and their energy costs. Once the choice is made, the remaining part of the design can be carried out as in case 2.

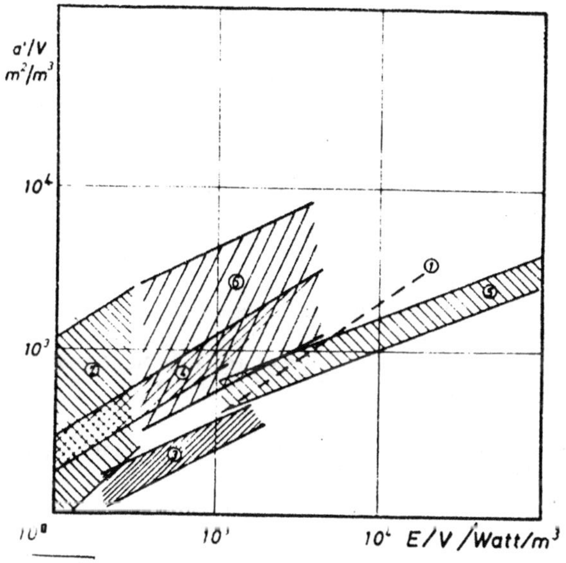

Fig.28. Comparison of gas/liquid interfacial area as a function of the energy dissipation rate (necessary power input) in sulphite oxidation systems. [1, stirred tank reactor S [Reith(118)];2,bubble column-porous plate G [authors]; 3, bubble column-perforated plate S[Reith(118)];4, ejector nozzle S [Nagel(118)]; 5, packed bubble column S [Nagel(117)]; 6, bubble column-injector nozzle G[authors]. G: geometrical area; S: interfacial area obtained by sulfite oxidation method].

LITERATURE CITED

(1) Charpentier, J.C., "Mass Transfer Rates in Gas-Liquid Absorbers and Reactors", Adv. Chem. Eng., Edt by Drew and Vermeulen, Academic Press, 1, 11 (1981) .
(2) Reith, T., S. Renken and B.A. Israel, Chem. Eng. Sci., 23, 619 (1968).
(3) Burgess, J.M. and P.H. Calderbank, Chem. Eng. Sci., 30, 743 and 1107 (1975).
(4) Linek, V. and J. Mayrhoferova, Chem. Eng. Sci., 24, 481, (1969).
(5) Vermeulen, T., G.M. William and G.E. Langlois, Chem. Eng. Prog., 51, 85 (1955).
(6) Cameron, J.F., "Fluid Density Measurements in Enclosed Systems", Internat. Confer. on Radio Isotopes in Scientific Research, 1, 426 (1957).
(7) Calderbank, P.H., Trans. Inst. Chem. Eng., 37, 443 (1958).
(8) Wawecki, W., T. Reith, J.W. Van Heuven and W.J. Beek, Chem. Eng. Sci., 22, 1519 (1967).
(9) Todtenhanft, E.K., Chem. Ing. Techn., 43, 336 (1971).
(10) Landau, J., H.G. Gomaa and A.M. Al Taweel, Trans. Inst. Chem. Engrs., 55, 212 (1977).
(11) Calderbank, P.H., F. Evans and J. Rennie, Proceed. Eur. Chem. Eng. Symp., Heidelberg.(1976).
(12) Kürten, H. and P. Zehner, German. Chem. Eng., 1, 347 (1978).
(13) Sridhar, T. and O.E. Potter, Chem. Eng. Sci., 33, 1347 (1978).
(14) Danckwerts, P.V., "Gas-Liquid Reactions", Mc Graw Hill (1970).
(15) Linek, V., Chem. Eng. Sci., 27, 627 (1972).
(16) Robinson, C.W. and C.R. Wilke, AIChE J., 20, 285 (1974).
(17) Beenackers, A.A. and P.W.M. Van Swaaij, Proceed. Eur. Chem. Eng. Symp., Heidelberg (1976).
(18) Prasher, B.D., AIChE J., 21, 407 (1975).
(19) Matheron, E.R. and O.C. Sandall, AIChE J., 25, 332 (1979).
(20) Levenspiel, O. and J.H. Godfrey, Chem. Eng. Sci., 29, 1723 (1974).
(21) Laurent, A., Thesis, INPL, Nancy, France (1975).
(22) Midoux, N., A. Laurent and J.C. Charpentier, AIChE J., 26, 157 (1980).
(23) Gal-Or, B. and W. Resnick, I.E.C. Proc. Des. Devel., 5, 15 (1966).
(24) Reith, T., Brit. Chem. Eng., 15, 1559 (1970).
(25) Hassan, I.M. and C.W. Robinson , Chem. Eng. Sci., 35, 1277 (1980).
(26) Sridharan, K. and M.M. Sharma, Chem. Eng. Sci., 31, 767 (1976).
(27) Ganguli, K. and H. Van Den Berg, Chem. Eng. Sci., 33, 27 (1978).
(28) Alvarez-Fuster, C., N. Midoux, A. Laurent and J.C. Charpentier, Chem. Eng. Sci., 35, 1717 (1980).

(29) Sridharan, K., Ph. D. Thesis, University of Bombay, India (1975).
(30) Alvarez-Fuster, C., Thesis, INPL, Nancy, France (1980).
(31) Morsi, B.I., A. Laurent, N. Midoux and J.C. Charpentier, Chem. Eng. Sci., 35, 1467 (1980).
(32) Morsi, B.I., N. Midoux, A. Laurent and J.C. Charpentier, Entropie, n° 91, 38 (1980).
(33) Joosten, G.E.H. and P.V. Danckwerts, Chem. Eng. Sci., 28, 453 (1973).
(34) Laurent, A., C. Fonteix and J.C. Charpentier, AIChE J., 26, 282 (1980).
(35) Veteau, J.M., "Mesure des aires interfaciales dans les écoulements diphasiques", Rapport C.E.A. - R. 5005, CENG, France (1979).
(36) Kan, K.M. and P.F. Greenfield, I.E.C. Proc. Des. Devel., 17, 482 (1978), 18, 740 (1979).
(37) Onda, K., H. Takeuchi and Y. Okumoto, J. Chem. Eng. Jap., 1, 56 (1968).
(38) Linek, V., Z. Krirsky and P. Hudec, Chem. Eng. Sci., 32, 323 (1977).
(39) Mahajani, V.V. and M.M. Sharma, Chem. Eng. Sci., 35, 941, (1980).
(40) Mangers, R.J. and A.B. Ponters, I.E.C. Proc. Des. Devel., 19, 530 (1980).
(41) Sherwood, T.K. and F. Holloway, Trans. Am. Inst. Chem. Eng., 36, 39 (1940).
(42) Mohunta, D., A.S. Vaidyanathan and G.S. Laddha, Indian Chem. Eng., 11, 39 (1969).
(43) Sridharan, K. and M.M. Sharma, Chem. Eng. Sci., 31, 767 (1976).
(44) Mashelkar, R.A. and M.M. Sharma, Trans. Inst. Chem. Engrs., 48, T 162 (1970).
(45) Sahay, B.N. and M.M. Sharma, Chem. Eng. Sci., 28, 2245 (1973).
(46) Pexidr, V. and J.C. Charpentier, Collect. Czech. Chem. Commun., 40, 3130 (1975).
(47) Chen, B.M. and R. Vallabh, Ind. Eng. Proc. Design Devel., 9, 121 (1970).
(48) Sawant, S.B., V.G. Pangarkar and J.B. Joshi, Chem. Eng. Journ., 18, 143 (1979).
(49) Shah, Y.T., "Gas-Liquid-Solid Reactor Design", Mc Graw Hill, New York, (1979).
(50) Germain A., G. L'Homme and A. Lefebvre, Chem. Engng. Symposium, Liège, March 1978, G.A. L'Homme Editor, CEBEDOC, Liège (1979).
(51) Charpentier, J.C., Chem. Eng. Journ., 11, 161 (1976).
(52) Gianetto, A., G. Baldi, V. Specchia and S. Sicardi, AIChE J., 24, 1087 (1978).
(53) Hofmann, H., Inter. Chem. Engng., 17, 19 (1977).
(54) Midoux, N., M. Favier and J.C. Charpentier, J. Chem. Eng. Japan, 9, 350 (1976).

(55) Talmor, E., AIChE J., 23, 858 (1977).
(56) Block, J.R., "Pulsing Flow in Trickle-Bed Columns", University of Groningen, (1981).
(57) Sicardi, S., H. Gerhard and H. Hofmann, Chem. Eng. Journ., 18, 173 (1979).
(58) Perez Sosa, J.A., "Ermittlung wichtiger fluidynamischer grössen bei rieselreaktoren mit den system wasser-luft und cyclohexen-kohlendioxid", Universität Erlangen-Nürnberg, (1981).
(59) Larkins, R., R. White and D. Jeffrey, AIChE J., 7, 231 (1961).
(60) Sato, Y., T. Hiroshe, F. Takahashi and M. Toda, J. Chem. Eng. Jap., 6, 147 and 315 (1973).
(61) Turpin, J. and R. Huntington, AIChE J., 13, 1196 (1967).
(62) Charpentier, J.C., Journ. Powder and Bulk Solids Technology, 2, 53 (1978).
(63) Satterfield, C.N., AIChE J., 21, 209 (1975).
(64) Reiss, L.P., I.E.C. Proc. Des. Devel., 6, 486 (1967).
(65) Gianetto, A., V. Specchia and G. Baldi, AIChE J., 19, 916 (1973).
(66) Shende, B.W. and M.M. Sharma, Chem. Eng. Sci., 29, 1763 (1974).
(67) Mahajani, V.V. and M.M. Sharma, Chem. Eng. Sci., 34, 1425 (1979).
(68) Morsi, B.I., N. Midoux, A. Laurent and J.C. Charpentier, CHISA 81, Praha, Czechoslovakia, sect. Distillation and Absorption, Aug. 31 - Sept. 4, (1981).
(69) Morsi, B.I. and J.C. Charpentier, NATO ASI on" Mass transfer with chemical reaction in multiphase systems", CESME, Izmir, Turkey, Aug. 10-21 (1981).
(70) Westerterp, K.R., L. Van Dierendonck and J.A. De Kraa, Chem. Eng. Sci., 18, 1157 (1963).
(71) Midoux, N. and J.C. Charpentier, Entropie, 5, 88 (1979).
(72) Dickey, D.S., AIChE Meeting, Paper 116d, 72d Annual Meeting, San Francisco, Nov. 1979.
(73) Hassan, I.T. and C.W. Robinson, AIChE J., 23, 48 (1977).
(74) Luong, H.T. and B. Volesky, AIChE J., 25, 893 (1979).
(75) Michel, B.J. and S.A. Miller, AIChE J., 8, 262 (1962).
(76) Kurten, K. and P. Zehner, Ger. Chem. Eng., 2, 220 (1979).
(77) Calderbank, P.H., Trans. Inst. Chem. Engrs., 36, 443 (1958).
(78) Sridhar, P. and O.E. Potter, Chem. Eng. Sci., 35, 683 (1980).
(79) Van Riet, K., I.E.C. Proc. Des. Devel., 18, 357 (1979).
(80) Joosten, G.E.H., J.G.M. Schilder and J.J. Janssen, Chem. Eng. Sci., 32, 563 (1977).
(81) Ranade, V.R. and J.J. Ulbrecht, AIChE J., 24, 796 (1978).
(82) Miller, D.N., AIChE J., 20, 445 (1974).
(83) Westerterp, K.R., Chem. Eng. Sci., 18, 495 (1963).
(84) Van Dierendonck, L., J.M.H. Fortuin and D. Venderbos, Proceed 4th Eur. Symp. Chem. Engng., 205 (1971).

85) Mehta, V.D. and M.M. Sharma, Chem. Eng. Sci., 26, 461 (1971).
86) Sridharan, K., "Studies in mass transfer", Ph. D. Thesis, University of Bombay, India, (1975).
87) Sankholkar, D.S. and M.M. Sharma, Chem. Eng. Sci., 30, 729 (1975).
88) Ganguli, K.L. and H.J. Van Den Berg, Chem. Eng. Sci., 33, 27 (1978).
89) Midoux, N. and J.C. Charpentier, Entropie, 101, 3 (1981).
90) Figueiredo, M. and P.H. Calderbank, Chem. Eng. Sci., 34, 1333 (1979).
91) Hughmark, G.A., I.E.C. Proc. Des. Devel., 19, 638 (1980).
92) Midoux, N. and J.C. Charpentier, Proceed. 3rd Eur. Conf. on Mixing, York, England, Vol. 1, 337 (1979).
93) Mashelkar, R.A., Brit. Chem. Eng., 10, 1297 (1970).
94) Mashelkar, R.A. and M.M. Sharma, Trans. Inst. Chem. Engrs., 48, 162 (1970).
95) Sharma, M.M., R.A. Mashelkar and V.D. Mehta, Brit. Chem. Eng. 1, 14, 70 (1969).
96) Botton, R., D. Cosserat and J.C. Charpentier, Chem. Eng. Journ., 16, 107 (1978).
97) Botton, R., D. Cosserat and J.C. Charpentier, Chem. Eng. Journ., 20, 87 (1980).
98) Hughmark, G.A., I.E.C. Proc. Des. Devel., 6, 218 (1967).
99) Akita, K. and F. Yoshida, I.E.C. Proc. Des. Devel., 1, 76 (1973).
100) Hikita, H., S. Asai, K. Tanigawe, K. Segawa and M. Kitao, Chem. Eng. Journ., 20, 59 (1980).
101) Koide et al., J. Chem. Eng. Jap., 12, 98 (1979).
102) Ueyama, K. et al., I.E.C. Proc. Des. Devel., 19, 593 (1980).
103) Bach, H.F. and T. Pilhoter, Germ. Chem. Eng., 270, 1 (1978).
104) Nicklin, D.J., J.O. Wilkes and J.F. Davidson, Trans. Inst. Chem. Eng., 40, 61 (1962).
105) Sharma, M.M. and R.A. Mashelkar, AIChE Chem. Eng. Symp. Ser., 28, 10 (1968).
106) Geddes, R.L., Trans. Inst. Chem. Engr., 42, 79 (1946).
107) Gestrich, W. and W. Kraus, Int. Chem. Eng., 16, 10 (1976).
108) Deckwer, W.D., R. Burckhart and G. Zoll, Chem. Eng. Sci., 29, 2177 (1974).
109) Kastanek, F., M. Rylek, J. Kratochvil and M. Hartman, CHISA, H3, Abs. Dist., Prague, (1975).
110) Oels, U., J. Lücke, R. Buchholz and K. Schügerl, Germ. Chem. Eng., 1, 115 (1978).
111) Hikita, H., S. Asai, K. Tanigawa, K. Segawa and M. Kitao, Chem. Eng. Journ., 22, 61 (1981).
112) Mehta, K.C. and M.M. Sharma, Brit. Chem. Eng., 15, 1440 and 1556 (1970).
113) Pigford, R.L. and C. Pyle, Ind. Eng. Chem., 43, 1649 (1951).
114) Ramm, V.M. "Absorption of Gases", Israel Program for Scientific Translations, Jerusalem, (1968).

[115] Kohl, A.L. and F.C. Riesenfeld, "Gas Purification", 3rd Edt Gulf Publ., Houston (1979).
[116] Zenz, F.A., "Design of Gas Absorption Towers", in Handbook of Separation Techniques for Chemical Engineers, P.A. Schweitzer Editor, Mc Graw Hill (1975).
[117] Nagel, O., H. Kurten and R. Sinn, Chem. Ing. Techn., 42, 474 and 921 (1970).
[118] Reith, T., Ph. D. Thesis, Delft University (1968).
[119] Virkar, P.D. and M.M. Sharma, Canad. J. Chem. Engr., 53, 512 (1975).
[120] Laurent, A., C. Fonteix and J.C. Charpentier, AIChE J., 26, 282 (1980).
[121] Stichlmair, J. and A. Mersmann, Int. Chem. Eng., 18, 223 (1978).
[122] Mehta, V.D. and M.M. Sharma, Chem. Eng. Sci., 26, 461 (1971)
[123] Laurent, A. and J.C. Charpentier, Chem. Eng. Journ., 9, 85 (1974).
[124] Sharma, M.M. and R.K. Gupta, Trans. Inst. Chem. Engrs., 45, T 69 (1967).
[125] Nagel, O., H. Kurten and R. Sinn, Chem. Ing. Techn., 44, 367 and 899 (1972).
[126] Van Landeghem, H., Chem. Eng. Sci., 35, 1912 (1981).

HEAT AND MASS TRANSFER IN EXOTHERMIC GAS ABSORPTION

Reginald Mann
Department of Chemical Engineering
University of Manchester Institute of Science and Technology
Manchester M60 1QD England

Gas absorption and any associated chemical reaction is always accompanied by the simultaneous release of heat of solution and heat of reaction. The micro-scale phenomena taking place close to the interface therefore involve the generation and diffusion of heat as well as the diffusion and reaction of material species. In developing a fundamental appreciation of simultaneous mass and heat transfer in gas-liquid reactions it is important for the heat effects to be incorporated into the analysis of diffusion and reaction because the rates and pathways of chemical reactions are usually enormously sensitive to temperature. In particular, for the case of gas-liquid reactor performance, if the heat effects are such that the mass transfer with reaction zone adjacent to the interface is at a temperature significantly different from the bulk, the yield and the selectivity performance will be erroneously interpreted if reaction is assumed to take place at the bulk liquid temperature. In consequence, the basic conceptual design of a commercial gas-liquid reactor could incorporate fallacious reasoning leading to inefficient operation at sub-optimal yield.

A PENETRATION THEORY APPROACH

According to the well-known penetration theory, the diffusion and reaction of a single species is described by the partial differential equations

$$D \frac{\partial^2 C}{\partial x^2} - r(C, T) = \frac{\partial C}{\partial t} \qquad (1)$$

$$\alpha \frac{\partial^2 T}{\partial x^2} + \frac{\Delta H_R}{\rho C_p} \cdot r(C, T) = \frac{\partial T}{\partial t} \tag{2}$$

In developing an understanding of the likely magnitude of heat release, the influence of reaction can be initially ignored, so that $r(C, T) = o$, and the processes taking place now simply involve physical dissolution and diffusion of the dissolved gas and the accompanying conduction of the heat of solution into the semi-infinite liquid phase. By manipulation of the two simultaneous unsteady diffusion/conduction equations, which obey the interfacial boundary condition

$$- \Delta H_s D \left.\frac{\partial C}{\partial x}\right|_{x = o} = -K \left.\frac{\partial T}{\partial x}\right|_{x = o} \tag{3}$$

it can be shown that the interface temperature and concentration adjust instantaneously on exposure of the liquid to the gas at $t = o$ to the equilibrium solubility values. Thereafter, irrespective of the time of exposure, the interfacial solubility and temperature are invariant, and are linked by the relationship

$$T^*(o) = C^*(o) \frac{\Delta H_s}{\rho C_p} \sqrt{\frac{D}{\alpha}} \tag{4}$$

In this case the penetration profiles appear as shown in Fig. 1 when they are scaled together. The figure indicates a very important feature of heat release acompanying gas absorption, which is that the extent of heat penetration as manifested by the temperature above the datum is very much greater than the depth of mass penetration. This circumstance arises because the thermal diffusivity α is much greater than the mass diffusivity. In fact the ratio of depth of penetration at a given fractional value of the interfacial condition is given by $\alpha \simeq \sqrt{\alpha / D}$. For many systems $\alpha \simeq 100D$, and therefore in physical absorption the temperature profile extends approximately ten times further into the liquid than the concentration profile.

If a simple first order reaction is now introduced, Eqns (1) and (2) become

$$D \frac{\partial^2 C}{\partial x^2} - k(T) C = \frac{\partial c}{\partial t} \tag{5}$$

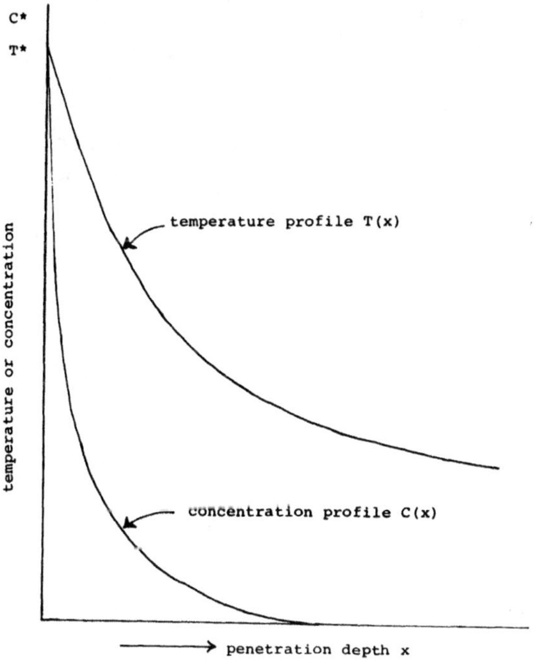

FIG. 1. RELATIVE DISPOSITION OF T AND C PROFILES

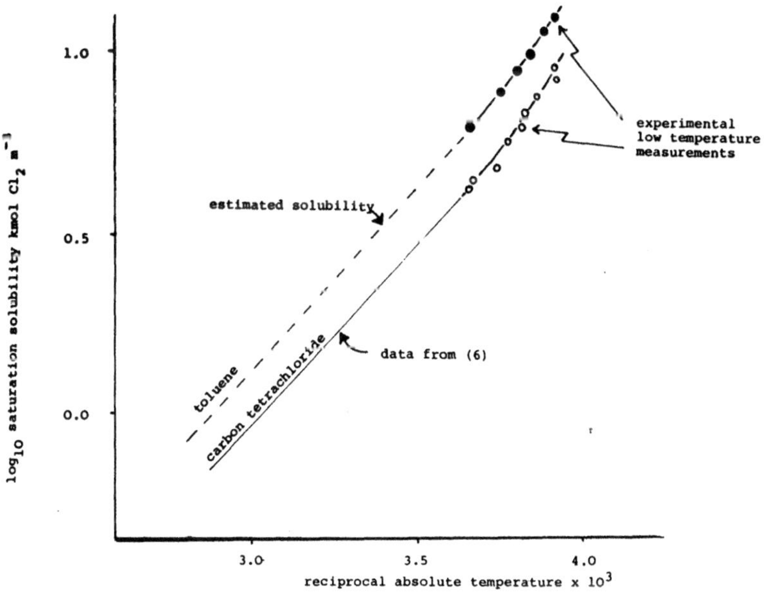

FIG. 2 ESTIMATION OF SOLUBILITY OF CHLORINE IN TOLUENE

$$\alpha \frac{\partial^2 T}{\partial x^2} + \frac{\Delta H_R \, k(T) \, C}{\rho \, C_p} = \frac{\partial T}{\partial t} \qquad (6)$$

The effect of chemical reaction is to increase the absorption rate, giving an increase in the heat released due to solution in comparison with the physical absorption case. Also, additional heat will now be released if the reaction is exothermic, and since usually $\Delta H_R > \Delta H_s$, absorption with reaction can be expected to offer substantially greater heat effects.

One important consequence of any increase in interfacial temperature due to the release of heat of solution and reaction is to decrease the interfacial solubility of the absorbing gas. This will result in a decrease in the absorption driving force and this in turn will lower the rate of absorption.

The potential magnitude of interfacial temperature increase can be estimated by first of all assuming that any temperature rise will be insufficient to change the interfacial solubility $C^*(o)$, and thus will have no influence upon absorption rate. This effectively uncouples the two Eqns (5) and (6). The rate of absorption is a well known analytical function from Eqn (5), and from it the rates of release of heat of solution and reaction are known. Referring again to Fig. 1, because the depth of the temperature profile is an order of magnitude greater than the concentration profile, and because the reaction reduces the penetration depth of the species being consumed by reaction, it appears as if the heat released due to reaction is being released at the interface. If the reaction is assumed to generate a heat flux at x = o, then the solution and reaction heat fluxes are solved in respect of the simplified heat diffusion equation

$$\alpha \frac{\partial^2 T}{\partial x^2} = \frac{\partial T}{\partial t} \qquad (7)$$

The result for interfacial temperature as a function of time is given by

$$(t) = C^{*}(o)\frac{\Delta H_s}{\rho C_p}\sqrt{\frac{D}{\alpha}}\exp\left[\frac{-kt}{2}\right]\left[I_o\left[\frac{kt}{2}\right] + (n+1)kt\left\{I_o\left[\frac{kt}{2}\right] + I_1\left[\frac{kt}{2}\right]\right\}\right] \quad (8)$$

ere the I_o and I_1 are hyperbolic Bessel Functions. This lysis was first performed by Danckwerts (1) using a series lution approximation, and later he gave the above result in ms of the analytical functions (2). The result above was ived on the basis that $\alpha/(\alpha - D) \simeq 1$, which is ivalent to considering the heat of reaction as an erfacial heat flux.

Eqn (8) predicts that the temperature increases as the time exposure increases. Danckwerts (1) used this approach to imate the heat effects for absorption of carbon dioxide in a bonate-bicarbonate buffer solution, and showed that an tial temperature increase of 0.02°C was expected, rising due reaction to 0.06°C after 0.5 s. He concluded that the orption process was essentially isothermal, and that the perature increase was certainly too small to have any effect n the rate of absorption.

Subsequently, Chiang and Toor (3) in considering the orption of pure ammonia into water in a laminar jet tactor, performed a slightly more complex analysis of sical absorption to take account of the volume change of the uid phase in addition to the heat release effect. Their ations for the physical absorption of ammonia into water e

$$D\frac{\partial^2 C}{\partial x^2} = \frac{\partial C}{\partial t} + v(t)\frac{\partial C}{\partial x} \quad (9)$$

$$D\frac{\partial^2 T}{\partial x^2} = \frac{\partial T}{\partial t} + v(t)\frac{\partial T}{\partial x} \quad (10)$$

re $v(t)$ is a velocity of the liquid phase arising from the ume change effect. In their theoretical analysis, they ived Eqn (4) via Duhamel's Theorem, and by neglecting $v(t)$, y predicted that the interfacial temperature during ammonia orption would be 15.0°C above a datum temperature of 22.2°C. effect of volume change was calculated to increase the rate of absorption by a small amount, giving rise to an interface temperature of 17.7°C. In any event, the system ammonia-water does appear to give quite appreciable interfacial temperature increases, sufficient to introduce a supression of the absorption rate due to solubility reduction.

The solution and reaction of chlorine in organic liquids where fast exothermic chlorination reactions are possible, presents another candidate system where heat release and thermal effects may be significant. Indeed, it was in an original set of experiments in 1965 (4), measuring the absorption of pure chlorine into a laminar jet of toluene at 25°C, where the mixing cup temperature after absorption in the jet was as high as 6°C, which prompted the speculation that quite high interfacial temperatures could be achieved in the absorption and reaction of chlorine in toluene (and of course in other organic liquids).

To interpret absorption experiments for this system it is essential to know the way in which the saturation solubility varies with temperature. This is experimentally impossible since the dissolution is always accompanied by reaction at normal temperatures. For chlorine-toluene an estimate has been made from solubility measurements performed in the range 0°C to - 20°C (5) which are presented in Fig. 2, along with some companion measurements using the same gas burette technique with the non-reactive system chlorine-carbon tetrachloride. The Cl_2 - CCl_4 results compare well with those published in International Critical Tables (6) which verifies the measurement technique. The extrapolation of the Cl_2 - $C_6H_5CH_3$ solubilities into the ambient temperature range shows that the data are not seriously different from the idealised solubility based on an application of Raoults Law, and the slope of the log solubility versus reciprocal absolute temperature indicates a heat of solution of 5,500 cal mol^{-1}. This estimated solubility data is presented also in Fig. 3 as a Bunsen coefficient <u>ie</u> as the solubility expressed in volumes of gas dissolved per unit volume of solvent toluene. The Bunsen coefficient of 70 at 25°C is a moderate solubility, though very much less than a value of around 440 for ammonia-water at the same temperature.

This solubility data is the basis for estimating the initial temperature adjustment from a datum of 25°C using Eqn (4). The values of C*(o) and T*(o) which obey Eqn (4) are shown in Fig. 4. Under purely physical absorption, or before significant reaction can take place, the interface temperature is 7.05°C above the datum of 25°C. Using data appropriate to this system (7), in particular for ΔH_R = 30,000 cal mol^{-1}, the predicted

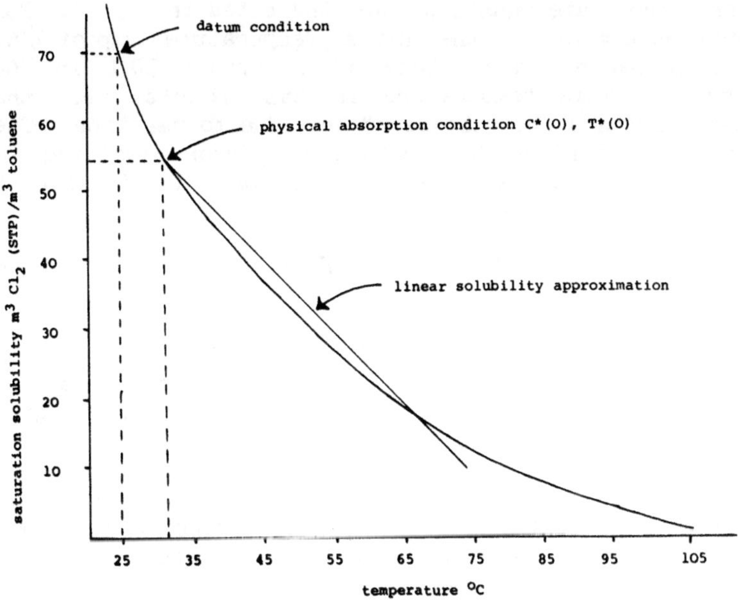

FIG. 3 LINEAR APPROXIMATION TO SOLUBILITY

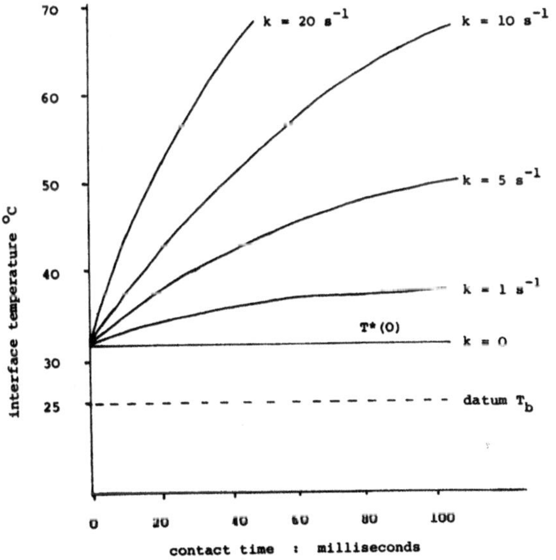

FIG. 4 SURFACE TEMPERATURE IGNORING SOLUBILITY REDUCTION DUE TO CHEMICAL REACTION

temperature increases according to Eqn (8) for various values of the first order rate constant k are indicated in Fig. 4. For a rate constant $k = 10\ s^{-1}$, the initial temperature jump of 7°C, has been followed by an increase of a further 30°C in 100 milliseconds. This temperature increase results from the extra dissolution and the release of heat due to reaction. The fact that the interface temperature has been predicted to increase to 37°C above the datum in 0.1 seconds is significant from several points of view:-

(i) the solubility will certainly be reduced at the higher temperatures

(ii) the chemical reaction, by reference to Fig. 4 is taking place at a temperature quite different from 25°C - competing reactions with differing activation energies would certainly have different yields

(iii) vaporisation of the toluene is possible at the higher temperatures giving a possibility of gas phase reaction, possibly again introducing an unanticipated gas phase mass transfer resistance.

The solubility reduction appears likely to severely curtail the absorption rate expected on the basis of isothermal absorption. From Fig. 3, the solubility of chlorine at 70°C is only 1/5th of that at the datum of 25°C, and the effective driving force would thereby be reduced by a factor of 5.

These observations prompted a theoretical analysis of the consequences of linking the interfacial solubility and temperature behaviour in solving Eqns (5) and (6). The normal logarithmic form of solubility relationship makes the solution of Eqns (5) and (6) intractable in the Laplace domain, but analytical results are possible using a linear solubility relationship shown in Fig. 4, where

$$T^*(t) - T^*(0) = \frac{1}{\mu_s}\,(C^*(0) - C^*(t)) \qquad (11)$$

If the heat release due to reaction is treated as an interfacial heat flux, then diffusion and reaction obey Eqn (5) but heat diffusion obeys Eqn (7). The evolution of the entire temperature profile T(x,t) for an arbitrary interfacial flux ϕ_H is a classical problem in heat flow and easily handled by Laplace transformation (8). In this case the interfacial heat flux is determined by the rate of absorption of the gas and its rate of reaction close to the interface. This heat flux is therefore given by

$$\phi_H = \Delta H_s \left[-D \frac{\partial C}{\partial x}\right]_{x=o} + \Delta H_R \left\{ \left[-D \frac{\partial C}{\partial x}\right]_{x=o} - \frac{\partial}{\partial t}\int_o^\infty C(x,t)dx \right\} \quad (12)$$

Heat flux due to solution — heat flux due to reaction

where

$$\text{heat flux due to reaction} = \Delta H_R \left[\text{total rate of absorption} - \text{rate of accumulation of unreacted solute}\right]$$

Using the linear solubility relationship (Eqn (11)) to link the interfacial concentration and temperature, $C^*(t)$ can be deduced by inverting the Laplace transformed Eqn (12), for a constant value of the reaction rate constant (ie assuming zero activation energy for the reaction). The algebraic manipulations are somewhat lengthy (7), but the result of allowing the solubility to reduce as the temperature at the interface rises is shown in Fig. 5. At a time of exposure of 0.1 seconds with $k = 10\ s^{-1}$ the interfacial temperature increase has been restricted to 29°C above the datum of 25°C.

This is reflected in the changed absorption rate behaviour shown as the enhancement factor in Fig. 6. In this case the enhancement factor E is defined by

$$E = \frac{\text{rate of absorption with reaction and heat effects}}{\text{physical absorption neglecting heat effects}}$$

Fig. 6 indicates an initial 22% reduction in the enhancement factor as the temperature and solubility adjust instantaneously at t = o. Thereafter, with $k = 10\ s^{-1}$ as the time of exposure increases, the enhancement factor continues to decline to a value of 0.65 after 100 milliseconds. The chemical reaction is therefore seen to significantly hinder the absorption process, rather than to enhance it. To some extent an enhancement factor of less than unity is a misnomer. In the complete absence of heat effects the enhancement factor would have become 1.25. The hindrance of the chemical reaction intensifies as the rate constant is made larger. In Fig. 6, the limit of reasonable applicability of the linear solubility approximation is shown by reverting to a broken line.

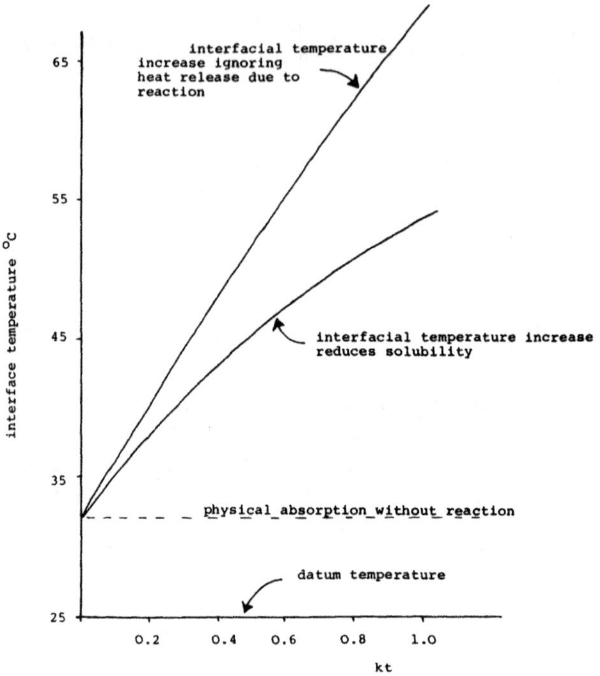

FIG. 5 EFFECT OF RELEASE OF HEAT OF REACTION

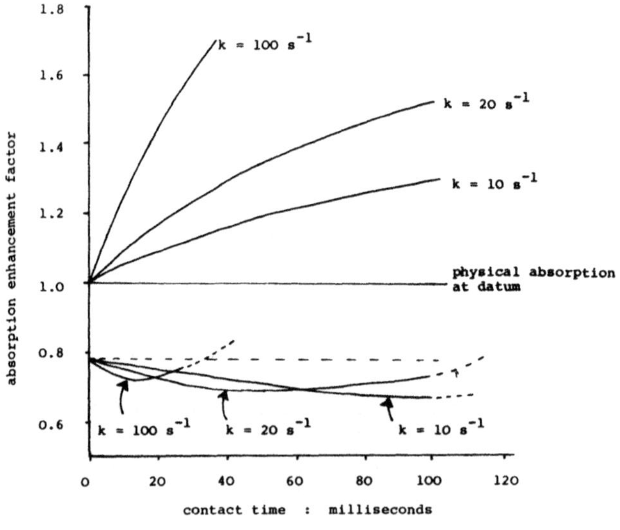

FIG. 6 ABSORPTION BEHAVIOUR WITH LARGE HEAT EFFECTS

This general analysis of the chlorine-toluene system based on first order reaction was developed in parallel with a ries of experimental measurements (9), in which chlorine was sorbed in toluene in a laminar jet. This absorption device ovides remarkable control of surface area and with a flat locity profile the penetration time is reasonably well fined so that the penetration theory can be directly applied thout any uncertainty concerning the complications of nvective transport. Experimental mixing cup temperatures r Cl_2-toluene ranged from 1^oC to 6^oC. These can be terpreted as the amount of heat accumulated per unit of jet rface as the jet plunges into the receiver via the equation

$$H(t) = \frac{1}{4} \rho C_p \Delta T d \tag{13}$$

A plot of the experimental results is shown in Fig. 7. The ojected heat accumulations for fixed interfacial temperatures 10^oC and 20^oC above the datum are also shown in Fig. 7. The rface temperature relationship which when solved in njunction with Eqn (7) exactly matches the experimental ttern of heat accummulation, is given by the empirical lationship $T^*(t) = 10.0 + 86\ t$. This would indicate that e temperature at the surface of the jet has increased to 28^oC ove the datum after 40 milliseconds. This interface mperature behaviour definitely verifies that large erfacial temperatures can occur in the absorption and action of chlorine. However, the observed behaviour does match the predicted behaviour since the reaction is rtainly not first order, and the mass transfer process oars to be complicated by some kind of interfacial bulence (9). Moreover, the experimental results could not at all explained by allowing for the rate constant to celerate with temperature according to the conventional henius type relationship. In fact, the intractability of e penetration theory equations when k(T) has an exponential mperature dependence, pointed the way to a simpler film eory based analysis.

.M THEORY ANALYSIS

The equation describing diffusion with (pseudo) first order action after dissolution at the interface is given by

$$D \frac{d^2C}{dx^2} = k(T)\, C \tag{14}$$

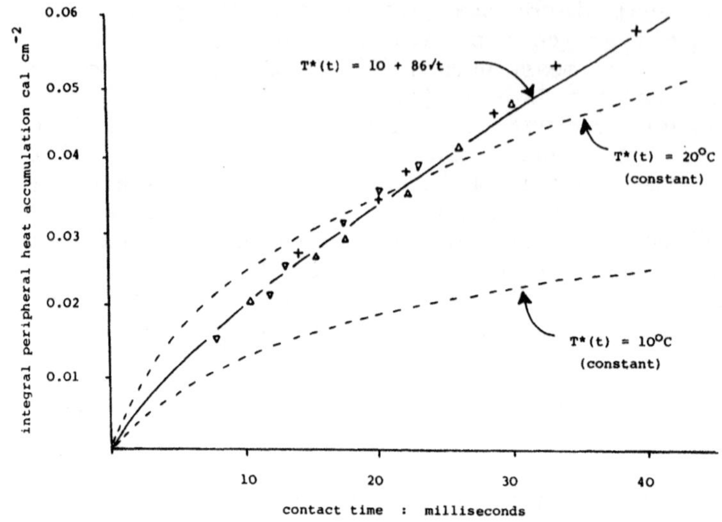

FIG. 7 ESTIMATION OF EXPERIMENTAL INTERFACE TEMPERATURE BEHAVIOUR

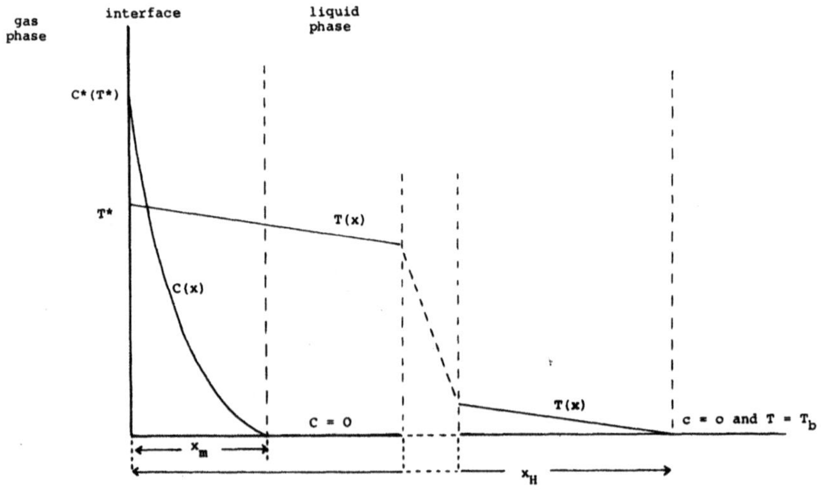

FIG. 8 CONCENTRATION AND TEMPERATURE PROFILES FOR FILM THEORY

ıd heat transfer is described by

$$\alpha \frac{d^2T}{dx^2} = - \frac{\Delta H_R}{\rho c_p} k(T) C \qquad (15)$$

ıese steady state equations of the film theory are to be solved th respect to the boundary conditions

$$C = C^*(T^*) \text{ and } T = T^* \text{ at } x = o \qquad (16a)$$

$$C = 0 \text{ at } x = x_M \text{ and } T = T_b \text{ at } x = x_H \qquad (16b)$$

.ere the interfacial concentration (solubility) and mperature are to be linked through some solubility lationship such that

$$C^*(T^*) = f(T^*) \qquad (17)$$

.e second set of boundary conditions imply that the mass 'ansfer film thickness and the heat transfer film thickness e not the same (10).

It is therefore assumed in the formulation of Eqns (14) and 5) that the potentially adverse effect of temperature crease on the absorption rate behaviour and the temperature pendence on the reaction rate constant are the most important fects in the analysis of absorption accompanied by gnificant heat release. Other factors, such as coupling of e heat fluxes and mass fluxes (the Dufour and Soret effects), lume change of the liquid phase and any bulk flow mplications are taken to be of secondary importance. The undary condition which assumes a zero concentration of agent in the bulk liquid phase has been widely adopted in lutions to the diffusion/reaction equations. It is likely be a good approximation as the reaction moves towards the so-lled fast reaction regime, and is most inappropriate for slow action when the liquid phase can approach a saturation ndition. Nevertheless it provides a good working hypothesis r characterising exothermic absorption using the film theory, d the use of this boundary condition has been examined in the ntext of gas-liquid reactor performance in a companion paper this ASI (11).

The boundary conditions of Eqns (16a) and (16b) are not sufficient to solve the two coupled differential eqns (14) and (15) and two further independent relationships are required, since four conditions are necessary to solve two simultaneous second order ordinary differential equations. Two further conditions are the interfacial heat balance equation

$$- \Delta H_s D \frac{dC}{dx}\bigg|_{x=o} = - K \frac{dT}{dx}\bigg|_{x=o} \tag{18}$$

and the overall heat balance

$$- \Delta H_S D \frac{dC}{dx}\bigg|_{x=o} + \Delta H_R \left[- D \frac{dC}{dx}\bigg|_{x=o} - \left(- D \frac{dC}{dx}\bigg|_{x=x_M} \right) \right] = - K \frac{dT}{dx}\bigg|_{x=x_M} \tag{19}$$

Eqns (14) and (15) are strongly coupled due to the fact that the variation of temperature through the mass transfer film induces a positional variation of reaction rate constant k(T). In this circumstance, analytical solutions to the basic equations are quite impossible.

However, as with the penetration theory analysis, the difference in magnitude of the mass and thermal diffusivities with $\alpha \simeq 100$ D, means that the heat transfer film is an order of magnitude thicker than the mass transfer film. This is depicted schematically in Fig. 8. The fall in temperature from T* over the distance x_M is (if $\alpha \simeq 100$ D) about 10% of the overall interface excess temperature above the datum temperature T_b. Furthermore, in considering the location of heat release due to reaction in the mass transfer film, this is bound to be greatest closest to the interface, and this is especially the case when the reaction becomes fast. Therefore, two simplifications can be introduced as a result of this: (i) the release of heat of reaction can be treated as an interfacial heat flux and (ii) the reaction can be assumed to take place at the interfacial temperature T*. The differential equation for diffusion and reaction can therefore be written

$$D \frac{d^2C}{dx^2} = k(T^*)\, C \tag{20}$$

subjected to $C^* = f(T^*)$ at $x = o$

$$C = 0 \quad \text{at } x = x_M \tag{21}$$

ıe rate constant to be utilised in solving the diffusion with ›action behaviour is therefore quite reasonably taken to be ıat at the interface temperature T*.

The appropriate value of T* is then found by applying the ›at balance relationships to the simplified heat transfer ıuation

$$\alpha \frac{d^2T}{dx^2} = 0 \qquad (22)$$

ıis means that whilst T* is approximately constant across the ıss transfer film x_M, it must be linear across the heat ›ansfer film x_H as shown in Fig. 8. T* is therefore found ›om

$$(T^* - T_b) = -\Delta H_s\, D \left.\frac{dC}{dx}\right|_{x=o} + H_R\left[- D \left.\frac{dC}{dx}\right|_{x=o} + D \left.\frac{dC}{dx}\right|_{x=x_M}\right]$$

(23)

Equation (14) has an analytical solution for C(x) in terms hyperbolic functions, and the result in terms of the ıhancement factor is given by

$$E = \frac{C^*(T^*)}{C^*(T_b)} \cdot \frac{\sqrt{M'}}{\tanh\sqrt{M'}} \qquad (24)$$

ıere $M' = D\, k(T^*)/k_L^{\,2}$

before, E is defined relative to physical absorption ınsidered to take place with negligible heat effects. The ffusion/reaction parameter M' is defined using the reaction .te constant evaluated at the interfacial temperature T*.

The enhancement factor defined by Eqn (24) is the product of ›o factors. The first term, which is the ratio of the lubility at the interface temperature T* to that at the datum mperature T_b, represents the effect of reduced driving force on the absorption behaviour. The second term represents the ıemical reaction contribution to the enhancement in absorption te. The competition between these two opposing effects for .ose cases where a large increase in interfacial temperature possible, gives rise to some quite complex phenomena in othermic absorption processes.

Determination of the interfacial temperatures achieved during absorption requires the solution of Eqn (22) in conjunction with the analytical results of Eqn (20). In the absence of reaction when

$\Delta H_R = 0$ and $\sqrt{M^1} = 0$, Eqn (23) gives

$$T^*(o) = T_b + \Delta H_s \, C^*(T^*(o)) \, k_L/h_L \quad (25)$$

The solution of Eqn (25) for a highly exothermic system can be illustrated for data relevant to the absorption of sulphur trioxide in dodecylbenzene (DDB). In this case, it is impossible to measure the solubility of SO_3 in DDB at moderate temperatures, because of the very high reactivity of SO_3. The method applied to the system Cl_2 -toluene cannot be used in this case, since SO_3 has a much higher condensation temperature. However, the ideal solubility can be estimated from an application of Dalton's and Raoult's Laws, whereby the liquid phase mol fraction of SO_3 is given by

$$x_A = Py_A/p_A^o \quad (26)$$

The saturation solubility, assuming negligible volume changes on mixing is then obtained from

$$C^* = \frac{1}{1 - x_A} \quad \frac{1}{(1 - x_A)\,\rho_A'} + \frac{1}{x_A\,\rho_L'} \quad (27)$$

The results for the ideal solubility of SO_3 in DDB are presented in Fig. 9 in terms of both concentrations and volumes (Bunsen coefficients). This is a compact graphical representation of the variation of solubility with temperature for various mol fractions of SO_3 in the gas phase.

In terms of solving Eqn (25) and illustrating the results for gas phase compositions of 3%, 10% and 30% of SO_3, the solubility data can be more conveniently visualised in Fig. 10. For a 30% gas phase composition of SO_3, an exact solution of Eqn (25), shows that for physical absorption, an interface temperature of 38°C will be achieved ie the interfacial temperature is 13°C above the datum of 25°C. The corresponding values for 10% SO_3 and 3% SO_3 are 6.1°C and 1.4°C. At the higher gas phase compositions, purely physical absorption gives a quite

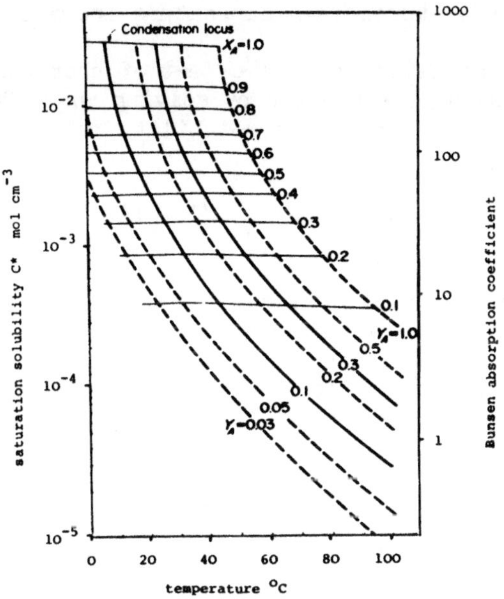

FIG. 9 SATURATION SOLUBILITY OF SO_3 IN DDB

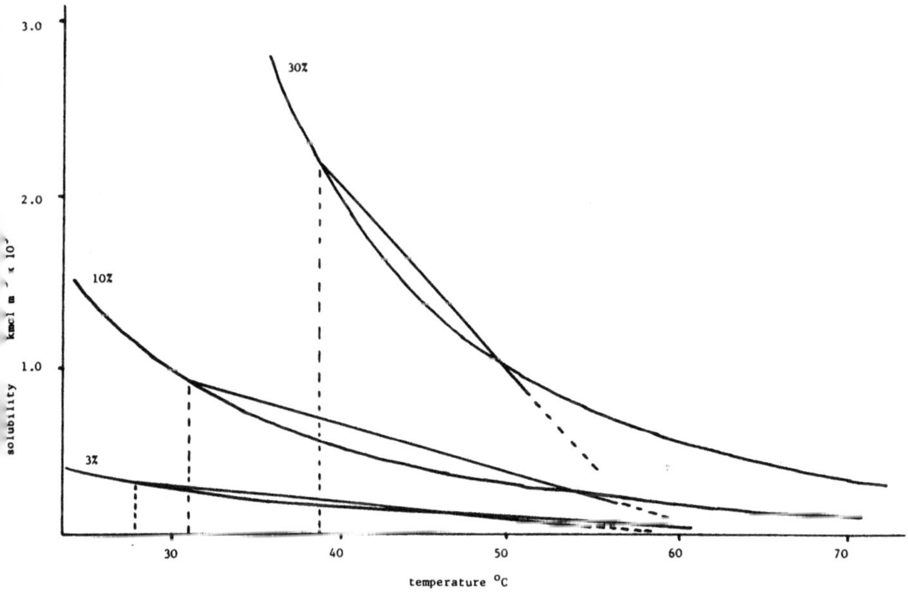

FIG. 10 IDEAL SOLUBILITY FOR SULPHUR TRIOXIDE IN DODECYLBENZENE

significant non-isothermality. The presence of reaction and the release of heat of reaction can be expected to give much more pronounced interfacial temperature increases.

The likely magnitude of the effect of chemical reaction can be very conveniently estimated by solving Eqns (20) and (22) in conjunction with a linearised solubility relationship given by

$$C^*(T^*) = C^*(T^*(o)) - \mu_s (T^* - T^*(o)) \tag{28}$$

where the linearisation commences from the conditions existing for purely physical absorption. Using this linearisation makes Eqn (23) explicit in T^* for the case where the reaction is in the fast regime and the activation energy is zero, so that $\tanh\sqrt{M'} \to \sqrt{M}$. Rearrangement of Eqn (23) then gives

$$T^* = T^*(o) + \frac{C^*(T^*(o))}{\mu_s + \dfrac{1}{\Delta H_R + \Delta H_s} \cdot \sqrt{\dfrac{1}{M}} \cdot \dfrac{h_L}{k_L}} \tag{29}$$

and the reduction in C^* follows directly so that

$$\frac{C^*(T^*)}{C^*(T^*(o))} = 1 - \frac{\mu_s}{\mu_s + \dfrac{1}{\Delta H_R + \Delta H_s}\sqrt{\dfrac{1}{M}}\,\dfrac{h_L}{k_L}} \tag{30}$$

Hence, the enhancement factor is given by

$$E = \frac{\sqrt{M}}{1 + \mu_s (\Delta H_R + \Delta H_s)\sqrt{M}\, k_L/h_L} \tag{31}$$

If the reaction is very fast, then at large $\sqrt{M}$, the enhancement factor reaches the asymptotic value

$$E = \frac{1}{\mu_s(\Delta H_R + \Delta H_s)} \frac{h_L}{k_L} \tag{32}$$

Therefore, further increases in reaction rate do not result increases in absorption rate and the enhancement factor mains constant. This asymptote reflects the balance between e tendency for increases in $\sqrt{M}$ to steepen the interfacial ncentration gradient, but at the same time to increase the mperature and thereby decrease the available driving force r the absorption. Eqn (29) therefore gives a measure of the gnitude of interfacial temperature achieved by the release of at of reaction. This measure of interfacial temperature is nservative since in Eqn (29) M refers to the case where the tivation energy is zero and the reaction rate is not celerated by the rise in interfacial temperature.

The full solutions of Eqn (23) over the entire range of $\sqrt{M}$ lues for gas phase composition of 3%, 10% and 30% SO_3 are ven in Fig. 11. The values of μ_s implied by the linear proximation to solubility shown in Fig. 11 when allied to the lubility reductions implied by the adjustment to T*(o) from under physical absorption or slow reaction conditions, gives complicated picture of the ultimate temperature asymptotes hieved at the three gas phase compositions. The fact that gher temperatures are predicted for the lower compositions of % and 3% SO_3 can be explained by the crossing over of the nearisation approximations in the temperature range between and 60°C. It is clear that above 50°C, the linearisation of e solubility curve becomes particularly inappropriate and the edictions above this temperature should be treated with ution. This is reflected by the use of broken lines in Fig. . However, it is very clear from Fig. 11 that quite high mperatures can be expected in the absorption with reaction of $_3$ in DDB, such that interfacial temperature increases in the esence of fast reaction in excess of 25°C are anticipated. e corresponding behaviour of the enhancement factor is esented in Fig. 12. The exothermic absorption behaviour is ite different from that expected on the basis of hypothetical othermal absorption with reaction.

The most severe aberrations can be observed for the 30% SO_3, the gas phase. The 13°C temperature increase for physical sorption reduces the absorption potential by a factor of most 6, giving an enhancement factor for values of $\sqrt{M} < 1.0$ 0.16. As $\sqrt{M}$ increases, the enhancement factor is predicted fall to an asymptote of 0.13. This is an example of an crease in the reaction rate giving rise to a decrease in the sorption rate due to the adverse impact of release of heat of action. That this behaviour is theoretically possible can be en from a combination of Eqns (32) and (25) whereby

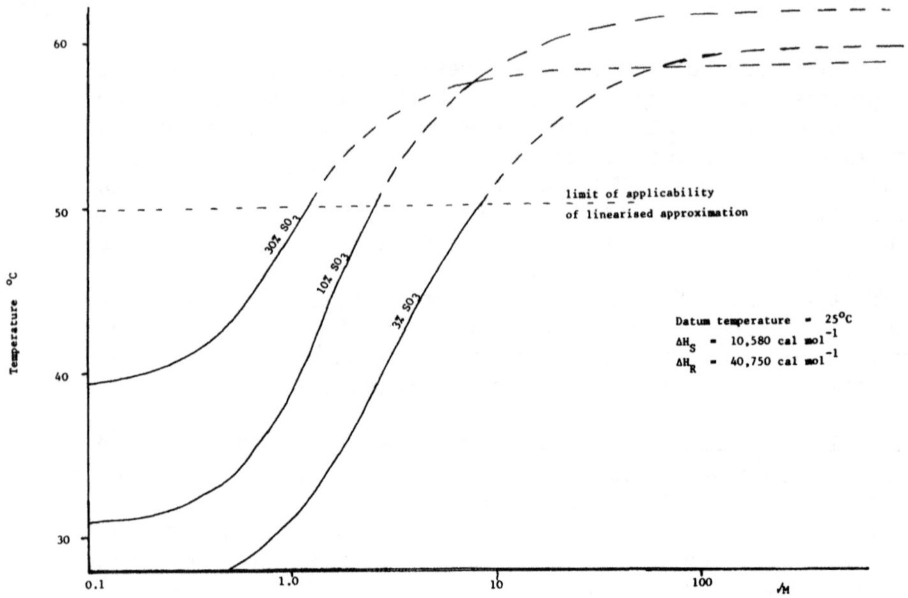

FIG. 11 INTERFACE TEMPERATURE PREDICTIONS FROM LINEARISED SOLUBILITY

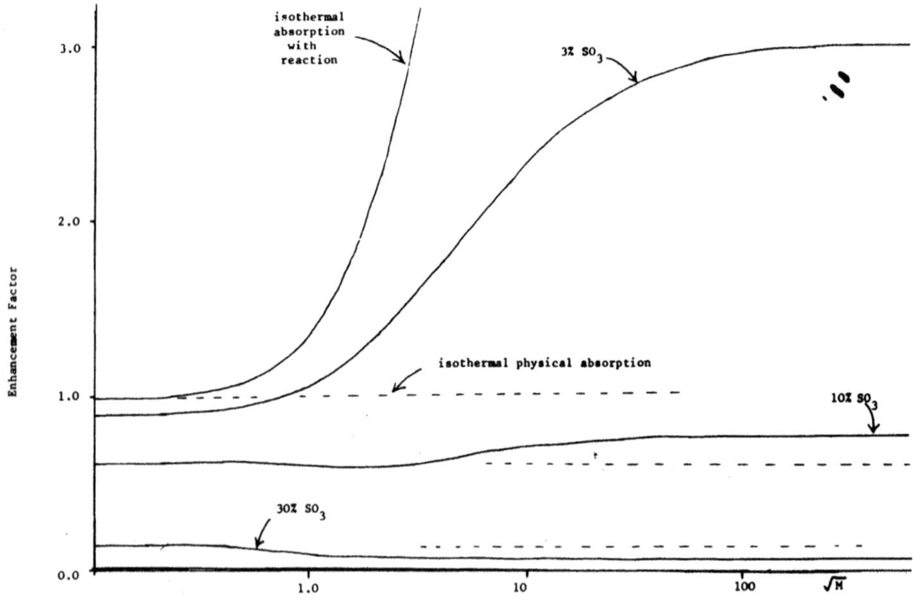

FIG. 12 ENHANCEMENT FACTOR PREDICTIONS FOR LINEARISED SOLUBILITY

$$* = T^*(o) + \left\{ \frac{\mu_s (\Delta H_s + \Delta H_R) k_L \sqrt{M}}{h_L \tanh\sqrt{M}} - \frac{\mu_s' \Delta H_R k_L \sqrt{M}}{h_L \sinh\sqrt{M}} \right\}^{\frac{1}{n+1}} \quad (35)$$

nd

$$= \frac{\mu_s' \sqrt{M}}{(T^* - T^*(o))^n \, C^*(T^*(o)) \tanh\sqrt{M}} \quad (36)$$

owever, whilst the explicit relationships simplify the ssessment of the potential magnitude of the heat effects in as absorption, their usefulness is still restricted to the ase where $E_R = 0$ and the rate constant is thereby considered to e always at its value corresponding to the liquid phase datum emperature T_b.

The effect of using the actual idealised solubility elationships and solving the two coupled equations (20) and 22) by trial and error to find the interfacial temperature and ence the absorption enhancement factor can be observed in Fig. 3. It is immediately evident that the enhancement factor no onger shows an asymptote under fast reaction conditions. This corrects a previous analysis (10), where a slowly ncreasing E was mistakenly identified as an asymptote). The symptotic enhancement factor would seem to be a result of the inearisation of the solubility curve. Nevertheless, Fig. 13 oes show a slightly intensified effect on E in the low/intermediate reaction regime where $\sqrt{M} \leq 1.0$ with an nitial decline in E in both the 10% and 30% cases. Thereafter, he enhancement factor always increases as $\sqrt{M}$ increases, lthough it never can reach the hypothetical values for sothermal absorption with reaction.

This conclusion has an interesting contrast with the redicted behaviour for finite values of the activation energy . As with ideal solubility, the incorporation of a finite tivation energy eliminates the possibility of obtaining onvenient analytical results. Even so, suitable trial and ror methods give a reasonably straightforward evaluation of he non-isothermal behaviour.

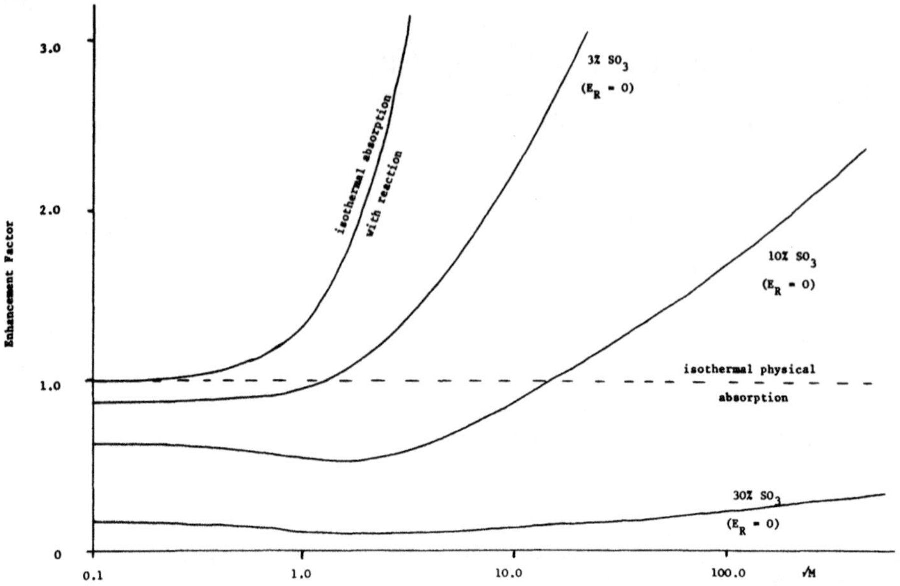

FIG. 13 ENHANCEMENT FACTORS FOR ABSORPTION OF SO_3 USING IDEALISED SOLUBILITY

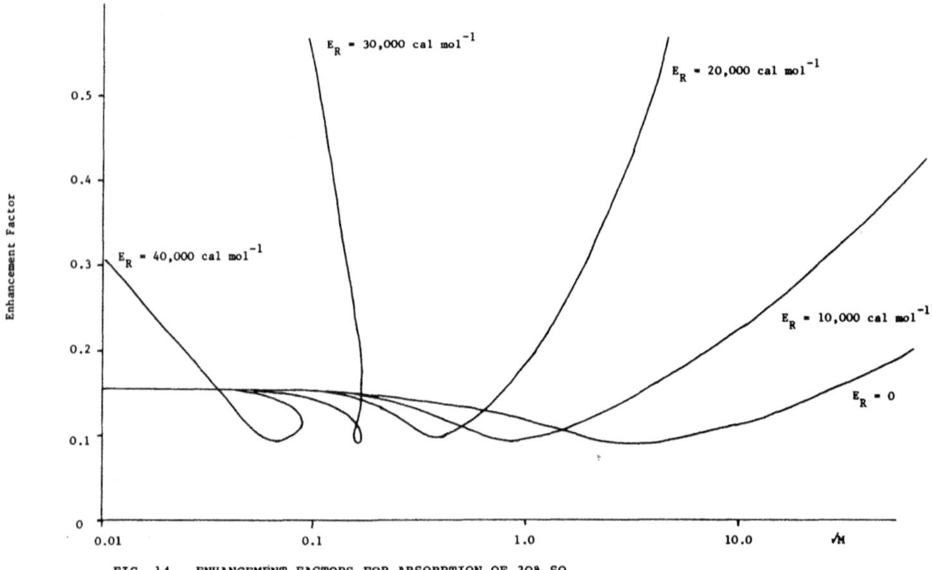

FIG. 14 ENHANCEMENT FACTORS FOR ABSORPTION OF 30% SO_3 (IDEALISED SOLUBILITY)

$$\frac{E\ (\text{fast reaction})}{(\text{physical absorption})} = 1 + \frac{1 - \mu_s \Delta H_R \frac{k_L}{h_L}}{\mu_s(\Delta H_R + \Delta H_s)\frac{k_L}{h_L}}$$

(33)

om this equation, it can be seen that the enhancement factor creases or increases depending upon whether $\mu_s \Delta H_R k_L / h_L$ is eater or less than unity (10).

Thus, in Fig. 12, for 10% SO_3 in the gas phase, the advent of st reaction results in an asymptote somewhat higher than that r physical absorption. In the case of 3% SO_3, the hancement factor initially increases relatively bstantially, but nevertheless still reaches an asymptote ich is much very less than expected if heat release were to be nored. For instance at $\sqrt{M} = 10$, the enhancement factor with at release due to reaction is about 2.3, whereas it would be for isothermal pseudo first order reaction.

It should of course be realised that the above enhancement ctor predictions are wholly inappropriate if the predicted terfacial temperatures are above the temperatures for which e linearised solubility can be considered a reasonable proximation to the actual solubility. Such a linearisation nds to be very reasonable over an initial range of mperature, but eventually fails to match the real behaviour predicting that the solubility actually reaches zero at some mperature. In reality the solubility is never predicted to zero, but merely continues to decline at a steadily reducing te.

This kind of behaviour can be represented by a solubility lationship of a hyperbolic form, so that

$$C^* = \frac{\mu'_s}{(T^* - T^*(o))^{n'}} \qquad (34)$$

is form of relationship has been shown to give convenient alytical solutions (12) whereby

In respect of Eqn. (24), the interactions of solubility reduction and the factor $\sqrt{M'}/\tanh\sqrt{M'}$ in Eqn. (24), now give rise to a further degree of complexity in absorption behaviour. Depending upon the size of the activation energy, the increase in the rate constant at high interfacial temperatures can offset the reduced solubility in a number of ways, which unexpectedly introduces an added level of complexity to these interactions.

This is clearly shown in Fig. 14 for a gas phase composition of 30% SO_3. For a given value of the activation energy, in Eqn (24) the value of the diffusion reaction parameter $\sqrt{M}$ which refers to the datum temperature has to be replaced by $\sqrt{M'}$ which is the value that $\sqrt{M}$ has at the interfacial temperature T*. The inter-relation between the two factors is

$$\sqrt{M} = \sqrt{M} \exp \left\{ \frac{-E_R}{2R} \left[\frac{1}{T^*} - \frac{1}{T_b} \right] \right\} \quad (37)$$

There is thus a direct mapping between $\sqrt{M'}$ and $\sqrt{M}$, such that the value at the interfacial temperature condition is related back to the diffusion/reaction parameter at the datum conditions. Each curve in Fig. 14 is of an identical shape, but displaced on the $\sqrt{M}$ abscissa according to Eqn. (37).

If E_R is set to 10,000 cal mol^{-1}, a particular value of the enhancement factor arises at a lower value of $\sqrt{M}$. Thus in Fig. 14 an enhancement factor of 0.2 corresponds to a $\sqrt{M}$ value of 7.4, whereas the interfacial temperature makes the actual $\sqrt{M'}$ value equal to 66.0. At this same interfacial value of 66.0, if the activation energy were 20,000 cal mol^{-1}, the corresponding datum $\sqrt{M}$ value is only 1.1. In other words, whilst the datum value, ignoring heat release, would indicate reaction taking place at the border between slow and fast reaction, the high heat release and associated interface temperature rise place the reaction far into the fast regime.

The conformal mapping of M^1 onto M then gives rise to a new complication at the higher values of the activation energy. Multiple steady states of the film concentration and temperature profiles are apparent at E_R = 30,000 and E_R = 40,000 cal mol^{-1} and the enhancement factor - $\sqrt{M}$ plot then gives rise to a curious loop formation. This effect has already been noted (13), and is analogous to the steady state multiplicity of the exothermic catalyst pellet (14). It seems that in certain parameter ranges, a given value of $\sqrt{M}$ at the datum temperature can give rise to more than one set of profiles

TABLE 1 : ABSORPTION OF SO_3 IN LAMINAR JET OF DDB

DDB flowrate = 146 cm^3 min^{-1}

jet length = 9.1 cm

contact time = 51.6 milliseconds

mass transfer coefficient = 5.94×10^{-3}

EXPERIMENT					THEORY		
SO_3 in Gas Phase	Gas Temp	DDB Outlet Temp	DDB Feed Temp	T^*-T_b	T^*-T_b	$\sqrt{M^1}$	SO_3 in Gas Phase
0.035	25.1	26.0	25.2	6.81	5.12	1.78	0.03
0.063	25.0	26.3	24.8	11.5	11.76	2.62	0.06
0.093	25.2	27.7	25.1	20.9	19.94	4.23	0.09
0.123	26.5	28.6	25.1	36.3	29.64	7.73	0.12
0.156	27.0	29.9	24.8	40.9	40.88	15.6	0.15
0.191	26.0	34.1	25.1	72.1	53.64	29.7	0.18
0.220	25.0	30.5	24.9	44.6	67.94	59.5	0.21
0.248	25.0	32.3	25.3	56.5	83.76	125	0.24
0.273	25.5	38.2	25.5	101.9	101.12	244	0.27
0.299	26.0	39.9	25.7	114.2	120.00	479	0.30
EXPERIMENT					THEORY		

across the mass and heat transfer films which obey the appropriate heat and mass diffusion and reaction relationships.

The interfacial temperature behaviour corresponding to Fig. 14 is shown in Fig. 15. At a $\sqrt{M}$ value of 0.165 with E_R = 30,000 cal mol^{-1} four steady states appear in the interfacial temperature range up to 300°C - at the values 49°C, 59C, 85°C and 112°C with enhancement factors corresponding to 0.092, 0.110, 0.125 and 0.190. In addition there must be a further steady state, but this would seem to correspond to very high temperatures in excess of 300°C. Such a steady state would stretch this theoretical analysis beyond reasonable credibility, since at such temperatures thermal decomposition of liquid phase DDB would be expected.

At a very high value of E_R = 40,000 cal mol^{-1} only two steady states are predicted in a reasonable interfacial temperature range, although at this value of E_R, the re-entrant sweep of the temperature -$\sqrt{M}$ curve is very pronounced. Altogether, there is a wide scope for complex heat and mass transfer behaviour and it is interesting to speculate upon the stability of the individual multiple steady states and the further possibilitites for their interaction with the overall characteristics of a continuous flow gas-liquid reactor (15).

EXPERIMENTAL RESULTS FOR THE SYSTEM SO_3 - DDB

A number of experiments have been performed absorbing sulphur trioxide into a laminar jet of liquid toluene (16). In these experiments, an estimate of the temperature achieved on the surface of the jet on entry into the jet receiver can be obtained by interpreting the mixing cup temperature of the jet as it becomes fully mixed in the receiver take-off line. The form of the heat transfer profile is as shown in Fig. 16. The depth of heat penetration into the jet is assumed to be given by

$$x_H = \sqrt{\pi \alpha t} \qquad (38)$$

so that the dimensions of heat film thickness x_H and the mass film thickness x_M have been assumed to be given by the penetration theory. The details of a full set of experimental results with gas phase composition varying from 3% to 30% are presented in Table 1. The contact times in the jet have been estimated from enlarged jet photographs, and the values refer to jet lengths of 48 mm and 91 mm with a nominal jet diameter of 14 mm. The interfacial temperature increase on entry into the receiver is calculated from

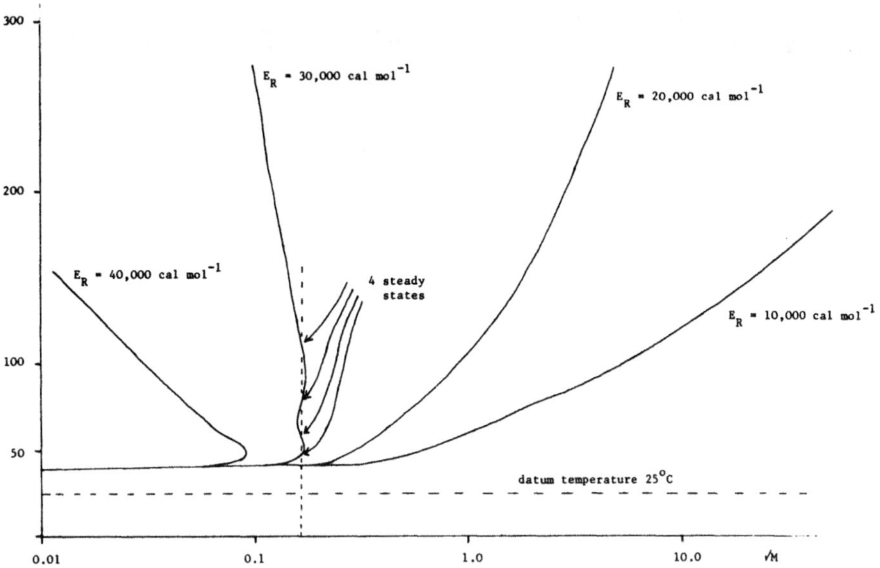

FIG. 15 INTERFACIAL TEMPERATURES FOR ABSORPTION OF 30% SO_3 (IDEALISED SOLUBILITY)

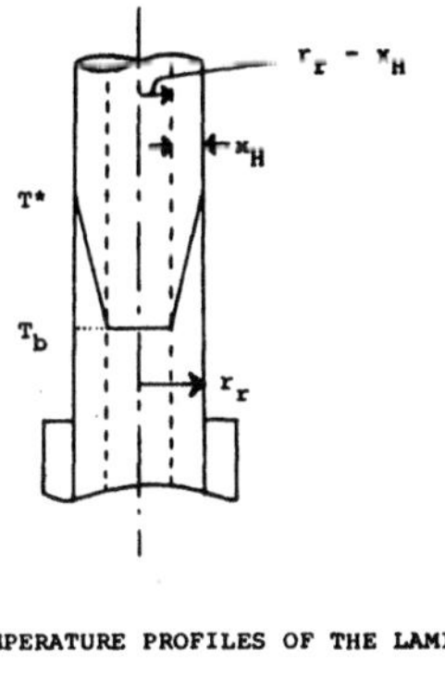

FIG. 16 TEMPERATURE PROFILES OF THE LAMINAR JET

$$T^* = 2\left\{ T' - \frac{(r_r - x_H)^2 T_b}{r_r^2} \right\} \left[\frac{r_r^2}{2r_r x_H - x_H^2} \right] - 2T_b \qquad (39)$$

where T' is the mixing cup temperature.

As Table 1 indicates, at the highest gas phase composition of 30% SO_3 the surface temperature after a contact time of 51.6 milliseconds is 114.2°C above the datum temperature of 25°C. These are remarkably high temperatures and the system SO_3 - DDB is seen to be exceptionally non-isothermal.

A plot of the experimentally inferred surface temperatures versus gas phase composition is given in Fig. 17. These data require to be interpreted through the Eqns (20) and (22), using

$$k_L = \sqrt{\frac{D}{\pi t}} \quad \text{and} \quad h_L = \sqrt{\frac{\alpha}{\pi t}}$$

so that the ratio of k_L/h_L is proportional to the ratio of the square roots of D and α. The diffusivity has been estimated from the Wilke-Chang correlation (17) to be 5.73×10^{-6} and the thermal diffusivity was calculated to be 4.13×10^{-4}. This interpretation has been carried out using the assumption that the absorption is in the fast reaction regime and that the kinetics are pseudo first order. Since M' is defined by

$$M' = k(T^*)\, D/k_L^2 \qquad (40)$$

the basic kinetic parameters A and E_R are given by the relationship

$$M^1 = \frac{D}{k_L^2} A \exp\left\{ - E_R/R\,(T^* + 298) \right\} \qquad (41)$$

The results of this analysis are shown as the final column in Table 1. With the possible exception of the first two values, all the results do appear to lie in the fast reaction regime, and the best fit result is shown in Fig. 17 with $A = 1.24 \times 10^{19}$ s^{-1} and $E_R = 24.7 \times 10^3$ kcal mol^{-1}. The upper and lower limits of the scatter of the data are also shown in Fig. 17.

This estimate of the reaction parameters now permits a prediction of the jet temperature profiles which are shown in

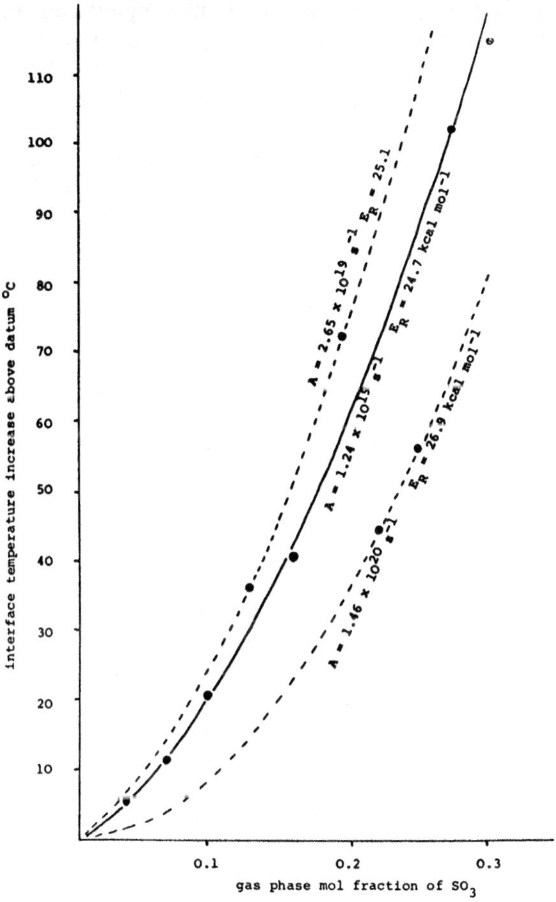

FIG. 17 INTERFACE TEMPERATURE RISE AT THE JET SURFACE

Fig. 18. With reference to Fig. 14 the interference of the reaction kinetic parameters is not complicated by possible multiplicity of the heat and mass transfer profiles, since an activiation energy of around 25 kcal mol^{-1} can only give rise to single solutions. The exothermicity of the absorption process results in a rapid increase in the reaction speed along the jet. This is presented in terms of the half-lives of the reacting sulphur trioxide at the jet surface in Fig. 19. For the nominal 10% SO_3 in the gas phase, the half-life has become less than 10 milliseconds at the end of the jet. The much greater surface temperature achieved in the 30% SO_3 case means that the half-life decreases along the jet surface from around 10 milliseconds close to the jet nozzle to less than a micro-second on entry to the receiver.

These experiments appear to establish that the absorption of SO_3 into DDB is significantly exothermic for gas compositions of 10% or greater, especially at long exposure times. Thus even for 10% SO_3, the boiling point of DDB of 296°C is calculated to be achieved in an exposure time of 1.2 seconds. The corresponding values for 3% and 30% are 12 seconds and 0.1 seconds respectively. These observations are relevant to operating practise for falling film sulphonation columns.

The operation of bubbling sulphonators with surface exposure times of 10 milliseconds for typical bubble size ranges do not appear to be prone to interfacial temperatures in the region of 100°C above datum. However, the industrial sulphonation of linear alkyl benzenes presents an interesting problem of associated severe discolouration and the formation of malodourous compounds when sulphonation is carried out at high gas composition of SO_3. The productivity of sulphonators is limited by these factors and the role of localised high temperatures in the absorption process has yet to be fully appreciated.

In the longer term, improved fundamental understanding of exothermic gas absorption will lead to new and possibly novel concepts of sulphonator reactor design which will be capable of high productivity at high selectivity with reduced by-product formation.

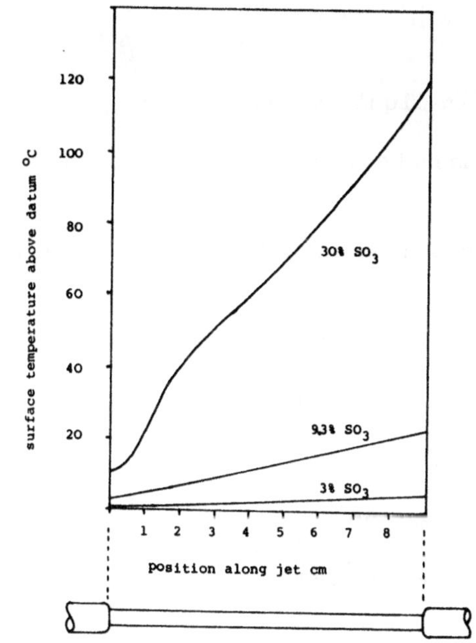

FIG. 18 TEMPERATURE PROFILE ALONG A JET

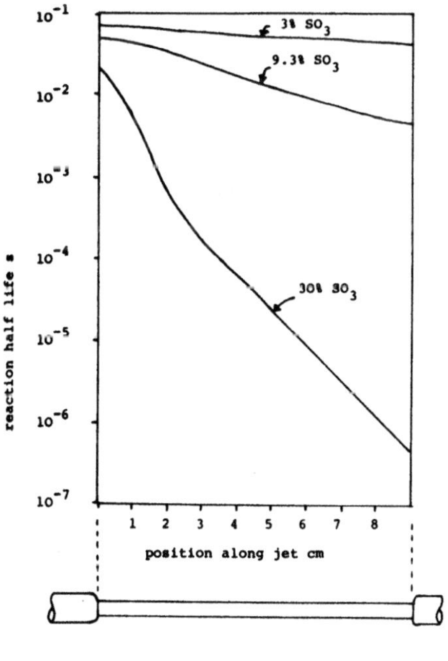

FIG. 19 REACTION HALF-LIFE OF SO_3 AT JET SURFACE

NOMENCLATURE

C	concentration in liquid phase
$C^*(o)$	interfacial concentration at $t = o$ or for physical absorption
$C^*(t)$	interfacial concentration after time t in penetration theory
C^*	interfacial concentration in film theory
c_p	specific heat of liquid phase
D	liquid phase diffusion coefficient
d	diameter of laminar jet
E	absorption enhancement factor
E_R	activation energy
$H(t)$	peripheral heat accumulation on jet surface after exposure time t
ΔH_R	heat of reaction
ΔH_s	heat of solution
h_L	heat transfer coefficient
K	liquid phase thermal conductivity
k_L	mass transfer coefficient
$k(T)$	reaction rate constant at temperature T
M	diffusion/reaction parameter evaluated at T_b
M'	diffusion/reaction parameter evaluated at T^*
n	ratio of $\Delta H_R / \Delta H_s$
n'	constant in hyperbolic solubility relationship
p_A^o	vapour pressure of pure absorbing gas
r_r	radius of laminar jet

$r(C, T)$	reaction rate at C and T
T	temperature
$T^*(o)$	interfacial temperature at t = o or for physical absorption
$T^*(t)$	interfacial temperature after time t in penetration theory
T^*	interfacial temperature in film theory
t	exposure time
$v(t)$	liquid phase velocity due to volume change
x	penetration depth
x_A	mol fraction of absorbing gas in liquid phase
x_M	thickness of mass transfer film
x_H	thickness of heat transfer film
y_A	mol fraction of absorbing gas in gas phase
α	thermal diffusivity
ρ	liquid phase density
ϕ_{II}	interface heat flux
μ_s	linear solubility coefficient
μ'_s	hyperbolic solubility coefficient

REFERENCES

(1) Danckwerts, P.V., Appl.Sci.Res., A3, 385, (1953)

(2) Danckwerts, P.V., Chem.Eng.Sci., 22, 472, (1967)

(3) Chiang, S.H., and Toor, H.L., A.I.Ch.E.Jl., 10, 398, (1964)

(4) Mann, R., M.Sc. Thesis, University of Manchester, (1965)

(5) Mann, R., Ph.D. Thesis, University of Manchester, (1968)

(6) Int.Crit.Tables, Vol.3

(7) Clegg, G.T., and Mann, R., Chem.Eng.Sci., 24, 321, (1969)

(8) Carslaw, H.S., and Jaeger, J.C., "Conduction of Heat in Solids", Oxford University Press, (1947)

(9) Mann, R., and Clegg, G.T., Chem.Eng.Sci., 30, 97, (1975)

(10) Mann, R., and Moyes, H., A.I.Ch.E.Jl., 23, 17, (1977)

(11) Mann, R., "Absorption with Complex Reaction in Gas-Liquid Reactors", NATO A.S.I., Izmir, (1981)

(12) Allan, J.C., and Mann, R., Chem.Eng.Sci., 34, 413, (1979)

(13) Allan, J.C., and Mann, R., Submitted to Can.Jl.Chem.Eng.

(14) Weisz, P.B., and Hicks, J.S., Chem.Eng.Sci., 17, 265, (1962)

(15) Hoffman, L.A., Sharma, S., and Luss, D., A.I.Ch.E.Jl., 21, 318, (1975)

(16) Allan, J.C., M.Sc. Thesis, University of Manchester, (1978)

(17) Wilke, C.R., and Chang, P., A.I.Ch.E.Jl., 1, 264, (1955)

ABSORPTION WITH COMPLEX REACTION IN GAS-LIQUID REACTORS

Reginald Mann
Department of Chemical Engineering
University of Manchester Institute of Science
and Technology
Manchester M60 1QD

INTRODUCTION

That reactions between gases and liquids are important in chemical engineering can be judged from the vast literature on this subject. Almost all of the published work up to 1970 has been thoroughly reviewed, drawn together and elegantly summarised in the very comprehensive "Gas-Liquid Reactions" by Danckwerts (1). Since 1970 a number of associated surveys and monographs have been produced, notably those of Juvekar and Sharma (2), Barona (3) and Charpentier (4). This present survey is an attempt to draw attention to the difficulties associated with the analysis of gas-liquid reactions in the context of the interpretation of gas-liquid reactor performance. It is restricted to consideration of reactors which do not involve volatile intermediates or products and those which do not involve a third solid phase. The subject of desorption with reaction has been thoroughly dealt with by Shah and Sharma (5) and a definitive account of gas-liquid-solid reactors has been undertaken by Shah (6).

A great deal of work in the field of gas absorption concentrates upon the performance of a gas-liquid contactor as an absorber. In considering the performance of a contactor as a reactor, it is necessary to appreciate the ways in which the processes of diffusion accompanied by chemical reaction influence the extents of reaction taking place in the film and bulk of the liquid phase. In seeking a general approach to the problem of quantifying reactor design for an arbitrary number of gas and liquid phase reagents reacting together through an arbitrarily numerous set of reactions, the literature is systematically reviewed for a classification of gas-liquid reactions in order of increasing complexity as indicated in Table 1.

In each case, the diffusion/reaction equations have to be solved in conjunction with boundary conditions at the film/bulk interface. These boundary conditions are determined by the operating mode of the reactor and the overall material balances. In general, the way in which this can be achieved relies upon resolving the diffusion/reaction differential equations to suitable approximating expressions for the fluxes involved. Whilst this can be satisfactorily achieved for the simpler reaction cases, the general case involving only a modest degree of complexity presents considerable difficulties. Thus in many instances the analysis of gas-liquid reactor performance and associated design problems relies upon extensive experience and intuition. A flexible and robust generalisation approach remains a difficult long term goal.

POTENTIAL COMPLEXITIES OF A GAS-LIQUID REACTOR

There are probably more gas-liquid reactors operating worldwide in the chemical industry than reactors of any other type. Since chemical transformations are the very backbone of chemical manufacture, it follows that gas-liquid reactors merit continued attention and research in the effort to improve product quality and raw material utilisation.

Moreover, gas-liquid reactors represent the simplest departure from reactors which involve purely homogeneous reactions, but nevertheless the interactions of the processes of simultaneous mass transfer and reaction in gas-liquid reactors have presented and continue to present a formidable challenge to the applied science of chemical reaction engineering. This is particularly true of the vigorous intense reactions which are involved in the production of many major chemical intermediates. For example, in many liquid phase hydrocarbon oxidation processes, partial oxidation competes with complete combustion to produce a wide range of oxidised hydrocarbons, where the kinetics may be summarised as

BLE 1: ABSORPTION WITH COMPLEX REACTION IN GAS-LIQUID REACTORS

mber of s phase actants	Number of Liquid phase Reactants	Number of Liquid phase Reactions	Nature of Liquid phase Reactions
1	0	1	Simple
1	0	1	Complex
1	0	2	Reversible
1	0	2	Consecutive
1	0	2	Parallel
1	1	1	Simple
1	1	1	Autocatalytic
1	1	2	Reversible
1	2	2	Parallel
1	2	2	Consecutive
1	3	5	Parallel/ Consecutive
2	0	1	Simple
2	1	4	Complex reversible

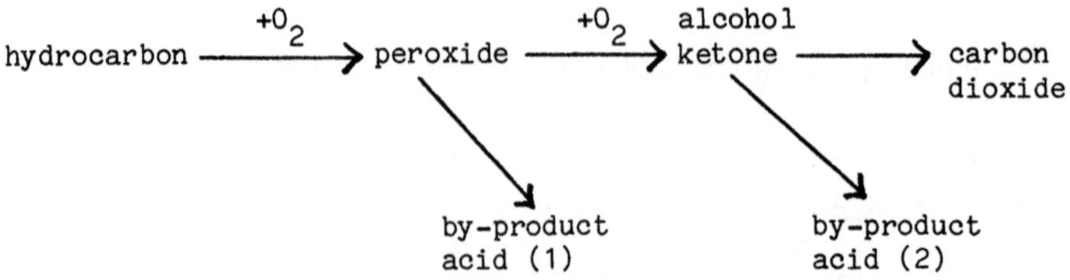

Spielman (7) has noted that in many cases overoxidation results in a host of useless and undesirable tars and condensation products. In many instances just one or two of this spectrum of products is desired and the elimination of by-products is necessary for profitable manufacture.

The same is true of chlorinated hydrocarbons and the chlorination of n-decane can be summarised as (8)

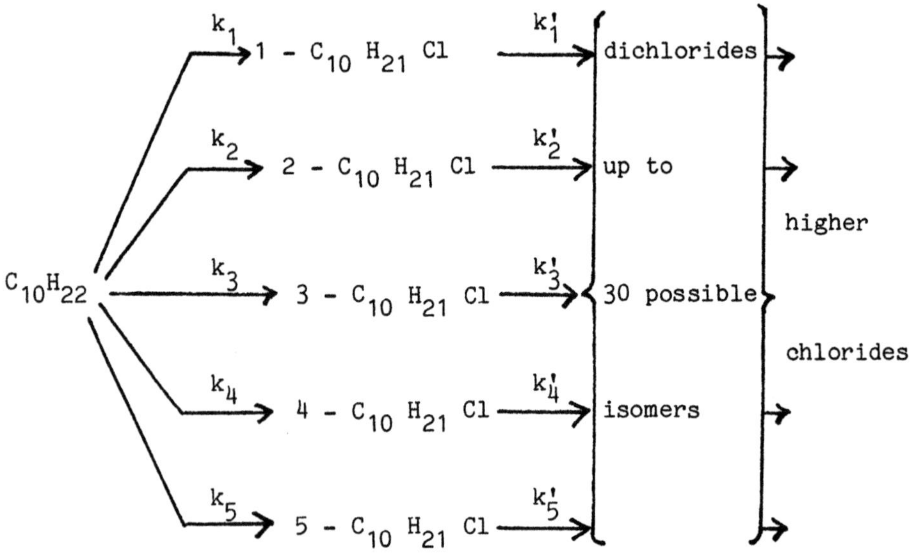

Occasionally particular isomers have specially desirable properties, but their manufacture by gas-liquid reaction gives rise to considerable difficulty when such a massive variety of associated products is possible.

Similar considerations apply to the hydrogenation of unsaturated fatty acids (9), which can be represented by

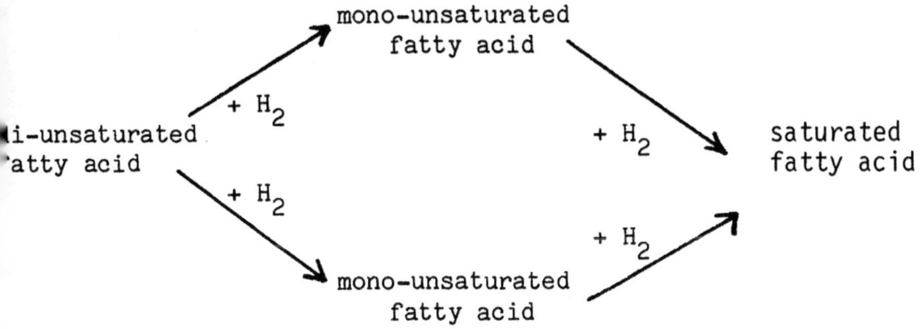

.s a typical example.

These problems of liquid product selectivity typified by ·xidation, chlorination and hydrogenation can be contrasted with .hose involving a requirement for selective absorption from a gas tream containing a number of soluble reactive gases. For ·xample, in the scrubbing of gases produced from the incineration f pvc type plastic waste with low-grade high-sulphur fuel oil, .t is desirable to selectively absorb the acid gases hydrogen hloride and sulphur dioxide whilst at the same time suppressing he absorption of carbon dioxide, which is necessarily present in .arge excess.

All of these gas-liquid reaction problems are characterised y a considerable reaction complexity because of the abundance of ossible reactions accompanying the necessary mass transfer and bsorption processes. This reaction complexity is compounded by he complexities inherent in the mixing and contacting of the ;ases and liquids in some suitable reaction vessel. Fig. 1 ortrays this schematically for a quite arbitrary contacting of ;as and liquid. Complex reactions are then often accompanied by as-liquid flow complexities and the overall difficulty is ntensified by the fact that the classical gas-liquid reactor .ypes

packed column
plate column
bubble column
stirred vessel

mply only a weak control over the gas-liquid flow processes and ontacting pattern, with a good deal of uncertainty over the istribution of surface area and mass transfer potential.

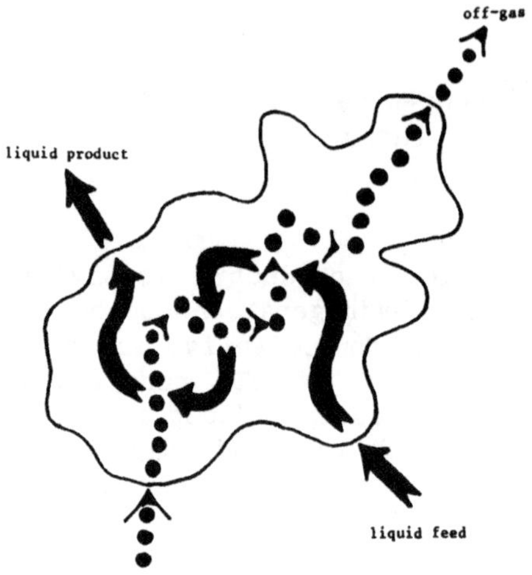

	FEEDS	REACTOR	PRODUCTS	
LIQUID	flow rates and compositions	reaction kinetics - number of reactions - types of reactions reactor properties	?	LIQUID
GAS	flow rates and compositions	- structure of liquid flow - structure of gas flow - contacting pattern - distribution of interfacial area - mass transfer coefficient	?	GAS

Fig. 1. GENERAL PROBLEM OF GAS-LIQUID REACTOR ANALYSIS AND DESIGN

Thus from Fig. 1, the general problem can be stated thus. If ne flow rates and composition of the gas and liquid flow streams re specified, the reaction kinetics are fully quantified and the as-liquid contacting pattern is accurately definable, what will e the flow rates and composition of any product streams? Even eaving aside the fact that reaction kinetic information is carce, and understanding of gas-liquid contacting is weak, the roblem remaining is still significantly severe, since etermining the overall performance of the reactor/contactor equires the resolution of the interactions of absorption, mass ransfer and reaction.

A general approach capable of handling reactions which are bitrarily numerous and complex in character is a long term oal. What follows is merely a preliminary evaluation of the ifficulties involved in doing this.

NALYSIS OF A BACKMIXED REACTOR FOR A SINGLE SIMPLE REACTION

The strong interactions and coupling between the mass ansfer and chemical reaction processes are the primary omponent in complicating the analysis of even a very simplified eactor. This can be illustrated with respect to some ualitative features of a simple single reaction in which a eagent 'A' is decomposed irreversibly in the liquid phase to a roduct P, such that $A \longrightarrow P$.

The simplest possible reactor configuration is that in which oth the liquid and gas phases are absolutely perfectly backmixed ith uniform internal properties as indicated in Fig. 2. The erfect backmixing assumption incorporates a uniform specific urface area per unit volume of liquid $\bar{a}$ and a uniform mass ransfer coefficient k_{LA} for the purposes of simplicity, although e assumption of perfect backmixing does not restrictively equire that k_{LA} and $\bar{a}$ should be uniform, or that the associated as bubbles or liquid drops which create $\bar{a}$ should all be of the ame size.

Qualitatively, four distinct regimes can be recognised oncerning the interaction between the mass transfer and reaction f A as shown in Fig. 3, in which it is assumed that the film eory is an adequate basis for considering diffusion and eaction phenomena in a gas-liquid reactor.

In regime I the reaction is so negligibly slow that the ntire film and bulk liquid phase are saturated at the oncentration C_A*. As the reaction speed increases to regime II, e rate of consumption of A in the bulk liquid phase is balanced

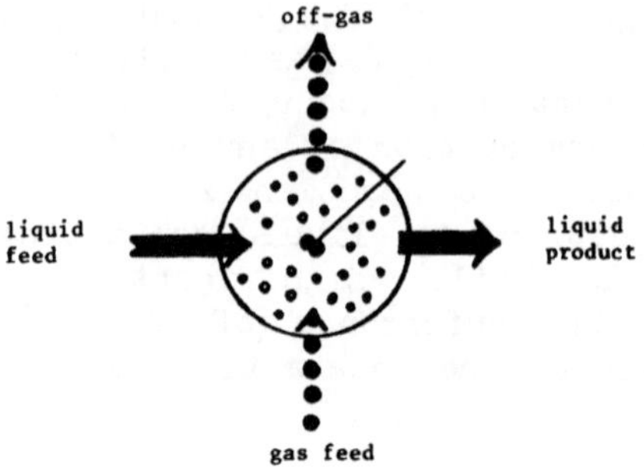

Fig. 2 A BACKMIXED GAS-LIQUID REACTOR ELEMENT

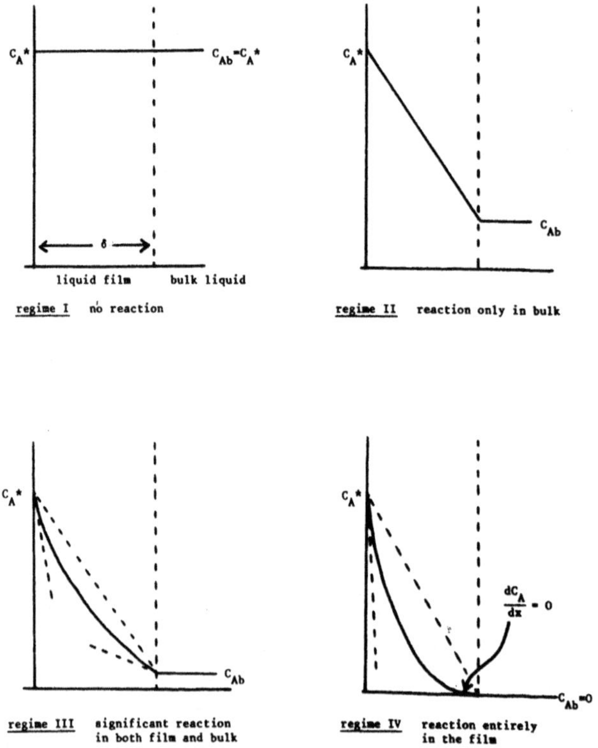

Fig. 3 REACTION REGIMES IN ABSORPTION OF A SINGLE GAS

y the rate of transfer of A through the liquid film. Negligible ate of reaction in the film results in a linear concentration radient. A further increase in reaction speed results ventually in comparable amounts of reaction taking place in the ilm and bulk and this is characterised by a curved concentration rofile as shown in regime III. Ultimately at very high reaction peed, virtually all the A is consumed within the film and this is haracterised by a zero gradient at the film/bulk junction as hown in regime IV in Fig. 3.

Danckwerts (1) has proposed characterising these regimes by eans of inequalities given by (for 1st order reaction of A)

egime I $$k_1 \tau >> 1$$

egime II $$D_A k_1/k_L^2 << 1$$

egime III $$\left[\frac{k_L}{D_A \bar{a}} \left[1 + \frac{1}{\tau k_1} \right] \right]^{\frac{1}{2}} >> 1$$

egime IV $$D_A k_1/k_L^2 >> 1$$

Regime IV is the so called fast reaction regime, and the riterion is derived on the basis that the volume of the bulk iquid phase is significantly greater than the film volume. Two ifficulties arise in relation to the use of criteria of this ype. Firstly, how can one handle the case of non-first-order eactions? Secondly, in respect of the problems arising when ore than one reaction is present, it becomes necessary in ualitatively establishing reactor performance to move beyond hese simple qualitative discriminants in such a way that the roportion of film and bulk reaction are directly determined. t is possible for example when two reactions are present for ne to take place largely in the film and the other redominantly in the bulk. In such circumstances direct alculation of performance needs to be initiated without prior ssumptions on the role of filmwise and bulkwise reaction.

In recognition of some aspects of these difficulties, Kulkarni and Doraiswamy (10) have proposed the use of an effectiveness factor concept to characterise reactor behaviour (by analogy with similar definitions used in diffusion/reaction analysis in catalytic phenomena). They proposed that an effectiveness factor be defined by

$$\eta = \frac{\text{rate of reaction evaluated at interfacial concentration}}{\text{overall rate of reaction in film and bulk.}}$$

For first order reaction, with

$$D_A \frac{d^2 C_A}{dx^2} = k_1 \ C_A \tag{1}$$

or with
$$M(1) = \frac{\delta^2 k_1}{D_A} \tag{2}$$

their results show that for only bulk reaction

$$\eta = \frac{1}{1 + M(1)\beta} \tag{3}$$

and for significant reaction taking place in the film

$$\eta = \frac{1}{1+\beta}\left[\frac{1}{\{\cosh\sqrt{M(1)}\}^2 \ (1 + \sqrt{M(1)}\,\beta \tanh \ \sqrt{M(1)}} + \frac{\tanh\sqrt{M(1)}}{\sqrt{M(1)}}\right] \tag{4}$$

The effectiveness factor depends upon $\sqrt{M(1)}$ (often called the Hatta number) and the parameter β given by

$$\beta = \text{volume per unit area of bulk/volume per unit area of film} \tag{5}$$

and this is an essential distinction between the gas-liquid reactor and the catalyst particle, since in the latter case there is no equivalent of reaction in the bulk.

However, the use of the term effectiveness factor is somewhat misleading since η becomes very small when the reaction becomes fast. In this circumstance the absorption effectiveness is increased, and so an increasing absorption effectiveness is described by a decreasing effectiveness factor. The definition of effectiveness factor is also nadequate since it does not directly distinguish the ontributions of film and bulk reaction to the overall erformance. This can be more satisfactorily achieved in the ollowing way.

The perfectly backmixed uniform property gas-liquid zone oncept illustrated in Fig. 2 can be envisaged alternatively as nown in Fig. 4.

The interaction of the processes of dissolution, mass ransfer with reaction and the overall input and output flows gas and liquid can be set out as follows. If G_i is the molar eed rate of gas containing a dissolving and reacting component ' at a mol fraction y_{Ai}, then the rate of absorption at the nterface $N_A|_{x=o}$ is given by

$$|_{x=o} = G_i y_{Ai} - G_o y_{Ao} \tag{6}$$

oreover, if there is no net change in the throughput of ertgases then

$$(1 - y_{Ai}) = G_o (1 - y_{Ao}) \tag{7}$$

or transport through the gas film

$$|_{x=o} = V k_{gA} \bar{a} (P_{Ag} - P_A^*) \tag{8}$$

d since the gas phase is fully backmixed, the outlet gas mposition equals the bulk composition and therefore

$$P_{Ag} = P y_{Ao} \tag{9}$$

so if the solubility of the reacting component A obeys nry's law

$$P_A^* = H_A C_A^* \tag{10}$$

ffusion and simultaneous reaction of A through the liquid lm is described by the differential equation

$$D_A \frac{d^2 C_A}{dx^2} = k_1 C_A \tag{11}$$

the reaction is simple first order. The solution of this uation is subject to the boundary conditions at each end of e liquid film

$$= 0, \quad C_A = C_A^* \quad ; \quad x = \delta \quad , \quad C_A = C_{Ab} \tag{12}$$

The flux through the gas film must equal the flux of A into the liquid phase at the interface, so that

$$N_A|_{x=o} = V_L \bar{a} D_A \left[\frac{-dC_A}{dx}\Big|_{x=o} \right] \quad (13)$$

The corresponding material balance on A over the bulk liquid phase is then given by

$$N_A|_{x=\delta} = V_L \bar{a} D_A \left[\frac{-dC_A}{dx}\Big|_{x=\delta} \right] \quad (14)$$

$$= k_1 V_{Lb} C_{Ab} + Q C_{Ab}$$

$$= k_1 V_{Lb} C_{Ab} \left(1 + \frac{1}{k_1 \tau}\right) \quad (15)$$

The rate of absorption is normally expressed in terms of an Enhancement Factor E_A, defined by

$$E_A = \frac{\text{rate of absorption with reaction}}{\text{rate of physical absorption}} = \frac{-D_A \frac{dC_A}{dx}\Big|_{x=o}}{k_{LA} (C_A^* - C_{Ab})}$$

which can be analytically expressed using the solution of Eqn (11) by

$$E_A = \frac{\left\{ C_A^* - \frac{C_{Ab}}{\cosh\sqrt{M(1)}} \right\} \frac{\sqrt{M(1)}}{\tanh\sqrt{M(1)}}}{C_A^* - C_{Ab}} \quad (16)$$

Hence

$$N_A|_{x=o} = V_L \bar{a} k_{LA} (C_A^* - C_{Ab}) E_A \quad (17)$$

Equating the rate of reaction in the bulk liquid to the rate of diffusion of A into the bulk less the rate at which A is carried away by the effluent stream, the bulk concentration of A is given by

$$C_{Ab} = \frac{C_A^*}{\left\{1 + \frac{k_1(1 + \frac{1}{k_1 \tau})}{k_{LA}\,\bar{a}\,\frac{\sqrt{M(1)}}{\tanh\sqrt{M(1)}}}\right\}\cosh\sqrt{M(1)}} \qquad (18)$$

combining this result with the absorption flux predicted by n (17)

$$N_A|_{x=o} = V_L \; \bar{a} \; k_{LA}^* \quad C_A^* \qquad (19)$$

ere

$$_A^* = k_{LA}\left\{1 - \frac{1}{\left[1 + \frac{k_1(1 + \frac{1}{k_1 \tau})}{k_{LA}\,\bar{a}\,\frac{\sqrt{M(1)}}{\tanh\sqrt{M(1)}}}\right](\cosh M(1))^2}\right\}\frac{\sqrt{M(1)}}{\tanh\sqrt{M(1)}} \qquad (20)$$

om the overall material balances expressed in Eqns (6) and), making use of the additivity of resistances

$$\frac{1}{K_{gA}} = \frac{1}{k_{gA}} + \frac{H_A}{k_{LA}^*} \qquad (21)$$

have

$$|_{x=o} = G_i \frac{y_{Ai} - y_{Ao}}{1 - y_{Ao}} = V_L \; \bar{a} \; K_{gA} \; P y_{Ao} \qquad (22)$$

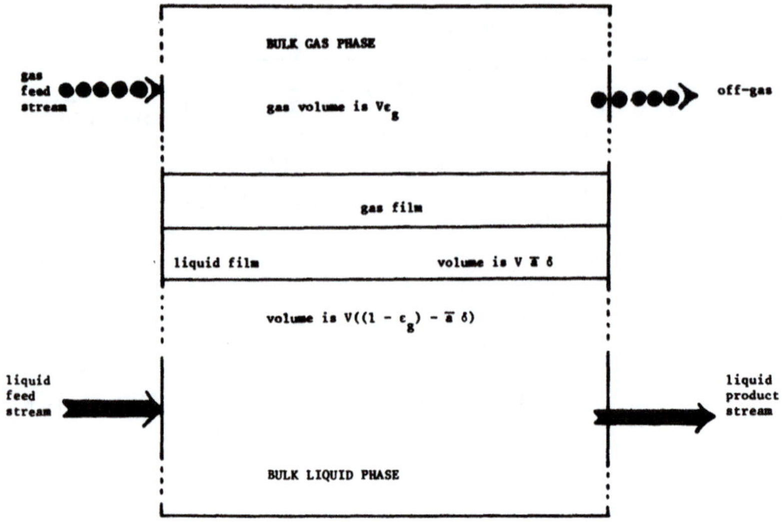

FIG. 4 SCHEMATIC REPRESENTATION FOR BACKMIXED GAS-LIQUID ELEMENT OF VOLUME V

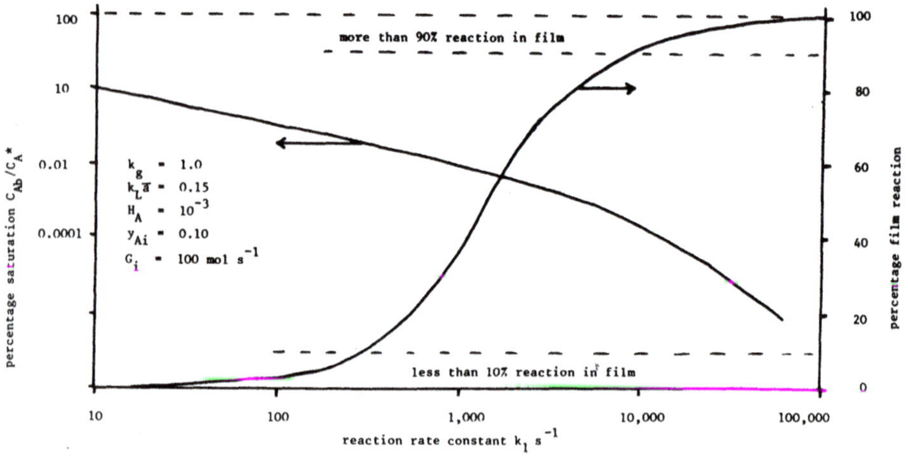

FIG. 5 Film and Bulk Reaction in Backmixed Gas-Liquid Element

he resulting quadratic in y_{Ao}

$$y_{Ao}^2 - [1 + \frac{G_i}{V_L \bar{a} K_{gA} P}] y_{Ao} + \frac{G_i y_{Ai}}{V_L \bar{a} K_{gA} P} = 0 \quad (23)$$

s then easily solved to obtain the output mol fraction A in the xit gas stream.

Therefore, given the input flows and compositions as ndicated by G_i and y_{Ai}, with the appropriate reactor arameters V_L and $\bar{a}$, the mass transfer coefficients k_{LA} and k_{gA} nd finally the reaction speed denoted by the rate constant k_1, he reactor output composition is directly determined.

The interface concentration C_A^* and the bulk concentration Ab then follow directly, as do the proportions of film and bulk eaction determined from $N_A|_{x=o}$ and $N_A|_{x=\delta}$. In this way, the elative contributions from film and bulk reaction are found ithout making any a priori assumptions.

Fig. 5 shows some calculations (11) performed for the case here Q = 0 ($\tau = \infty$), and the liquid phase consists of the eaction product. This liquid phase is assumed to overflow rom the backmixed zone at a rate corresponding to its ormation from A by liquid phase chemical reaction. In Fig. 5 negligible amount of A has been assumed to be carried away ith this product overflow.

Fig. 5 indicates clearly that the range of 10% bulk eaction to 90% bulk reaction involves k_1 increasing from round 500 s^{-1} to 10,000 s^{-1} and a significant fraction of all as-liquid reactions can be expected to occur in this range.

The corresponding $\sqrt{M(1)}$ values for the example in Fig. 5 re 0.5 and 3.0. The values used in Fig. 5 are typical of ubbling gas liquid systems for which the volume of the bulk is etween three and four orders of magnitude greater than the ilm volume. The 50% bulk/50% film reaction therefore occurs a fractional saturation of less than 10^{-2}% for the articular parameter values of Fig. 5.

Basing the analysis upon the idealised gas-liquid ackmixed element of Fig. 2 is not meant to be at all estrictive in the development of a truly generalised approach the analysis and design of gas-liquid reactors. Such idealised backmixed elements form the basic building 'block' from which any type of gas-liquid reactor can be constructed. This is depicted in Fig. 6.

A vertical assembly of elements, in conjunction with a high rate of liquid circulation, can represent plug flow with dispersion of the upward flowing gas and the high level of liquid backmixing typical of a bubble column. Plate columns and stirred vessels can be likewise represented by suitable assemblies as shown in Fig. 6. If a generalised sub-routine for the gas-liquid element were available, assemblies like those in Fig. 6 can be readily solved with a standard flowsheeting system.

ABSORPTION OF A SINGLE GAS WITH MORE COMPLEX KINETICS

For a first order reaction, the direct resolution of the proportions of film and bulk reaction can be ascertained for a single backmixed gas-liquid reactor element without trial and error or a priori assumptions. This is due to the fact that analytical expressions for the concentration profile and its gradients are obtained from the diffusion reaction equation. Such analytical expressions do not exist if A is consumed by other than a first order reaction, but it is possible to calculate the performance a priori even so.

For the general mth order reaction

$$D_A \frac{d^2C_A}{dx^2} = k_1 C_A^{\,m} \tag{24}$$

the boundary conditions are exactly as previously, so that

$$C_A = C_A^* \text{ at } x = o \quad \text{and} \quad C_A = C_{Ab} \text{ at } x = \delta.$$

If each side of Eqn. 24 are multiplied by $2D_A \frac{dC_A}{dx}$

and the equation integrated between 0 and δ, the result is

$$D_A \int_0^{\delta} 2 \frac{dC_A}{dx} \frac{d^2C_A}{dx^2} dx = 2 \int_{C_A^*}^{C_{Ab}} k_1 C_A^{\,m} dC_A \tag{25}$$

from which

$$D_A \left[\left[\frac{dC_A}{dx} \right]^2 \right]_o^{\delta} = 2 \left[\frac{k_1}{m+1} C_A^{m+1} \right]_{C_A^*}^{C_{Ab}}$$

hence

$$- \left. \frac{dC_A}{dx} \right|_{x=o} + \left. \frac{dC_A}{dx} \right|_{x=\delta} = \left[\frac{2}{D_A} \frac{k_1}{m+1} \{ C_A^{*\,m+1} - C_{Ab}^{m+1} \} \right]^{\frac{1}{2}}$$

he enhancement factor E_A is defined as before by

$$E_A = \frac{-D_A \left. \frac{dC_A}{dx} \right|_{x=o}}{k_{LA} (C_A^* - C_{Ab})}$$

o that

$$E_A = \frac{\left[\frac{2}{D_A} \frac{k_1}{m+1} \{ C_A^{*\,m+1} - C_{Ab}^{m+1} \} \right]^{\frac{1}{2}}}{k_{LA} (C_A^* - C_{Ab})} \tag{26}$$

or very fast mth order reaction, $C_{Ab} \rightarrow 0$ and the asymptotic alue for E_A is then given by

$$E_A = \frac{1}{k_{LA}} \left[\frac{2}{m+1} D_A k_1 C_A^{*\,m-1} \right]^{\frac{1}{2}} \tag{27}$$

f the right hand side of the above equation is designated by (m), then

$$E_A = \sqrt{M(m)} \tag{28}$$

hich is the Hatta number for a mth order reaction. Thus if he enhancement factor is represented by an expression nalogous to that for the 1st order reaction case, then

$$E_A = \frac{\left(C_A^* - \frac{C_{Ab}}{\cosh\sqrt{M(m)}}\right)}{C_A - C_{Ab}} \frac{\sqrt{M(m)}}{\tanh\sqrt{M(m)}} \tag{29}$$

his expression for E_A is exactly asymptotically correct for

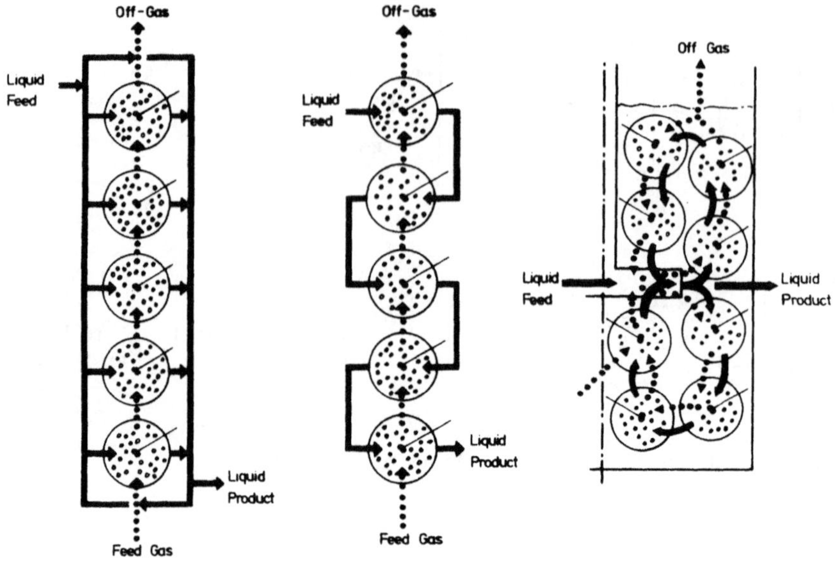

FIG. 6 ASSEMBLING BACKMIXED GAS-LIQUID ELEMENTS INTO CLASSICAL REACTOR TYPES.

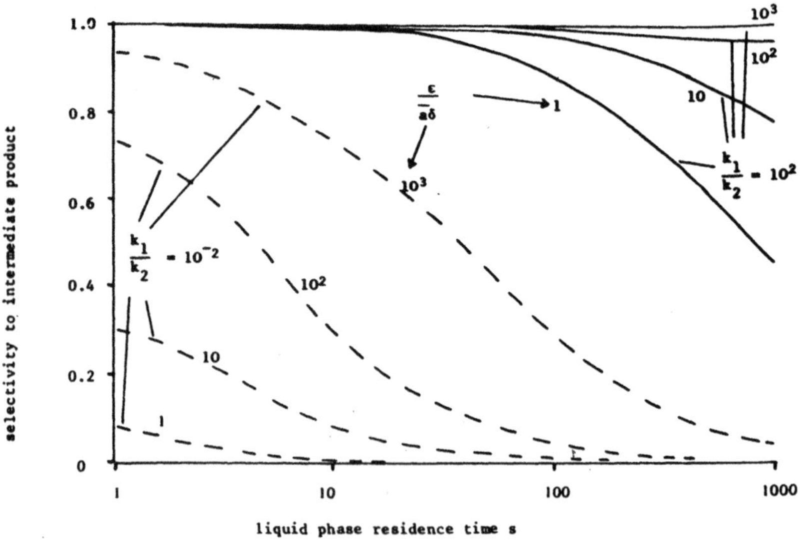

FIG. 7 SELECTIVITY OF INTERMEDIATE R FOR CONSECUTIVE GAS-LIQUID REACTION A→R→S [Ref(16)]

all values of M(m) when E_A approaches unity at the slow action condition and also under fast reaction conditions when n) is large and $E_A \rightarrow \sqrt{M(m)}$. As Brian has shown (12), this proximation for E_A will involve only small errors in the tervening range of reaction speeds. Also, by analogy with e first order reaction case, slow reaction involving less en 10% reaction in the mass transfer film should occur for $\overline{M(m)}<0.5$, with the converse case of fast reaction involving ss than 10% reaction in the bulk of the liquid phase should cur for $\sqrt{M(m)} > 3.0$. Otherwise the treatment for mth order action exactly follows the treatment for first order action. The exact proportions of film and bulk reaction are antitatively determined without resorting to 'A' priori sumptions. The only additional difficulty, in comparison th the 1st order reaction, is that a simple one variable eration is required upon C_A*.

It should be noted however, that care needs to be taken th zero order reaction since the reagent 'A' can be fully nsumed in the film, in which case there is zero bulk reaction d $C_{Ab} = 0$. This is in contrast to the mth order reaction when is never reduced to zero though it may be extremely small. kewise care should be exercised in using the approximation E_A if m < o in which case errors are likely in the range $5 < \sqrt{M(m)} < 3.0$.

The mth order reaction case is straightforwardly extended quite arbitrarily complex reaction rate expressions such as

$$(-r_A) = \frac{k_1 C_A^m}{(1 + KC_A)^n} \tag{30}$$

such cases, the diffusion reaction equation is

$$D_A \frac{d^2 C_A}{dx^2} = (-r_A) \tag{31}$$

which a general reaction independent modulus (Hatta number) signated by M(g) is given by

$$\sqrt{M(g)} = \frac{\left[2 D_A \int_o^{C_A^*} (-r_A)\, dC_A\right]^{\frac{1}{2}}}{k_{LA} C_A^*} \tag{32}$$

The proportions of film and bulk reaction in the overall backmixed gas-liquid reactor element can then be directly obtained, again without resorting to any a priori assumption.

Only the feed rates and composition, the reactor parameters and the reaction kinetics need to be specified, after which the outputs of the reactor element are directly determined.

ABSORPTION OF A SINGLE GAS ACCOMPANIED BY TWO REACTIONS

The simplest example of two simultaneous reactions is the reversible first order reaction

$$A \underset{k_2}{\overset{k_1}{\rightleftharpoons}} P$$

The diffusion reaction equations are

$$D_A \frac{d^2C_A}{dx^2} - k_1 \left(C_A - \frac{C_P}{K}\right) = 0 \qquad (33)$$

$$D_p \frac{d^2C_P}{dx^2} + k_1 \left(C_A - \frac{C_P}{K}\right) = 0 \qquad (34)$$

subject to the boundary conditions

$$C_A = C_A^* \quad , \quad \frac{dC_p}{dx} = 0 \text{ at } x = 0 \text{ ; } C_A = C_{Ab} \text{ and } C_p = C_{pb} \text{ at } x = \delta$$

By adding Eqns (33) and (34)

$$D_A \frac{d^2C_A}{dx^2} + D_P \frac{d^2C_P}{dx^2} = 0 \qquad (35)$$

By integrating once and applying the boundary conditions

$$D_A \frac{dC_A}{dx} + D_p \frac{dC_p}{dx} = c' \qquad (36)$$

with a further integration giving

$$D_A \; C_A \; + \; D_p \; C_p \; = \; c'x + c'' \tag{37}$$

pplying the boundary conditions, and the definition of the nhancement factor

$$E_A = \frac{-D_A \left.\frac{dC_A}{dx}\right|_{x=o}}{k_{LA} \; (C_A{}^* - C_{Ab})}$$

he result is

$$E_A = 1 + \frac{D_P}{D_A} \left\{ \frac{C_P{}^* - C_{Pb}}{C_A{}^* - C_{Ab}} \right\} \tag{38}$$

y further manipulation (13), the analytical result with equal iffusivities is given by

$$E_A = \frac{1 + K}{1 + \{K \frac{\tanh\sqrt{M_e(1)}}{\sqrt{M_e(1)}}\}} \tag{39}$$

here

$$M_e(1) = \frac{D_A k_1}{k_{LA}{}^2} \left\{\frac{1 + K}{K}\right\} \tag{40}$$

he solution of the perfectly backmixed element can then roceed as with the first order irreversible reaction, by ssuming a value for C_{Pb}, the bulk concentration of the product . The bulk reaction condition for A given by Eqn (15) is asily modified to account for the reverse reaction in the ulk. A suitable trial and error approach will converge to the orrect value for C_{Pb} and the overall performance of the gas-liquid reactor element will be known. However detailed esults for this case, equivalent to Fig. 5 have not so far been resented in the literature. The approach is easily eneralised to the arbitrary stoichiometry $A \rightleftharpoons \nu_p P$ and the eversible mth order reaction.

The case of two independent parallel decompositions of the as 'A' in a liquid phase can be represented by

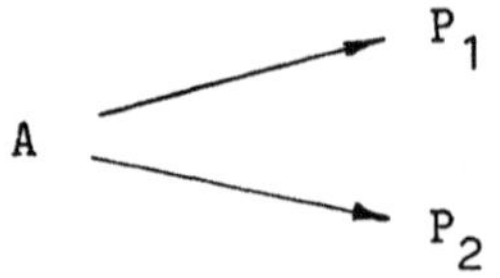

for which the diffusion equation is

$$D_A \frac{d^2 C_A}{dx^2} = k_1 C_A^{\ p} + k_2 C_A^{\ q} \tag{41}$$

In considering the consumption of the dissolving and reacting gas 'A', the treatment follows that indicated for absorption with arbirtrarily complex reaction, so that

$$(- r_A) = k_1 C_A^{\ p} + k_2 C_A^{\ q} \tag{42}$$

and the generalised reaction modulus M(g) is defined by Eqn. (32). The treatment for determining the exit composition y_{Ao} from a perfectly backmixed gas-liquid element then proceeds as for the first order reaction with M(g) replacing M(1), provided that the rate of reaction in the bulk is represented by the above kinetics.

The yield performance of the backmixed element is then determined by concentrating on one of the parallel reaction paths and calculating the filmwise and bulkwise rates of reaction of that component. The other component can then be found by a simple overall material balance. This can be illustrated by considering the component P_1. The rate of reaction of 'A' to 'P_1' in the bulk will be given by $k_1 C_{Ab}^{\ p} V_{Lb}$ and the rate of production of P_1 in the film is obtained from $V \bar{a} \int_0^{\delta} C_A(x)\,dx$, where $C_A(x)$ is given to a good approximation by

$$C_A(x) = \frac{1}{\sinh\sqrt{M(g)}} \left\{ C_A^{*} \sinh\left(1 - \frac{x}{\delta}\right)\sqrt{M(g)} + C_{Ab} \sinh\left(\frac{x}{\delta}\right)\sqrt{M(g)} \right\} \tag{43}$$

Exactly as with the first order reaction, the backmixed gas-liquid element performance is evaluated without prior assumptions about the relative contribution of film and bulk reaction.

One interesting point that should be noted concerns the se of negligible reaction in the film, when the relative elds of P_1 and P_2 will be influenced by the magnitude of the ss transfer coefficient k_{LA} and/or the magnitude of the terfacial area $\bar{a}$, since these both govern the value of C_{Ab}. other words, even when the reaction is influenced only by lk reaction, the chemical yield performance will be sensitive the mass transfer parameters.

However, an efficient reactor configuration and design for rallel reactions of different order is complicated by the nflicting requirements of productivity and yield. Depending the magnitude of p and q, high and low rates of absorption y produce either high or low yields. The general aracteristics of reactor performance remain to be established r this case. Once again, care should be exercised with zero der reactions in case 'A' is fully depleted in the film.

The remaining possibility for two reactions accompanying e absorption and reaction of a single gas is the case where e first product of the decomposition of 'A' can itself rther decompose. This consecutive reaction sequence can be signated by

$$A \xrightarrow{k_1} R \xrightarrow{k_2} S$$

e diffusion reaction equations for this case are

$$D_A \frac{d^2C_A}{dx^2} = k_1C_A \tag{44}$$

$$D_R \frac{d^2C_R}{dx^2} = k_2C_R - k_1C_A \tag{45}$$

oject to the boundary conditions

$$\begin{aligned} C_A &= C_A^* \quad \frac{dC_R}{dx} = 0 \text{ at } x = 0 \\ C_A &= C_{Ab} \quad C_R = C_{Rb} \text{ at } x = \delta \end{aligned} \tag{46}$$

It is immediately apparent from these equations that the behaviour of component A in its decomposition to R is absolutely independent of the further decomposition of $R \longrightarrow S$. In calculating the performance of the backmixed gas-liquid element therefore the procedure to determine the reaction of A is exactly identical to the treatment for a single reaction. Thus the proportions of film and bulk reaction and the overall absorption of A are not influenced by the second reaction.

The solution of the diffusion/reaction equation for the intermediate R can be directly solved to provide an expression for $C_R(x)$, with the result that

$$C_R(x) = \left(C_R^* + \frac{\nu_s}{\nu_{s-1}} C_A^* \right) \frac{\cosh \sqrt{M_2(1)}(1-x)}{\cosh \sqrt{M_2(1)}} +$$

$$\sqrt{M_2(1)} \left[(C_{Rb} - \nu_s C_{Ab}) - \frac{\nu_s}{\nu_{s-1}} \frac{M_1(1) \beta C_{Ab}}{\sqrt{M_2(1)}} \right] \frac{\sinh \sqrt{M_2(1)}\, x}{\cosh \sqrt{M_2(1)}}$$

$$- \frac{\nu_s}{\nu_{s-1}} C_A(x) \qquad (47)$$

By adding the two diffusion reaction relationships Eqns (44) and (45)

$$D_A \frac{d^2 C_A}{dx^2} + D_R \frac{d^2 C_R}{dx^2} = k_2 C_R \qquad (48)$$

and integrating from 0 to δ across the film, the result is

$$D_A \left. \frac{dC_A}{dx} \right|_{x=\delta} + E_A k_1 (C_A^* - C_{Ab}) - D_R \left. \frac{dC_R}{dx} \right|_{x=\delta} = k_2 C_R(x) dx$$

(49)

his equation contains the unknowns C_{Rb} and C_R*. A further elationship between these two unknowns is obtained by onsidering a material balance on R across the bulk liquid hase. Thus,

$$V \bar{a} \; D_R \left. \frac{dC_R}{dx} \right|_{x=\delta} = k_1 \, C_{Ab} \, V_{Lb} - k_2 \, C_{Rb} \, V_{Lb} - Q \, C_{Rb} \tag{50}$$

$$\begin{bmatrix}\text{ate at which R} \\ \text{nters the bulk}\end{bmatrix} = \begin{bmatrix}\text{net rate at which R} \\ \text{is produced in bulk}\end{bmatrix} - \begin{bmatrix}\text{rate at which R} \\ \text{leaves in liquid stream}\end{bmatrix}$$

he overall backmixed reactor characteristics of this gas-iquid consecutive reaction have not been presented in the iterature so far, though a number of qualitative aspects worth oting are:

i) If the component R is subjected to strong diffusional influence the yield of R is always lowered, since the concentration of R will be higher in the liquid film than in the bulk. A faster decomposition to product S will thus occur in the film than would occur if its reaction were confined to the bulk. Hence there is a corresponding yield taxation.

ii) It seems possible that in many cases of practical interest where R is a desired intermediate, it will be the case that $k_2 << k_1$. It would then be likely that even if a large proportion of A were consumed by fast reaction in the film, in contrast the reaction of R to product S could be significant only in the bulk.

iii) In the circumstances where bubbles of different size have different values of k_{LA}, for fast film reaction bubbles of different sizes could have different yields. This has been noted by Steeman (14), and has an analogy with the catalysis of consecutive reactions where small and large pores have differing chemical yields (15).

The detailed behaviour of the backmixed element still emains unreported, and the solutions of Nagel et al (16) isregard the material balance on A, and use the boundary ondition on C_A* as a constant. Their results do show however, hat the yield of R can be reduced even when $k_1 > k_2$. Also the yield of R is finite even though the conversion of A is completed in the liquid phase of the reactor. This contrasts with the homogeneous reactor, when the yield is zero at 100% conversion of A. An example of results calculated at a fixed C_A^* (16) are shown in Fig. 7.

REACTION BETWEEN AN ABSORBING GAS AND A LIQUID PHASE REAGENT

For the case where $A + \nu_B B \rightarrow$ products, considerations of the interactions of mass transfer and chemical reaction require additional material balances for the reactive component 'B' contained in the liquid phase. The diffusion reaction equations for mth-nth order reaction are

$$D_A \frac{d^2 C_A}{dx^2} - k_1 C_A^m C_B^n = 0 \tag{51}$$

$$D_B \frac{d^2 C_B}{dx^2} - \nu_B k_1 C_A^m C_B^n = 0 \tag{52}$$

As Levenspiel has pointed out, there are now eight distinct regimes of behaviour summarised in Fig. 8. Those cases which involve a sufficiently large excess of 'B' present in the bulk and at the film/bulk junction reduce to pseudo mth-order behaviour, and there will exist a negligible concentration gradient of B in the film. These cases are exactly equivalent to the single reaction case detailed previously. The four new additional behaviour regimes each involve the existence of concentration gradients of the liquid phase reactant B.

In providing a general approach to the solution of a backmixed zone, without 'a priori' assumptions on the relative contributions of film and bulk reaction or prior knowledge of the values of the two bulk concentrations C_{Ab} and C_{Bb}, it is again useful if the direct solution of the two second order non-linear ordinary differential equations which are embedded in the algebraic equations governing the overall material balances, can be avoided.

In order to do this, it is necessary to provide a suitable approximation to the enhancement factor E_A and from it to deduce the proportion of component A reacting in the mass transfer film.

The low reaction speed approximation is of course $E_A = 1$, and

	reaction speed	mass transfer character	concentration profiles	
I'	negligible	no mass transfer resistance	no gradient of A through film no gradient of B through film	negligible reaction in film
II'	slow	diffusional resistance to A	straight profile of A through film no gradient of B through film	negligible reaction in film
III'	intermediate	diffusion of A with reaction	curved profile of A through film no gradient of B through film	reaction in both film and bulk
IV'	intermediate	diffusion of A and B with reaction	curved profile of A through film curved profile of B through film	reaction in both film and bulk
V'	fast	reaction restricted to a zone adjacent to interface	curved profile of A in film no gradient of B through film	reaction confined to film
VI'	fast	reaction restricted to a zone within the film	curved profile of A in film curved profile of B in film	reaction confined to film
VII'	instantaneous	reaction at a plane within the film	linear profile of A to reaction plane linear profile of B to reaction plane	reaction confined to film
VIII'	instantaneous	surface reaction	linear profile of B to interface	reaction confined to film

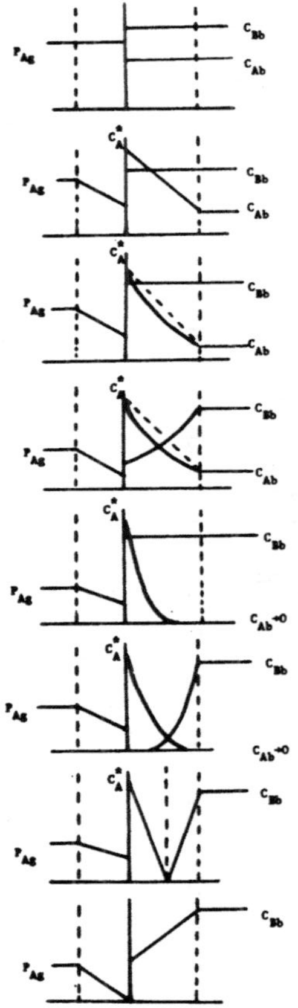

FIG. 8 REACTION REGIMES WHEN GAS REAGENT 'A' REACTS WITH LIQUID REAGENT B

under infinitely fast reaction conditions, the overall reaction rate is governed by the diffusion of A and B to the reaction plane. At this condition, the nature of the kinetics is irrelevant, and the enhancement factor has reached an ultimate asymptotic value. This is given by

$$E_{Ai} = 1 + \frac{C_{Bb}}{\nu_B C_A^*} \frac{D_B}{D_A} \tag{53}$$

and this condition can be expected to apply when

$$\sqrt{M(g)} >> \frac{C_{Bb}}{\nu_B C_A^*} \frac{D_B}{D_A} \tag{54}$$

As long ago as 1948 Van Krevelyn and Hoftizjer proposed that approximations based upon

$$E_A = \frac{\sqrt{M(m)} \left\{ \frac{E_{Ai} - E_A}{E_{Ai} - 1} \right\}}{\tanh \sqrt{M(m)} \left\{ \frac{E_{Ai} - E_A}{E_{Ai} - 1} \right\}} \tag{55}$$

would be asymptotically exact and moreover provided an excellent approximation (with maximum error of a few percent) in the intermediate range (17).

Setting

$$\sqrt{M(m,n)} = \sqrt{M(m)} \left\{ \frac{E_{Ai} - E_A}{E_{Ai} - 1} \right\} \tag{56}$$

the procedure for determining the rate of absorption of A is to use

$$E_A = \frac{\left\{ C_A^* - \frac{C_{Ab}}{\cosh M(m,n)} \right\} \frac{\sqrt{M(m,n)}}{\tanh \sqrt{M(m,n)}}}{(C_A^* - C_{Ab})} \tag{57}$$

Typical computations for E_A based upon the assumption that $C_{Ab} \rightarrow 0$ are shown in Fig. 9. The pseudo first order case with C_{Bb} constant through the film is the upper limit of behaviour. Eqn (57) is implicit in E_A and contains the unknown bulk concentration C_{Bb}. The performance of a backmixed reactor

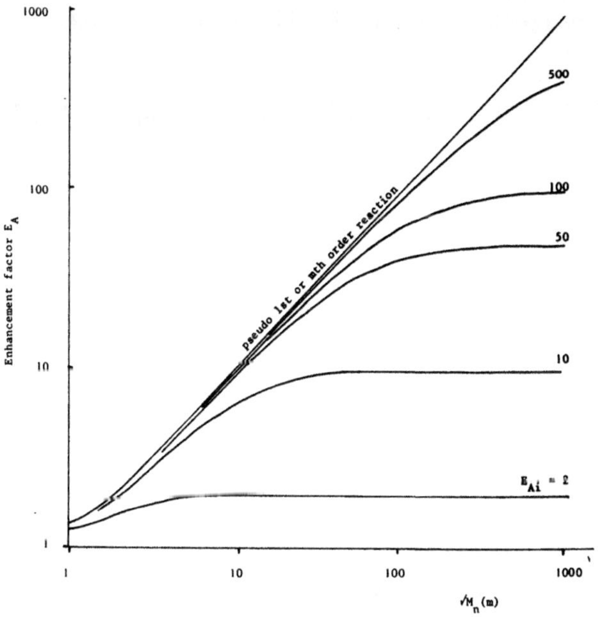

FIG. 9 ENHANCEMENT FACTORS FOR mth-nth ORDER REACTION BETWEEN GAS PHASE REACTANT A AND LIQUID PHASE REAGENT B

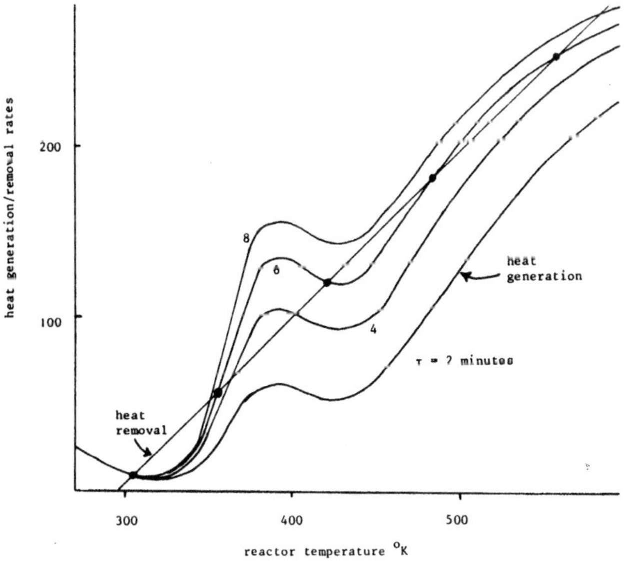

FIG. 10 MULTIPLICITY OF OPERATING STATES FOR A GAS-LIQUID REACTOR PROCESSING A SIMPLE SINGLE REACTION [Ref(18)]

element is therefore determined by the stoichiometrically balanced consumption of 'A' and 'B', where A is treated exactly as in the single reagent case and the liquid phase reagent 'B' should satisfy

$$Q\,(C_{Bo} - C_{Bb}) = V_L\,\bar{a}\,\left.\frac{dC_B}{dx}\right|_{x=\delta} + k\,V_{Lb}\,C_{Ab}\,C_{Bb}\left\{1 + \frac{1}{k\,C_{Ab}\,\tau}\right\} \quad (58)$$

overall utilisation of B	=	rate of reaction in film	+	rate of reaction in bulk	+	rate efflux of B

This will be satisfactory for the regimes I', II', III', IV' and V' in Fig. 8.

However for the case where C_B^* just falls below zero at the interface, so that reaction appears to be confined to a zone within the mass transfer film - this demarcates case VI' and those beyond it (i.e. those for even faster reaction) -the proposed treatment becomes suspect, since by its very analogy to the simple single first order reaction, the concentration C_{Ab} whilst it may become extremely small is never calculated to be zero. In other words the proposed approach presents some difficulties in the instantaneous regime since reaction at a plane can never take place. It may well be that it is never possible to arrive a priori at the "reaction at a plane" condition for the very reason that no reaction can be truly instantaneous unless the reaction rate constant is infinite. All analyses that proceed from a finite reaction rate constant, no matter how large, will always give finite values for C_{Ab} and C_{Bb}. This case therefore has yet to be fully explored and resolved.

The condition at which C_B^* falls below zero at the interface which gives the limiting condition for reaction not to take place throughout the film, can be deduced by a suitable manipulation of Eqns (51) and (52). The appropriate boundary conditions are

$$C_A = C_A^*\,,\quad \frac{dC_B}{dx} = 0 \text{ at } x = 0;\quad C_A = C_{Ab},\ C_B = C_{Bb} \text{ at } x = \delta$$

and these hold up until the point that C_B^* falls below zero.

Subtracting Eqns (51) and (52) gives

$$D_A \frac{d^2C_A}{dx^2} - \frac{D_B}{\nu_B} \frac{d^2C_B}{dx^2} = 0 \tag{59}$$

ich by integrating twice and using the boundary conditions
ves

$$- C_{Bb} = \nu_B \frac{D_A}{D_B} (C_A - C_{Ab}) - \frac{\nu_B}{D_B} D_A \left.\frac{dC_A}{dx}\right|_{x=o} (x - \delta) \tag{60}$$

d from the definition of E_A,

$$\frac{C_{Bb} - C_B^*}{C_A^* - C_{Ab}} = \nu_B \frac{D_A}{D_B} (E_A - 1) \tag{61}$$

The value of E_A which gives rise to $C_B^* = 0$ can therefore be und, subject to the correctness of C_{Bb}, which can be determined a simple single variable iteration, in conjunction with an eration on the implicit value for E_A itself, as indicated by Eqn 5).

An analysis equivalent to that proposed above has been rried out by Hoffman et al (18). They used data relevant to e chlorination of n-decane with m = n = 1 i.e. the reaction is rst order with respect to each component. For a single ckmixed gas-liquid reactor (equivalent to the element of Fig. , it was demonstrated that the interaction of mass transfer and emical reaction gave rise to the possibility of up to 5 steady ates for a single overall second order reaction. In their antitative treatment, they made use of a reaction factor E_A^*, ich is related to the normal enhancement factor by

$$E_A^* C_A^* = E_A (C_A^* - C_{Ab}) \tag{62}$$

is reaction factor has an analytical solution for the general n order case given by

$$* = M(m,n) \left\{ \frac{M(m,n)(\alpha-1) + (1/\theta) + \sqrt{M(m,n)} \tanh \sqrt{M(m,n)}}{M(m,n)(\alpha - 1) + (1/\theta) \tanh \sqrt{M(m,n)} + \sqrt{M(m,n)}} \right\} \tag{63}$$

ere for pseudo first order reaction

$$\sqrt{M(1,0)} = \sqrt{k_1 D_A C_{Bb}}/k_{LA}$$

with $\alpha = \varepsilon/\bar{a}\,\delta$ and $\theta = k_{LA}\,\bar{a}\,\tau\,/\varepsilon$

In the instantaneous regime the enhancement factor and reaction factor are identical and Eqn (53) applies. The authors (18) used the following generalised expression

$$E_A^* = \left[\frac{M(1,0)(\alpha - 1) + (1/\theta) + \sqrt{M(1,0)\,C_B^*}\ \tanh\sqrt{M(1,0)C_B^*}}{M(1,0)(\alpha - 1) + (1/\theta)\tanh\sqrt{M(1,0)\,C_B^*} + \sqrt{M(1,0)\,C_B^*}}\right]$$

$$\times\ M(1,0)C_B^* \qquad (64)$$

which hold approximately from slow through to instantaneous reaction. In their calculations of adiabatic performance, the number of steady states can be visualised on a classical heat generation/removal plot of which an example for the system chlorine-n-decane is given in Fig. 10. At low temperatures, heat is generated principally by dissolution and the interaction of solubility with temperature makes the heat generation curve fall initially. At higher temperatures, the interactions of solubility reduction and reaction rate constant acceleration produce a second trough in the heat generation curve, as reaction passes from being mainly in the bulk to mainly in the film. A sustained rise in heat generation rate is evidenced at higher temperatures when enhancement due to fast reaction in the film overrides the tendency for the solubility to be reduced.

The net result is that up to five steady states are possible dependent upon the residence time. This steady state multiplicity is accompanied by the customary extinction and ignition hysteresis of the steady states. However, the added complications of the interaction of mass transfer and reaction produce additional pecularities in the ignition/extinction behaviour. Fig. 11 (adapted from (18)), indicates the influence of the variation of liquid phase residence time τ . On increasing the residence time, an ignition to a high temperature steady state occurs at τ = 55 minutes. On reducing τ , the extinction behaviour is quite different in character and shows two stages. An extinction at τ = 22 minutes gives a first extinction to an intermediate steady state. A further reduction to τ = 9 minutes gives a second extinction.

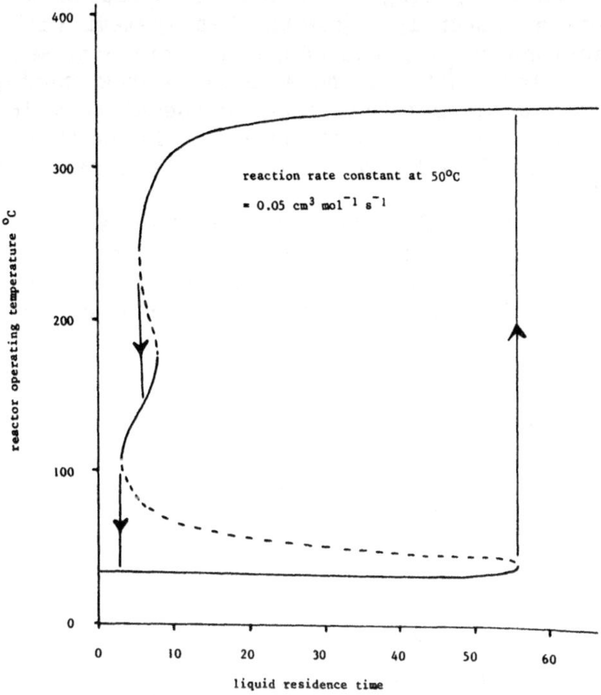

FIG. 11 HYSTERESIS OF GAS-LIQUID REACTOR OPERATING STATES AS LIQUID PHASE RESIDENCE TIME VARIES [Ref(18)]

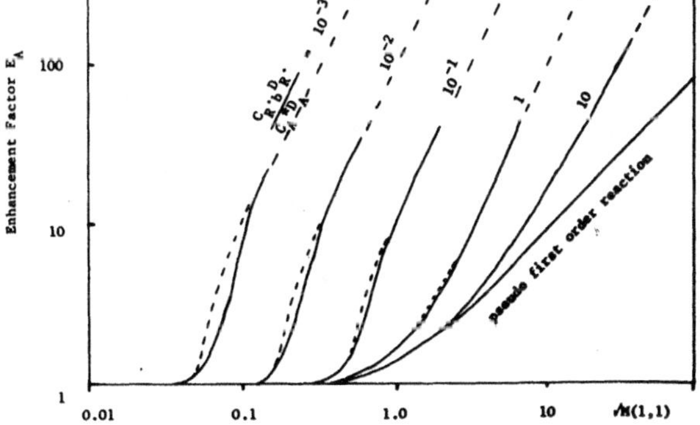

FIG. 12 ENHANCEMENT FACTORS FOR AUTOCATALYTIC REACTION

That such multiplicity complexities are available from a reaction involving a single gas phase reagent and a single liquid phase reagent participating in a single reaction with simple reaction kinetics, strongly suggests that systems with a higher number of reactions or reagents could show still more pronounced peculiarities. Indeed, the authors go on to show exactly this in respect of two reactions taking place consecutively in a single backmixed factor (28). This example is discussed in the next section where examples with two reactions are considered.

One remaining case of a reaction type where a gas A dissolves and reacts with a liquid phase reagent is the autocatalytic reaction in which the product can catalyse the decomposition of A. Radical mechanisms are of this type and can be represented by the stoichiometry

$$A + R^{\bullet} \xrightarrow{k_1} R^{\bullet} + R^{\bullet}$$

with diffusion and reaction equations

$$D_A \frac{d^2 C_A}{dx^2} = k_1 C_A C_{R^{\bullet}} \qquad (65)$$

$$D_R \frac{d^2 C_{R^{\bullet}}}{dx^2} = - k_1 C_A C_{R^{\bullet}} \qquad (66)$$

subject to the boundary conditions

$$C_A = C_A^{*} \quad \frac{dC_{R^{\bullet}}}{dx} = 0 \quad \text{at} \quad x = o \qquad (67)$$

$$C_A = C_{Ab} \quad C_{R^{\bullet}} = C_{R^{\bullet}b} \quad \text{at} \quad x = \delta \ .$$

Numerical solution of these non-linear equations (assuming $C_{Ab} \rightarrow 0$)(19) have indicated that the Enhancement Factor increases greatly above the limiting pseudo first order case due to the accumulation of autocatalysing radicals in the mass transfer film. Fig. 12 shows a typical plot of E_A (at negligible C_{Ab}) versus the diffusion/reaction parameter $\sqrt{M(1,1)}$. It can be shown that these solutions can be approximated by

$$E_A = \frac{\sqrt{M(1,1)}\{1 + (E_A - 1)C_A{}^*D_A/C_{Rb}^{\bullet}D_R^{\bullet}\}}{\tanh \sqrt{M(1,1)\{1 + (E_A - 1)C_A{}^*D_A/C_{Rb}^{\bullet}D_R^{\bullet}\}}} \tag{68}$$

A more detailed understanding of reactions of this kind will ɜ necessary to appreciate the performance of oxidation and ılorination gas-liquid reactors, which are usually supposed to ·oceed by radical mechanisms. It is possible that a variety of ıdical species may occur and their many possibilities for ɜcombination may lie behind the wide spectrum of products ıaracteristic of oxidations and chlorinations. A related case ıere product HCl in a chlorination substitution acts ıtocatalytically has been reported by Joosten et al (20). The ɜneral overall reaction characteristics have so far not been ɜported in the literature. However, the inflected character of ıe reaction rate/concentration behaviour of these autocatalytic ɜactions makes the occurrence of unusual patterns of reactor ıltiplicity highly probable.

⁝TENSION TO TWO REACTIONS IN THE LIQUID PHASE

The simplest case of a gas reagent A and liquid phase reagent accompanied by two reactions can be represented by

$$+ \quad \nu_B B \quad \underset{k_2}{\overset{k_1}{\rightleftharpoons}} \quad \nu_E \ E + \nu_F \ F \tag{69}$$

ıd this case has been studied by Onda et al (21).

ıe desorbing diffusion reaction equations for the general case ' m,n - p,q order reactions are

$$D_A \frac{d^2C_A}{dx^2} = k_1 C_A{}^m C_B{}^n - k_2 C_E{}^p C_F{}^q \tag{70}$$

$$D_B \frac{d^2C_B}{dx^2} = \nu_B (k_1 C_A{}^m C_B{}^n - k_2 C_E{}^p C_F{}^q) \tag{71}$$

$$D_E \frac{d^2C_E}{dx^2} = -\nu_E (k_1 C_A{}^m C_B{}^n - k_2 C_E{}^p C_F{}^q) \tag{72}$$

$$D_F \frac{d^2C_F}{dx^2} = - \nu_F (k_1 C_A{}^m C_B{}^n - k_2 C_E{}^p C_F{}^q) \quad (73)$$

Subject to the boundary conditions

$$C_A = C_A{}^* , \frac{dC_B}{dx} = 0, \frac{dC_E}{dx} = o , \frac{dC_F}{dx} = o \text{ at } x = o \quad (74)$$

$$C_A = C_{Ab} , C_B = C_{Bb} , C_E = C_{Eb} , C_F = C_{Fb} \text{ at } x = \delta$$

Equations (70) and (72) are rearranged to

$$D_A \frac{d^2C_A}{dx^2} + \frac{D_E}{\nu_E} \frac{d^2C_E}{dx^2} = 0 \quad (75)$$

so that, by integration with the boundary conditions

$$C_E(x) = \{ C_{Eb} + - D_A \frac{dC_A}{dx} \} \frac{\nu_E}{D_E} (\delta - x) - \nu_E \frac{D_A}{D_E} (C_A(x) - C_{Ab}) \quad (76)$$

Two similar expressions give the relationships between B, E and F such that

$$C_B{}^* = \frac{C_{Bb}}{C_A{}^*} - \frac{\nu_B}{\nu_E} \frac{D_E}{D_B} \{ C_E{}^* - \frac{C_{Eb}}{C_A{}^*} \} \quad (77)$$

$$*\ = \frac{C_{Fb}}{C_A^*} + \frac{\nu_F}{\nu_E} \frac{D_E}{D_F} \{ C_E^* - \frac{C_{Eb}}{C_A^*} \} \qquad (78)$$

da et al (21) then used the linearisation method of Hikita and ai (22) (which is analogous to the Van Krevelen and Hoftijzer eatment (17)), so that Eqn (70) becomes

$$\frac{d^2C_A}{dx^2} = \{ \frac{2}{m+1} k_1 C_A^{*m-1} C_B^{*n} \} C_A - \{ \frac{2}{p+1} k_2 C_E^{*p-1} C_F^{*q} \} C_E$$

$$= M_n(m)\, C_A - M_q(p)\, C_E \qquad (79)$$

analytical result for the reaction factor E_A^* defined by

$$E_A^* = \frac{-D_A \left.\frac{dC_A}{dx}\right|_{x=o}}{k_{LA}\ C_A^*}$$

then obtained and given by

$$* = \frac{\left(1 - \frac{C_A^*}{C_{Ab}}\right)\left[1 + \frac{p+1}{m+1}\frac{D_E}{D_A}\frac{1}{\nu_E}\frac{1}{T} + \frac{1}{\nu_E}\frac{D_E}{D_A}\frac{p+1}{m+1}\frac{1}{T}\frac{C_{Ab}}{C_A^*} - \frac{C_{Eb}}{C_A^*}\right]}{1 + \frac{1}{\nu_E}\frac{D_E}{D_A}\frac{1}{T}\frac{\tanh\sqrt{M^1}}{M^1}}$$

$$\times [\, 1\text{-sech}\ M^1\,] \qquad (80)$$

ere

$$M^1 = M_n(m)\left(\frac{C_A{}^*C_B{}^*}{C_{Bb}}\right)^n\left(1 + \nu_E\,\frac{D_A}{D_E}\,T\right)$$

with

$$T = K_2\left(\frac{C_E{}^*}{C_{Eb}}\right)^{p-1}\left[\frac{C_F{}^*}{C_{Fb}}\right]^q\left(\frac{C_B{}^*}{C_{Bb}}\right)^{-n}$$

and $$K_2 = C_A{}^{*\,p+q-m-n}/K$$

By numerical integration of the diffusion and reaction equations, Onda et al (21) showed that the Hikita and Asai linearisation is always accurate to within a few percent.

However the general performance characteristics of a backmixed gas-liquid reactor element, determined without a priori assumptions, and governed by the coupling of the diffusion reaction equations to the gas and liquid phase material balances, has not been reported in the literature. Whilst the treatment is simplified by the analytical expressions of the type in Eqn (80), it would appear that the complexity of the trial and error solutions for the interface concentrations $C_A{}^*$ and $C_B{}^*$ and the bulk concentrations C_{Ab}, C_{Bb}, C_{Eb} and C_{Fb} will present some difficulty.

The case where a single gaseous species 'A' reacts with two liquid phase reagents 'B_1' and 'B_2' in parallel can be represented by

$$A + \nu_{B_1} B_1 \xrightarrow{k_1} (\text{product})_1$$

$$A + \nu_{B_2} B_2 \xrightarrow{k_2} (\text{product})_2$$

This case has also been considered in detail by Onda et al (23). The diffusion reaction equations for general reaction orders can be written

$$D_A \frac{d^2C_A}{dx^2} = k_1 C_A^m C_{B_1}^n + k_2 C_A^p C_{B_2}^q \quad (81)$$

$$D_{B_1} \frac{d^2C_{B_1}}{dx^2} = \nu_{B_1} k_1 C_A^m C_{B_1}^n \quad (82)$$

$$D_{B_2} \frac{d^2C_{B_2}}{dx^2} = \nu_{B_2} k_2 C_A^p C_{B_2}^q \quad (83)$$

th boundary conditions

$$= C_A^* , \quad \frac{dC_{B_1}}{dx} = 0 , \quad \frac{dC_{B_2}}{dx} = 0 \quad \text{at} \quad x = 0 \quad (84)$$

$$= C_{Ab} , \quad C_{B_1} = C_{B_1b} , \quad C_{B_2} = C_{B_2b} \quad \text{at} \quad x = \delta$$

uation (81) is linearised as in the previous general reversible eaction case, so that

$$\frac{d^2C_A}{dx^2} = \{ \{\frac{2}{m+1} k_1 C_A^{*m-1} C_{B_1}^{*n}\} + \{\frac{2}{p+1} k_2 C_A^{*p-1} C_{B_2}^{q}\} \} C_A \quad (85)$$

e reaction factor E_A^* (defined by Eqn (62)) is obtained very mply for the cases where there is negligible depletion of the quid phase reagents B_1 and B_2, and is given by

$$E_A^* = \frac{\sqrt{M''}\{1 - \frac{C_{Ab}}{C_A^*} \text{sech} \ M''\}}{\tanh\sqrt{M''}} \quad (86)$$

where

$$M'' = M^1(m)\ \{\ 1 + [\frac{m+1}{p+1}][\frac{k_2}{k_1}]\ C_A{*}^{p-m}\ \} \qquad (87)$$

where $M^1(m)$ refers to diffusion reaction parameters evaluated at the bulk concentrations of the liquid phase reagents C_{B_1b} and C_{B_2b}.

For the case where significant depletion of the liquid phase reagents takes places, the treatment parallels that for A reacting with a single liquid phase component. Manipulation of Eqns (81), (82) and (83) gives

$$E_A{*} = [1 - \frac{C_{Ab}}{C_A{*}}] + \frac{C_{Bb}}{C_A{*}}\ \{\frac{D_{B_1}/D_A}{\nu_{B_1}}[\ 1 - \frac{C_{B_1}{*}}{C_{B_1b}}\ ] + \frac{D_{B_2}/D_A}{\nu_{B_2}}\ [\frac{C_{B_2b}}{C_{B_1b}} - \frac{C_{B_2}{*}}{C_{B_1b}}]\ \} \qquad (88)$$

$$\text{with} \quad E_A^* = \frac{\sqrt{M'''}\ \{\ 1 - \frac{C_{Ab}}{C_A{*}}\ \text{sech}\sqrt{M'''}\ \}}{\tanh\sqrt{M'''}}$$

where

$$M''' = M_n(m)\ \{\frac{C_{B_1}{*}}{C_{B_1b}} + \frac{m+1}{p+1}\cdot\frac{k_2}{k_1}\ C_A{*}^{p+q-m-n}\ [\frac{C_{B_1b}}{C_A{*}}]^{q-n}\ [\frac{C_{B_2}{*}}{C_{B_1b}}]^{q}\ \}$$

where $M_n(m)$ refers to component B_1.

The analysis can only be completed by making some assumptions about the concentration profiles of the liquid reagent. The authors (21) assume that these can be expressed as quadratic functions of the dimensionless film thickness, so that

$$\frac{C_{B_1}(x)}{C_{B_1 b}} = \left[1 - \frac{C_{B_1}^*}{C_{B_1 b}}\right]\left[\frac{x}{\delta}\right]^2 + \frac{C_{B_1}^*}{C_{B_1 b}} \qquad (89)$$

with a similar relationship for C_{B_2}.

The general performance features of the backmixed gas-liquid reactor element requires the incorporation of the above expressions into the usual gas and liquid material balance relationships and this remains to be undertaken. Also, the elucidation of optimal gas-liquid element configurations as envisaged in Fig. 6, so that gas-liquid reactor designs can give good yield performance, appears to require considerable further effort.

The case of two reactors in which the product of the first reaction between the gas and liquid phase reagents can react further with the absorbing gas is represented by

$$A + B \xrightarrow{k_1} R$$

$$A + R \xrightarrow{k_2} S$$

If the reactions are all first order in each of the reacting components, the diffusion reaction equations are

$$D_A \frac{d^2 C_A}{dx^2} = k_1 C_A C_B + k_2 C_A C_R \qquad (90)$$

$$D_B \frac{d^2 C_A}{dx^2} = k_1 C_A C_B \qquad (91)$$

$$D_R \frac{d^2 C_R}{dx^2} = - k_1 C_A C_B + k_2 C_A C_R \qquad (92)$$

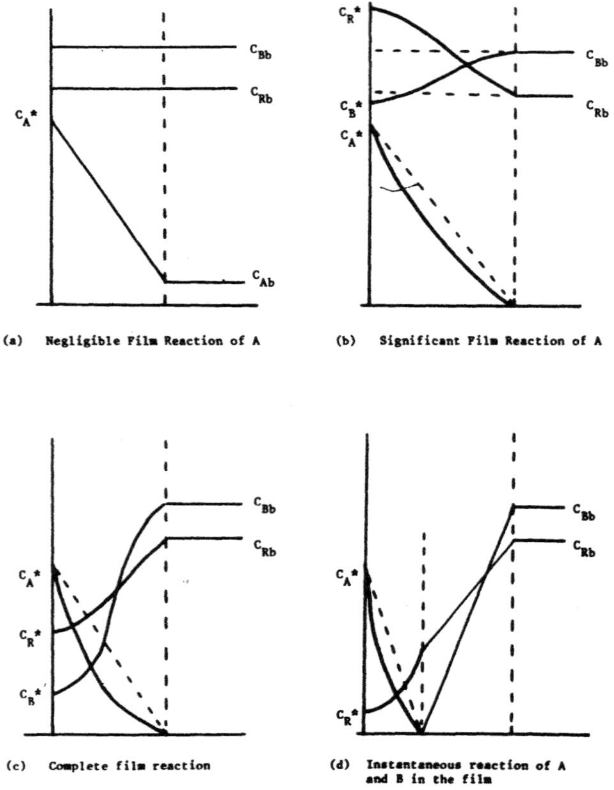

FIG. 13 SOME REACTION REGIMES FOR THE (1,1) - (1,1) CONSECUTIVE GAS-LIQUID REACTION

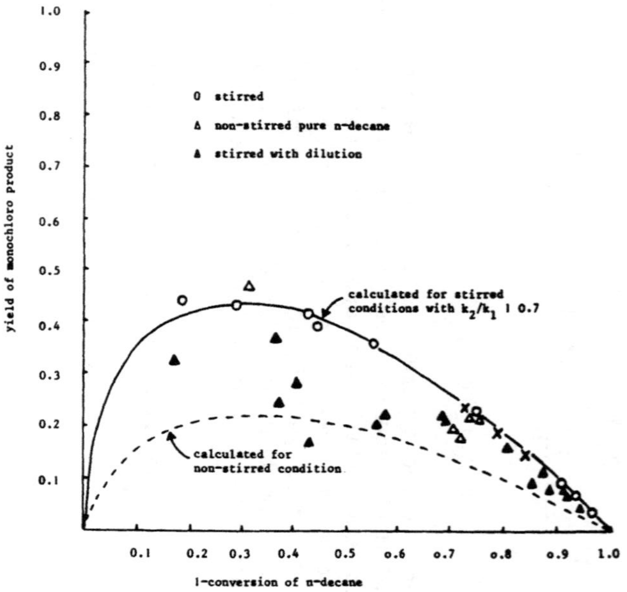

FIG. 14 INFLUENCE OF STIRRING ON YIELD OF MONOCHLORO INTERMEDIATE [Ref(24)]

th boundary conditions

$$x = o \qquad C_A = C_A^* \qquad \frac{dC_B}{dx} = 0 \qquad \frac{dC_R}{dx} = 0 \qquad (93)$$

$$x = \delta \qquad C_A = C_{Ab} \qquad C_B = C_{Bb} \qquad C_R = C_{Rb}$$

corporation of the consecutive reaction possibility multiplies e number of reaction regimes discussed in the context of the ngle reaction between A and B.

Some of the more significant possibilities are presented in g. 13. In Fig. 13 (a), if all the reaction takes place in the lk liquid phase, then no reduction occurs in the potential eld of the intermediate product R. If, however, the reaction significantly fast in terms of A, then it is possible epending upon the current magnitude of C_{Bb} and C_{Rb} and the rate nstant ratio k_1/k_2) for some depletion of B to take place rough the film, with the net reaction to R providing an creasing concentration of R through the film from C_{Rb} to C_R^*, ere is thus a net production of R within the film. However, is net production of R in the film results in a lower yield of R th respect to A, since the relative increase in $C_R(x)$ compared the relative decrease in $C_B(x)$ lowers the instantaneous yield nction at every point in the film. It is therefore possible r the case of diffusion and fast reaction to see a 'taxation' of e yield of a desired intermediate which accumulates gnificantly in the reaction film. Any accompanying reaction in e bulk gives a high yield of R.

If the intensity of the film reaction increases further, en it is possible for R to be consumed within the film and the eld is lowered still further, relative to that obtainable at e bulk condition. In the profiles of Fig. 13(c), net overall nsumption of R is taking place.

Ultimately, the reaction between A and B can become stantaneous with reaction between them occurring at a plane. y intermediate R can then be reacted with A before it can ffuse to react with B. At this condition there is potentially serious loss of R if it is the desired termediate. It needs to be stressed that the examples of Fig. (a), (b), (c), (d) are chosen to emphasise the qualitative ssibilities of regimes which influence the yield of termediate product R with respect to the final product S.

ıne earliest analyses of the diffusion/reaction equation was by Van de Vusse (24) for a semi-batch reactor in which all reaction took place in the mass transfer film, so that the boundary condition on A at $x = \delta$ is given by

$$\left.\frac{dC_A}{dx}\right|_{x=\delta} = 0$$

Use of the boundary condition with C_A^* constant is somewhat artificial and neglects the material balance on the gas stream, and therefore avoids a consideration of the yield of liquid phase product R with respect to the utilisation of gas phase reagent A. Van de Vusse (24) solved the diffusion reaction equations numerically, and showed how the theory could be used to explain the experimentally dissolved influence of stirring on the chlorination of n-decane. Fig. 14 shows yield of R with respect to B as a function of the conversion of B, as determined experimentally. A serious loss of yield occurs in the absence of stirring, due to the reduction in the mass transfer coefficient. Van de Vuuse (22) concluded that selectivity limitations would occur if $\sqrt{M(1)} > 2$ and moreover if $\frac{C_{Bb}}{C_A^*} \lesssim M(1)$.

From this second criterion, it may be concluded that for those cases where a gas reacts with a pure liquid over a restricted range of conversion, so that C_{Bb} is always fairly high, a gas of low solubility is unlikely to react so as to produce a selectivity limitation.

In subsequent treatments, Teramoto et al (25, 26) developed an analysis of both semi-batch and continuous reactor performance which incorporated a quantitative discrimination of the role of film and bulk reaction. The intractability of the non-linear product terms in the diffusion/reaction equations was ultimately avoided by a linearisation method, identical to that proposed by Hikita and Asai (22). In this approach the profiles $C_B(x)$ and $C_R(x)$ are replaced by their interfacial values, so that the diffusion/reaction equations become

$$D_A \frac{d^2C_A}{dx^2} = k_1 \{ 1 + \frac{k_2 C_R^*}{k_1 C_B^*} \} C_B^* C_A \qquad (94)$$

$$D_B \frac{d^2C_B}{dx^2} = k_1 C_B^* C_A \qquad (95)$$

$$D_R \frac{d^2C_R}{dx^2} = (- k_1 C_B^* + k_2 C_R^*) C_A \qquad (96)$$

From Eqn. (94), now linear in C_A, an analytical solution is tained. Manipulation of the equations for C_A and C_B yields

$$\frac{d^2C_A}{dx^2} - \frac{D_B}{D_A} \{ 1 + \frac{k_2 C_R^*}{k_1 C_B^*} \} \frac{d^2C_B}{dx^2} = 0 \qquad (97)$$

ich is integrated with the appropriate boundary conditions to ve both the interfacial value C_B^* and the flux of B out of the lm given by

$$- D_B \left. \frac{dC_B}{dx} \right|_{x=\delta} .$$

The reactions of B and R in the film are thereby determined rectly, so that the instantaneous fractional yield of termediate R with respect to the liquid phase reagent B is given

$$\psi \left(\frac{R}{B} \right) = \frac{\text{rate of production of R in film and bulk}}{\text{rate of consumption of B in film and bulk}}$$

assuming pseudo steady state with respect to C_A^* and C_B^*, ramoto et al (26), numerically integrated the instantaneous eld to give the overall yield at various times (or conversions) the relation

$$\phi\left(\frac{R}{B}\right) = \int_0^\theta \psi \;\; dX_B.$$

In this way they constructed a time-conversion-yield plot similar to that shown in Fig. 15.

In a subsequent analysis, Teramoto et al (27) extended the treatment to the general case of (m,n) - (p,q) order reaction. In this later work, they showed how the use of generalised moduli reduced the (m,n) - (p,q) case to the (1,1) - (1,1) case. Typically they presented time-conversion-overall yield plots of the type shown in Fig. 15. This figure clearly shows the influence of increasing reaction speed on the selectivity. The case of M(1) = 0 corresponds to infinitely slow bulk reaction at the interfacial concentration of A which gives the maximum intrinsic yield. As the modulus M(1) moves into the diffusion influenced regime, the yield at a given conversion level declines. The results in Fig. 15 refer to a selectivity ratio of 4. To achieve a conversion of 50% of B as $\sqrt{M(1)}$ increases from 1.0 through to 5.0, the batch time decreases (ie the rate of conversion of B increases) and the yield falls. The productivity of intermediate R can therefore be expected to be fairly insensitive to reaction speed at these conditions. For instance at 50% conversion as $\sqrt{M(1)}$ increases from 1.0 to 5.0, the batch time decreases from 3 to 1, and the yield decreases from 0.38 to around 0.17.

These simulations, as with Van de Vusse, neglect the influence of the utilisation of the gas phase reagent (since a constant value of C_A* is assumed) and the authors do not explicitly indicate the proportions of film and bulk reaction taking place. Although the authors demonstrated that their linearisation approximation is accurate, their results probably are restricted to the range before the reaction between A and B approaches the instantaneous condition, since at a circumstance as depicted in Fig. 13(d), the use of C_B* and C_R* could hardly be expected to be appropriate.

Following on the work of Teramoto, Hoffman, Sharma and Luss (28) have performed an analysis of the adiabatic gas-liquid reactor operating in continuous backmixed flow of the liquid phase for this consecutive (1,1) - (1,1) reaction. They used data relative to the system chlorine/n-decane with a selectivity ratio of k_1/k_2 = 1.414. The boundary conditions were formulated in terms of overall material balances on the gas and liquid phases, so that for component A, the boundary condition at the film-bulk junction is given by

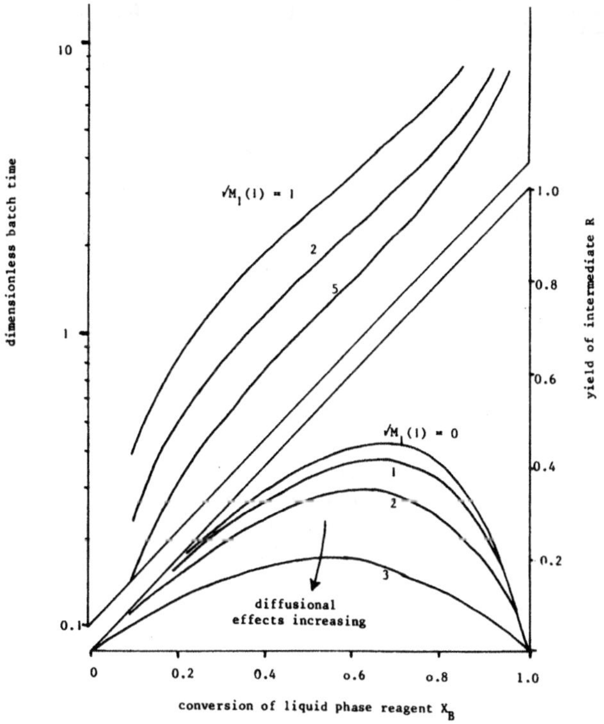

FIG. 15 YIELD PERFORMANCE OF A SEMI-BATCH GAS-LIQUID REACTOR FOR (1,1)-(1,1) CONSECUTIVE REACTION [Ref(27)]

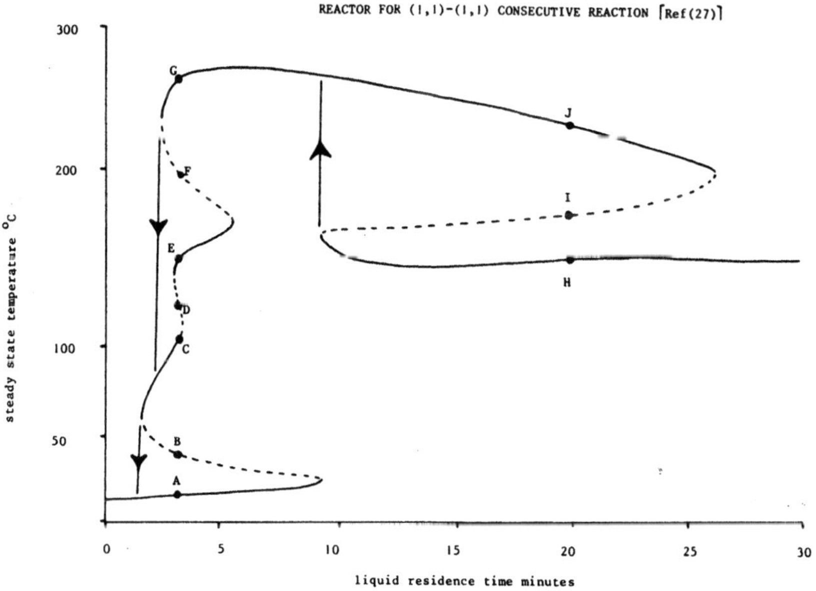

FIG. 16 UNUSUAL EXTINCTION BEHAVIOUR FOR GAS-LIQUID REACTOR PROCESSING (1,1)-(1,1) CONSECUTIVE REACTION [Ref(28)]

$$- D_A \left. \frac{dC_A}{dx} \right|_{x=\delta} = (k_1 C_{Ab} C_{Bb} + k_2 C_{Ab} C_{Rb})V + Q C_{Ab}$$

which states that the rate of transfer of dissolved gas A to the bulk of the liquid must at steady state be equal to the amount consumed by reaction in the bulk plus that which leaves the reactor in the effluent stream. Similar balances were written for each component without making any a priori assumptions on the extents of film and bulk reaction. Teramoto's (26) approximate analytical results were employed whereby the reaction factors defined for each component, E_A*, E_B* and E_R* enabled the rate of absorption of A and its extent of reaction in the film, as well as the corresponding consumptions of B and R in the film to be determined, starting only from an assumption for the concentration of A in the bulk liquid phase ie C_{Ab}. A simple iteration on C_{Ab} gives a complete definition of the reactor state for any feed conditions. The total number of steady states of such an adiabatic gas-liquid reactor are then found from the intersections of heat generation and heat removal lines when drawn versus temperature (as in Fig. 10 for the single reaction case). The resulting steady states can be as many as seven in number (cf with 5 for a homogeneous reactor processing a consecutive reaction), with an impressive degree of complexity being predicted, which is seemingly out of all proportion to the apparent simplicity of two reactions taking place in a gas-liquid reactor. As before, these multiplicities of steady states are caused by subtle interactions amongst the rates of the two reactions, the rates of interphase transfer and the solubility of the gaseous reactant.

Some very curious extinction and ignition phenomena are predicted to occur, as can be seen from Fig. 16 which shows how the steady-state temperature changes as the liquid phase residence time is varied. For instance, as the residence time is decreased along the branch H-K, ignition to the branch G-J can occur. This is the inverse to the usual ignition expected from an increase in residence time.

At a $\tau = 3$, the operating temperature shows a number of increasing temperature steady states, which exhibit an increasing conversion with a decreasing selectivity to R. The diffusion/reaction parameter $\sqrt{M(1)}$ increases from 0.003 to 113.1 at each of the steady states. The corresponding values of E_A*, E_B* and E_R* are given in Table 2.

Sharma et al (28) point out that the simple mass and chemical ries resistance model previously used by Schmitz and Amundsen :9) is completely inadequate for describing such a system, since is precisely the complex nature of the interactions of ffusion and reaction which give rise to the complex reactor ltiplicity phenomena. Sharma et al's analysis is somewhat complete since they do not evaluate the film/bulk interactions, r do they make an appraisal of the validity of the Teramoto nearisation approximation. Again, this particular form of proximation could be highly inappropriate if the concentration reagent B falls to zero in the mass transfer film. This obably occurs at their highest temperature steady states at a $\sqrt{M(1)}$ value of around 100.

The associated problem of understanding the complexities ssible in the performance of an isothermally operating rfectly backmixed reactor was not undertaken by Hoffman et al :8), though the problem has been tackled more recently by Ho and e (30). Instead of using a linearisation approximation, they ed the collocation method to solve the diffusion/reaction eqns 0), (91) and (92) for general (m,n) - (p,q) order reaction netics when they are embedded in the overall material balances the gas and liquid phases. Whilst the authors have tackled the oblem directly without a priori assumptions, it is difficult to rceive any generalities from the two specific cases they nsidered. The authors do not detail the overall aracteristics of their two reactor examples, and they do not ntion the yield/selectivity behaviour, nor the absorption ficiency of A, nor the utilisation of the liquid phase mponent B. Fig. 17 gives example profiles of A, B and C at rious values of k_L and k_1/k_2. The figure shows that the riation of $C_R(x)$ through the film depends upon the size of k_L lative to k_2. For $k_2 = 10$, increasing k_L from 0.005 to 0.02 cm ^{1}, flattens the profile. However, from Fig. 17 this then duces the bulk concentration of intermediate R. This does not an that the yield of R is reduced, because changes in k_L also ange the conversion of A and B. At a fixed $k_L = 0.02$, creasing k_2 from 10 through to 1000 steepens the profile in R d lowers C_{Rb}, although only from 0.094 to 0.092, which would ggest a 2% yield loss if the conversion of A were approximately nstant.

Huang, Carberry and Varma (31) have revisited this case most cently. They also solved the diffusion reaction equations ing the method of orthogonal collocation, which they showed to computationally more accurate than the conventional direct merical solution of the set of three simultaneous second order

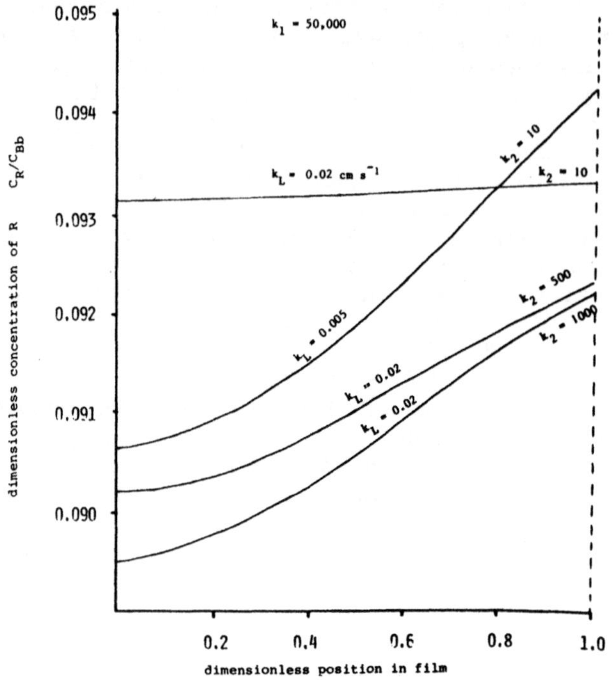

FIG. 17 FILM PROFILES OF INTERMEDIATE R FOR ISOTHERMAL GAS-LIQUID REACTOR PROCESSING (1,1)-(1,1) REACTION [Ref(30)]

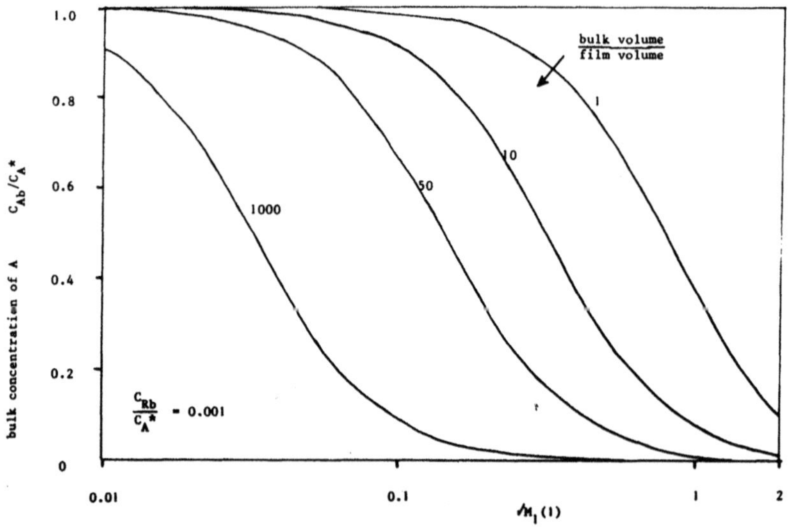

FIG. 18 EFFECT OF FILM/BULK VOLUME RATIO ON BOUNDARY CONDITION ON 'A' FOR (1,1)-(1,1) CONSECUTIVE REACTION IN A GAS-LIQUID REACTOR [Ref(31)]

ES in Eqns (90), (91) and (92). However, they have not nsidered the way in which the diffusion and reaction interact th the overall material balances on the gas and liquid phases, d the equations were solved on the basis of arbitrary film/bulk undary conditions. Instead, they compared reaction factors * as calculated using the film and penetration theories to scribe the diffusion and reaction. Also, they calculated film elds according to both the film and penetration theories and owed that the differences in film yields are somewhat greater an the differences in reaction factors. The authors also ressed that their computed yields were point yields, and that fferences between film and penetration theories for an overall actor yield would indeed be magnified.

The authors also investigated the results of properly counting for C_{Ab} (at an assumed value of C_{Rb}/C_A* of 0.001) which ey showed to be important in the $\sqrt{M(1)}$ range of 0.01 to 2.0. is can be seen from Fig. 18. They also demonstrated that E_A* significantly in error in this range of $\sqrt{M(1)}$ if C_{Ab} is sumed to be zero instead of finite, and this is shown in Fig. .

Whilst Huang et al (31) considered film yield problems, cluding the philosophically interesting distinctions between lm and penetration theory yields, they did not address emselves to the reactor overall yield performance as influenced the relative contributions of reaction in the film and bulk. e yield differences between the two mass transfer theories uld moreover be expected to be significant for an operating actor where a few percent of yield improvement could make the fference between profit and loss for a reactor. Neither did e authors consider describing reactor performance in terms of e utilisation of gas phase reagent A and co-utilisation of quid phase reagent B.

RE THAN TWO REACTIONS IN THE LIQUID PHASE

The previous scheme represents the simplest case where the oduct formed can react with the absorbing gas. In many stances, as in oxidations and chlorinations, each of a ccession of products can often react further with the absorbing s.

An example scheme involving both consecutive and parallel mponents with five reactions accompanying the dissolution of e gas A is (32)

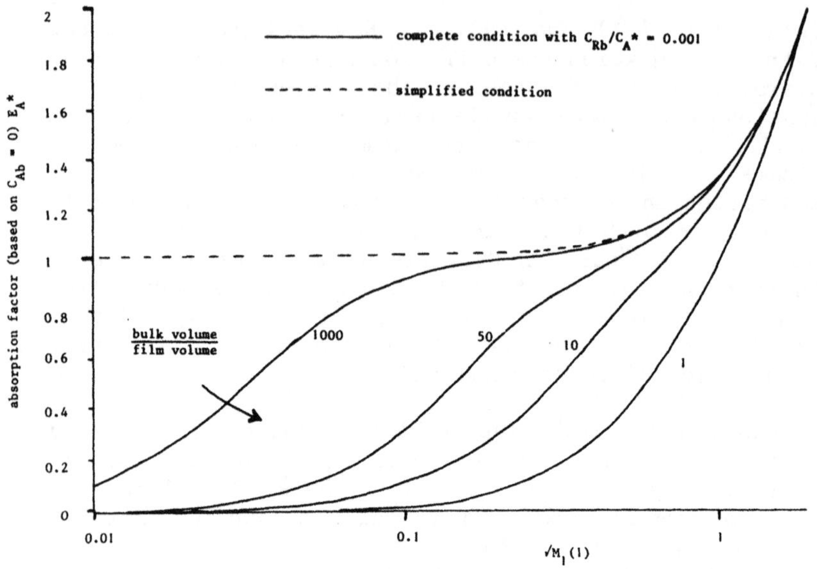

FIG. 19 INFLUENCE OF SIMPLIFIED AND COMPLETE BOUNDARY CONDITION ON ABSORPTION FACTOR E_A^* FOR (1,1)-(1,1) REACTION [Ref(31)]

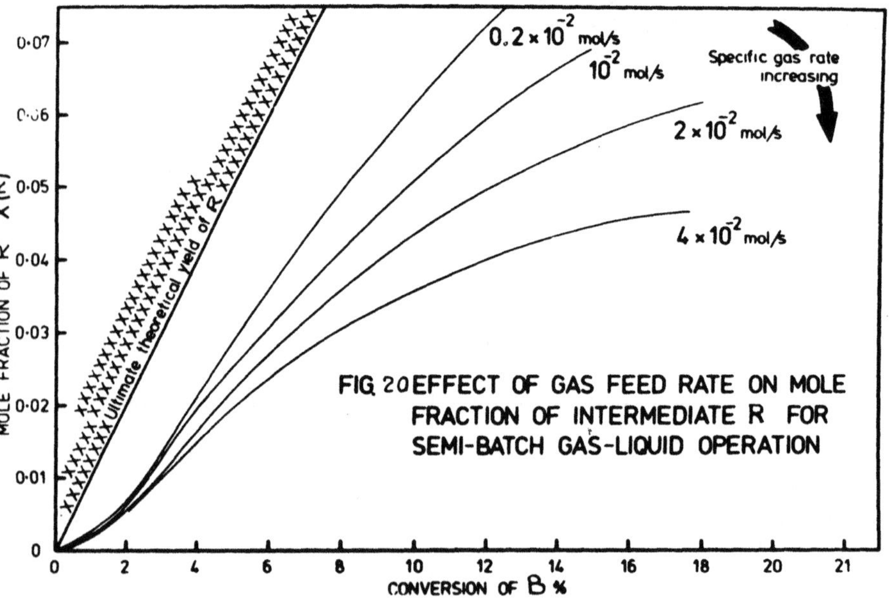

FIG. 20 EFFECT OF GAS FEED RATE ON MOLE FRACTION OF INTERMEDIATE R FOR SEMI-BATCH GAS-LIQUID OPERATION

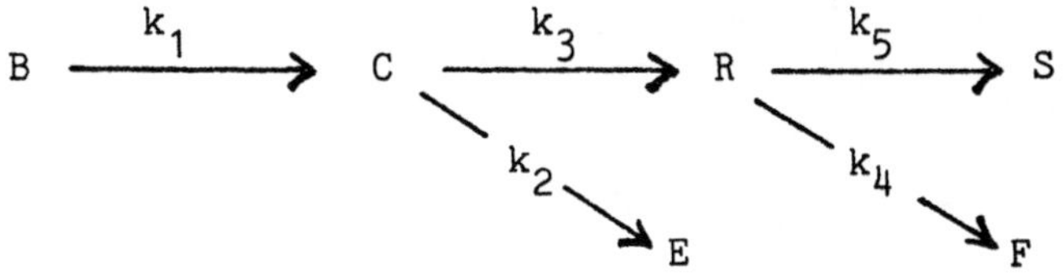

A series of differential equations describes the evolution ' liquid phase products with time in a semi-batch reactor with ntinuous feed of the gas. If the formations of the by-products and F are second order in the reacting gas A and all other actions are first order, and moreover if A absorbs and reacts gnificantly only in the bulk liquid phase, so that film action is negligible and the 'slow' regime applies, then a ries of ordinary differential equations describe the ncentration trajectories of liquid phase products with time.

ese were:

$$\frac{dC_{Bb}}{dt} = - k_1 C_{Ab} C_{Bb} \quad (99)$$

$$\frac{dC_{Eb}}{dt} = k_1 C_{Ab} C_{Bb} - (k_3 + k_2 C_{Ab}) C_{Ab} C_{Cb} \quad (100)$$

$$\frac{dC_{Rb}}{dt} = k_3 C_{Ab} C_{Cb} - (k_5 + k_4 C_{Ab}) C_{Ab} C_{Rb} \quad (101)$$

$$\frac{dC_{Sb}}{dt} = k_5 C_{Ab} C_{Rb} \quad (102)$$

$$\frac{dC_{Eb}}{dt} = k_2 \, C_{Ab}^2 \, C_{Cb} \quad (103)$$

$$\frac{dC_{Fb}}{dt} = k_4 \, C_{Ab}^2 \, C_{Rb} \quad (104)$$

Since the decompositions to by-products E and F are of a higher order in the dissolved gas A, in regions of any reactor where high dissolved gas concentrations occur, the rate of reaction to form by-products will be accelerated and the yield of desired product R will be reduced on two counts. Firstly because the precursor C will be diverted to E and secondly because the desired product R will be diverted to F.

If the performance of a batch reactor is to be described by solving the differential equations for liquid phase components B, C, R, S, E and F, it is necessary to determine C_{Ab} by solving the material balance on A across the semi-batch reactor.

The unsteady behaviour of dissolved gas concentration, assuming quasi-stationary behaviour of the gas-liquid phases, is then given by

$$\frac{dC_{Ab}}{dt} = \frac{G_i y_{Ai} - G_o y_{Ao}}{V_L} - \Sigma \, \alpha \, (I) \, \frac{dC_{Ib}}{dt} \quad (105)$$

where B + α (I) A $\longrightarrow$ I for I = C, R, S, E and F. In this last relationship α (I) is the number of mols of reacting gas A contained in a mol of B when converted into I. The material balance on A across the gas phase requires that

$$G_i y_{Ai} - G_o y_{Ao} = k_L \, \bar{a} \, V_L \, (C_A^* - C_{Ab})$$

d the treatment parallels that outlined for the single reaction th a quadratic in y_{Ao}, from which C_A* and C_{Ab} can be found as e set of differential equations are integrated through time.

For the parameters shown in Table 3, results showing the mol action of desired intermediate R as a function of the nversion of B are presented in Fig. 20. As the gas input rate creases, the mols of R produced in the batch decline. The rresponding yield behaviour is shown in Fig. 21.

In this way, the reaction rate constants k_1 to k_5 can be und by matching semi-batch product trajectories to experimental sults. This kinetic information can then be utilised to aluate various forms of reactor design, particularly for plant ale continuous flow reactors.

In carrying out such calculations, it is unwise to make too ny assumptions concerning the ideality of mixing achieved side a reactor. This is especially the case for very large lume reactors used in large tonnage production. A series of ealised backmixed elements can be assembled into a stirred tank nfiguration as shown in Fig. 6 which is intended to allow for e mass transfer, absorption and reaction to be incorporated to reasonably realistic descriptions of the internal gas-liquid ow phenomena. The calculation of the overall performance of ch a reactor, with backmixed zones assembled into loops, given ly the input gas and liquid flow rates, involves the solution of rge sets of algebraic equations. In the example given with re reactions, the solution of a single backmixed zone contains ven equations with seven unknowns. When constructed into two ops with four elements in each loop, the problem comprises rty six simultaneous equations. These can nevertheless be lved for the case where mass transfer resistance is nificant, but there is negligible reaction in the film i.e. e slow reaction regime.

Some results relating to the distribution of gas phase nposition of A (for a feed mol fraction y_{Ai} = 0.21) for the nal reactor in a train of three reactors are presented in Fig. Whilst small scale reactors are almost perfectly mixed, the culations show that scaling up at equal tip speed gives ogressively worse mixing quality as the scale increases. Plant le in this example represents two orders of magnitude increase linear length dimension. The regions of high reactant gas nposition in the gas phase give rise to high dissolved ncentration C_{Ab} in the associated bulk phase, with a reponding high rate of by-product formation locally. A poor ld is therefore obtained at the large scale when compared with ll scale operation.

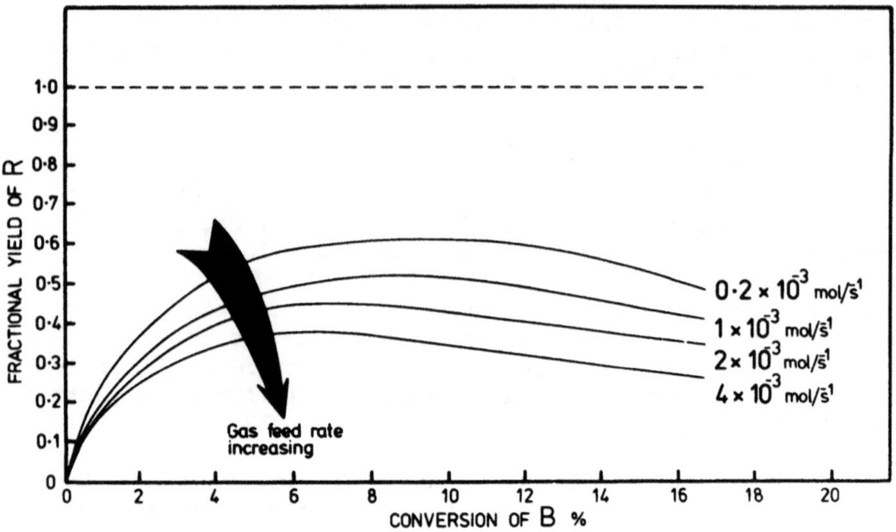

FIG. 21 EFFECT OF GAS FEED RATE ON YIELD OF R FOR SEMI-BATCH OPERATION

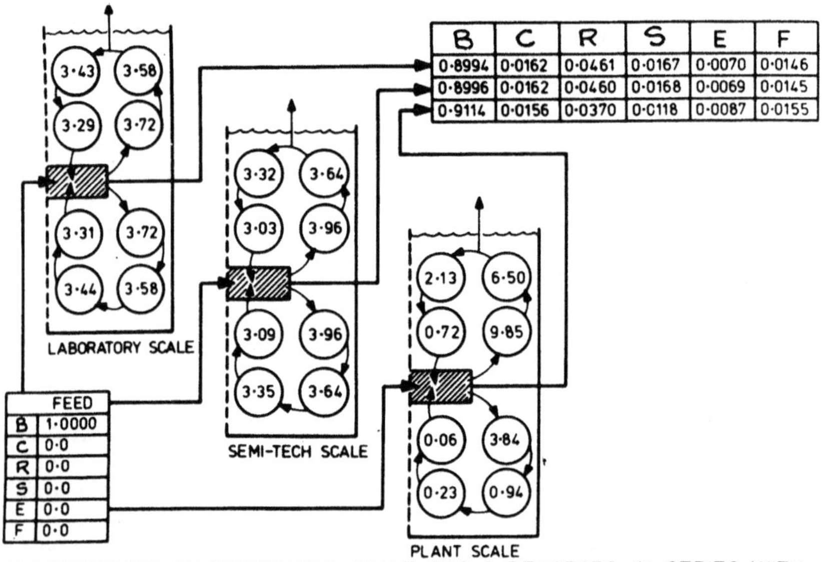

FIG. 22 EFFECT OF INCREASING SCALE FOR 3 REACTORS IN SERIES WITH GAS FED TO THE IMPELLER

It is important to appreciate this effect, since even in the ow reaction regime, as Fig. 5 has shown, bulk phase reaction can rrespond to relatively large rate constants. In Fig. 5 bulk action takes place with a first order rate constant of up to 200 ^{1}. The rates of mixing processes close to the interface are fficiently fast to give negligible depletion by reaction. wever, in a reactor for which overall mixing processes are in e range of seconds, as in the example just described, croscopic depletion of the reagent gas does take place. As the ample showed, for complex reactions the consequences can be ghly significant.

Moreover, this effect has only been considered in respect of e slow reaction regime. The problem of handling say five actions with diffusion and reaction taking place in the mass ansfer film has so far proved too difficult to resolve, though is certainly worthy of continued effort.

SORPTION OF TWO GASES REACTING TOGETHER IN THE LIQUID

The case where two gases A_1 and A_2 dissolve in a liquid phase d react together is a case for which a large number of mmercial examples exist (33). A general stoichiometry can be itten

$$A_1 + \nu_{A_2} A_2 \xrightarrow{k_1} \text{products}$$

Some general qualitative features of such cases are esented in Fig. 23. If the liquid phase reaction is so fast as be considered instantaneous, the transport of the least luble gas to the interface through the gas film will be rate ntrolling. If A_1 is the least soluble gas, profiles will be as Fig. 23(a) as Pangarkar and Sharma (34) have shown. On the ner hand, if the reaction is slower and the solubility of A_2 is ry much greater than for A_1, then A_2 may be readily absorbed to the liquid phase giving a large excess of A_2 with rrespondingly negligible consumption of A_2 in the liquid film. might then be consumed by pseudo first order reaction in either e slow, intermediate or fast reaction regimes.

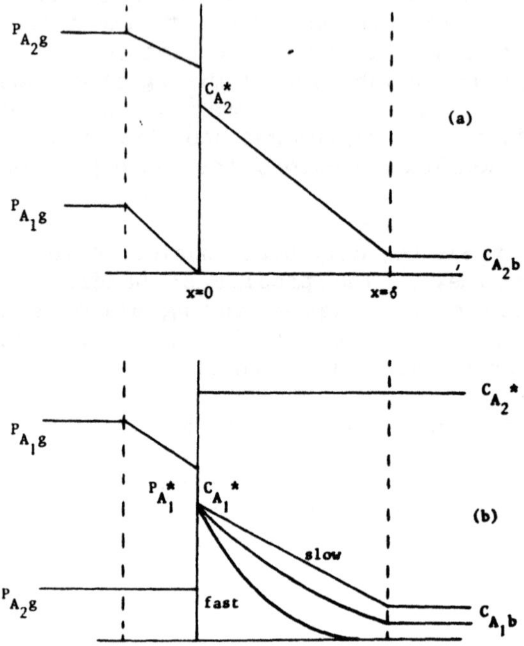

FIG. 23 SIMPLIFIED CASES FOR ABSORPTION AND REACTION OF TWO GASES REACTING TOGETHER [Ref(34)]

In some respects, however, these specific instances are no asier to resolve than the general case. This is because, as ith our previous analyses of a single absorbing gas, the bulk hase boundary conditions can only be determined by a full valuation of the interactions of film and bulk reactions with he overall gas phase transfer and material balance elationships.

The use of profile approximations has been proposed by Sada t al (35), and recently, Zarzycki et al have proposed some nalytical (36) and approximate (37) solutions to the enhancement actors which might assist in the solution of the general case.

The extended case where A_1 and A_2 can react together and one f the gas phase reagents can react with a liquid phase reagent as also considered by Pangarkar and Sharma (34). toichiometrically this case can be represented by

$$A_1 + \nu_{A_2} A_2 \xrightarrow{k_1} P_1$$

$$A_1 + \nu_B B \xrightarrow{k_2} P_2$$

practical example of this is the absorption of CO_2 and NH_3 into lkanolamines, when it is desired to absorb one of the species electively.

As far as selective absorption of A_1 is concerned, this is ound to be greatest when A_1 reacts instantaneously with B and here is a large excess of B. In this case the instantaneous eaction plane is close to the interface, so that the opportunity or A_1 and A_2 to react together is suppressed. From this point f view, co-current flow of gas and liquid would be most ppropriate, since in this contacting mode A_1 and B will deplete ogether and the possible reaction between A_1 and A_2 in the egion between $x = o$ and $x = \delta$ will thereby be reduced to a inimum.

However, as might be expected from Fig. 8, the overall number of reaction regimes is formidably large and a general approach to the solution of the backmixed gas-liquid reactor element remains daunting.

ABSORPTION OF TWO GASES ACCOMPANIED BY SEVERAL REACTIONS

As the number of absorbing components and the number of liquid phase reactions increases the feasibility of a generalised approach seems to recede, and the nature and type of approximation and the strategy for incorporating them into a generalised approach can hardly be discerned. Each example becomes an individual case, and each individual case requires a depth and understanding of the specific complications which can only be realised by those with considerable experience and insight into the special features of the individual case.

This can be illustrated by the work of Cornelisse et al (38) who have considered the selective absorption of H_2S into secondary or primary amines in the presence of CO_2. This is of commercial importance in the selective scrubbing of a number of industrial gas streams. The reactions involved can be written (following previous notations)

$$A_1 + B \underset{k_2}{\overset{k_1}{\rightleftharpoons}} C + D \quad - \text{ infinitely fast}$$

$$A_2 + 2B \underset{k_4}{\overset{k_3}{\rightleftharpoons}} C + E \quad - \text{ slower}$$

A_1 = H_2S

A_2 = CO_2

B = amine

C = ammonium ion

D = bisulphide ion

E = carbonate ion

In this reaction scheme, the formation of bicarbonate is neglected. This is therefore an example of two gas phase reagents reacting with a liquid phase reagent with two reversible reactions incorporating four reactions overall. Reversible reactions are essential to such systems, since it is the reversibility that is the basis of regeneration and reuse of the absorbing liquor. In this example, not only are there four liquid phase reactions taking place, but also the consumption of A_2 (identified with CO_2) is described by a complex rate law given by

$$r(A_2)) = k_3 C_{A_2} C_B - \frac{1}{K_{A_2}} \frac{C_E C_c}{C_B} \qquad (106)$$

ich adds further complexity to the diffusion/reaction enomena.

The authors analysis (38) uses film theory on the gas-side d penetration theory on the liquid side. The penetration eory was adopted as offering a more realistic basis to describe e diffusion and reaction. The set of parabolic partial fferential equations which describe this diffusion and reaction re solved in conjunction with the customary boundary nditions, plus a number of subsidiary relationships which sure electrical neutrality and are compatible with the initial adings of reagent amine. Equilibrium was assumed to be tablished in the bulk, but otherwise the fluxes were not inted to the overall material balances on the gas and liquid ases. Even so, the computation appears quite formidable.

In a very similar example, Cornelissen (39), examined the me selective absorption problem using a tertiary amine for ich case the reaction of CO_2 is slower. The overall scheme is

$$A_1 + B \underset{k_2}{\overset{k_1}{\rightleftharpoons}} C + D \qquad \text{infinitely fast}$$

$$A_2 + B + H_2O \xrightarrow{k_3} C + F \qquad \text{slower}$$

$$A_2 + B + OH^- \xrightarrow{k_4} E + H_2O \qquad \text{slower}$$

In this instance there are now two parallel irreversible actions consuming CO_2. Cornelissen assumed that A_1/B uilibrium is established instantaneously everywhere (in the lm and the bulk), but the reaction speed of the A_2 (carbon oxide) is slower and whilst some reaction took place within the lm, significant amounts of reaction also occurred in the bulk.

The reaction of A_2 was taken to be pseudo first order in the film. The curved concentration profiles within the mass transfer film were approximated by straight lines, such that at either end of zones within the film, the linearised concentration gradients exactly matched the gradients of the curved profile. This approach is equivalent to approximating reaction throughout a zone by the equivalent amount of reaction taking place at a properly positioned plane within the zone. The replacement of the curved profiles by straight lines preserves unchanged the diffusion and reaction balances across the whole mass transfer film. By this ingenious method, the set of ordinary differential equations describing the complexities of diffusion and reaction becomes replaced by a set of algebraic equations.

Unlike the previous analysis (38), the boundary conditions were properly treated as dependent variables found by taking account of film and bulk reaction balanced with the input and output flows and mass transfer fluxes across a tray. The approach (39) involved perfect backmixing of the liquid, but with provision of the possibility of axial mixing in the gas phase being represented by staging of backmixed zones. In practise though, the authors claimed that a single gas zone was sufficiently accurate.

For an individual tray in a tray column, the problem resolves to 27 non-linear algebraic equations in 27 unknowns. Solution of each tray has to be incorporated into a separate algorithm for the iterative tray-to-tray calculations. The performance of a 0.11m valve tray column was successfully simulated, and the method has been adopted as a basis for the design of commercial scale absorbers up to 8.5m in diameter used for the selective absorption of H_2S in the presence of CO_2.

The possibility of significant errors developing due to the use of approximating straight line profiles was not considered. However, the method offers a promising basis for avoiding the computational difficulties encountered when the diffusion/reaction equations are embedded into the overall mass transfer and flow material balances.

A general adoption of the technique requires a framework for realistic replacement of the diffusion reaction equations given only the diffusion and reaction kinetic parameters. As with many of the cases described so far, there is still a great deal to be done before gas-liquid reactors can be designed or analysed from first principles using only physico-chemical data and knowledge of the input flow rates and compositions.

)MENCLATURE

	gas-liquid interfacial area
	interface concentration of component I
b	bulk concentration of component I
	diffusion coefficient of component I
	absorption enhancement factor for component I
	absorption factor for absorption of I
i	asymptote of absorption enhancement factor for I
	total molar feed rate of gas
	total molar efflux rate of gas
	Henry's Law coefficient for solubility of I
	reaction rate constant for ith reaction
	liquid mass transfer coefficient for I
	equilibrium constant
I	overall mass transfer coefficient for I
I	gas phase mass transfer coefficient for I
m)	diffusion/reaction factor for (pseudo) mth order reaction
g)	diffusion/reaction factor for general reaction kinetics
(m)	diffusion/reaction factor for reversible mth order reaction
(m)	diffusion/reaction factor for mth-nth order reaction
n,n)	diffusion/reaction factor modified according to Eqn (56)

N_I	molar flux of I
P_{Ig}	partial pressure of I in bulk gas phase
P_I^*	partial pressure of I at gas-liquid interface
Q	volumetric flow rate of liquid phase
V_L	volume of liquid phase
X_I	conversion of component I
x	position in mass transfer film
y_{Ii}	inlet mol fraction of I in gas phase
y_{Io}	outlet mol fraction of I in gas phase
β	volume per unit area of bulk/volume per unit area of film
δ	mass transfer film thickness
η	effectiveness factor for a gas-liquid reactor
τ	liquid phase residence time
ψ	instantaneous yield function
ϕ	overall yield function

subscripts

I	is a general subscript referring to components A, B, C, D,... etc
A	refers to gas-phase reactants, subscripted 1, 2, ... for more than one gas-phase reagent
B	refers to liquid-phase reactant, subscripted 1, 2 .... for more than one liquid phase reagent
P	usually refers to products of reaction
R	usually refers to a desired intermediate product.

EFERENCES

1) Danckwerts, P.V., "Gas-Liquid Reactions", McGraw-Hill, (1970).

2) Juvekar, V.A., and Sharma, M.M., Tran.I.Chem.Engrs., 55, 77, (1977).

3) Barona, N., Hydrocarbon Processing, 58, 179, (1979).

4) Charpentier, J.C., A.C.S. Symposium Series, 72, 223, (1978).

5) Shah, Y.T., and Sharma, M.M., Trans.I.Chem.Engrs, 54, 1, (1976).

6) Shah, Y.T., "Gas-Liquid-Solid Reactor Design", McGraw-Hill, (1979).

7) Spielman, M., A.I.Ch.E.Jl., 10, 496, (1964).

8) Ding, S.Y.J., Sharma, S., and Luss, D., Ind.Eng.Chem.Fund., 13, 76, (1974).

9) Hashimoto, K., Teramoto, M., and Nagata, S., Jl.Chem.Eng.Japan, 4, 150, (1971).

0) Kulkarni, B.D., and Doraiswamy, L.K., A.I.Ch.E.Jl., 21, 501, (1975).

1) Mavros, P.P., M.Sc. Dissertation, University of Manchester, (1978).

2) Peterson, E.E., "Chemical Reaction Analysis", Prentice-Hall, (1965).

3) Danckwerts, P.V. and Kennedy, A.M., Trans.I.Chem.Engrs., 32, 549, (1954).

4) Steeman, J.W.M., Kaarsemaker, S., and Hoftijzer, P.J., Chem.Eng.Sci., 14, 139, (1961).

5) Carberry, J.J., Chem.Eng.Sci., 17, 675, (1962).

6) Nagel, O., Hegner, B, and Kurten, H., Chem.Ing.Tech., 50, 934, (1978).

(17) Van Krevelen, D.W., and Hoftijzer, P.J., Rec.Trav.Chim., 67, 563 (1948).

(18) Hoffman, L.A., Sharma, S., and Luss, D., A.I.Ch.E.Jl., 21, 318, (1975).

(19) Sim, M.T., and Mann, R., Chem.Eng.Sci., 30, 1215, (1975).

(20) Joosten, G.E.H., Maatman, H., Prins, W., and Stamhuis, E.J., Chem.Eng.Sci., 35, 223, (1980).

(21) Onda, K., Sada, E., Kobayashi, T., and Fujine, M., Chem.Eng.Sci., 25, 753, (1970).

(22) Hikita, H., and Asai, S., Kagaku Kogaku, 27, 823, (1963).

(23) Onda, K., Sada, E., Kobayashi, T., and Fujine, M., Chem.Eng.Sci., 25, 1023, (1970).

(24) Van de Vusse, J.G., Chem.Eng.Sci., 21, 631, (1966).

(25) Hashimoto, K., Teramoto, M., Nagayasu, T, and Nagata, S., Jl.Chem.Eng.Japan, 1, 132, (1968).

(26) Teramoto, M., Nagayasu, T., Matsui, T., Hashimoto, K., and Nagata, S., Jl.Chem.Eng.Japan, 2, 186, (1969).

(27) Teramoto, M., Hashimoto, K., and Nagata, S., Jl.Chem.Eng.Japan, 6, 522, (1973).

(28) Sharma, S., Hoffman, L.A. and Luss, D., A.I.Ch.E.Jl., 22, 324, (1976).

(29) Schmitz, R.A., and Amundson, N.R., Chem.Eng.Sci., 18, 265, (1963).

(30) Ho, S-P., and Lee, S-T., Chem.Eng.Sci., 35, 1139, (1980).

(31) Huang, D.T-J., Carberry, J.J., and Varma, A., A.I.Ch.E.Jl., 26, 832, (1980).

(32) Mann, R., Middleton, J.C., and Parker, I.B., Proc.2nd.Europ.Conf.Mixing, F3, (1977).

(33) Ramachandran, P.A., and Sharma, M.M., Trans.I.Chem.Engrs., 49, 253, (1971).

(34) Pangarkar, V.G., and Sharma, M.M., Chem.Eng.Sci., 29,

2297, (1974).

5) Sada, E., Kumazawa, H., and Butt, M.A., Can.Jl.Chem.Eng., 54, 97, (1976).

6) Zarzycki, R., Ledakowicz, S., and Starzak, M., Chem.Eng.Sci., 36, 105, (1981).

7) Zarzycki, R., Ledakowicz, S., and Starzak, M., Chem.Eng.Sci., 36, 113, (1981).

8) Cornelisse, R., Beenackers, A.A.C.M., Van Beckum, F.P.H., and Van Swaaij, W.P.M., Chem.Eng.Sci., 35, 1245, (1980).

9) Cornelissen, A.E., Trans.I.Chem.Engrs, 58, 243, (1980).

PROCESS DESIGN ASPECTS OF GAS ABSORBERS

Erdogan Alper

Department of Chemical Engineering
Ankara University, Besevler, Ankara, Turkey

1. INTRODUCTION

Equipment which is used in contacting a gas with a reactive liquid can be gas absorber or a gas-liquid reactor. This terminology itself shows the interdisciplinary nature of the process which involves both chemical (i.e. reaction kinetics) and physical (molecular diffusion, fluid mechanics etc.) phenomena. Thus the subject does not fall entirely within the province of either the chemist or the conventional engineer. The classical literature on this area (Astarita (1), Danckwerts (2), Sherwood et al. (3) etc.) has mainly dealt with gas absorption, in which the reaction is applied merely to enhance the rate of mass transfer. In such cases, there is also always a physical gas absorption process to refer to and the reactions are usually "fast". On the other hand, many industrial reactions in organic chemistry such as oxidations and chlorinations (4), are relatively slow and the main emphasis is the conversion of the liquid phase product. Therefore, two approaches may be used to characterize the interaction of mass transfer and chemical reaction between components of a gas and a liquid, one expressing the enhancement effect of a relatively fast reaction on the physical mass transfer leading to the classical concept of the "enhancement factor" (1-3) and a second, a relatively new one, expressing of slowing down of the already slow reaction rate by mass transfer and leading to the "utilization factor" (5,6). Consequently equipment may respectively be called a gas absorber or a gas-liquid reactor. Although, the treatment in this review has a general approach, the emphasis is on the enhancement of gas absorption rate; hence it deals with the process design aspects of gas absorbers in which relatively fast reactions are occuring.

The design of gas absorbers when only physical absorption is involved, is relatively a simple matter (provided that the necessary data are available) and need not be described here. Apart from the hydrodynamical data such as flooding, only values of k_Ga and k_La under the prescribed conditions and the parameters of a proper two-phase contactor model are required. The latter will be discussed in some detail in Section 3; in many cases even the simple generalised design procedures, such as those reviewed by Pavlica and Olson (7) are not necessary and ideal flow patterns - for instance, plug flow - may be adequately used.

On the other hand, in the presence of a reaction, the rational process design is usually a complicated matter. The main stages of the design procedure is as shown schematically in Figure 1. First, the specific design problem should be defined. This leads to a number of independent parameters, such as flow-rates of each phase, the choice of equipment and its details (for instance, packing material) temperature and pressure. The choice of such parameters for an indivual task often involves very delicate economic balances and is, to a large extent, well beyond the scope of this review. Then a number of other parameters, such as mass transfer coefficients, gas-liquid interfacial area and liquid hold-up, depend on these independent parameters and should be known or estimated prior to any rational design approach. Such data are often available for some standard systems and conditions and the estimation for precribed conditions, often causes a major problem.

The second stage, which involves the determination and the estimation of "process specific data", such as solubilities and diffusivities, reaction kinetics and rate constants, is highly specific and has to be obtained for each system at the prescribed conditions. Some of these data, such as relevant kinetics and the equilibrium constants etc., may in fact be measured but this is as laborious task. Even more difficult is the estimation of quantities which cannot be measured directly, particularly the solubility and diffusivity of the dissolved gas in a solution with which it reacts.

The final stage of the design consists of both "microscopic" scale (or local) modeling (or absorption-reaction model) and "macroscopic" scale (or integral) modeling (or two-phase contactor model) in order to compute the capacity of the equipment from first principles -for instance, in the case of a packed column, the required height). Under certain circumstances, it is possible to use laboratory models instead of theoretical modeling either at microscopic or both microscopic and macroscopic scales. By doing so, it is often possible to avoid most or all of the Stage 2, which is of course highly specific and requires normally a laborious task. However, it may be pointed out that all the information involved in Stage 1, -with the exception of separate

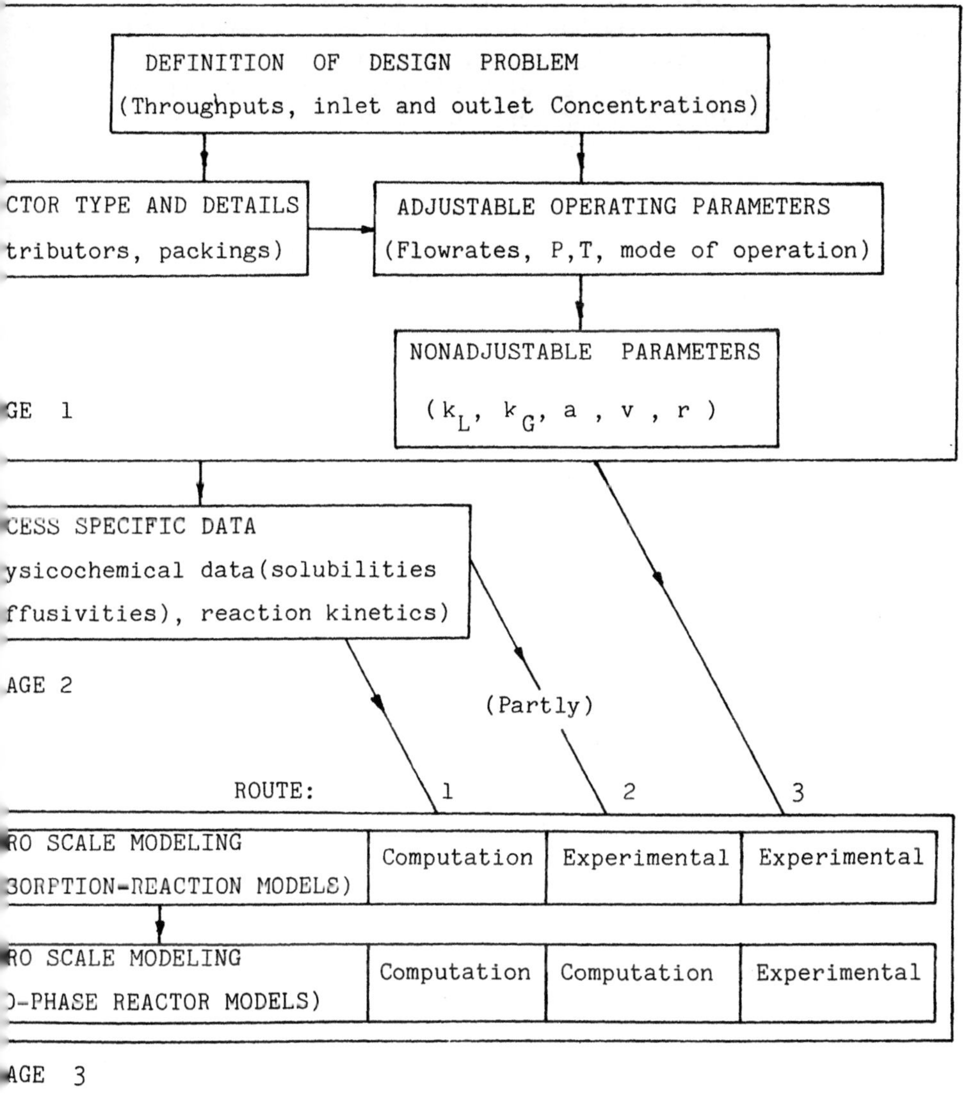

gure 1. Main stages of gas absorber (gas-liquid actor) design.

values of k_L and a rather than k_La all these data are also required for rational design of physical absorbers -, should be available no matter whether theoretical or laboratory models are used. In this review, theoretical modeling both at microscopic scales will only be very briefly discussed and the main emphasis will be given to the laboratory models and their use with special reference to packed columns.

2. MICROSCOPIC SCALE MODELING (ABSORPTION-REACTION MODELS)

Microscopic modeling considers a small but statistically representative volume element of the absorber (or reactor), that is, a "point" in the equipment. Recently, Thoenes (8) grouped such considerations as "volume element modeling". It is necessary to make energy and component mass balances in the reactive liquid-phase (it is assumed that no reaction takes place in the gas phase). Fortunately for most systems, the isothermal approximation is often justified. Thus, the components mass balance yields:

$$(D_i\nabla - \vec{v})\nabla A_i = \frac{\partial A_i}{\partial t} + z_i\ r_i \qquad (1)$$

Solution of this equation requires not only various physicochemical and reaction kinetics data, but also a detailed knowledge of the fluid mechanics near the interface which is however not available in any but artificially simplified agitated system (see, for instance, reference (9) for a detailed review of interphase mass transfer models and description of interfaces). However, the concentration gradient ∇A_i and the velocity vector $\vec{v}$ are approximately perpendicular to each other, hence the scalar product of them is always negligible. Furthermore, diffusion is usually unidimensional. Therefore Eqn (1) reduces to:

$$D_i \frac{\partial^2 A_i}{\partial x^2} = \frac{\partial A_i}{\partial t} + z_i\ r_i \qquad (2)$$

It is now justifiable to solve this equation using the boundary conditions of the simplified physical absorption models (1,2). Solution of Eqn (2) then enables us to calculate the absorption rate at a particular "point" in the absorber. These results are usually expressed in terms of an enhancement factor, i.e. a factor by which the rate of absorption is increased by the chemical reaction. It is well known that this enhancement factor differs little in value whether film or the Higbie or Danckwerts surface renewal models are used as the basis of calculation. Figure (2) shows the typical representation of the effect of chemical reaction for a second order ($r=k_2A\ B$) irreversible reaction. With the exception of Region IV, all regions are amenable to analytical solutions. In fact, the enhancement factor predic-

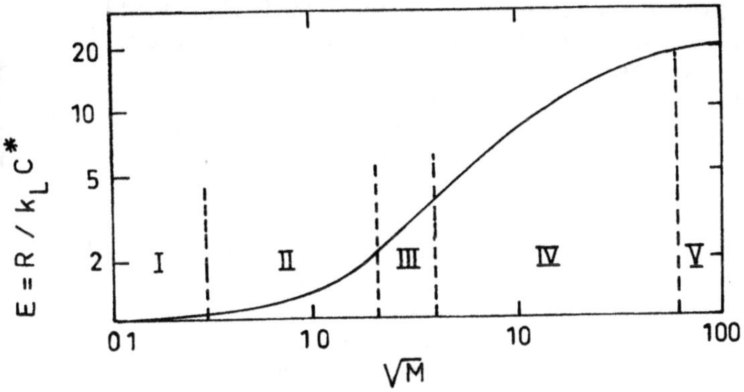

Figure 2. Effect of second order reaction on absorption rate

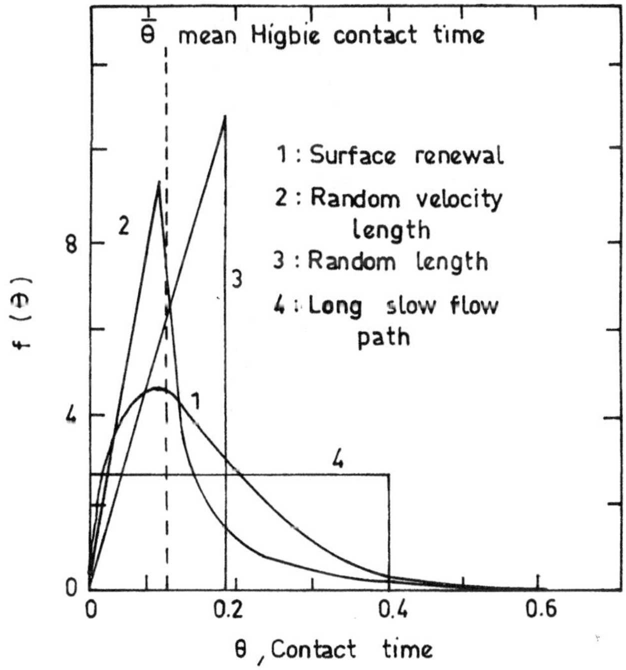

Figure3. Different contact-time distributions compared at the same overall physical absorption rate corresponding to a mean Higbie contact time of 0.1 s (11).

tions are identical for Regions I and III for all three models. In Region V, predictions based upon the three models are similar -indeed they are identical if the diffusivities are equal. In Region II, the maximum difference is often much less than 5%. Finally, the the transition Region IV, no analytical solution is possible. However, numerical solutions indicate little dependency of the enhancement factor E on the chosen model (10). Further, in this regime, the effect of contact time distributions has been examined in some detail by many workers (11-14). For instance, for arbitrarily chosen models, including those of Higbie and Danckwerts, Porter (11) derived the contact time distribution functions which are shown in Figure (3). These distribution functions (11-15) are based on considerations which bear some relation to possible flow mechanisms in packed columns and they all lead to numerically almost the same predictions (2). Similar types of conclusions may also be drawn for other kinetics such as (m,n) th order reactions (14); thus it appears that the effect of chemical reaction on the relative increase in the rate of mass transfer does not depend strongly on the theoretical models (thus the actual flow pattern near the interface). It means that it is not of crucial importance which sort of a theoretical model we use if we want to predict the effect of chemical reaction on gas absorption. Hence, for instance, explicit approximate equations which agree with predictions from one of the theoretical models may be used for design purposes (16). One of the earliest and the most accurate equation for second order irreversible reaction is (16,17):

$$E = \left(1 + \frac{\sqrt{M}}{\alpha}\right) \left(1 - \exp(-0.65\sqrt{M}\,\alpha \right) \tag{3.a}$$

$$\alpha = \sqrt{M/\ (E_i - 1)} + \exp \left(0.68/\sqrt{M} - 0.45\sqrt{M/(E_i - 1)}\right) \tag{3.b}$$

Recently other simple explicit equations were also proposed (18, 22) and the accuracy was tested by Wellek et al. (11). In general the overall accuracy is reduced with simplicity and Wellek et al recommend the following equation for general use:

$$(E - 1)^{-1.35} = (E_i - 1)^{-1.35} + (E_1 - 1)^{-1.35} \tag{4.a}$$

where

$$E_1 = \sqrt{M} / \tanh\sqrt{M} \tag{4.b}$$

A second but more important conclusion in the above considerations is that it suggests that it does not matter much what sort of a laboratory model (as opposed to a theoretical model) we use if we want to experimentally simulate an absorber -provided it has the right value of k_L, it is of no importance how it is generated. Further consideration of this aspect is given in

OPERATION	APPROXIMATE FLOW PATTERN Gas Phase	Liquid Phase	EXAMPLES
Semi-batch	Plug flow	Completely mixed	Gas sparged reactor
	Completely backmixed	Completely backmixed	Mechanically agitated
Continuous, Cocurrent, Counter-current	Plug flow	Plug flow	Packed column,packed bubble column
	Plug flow	Completely backmixed (?)	Bubble column
Continuous, Cocurrent, Countereurrent Cross-flow	Completely backmixed (?)	Plug flow	Spray column
Continuous, (stagewise) counter-current, Cross-flow	Plug flow	Completely backmixed (?)	Each plate of a plate column Sectionalized bubble column
Continuous, Cocurrent, Counter-current, Cross flow	(Not studied in sufficient detail)		Miscellaneous (Wetted-wall column, venturiscrubber, turbulent bed contactor etc.

BLE 1. DIVERSE GAS ABSORBERS (GAS-LIQUID REACTORS)

Section 5.

The above treatment assumes a known value of interfacial concentration of the dissolved gas, which, of course, depends on the gas phase resistance, if any. As in the case of liquid-side phenomena, the exact nature of processes on the gas-side is also not clearly known. However, the situation is simpler as there is usually no reaction to be considered and it is usual to employ the film model approach. The addition of resistances was studied by many workers (23-27) and mathematical investigations (25,26) showed that it does not matter which liquid-side model one adopts.

3. MACROSCOPIC SCALE MODELING (TWO-PHASE CONTACTOR MODELS)

The above treatment considered only the modeling of the local process. The design of an absorber/gas-liquid reactor requires also an examination of global issues. In the terminology of Thoenes (8) this covers both "partial and overall reactor" models. Table 1 shows some of the diverse gas absorbers/reactors. Each of the two phases of gas and liquid may be either in plug flow or completely backmixed as the two extreme cases of macromixing. In plug flow, longitudinal mixing is nonexistent; but due to complete radial mixing, all fluid elements within the system have identical residence time. In a completely mixed system, the residence time distribution of fluid follows an exponential decay, with the exit stream composition being identical to that within the system. It is a well known fact, that, in general, the flow of one or both of the phases may deviate considerably from the above extreme cases and the backmixing lies in between. These deviations may be the combined results a number of different phenomena; these may be nonuniform velocity profiles, short circuiting, bypassing and channeling, velocity fluctuations due to molecular and turbulent diffusion, effects of contactor shape and internals, backflow of fluids due to velocity differences between phases and recycling due to agitation. Hartland and Mecklenburgh (28) and Mecklenburgh (29) discussed in some detail, these so called nonideal flow patterns and unlike axial mixing phenomena such as channeling,recirculation, wall flow etc. cannot be considered as random processes. Figure (4) shows possible transverse -that is, the direction perpendicular to flow-nonuniformity of the velocity distribution in countercurrent absorbers (for example, a packed column) (30). Figure (5) shows nonuniformities in plates of a plate-column absorber which employs cross-current flow (30).

One of the most simple models is known as the "axial dispersion model". Here a one dimensional Fick's law type of diffusion equation is accepted and the constant of proportionality is commonly termed the axial dispersion coefficient. In the model, complete mixing in the radial direction is assumed. Although the

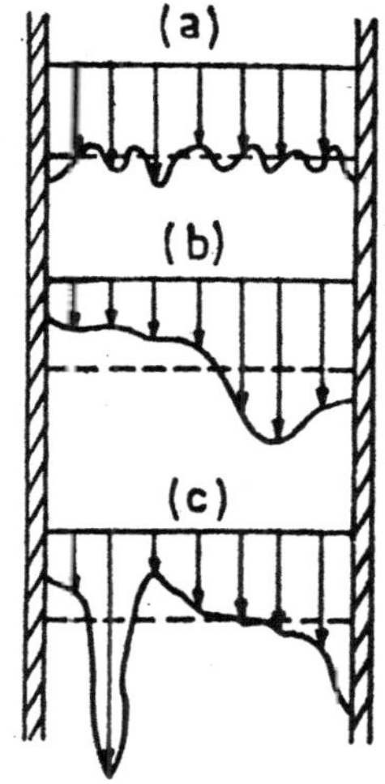

Figure 4. Transverse non-uniformity of the velocity distribution :(a) random non-uniformity, (b) considerable transverse non-uniformity, (c) channeling.

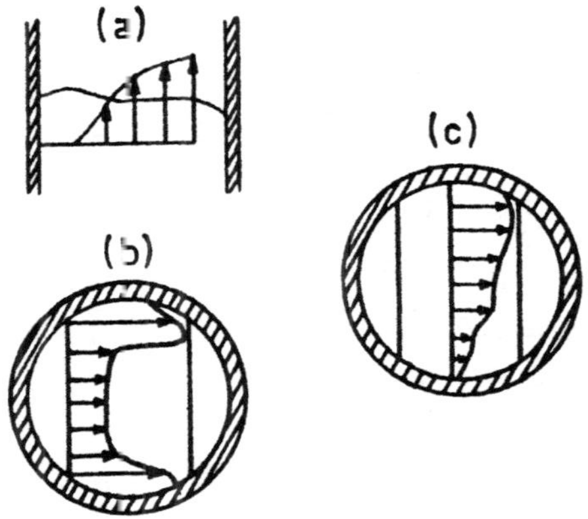

Figure 5. Non-uniformity of plate operation with cross-current flow: (a) longitudinal non-uniformity in gas flow, (b) transverse non-uniformity in liquid flow, (c) channeling (by passing).

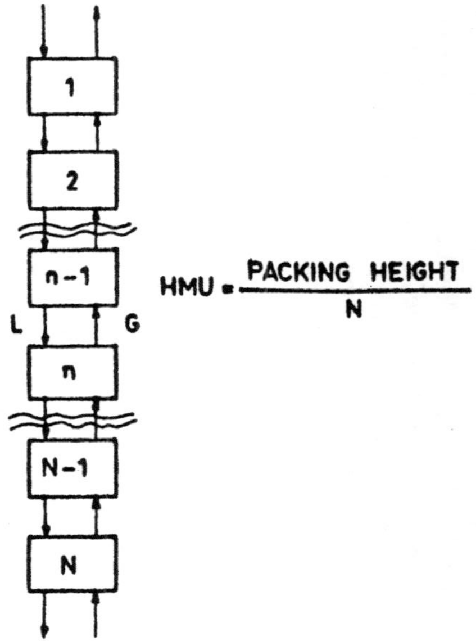

Figure 6. Schematic representation of series of stirred tanks model (cell model).

assumption that all mixing processes follow Fick's law type of diffusion equation is a gross oversimplification, it is widely used as it involves only one parameter, the dispersion coefficient (E_z) expressed as the Peclet number ($Pe=UL_c/E_z$) in dimensionless form where U is the interstitial velocity. In bubble or spray columns, L_C could be either the diameter of the column or the diameter of the bubble. For packed columns, L_C is usually the characteristic diameter of the packing. Under this situation, Pe is often denoted as the Bodenstein number. The value of the Pe (or Bodenstein) number denotes the degree of backmixing. For complete mixing Pe=0 and for $Pe \to \infty$ the plug flow prevails. Sherwood et al. (3) have reviewed and outlined the use of the "axial dispersion model" in two-phase contactor design.

Another one parameter model is the series of stirred tank model (often referred to as the cell model). In this model, the equipment is represented by a series of perfectly mixed stages (31-34) and the number of cell is a measure of the degree of backmixing.

Considerable effort has also been devoted to many multiparameter models. One of the simplest is the "two zone model" which is largely applied to packed columns (35). The underlying idea for such a model is that only a fraction of liquid flows through the packing, while at each height there is a stagnant zone in which the liquid is well mixed and which exchanges mass with the flowing fraction. The general problem of backmixing in gas-liquid systems for both simple and complex models of gas-liquid reactors were recently reviewed by Shah, Stiegel and Sharma (36) and Calo (37).

4. PROCESS DESIGN CALCULATIONS FOR GAS ABSORBERS

Design procedures of contactors for simultaneous gas absorption with chemical reaction require all the data -such as floodin hold-up, k_La and k_Ga and axial dispersion coefficients whenever they are relevant-which are normally required for the design of physical gas absorbers too. Further to these data, separate values of k_L and a are also required in order to estimate the enhancement factor using one of the absorption-reaction models.

The quantities k_Ga and k_La can easily be measured but special care must be paid to the validity of the macroscopic model employed. The value of specific interfacial area (a) may, for example, be obtained by using so called chemical methods (38-41). Other required data, such as liquid and/or gas hold-ups etc., present few difficulties. However all these data must be obtained from large scale equipment which are representative of industrial absorbers. The work is thus expensive and laborious but is worth doing once and for all to establish the essential characteristics of the absorber under consideration, without which no rational de

ign as outlined in Figure (1) can be undertaken. In these re-
pects, special attention has to be devoted to non-aqueous solu-
ions of industrial importance which have been studied only
rarely (42-48).

There is also a need for caution about the interpretation
f such measurements as the precise fulfillment of the required
onditions is not usually easy. For instance, chemical methods of
easuring interfacial area (a) purport to measure the area which
s effective under the chosen conditions. There is however good
vidence that the effective interfacial area may well depend on
he type of reaction proceeding in the liquid as illustrated for
acked columns (49).

Finally, the theoretical predictions require data which are
xtremely specific -that is, Stage 2 of Figure 1. It is essential
hat such data should be known both quantitatively and accurately.
btaining them is not only laborious, in addition some of them
an only be estimated in any case. However, if all these data are
vailable (i.e. Stage 1 and 2 of Figure 1), we can proceed and
alculate the capacity by coupling microscopic and macroscopic
cale models. For the former, any one of the well known models
s sufficient; particularly such explicit expressions as Eqns. (3)
nd (4) are most suitable for design purposes (16). For macrosco-
ic scale modeling, it seems, in many cases, ideal flow patterns
uffice (50). In any case, so far only the axial dispersion model
or the cell model) has been used for improvement and the ex-
erience shows that any model containing more than two parameters
ill not find extensive use for design purposes.

Recently, Juvekar and Sharma (50) considered the reaction
-zb → products ($r=k_{mn}\ A^m\ B^n$) and many cases of different con-
itions -but all assuming one of the two extreme cases of plug
ow and/or complete mixing - and derived analytical equations
hich can readily be used for design purposes. Table 2 shows the
ases examined by them and the underlying assumptions and the
asis of derivations are outlined below.

.1. Packed columns

Many workers (51-56) have considered the design of counter-
urrent packed columns. The equipment can effectively be used in
o-current operation since there is no disadvantage due to dri-
ing force as in the case of physical absorbers (57). Figure 7
hows schematically such a column and Table 3 gives the simpli-
ying assumptions of Juvekar and Sharma (50).

A material balance of the solute over a differential height
h of the column can now be written as:

$$-G'\ dy = a\ (\ He\ )P\ [Y/(1+Y)]\ R\ S\ dh \qquad (5)$$

$$= (\ L/z\ )\ dB^o$$

CASE NO.	REACTION ORDER m	n	CONDITIONS TO BE SATISFIED	ABSORPTION RATE R_a, kmol/m^3s
1	1	1	$k_L a \gg v\, k_2\, B^\circ$	$k_2\, A^*\, B^\circ$
2	2	1	$k_L a \gg v\, k_3\, A^*\, B^\circ$	$k_3 (A^*)^2\, B$
3	1	1	$k_L a \approx v\, k_2\, B^\circ$	$k_L\, a (A^* - B^\circ)$
4	m	n	$k_L a \ll v\, k\, (A^*)^{m-1} (B^\circ)^n$	$k_L\, a\, A^*$
5	1	1	$E_i \gg \sqrt{M} \gg 1$ $k_g \gg He\, k_L \sqrt{M}$	$a\, A^* \sqrt{D_A k_2 B^\circ}$
6	0	1		$a \sqrt{2 D_A k_1 A^* B^\circ}$
7	1	0		$a\, A^* \sqrt{D_A k_1}$
8	0	2		$a\, B^\circ \sqrt{2 D_A k_2 A^*}$
9	1	2		$a\, A^*\, B^\circ \sqrt{D_A k_1}$
10	1	1	$E_i \gg \sqrt{M} \approx 1$ $k_G \gg He\, k_L \sqrt{M+1}$	$a\, A^* \sqrt{D_A k_2 B + k_L^2}$
11	1	2		$a\, A^* \sqrt{D_A k_3 (B^\circ)^2 + k2}$
12	1	1	$E_i \gg \sqrt{M} \gg 1$ $k_G \gg He\, k_L \sqrt{M}$	$\dfrac{k_G a A^* \sqrt{D_A k_2 B^\circ}}{k_G + He \sqrt{D_A k_2 B^\circ}}$
13	1	1	$E_i \approx \sqrt{M} \gg 1$	$\dfrac{k_L a A^*}{2} \left\{ -\dfrac{M}{E_i} + \sqrt{\left(\dfrac{M}{E_i}\right)^2 + 4M} \right\}$
14	m	n	$\sqrt{M} \gg E_i \gg 1$ $k_G \approx He\, k_L\, E_i$	$\dfrac{A^* k_G k_L E_i}{k_G + He\, k_L E_i}$

TABLE 2. ANALYTICAL EXPRESSIONS FOR THE RATE OF ABSORPTION OF A INTO A SOLUTION CONTAINIG REACTANT B (Reaction is irreversible (m,n)th order; $E_i = 1 + \sqrt{[D_B/D_A]}\,[B^0/zA^*]$ $M = \dfrac{\sqrt{2/(m+1)\, D_A k_{mn} [A^*]^{m-1} [B^0]^n}}{k_L}$

where G' is the molar flowrate of inerts, Y is the mole ratio of solute A to inert gas, He is the Henry's law constant, P. is the total pressure, R is the specific absorption rate and S is the cross sectional area. Expressions for R can be obtained from microscopic modeling (that is, absorption-reaction models).

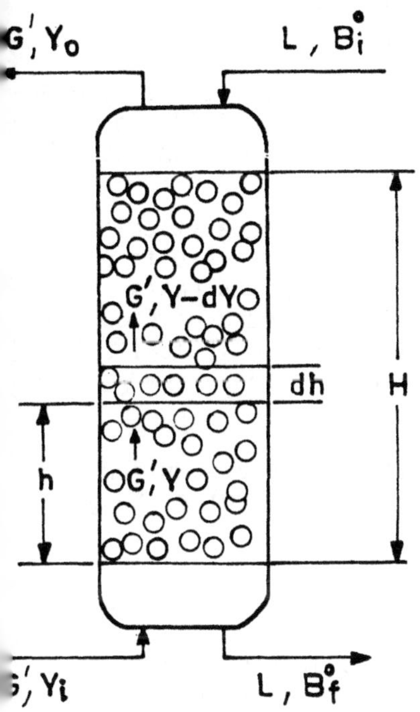

Figure 7. Schematic representation of a packed-column absorber

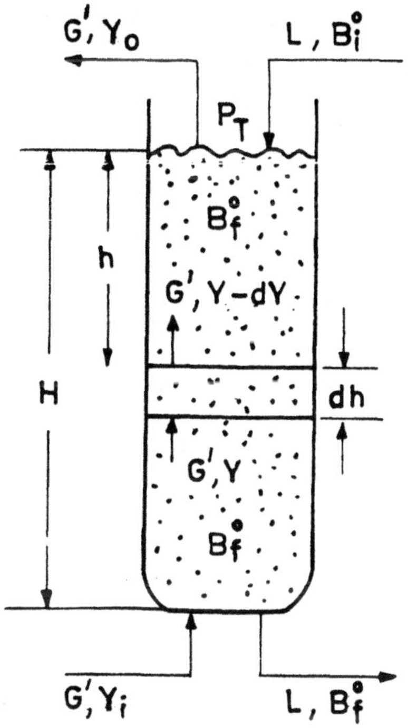

Figure 8. Schematic representation of a bubble column

TABLE 3. ASSUMPTIONS FOR PACKED-COLUMN ABSORBER DESIGN (50)

1. Plug flow of both gas and liquid phases
2. Isothermal operation ($T = \text{constant}$)
3. Constant total pressure in the column
4. Constant physical properties of the liquid
5. Negligible gas-side resistance; in some cases it can however be accounted for

TABLE 4. ASSUMPTIONS FOR BUBBLE COLUMN DESIGN (50)

1. Plug flow of gas and completely backmixed liquid
2. Linear pressure variation with height:

$$P = P_T\,(1 + \Psi h) \quad \text{where} \quad \Psi = \frac{1}{P_T}\left(\frac{dP}{dh}\right)$$

3. Constant k_L, k_G and a

For instance, for the case 14 of Table 2, the integration yields:

$$H = \frac{1}{\beta}\left[\alpha\,\delta \ln\left(\frac{\gamma - Y_o}{\gamma - Y_o}\right) + \ln\left(\frac{Y_i}{Y_o}\right) + (Y_i - Y_o)\right] \qquad (6)$$

where

$$\alpha = \frac{L}{zG}\ ;\ \beta = k_G a\ P\ S/G'\ ;\ = \frac{zk_G\ P}{k_L}\sqrt{D_A/D_B} \qquad (7)$$

Similar equations, often with more complicated terms, are also derived by Juvekar and Sharam (50) for cases 3,5,10,11,12 and 14 of Table 2 for both co-current and countercurrent operations as well as simplifications for lean gases.

Juvekar and Sharma (50) have compared their analytical solutions with either the actual packing height or the results of numerical solutions of other workers for the limited cases where all the required information -that is, Stage 1 and 2 of Figure 1- were available. These results indicate that the analytical solutions are sufficiently accurate for design purposes.

4.2. Bubble Column Contactors

Figure 8 shows the pertinent details and the assumptions are given in Table 4. A material balance over a differential height gives:

$$\frac{dY}{dh} = R\ a\left[S/G'\right] \qquad (8)$$

Inserting the expression for R for a particular case and integrating gives the desired total dispersion height. For instance, for case 5 of Table 2, we obtain:

$$H = \frac{-1}{\Psi} + \frac{1}{\Psi}\left[1 + \frac{2\Psi}{\beta}\{(Y_i - Y_o) + \ln(Y_i/Y_o)\}\right]^{1/2} \qquad (9)$$

where

$$\beta = \frac{a\ H\ P\ S\sqrt{D_A\ k_2\ B^o}}{G'}$$

and Ψ is as defined in Table 4.

Juvekar and Sharma (50) only compared their results with those of Mashelkar (58) but found good agreement. However, unlike packed column, it is exceedingly difficult to accept the ideal flow pattern of Table 4 for many bubble column applications. Indeed, the backmixing in the liquid phase is unavoidable and it usually reduces the contactor performance. Many authors (7,59-64) have employed the axial dispersion model for the liquid phase -in some cases the gas phase too. One of the most sophisticated design procedure was described by Deckwer (63,64) who showed that

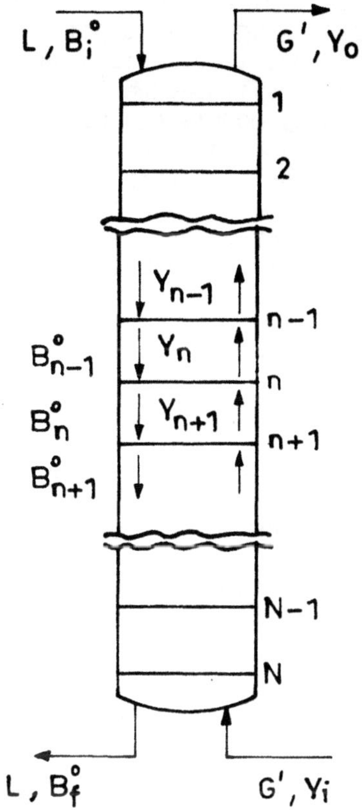

Figure 9.Schematic representation of a plate column absorber

TABLE 5.ASSUMPTIONS FOR PACKED BUBBLE COLUMN DESIGN(50)

1.Plug flow of both gas and liquid phases
2.Isothermal operation (T- constant)
3.Linear pressure variation with height
4.Constank k_L, k_G and a

TABLE 6.ASSUMPTIONS FOR PLATE-COLUMN ABSORBER DESIGN(50)

1.Lean gas with respect to solute
2.Plug flow of gas and completely backmixed liquid on each plate
3.The solute concentration on a plate is the arithmetic mean of the concentration of inlet and outlet gases
3.Negligible pressure drop;or a linear variation of pressure with plate number n :

$$P = P_T\ (1+n\theta) \quad \text{where} \quad \theta = \frac{1}{P_T}\left(\frac{dP}{dn}\right)$$

---

axial variations in pressure and gas flowrate should be also accounted for. Even under isobaric conditions, the gas flow decreases owing to absorption which in turn leads to increased gas residence time; this higher conversions are obtained. Based on his analysis, Deckwer concluded simple isobaric, constant gas velocity models can be used without serious errors if the column operates at elevated pressures, say 20 atm, and if the gas shrinkage by absorption is small. He also pointed out that in large diameter bubble columns (diameter > 0.5 m), gas phase dispersion may be very important. In a subsequent study, Deckwer applied his model successfully to the case of absorption and reaction of isobutene in sulphuric acid.

4.3. Packed Bubble Columns

By noting the observation (42) that the addition of packings in a bubble column considerably reduces the backmixing in the liquid phase, Juvekar and Sharma (50) assumed plug-flow in both phases. Then, for short columns or columns operated at high pressures, the height can be obtained from the expressions for packed columns. Otherwise, a linear pressure variation with the height may be assumed and Juvekar and Sharma (50) presented analytical expressions for the cases 3,5,8,10,11 and 12 of Table 2 for both co-current and countercurrent operations.

4.4. Plate Column Absorbers

Figure 9 shows the plate column schematically and Table 6 shows the simplifying assumptions of Juvekar and Sharma (50). The material balance for the solute over the n th plate can be written as:

$$G' \left[Y_{n+1}-Y_n\right] = R\left[B_n^o,\ (Y_n+Y_{n+1})/2\right] a\ V_d \qquad (11)$$

$$= (L/z)\left[B_{n-1}^o - B_n^o\right]$$

where the absorption rate R is evaluated at the outlet concentration B_n^o and at the arithmetic mean of solute concentrations in the gas at the inlet and exit of a plate. The material balance for the solute between the (n+1) th and n th plates gives:

$$Y_i - Y_{n+1} = (L/zG')\left[B_n^o - B_f^o\right] \qquad (12)$$

If the specific form of the absorption rate equation is inserted into Eqn. (12), these two equations may be arranged to yield a relationship between B_{n-1}^o and B_n^o. Now it is possible to plot B_{n-1}^o against B_n^o and the to construct McCabe-Thiele type steps, the actual number of plates required for the operation may then be found graphically (see, for example, (65)). Juvekar and Sharma (50) proposed, however, that the plot of B_{n-1}^o versus B_n^o may be approximated to a straight line by a least square fit. Thus:

$$B^o_{n-1} = m' \, B^o_n + C \tag{13}$$

Here m' and C are obtained from the least square fittings The number of plates can now be obtained by the following Fenske type equation:

$$N = \frac{\ln \left\{ \dfrac{B^o_i - C/(1-m')}{B^o_f - C/(1-m')} \right\}}{\ln m'} \tag{14}$$

Under some circumstances it is also possible to use a modified Lewis method (50). This assumes a continuous function instead of a discrete realtion between B^o_n and n. Then if:

$$\left| \frac{d^2 \, B^o_n / \, dn^2}{d \, B^o_n / \, dn} \right| << 1 \tag{15}$$

The Taylor expansion yields:

$$B^o_n = B^o_{n-1} + \frac{d \, B^o_n}{d \, n} \tag{16}$$

The number of plates can then be found from the following equation:

$$N = \int_0^N d \, n = \int_{B^o_f}^{B^o_i} \frac{d \, B^o_n}{(-d \, B^o_n / d \, n)} \tag{17}$$

The modified Lewis method can easily be employed once the relationship between B^o_{n-1} and B^o_n is obtained provided that the condition in Eqn. (15) is satisfied. Juvekar and Sharma (50) listed expressions which were obtained from the modified Lewis method, for calculating the number of plates for cases 3,5,10,11 and 12 of Table 2. They compared the graphical results of Kawagoe et al. (65) for the case 12 of Table 2, and obtained excellent agreements both with a Fenske type equation and with the modified Lewis method.

Juvekar and Sharma (50) also took into consideration the pressure variation along the column. However, the validity of these methods is far from beeing tested experimentally verified and in view of the great industrial importance of plate columns, experimental evidence in support of them is certainly required.

5. USE OF LABORATORY MODELS IN PROCESS DESIGN

Computational design methods from first principles -that is, the Route 1 of Figure 1) require not only all the physical gas absorber data (such as k_L a,k_Ga, a etc.) but also the process specific data of reaction kinetics, solubilities and diffusivities. Some of these data (i.e. Stage 2 of Figure 1) can only be estimated using methods which are generally speaking not very reliable. Therefore, Danckwerts and his coworkers (66-69) thought that it might be more statisfactory to build a laboratory model of the absorber, which simulates the essential features; and to make measurements on that rather than the computation from first principles. By "essential" features, it is not suggested to make a geometrically accurate model of the absorber, and to try to get complete dynamical similarity. This would, of course, be impossible, partly because of the large number of dimensionless groups involved and partly because important data, such as the kinetics and the rate of reaction, might be ignored (69).

The model building of the Cambridge School was however based on the fact that the rate of absorption per unit interfacial area (R) depends only on the interfacial partial pressure of the gas, the liquid composition and the value of k_L, the liquid-side mass transfer coefficient in the absence of reaction. It makes no difference how k_L is generated -by turbulence promoted by stirring by flow over a packing, or in other ways- equality of k_L will give equality of specific rate of absorption. Equality of partial pressure at the interface will be assured by having the right bulk composition and value of k_G. Another very important characteristic of the model is that it should have a definitely known contact area.

It is convenient to describe two different types of model. The first one may be called a "point" model (67,69). Here only the absorption-reaction interaction (that is, microscopic scale phenomena) is simulated so that gas absorption rate per unit interface (R) may be measured for a variety of combinations of bulk gas and bulk liquid compositions. Then, these results are inserted into an appropriate two-phase contactor model (for example, those listed in Tables 3-6), to yield the required capacity. It is clear that the method eliminates theoretical modeling at the microscopic level and none of the quantitative process specific data (i.e. Stage 1 of Figure 1) is needed. However, some qualitative data are required as the model is applicable only to reactions which are fast enough to take place in the diffusion film near the interface so that there is no unreacted dissolved gas, and no reaction in the bulk of the liquid (inappropriate considerations of bulk reactions may result in vast design errors (70)). It is also confined to the case where a single gas is being absorbed. The reason for these limitations is mainly that, in these cases, there is a simple stoichiometric relationship between the bulk composition of the gas and the bulk composition of the liquid at any point (or level) in the absorber.

The second method of modeling an absorber is to make what might be called an "integra" or "complete" model (68,69). This would then consist of a laboratory scale absorber which simulates the industrial absorber with regards to microscopic as well as macroscopic modeling. The rules for such an "integral" or "complete" modeling were first established by Alper and Danckwerts (68,69). It is interesting to note that, in this type of modeling, none of the process specific data are needed; thus the Stage 2 of Figure 1 is completely eliminated. Alper and Danckwerts (68, 69) have applied such modeling to packed columns and have shown that a special type of absorber could barely satisfy all the necessary conditions (see, Section 6.4). Recently, Charpentier and his coworkers (71,72) tried to apply such a modeling to venturi scrubbers but could not satisfy all the necessary conditions simultaneously. They were however able to simulate the absorber for some limiting cases, such as the case of only liquid phase resistance (71,72).

5.1. Laboratory Absorbers

Laboratory absorbers for studying absorption into liquids may be divided into two groups. Some of them have effectively a quiescent liquid which comes into contact with gas for a desired time which can also be changed with relative ease. Strictly speaking these absorbers simulate the conditions which are foreseen by the Higbie model. Table 7 shows the main types and Danckwerts (2) discusses in some detail design characteristics and proper operation of many of these absorbers. These absorbers may also be used to obtain "process specific data" -such as, reaction kinetics, diffusivities etc. (2,73-75).

The remaining laboratory absorbers (see Table 8) involve random or regular movements which tend to bring about mixing between liquid near the surface and in the bulk or replacement of one by the other, thus simulating agitated liquids. Here the essential feature is, in each case, a well defined and known interface and relatively easy adjustment of the mass transfer coefficients k_L and k_G.

Although, the use of all the types in Table 7 and 8 are advised repeatedly (71,72,76), it is our opinion that some have only a historical value. For instance, a conical wetted wall column (77) has no advantage over that of a cylindrical one, but it has the disadvantage that the end effect due to a rigid film at the liquid exit will be increased. For instance, the results from a disk column are difficult to reproduce (78) thus a string of spheres column may be preferred. Other types which have only historical value include the rotating drum and moving band absorbers

TABLE 7. MAIN LABORATORY ABSORBERS WITH AN EFFECTIVELY QUIESCENT LIQUID (2).

APPARATUS	CONTACT TIME	REMARKS
Rotating drum	0.01-0.25 s	Significant exit end effect; it has only historical value
Falling films:		Use of surface active agents is necessary
1. Cylindrical	0.01-2 s	Very convenient, end effects may be eliminated by a special design of exit stream collar
2. Conical (77)	0.2-1 s	No particular advantage over cylindrical one; enhanced end effects. It has only historical value.
3. Spherical	0.1-0.5 s	Reduced exit end effect, useful but it has limited contact time
Laminar jet	0.001 - 0.1 s	Very useful, departures from the ideal flow may be eliminated by proper design of nozzle or orifice
Moving band	0.001 s	Difficult to construct and operate. It has historical value

TABLE 8. MAIN LABORATORY ABSORBERS SIMULATING AGITATED LIQUIDS (69)

APPARATUS	RANGE OF k_L (cm/s)	RANGE OF k_G (mol/cm^{2}s atm)	REMARKS
String of discs	$0.5 - 2.5\times10^{-3}$	$3 - 30\times10^{-5}$	Not reproducible, k_L depends on the number of discs (78). Considerable concentration gradient, hence not suitable as a "point" model
String of spheres	$3 - 30\times10^{-3}$	$2 - 25\times10^{-5}$	Reproducible and partially amenable to theoretical analysis. Considerable concentration gradients
Stirred cell	$2 - 15\times10^{-3}$	$3 - 20\times10^{-5}$	Reasonably reproducible, concentration gradientless. hence a good "point" model

6.1. Deviation from Plug-Flow Behaviour

Previous considerations (see Table 3) indicate that the assumption of plug flow of both phases seems to be reasonable. It is however known that this is not exactly the case, and that there is some mixing or exchange between elements of fluid which enter at different times; measurements of the extent of mixing of the liquid and of the gas have been made by many workers (80-83).

The longitudinal dispersion can be either described simply by the "cell method" or by the "axial dispersion" model. In the former (see, Figure 6), a quantity characteristic of the type of packing and flow, is the height of a mixing unit, HMU, which is the height of the packing divided by the number of completely mixed cells required to simulate the dispersion. The weight of the evidence shows that for the liquid the HMU is about two-packing particle diameter; for the gas it is one particle diameter in a dry packing, increasing to about five diameters at high gas and liquid rates (2).

The "Axial dispersion" model is also used by many workers and the Pe numbers for both phases are often reported. In this respect, Dunn et al. (83) have studied the characteristics of a 60 cm in diameter column using the tracer method; their main conclusions agree with the above. Recently, Van Landeghem stressed the fact that "poor" distribution is a much more serious problem and he also claimed that "axial dispersion" would not be a good model because it implicitly assumes a uniform flow of liquid through the entire cross section of the column. In reality, the liquid, of course, flows in a very heterogeneous way in the form of a film, or of small streams or drops advancing at very different velocities (84). Also, channeling caused by inhomogeneities in the density of packing, or by flow down the wall, may cause a much more marked dispersion. Although, the latter has received considerable attention (85,86), the information available is still scanty.

Unlike some other absorbers (for example, bubble columns), by proper design, harmful deviations from plug flow behaviour may be avoided although many different parameters have adverse effects. For instance, axial mixing is reduced by having a column with a height many times the diameter of one particle, but this will lead to considerable wall flow (unless redistributors at intervals of 2.5-3 column diameters are employed). Then increasing the ratio of column diameter to particle diameter reduces wall flow and Porter and Templemann (85) concluded that wall flow is of negligible importance in large industrial packed columns. On the other hand, it is generally believed that the effectiveness of packed columns decreases as the diameter is increased. This phenomena is not understood, but may be due to segregation of gas and liquid flows, so that parts of the column receive more and others less than the average flow rates of gas and especially liquid (87).

6.2. Mass Transfer Coefficients (k_La and k_Ga), Effective Interfacial Area (a) and Liquid Hold-up

Measured values of these parameters as functions of flowrates, liquid properties etc. for various packings are unfortunately still very few-especially for those of large dimensions. Tables 9 and 10 show some of the recent relevant literature. Unfortunately, there still seems to be no completely satisfactory correlations relating various data. Empirical correlations, such as those of Norman (100) and Sherwood and Holloway (101) for k_La are not always applicable (see, for example reference (107) for discussions) and should only be used with some care. Furthermore most of the correlations on interfacial area (e.g. Onda et al. (108)) give an upper limit, which is equal to the geometrical surface area of packing. On the other hand, recent works show clearly that for some packings, the effective area may well exceed the geometrical surface substantially at high liquid flowrates.

It seems that, unlike gas-in-liquid dispersion systems, interfacial areas do not depend strongly on the various properties of the chemical systems employed and, for a given packing, depend mainly on the liquid flowrate. For instance, Figure 10 shows the data of Sharma and coworkers (47,110,111) which has been checked by several workers over a span of several years with good agreement; Table 11 shows the characteristics of these systems. It appears that ionic strength (varied from 1 to 34.5 ion/l) and viscosity (varied from 1 to 9 cP) have little effect on interfacial area (111). Other published information with aqueous solutions (95,96) support this view. Information with nonaqueous solutions is scanty but may well be different (47).

One point of dispute may however be the effect of the liquid viscosity. Recently, Rizzuti and his coworkers (112) found a strong dependency of viscosity on the effective interfacial area in sugar solutions (γ=0,9 - 1.55 x 10^{-2} cm^2/s) and they proposed:

$$a = 39 \quad \gamma^{0.70} \quad L^{0.326} \tag{18}$$

Thus it appears that the viscosity has a strong effect. One disadvantage of their experiments was that the ration of column diameter to that of packing was unfortunately less than 4. Recently Alper (113) has measured the interfacial areas by using the sulphite oxidation method in the absence and presence of CMC. The addition of CMC hardly affects the specific absorption rate, but hydrodynamical properties are of course strongly affected. Figure 11 shows the summary of his results. It is seen that the interfacial area increases considerably by adding CMC, the effect being much less at high liquid flowrates. It seems further work is needed in this area.

ABLE 9. SOME OF THE RECENT LITERATURE ON THE EFFECTIVE GAS-LIQUID NTERFACIAL AREA AND THE MASS TRANSFER COEFFICIENTS IN ACKED COLUMNS

REFERENCES	REMARKS
harma and anckwerts (38)	A critical literature survey of chemical methods of measuring k_L, k_G and a
anckwerts nd Sharma (52)	A critical review with data of (27) and (106)
uranik and ogelpohl (88)	A generalised correlation of the interfacial area in terms of a reaction factor
harpentier (89)	A critical survey
olev (90)	Literature and original data about k_L and a covering German and Eastern European countries
eichelt nd Blass (91)	A critical survey, limited to physical absorption
lper (92)	A critical survey of physical and chemical methods of measuring a
nda (55)	Review of Japanese work on packed columns
ridharan nd Sharma (47)	Limited k_La and a data as well as discussions of nonaqueous chemical systems
lper (40)	k_L and a data for plastic rings of different legth to diameter showing its effect
lper (95)	Data for porous and nonporous packings suggesting a is not likely to be influenced by porosity
nende and narma (57)	k_Ga and a values of various packings for cocurrent operation
ahay and narma (96)	Values of k_Ga, k_La and a for several 1 in. packing materials
inek et. al. 97-99)	Increased interfacial area due to application of a hydrophilic layer and comparison of various correlations
angers and ontner (94)	Effect of viscosity on k_La and a proposed correlation
iscellaneous 100-105)	Experimental data and correlations for k_La

TABLE 10. SELECTED LITERATURE ON THE LIQUID HOLD-UP IN PACKED COLUMNS

REFERENCES	REMARKS
Tichy (120-121)	Data for spheres. A method to predict the effect of gas flowrate is given.
Buchanan (122)	A correlation for Raschig rings
Hofmann (123)	A critical review of fundamental aspects
Shulman et.al. (124-125)	The most comprehensive data for ceramic and carbon rings. Both aqueous and non aqueous systems
Mohunta and Laddha (126)	New data and correlations for small packings
Charpentier and Favier (127)	Data for small particles using foaming and nonfoaming hydrocarbons

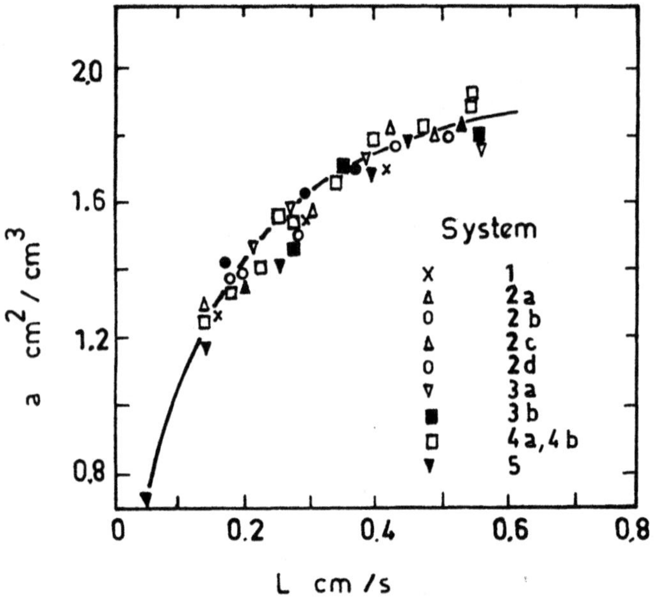

Figure 10. Effective interfacial area of 0.95 cm (3/8 in.) ceramic Raschig rings

TABLE 11. CHARACTERISTICS OF SYSTEMS IN FIGURE 10

No.	Solute Gas	Absorbent	Viscosity (cP)	Ionic St. (ion/l)	Ref.
1	i-butylene	aq. H_2SO_4	8.900	34.50	(111)
a	Oxygen	Various combinations of CuCl, $CuCl_2$ and HCl	2.400	8.00	(111)
b			2.460	9.15	
c			0.905	3.50	
d			1.200	5.10	
a	Oxygen	aq. dithionite and NaOH	0.900	0.62	(110)
b			1.000	1.31	
a	CO_2	aq. NaOH	1.190	1.80	(110)
b		monoethanolamine	1.200	-	
5	CO_2	Cuprousamine complex	1.280	-	(47)

It is known that for a given packing, k_L depends mainly on the liquid flowrate. Viscosity of the solution will also have a profound effect on the value of k_L. This will be both through the reduction in diffusivity and the effect of viscosity on the hydrodynamics. It is possible that the form $k_L \propto$ D (which includes the effect of viscosity through diffusivity) may be sufficient for practical purposes for moderate changes in viscosity. Experimental evidence is however very limited. Figure 12 and 13 show some data of Alper (114) for two different columns containing either ceramic 1 cm spheres or plastic 5/8 in. Pall rings. These are obtained by sugar solutions, the degree of dependency of k_La/ D on viscosity clearly depends on also the type of packing. That is, while the dependency for ceramic spheres is weak, it is much more pronounced for plastic Pall rings. Figure 12 and 13 show also the combined results of two adverse effects; at a fixed liquid flowrate, an increase in viscosity increases a, while decreasing k_L/ D due to probably decreased turbulence. At low viscosities, the first effect is more pronounced than the other, but the latter overtakes it at high viscosities.

Sharma and coworkers (57,96,110) have shown repeatedly that, for a given packing, k_G depends only on the gas flowrate even though theoretical considerations and some experimental evidence in model absorbers (such as string of discs or spheres columns) (69,115) suggest some dependency on the liquid flowrate too. Measured values of k_G are, of course, specific to the soluble gas but a suitable correction can be made. Sharam and corworkers (110, 116,119) have reported data, covering an eight-fold variation in

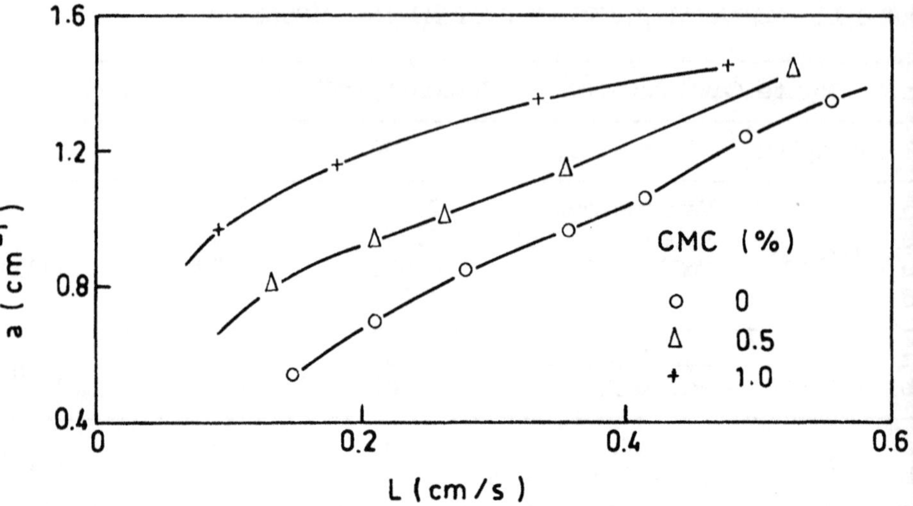

Figure 11.Effect of CMC addition on a (5/8 in. polyethylene Pall rings (Results of sulphite oxidation experiments at 25°C)

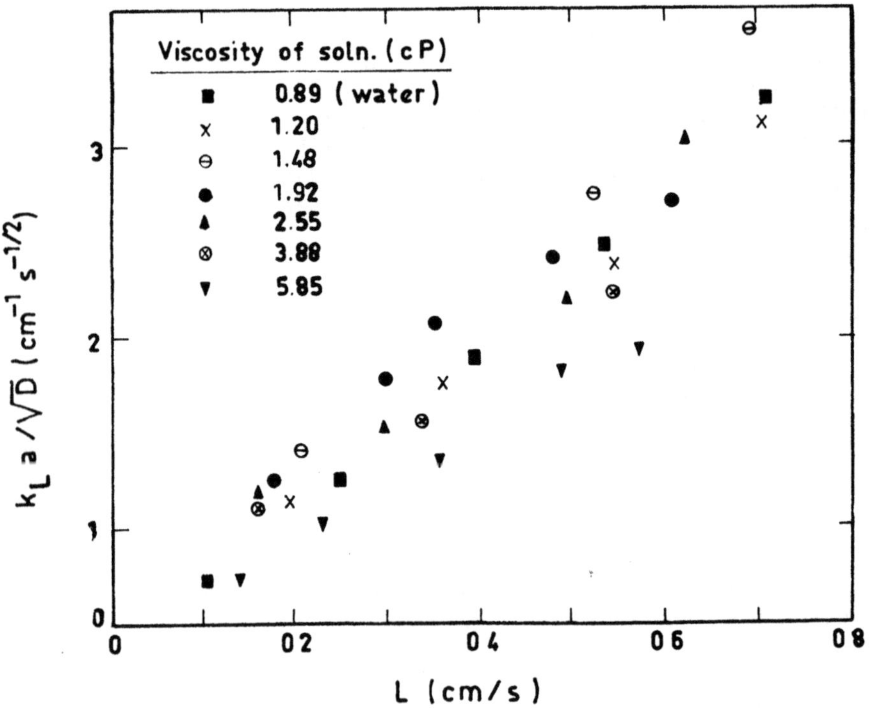

Figure 12.Effect of viscosity on $k_L a/\sqrt{D}$(1 cm ceramic spheres)

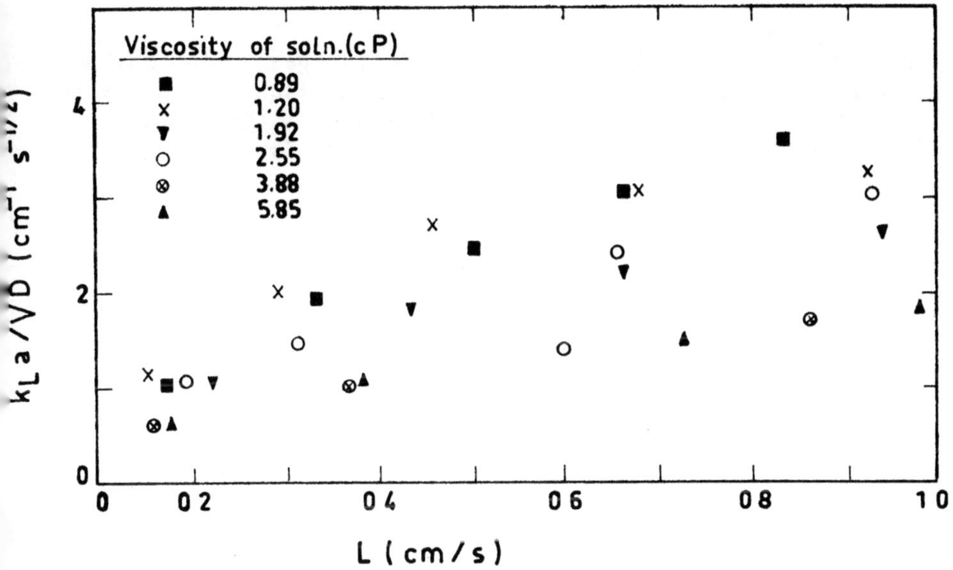

'igure 13.Effect of viscosity on $k_L a/\sqrt{D}$ (5/8 in.Pall rings)

; diffusivity, which show not only that k_G is proportional to .5 but also that the Schmidt number is not the proper correla- on factor in contrast to many typical correlations (119). The :ter point of view was also confirmed by Merchuk and Tamir (118)) however favoured a relationship of $k_G \propto D_G^{2/3}$ (see, Sharma and lav (119) for possible explanation of the errors in the data (118)).

Finally, it may be pointed out that most of the reported .a on k_L, a and liquid hold-up were obtained without any flo- g gas or at low gas flowrates. Tichy (120) gives a method of :imating the effect of gas flowrate from the liquid hold-up at 'o gas flowrate and the gas flowrate at the flooding point. How- 'r, it can be assumed that the dependency of all these quan- ies on gas flowrate is negligible if operating conditions are l below the loading point.

. Modeling of a Packed Column

If we assume "plug flow" for both phases, setting up a ma- ial balance on a differential height of a packed column for ab- ption of solute A into a liquid containing dissolved reactant es:

$$R\left[p^o, A^o, B^o, k_L, k_G\right] a\, dh = L\, dC^o + v\, r(A^o, B^o) dh \qquad (19)$$

$$-L\,dB^{\circ} = zvr(A^{\circ}, B^{\circ})\,dh \qquad (20)$$

where p^{o} is the partial pressure of soluble gas in the bulk gas stream, A^{o} is the bulk concentration in the liquid of unreacted dissolved gas, B^{o} is the bulk concentration of the reactant. These equations can be rearranged and integrated to give:

$$\left(\frac{aH}{L}\right) = \int_{i}^{o} \frac{dA^{\circ}}{R(p^{\circ}A^{\circ}B^{\circ}k_L k_G)} - \int_{i}^{o} \frac{dB^{\circ}}{zR(p^{\circ}A^{\circ}B^{\circ}k_L k_G)} \qquad (21)$$

$$\left(\frac{vH}{L}\right) = \int_{i}^{o} \frac{dB^{\circ}}{zr(A^{\circ}, B^{\circ})} \qquad (22)$$

Similar equations with the same R.H.S.'s may be derived when various gases are absorbed simultaneously or when a gas is absorbed into a solution which contains various reactants (68,69).

6.4. Stirred Cell as a "Point" Model

When there is no reaction in the bulk of the liquid, the first term in the R.H.S. of Eqn. (21) is zero. Thus the required packing height, H, of the column is calculated by taking measurements of R, that is absorption rate per unit interface, under a number of conditions representing different points in the column (see, Figure 14) then integrating (1/R) with respect to B^{o}. The "point" model can take various forms. The essential feature in

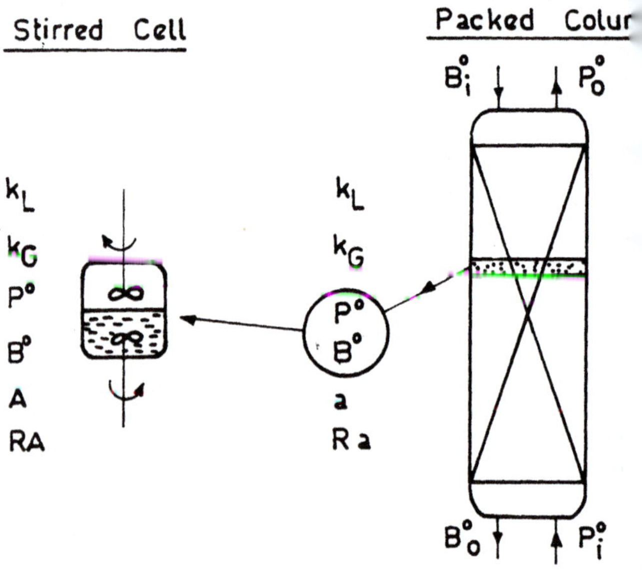

Figure 14. The schematic representation of the principle of "point" modeling

each case is that the area of interface between gas and liquid is well defined and known. The required values of mass transfer coefficients (k_L and k_G) can be achieved and easily adjusted. There should preferably be no significant concentration gradients in either phase and the absorption rate should be measured easily. Alper (69) examined various possibilities critically and found that a stirred cell such as shown in Figure 15 satisfies all these requirements. A stirred cell was first used by Danckwerts and Gillham (66); later Sharma and Jhaveri (128), Shafer and Yano (129) used similar devices. These workers considered only the liquid side phenomena (hence matching k_L) and not gas side phenomena. Alper extended this approach so that the value of k_G in the cell could be varied to match the value in a packed column as well (67,69).

In his stirred cell (67,69), the stirres were on coaxial shafts driven by synchronous motors through variators. The bulk of the liquid was kept uniform by a turbine and another stirrer on the same shaft just skimmed the liquid surface, producing a value of k_L which depended on the stirrer speed. Likewise on the gas side, there was one stirrer, close to the surface but also another stirring the bulk, and the rate of rotation of this stirrer determined k_G. Recently, somewhat similar devices have been described by Godfrey and Levenspiel (130) and Sridharan and Sharma (47). The measured value of k_G and k_L for specific systems as functions of liquid and gas side stirrer speeds are given in Figures 16 and 17. They cover the range -albeit on the limit- of those obtained in packed columns.

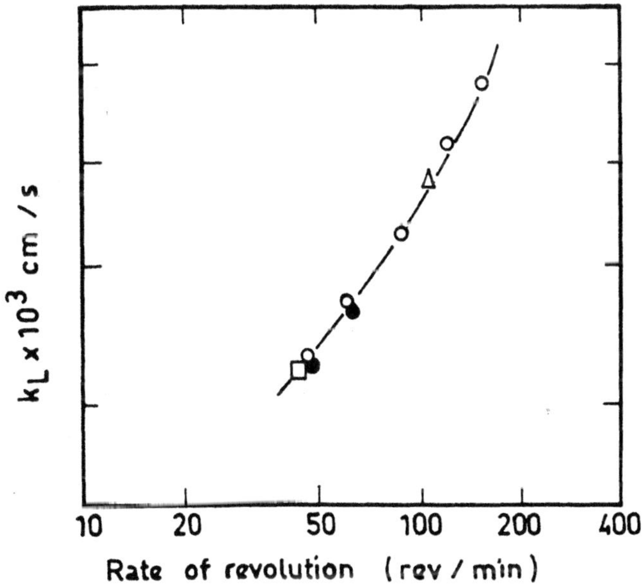

Figure 16.Liquid-side mass transfer coefficient in the stirred cell (Carbondixide- water system at 25 C)

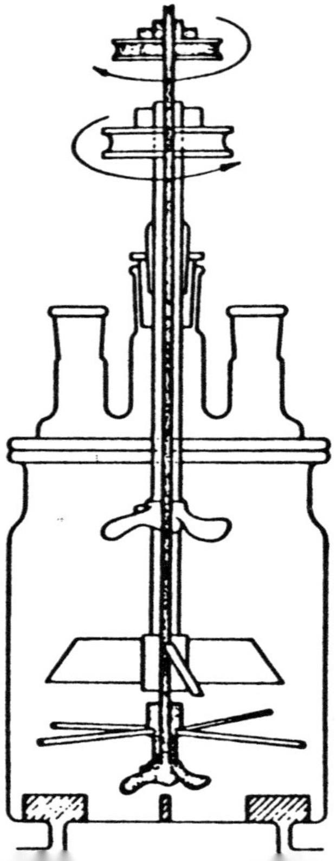

Figure 15. The stirred cell (67)

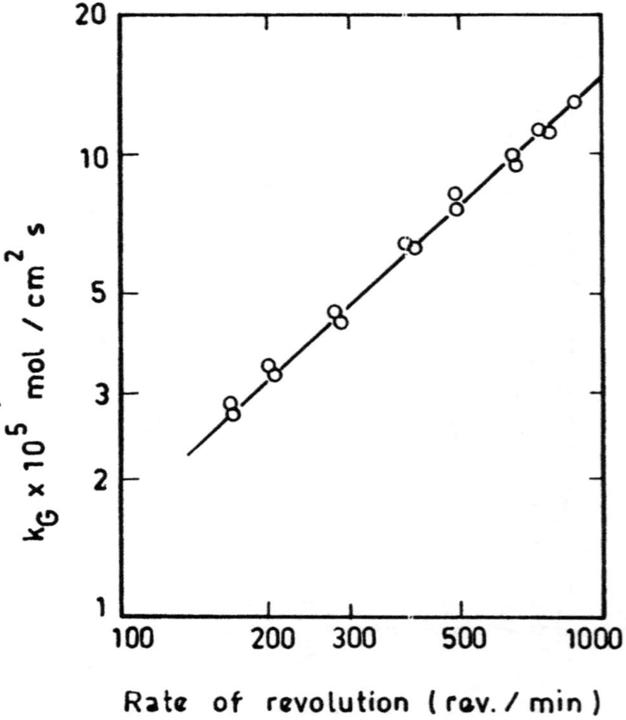

Figure 17. k_G in the stirred cell (SO_2 -water)

The technique discussed here was tested by carrying out absorption experiments in a packed column of 10.2 cm inside diameter and upto 183 cm high. The packing consisted of 1.27 cm ceramic Raschig rings and it was possible to easily change the height of packing exposed to the gas. Other details of this equipment can be found elsewhere (67,69).

The values of k_L and k_G were made the same in the stirred cell by adjusting rated of stirring and absorption rates at various levels in the column, were also measured in the stirred cell (combinations of B^o and p^o were obtained from total material balances) -see, Figure 14. For each experiment in the packed column the relevant value of R from the stirred cell measurements were used to integrate Eqn. (21) numerically.

This procedure has been carried out for absorption with and without gas-side resistance and a good agreement between predictions and the actual heights was found for a given task. Table 12 shows some typical results for CO_2/air-NaOH-Na_2CO_3 solutions system (67,69). Similar tests were made later on by Laurent (131,132) using the same chemical systems.

TABLE 12. RESULTS OF CO_2-AIR-AQ. (NaOH + Na_2CO_3) EXPERIMENTS (67,69)

OH^- Inlet	OH^- Outlet	p_{CO_2} x10^2 Inlet	p_{CO_2} x10^2 Outlet	Packing Height	Predicted Height	Diff.	Error in material balance
mol/l		atm		cm	cm	%	%
0.39	0.29	4.6	3.0	48.0	46.3	3.5	8.4
0.52	0.42	4.1	3.3	48.0	44.1	8.1	4.2
0.58	0.46	4.6	3.6	48.0	46.3	3.5	3.6
0.28	0.10	5.2	3.6	108.0	104.0	3.7	3.5
0.62	0.19	7.6	4.0	163.0	154.0	5.5	0.5
0.60	0.08	10.3	6.2	163.0	154.0	5.5	0.8
0.55	0.19	6.0	3.1	163.0	153.0	6.1	7.0
0.52	0.20	5.0	2.2	163.0	154.0	5.5	0.0
0.55	0.08	10.6	7.0	143.0	142.0	0.7	5.3
0.46	0.10	7.8	5.0	143.0	141.0	1.4	6.7
0.46	0.13	7.8	5.2	123.0	121.0	1.6	5.4
0.58	0.13	10.2	6.7	123.0	122.0	0.8	2.6
0.53	0.26	5.3	3.0	123.0	121.0	1.6	6.0

6.4. String of Spheres Column as a "Complete" Model

The second method of experimentally simulating an absorber is to make what might be called an "integral" or "complete" model. Here not only the microscopic scale, but also the macroscopic scale issues are considered. Such a model would have the same mode of operation (for packed columns, countercurrent for instance) into which feeds of gas and liquid at the right composition, pressure and temperature are fed in the right ratio of feed-rates and the compositions of the emerging streams have the same values as would be obtained from the industrial equipment which is being modeled.

The idea of such a model is an attractive one, particularly if it can deal with the type of systems with which the "point" model cannot -e.g. simultaneous absorption of two gases, slow reaction leading to reaction in the bulk of the liquid etc. In addition, the integral model may prove much less laborious than the "point" model, and it has the psychological advantage of simulating the observable featured of the industrial plant directly. The question remains of how to make a model which reproduces the essential features of the plant.

One feature which must be retained is the same mode of operation (i.e. countercurrent in our case); thus for packed column simulation one has to think in terms of the liquid running down over some kind of surface, and gas flowing upwards. If the model has also the same flow patterns, that is plug flow of both phases it is easy to show that (69):

$$\frac{A_m\ H_m}{Q_m} = \int_i^o \frac{d\ A^o}{R\ \ p^o,A^o,B^o,k_L,k_G} - \int_i^o \frac{d\ B^o}{zR\ p^o,A^o,B^o,k_L,k_G} \qquad (23)$$

$$\frac{V_m H_m}{Q_m} = \int_i^o \frac{d\ B^o}{z\ r\ A^o,B^o} \qquad (24)$$

where Q is the total liquid flowrate, A_m is the wetted area per unit height of the model and V_m is the volume of liquid per unit height.

In general, we do not know how the specific rate of absorption varies with p^o and B^o; however we can make the relationship between specific absorption rate and bulk concentrations in the model the same as that in the industrial plant by using the same values of k_G and k_L (see Figure 18). To get the same inlet and outlet concentrations in both, the integrals on the right hand sides of Eqns. (23) and (24), must be equal. This follows:

$$\frac{a\ H}{L} = \frac{A_m\ H_m}{Q_m} \tag{25}$$

$$\frac{H}{H_m} = \frac{(L/a)}{(Q_m/A_m)} = \text{Scaling ratio} \tag{26}$$

Here H and H_m are the heights of the industrial column and the model respectively. (L/a) and (Q_m/A_m) are the corresponding "wetting rates" -that is, flow per unit wetted perimeters. Since we want the height of the model to be a good deal less than that of the industrial column, for instance 1/5 or 1/10, we have to make the wetting rate in the model 1/5 or 1/10 that of in the industrial column. This is one of the scaling rules, and makes for very low wetting-rates in the model. From Eqns. (22) and (24), we obtain:

$$\frac{v\ H}{L} = \frac{V_m\ H_m}{Q_m} \tag{27}$$

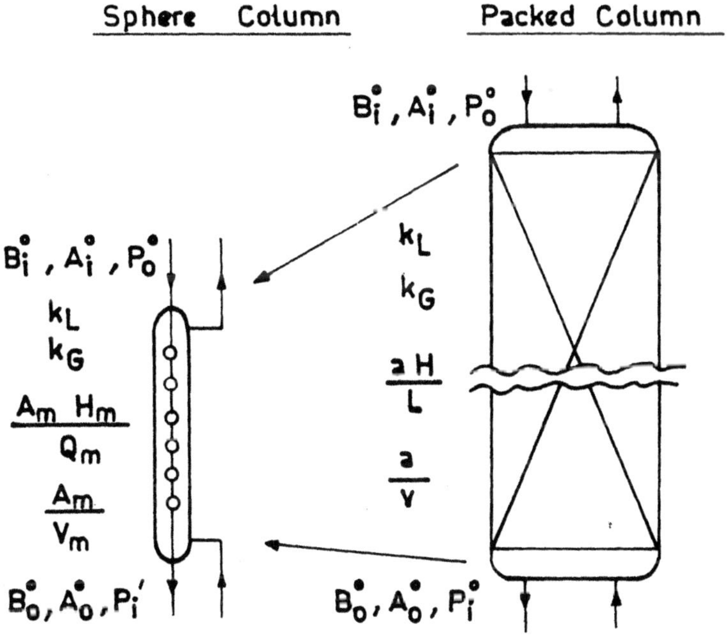

Figure 18. Schematic representation of "complete" (integral) modeling of a packed column

Rearranging Eqn (27) by using Eqn. (26) gives:

$$a/v = A_m/V_m \qquad (28)$$

This is of course, the rule which must be obeyed by the model if any slow reaction takes place in the bulk. It can easily be shown that the condition of Eqn. (28) means the average residence times of the liquids are the same in the model as in the industrial column, so that the bulk reaction is present to the same extent in each. The rules which have to be obeyed by the model vis à vis the industrial column are summarised in Table 13. Table 14 gives the values assumed for the packed column in order to see whether the conditions in Table 13 can be achieved by a model.

One problem which arises immediately is that, for a given geometry of the wetted surface, k_L is a function of the wetting rate, since it depends on the liquid velocity on the surface, the degree of turbulence, the rate of liquid mixing and turnover, and so on. If we want to use a 1/10 scale model, we have to make the wetting rate in the model 1/10 of that in the industrial equipment, but at the same time we have got to keep k_L the same. Clearly, a different geometry in the model must be used. Simply scaling-down the size of the packing does not work; the value of k_L for a small packing is not much different than that for large packing at the same wetting rate (2); thus it is less than for a large packing if one takes the diminished wetting rate into account. Hence, we must think in terms of something with a very different appearance from the packings in the industrial columns.

Alper and Danckwerts (68,69) tried wetted-wall columns, with various types of flow mixer fitted at intervals to give the required value of k_L, in spite of the very low Reynolds number. However, none of these attempts were successful. They also tried other configurations, but eventually came down in favour of a string of spheres (Figure 19), threaded on to a central metal rod. They could produce liquid hold-ups approximating those found in industrial equipment by making a depression in the top of each sphere. This configuration has also the advantage that k_L is the same on each sphere, instead of increasing downard from sphere to sphere as it does if we simply use a string of contiguous spheres with no depressions (69,78). The value of k_L depends on the sphere diameter and the wetting rate, and was measured for various conditions. Figure 20 shows the experimental results (68). The value of k_G depends on the velocity of the gas in the tube enclosing the spheres, which can be varied to get the desired value. Alper and Danckwerts (68,69) have found with representative wetting rates and gas velocities they could obtain values of k_G and k_L characteristic of industrial packed columns.

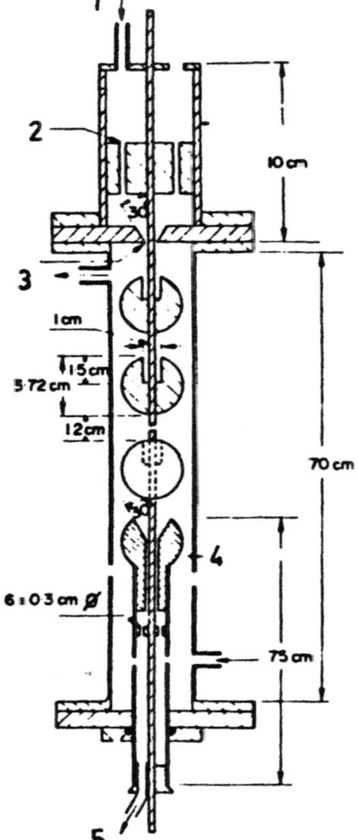

Figure 19.The string of spheres column(68,69)

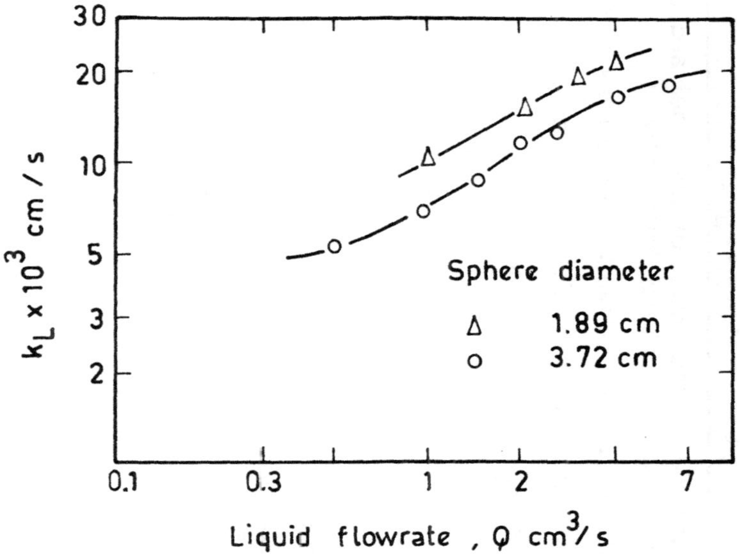

Figure 20.Liquid-side mass transfer coefficient in the 1.85 cm and 3.72 cm dia.spheres column (68,69)

TABLE 13. RULES TO BE OBEYED BY THE "COMPLETE" MODEL (69)

No.	PARAMETER
1	Same mode of operation in the model and the industrial column (e.g. countercurrent flow)
2	Feeds at right temperature, pressure and composition
3	Same liquid to gas flowrate ratio
4	Same k_L and k_G values
5	Same wetting rate in proportion to height (i.e. same(aH/
6	Same liquid hold-up per unit interface (i.e. same (v/a)

TABLE 14 TYPICAL VALUES OF CHARACTERISTICS OF A COLUMN PACKED WITH 3.8cm RASCHIG RINGS (69)

PARAMETER	UNITS	TYPICAL VALUES
L, superficial liquid velocity	cm/s	1.0
k_L, physical mass transfer coefficient	cm/s	1.8×10^{-2}
a, effective interfacial area per unit packed space	cm^2/cm^3	1.0
$\overline{(G/L)}$, ratio of gas to liquid superficial velocity	-	40.0
v, liquid hold-up	cm^3/cm^3	6×10^{-2}
k_G, gas-side mass transfer coefficient	mol/atm cm^2s	12×10^{-5}
(a/v), interfacial area per unit volume of liquid	cm^2/cm^3	17.0

6.5. Tests of the "complete" model

A number of tests were made by Alper and Danckwerts (68,69) to see whether the model could be successfully used to predict gas absorption behaviour in a packed column. For economical reasons they were unable to work with a packing larger than 1/2 in., and the scaling factor was about 3. It was possible to make the value of k_L the same in the sphere column and the

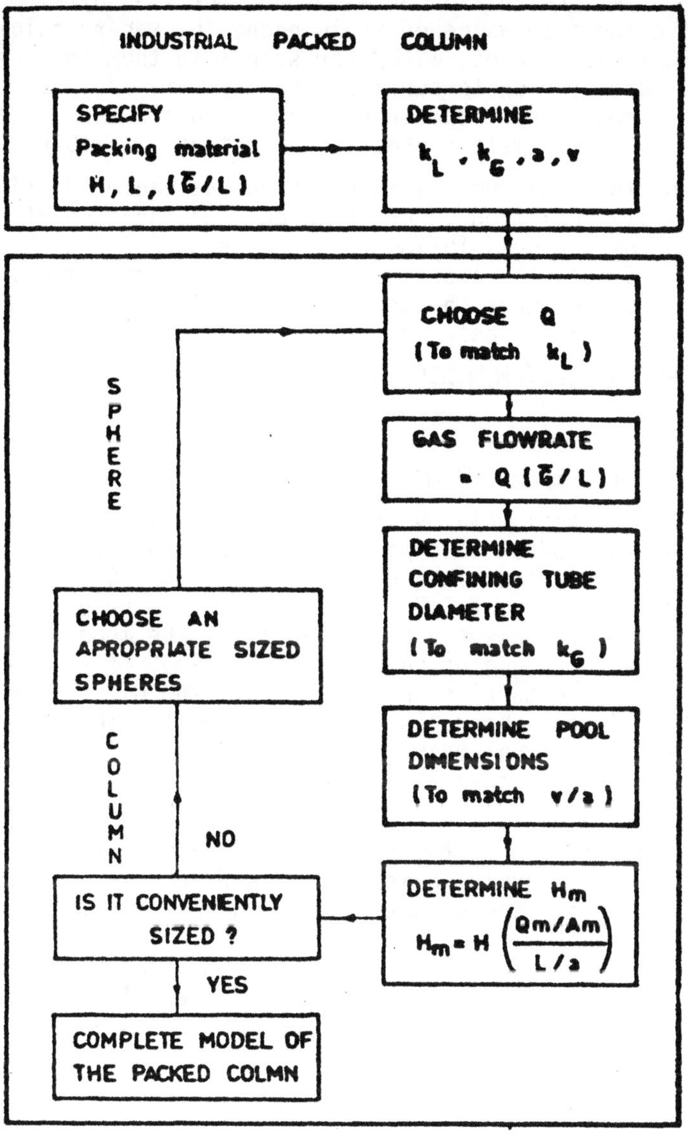

FIGURE 21. THE SCHEMATIC REPRESENTATION OF THE USE OF THE STRING OF SPHERES TO SIMULATE AN INDUSTRIAL PACKED COLUMN

packed column, and at the same time satisfy the wetting rate criterion. As for the gas side resistance (where relevant), the gas-liquid ratio was kept the same in the sphere column as in the packed column, and the diameter of the tube enclosing the sphere column was calculated by a method, outlined by Alper (68, 69), to give the same value of k_G in each. The sphere column contained upto 10 spheres, with depressions in them to make the ratio of liquid hold-up to surface area the same as in the packed column. They could thus achieve similarity in all essential respects and the test systems used included those in Table 15.

Each of these systems is interesting but causes difficulties in theoretical predictions. For instance, for the case of absorption of CO_2 into amine solutions containing arsenite at high carbonation ratios, the experiments correspond to a situation where fast reaction near the interface is followed by another reaction in the bulk. These are respectively:

$$CO_2 + 2RR'NH \rightleftharpoons RR'NCOO^- + RR'NH_2^+ \qquad (29)$$

and

$$RR'NCOO + H_2O \rightleftharpoons RR'NH + HCO_3^- \qquad (30)$$

Danckwerts and McNeil (133) have given some simplified approximate methods to predict absorption rate for carbonation ratios either considerably less than 0.5 or greater than 0.5; when this ratio is in the region of 0.5-0.6 neither method predicts the absorption rate sufficiently accurate. Thus, the model experiments can take the place of calculation. Table 5 of reference (68) shows the results of such experiments; the difference between the "predicted" absorption rate from the sphere column experiments and the actual measured absorption rate in the packed column is always less than 7%.

Results of experiments with other systems are also in quite good agreement and can be found elsewhere (68,69). In general, these experiments, involving as they do a number of very different systems governed to varying extents by k_G, k_L and film and bulk reactions, and including several which could not be dealt by the point method, are reasonably encouraging, and some industrial application for the method is foreseen. Without alteration, in principle it should be possible to use it at elevated temperatures and pressures.

The procedure, then for the use of the model to simulate an industrial packed column is outlined in Figure 21. First, we assume that we have specified the nature of packing material, the height of the column, the superficial liquid velocity and the gas to liquid ratio; from these follow the values of k_L, k_G the effective interfacial area and the liquid hold-up (that is, Stage 1 of Figure 1). It is assumed that these latter quantities

TABLE 15. CHEMICAL SYSTEMS USED TO TEST THE "COMPLETE" MODEL (69)

CHEMICAL SYSTEM	CHARACTERISTICS
Absorption of CO_2 into carbonated monoethanolamine solutions containing arsenite	A fast reaction near the interface is followed by a slow reaction in the bulk
Simultaneous absorption of CO_2 and SO_2 into monoethanolamine solutions	Two gases react simultaneously with the same reactant, hence the effect of the presence of one gas is to reduce the rate of absorption of the other
Simultaneous absorption of NH_3 and CO_2 into water	Two gases react in solution,hence absorption of second gas increases the absorption rate of the other gas
Absorption of CO_2 from air into 2,6 dimethylmorpholine solutions	Approximately upto 35% gas side resistance
Absorption of CO_2 into a solution which contains two amines : 1.MEA,di-isopropanolamine 2.MEA,2-methylaminoethanol 3.MEA,monoisopropanolamine 4.MEA,diethanolamine 5.Diethanolamine,di-isopropanolamine	Both amines compete for the same dissolved gas. Reaction rates are such that various mass transfer regimes are covered

have been measured once and for all for different packings at different gas and liquid flowrates.

The liquid flowrate Q in the sphere column is chosen to make k_L the same. This gives the gas flowrate in the model, that is, Q x (G/L). Then the confining tube diameter is determined to adjust the linear gas velocity to the point at which k_G is the same in both absorbers. The depression on top of the spheres is sized to make the ration of liquid hold-up to surface area the same as that in the packed column. Then the number of spheres, N, required to make the ratio of heights of the columns the same as the ratio of wetting rates. Finally, if the height of sphere column so calculated is not convenient, it will be necessary to repeat the calculations using data for different sized spheres. The data of Alper (68,69) for two sizes of spheres should help in the choice, but it is necessary ideally to have laboratory data giving k_L and k_G for a range of sphere sizes; this again should be done once and for all.

7. DISCUSSIONS AND CONCLUSIONS

It has been established that any rational process design method of gas absorbers (gas-liquid reactors) may conveniently be divided into three main stages (see, Figure 1). The first stage which is essential for any rational process design approach seems to be still one of the main problems. That is, the values of the mass transfer coefficients, interfacial areas, hold-ups etc. are difficult to estimate under the conditions of the reaction system. There is a real dearth of such data in all absorbers where liquid other than water or aqueous solutions (for instance, nonaqueous, nonnewtonian, viscous solvents etc.) are involved and various reported data and correlations are not all reliable when employed under different experimental conditions and they should therefore be used cautiously. In this respect, although there is some dispute, the packed column seems to be one of the gas absorbers whose characteristics (i.e. k_La, a etc.) can be predicted with a fair degree of confidence.

The second stage which involves obtaining "process specific data" (such as relevant kinetics, solubilities and diffusivities) is only necessary if theoretical modeling at the microscale is to be used. If however, such data are available, theoretical predictions -for many cases, analytical expressions- are possible provided that a satisfactory reactor model is also available. On the other hand, obtaining such data is not only a laborious task, but in many cases they have to be estimated by methods which are, generally speaking, not very reliable. This stage may therefore conveniently be avoided by making use of laboratory models. A number of workers (66,67,69,128-132,134-137) have simulated essential features at the microscopic scale (i.e. "point" modeling) and illustrated that this method can be used for design purposes satisfactorily.

In certain cases, when the chemical system is more complicated a "point" model may not be applicable (67,69). Under these conditions the theoretical predictions would also have been eithe too complicated -if indeed possible- or would lead to large errors in several instances. On the other hand, it appears that if one builds a "complete" (or integral) model of the gas absorber, the behaviour of the industrial equipment (for instance, total absorption rate or the composition of gas or liquid leaving the column etc.) can be obtained with sufficient accuracy from the results of an appropriate sphere column. The rules to be obeyed by the "complete" model are as given in Table 13 if both phases are essentially in plug flow. Recently Charpentier and his coworkers (71,72) tried to simulate a venturi scrubber by a laminar jet. They have assumed the validity of Eqns. (21) - (24), in other words the rules given in Table 13. Strictly speaking, these rules are derived from "plug flow" considerations for both phases; they may not necessarily be the same if the industrial absorber under consideration has other flow patterns. Charpentier and his coworkers (71,72) did not consider this aspect and implicitly assumed plug flow in both absorbers. Further, their study showed how stringent the simulation rules are; it appears they could either attain k_G or k_L (but not both simultaneously) and the same (v/a) could not be realised at all.

For processes which occur at high temperatures and pressures these design methods from laboratory models could - although not tested- probably be used. It must however be noted that these design methods do not consider non-isothermal systems. Although there has been considerable effort directed at such systems (138), theoretical predictions are not all that refined. Experimental modeling of such systems will probably be not possible due to the further requirement of matching the heat transport phenomena. However due to its operating characteristics -that is, poor heat removal- a packed column (which was the main concern here) is not suitable for such operations.

ACKNOWLEDGEMENTS

Most of the earlier experimental data, as well as the general philosophy have been obtained whilst the author was at the Chemical Engineering Department of Cambridge University, England. The Author is grateful to Prof. Peter V.Danckwerts for his support; he also thanks the Foundations of Alexander von Humboldt and Volkswagen of F.R.Germany for their generous financial support for the continuation of the research.

REFERENCES

1. Astarita,G. Mass Transfer With Chemical Reaction (Elsevier Publishers, Amsterdam, 1967).

2. Danckwerts,P.V. Gas-Liquid Reactions (Mc Graw Hill Co., New York, 1970)

3. Sherwood,T.K., Pigford,R.L. and C.R.Wilke. Mass Transfer (Mc Graw Hill Co., New York, 1975).

4. Barona,N. and H.W.Prengle. "Reactor Design". Hydrocarbon Processing. 52 No: 3 (1973) 63.

5. Carberry,J.J. Chemical and Catalytic Reaction Engineering (Mc Grwa Hill Co., New York, 1976).

6. Froment,G.F. and K.B.Bischoff. Chemical Reactor Analysis and Design (John Wiley and Sons, New York, 1979).

7. Pavlica,R.T. and J.H.Olson. "Unified design method for continous-contact mass transfer operations" Ind.Engng.Chem. 62 No: (1970) 45.

8. Thoenes,D. "Current problems in the modelling of chemical reactors". Chem.Engng.Sci. 35 (1980) 1840.

9. Krishna,R. "Interphase Mass Transfer Models" (Proceedings NATO ASI on "Multiphase Chemical Reactors", Portugal, 1981).

10. Brian,P.L.T., Hurley,H.F. and E.H.Hasseltine. "Gas absorption accompanied by second order chemical reaction". AIChE J. 7 (1961) 226.

11. Porter,K. "The effect of contact time distribution on gas absorption with chemical reaction".Trans.Instn.Chem.Engrs. 44 (1966) T25.

12. Alper,E, "Effect of contact time distribution on gas absorption with chemical reaction" (To be published).

13. Merchuk,J.C. "Further considerations on the Danckwerts-Gilham method for design of gas absorbers". AIChE J. 21 (1975) 815.

14. Porter,K. and D.Roberts. "Similarities between the effect of different flow pattern on diffusion with chemical reaction near an interface". Chem.Engng.Sci. 24 (1969) 695.

15. Davidson,S.F. "The hold-up and liquid film coefficient at packed towers. Part II. Statistical models of the random packing Trans.Instn.Chem.Engrs. 37 (1959) 2092.

16. Alper,E. "Gas absorption with second order reaction: Comparison with approximate enhancement factor equations". Chem. Engng.Sci. 28 (1973) 2092.

17. Kishinevskii,M.K., Kornekko,T.S. and T.M.Popa. "Kinetics o gas absorption". Theor.Found.Chem.Eng. 4 (1971) 641.

18. De Coursey,W.J. "Absorption with chemical reaction: Develo ment of a new relation for the Danckwerts model". Chem.Engng. Sci. 29 (1974) 1867.

19. Santiago,M. and I.H.Farina. "Mass transfer with second ord chemical reaction. Numerical solution". Chem.Engng.Sci. 25 (1970 744.

20. Yeramian,A.A., Gottifredi,J.A. and J.J.Ronco. "Mass transf with homogeneous second order irreversible reaction. A note on

an explicit expression for the reaction rate". Chem.Engng.Sci. 25 (1970) 1622.

21. Baldi,G. and S.Sicardi. "A model for mass transfer with or without chemical reaction in packed columns". Chem.Engng.Sci. 30 (1975) 617.

22. Wellek,R.M., Brunson,R.J. and F.H.Law. "Enhancement factors for gas absorption with second order irreversible chemical reaction". Can.J.Chem.Engng. 56 (1978) 181.

23. Abramzon,A.A. and M.V.Ostrovskii. Zh.Prikl.Khim. 36 (1963) 789.

24. Goodgame,T. and Sherwood,T.K. "The additivity of resistances in mass transfer between phases". Chem.Engng.Sci. 3 (1954) 37.

25. King,J.C. "The additivity of individual phase resistances in mass transfer operations". AIChE Jl. 10 (1964) 671.

26. Szekely,J. "On the addition of phase resistances". Chem. Engng.Sci. 20 (1965) 141.

27. Mannford-Doble,M. "Gas absorption in packed columns". Ph.D. Thesis, Cambridge University (1966)

28. Hartland,S. and J.C.Mecklenburgh. "The concept of backmixing". Chem.Engng.Sci. 23 (1968) 186.

29. Mecklenburgh,J.C. "Backmixing and Design: A Review". Trans. Instn.Chem.Engrs. 52 (1974) 180.

30. Rosen,A.M. and Krylov,V.S. "Theory of scaling-up and hydrodynamic modeling of industrial mass transfer equipment". Chem. Engng.J. 71 (1974) 85.

31. Deans,H.A. and Lapidus,L."A computational model for predicting and correlating the behaviour of fixed bed reactors: I. Derivation of model for nonreactive systems". AIChE Jl. 6 (1960) 656.

32. Deans,H.A. and L.Lapidus. "A computational model for predicting and correlating the behaviour of fixed bed reactors: II. Extension for chemically reactive systems". AIChE Jl. 4 (1960) 563.

33. Nishiwaki,A. and Y.Kato. "Relations between the dispersion model and other models for first order reactions". Can.J.Chem.Engng. 52 (1974) 276.

34. Ham,A. and H.S.Coe. "Calculation of extraction in continous agitation". Chem.Met.Engng. 19 (1918) 663.

35. Schwartz,J.G. and G.W.Roberts. "An evaluation of models for liquid backmixing in trickle bed reactors". Ind.Engng.Chem.Proc. Des.Dev. 12 (1973) 262.

36. Shah,Y.T., Stiegel,G.J. and M.M.Sharma. "Backmixing in gas-liquid reactors". AIChE Jl. 24 (1978) 369.

37. Calo,J. "Multiphase reactor models" (Proceedings of NATO ASI on "Multiphase Chemical Reactors", Portugal, 1980).

38. Sharma,M.M. and P.V.Danckwerts. "Chemical methods of measuring interfacial areas and mass transfer coefficients in two-fluid systems". Brit.Chem.Engng. 15 No: 4 (1970) 206.

39. Alper,E, and W.-D.Deckwer. "Absorption with reaction: Design methods and effect of viscosity for packed columns". (IChemE Jubilee Symposium, EFCE Pub.Series No: 21. London, 1982) p.D129.

40. Alper,E. "Optimum length to diameter ratio of plasctic rings for gas absorption with irreversible chemical reaction". Chem.Engng.Sci. 34 (1979) 1091.

41. Alper,E., Deckwer,W.-D. and P.V.Danckwerts. "Comparison of effective interfacial areas with the actual area for gas absorption in a stirred cell". Chem.Engng.Sci. 35 (1980) 1263.

42. Mashelkar,R.A. and M.M.Sharma. "Mass transfer in bubble and packed bubble columns". Trans.Instn.Chem.Engrs. 48 (1970) TI62.

43.Mehta,V.D. and M.M.Sharma. "Mass transfer in mechanically agitated gas-liquid contactors." Chem.Engng.Sci. 26 (1971) 461.

44. Ganguli,K.B. "Measurement of H_2/edible oil interfacial area in an hydrogenator using a Ziegler-Natta catalyst". Ph.D. Thesis, Technische Hogeschool, Delft (1978).

45. Shende,B.W., Ph.D. Thesis, University of Bombay, India(1975)

46. Bossier,J.A., Farritor,R.E., Hughmark,G.A. and J.T.G.Kao. "Gas-liquid interfacial area determination for a turbine agitated reactor". AIChE Jl. 19 (1973) 1065.

47. Sridharan,K. and M.M.Sharma. "New systems and methods for the measurement of effective interfacial area and mass transfer coefficients in gas-liquid contactors". Chem.Engng.Sci. 31 (1976) 767.

48. Morsi,B.E., Midoux,N., Laurent,A. et J.C.Charpentier. "Hydrodynamique et aire interfacials des écoulements gaz-liquide à co-courant vers le bas en lit fixe. Influence de la nature du liquide". Entropie 16 No:91 (1980) 39.

49. Joosten,G.E.H. and Danckwerts,P.V. "Chemical reaction and effective interfacial areas in gas absorption". Chem.Engng.Sci. 28 (1973) 453.

50. Juvekar,V. and M.M.Sharma. "Some spects of process design of gas-liquid reactors". Trans.Instn.Chem.Engrs. 55 (1977_ 72.

51. van Krevelen,D.V. and P.J.Hoftijzer."Graphical design of gas-liquid reactors". Chem.Engng.Sci. 2 (1953) 145.

52. Danckwerts,P.V. and Sharma,M.M. "The absorption of carbon dioxide into solutions of alkalis and amines (with some notes on hydrogen sulphide and carbonyl sulphide)". Chem.Engr. (London) Oct. (1966) CE264.

53. Kafarov,U.V. and U.A.Reutskii. "New methods for chemical absorber design calculations". J.Appl.Chem.USSR 45 (1972) 1054.

54. Jhaveri,A.S. "Kinetics of gas absorption". Ph.D.Thesis, University of Bombay, India (1968).

55. Onda,K. "Gas absorption in packed columns". Mem.Fac.Eng., Nagoya University, 24 No:2 (1972) 165.

56. Alper,E. "Auslegung von Füllkörperkolonnen als chemische Absorber durch Simulation in Labormodellen". Chemie-Ing.Tech. 51 (1979) 1136.

57. Shende,B.W. and M.M.Sharma. "Mass transfer in packed columns: Cocurrent operation". Chem.Engng.Sci. 29 (1974) 1763.

58. Mashelkar,R.A. "Bubble columns". Brit.Chem.Engng. 15 (1974) 1763.

59. Todt,J., Lücke,J., Schügerl,K. and A.Renken. "Gas hold-up and longitudinal dispersion on different types of multiphase reactors and their possible applications for microbial processes". Chem.Engng.Sci. 32 (1977) 369.
60. Hagberg,C.G. and F.X.Krupa. Proceedings of the 4th Int. 6th European Symposium on Chemical Reaction Engineering. p.408. Dechema, Frankfurt (1976).
61. Mhaskar,R.D. "Effects of backmixing on the performance of bubble column reactors". Chem.Engng.Sci. 29 (1974) 897.
62. Szeri,A., Shah,Y.T. and Madgavkar,A. "Axial dispersion in two phase cocurrent flow with fast and instantaneous reactions". Chem.Engng.Sci. 31 (1976) 225.
63. Deckwer,W.-D. "Non-isobaric bubble columns with variable gas velocity". Chem.Engng.Sci. 31 (1976) 309.
64. Deckwer,W.-D. "Absorption and reaction od isobutene in sulfuric acid". Chem.Engng.Sci. 32 (1977) 51.
65. Kawagoe,M., Nakao,K. and T.Otake. "Design of multistage gas-liquid reactor". Chem.Engng.Japan 5 (1972) 149.
66. Danckwerts,P.V. and A.J.Gillham. "The design of gas absorbers. I.Methods for predicting rates of absorption with chemical reaction in packed columns and tests with 1 1/2 in Raschig rings". Trans.Instn.Chem.Engrs. 44 (1966) T42.
67. Danckwerts,P.V. and E.Alper. "Design of gas absorbers: Part III. Laboratory point model of a packed column absorber". Trans.Instn.Chem.Engrs. 53 (1975) 54.
68. Alper,E. "Laboratory scale-model of a complete packed column absorber". Chem.Engng.Sci.31 (1976) 599.
69. Alper,E. "Laboratory models of a packed-Column absorber" Ph.D.Thesis, Cambridge University, England (1972).
70. Andrew,S.P.S. "Scale up in distillation and absorption". Proc. of Symposium on "Scale-up", Instn.Chem.Engrs. (1968).
71. Laurent,A., Fonteix,C. and J.C.Charpentier. "Simulation of a pilot scale, liquid motivated venturi jet scrubber by a laboratory scale model". AIChE Jl. 26 (1980) 282.
72. Laurent,A. and J.C.Charpentier. "Anwendung experimenteller Labormodelle bei der Voraussage der Leistung von Gas/Flüssigkeits-Reaktoren". Chemie-Ing.Tech. 53 (1981) 244.
73. Alper,E. and W.-D.Deckwer. "Kinetics of absorption of CO_2 into buffer solutions containing carbonic anhydrase". Chem. Engng.Sci. 35 (1980) 2147.
74. Alper,E., Lohse,M. and W.-D.Deckwer. "On the mechanism of enzyme catalyzed gas-liquid reactions: Absorption of CO_2 into buffer solutions containing carbonic anhydrase". Chem.Engng.Sci. 35 (1980) 549.
75. Lohse,M., Alper,E., Quicker,G. and W.-D.Deckwer. "Diffusivity and solubility of carbon dioxide in polymer solutions". Phase Equilibria and Fluid Properties in the Chemical Industry. EFCE Pub. Series No. 11, p. 110 (1980).
76. Charpentier,J.C. and Laurent A. "Some considerations and recalls on the design of gas absorbers: Laboratory apparatus to

simulate absorber and to predict the effect of chemical reaction". AIChE Jl. 20 (1974) 1029.

77. Volgin,B.P., Efimova,T.F. and Gofman,M.S. "Absorption of sulfur dioxide by ammonium sulfite-bisulfite solution in a venturi scrubber". Int.Chem.Engng.8 (1968) 113.

78. Alper,E. "Physical absorption of a gas in laboratory models of a packed column". AIChE Jl. 25 (1979) 545.

79. Sherwood,T.K. Proceedings of the 20th Anniversary Symposium "Mass Transfer and Diffusion", p. 4. University of Houston, Texas, U.S.A. (1973).

80. de Maria,K. and R.R.White. "Transient response study of gas flowing through irrigated packings". AIChE Jl. 6 (1960) 473.

81. Sater,U.E. and O.Levenspiel. "Two-phase flow in packed beds". Ind.Engng.Chem.Fund. 5 (1966) 86.

82. de Waal,K.J.A. and van Mameren,A.C. Proceedings of Symposium on "Transport Phenomena", Rotenburg,P.A. (Ed.), London, The Instn.Chem.Engrs., p. 60 (1965).

83. Dunn,W.E., Vermeulen,T., Wilke,C.R. and T.T.Ward. "Longitudinal dispersion in packed gas-absorption columns". Ind.Engng. Chem.Fund. 16 (1977) 116.

84. Lespinasse,B. Rev. IFP, Génie Chimique 11 (1962) 41.

85. Porter,K.E. and Templemann,J.J. Trans.Instn.Chem.Engrs. 46 (1968) T86.

86. Gillham,A.J. "Kinetics of gas absorption". Ph.D.Thesis, Cambridge University, England (1963).

87. Porter,K.E. Trans.Instn.Chem.Engrs. 46 (1968) T69.

88. Puranik,S.S. and Vogelpohl,A. "Effective interfacial area in irrigated packed columns". Chem.Engng.Sci. 29 (1974) 501.

89. Charpentier,J.C. "Recent progress in two-phase gas-liquid mass transfer in packed columns". Chem.Eng.J. II (1976) 161.

90. Kolev,N. "Wirkungsweise von Füllkörpernschüttungen". Chemie-Ing.Tech. 48 (1976) 1105.

91. Reichelt,W. and Blass,F. "Die Berechnung von Füllkörperapparaten bei Gegenstrom von Gas und Flüssigkeit". Chem.-Ing. Tech. 46 (1974) 171.

92. Alper,E. "Chemical methods of measuring interfacial areas in two-phase systems". (Proceedings of NATO ASI on "Two-phase flow and heat transfer", Turkey, 1976)

93. Alper,E, "Comments on "Gas-liquid reactions: Formulation as initial value problems". Chem.Engng.Sci. 34 (1979) 1076.

94. Mongers,R.J. and Ponter,A.B. "Effect of viscosity on liquid film resistance to mass transfer in a packed column". Ind.Engng.Chem.Process Des.Dev. 19 (1980) 530.

95. Alper,E. "Measurement of effective interfacial area in a packed-column absorber by chemical methods". Trans.Instn.Chem. Engrs. 57 (1979) 64.

96. Sahay,B.N. and Sharma,M.M. "Effective interfacial area and liquid and gas-side mass transfer coefficients in a packed column". Chem.Engng.Sci. 28 (1973) 41.

97. Linek,V., Krivsky,Z. and P.Hudec. "Effective interfacial area in plastic -packed absorption columns". Chem.Engng.Sci. 32 (1977) 323.
98. Linek,V., Petricek,P., Bense,P. and Z.Krivsky. "Spezifische Phasengrenzfläche und die Stoffübergangskoeffizienten in Absorptionskolonnen mit Füllkörpern aus Kunststoffe". Verfahrenstechnik 14 (1980) 733.
99. Linek,V., Stoy,V., Machon,V. and Z.Krivsky. "Increasing the effective interfacial area in plastic absorption columns". Chem.Engng.Sci. 29 (1974) 1955.
100. Norman,W.S. Distillation, Absorption and Cooling Towers. (Longmans and Green and Co. ltd., 1960).
101. Sherwood,T.K. and F.A.L.Holloway. "Performance of packed towers: Experimental studies of absorption and desorption" Trans.Am.Instn.Chem.Engrs. 34 (1940) 21.
102. Mika,V. "Gas absorption in packed columns". Collectn.Czech. Commun. 32 (1967) 2933.
103. Copp,D. and A.B.Pontner. Waerme Stoffübertrag. 5 (1972) 129.
104. Reiss,C.P. "Cocurrent gas-liquid contacting in packed-columns". Ind.Engng.Chem.Proc.Des.Dev. 6 (1967) 846.
105. Mohunta,D., Vaisyanathan,A. and G.Laddha. Indian Chem. Engng. 11 (1965) 73.
106. Richards,G.M., Ratcliff,G.A. and P.V.Danckwerts. "Kinetics of CO_2 absorption. III. First order reaction in a packed column". Chem.Engng.Sci. 19 (1964) 325.
107. Sahay,B.N. and M.M.Sharma. "Letters to the editor". Chem. Engng.Sci.30 (1975) 325.
108. Onda,K., Sada,E. and Y.T.Takeuchi. Chem.Engng.Japan 1 (1968) 56.
109. Danckwerts,P.V. and Rizvi,S.F. "The design of gas absorbers. Part II. Effective interfacial areas for several types of packings" Trans.Instn.Chem.Engrs. 49 (1971) 124.
110. Vidwans,A.D. and Sharma,M.M. "Gas-side mass transfer coefficient in packed columns". Chem.Engng.Sci. 23 (1968) 669.
111. Jhaveri,A.S. and M.M.Sharma. "Effective interfacial area in a packed column". Chem.Engng.Sci. 23 (1968) 669.
112. Rizzuti,L., Augugliaro,V. and Cascio,G.L. "The influence of the viscosity on the effective interfacial area in packed columns". Chem.Engng.Sci. 36 (1981) 973.
113. Alper,E. "The influence of CMC addition on the effective interfacial Area in packed columns" (To be published).
114. Alper,E. "The influence of viscosity on the volumetric mass transfer coefficient" (To be published).
115. Stephens,E.J. and G.A.Morris. "Determination of liquid-film absorption coefficients". Chem.Engng.Prog. 47 (1951) 232.
116. Mehta,V.D. and M.M.Sharma. "Effect of diffusivity on gas side mass transfer coefficient". Chem.Engng.Sci. 21 (1966) 361.
117. Taecker,R.G. and G.A.Hougen. "Heat, mass transfer of gas film in flow of gases through commercial tower packings". Chem.

Engng.Prog. 45 (1949) 188.
118. Tamir,A. and J.C.Merchuk. "Effect of diffusivity on gas side mass transfer coefficient". Chem.Engng.Sci. 34 (1979) 1423.
119. Yadav,G.D. and M.M.Sharma. "Effect of diffusivity on true gas-side mass transfer coefficient in a model stirred contactor with a plane liquid interface". Chem.Engng.Sci. 34 (1979) 1423.
120. Kolar,V., Broz,Z. and J.Tichy. "Liquid hold-up in gas-liquid countercurrent flow through a bed of packing". Coll.Czech. Chem.Comm. 35 (1970) 3344.
121. Tichy,J. "Liquid hold-up in gas-liquid coutercurrent flow through a bed of packing". Chem.Engng.Sci. 28 (1973) 665.
122. Buchanan,J.E. "Hold-up in irrigated ring packed towers below the loading point". Ind.Engng.Chem.Fund. 6 (1967) 400.
123. Hofmann,H. "Hydrodynamik, Transportvorgänge und mathematische Modelle bei Rieselreaktoren". Chemie-Ing.Tech. 47 (1975)8
124. Shulman,H.L., Ullrich,C.F. and N.Wells. "Performance of packed columns. I. Total, static and operating hold-ups". AIChE Jl. 1 (1955) 259.
125. Shulman,H.C., Ullrich,C.F., Wells,N. and A.Z.Prouix. "Performance of packed columns. 3. Hold-up for aqueous and nonaqueou systems". AIChE Jl. 1 (1955) 259.
126. Mohunta,D.M. and G.S.Laddha, "Prediction of liquid hold-up in random packed beds". Chem.Engng.Sci. 20 (1965) 1069.
127. Charpentier,J.C. and M.Favier. "Some liquid hold-up experi mental data in trickle bed reactors for foaming and nonfoaming hydrocarbons". AIChE Jl. 21 (1978) 1213.
128. Jhaveri,A.S. and M.M.Sharma. "Absorption with fast chemica reaction". Chem.Engng.Sci. 24 (1969) 189.
129. Shafer,D.L., Jones,J.H. and T.E.Daubert. "Simultaneous absorption and chemical reaction of butenes". Ind.Engng.Chem.Proc. Des.Dev. 13 (1974) 14.
130. Levenspiel,O. and J.H.Godfrey. "A gradientless contactor for experimental study of interphase mass transfer with/without reaction". Chem.Engng.Sci. 29 (1974) 1723.
131. Laurent,A. Ph.D.Thesis, Nancy University, France (1975).
132. Laurent,A. and J.C.Charpentier. "Le role et utilité des modè les experimentaux de laboratoire dans la prévision des performances d'un réacteur gaz-liquide industriel". J.Chim.Phys. No: 11 (1977) 1001.
133. Danckwerts,P.V. and K.M.McNeil. "Kinetics of CO_2 absorptio into amine solutions". Trans.Instn.Chem.Engrs. 45 (1967) T32.
134. Ouwerkerk,C. "Design for selcetive H_2S absorption". Hydrocarbon Process. April (1978) 89.
135. Hatcher,W.J. and D.R.Hart. "Reaction and mass transport in two-phase reactors: Sulfonation of benzene". Chem.Engng.Sci. 35 (1980) 90.
136. Krötzsch,P. "Abgasreinigung durch Chemiesorption mittels Strahlwäschern am Beispiel der Cl_2-Absorption in wässrige Natriu Thiosulfat-Lösung". Chemie-Ing.Tech. 47 (1975) 213.

137. Kastanek,F., Zahradnik,J., Rylek,M. and J.Kratochvil. Scaling-up of bubble column reactors on basis of laboratory ata". Chem.Engng.Sci. 35 (1980) 456.
138. Mann,R. "Heat and mass transfer in exothermic gas absorpion". (Proceedings of NATO ASI on "Mass transfer with chemical eaction in multiphase systems", Turkey, 1981).

GAS-LIQUID RATE CONSTANT MEASUREMENTS BY CHROMATOGRAPHY

J. ANDRIEU* and J.M. SMITH **

**University of California,Davis,95616,California USA
* Université des Sciences et Techniques du Languedoc, place Eugène Bataillon, 34060 Montpellier (France

ABSTRACT
We present an analysis for absorption and reaction of a pulse of reactant gas moving along a column containing a stationary liquid phase. For first order homogeneous reaction in the liquid film , measured moments of the effluent curve can be used to evaluate rate constants for gas-liquid reactions. This model has been applied to experimental data obtained for the absorption and reaction of carbon dioxide in aqueous Na_2CO_3 - $NaHCO_3$ solutions .

1 INTRODUCTION

In the past, steady-state methods have been used for determining equilibrium and rate parameters (solubilities, rate constants, diffusivities, ...) for gas-liquid process design (1); of recent industrial interest is the selective separation of acid gases by aqueous amine solutions (SO_2, H_2S or CO_2) . On the other hand, pulse response techniques have been useful for obtaining these parameters for first order and reversible adsorption process - SMITH et col (2), (3) . The objective of this communication is to apply this moment method to a chromatographic - type column in which the reacting gas flows and reacts in a stationary liquid phase .

2 THEORY

Consider a column packed with inert non porous particles uniformly coated with absorbent liquid of thickness δ much less than the diameter, d_p of the particles so that the liquid layer may be considered to be flat. At the interface gas and liquid phase concentrations are assumed to be in equilibrium :

$$\frac{c_{g,i}}{c_i} = H \qquad (1)$$

In the liquid, diffusion of gaseous reactant occurs simultaneously with first order irreversible reaction

$$A + M \xrightarrow{k_1} AM \qquad (2)$$

where AM is a non-volatile product

Concentration profile is shown in Figure 1 ; the column is assumed isobaric and isothermal and the gas flow through the bed is represented by the axial dispersion model (dispersion coefficient E_d)

2.1 BASIC EQUATIONS

Mass conservation of A in the gas phase leads to :

$$E_d \cdot \frac{\partial^2 c_g}{\partial z^2} - u \frac{\partial c_g}{\partial z} - D_A \left(\frac{\partial c}{\partial x}\right)_{x=\delta} a = \varepsilon_g \frac{\partial c_g}{\partial t} \qquad (3)$$

where a is the gas-liquid interfacial area per unit volume of empty column .

Mass conservation of A in the liquid film where diffusion and reaction occur simultaneously, gives :

$$D_A \frac{\partial^2 c}{\partial x^2} - r = \frac{\partial c}{\partial t} \qquad (4)$$

The reaction is assumed irreversible and first order so that the reaction rate is :

$$r = k_1 . c \qquad (5)$$

where c is the concentration of reactant A in the liquid

The boundary and initial conditions for a pulse input (injection time = t_m) are :

$$x = \delta \qquad D_A \left(\frac{\partial c}{\partial x}\right)_{x=\delta} = k_f (c_g - c_{g,i}) \qquad (6)$$

$$x = 0 \qquad c = 0 \qquad (7)$$

$$z = 0 \qquad c_g = 0 \text{ for } t > t_m > 0, \; c_g = c_{g,0} \text{ for } 0 < t < t_m \qquad (8)$$

$$t = 0 \qquad c_g = c = 0 \qquad (9)$$

This system of linear equations can be solved in the Laplace domain to give $\bar{c}_g(L,s)$ (4) . Then moments, m_n of the response curve at the bed exit (z = L) may be obtained from $\bar{c}_g(L,s)$ and the limit equation :

$$m_n = (-1)^n \lim_{s \to \infty} \frac{d^n \bar{c}_g(L,s)}{ds^n} \qquad (10)$$

'he n^{th} moment at the bed exit is defined in terms of the 'esponse curve by the expression :

$$m_n = \int_0^\infty t^n . \; c_g(L,t)dt \qquad (11)$$

;o, the zeroth reduced moment is defined by :

$$\mu_0 = \frac{1}{c_{g,0} . \; t_m} \int_0^\infty c_g(L,t) \; . \; dt \qquad (12)$$

ınd, in the case of an irreversible reaction, is equal to the 'raction of A which has not reacted at the bed outlet

'ith equation (11) the first absolute moment is

$$\mu_1 = \frac{m_1}{m_0} = \frac{\int_0^\infty t.c_g(L,t)\;dt}{\int_0^\infty c_g(L,t)dt} \qquad (13)$$

.2 MOMENT EQUATIONS

olution of equations (3-9) in the LAPLACE domain may be ubstituted in equation (11) to evaluate μ_0 and μ_1 (4). f axial dispersion is neglected, the reduced zero moment is iven by :

$$\mu_0 = \exp\left[-\,\alpha . \; R(0)\right] \qquad (14)$$

here:

$$R(0) = \frac{\Phi . \; Sh. \; \sinh \Phi}{\Phi \;\; \sinh \Phi + Sh. \; \cosh \Phi} \qquad (15)$$

$$\alpha = \frac{D_A . L . a}{H. \; \delta . \; u} \qquad (16)$$

$$\Phi^2 = \frac{k_1 \; . \; \delta^2}{D_A} \qquad (17)$$

$$Sh = \frac{k_f . \; \delta . \; H}{D_A} \qquad (18)$$

or $\Phi \to 0$ from equations (14) and (15), we obtain $\mu_0 \to 1$, onsistent with the fact that all reactant in the pulse would ppear in the response .

An interesting limiting case correspond to negligible resistance from gas to liquid (large Sherwood number values) and negligible diffusion resistance within the liquid with respect to reaction resistance; then equations (14 - 18) reduce to :

$$\mu_0 = \exp- \alpha . \Phi^2 = \exp- \frac{k_1 . \varepsilon_L . L}{H . u} = \exp- \frac{k_1 \varepsilon_L}{H . \varepsilon_g} . \tau \quad (19)$$

This expression is the same as that derived by LANGER (5) for a first order irreversible reaction in a chromatographic column. Thus a plot of experimental values of μ_0 as a function of τ provides a method for evaluating the kinetic constant if H is known from literature data. At the other extreme when the process is controlled by mass transfer from gas to liquid (fast reaction) $\Phi \to \infty$ the zero moment becomes :

$$\mu_0 = \exp- (\alpha . Sh) = \exp- \left(\frac{k_f . L . a}{u}\right) \quad (20)$$

Besides, for the first moment we obtain :

$$\mu_1 = \tau \left\{ 1 + \frac{1}{2} \frac{Sh^2}{H} . \left(\frac{\varepsilon_L}{\varepsilon_g}\right) \frac{1 + \frac{\sinh . 2\Phi}{2\Phi}}{[\Phi \sinh \Phi + Sh . \cosh \Phi]^2} \right\} \quad (21)$$

Thus, we obtain a more complicated expression that in the case of a reversible first order reaction that is (4) :

$$\mu_1 = \tau \left\{ 1 + \frac{(K+1)}{H} \quad \frac{\varepsilon_L}{\varepsilon_g} \right\} \quad (22)$$

This expression shows that for a non reacting gas (K = 0), the first moment of the response peak can be used to determine solubility of A in the liquid .

In figure 2 the reduced zero moment is plotted as a function of Thiele modulus, Φ, for α ranging from 0.1 to 10 and for $Sh = 10^3$. This value of Sh corresponds, approximately to experimental conditions given in Table 1 . This plot indicates that systems and conditions for which α is in this range should be suitable for determining k_1 from zeroth moments .

3 EXPERIMENTAL

Suitability of this theory for evaluating rate constants by chromatography was checked with the absorption and reaction of CO_2 in an aqueous Na_2CO_3 - $NaHCO_3$ solution . The kinetic constant of this irreversible , first-order reaction has been measured by classical methods by ROBERTS and DANCKWERTS (6).

In order to verify hypothesis of derivation of equation (14) - that is, contributions to zero moments of axial dispersion, gas-to-liquid mass transfer and liquid-phase diffusion is negligible- our experiments were carried out with high gas flow rates relatively long columns and thin liquid films .

Inert support consist of glass beads (1 mm nominal diameter) ; coating ratio (mass of liquid/mass of dry particles) was about 10 % for all experiments. Column was made of soft copper (ID = 0.92 cm ; L = 0.27 m or 0.56 m) and was packed by adding particles slowly while vibrating the tube. Then the tube was coiled for insertion in the chromatograph oven. The major experimental problem was the control of column preparation which gave reproducible data. We believe that variations are primarily due to non uniform liquid coating and partial breakdown of this liquid film during the packing. For the experimental conditions given in Table 1 Peclet and Sherwood numbers and Thiele modulus were estimated Pe = 250, Sh = 3000 and $\Phi = 0.2$. After numerous runs, reasonably reproducible moments were obtained from Equation (12). Final results for two columns are plotted as ($10.\ \mu_0$) vs L/v in figure 3 . The data show a linear behaviour as indicate by equation (19). The slope of these lines is equal to $k_1 . \varepsilon_L / H.\varepsilon_g$. The proper value of H, calculated by the method of VAN KREVELEN and HOFTIJZER (7) at 25°C is H=2.2 (moles/cm^3 of gas)/(moles/cm^3 of liquid). With this value and the data of Table 2, the slopes of the lines in figures 4 give the mean value for the rate constant :

$$k_1 = 0.43 \text{ sec}^{-1}$$

For the same conditions DANCKWERTS and ROBERTS reported $k_1 = 1.0 \text{ sec}^{-1}$, value obtained with a wetted wall column

4 CONCLUSION

Equations have been derived for zero and first moments of the response curve for a pulse input of absorbable gas (in an inert carrier) which reacts homogeneously inside a layer of stationary liquid. The results are restricted to first order irreversible reactions, isothermal and isobaric plug flow conditions.

As predicted by the theory, experimental results show that the zero moment is a decreasing fonction of the space time .

Kinetic constant given by this simplified theory - by plotting $\ell n\ \mu_0$ as a function of the residence time L/v - is of the same order of magnitude as the literature values obtained with steady state method in laboratory gas-liquid contactors .

Improvements in chromatographic method of measuring rate constants will depend upon improved technique of particle coating and column packing. We think, that for convenient systems capillary column, could also be advantageously used .

NOMENCLATURE

a	gas-liquid interfacial area per unit volume of empty column, cm^{-1}
c_g	concentration of absorbable reactant A in the gas phase, $g.mole/cm^3$; $c_{g,o}$ = concentration in injection pulse
c	concentration of A in the liquid phase
$\bar{c}$	Laplace transform of concentrations
D_A	diffusivity of dissolved A in the liquid layer cm^2/sec
d	column diameter
d_p	particle diameter
E_d	axial dispersion coefficient cm^2/sec
H	Henry's law constant
K	equilibrium constant
k_f	gas side mass transfer coefficient cm/sec
k_1	first-order, forward reaction rate constant sec^{-1}
L	column length
m_n	n-th moment at column outlet
Pe	axial Peclet number in the gas $Pe = \frac{L.u}{E_d}$
$R(0)$	dimensionless group defined by equation (15)
r	reaction rate per unit volume of liquid g. mole/ cm^3.sec
Sh	Sherwood number defined by equation (18)
s	Laplace variable
t	time, sec
t_m	pulse injection time, sec
u	superficial velocity, cm/sec
v	interstitial velocity in the packed bed, $v = \frac{u}{\varepsilon_g}$ cm/sec
x	coordinate normal at gas-liquid interface
z	axial coordinate from bed entrance

GREEK

α	dimensionless liquid-diffusion group defined by equation(16)
δ	thickness of liquid layer
ε_o	porosity of bed without liquid $\varepsilon_o = \varepsilon_L + \varepsilon_g$
ε_L	liquid hold-up in column $\varepsilon_L = a.\delta$
ε_g	gas hold-up in column

μ_o zero moment defined by (12)

μ_1 first absolute moment defined by equation (13)

Φ Thiele modulus defined by equation (17)

τ residence time of gas $\tau = L/v = \varepsilon_g . L/u$

REFERENCES

1. Danckwerts P.V. and M.M. Sharma. The Chemical Engineer, 10 , (1966), 244-280

2. Schneider P. and J.M. Smith. A.I.Ch.E Jl, 14 (1968), 262

3. Hashimoto N. and J.M. Smith. Ind.Eng.Chem. Fundamen., 12 (1973), 351

4. Andrieu J. and J.M. Smith . The Chem. Eng. Jl., 20 ,(1980),211

5. Langer J.H., H.R. Melton and T.D. Griffith. Jl of Chromatography, 122 (1976),487

6. Roberts D. and P.V. Danckwerts . Ch. Eng.Sci. , 17 (1962),961

7. Van Krevelen D.W. and Hoftijzer . Chimie et Industrie, Congrès de Chimie Industrielle, Bruxelles, sep. 1968 , p. 168

TABLE 1 : OPERATING CONDITIONS FOR A LIQUID-COATED PACKED COLUMN (Solute gas in helium carrier)

Temperature =	25°C
Pressure =	1 atm
Column diameter, d =	0.82 cm
Column length, L =	30 cm
Packing : glass beads, d_p =	0.1 cm
Total bed porosity, ε_0 =	$\varepsilon_L + \varepsilon_g = 0.40$
Liquid film thickness, δ =	10^{-3} cm
Liquid retention, ε_L =	0.036
Retention ratio $\varepsilon_L / \varepsilon_g$ =	0.09
Diffusivity of A in liquid, D_A =	2×10^{-5} cm^2/sec
Reaction rate constant, k_1 =	100 $(sec)^{-1}$
Diffusivity of A in gas phase =	0.60 cm^2/sec
Kinetic viscosity of gas phase =	1.18 cm^2/sec
Gas flow rate,	0.83 to 5.5 cm^3/sec

TABLE 2 : EXPERIMENTAL CONDITIONS

COLUMN	
temperature	25°C
pressure	1 atm
diameter (ID)	0.92 cm
length, column I, L_1	0.27 m
column II, L_2	0.56 m
PACKING, GLASS BEADS, diameter	0.092 cm
BED POROSITY (liquid + gas)	0.40
LIQUID	
hold-up ε_L	~0.06
film thickness δ	12-14 microns
GAS	
hold up ε_g	~0.34
flow rate (25°C , 1 atm)	41.7 to 117 cm^3/min
pulse composition	1.5 % CO_2 in He
pulse injection volume	1.0 cm^3
COATING SOLUTION	
Na_2CO_3 conc.	0.625 molal
$NaHCO_3$ conc.	0.530 molal
CARRIER GAS	pure helium

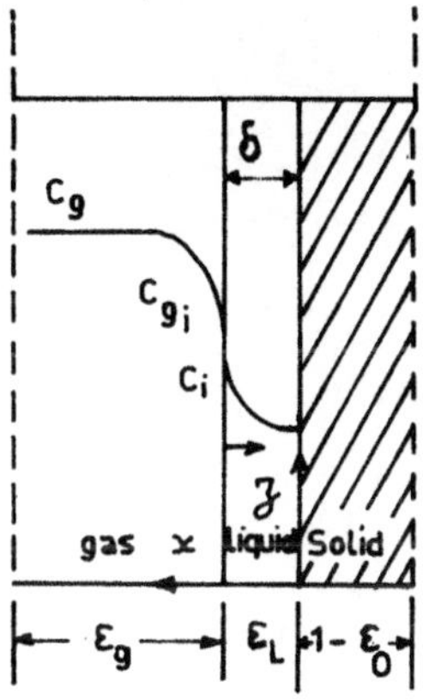

Figure 1 Concentration Profile For Absorbable Reactant

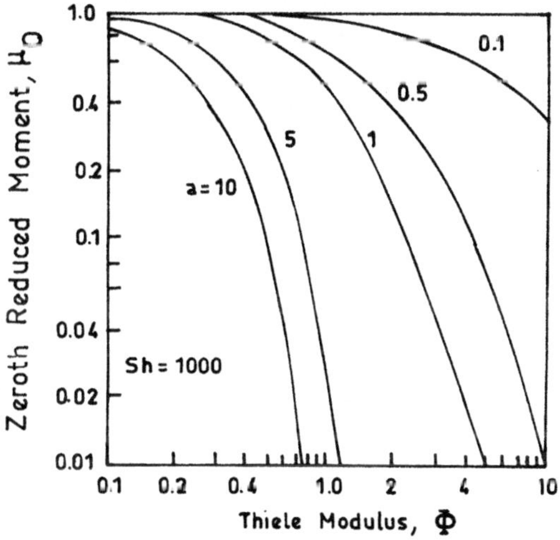

Figure 2 Effect of liquid Phase Diffusion (a) on Zeroth Moment for Irreversible Reaction and Sh = 1000

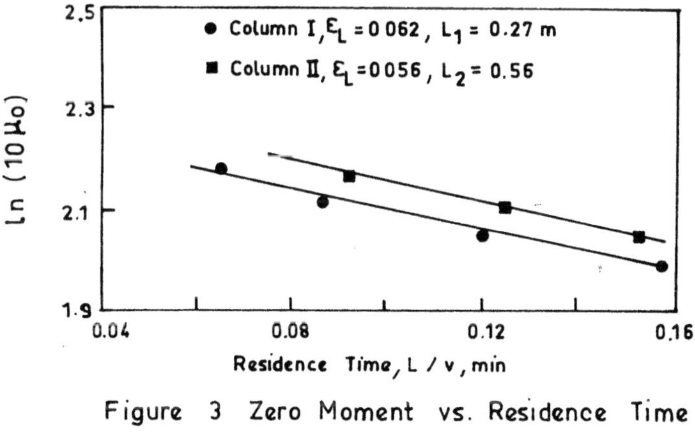

Figure 3 Zero Moment vs. Residence Time

DETERMINATION OF GAS-LIQUID MASS TRANSFER BY OXIDATION OF HYDRAZINE

R. Sick, P. Weiland and U. Onken

Lehrstuhl für Technische Chemie B, Universität Dortmund, F.R.G.

Summary
It is shown that the oxidation of hydrazine in the presence of homogeneous catalysts is a suitable model reaction for the determination of mass transfer in gas/liquid systems.

1. Introduction

For the determination of volumetric mass transfer coefficients $\beta_L a$ in gas/liquid systems both physical and chemical methods are used. Commonly used chemical methods are the sulfite oxidation and the reaction of carbon dioxide in alkali hydroxide or ethanolamine solutions. The disadvantage of these methods is, that the measured $\beta_L a$-values depend on the type of model medium. All these systems are combined with high salt concentrations, which inhibit coalescence. A change of the coalescence behaviour means a change in gas hold-up, interfacial area, and gas residence time. In systems with coalescence restraining, $\beta_L a$-values are 1.5 to 7 times higher than in coalescing systems. As a suitable reaction for the determination of $\beta_L a$ in coalescing systems, Zlokarnik |1| proposed the oxidation of hydrazine in aqueous solution by atmospheric oxygen. According to eq. (1) the reaction products are not accumulating and do not change the coalescence behaviour:

$$N_2H_4 + O_2 \xrightarrow{\text{cat.}} N_2 + 2H_2O \qquad (1)$$

Up to now only precipitated copper hydroxide has been used as catalyst. In the present work we tested the application of several soluble compounds for homogeneous catalysis. Besides that we also investigated the application and reliability of the system containing the suspended catalyst.

2. Hydrazine Methode
For the determination of $\beta_L a$, the solution or the suspension of the catalyst is prepared in an aerated 4.5 l vessel, which is agitated by a six-blade turbine. After adjusting pH, aeration rate and stirring speed, a constant rate of hydrazine is fed into the reactor. At steady state the feed rate of hydrazine $\dot{n}_{Hy}$ is equal to the absorption rate of oxygen $\dot{n}_{O_2}$, as written in eq. (2):

$$\dot{n}_{Hy} = \dot{n}_{O_2} = \beta_L a \, V_L \, \Delta c_L \tag{2}$$

V_L volume of liquid

Δc_L liquid side concentration gradient

Our experiments were carried out at 25 °C. In all cases the ionic strength of the system was low enough to avoid coalescence inhibition.

3. Suspended copper catalyst
According to Zlokarnik |1| the system copper sulfate/sodium hydroxide was used for catalyzing the hydrazine oxidation. Addition of sodium hydroxide to the solution of copper sulfate causes precipitation of copper ions as copper hydroxide and cupric oxide. After feeding hydrazine into the reactor, also copper(I)-oxide is formed, which is reoxidized by oxygen.

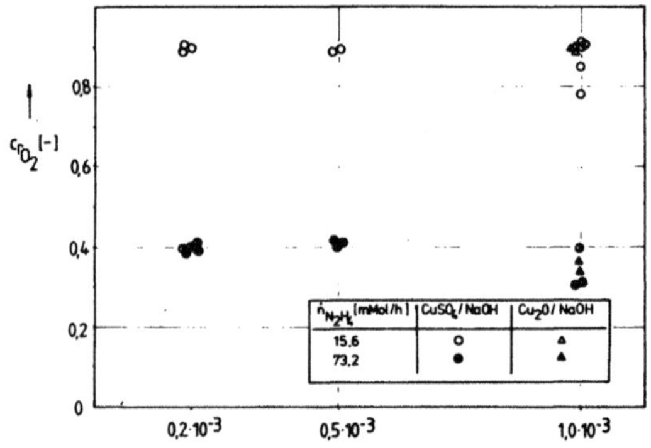

copper content |gmol/l|

Fig. 1: Relative oxygen saturation as function of the copper content

In the experiments reported here, the total copper content was varied. Fig. 1 shows steady state oxygen concentration at two different feed streams of hydrazine. As can be seen, there is no influence of the copper content on the consumption of oxygen. Suspensions, prepared by adding Cu_2O to NaOH-solution instead of precipitating of hydroxide from $CuSO_4$, yielded the same absorption rate (see fig. 1).

From these results, it can be assumed that the reaction is catalyzed homogeneously by dissolved copper ions, which are present in concentrations of about 10^{-6} gmol/l at the conditions of our experiments. This agrees with results of Gaunt et al. |2|. The reaction could be stopped by complexing the copper ions using ethylenediamine tetraacetic acid.

In suspensions of freshly precipitated $Cu(OH)_2$ a higher conversion of hydrazine was observed; reproducible values could only be measured in suspensions older than one day. Another restriction of the determination of $\beta_L a$ by the hydrazine method concerns the concentration of dissolved oxygen. Chemically determined values only agreed with those from dynamic measurements by a physical method, when the oxygen content was between 40 and 60% of saturation (air, 1 bar). $\beta_L a$-values versus superficial gas velocity are shown in fig. 2.

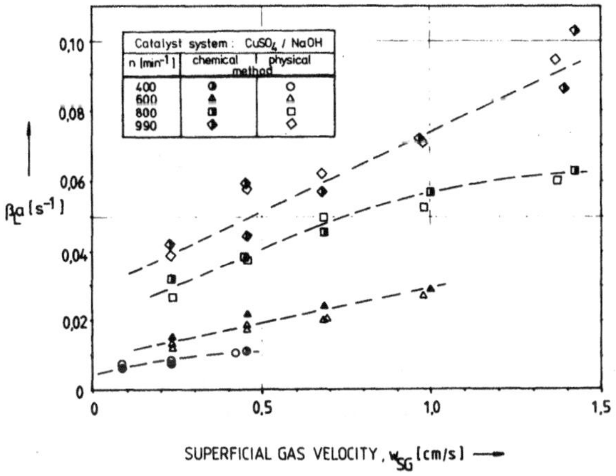

Fig. 2: Comparison of volumetric mass transfer coefficients in a stirred tank from chemical and physical method

pH of the reaction must be above 10. In the range of pH = 10.8-12, the reaction is fast enough, to consume hydrazine immediately |3|; this means that $\beta_L a$ is not affected.

4. Homogeneous Catalysis

Though $\beta_L a$-values determined in the suspension agree well with those from physical measurements, the method shows following disadvantages:

- long time (24 h) up to stable catalytic conditions;
- inhomogeneous dispersion of solids, especially in tall reactors;
- solids deposited on probes may cause experimental errors;
- $\beta_L a$-values may depend on relative oxygen concentration;
- bubbles and flow pattern cannot be observed visually because of turbidity.

By replacing the suspended copper catalyst by homogeneous catalysts, an essential improvement of the method is to be expected.

It has been found, that ions of elements of the 1st and 8th subgroup of the periodic system are able to catalyze the hydrazine oxidation |2|. In order to protect the ions from reduction into insoluble compounds, complexing agents must be used. According to stereospecifity, possible redox mechanisms, and stability constants, various systems were selected |3|.

Tab. 1: Complexes, investigated in respect to application as homogeneous catalysts

Ligand	Central Ion	pH-agent	Range of pH	Concentration |gmol/1|
Ammonia	Cu^{2+}	NH_4OH	10.5–11.0	$1\cdot10^{-4}$–$1.6\cdot10^{-4}$
	Co^{2+}		10.5–12.0	$1.6\cdot10^{-6}$ – $1.6\cdot10^{-4}$
Sulfonated Phthalocyanines	Cu^{2+}	NaOH	12.0–13.0	$1\cdot10^{-5}$–$1\cdot10^{-4}$

Tab. 1 shows a few of these systems. Most suitable for the determination of $\beta_L a$ were cupra ammines and copper sulfo-phthalocyanines.

4.1 Cupra Ammines

The cupra ammines were prepared by adding ammonia to a solution of $CuSO_4$. Above pH = 10 the solution was completely clear. The relative oxygen content was varied between 37 and 70%. Within the range of our experiments no effect of catalyst concentration and of pH on $\beta_L a$ was observed. Some results are shown in fig. 3.

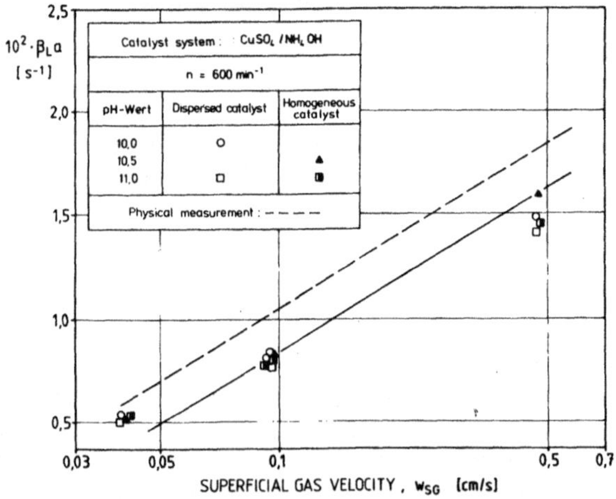

Fig. 3: $\beta_L a$ determined in presence of cupra ammines

This figure also shows results obtained with precipitated Cu_2O and from the physical dynamic method. The suspension was generated by reduction of dissolved copper(II) to solid cuprous oxide by an excess of hydrazine. There is no difference of the β_La-values determined in homogeneous solution and in suspension. But the results from the hydrazine method are about 13% smaller than from the physical method. This may be due to a gas phase gradient caused by desorption of ammonia.

The method employing cupra ammines is simple in application and reliable because the results are independent of the conditions of the system in a wide range. Losses of ammonia cannot be avoided, even by cooling the effluent gas. Therefore further installations, such as an absorber for ammonia, are necessary.

.2 Copper Sulfophthalocyanines
Copper phthalocyanines are very stable chelates with a planar-quadratic structure. Their sulfonated derivatives are well soluble in water. A sodium salt of copper tetrasulfophthalocyanine of high purity (CuTSP) and an industrial dyestuff, containing 2.7-2.8 sulfogroups per phthalocyanine, were investigated. Experimental conditions are shown in tab. 1.

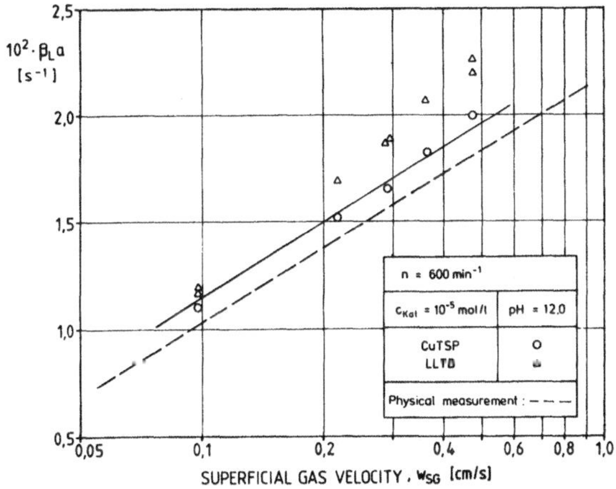

Fig. 4: Comparison of β_La from dynamic physical and chemical (phthalocyanine catalysts) measurements

Fig. 4 shows β_La as a function of superficial gas velocity. The results of the system containing CuTSP are very similar to those from the physical method. But on an average, the β_La-values of the hydrazine method are 8% higher. Within the investigated range both methods agree excellently. The results with the industrial product, Lurantinlichttürkisblau (LLTB) supplied by BASF, show an increasing deviation on increasing aeration rate.

The results show that the oxidation of hydrazine can be catalyzed homogeneously by copper complexes. So the disadvantages combined to the suspended catalyst can be avoided. The ionic strength and therefore the coalescence behaviour can be adjusted individually.

In semi-industrial scale, the limits of application and the feasibility of the new method are still to be investigated. At presence other phthalocyanines are tested with respect to replace the expensive tetrasulfonate by cheaper mass products.

Acknowledgment
The authors have to acknowledge Mrs. Gadooni and Mr. Materne, who have carried out a part of the experiments.

References

|1| Zlokarnik, M.: Adv. Biochem. Eng. 8 (1978) 133
|2| Gaunt, H. and Wetton, B.A.M.: J. Appl. Chem. 16 (1966) 171
|3| Weiland, P.; Sick, R. and Onken, U.: Chem.-Ing-Tech. 53 (1981) 580

SIMULTANEOUS MASS TRANSFER OF TWO GASES WITH COMPLEX REVERSIBLE REACTIONS:AN EXAMPLE BEING THE SIMULTANEOUS ABSORPTION OF H_2S AND CO_2 INTO AQUEOUS AMINE SOLUTIONS

P.M.M.Blauwhoff,G.J.B.Assink,W.P.M. van Swaaij

Department of Chemical Engineering
P.O.Box 217,7500 AE ENSCHEDE
The Netherland

INTRODUCTION

Though scrubbing of gases containing H_2S and CO_2 by alkanolamines in aqueous or mixed solutions is a well established process,the calculation of mass transfer fluxes, taking into account interactive chemical reactions , is still a very complex problem.

The reaction of H_2S with primary and secondary amines is reversible and instantaneous while for CO_2 the reaction is also reversible ,but has a finite rate [6].Several (approximate) analytical models are available in literature describing either mass transfer of a single gas component with instantaneous reversible reactions[13,14,16,17] and irreversible reaction of finite rate [9,11,12,14,18,19] or mass transfer of two gases with one or two irreversible reactions[3,5,10,15].Neither of these models nor combination of models is however able to describe the simultaneous mass transfer of H_2S and CO_2 and liquid phase reactions sufficently accurate under all relevant conditions.

Therefore we developed a more complete model and solved it with numerical methods[4].In our previous work [4] we demonstrated the numerical stability of the solution method as well as the agreement between most of the analytical models mentioned above and the corresponding limiting cases of our model. We also demonstrated some of the features of simultaneous absorption accompanied by complex reactions of hydrogen sulphide and carbon dioxide as an example.

One very interesting feature of simultaneous mass transfer, previously defined as forced desorption [4], will be the main issue of this work. In this case net desorption of one of the gaseous species will occur against the direction of its overall driving force due to extreme interaction of the liquid-phase reactions and consequently the enhancement factor will be negative. We selected this specific item out of numerous possibilities because we were interested to see if this phenomenon could be realized in practice and because it reflects the ultimate degree of interaction which cannot be described by any other model.
The essence of forced desorption has been reported in earlier work [4] without experimental results and for very high mass transfer coefficients which require the use of special equipment [1].
Therefore we studied the absorption of H_2S and CO_2 into an aqueous DIPA solution using a stirred cell reactor and the associated physico-chemical parameters and mass transfer properties were used as a starting point for calculations.

The model

In our previous work [4] the model for simultaneous absorption of H_2S and CO_2 with complex reactions and its refinements is described extensively. Here we confine ourselves to a recapitulation of the starting points.

The reaction between H_2S and an amine R_2NH is reversible and instantaneous and is given by [6]:

$$H_2S + R_2NH \rightleftharpoons HS^- + R_2NH_2^+ \qquad (1)$$

$$\text{where } K_{H_2S} = \frac{[HS^-][R_2NH_2^+]}{[H_2S][R_2NH]} \qquad (2)$$

Throughout the liquid equilibrium (1) is established due to infinitely fast forward and backward reaction rates.
CO_2 reacts reversibly with primary and secondary amines according to [6]:

$$CO_2 + 2R_2NH \rightleftharpoons R_2NCOO^- + R_2NH_2^+ \qquad (3)$$

$$\text{where } K_{CO_2} = \frac{[R_2NCOO^-][R_2NH_2^+]}{[CO_2][R_2NH]^2} \qquad (4)$$

Reaction (3) takes place at a finite rate and the net overall reaction rate is expressed by [7].

$$r = k_2 \left\{[CO_2][R_2NH] - \frac{1}{K_{CO_2}} \frac{[R_2NCOO^-][R_2NH_2^+]}{[R_2NH]}\right\} \qquad (5)$$

Although more complicated rate expressions for reaction (3) are suggested in literature [8] and also our recent findings suggest a slightly different rate equation [2], equation (5) was found to be sufficiently accurate for the purpose of this work. Other CO_2 consuming reactions as well as the hydrolysis of the carbamate ion, R_2NCOO^-, are slow compared to reaction (3) and are therefore not incorporated in the model.

The gas-phase mass transfer is described by the stagnant film model whilst for the liquid phase Higbie's penetration model was used. The process of diffusion and simultaneous reaction in the liquid-phase penetration zone is given by the following balances [4]:

the carbon dioxide reaction balance:

$$\frac{\partial[CO_2]}{\partial t} = D_{CO_2}\frac{\partial^2[CO_2]}{\partial x^2} - k_2\,[CO_2][R_2NH] + \frac{k_2}{K_{CO_2}}\frac{[R_2NCOO^-][R_2NH_2^+]}{[R_2NH]} \qquad (6)$$

the total carbon dioxide balance:

$$\frac{\partial[CO_2]}{\partial t} + \frac{\partial[R_2NCOO^-]}{\partial t} = D_{CO_2}\frac{\partial^2[CO_2]}{x^2} + D_{R_2NCOO^-}\frac{\partial^2[R_2NCOO^-]}{\partial x^2} \qquad (7)$$

total sulphur balance:

$$\frac{\partial[H_2S]}{\partial t} + \frac{\partial[HS^-]}{\partial t} = D_{H_2S}\frac{\partial^2[H_2S]}{\partial x^2} + D_{HS^-}\frac{\partial^2[HS^-]}{\partial x^2} \qquad (8)$$

total amine balance

$$\frac{\partial[R_2NH]}{\partial t} + \frac{\partial[R_2NH_2^+]}{\partial t} + \frac{\partial[R_2NCOO^-]}{\partial t} = D_{R_2NH}\frac{\partial^2[R_2NH]}{\partial x^2} + D_{R_2NH_2^+}\frac{\partial^2[R_2NH_2^+]}{\partial x^2} + D_{R_2NCOO^-}\frac{\partial^2[R_2NCOO^-]}{\partial x^2} \qquad (9)$$

the acid balance:

$$\frac{\partial[H_2S]}{\partial t} + \frac{\partial[CO_2]}{\partial t} + \frac{\partial[R_2NH_2^+]}{\partial t} = D_{H_2S}\frac{\partial^2[H_2S]}{\partial x^2} + D_{CO_2}\frac{\partial^2[CO_2]}{\partial x^2} + D_{R_2NH_2^+}\frac{\partial^2[R_2NH_2^+]}{\partial x^2} \qquad (10)$$

Equilibrium equation (2) is used as reaction balance for H_2S. After discretisation of the set of partial differential equations (6)-(10) and linearization by the Newton-Raphson technique the system is solved by an iterative numerical procedure. Boundary conditions and more detailed information are given elsewhere [4].

A qualitative view on simultaneous absorption of H_2S and CO_2 into amine solutions.

Consider the absorption process schematically given in figure 1. The species H_2S and CO_2 diffuse from the bulk of the gas phase to the interface. Here the gas concentrations are in dynamic equilibrium with their liquid-phase concentrations. In the liquid, H_2S and CO_2 diffuse towards the bulk and both react simultaneously with the amine according to equations (1) and (3) respectively. As long as the mass-transfer rates of the commonly involved species, R_2NH and $R_2NH_2^+$, are high in comparison with the net conversion rates, their concentrations in the penetration zone equal the respective bulk concentrations. Hence reactions (1) and (3) can be regarded as totally independent of each other (see figure 2a).

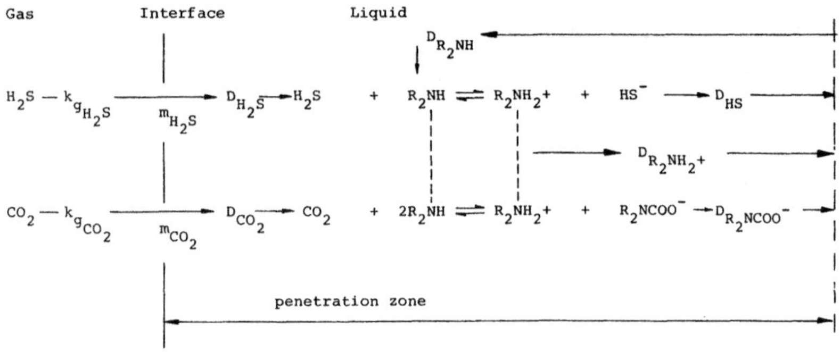

figure 1: Scheme of the absorption proces.

Increasing the two gas-phase concentrations to the same extent enhances the overall net reaction rates and thereby the amine consumption in the penetration zone. If the amine supply from the bulk and the coupled transport of reaction products to the bulk are slow compared to the reaction rate, the amine concentration will fall sharply while the product concentrations increase. At this point the H_2S and CO_2 compete for the amine present in the penetration zone and reactions (1) and (3) have become interdependent (see figure 2b).
If we start again with the first non-interactive situation and increase for example the CO_2 gas-phase concentrations sharply, the transport phenomena become quite different as only the CO_2 conversion rate is increased. The amine for this increased reaction will now be supplied by two mechanisms: firstly through diffusion from the bulk to the penetration zone and secondly by means of a reversal of the overall net rate of the H_2S-amine reaction. This latter mechanism is enforced by the increased production of $R_2NH_2^+$ which in combination with the amine consumption causes a shift in the H_2S-amine reaction and produces both amine and free H_2S. If the transport rate of this free gaseous H_2S is small in proportion to its production rate, a high local H_2S concentration can be obtained. This local concentration can exceed the interfacial concentration and consequently leads to diffusion of part of the free H_2S towards the gas phase. The net result will be desorption although based on the overall driving force absorption of H_2S would have been

figure 2a:

Concentration profiles
I without interaction

$[H_2S]_g = 4$ moles/m^3
$[CO_2]_g = 4$ moles/m^3

figure 2b:

Concentration profiles
II with interaction

$[H_2S]_g = 20$ moles/m^3
$[CO_2]_g = 20$ moles/m^3

figure 2c:

Concentration profiles
III forced desorption

$[H_2S]_g = 4$ moles/m^3
$[CO_2]_g = 80$ moles/m^3

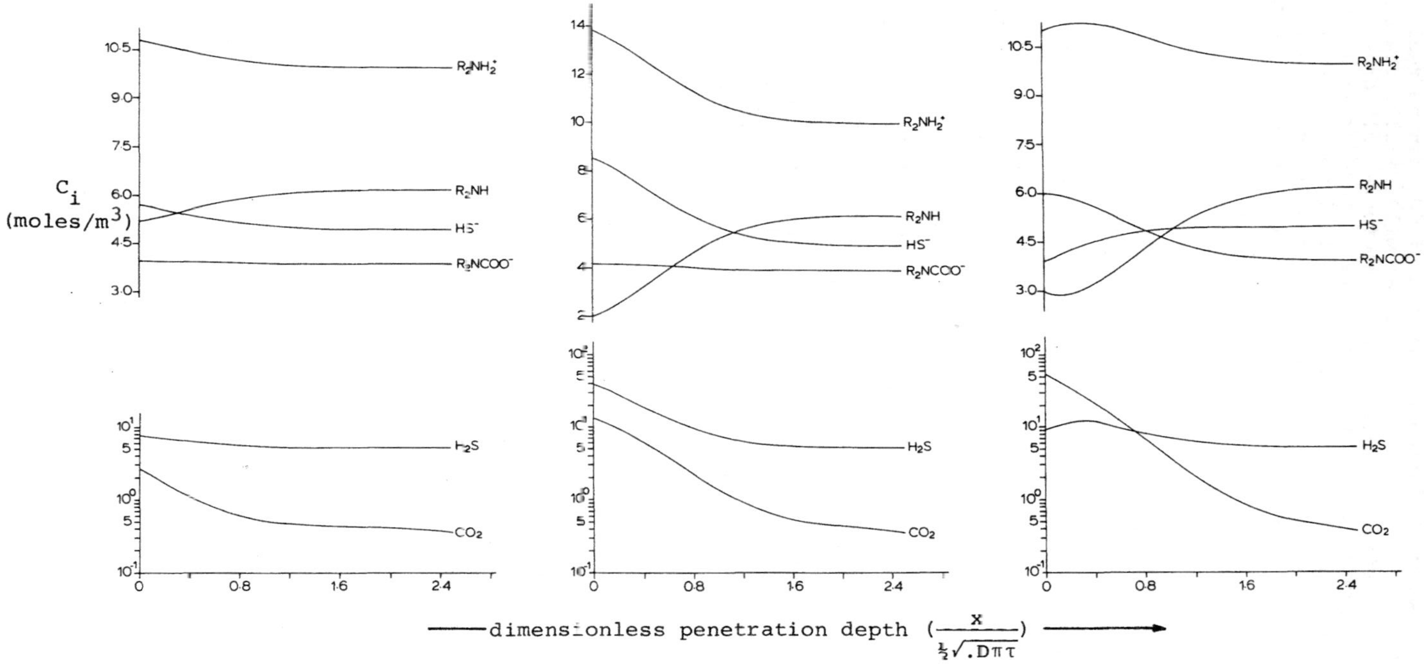

expected (see figure 2c). Consequently the enhancement factor is negative. We defined this phenomenon as forced desorption [4]. Analogously forced desorption of CO_2 can be realized at high H_2S gas-phase concentrations as will be illustrated in the next section. Although in the above we only presented a qualitative view on simultaneous mass transfer with complex reactions this approach can be a usefull tool in estimating the effects of physico-chemical and mass-transfer parameters on transport phenomena. This will be illustrated with the help of various calculations.

Illustrative calculations

In this section the influence of various parameters on simultaneous mass transfer in general and forced desorption in particular will be demonstrated.
Of course neither these illustrations nor the number of parameters varied are limitative.
In the calculations the larger part of the physico-chemical parameters is fixed (see table 1) and only mass-transfer coefficients, k_g and k_1, and gas-phase concentrations have been varied. We selected these particular parameters because they can be easily varied in industrial and laboratory mass-transfer equipment.

Figure 3 shows the influence of the CO_2 gas-phase concentration on H_2S and CO_2 mole fluxes, J_{H_2S} and J_{CO_2} respectively. The charged DIPA solution ($\alpha_{H2S} = \alpha_{CO_2} = 0.25$) is in equilibrium with a CO_2 gas-phase concentration of 0.55 mole/m^3, indicated by the dotted line. Below this concentration a negative driving force is present and desorption of CO_2 will occur. Increasing $[CO_2]_g$ to slightly above the equilibrium concentration yields a positive driving force but desorption is still found.

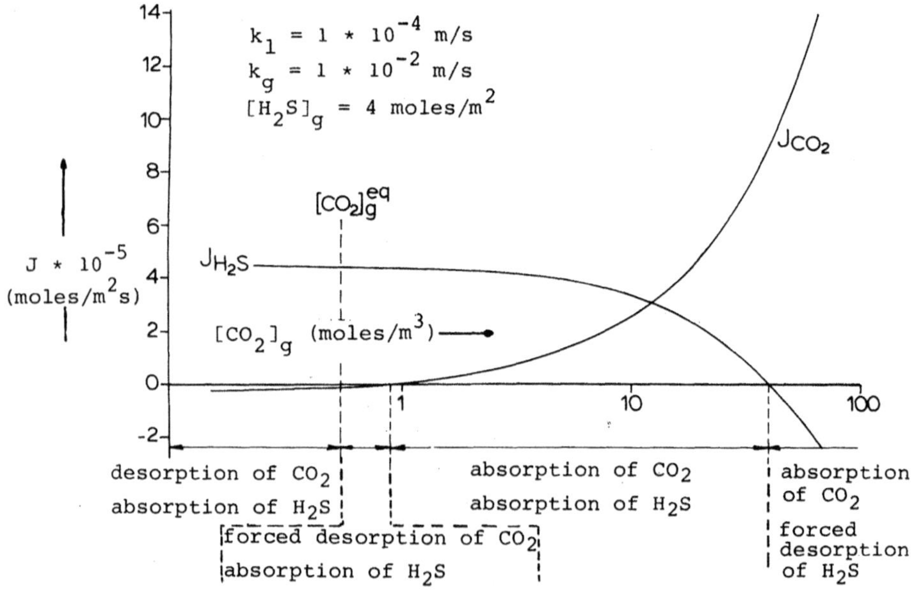

figure 3: Influence of CO_2 gas-phase concentration on J_{H_2S} and J_{CO_2}

This forced desorption originates from the comparatively high amine consumption by the H_2S absorption reaction and the induced shift in the net rate of reaction (3). At a CO_2 concentration of approximately 0.80 mole/m^3 the positive driving force balances the high local CO_2 concentration in the penetration zone and results in a net zero mole flux. Above this point the CO_2 mole flux increases steadily with the CO_2 gas-phase concentration.
The H_2S mole flux is hardly affected by CO_2 up to gas-phase concentrations of about 4 moles/m^3. Calculated free DIPA concentration profiles in the penetration zone correspondingly show no depletion. At higher $[CO_2]_g$ the amine becomes depleted and leads to a decrease of the H_2S flux at a constant H_2S driving force. CO_2 concentrations exceeding 40 moles/m^3 yield forced desorption of H_2S according to the mechanism mentioned earlier.
In figure 4 the analogous results for variations in H_2S gas-phase concentrations are shown.

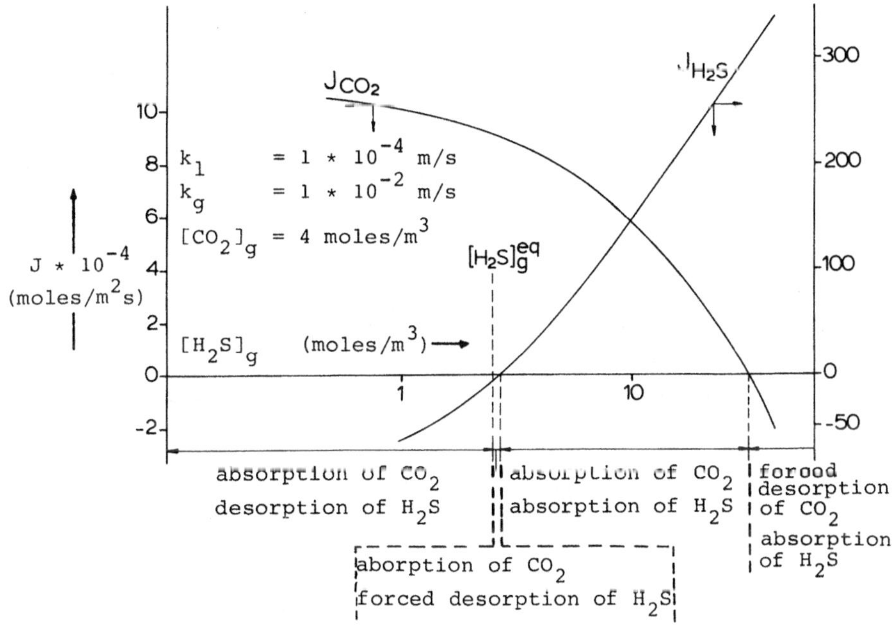

figure 4: Influence of H_2S gas-phase concentration on J_{CO_2} and J_{H_2S}

The effects of mass-transfer coefficients and CO_2 gas-phase concentrations are briefly indicated in figure 5. Doubling of the gas-phase mass-transfer coefficient k_g hardly affects the largely liquid-phase controlled CO_2 absorption but increases the substantially gas-phase limited H_2S mole flux. In the forced desorption concentration region however, the doubled k_g causes a higher desorption flux due to a facilitated H_2S transport to the gas phase.
A decrease of the liquid-phase mass-transfer coefficient k_l (see figure 5) impedes the transport of amine to the penetration zone and the removal of reaction products. Due to this hampered transport H_2S and CO_2 fluxes decrease and the forced desorption region consequently shifts to lower CO_2 concentrations.

Table 1.Fixed parameters in mass transfer calculations

$[DIPA]_{total}$	= 2.0 kmol/m^3
T	= 298 K
k_2	= 0.350 m^3/mole sec
D_{R_2NH}	= $1.93*10^{-10}$ m^2/s
$D_{R_2NCOO^-}$	= $1.93*10^{-10}$ m^2/s
$D_{R_2NH_2^+}$	= $1.93*10^{-10}$ m^2/s
D_{H_2S}	= $5.94*10^{-10}$ m^2/s
D_{HS^-}	= $1.93*10^{-10}$ m^2/s
D_{CO_2}	= $7.70*10^{-10}$ m^2/s
m_{H_2S}	= 2.05
m_{CO_2}	= 0.672
K_{H_2S}	= 155
K_{CO_2}	= 2.38 m^3/mole
$\alpha_1 = \frac{[H_2S]_{total\ in\ liquid}}{[DIPA]_{total}}$	= 0.25
$\alpha_2 = \frac{[CO_2]_{total\ in\ liquid}}{[DIPA]_{total}}$	= 0.25

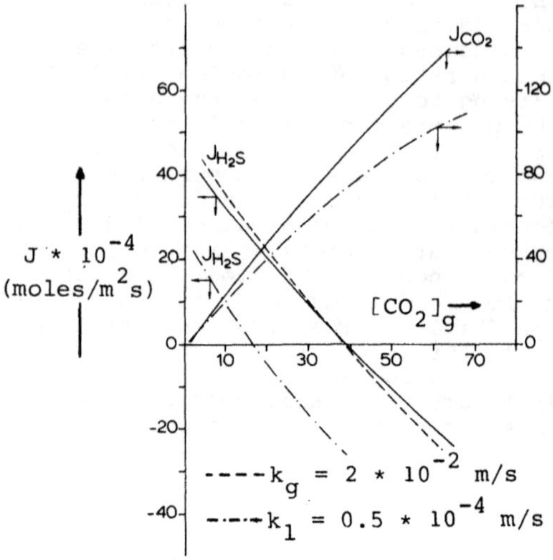

figure 5: Influence of k_g, k_1 and $[CO_2]_g$ on J_{H_2S} and J_{CO_2}

Experimental

The experimental set-up is given in figure 6. A closed reactor-detector system was used to enable detection of small mole fluxes. The stirred cell reactor is 0.10 m in diameter and was filled before each experiment with ~720 ml of charged ~2.0 M DIPA solution. The gas phase in the system was circulated by means of a flexible tube pump over a flow-through cell in a Perkin Elmer model 257 Infrared Grating Spectrophotometer for CO_2 detection. Although spectrophotometers are not exceptionally well-suited for quantitative measurements, we preferred this type of analysis compared to gas chromatography for example because it does not influence the gas phase.

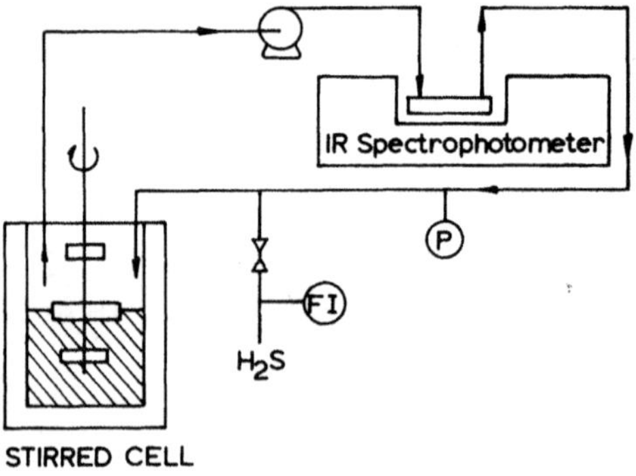

figure 6: Experimental set-up

The experiments started with equilibration of a solution containing ~0.35 moles CO_2/mole DIPA at 25°C. The amount of CO_2 charged was determined by the detection limits and sensitivity of the infrared spectrophotometer and the expected CO_2 gas-phase concentration increase. Equilibration of the solution was checked by the infrared spectrophotometer and took some 2 hours.
After this period pure H_2S was introduced into the gas phase at a constant flow rate. The changing concentrations of H_2S plus CO_2 were recorded by a manometer whilst the CO_2 concentration alone was determined by the spectrophotometer.
Directly on admittance of H_2S the CO_2 desorbed from the solution into the gas phase which was unambiguously recorded by the spectrophotometer. The CO_2 desorption increased the CO_2 gas-phase concentration immediately to above its equilibrium situation and forced desorption of CO_2 was obtained.
After about 10 minutes the H_2S flow was stopped to enable re-equilibration. It was then observed that under our experimental conditions CO_2 was again absorbed into the solution to almost the initial equilibrium (see figure 7 for a typical example). This proves that the recorded concentration curves in the gas phase are due to reaction processes in the penetration zone alone and have nothing to do with the bulk equilibrium condition. The total amount of H_2S introduced into the system was always less than 0.02 mole/mole DIPA and is negligible compared to the CO_2 charge nor dit it affect the original equilibrium.

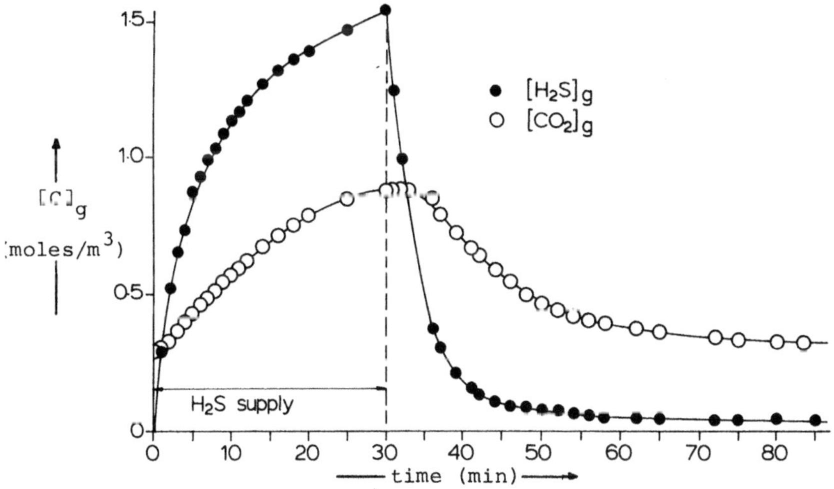

figure 7: A typical example of measured concentration curves during forced desorption experiments.

Experimental results and conclusions

Four experiments were caried out at different stirrer speeds under conditions as given in table 2. Mole fluxes were obtained from measured concentration curves and plotted in figure 8 together with calculated fluxes for both penetration and film theory models. The latter model was derived from our numerical penetration model by zeroing time derivatives and equalizing the penetration depth to the film thickness.

Table 2. Experimental conditions

Experiment no.	1	2	3	4
stirrer speed rpm	35	48	55	69
k_1 $* 10^{-5}$ m/s	0.79	0.94	1.00	1.15
k_g $* 10^{-3}$ m/s	2.0	2.4	2.6	3.0
α_{H_2S} $* 10^{-3}$ start	7.7	0.4	4.9	9.3
α_{H_2S} $* 10^{-3}$ end	14	4.0	9.3	18.2
α_{CO_2}	0.357	0.304	0.30	0.304
$[DIPA]_{total}$ $\frac{moles}{m^3}$	2140	1970	1970	1970
V_{liquid} $* 10^{-6}$ m^3	730	720	720	720
V_{gas} $* 10^{-6}$ m^3	1094	1104	1104	1104
ϕ_{H_2S} $* 10^{-6}$ moles/s	6.48	10.4	10.3 (t≤8 min)	10.3 (t≤18 min)
			15.1 (t>8 min)	12.6 (t>18 min)

From figure 8 it can be concluded that forced desorption of CO_2 can easily be realized under practical conditions and can be also predicted by the models. Measured H_2S mole fluxes fall between penetration and film theory calculations. The forced desorption of CO_2 agrees better with the film theory than with the penetration theory. It should be kept in mind however that the calculations are extremely sensitive to mass-transfer coefficients, diffusion and equilibrium constants which were obtained from separate experiments and open literature.

figure 8: Experimental and calculated results.

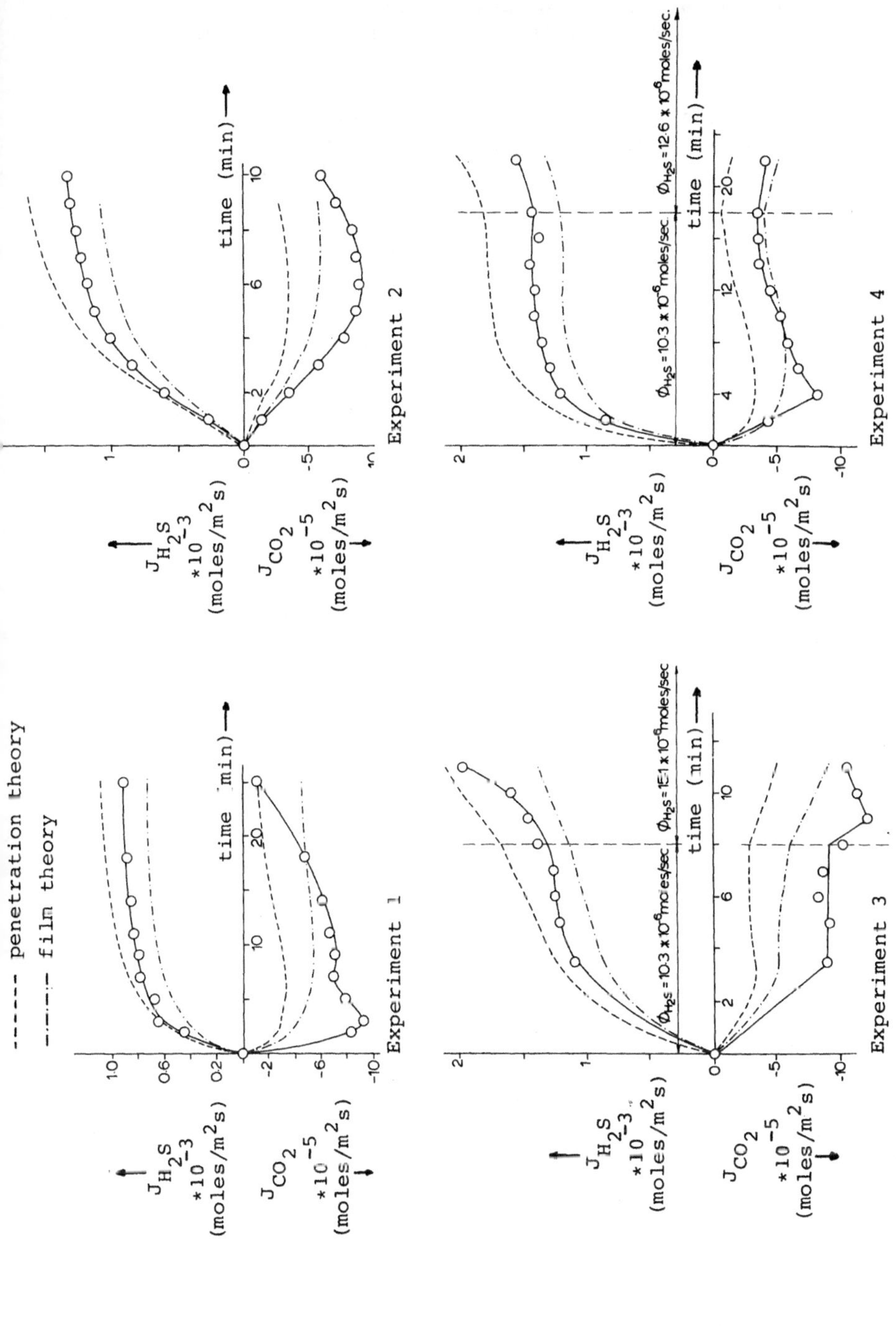

Literature

1. Beenackers, A.A.C.M.; Ph.D. Thesis, Twente University of Technology, The Netherlands, 1977.
2. Blauwhoff, P.M.M., Versteeg, G.F., van Swaaij, W.P.M.; to be published.
3. Cornelisse, R., Beenackers, A.A.C.M., van Swaaij, W.P.M.; Chem. Eng. Sci. 33,1532 (1977).
4. Cornelisse, R., Beenackers, A.A.C.M., van Beckum, F.P.H., van Swaaij, W.P.M.; Chem. Eng. Sci. 35, 5, 1245 (1980).
5. Cornelissen, A.E.; Trans. Inst. Chem. Eng. 58, 4, 242 (1980).
6. Danckwerts, P.V., Sharma, M.M.; Chem. Eng. 10, CE 244 (1966).
7. Danckwerts, P.V.; "Gas-Liquid Reactions", McGraw-Hill, New York, 1970.
8. Danckwerts, P.V.; Chem. Eng. Sci. 34, 4, 443 (1979).
9. Decoursey, W.J.; Chem. Eng. Sci 29, 1867 (1974).
10. Goetler, L.A., Pigford, R.L.; I. Chem. E. Symp. Sci. 28 (1968).
11. Hikita, H., Asai, S.; Kagaku Kogaku 27, 823 (1963).
12. Van Krevelen, D.W., Hoftijzer, P.J.; Rec. Trav. Chim. 67, 563 (1948).
13. Olander, O.R.; AICHE J., 6, 233 (1960).
14. Onda, K., Sada, E., Kobayashi, T., Fujne, M.; Chem. Eng. Sci. 25, 753 (1970).
15. Ouwerkerk, C.; I. Chem. E. Symp. Ser 28, 39 (1968).
16. Ramachandran, P.A., Sharma, M.M.; Chem. Eng. Sci. 26, 349 (1971).
17. Secor, R.M., Beuttler, J.A.; AICHE J. 13, 365 (1967).
18. Wellek, R.M., Brunson, R.J., Law, F.H.; Can. J. Chem. Eng., 56, 181 (1978).
19. Yeramian, A.A., Gottifredi, J.C., Ronco, J.J.; Chem. Eng. Sci. 25, 1622 (1970).

FACILITATED GAS TRANSPORT IN LIQUIDS

Jerry H. Meldon

Chemical Engineering Department, Tufts University,
Medford, Massachusetts 02155, USA

1 INTRODUCTION

The subject of carrier-mediated or "facilitated" gas transport in liquid film membranes has been an active field of research the past 15-20 years. It refers to the enhancement of gas transport attributable to reversible reaction of physically dissolved gas with non-volatile, diffusible solutes. The underlying physics are analogous to those governing gas absorption with chemical reaction. However, the two processes differ insofar as the membrane process has been primarily operated in the steady-state, and without the occurrence of irreversible reactions.

The partial pressure of the transported species is maintained constant in the two gas phases separated by the membrane, by virtue of either the velocity of a sweep gas or the volume of closed gas phases. The membranes are generally in the form of a soaked filter paper, or of homogeneous solution supported by highly permeable polymer membranes. At one membrane face there is absorption with chemical reaction; at the other there is stripping. Steady-state operation is ensured by zero net conversion within the membrane as a whole.

Because of their possible physiological ramifications, the systems which long attracted the widest interest were those involving oxygen transport in solutions containing either the red blood cell protein hemoglobin or its cousin from muscle, myoglobin. Authoritative reviews of these subjects have been written by Kreuzer (1) and Wittenberg (2). Much of the ensuing work on these and other chemical systems, both experimental and theoretical, was stimulated by the early reports of facilitated oxygen transport by Wittenberg (3), Scholander (4) and Hemmingsen and Scholander (5).

A comparable degree of interest has been shown in the transport of the other key physiological gas, carbon dioxide, as enhanced by reaction to form bicarbonate ion. The pioneering work of Longmuir et al. (6), Enns (7) and Ward and Robb (8) made it clear that fruitful theoretical analyses required consideration of the kinetics of CO_2 hydration and its catalysis by, in particular, the red blood cell enzyme carbonic anhydrase and arsenite ion. A review of this subject is forthcoming (9).

In addition to oxygen and carbon dioxide, gases for which carrier-mediated transport has been demonstrated include carbon monoxide (10), nitric oxide (11), sulfur dioxide (12) and ethylene (13). A system involving simultaneous transport of carbon dioxide and hydrogen sulfide, with potential applications to sour gas treating (14) is discussed in some detail below.

The fundamental theoretical aspects of carrier-mediated transport have been reviewed, from a chemical engineering standpoint, by Schultz et al. (15,16) and Smith et al. (17), and the interested reader is encouraged to refer to them. As such, the remainder of this paper will touch on the fundamentals only in brief, while focussing instead upon a number of interesting developments of the past several years.

2 THEORETICAL FUNDAMENTALS

Much theoretical effort has been devoted to systems in which the permeant gas is presumed to undergo the following reaction with non-volatile solute B:

$$A + B \rightleftarrows AB$$

The equations governing steady-state transport, assuming no electrically driven fluxes, are:

$$D_A d^2C_A/dx^2 = D_B d^2C_B/dx^2 = -D_{AB} d^2C_{AB}/dx^2 = k_1 C_A C_B - k_2 C_{AB} \quad (1)$$

where the symbology follows convention.

Assuming $D_B = D_{AB}$, the analysis is simplified by the fact that:

$$C_B + C_{AB} = C_T \quad (2)$$

where C_T is the overall total carrier concentration.

Boundary conditions are:

$C_A = C_A^o$ at $x = 0$

$C_A = C_A^L$ at $x = L$

$dC_B/dx = dC_{AB}/dx = 0$ at $x = 0,L$

where L is the film thickness, and C_A^o and C_A^L are determined by the respective gas phase partial pressures and a solubility coefficient.

General analytical solution is not possible. A variety of numerical solutions have been developed, as well as analytical approximations, as described in the aforementioned reviews. The analytical solutions are generally applicable to either the "thin film" regime - i.e., in which diffusion times are comparable to or less than reaction times, and minimal flux enhancement occurs, or the "thick film" regime in which perturbations from local reaction equilibrium are small (18). Friedlander and Keller (19) showed that there is a characteristic length scale, λ, a function of the reaction and diffusion constants, such that L/λ is a measure of the approach to local reaction equilibrium. It is thus similar to the Thiele modulus of porous catalysts.

As $L/\lambda \to \infty$, the facilitation factor F defined by:

$$\text{Net gas flux} = D_A(C_A^o - C_A^L)(1 + F)/L \tag{3}$$

approaches asymptotically the value F_{eq}, corresponding to local reaction equilibrium. For the stoichiometry assumed above,

$$F_{eq} = \frac{D_{AB}K_{eq}C_T}{D_A(C_A^o - C_A^L)}\left\{\frac{C_A^o}{1 + K_{eq}C_A^o} - \frac{C_A^L}{1 + K_{eq}C_A^L}\right\} \tag{4}$$

where $K_{eq} = k_1/k_2$.

The mathematical approaches developed for the simple kinetic model can be applied in straightforward fashion to systems of arbitrary kinetics. Complications arise, however, in the event that one cannot neglect electrically driven fluxes. Such is the case, obviously, when the field is applied externally. Ward (11) showed that nitric oxide can be transported in the absence of a partial pressure gradient, by the electrically driven flux of its complex with ferrous ion. Similar effects have been demonstrated in the carbon dioxide/bicarbonate system (20).

In the presence of an appreciable electrical field, Fick's law of diffusion is no longer adequate, and must be replaced by the

Nernst-Planck relation:

$$N_j = -D_j(dC_j/dx + z_j C_j \frac{\overline{F}}{RT} dV/dx) \quad (5)$$

where N is the flux and z the charge of species j, $\overline{F}$ is Faraday's constant and V is electrical potential.

Even in the absence of an externally applied field, the zero current requirement and electroneutrality constraint can only be satisfied if the diffusion coefficients of all transported ionic species are equal, or else a potential gradient (the "diffusion potential") prevails (21). The consequences of the diffusion potential in the transport of CO_2 in alkaline solutions are analyzed in a companion paper in this symposium (22). The particularly interesting results for CO_2 transport in protein solutions are reviewed in the section that follows.

3 CO_2 TRANSPORT IN BUFFER SOLUTIONS

An understanding of bicarbonate-mediated carbon dioxide transport in physiological systems requires consideration of the coupling of the reactions intrinsic to CO_2 in water:

$$CO_2 + H_2O \rightleftarrows H_2CO_3 \rightleftarrows HCO_3^- + H^+$$

$$CO_2 + OH^- \rightleftarrows HCO_3^-$$

$$HCO_3^- \rightleftarrows CO_3^{2-} + H^+$$

$$H_2O \rightleftarrows OH^- + H^+$$

to the dissociation reactions of any buffers which are present:

$$HB \rightleftarrows B^- + H^+$$

The buffer reaction above can have a profound effect upon net CO_2 transport, and does so via a mechanism which is perhaps more complex than it might appear at first glance. Clearly, because of the coupling via hydrogen ions, the dissociation of the weak acid affects the extents of formation of carbonate and bicarbonate ions. In moderately alkaline solutions, the CO_2 flux is enhanced by that of bicarbonate, and diminished by the counter-transport of carbonate ion. The bicarbonate concentration gradient exceeds that of carbonate, and that difference is enhanced by the presence of buffer (23). Thus, simple inorganic weak acids have a positive effect upon CO_2 transport in alkaline solutions, as illustrated in Figure 1, and confirmed in a later study (24).

However, as pointed out by Gros et al. (25), the dissociation

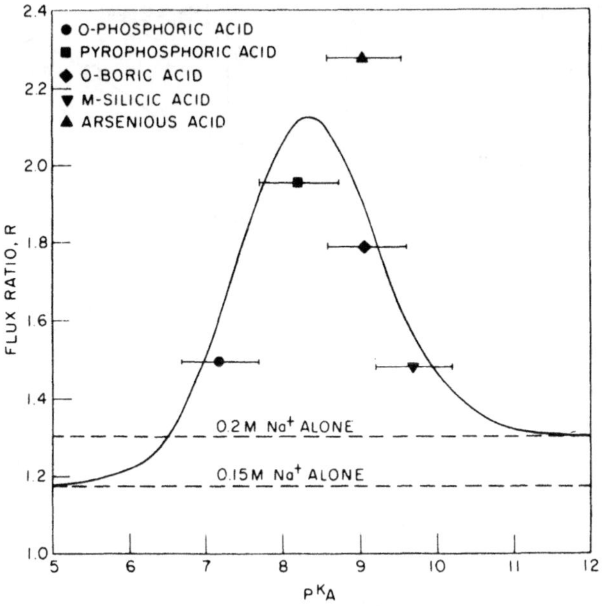

Fig. 1: Flux of carbon dioxide through a thin film of alkaline solution relative to the flux through water, as a function of the pK of the weak acid added to 0.2 M bicarbonate solution, to an effective extent of 0.05 M. Bell curve peaks in the pH range prevailing in solution. Taken from ref. 23.

reaction of the buffer alone is insufficient to ensure enhancement. The buffer must also be mobile. A highly diffusible buffer not only consumes hydrogen ions generated at the upstream side of the film, but also transports them to the downstream side, where it promotes the formation of carbonic acid, and hence CO_2, from bicarbonate ions. This phenomenon can be interpreted in terms of the diffusion potential caused by the difference between the mobilities of bicarbonate, carbonate and buffer species.

In the particular case in which the buffer is a protein, e.g., the hemoglobin in red cells or albumin in blood plasma, the discrepancy between the mobility of the buffer and that of the smaller anions, HCO_3^- and CO_3^{2-} is quite large. In such instances, one can predict with reasonable accuracy the prevalence of appreciable voltage differences across the liquid membrane (26), as well as a significant reduction in the facilitation factor, compared to what one would estimate in the hypothetical event that the protein's mobility were equal to that of the smaller ions. This is demonstrated by the experimental results of Stroeve and Ziegler

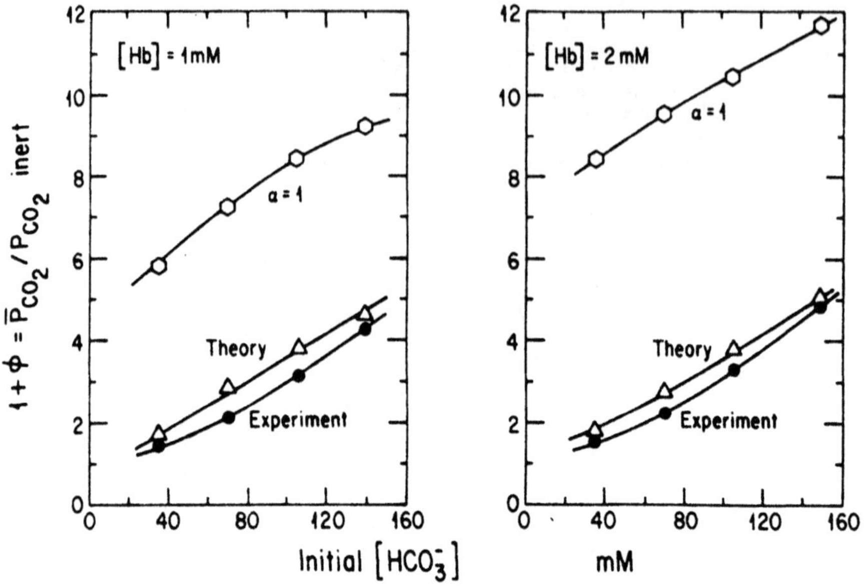

Fig. 2: Ratio of CO_2 permeability in a solution of hemoglobin and sodium bicarbonate to the measured physical permeability of CO_2 in the same solution. The presence of carbonic anhydrase ensured local equilibrium of the CO_2 hydration reaction. α=1 denotes theoretical result assuming equality of protein and bicarbonate diffusivities. Theory line based on analysis in ref. 27. Results taken from ref. 28.

(28), shown in Figure 2.

The transport of oxygen in hemoglobin solutions is similarly a more complex process than diffusion with the reaction A + B = AB as it has most often been modelled. The extent of oxyhemoglobin formation varies sigmoidally, rather than hyperbolically with the equilibrium partial pressure of oxygen, a consequence of "cooperative" oxygen binding at the four sites per protein molecule (29). In addition, diffusion potential effects arise from the difference in the electrical charges of oxy- and deoxy-hemoglobin at a given pH - i.e., oxyhemoglobin is a stronger acid. Thus, O_2 transport in purified hemoglobin solutions involves the simultaneous diffusion of protein, hydrogen and hydroxyl ions, and the mobilities of the latter two species far exceed that of the macromolecule (30). Analysis of the complex, but physiologically relevant case of simultaneous O_2 and CO_2 transport in hemoglobin solutions, has only bee approached in very approximate fashion (31).

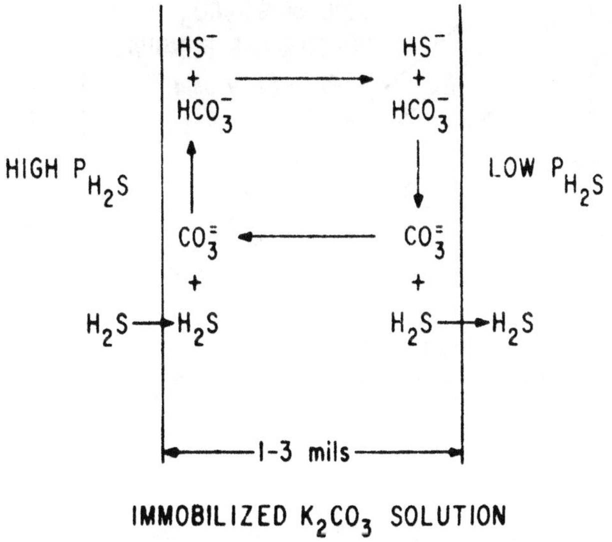

Fig. 3: Schematic diagram of facilitated H_2S transport in a carbonate/bicarbonate solution. Taken from ref. 14. (Copyright American Chemical Society)

4 H_2S-SELECTIVE MEMBRANES

Transport of hydrogen sulfide in aqueous systems is facilitated by its acid dissociation reaction:

$$H_2S \rightleftarrows HS^- + H^+$$

The reaction is effectively instantaneous, which justifies the assumption of local equilibrium. Facilitation, though, is limited by the availability of a counter-ion. The dissociation is weak enough (pK ca. 7) that hydrogen ions cannot serve as a co-ion, as they do in the case of sulfur dioxide (12). On the other hand, the second dissociation to form sulfide ion is too weak for the latter to serve the same role as does carbonate ion in CO_2 systems. However, as illustrated in Figure 3, the presence of carbonate/bicarbonate promotes H_2S transport, and in a manner analogous to the effect of buffers upon carbon dioxide transport.

Furthermore, when H_2S and CO_2 are each present in the gas phase (which is also an inevitable outcome of the presence of carbonate/bicarbonate in solution), there is selective transport of hydrogen sulfide by virtue of the kinetic limitations upon CO_2 transfer. As reported by Matson et al. (14), selectivity is further promoted by use of several liquid membranes in series, with gas-

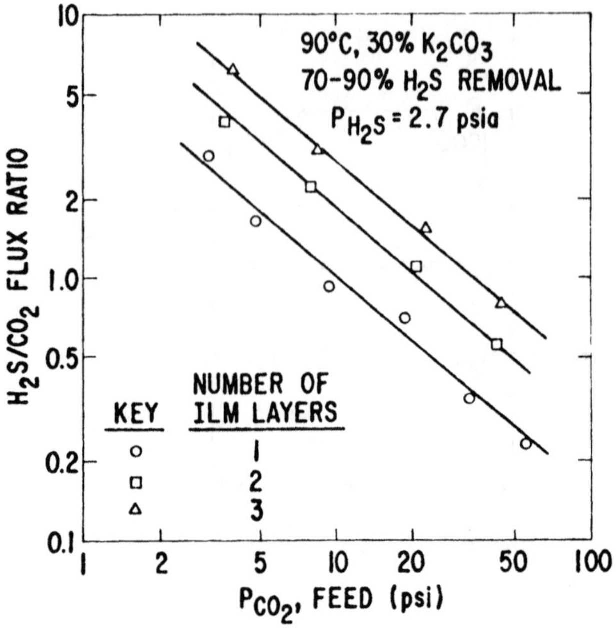

Fig. 4: Enhancement of membrane H_2S/CO_2 selectivity by introduction of gas gaps between membranes placed in series. Taken from ref. 14. (Copyright American Chemical Society)

filled gaps between them. The effect is due to the introduction of additional boundary zones in which the perturbations from reaction equilibrium are greatest, an effect which is limited to the reactions of CO_2. As illustrated in Figure 4, selectivity with respect to hydrogen sulfide increases in rough proportion with the number of immobilized liquid membrane (ILM) layers, as it is increased from one to three.

5 FACILITATED GAS TRANSPORT IN TISSUE

Among the more interesting recent developments have been in the fields of oxygen and carbon dioxide transport in living tissue. The question of whether myoglobin actually facilitates steady-state O_2 transfer from capillary blood to respiring muscle remains a controversial one, compared with the generally accepted role of myoglobin as a short-term source of oxygen in periods of temporary cut-off of oxygen supply - e.g., in diving mammals. However, Wittenberg et al. (32) demonstrated distinctly reduced steady-state rates of oxygen uptake by bundles of pigeon breast muscle fibers, when the myoglobin was poisoned by a number of reagents known not to impair oxidative metabolism. If myoglobin does indeed transport

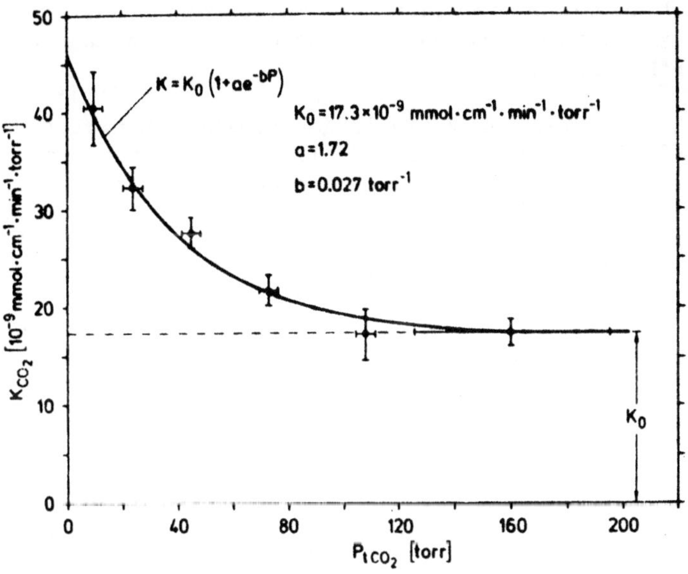

Fig. 5: Experimentally determined permeability coefficient for CO_2 transport in rat muscle tissue as a function of the CO_2 partial pressure level. Taken from ref. 34.

it can be said to have the effect of reducing the capillary blood oxygen partial pressure necessary to sustain tissue respiration, and therefore of reducing the required cardiac output (33).

As shown by Kawashiro and Scheid (34), and illustrated in Figure 5, the effective permeability of carbon dioxide in rat muscle tissue increases markedly as the CO_2 partial pressure is reduced. This is justifiably interpreted as evidence of facilitated CO_2 transport by bicarbonate, the concentration gradient of which becomes negligible at high CO_2 partial pressures. The significance of such facilitation remains to be seen, however, since the CO_2 partial pressure gradient between tissue and capillary is very small, even at elevated metabolic rates, and CO_2 elimination from tissue is not limited by the flow of blood.

6 FUTURE DIRECTIONS

In addition to the research efforts described above, there has been significant progress on a number of other frontiers of carrier-mediated gas transport. Among the more noteworthy developments have been the use of tracer flux measurements by Quinn and coworkers (see ref. 35, for example) to determine reaction kinetic constants, and the analysis of transport and chemical reaction in heterogeneous media by Stroeve and coworkers (e.g., ref. 36) and its application to, among other systems, oxygen transfer in whole blood. The techniques developed by these two groups are expected to bear further fruit.

An actual commercially viable gas separation process based on carrier-facilitated transport in liquid membranes remains to be synthesized, although that goal has come close to attainment. The fact that silver ions form complexes with olefins has led to at least one patent for separation of unsaturated from saturated hydrocarbons (13). Very recently, however, Amoco was forced by economics to shelve plans to replace distillation columns in olefin plants with membrane-based separators involving hollow fibers of cellulose acetate saturated with solutions of silver nitrate (37).

There remain many other unanswered questions. For example, despite the investigative effort devoted to unravelling the mechanism of oxygen transport through hemoglobin solutions, it is still a matter of debate whether or not diffusion of hemoglobin enhances O_2 transport from blood to tissue.

REFERENCES

1. Kreuzer, F. "Facilitated diffusion of oxygen and its possible significance: A review." Respiration Physiology 9 (1970) 1-30.
2. Wittenberg, J.B. "Myogoblobin-facilitated oxygen diffusion: Role of myoglobin in oxygen entry into muscle." Physiological Reviews 50 (1970) 559-636.
3. Wittenberg, J.B. "Oxygen transport: a new function proposed for myoglobin." Biological Bulletin 117 (1959) 402.
4. Scholander, P.F. "Oxygen transport through hemoglobin solutions." Science 132 (1960) 368.
5. Hemmingsen, E. and P.F. Scholander. "Specific transport of oxygen through hemoglobin solutions." Science 132 (1960) 1379-1381.
6. Longmuir, I.S., Forster, R.E. and C.-Y. Woo. "Diffusion of carbon dioxide through thin layers of solution." Nature 209 (1966) 394-395.
7. Enns, T. "Facilitation by carbonic anhydrase of carbon dioxide transport." Science 155 (1967) 44-47.
8. Ward, W.J. and W.L. Robb. "Carbon dioxide - oxygen separation: facilitated transport of carbon dioxide across a liquid film." Science 156 (1967) 1481-1484.
9. Meldon, J.H., Stroeve, P. and C.E. Gregoire. "Facilitated transport of carbon dioxide: A review." Chemical Engineering Communications, in press.
10. Smith, D.R. and J.A. Quinn. "The facilitated transport of carbon monoxide through cuprous chloride solutions." American In-

stitute of Chemical Engineers Journal 26 (1980) 112-120.
11. Ward, W.J. "Electrically induced carrier transport." Nature 227 (1970) 162-165.
12. Roberts, D.L. and S.K. Friedlander. "Sulfur dioxide transport through aqueous solutions." American Institute of Chemical Engineers Journal 26 (1980) 593-610.
13. Steigelman, E.F. and R.D. Hughes, U.S. Patent 3,758,603, September 11, 1973, cited in Kimura, S.G., Matson, S.L. and Ward, W.J. "Industrial applications of facilitated transport." Recent Developments in Separation Science 5 (1979) 11-25.
14. Matson, S.L., Herrick, C.S. and W.J. Ward. "Progress on the selective removal of H_2S from gasified coal using an immobilized liquid membrane." Industrial & Engineering Chemistry, Process Design & Development 16 (1977) 370-374.
15. Schultz, J.S., Goddard, J.D. and S.R. Suchdeo. "Facilitated transport via carrier-mediated diffusion in membranes. Part I: Mechanistic aspects, experimental systems and characteristic regimes." American Institute of Chemical Engineers Journal 20 (1974) 417-445.
16. Goddard, J.D., Schultz, J.S. and S.R. Suchdeo. "Facilitated transport via carrier-mediated diffusion in membranes: Part II: Mathematical aspects and analyses." American Institute of Chemical Engineers Journal 20 (1974) 625-648.
17. Smith, D.R., Lander, R.J. and J.A. Quinn. "Carrier-mediated transport in synthetic membranes." Recent Developments in Separation Science 3B (1977) 225-241.
18. Smith, K.A., Meldon, J.H. and C.K. Colton. "An analysis of carrier-facilitated transport." American Institute of Chemical Engineers Journal 19 (1973) 102-111.
19. Friedlander, S.K. and K.H. Keller. "Mass transfer in reacting systems near equilibrium. Use of the affinity function." Chemical Engineering Science 20 (1965) 121-129.
20. Lin, C.H. and J. Winnick. "An electrochemical device for carbon dioxide concentration. II." Industrial & Engineering Chemistry, Process Design & Development 13 (1974) 63-70.
21. Vinograd, J.R. and J.W. McBain "Diffusion of electrolytes and of the ions in their mixtures." Journal of the American Chemical Society 63 (1941) 2008-2015.
22. Meldon, J.H. and J.E. Roberts. "Theory of membrane CO_2 transport with equilibrium reaction." This volume.
23. Meldon, J.H., Smith, K.A. and C.K. Colton. "The effect of weak acids upon the transport of carbon dioxide in alkaline solutions." Chemical Engineering Science 32 (1977) 939-950.
24. Lander, D.R., Smith, D.R. and J.A. Quinn. "The effect of buffers and pH gradients on CO_2 transport through liquid films." Chemical Engineering Science 34 (1979) 745-747.
25. Gros, G., Moll, W., Hoppe, H. and H. Gros. "Proton transport by phosphate diffusion - a mechanism of facilitated CO_2 transfer." Journal of General Physiology 67 (1976) 773-790.
26. DeKoning, J., Stroeve, P. and J.H. Meldon. "Electrical poten-

tials during carbon dioxide transport in hemoglobin solutions." Advances in Experimental Medicine and Biology 94 (1978) 183-188.
27. Meldon, J.H. "Theory of the effect of diffusion potentials on the transport of carbon dioxide in protein solutions." Abstract 413, Proceedings of the Fifth International Biophysics Congress (1975) Copenhagen.
28. Stroeve, P. and E. Ziegler. "The transport of carbon dioxide in high molecular weight buffer solutions." Chemical Engineering Communications 6 (1980) 81-103.
29. Meldon, J.H., Smith, K.A. and C.K. Colton. "Analysis of 2,3-diphosphoglycerate-mediated, hemoglobin-facilitated oxygen transport in terms of the Adair reaction mechanism." Advances in Experimental Medicine and Biology 37A (1973) 199-205.
30. Bright, P.B. "The basic flow equations of electrophysiology in the presence of chemical reactions: II. A practical application concerning the pH and voltage effects accompanying the diffusion of O_2 through hemoglobin solution." Bulletin of Mathematical Biology 29 (1967) 123-138.
31. Ulanowicz, R.E. and G.C. Frazier. "The transport of oxygen and carbon dioxide in hemoglobin systems." Mathematical Biosciences 7 (1970) 111-129.
32. Wittenberg, B.A., Wittenberg, J.B. and P.R.B. Caldwell. "Role of myoglobin in the oxygen supply to red skeletal muscle." Journal of Biological Chemistry 250 (1975) 9038-9043.
33. Meldon, J.H. "The theoretical role of myoglobin in steady-state oxygen transport to tissue and its impact upon cardiac output requirements." Acta Physiologica Scandinavica Suppl. 440 (1976) 93
34. Kawashiro, T. and P. Scheid. "Measurement of Krogh's diffusion constant of CO_2 in respiring muscle at various CO_2 levels: Evidence for facilitated diffusion." Pflugers Archiv 362 (1976) 127-133.
35. Donaldson, T.L. and J.A. Quinn. "Kinetic constants determined from membrane transport measurements: Carbonic anhydrase activity at high concentrations." Proceedings of the National Academy of Sciences, USA 71 (1974) 4995-4999.
36. Stroeve, P., Smith, K.A. and C.K. Colton. "An analysis of carrier facilitated transport in heterogeneous media." American Institute of Chemical Engineers Journal 22 (1976) 1125-1132.
37. R.I. Berry. "Membranes separate gas." Chemical Engineering, July 13, 1981, 63-67.

THEORY OF MEMBRANE CO_2 TRANSPORT WITH EQUILIBRIUM REACTION

Jerry H. Meldon and John E. Roberts

Chemical Engineering Department, Tufts University,
Medford, Massachusetts 02155, USA

1 INTRODUCTION

In the past two decades there has been considerable interest in CO_2-selective liquid membranes (1-9). Over fifteen years ago prototypes thousands of times more permeable to carbon dioxide than to oxygen were reported (10). Systems have been explored more recently which employ an electrical field to pump CO_2, in the form of bicarbonate ion, up its chemical potential gradient (11). Still, no commercial applications are known to the authors.

Perhaps one reason for the non-competitiveness of liquid films as gas separators - besides the difficulties of fabricating ultra-thin porous membranes and preventing their dessication - is that the theoretical aspects of CO_2 transport in alkaline media have not been fully explored. Solutions to the differential equations governing steady-state CO_2 diffusion with non-equilibrium chemical reaction are available, and the appreciable effects of catalysts and buffers have been elucidated. However, noteworthy aspects of the equilibrium (fast reaction) regime in simple alkaline solutions have not been fully examined.

We refer in particular to cases in which the downstream CO_2 partial pressure is so low as to ensure that carbonate and hydroxyl, rather than bicarbonate ions, are the dominant diffusing species. Under such conditions there are two notable phenomena:
i) enhancement factors are exceedingly large because of the ratio of the concentrations of carrier and transported species, and
ii) electrostatic effects, due to the greater mobility of hydroxyl than of carbonate ion, endow carbonate with an effectively enhanced diffusivity, and thereby afford additional CO_2 flux enhancement.

In this paper we consider steady-state transport of CO_2 across a semi-infinite flat sheet of liquid which separates gas phases of CO_2 partial pressures p^o and p^L, respectively (where L is the thickness of the membrane). The liquid membrane contains an alkali metal ion, M^+, at an average concentration $\overline{[M^+]}$.

As CO_2 diffuses it undergoes the following reversible chemical reactions:

$$CO_2 + H_2O \rightleftarrows H_2CO_3 \rightleftarrows HCO_3^- + H^+ \quad (A)$$

$$CO_2 + OH^- \rightleftarrows HCO_3^- \quad (B)$$

which are coupled to:

$$HCO_3^- \rightleftarrows CO_3^{2-} + H^+ \quad (C)$$

$$H_2O \rightleftarrows OH^- + H^+ \quad (D)$$

The effective overall reaction in moderately alkaline solutions is:

$$CO_2 + CO_3^{2-} + H_2O \rightleftarrows 2HCO_3^- \quad (E)$$

while in highly alkaline solutions it becomes:

$$CO_2 + 2OH^- \rightleftarrows CO_3^{2-} + H_2O \quad (F)$$

We assume that reaction equilibrium effectively prevails locally throughout the membrane; i.e., that either the film thickness or the concentration of a catalyst for the CO_2 hydration step in reaction A (e.g., the enzyme carbonic anhydrase) are sufficiently large to ensure that the reaction time scale is much less than the diffusion time scale. Note also that equilibration of the only other slow reaction, B, is effected by the equilibria of A and D. This greatly simplifies the mathematics and allows us to focus on electrostatic effects which would also prevail in the absence of reaction equilibrium.

In the moderately alkaline regime (I), bicarbonate ion diffuses co-currently with CO_2, and carbonate serves as a counter-ion. At high pH (i.e., extremely low CO_2 partial pressures in the steady state), carbonate ion becomes the CO_2 carrier, hydroxyl becomes the counter-ion, and this defines régime II. Because $D_{OH^-} > D_{HCO_3^-} > D_{CO_3^{2-}}$ (the ratios at 25°C and infinite dilution are 5.7:1.3:1) diffusion potentials prevail such that in regime I, $dV/dx < 0$ and in regime II, $dV/dx > 0$ (where V denotes electrical potential and x the distance into the membrane from the surface exposed to p^o, the higher of the boundary CO_2 partial pressures).

The diffusion potential affects carrier-mediated CO_2 transport in two ways : i) electrically driven fluxes of carbonate and bicarbonate ions, and ii) a gradient in the concentration of M^+ that parallels the electrical field. Where M^+ accumulates, the concentrations of carbonate, bicarbonate and hydroxyl ions are each higher than those which would prevail in the absence of the diffusion potential. Consequently, in regime I bicarbonate-mediated CO_2 transport is diminished by both a reduced bicarbonate concentration gradient and an adverse electrical field. In regime II, carbonate-mediated CO_2 transport is enhanced by the opposite effects.

Regime I has been examined in a previous study (12) through an approximate analytical solution to the governing equations. We present here the results of numerical analysis of the equations describing CO_2 transport in alkaline solutions in general. These confirm the earlier calculations. Furthermore, they demonstrate the analogous physics of regime II, and reveal the particularly interesting behavior in cases which span the alkaline pH range.

2 MATHEMATICAL ANALYSIS

Steady-state transport of CO_2 is governed by the following constraints:

i) Local species balances:

$$dN_j/dx = -r_j \tag{1}$$

where N_j is the flux and r_j the rate of consumption by reaction of species j.

ii) The Nernst-Planck equation:

$$N_j = -D_j(dC_j/dx + z_j C_j \frac{\bar{F}}{RT} dV/dx) \tag{2}$$

where C_j is the concentration and z_j the charge of species j, $\bar{F}$ is Faraday's constant, R the ideal gas constant and T the absolute temperature.

iii) Zero electrical current:

$$\Sigma z_j N_j = 0 \tag{3}$$

iv) Effective electroneutrality, locally, as follows from the Poisson equation for systems of practical interest (12):

$$\Sigma z_j C_j = 0 \tag{4}$$

v) Global conservation of alkali metal ion:

$$\int_0^L [M^+]\,dx = \overline{[M^+]}L \qquad (5)$$

vi) The reaction equilibria:

$$[HCO_3^-][H^+]/[CO_2] = K_1 \qquad (6)$$

$$[CO_3^{2-}][H^+]/[HCO_3^-] = K_2 \qquad (7)$$

$$[H^+][OH^-] = K_w \qquad (8)$$

vii) The enforceable boundary conditions:

$$[CO_2] = \alpha p^o \text{ at } x = 0 \qquad (9)$$

$$[CO_2] = \alpha p^L \text{ at } x = L \qquad (10)$$

$$N_{HCO_3^-} + N_{CO_3^{2-}} = 0 \text{ at } x = 0 \text{ and } L \qquad (11)$$

$$N_{M^+} = 0 \text{ at } x = 0 \text{ and } L \qquad (12)$$

(Boundary conditions (11) are all that may be applied to HCO_3^-, CO_3^{2-} and OH^-. The more general conditions, $N_j = 0$, $j \neq CO_2$, are preëmpted by the assumptions of reaction equilibrium.)

From equation (1) and the stoichiometry it follows that:

$$d(N_{CO_2} + N_{HCO_3^-} + N_{CO_3^{2-}})/dx = 0 \qquad (13)$$

Upon integration of equation (13) and application of conditions (11) one obtains:

$$N_{CO_2} + N_{HCO_3^-} + N_{CO_3^{2-}} = \Phi \qquad (14)$$

where Φ is the net flux of carbon dioxide.

Equation (1) and conditions (12) require that:

$$N_{M^+} = 0 \qquad (15)$$

and, therefore, that:

$$d[M^+]/dx = -[M^+]\frac{F}{RT}\,dV/dx \qquad (16)$$

Equations (2), (3), (4), (15) and (16) may then be combined to show that:

$$dV/dx =$$

$$\frac{RT}{F}\left\{\frac{(D_{HCO_3^-} - D_{CO_3^{2-}})d[HCO_3^-]/dx + (D_{OH^-} - D_{CO_3^{2-}})d[OH^-]/dx}{D_{HCO_3^-}[HCO_3^-] + D_{CO_3^{2-}}(4[CO_3^{2-}] + [M^+]) + D_{OH^-}[OH^-]}\right\} \quad (17)$$

and thus the sense of dV/dx in regimes I and II as stated in the introductory section.

We have solved the equations numerically. The method (13) involves a trial-and-error search for the M^+ concentration at x = 0 which leads ultimately to satisfaction of equation (5).

3 RESULTS AND DISCUSSION

Calculations were performed with $\overline{[M^+]}$ fixed at 1N, using physicochemical parameters at 25°C from the published literature, and neglecting ionic strength corrections. Thus, the solubility coefficient of CO_2, α, was set at 4.4×10^{-5} M./mm Hg, K_1 at 4.37×10^{-7} M., K_2 at 4.68×10^{-11} M., K_w at 10^{-14} (M.)^2.[21] Diffusivities in cm^2/sec x 10^{-5} were 1.94 for CO_2, 1.09 for HCO_3^-, 0.804 for CO_3^{2-} and 3.88 for OH^-.

Given values of p^o and p^L the computer generated dimensionless concentration profiles within the liquid membrane, with the concentration of CO_2 normalized by αp^o, and those of M^+, HCO_3^-, CO_3^{2-} and OH^- normalized by the average metal ion concentration (see Figure 1).

Also calculated were the enhancement factor F, defined by:

$$\Phi = D_{CO_2}\alpha(p^o - p^L)(1 + F)/L \quad (18)$$

and the potential difference across the film, $V^L - V^o$. Since equilibrium was assumed to prevail locally, both F and $V^L - V^o$ are independent of L.

In addition, F* was defined as the hypothetical F value calculated when the carbonate and hydroxyl ion diffusivities are arbitrarily set equal to that of bicarbonate ion. Electrical effects then vanish, the alkali metal ion concentration is constant, and it follows that:

$$F^* = \frac{D_{HCO_3^-}([HCO_3^-]^o - [HCO_3^-]^L + [CO_3^{2-}]^o - [CO_3^{2-}]^L)}{D_{CO_2}\alpha(p^o - p^L)} \quad (19)$$

Table 1

Numerical Results

Case	p^o (mm Hg)	p^L/p^o	pH^o	pH^L	F*	F/F*	V^L-V^o (mv)
I	100	10^{-4}	8.67	11.38	58.2	0.884	-2.85
II	10^{-7}	10^{-2}	13.75	13.94	2.99×10^{12}	1.07	6.09
III	100	10^{-9}	8.70	13.61	93.8	1.37	9.10
IV	10^{-4}	10^{-2}	12.39	13.30	1.22×10^{7}	2.84	5.14

The boundary values of the carrier concentrations, in this limiting case, may be calculated in straightforward fashion from the corresponding partial pressures of CO_2, the reaction equilibria, and the electroneutrality constraint.

Table 1 lists the numerical results calculated with four different combinations of p^o and p^L. Case I is entirely in regime I, with the hydroxyl ion concentration never exceeding 1/20 of the bicarbonate concentration. The value of $V^L - V^o$ is therefore negative and F/F* is less than unity. Cases II and IV are in regime II. Case III spans the two regimes but is predominantly in regime II. Thus the sign of $V^L - V^o$ is positive in the last three cases.

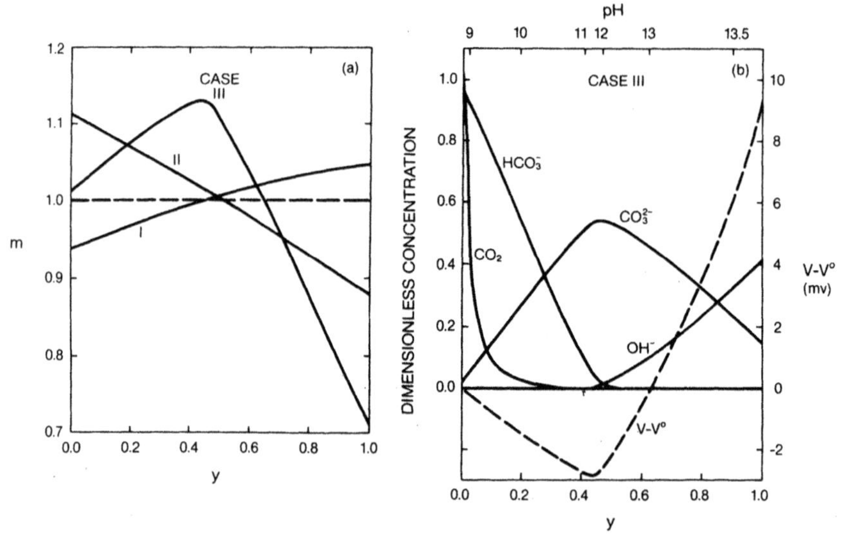

Fig. 1: Normalized concentration and voltage profiles (y=x/L).

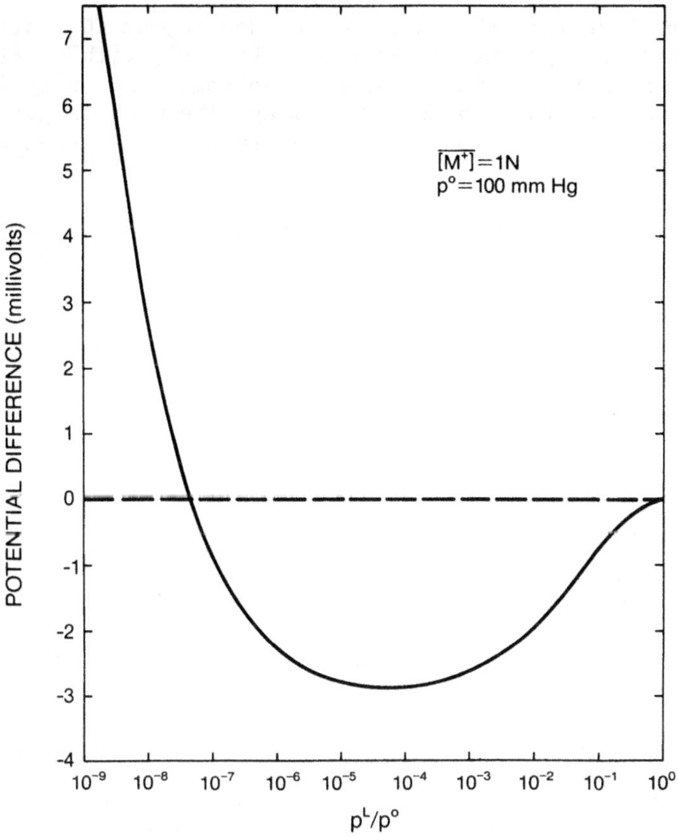

Fig. 2: Overall potential difference as a function of CO_2 partial pressure driving force.

As noted previously (12), ionic diffusion rates are determined largely by the mobility of the more dilute of two predominant species. In case I the major species switches within the film from bicarbonate to carbonate, and the effective ionic diffusivity is betweeen the two intrinsic values, i.e., $D_{CO_3^{2-}}/D_{HCO_3^-} < F/F^* < 1$. In cases II and IV, $D_{CO_3^{2-}}/D_{HCO_3^-} < F/F^* < D_{OH^-}/D_{HCO_3^-}$. Since carbonate predominates in case IV, the effective ionic diffusivity is close to D_{OH^-} and F/F^* attains its largest value in the table.

Figure 1a shows how the dimensionless M^+ concentration varies with y (=x/L) in case I, II and III. Consistent with equation (16), the alkali metal ion distributes along the electrical field, which in case III switches sign midway through the membrane. Figure 1b examines in detail the concentration and potential profiles in case III and reveals the switch from regime I to regime II.

Figure 2 depicts the influence of downstream CO_2 partial pressure upon the potential difference $V^L - V^o$, with p^o fixed at 100 mm Hg. Note the rapid increase in voltage upon entry into regime II when p^L falls below 10^{-5} mm Hg. The much larger potentials in the latter regime are due the fact that $D_{OH^-}/D_{CO_3^{2-}} \gg D_{HCO_3^-}/D_{CO_3^{2-}}$.

These results demonstrate the appreciable electrostatic effects - as well as extremely high enhancement factors - in CO_2 transport in highly alkaline solutions. It is interesting to note that the importance of the same phenomenon in CO_2 scrubbing with caustic solutions was recognized long ago by several investigators (14,15,16).

REFERENCES

1. Longmuir, I.S., Forster, R.E. and C.-Y. Woo. "Diffusion of carbon dioxide through thin layers of solution." Nature 209 (1966) 394-395.
2. Enns, T. "Facilitation by carbonic anhydrase of carbon dioxide transport." Science 155 (1967) 44-47.
3. Otto, N.C. and J.A. Quinn. "The facilitated transport of carbon dioxide through bicarbonate solutions." Chemical Engineering Science 26 (1971) 949-961.
4. Gros, G. and W. Moll. "The diffusion of carbon dioxide in erythrocytes and hemoglobin solutions." Pflugers Archiv 324 (1971) 249-266.
5. Suchdeo, S.R. and J.S. Schultz. "The permeability of gases through reacting solutions: the carbon dioxide-bicarbonate membrane system." Chemical Engineering Science 29 (1974) 13-23.
6. Donaldson, T.L. and J.A. Quinn. "Carbon dioxide transport through enzymatically active synthetic membranes." Chemical Engineering Science 30 (1975) 103-115.
7. Meldon, J.H., Smith, K.A. and C.K. Colton. "The effect of weak acids upon the transport of carbon dioxide in alkaline solutions." Chemical Engineering Science 32 (1977) 939-950.
8. Lander, R.J., Smith, D.R. and J.A. Quinn. "The effect of buffers and pH gradients on CO_2 transport through liquid films." Chemical Engineering Science 34 (1979) 745-747.
9. Stroeve, P. and E. Ziegler. "The transport of carbon dioxide in high molecular weight buffer solutions." Chemical Engineering Communications 6 (1980) 81-103.
10. Ward, W.J. and W.L. Robb. "Carbon dioxide - oxygen separation: facilitated transport of carbon dioxide across a liquid film." Science 156 (1967) 1481-1484.

11. Lin, C.H. and J. Winnick. "An electrochemical device for carbon dioxide concentration. II." Industrial & Engineering Chemistry, Process Design & Development 13 (1974) 63-70.
12. Meldon, J.H., Smith, K.A. and C.K. Colton. "An analysis of electrical effects induced by CO_2 transport in alkaline solutions." Recent Developments in Separation Science 5 (1979) 1-10.
13. Meldon, J.H. and J.E. Roberts, manuscript in preparation.
14. Danckwerts, P.V. and A.M. Kennedy. "The kinetics of absorption of carbon dioxide into neutral and alkaline solutions." Chemical Engineering Science 8 (1958) 201-215.
15. Nijsing, R.A.T.O., Hendriksz, R.H. and H. Kramers. "Absorption of CO_2 in jets and falling films of electrolyte solutions, with and without chemical reaction." Chemical Engineering Science 10 (1959) 88-104.
16. Astarita, G. Mass Transfer with Chemical Reaction (Amsterdam: Elsevier, 1967).

DISTILLATION WITH CHEMICAL REACTION

H. Sawistowski

Department of Chemical Engineering and Chemical Technology
Imperial College of Science and Technology
London SW7

INTRODUCTION

In process engineering mass transfer with chemical reaction is encountered in separation processes and in chemical reaction engineering. In separation processes a reactant is usually introduced to react with the transferring solute and thus speed up the rate of mass transfer. The effect of mass transfer in reaction engineering is entirely different. It may affect the supply or removal of reactants or products, particularly in heterogeneous systems, from the reaction zone and thus limit the rate of reaction, so that the reaction can become mass-transfer controlled. This distinction is not always clear-cut. In general, however, the generic term 'mass transfer with chemical reaction' is employed when the aim of the process is the separation of a mixture or isolation of a particular component. Thus, absorption and extraction with chemical reaction fall into this category. Conversely, if the aim of the process is the conversion of a substance or formation of a product, one talks about mass-transfer controlled chemical reaction and treats it under the heading of chemical reaction engineering.

Distillation with chemical reaction does not fall into either of these categories as it aims at simultaneous attainment of both objectives, i.e. product formation and separation. It is therefore an operation of particular interest since it combines two processes in a single unit so that capital and operating costs can be reduced considerably. Furthermore, if the reaction is reversible, e.g.

$$aA + bB \rightleftarrows qQ + sS$$

where Q, the desirable product, is also the most volatile componen its removal from the liquid phase increases the yield. Distillation is also usually conducted at elevated temperatures. This generally speeds up the rate of reaction. However, higher temperatures may also lead to differential enhancement of side reactions. Consequently, distillation with chemical reaction is not a generally applicable process. It is highly system-specific and its suitability should be assessed separately for each process.

PREVIOUS WORK

Esterification has been the most popular reaction to be conducted in a distillation column. Patents were awarded as early as 1921 to Backhaus (1) for combining distillation with chemical reaction of alcohol and acetic acid using sulphuric acid as catalyst. No details of column design or calculation were, however, given. Subsequently, in 1932 Keyes (2) reviewed the various esterification processes used in industry, mainly summarizing material from patent literature. Further patents were granted to McKeon (3) and Brunn and Grubb (4).

Schnipp (5) produced in a continuous process 2,3-butylene glycol diacetate from butylene glycol and acetic acid with sulphuric acid as a catalyst. The column consisted of an interconnected series of reaction vessels each containing a single bubble-cap. The glycol and the catalyst were continuously supplied to the top of the column with glacial acetic acid introduced at the bottom. The distillate, consisting mainly of acetic acid and water, was separated and the acetic acid recycled. The bottom product was separated by vacuum distillation and a 97% yield of ester was obtained experimentally.

Leyos and Othmer (6) studied the continuous esterification of butanol and acetic acid in a five-stage apparatus where a hot feed was introduced to the top plate of the esterification column. They showed that long times of contact (produced by larg hold-ups or large number of plates) were not needed for high yields. However, overall conversion was increased by an increase in catalyst concentration.

Belk (7) describes a plate-to-plate calculation for hypothetical, two and three-component liquid-phase reversible reactions, carried out continuously in a single distillation column. He disregarded deviations of the liquid mixture from ideality, assumed adiabatic operation of the column and 100% plate efficien Heat of reaction was assumed to be negligible and all parameters to be independent of temperature. He considered specific examples of a two-component reversible reaction A $\rightleftharpoons$ B as well as hypothetical reactions 2A $\rightleftharpoons$ B and 2C $\rightleftharpoons$ A + B.

Marek (8) derived a design procedure for plate columns based on material and enthalpy balances which included the presence of a chemical reaction. For a ternary mixture a combined numerical and graphical method was suggested using a modified McCabe-Thiele construction. The problem was simplified by assuming negligible heat of reaction and 100% stage efficiency. The hypothetical reactions employed were A + B → 2Q and A + B → Q.

Subsequently, Marek (9) used his procedure to predict conditions in a 30-plate bubble-cap column used for simultaneous reaction and distillation of the mixture water-acetic acid-acetic anhydride at a pressure of 400 mm Hg. Additional assumptions included absence of vapour-phase reaction, introduction of reflux and feed at their boiling points and constant molar overflow. Provided plate efficiency was around 50% for each component, reasonable agreement was claimed between theory and experiment.

Costa (10) studied both theoretically and experimentally the continuous hydrolysis of acetic anhydride in a ten plate distillation column also operating under a pressure of 400 mm Hg. Data of only a single experiment are reported.

Matz (11) investigated the production of hydrocarbon halide in a packed column using phosphorus and methanol. The resulting methanol-methyl halide azeotrope was separated by liquid-liquid extraction using water as the continuous phase.

Parker (12) recommended the use of a distillation reactor for hydrolyzation of ethylene oxide to ethylene glycol. Miller (13), and subsequently Corrigan and Miller (14), analysed this process using a crude plate model and concluded that increased temperature in the distillation reactor adversely affected selectivity of the process as compared to the two-stage Shell process. However, this was disproved by Sive (15) who found no effect on selectivity of operating pressure or feed composition when modelling a packed distillation reactor for this process.

Lopez (16) investigated esterification of acetic acid by methanol in a wetted-wall column under distillation conditions. He found experimentally that conversion was enhanced by an increase in feed rate of the acid or the concentration of the catalyst but decreased by an increase in reflux ratio. He attributed differences between predicted and experimental values to inadequacy of the predicted vapour-liquid equilibrium data. The same system was studied experimentally by Corrigan and Ferris (17) in an Oldershaw sieve-tray column. Batch and continuous experiments were conducted and the latter were claimed to yield a 90.4% methyl acetate composition in the top product.

Jeffreys (18) studied the hydrolysis of acetic anhydride to acetic acid in a six-tray sieve distillation unit. The process was investigated both theoretically and experimentally. Theoretical analysis comprised the solution of mass and enthalpy balances with an additional term allowing for the effect of chemical reaction. Good agreement was obtained between calculated and experimental results.

Geelen and Wijffels (19) investigated the reaction of vinyl acetate with stearic acid in a distillation column to form vinyl stearate and acetic acid. A modified form of the McCabe-Thiele diagram was employed to obtain the number of theoretical plates for a given conversion. This was tested experimentally using a bubble-cap Oldershaw column for the distillation reactor followed by a packed column for separation of top products (vinyl acetate, acetic acid). Theoretical considerations agreed reasonably well with experimental findings.

Zyryanov (20) proposed a calculation method for rectification of a binary mixture accompanied by a first-order isomerisation $A \rightarrow B$. However, the method has not been tested.

Hirata (21) discussed an iterative method for the determination of stage temperatures, stage reaction rates and interstage flow rates. This was conducted in the context of multicomponent distillation with a simultaneous chemical reaction and the use of Muller's modified method for the convergence of the column temperature profile. He assumed the existence of one feed stream F_j, one vapour stream G_j, one liquid stream L_j and one intercooler or interheater Q_j on each stage except the condenser and the reboiler. Each stage was regarded as an equilibrium stage. A tridiagonal matrix algorithm was employed for the solution of the linearized material balance equations.

Hirata (22) also studied the behaviour of an Oldershaw column, a packed column and a sieve-tray column as distillation reactors for the esterification of ethanol and acetic acid. He found reflux ratio to be an important factor affecting not only separation but also conversion.

Saunders (23) investigated both theoretically and experimentally the sulphonation of toluene in a packed distillation column. The mathematical approach consisted of numerical integration of a set of simultaneous differential equations. The problem was complicated by the existence of two liquid phases.

Nelson (24) studied theoretically the general case of countercurrent equilibrium stage separation with chemical reaction and applied his technique to describe distillation reactors. His model relied on the assumption of each stage being a perfectly mixed reactor and also an equilibrium stage.

Davies (25, 26) studied the continuous transesterification of ethanol and butyl acetate in a column of six sieve plates. The approach was similar to that adopted by Jeffreys (18) and again good agreement was reported between predicted and observed results. Conversion was observed to depend on stage hold-up, reflux ratio and boil-up rate.

Davies et al. (27) also extended their studies to distillation of formalin solutions, that is the analysis of a distillation column involving reactions of formaldehyde with water and methanol. The problem was complicated by the estimation of equilibrium data for a system of five-components of which one (hemiformal) does not exist in a pure state. Although good agreement is claimed between theory and experiment, some of the assumptions are rather doubtful, e.g. the use of the A.I.Ch.E. method for prediction of plate efficiencies.

Pilavakis (28, 29) investigated the esterification of methanol by acetic acid in a packed column. He assumed the reaction to be pseudo-first-order with respect to either methanol or acid over certain specified concentration ranges and incorporated the effect of heat of reaction not only in the enthalpy balances but also in the flux equations. The column was calculated by numerical solution of a set of differential equations. The top product was an azeotropic mixture of methanol and ester which could, however, be broken by introduction of acetic acid high up in the column rather than further down as a mixed feed with methanol. Consequently, in practice such a column will consist of a rectifying section, an extractive distillation section with acetic acid as the extractive solvent and a distillation reactor section. Good agreement was obtained between theory and experiment which, however, suffered from the fact that the hold-up of liquid in the column was small in comparison to the reboiler hold-up so that most of the reaction occurred in the latter location.

Hegner and Molzahn (30) suggested the extension of their method for calculation of countercurrent separation processes accompanied by chemical reaction to differential-contact (packed column) distillation reactors. However, the method does not incorporate the constraint that liquid should stay at its boiling point and vapour at the condensation point and hence requires some modifications.

Kaibel et al. (31) present a mathematical model for the calculation of stagewise distillation reactors both for continuous and batch operations. The assumptions made are similar to those of Hirata (21) and calculations are again conducted in terms of theoretical plates. The behaviour of distillation reactors is illustrated with reference to hypothetical cases of a fast reaction with an unfavourable equilibrium position, an esterification column and a saponification column. In the first case four configurations were considered: steady-state or batch operati with the reaction confined only to the reboiler or taking place in the reboiler and in the column. Only the latter will be referred to in this text as a distillation reactor, whereas the former will be termed reboiler reactor. The authors have found that for the same fractional conversion the energy requirements of distillation reactors are significantly lower than those of reboiler reactors and that in each case they were also relatively lower for continuous than batch operation. The first statement confirmed previous findings of Block (32) who used a quasi-McCabe-Thiele diagram to analyse the esterification reaction between acetic acid and butanol. The superiority of the distillation reactor over the reboiler reactor was quite striking. Kaibel et al. also observed a strong effect of feed plate location The analysis of the esterification column indicates the existence in distillation reactors of concepts of minimum number of theoretical stages and minimum reflux ratio for a specified conversion. The number of stages is strongly affected by the residence time of liquid on the plate. It is claimed that the BASF simulation method described by the authors is supported by experimental evidence although no such data are presented.

The BASF model presented by Kaibel et al. (31) was further analysed by Mayer and Worz (33). They considered the reaction $A + B \rightleftharpoons C + D$ and investigated the effect on conversion of feed plate location, number of plates and mutual variation in relative volatilities of A, B and D, all as a function of heat input in continuous and batch operations. For C as the most volatile component, it was found that a batch reboiler reactor is energetically a more attractive solution if D is more volatile than A. Otherwise a distillation reactor is preferable.

As will be noticed, the reactions investigated for potential use in distillation reactors were mainly esterification, saponification and transesterification. In addition to the work already reported, esterification of acetic acid with ethanol and with ethylene glycol has also been studied by Hartig and Regner (34) and of acetic acid with butanol by Block and Hegner (35). Description of esterification of terephthalic acid with ethylene glycol in a distillation reactor is contained in a Monsanto patent (36). Other studies of transesterification reactions included butyl acetate with ethanol by Koban and Wagner (37) and of dimethylterephthalate with ethylene glycol by Baratella et al. (38). Other types of reactions are less frequently encountered but investigations are described of isomerisation (39, 40) and amide formation (41). There is now little doubt that, for specific reactions, the use of distillation reactors represents substantial savings in both operating and capital costs.

PROBLEM FORMULATION

Distillation with chemical reaction is a process of great complexity. Not only does it impose simultaneous consideration of the rate equations of mass transfer and chemical change but also amplifies their interaction through the presence of heat of reaction. Further, it poses questions regarding equilibrium data and questions the validity of such concepts as plate efficiency.

Full description of distillation reactors requires the knowledge of:

(a) vapour-liquid equilibrium data,
(b) rate equation of chemical change,
(c) thermal effects of the reaction,
(d) rate equation of mass transfer as modified by the presence of chemical change and of thermal effects,
(e) performance data of column, e.g. mass transfer coefficients or plate efficiencies (if applicable),
(f) column calculation procedure.

Each step will be discussed in turn with the final calculation procedure being presented separately for packed and plate columns. It is at this stage that any shortcomings in the constituent data become compounded on synthesis of the individual elements into column calculation.

VAPOUR-LIQUID EQUILIBRIA

There is no method for theoretical prediction or correlation of VLE data in reacting systems. It has to be remembered that equilibrium in such systems means not only concentration and thermal equilibrium but also absence of chemical change. In other words, the chemical reaction must also be at equilibrium or, if it is irreversible, it must have reached completion. This poses the question of how to express the interphase concentration driving force in, e.g., a packed column in the presence of, say, a slow reaction. Should it be the difference between the local concentration of a component in the vapour phase and a 'true' equilibrium value corresponding to a given liquid concentration of the component or should a 'pseudo' equilibrium value, which would treat the reaction as frozen in time, be used for this purpose? It is the latter approach which is generally employed, that is multicomponent VLE data are predicted as if the reaction did not exist. Prediction is normally based on binary data using the Wilson equation. In theoretical treatments constant K-values or constant relative volatilities were often assumed. Neither of these procedures is very satisfactory since the fact that a reaction is taking place already indicates high non-ideality of the system and, in addition, frequent presence of carboxylic acids indicates a non-accounted for lack of ideality in the vapour phase. For the same reasons, the suggested use of the UNIFAC method is also unacceptable.

Experimental determination of VLE data depends very much on the choice of still. The traditional type of a circulation still is unsatisfactory for a reacting system since the compositions of the phases are constantly changing. This necessitates the use of a flow still. Pilavakis (28) used for this purpose a suitably modified version of the Cathala ebulliometer (42) which was first successfully tested on a number of well known binary mixtures. For the system methyl acetate-acetic acid-methanol-water he determined VLE data for all six binary systems, unless thermodynamically consistent values were already available from literature, for the four ternary systems and for the quaternary system with and without the catalyst. It was found that the presence of sulphuric acid in the composition range of up to 4 wt-%, as normally encountered in homogeneous catalysis, had negligible effect on the equilibrium data calculated on acid-free basis. This has also been found by Teshima (43).

The experimental data were correlated by third-order Margules equation using Marek's equation (44, 45) to take into account non-ideality in the vapour phase. An attempt was also made to correlate the data using the NRTL method but this proved to be successful for the binary systems but the quaternary predictions were not as good as those given by the Margules equation. The same quaternary system was investigated by Hirata (46). He presented four equations to predict vapour-liquid equilibrium ratios as function of temperature without quoting the actual VLE data. His results did not agree with those of Pilavakis and were also found to be thermodynamically inconsistent. Subsequently, Hirata (47) also presented an equation for the calculation of activity coefficients for this quaternary system. The equation was based on the Margules equation and consisted of 64 constants rearranged as a polynomial series in terms of mole fractions. It was tested by Pilavakis and the results found inconsistent even for binary mixtures.

From the scant information available in literature it seems, therefore, that the use of pseudo-equilibrium data is justified. Further, with the present state of knowledge of VLE determinations, they have to be established by laborious experimentation using a flow still and, unless a better method is tested, correlated by third-order Margules equation corrected for non-ideality in the vapour phase.

REACTION KINETICS AND REACTION EQUILIBRIUM

Rate equations of chemical change are usually available at moderate temperatures for most of the reactions of interest. The same applies to reaction equilibrium constants. Sufficient information is needed to extrapolate both the equilibrium and rate constants to the boiling point. This is normally performed by an Arrhenius-type equation for which data are not difficult to acquire. It should be noted that for work on stagewise distillation reactors no reaction kinetic data were needed, since theoretical plates were assumed throughout. When such information was required, the reaction was assumed to be of pseudo-first order.

THERMAL EFFECTS

In absorption with chemical reaction thermal effects, that is heat of reaction, result only in temperature changes of the phases. This is no longer the case in distillation where the liquid is at its boiling point and vapour at the condensation point. An exothermic reaction will produce surface evaporation (presence of nucleate boiling is considered unlikely) and thus introduce a 'thermal' distillation effect superimposed on the normal or so-called 'contact' distillation process. Similarly, an endothermic reaction will result in surface condensation, again resulting in a thermal distillation contribution. The situation is similar to non-adiabatic distillation which, for instance, could occur in a wetted-wall column by supplying or extracting heat through the column wall. Such non-adiabatic operation of a distillation column was first considered by Byron, Bowman and Coull (48) whose qualitative predictions are presented in Fig. 1.

They considered contact distillation, i.e. concentration-driven diffusion process, to possess a maximum at adiabatic conditions (curve 1) and thermal distillation to increase rapidly with both heat supply and heat removal (curve 2) as the result of fractional vaporization or condensation respectively. The total effect (curve 3) was regarded as a sum of the two separate effects.

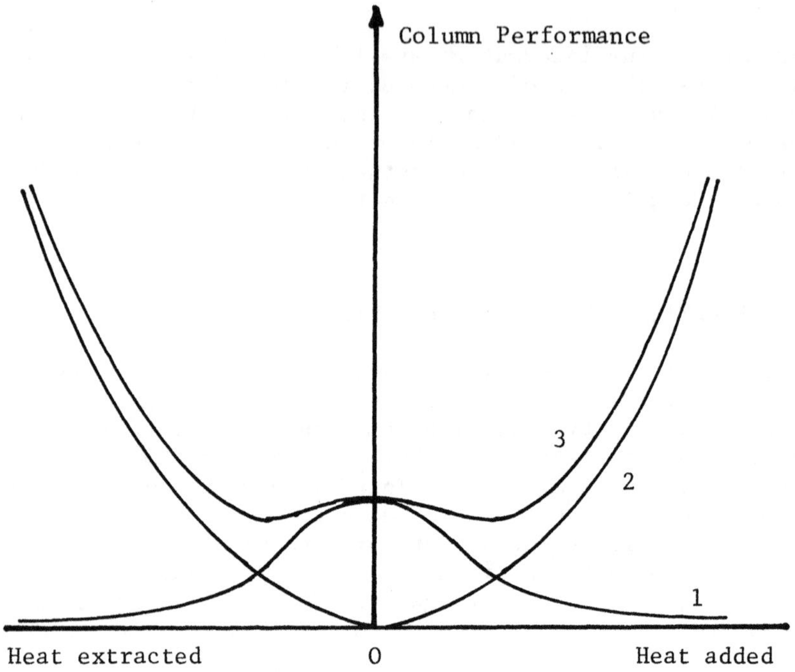

Fig. 1. Effect of non-adiabatic operation on column performance according to Byron, Bowman and Coull (48).

Bainbridge (49, 50) tried to confirm the predictions of Byron et al. using a sieve-tray column but the results were inconclusive. Further work was, therefore, conducted by Wildey (51), Oyekan (52), Teja and Thong (53) and independently by Hochgesand (54) using wetted-wall columns. The relevant theory will be summarized here briefly, since it affects the rate equation for distillation with chemical reaction.

Consider a binary, gas-phase controlled distillation process of components A and B. If N_A, N_B are the molar fluxes normal to the interface and λ_A, λ_B the relevant molar latent heats of evaporation, it is customary to write the heat balance at the interface as

$$N_A \lambda_A + N_B \lambda_B = 0 \tag{1}$$

However, if a heat flux q_s resulting in surface vaporization is supplied through the wall

$$N_A + N_B = q_s/\lambda \qquad (2)$$

if $\lambda_A = \lambda_B$.

But, from the Stefan-Maxwell equation for binary mixtures

$$N_A = (N_A + N_B)\, y'_A + n_A \qquad (3)$$

where n_A is the diffusional component of the total flux and y'_A the local mole fraction of A (say the more volatile component) in the vapour phase. Since

$$n_A = - D_{AB}\, C_G\, (dy'_A/dz) \qquad (4)$$

where D_{AB} is molecular diffusivity, C_G molar density and z direction normal to the interface, combination of (2), (3) and (4) gives

$$N_A = (q_s/\lambda)\, y'_A - D_{AB}\, C_G\, (dy'_A/dz) \qquad (5)$$

Assuming applicability of the film theory and hence of boundary conditions: $y_A' = y_{As}$ at $z = 0$ and $y'_A = y_A$ at $z = \ell$, the flux at the interface N_{As} becomes

$$N_{As} = (q_s/\lambda)(y_{As} \exp \varepsilon_G - y_A)/(\exp \varepsilon_G - 1) \qquad (6)$$

where $\varepsilon_G = q_s/\lambda\, K^o_G$ and K^o_G is the mass transfer coefficient for adiabatic distillation ($K^o_G = D_{AB}C_G/\ell$, according to the film theory). It can be seen from eqn (6) that for small values of ε_G, e.g. $\varepsilon_G < 0.1$,

$$N_{As} = K^o_G\, (y_{As} - y_A) + (q_s/\lambda)\, y_{As} \qquad (7)$$

and hence only under these conditions are the 'contact' and 'thermal' fluxes additive.

From conservation equation for A over a differential section of column height dh

$$d(Gy_A) = N_{As}\, a_h\, dh \qquad (8)$$

where G is the molar flow rate of the vapour and a_h the interfacial area per unit height of column.

But $$dG = (q/\lambda)\, a_h\, dh \qquad (9)$$

and $G = G^{o} + (q/\lambda)\ a_h\ h$ (10)

where h is measured from column bottom where the vapour rate is G^{o}. On combining eqns (6), (8), (9) and (10) as well as assuming that $(q/\lambda)\ a_h\ h << G^{o}$ the column performance, expressed in terms of number of transfer units N_G, is

$$N_G = \int_{y_{A1}}^{y_{A2}} dy_A/(y_{As}-y_A) = (q_s/\lambda)\ a_h\ H/G^{o}(1-\exp(-\varepsilon_G)) \quad (11)$$

Since, for adiabatic operation,

$$N_G^{o} = K_G^{o}\ a_h\ H/G^{o} \quad (12)$$

therefore $N_G/N_G^{o} = \varepsilon_G/(1-\exp(-\varepsilon_G))$ (13)

Thus, heat supply to the column increases its performance, whereas heat extraction (negative ε_G) results in lower separating power. These findings have been confirmed experimentally.

In the presence of liquid phase resistance, an equation analogous to eqn (6) can be established for the liquid phase containing x_{As}, the mole fraction of liquid at the interface, and the parameter $\varepsilon_L = q_s/\lambda K_L^{o}$, where K_L^{o} is the liquid phase mass transfer coefficient for adiabatic distillation. Assuming equilibrium at the interface, y_{As} and x_{As} can be eliminated from eqn (6) and the corresponding equation for the liquid phase. This would give a flux equation in terms of y_A, x_A, K_G^{o}, K_L^{o} and q_s. Values of the relative performance for non-adiabatic distillation, expressed as the ratio of overall number of transfer units N_{OG}/N_{OG}^{o}, have been calculated for such a case (53) and are presented in Fig. 2 for various ratios of K_G^{o}/K_L^{o}. It can be seen that, should distillation be a liquid-phase controlled process, heat extraction and not heat supply would produce an increase in performance over the adiabatic case.

Strictly speaking, eqn (1) is only an approximation as it neglects the existence of heat transfer arising from the fact that the boiling point of the liquid phase is not equal to the condensation point of the vapour phase. This, in the case of adiabatic distillation, produces the 'thermal' distillation effect first discussed by Danckwerts, Smith and Sawistowski (55). As the effect is usually not large, an additive form, analogous to eqn (7), can be employed, i.e.

$$N_{As} = K_G^{o}\ (y_A^* - y_A) + h_G\ (T_G - T_L)/\lambda \quad (14)$$

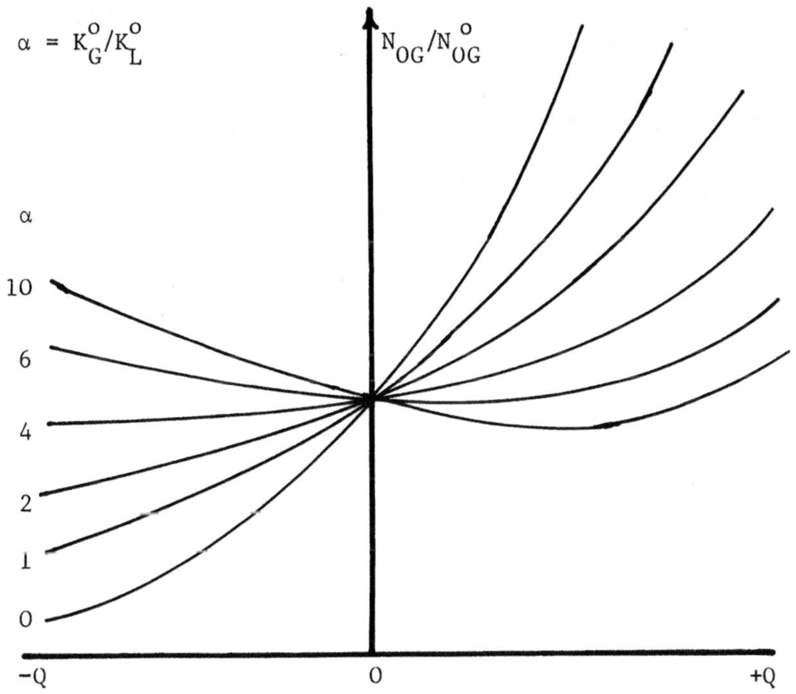

Fig. 2. Variation of relative column performance for non-adiabatic operation of a distillation column at different K_G^o/K_L^o ratios.

The distillation process is again considered as gas-phase controlled, y_A* being the equilibrium value with respect to x_A, h_G the heat transfer coefficient in the vapour phase and T_G, T_L denoting the temperatures of the two phases.

From the Chilton-Colburn analogy

$$h_G = K_G^o\, M_G c_G\, Le_G^{2/3} \tag{15}$$

where M_G is the molar mass, c_G the specific heat capacity and Le the Lewis number. Consequently, from eqns (14) and (15)

$$N_{As} = K_G^o\{(y_A^* - y_A) + Le_G^{2/3} M_G c_G (T_G - T_L)/\lambda\} \tag{16}$$

For the system methanol-water operated at total reflux the following values apply at $x_A = 0.5$: $y_A = 0.5$, $y_A^* = 0.784$, $T_G = 86.0^oC$ and $T_L = 73.2^oC$. The contribution of thermal distillation to the molar flux for this particular case is, therefore, of the order of 12%. However, the effect is normally neglected in distillation calculations and will also be neglected in subsequent considerations. Only the thermal effect arising from the heat of reaction will be taken into account.

MASS TRANSFER KINETICS

In the vapour phase, where no chemical reaction takes place, the flux equation is given by eqn (6) which, for component j, is given by

$$N_{js} = (q_s/\lambda)(y_{js} \exp \varepsilon_{Gj} - y_j)/(\exp \varepsilon_{Gj} - 1) \tag{17}$$

where $\varepsilon_{Gj} = (q_s/\lambda)/K^o_{Gj}$.

For the liquid phase, conservation equation of component j leads to the following differential equation

$$-(dN_j/dz) + r_j = 0 \tag{18}$$

where r_j is the rate of formation of component j in a reaction $A + B \rightleftharpoons C + D$, i.e. absolute value of all stoichiometric coefficients is equal to one.

Assuming the reaction to be of pseudo-first order with respect to component j, the rate equation per unit volume of liquid is

$$r_j = - kC_L x_j \tag{19}$$

where k is the pseudo-first order rate constant, C_L the molar density of the liquid and x_j the mole fraction of component j in the liquid phase.

The reaction-induced thermal component of the molar flux in the gas phase induces a similar component in the liquid phase. Hence, the combination of equations (18), (19) and the liquid phase equivalent of eqn (5) leads to the following differential equation

$$(d^2x^l_j/d\zeta^2) - \varepsilon_{Lj}(dx^l_j/d\zeta) - \beta^2_j x^l_j = 0 \tag{20}$$

where x^l_j denotes the local mole fraction in the film, $\zeta = z/\ell_L$ with ℓ_L being the thickness of the liquid film, $\varepsilon_{Lj} = (q_s/\lambda)/K^o_{Lj}$ and $\beta^2_j = kD_{jm}C^2_L/(K^o_{Lj})^2$. D_{jm} is the effective diffusivity of j in the liquid and, according to the film theory, $K^o_{Lj} = D_{jm}C_L/\ell_L$. By assuming the direction from the liquid to the gas phase as positive, the boundary conditions are: $x^l_j = x_j$ at

$\zeta = 0$ and $x_j^1 = x_{js}$ at $\zeta = 1$.

The solution of eqn (20) subject to the above boundary conditions yields the following expression for the molar flux of j at the interface

$$N_{js} = K_{Lj}^{o} \frac{b}{\tanh b}\left[\frac{x_j \exp(\varepsilon_{Lj}/2)}{\cosh b} - x_{js}\left(1 + \frac{\varepsilon_{Lj}}{2b} \tanh b\right)\right] \quad (21)$$

where $b^2 = (\varepsilon_{Lj}^2 + 4\beta_j^2)/4$.

By assuming equilibrium at the interface ($y_{js} = m_j x_{js}$) and thus eliminating y_{js} and x_{js} from eqns (17) and (21) the final expression for the flux is obtained as

$$N_{js} = \left[\frac{\exp(\varepsilon_{Gj})-1}{K_{Gj}^{o}\, \varepsilon_{Gj}} + \frac{m_j \exp(\varepsilon_{Gj}) \tanh b}{b\, K_{Lj}^{o}\{1+(\varepsilon_{Lj}/2b) \tanh b\}}\right]^{-1} x$$

$$\left[\frac{\exp(\varepsilon_{Gj} + \varepsilon_{Lj}/2)}{\cosh b + (\varepsilon_{Lj}/2b) \sinh b} y_j^* - y_j\right] \quad (22)$$

The first term represents the reciprocal of an overall mass transfer coefficient and incorporates the enhancement factors for both the gas-phase and liquid-phase mass transfer coefficients. The second term represents the effective driving force which is modified by the effect of the reaction on the equilibrium value.

By analogy to absorption with chemical reaction, eqn (22) can be simplified for small values of ε and small or large values of b.

The parameter q_s is calculated as follows

$$q_s = - r_j(-\Delta H)\phi_L/a_s \quad (23)$$

where $-\Delta H$ is the enthalpy of reaction per mole of j, ϕ_L is the volumetric liquid hold-up per unit volume of packed bed or dispersion and a_s the corresponding interfacial area.

On combining eqns (23) and (19)

$$q_s = \{kC_L(-\Delta H)\phi_L/a_s\}\, x_j \quad (24)$$

It should be noted that in this equation j represents a component with respect to which ΔH is defined and hence is not a general subscript.

First-order chemical kinetics have been assumed throughout this treatment for two reasons:

(a) it provides an analytical solution for the molar flux, and hence
(b) highlights the various problems specific to a distillation situation.

The assumptions underlying the choice of this type of kinetics are negligible effect of the reverse reaction as well as pseudo-constant values of concentrations of some of the other components and also of the homogeneous catalyst. Only the latter is generally justified. Hence, first-order kinetics is a crude approximation which can only be applied over certain ranges of concentration with different rate constants applying to different locations in the column.

Higher order or more complex kinetics leads to solution of a set of simultaneous differential equations so that the value of the interfacial flux can only be obtained by a numerical method.

PERFORMANCE DATA

As indicated by eqn (22) the performance data which are required for column calculation are the mass transfer coefficients K^o_{Gj} and K^o_{Lj}. Each phase is a multicomponent mixture and consequently the mass transfer coefficients are multicomponents mass transfer coefficients. In absorption with chemical reaction the consequences of this problem are usually avoided by the assumption of dilute solutions and hence of mutually independent binary diffusivities. Such an assumption is not valid for distillation systems, where effective binary diffusivities are functions of concentration of all the components and of their binary diffusivities. Consequently, mass transfer coefficients of the different species are not equal either. Further, they are concentration dependent and do not represent an independent set of values. This problem is not specific to distillation reactors but applies to multicomponent distillation in general. Although a method of calculation of effective mass transfer coefficients K^o_j as a function of binary mass transfer coefficients K_{ij} has been developed by Krishna and Standart (56, 57) it is seldom employed. Instead, in packed columns, equal mass transfer coefficients are used for all components with consequent distortion of the concentration profile along the column. Such an approach is perhaps not surprising if one considers that even the best available correlations for volumetric mass transfer coefficients require a safety factor of 1.70 for confidence limit of 95% (58). For distillation reactors this was best exemplified by the work of Pilavakis who obtained such widely different and often obviously unacceptable values of volumetric mass transfer coefficients from existing correlations that he was forced to obtain them from direct experimentation on his pilot-plant column using the EFCE recommended mixture methanol-ethanol. It seems, therefore, that prediction of basic mass transfer data lags considerably behind the refinement of theory.

In plate column distillation the basic performance concept is that of plate efficiency, usually the Murphree type, although the Hausen and Standart variations have occasionally been recommended and used. The mass transfer process enters the plate efficiency concept through the relation between the flux, and hence mass transfer coefficients, and the point efficiency which, in turn, is related to the plate efficiency by flow behaviour on and between the plates. Since in multicomponent mixtures the mass transfer coefficients of the components are different, the same must apply to the point efficiencies and also to the plate efficiencies. Again, the method of Krishna and Standart can be used to predict plate efficiencies of individual components and it has been successfully tested for ternary mixtures. However, it has not as yet been accepted in general distillation practice. Instead, a plate efficiency based on the key components is calculated and applied to all components with consequent loss in accuracy of prediction of concentration profiles.

Although the percentage variation in mass transfer coefficients is significantly reduced on converting them into plate efficiencies, nevertheless the application of the plate efficiency concept to distillation reactors suffers from another, more serious, disadvantage. Plate efficiency, in its Murphree formulation, is defined as the ratio of the actual progress of a process to that attained on a theoretical plate. A theoretical plate is, by definition, a stage in which the two streams leaving it are in thermodynamic equilibrium, and this applies not only to concentrations and temperatures but also to chemical reactions. It is useful in this context to consider the extreme case of a slow irreversible reaction. A theoretical stage would imply the complete conversion of one of the reactants, that is a 100% yield of the product. In practice this means that the complete distillation reactor is equivalent to less than a single theoretical stage if the yield is, for instance, 97%. With reference to a single plate it means that the plate will have a low reaction plate efficiency but its mass transfer plate efficiency as a measure of approach to the pseudo-equilibrium values may be quite high. Quite apart, therefore, from the fact that concentration plate efficiencies of individual components may be different, the reaction plate efficiency will also, in general, be different. Hence, any values of single plate efficiencies for mass transfer and chemical reaction quoted in literature have to be viewed with suspicion and may represent no more than a fitted parameter.

From general considerations it follows that concentration plate efficiencies for distillation reactors will have values similar to those for ordinary distillation columns modified slightly by thermal distillation effects. Thus, they will be higher for exothermic than for endothermic reactions under otherwise similar conditions. Prediction of reaction efficiencies is an uncharted field. It requires the knowledge of residence time distributions with modification of reactor behaviour by continuous exchange of components with the vapour phase.

COLUMN CALCULATION PROCEDURES

Packed Columns

The method of calculation of packed columns is based on stepwise calculation of multicomponent distillation columns with variable molar overflow. The introduced modifications represent simply the presence of the reaction.

Thus, the equations for the component balances are:

$$d(Gy_j) = - N_{js}\ a_s\ A_c\ dh \tag{25}$$

$$= d(Lx_j) - r_j\ \phi_L\ A_c\ dh \tag{26}$$

where A_c is the cross-sectional area of the column and h is the variable column height measured from the top of the column. Again the stoichiometric coefficients in eqn (26) are assumed to be equal to one.

Addition of eqn (25) for all components gives

$$dG/dh = - \sum_j N_{js}\ a_s\ A = - (q_s/\lambda)\ a_s\ A \tag{27}$$

where λ is the mean molar latent heat of evaporation.

$$\text{Thus } \sum_j N_{js} = q_s/\lambda \tag{28}$$

and values of N_{js} obtained from the flux equations must be normalized prior to use in eqn (25).

By analogy to eqn (27)

$$dL/dh = - (q_s/\lambda)\ a_s\ A \tag{29}$$

The liquid and vapour temperatures in any section of the column can be calculated from the respective liquid and vapour compositions, since the liquid is assumed to be always at its boiling point and the vapour at the condensation point. An energy balance is, however, required for the calculation of q_s. Such a balance over a differential section of the column, dh, gives

$$d(G\ C_G\ T_G) = -\ q_s\ a_s\ A\ dh \tag{30}$$

$$= d(L\ C_L\ T_L) - r_j\ \Delta H\ \phi_L\ A\ dh - dH_D - dH_L \tag{31}$$

where H_D, H_L represent the heat of dilution and heat loss from the column. If these two terms are negligible, q_s is given by eqn (23) and the incorporation of an energy balance is not necessary.

Assuming that the values of G, L, y_j, x_j, T_G and T_L are known at a particular plane h in the column, the values of physical properties can be calculated there as well as the values of the derived parameters r_j, N_{js} and q_s. However, in order to obtain the values of the primary parameters at a plane h + Δh, the local gradients of G, L, y_j, x_j at h must also be known. The first two are already given by eqns (27) and (29), while the last two are obtained from eqns (25) and (26) as

$$dx_j/dh = \{r_j\ \phi_L\ A - x_j(dL/dh) - N_{js}\ a_s\ A\}/L \tag{32}$$

$$dy_j/dh = \{-\ y_j(dG/dh) - N_{js}\ a_s\ A\}/G \tag{33}$$

All the equations are thus available to calculate G, L, y_j and x_j in plane h + Δh using, e.g. the Runge-Kutta routine. In this procedure, the derivatives of y_j and x_j are calculated, however, without taking into consideration the value of, or change in, other mole fractions. For this reason the mole fractions must be normalized at each step of the calculation. Hence

$$x_j = (x_j)_{calc}/\Sigma_j(x_j)_{calc} \tag{34}$$

and $$y_j = (y_j)_{calc}/\Sigma_j(y_j)_{calc} \tag{35}$$

It should be noted that separate material and energy balances have to be set up for the bottom and top of the column and around the feed points. Alternatively, corresponding terms could be included in the basic equations and set to zero at all other places in the column.

A full set of equations and a basic procedure is thus available for preparation of an algorithm for column calculation. This will be similar to an algorithm for calculation of a multicomponent packed distillation column. Its use in practice (28, 29) has shown that cases exist when Euler's method can be used instead of the Runge-Kutta routine with considerable saving in computer time. On the other hand, some problems are stiff and convergence is only attained by application of Gear's predictor-corrector technique.

Plate Columns

A theoretical model for the calculation of the number of theoretical plates using the Newton-Raphson method is presented by Kaibel et al. (31). However, it does not incorporate a constraint on T so that temperature becomes an independent variable. Such an assumption is obviously highly questionable. Nevertheless this difficulty can be overcome by incorporating such a constraint into the equations. The problem of different plate efficiencies for concentration and reaction equilibrium is, however, considerably more difficult to handle. It would appear that the best approach will be to abandon completely the concept of theoretical plates and efficiencies and develop instead a plate-to-plate calculation method based on real plates. Here the extension of the differential equations for packed columns into difference equations and their subsequent modification to apply to each individual plate offers the best chance of success.

ECONOMIC ANALYSIS

Economic feasibility studies presented so far in literature refer primarily to energy requirements. In most cases they are based on hypothetical systems employing the concept of theoretical plates. Although actual values may therefore be questionable, nevertheless the trends will remain significant. Thus, Fig. 3 represents percentage conversion as a function of energy requirement for steady-state operation of a distillation reactor and a reboiler reactor according to the calculations of Kaibel et al. (31). The reaction in question is $A + B \rightleftharpoons C + D$ with chemical equilibrium placed on the far left. Product C is volatile and D non-volatile, whereas A has a boiling point $5^{o}C$ above B. In such a case for 90% conversion the distillation reactor requires only 0.5% energy needed for the reboiler reactor. This is due entirely to the much faster removal of C from the system.

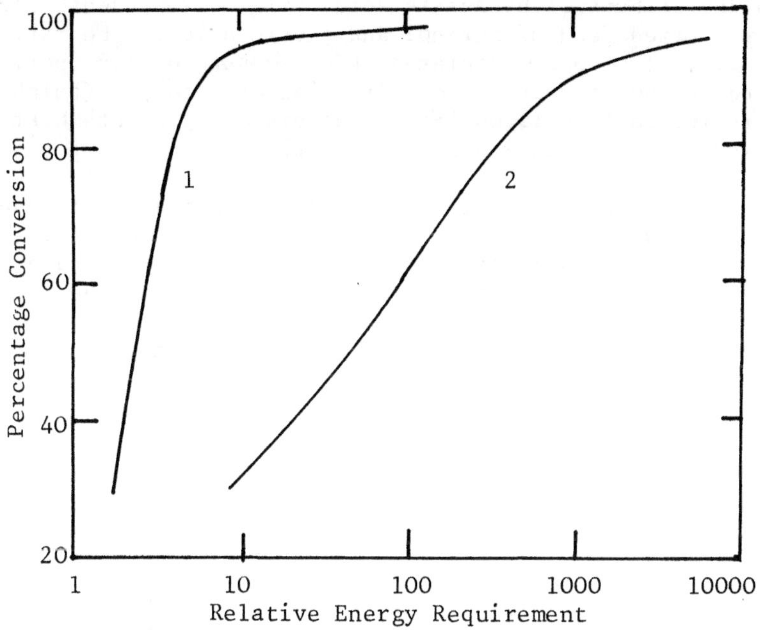

Fig. 3. Comparison of energy requirements for a distillation reactor (curve 1) and a reboiler reactor (curve 2). Mixed feed is employed in both cases.

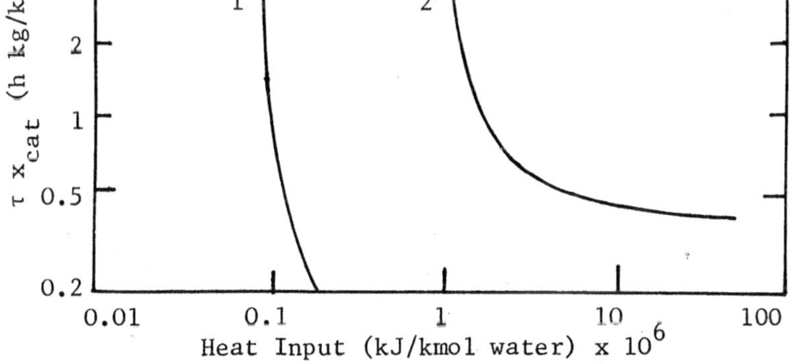

Fig. 4. Energy requirements of a distillation reactor (curve 1) and reboiler reactor (curve 2) in the production of butyl acetate for acetic acid conversion of 97%.

Similar findings are reported by Block (32) in the study of esterification of butanol by acetic acid. Fig. 4 represents the results for a mixed feed of butanol and acetic acid in the ratio of 1.14 to 1.0. The column operated at a pressure of 1.25 bar. The ordinate is the product of mole fraction of catalyst (sulphuric acid) and residence time τ, whilst the abscissa represents heat input per kmol of water formed in the reaction.

Since a reboiler reactor requires a lower capital cost than a conventional reactor followed by a distillation column, this must also apply to a distillation reactor. In the latter case this cost can be further reduced by a correct choice of the reflux ratio and residence time on plate. This is illustrated in Fig. 5 taken again from the work of Kaibel et al. (31) for the operation of a hypothetical esterification column.

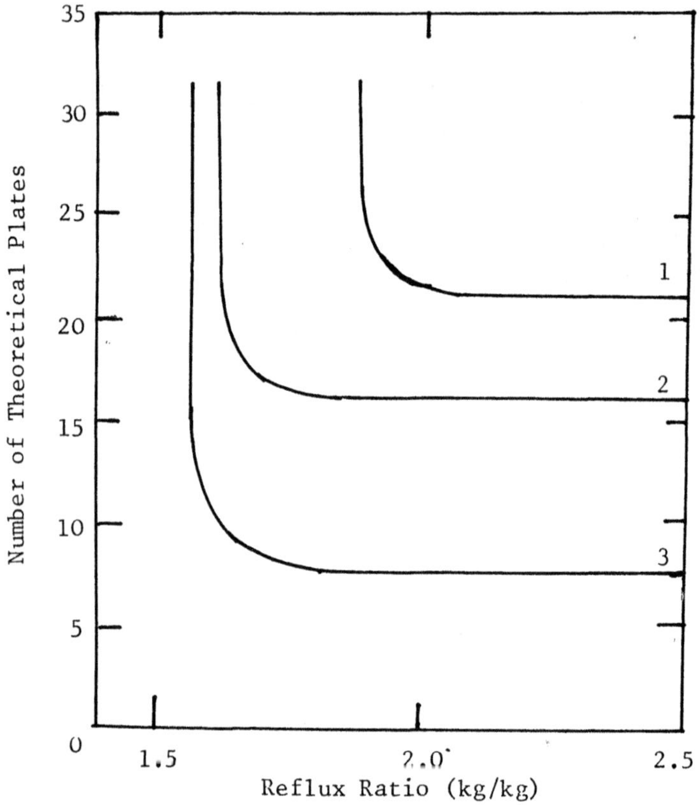

Fig. 5. Variation of number of theoretical plates with reflux ratio for a 97.5% conversion with residence time on plate as parameter (curves 1 - τ, 2 - 2τ, 3 - 3τ).

REFERENCES

1. Backhaus, A.A., U.S. Patents 1400849, 1400850, 1400851 (1921); 1403224, 1403225, 1425624, 1425625 (1922); 1454462, 1454463 (1923).
2. Keyes, D.B., Ind. Engng Chem., 24 (1932) 1096.
3. McKeon, T.J., U.S. Patent 2208769 (1940).
4. Brunn, J.H. and Grubb, H.W., U.S. Patent 2384793 (1945).
5. Schniepp, L.E., Ind. Engng Chem., 37 (1945) 872.
6. Leyes, C.E. and Othmer, D.F., Trans. Am. Inst. Chem. Engrs, 41 (1945) 157.
7. Belk, L.H., A.I.Ch.E.Jl., 1 (1955) 467.
8. Marek, J., Coll. Czech. Chem. Commun., 19 (1954) 1055.
9. Marek, J., Coll. Czech. Chem. Commun., 21 (1956) 1561.
10. Costa, P. and Canepa, B., Quad. Ing. Chim. Ital., 5 (1929) 113.
11. Matz, G., Chem-Ingr-Tech., 33 (1961) 653.
12. Parker, A.S., U.S. Patent 2839588 (1958).
13. Miller, J.H., MSc thesis, Ohio State University (1965).
14. Corrigan, T.E. and Miller, J.H., Ind. Engng Chem. Process Design & Development, 7 (1968) 383.
15. Sive, C.D., MSc thesis, Imperial College (1969).
16. Lopez Merono, J., MSc thesis, Imperial College (1969).
17. Corrigan, T.E. and Ferris, W.R., Can. J. Chem. Engng, 47 (1969) 334.
18. Jeffreys, G.V., Genie Chimique, 101 (1969) 1111.
19. Geelen, H. and Wijffels, J.B., Proc. 3rd European Symp. on Chemical Reaction Engng, Amsterdam (1964) 125.
20. Zyryanov, V.V., Zh. prikl. Khim., 39 (1966) 1070; 40 (1967) 2059.
21. Suzuki, J., Yagi, H., Komatsu, H. and Hirata, M.J., Chem. Engng of Japan, 4 (1971) 26.
22. Hirata, M.J., Kagaku Kogaku, 34 (1970) 45.
23. Saunders, A., PhD thesis, Univ. of London (1971).
24. Nelson, P.A., AIChE Jl, 17 (1971) 1043.
25. Davies, B., PhD thesis, Univ. of Aston (1972).
26. Davies, B., Jeffreys, G. and Jenkins, J.D., Trans. Inst. Chem. Engrs, 51 (1973) 267.
27. Davies, B. Jenkins, J.D. and Dilfanian, S., Distillation, IChemE Symp. Series 56 (1979) 4.2/65.
28. Pilavakis, P.A., PhD thesis, Univ. of London (1974).
29. Pilavakis, P.A. and Sawistowski, H., Distillation, IChemE Symp. Series 56 (1979) 4.2/49.
30. Hegner, B. and Molzahn, M., Distillation, IChemE Symp. Series 56 (1979) 4.2/81.
31. Kaibel, G., Mayer, H.H. and Seid, B., German Chemical Engng 2 (1979) 180; also Chem-Ingr-Tech., 50 (1978) 586.
32. Block, U., German Chemical Engng, 1 (1978) 79.
33. Mayer, H.H. and Worz, O., German Chemical Engng 3 (1980) 252.
34. Hartig, H. and Regner, H., Chem-Ingr-Tech., 43 (1971) 1001.

35. Block, U. and Regner, H., Verfahrenstechnik (Mainz), 11 (1977) 157.
36. U.S. Patent 3590072 (1971).
37. Koban, H. and Wagner, H., Chem-Ingr-Tech., 50 (1978) 313.
38. Baratella, P., Carra, S., Giardini, G. and Santi, R., Quad. Ing. Chim. Ital., 10 (1974) 1.
39. Danov, S., Polyakov, V., Dozorov, V. and Sibiryakova, L., Teor. Osn. Khim. Tekhnol., 8 (1974) 781.
40. Danov, S., Dozorov, V. and Polyakov, V., Teor. Osn. Khim. Tekhnol., 9 (1975) 615.
41. Solodnikov, V., Zadorski, V., Khokhlov, S. and Zherts, A., Konf. Teor. Prakt. Rektifikatsii, (1973) 173.
42. Cathala, M.E.J., Intl Symp. on Distillation, Inst. Chem. Engrs, London (1960) 115.
43. Teshima, T., J. Chem. Soc. Japan, Ind. Chem. Section, 55 (1952) 492.
44. Marek, J., Coll. Czech. Chem. Commun., 19 (1954) 1074.
45. Marek, J., Coll. Czech. Chem. Commun., 20 (1955) 1490.
46. Hirata, M., Kagaka Kogaku, 31 (1967) 1184.
47. Hirata, M., Chem. Engng of Japan, 3 (1970) 152.
48. Byron, E.S., Bowman, J.R. and Coull, J., Ind. Engng Chem., 43 (1951) 1002.
49. Bainbridge, G.S., PhD thesis, Univ. of London (1964).
50. Sawistowski, H., Bainbridge, G.S., Stacey, M.J. and Theobald, A., Distillation, ABCM/BCPMA, London (1964) 143.
51. Wildey, B.R., MPhil thesis, Univ. of London (1964).
52. Oyekan, A.J.M.A., MSc thesis, Univ. of London (1967).
53. Teja, A. and Thong, K.C., BSc project, Imperial College (1968).
54. Hochgesand, G., VDI-Forschungsheft 498 (1963).
55. Danckwerts, P.V., Sawistowski, H. and Smith, W., Intl. Symp. on Distillation, Instn. of Chemical Engineers (1960) 7.
56. Krishna, R. and Standart, G.L., AIChE Jl, 22 (1976) 383.
57. Krishna, R., Chem. Engng Sci., 32 (1977) 1197.
58. Bolles, W.L. and Fair, J.R., IChemE Symp. Series 56 (1979) 3.3/35.

INTRODUCTION TO BIOCHEMICAL SYSTEMS. FORMAL TREATMENT OF BIOCHEMICAL REACTIONS AND CHARACTERISATION OF FERMENTATION SYSTEMS

K. Schügerl

Institut für Technische Chemie
Universität Hannover
D-3000 Hannover 1, FRG

1. FORMAL TREATMENT OF BIOCHEMICAL REACTIONS

Bioconversion of raw materials into desired products can be carried out by microorganisms or cell-free biocatalysts (enzymes). In the following, only the conversion by microorganisms will be considered.

Bioconversion can have different aims:

- production of cell mass (conversion of substrates into biomass)
- production of metabolites (conversion of substrates and/or precursor into metabolites)
- conversion of chemicals into other compounds (strictly speaking biotransformation)

Bioconversion always begins with cultivation of the cells. For metabolite production the cell growth phase is followed by a production phase and for transformation of chemicals by the biochemical alteration of chemical compounds.

Only cell growth and secondary metabolite production will be considered here.

GROWTH OF THE CELL POPULATION

In unstructured models the growth of the cell population is measured by means of the variation of (dry) cell mass or biomass. In structured models the variation of the number of cells is the measure for the cell population growth. If the average weight of the cells is constant there is a simple relationship between these two population properties. Since the determination of (dry) cell mass is easier than that of the number of cells, with the exception of plant and animal tissue cultures the (dry) cell mass, DCM, is used to characterize growth. In the following, we will use the (dry) cell mass concentration, X, to describe cell growth.

The cells are usually cultivated in a 500 ml Erlenmeyer flask to produce a preseeding of about 200 ml volume. After a given cell concentration has been attained a larger reactor is inoculated by these living cells which are in the nonlimited (exponential) growth phase. This process is repeated until the production reactor is inoculated. The ratio of inoculum cell mass to final cell mass of a stage is usually 1/10. Thus, for a 100 m^3 reactor 5 to 6 preculture stages are necessary.

During inoculation of a reactor by the correspondin preculture the cells are exposed to a strong alteration of their environment. They need more or less time to adapt to the new conditions, e.g. to the new substrate type, substrate and oxygen concentration etc. During this so-called lag time no growth can be observed. After a transition range, where the cell growth is accelerated the cells attain the exponential growth range. Cell growth in this range can be described by a simple relationship which generally holds. No general relationships are known for the lag and transition ranges.

1.1 Nonlimited Cell Growth

After the transition range the cell population growth rate can be described by eq. (1).

$$R_X = \frac{dX}{dt} = \mu_m X \qquad (1)$$

where X is the (dry) cell mass concentration and
μ_m is the maximum specific growth rate.
Eq. (1) holds true only for nonlimited growth, i.e. when growth is not limited by the lack of substrate or not inhibited by substrate concentrations which are too high. With the initial conditions (2)

$$X = X^o \quad \text{at } t = 0 \qquad (2)$$

the integration of eq. (1) yields (3)

$$\frac{X}{X^o} = \exp(\mu_m t) \qquad (3)$$

Therefore, this growth range is called "exponential growth".

Since R_X depends neither on the substrate concentration, S, nor on the dissolved oxygen concentration, O, the growth is formally a reaction of zeroth order with regard to S and O.

1.1.1 Cell population growth in stirred tank reactors. In perfectly mixed batch reactors the cell mass concentration is constant everywhere, hence eqs. (1) and (3) can be used.

For perfectly mixed continuous stirred tank reactors, CSTR, eq. (4) holds true:

$$\frac{dX}{dt} = \mu_m X_1 - D(X_1 - X_o) \qquad (4)$$

If sterile feed is used, $X_o = 0$ holds.

In eq. (4) $D = \frac{1}{\tau}$ is the dilution rate $= \frac{F_V}{V_r}$,

τ is the mean residence time of the medium in the reactor $= V_r/F_V$,

X_1 is the (dry) biomass concentration in the reactor,

F_V is the volumetric feed rate and

V_r the medium volume in the reactor.

Under steady-state conditions: $\frac{dX}{dt} = 0$ and with $X_o = 0$ eq. (5) holds:

$$\mu_m = D \qquad (5)$$

The substrate and oxygen utilization rates $-R_S$ and $-R_O$ are given by eqs. (6) and (7):

$$-R_S = -\frac{dS}{dt} = \mu_m X \frac{1}{Y_{X/S}} \qquad (6)$$

$$-R_O = -\frac{dO}{dt} = \mu_m X \left(\frac{1}{Y_{X/O}} \right) \qquad (7)$$

if no aeration is carried out.

S is the substrate concentration,
O the dissolved oxygen concentration in the medium
$Y_{X/S}$ the substrate yield coefficient and
$Y_{X/O}$ the oxygen yield coefficient, the latter two being defined by eqs. (8) and (9):

$$Y_{X/S} = \frac{\text{biomass produced}}{\text{substrate utilized}} \quad (8)$$

$$Y_{X/O} = \frac{\text{biomass produced}}{\text{oxygen utilized}} \quad (9)$$

In well mixed batch reactors eqs. (6) and (7) hold. For the CSTR the following mass balances are valid if sterile feed is used:

$$\frac{dS}{dt} = D(S_o - S_1) - \frac{1}{Y_{X/S}} \mu_m X_1 \quad (10)$$

$$\frac{dO}{dt} = D(O_o - O_1) - \frac{1}{Y_{X/O}} \mu_m X_1 - k_L a(O^* - O_1) \quad (11)$$

where S_o and O_o are the substrate and dissolved oxygen concentrations in the feed,
S_1 and O_1 the substrate and dissolved oxygen concentrations in the reactor medium,
$k_L a$ is the volumetric gas/liquid mass transfer coefficient with regard to medium volume,
O^* the dissolved oxygen concentration in the medium at saturation

Under steady-state conditions

$$\frac{dS}{dt} = \frac{dO}{dt} = 0 \quad (7)$$

In the nonlimited growth range, the steady state is nonstable: If the cell concentration, X, exceeds the steady state cell concentration, $\tilde{X}$, X increases until it reaches a substrate or oxygen transfer limited state. If X drops below $\tilde{X}$, the cells are washed out. A continuous operation is only possible if the dilution rate, D, is varied to keep the cell concentration constant: if X exceeds $\tilde{X}$, D is increased, if X drops below $\tilde{X}$, D is diminished. A reactor operated in this manner is called turbidostat.

1.1.2 Cell population growth in tower reactors.
Tower reactors are usually described by means of dispersion models. According to Chen (1) the cell mass balance yields for nonsterile feed $X_o > 0$:

$$\frac{1}{Bo}\frac{d^2\tilde{X}^*}{dz^2} - \frac{d\tilde{X}^*}{dz} + Da\tilde{X}^* = 0 \qquad (12)$$

with Danckwerts' boundary conditions:

$$\tilde{X}^* - \frac{1}{Bo}\frac{d\tilde{X}^*}{dz} = 1 \qquad \text{at } z = 0 \qquad (12a)$$

$$\frac{d\tilde{X}^*}{dt} = 0 \qquad \text{at } z = 1 \qquad (12b)$$

where $Bo = \frac{uL}{D_{ax}}$ is the Bodenstein number,

$Da = \frac{\mu_m L}{u} = \mu_m \tau$ is the Damköhler number,

$\tilde{X}^* = X/X_o$ the dimensionless cell mass concentration in steady state,

$z = x/L$ the dimensionless tower length,

x the length coordinate,

L the length of the tower,

u the flow velocity and

D_{ax} the axial dispersion coefficient.

The solutions of eq. (12) are given (1):
for $Bo > Da$ or $q > 0$

$$\tilde{X}^*(z) = \frac{2\exp(\frac{Boz}{2})\left[\sinh\frac{qBo}{2}(1-z) + 2q\cosh\frac{qBo}{2}(1-z)\right]}{(1+q^2)\sinh\frac{qBo}{2} + 2q\cosh\frac{qBo}{2}} \qquad (13)$$

where $q = \left(1 - \frac{4Da}{Bo}\right)^{1/2}$

for $Bo = 4Da$ or $q = 0$

$$\tilde{X}^*(z) = \frac{1 + \frac{Bo}{2}(1 - z)}{1 + \frac{Bo}{4}}\exp(\frac{Boz}{2}) \qquad (14)$$

and for $Bo < 4Da$ or $b > 1$

$$\tilde{X}^*(z) = \frac{2\,\exp(\frac{Boz}{2})[\sin\frac{bBo}{2}(1-z) + \cos\frac{bBo}{2}(1-z)]}{(1-b^2)\sin\frac{bBo}{2} + 2b\cos\frac{bBo}{2}} \tag{15}$$

where $b = \left(\frac{4\ Da}{Bo} - 1\right)^{1/2}$

Eq. (15) is only valid when Bo and Da are related in such a way that

$$\frac{Bo}{4} \le Da \le \frac{Bo}{4} + \frac{4\,\alpha_1^2}{Bo} \tag{16}$$

where α_1 is the smallest positive root of the following equation

$$\alpha \tan \alpha = \frac{Bo}{4} \tag{17}$$

As can be seen from Fig. 1 in region I there is no solution of eq. (12) for steady state operations: microbial cells grow faster than can be carried away by the exit flow stream. Thus X increases until it reaches a substrate or oxygen transfer limited state at the reactor exit. In regions II and III the finite solutions exist as presented by eq. (15) and (13), respectively.

In Fig. 2, $\tilde{X}^*(1)$ is plotted as a function of Da for different Bo numbers. It can be recognized that for Da = const. the highest dimensionless exit cell mass concentrations, $\tilde{X}^*(1) = X(1)/X_o$, can be achieved at Bo = 0.

In general, the cells are seldom fed to the reactor continuously. The feed is usually sterile (cell free): $X_o = 0$.

The start-up consists of (1) cell inoculation, (2) cell growth in batch operation, (3) when the desired cell mass concentration is attained and cells are still in the nonlimited growth range the batch operation is switched to a continuous one, (4) the reactor is operated with sterile feed in the steady state.

To investigate whether it is possible to switch from batch to continuous operation the nonsteady state cell mass balance equation must be solved:

$$\frac{\partial X^*}{\partial \theta} = \frac{1}{Bo}\frac{\partial^2 X^*}{\partial z^2} - \frac{\partial X^*}{\partial z} + Da X^* \tag{18}$$

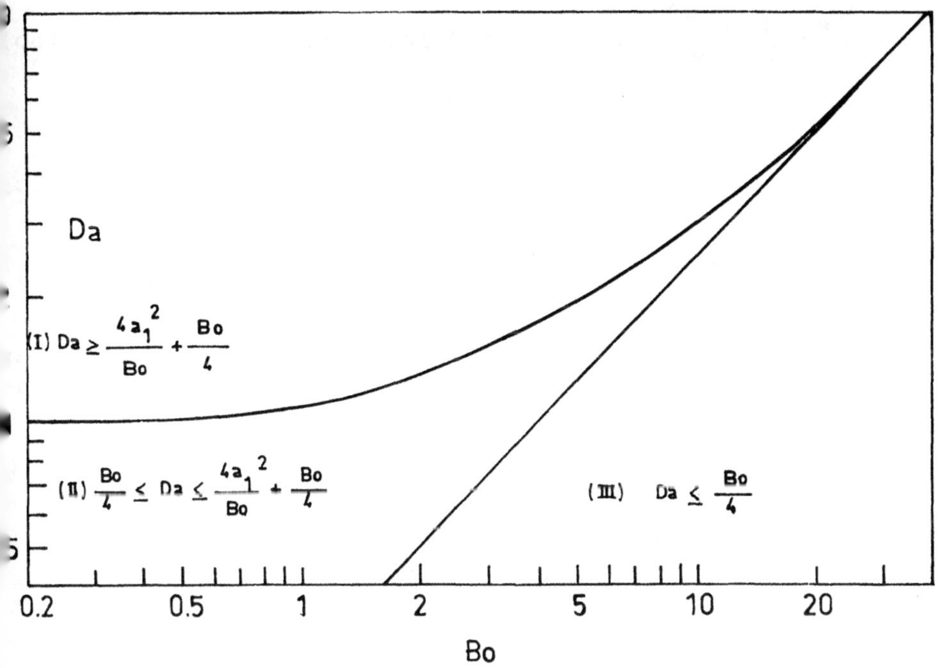

Fig.1:Da-Bo Diagram

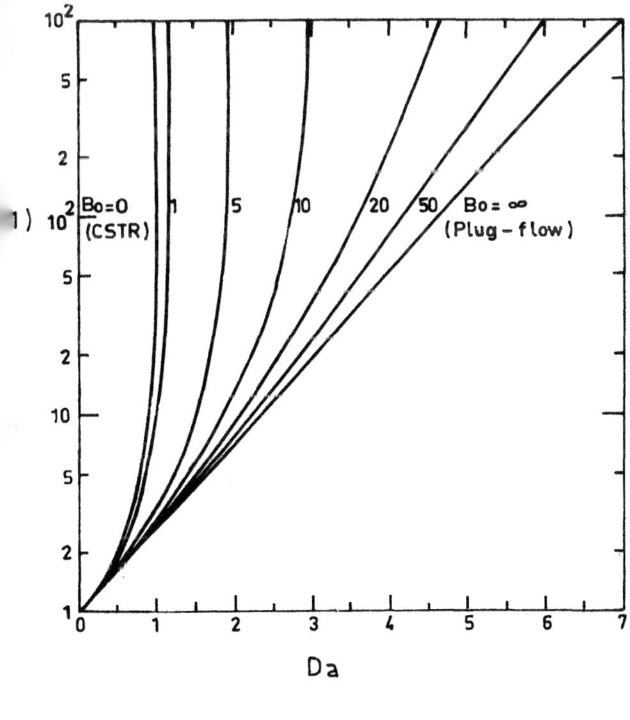

Fig.2.Dimensionless cell mass concentration at reactor outlet $X^*(1)=X/X_o$ as a function of Da for different Bo-numbers. Steady state operation. Nonsterile feed ($X_o > 0$) (1).

with $X^* = 1$ for all z at $\theta = 0$ (18a)

$$X^* - \frac{1}{Bo}\frac{\partial X^*}{\partial z} = 0 \quad \text{at } z = 0 \text{ for } \theta > 0 \tag{18b}$$

and

$$\frac{\partial X^*}{\partial z} = 0 \quad \text{at } z = 1 \text{ for } \theta > 0 \tag{18c}$$

where $X^* = X/X^o$
X^o is the cell mass concentration at $\theta = 0$
$\theta = \frac{t}{\tau}$ the dimensionless time and
t the real time measured from the moment of switching from batch to continuous operation

The solution of eq. (18) is given by (1):

$$X^*(\theta,z) = \exp\left(\frac{Boz}{2}\right) \sum_{n=1}^{\infty} A_n \exp(-\lambda_n \theta) \left[\frac{Bo}{4\alpha_n} \sin(2\alpha_n z) + \cos(2\alpha_n z)\right] \tag{19}$$

where α_n, n=1,2,... are real roots of $\alpha \tan \alpha = \frac{Bo}{4}$

$$\text{or } \alpha \cot \alpha = -\frac{Bo}{4} \tag{20}$$

$$\lambda_n = \frac{Bo}{4} + \frac{4\alpha_n^2}{Bo} - Da \tag{21}$$

and

$$A_n = \frac{\frac{8\alpha_n^2}{Bo}\left[\left(\frac{Bo}{4}\right)^2 + \alpha_n^2\right]}{\frac{Bo}{4} + \left(\frac{Bo}{4}\right)^2 + \alpha_n^2} \tag{22}$$

The following particular cases should be considered

1. If all eigen values α_n, n=1,2,... are positive for given Bo and Da washout will occur as $\theta \to \infty$. The washout conditions follow from eq. (21):

$$Da < \frac{4\alpha_1^2}{Bo} + \frac{Bo}{4}, \tag{23}$$

where α_1 is the smallest positive root satisfying eq. (20). This condition coincides with regions II and III in Fig. 1. In these regions steady state solutions do not exist for sterile feed but they do for nonsterile feed.

2. If one or more of λ_n are negative, the cells will grow as $\theta \to \infty$ and attain a substrate or oxygen transfer limited growth state at the reactor exit.
3. If the smallest λ_n is zero the steady state solution is possible. However, this is an unstable steady state solution because any flow fluctuation will make the smallest λ non-zero and the cells will either grow and reach the limited state or be washed out.

Steady state nonlimited growth can only be maintained in a tower reactor if growth is limited at the reactor exit.

1.1.3 Cell population growth in tower loop reactors. Air lift tower loop reactors are often used to carry out cultivation of microbial cells (2,3,4). In these reactors the medium is recirculated from the top of the tower to its bottom.

When assuming that (a) the tower can be described by a one-dimensional dispersion model, (b) the liquid residence time in the loop is negligible, and (c) the axial dependence of the local relative gas holdup is negligible, the following cell mass balance is obtained (5,6):

$$\frac{\partial X^*}{\partial \theta} = \frac{1}{Bo_R} \frac{\partial^2 X^*}{\partial z^2} - \frac{\partial X^*}{\partial z} + Da_R X^* \qquad (24)$$

where $Bo_R = Bo(1+\gamma)$ is the modified Bodenstein number

$Da_R = \frac{Da}{1+\gamma}$ is the modified Damköhler number

$\gamma = \frac{u_R}{u}$ is the liquid recirculation ratio

$\theta = \frac{t(u+u_R)}{L}$ is the dimensionless time,

$X^* = \frac{X}{X^o}$ for sterile feed

$X^* = \frac{X}{X_o}$ for nonsterile feed.

$u = \dot{V}_L/Q$ is the superficial liquid velocity,

$u_R = \dot{V}_R/Q$ is the superficial liquid velocity due to recirculation,

$\dot{V}_L$ is the volumetric feed rate,
$\dot{V}_R$ is the volumetric flow rate in the loop,
Q is the cross-sectional area of the tower and
t is the time measured from the moment of switching from batch to continuous operation

Under steady state conditions eq. (24) is reduced to (25):

$$\frac{1}{Bo_R} \cdot \frac{d^2\tilde{X}^*}{dz^2} - \frac{d\tilde{X}^*}{dz} + Da_R\tilde{X}^* = 0 \qquad (25)$$

When no cells are in the feed, the boundary conditions are

$$\tilde{X}^*(0) = \frac{\gamma}{1+\gamma}\tilde{X}^*(1) + \frac{1}{Bo_R}\frac{d\tilde{X}^*(0)}{dz} \qquad (26)$$

and

$$\frac{d\tilde{X}^*(1)}{dz} = 0 \qquad (27)$$

Solution of eq. (25-27) results in $\tilde{X}^*(z) = 0$, i.e. washout.

When the feed contains some biomass the boundary conditions are:

$$\tilde{X}^*(0) = \frac{1}{1+\gamma} + \frac{\gamma}{1+\gamma}\tilde{X}^*(1) + \frac{1}{Bo_R}\frac{d\tilde{X}^*(0)}{dz} \qquad (28)$$

and

$$\frac{d\tilde{X}^*}{dz} = 0 \qquad (29)$$

Eq. (25), subject to (28) and (29), were solved analytically (5,6). Again three cases can be distinguished:

I. For $Da_R < \frac{Bo_R}{4}$, i.e. $(A > 0)$, the solution is given by

$$\tilde{X}^*(z) =$$

$$= \frac{\sinh\left(\frac{Bo_R}{2} A(1-z)\right) + A \cosh\left(\frac{Bo_R}{2} A(1-z)\right)}{\boxed{(1+\gamma)} \left[(1+A^2)\sinh\left(\frac{Bo_R}{2} A\right) - \boxed{2A \frac{\gamma}{1+\gamma} \exp\left(\frac{Bo_R}{2}\right)} + 2A \cosh\left(\frac{Bo_R}{2} A\right)\right]} \cdot 2 \exp\left(\frac{Bo_R}{2} z\right) \quad (30)$$

where $A = \left(1 - \frac{4Da_R}{Bo_R}\right)^{1/2}$

Since only positive X^*s' are reasonable, the denominator in eq. (30) must always be positive. This leads to

$$\gamma < \frac{x'}{1 - x'} \quad (31a)$$

$$\text{with } x' = \left[\frac{1 + A^2}{2A} \sinh\left(\frac{Bo_R}{2} A\right) + \cosh\left(\frac{Bo_R}{2} A\right)\right] \exp\left(-\frac{Bo_R}{2}\right) \quad (31b)$$

II. For $Da_R = \frac{Bo_R}{4}$, i.e. $A = 0$, the solution is given by

$$\tilde{X}^*(z) = \frac{2 + Bo_R (1-z)}{\boxed{(1+\gamma)} \left(2 + \frac{Bo_R}{2}\right) - \boxed{2 \gamma \exp\left(\frac{Bo_R}{2}\right)}} \exp\left(\frac{Bo_R}{2} z\right) \quad (32)$$

The condition for positive cell mass is given by

$$\gamma < \frac{2 + \frac{Bo_R}{2}}{2 \exp\left(\frac{Bo_R}{2}\right) - \left(2 + \frac{Bo_R}{2}\right)} \quad (33)$$

III. For $Da_R > \frac{Bo_R}{4}$, i.e. A becomes imaginary, and the solution is given by

$$\tilde{X}^*(z) = \frac{Bo_R \sin(\chi(1-z)) + 2\chi \cos(\chi(1-z))}{\boxed{(1+\gamma)}\ (Bo_R - \frac{4}{Bo_R}\chi^2)\sin\chi - \boxed{4\gamma\,\chi\,\exp(\frac{Bo_R}{2})} + 4\boxed{(1+\gamma)}\chi\cos\chi}$$

$$2\exp(\frac{Bo_R}{2}z) \quad (34)$$

With:

$$B = \sqrt{\frac{4Da_R}{Bo_R} - 1} \quad (35)$$

and

$$\chi = \frac{Bo_R}{2} B \quad (36)$$

The requirement of positive cell mass concentration is fulfilled only, if

$$\frac{Bo_R}{4} < Da_R < \frac{Bo_R}{4} + \frac{4\alpha_1^2}{Bo_R} \quad (37)$$

The value of α_1 results from the condition that both numerator and denominator must be positive (5,6).

In analogy to tower reactors without a loop (1.1.2) Fig. 3 shows the ranges of Bo_R and Da_R, where the different cases of the solution apply. It can be shown (5,6) that in case III an upper limit of Da_R exists which leads to a cell mass concentration increase and to substrate or oxygen transfer limited growth at the reactor exit. This critical limit diminishes with increasing γ (Fig. 3). With an increasing recirculation ratio, γ, one moves from region III via the limiting case II to region I. In region I, γ can only be raised to γ_{max}, which is defined by eq. (31). In Fig. 4, γ_{max} is plotted as a function of Da for various Bo. Only for Da $\geq$ 1 is γ_{max} finite. At Bo = const, γ_{max} increases with decreasing Da, as long as Bo is high enough

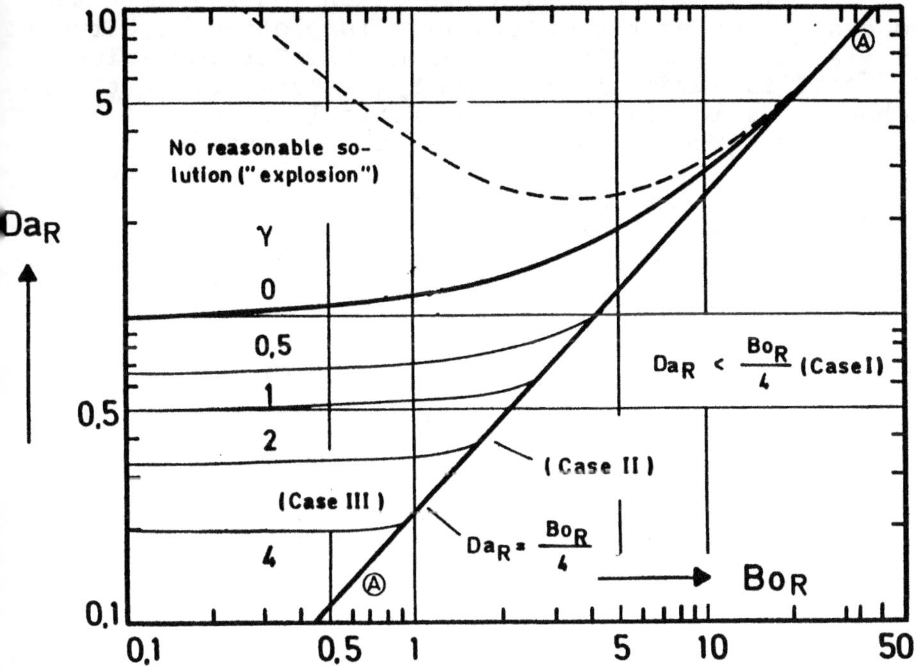

ig.3: Ranges of physically correct solutions(5) (Case I: eft from the line A, Case II: equal to the line A , ase III: right from the line A)

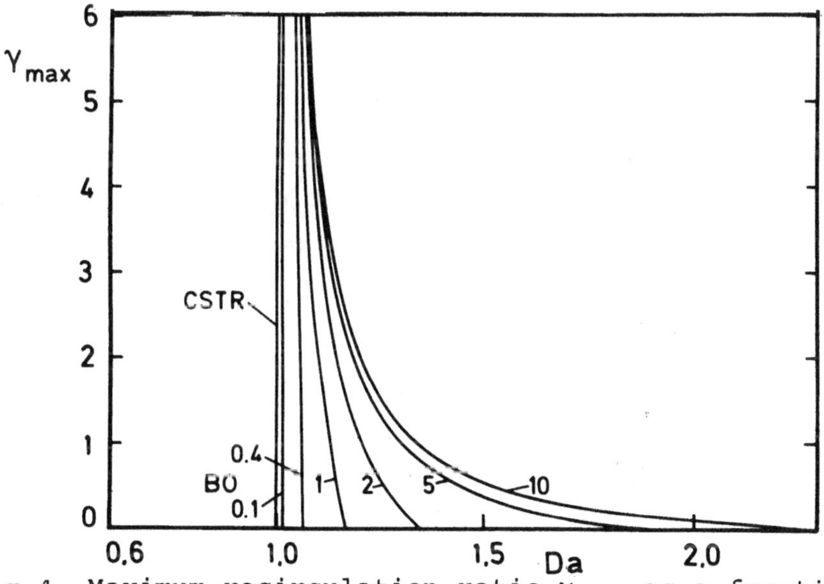

ig.4. Maximum recirculation ratio γ_{max}, as a function f Da with Bo as parameter(5)

In the CSTR, $\gamma_{max} \to \infty$.

In Fig. 5, the attainable dimensionless cell mass concentration at the exit, $\tilde{X}^*(1)$, is plotted as a functi of γ at Bo = 1 and Da as a parameter. For low values of Da, $\tilde{X}^*(1)$ is hardly influenced by γ. However, if the Da number is slightly below the critical value of 1.17 (Fi 4), the recirculation ratio reveals a considerable influence on $\tilde{X}^*(1)$. At Da = 1.1 and γ = 2, for instance, $X^*(1)$ is ten times as high as for γ = 0.

Under nonsteady state conditions and with cells in the feed, eq. (24) has to be solved with the boundary conditions:

$$X^*(0,\theta) = \frac{1}{1+\gamma} + \frac{\gamma}{1+\gamma} X^*(1,\theta) + \frac{1}{Bo_R} \frac{\partial X^*(0,\theta)}{\partial z}, \quad \theta > 0 \tag{38}$$

$$\frac{\partial X^*(1,\theta)}{\partial z} = 0 \qquad \theta > 0 \tag{38a}$$

and the initial condition

$$X^*(z,\theta) = 0 \qquad \theta \leq 0 \tag{38b}$$

The following closed solution was obtained (5,6):

$$X^*(z,\theta) \quad X^*(z,\infty) + \exp\left(\frac{Bo_R}{2} z\right) \sum_{k=1}^{\infty} R_K(z) \exp(S_k\theta) \tag{39}$$

where

$$S_k = Da_R - \frac{Bo_R}{4} - \frac{4a_k^2}{Bo_R} \tag{39a}$$

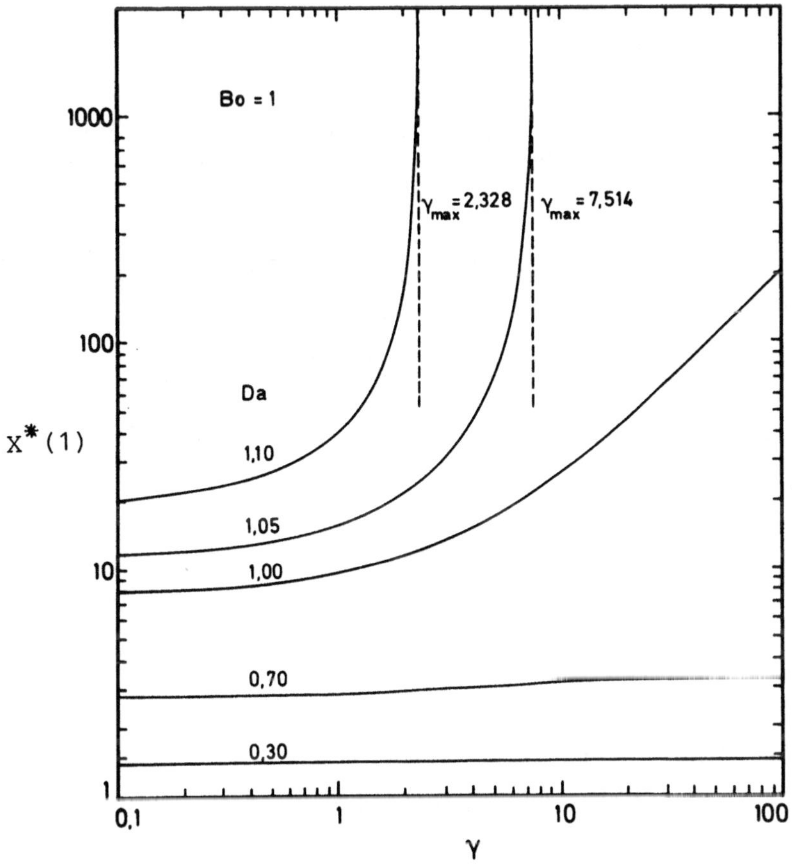

Fig. 5: Outlet biomass concentration in dependence of γ for various Da- numbers (Bo = 1) (5).

For the roots see (6) and for the solution technique see (5).

The nonsteady-state behavior depends only on the exponential terms of eq. (39). To attain a steady-state solution all these terms must disappear for $\Theta \to \infty$, i.e. all values of S_K must be negative. The condition for $S_K < 0$ follows from eq. (39a):

$$Da_R < \frac{Bo_R}{4} + \frac{4\alpha_1^2}{Bo_R} \tag{40}$$

If this condition is fulfilled and $\Theta \to \infty$, eq. (39) reduces to the steady-state solution $X(z,\infty)$: eq. 30, 32 or 34 depending on the value of A.

Under <u>nonsteady state conditions</u> and with <u>cell-free feed</u> eq. (24) has to be solved with the boundary conditions:

$$X^*(0,\Theta) = \frac{\gamma}{1+\gamma} X^*(1,\Theta) + \frac{1}{Bo_R} \frac{\partial X^*(0,\Theta)}{\partial z} \tag{41a}$$

$$\frac{\partial X^*(1,\ )}{\partial z} = 0 \qquad \Theta > 0 \tag{41b}$$

and with the initial condition:

$$X^*(z,\Theta) = 1 \qquad \text{at } \Theta = 0 \tag{41c}$$

$\Theta = 0$ is the moment when the batch is switched to continuous operation.

The solution of eq. (24) and (41) is

$$X^*(z,\Theta) = \exp(\frac{Bo_R}{2}, z) \sum_{k=1}^{\infty} R_K(z) \exp(S_K \Theta) \tag{42}$$

The $R_K(z)$ values are given in ref. (5,6). S_K is given by eq. (37)

If $S_K < 0$, i.e. condition (40) is valid, $X^*(z,\Theta)$ approaches zero for $\Theta \to \infty$, which corresponds to washout. If only one $S_K > 0$, the cell mass concentration increases as long as substrate or oxygen transfer limited growth is attained at the reactor exit. Steady-state operation is only possible if one $S_K = 0$, all other $S_K < 0$, i.e.

$$Da_R = \frac{Bo_R}{4} + \frac{4K^2}{Bo_R} \qquad (43)$$

However,this solution is not stable and small deviations from the steady state lead to either washout or exponential cell growth and limited growth.All solutions for tower loop reactors yield the corresponding solutions for tower reactors if γ is set to zero.

A comparison of stirred tank reactors (1.1.1) with tower reactors (1.1.2) and tower loop reactors (1.1.3) indicate that all of them behave similar if cell-free feed is used: at nonlimiting growth there is no stable steady state.

With cells containing feed stable steady states exist which differ from washout,and cell "explosion". The ranges of Bo and Da values which correspond to these states are similar in tower and tower loop reactors. However, application of a recycling process may increase the cell mass production considerably.

1.2 Substrate limited growth

Cell cultivation is usually carried out in the substrate limited growth range.

1.2.1 Substrate limited growth in stirred tank reactors. Several relationships have been recommended for the dependence of the specific growth rate, μ , on the substrate concentration. The most popular is the relationship of Monod (7):

$$\mu = \mu_m \frac{S}{K_S + S} \qquad (44)$$

where K_S is the saturation constant with regard to the substrate

Two limiting cases can be distinguished:

$$\text{for } S \gg K_S , \qquad \mu \simeq \mu_m \qquad (45)$$

the growth is of zeroth order with regard to the substrate (nonlimited growth) and

$$\text{for } S \ll K_S , \qquad \mu \simeq \frac{\mu_m}{K_S} S \qquad (46)$$

growth is of first order with regard to the substrate (strong substrate limited growth).

In a batch reactor, eq. (47) holds true:

$$R_X = \frac{dX}{dt} = \mu_m \frac{SX}{K_S+S} \tag{47}$$

The corresponding substrate and oxygen consumption rates are:

$$-R_S = -\frac{dS}{dt} = \frac{\mu_m S}{K_S+S} \quad X \frac{1}{Y_{X/S}} \tag{48}$$

and

$$-R_O = -\frac{dO}{dt} = \frac{\mu_m S}{K_S+S} \quad X \frac{1}{Y_{X/O}}\ , \tag{49}$$

if no aeration is carried out.

For the CSTR the following balance equations are valid:

$$\frac{dX_1}{dt} = \frac{\mu_m S_1}{K_S+S_1} X_1 + D(X_o-X_1) \tag{50}$$

$$\frac{dS_1}{dt} = -\frac{\mu_m S_1}{K_S+S_1} X_1 \frac{1}{Y_{X/S}} + D(S_o-S_1) \tag{51}$$

$$\frac{dO}{dt} = -\frac{\mu_m S_1}{K_S+S_1} X_1 \quad \frac{1}{Y_{X/O}} + D(O_o-O_1) + k_L a(O_1^*-O_1) \tag{52}$$

where O_1^* is O_1 at saturation.
$k_L a$ is the volumetric mass transfer coefficient for O_2 across the gas/liquid interface.

Under steady-state conditions eq. (53) holds:

$$\frac{dX_1}{dt} = \frac{dS_1}{dt} = \frac{dO_1}{dt} = 0 \tag{53}$$

This substrate limited cell growth in a steady-stat CSTR is called chemostat culture, if $X_o = 0$, because it has a self-regulating character (8,9), i.e. it is a stable state:

If X_1 exceeds its steady state value $\tilde{X}_1$, S_1 drops below $\tilde{S}_1$ according to eq. (51), this causes a reduction of X_1 to $\tilde{X}_1$.

If X_1 drops below $\tilde{X}_1$, S_1 exceeds $\tilde{S}_1$ and this causes an increase in X_1 to $\tilde{X}_1$.

With $X_o = 0$, eqs (50) and (53) yield:

$$\mu_1 \tilde{X}_1 - D\tilde{X}_1 = (\mu_1 - D)\,\tilde{X}_1 = 0 \qquad (54)$$

where $$\mu_1 = \mu_m \frac{S_1}{K_S + S_1} \qquad (55)$$

or $$\mu_1 = D \qquad (56)$$

Eqs. (51) and (53) yield:

$$\mu_1 \tilde{X}_1 \frac{1}{Y_{X/S}} = D(S_o - \tilde{S}_1) \qquad (57)$$

When substituting eq. (56) into eq. (55) we obtain

$$\tilde{S}_1 = \frac{K_S D}{\mu_m - D} \qquad (58)$$

and eq. (56) into eq. (57):

$$\tilde{X}_1 = Y_{X/S}(S_o - \tilde{S}_1) = Y_{X/S}\,[\,S_o - K_S D/(\mu_m - D)\,] \qquad (59)$$

In Fig. 6, $\tilde{X}_1$ is plotted as a function of D for two different S_o values.

The cell production rate, R_x, is given by

$$R_x = D\tilde{X}_1 = DY_{X/S}[S_o - K_S D/(\mu_m - D)] \qquad (60)$$

In Fig. 7, R_x is plotted as a function of D. It can be recognized that R_x passes a maximum. To evaluate the dilution rate, D_m, at which this maximum occurs eq. (60) is differentiated with respect to D and the derivative is equated to zero. This results in (12):

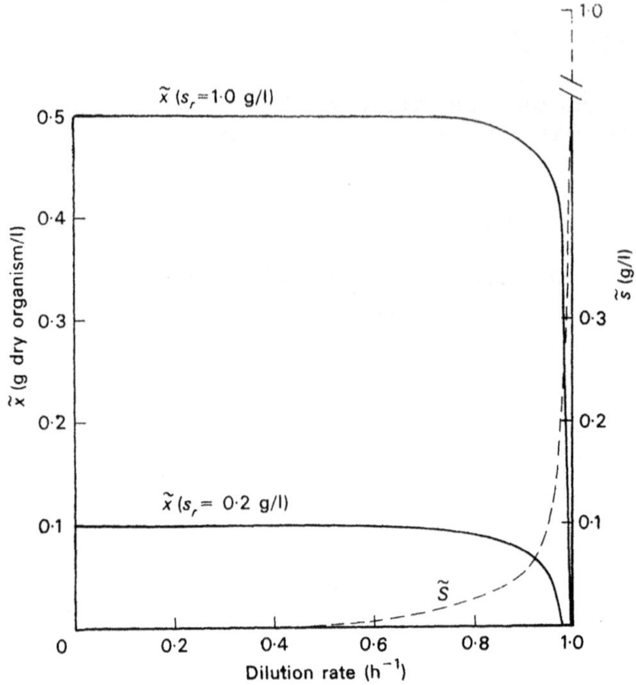

Fig. 6: Steady-state values of cell mass X, and growth limiting substrate, S, concentrations in stirred tank (Chemostat). $\mu_m = 1.0\ h^{-1}$, $K_S = 0.005\ g\ l^{-1}$, $Y_{X/S} = 0.5$ (12).

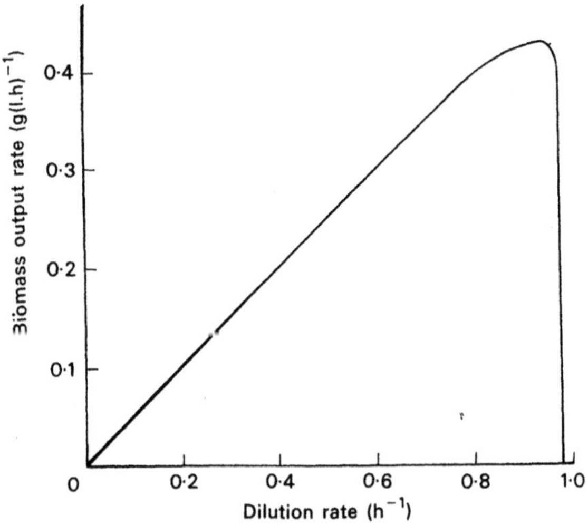

Fig. 7: Steady-state cell mass productivity in stirred tank (Chemostat). $\mu_m = 1.0\ h^{-1}$, $K_S = 0.005\ g\ l^{-1}$, $Y_{X/S} = 0.5$ (12).

$$D_m = \mu_m [1 - (\frac{K_S}{S_o + K_S})^{1/2}] \quad (61)$$

Inserting eq. (61) into eq. (59) yields $\tilde{X}_m$

$$\tilde{X}_m = Y_{X/S} [S_o + K_S - \{ K_S(S_o + K_S) \}^{1/2}] \quad (62)$$

1.2.2 Substrate limited growth in tower reactors.
For steady-state cell mass balance eq. (63) is valid:

$$\frac{1}{Bo} \frac{d^2\tilde{X}^*}{dz^2} - \frac{d\tilde{X}^*}{dz} + \frac{\tilde{S}^*\tilde{X}^*}{K + \tilde{S}^*} Da = 0 \quad (63)$$

and for a steady-state substrate balance eq. (64) holds true

$$\frac{1}{Bo} \frac{d^2\tilde{S}^*}{dz^2} - \frac{d\tilde{S}^*}{dz} - \frac{\tilde{S}^*\tilde{X}^*}{K + \tilde{S}^*} \frac{Da}{Y_{X/S}} = 0 \quad (64)$$

The corresponding boundary conditions are:

$$\frac{1}{Bo} \frac{d\tilde{X}^*}{dz} = \tilde{X}_o^* - \tilde{X}^* \qquad \text{at } z = 0 \quad (63a)$$

$$\frac{d\tilde{X}^*}{dz} = 0 \qquad \text{at } z = 1 \quad (63b)$$

and

$$\frac{1}{Bo} \frac{d\tilde{S}^*}{dz} = \tilde{S}_o^* - \tilde{S}^* \qquad \text{at } z = 0 \quad (64a)$$

$$\frac{d\tilde{S}^*}{dz} = 0 \qquad \text{at } z = 1 \quad (64b)$$

where $S^* = \frac{S}{S_o}$ is the dimensionless substrate concentration

$S_o^* = 1$

$X^* = \frac{X}{S_o Y_{X/S}}$ is the dimensionless cell concentration

$X_o^* = \frac{X_o}{S_o Y_{X/S}}$

$K = \frac{K_S}{S_o}$ is the dimensionless saturation constant with regard to the substrate.

Eqs. (63) and (64) were solved numerically (10,11). In Fig. 8, S* is plotted as a function of Bo for different Da as parameters. It can be recognized that for small Da numbers the CSTR (Bo = 0) yields the lowest S^*, i.e. the highest substrate conversion $U_S = 1-S^*$. With an increasing Da number (mean residence time τ) the course of the S^*(Bo) function changes and for Da $\geq$ 3.0 this function passes a minimum. With an increasing Da number this minimum shifts to higher Bo numbers. In this range an optimum Bo number, Bo_{opt}, exists at which the minimum of S* and/or maximum of U_S prevails. However, with an increasing Bo number the cells are washed out, e.g. at Da = 3.0 and Bo$\geq$ 6. At Bo_{opt} higher cell mass concentrations can be obtained than in the CSTR.

1.2.3 Substrate limited growth in tower loop reactors. When assuming that (a) the tower reactor can be describe by a one-dimensional dispersion model, (b) the liquid residence time in the loop is negligible, (c) the spacial dependence of the gas holdup is negligible and (d) the growth rate can be described by Monod kinetics, one obtains for the cell mass balance:

$$\frac{1}{Bo_R}\frac{d^2X^*}{dz^2} - \frac{dX^*}{dz} + Da_R\frac{X^*(1-X^*)}{K+1-X^*} = 0 \qquad (65)$$

with the boundary conditions

$$X^*(0) = \frac{1}{1+\gamma}X_o^* + \frac{\gamma}{1+\gamma}X^*(1) + \frac{1}{Bo_R}\frac{dX^*(0)}{dz} = 0 \qquad (65a)$$

and

$$\frac{dX^*}{dz} = 0 ,$$

here $X^* = \frac{X}{X_o + Y_{X/S}S_o}$ is the dimensionless cell mass condentration and

$K = \frac{K_S}{S_o + X_o/Y_{X/S}}$ is the dimensionless saturation con stant

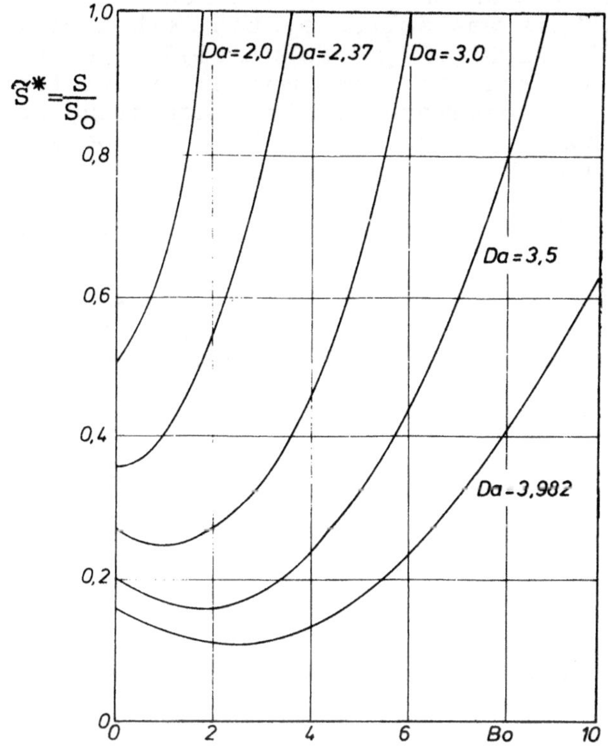

Fig. 8: Dimensionless substrate concentration at the reactor exit S/So as a function of Bo-numbers for different Da-numbers. Cell-free feed (10) K = 0.5

Eq. (65) is a nonlinear ordinary D.E. It was solved by quasi-linearisation which yields a linear boundary value problem. This was solved by the method of finite differences (6). In Fig. 9, $\tilde{S}^*$ is plotted as a function of Bo at Da = 3.0 and $\gamma = 0$ (no medium circulation) and for different K as parameters. With increasing Bo the dimensionless substrate concentration, $\tilde{S}^*$, passes a minimum as long as K is low enough.

Above a critical Da number, Da_{crit}, $\tilde{S}^*(Bo)$ passes a minimum:

$$Da_{crit} = \frac{\left[(K+1) - \sqrt{K(K+1)} - x_o^*\right] \sqrt{K(K+1)}}{\left[(K+1) - \sqrt{K(K+1)}\right] \left[\sqrt{K(K+1)} - K\right]} \tag{66}$$

For sterile feeding $X_o^* = 0$, eq. (66) reduces to eq. (67):

$$Da_{crit} = \frac{1}{1 - \frac{K}{\sqrt{K(K+1)}}} \tag{67}$$

From eq. (67) the critical K value, K_{crit}, can be evaluated at which $\tilde{S}^*(Bo)$ passes a minimum (Fig. 9):

$$K_{crit} = \frac{1}{\frac{1}{(1-1/Da)^2} - 1} \tag{68}$$

In Fig. 10, $\tilde{S}^*$ is plotted as a function of γ for K = 0.5 and Da = 2 as well as for Da = 5. For Da > Da_{crit} $\tilde{S}^*$ passes a minimum as a function of γ, if Bo > Bo_{opt}. For Bo < B_{opt}, $\gamma_{opt} = 0$.

In Fig. 11, γ_{opt} and $\tilde{S}^*$ are plotted as functions of the Bo number at Da = 5 and K = 0.5. Near Bo_{opt}, the dependence of γ_{opt} on the Bo number is considerable. The variation of $\tilde{S}^*$ on Bo is less significant. For comparison, $\tilde{S}^*$ is also plotted for the CSTR as well as for the tower reactor without a loop and at Bo_{opt}. The transitic from the CSTR (Bo = 0) to Bo_{opt} causes an improvement of $\tilde{S}^*$ by 64 %. When employing γ_{opt} at Bo = 15 a further improvement of 33 % is possible (Fig. 11). Under these con ditions, with $\gamma = 0$, washout of the cells occurs.

Figs. 12 and 13 show the dependence of γ_{opt} on the Da number and K at a constant Bo number (Bo = 7). Above

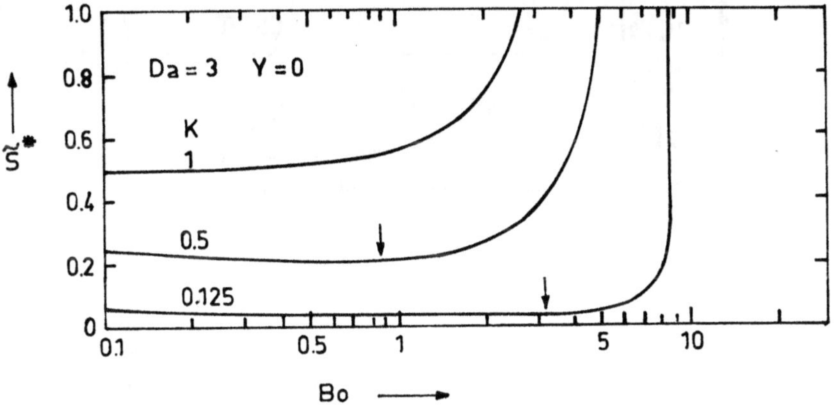

Fig.9: Dimensionless substrate concentration, $\tilde{S}^*$, as a function of the Bo number as a parameter (5).

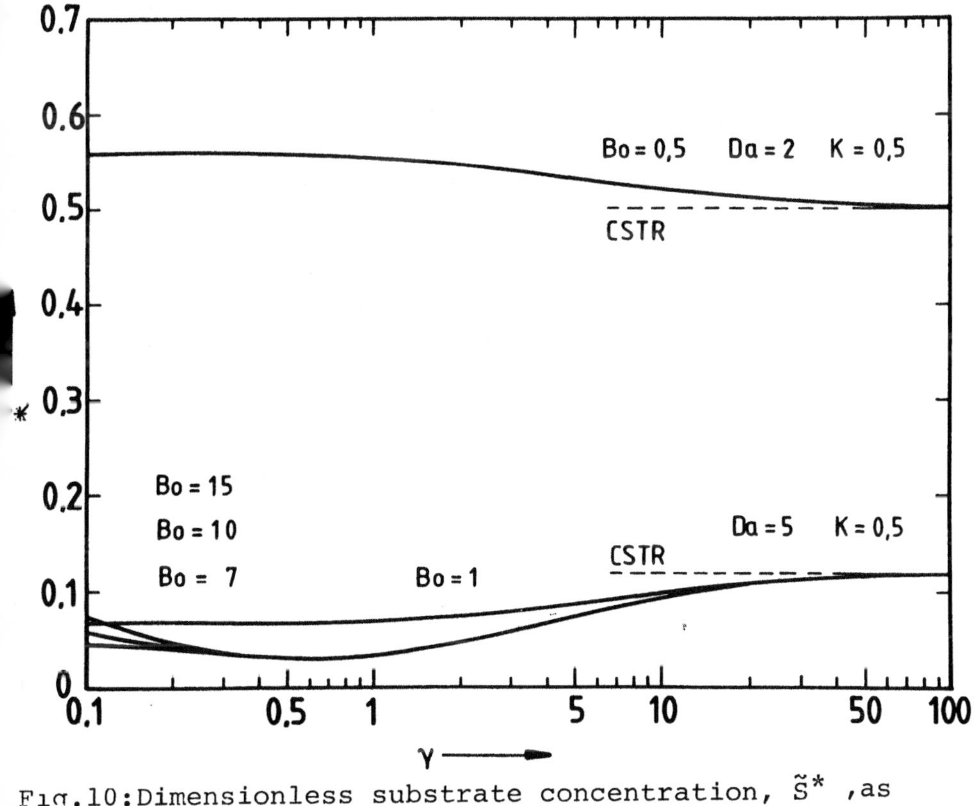

Fig.10: Dimensionless substrate concentration, $\tilde{S}^*$, as a function of the recyling ratio, γ. Comparison (5).

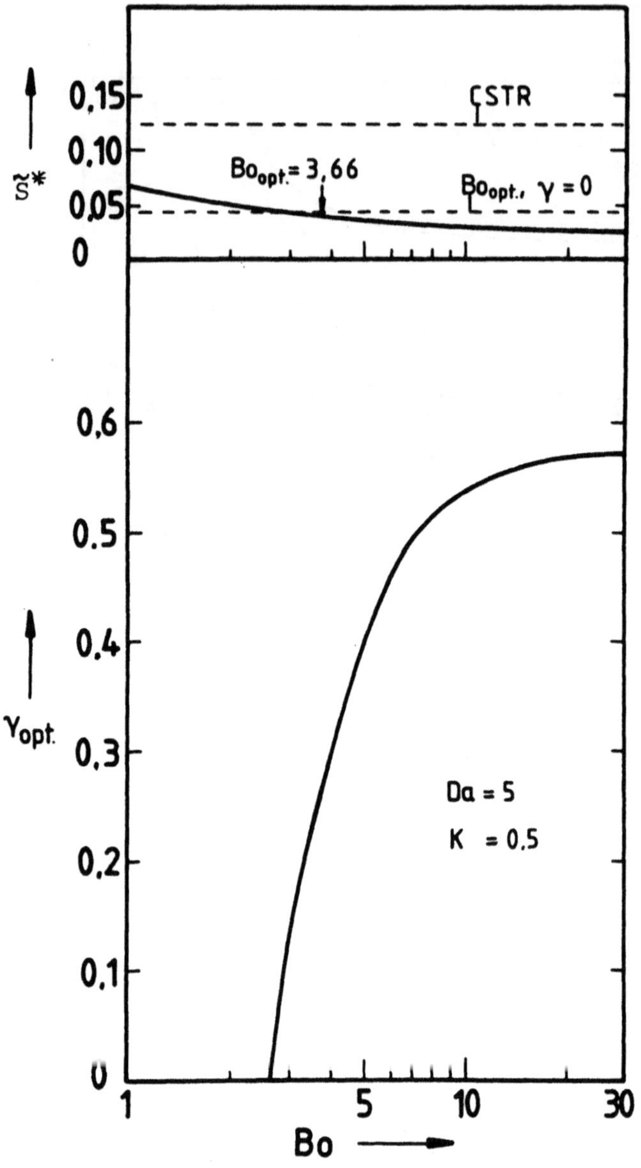

Fig.11: Optimum recycling ratio, γ_{opt}, and dimensionless substrate concentration, $\tilde{S}^*$, as a function of the Bo number. Comparison of continuously stirred tank reactor, CSTR, tower reactor without liquid recycling and at optimum Bo-number, Bo_{opt}, $\gamma = 0$, and tower reactor with optimum liquid recycling and Bo-number, Bo_{opt} (5).

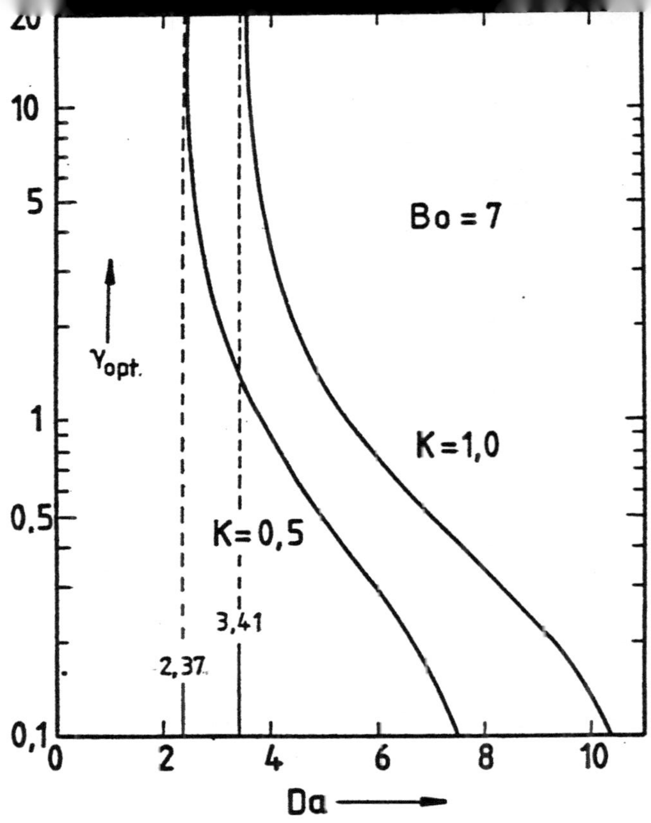

Fig.12: Optimum recycling ratio, γ_{opt}, as a function of Da no. K as a parameter (5).

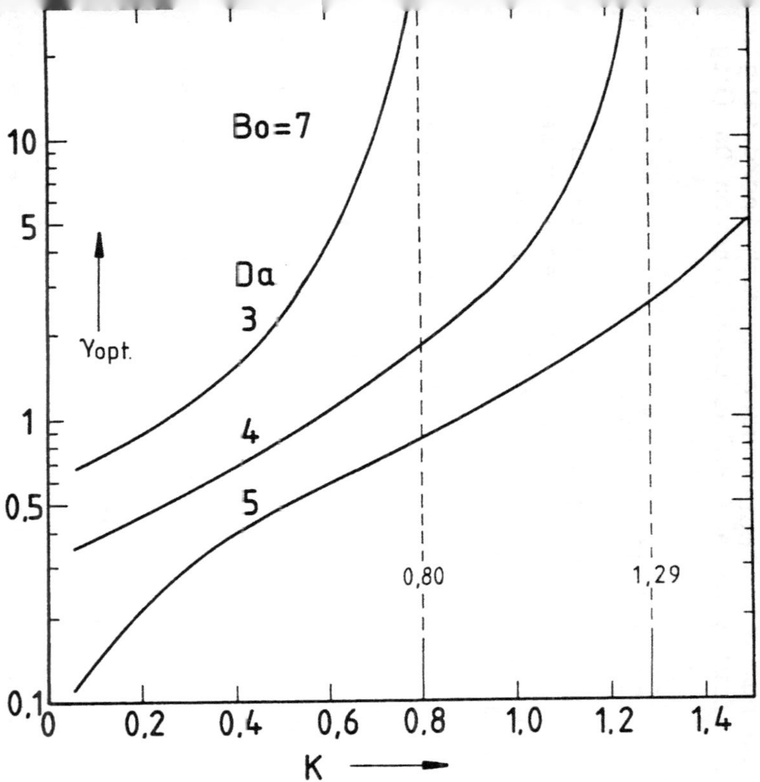

Fig.13: Optimum recycling ratio, γ_{opt}, as a function of K. Da number as a parameter (5).

critical Da number, with a value depending on K, γ_{opt} exists and diminishes with increasing Da (Fig. 12). At a constant Da number, γ_{opt} increases with increasing K and reaches infinity at a K value depending on Da (Fig. 13).

1.3 Oxygen Transfer Limited Growth

Cell cultivation under oxygen transfer limitation is usually not desired. However, often oxygen transfer limited growth prevails due to the insufficient oxygen transfer rate into the reactor.

1.3.1 Oxygen transfer limited cell growth in stirred tank reactors. It is usually assumed that the influence of the dissolved oxygen concentration on the specific growth rate can be described by Monod kinetics:

$$\mu = \mu_m \frac{O}{K_O + O} \tag{69}$$

where K_O is the saturation constant with regard to the dissolved oxygen

Two limiting cases can be distinguished:

1) For $C \gg K_O$, $\mu \simeq \mu_m$, no oxygen transfer limitation. The growth is of zeroth order with regard to the dissolved oxygen concentration
2) For $O \ll K_O$, $\mu \simeq \frac{\mu_m}{K_O} O$, (70)

i.e. strong oxygen transfer limitation. The growth is of first order with regard to the dissolved oxygen concentration.

In the batch reactor eqs.(71) to (72) hold true:

$$R_X = \frac{dX}{dt} = \mu_m \frac{OX}{K_O+O} \tag{71}$$

$$-R_S = -\frac{dS}{dt} = \mu_m \frac{OX}{K_O+O} \frac{1}{Y_{X/S}} \tag{72}$$

$$-R_O = -\frac{dO}{dt} = \mu_m \frac{OX}{K_O+O} \frac{1}{Y_{X/O}} , \tag{73}$$

if no aeration is carried out.

In the CSTR the following balance equations are valid:

$$\frac{dX_1}{dt} = \mu_m \frac{O_1}{K_O+O_1} X_1 + D(X_o-X_1) \tag{74}$$

$$\frac{dS_1}{dt} = -\mu_m \frac{O_1}{K_O+O_1} X_1 \frac{1}{Y_{X/S}} + D(S_o-S_1) \tag{75}$$

$$\frac{dO_1}{dt} = -\mu_m \frac{O_1}{K_O+O_1} X_1 \frac{1}{Y_{X/O}} + D(O_o-O_1)+k_La(O_1^*-O_1) \tag{76}$$

Under steady-state conditions

$$\frac{dX_1}{dt} = \frac{dS_1}{dt} = \frac{dO_1}{dt} = 0 \tag{77}$$

Since under oxygen transfer limited growth $O_1^* \gg O_1$ is valid and because of the low solubility of oxygen the term $D(O_o-O_1)$ is small in comparison with the others, eq. (76) can be reduced to eq. (78) is steady state prevails:

$$\mu_m \frac{O_1}{K_O+O_1} \tilde{X}_1 \simeq k_La\, O_1^*\, Y_{X/O} \tag{78}$$

The cell growth rate, μX_1, can mainly be influenced by the volumetric mass transfer coefficient, k_La, and yield coefficient, $Y_{X/O}$. Since the oxygen transfer rate, $k_LaO_1^*$, is independent of D, one can reduce eq. (74) for cell-free feed and steady-state operation:

$$\mu \tilde{X}_1 - D\tilde{X}_1 = 0 \tag{79}$$

or $\quad \mu = D \qquad (80)$

i.e. a behavior analogous to chemostat is expected: if X_1 exceeds $\tilde{X}_1$, O_1 will drop below $\tilde{O}_1$, this reduces μ and X_1 to $\tilde{X}_1$. If X_1 drops below $\tilde{X}_1$ O_1 will exceed $\tilde{O}_1$, this increases μ as well as X_1 to $\tilde{X}_1$. The steady-state is stable.

1.3.2 Oxygen transfer limited cell growth in tower reactors. If one assumes that the tower can be described by a one-dimensional dispersion model, the gas holdup variation along the tower is negligible and the longitudinal dispersion coefficients in the liquid phase, D_F, and in the gas phase, D_G, are constant in the tower, the following steady state oxygen balance can be obtaine (13,14) in the liquid phase:

$$D_F \frac{d^2\tilde{O}_F(x)}{d\,x^2} - u_F \frac{d\,\tilde{O}_F(x)}{d\,x} - R_{OF}(\tilde{X}_F, \tilde{S}_F, \tilde{O}_F x) + k_L(x)\; a(x)\; [\tilde{O}_F^*(x) - \tilde{O}_F(x)] = 0 \qquad (81)$$

and in the gas phase:

$$D_G \frac{d}{d\,x}[P(x) \frac{d\,\tilde{x}_{OG}(x)}{d\,x}] - \frac{d}{d\,x}[(P(x)\tilde{X}_{OG}(x)), U_G(x)] - \frac{R_G\; T\; E_F}{M_{O2}\; E_G}\; k_L(x)\; A(x)\; [\tilde{O}_F^*(x) - \tilde{O}_F(x)] = 0 \qquad (82)$$

The Danckwerts boundary conditions are assumed; in the liquid phase

$$\frac{d\,\tilde{O}_F(0)}{dx} = \frac{u_F}{D_F}\left[\tilde{O}_F(0) - \tilde{O}_F^E\right] \quad \text{at } x = 0 \qquad (81a)$$

and

$$\frac{d\,\tilde{O}_F(L)}{d\,x} = 0 \quad \text{at } x = L \qquad (81b)$$

and in the gas phase

$$\frac{d\,\tilde{x}_{OG}(0)}{dx} = \frac{u_G^E}{D_G}\left[\tilde{x}_{OG}(0) - \tilde{x}_{OG}^E\right] \quad \text{at } x = 0 \qquad (82a)$$

and

$$\frac{d\,\tilde{x}_{OG}(L)}{d\,x} = 0 \quad \text{at } x = L \qquad (82b)$$

where $\tilde{x}_{OG}$ is the mole fraction of O_2 in the gas phase,

$\tilde{O}_F^E$ is $\tilde{O}_F$ at the entrance, at $x = O^-$

$\tilde{x}_{OG}^E$ is $\tilde{x}_{OG}$ at $x = o^-$

E_G and E_F are the gas and liquid holdup, respectively

The D.E. system (81)(82) can only be solved numerically.

Since, in general, it is assumed that $\tilde{X}$, $\tilde{S}$ and $\tilde{O}$ are space dependent, the oxygen consumption rate also depends on x. Thus, for $\tilde{X}$ and $\tilde{S}$ similar balance equations are obtained.

1.3.3 Oxygen transfer limited cell growth in tower loop reactors. In addition to the assumptions in 1.3.2 it can be assumed that the medium is completely free of a gas phase in the loop, furthermore, the longitudinal liquid dispersion coefficient in the loop can be neglected. These two assumptions hold true for tower reactors with an outer loop which has a considerably smaller cross-sectional area than the tower cross-sectional area (13, 14).

The following D.E.'s are needed in addition to eq. (81) and (82): oxygen mass balance in the loop liquid:

$$-u_B \frac{d\tilde{O}_B(x^*)}{dx^*} - R_{OB}(\tilde{X}_B, \tilde{S}_B, x^*) = 0 \tag{83}$$

where index B refers to the loop,
x^* is the longitudinal coordinate in the loop.

To solve eq. (83) the boundary condition at the lower end of the loop

$$\tilde{O}_B(O) = \tilde{O}_F(L) = O_F^A \tag{84}$$

is used. In eq. (84), O_F^A is the exit concentration of the tower.

The solution of the oxygen balance is only possible if the conditions at the lower end of the tower are defined by

$$\frac{u_F}{D_F}\,[\tilde{O}_F(0) - \gamma O_B^A - \{1-\gamma\}\tilde{O}_o\,] = 0 \qquad (85)$$

By means of the CO_2 balance $u_G(x)$ can be calculate in the tower (13,14). The substrate balance for a stead state in the tower is given by

$$D_F \; \frac{d^2\tilde{S}_F(x)}{dx^2} \; u_F \; \frac{d\,\tilde{S}_F(x)}{d\;x} - R_{SF}(\tilde{X}_F,\tilde{S}_F,\tilde{O}_F,x) = 0 \qquad (86)$$

and in the loop by

$$-u_B \; \frac{d\tilde{S}_B(x^*)}{d\;x^*} - R_{SB}(\tilde{X}_B,\tilde{O}_B,x^*) = 0 \qquad (87)$$

with the boundary conditions

$$\frac{d\tilde{S}_F(0)}{dx} = \frac{u_F}{D_F}\,[\tilde{S}_F(0) - \gamma S_B^A - \{1-\gamma\}\tilde{S}_o] \qquad (86a$$

$$\frac{d\,\tilde{S}_F(L)}{d\;x} = 0 \qquad (86b$$

$$\tilde{S}_B(0) = S_F^A \qquad (87a$$

The cell mass balance for a steady state in the tower i given by

$$D_F \; \frac{d^2\tilde{X}_F(x)}{d\;x^2} - u_F \; \frac{d\tilde{X}_F(x)}{d\;x} + R_{XF}(\tilde{X}_F,\tilde{S}_F,\tilde{O}_F,x) = 0 \qquad (88)$$

and in the loop by

$$-u_B \; \frac{d\tilde{X}_B(x^*)}{d\;x^*} + R_{XB}(\tilde{X}_B,\tilde{S}_B,\tilde{O}_B,x^*) = 0 \qquad (89)$$

Again boundary conditions link these two D.E.'s

$$\frac{u_F}{D_F}\,[\,\tilde{X}_F(0) - \gamma X_B^A - \{1-\gamma\}\;\tilde{X}_o\,] = 0 \qquad (88a$$

$$\frac{d\tilde{X}_F(L)}{dx} = 0 \tag{88b}$$

$$\tilde{X}_B(0) = X_F^A . \tag{89a}$$

For sterile feed $X_o = 0$.

For R_{XF}, R_{XB}, R_{SF}, R_{SB}, R_{OF} and R_{OB} the validity of the corresponding Monod kinetics is assumed, e.g.

$$R_{XF}(x) = \{\mu(\tilde{S}_F, \tilde{O}_F, x) - \mu_T\}\tilde{X}_F(x) \tag{90}$$

and

$$\mu(\tilde{S}_F, \tilde{O}_F, x) = \mu_m \frac{\tilde{S}_F(x)}{K_S + \tilde{S}_F(x)} \frac{\tilde{O}_F(x)}{K_O + \tilde{O}_F(x)} \tag{91}$$

where μ_T is the specific death rate.

These D.E. systems were numerically solved and the calculated data was fitted to the measured one. These measurements and simulations will be considered in the article "Biochemical Reactions and oxygen transfer into different fermentation broths" by the author in this book.

2 PRODUCT FORMATION

Different models were recommended for product formation. In general, a product can be linked with cell growth or not.

In general, the product formation rate is given by

$$\frac{dP}{dt} = q_p X \tag{92}$$

When the product is growth linked, the amount of product formed is directly proportional to the biomass formed:

$$dP = Y_{P/X}\, dX \tag{93}$$

where $Y_{P/X}$ is the product yield referred to biomass formed.

It follows that

$$\frac{dP}{dt} = Y_{P/X} \frac{dX}{dt} = Y_{P/X} \mu X \tag{94}$$

Thus the specific rate of product formation is given by

$$q_p = Y_{P/X} \ \mu \tag{95}$$

When the product is <u>non-growth-linked</u> it can be a complex function of μ. In the simplest case

$$q_p = k_1 = \text{constant} \tag{96}$$

Sometimes the product formation is partly growth linked and partly independent of a growth rate:

$$q_p = k_1 + k_2 \mu \tag{97}$$

where k_1 is the rate constant of the non-growth-linked product formation,
$k_2 = Y_{P/X}$ constant of the growth linked product formation.

In the following, eq. (97) will be used to treat product formation.

<u>2.1 Product Formation in the Stirred Tank Reactor.</u>

The mass balance of the product for nonlimiting growth is given by

$$\frac{dP_1}{dt} = (k_1 + Y_{P/X} \mu_m) \ X_1 + D(P_o - P_1) \tag{92}$$

where P_o and P_1 are the product concentrations in the feed and in the reactor.

With substrate and oxygen transport limited growth the following product balances hold true:

$$\frac{dP}{dt} = (k_1 + Y_{P/X} \frac{\mu_m S_1}{K_S + S_1} X_1 + D(P_o - P_1) \tag{93}$$

and

$$\frac{dP}{dt} = (k_1 + Y_{P/X} \frac{\mu_m O_1}{K_O + O_1}) X_1 + D(P_o - P_1) \tag{94}$$

Under steady state

$$\frac{dP}{dt} = 0 \qquad \text{is valid.}$$

2.2 Product Formation in Tower Reactors

No general solutions of the product mass balance equation in tower reactors are known. However, one can draw some conclusions from the longitudinal cell, substrate and dissolved oxygen concentration profiles with regard to the longitudinal product concentration profile.

When employing nonflocculating bacteria or yeasts or fungi without pellet formation in a laboratory bubble column only slight longitudinal cell mass concentration profiles can be expected (5). In non-substrate-limited growth and in laboratory bubble column reactors the longitudinal substrate concentration is uniform (13). When substrate limited growth prevails longitudinal substrate concentration can be expected (13).

Longitudinal dissolved oxygen concentration profiles always exist. They are most considerable at the end of the nonlimiting growth phase (15).

In case of non-growth-linked product formation uniform product concentration should prevail. But also for growth linked product and in the presence of nonuniform longitudinal dissolved oxygen and/or substrate concentration profiles no or only slight longitudinal product concentrations can be expected.

Hence in laboratory bubble column reactors, eqs. (92), (93) and/or (94) can be used, where X_1, S_1 and O_1 should be replaced by their mean values in the tower.

2.3 Product Formation in Tower Loop Reactors

In tower loop reactors with increasing liquid recycling the longitudinal concentration profiles become more and more uniform. Therefore the use of eqs. (92), (93) and/or (94) is recomended.

In the presence of flocculating cells or pellets a considerable longitudinal cell mass concentration profile usually prevails. This can cause very significant nonuniformity of dissolved oxygen concentration profiles. It is unlikely that under these conditions the longitudinal profile of product concentration is uniform in tower loop reactors, especially in tower reactors without liquid recycling.

B. CHARACTERIZATION OF FERMENTATION SYSTEMS

The performance of bioreactors is considerably influenced by the medium properties. Therefore, knowledge of medium properties is necessary for the construction of bioreactors.

1 MEDIUM PROPERTIES

Several properties influence the performance of bioreactors. However, in this lecture only a few selected ones should be considered.

1.1 Viscosity

Yeast and bacterium suspensions at concentrations which are used in bioreactors exhibit Newtonian properties. Reuss et al. (16) recommend the following relationship for the dynamic viscosity, η , of yeast (Saccharomyces cerevisiae and Candida utilis) suspensions:

$$\frac{\eta_s}{\eta_o} = \frac{1}{1-(h_s \varepsilon_x)^a} \tag{95}$$

where η_s is the viscosity of the suspension,
η_o the viscosity of the supernatant,
ε_x the volume fraction of the cells,
h_s the packing factor:
$h_s = 0.0487\ P_{osm} + 1.59$ and
P_{osm} the osmotic pressure (bar = 10^5 Nm^{-2})

The rheology of highly viscous fermentation moulds of fungi cultivations was investigated by Metz et al. (17). However, these systems will not be considered here.

1.2 Coalescence Suppressing or Promoting Character.

The interfacial properties considerably influence the bubble coalescence process in liquids.

At the same concentration and chain length, fatty acids have the strongest, alcohols intermediate, polyalcohols and ketones the slightest influence. The concentration, C_{CO}, at which coalescence suppression begins is inversely proportional to the number of carbon atoms, n_C (18):

$$C_{CO} \propto n_C^{-1.5} \tag{96}$$

Interfacial tension suppression, $d\sigma/dc$, is also influenced by C_{CO} (18):

$$\ln \frac{d\sigma}{dc} = -(1.5 + 0.5 \ln C_{CO}) \tag{97}$$

There is a definitive relationship between the coalescence suppressing effect of salt solutions, their ion strength and their position in the lyotropic (Hofmeister) series (19). At the same ion strength, the salt has a stronger effect which exhibits a greater tendency to flocculate proteins. Since no relationship is known for complex cultivation media the bubble coalescence behavior of media was experimentally determined by means of the volumetric mass transfer coefficients, k_La, which were measured in a standard bubble column reactor. These k_La values were compared with the k_La values measured in the same equipment under the same operational conditions with water: $(k_La)_{H_2O}$. The ratio, m:

$$m = \frac{k_La}{(k_La)_{H_2O}} \tag{98}$$

was employed to characterise the cultivation medium for a fixed superficial gas velocity, w_{SG} (20). In the presence of antifoam agents a modified relationship was used:

$$m = \frac{(k_La)_{corr}}{(k_La)_{ref}} \tag{99}$$

where $(k_La)_{corr} = k_La - \Delta(k_La)$

$$\Delta(k_La) = \left[k_La - (k_La)_{ref}\right]_{w_{SG}=2\,cm\,s^{-1}}$$

As a reference a nutrient salt solution with an antifoam agent was employed.

1.3 Foam Formation

Cultivations are often accompanied by foam formations due to the high foaming capacity of protein solutions. This capacity results from the stabilization of the gas liquid interface caused by the denaturation and strong adsorption of the surface proteins (21).

Since mechanical foam breaking with its high input power requirements is expensive, antifoam agents are usually preferred in the fermentation industry. The presence of antifoam agents, however, deteriorates the efficiency of gas dispersion by increasing the bubble coalescence rate.

The influence of salts (22) and alcohols (23) on the foaminess were investigated and simple relationships were found. The salt effect can be explained by the interaction between the water and the salt, i.e. by its influence on the water structure. The alcohol effect is due to the change of the water structure and the direct interaction between alcohols and proteins, where the direct interaction overcompensates the water structure effect. The influence of antifoam agents on foam formation seems to be a complex process. With increasing antifoam concentration the foam formation is diminished step by step (24).

1.4 Cell Sedimentation and Flocculation

There is a significant disagreement among different research groups with regard to the cause of cell flocculation (25). The flocculation of yeast seems to be a genetic property. However, there is no flocculation during the nonlimited exponential growth phase. Ions, especially Ca ions, phosphates, the properties of cell membranes as well as the glycogen content of the cells, can influence the flocculation of yeasts during fermentation. Additives, which influence the water structure and protein solubility can alter the flocculation properties.

Since the influence of different additives on cell sedimentation based on literature data is contradictory it is necessary to determine the cell sedimentation rate experimentally. For cells with low sedimentation rates batch runs were carried out and the displacements of the interfaces between layers of different cell concentrations along the column were determined optically.

For flocculating cells with high sedimentation rates continuous runs were used. The cell suspension was continuously fed into the sedimentation tank and, from the top, the clear liquid and, from the bottom, the concentrated cell suspension were continuosuly removed. Cell mass enrichment was measured as a function of the mean cell residence time in this sedimentation tank (26).

2 PROPERTIES OF BIOLOGICAL TWO-PHASE SYSTEMS

Two-phase system properties can strongly influence cultivation conditions, especially if growth is oxygen transfer limited. To treat the oxygen transfer rate quantitatively it is necessary to determine the volumetric mass transfer coefficient, k_La, the dissolved oxygen concentration in the liquid bulk, O_F, and at the gas liquid interface, O_F^*. The specific interfacial area is influenced by the Sauter bubble diameter, d_s, and the relative gas holdup, ε_G, according to eq. (100):

$$\underline{a} = \frac{6E_G}{d_s(1-E_G)} , \quad (100)$$

if the bubbles have a spherical shape, which holds true for small bubbles. The mass transfer coefficient, k_L, is a complex function of several parameters such as d_s, interfacial properties and turbulence.

2.1 Relative Gas Holdup, E_G

The mean relative gas holdup, E_G, is defined by eq. (101)

$$E_G = \frac{V-V_L}{V} \quad (101)$$

where V is the volume of the bubbling layer,
V_L the volume of the bubble-free layer.

In bubble column reactors the determination of E_G is carried out by means of the height of the bubbling layer, H and bubble-free layer, H_L, or of the corresponding hydrostatic pressures.

The local relative gas holdup was measured by electrical conductivity probes (see 2.2).

2.2 Bubble Size Distributions

The Sauter bubble diameter, d_s, can be calculated from the bubble size distribution by eq. (102):

$$d_s = \frac{\sum_1^N n_i d_i^3}{\sum_1^N n_i d_i^2} \qquad (102)$$

where n_i is the frequency of the bubbles with the diameter d_i.

The most bubble size distributions were measured with flash photography and semiautomatic evaluation of the photographs (27,29). Miniaturized electrical conductivity probes with two sensors and on-line computer evaluation are also popular (27,28,29).

The measurements with these probes also yield the local gas holdup in tower reactors (27,29).

2.3 The Specific Interfacial Area

The specific interfacial area can be measured by means of chemical reaction (e.g. sulphite oxydation) if a model medium is used (e.g. 30). However, chemical methods cannot be employed for fermentations. Therefore they are usually calculated by means of eq. (100), if d_s and E_G are available. Specific interfacial areas were determined during the cultivation of *Candida boidinii* (31), *Hansenula polymorpha* (32) and *Escherichia coli* (5) in tower loop reactors.

2.4 Volumetric Mass Transfer Coefficients

In reactors with uniform concentration of substrate cells and dissolved oxygen (reactors with lumped parameters), k_La can be calculated in batch operation by eq. (103):

$$k_La = \frac{OTR}{O_1^* - O_1} \qquad (103)$$

and in continuous operation by eq. (104):

$$k_La = \frac{OUR - D(O_o - O_1)}{O_1^* - O_1} = \frac{OTR}{O_1^* - O_1} \tag{104}$$

The oxygen transfer rate, OTR, can be calculated by the O_2 balance of the gas phase by means of the gas compositions (O_2, CO_2, N_2) at the gas reactor inlet and outlet.

The oxygen utilization rate, OUR, can be calculated by eq. (105):

$$OUR = \frac{1}{Y_{X/O}} \mu (X_1 - X_o) \tag{105}$$

The saturation concentration of dissolved oxygen, O_1^*, is calculated by means of the measured gas composition in the reactor and the measured solubility of O_2 in the medium (33). The dissolved oxygen concentrations in the feed, O_o, and reactor, O_1, are measured by oxygen electrodes and calculated by the corresponding O_2 solubility. In tower and tower loop reactors k_La was determined by fitting the calculated longitudinal dissolved oxygen profiles to the measured ones (13). Such data were evaluated during cultivation of Candida boidinii (31) and Hansenula polymorpha (13,15).

Flow pattern and turbulence properties will not be discussed here, because they will be treated in the article of A. Lübbert in this book.

REFERENCES

1. Chen, M.S.K. AIChE J. 18 (1972) 849.

2. Seipenbusch, R. and H. Blenke. "The loop reactor for cultivating yeast on n-paraffin substrate". Adv. in Biochem. Eng. 15 (1980). Ed. A. Fiechter, Springer Verl. p. 1

3. Cow, J.S., J.D. Littlehailes, S.R.L. Smith and R.B. Walter. "Single Cell Protein" II. Eds. S.R. Tannenbaum, D.I.C. Wang, MIT Press (1975) 370

4. Faust, U. and W. Sittig. "Methanol as Carbon Source for Biomass Production in a Loop Reactor". Adv. in Biochem. Eng. 17 (1980) 63, ed. A. Fiechter, Springer Verlag

5. Adler, I. Dissertation, University Hanover 1980
6. Adler, I., W.-D. Deckwer and K. Schügerl. Part I. Chem. Eng. Sci. (in the press)
7. Monod, J. Recherches sur la Croissance des Cultures Bacteriennes. 2nd edn (A942) Hermann Paris
8. Monod, J. Ann. Inst. Pasteur 79 (1950) 715
9. Novick, A. and L. Szilard. Science 112 (1950) 715
10. Todt, J., J. Lücke, K. Schügerl and A. Renken. Chem. Eng. Sci. 32 (1977) 369
11. Chen, G.K.C., L.T. Fan, L.E.Erickson, Can. J. Chem. Eng. 50 (1972) 157
12. Pirt, S.J. "Principles of Microbe and Cell Cultivation". Blackwell Scientific Publ. Oxford 1975
13. Luttmann, R. Dissertation, University of Hanover 1980
14. Luttmann, R., M. Thoma, H. Buchholz and K. Schügerl. Computer and Chem. Eng. Part II (submitted)
15. Buchholz, H. Dissertation, University of Hanover 1979
16. Reuss, M. D. Josic, M. Popovic, W.K. Brown. European J. Appl. Microbiol. Biotechnol. 8 (1979) 167
17. Metz, B., N.W. Kossen and J.C. van Suijdam. "The Rheology of Mould Suspensions". Adv. in Biochem. Engng. Springer Verlag Vol. 11 (1979) 103.
18. Keitel, G. Dissertaion, University of Dortmund 1978
19. Zlokarnik, M. "Sorption Characteristics for Gas-Liquid Contacting in Mixing Vessels". Adv. in Biochem. Engng. Springer Verlag. 7 (1978) 113.
20. Adler, I, J. Diekmann, W. Hartke, V. Hecht, F. Rohn and K. Schügerl. European J. Appl. Microbiol. Biotechnol 10 (1980) 171
21. Cumper, C.W.N., A.E. Alexander, Trans. Farad. Soc 46 (1950) 235
22. Bumbullis, W., K. Kalischewski and K. Schügerl. European J. Appl. Microbiol. Biotechnol. 7 (1979) 147
23. Bumbullis, W. and K. Schügerl. European J. Appl. Microbiol. Biotechnol. 8 (1979) 17
24. Müller, B. , V. Pfanz and K. Schügerl (in preparation)
25. Windish, W. Monatsschr. f. Brauerei 22 (1969) 69
26. Kuhlmann, W., A.Gebauer and , Schmidt. (In preparation)
27. Buchholz, R. and K. Schügerl . European J. Appl. Microbiol. Biotechnol. 6 (1979) 301
28. Buchholz, R. and K. Schügerl. European J. Appl. Microbiol. Biotechnol. 6 (1979) 315
29. Buchholz, R., W. Zakrzewski and K. Schügerl. Chem. Ing. Techn. 51 (1979) 568

30. Schumpe, A. and W.-D. Deckwer. Chem. Ing. Techn. 52 (1980) 468
31. Schügerl, K., J. Lücke, J. Lehmann, and F. Wagner. Adv. in Biochem. Engng. 8 (1978) 63
32. Zakrzewski, W. Dissertation University of Hanover 1980
33. Deckwer, W.-D."Physical Transport Phenomena in Biological Tower Reactors"(Proceedings of NATO ASI on "Mass transfer with chemical reaction in multiphase systems".İzmir,Turkey,1981)

PHYSICAL TRANSPORT PHENOMENA IN BIOLOGICAL TOWER REACTORS

W.-D. Deckwer

Institut für Technische Chemie
Universität Hannover (TH)
D-3000 Hannover 1, FRG

1 INTRODUCTION

It is well known that the performance of biological processes may be influenced significantly by physical transport phenomena. In general, physical transport processes concern the transfer of mass, momentum and various kinds of energy. One can suspect that until now only few of the physical transport phenomena have been fully recognized and understood.

The present paper will be mainly confined to mass transfer phenomena in a special type of biological process, i.e. aerobic fermentations. Various reactors have been developed to carry out aerobic fermentations. Among them the more popular are stirred aerated vessels, bubble columns (with external circulation of liquid) and various kinds of loop reactors. The present lecture will only deal with gas-in-liquid dispersions generated by various kinds of spargers and processed without mechanical agitation at moderate liquid flow rates. In chemical reaction engineering such gas-liquid reactors are called bubble columns and in biotechnology tower bioreactors. Applications of these reactors are found in waste water treatment (1-3) and aerobic fermentations (4,5). Recent examples concern the production of yeasts, i.e. Candida boidinii (6) and Hansenula polymorpha (7-9), the fermentation of Penicillium chrysogenum (10), and the production of animal cells (11).

A decisive mass transfer problem in the majority of fermentations constitutes the transport of oxygen from the air phase to the locale of the reaction, i.e. the biomass phase which, in accordance with chemical engineering, will be conveniently referred to as solid phase. The major reason that oxygen transfer may play an important role in biological processes is the limited oxygen capacity of the fermentation broth due to the low solubility of O_2.

The oxygen transfer path in a typical three-phase biological system is schematically shown in Fig. 1. The situation is completely equivalent to the one encountered in catalytic slurry reactors. It is generally accepted that before oxygen can be consumed by biomass particles several physical resistances have to be overcome, in principle. In terms of the simple film theory possible resistances may be

- diffusion through the liquid films around the bubble and the biomass particles (pellets, flocculant, fibrous or filamentous material)

- effective diffusion in parallel with the consumption by reaction in the interior of the biomass particles.

The latter phenomenon, i.e. effective diffusion and reaction in the biomass, may be totally or almost completely free of external influence. Several clear expositions are now available which treat the interaction of diffusion and reaction in analogy to heterogeneous catalysis by introducing the concept of effectiveness factors (12-14). On the other hand, operational conditions and hydrodynamic flow behavior exert a strong impact on the transfer of oxygen through the liquid films around both the bubble and biomass particle. These possible transport resistances are characterized by volumetric mass transfer coefficients and among others they will be the major subject of this paper.

2 MODEL EQUATIONS OF FERMENTER

Today, there is no doubt that the design and scale-up of chemical reactors should be based on mathematical models and computational optimization procedures. This is certainly also true of biological reactors though they are definitely more complex. The use of mathematical models requires to embed the microscopic picture of mass transport and reaction into the governing macro-

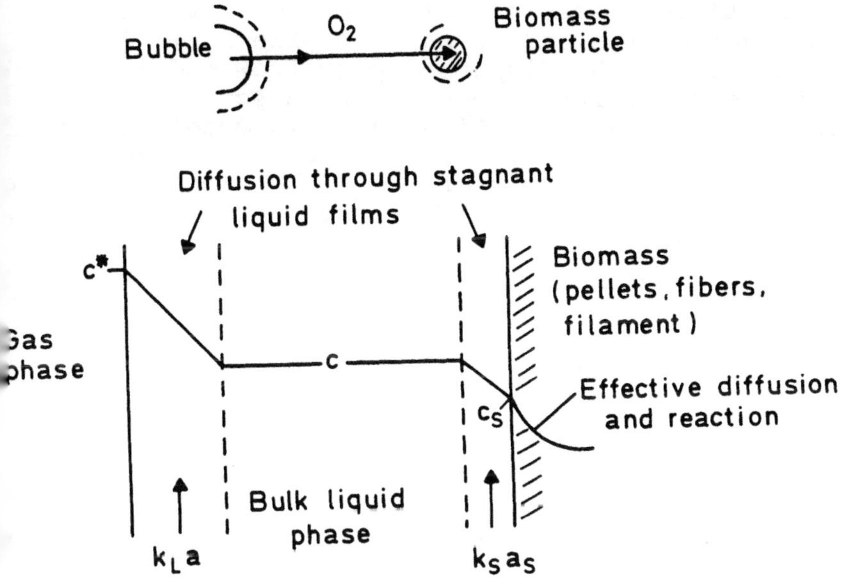

Fig. 1: Mass transfer resistances in biological reactor

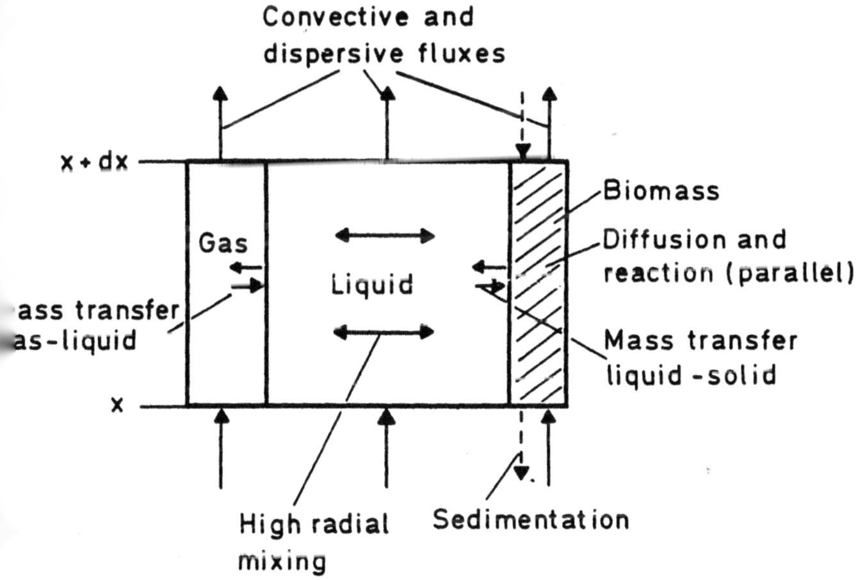

Fig. 2: Schematic volume element of bioreactor

scopic balance equations of the fermenter. With the exception of stirred vessels, biological reactors usually have some degree of slenderness, i.e. the ratio $L/d_C > 1$, therefore the dispersed plug flow model can be assumed to be a pertinent approach to describe fermenters. The differential equations of this model are obtained in the usual way by balancing over a volume element under consideration of those phenomena which are thought to be of influence. In view of Fig. 2 the balance equations for oxygen are as follows:

Gas phase

$$\frac{d}{dx}\left(\varepsilon_G E_G \frac{dc_G}{dx}\right) - \frac{d}{dx}(u_G c_G) - k_L a(c_L^* - c_L) = 0 \qquad (1)$$

Liquid phase

$$\frac{d}{dx}\left(\varepsilon_L E_L \frac{dc_L}{dx}\right) - u_L \frac{dc_L}{dx} + k_L a(c_L^* - c_L) - k_s a_s(c_L - c_S) = 0 \qquad (2)$$

Biomass phase (external surface)

$$k_s a_s(c_L - c_s) - R(c_S, c_C, c_B, D_{eff}, \ldots) = 0 \qquad (3)$$

Of course, the rate term R in the oxygen balance depends additionally on the local concentrations of the C source and the biomass (c_C and c_B), and equivalent balances have to be formulated for both of them. In the case of biomass it may be necessary to take into account sedimentation, which gives the following balance equation:

$$\frac{d}{dx}\left(\varepsilon_L E_S \frac{dc_B}{dx}\right) + (u_S - u_L)\frac{dc_B}{dx} + R' = 0 \qquad (4)$$

where R′ is the generation term for biomass which needs no further specification for the present purpose. u_S is the settling velocity of the biomass particles in the swarm.

The above model equations, the structure of which is representative for distributed reactor models involve a series of parameters which characterize the physical transport phenomena. These are mixing (expressed by the dispersion coefficients E_G, E_L, E_S) and mass transfer properties (ε_G, $k_L a$, $k_S a_S$). These physical or hydrodynamic transport parameters depend generally on physico-chemical properties (viscosity, surface tension geometrical sizes, the kind of gas distribution, the phase velocities and the flow regime. Due to the various influences the calculation or estimation of physical transport parameters in nonagitated gas-in-liquid dispersions is often subject to serious unreliabilities, and their availability is an important problem in design and scaleup of fermenters. In the following, the utility of hydrodynamic parameters which characterize the physical transport phenomena will be reviewed in light of recent findings.

3 GAS-LIQUID MASS TRANSFER

3.1 Gas Holdup

The fractional gas holdup ε_G is an important parameter to characterize gas-in-liquid dispersions. ε_G depends mainly on the gas throughput, the sparger design and on physico-chemical properties. If the column diameter is large compared to the bubble diameter, say larger by a factor of about 40, the column diameter has no significant effect. This is commonly valid if $d_c \geqslant 10$ cm.

The influence of the gas velocity on ε_G can conveniently be expressed by

$$\varepsilon_G \propto u_G^{\,n} \tag{5}$$

At low gas velocities and if porous spargers are used bubbly flow prevails. Then the exponent n may vary from 0.7 to 1.2. In churn-turbulent (heterogeneous) flow, which occurs at higher gas velocities, and if single and multinozzle spargers ($d_o \geqslant 1$ mm) are used, n is in the range of 0.4 to 0.7. Fig. 3 presents data for water and aqueous systems including some from large-scale equipment. The flow is heterogeneous and the exponent of eq (5) is about 0.6.

Akita and Yoshida (15), among others, proposed the following empirical correlation

$$\frac{\varepsilon_G}{(1-\varepsilon_G)^4} = 0.2\left(\frac{g d_c^{\,2} \rho_L}{\sigma}\right)^{1/8} \left(\frac{g d_c^{\,3}}{\rho_L^{\,2}}\right)^{1/12} \frac{u_G}{\sqrt{g d_c}} \tag{6}$$

The column diameter d_c is only included to present the

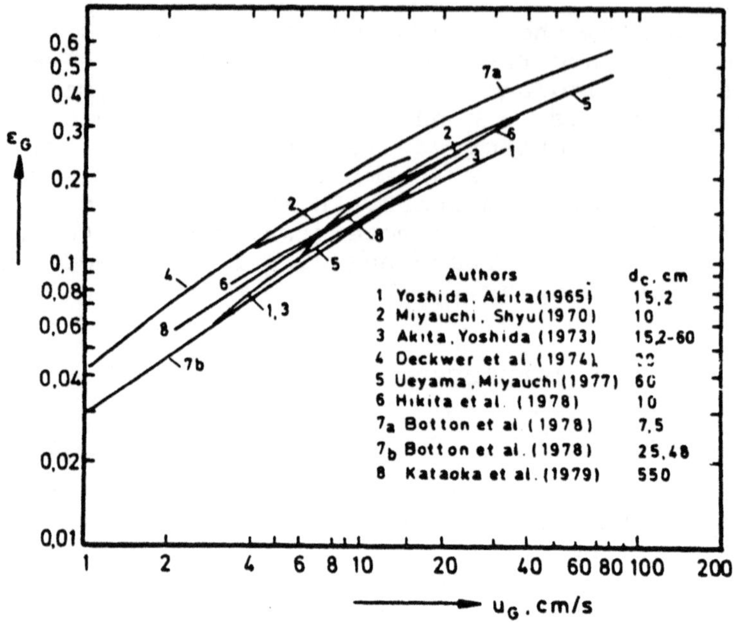

Fig. 3: Holdup vs. gas velocity (single & multi-orifice spargers)

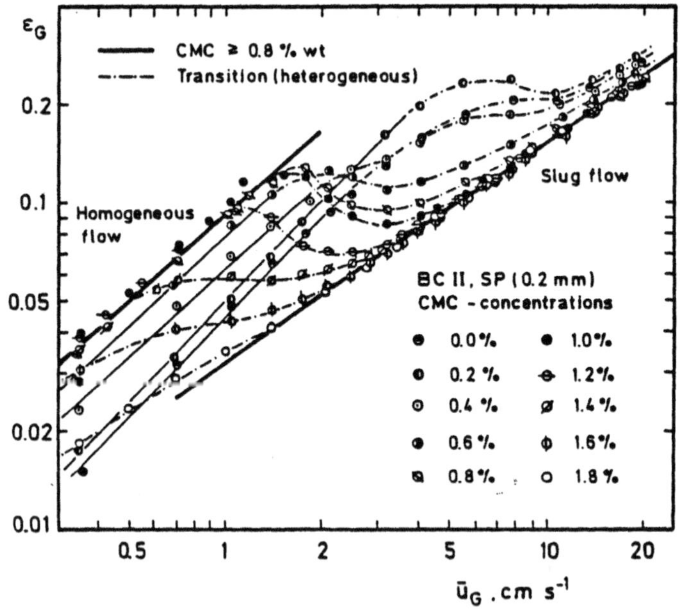

Fig. 4: Gas holdup in CMC solutions - bubble column aerated with sintered plate

dependencies in dimensionless form. Generally, the correlation of Akita and Yoshida gives a conservative estimate, see also Fig. 3. Considerably higher gas holdup values for nutrition and fermentation media have been reported by Schügerl et al. (16), particularly, if more effective spargers, such as porous plates and two-phase nozzles were used. However, it has not yet been possible to summarize the data measured for such systems into utilizable correlations.

Aqueous solutions of CMC (Na-carbomethoxy cellulose) are often used as model media antibiotica fermentation broths. The gas holdup in such solutions shows a complex behavior as shown in Fig. 4 for 14 cm ID bubble column aerated with a sintered plate (17). At low gas velocities a homogeneous flow can be realized. By increasing the gas velocity the final regime will be slug flow after a transition range where the flow is heterogeneous. Fig. 5 shows that in homogeneous flow, which can only be realized with porous spargers like sintered plates, the holdup does not depend on CMC concentrations and hence on viscosity. In this flow regime the exponent n of u_G in eq. (5) is 0.85. If perforated plates are used slug flow exists for all gas velocities and eq. (5) applies also at $n = 0,67$. Once again, for fully established slug flow ε_G is independent of the viscosity. At higher gas velocities the ε_G data of various spargers fall together. It is noticeable that in homogeneous flow ε_G values can be obtained at low gas velocities which in slug flow regime can only be realized at considerably higher gas velocities.

As the holdup can be measured easily by various methods it seems therefore always better to carry out ε_G measurements in a labscale column (with $d_c \geq 10$ cm). This will certainly give a better estimate for a specific design system than any correlation.

3.2 Physical Gas-Liquid Mass Transfer Coefficient

Various models are available to calculate liquid side mass transfer coefficients k_L. The value of this hydrodynamic parameter and the equations that apply to its calculation largely depend on bubble size and the constitution of the bubble surface. Fig. 6 presents some recent measurements on mass transfer from single bubbles (19) which demonstrate the above influences. The evaluated k_L values are plotted as Sherwood numbers vs. Peclet numbers. Large circulating bubbles with mobile surface yield k_L values which approach the

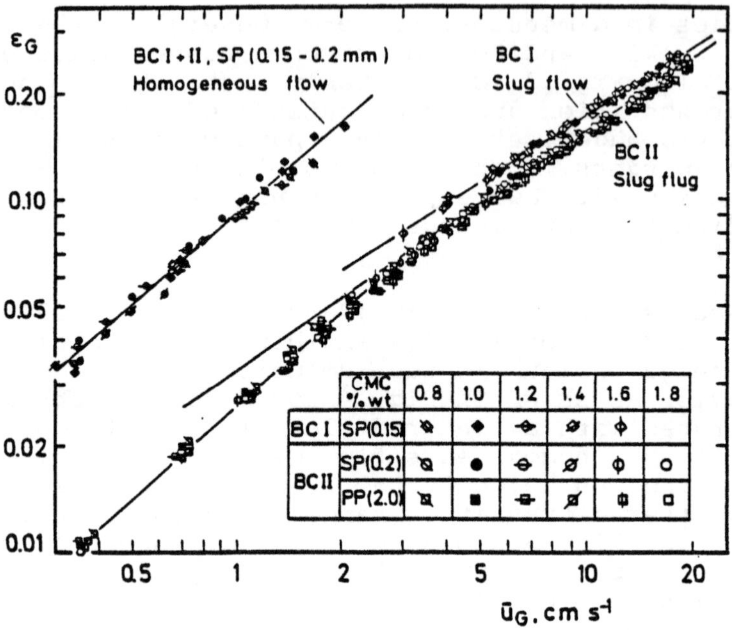

Fig. 5: Gas holdup vs. gas velocity - CMC solutions

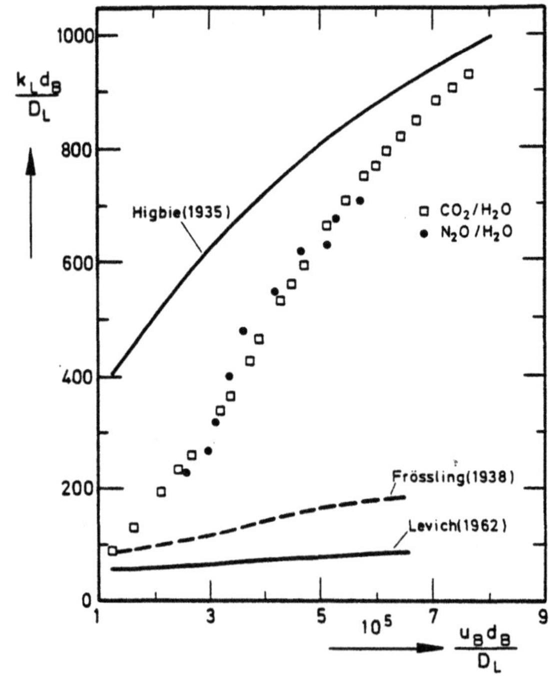

Fig. 6: Sherwood number vs. Peclet number

predictions of the model of Higbie (20)

$$\frac{k_L d_B}{D_L} = 1.13 \sqrt{\frac{u_B d_B}{D_L}} \tag{7}$$

where the contact time Θ is calculated from the bubble rise velocity u_B and the diameter d_B. The Higbie model i.e. eq. (7), is in accordance with the solution given by Boussinesq (21) for potential flow around a circulating sphere. If the bubbles are small the surface mobility is decreased and they behave as rigid spheres. Therefore, the k_L values of small bubbles approach the limiting solution given by Levich (22) and Frößling (23). The Levich solution

$$Sh = 0.997\ Pe^{1/3} \tag{8}$$

is valid if $Pe \gg 1$ and $Re < 1$.

For practical purposes and if only estimates are needed, the correlations provided by Calderbank and Moo-Young (24) can be recommended

$d_B < 2.5$ mm

$$k_L (Sc)^{2/3} = 0.31 \left(\frac{\Delta \rho u_L g}{\rho_L^2}\right)^{1/3} \tag{9}$$

$d_B > 2.5$ mm

$$k_L S_c^{1/2} = 0.42 \left(\frac{\Delta \rho u_L g}{\rho_L^2}\right)^{1/3} \tag{10}$$

Thus, provided it is known whether the bubble diameter is greater or smaller than 2.5 mm the k_L value can be calculated from physico-chemical properties alone.

Hallensleben (19) has shown recently that liquid-side mass transfer coefficients obtained from measurements with single bubbles apply with good accuracy to bubble swarms provided the bubbles do not interfere. This is the case if the gas-in-liquid dispersion is operated in the bubbly flow regime, i.e. at gas velocities less than about 5 cm/s. Therefore the models and correlations for single bubbles can be utilized to estimate k_L values in bubble columns at low gas flow rates, at least.

Empirical correlations for k_L in bubble columns have been developed by Hughmark (25) and Akita and Yoshida (26). The equation given by Hughmark is often a good compromise, while, according to our experience, Akita's and Yoshida's correlation predicts a rather conservative estimate of k_L.

3.3 Bubble Diameter

Together with the fractional gas holdup the bubble diameter (volume-to-surface mean value) decisively determines the gas-liquid interfacial area which is given by

$$a = \frac{6\,\varepsilon_G}{d_s} \quad . \tag{11}$$

In addition, the discussion of k_L for single bubbles has indicated the importance of the bubble diameter which essentially influences the constitution of the gas-liquid interface and the mean bubble rise velocity. Initial bubble size d_s generated from single-orifice spargers can be calculated from the empirical equation

$$\frac{d_s^*}{d_o} = 1.88\left(\frac{u_o}{gd_o}\right)^{1/3} \tag{12}$$

given by Akita and Yoshida (26). Here d_o is the orifice diameter and u_o the gas velocity in the orifice. The volume-to-surface mean bubble diameters d_s in gas-in-liquid dispersions can be calculated by a correlation developed by the same authors:

$$\frac{d_s}{d_c} = 26\left(\frac{gd_c^2\,\rho_L}{\sigma}\right)^{-0.5}\left(\frac{gd_c^3}{\nu_L}\right)^{-0.12}\left(\frac{u_G}{\sqrt{gd_c}}\right)^{-0.12} \tag{13}$$

Eq. (13) was established from experimental data with water, glycol (30 % and 100 %), methanol, and carbon tetrachloride in columns of 7.7, 15 and 30 cm diameter. Curiously, the correlation involves the column diameter giving $d_s \propto d_c^{-0.3}$ which is unexpected and probably due to the churn-turbulent flow.

Koide and coworkers (27) studied bubble sizes in coalescing media (water) and noncoalescing solutions of alcohols and electrolytes and proposed empirical correlations which involve modified Froude and Weber numbers. However, Schügerl et al. (16) have shown that the agreement is only sufficient for water and methanol solutions. Solutions of ethanol show large deviations which cannot be explained by surface tension veriation.

Comprehensive data on mass transfer and bubble sizes in columns equipped with various gas spargers have been reported by Schügerl et al. (16). These authors used liquid model media for simulating yeast fermentation, i.e. solutions of alcohols and glucose in the presence of salt mixtures typical in yeast fermentation. Due to the large variability of their findings, Schügerl et al. (16) did not develop empirical correlations but discussed their results with respect to coalescence promoting and hindering properties of the liquid media and with respect to the bubble size stability diagram derived by Berghmans (28). For a given Bond number defined by

$$Bd = \sqrt{\frac{\rho_L g d_B{}^2}{4\sigma}} \tag{14}$$

Berghmans' analysis gives a critical Weber number

$$We = \sqrt{\frac{\rho_L u_B{}^2 d_B}{2\sigma}} \tag{15}$$

which devides regions of stable and unstable bubble sizes as shown in Fig. 7. The coalescence properties are mainly dependent on the added salts and the organic substances present. Let us first consider the effect of salts. If a bubble is generated, the concentrations of electrolyte at the interface and in the bulk liquid are equal at first. The ions have the tendency to move away from the interface, giving an enrichment of water there accompanied by an increase in surface tension. Since the transport of ions in the bulk liquid requires some time, the coalescence hindering action is only pronounced at short residence times of the bubbles. Therefore large effects of added electrolytes on d_S (and hence on interfacial area a and volumetric mass transfer effects) can be observed in stirred vessels and multistage columns

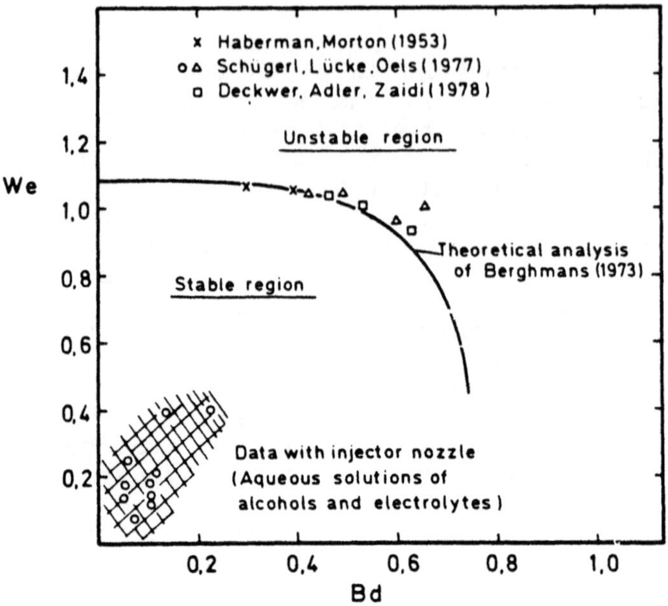

Fig. 7: Regions of stable and unstable bubble sizes

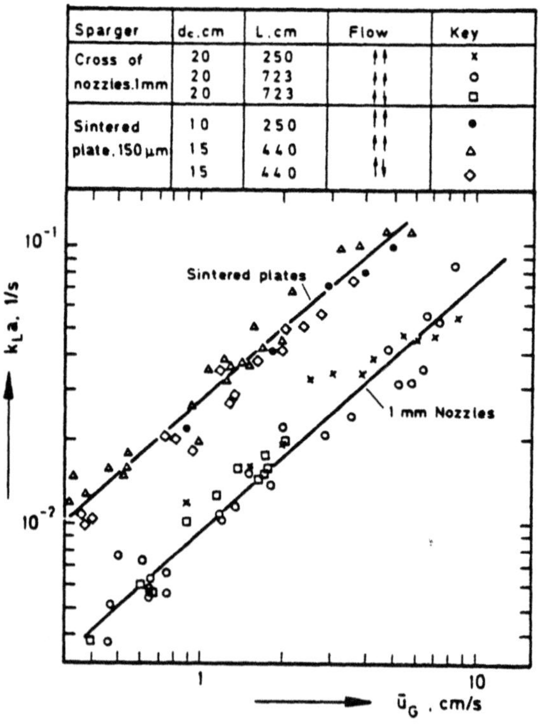

Fig. 8: Volumetric mass transfer coefficients in
tall bubble columns with different spargers

where the life expectancy of the individual bubbles is short. For instance, Zlokarnik (29) reports a 700 % increase in k_La in stirred aerated vessels. If, on the other hand, the bubble residence time is large, as in single-stage tower fermenters, the influence of added electrolytes is much less pronounced. Therefore, Deckwer, Adler, and Zaidi (30) did not find significant differences in bubble diameters for water and electrolytes in high bubble columns. Under such conditions the analysis of Berghmans should apply, and indeed measured bubble diameters are located in the vicinity of the curve for the dynamical equilibrium diameter (Schügerl et al. (16) (Fig. 7).

The addition of alcohols lowers the surface tension if these substances have enough time to accumulate at the surface. As they orient with their hydrophobic group towards the gas phase, the generated dipole layer suppresses coalescence. The effect largely depends on the alcohol: the influence of methanol is small, while that of ethanol is medium and propanol gives the largest effect. In addition, the gas distributor plays a major role. If sintered plates of high porosity and two-phase nozzle spargers of injector or ejector type are applied as spargers, small bubbles of low rise velocity are produced. Hence, there is enough time for alcohols (and other surfactants) to accumulate at the bubble interface. Therefore coalescence is largely suppressed and small bubbles are observed through the entire column. As shown in Fig. 7, the bubble sizes are in the stable region but considerably below the curve corresponding to dynamic equilibrium between bubble coalescence and break-up.

The above considerations and the conclusions of Schügerl et al. (16) and of Voigt and Schügerl (31) present some useful guidelines for producing small bubble diameters in gas-liquid processes. Usually small bubbles are desirable as they yield large interfacial areas, and this effect usually overbalances the decrease in k_L with decreasing diameter. The findings of Schügerl et al. (16) with synthetic fermentation media and the guidelines drawn therefrom could successfully be applied to explain qualitatively, at least, observed bubble sizes and mass transfer properties in yeast fermentations (7,9,10)

3.4 Volumetric Mass Transfer Coefficients

Although it is most useful to know the influence of various physical and operating variables on the indiv dual values of k_L and a the model equations (1) to (4) of a biological reactor involve only k_La since an absorp tion enhancement caused by fast reaction in the liquid film can be neglected. However, the possibility of absor tion enhancement has been discussed controversely at the 5th International Fermentation Symposium (32-35). Theoretical considerations (36,37) indicate that any appreciable absorption enhancement cannot be expected in biological reactors of practical importance. New experimental results with enzyme catalyzed absorption reaction systems (38) did not confirm the two-zone model of mass transfer postulated by Tsao (39) for biological reaction systems.

With respect to oxygen transfer the volumetric mass transfer coefficient k_La is the most important design parameter and many data are available now. The oper ating variable which strongly affects k_La in gas-liquid dispersions is the gas flow rate. Kastanek (40) used Hig bie's penetration theory, and by substituting the contac time by an expression derived from Kolmogoroff's theory of isotropic turbulence (41) he arrived at the following equation for k_La

$$k_La \propto \left(\frac{u_G(u_G+c)}{2u_G+c}\right)^{13/20} \quad (16$$

which can be reduced to

$$k_La = bu_G^{\,n} \quad (17$$

where the exponent n may vary from 0.8 to 1.2. From his own measurements and those of Deckwer, Burckhart and Zol (42) Kastanek recommended a value of n of about 0.8 for water and electrolyte solutions if the gas velocity is less than 25 cm/s. The exponent is not (40,42) or only little (45) affected by liquid flow rate and the kind of gas sparger. However, the constant b is largely dependent on the sparger design and liquid media (43), Schügerl et al. (16). Fig. 8 shows k_La data in tall bubb columns for tap water and cocurrent and countercurrent flow using porous plates and orifice spargers. The corre lation lines for the two sparger types are practically i parallel (same n) but the k_La data in the column with orous sintered plates are considerably larger than the nes measured with orifice spargers even though the colmns are high and one should expect the gas sparger inluence to be negligible. The strong influence of the parger design on k_La has been confirmed by other inestigations (16).

It should be pointed out that in the vicinity of ne gas spargers of porous and plate type higher turbuence intensities prevail and therefore higher k_La values re observed. This was shown for the absorption of CO_2 30,44). Recently, increased k_La values near the sparger ave been reported for oxygen transfer as well (45). nen describing oxygen concentration profiles measured uring yeast fermentations in a bubble column Buchholz 7) also introduced a coalescence function which accounts or the decrease in k_La with increasing distance from the parger.

In industry, porous plates are commonly not used nough the achievable mass transfer rates are favorable. ne gas is either sparged by single or multiorifice disributors or by two-component nozzles of various types. or the case of the less effective single and multiorice spargers the correlation of Akita and Yoshida (15) an be recommended. This correlation which was estabshed for oxygen transfer in water and various aqueous olutions is given by

$$
{}_La\,\frac{d_c^{\,2}}{D_L} = 0.6\,\varepsilon_G^{\,1.1}\left(\frac{\nu_L}{D_L}\right)^{0.5}\left(\frac{g d_c^{\,2}\rho_L}{\sigma}\right)^{0.62}\left(\frac{g d_c^{\,3}}{\nu_L^{\,2}}\right)^{0.31} \qquad (18)
$$

in dimensionless numbers

$$
Sh\,(ad_c) = 0.6\,\varepsilon_G^{\,1.1} Sc^{0.5} Bd^{0.62} Ga^{0.31} \qquad (19)
$$

Fig. 9 demonstrates that the empirical equation of ita and Yoshida (15) describes fairly well k_La data asured by various authors for equipment of varying zes. Of course, in large diameter reactors the influence the diameter levels off, therefore the calculation for e 1 m and 5.5 m columns were done with d_c = 0.6 m as oposed by Kataoke and coworkers (46).

Numerous k_La data of aqueous solutions of alcohols d glucose with and without inorganic electrolytes in bble columns equipped with various spargers (porous

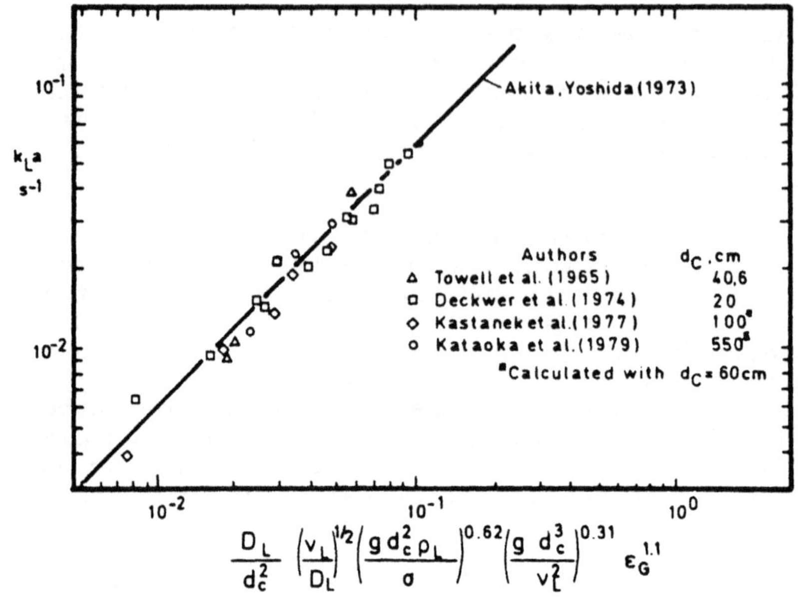

Fig. 9: k_La values plotted as to the correlation of Akita and Yoshida (1973)

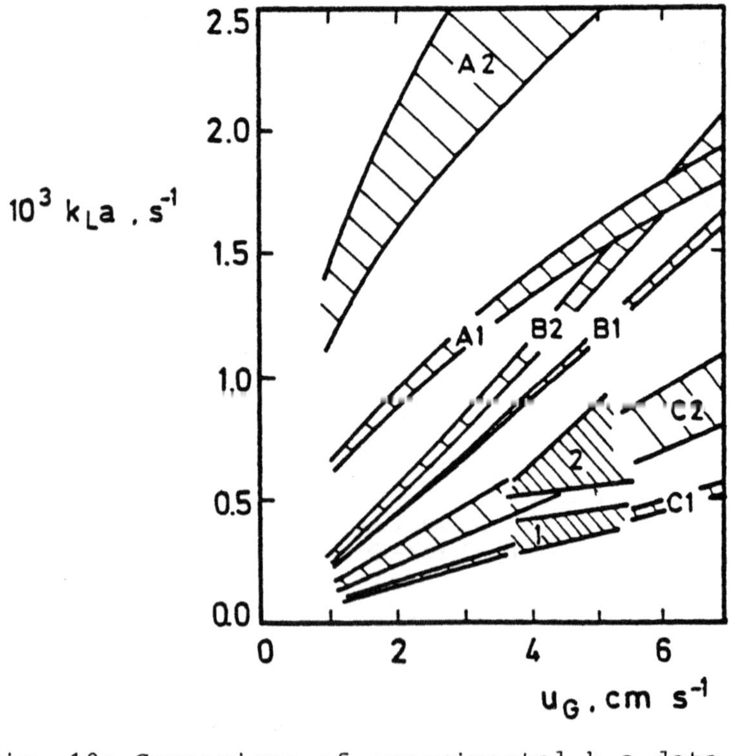

Fig. 10: Comparison of experimental k_La data in Penicillium chrysogenum (1,2) with correlations of Henzler (A), Nakanoh and Yoshida (B), and Schumpe et al. (C)

›lates and multiorifice plates) have been published by ;chügerl et al. (16). Only few of these findings are in ıgreement with the equation of Akita and Yoshida. In adition, results in tall bubble columns with porous ;pargers are considerably larger than the predictions •f this correlation (42). The various experimental re-ults can be interpreted qualitatively by assuming that :oalescence affected properties of the studied liquid ıedia, an effect which was already mentioned above in :he discussion of bubble diameters. The findings lead :o reasonable guidelines for estimating the effect of 'arious substances on k_La which are in general accordance 'ith the results reported for aerated vessels (29) and .arger tower reactors (3).

Nakanoh and Yoshida (47) studied oxygen transfer with •oth Newtonian as well as non-Newtonian liquids (aqueous olutions of sucrose, sodium carboxylmethyl cellulose CMC) 0.3 - 1 %, and of sodium polyacrylate (PA), 0.01 - •.1 % in a 14.5 cm diameter bubble column of 190 cm in .eight. For gas velocities less than 10 cm/s the authors orrelate all their data by the following equation

$$k_L a' \frac{d_c^2}{D_L} = 0.09\ Sc^{1/2} Bd^{0.75} Ga^{0.39} Fr (1 + c\ De^m)^{-1} \qquad (20)$$

here for unelastic liquids c = 0, and for elastic liuids c = 0.13 and m = 0.55. The Deborah number De acounts for the elastic properties of PA solutions. The pparent viscosity applied in eq. (20) was obtained from he viscosity vs. shear rate curve, and the shear rate as calculated from

$$\gamma = 50\ u_G \qquad (21)$$

γ in s^{-1}) and u_G in cm/s) which was proposed by Nishika-a et al. (48) from heat transfer measurements in bubble olumns with non-Newtonian liquids.

Volumetric mass transfer coefficients in CMC solu-ions were also reported by Buchholz et al (48) for a ingle-stage bubble column (14 cm ID by 3.9 m height) and y Voigt et al. for a multistage bubble column. k_La data or PAA solutions can be found by Hecht et al. (50). The esults with the CMC solutions were correlated by Henz-er (51). This correlation predicts an unusually low de-endency on u_G, i.e.

$$k_La \propto u_G^{0.1} \tag{2}$$

and the predictions differ largely from eq. (20). Ther fore, oxygen mass transfer in CMC solutions was remeasured in a 14 cm ID bubble column equipped with differ spargers by Nguyen-tien (52). The course of the measur k_La data in dependence of the gas throughput is comple and similar to that of the gas holdup, see Fig. 4 (18) However, in the most important flow regime namely slug flow all measured data can be simply correlated by (18

$$k_La = 2{,}08\times10^{-4}\ \bar{u}_G^{0.59}\ \mu_{eff}^{-0.84} \tag{2}$$

(s^{-1}) (cm/s) (Pas)

The effective dynamic viscosity in CMC solutions calculated by the procedure outlined by Nishikawa et a (48). Use of eq. (21) and the viscosity vs. shear rate curve gives

$$k_La = 2.08\times10^{-4}\bar{u}_G^{0.59}\left(K(50\ \bar{u}_G)^{n-1}\right)^{-0.84} \tag{2}$$

Eq. (23) describes the measured data of Nguyen-tien (5 with striking agreement. König (10) has measured k_La values in fermentation broths of Penicillium chrysogen and also reported their rheological properties. Though the fermentation broths show a larger deviation from N tonian behavior than the CMC solution used by Nguyen-t Fig. 10 shows that eq. (23) describes the data fields König (10) with excellent agreement while the correlat of Nakanoh and Yoshida (47) and above all that of Henz (51) deviate considerably.

More effective gas distributors than single and m tiorifice spargers (i.e. perforated plates) are porous plates and nozzles of ejector and injector type. This demonstrated in Fig. 11 where, as an example, measured interfacial areas for various spargers are plotted vs. gas velocity for 1 % ethanol solution (16). The interfa cial area achievable with two-component nozzles is as large as 50 cm^{-1} while with methanol (1 % solution) va of a in the range of 20 cm^{-1} can be obtained. Interfac areas generated from various nozzles have been termined by Nagel and coworkers (53,54) who used the s fite oxydation system and correlated their data with t energy input. The authors presented useful design char from which the interfacial areas generated in differen

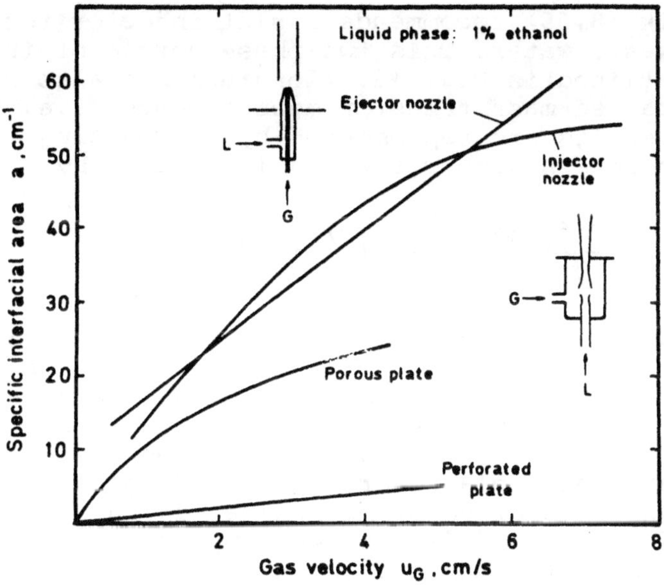

ig. 11: Interfacial areas generated by different spargers

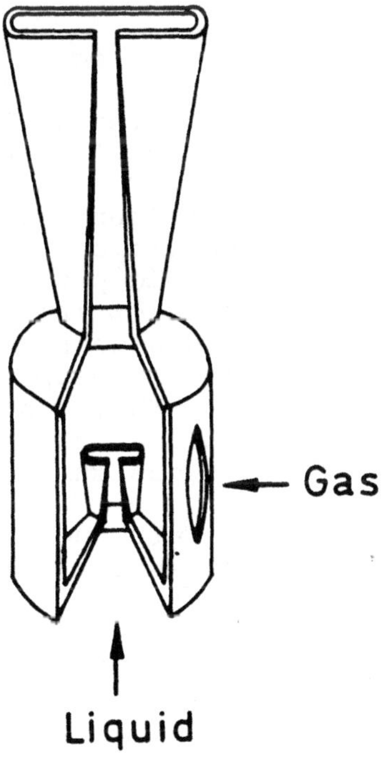

Fig. 12: Slot injector

gas-liquid contacting apparatuses can be obtained.

Zlokarnik (3,55) recommends a slot injector for the aeration of waste water. This two-phase nozzle of injector type is outlined in Fig. 12. Zlokarnik has also reported on detailed mass transfer measurements, i.e. k_La values. The results are represented by introducing a dimensionless sorption number Y which is defined by

$$Y = \frac{G}{Lq\,\Delta c}\left(\frac{\nu^2}{g}\right)^{1/3} = \frac{k_La}{u_G}\left(\frac{\nu^2}{g}\right)^{1/3} \tag{25}$$

where G is the oxygen mass flow rate through the interface (kgO_2/s), L the contactor height (m), q the gas throughput (m^3/s), and Δc the mean driving concentration difference. If one assumes as a rough approximation $k_La \propto u_G$, then Y should be constant for a given system if porous and perforated plates are used to disperse the gas. If the gas is sparged by mechanical agitation or by the energy of the liquid jet in two-component nozzles Y is a function of the dispersion number X which is given by

$$X = \frac{P_L/q}{\rho(\nu g)^{2/3}} \tag{26}$$

In this equation P_L is the power of the liquid jet. The sorption characteristics of the slot injector for different directions are represented in Fig. 13. High values of Y and hence k_La are obtained if the slot injector is directed with an angle of 35 °C towards the bottom. As pointed out by Zlokarnik (3) bubble coalescence is strongly suppressed under this condition and the energy of the jet utilized for the bubble production is highly preserved. A comparison of the slot injector with other volume aerators of various design shows that the slot injector is a very efficient nozzle sparger reaching efficiencies of about 3.8 kg O_2/kWh if the liquid height L is the range of 10-18 m (55). At such liquid heights the O_2 concentration in the off gas can be reduced to 4 %.

The sorption number Y introduced by Zlokarnik (3) and defined by eq. (25) presents a reasonable measure to characterize mass transfer properties of tower bioreacto[rs]. Table 1 summarizes experimental studies in bubble columns of various geometry and gives sorption numbers calculated for u_G = 5 cm/s. Besides physico-chemical properties it is particularly the gas distributor which affects the orption number. This can clearly be discerned from Table . For single and multiorifice (perforated plates, spider ype) sparger with hole diameters of 1 mm or larger Y aries from 2.4 to 5.4 the mean value being about 4. In he case of porous plate distributors Y is about 10 for ater but for noncoalescing media considerably larger alues can be obtained, especially, if the pore diameter s small. The high k_La values and hence sorption numbers ttainable in such liquids can be entirely attributed to nall bubble sizes, see above discussion on bubble diaeter. High sorption numbers can be obtained by applicaion of two-phase nozzles, of course, where additional nergy is introduced by the liquid jet.

Though porous spargers give favorable sorption umbers they are not used in industrial applications beause of their tendency for blocking. In addition, porous lates are rather limited in their diameter and do not ead to gas-in-liquid dispersions which are uniform over he entire reactor volume. In industry, multiorifice pargers and recently two-phase nozzles of injector and ector type are always preferred. The latter spargers eem to provide for a rather uniform gas distribution and ield high sorption numbers even in coalescing media ke tap water. Though these two-phase nozzle spargers equire additional energy for the liquid jet their appliation in large scale equipment like biological waste ater treatment appears very feasible (3,54,55).

LIQUID-SOLID MASS TRANSFER

If respiring biomass particles are relatively large ass transfer resistances from the bulk fermentation oth to the outer surface of the biomass particles may ccur. In their paper on physical rate processes in inustrial fermentation, Calderbank and Jones (58) have ported on liquid-solid mass transfer coefficients k_S th ionic resin beads suspended in a stirred vessel. ey recommend the following equation

$$\frac{k_s d_p}{D_L} = 2 + 0.31 \left(\frac{d_p{}^3 \Delta \rho g}{\mu_L D_L}\right)^{1/3} \tag{27}$$

$$Sh = 2 + 0.31\ Ra^{1/3} \tag{28}$$

If $Sh \gg 2$ eq. (28) reduces to eq. (9) proposed by

TABLE 1: Typical experimental values of sorption number for various columns (evaluated for $u_G = 5$ cm/s)

Gas sparger (mean orifice or pore diameter in mm)	Column sizes d_c, cm	Column sizes L, cm	Liquid system	Sorption number 10^5 Y	Reference
Multiorifice, 30	550	900	Tap water	2.4	Kataoke and coworkers (46)
Multiorifice, 1.6	100	120	Sulfite soln	3.7	Kastanek, Kratochvil, Rylek (56)
Single and multiorifice, 3.8 - 8.6	7.6 183, 762	600-1600	Tap water	5.4	Jackson, Shen (57)
Multiorifice, 1	20	720	Tap water Salt solns	3.6 4.3	Deckwer, Burckhart, Zoll (42)
Multiorifice, 0.5 (perforated plate, 180 bores)	14	400	Demineralized water solns of alcohols (0.5-2%) and salts	8.7 10.5	Schügerl, Lücke, Oels (16)
Porous plate, 0.150	15	440	Tap water Salt solns	3.6 15.4	Deckwer, Burckhart, Zoll (42)
Porous plate, 0.0175	14	400	Demineralized water solns of alcohols (0.5-2%) and salts	11.2 >33.0	Schügerl, Lücke, Oels (16)
Two-phase nozzle	30	230	Sulfite soln	9 - 18*	Nagel, Hegner, Kürten (54)
Slot injector (two-phase)	280	700	Tap water	8 - 16*	Zlokarnik (3, 55)

* Dependent on energy provided by liquid jet

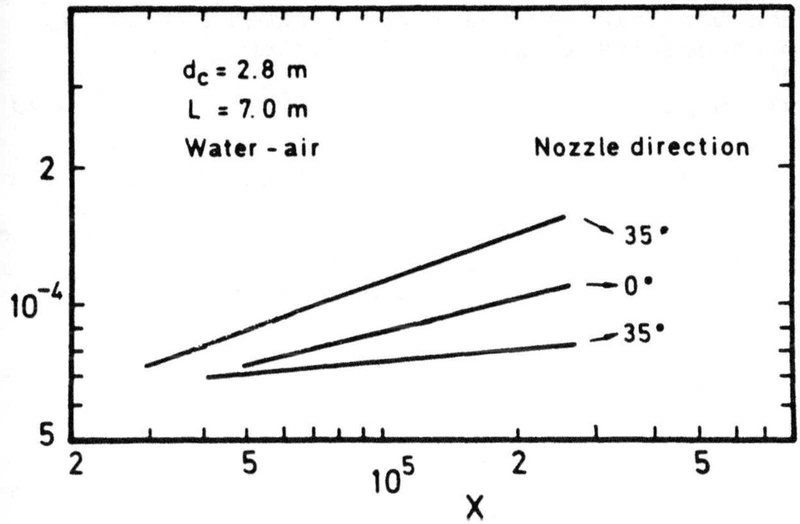

g. 13: Sorption characteristics of slot injector Zlokarnik (1979)

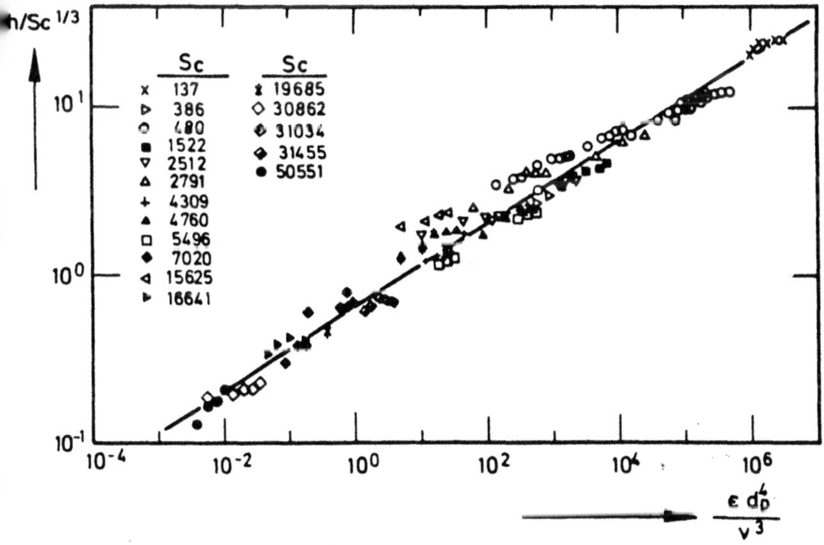

g. 14: Correlation of liquid-solid mass transfer coefficients

Calderbank and Moo-Young (24) for gas-liquid mass tran fer from small rising bubbles. If the density difference between the particle and the fluid becomes nil the k_s value approaches its lower limiting value

$$k_s = 2\, D_L/d_p \,. \quad (29)$$

According to Calderbank and Jones (58) the liquid-solid transfer depended only on physico-chemical properties did not consider the influence of hydrodynamics, impelle speed, and geometry. Other investigations (see, for instance, Boon-Long et al. (59))have shown that such parameters may not be negligible.

In general, liquid-solid mass transfer coefficients are correlated by expressions like

$$\frac{k_s d_p}{D_L} = 2 + \alpha \left(\frac{\nu}{D}\right)^n \left(\frac{u_s d_p}{\nu}\right)^m \quad (30)$$

or

$$Sh = 2 + \alpha Sc^n Re^m \quad (31)$$

The slip velocity u_s is usually difficult to estim (60). Therefore, it is common practice now to compute the Reynolds number on the basis of Kolmogoroff's theory which gives

$$Re = c \left(\frac{\epsilon d_p^4}{\nu^3}\right)^p \quad (32)$$

Here the exponent p is dependent on the ratio of the particle size to the microscale of the eddies. In the case of tower bioreactors, the energy dissipation rate per unit mass ϵ can simply be calculated from

$$\epsilon = u_G g \,. \quad (33)$$

While many studies on k_s in the two-phase system liquid-solid were carried out only few have been reported for aerated suspensions in bubble columns (61,62). Most recently, liquid-solid mass transfer on suspended ionic resin beads was investigated by Sänger and Deckwer (63) in a bubble column. Aqueous solutions of polyethylene glycol (PEG) of various molecular weight were applied. With PEG solutions which show Newtonian flow behavior he viscosities the diffusivities decreased only slightly (64). Owing to this behavior and the small density ifference between the resin beads and the liquid (i.e. .1 to 0.2 g/cm^3) the results are thought to be particuarly relevant to fermentations. Sänger and Deckwer (63) orrelated their findings by

$$Sh = 2 + 0.545\ Sc^{1/3} \left(\frac{\epsilon d_p^4}{V^3}\right)^{0.264} \qquad (34)$$

Figure 14 shows that eq. (34) describes the measured ata fairly well. The proposed correlation is also in easonable agreement with the results reported by Sano t al. (62).

The relative importance of mass transfer resistances t the liquid-biomass interface (i.e. $1/k_s a_s$) depends ainly on the size of the biomass particles. For instance, n yeast fermentations the particles are in the µm range 1-3 µm). One can show that for typical fermentation conitions (7) $k_s a_s \gg k_L a$ hence the oxygen transfer resistance s located completely at the gas-liquid interface. On the ther hand, König (10) carried out fermentations of Penicillium chrysogenum in bubble columns and found, under pecial conditions, biomass pellets of 0.3 to 2 mm diaeter. Here oxygen mass transfer at the liquid-solid inerface (and possibly pore diffusion) should be considered major resistance since $k_s a_s \ll k_L a$.

MIXING

The global mixing effects in tower bioreactors can onveniently be described by the dispersion coefficients both phases which are defined in analogy to Fick's w for diffusive transport.

Dispersion in liquid phase has been the subject of ny investigations which were recently reviewed by ah et al. (65). In general, the liquid phase dispersion epends mainly on gas velocity and column diameter. The erall liquid flow rate, the kind of gas sparger, and so physico-chemical properties like viscosity and denty do not significantly effect liquid phase dispersion. e majority of all the measured data in various bubble lumns can be described empirically by a correlation ven by Deckwer et al. (42)

$$E_L = 2.7\ d_c^{1.4}\ \bar{u}_G^{0.3} \tag{35}$$

Here E_L is in cm^2/s, d_c in cm and u_G in cm/s. A dimensionless form of eq. (35) under consideration of additional experimental data was recently proposed (66)

$$\frac{u_G d_c}{E_L} = 2.83\ \left(\frac{u_G^2}{g d_c}\right)^{0.34} \tag{36}$$

or

$$Pe_L = 2.83 \times Fr^{0.34} \tag{37}$$

Baird and Rice (67) have given a theoretical interpretation of dispersion in multiphase reactors on the basis of Kolmogoroff's theory of isotropic turbulence. Their results agree with eq. (3) only the exponent of Fr being 1/3. The predictions of eq. (37) also agree with the empirical correlations proposed by Kato and Nishiwaki (68) and Akita (69).

Joshi and Sharma (70) have developed a circulation cell model for bubble columns. This model gives a circulation velocity u_c which correlates well measured dispersion coefficients. The expression of Joshi and Sharma (70) is given by

$$E_L = 0.31\ u_c\ d_c \tag{38}$$

where u_c follows from

$$u_c = 1.31\ \left[g d_c (u_G - \varepsilon_G u_{B\infty})\right]^{1/3} \tag{39}$$

Eq. (38) also gives a good description of experimental results and shows approximately the same dependencies on u_G and d_c as eq. (35) and (36).

An unusual dispersion behavior was observed by König et al.(71) who reported on measurements in bubble columns equipped with porous spargers and employing diluted solutions of alcohols as liquid phase. The dependency of E_L on the gas velocity revealed steep changes and passed through maximum values. The authors interpreted their results as being affected by bubble size distribution, i.e. the Sauter diameter of the bubbles, which were measured simultaneously. Figure 15 shows E_L as a function of d_S. At very small diameters the entire liquid is attached to

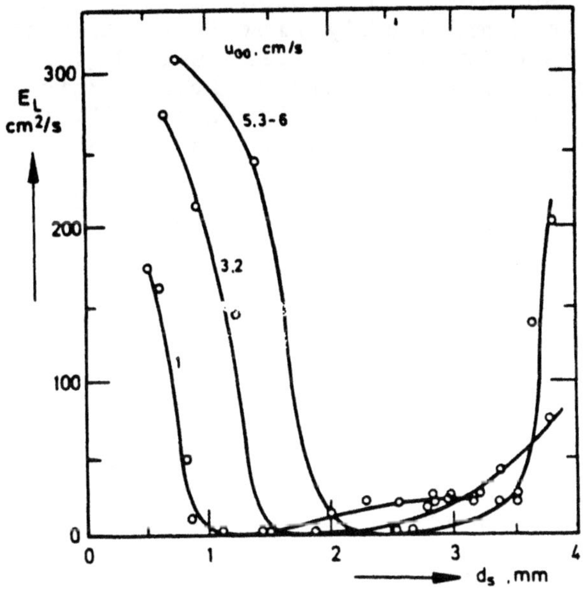

ig.15: E_L vs. d_S in diluted alcohol solutions, porous sparger (König and coworkers, 1978)

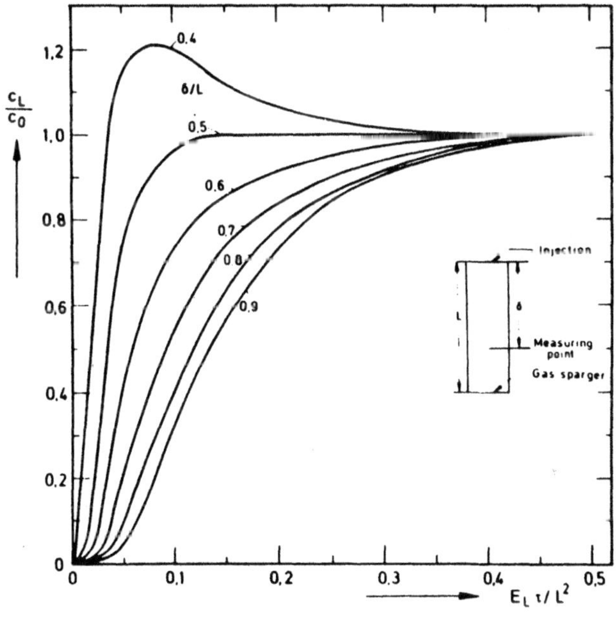

ig. 16: Homogeneity vs. mixing time

the gas-liquid interface and carried upwards, which cause an underpressure and leads to violent eddies. At medium values of d_S (1 to 3 mm) bubbly flow is assumed to prevail, which yields extremely low values of E_L. For larger bubbles the flow is churn-turbulent giving again high values of E_L. The findings of König et al (71) indicate that the structure of the gas-in-liquid dispersion may play an important role in the mixing behavior. However, the interdependence appears rather complex and cannot yet be fully understood.

The influence of viscosity on E_L was studied by Hikita and Kikukawa(73). Only a small effect was found, i.e. $E_L \alpha \mu^{-0.12}$. Liquid phase dispersion coefficients in aerated non-Newtonian media were not yet measured directly. When matching experimental profiles of liquid phase oxygen concentrations with the predictions of the dispersion model, Schumpe et al. (18) obtained higher E_L values than for low viscous media. However, these preliminary results need further clarification by direct measurements

In fermentation technology it is the mixing time rather than the dispersion coefficient E_L which is used to characterize the global mixing effects. The mixing time can be obtained from the transient solution of the dispersion model and is usually defined for 90 % homogeneity. In case a tracer is injected at the column top or bottom the solution of Ohki and Inoue (72) can be used. This solution is plotted for various values of δ/L (dimensionless distance between injection and measuring point) in Fig. 16. For a desired homogeneity and a given value of δ/L the value of

$$a(c/c_o, \delta/L) = \frac{E_L \tau}{L^2} \tag{40}$$

can be read from the abscissa. Hence, by introducing the correlation for E_L, i.e. eq. (36) the mixing time is given by

$$\tau = \frac{a(c/c_o, \delta/L)}{3.676} \frac{L^2}{d_c^{1.34} u_G^{0.32}} \tag{41}$$

In Fig.17, τ is plotted vs. the tower diameter for three different heights L. It is particularly the slenderness (L/d_C) which largely influences the mixing time.

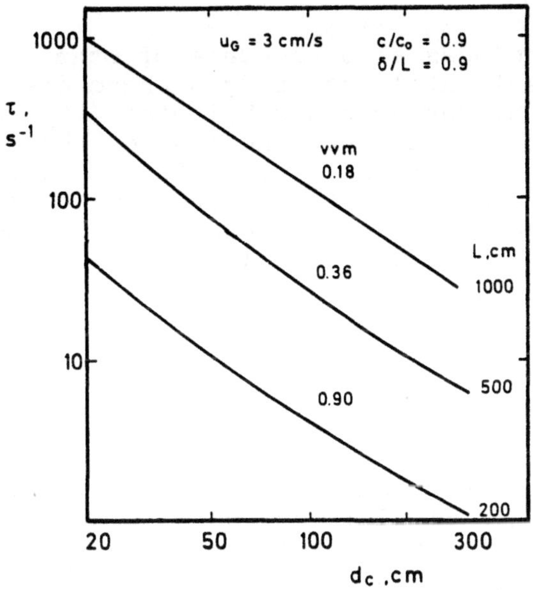

.g. 17: Mixing time vs. tower diameter

Owing to slight density differences between fermenta-
on liquid and biomass the particles have the tendency to ttle. Thus a biomass concentration profile along the wer may result. The pertinent model to account for bio- ss concentration profiles is the sedimentation-disper- on model (74,75). This model involves two parameters, mely, the solid dispersion coefficient E_S and the mean ttling velocity u_S of the biomass particles in the swarm. th parameters were determined by Kato et al. (75) in bble columns for glass beads of 75 and 163 µm diameter. e authors presented their results by empirical cor- lations for both E_S and u_S. Until other data for smaller nsity differences are available the application of the rrelations of Kato and coworkers is recommended for ological systems also. It should be pointed out that the lid phase dispersion coefficient E_S almost completely rees with E_L, i.e. the liquid phase dispersion coeffi- ent.

In contrast to liquid phase dispersion coefficients, ta on gas phase dispersion are sparse, and, in general, e measurements reveal considerable scatter. Towell and kerman (76) proposed the following empirical equation r the gas phase dispersion coefficient

$$E_G = 0.2\ d_c^{\ 2} u_G \qquad (42$$

This correlation includes also the data of other author Most recently, experimental data on gas phase dispersio have been reported by Mangartz and Pilhofer (79). On th basis of their findings with various liquids these auth conclude that the bubble rise velocity in the swarm ($u_G = u_G^* / \varepsilon_G$) is a characteristic variable which mainly influences gas phase dispersion. Mangartz and Pilhofer (79) recommend the subsequent correlation:

$$E_G = 5 \times 10^{-4}\ u_G^{*3}\ d_c^{\ 1.5} \qquad (43$$

Though gas phase dispersion coefficients are high, and often considerably larger than those of the liquid phase the impact of gas phase dispersion on reactor per formance was seldom taken into account (80), particular no experimental or numerical investigations have been r ported for biological reactors though, owing to their usually large scale, gas phase dispersion may be of sig nificant influence.

6 HEAT TRANSFER

Fermentation processes are accompanied by heat effects. Luong and Volesky (81) measured heat generation 12 aerobic fermentation systems. They correlated the heat production with the oxygen uptake rate (OUR) and p posed

$$Q_{ferm} = 0.465\ (MJ/mol\ O_2) \times OUR$$

where OUR is in mol O_2/m^3 h. One can estimate from this figure that heat generation in bioprocesses has the sam order of magnitude as chemical reactions. For instance, in 3000 m^3 fermenter of the Hoechst-Uhde SCP process th UOR is 10 kg O_2/m^3h giving a heat generation of 4.36x10 kJ/h. One has to consider that in case of bioprocesses the heat is deliberated in vessels which are larger by one or two orders of magnitude than conventional chemic reactors. In addition, the temperature level is low, therefore latent heats cannot be used. Owing to the low temperature difference between reaction and cooling med the heat transfer area must be large. Therefore the cal culation of heat transfer coefficients is important for biological processes also.

Heat transfer coefficients in two-phase and three-
ıase (i.e. slurry) column reactors were recently re-
iewed by Deckwer (82) and Deckwer et al.(83). The avail-
ɔle data can be excellently described on the basis of a
ıeoretical model which gives (82)

$$\frac{h}{\rho c_p u_G} = 0.1 \left[\frac{u_G^3}{\nu g} \left(\frac{\nu \rho c_p}{k}\right)^2\right]^{-1/4} \quad (44)$$

ıd in dimensionless numbers

$$St = 0.1\ (Re\ Fr\ Pr^2)^{-1/4} \quad (45)$$

Eq. (45) is also in good agreement with the findings
ɛported by Nishikawa, Kato and Hashimoto (84) for low
ıd higher viscous Newtonian media. These authors also
ɛasured heat transfer coefficients in CMC solutions.
' assuming that heat transfer data in aerated Newtonian
ıd non-Newtonian fluids follow the same dependencies
.shikawa, Kato, and Hashimoto (84) have proposed valuable
ɔrrelations for the average shear rate as a function of
ıe gas velocity. With the shear rate known, the effective
.scosity of non-Newtonian media in the tower reactor can
ː obtained from the shear stress vs. shear rate curve.

In general, use of eq. (45) can be recommended for
.bble column bioreactors. However, one should consider
.e possibility of slime formation and adherence of bio-
.ss particles at the heat transfer area which might re-
.ce considerably heat transfer.

SUMMARY

From the lot of experimental data on physical trans-
rt parameters in aerated nonagitated tower bioreactors
number of correlations have been developed. These are
ought to give rather reliable estimates for the dis-
rsion parameters, the heat transfer coefficient and the
quid-solid mass transfer coefficients. Gas-liquid mass
ansfer data are particularly sensitive to the composi-
on of the liquid media and the sparger design, i.e.
e initial state of the gas-in-liquid dispersion. Corre-
tions for k_La are based on data from water and aqueous
lutions of simple composition. In simulated nutrition
d fermentation media large variations in k_La can be ob-
rved which are not yet describable by correlations of
oader applicability. However, some guidelines to esti-

mate the qualitative behavior have been proposed.

Future investigations should place emphasis on measurements of physical transport parameters under "in sit" conditions, i.e. during fermentations. Under such circum stances only few data are available.

OTATION

specific gas-liquid interfacial area referred to dispersion volume

' specific gas-liquid interfacial area referred to liquid volume

s specific liquid-solid interfacial area referred to volume of nonaerated suspension

constant in eq. (17)

d Bond number, eq. (14) or $gd_c^2\rho_L/\sigma$

constant in eq. (16)

concentration

equilibrium solubility in liquid phase

L heat capacity of gas-free liquid or suspension

p diffusivity

effective diffusivity in biomass

eff

e Deborah number, $u_B/(\lambda d_s)$

mean bubble diameter

B column diameter

c orifice diameter

o particle diameter

p Sauter (volume-to-surface) diameter

s initial Sauter diameter

s dispersion coefficient

r Froude number, $u_G/(gd_c)^{1/2}$ or $u_G^2/(gd_c)$

oxygen mass flow rate through interface

a Galilei number, gd_c^3/ν_L^2

gravitational acceleration

heat transfer coefficient

thermal conductivity

gas-liquid mass transfer coefficient

L liquid-solid mass transfer coefficient

s tower length

power of liquid jet

L Peclet number, u_Bd_B/D_L

e modified Peclet number, u_Gd_c/E_L

L Prandtl number, $\nu\rho c_p/k$

r gas throughput

, R' general rate terms

a Raleigh number, $d_p^3\Delta\rho g/(\mu_LD_L)$

Re	Reynolds number, $u_s d_p / \nu$ or eq. (32)
Sc	Schmidt number, ν_L / D_L
Sh	Sherwood number, $k_L d_B / D_L$ or $k_s d_p / D_L$
u_G	superficial gas velocity
u_B	bubble rise velocity
u_o	gas velocity in orifice
u_s	slip or settling velocity in particle swarm
We	Weber number, eq. (15)
X	power number, eq. (26)
x	axial coordinate
Y	sorption number, eq. (25)

Greek

α	constant
γ	shear rate
ε	fractional holdup
ϵ	energy dissipation rate, eq. (33)
λ	relaxation time
μ	dynamic viscosity
ν	kinematic viscosity
$\bar{\nu}$	effective (apparent) kinematic viscosity in non-Newtonian media
σ	surface tension
τ	mixing time

Indices

B	biomass or bubble
G	gas
L	liquid
S	surface of solid

1. Bayer Turmbiologie (Bayer Tower Biology) (1978)
yer Prospects D 991-7127/89 7797 and E 589-777/68 619
2. Leistner, G., G. Müller, G. Sell and A. Bauer.
em.-Ing.-Tech., 51 (1979) 288
3. Zlokarnik, M. Chem. Eng. Sci., 34 (1979) 1265
4. Smith, E.L. and R.N. Greenshields. Chem. Engng.
nuary 1974 (1974) 28
5. Schügerl, K. Chem.-Ing.-Tech. 49 (1977) 605
6. Schügerl, K., J. Lücke, J. Lehmann and F. Wagner.
v. Biochem. Eng. 8 (1978) 63
7. Buchholz, H. Dr. thesis, University of Hanover
RG) (1979)
8. Buchholz, H., R. Luttmann, W. Zakrzewski and K.
hügerl. Chem. Eng. Sci. 35 (1980) 111
9. Voigt, J. Dr. thesis, University of Hanover (FRG)
980)
10. König, B. Dr. thesis, University of Hanover (FRG)
980)
11. Katinger, H.W.D., W. Scheirer and E. Kröner (1979).
r. Chem. Eng. 2 (1979) 31
12. Kobayashi, T., G. van Dedem and M. Moo-Young.
otechn. Bioengng. 15 (1973) 27
13. Atkinson, B. "Biochemical Reactors," chapter 4,
974), Pion Ltd., London
14. Bailey, J.E. and D.F. Ollis. "Biochemical Engi-
ering Fundamentals," chapter 7, (1977), McGraw-Hill
. New York
15. Akita, K. and F. Yoshida. Ind. Eng. Chem. Proc.
s. Dev. 12 (1973) 76
16. Schügerl, K., J. Lücke and U. Oels. Adv. Biochem.
g. 7 (1977) 1
17. Schumpe, A. and W.-D. Deckwer. Paper presented
Int. Symp. on "Mass Transfer with Chemical Reaction
Two-Phase Systems" ACS meeting, March 29 - April 3,
81, Atlanta, USA
18. Schumpe, A., K. Nguyen-tien and W.-D. Deckwer.
em.-Ing.-Tech., submitted
19. Hallensleben, J. Dr. thesis, University of Hanover
RG) (1980). Chem.-Ing.-Tech., to be published
20. Higbie, R. Trans. AIChE 31 (1935) 365
21. Boussinesq, J. J. Mathematiques, 6e serie 1
905) 285
22. Levich, V.G. "Physicochemical Hydrodynamics," Pren-
ce Hall, Englewood Cliffs (1962)
23. Frössling, N. Beitr. Geophys. 32 (1938) 170
24. Calderbank, P.H. and M.B. Moo-Young. Chem. Eng.
. 16 (1961) 39
25. Hughmark, G.A. Ind. Eng. Chem. Proc. Des. Dev. 6
967) 218

26. Akita, K. and F. Yoshida. Ind. Eng. Chem. Proc. Des. Dev. 13 (1974) 84
27. Koide, K., K. Kato, Y. Tanaka and H. Kubota. J. Chem. Eng. Japan 1 (1968) 51
28. Berghmans, J. Chem. Eng. Sci. 28 (1973) 2005
29. Zlokarnik, M. Adv. Biochem. Eng. 8 (1978) 133
30. Deckwer, W.-D., I. Adler and A. Zaidi. Can. J. Chem. Eng. 56 (1978) 43
31. Voigt, J. and K. Schügerl. Chem. Eng. Sci. 34 (1979) 1221
32. Tsao, G.T. Abstracts of papers, Session 4, 5th Int. Ferm. Symp., Berlin (1976)
33. Tsao, G.T. Biotechn. Bioeng. 19 (1977) 557
34. Yoshida, F. and H. Yagi. Biotechn. Bioeng. 19 (1977) 561
35. Linek, V. and P. Benes. Biotechn. Bioeng. 19 (1977) 565
36. Merchuk, J.C. Biotechn. Bioeng. 19 (1977) 1889
37. Alper, E., Y. Serpemen and W.-D. Deckwer. Poster paper 6th Int. Ferm. Symp., London, Ontario (1980)
38. Alper, E., M. Lohse and W.-D. Deckwer. Chem. Eng. Sci. 35 (1980) 2147
39. Tsao, G.T. Chem. Eng. Sci. 27 (1972) 1593
40. Kastanek, F. Collect. Czechoslov. Chem. Commun. 42 (1977) 2491
41. Hinze, J.O. "Turbulence" (1975), McGraw-Hill, New York
42. Deckwer, W.-D., R. Burckhart and G. Zoll. Chem. Eng. Sci 29 (1974) 2177
43. Deckwer, W.-D. and R. Burckhart. Abstract of Papers, Session 4, 5th Int. Ferm. Symp., Berlin (1976)
44. Deckwer, W.-D., J. Hallensleben and M. Popovic. Can. J. Chem. Eng. 58 (1980) 190
45. Alvarez-Cuenza, M., C.G.J. Baker and M.A. Bergougnou. Chem. Eng. Sci 35 (1980) 1121
46. Kataoke, H., H. Takeuchi, K. Nakao, H. Yagi, T. Tadaki, T. Otake, T. Miyauchi, K. Washime, K. Watanabe and F. Yoshida. J. Chem.Eng. Japan 12 (1979) 105
47. Nakanoh, M. and F. Yoshida. Ind. Eng. Chem. Proc Des. Dev. 19 (1980)
48. Buchholz, H., R. Buchholz, J. Lücke and K. Schügerl. Chem. Eng. Sci 33 (1978) 1061
49. Voigt, J., V. Hecht and K. Schügerl, Chem. Eng. Sci 35 (1980) 1317
50. Hecht, V., J. Voigt and K. Schügerl. Chem. Eng. Sci. 35 (1980) 1325
51. Hikita, H., S. Asai, K. Tanigawa and K. Segawa. Paper presented at CHISA '78, Prague 1978
52. Nguyen-tien, K., Diplomarbeit, University of Hanover, 1981

53. Nagel, O., H. Kürten and B. Hegner. Chem.-Ing.-
ech. 45 (1973) 913
54. Nagel, O., B. Hegner and H. Kürten. Chem.-Ing.-
ech. 50 (1978) 934
55. Zlokarnik, M. Korrespondenz Abwasser 27 (1980)
94
56. Kastanek, F., J. Kratochvil and M. Rylek. Collect.
zechoslov. Chem. Commun. 42 (1977) 3549
57. Jackson, M.L. and C.-C. Shen, AIChE-J. 24 (1978)
3
58. Calderbank, P.H. and S.J.R. Jones. Trans. Instn.
hem. Engrs. 39 (1961) 363
59. Boon-Long, S., C. Laguerie and J.P. Couderc.
hem. Eng. Sci. 33 (1978) 813
60. Satterfield, C.N. "Mass Transfer in Heterogeneous
atalysis" (1970). MIT Press, Cambridge (Mass.)
61. Kamawura, K. and T. Sasano. Kagaku Kogaku 29
1965) 693
62. Sano, Y.N. Yamaguchi, and T. Adachi. J. Chem. Eng.
apan, 7 (1974) 255
63. Sänger, P. and W.-D. Deckwer. Chem. Eng. J. (1981)
o be published
64. Lohse, M., E. Alper, G. Quicker and W.-D. Deckwer.
FCE Publ. Ser. 11 (1980) 116
65. Shah, Y.T., G.J. Stiegel and M.M. Sharma. AIChE-J.
4 (1978) 369
66. Shah, Y.T., and W.-D. Deckwer, in "Scaleup in the
hemical Process Industries" (1981), Ed. by R. Kabel and
. Bisio, J. Wiley & Sons, New York
67. Baird, M.H.I. and R.G. Rice. Chem. Eng. J. 9 (1975)
7
68. Kato, Y. and A. Nishiwaki. Int. Chem. Eng. 12
972) 182
69. Akita, K. Dr. thesis, Kyoto University (1973)
70. Joshi, J.B. and M.M. Sharma. Trans. Instn. Chem.
ngrs. 57 (1979) 244
71. König, B., R. Buchholz, J. Lücke and K. Schügerl.
er. Chem. Eng. 1 (1978) 199
72. Ohki, Y. and H. Inoue. Chem. Eng. Sci 25 (1970) 1
73. Hikita H. and H. Kukikawa. Chem. Eng. J. 8 (1974) 7
91
74. Cova, D.R. Ind. Eng. Chem. Proc. Des. Dev. 5 (1966)

75. Kato, Y. , A. Nishiwaki, T. Fukuda and S. Tanaka.
hem. Eng. J. Japan 5 (1972) 112
76. Towell, G.D. and G.H. Ackerman. Proc. 2nd Int.
ymp. Chem. React. Engng. (Amsterdam), B 3-1 (1972)
77. Kölbel, H., H. Langemann and J. Platz. Dechema-
onogr. 49 (1964) 253

78. Carleton, A.J., R.J. Flain, J. Rennie and F.H.H. Valentin. Chem. Eng. Sci. 22 (1967) 1839
79. Mangartz, K.-H. and T. Pilhofer. Verfahrenstechnik (Mainz) 14 (1980) 40
80. Deckwer, W.-D. Chem. Eng. Sci. 31 (1976) 309
81. Luong, J.H.T. and B. Volesky. Can. J. Chem. Eng. 58 (1980) 497
82. Deckwer, W.-D. Chem. Eng. Sci. 35 (1980) 1341
83. Deckwer, W.-D., Y. Louisi, A. Zaidi and M. Ralek. Ind. Eng. Chem. Proc. Des. Dev. 19 (1980) 198
84. Nishikawa,M.,H.Kato and K.Hashimoto.Ind.Engng. Chem.Proc.Des.Dev. 16 (1977) 1.

BIOCHEMICAL REACTIONS AND OXYGEN TRANSFER INTO DIFFERENT FERMENTATION BROTHS AND REACTORS

K. Schügerl

Institut für Technische Chemie
Universität Hannover
Callinstr. 3, D-3000 Hannover 1

1 INTRODUCTION

Few biochemical reaction systems are known which are thoroughly analysed. Since in the author's laboratory detailed investigations have been carried out in tower loop reactors, only these reactors will be considered here.

Four different tower loop reactor types were used for the investigations: Two concurrent air lift tower loop systems (a single-stage and a ten-stage reactor) and two countercurrent tower loop systems (a single-stage and a three-stage reactor).

A stainless steel single-stage concurrent bubble column air lift loop reactor, 15 cm in diameter, with a bubbling layer, 275 cm high, and a stainless steel porous plate were used for the cultivation of Hansenula polymorpha (1) and Escherichia coli (2) (Fig. 1)

The same tower was also used as a ten-stage reactor. Nine perforated plates were installed, which separated the tower into 10 sections (Fig. 2). Each of the perforated plates had holes, 3 mm in diameter, and a relative free cross-sectional area of 6.5%. E. coli was cultivated in this reactor (2).

A stainless steel, 254 cm high countercurrent tower loop reactor, 20 cm in diameter, was operated in a single- or three-stage mode (3) (Fig. 3).

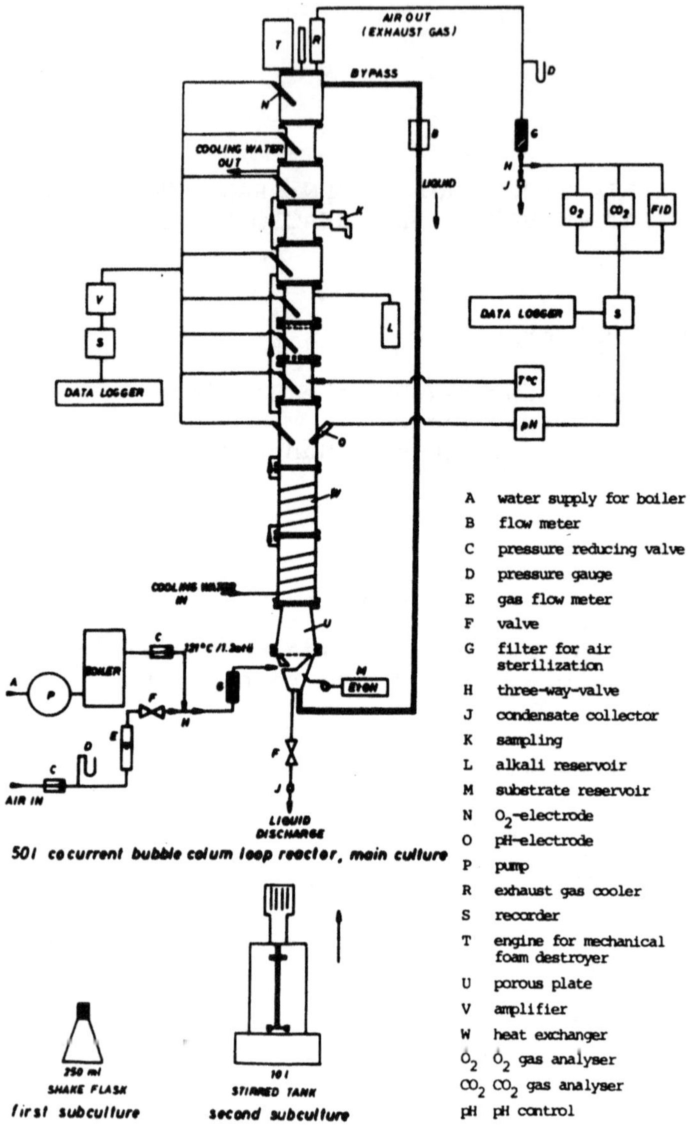

Fig.1 Schematic view of the single-stage concurrent (air lift) tower loop reactor (Reactor A)

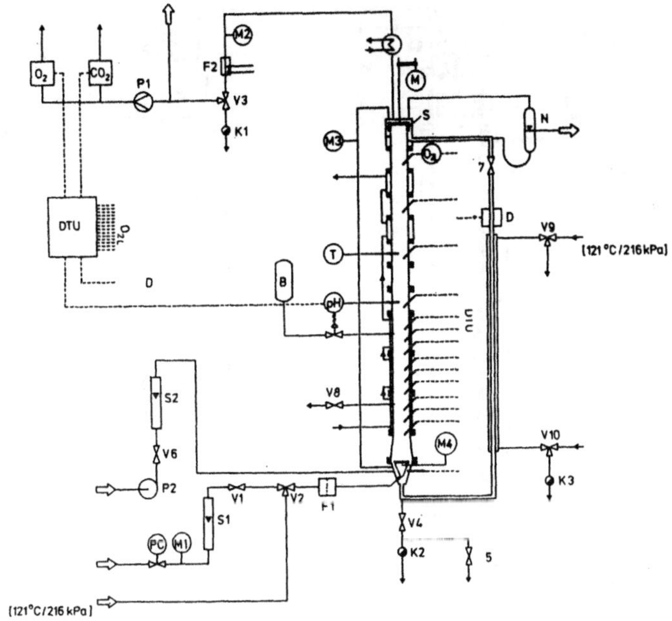

Fig. 2 : Schematical view of the ten-stage concurrent (air lift) tower loop reactor (Reactor B)

PC	pressure-reducing valve	M3	difference pressure gauges
M1,M2,M4	pressure gauge	O_2	paramagnetic O_2 analyser
S1,S2	gas flow meter	CO_2	infrared CO_2 analyser
V1,V6,V7	throttle valves	DTU	data transfer and storage unit
V2,V3,V9,V10	three/two way valves		
F1	sterile filter	(1)	exit gas
S M	mechanical foam destroyer	(2)	condensed water
F2	heated exit gas section	(3)	sampling
K1-K3	condenser trap	(4)	feed
P1	membrane compressor	(5)	air in
V4,V5,V8	shut-off valves	(6)	steam
P2	centrifugal pump	(7)	medium exit
N	level control and overflow	(8)	cooling water
D	inductive flow meter	(9)	electrical heating
O_{2L}	dissolved oxygen measuring electrodes	(10)	drain
pH	pH meter and control		
B	supply tank		
T	temperature meter (Pt 100)		

2: Schematic view of the ten stage concurrent
r lift) tower loop reactor (Reactor B)

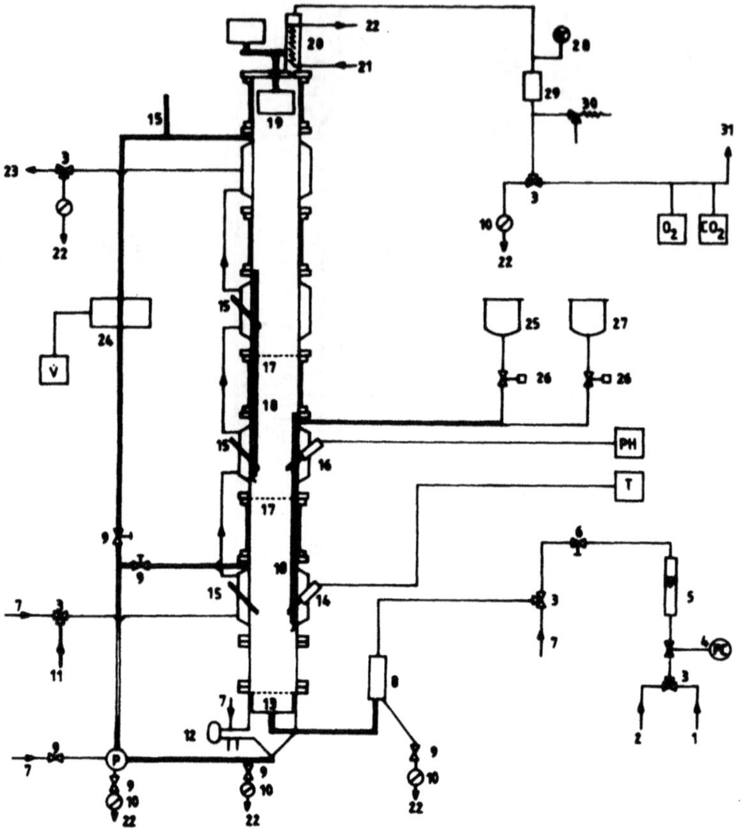

1 air supply	19 mechanical foam destroyer
2 N_2 supply	20 exhaust gas cooler
3 three-way valve	21 cooling water
4 pressure reduction valve	22 waste water
5 gas flow meter	23 thermostatized water backflow
6 needle valve	24 inductive liquid flow meter
7 water vapor (121°C)	25 alkali reservoir
8 sterile filter	26 magnetic valve
9 valve	27 substrate reservoir
10 condensation collector	28 pressure gauge
11 thermostatized water inlet	29 heated exhaust gas tube sterile gas outlet
12 sampling valve	30 safety valve
13 gas distributor	31 exhaust gas
14 resistance thermometer	CO_2 CO_2-analyzer (exhaust gas)
15 O_2 electrode	O_2 O_2-analyzer (exhaust gas)
16 pH electrode	T temperature measure and control
17 compartment separating tray	P liquid pump
18 overflow tube	V̇ inductive flow meter

Fig.3 : Schematical view of the three stage counter-current tower loop reactor (Reactor D)

At the bottom of the column a perforated plate aerator, 12.4 cm in diameter, and with holes, 0.5 mm in diameter, were installed. In the three-stage column separating trays were employed, which consisted of a perforated plate with a hole, 0.5 mm in diameter, and a free cross-sectional area of 0.34% and a 20 and/or 40 cm high overflow. Hansenula polymorpha was cultivated in these reactors (3). All of these reactors were provided with pH- and temperature control, 13 oxygen electrodes along the concurrent columns and/or four oxygen electrodes along the countercurrent columns, a flow rate meter for recycling medium velocity, CO_2, O_2, and ethanol concentration in exhaust gas and data loggers to store these data. Biomass, ethanol, and glucose concentrations, gas flow rate, and substrate feed rate were determined off-line.

For seeding, composition of sub- and main cultures, see Ref. 2 and 3.

2 CONCURRENT SINGLE- AND TEN-STAGE TOWER LOOP REACTORS

The measurements of the local properties of two-phase systems during cultivation indicate that radial profiles of d_S are fairly uniform. Also, their longitudinal variations are fairly moderate, except in the neighborhood of the aerator (1, 4). The same holds true for the spacial variations of the local relative gas holdups. At low superficial gas velocities the specific interfacial area, a, is fairly uniform also. At high superficial gas velocities (turbulent or heterogeneous flow range) the radial profile of a has a shape of an error function, with its maximum in the column center (5). The behavior of these parameters near the aerator depends on the aerator itself and on the medium character.

If the diameter of the primary bubbles, d_P, (at the aerator) is larger than the corresponding dynamic equilibrium bubble diameter, d_e, or if they are equal, d_S is constant in the aerator vicinity also. Only if $d_P \ll d_e$, does the bubble size increase with increasing distance from the aerator, x, due to coalescence until d_e is attained. How quickly $d_S \approx d_e$ is reached depends on the medium property. In these systems a depends considerably on the longitudinal position, x. a attains a constant value as soon as $d_S \approx d_e$ has been established. Also, k_La exhibits a strong dependency on x in these systems in the vicinity of the aerator, if $d_P < d_e$. Since in the investigated systems $d_p \ll d_e$, the spacial dependency of k_La was to be considered by

$$k_La(x) \begin{cases} k_La^E \exp\left[-K_{ST} \frac{x}{\alpha L}\right] \text{ for } 0 \leq x \leq \alpha L & (1a) \\ k_La^E \exp\left[-K_{ST}\right], \text{ for } \alpha L \leq x \leq L, & (1b) \end{cases}$$

where k_La^E is the volumetric mass transfer coefficient at the gas entrance, x = 0,
K_{ST} the "coalescence factor".

The simulations of longitudinal concentration profiles of dissolved oxygen indicate that all profiles could be fitted by assuming a constant validity range of eq.(1): α = 0.1 L, where L = 276 cm. Thus, in these tower reactors the variation of k_La is only taken into account in the range x = 0 to x = 27.6 cm. In this range k_La diminishes exponentially with x according to eq.(1a). In the range x = 27.6 to 276 cm, k_La is constant and given by eq.(1b). This k_La is called k_La^α here:

$$k_La^\alpha = k_La^E \exp\left[-K_{ST}\right] \tag{2}$$

The reduction of k_La^E at the reactor entrance to its spacial independent value, k_La^α, is characterized by the coalescence function, :

$$\psi^\alpha = \frac{k_La^\alpha}{k_La^E}, \tag{3}$$

which is influenced considerably by the medium properties and operation conditions.

In the following only the k_La^α-values will be considered. Since the cultivations were carried out in the homogeneous or laminar flow range, the specific interfacial area, a, was also constant for $0.1 \leq z = \frac{x}{L} \leq 1$. Thus, the mass transfer coefficient, k_L, was calculated by

$$k_L = \frac{k_La^\alpha}{\underline{a}}, \tag{4}$$

where $\underline{a}$ is the geometric specific interfacial area, which was calculated by eq.(5):

$$\underline{a} = \frac{6\varepsilon_G}{d_S(1-\varepsilon_G)} \qquad (5)$$

The Sauter bubble diameter, d_S, and the local relative gas holdup, ε_G, were determined in three longitudinal positions in the tower. Since they were nearly identical, it was assumed that they were uniform in the range $0.1 \leqslant z \leqslant 1$.

At first let us consider nonlimited and oxygen-limited growths. At the high medium recycling rates (1000 to 2000 l h^{-1}) the tower reactor exhibited CSTR-behavior with regard to the cell mass, X, and substrate, S, concentrations. The longitudinal concentration profiles of dissolved oxygen were nonuniform and were described by a dispersion model (2, 6) with particular Bo-numbers ($Bo = \frac{u \cdot L}{D_{ax}}$) and space dependent k_La (6).

The profiles were fairly uniform in nonlimited growth due to the low oxygen uptake rates (curves 1 to 5 in Fig. 4), and in the strongly oxygen-transfer-limited range due to the high oxygen uptake rate (curve 8). At the beginning of oxygen transfer limitation (curve 7 in Fig. 4 and curves 1 to 3 in Fig. 5) the nonuniformity of the profiles is most significant.

The dissolved oxygen was not recycled in the oxygen-transfer-limited growth range, because it was consumed in the loop, as can be observed in the upper part of Fig. 6, in which the dissolved oxygen saturation is plotted as a function of the cultivation time, t, measured for *Hansenula polymorpha* on ethanol substrate at three different positions. At the end of the loop (z = -0), no oxygen could be detected even at t = 14 h, that is, before the oxygen transfer limitation begins at t = 15 h (see the lower part of Fig. 6). By using the dispersion model, the dissolved oxygen concentration profiles were calculated and fitted to the measured ones, thus k_La^E, k_La^α and K_{ST} were identified. In Fig. 7 k_La^α is shown as a function of the cultivation time for *H. polymorpha* and ethanol substrate (1). After inoculation k_La^α drops to low values, then quickly increases at $t \simeq 8$ h, passes a maximum and diminishes at first rapidly and after 14 h, gradually. In Fig. 8 the specific interfacial area, $\underline{a}$,

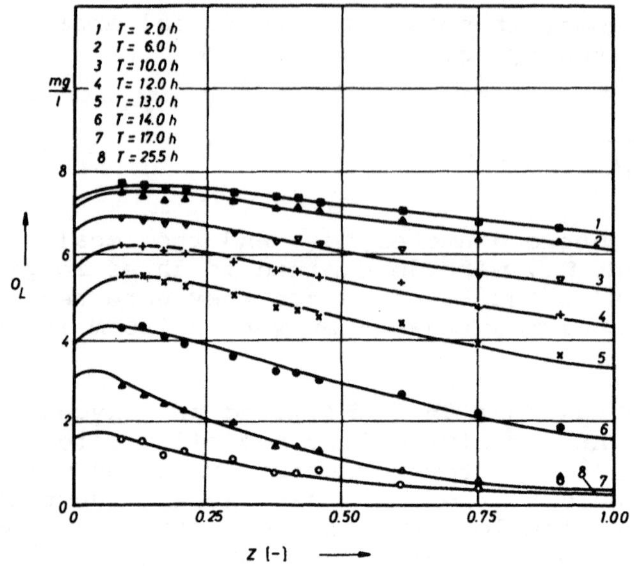

Fig.4: Longtidunal concentration profiles of dissolved oxygen, DOC (mg l^{-1}) in Reactor A during H.polymorpha cultivation at different cultivation times,t.Substrate concentration: S = 5 g l^{-1} ethanol.

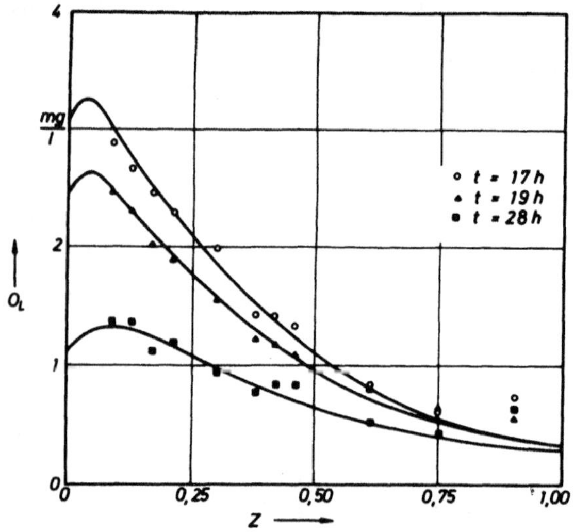

Fig.5 :Longitudinal concentration profiles of dissolved oxygen, DOC (mg/l) in Reactor A during H.Polymorpha cultivation at different cultivation times,t.Substrate concentration: 5 g/l ethanol (under strong oxygen transfer limited growth).

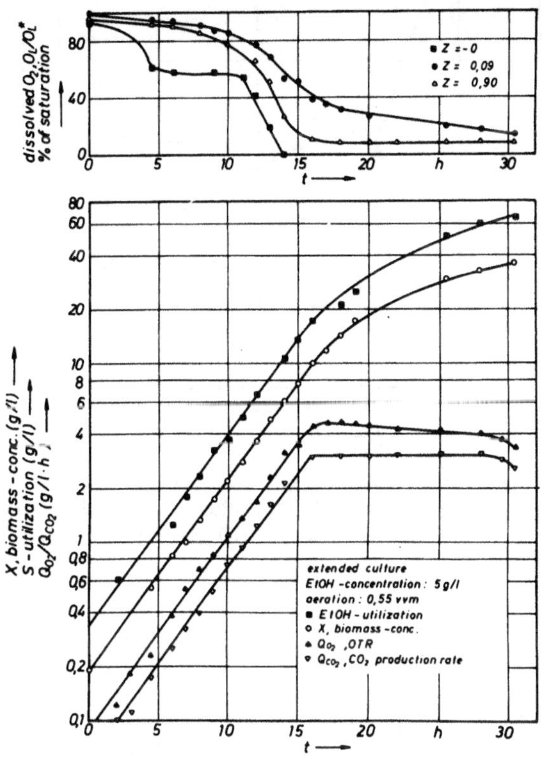

ig.6: H.polymorpha cultivation in Reactor A employing substrate thanol in extende culture operation. Substrate concentration : g/l kept constant by substrate feed. Aeration rate 0.55 vvm.

oper part of figure: relative saturation of dissolved oxygen as a unction of the cultivation time. Longitudinal position of the O_2 robes:

■ z= -0 (just below the aerator)
● z= 0.90 (at the aerator)
Δ z= 0.90 (at the tower head)

ower part of figure : variation of the cultivation with time

■ substrate uptake rate (g/l.h)
o (dry) cell mass concentration, X (g/l)
▲ oxygen uptake rate (g/l.h)
∇ CO_2 production rate (g/l.h)

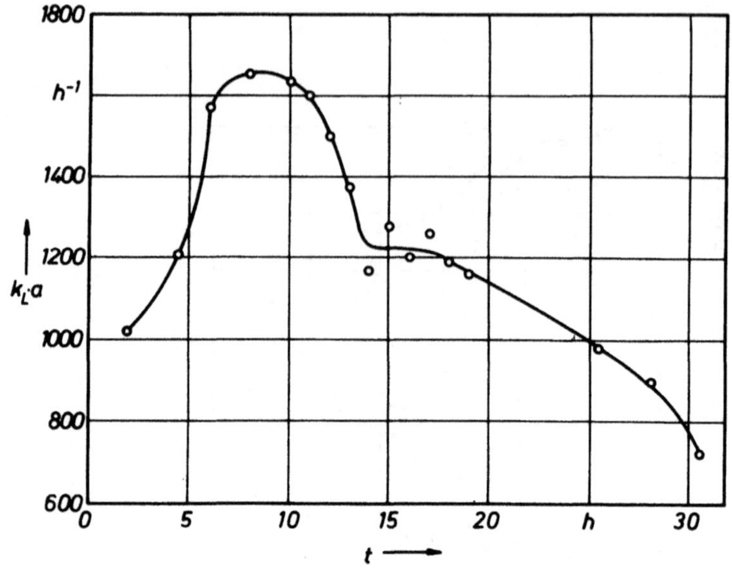

Fig.7:Volumetric mass transfer coefficient,$k_L a$,as a function of the cultivation time in Reactor A during the cultivation of H.polymorpha on ethanol substrate,S= 5 g/l.Aeration rate:0.55vvm.

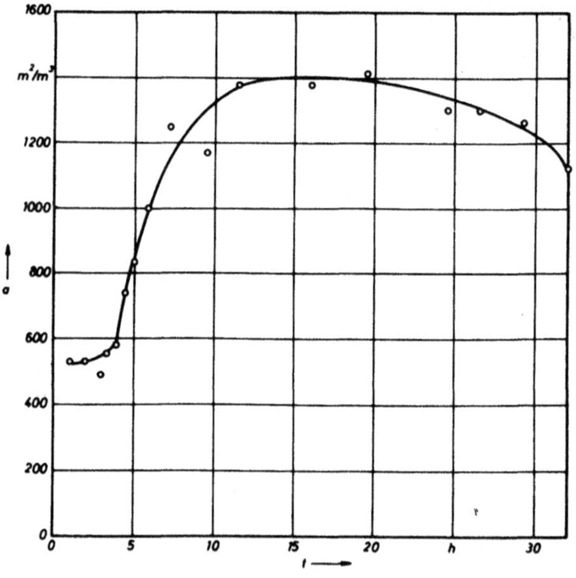

Fig.8: Specific interfacial area,a, as a function of the cultivation time,t, in Reactor A during the cultivation of H.polymorpha on ethanol substrate , S= 5 g/l. Aeration rate : 0.55 vvm.

calculated by eq.(5), is plotted as a function of t for the same run. An initial drop, then a quick increase is common for k_La^α and a. However, k_La^α passes a maximum earlier and drops more rapidly, while a attains its maximum later and diminishes only slightly. The mass transfer coefficient, k_L, was calculated by eq.(4) (Fig. 9). For $t \leqslant 4$ h it is constant, but soon after it gradually diminishes for $t \leqslant 15$ h somewhat quicker, then slighter(1).

The courses of k_La^α and a as a function of the cultivation time are rather different during cultivations on ethanol and glucose substrate (1). When using glucose, the dissolved oxygen concentration diminishes with increasing t as usual, but at 3.5 h it passes a minimum and gradually increases (upper part of Fig. 10). However, the deviation from exponential growth was not caused by growth limitation, because sufficient substrate and dissolved oxygen were present (Fig. 10 and upper part of Fig. 11). The reduction of the growth rate is accompanied by diminutions of the yield coefficients $Y_{X/O}$ and $Y_{X/S}$ and the cell mass productivity (middle and lower parts of Fig. 11). Also, ethanol is produced (upper part of Fig. 11). Obviously some kind of repression has occurred due to glucose. The longitudinal dissolved oxygen profiles are similar in this range to those measured in systems with ethanol substrate at the end of the exponential growth phase (Fig. 5 and Fig. 12). By fitting the calculated profiles to the measured ones, the corresponding k_La^α-values were determined. Fig. 13 shows that k_La^α increases with t. The same holds true for a (Fig. 14). This is caused by the increasing ethanol concentration. The k_L-value is nearly constant (Fig. 15). A comparison of k_La^α-values with ethanol and/or glucose substrate show that the former k_La^α is much higher ($1400 \rightarrow 800$ h^{-1}) than the latter ($150 \rightarrow 700$ h^{-1}) (Fig. 7 and 13). With increasing alcohol concentration in the glucose system the k_La -values also approach 800 h^{-1}. The same is true for the specific interfacial area, a, with ethanol ($600 \rightarrow 1200$ m^{-1}) (Fig. 8) and with glucose ($200 \rightarrow 1000$ m^{-1}) (Fig. 14). k_L-values are also higher in the ethanol (Fig. 9) than in the glucose (Fig. 15) system, but with increasing time and ethanol concentration they approach the same value.

The influence of ethanol concentration on k_La^α is shown in Fig. 16. k_La^α increases with increasing ethanol concentration. Also, the operation mode influences k_La^α. In media with oxygen-transport-limited growth k_La^α is the smallest. In Fig. 17 k_La^α is plotted as a function of the superficial gas velocity, W_{SG}. In tower loop reac-

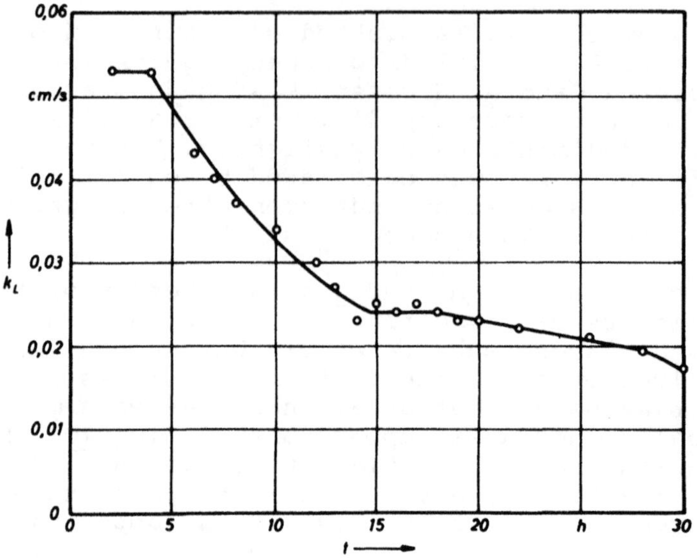

Fig.9:Mass transfer coefficient,k_L ,as a function of the cultivation time,t,in Reactor A during the cultivation of H.polymorpha on ethanol substrate, S=5 g/l. Aeration rate:0.55 vvm.

Fig.11.Variation of yield coefficients,$y_{X/O}$ and $y_{X/S}$ respiratory quotient,RQ, cell mass concentration, X,cell productivity,Pr, and produced ethanol concentration as a function of cultivation time in Reactor A during the cultivation of H.polymorpha at high glucose concentrations (S=9.2 g/l).

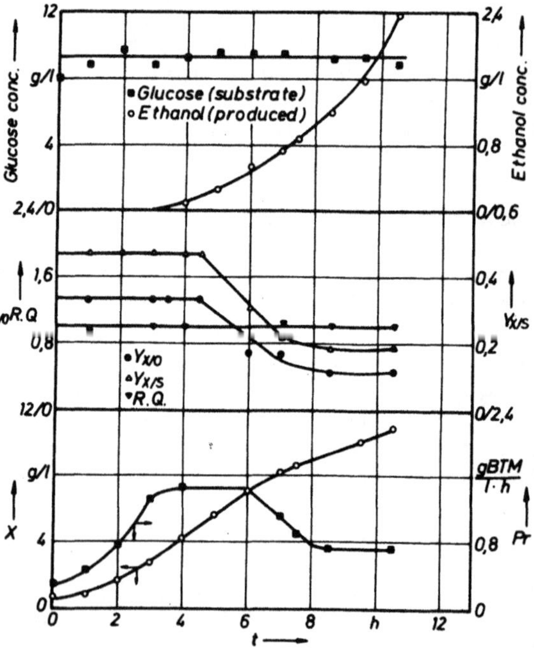

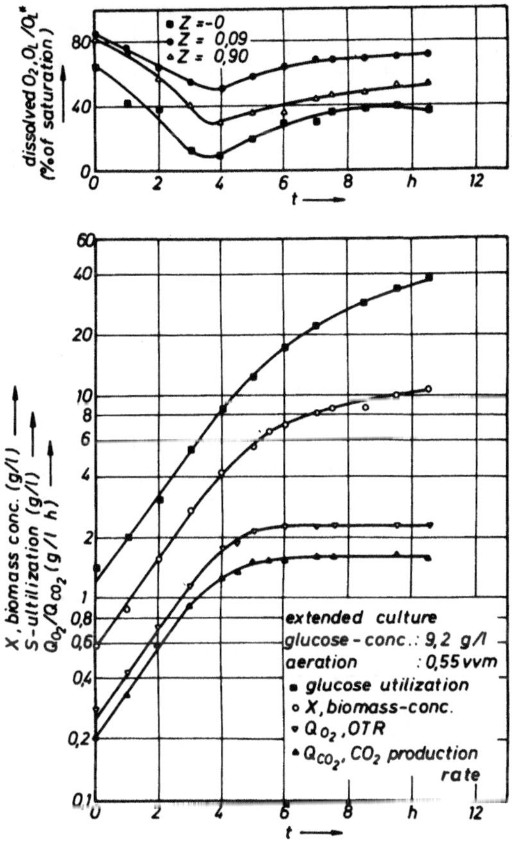

Fig.10 : H.polymorpha cultivation in Reactor A employing glucose substrate in extended culture operation. Substrate concentration S = 9.2 g/l , kept constant by substrate feed. Aeration rate: 0.55 vvm.

Upper part of the figure: dissolved oxygen concentration as a function of the cultivation time. Longitudinal position of the oxygen probes:

z= -0	(just below the aerator)
z= 0.09	(at the aerator)
z= 0.90	(at the column head)

Lower part of the figure: variation of the cultivation with time.

glucose uptake rate (g/l h)
(dry) cell mass concentration, X , (g /l)
oxygen uptake rate (g/l.h) ; CO_2 production rate (g/l h)

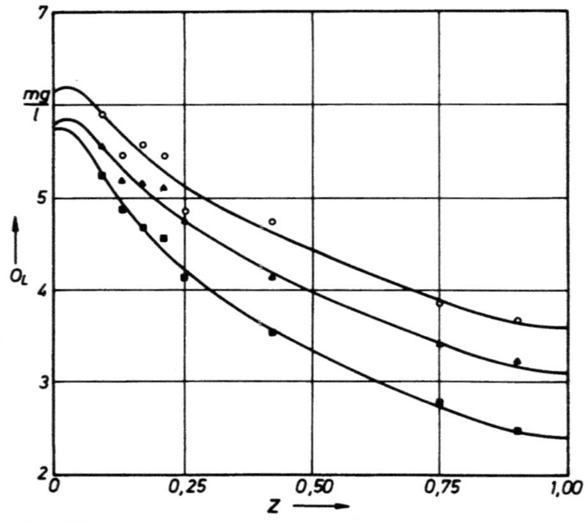

Fig.12: Longitudinal profiles of dissolved oxygen (mg/l) in Reactor A during the cultivation of H.polymorpha on glucose substrate (S = 9.2 g/l) at different cultivation times ,t

1 t = 6.0 h X = 7.10 g/l
2 t = 7.0 h X = 9.65 g/l
3 t = 9.5 h X =10.05 g/l

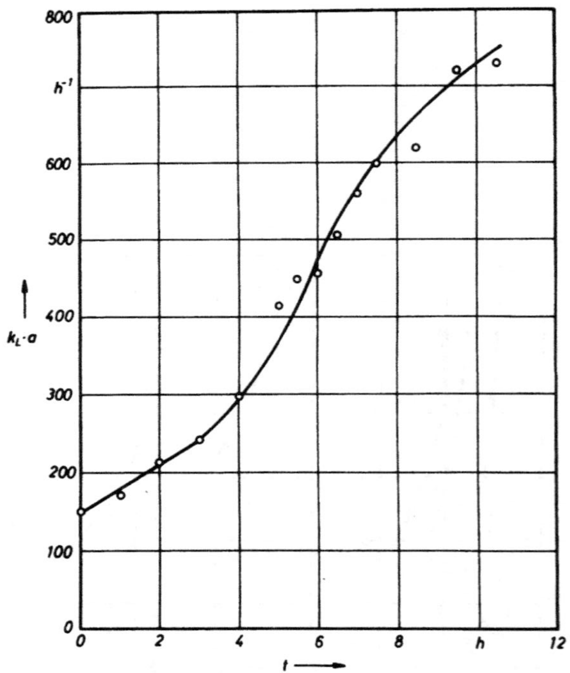

Fig.13:Volumetric mass transfer coefficients,k_La, as a function of the cultivation time,t, in Reactor A during the cultivation H.polymorpha employing glucose substrate, S = 9.2 g/l. Aeration rate : 0.5 vvm.

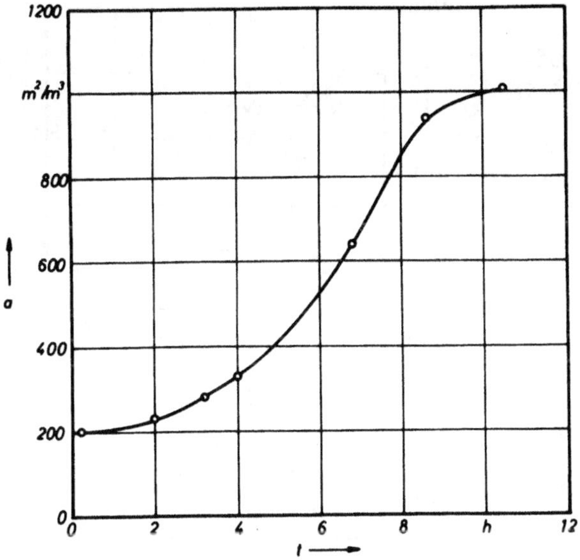

Fig.14: Specific interfacial area,a,as a function of the culti-vation time,t, in Reactor A during the cultivation of H.polymorpha employing glucose substrate, S = 9.2 g/l. Aeration rate=0.5 vvm.

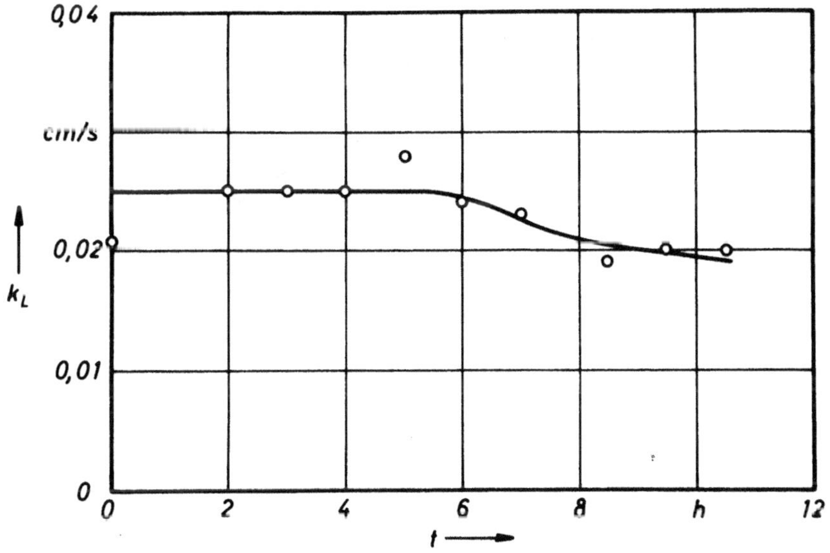

Fig.15: Mass transfer coefficient,k_L, as a function of the culti-vation time,t, in Reactor A during the cultivation of H.polymorpha on glucose substrate, S=9.2 g/l ,Aeration rate: o.5vvm.

Fig.16: Influence of ethanol concentration tion on the mass transfer coefficient $k_L a$, during cultivation of *H.polymorpha*

$k_L a^{\alpha} = 235\ h^{-1}$

$S_D = 1.25\ g/l$

▽△ batch cultivation
□ substrate limited cultivation
○ extended culture
● oxygen transfer limited cultivation
— lowest limit for all $k_L a$ values
I $k_L a$ variation range without cells

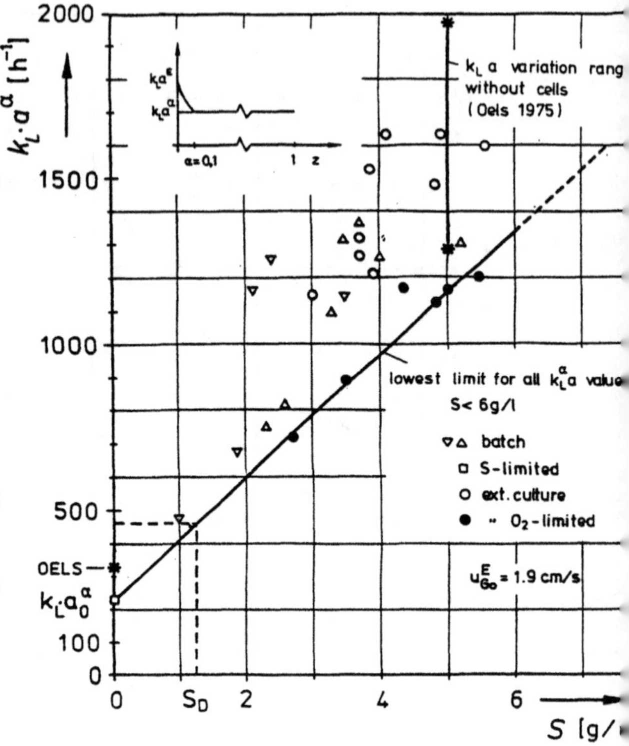

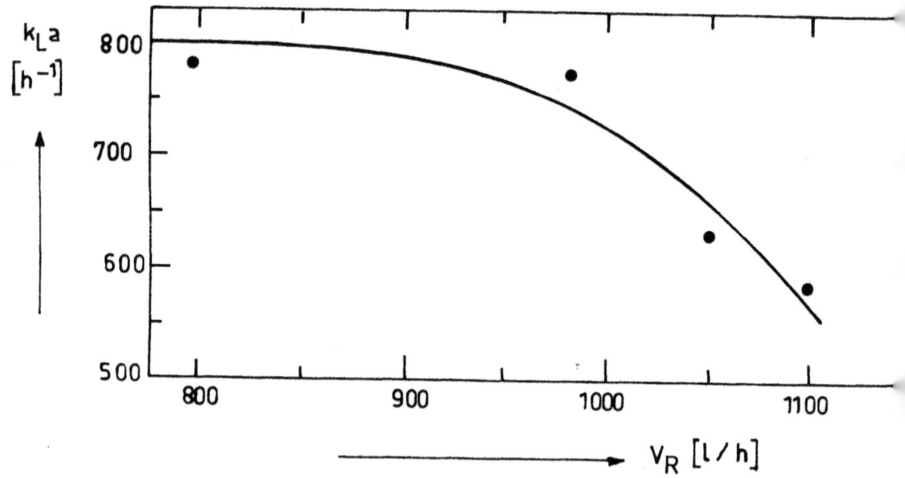

Fig.18: Volumetric mass transfer coefficient, $k_L a$, as a function of the medium recycling rate, $\dot{V}_R$ in Reactor A during the cultivation of *E.coli* in continuous operation.

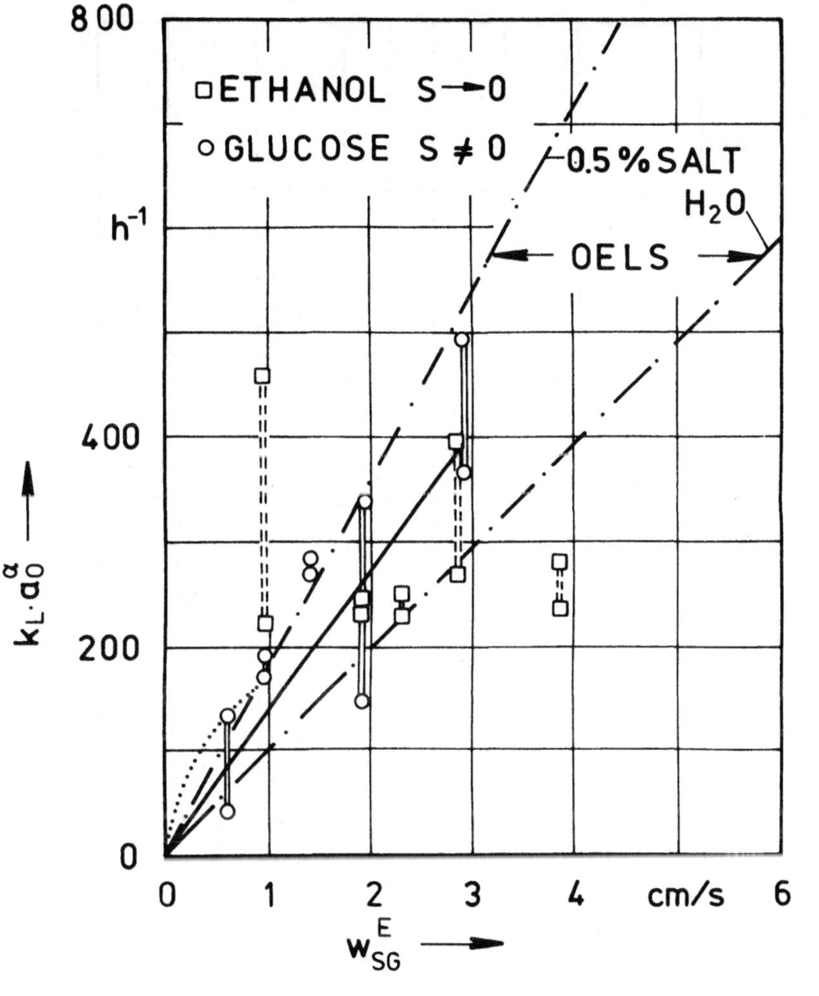

Fig.17: Influence of the superficial gas velocity, u_{Go}^{E}, on the volumetric mass transfer coefficient in the absence of ethanol during the cultivation of H.polymorpha

□ substrate limited growth,ethanol substrate, S = 0 g/l
○ glucose substrate S ≠ 0 g/l
-·-·- range for 0.5 % nutrient salt solution in tower reactor

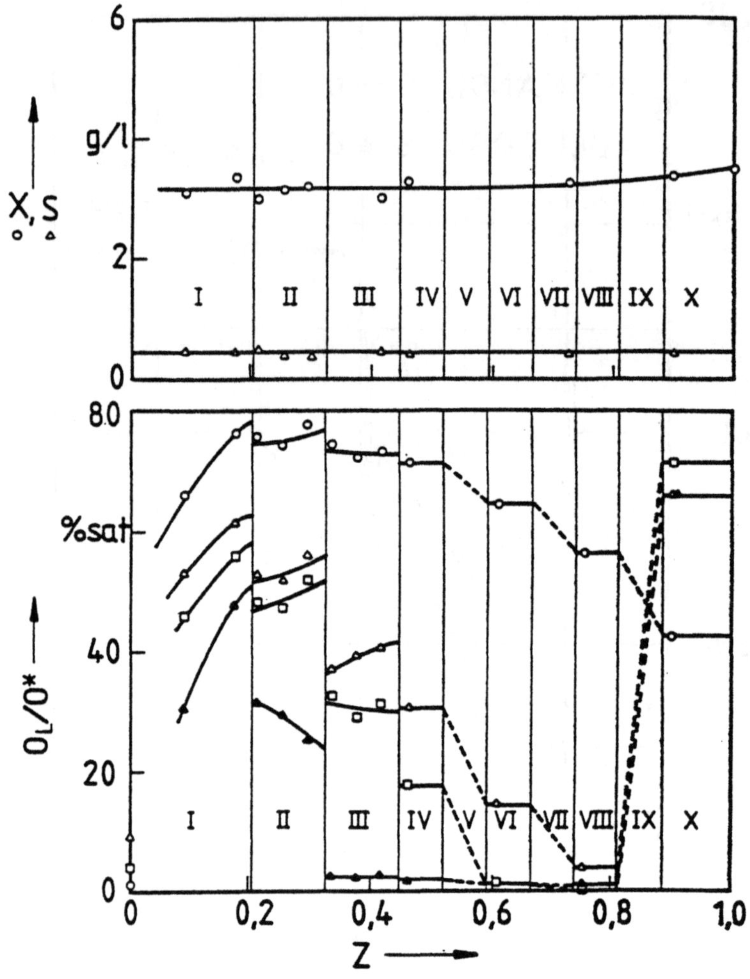

Fig.19: Longitudinal profiles of cell mass concentration, X, substrate concentration, S, (upper part) during the cultivation of E.coli in Reactor B in batch and continuous operation

o x , △ S (continuous) upper part

- o DOS at t = 40 min (batch)
- ▲ DOS at t = 220 min (continuous)
- □ DOS at t = 270 min (continuous)
- ▲ DOS at t = 400-820 min (continuous)

lower part

I-X are the stages counted from the bottom.

tors without loop (measurements of Oels (7)) k_La increases with w_{SG} up to 6 cm s^{-1}. In tower loop reactors k_La^{α} increases with w_{SG}, passes a maximum at about 3 cm^{-1}, then diminishes (4). This is due to the increasing medium recycling rate, $\dot{V}_R$. In Fig. 18 k_La is shown as a function of $\dot{V}_R$ during *E. coli* cultivation (2). The reduction of k_La with increasing $\dot{V}_R$ is significant. This is mainly due to the reduction of ε_G.

If the nonlimited growth along the tower turns into substrate-limited growth, the longitudinal dissolved oxygen concentration profiles pass a minimum at this position. These longitudinal profiles can only be simulated with a distributed parameter model with regard to the substrate (4). In Fig. 19 the longitudinal cell mass, X, substrate, S, and relative oxygen, O_L/O^*, concentrations are shown in the ten-stage concurrent tower loop reactor with *E. coli* (2). Whereas X and S are uniform in the tower, O_L/O^* diminishes with increasing z, but remains higher than 0.5 at high liquid recycling rates and in batch operation. Thus, no oxygen transfer limitation occurs. In continuous cultivation O_L/O^* drops to zero. With increasing cultivation time the numbers of stages increase, in which the oxygen supply of bacteria is not sufficient (Fig. 19). (The increase of O_L/O^* in the 10th stage is caused by the mechanical foam destroyer at the top of the column, which also acts as an aerator). The inadequate oxygen transfer rate in the ten-stage tower is due to the formation of liquid-free layers below the perforated trays, which acted as very ineffective gas distributors, because of their large holes (3 mm in diameter) and free cross sectional area (6.52%). In addition, the efficiency of the stage separating trays probably deteriorated because of foam formation below the trays.

3 COUNTERCURRENT SINGLE- AND THREE-STAGE TOWER LOOP REACTORS

In Fig. 20 the volumetric mass transfer coefficients are shown as a function of the cultivation time with *H. polymorpha* and ethanol substrate (3). In the three-stage tower with a bubbling layer height of 40 cm considerably higher k_La-values can be attained than in the single-stage one with a bubbling layer of 160 cm height. However, whereas in the single-stage tower reactor the cultivation could be carried out without any difficulty, the multistage reactor could not be operated in absence of antifoam agents, since in this case the free volume

Fig.20: Comparison of the volumetric mass transfer coefficients k_La's in Reactor C (single stage tower loop) and Reactor C and D (single- and three-stage tower loops)

o single-stage

● three-stage (h = 40 cm)

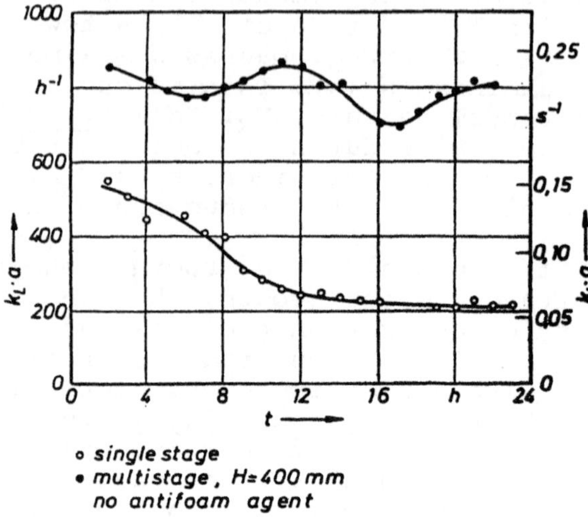

Fig.21: Cell mass concentration X, as a function of the cultivation of H.polymorpha in Reactor D (three-stage tower) in the absence of antifoam agents employing substrate ethanol

o 1. stage (at the top)

● 2. stage

△ 3. stage (at the bottom)

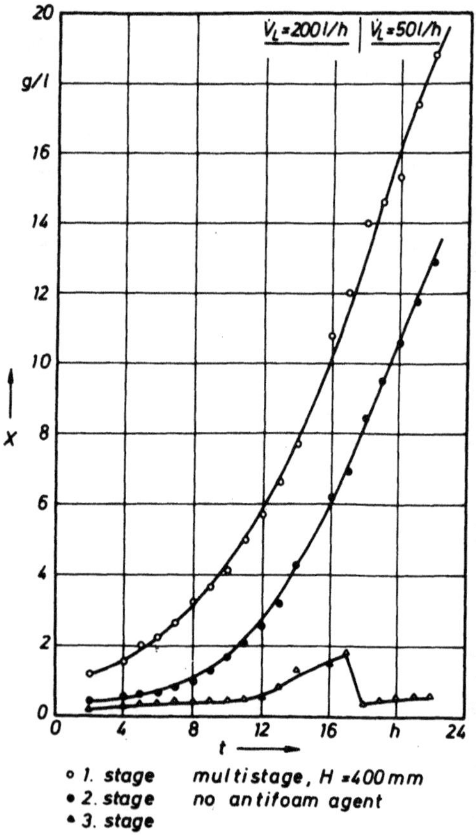

between the bubbling layer and the tray above the layer became filled by foam. By microflotation, the cells which were enriched in foam, passed through the tray into the stage above. Here they were enriched in the foam again and so on. Through this microflotation effect, the cells were enriched in the upper stage; in the lower stages the cell concentration was diminished considerably (Fig. 21). It was not possible to reduce this cell segregation by increasing the liquid circulation rate, since the liquid transport capacity of the overflow was reduced considerably when foaming broth was present. To avoid this segregation antifoam agent was used.

These runs with antifoam agent were carried out with two different bubbling layer heights: H = 20 and 40 cm in the three-stage tower. During the cultivations no foam formation was observed. Also, no differences were found in the cell mass concentrations of the three different stages. In Fig. 22 the volumetric mass transfer coefficients are plotted as functions of the cultivation time for the single-stage and the three-stage columns at H = 40 and 20 cm. The lowest k_La-values were found in the single-stage tower and the highest ones in the three-stage tower with H = 20 cm. The installation of trays improved the k_La-values. This improvement increased with diminishing bubbling layer height.

COMPARISON OF THE SYSTEMS

Because of the low liquid velocity, the difference in E_G in the concurrent and the countercurrent tower reactors is slight. Therefore, they can be compared directly. In Table 1 the concurrent single-stage tower reactor with a porous plate gas distributor (Reactor A) and the countercurrent one with a perforated plate gas distributor (Reactor C) are compared. One recognizes that OTR, k_La as well as the productivity, Pr, is much higher in Reactor A than in Reactor C due to the more efficient aerator.

In Table 2 the single-stage and ten-stage tower loop reactors are compared. Because the stage-separating trays act as gas distributors and their efficiency is very low (3 mm hole-diameter and foam formation), the single-stage tower with an efficient aerator yields a much higher productivity than the ten-stage tower.

In Table 3 the single- and three-stage countercurrent tower reactors are compared in absence and pre-

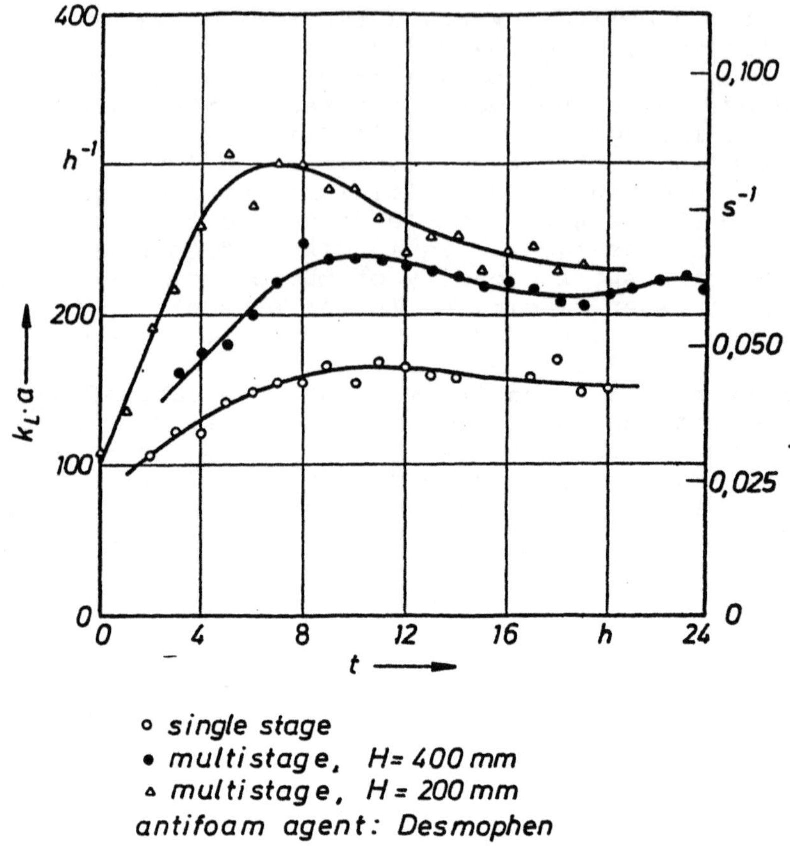

Fig.22:Volumetric mass transfer coefficients , k_La's, as function of the cultivation time,t, during the cultivation of H.polymorpha in Reactors C and D employing substrate ethanol and Desmophen 3600 as antifoam agent.

- o single stage
- ● three-stage, h = 40 cm
- Δ tree-stage, h = 20 cm

Table 1: Comparison of the performance of Reactors A and C employing H.polymorpha on ethanol substrate. $Y_{X/O}$, k_La, OTR and P_r given for oxygen transfer limited growth range.

	Reactor A	Reactor C
	porous plate 17.5 μ_m pore dia.	perforated plate 0.5 mm hole diam.
$\mu_m (h^{-1})$	0.26	0.25
$Y_{X/O}$ (-)	0.53	0.55
OTR ($gl^{-1}h^{-1}$)	4.7	1.28
k_La (h^{-1})	1200	250
Pr ($gl^{-1}h^{-1}$)	2.5	0.71

Table2: Comparison of the performances of Reactor A (single-stage with porous plate aerator) and Reactor B (ten-stage ,with porous plate aerator and 9 trays , perforated plates 3.0 mm in hole diameter) employing E.coli and antifoam agent (Desmophen) in batch and continuous operation.

operation	batch, $\dot{V}_R$ = 340 lh^{-1}		continuous, $\dot{V}_R$ = 1400 lh^{-1}	
stages	single	ten	single**	ten**
gas distribution	porous plate	perforated plate trays	porous plate	perforated plate trays
$\mu_m (h^{-1})$	0.54	0.52	0.36-0.44	0.36
$Y_{X/S}$ (-)	0.80	0.75	1.16	0.53
RQ (-)	0.88	0.93	0.93	0.86
Pr,Pr* ($gl^{-1}h^{-1}$)	2.33	1.66	2.64	1.15

* Pr = DX in continuous culture

Pr = $\frac{\Delta x}{\Delta t}$ in batch culture

** $\mu \simeq D$

Table 3 : Comparison of single- and tree-stage countercurrent tower loops (Reactors C and D) by employing *H.polymorpha* in antifoam agents

Antifoam agent	AA absence		Desmophen 3600			soy oil		
	single-stage	three-stage	single-stage	three-stage		single-stage	three-stage	
H cm	(160)[1]	40[3]	(160)[1]	40	20	(160)[1]	40	20
μ_m (h^{-1})	0.25	0.17	0.21	0.21	0.21	0.24/0.26[2]	0.11/0.26[2]	0.09/0.27[2]
k_La[4] (h^{-1})	250	800	160	220	250	290	440	650
(s^{-1})	0.07	0.22	0.04	0.06	0.07	0.08	0.12	0.18
$Y_{X/O}$ $\left[\frac{g\ biomass}{g\ O_2}\right]$[5]	0.55	0.36	0.69	0.58	0.74	1.06	1.16	1.17
Pr [g/l,h][5]	0.71	0.80	0.68	0.75	1.09	1.69	2.43	2.93
Q_{O_2m} [g/l,h]	1.28	2.20	0.98	1.30	1.48	1.60	2.10	2.50
spec.energy requirement $\left[\frac{kW\cdot h}{kg\ O_2}\right]$	0.20	0.14	0.27	0.23	0.20	0.16	0.14	0.12
spec.energy requirement $\left[\frac{kW\cdot h}{kg\ biomass}\right]$	0.37[6]	0.38[6]	0.38	0.40	0.28	0.15	0.12	0.10

1) mean bubbling layer height
2) first exp. growth phase/sec. exp. growth phase
3) large error due to foam formation
4) at maximum oxygen transfer rate
5) in the oxygen transfer limited linear growth phase
6) without the energy requirement of the mechanical foam destroyer

sence of antifoam agents. The primary gas distributors and the stage-separating trays consisted of a perforated plate with a hole, 0.5 mm in diameter. In absence of antifoam agents, k_La is much higher in the three-stage tower than in the single-stage one. In spite of this the corresponding cell productivity is only slightly higher due to the cell enrichment in the upper stage.

In the presence of antifoam agent (Desmophen 3600) the highest k_La-value and productivity are attained in the three-stage tower with H = 20 cm, and the lowest ones in the single-stage tower.

In Table 4 the biological parameters of H. polymorpha are compared with ethanol and/or glucose substrate. The maximum specific growth rate, μ_m, is higher for glucose than for ethanol. Also, the oxygen yield coefficient is higher with the former than with the latter. This means that H. polymorpha has a higher growth rate with glucose than with ethanol and needs less oxygen.

In Table 5 specific interfacial area, a, volumetric mass transfer coefficient k_La, OTR, and productivity, Pr, are compared. They were evaluated with ethanol substrate as well as glucose substrate.

Although a, k_La, and OTR are higher with ethanol than with glucose, the productivity for the latter is higher than for the former. This is due to the higher μ_m and $Y_{X/O}$ for glucose substrate than for ethanol.

5 CONCLUSION

The reactor construction, medium properties, and biological parameters of the microorganism influence the oxygen transfer and cell productivity considerably. Efficient aerator, low bubbling layer height, high maximum specific growth rate, μ_m, and high oxygen yield coefficient, $Y_{X/O}$, improve the cell productivity considerably.

Table 4. Comparison of single-stage concurrent tower loop reactors by employing *H.polymorpha*, ethanol and glucose as substrate. Biological parameters of the investigated systems

substrate	growth phase	μ_m	$Y_{X/S}$ $\frac{g\ X}{g\ substrate}$	$Y_{X/O}$ $\frac{gX}{g\ O_2}$	R.Q. $\frac{mol\ CO_2}{mol\ O_2}$
ethanol	exponential	0.26	0.25 - 0.55	0.32-0.53	0.48
	O_2-limited		0.50 - 0.75	0.25-0.53	0.48
	S-limited		0.55 - 0.75	0.32-0.56	0.48
glucose	exponential	0.58	0.47	1.16-1.36	1.0
	O_2-limited		0.46	0.65-1.24	1.0
	S-limited		0.46	0.98-1.21	1.0

Table 5. Comparison of single-stage concurrent tower loop reactors by employing *H.polymorpha*, ethanol and glucose as sustrate. Maximum values of productivity, OTR, k_La and a in the investigated systems

substrate	growth phase	Pr g/l h	Q_{O_2} g/l h	k_La h^{-1}	a m^{-1}
ethanol	O_2-limited	2.5	4.7	1200	1400
	S-limited	0.95	1.7	300	400
glucose	O_2-limited	3.3	2.8*)	540*)	650*)
	S-limited	1.7	1.4	300	300

*) with ethanol production

REFERENCES

1. Buchholz, H. Dissertation, University of Hanover, 1977
2. Adler, I. Dissertation, University of Hanover, 1980
3. Voigt, J. Dissertation, University of Hanover, 1980
4. Luttmann, R. Dissertation, University of Hanover, 1980
5. Zakrzewski, W. Dissertation, University of Hanover, 1980
6. Schügerl, K. "Introduction to biochemical systems. Formal treatment of biochemical reactions and characterization of fermentation systems". (Proceedings of the NATO ASI on " Mass transfer with chemical reaction", Turkey, 1981)
7. Oels, U. Dissertation, University of Hanover, 1975.

PROCESS DESIGN ASPECTS AND COMPARISON OF DIFFERENT BIOREACTORS

K. Schügerl

Institut für Technische Chemie
Universität Hannover
D-3000 Hannover 1, FRG

1 INTRODUCTION

The aim of a bioreactor is to ensure the optimal environment for the particular microorganism with respect to cell growth and/or product formation. Therefore the optimum reactor construction and operation is strongly influenced by the microorganism directly as well as indirectly due to medium properties and the biochemical process itself.

2 GENERAL CONSIDERATIONS TO REACTOR SELECTION

During the cultivation of aerobic organisms a sufficient amount of dissolved oxygen must be present in the broth. The oxygen transfer rate, OTR, must be high enough to satisfy the oxygen demand of the cells. Especially for microorganisms with a high specific growth rate, μ, and a low oxygen yield coefficient, $Y_{X/O}$, with regard to the substrate employed, reactors with very efficient aerators are needed. Spherical cells are usually smaller and less sensitive to shear stresses than filamentous microorganisms. However, they usually have a high oxygen demand. Filamentous microorganisms grow only on the ends of the mycelia. This results in low growth and low oxygen consumption rates. Since such organisms are usually sensitive to shear stresses gas dispersing facilities or operational conditions with low shear stresses are used which are less effective in gas dispersion than the high performance aerators.

Floc formation increases microbial holdup, facilitates the separation of the biomass from the fermentatio broth, makes it possible to use particular fermenter con figurations (e.g. tower reactors) for continuous culture or to feed back a fraction of the microorganisms to the reactor.

In case of film forming microorganisms the use of film fermenters (trickle bed, rotating disc, rotating drum, etc.) is possible.

Microbial flocs can have different morphologies. They depend on the strain as well as on the operational conditions in the reactor. The filamentous form is sensitive to shear stress and usually causes highly viscous non-Newtonian behavior of the mould. The pellet form is less sensitive to shear stress and the fermentation brot behaves like a three-phase system with a relatively low liquid viscosity. The OTR into filamentous mould is usually lower than the OTR into a pellet broth.

Reactor operation is strongly influenced by the strain stability. Only strains which are sufficiently stable can be used in continuous cultures.

3 GENERAL CONSIDERATIONS TO REACTOR OPTIMIZATION

3.1 Optimization with Regard to Substrate Conversion

Since substrate costs generally make up about 50 % of production costs substrate conversion is a key parameter for the economy of bioreactions. The lower the intensity of longitudinal medium dispersion, the higher the substrate conversion in continuous bioreactors under corresponding operational conditions. However, at a low dispersion intensity, cell washout occurs. To avoid washout and to achieve high substrate conversion, tower reactors with optimum longitudinal dispersion or tower loop reactors with an optimum recycling rate can be used (1).

It is possible to decouple the longitudinal dispersions of the cells and the substrate, if the substrate concentration at the tower exit, S_e, is very low (2). Since the substrate passes through the tower only once the effective Bodenstein number, $Bo_R = Bo(1+\gamma)$, increases with the medium recirculation ratio, γ . On th other hand, a high percentage of cells is recycled. Thus the cell dispersion intensity increases with increasing

At a high medium recirculation ratio ($\gamma > 20$), the tower loop behavior approaches that of a CSTR with regard to the cells but the behavior of a PFR with regard to the substrate (2). This decoupling holds true as long as $S_e \simeq 0$.

When the γ values are too high, $S_e > 0$, the substrate is recycled as well and the tower loop system also behaves as a CSTR with regard to the substrate. Thus there is an optimum γ value at which $C_S = S_e/S_o$ is at a minimum and the substrate conversion, $U_S = 1 - C_S$, is at a maximum as long as the Bo number is high enough (2).

This problem was quantitatively investigated by Adler (3) (see also in (1)).

If the recirculation ratio, γ, is zero, there is an optimum Bo number, Bo_{opt}, at which C_S is at a minimum and U_S at a maximum. The position of this C_S minimum strongly depends on $Da = \mu_m \tau$ where τ is the mean residence time in the reactor, as well as on the dimensionless substrate saturation constant, K

$$K = \frac{K_S}{S_o + X_o/Y_{X/S}} \qquad (1)$$

where X_o is the cell mass concentration in the feed. Below a critical Da number, Da_{crit}, or above a critical K, K_{crit}, the optimum reactor is the CSTR.

For cell-free feed ($X_o = 0$) (3):

$$Da_{crit} = \frac{1}{1 - \frac{K}{\sqrt{K(K+1)}}} \qquad (2)$$

and

$$K_{crit} = \frac{1}{\frac{1}{(1 - \frac{1}{Da})^2} - 1} \qquad (3)$$

Introduction of medium recirculation can improve C_S and U_S if $Bo > Bo_{opt}$, $Da > Da_{crit}$ and $K < K_{crit}$, i.e. $\gamma_{op} > 0$. In this case γ_{opt} diminishes with increasing Da and K. (For $Da < Da_{crit}$ and/or $K > K_{crit}$ $\gamma_{opt} = \infty$ (CSTR). On the other hand, for $Bo < Bo_{opt}$, $\gamma_{opt} = 0$.

3.2 Optimization with Regard to Cell Mass Productivity

It is obvious that maximum cell productivity can be attained if the cells are cultivated in the nonlimited growth range. However, when using cell-free feed, nonlimited growth systems are unstable (1). Furthermore, the substrate loss is considerable for nonlimited growth operations. Because of this, nonlimited growth is an uneconomical operation. However, it is possible to maintain nonlimited growth at the entrance of a tower reacto and substrate limited growth at its exit and by that to achieve high productivity and high substrate conversion in a stable steady state operation (2). In this case the reactor behavior can be described by the substrate limited rate equation.

There is formal analogy between autocatalytic reactions and cell growth with substrate limitation. Bisho has shown (4) that the maximum growth rate can be attained, if one uses a combination of a CSTR and a PFR. The size of the CSTR has to be chosen so that the growth rate has a maximum in it. The size of a CSTR can be evaluated graphically by plotting the inverse growth rate, R_X^{-1}, as a function of the cell mass concentration, X. Figures 1 and 2 show such plots where the dimensionless growth rate is

$$R_X = \frac{R_X^*}{X_o + Y_{X/S} S_o} \tag{4}$$

the growth rate according to Monod

$$R_X^* = \mu_m \frac{XS}{K_S + S} \tag{5}$$

and the dimensionless cell mass concentration

$$C_X = \frac{X}{X_o + Y_{X/S} S_o} \tag{6}$$

Figs. 1 and 2 show that R_X^{-1} passes a minimum at a particular C_X which is called C_{Xcrit}. For $C_X \leqslant C_{Xcrit}$ the CSTR is the optimum reactor. C_{Xcrit} is given by

$$C_{Xcrit} = (K+1) - \sqrt{K(K+1)} \geqslant 0.5 \tag{7}$$

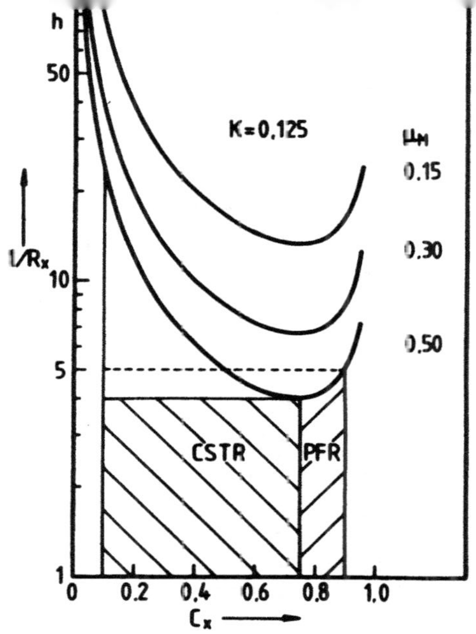

Fig.1: Reciprocal related growth rate, R_x^{-1}, as function of the dimensionless substrate concentration, C_x, at K = 0.125 for different μ_{max}. Combination of CSTR and PFR (3).

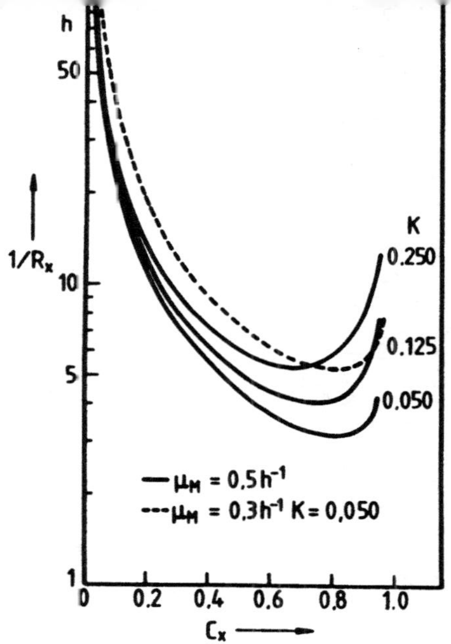

Fig.2: Reciprocal related growth rate, R_x^{-1}, as a function of the dimensionless substrate concentration, C_x, at $\mu_{max} = 0.5\ h^{-1}$ (———) and variable K, as well as $\mu_{max} = 0.3\ h^{-1}$ and K = 0.05 (- - - -) (3).

i.e. for $C_X \leq 0.5$, the CSTR is always the optimum reacto

If a tower reactor with negligible longitudinal dispersion (PFR) is used, mixing can be controlled by employing medium recycling. There is an optimum recycling ratio, γ_{opt}, at which R_X reaches its maximum (3). If the dimensionless cell concentration is fixed at the exit of the reactor, C_{XF}, and a reactor is used which yields the smallest necessary volume, V_R, and a mean residence time of the medium, τ or Da, it turns out that there is a PFR loop combination with γ_{opt} at which Da has a minimum (Fig. 3), if $C_{XF} > C_{XFcrit}$ where

$$C_{XFcrit} = (K+1) - \sqrt{K(K+1)} \tag{8}$$

For $C_{XF} \leq C_{XFcrit}$, $\gamma_{opt} = \infty$ (CSTR)

For $C_{XF} > C_{XFcrit}$

$$Da = (1+\gamma)\left[\ln\frac{1+\gamma}{\eta+\gamma} + K\ \ln\frac{\xi+\gamma}{\eta+\gamma}\right] \tag{9}$$

where $\eta = \frac{C_{Xo}}{C_{XF}}$

$$\xi = \frac{1-C_{Xo}}{1-C_{XF}}$$

γ_{opt}, to a considerable degree, depends on the exit cell mass concentration, C_{XF} (Fig. 4). With decreasing C_{XF}, γ_{opt} increases and for $C_{XF} \leq C_{XFcrit}$, $\gamma_{opt} = \infty$. γ_{opt} also depends on K: with increasing K, γ_{opt} diminishes (3)

When using a tower reactor with longitudinal dispersion and medium recycling the relationships for substrate limitations (chapter 3.1) can be employed since the exit cell mass concentration, C_{XF}, can also be written as

$$C_{XF} = \frac{X_F}{X_o+Y_{X/S}S_o} = \frac{Y_{X/S}(S_o-S_e)}{X_o+Y_{X/S}S_o} \doteq \frac{S_o-S_e}{S_o} = U_S \tag{10}$$

Thus, reactor optimization with regard to U_S (chapter 3.1) is also optimization with regard to C_{XF}.

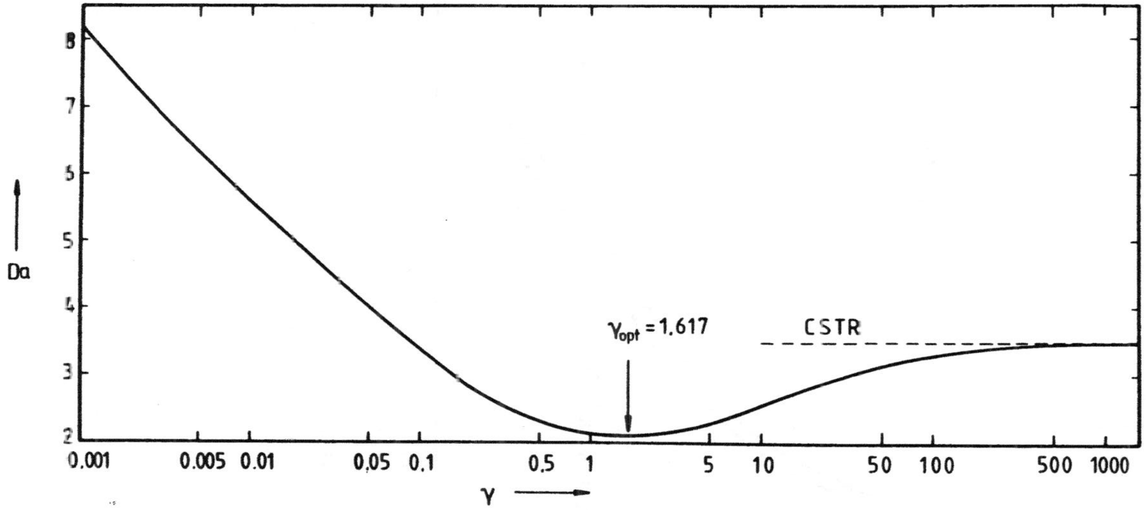

Fig.3: Da - number as a function of the medium recycling ratio for PFR with loop at $K = 0.125$, $C_{Xo} = 0$, $C_{XF} = 0.95$ (3).

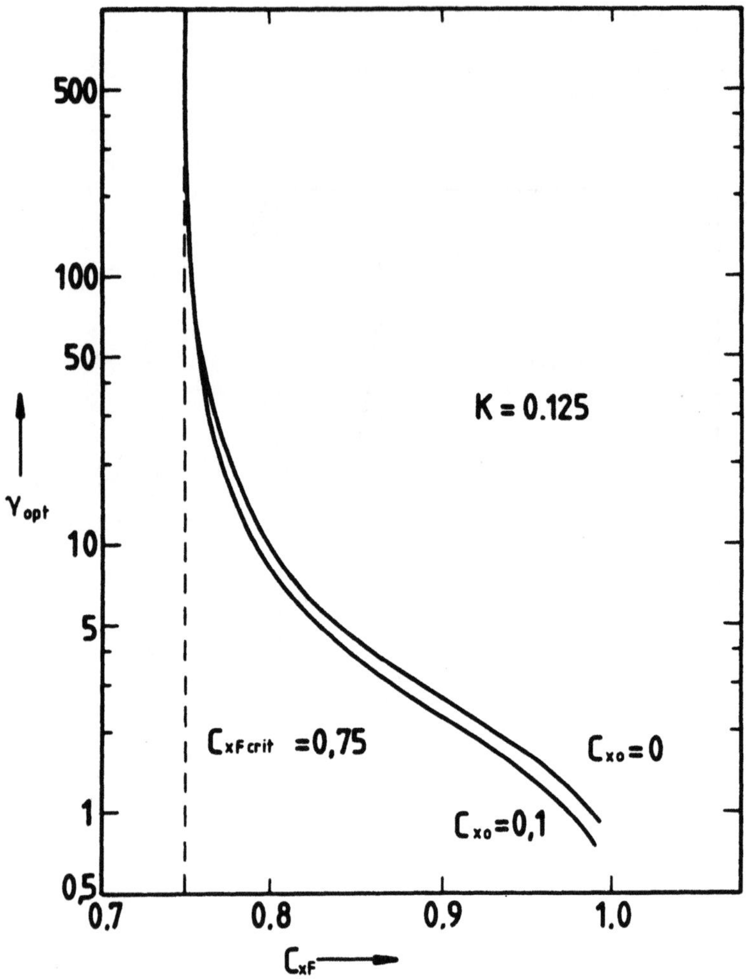

Fig.4: Optimum recycling ratio γ_{opt}, as a function of the exit cell mass concentration, C_{XF}, for PFR with loop at K = 0.125, C_{XO} = o.1 (3).

3.3 Optimization with Regard to the Oxygen Transfer and Power Input

To maximize the oxygen conversion, U_O

$$U_O = \frac{OUR}{DO_{LO}+OTR}, \tag{11}$$

where O_{LO} is the dissolved oxygen concentration in the medium at the medium entrance,

$OUR = \int_{z=0}^{z=1} \frac{1}{Y_{X/O}} \mu X dz$, the oxygen utilization rate

$OTR = \int_{z=0}^{z=1} k_L a(O_L^* - O_L)\, dz$, the oxygen transfer rate,

growth must be oxygen transfer limited, at least at the medium exit. This leads to a problem similar to the one considered in chapter 3.1. However, the theoretical treatment of this system is more complex, since oxygen is transferred into the medium along the tower. This treatment is not considered here.

In the oxygen transfer limited range, cell productivity is controlled by the OTR. In the CSTR, for maximum productivity, Pr_m, eq. (12) holds:

$$Pr_m \simeq Y_{X/O}\, k_L a(O_L^* - O_L^c); \tag{12}$$

since usually $D(O_{LO} - O_L^c) \ll k_L a(O_L^* - O_L^c)$. Here, O_L^c is the critical dissolved oxygen concentration in the medium below which the cells cannot utilize oxygen and O_L^* is the saturation value of O_L. Because of the low oxygen solubility the maximum driving force $(O_L^* - O_L^c)$ is low. $Y_{X/O}$ is determined by the microorganism and the substrate. Thus, Pr_m can usually be controlled by $k_L a$. Since the variation range of k_L is narrow the specific gas liquid interfacial area, a, is the main controlling parameter. a is given by:

$$a = \frac{6\, E_G}{d_S(1-E_G)}, \tag{13}$$

if d_S is small. Since E_G is also a function of d_S the Sauter bubble diameter, d_S, is the primary variable.

d_S is a function of the energy dissipation rate in the reactor and also depends on the medium properties with regard to bubble coalescence. It has been shown (5) that in coalescence suppressing media aerators with a spacially concentrated energy dissipation rate and in coalescence promoting media aerators with a spacially distributed energy dissipation rate give the highest OTR at the same specific power input, P/V_R. However, in the same medium and with the same type of gas distributor the OTR also depends on the efficiency of the aerator. The higher the fraction of microeddies in the turbulence, the higher the efficiency of the aerator. To minimize the specific power input the aerator efficiency has to be maximized. This holds true for primary dispersion. To minimize the specific power input P/V_R, the coalescence rate has to be minimized as well. Different strategies for minimizing the coalescence rate were considered in ref. 5.

4 PARTICULAR BIOREACTORS

Because of the wide variety of bioreactors it is difficult to classify them according to a physically acceptable system. They can be classified according to their construction or operation. However, here the power input mode is chosen as a decisive criterion. According to this criterion three different types of bioreactors can be distinguished:

1. Power input with mechanically agitated insertions (Fig. 5, Table 1),
2. Power input with a liquid pump in the outer loop (Fig. 6, Table 2),
3. Power input with gas compression (Fig. 7, Table 3)

Before the different reactors are briefly considered some general statements with regard to gas dispersion should be made.

It is useful to distinguish between primary and secondary gas dispersing facilities. In a stirred tank, the gas is introduced into the medium through a simple tube, ring nozzle, etc. and dispersed only coarsely. The fine dispersion is carried out with a secondary gas dispersing facility (impeller). In a multistage tower reactor, the primary gas dispersion is usually carried out with a perforated plate or porous plate, the secondary gas dispersion facilities (stage separating trays) redisperse the large bubbles formed by coalescence from

'ABLE 1

ioreactors with power input due to mechanically agitated
nsertions (Fig. 5)(20)

.1 rotating impeller
.2 rotating impeller and loop (H/D $\leqslant$ 2)
.3 rotating impeller and loop (H/D $>$ 2)
.4 rotating self-aerator impeller
.5 rotating self-aerator impeller and loop
.6 rotating self-aerator impeller horizontal loop
.7 cascade reactors with rotating impellers
.8 cascade reactors with axially oscillating mixing elements
.9 cascade reactors with pulsed liquid
.10 rotating film reactor
.11 rotating disc reactors
.12 rotating vane wheel reactors

ABLE 2

eactors with power input due to a liquid pump in the
uter loop (Fig. 6)(20)

.1 plunging jet reactor
.2 jet loop reactor
.3 plunging channel reactor
.4 nozzle loop reactor
.5 multistage countercurrent tower reactor
.6 tubular loop reactor
.7 reactor with rotating injector
.8 counter current packed tower reactor
.9 concurrent down flow tower reactor

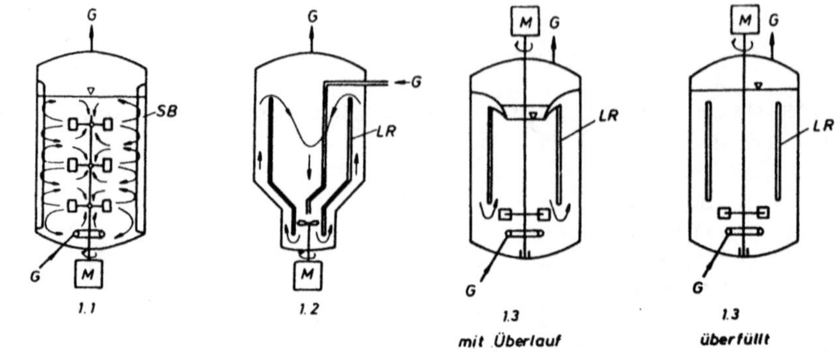

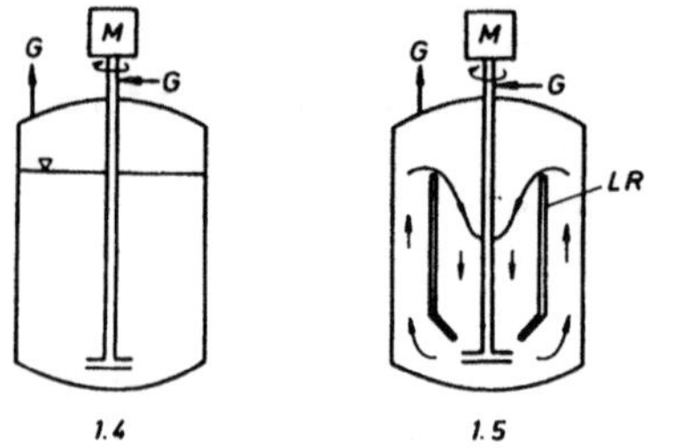

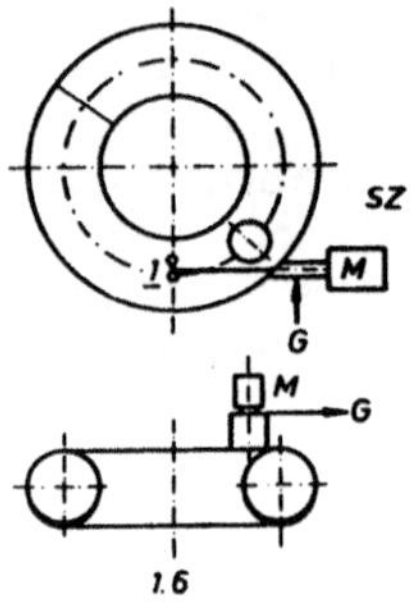

M engine
G gas (air)

SB baffles
LR draft tube

M engine
SZ foam destroyer
G gas (air)

LR draft tube
1 self-aerator impeller

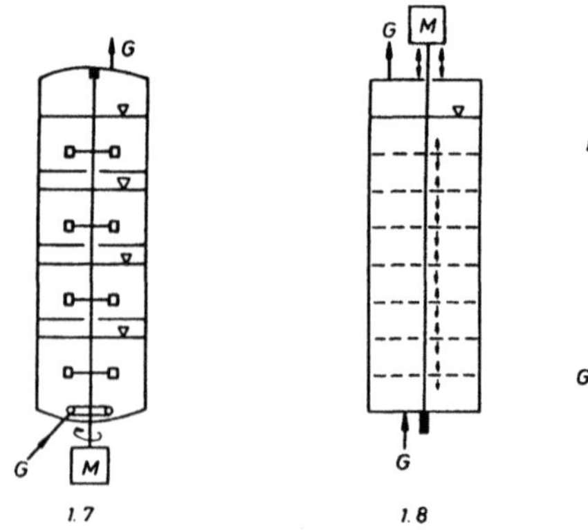

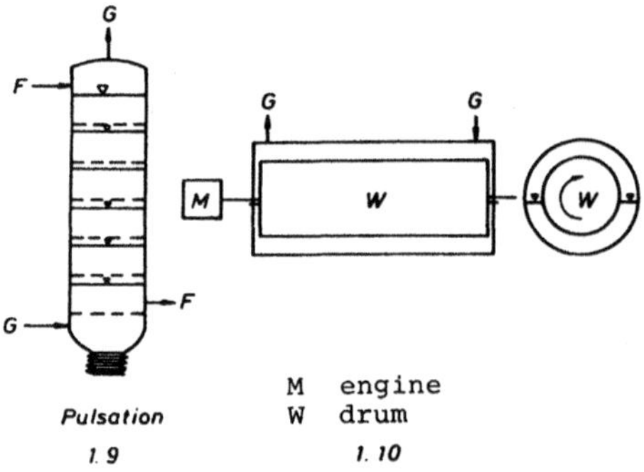

M engine
W drum

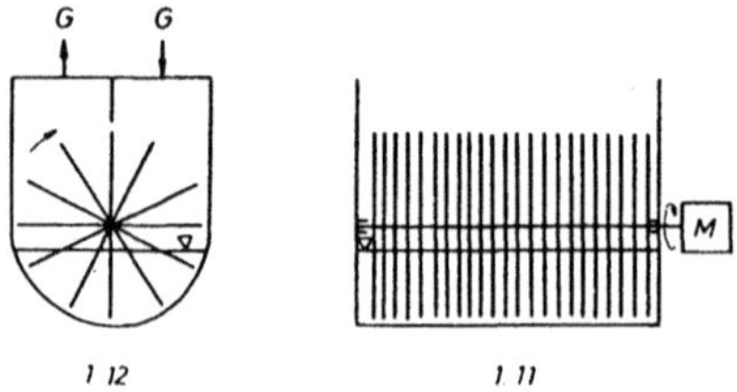

Fig. 5:
Bioreactors with power input due to mechanical insertions (20).

TABLE 3

Reactors with power input due to gas compression (Fig.7) (20)

3.1 Single-stage tower reactor
3.2 single-stage air lift with a draft tube loop
3.3 single stage air lift with an outer loop
3.4 single stage air lift with a partition wall
3.5 concurrent down flow loop reactor (deep shaft reactor)
3.6 concurrent multistage tower
3.7 concurrent tower with static mixer
3.8 tower loop with trays
3.9 multistage tower loop with liquid pulsation and static mixer

TABLE 4

Sorption characteristics of aerators according to Zlokarnik (21)

Aerator	$Y_3 \cdot 10^5$	$X \cdot 10^{-5}$	
single orifice*	6.2	independent	
perforated plate* (3 mm in hole diameter)	6.4	independent	
perforated plate* (1 mm in hole diameter)	8.8	independent	
porous plate *	14.0	independent	
submerge aerator of Frings**	9.5	independent	
static mixer (Kenics)	6.0	independent	
injector nozzle *	10.0	1.0	H = 7 m
	13.0	2.0	H = 7 m
ejector (slot) nozzle	10.0	1.0	H = 7 m
	15.0	2.0	H = 7 m

* in bubble column
** in bubble column and with sulphite solution

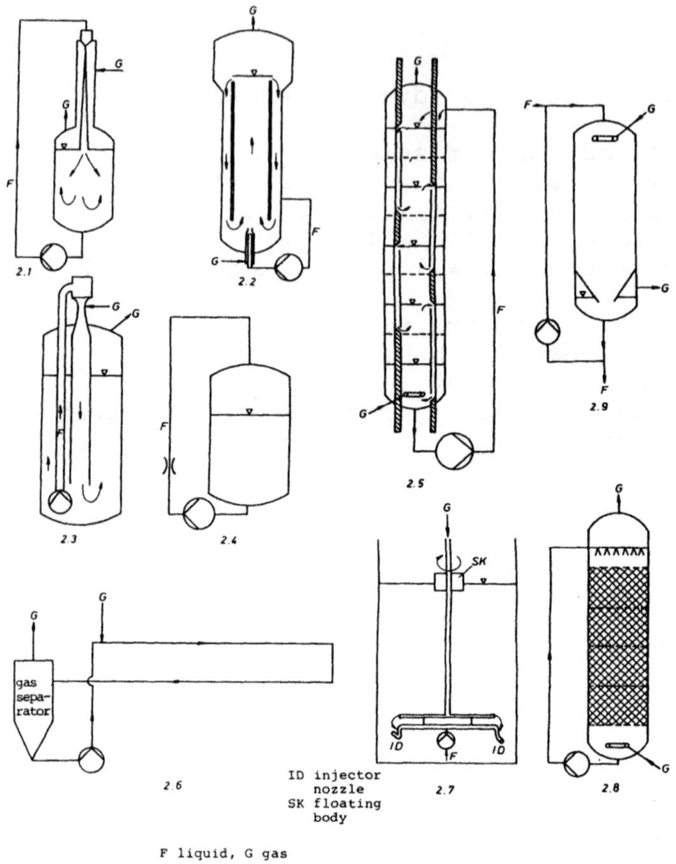

Fig. 6: Bioreactors with power input due to liquid pump in the loop (20).

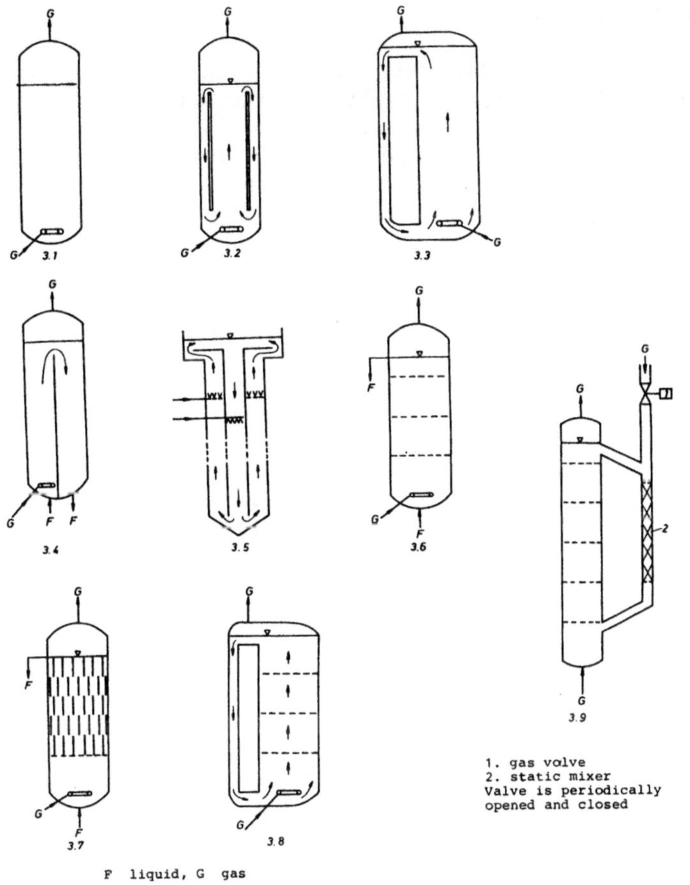

'ig. 7: Bioreactors with power input due to gas compression (20).

the small primary bubbles. In the latter case the degree of secondary gas dispersion is much lower than that of primary dispersion.

It is useful to distinguish between coalescence suppressing and promoting media. In coalescence suppressing media the gas dispersion should be spacially concentrated (usually to the bottom of the reactor). The high dispersion degree of the gas phase which is achieved in this range is largely preserved also if the bubbles arrive in ranges in which the dynamic equilibrium bubble sizes are much larger than the primary bubble sizes. The bubble diameter is mainly influenced by the primary dispersion degree and only slightly by the conditions which prevail in their actual surroundings. In contrast to this, in coalescence promoting media the gas dispersion should be uniformly distributed within the reactor since the actual bubble size is controlled by the dynamic equilibrium bubble size in their momentary surroundings. For the optimal strategy for the spacial distribution of the energy dissipation rate in the reactor, see ref. 5.

4.1 Reactors with Power Input due to Mechanically Agitated Insertions

In the following, only the physical action of a turbine stirrer should be considered because it is the most popular and best investigated stirrer type. The movement of the fluid in a stirred tank reactor enforced by a rotating impeller can be divided into primary and secondary flow patterns. The primary movement is the rotational or tangential flow of the fluid. Most of the energy transferred from the stirrer to the fluid is consumed by this movement. This controls the gas dispersion. However, it is of minor importance for the mixing process. The secondary movement consists of radial and axial components (Fig. 8). In mixing processes the secondary movement is the most important part of the flow pattern. Only a small fraction of the energy transferred to the fluid is contained in this secondary movement (6). Behind the upper and lower edges of the impeller gas is sucked into the wakes by the underpressure which prevails within them and dispersed by the high local shear stress. The bubbles formed in the wakes are collected in the core of the vortex. Because of the high bubble concentration in the vortex core, bubble coalescence is promoted here. The vortices lose their energy and thus their rotational speed with an increasing distance from the impeller due to internal friction. As a consequence, the

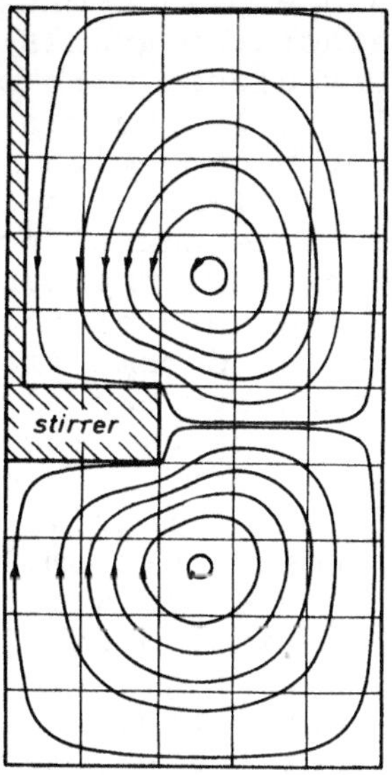

Fig.8: Streamlines of secondary fluid movement(6)

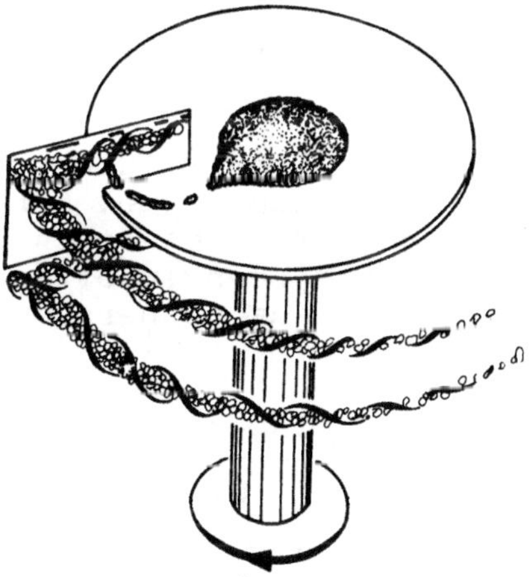

Fig.9: Bubble generation in vortex threads in the wake of a turbine stirrer (6).

underpressure reduces in the vortex core and the bubble gradually become free. Their size depends on the local energy dissipation rate which prevails at this position if a coalescence promoting medium is used. In the vicinity of the impeller blade pseudo-macroturbulence with a strong periodical character prevails which gradually transfers into true turbulence with an increasing distance from the impeller (7). One can assume that on the vortex surface free turbulence with microscales is formed which is very efficient with regard to gas dispersion. However, this free turbulence can hardly be used for gas dispersion since the bubbles are held together in the tube-like core of the vortex. At the position where the bubbles become free from this core, the free turbulence intensity has already been reduced considerably. Hence, the energy dissipation in the vort cannot be used very efficiently for gas dispersion. Onl at longer distances from the impeller where the vortex has already been decomposed into small turbulent eddies, can microturbulence be detected which is marked by a high energy dissipation rate, a nearly isotropic proper and a high efficiency with regard to gas dispersion (8) Because of the high shear forces in the range of the impeller blades the size of the primary bubbles is much smaller than the size of the bubbles which are formed at the vortex end by microturbulence. Therefore the efficiency of the power input with regard to gas dispersion in stirred tank reactors is rather low.

4.2 Reactors with Power Input due to a Liquid Pump in the Outer Loop

The utilization of the power input can be very different depending on the aerator type. Since gas dispersion with injector and/or ejector nozzles and plunging jets have recently been investigated fairly thoroughly, only these two gas dispersion organs will b considered.

The mechanism of gas dispersion with ejector nozzl was investigated by Kürten and Maurer (9) and Hallensleben et al. (10). Through the internal nozzle a fast liquid jet is formed. From the outer annulus (nozzle) gas flows into the momentum exchange tube in which a large bubble is formed at first. The fast liquid jet pulsates and decomposes into fine droplets which hit th gas/liquid interface of the large bubble and destroy it The small bubbles formed during this process are redispersed into much smaller bubbles in the momentum exchange tube due to the high dynamic pressure of the turbulence which prevails in it.

The mechanism of gas dispersion by means of plunging jets was investigated by Smith et al. (11), Burgess et al. (12) and Suicu et al. (13). By means of a nozzle a downcoming liquid jet is produced which hits the liquid pool surface and penetrates into it (2.1 in Fig. 6). The jet identity is preserved until a gas envelope of the jet exists. The disintegration of this envelope yields small bubbles of high density which move downwards in the liquid and in a transverse direction. Hence the bubble swarm forms a cone. This two-phase jet drags the surrounding liquid downwards due to its momentum exchange and mixes with it. Due to this the jet gradually decelerates. At the boundary of the two-phase jet and the surrounding liquid free turbulence develops. Its thickness increases in the direction of the flow. The final gas dispersion is again caused by turbulence. Large bubbles which are formed by coalescence from small bubbles rise and leave the pool. The small bubbles are dragged with the liquid and separated from it in a two-phase separator pump. The liquid is pumped to the top of the reactor and the loop is closed by forming the liquid jet by the nozzle.

The properties of the liquid jet significantly influence gas dispersion (11).

4.3 Reactors with Power Input due to Gas Compression

Three ranges of gas dispersion can be distinguished: single bubble formation at low aeration rates, gas jet formation at intermediate aeration rates and dispersion in a turbulent field at high aeration rates. The range of single bubble formation was investigated most thoroughly (14). However, this range is not interesting in practice. The bubble formation in the second range is caused by the nonstable interface between the gas jet and the surrounding liquid. The bubble formation mechanism is analogous to the droplet formation from a liquid jet according to Rayleigh (15). Meister and Scheele developed a theory for this bubble formation (16). Hallensleben et al. (10) investigated this bubble formation experimentally and confirmed the theory of Meister and Scheele. If the aeration rate is increased the primary bubble size increases at first, then passes a maximum and diminishes. In the increasing size range, single bubble formation occurs, at the maximum, gas jet decomposition prevails and in the decreasing size range, a turbulence mechanism is responsible for bubble formation (10). The latter bubble formation mechanism prevails in industrial reactors.

4.4 Comparison of the Different Reactor Types

The considerations in 4.1 to 4.3 show that in the operation ranges which are important for the industry gas dispersion is always caused by turbulence, thus the comparison of different reactors with regard to the oxygen transfer rate based on the turbulence theory should be possible.

Kolmogoroff (17) postulates that, at large Reynolds numbers of turbulent motion, the local property of turbulent motion should have a universal character described by the following concepts. First, it is locally isotropic whether the large scale motions are isotropic or not. Second, the motion at the very small scales is chiefly governed by the viscous forces and the amount of energy which is passed down to them from the larger eddies. The large eddies tend to break down into smaller eddies due to inertial forces. These in turn break down into still smaller eddies and so on. At the same time, viscous forces dissipate these eddies into heat at very small scales. In the long series of cascading processes of reaching the smallest eddies, the turbulent motion adjusts itself to some definite state. The further down the scale, the less the motion is dependent on the large eddies. Furthermore, in line with Taylor's experimental findings, Kolmogoroff essentially postulates that practically all the dissipation of energy occurs at the smallest scales when Reynolds number of turbulent motion is sufficiently high (18). When assuming the validity of this theory, the microscale, l, depends only on the power input, P/V_R and the liquid porperties.

$$l = \frac{\eta_F^{3/4}}{\rho_F^{1/2}} \left(\frac{P}{V_R}\right)^{-1/4} \qquad (14)$$

where η_F is the dynamic viscosity of the liquid and ρ_F the density of the liquid.

If one assumes that the local structure of turbulence does not change in the presence of bubbles, relationship (15) holds true for the dynamic equilibrium (maximum stable) bubble size, d_e, which must be considerably larger than l (19).

$$d_e = C_1 \frac{\tau^{0.6}}{\rho_F^{0.2}} \left(\frac{P}{V_R}\right)^{-0.4} \tag{15}$$

where τ is the surface tension of the liquid

Several relationships analogous to eq. (15) were developed (20). However, if d_e were only the function of τ , ρ and P/V_R, all of the aerators would result in the same d_e value, in the same medium and at the same specific power input, P/V_R. This is obviously not the case. Of course, for coalescence suppressing media the deviation from this expectation could be caused by the difference between the actual bubble diameter, d_S, and the dynamic equilibrium diameter, d_e. Therefore, in the following, coalescence promoting media are considered for which it can be assumed that $d_S \simeq d_e$ holds true. Zlokarnik (21) recommended for stirred tank reactors that instead of the volumetric mass transfer coefficient

$$k_L a = \frac{G}{V_R \Delta C} \tag{16}$$

a dimensionless group

$$Y_2 = \frac{G}{q \Delta C} \tag{17}$$

should be plotted as a function of a dimensionless group

$$X = \frac{P}{q} \frac{1}{\rho_F (\gamma g)^{2/3}} \tag{18}$$

(instead of the specific power input P/V_R). In eqs. (16 to 18) G is the oxygen transferred into the medium per unit of time,

$\Delta C = C_F^{*} - O_F^{C}$ is the driving force for oxygen transfer,

q the volumetric gas flow rate and

γ the kinematic viscosity of the liquid.

He found simple relationships between Y_2 and X for coalescence promoting systems:

$$Y_2 = 1.5 \cdot 10^{-2} X^{0.5} \tag{19}$$

For ejector-injector nozzles and static aerators the use of the following dimensionless groups was recommended (21):

$$\frac{k_L a}{w_{SG}} = \frac{G}{\Delta CqH}, \quad Y_3 = \frac{G}{\Delta CqH}\left(\frac{\gamma^2}{g}\right)^{1/3} \qquad (20)$$

In eq. (20) w_{SG} is the superficial gas velocity and H the height of the aerated layer.

For perforated plates, single orifice, porous plates submerse aerators of Frings and static mixers Y_3 does not depend on X (Table 4). For injector and ejector nozzles Y_3 increases with increasing X from $Y_3 = 7 \cdot 10^{-5}$ at $X = 2 \cdot 10^4$ to $Y_3 = 1.5 \cdot 10^{-4}$ at $X = 5 \cdot 10^5$ (21). For the economy of the process the efficiency of the aerator is important. Some efficiencies, $E = G/P$, for different aerators at different aerated layer heights are given in Table 5. The injector nozzle attains its optimum efficiency at H = 14 m: E = 3.9 kg O_2/kWh. Only the porous plate and the porous filter body exceed this high value at lower liquid layer heights.

The efficiency of different aerators differs considerably also in coalescence promoting media in which $d_S \simeq d_e$ can be set.

What is the cause of this difference? According to the theory of Kolmogoroff and eq. (15), d_e should depend only on the liquid properties (in Tables 4 and 5, with one exception only water was used) and on P/V_R. Thus, the Y_3 values for the same P/V_R should be the same. This is not the case.

Since for all dispersion organs in the ranges which are important for the practice the gas dispersion occurs due to turbulence mechanisms it is to be expected that there is a relationship between the turbulence properties and the efficiency of gas dispersion. It has already been pointed out that efficient gas dispersion is only possible, if the microeddy size is smaller than d_e and if the percentage of microturbulence is high enough. Therefore the power spectrum of turbulence should influence the efficiency of gas dispersion. Fig. 10 shows one-dimensional power spectra in stirred tank reactors of different sizes according to van der Molen and van Maanen (22). As usual, the energy content of turbulent eddies was plotted as a function of the wave number, K, which is inversely proportional to the diameter of the

TABLE 5

Efficiency of different aerators according to Zlokarnik (21)

Aerator	E(kgO_2/hWh)	aerated liquid height H(m)
tube stirrer	1.4	2
turbine stirrer	2.0 to 2.8	3
propeller	0.8 to 1.1	3
plunging jet	0.88	10
single orifice**	3.32	0.9 to 3.5
perforated plate (3 mm in hole diameter)	3.39	0.3
static mixer (Kenics)	2.5 to 3.5	8.25
slot nozzle *	{3.6 3.8	7 15
injector nozzle*	3.6	7
perforated plate* (1 mm in hole diameter)	⩾ 4	0.3
porous plate*	⩾ 4	0.3
submerse aerator of Frings	⩾ 4	2 to 5

* in bubble column
** in bubble column and with sulphite solution

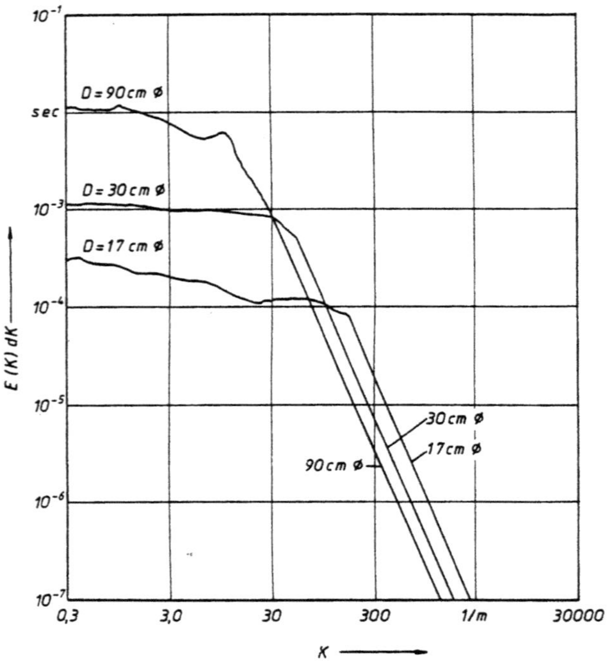

Fig. 10: One-dimensional power spectra in stirred tank reactors with different tank diameters, D (23)

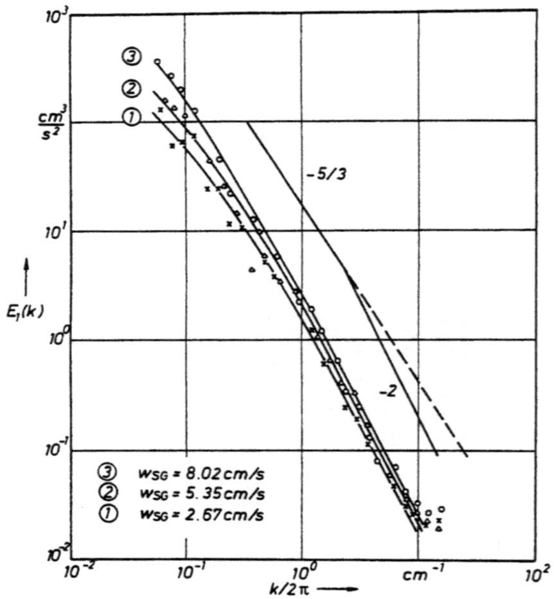

Fig. 11: One-dimensional power spectra in a tower reactor with perforated plate aerator 3 mm in hole diameter. 1 % methanol (23)

eddies. From this diagram the well-known phenomenon can be recognized that large eddies have a high energy content. With a diminishing eddy diameter, i.e. an increasing wave number, K, the energy content diminishes. The energy is mainly dissipated at very high wave numbers. The gas dispersion occurs exclusively at high wave numbers. While the energy cascades down from large eddies to microeddies, a considerable amount of energy is dissipated already. In bubble columns with perforated or porous plates primary eddies are produced which are much smaller than in stirred tank reactors. Therefore, in bubble columns the low wave number range of the power spectrum is missing in which the energy is uselessly dissipated with regard to the gas dispersion. But also the same bubble column with the same medium yields different power spectra, if different aerators are used (Figs. 11 and 12). With a diminishing hole diameter of the perforated plate the power spectra are shifted to higher wave numbers.

The relations can be illustrated even more clearly if the energy dissipation rate is plotted as a function of the wave number (Fig. 13). It can be noted that each energy dissipation spectrum exhibits a maximum. The shape of the spectra and the location of the maxima are different. Since gas dispersion occurs at high wave numbers it can be expected that the higher the wave number at which the maximum appears the higher is the efficiency of the aerator.

When using a porous plate the maxima appear at higher wave numbers than for perforated plates (Fig. 13). With an increasing hole diameter and a free cross-sectional area of perforated plates the maximum is shifted to lower wave numbers. A comparison of the sequence: porous plate, perforated plate 0.5 mm, 1.0 mm and 3.0 mm in hole diameter (with regard to the position of the maximum of their energy dissipation spectra) with the sequence for their efficiency shows a good agreement. It seems that when shifting the maximum of the energy dissipation spectrum to higher wave numbers by using different aerators an efficiency increase of the aerator can be expected. In addition, Fig. 13 shows that the increase in the aeration rate shifts the energy dissipation spectrum to higher energy dissipation rates but it does not influence its shape or the position of its maximum. This is consistent with the observation that the efficiency of porous plates is independent of the aeration rate, i.e. of X (Table 5). The same independence of the shape and the position of the maximum of the ener-

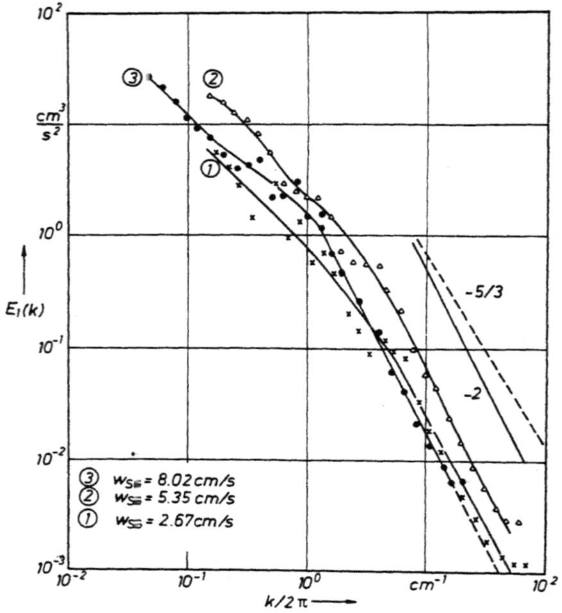

Fig. 12: One-dimensional power spectra in a tower reactor with a porous plate aerator 5 µm in pore diameter. 1 % methanol (23)

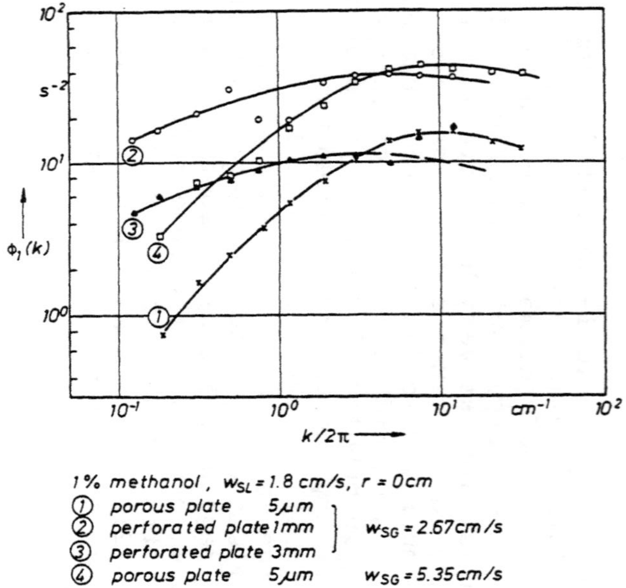

Fig. 13: Energy dissipation spectra in tower reactors (23)

gy dissipation spectrum can be expected also for perforated plates, for the single-orifice aerator, the static mixer and the submerse aerator of Frings. However, more measurements are needed to establish quantitative relationships.

REFERENCES

1. Schügerl, K. Introduction to biochemical systems. Formal treatment of biochemical reactions and characterization of fermentation systems (Proceedings of NATO ASI on "Mass transfer with chemical reaction", Turkey, 1981)
2. Schügerl, K. Chem. Ing. Techn. 49 (1977) 605
3. Adler, I. Dissertation, University of Hanover 1980
4. Bishoff, K.B. Canad. J. Chem. Engng. 44 (1966) 281
5. Schügerl, K. Paper on the 3rd European Conference on Biotechnology. Eastbourne 1980
6. Brauer, H. "Power consumption in aerated stirred tank reactor systems". Adv. in Biochem. Eng. Ed. T.K. Ghose, A. Fiechter, N. Blakebrough, Springer Verlag. 13 (1979) 87
7. Riet, K. van't, W. Bruin and J.M. Smith. Chem. Eng. Sci. 31 (1976) 407
8. Mujumdar, A.S., B. Huang, D. Wolf, M.E. Weber and W.J.M. Douglas. Can. J. Chem. Eng. 48 (1970) 475
9. Kürten, H. and B. Maurer. Partikel Technologie Nürnberg 1977. Ed. H. Brauer. O. Molerus H. 47
10. Hallensleben, J. R. Buchholz, J. Lücke and K. Schügerl. Chem. Ing. Techn. 49 (1977) 663
11. E. van de Sande and J.M. Smith. Chem. Eng. Sci. 31 (1976) 219
12. Burgess, J.M. and N.A. Molloy. Chem. Eng. Sci. 28 (1973) 183
13. Suicu, G.D. and O. Smigelschi. Chem. Eng. Sci. 31 (1976) 1217
14. Kumar, R. and N.R. Kuloor. Adv. Chem. Eng. 8 (1970) 225. Ed. Drew et al., Academic Press
15. Lord Rayleigh. Phil. Mag. 34 (1892) 177
16. Scheele, G.I. and B.J. Meister. AIChE J. 14 (1968) 9, 15.
17. Kolmogoroff, A.N. Compt. rend. Acad. Sci. U.R.S.S. 31 (1941) 538, 32 (1941) 16
18. Lin, C.C. Statistical Theories of Turbulence, Princeton University Press 1961.

19. Calderbank, P.H. Biochem. Biological Eng. Sci. 1 (1967) 101

20. Schügerl, K. Chem. Ing. Techn. 52 (1980) 951

21. Zlokarnik, M. Korrespondenz Abwasser 27 (198/) No.3 194, Chem. Ing. Techn. 52 (1980) 330

22. Molen, K. van der and H.R. E. van Maanen. Chem. Eng. Sci. 33 (1978) 1161

23. Zakrzewski, W. Dissertation, University of Hanove

TURBULENCE MEASUREMENTS IN BUBBLE COLUMNS

A. Lübbert

Institut für Technische Chemie
Universität Hannover
Callinstr. 3, D-3000 Hannover 1, FRG

1 INTRODUCTION

Most problems in the design and performance prediction of bubble column reactors appear, because it is - up to now - not possible to control the fluid dynamics in such reactors. Especially the parameters of the turbulent flow in these reactors are of major importance.

One interesting aspect is the breakup of bubbles in turbulent liquid motions. This is one of the main processes which determine the bubble size distribution and hence the specific interfacial area a.

It is the dynamic pressure force which has mostly been made responsible for the bubble breakup (1,2). Dynamic pressure forces which can influence the bubbles' sizes or forms result from changes in the liquid velocity over distances in the range of bubble diameters. This means that the bubble breakup is dependent on the spatial dimensions of the flow elements of the turbulent motions in the bubble column. Information on turbulent flows can be obtained by the methods of the statistical turbulence theory (3).

Systematic experimental studies on the spatial structure of the turbulent flow in bubble columns are missing. Therefore, prerequisite to further modelling, there is a need of accurate experimental data on this flow.

In this contribution we will report on turbulence measurements in the bubble column. Here we restrict ourselves to the determination of auto- and cross-correlations from measured velocity data.

2 EXPERIMENTAL

The liquid phase velocities in a laboratory scale bubble column of 140 mm in diameter have been measured with a constant temperature anemometer. Data acquisiti and analysis have been done by means of a modern proce computer.

The central part of this system - as shown schematically in figure 1 - is a "Digital Equipment" PDP 11/40 minicomputer. For fast computations of signa analysis routines a block floating point array process has been installed at the host computer.

Anemometer signals can be transferred to the proc computer in two ways: In the conventional method the output of the anemometer electronics is first recorded on an analog tape and then digitized by the computer's analog/digital converter (ADC). The second method uses a transient recorder, which permits very high sampling rates. Transient recorders are fast digital recording units.

3 DATA ANALYSIS

To calculate the correlation functions from the measured velocity-time histories u(t), data have been segmented in records of 1024 sampling points.

These records have been transformed into the frequency domain by a "Fast-Fourier Transformation" (FFT) and ensemble averaged there. The Fourier Transformatio of the resulting array back into the time domain gives the correlation function (4) by the Wiener Khintchine theorem.

The time critical part of this analysis is the segment averaging in the frequency domain. Figure 2 shows a flow diagram of the computer program for the determination of cross-correlations. Data have been sampled simultaneously from two anemometer probes and stored in alternating elements of an input buffer.

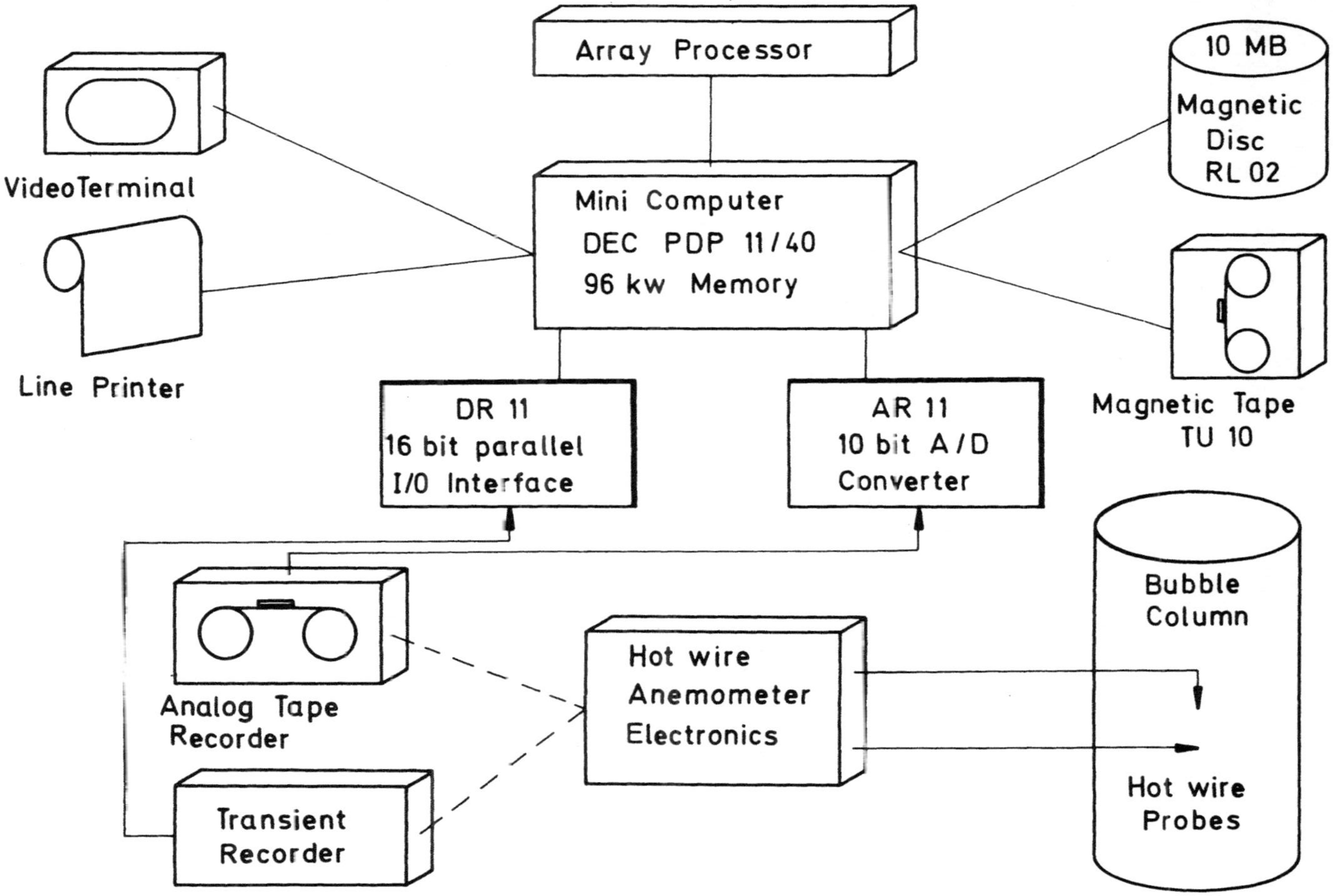

Fig. 1: Schematical representation of the experimental setup

READ FIRST DATA

DEMULTIPLEX INPUT
DATA INTO TWO
BUFFERS

FFT FORWARD CHANNEL I
FFT FORWARD CHANNEL II

COMPLEX CONJUGATE OF
RESULT OF CHANNEL II

READ NEXT
DATA SEGMENT
INTO THE
INPUT BUFFER

COMPLEX MULTIPLY
CHANNEL I AND II

ACCUMULATE RESULT
IN BUFFER III

REPEAT THIS CYCLE
N TIMES

Fig. 2: Flow diagram of the computer routine for the segment averaging in the frequency domain

First of all, these data have to be demultiplexed into data arrays $u_1(t)$ and $u_2(t)$, corresponding to the two input channels. Both arrays then have to be Fourier transformed yielding the raw spectra $X_1(f)$ and $X_2(f)$. After conjugating $X_2(f)$ the complex product $X_1(f) \cdot X_2^*(f)$ has been accumulated in a third buffer array.

This function chain has been carried out by the array processor, while the host computer reads the next data segment from the storage device.

The time needed for the calculation of one such cycle is about 20 ms, including accumulations for the determination of the mean velocity and the root mean square values - the latter operations are not shown in the figure -. This time is in the same order of magnitude as the time needed to process this data analysis on a large-scale computer like CDC's CYBER 76, and it is sufficient for an on-line analysis of turbulence data.

4 AUTO-CORRELATION FUNCTION

When the two data channels are identical - sampled from the same probe - the analysis results in the auto-correlation function of the time history. A typical example is shown in figure 3. The auto-correlation function R of the velocity data taken in the middle of the bubble column is plotted against the time displacement τ .

The most prominent feature of this auto-correlation function is that it does not show the exponential decay as one expects from single-phase experiments (3). The curve is divided into two parts, reflecting two mechanisms influencing the signal. The decrease at higher time displacements reflects the correlation within the velocity data of the liquid bulk, and from the steep decrease at small time displacements we believe that it is due to the bubbles' influence on the velocity signal. One reason for this assumption is that the time at which the turn of the curve occurs corresponds well with the time needed by bubbles of mean diameter to pass the probe with mean velocity. Other indications result from measurements of space-time correlations.

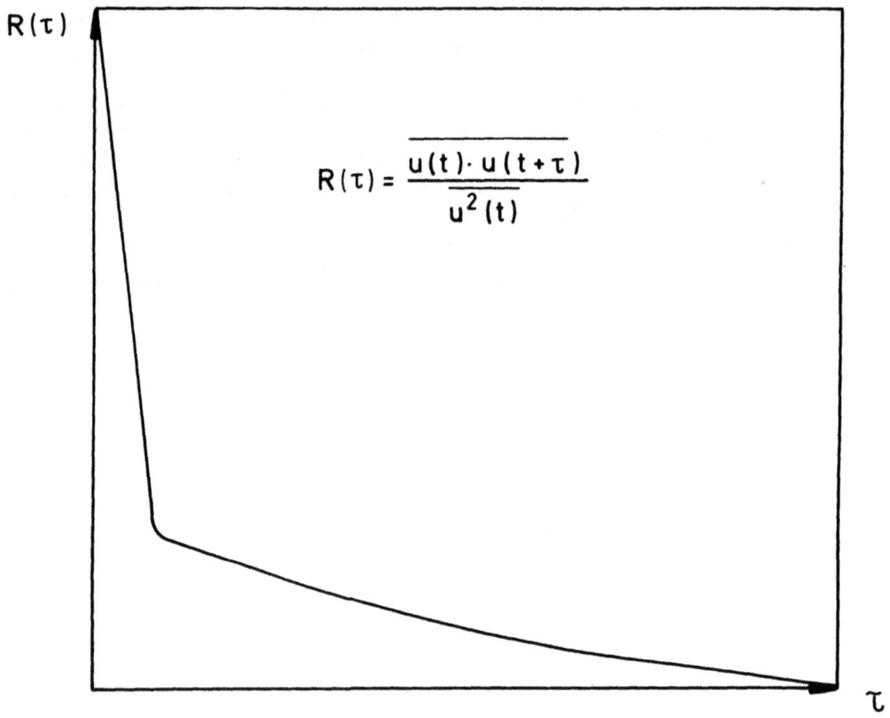

Fig. 3: Typical example for an autocorrelation function measured with a hot wire anemometer in an air/water bubble column

5 SPACE-TIME CORRELATIONS

The auto-correlation function in figure 3 contains information on the structure of the velocity signal in time. In chemical reaction engineering, however, one is mostly interested in spatial structures of the flow. Usually the link between the measured time histories u(t) and the spatial flow field u(X) were obtained by application of Taylor's hypothesis (5). Taylor's hypothesis, however, only holds when the absolute value of the fluctuating velocity component u(t) is small when compared with the mean flow velocity $\bar{U}$. Normally, this is not fulfilled in bubble columns operated in the heterogeneous mode.

The most direct way of overcoming this difficulty is the measurement of space-time correlations (6). Two hot wire anemometers were set up and placed on the axis of the bubble column. The two wires were parallel to one another, perpendicular to the column axis, a distance ΔX apart. The lower probe was held at a fixed position - 700 mm above the gas distributor - the upper one movable along the axis of the system.

Data were sampled from both probes simultaneously with the axial displacement ΔX of the probes as parameter. The results for several parameter values are plotted in figure 4. All these curves were normalized on their maximal value.

One can see immediately that the time displacements at which the maxima occur increase with the axial displacement. More interesting, however, are the shapes of the curves. They change significantly with increasing axial displacements of the probes. This fact speaks against the applicability of Taylor's hypothesis, which would require that the curves keep their shape and are only shifted.

From its maximum to higher time displacements the top curve in figure 4 shows a shape which is very similar to the one of the auto-correlation function in figure 3. At larger space displacements ΔX, however, the sharp peak in the region of the maxima is smeared out. This is another indication of the bubbles' influence on the correlation functions, because it is not very probable that one and the same bubble would hit both probes when they are far apart.

In figure 5 the time displacements Δt at which

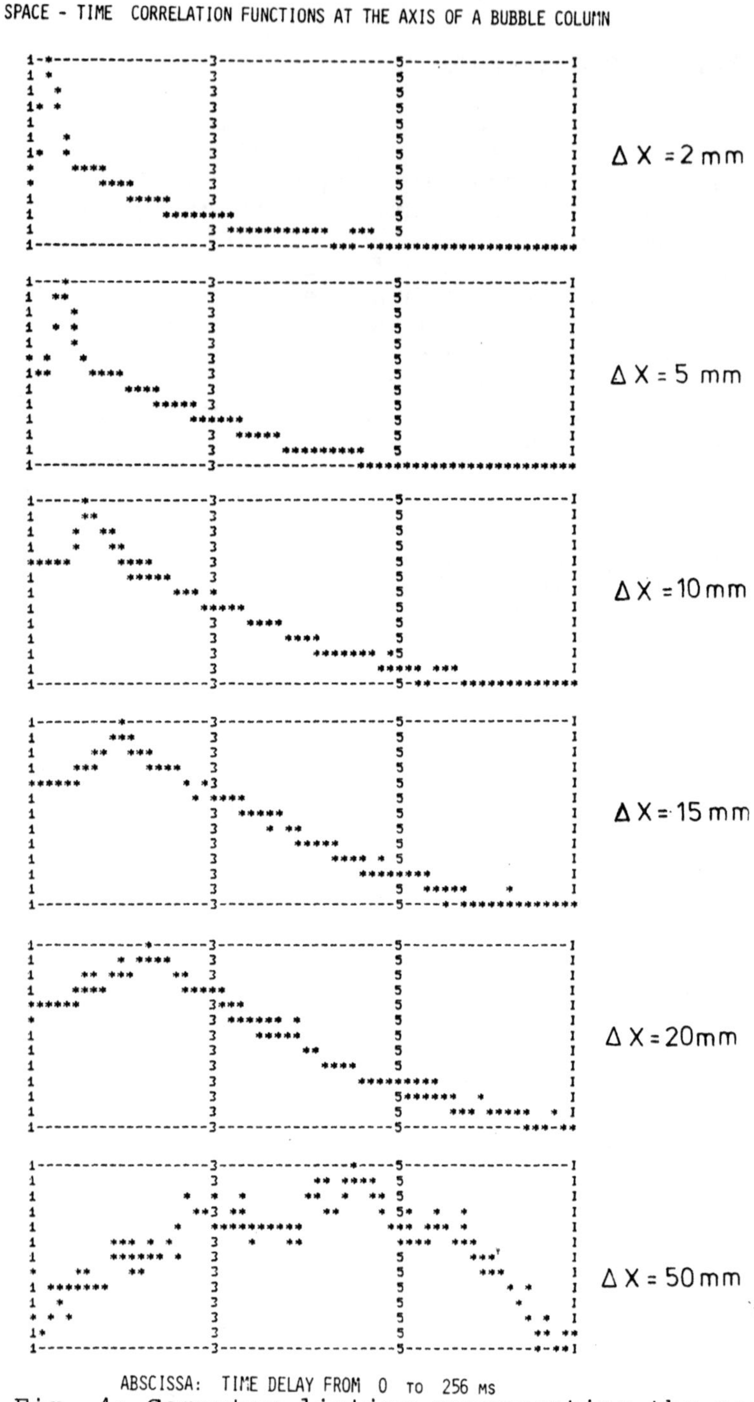

Fig. 4: Computer listing representing the results of a set of space-time correlations

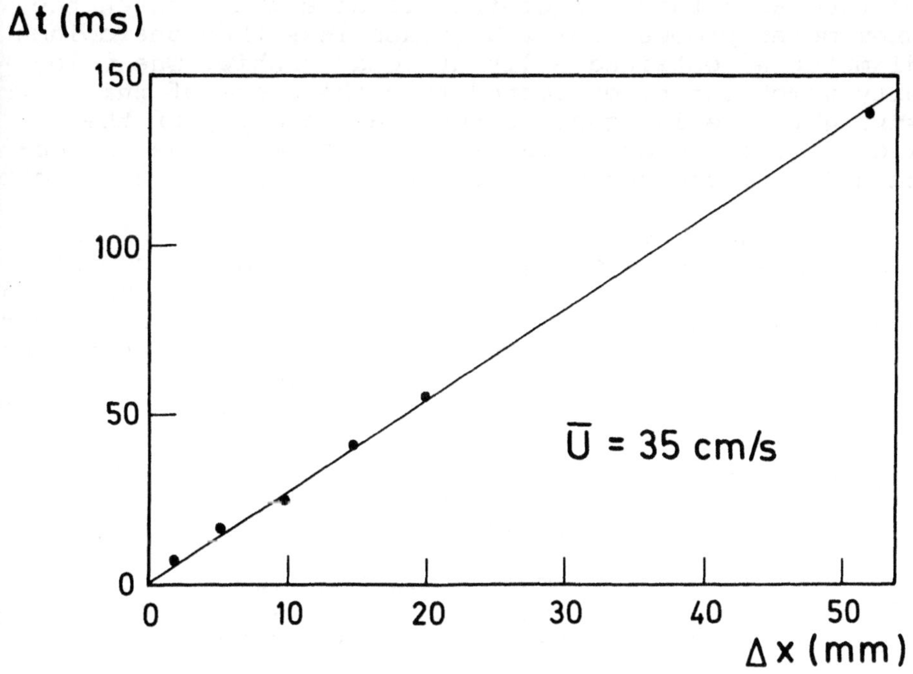

Fig. 5: Times of maximal correlation

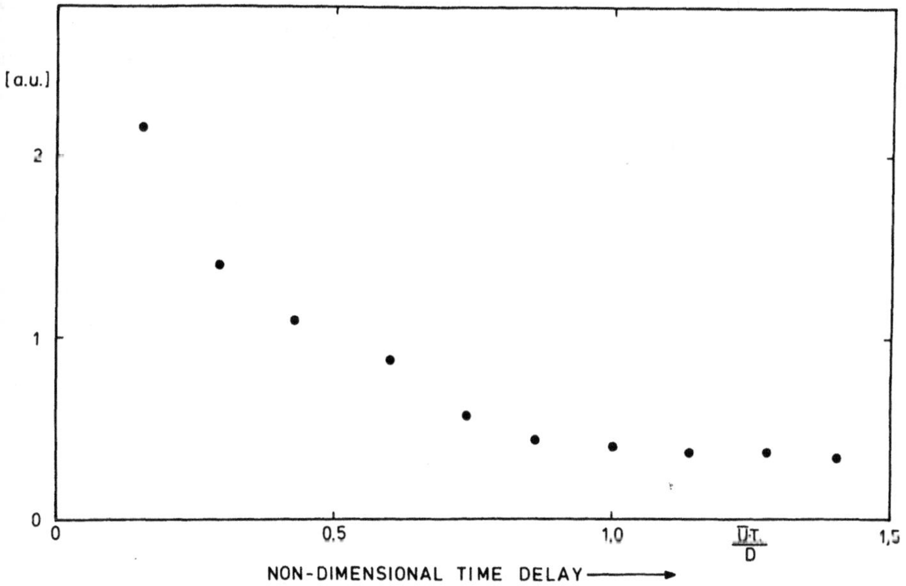

Fig. 6: Time-correlation following the mean motion of the flow

the maxima of the space-time-correlations occur are plotted as a function of the distance ΔX between the anemometer probes. For ΔX values less than one column diameter we obtained a linear relationship. The velocity which can be extracted from the slope of this straight line is equal to the mean velocity of the liquid. This velocity value is the same as the one obtained from the anemometers for the mean liquid velocity.

At space displacements greater than one column-diameter we observed significant deviations from this linear relation between the time of maximal correlation and the distance between the probes. Sharp peaks appear at those space displacements which are multiples of the column diameter D; however, they disappear after introducing some disturbing bodies into the flow (e.g. plexiglass rods). These findings lend some support to the assumption of convective flow cells within the bubble column. These coherent structures could be the circulation cells proposed by Joshi and Sharma (7).

More information can be derived from the heights of the maxima of the curves in figure 4, plotted in figure 6 in a nondimensional representation.

This curve represents the correlation as seen by an observer who follows the flow with the mean liquid velocity. This representation shows that the correlation is significantly greater than zero only within time displacements which are in the order of the times needed by the mean flow with the velocity $\bar{U}$ to move along one diameter D of the bubble column. Consequently, the observer does not see a frozen flow pattern.

6 SPACE CORRELATION

The most interesting descriptive function is the space-correlation function $R(\Delta X)$, which can be extracted from the space-time correlations of figure 4 by plotting the values at $\tau = 0$ as a function of the spatial displacement ΔX. The result is shown in figure 7. This space-autocorrelation decreases monotonously to zero within ΔX value in the order of one column diameter D.

It is interesting to compare this result with the autocorrelation function, which results from the time-autocorrelation by transformation using Taylor's hypothesis. We found large differences between these two curves, as shown in the figure.

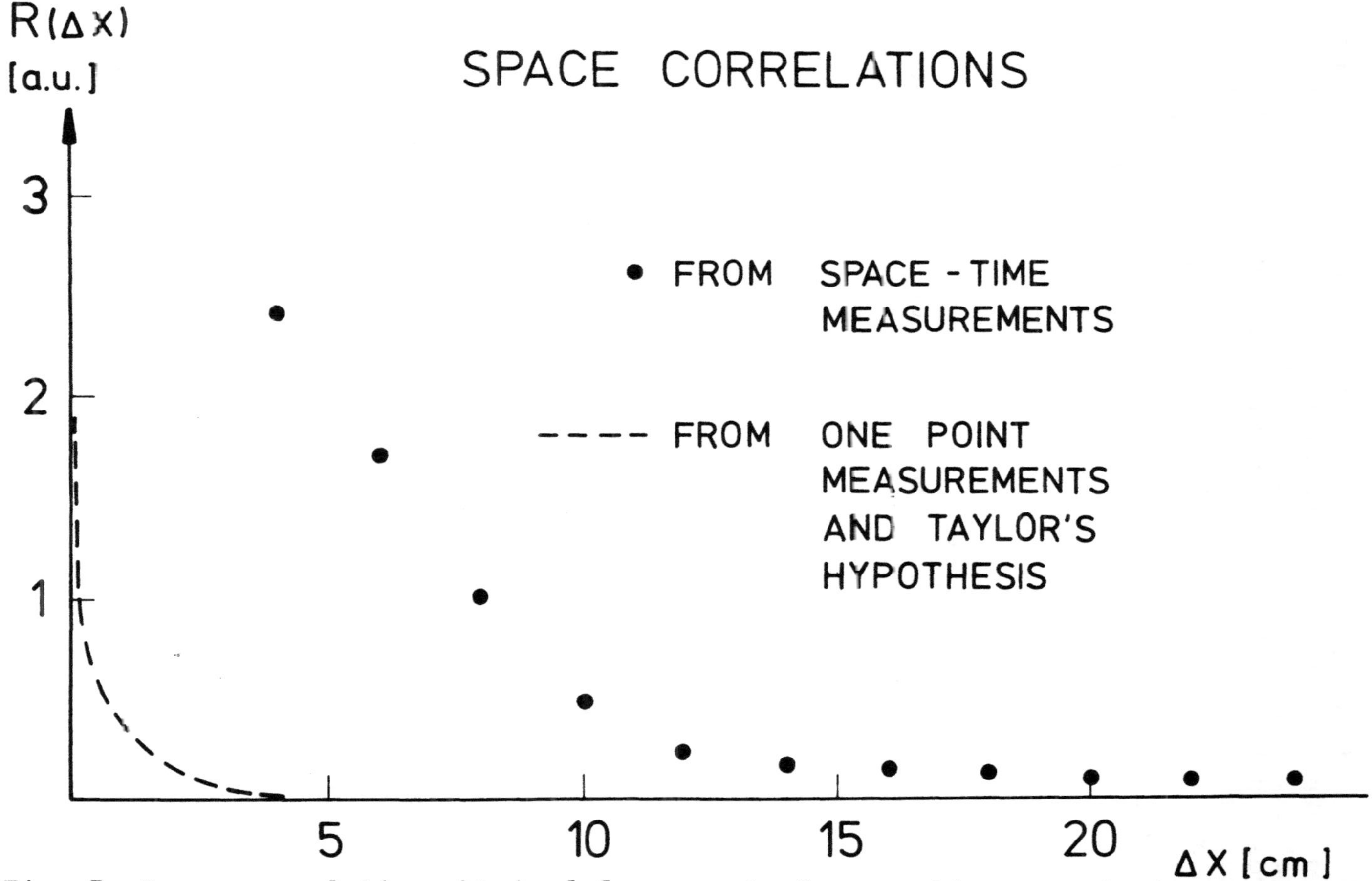

Fig. 7: Space-correlation obtained from a set of space-time-correlation measurements

7 CONCLUSION

It is possible to investigate the flow structure in bubble column reactors by means of hot wire anemometers. The analysis of the liquid-phase velocity data results in consistent descriptive functions on the turbulent motion. The large-scale structures of these flows are determined by the column diameter, and a coherent circulation cell structure must be taken into account. Further measurements are required to establish quantitative relations between the flow structures and disperging properties of turbulent flows.

8 ACKNOWLEDGMENTS

I want to thank Mr. J. Lippert, who has carried out the measurements in the laboratory, and Prof.K.Schügerl for his encouragement with this research project.

REFERENCES

1. Hinze, J.O. "Fundamentals of the Hydrodynamic Mechanism of Splitting in Dispersion Processes." A.I.Ch.E. Journal, vol. 1, pp. 289-295, 1955
2. Levich, V.G. "Physicochemical Hydrodynamics." English translation (Prentice-Hall, New York, 1962)
3. Hinze, J.O. "Turbulence." 2nd ed.(McGraw-Hill, New York, 1975)
4. Bendat, J.S. and Piersol, A.G. "Random Data." (Wiley-Interscience, New York, 1971)
5. Taylor, G.I. "The Spectrum of Turbulence," vol.A164 (Proc.Roy.Soc.) pp. 476-490, 1938
6. Frenkiel, F.N. and Klebanoff, P.S. "Space-Time Correlations in Turbulence," in "Dynamics of Fluids and Plasmas," S.I.Pai, ed. (Academic Press, New York, 1966) pp. 257-274
7. Joshi, J.B. and Sharma, M.M. "A Circulation Cell Model for Bubble Columns." Trans. I. Chem. E., vol. 57 (1979) pp. 244-251

GAS/LIQUID MASS TRANSFER PARAMETERS (ε_G, k_La, a) FOR NON-NEWTONIAN TWO-PHASE FLOW IN A BUBBLE COLUMN

A. Schumpe, K. Nguyen-tien and W.-D. Deckwer

Institut für Technische Chemie, Universität Hannover, Callinstr. 3, D-3000 Hannover 1
FR Germany

ABSTRACT

Gas/liquid mass transfer at non-Newtonian flow occurring, e.g., in aerobic fermentation is studied using highly viscous pseudoplastic solutions of sodium carboxymethylcellulose (CMC) as a model medium. Aeration measurements have been carried out in a 14 cm (I.D.) bubble column, 270 cm in height, operated cocurrently at bubbly and slug flow conditions. Volumetric mass transfer coefficients are evaluated by fitting the measured profiles of the oxygen liquid phase concentration to the predictions of the dispersion model. Interfacial areas have been measured by the sulfite oxidation method. Simple correlations describe ε_G, k_La and a in the slug flow regime. Since the k_La correlation holds also for a CMC/Na_2SO_4 solution the results on the interfacial area make it possible to evaluate k_L. Volumetric mass transfer coefficients reported for fermentations of penicillium chrysogenum are in striking agreement with the suggested correlation.

INTRODUCTION

Fermentation media may exhibit a highly viscous non-Newtonian flow behavior (1,2) resulting in low aeration efficiencies (3). For studying the volumetric mass transfer coefficients (k_La) under such conditions solutions of sodium carboxymethylcellulose (CMC) are often used as model media. In bubble columns such investiga-

tions have been carried out by Buchholz et al. (4) usi[n]
a stationary method for k_La measurement and by Nakanoh and Yoshida (5) using a transient technique. The correlation developed by Henzler (6) from the data of Buchholz et al. and the correlation proposed by Nakanoh and Yoshida predict very different dependencies on the gas velocity and the effective kinematic viscosity:

$$\text{Henzler} \qquad : k_La' \propto u_G^{0.1} \, \nu_{eff}^{-0.867} \tag{1}$$

$$\text{Nakanoh, Yoshida}: k_La' \propto u_G^{1.0} \, \nu_{eff}^{-0.28} \tag{2}$$

Furthermore the k_La values calculated from the cor relations may differ by a factor of about 5.

SCOPE AND PURPOSE

In order to investigate whether the different concentrations and the different sparger types applied contributed to this discrepancy further measurements of k_La in CMC solutions have been carried out.

Additionally, it seemed desirable to obtain some information on the individual values of a and k_L. The effective interfacial areas determined by chemical methods are smaller than the geometrical areas, in principle, due to the nonlinear dependency of the conversion on the interfacial area of a bubble (7,8). The measured overall conversion is an incorrect mean with respect to the interfacial area in as far as the dispersion structure is nonuniform and, hence, different conversions prevail in the bubbles at disengagement. The sulfite oxidation method, however, can be expected to give approximately correct results even in rather heterogeneous dispersions because of the low gas phase conversions involved (8,9). Furthermore, the reaction kinetics is almost unaffected by CMC (10,11,12). The photographic method, on the other hand, has been shown to overestimate the interfacial area in sulfite solutions both with and without CMC (8,9). Therefore, in this study, the sulfite oxidation method is used for interfacial area determination. The influence of the added salt (0.8 mole l^{-1} Na_2SO_3) is checked by additional k_La measurements in a CMC solution containing 0.8 mole l^{-1} Na_2SO_4.

EXPERIMENTAL

Aeration measurements have been carried out in a 14 cm (I.D.) bubble column (BC II), 270 cm in height which could be equipped with different gas spargers:

sintered plate	(SP):	0.2 mm (mean pore size)
perforated plate	(PP):	421 x 0.5 mm
perforated plate	(PP):	73 x 1.0 mm
perforated plate	(PP):	19 x 2.0 mm

Hydrodynamic investigations have also been carried out in a 10.2 cm (I.D.) bubble column, 236 cm in height (BC I, SP 0.15 mm).

The gas holdup was determined by the manometric method. The liquid flowed cocurrently to the gas phase and was recycled. For k_La determination (8,15) the stationary method was used, the liquid being regenerated in a second bubble column sparged with nitrogen. The profile of the oxygen liquid phase concentration was measured by withdrawing continuously small gas-free liquid flows from different locations along the column. The oxygen concentrations were measured in a stirred cell with a polarographic oxygen probe (Orion). The k_La values are evaluated by fitting a closed solution of the axial dispersion model (13) to the measured profiles. Either both the dispersion coefficient E_L and k_La were optimized or E_L was calculated from the empirical correlation of Deckwer et al. (14):

$$E_L = 2.7\, d_{BC}^{1.4}\, u_G^{0.3} \tag{3}$$

The sulfite oxidation measurements were carried out at 25 °C, a pH of about 8.7 and $8 \cdot 10^{-4}$m Co^{2+}. The oxygen concentration in the outlet gas was followed continuously by a paramagnetic oxygen analyzer (Maihak). The particular values of the absorption coefficient K_2 (7-9)

$$K_2 = (2\, k_2\, D\, /\, 3H^3)^{0.5} \tag{4}$$

were determined by absorption experiments in a stirred cell. Experimental details and the model equations are given in (8,9).

RHEOLOGICAL PROPERTIES

The rheological properties were measured with a Couette viscosimeter (Haake) at shear rates from 7 to 1139 s^{-1}. The pseudoplastic flow behavior follows the power law:

$$\tau = k\dot{\gamma}^{n} \tag{5}$$

$$\eta = k\dot{\gamma}^{n-1} \tag{6}$$

The fluid consistency indices ranged from 0.04 to 0.24 Pa s^n and the flow behavior indices from 0.92 to 0.68 (8,15). For estimation of the effective viscosity from eq. (6) the effective shear rates predicted by the relation of Nishikawa et al. (16) are introduced:

$$\dot{\gamma}_{eff} = 50\, u_G \tag{7}$$

where $\dot{\gamma}_{eff}$ is in (s^{-1}) and u_G in (cm s^{-1}).

GAS HOLDUP

The gas holdup measured at 0.8-1.8 %wt. CMC is plotted vs. the gas velocity in Figure 1. With the perforated plates large bubbles in rapid coalescence are observed even at low gas velocities. With the sintered plates homogeneous flow prevails resulting in much higher gas holdups which are independent of the column diameter. At higher gas velocities slug flow develops and ε_G becomes independent of the sparger type used in BC II:

$$\varepsilon_G = 3.22\cdot 10^{-2}\, \bar{u}_G^{\,0.674} \tag{8}$$

With the smaller bubble column (BC I) in the slug flow regime lower bubble rise velocities result in a higher gas holdup:

$$\varepsilon_G = 4.04\cdot 10^{-2}\, \bar{u}_G^{\,0.627} \tag{9}$$

In the case of CMC/Na_2SO_3 solutions ε_G closely resembles these dependencies.

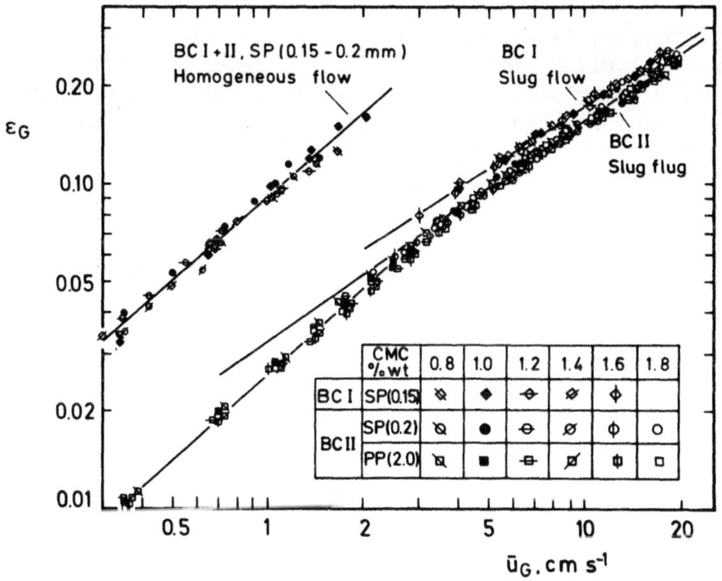

Fig. 1: Gas holdup

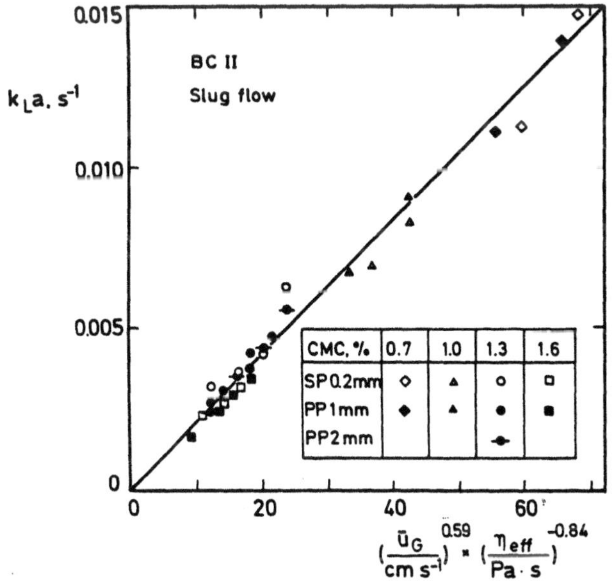

Fig. 2: Correlation of volumetric mass transfer coefficients at slug flow

VOLUMETRIC MASS TRANSFER COEFFICIENTS

Optimizing both k_La and E_L results in dispersion coefficients which are in good agreement with eq. (3) in case of water but tend to be higher, surprisingly, for CMC solutions (15). The k_La values, however, differ only slightly if the E_L predictions of eq. (3) are introduced instead of the optimized values.

If the sintered plate is used for aeration k_La shows a pronounced maximum at low $\bar{u}_G$ (homogeneous flow). At higher gas velocities ($\bar{u}_G > 2$-3 cm/s) in the slug flow regime k_La becomes independent of the sparger type and can be correlated by:

$$k_La = 2.08 \cdot 10^{-4}\, \bar{u}_G^{\,0.59}\, \eta_{eff}^{\,-0.84} \qquad (10)$$

$$s^{-1} \qquad\qquad cm\ s^{-1} \quad Pa\ s$$

for all applied CMC concentrations (0.7-1.6 %wt.)(Figure 2).

APPLICATION TO FERMENTATION MEDIA

König (3) reports volumetric mass transfer coefficients measured during fermentations of penicillium chrysogenum in a 20 cm (I.D.) bubble column. Two typical sets of rheological parameters are given:

Medium	n	k, Pa s^n
1	0.49-0.52	0.70-0.87
2	0.42-0.44	2.05-2.44

The experimental k_La values and the values calculated from eq. (10) agree fairly well (Figure 3). The predictions of the two other correlations are too high.

VOLUMETRIC MASS TRANSFER COEFFICIENTS OF A CMC/Na_2SO_4 SOLUTION

To investigate the effect of the sulfite added for interfacial area determination k_La has also been measured in a CMC solution containing 0.8 mole l^{-1} Na_2SO_4. The effective viscosity of the solution is lower than with a salt-free solution but if the actual rheological data are introduced eq. (10) holds very well (Figure 4).

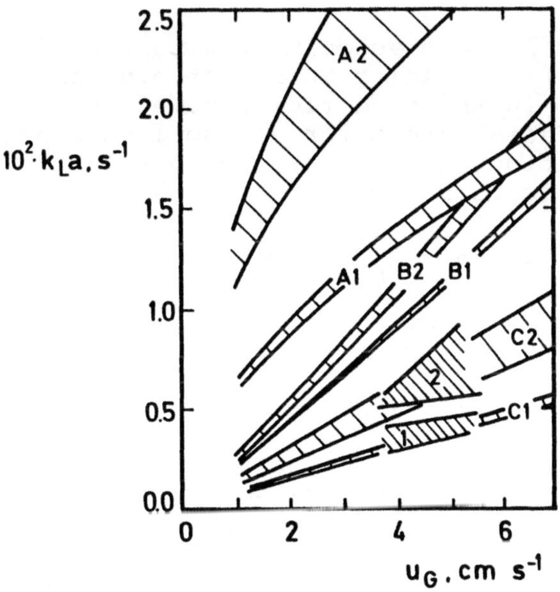

Fig. 3: Volumetric mass transfer coefficients (1,2) during fermentations of penicillium chrysogenum reported by König (3) and predictions after (A) Henzler (6), (B) Nakanoh, Yoshida (5) and (C) this work (eq. 10)

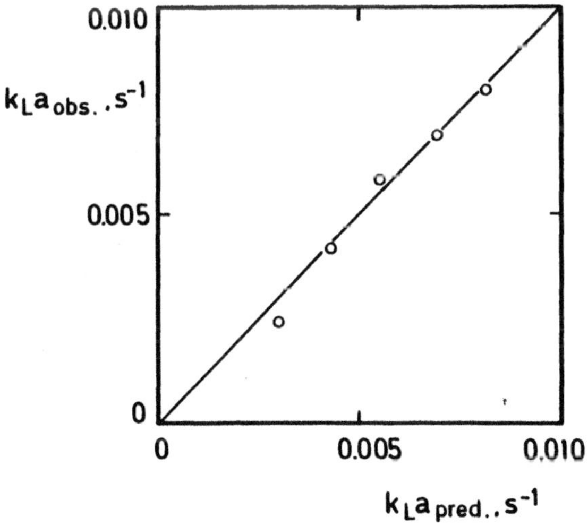

Fig. 4: $k_L a$ measured in a (1.5%wt CMC/0.8m Na_2SO_4) solution vs. the predictions of eq.(10)

SPECIFIC INTERFACIAL AREA

At very low gas velocities with the sintered plate homogeneous flow results in rather high interfacial areas (8,9). In the slug flow regime, however, the specific interfacial areas for all applied sparger types and CMC solutions of 1-2 %wt are correlated in striking agreement (Figure 5) by:

$$\underset{cm^{-1}}{a} = 4.65 \cdot 10^{-2} \, \underset{cm\ s^{-1}}{\bar{u}_G^{0.51}} \, \underset{Pa\ s}{\eta_{eff}^{-0.51}} \qquad (11)$$

MASS TRANSFER COEFFICIENT

Applying eq. (11) to the k_La values measured in the CMC/Na_2SO_4 solution mass transfer coefficients k_L are obtained which represent mean values for the total interfacial area. Figure 6 reveals that k_L becomes almost constant at high gas velocities at about 0.013 cm s^{-1}.

Since eq. (10) also held for the CMC/Na_2SO_4 solution eq. (11) may be introduced to obtain the following relation for k_L in the slug flow regime:

$$\underset{cm\ s^{-1}}{k_L} = 4.5 \cdot 10^{-3} \, \underset{cm\ s^{-1}}{\bar{u}_G^{0.08}} \, \underset{Pa\ s}{\eta_{eff}^{-0.32}} \qquad (12)$$

SUMMARY

In a bubble column of 14 cm I.D. gas holdups, volumetric mass transfer coefficients and specific interfacial areas have been measured in CMC solutions of highly viscous pseudoplastic flow behavior. In the slug flow regime ε_G, k_La and both a and k_L follow simple relations (eqs. 8,10-12) which consider the influence of the gas velocity and the effective viscosity. The predictions of eq. (10) are in close agreement with the k_La values observed in a tower fermenter with fermentation broths of pronounced non-Newtonian flow behavior.

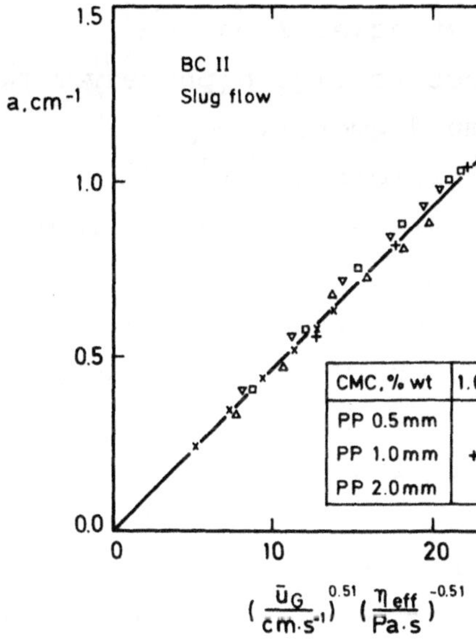

Fig. 5: Correlation of specific interfacial areas at slug flow

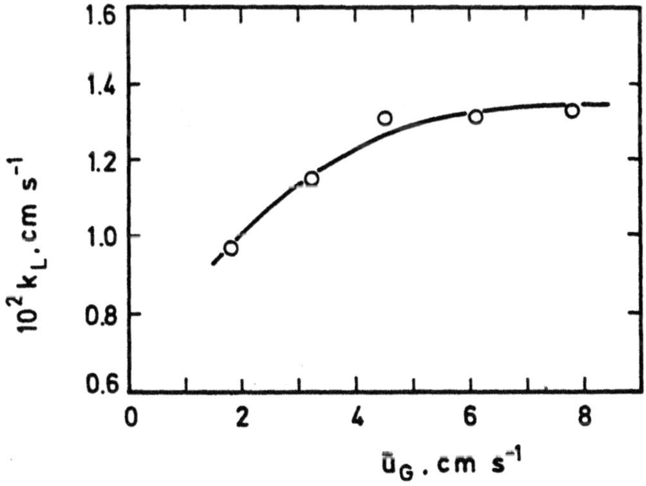

Fig. 6: Liquid-side mass transfer coefficients in the CMC/Na_2SO_4 solution

NOTATION

a	specific interfacial area, cm^{-1}
a'	a with respect to liquid phase volume, cm^{-1}
d_{BC}	bubble column diameter, cm
D	diffusion coefficient, $cm^2\ s^{-1}$
E_L	liquid phase dispersion coefficient, $cm^2\ s^{-1}$
H	Henry's constant, Pa $cm^3\ mol^{-1}$
k	fluid consistency index, Pa s^n
k_2	pseudo-second order rate constant, $cm^3 mol^{-1} s^{-1}$
k_L	liquid side mass transfer coefficient, cm s^{-1}
K_2	overall absorption coefficient (eq. 4), mol $cm^{-2} s^{-1} Pa^{-3/2}$
n	flow behavior index
$\bar{u}_G$	superficial gas velocity u_G at half the column height, cm s^{-1}
u_L	superficial liquid velocity, cm s^{-1}

Greek symbols:

$\dot{\gamma}$	shear rate, s^{-1}
$\dot{\gamma}_{eff}$	effective shear rate (eq. 7), s^{-1}
ε_G	gas holdup
η	apparent dynamic viscosity, Pa s
η_{eff}	effective dynamic viscosity (eqs. 6,7), Pa s
τ	shear stress, Pa

REFERENCES

1. Blanch, H.W. and S.M. Bhavaraju. Biotech. Bioeng. 18 (1976), 745.
2. Margeritas, A. and J.E. Jajic. Biotech. Bioeng. 20 (1978), 939.
3. König, B. Doctoral Thesis, Universität Hannover 1980.
4. Buchholz, H., R. Buchholz, J. Lücke and K. Schügerl. Chem. Eng. Sci. 33 (1978), 1061.
5. Nakanoh, M. and F. Yoshida. Ind. Eng. Chem. Proc. Des. Dev. 19 (1980), 190.
6. Henzler, H.-J. Chem.-Ing.-Tech. 52 (1980), 643.
7. Schumpe, A. and W.-D. Deckwer. Chem. Eng. Sci. 35 (1980), 2221.
8. Schumpe, A. Doctoral Thesis, Universität Hannover 1981.
9. Schumpe, A. and W.-D. Deckwer. Paper presented at the Int. Symp. on Mass Transfer with Chemical Reactions in Two-Phase Systems, ACS-Meeting, Atlanta 1981.
10. Wesselingh, J.A. and A.C. van't Hoog. Trans. Instn. Chem. Engrs. 48 (1970), T69.
11. Onken, U. and W. Schalk. Ger. Chem. Eng. 1 (1978), 191.
12. Ruchti, G., I.J. Dunn and J.R. Bourne. First European Congress on Biotechnology, Interlaken 1978.
13. Langemann, H. Brennstoff-Chemie 49 (1968), 22.
14. Deckwer, W.-D., B. Burckhart and G. Zoll. Chem. Eng. Sci. 29 (1974), 2177.
15. Deckwer, W.-D., K. Nguyen-tien, A. Schumpe and Y. Serpemen. Biotech. Bioeng., accepted for publication.
16. Nishikawa, M., H. Kato and K. Hashimoto. Ind. Eng. Chem. Proc. Des. Dev. 16 (1977), 133.

:NTRODUCTION TO LIQUID-LIQUID EXTRACTION WITH CHEMICAL REACTION

Erdogan Alper

Chemical Engineering Department
Ankara University, Besevler, Ankara, Turkey

. INTRODUCTION

Liquid-liquid reactions are very common in the rganic and related chemical industries. Applications re also found in process metallurgy. Indeed, the proress in hydrometallurgy, in the form of solvent extracion of metals, has been very rapid in the last two ecades. Although, fundamental studies on physical iquid-liquid extraction (or solvent extraction as it s normally called) may well be related to the modern irth of process engineering (together with other ommon separation processes of gas absorption and istillation) in the early 1920's, its potential, specially in reactive systems, has not yet been fully ealized. Moreover the number of extensive works is till limited. This may be linked to many factors; mong them, two are probably the main causes: firstly, arious fundamental aspects have not yet been properly nderstood and in the accompanying reviews, a number f important phenomena concerning physical aspects ave been authoritatively highlighted by Sawistowski 1) and Hartland (2). Secondly, many industrial liquidiquid reactions are carried out in reactors which perate batch- or semibatchwise and are not of very arge size. Hence they have not been in need of such ery precise process design methods. Indeed, although arious laboratory techniques and systems are well nown in organic and analytical chemistry, the industri1 exploitation of them has not yet been fully realised. able 1 gives some liquid-liquid reactions of industri-

Table 1. Selected examples of liquid-liquid reactions

No.	SYSTEMS
1.	Hydrolysis and/or saponification of a variety of esters
2.	Hydrolysis of organic compounds.
3.	Removal of free acid and esters from the products of oxidation of cyclohexane
4.	Removal of COS from liquefied C_3/C_4 fractions by treatment with aqueous NaOH and alkanolamine solutions
5.	Sulphonation of aromatics
6.	Nitration of aromatics
7.	Ammonolysis of nitrochlorobenzenes
8.	Manufacture of dithiocarbamates by reaction between aqueous solutions of amines and CS_2
9.	Alkylation of aromatics
10.	Epoxidation of oils by peracetic acid
11.	Oligomerisation of 1-butene in sulphuric acid; dimerisation of isobutylene and isoamylene
12.	Reduction of nitroaromatics by aqueous Na_2S, $NaHSO_3$/Na_2SO_3 and $Na_2S_2O_4$
13.	Oximation of cyclohexanone
14.	Production of ethylenediamine from ethylene-dichloride and aqueous NH_3
15.	Extraction of metals from aqueous solutions
16.	Pyrometallurgical operations to remove impurities from molten metals
17.	Caustic sweetening for removing mercaptan sulphur from Hydrocarbons

18. Recovery and separation of lower aliphatic acids and HF etc. from aqueous solutions with amines

19. Separation of a number of organic compounds by dissociation extraction

20. Interfacial polycondensation reactions

21. Reactions involving phase transfer catalysis

22. Reactions involving micellar and reverse micellar catalysis

23. Purification of molten phosphorous with concentrated sulphuric acid

24. Separation by liquid membranes

25. Waste water treatment by liquid ion exchange in liquid membrane systems

26. Extraction of acetic acid from water by solutions of cyclohexylamine in isobutanol

27. Extraction of n-butyric acid from benzene solution by aqueous NaOH

28. Solvent extraction of acetic acid from dilute aqueous industrial streams by commercial organic bases(such as various amines(Amberlite's,Adogen's,Alamine's etc.) and tri-octyl phosphine oxide(TOPO))

al importance and further details may be found in the excellent reviews of Sharma and coworkers (3,4).

Equipment customarily used for physical liquid-liquid extraction, may also be employed for reactive systems. They may be operated batch or semibatchwise for small scale productions (in the form of the usual mechanically agitated vessels) and continuously for large scale productions. The latter may involve both stagewise and differential contactors and Table 2 shows the classification of them according to Laddha and Degaleesan (5).

2. COMPARISON WITH GAS-LIQUID SYSTEMS

In many respects, liquid-liquid extraction with reaction resembles gas absorption accompanied by chemical reaction. For instance, both operations may be carried out either for separation or product formation purposes. In general, a similar interaction between diffusional factors and kinetics is encountered Hence many ideas, which are, strictly speaking, developed for gas-liquid operations, may also be applied to liquid-liquid systems. However, there are facets of liquid-liquid extraction which are distinctly different these differences occur both in process and theoretical aspects.

2.1. Process Aspects

Liquid-liquid reactions often involve a conversion or removal level of more than 99 %. Indeed, some gas-liquid reactions, where the boiling point of the solute is in the range of -20 to 10°C, may sometimes be advantageously carried out as liquid-liquid reactions (3). The removal of COS from a C_3-stream by aqueous amine solutions is a typical example.

Firstly, the solubility limitation of gases prevents the use of a truly batch gas-liquid reactor, so that semi-batch operation is required; on the other hand, batch liquid-liquid reactors are very common. In semi-batch operations, the outlet concentration of the solute component may vary significantly with time in gas-liquid systems whereas the analogous situation is varely encountered in liquid-liquid systems. The volumetric flowrate of gas may be very large resulting in very large diameter columns; in some cases, the gas flowrate may reduce drastically along the column if it is pure. In contrast, extraction equipment is usually not very large and vanishing flowrates of one of the phases is not normally encountered. Furthermore, the

INDUSTRIAL LIQUID-LIQUID EXTRACTORS						
DIFFERENTIAL				STAGEWISE		
GRAVITY	*PULSED AGITATED*	*MECHANICALLY AGITATED*	*CENTRIFUGAL*	*GRAVITY*	*PULSE AGITATED*	*MECHANICALLY AGITATED*
Spray column	Pulse packed column	RDC	Podbielniak-extractor	Plate column	Pulsed agitated	Scheibel column
Packed column		Mixco Lightnin-CM contactor (Oldshue-Rushton)	Luwesta-extractor		Pulsed sieve-plate column	Treybal contactor
		Graesser raining bucket contactor	westfalia-extractor		Pulsed mixer-settler	ARD-extractor
		Morris contactor	De Laval-extractor			Mixer-settlers

Table 2. Typical classification of industrial, conventional liquid-liquid extraction equipment (5).

orders of magnitude of the values of interfacial area and mass transfer coefficients may be quite different. Although, the interfacial areas in liquid-liquid systems are very large, and can be increased enormously by emulsifiers without introducing intense foaming as encountered in gas-liquid systems, the mass transfer coefficients are much smaller. Hence the "enhancement" of mass transfer rates in liquid-liquid systems is more easily manifested. However, many liquid-liquid reactions are conducted at relatively low temperatures. Liquid-phase alkylation, for example, is carried out at temperatures well below 273 K. The comparatively low boiling points of many organics naturally limit their liquid-phase reaction environment to temperatures which are low compared with most reaction systems and this usually results in rather low reaction rates.

It is therefore advantageous to switch from gas-liquid to liquid-liquid reactions if the overall rate is more pronouncedly increased by the interfacial area rather than the mass transfer coefficient. Also, unlike gas-liquid systems, the heat removal in exothermic liquid-liquid reactions may be carried out under refluxing conditions so that heat transfer is very efficiently carried out in an overhead condenser.

In general, gas-liquid systems have no phase separation problems; on the other hand, this causes some difficulties in liquid-liquid systems since the densities of two phases may be very similar and the interfacial tension may be rather low.

The locale of the reaction may also be quite different for these two systems; that is, while the reaction is generally confined to the liquid phase in a gas-liquid system, it is possible to anticipate the process of diffusion-reaction occuring in both phases in a liquid-liquid system. Indeed, the oximation reaction to manufacture cyclohexanone oxime as part of the process for making Nylon66, is an industrial example (3). In some cases, the locale of the reaction in an organic phase-aqueous phase reaction system may deliberately be changed from the aqueous phase to the organic phase; this process is now known as "phase transfer catalysis" and has made rapid progress during the last decade (6,7).

2.2 Theoretical Aspects

There have not been too many studies on the theoretical aspects of liquid-liquid extraction accompanied by a chemical reaction. Indeed, since similar in-

:eractions between diffusional factors and kinetics
)ccur, it has been common to use theories (or models)
which are, strictly speaking, developed specifically
for gas-liquid systems. However, it is not immediately
obvious that the models suitable for the gas-liquid
interface are directly applicable to the liquid-liquid
interface. This is because although an assumption of
ero shear stress in the liquid at the interface, hence
no velocity variation in the liquid as assumed in all
surface-renewal models, is reasonable for gas-liquid
contacting, it is unlikely to be valid at the interface
of liquid-liquid systems, where the density and vis-
cosity of the two phases are of the same order of
magnitude. On the other hand, Porter and Roberts (8)
have shown that the effect of a chemical on the rela-
tive increase in the rate of mass transfer does not
depend strongly on the flow pattern near the interface
indicating that the theories developed for gas-liquid
systems might safely be used for liquid-liquid reac-
tions. However, practical experience is still needed
where the highest discrepancy is expected, that is at
the transition regime between fast pseudo m th order
nd instantaneous reactions.

Although the phenomena of transport of various
pecies to (or from) the interface and the simultaneous
eaction can be accounted for by the theories of gas-
iquid systems, the third phenomenon crossing of the
nterface may require a different treatment. Firstly,
n cases of heavily contaminated liquid-liquid systems,
nterfacial resistance may no longer be negligible.
econdly interfacial turbulence, which is produced by
he interaction of mass transfer with interfacial ten-
ion, is in many cases very important. Indeed, there is
ery limited information in the role of this and
ertain other secondary phenomena in extraction with
eaction (9).

A detailed and very fundamental analysis of mass-
ransfer with chemical reaction in liquid-liquid disper-
ions has been published recently by Tavlarides and
tamatoudis (10). These authors point out that the de-
ign and analysis problem depends on the phase in
hich the reaction occurs, whether multiple reactions
re involved, the relative magnitudes of the rates of
ass transfer and reaction, and upon macromixing pro-
esses of the dispersed and continuous phases. Hence,
hey first examine dispersion phenomena such as
oalescence and breakage of droplets, and drop size
istribution. The topics discussed by the authors (10)

are schematised in Table 3 and further details can be found in various references quoted in (10).

Concerning the most relevant part of mathematical models for mass transfer with reaction in liquid-liquid dispersions, they review a variety of models to predict the extent of reaction and the selectivity for complex reactions. They refer to the simplest approaches as "effective interfacial surface area models"; here an effective interfacial surface area between the two liquid phases is defined (or estimated and mass transfer with reaction is then described usin models such as film, penetration, film-penetration or surface renewal for each of the two phases. Another type of model includes the use of drop size distributions and residence time distributions of the droplets in the agitated medium. These models assume that the number of drops in the vessel is constant and that the drops may have a distribution of sizes. The continuous phase is divided into an equal number of fluid element assigned to each droplet. Expressions are developed for mass transfer with reaction between the drop and the fluid element. The average moles transferred per unit volume of dispersion is obtained by integration of the flux over the distribution of droplet sizes for an average residence time of the drops in the reactor. These models do not however consider coalescence and breakage which can be accounted for by "dispersed phase interaction" models. Indeed, there are a variety of reaction systems where dispersed phase mixing has significant effects. These include (a) reactions in the drop phase with time scale of the order of coalescence and breakage time scale, (b) bimolecular reactions of intermediate rate in the dispersed phase with unmixed feed stream, (c) reaction in the continuous phase with diffusing species from the drop phase whose reaction and diffusion time scales are of the same order as the coalescence and breakage time scales, (d) bimolecular interfacial reactions where each reactant is soluble only in one phase (e.g. metal chelation and single-cell protein fermentation, (e) complex reaction in the dispersed phase of intermediate rate and (f) instantaneous bimolecular reactions in the dispersed phase with unmixed feeds. In general, coalescence and redipersion models can be applied to these reaction systems and they include population balance equations, Monte Carlo simulation techniques and a combination of macromixing and micromixing concepts with Monte Carlo simulations (10), which are however increasingly difficult to handle. Indeed, although the simple models of

:able 3. The analysis of interphase reactions and mass :ransfer in liquid-liquid dispersions (adapted from :avlarides and Stamatoudis (10)).

:. DISPERSION PHENOMENA

- Flow field in agitated dispersions(laminar/transitional dispersions,turbulent dispersions)
- Behaviour of liquid-liquid dispersions(maximum and minimum drop size in dispersions,phenomenological models for drop breakage and drop coalesence rates)
- Measurements and analysis of the properties(interfacial surface area,drop size and coalesence frequency measurements)

.I. MATHEMATICAL MODELS FOR MASS TRANSFER WITH REACTION IN LIQUID-LIQUID DISPERSIONS

- Effective interfacial surface area models
- Drop size and residence time distribution models
- Dispersed phase interaction models (1.Population balance techniques,2.Monte Carlo simulation models and 3.Models using macromixing and micromixing concepts)

he first type dismiss completely the fact that the ispersed phase consists of a distribution of droplet izes which may coalesce and redisperse, leading o different concentrations of the reacting species in ach drop and take no account of effects of the operaing variables on the dispersion parameters, they are evertheless still widely used in the design of contacing equipments.

. CONTROLLING REGIMES

If an interfacially stable system also has neglible interfacial resistance, the various regimes and e procedure for discerning them are the same as those gas-liquid systems. That is, depending on the relaive rates of diffusion and chemical reaction the stem may conform to one of the following regimes: very slow reactions, 2. slow reactions, 3. very fast actions and 4. instantaneous reactions. These well fined regimes are already discussed in connection th gas-liquid systems (11) and they are mostly anatically negotiable so that expressions to calculate the specific rate of extraction are readily available from the literature on gas-liquid systems (11). The various intermediate regimes can also be accounted for by expressions which are however only approximate in many cases (12). A variety of complex reactions, such as reversible reactions, two dissolved reactants, simultaneous extraction of two solutes, are industrially relevant. These include, among others, nitration/ sulphonation of mixed aromatics alkylation/oligomerisation in isobutene-isobutylene-butenes system and various aspects, for instance, selectivity problems etc., can be solved by extending analyses which are already available for gas-liquid systems (4,13).

In the case of accompanying interfacial turbulence, the local value of the true mass transfer coefficient is increased and this may well affect the controlling regime. Recently, Grosjean and Sawistowski (9) examined an instantaneous reaction and obtained good agreement between experimental and predicted results. They have also indicated the important part played by secondary phenomena and the need for their incorporation into any meaningful prediction of mass transfer coefficients; the phenomena referred to include the Marongoni effect, gravitational instability, interfacial resistance and the effect of ions on equilibrium and diffusion coefficients.

The procedure for discerning the controlling mechanism is similar to gas-liquid systems and a variety of model contactors can be employed for this purpose. However, a stirred cell with a definite interfacial area and two independent stirres is normally sufficient. The oldest design of such an equipment is known as a Lewis cell which has been adapted in modified forms by many workers (4). Hanson (14) and Sharma (4) discuss methods of discerning the controlling regimes for liquid-liquid systems together with very clear illustrative examples.

4. PROCESS DESIGN ASPECTS

Rational process design of liquid-liquid extractors (or reactors) is in many ways similar to those of equipment used in gas-liquid operations. The latter is discussed in some detail by Alper in this Proceedings (15). The important design parameter is the time of reaction for a specified level of conversion in the case of batch and semi-batch operations and volume (or height) for the continuous mode of operation. The required data include firstly hold-ups, interfacial areas and physical mass transfer coefficients and secondly a number of "process specific data" such as, reaction rate constants, diffusivities and solubilities. The results obtained from physical methods for the former have been reviewed recently by Van Landeghem (16). Although very few, some data are also available from reactive systems (17-19). Once these data are available, rate expressions may be incorporated with a proper extractor (reactor) model to obtain the desired parameter. For the latter, Vasudevan and Sharma (20) considered a number of idealised cases of combination of plug flow and/or completely mixed flow and obtained analytical expressions for a number of cases of practical importance. These include the reaction time for batch or semi-batch operated mechanically agiated contactors, the height/length of packed columns/plugflow pipeline contactors, operated cocurrently and countercurrently, the number of plates in plate columns and the volume of a continuously operated stirred tank (20). These design equations, which are similar to those of Alper (15) in essence, do not however consider backmixing in the continuous/dispersed phase, but this aspect can also be considered with suitable modifications of constituent differential equations in the case of differential and multistage contactors (21). Little is known about the deviation from complete mixing in mechanically agitated contactors when highly viscous liquids are involved.

Another aspect which has not received any detailed examination is the use of laboratory models in process design. Indeed, unlike gas-liquid systems, there is hardly any extensive studies here and the subject certainly deserves some attention.

5. SELECTED EXAMPLES OF LIQUID-LIQUID REACTIONS

Although liquid-liquid reactions are widespread systematic studies are still very few. Following are some brief accounts of conventional reactions, metal extraction and some novel aspects; the first and the last items have been investigated mainly by Sharma and his coworkers and the details can be found in their excellent reviews (3,4).

5.1. Conventional Reactions

Alkaline hydrolysis of a variety of compounds, such as esters of acids, alkyl halides and nitrochlorobenzenes has some technical relevance. Various specific systems have been recently examined by Sharma and coworkers (22,23); most of the reactions are very slow.

However, alkaline hydrolysis of formates conforms to the fast pseudo first order reaction regime, hence it may be employed to measure interfacial areas.

Nitration of aromatics has been studied by many workers (25-28) and complied by many workers (25). In general, all nitration reactions involve two steps: formation of nitronium ion (NO_2^+) and reaction between nitronium and the aromatic substance. Normally, the second step is the controlling one and the nitration of many aromatics, such as chlorobenzenes, is too slow to involve any diffusional resistance. On the other hand, depending on the conditions, the nitration of benzene and toluene may well fall into the fast reaction regime. Indeed, in the case of toluene nitration at high concentrations of toluene, the reaction may be controlled by the formation of the nitronium ion, leading to pseudo zero order reaction.

Alkylation of organic compounds with olefins is very important industrially (30-34). It appears that most acid-catalysed alkylations are very slow, hence kinetically controlled. However, Lee and Harriot (33) found that H_2SO_4-catalysed alkylation of isobutane with butenes to be rather fast. Dixon and Saunders (34) studied the H_2SO_4-catalysed alkylation of o-xylene with acetaldehyde and found that the reaction may be too fast to be kinetically controlled.

Reduction of aromatic compounds to corresponding aromatic amines with aqueous Na_2S and Na_2S_2 has been studied by Bhave and Sharma (35) who found that, in most cases, the process is kinetically controlled.

Removal of COS from liquefied petroleum fractions (C_3/C_4) whose boiling point is close to that of propylene can be accomplished by treatment with aqueous solutions of sodium hydroxide or alkanolamines. The process may be carried out either by gas-liquid or liquid-liquid operation and the latter seems to have many potential advantages (3).

The reaction between CS_2, and amines has been studied by Kothari and Sharma (36). Although there is no advantage, it is possible to carry out this process also as a gas-liquid operation. The process is important for the manufacture of pesticides and rubber chemicals.

Phosphorous chemistry provides also a number of interesting examples of liquid-liquid extraction with chemical reaction (37). Carr and Shah (38) studied the extraction of mercaptan sulphur from pentane by caustic soda solution and established that the system represents a case of mass transfer with an equilibrium reaction occurring in one phase. Table 1 gives a number of other liquid-liquid reactions, some of which however are only model reactions.

5.2. Solvent Extraction of Metals

The solvent extraction of metal salts from aqueous solutions has gained great commercial importance only during the last few decades, especially for low grade ores and for the recovery of metals from effluents (39,40). The process, which involves extracting the metal ion into an organic solvent by a water insoluble complexing agent in the form of a metal complex, has been known for a long time in nuclear material processing, mainly as uranium extraction. However, since these operations do not have the normal economic considerations, it is only after the development of copper extraction on a large scale that many aspects have been studied and clarified. The solvent extraction systems studied here differ considerably from those used, for example, in the petroleum and petrochemical fields, where mixtures of organics are separated into their components on a considerable scale. In metal extraction, the problems are associated with recovery of relatively small concentrations of metals from large volumes of solutions on a relatively small scale.

Lloyd (41) suggests that there is a cost involved in driving a mole of material from one phase to another, the cost per mole being relatively constant for different metals. Therefore, those, high in the periodic table are at an advantage (41). Warner (42) points out that the economics of many metals extraction processes are dominated by the cost of reagents necessary for chemical conditioning to provide reversal of the process for solvent recovery. Indeed, for example, hydroximes have been commercially successful as copper extractants despite their comparatively high price, while much cheaper carboxylic acids have not so far been adopted because of the greater reagent cost for chemical conditioning to achieve a reversible cycle (43).

Various aspects of metal extraction (including the application of liquid membranes which turn out to be a form of "facilatated transport") are discussed by Sawistowski (44) in this book; therefore the discussion is limited to certain chemical and descriptive aspects which are however not only of crucial interest, but are often too complicated to be discussed in full detail here.

5.2.1. Types of extractants. The most important properties of any metal extractant may be summarised as (a) favourable distribution of metal between the organic and aqueous phase, (b) selective Extraction of the desired metal in preference to other metals present in the solution, (c) rapid transfer of metal between phases in both extraction and stripping operations and (d) high solubility of both the reagent and its metal complex in the organic diluent. Table 4 shows a general classification of extractants according to Sharma (45).

Solvating extractants include the substances such as trialkyl phosphates, ketones and carbitols; the mos well known example being tributyl phophate which was originally used for the extraction of uranyl nitrate from leach liquors containing nitric acid. A number of other organophosphates have proved of value in the concentration and separation of the lantanide and actinide elements (46).

A rather outstanding example of acidic extractant is that of alkyl phosphoric acid, in particular die-2-ethyl hexylphosphoric acid (DEHPA) which has been used in the extraction of uranium, nickel, zinc, cobalt, chromium and many other metals (46). Dodecyl phosphori acidics also used in uranium extraction (46). A large number of carboxylic acids may be used to extract a variety of metals; these include naphtenic acids for base metals and rare earth metals, α-bromolauric acid, pivalic acid etc. Versatic acids (which is a trade mar of Shell Co.) have a general formulae of:

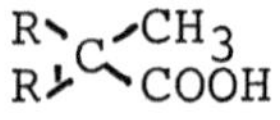

and they (e.g. Versatic 911) are used for the extraction of many metals, in particular cobalt and nickel. Recently, Rice (47) has reviewed very authoritatively developments and potential uses for carboxylic acid extractants.

Ring formation, resulting when a given coordinati agent can occupy more than single coordination positions has great importance in coordination chemistry. Such ring-forming groups are referred to as chelating groups. Chelating extractants have gained great importance due to successful applications, mainly in copper extraction from sulphuric acid as well as ammonia

Table 4. A general classification of extractants which are employed in the solvent extraction of metals (45).

EXTRACTANT	EXAMPLES
1. Solvating extractants	Trialkyl phosphates, ketones carbitols etc.
2. Acidic extractants	Naphtenic acids, DEHPA, ethyl hexyl phosphoric acid, versatic acid etc.
3. Chelating extractants	Aliphatic and aromatic oximes (such as LIX63 and LIX65N), quinolines (for instance, KELEX 100), polyols (e.g. phenyl glycol)
4. Ionic extractants	Higher molecular weight amines (primary, secondary and tertiary), higher molecular weight quaternary ammonium compounds, quaternized sulphonium compounds
5. Crown ethers	Cyclic polyether compounds such as 18-crown-6, dibenzo-18-crown-6 etc.

leach liquors. Here, ring structures are formed with molecules which may be neutral or charged, involving the extractant molecule as a ligand to the metal ion Detailed studies on the comparison of different chelating extractants and the kinetics of solvent extraction of metal chelates can be found respectively in recent papers (48,49).

In the case of copper, both aliphatic and aromatic oximes have been used and the patent literature is indeed extensive and receiving tremendous attention. The most well known aliphatic and aromatic oximes have the trade names LIX63(5,8-diethyl-7-hydroxy-dodecan-6-one-oxime) and LIX65N(2-hydroxy-5-nonylbenzophene oxime) respectively (see Figure 1. a,b and c). Another famous oxime is known as LIX64N, which is however only a mixture of LIX63 and LIX65N. In the case of LIX65N, both anti- and syn- isomers exist, but only the former is active. o-hydroxyaryloximes with general formulae of Figure 1.d, have been suggested by Price and Tumilty (50), and the Shell Company has been developing a class of selective copper extractants with the general formulae of Figure 1.e (50). Modifications, such as variations of R, R', X etc., or the substituent in the aromatic ring of LIX65N, have been suggested for improved selective performance (50,51).

Derivatives of quinolines have been used as chelating agents for copper extraction. The most well known is 7-alkenyl-8-hydroquinoline and has the trade name KELEX 100 (see Fiure 2) (52). Substituted 8-sulphonamidoquinolines have been suggested for Cu, Ni and Co extraction (53); β-dodecenyl 8-hydroxyquinoline was used to extract germanium (IV) (54). In many cases, chelating or stripping characteristics of the metal may vary markedly with the oxidation state; in the case of Co extraction by hydroxyquinoline, it is difficult to strip the chelate of higher oxidation state. Therefore, air oxidation of the cobaltous form, which is relatively fast, needs to be prevented (55).

A variety of chelates of B-compounds are formed by polyols. For instance, phenyl glycol may be used to extract sodium borate from dilute aqueous solutions. Recently, 2-ethyl-hexanol in kerosene was used to extract boric acid from aqueous solutions (57).

Ion extractants are sometimes called liquid ion exchangers as they carry within an ion pair a labile cation or anion which is capable of exchanging with the metal species in the aqueous phase. These extractants are predominantly higher molecular weight amines

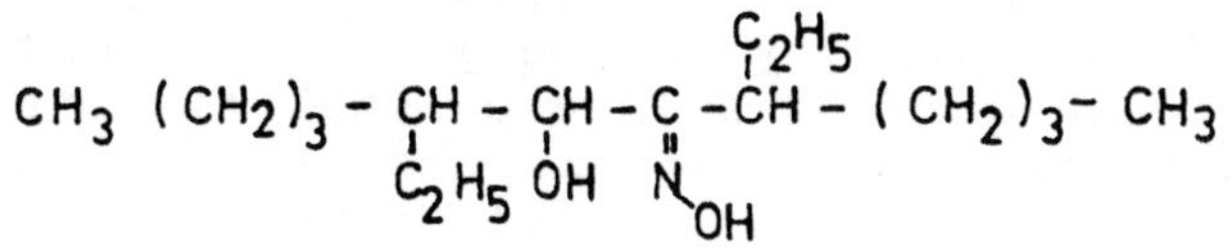

(a) 5,8-diethyl-7-hydroxy-dodecan-6-one-oxime
(Trade name LIX63)

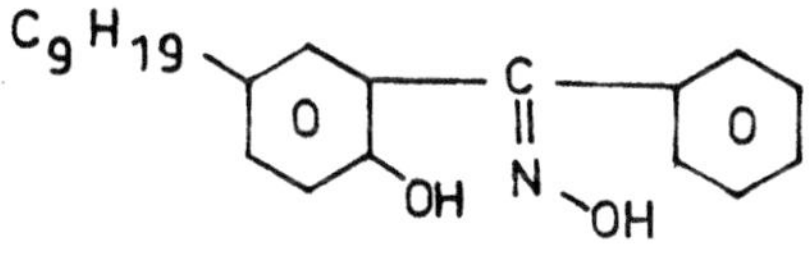

(b) 2-hydroxy-5- nonylbenzophene oxime
(Trade name LIX65N) (anti-form)

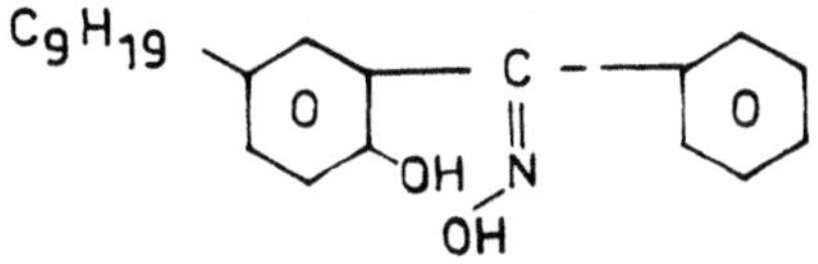

(c) LIX65N (syn-form)

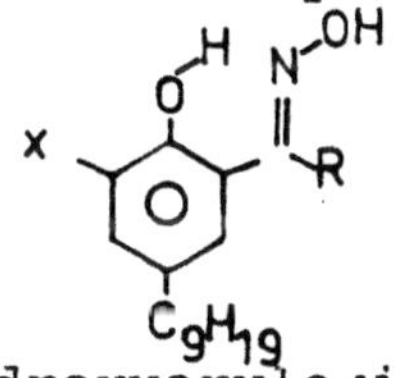

(d) o-hydroxyaryloximes
(X may be C_5H_{11}, H, Br etc.)

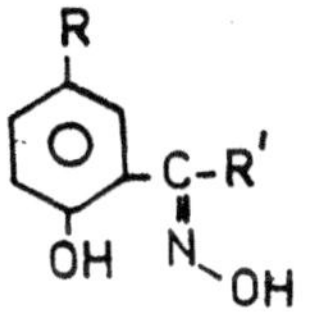

(e) 5-alkyl-2-hydroxyphenyl alkyl ketone oxime

Figure 1. Miscellaneous oxime-type extractants for the solvent extraction of metals

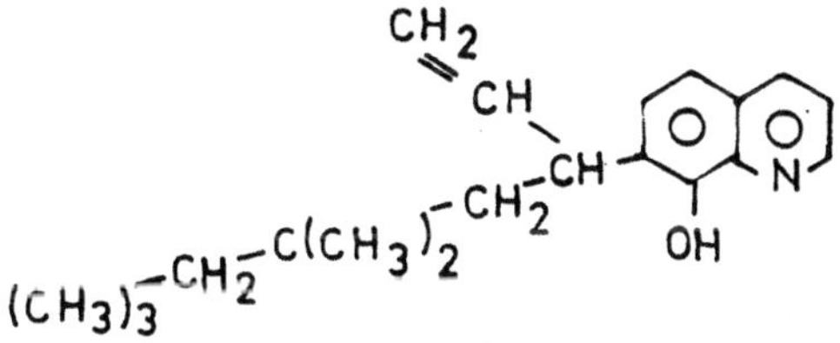

Figure 2. 7-alkenyl-8-hydroquinoline
(Trade name KELEX 100)

(primary, secondary or tertiary) and higher molecular weight quaternary ammonium compounds. Tertiary amines have proved of more utility in the particular case of extraction of uranium from sulphate leach liquors (58). Secondary and tertiary amines were used to extract molybdenum selectively. Tricaprylyl amine (trade name Alamine 336) and methyltrioctylammonium salt (Aliquat 336) were used, among others, in vanadium extraction (46).

Crown ethers, cryptates and other chelating agents may also be used as extractants. Crowns are defined as macroheterocycles, usually containing the basic unit $(-Y-CH_2-CH_2)_n$ where Y is O,S or N. They have the unique property that, depending on the basic structure, one can get different cavity sizes. They can not only provide simple and efficient means for solubilising metal salts in nonpolar and dipolar aprotic solvents, but also can distinguish between different metal ions depending on the cavity size. Because the systematic nomenclature of these compounds is very clumsy, common names are used. These are exemplified by 18-crown-6 (Figure 3.a) in which 18 indicates the number of atoms in the ring, crown the calss, in this case ether, and 6 the number of oxygen atoms in the relation 1,4,7 etc. Other common commercially available crown ethers, as shown in Figure 3.b,c and d respectively are dibenzo-18-crown-6, dicyclohexano-18-crown-6 and 15-crown-5. Now crown ethers, azo-analogues, polyoxa-polyazo-macrocycles, analogues containing annelated hetero-cycles, and bi- and polycyclic analogues are being reported. Figure 3.e shows a compound designated as cryptate 2.2.2. (the numbers indicate the three two-oxygen-atom bridges) which can be found under the trade name Kryptofix 222 (59). Their use in phase transfer catalysis (PTC) has been already known for the last few decades (6,7) and a simplistic view is shown in Figure 4. Recently, they have been considered to extract alkali metals.

Mc Dowell et al. (60) have suggested a combination of organic soluble cation exchangers, such as DEHPA, with the crown ethers in the organic phase to extract K^+. Mühl and Gloe (49) have shown that dibenzo-18-crown-6 shows a very selective behaviour depending on the metal ion diameter and with nitrates the following behaviour was observed $K^+ > Rb^+ > Cs^+ > Na^+$. Marcus et al. (61) have shown that crown ethers dissolved in solvents such as creosols, xylenols etc. can selectively extract KCl from brine containing Na, Mg and Ca chlorides. They can also be used in conjunction with liquid membranes.

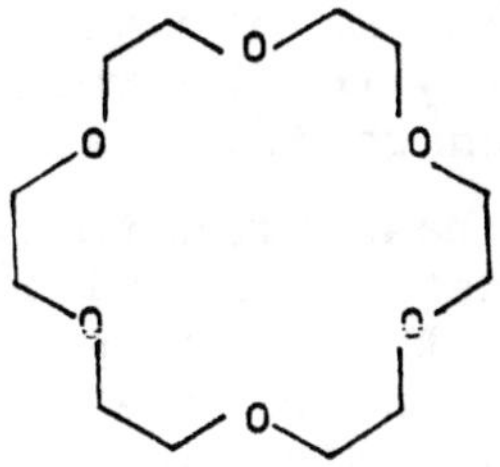

(a) 18-crown-6

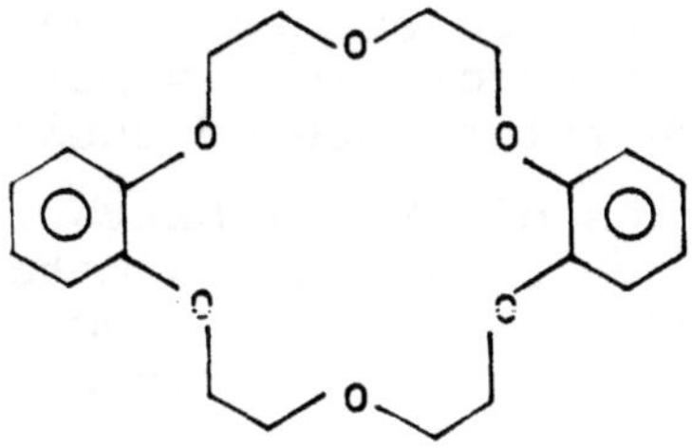

(b) dibenzo-18-crown-6

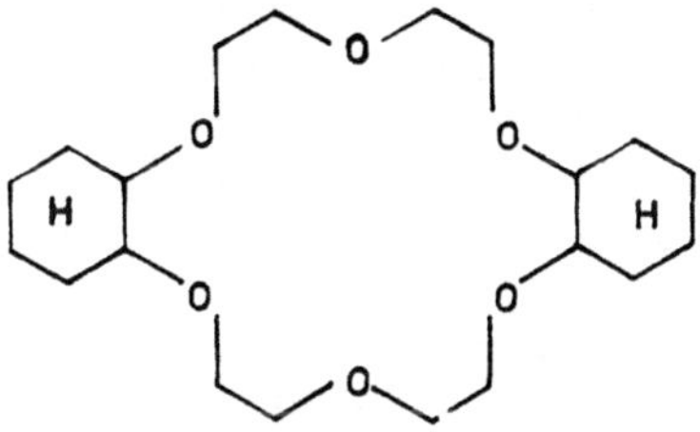

(c) dicyclohexano-18-crown-6

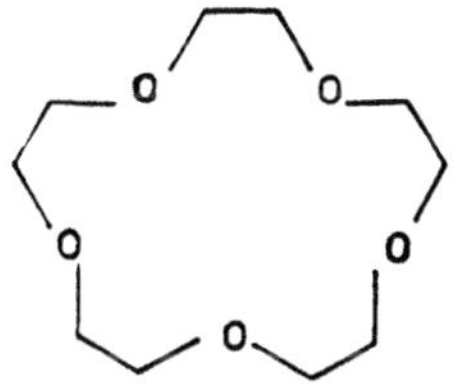

(d) 15-crown-5

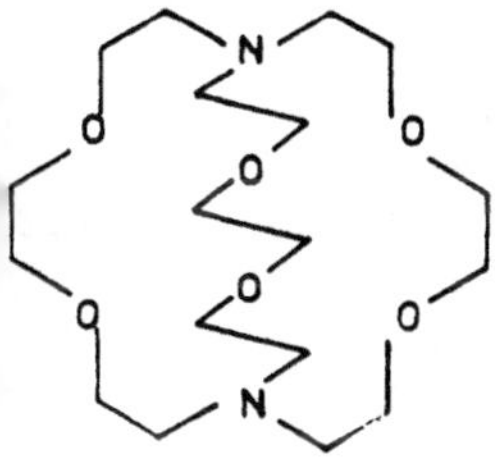

(e) Cryptate[2.2.2]
(Trade name Kryptofix222)

Figure 3. The structure of some of the simple Crown Ethers and Cryptates.

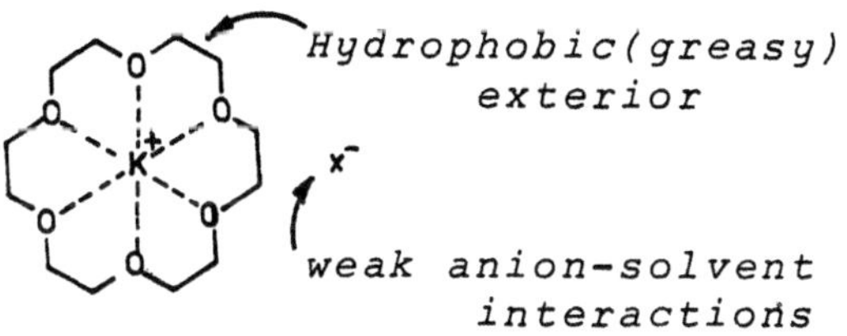

Figure 4. A simplistic view of the solubilization process.Solubilization of potassium salts in nonpolar and dipolar aprotic solvents.

However it must be stressed that the commercial explotation of crown ethers is not likely to be realised in immediate future due to their high costs.

5.2.2.Kinetics of Metal Extraction:Mass Transfer with reaction. Investigations on metal extraction kinetics were carried out on the assumption that the reaction was very slow and proper considerations of mass transf effects were often neglected.Thus,the information on relevant properties,such as contact area etc. were oft missing leading to contradictions.Recent studies (62,63 however, indicate reasonably clear that mass transfer play an important role and that there is interaction between mass transfer factors and chemical kinetics at least in the case of copper extraction with oximes However,the process differs probably from the usual two-phase systems considered in the main part of this book; that is; in normal systems,the sparingly sol solute dissolves in a phase (usually aqueous) and react there with an already dissolved reactant, the products being also soluble in the same phase.On the other han in metal systems,ions are insoluble in the organic phase and insolubility in water is one of the main selection criteria of complexing agents and the metal complex must also be insoluble in the aqueous phase. There is however still apparent controversy over the actual locale of the reaction, some workers (44,62) adopt the view of a truely interfacial reaction. Other believe that the reaction takes place in a zone in the aqueous phase adjacent to the interface. This is, however,partly due to the different views of the interface as some appear to view it as a pseudo-crysalline boundary,while others consider it as a region of finite thickness(64). Considering the solubility of commercial extractants LIX63 and LIX65N (anti-form) in water are respectively 15.5 ppm and 1 ppm, Hanson an coworkers (67) postulated a reaction immediately adjacent to the interface but with a significant (and poss ly sometimes dominating) diffusional resistance, in a zone (likely to be of greater thickness) on the organic side of the interface. Kondo and coworkers (65,66 have studied the extraction of copper by benzylacetone and have found the customary interactions between mass transfer and reaction. Here, the partition constan is 7.24×10^{-4} and the enhancement factors of as high as 1000 were observed in a Lewis type stirred cell. Indeed, at low pH range of 4.5, experiments appear to fall into the regime of pseudo first order. At high pH of 7.5, the regime is close to that of an instantaneous reaction so that the process of solvent extraction can be controlled by diffusion of metal ions and the agent to the reaction zone in the aqueous phase. However, often the process is extremely complicated great caution must be exercised in interpreting experimental results which in turn should be planned very carefully. Since the chemical reaction has usually a crucial role, together with mass transfer factors, the followings should also be carefully considered (45) : (a) the difference in the reaction rate of possible stereospecific isomers (68), (b) nature of species (monomer, dimer etc.) (69), (c) role of impurities; synergism or inhibition (70), and (d) effect of diluents, including nature of specied, effect on interfacial area and interactions with extractants (71).

5.3. Selected Novel Aspects

5.3.1. Reactions in organic phase. In the case of organic-aqueous phase reactions, the locale of the reaction is often assumed to be in the aqueous phase. This is, of course, true for most of the liquid-liquid reactions. There are however cases where the reaction occurs entirely in the organic phase. Sharma (3,4) points out a number of examples. For instance, the rates of nitration of 1-dodecene and 1-octadecene under identical experimental conditions, are very similar even though the solubilities of them in aqueous phase are very different. This argument and some other related arguments indicate clearly that the reaction occurs in the organic phase (Gregory et al. (72)).

Another interesting example is polycondensation reactions to procedure polyesters and polyamides (73). Sharma (3,4) discusses also some aspects of interfacial polycondensation; here, for instance, the diacyl chloride, dissolved in a water immiscible organic solvent, may be contacted with hexamethylene diamine to procedure the polyamide, Nylon 66. Since diamine is very soluble in the organic phase and the reaction is very fast, the reaction occurs probably in organic phase, therefore acyl chloride cannot get an opportunity to get into aqueous phase where concurrent hydrolysis would have occured. Usually, a film of polymer is formed on the organic phase droplets disposed in water and thus droplets are encapsulated and interfacial area measurements can well be envisaged.

5.3.2. Phase transfer catalysis. This is an outstanding example where the locale of the reaction is deliberately switched from aqueous to organic phase. The basic premise of phase transfer catalysis of two-phase reaction is that one can select a phase transfer agent that, used in catalytic quantities, can bring one of the reactants into the normal phase of the other reactant in such form that high reaction rates are observed (6,7,76) The most common example, and the one for which a large amount of data is available, is simple cyanide displacement on alkyl chloride or bromides:

$$(\text{R-Cl})_{\text{org}} + (\text{NaCN})_{\text{aq}} \rightarrow (\text{R-CN})_{\text{org}} + (\text{NaCl})_{\text{aq}} \quad (1)$$

Simply heating and stirring of a two-phase mixture of 1-chlorooctane with aqueous sodium cyanide leads to essentially zero yield of 1-cyanooctane. However, if a small amount of an appropriate quaternary ammonium salt is added, then very rapid formation of 1-cyanooctane is observed. The general outline of the catalysis sequence is represented by Figure 5.

Alternate types of phase transfer catalysts for two-phase reactions involving salts are crown ethers, cryptates and dialkylpolyethylene oxides, which form reversible complexes with many cations. For example, crown ether 18-crown-6, also strongly catalyzes reaction (1). In this case, the crown ether transfers the entire KCN molecule into the organic phase by complexation.

In principle, diffusional resistance may be associated with the transfer of the reactant and the catalyst in Figure 5. Anion transfer by the catalyst into the organic phase is considered to be an equilibrium process in which two different ions in the aqueous phase associate with a quaternary cation in the organic phase:

$$(\text{QCl})_{\text{org}} + (\text{CN}^-)_{\text{aq}} \rightleftarrows (\text{QCN})_{\text{org}} + (\text{Cl}^-) \quad (2)$$

The rate determining step of the reaction is in the organic phase of the reaction mixture rather than in the aqueous phase or at the interface or in the micelles.

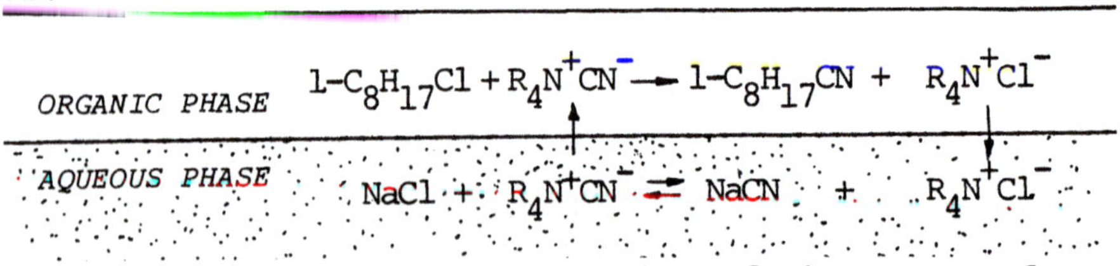

Figure 5. Schematic representation of phase transfer catalysis (Quaternary cation, $Q^+ = R_4N^+$ is ammonium or phosphonium.

In these studies (6,7,76), the intrinsically slow two-phase reactions have been studied. However, Sharma and coworkers (77) have recently studied the alkaline hydrolysis of a variety of formate esters, which are normally also fast, catalysed by PTC agents, such as cetyltrimethyl-, tricaprylmethyl-ammonium bromide etc. They have carried out experiments with this system in a fully baffled mechanically agitated contactors as well as in a constant area cell. The results are truly remarkable so that the presence of PTC results in an increase in the mass transfer rate ranging from 20 to over 200.

It is interesting to note that much of the phase transfer catalysis has used a quaternary ammonium reagent (trade name Aliquat 336), which is also widely used in solvent extraction of metals. Recently, it has been suggested that PTC (as well as micellar catalysis) may play an important role in mixed reagents systems involving LIX63, DEHPA, Lauric acid, Aliquat 336 and HDDNS (78). Indeed, these authors (78) were able to offer an explanation for the catalytic effect of LIX63 on copper extraction by LIX65N. It seems probable that metal extraction with mixture of chelating reagents and organic acid mixtures may well proceed via a phase transfer catalysis type mechanisms.

Phase transfer catalysis may function not only through liquid-liquid systems, but also with liquid-gas, liquid-solid, solid-gas and presumably solid-solid systems (6). Since the subject is actively pursued from many sides and since many of its potential fruits are far from being fully harvested, the industrial explotations have probably not yet been realised to any noticeable extent.

5.3.3. Micellar catalysis. A somewhat related phenomena to phase transfer catalysis occurs in Micellar Catalysis (79), which is however mechanistically and preparatively different. When good surfactants are added to a two-phase aqueous-organic system, normally micelles are produced. These micelles usually take the form of small aggregations of 20-100 organic molecules dispersed in the aqueous phase, wherein the nonpolar parts of the surfactant and other nonpolar organic molecules occupy the internal hydrophobic volume of the micelle, while the highly polar groups, which are referred as heads, occupy the highly hydrophilic outer surface. Figure 6 illustrates the situation schematically. The micelles formed in aqueous surfactant solutions behave as a separate (pseudo phase) medium with unique

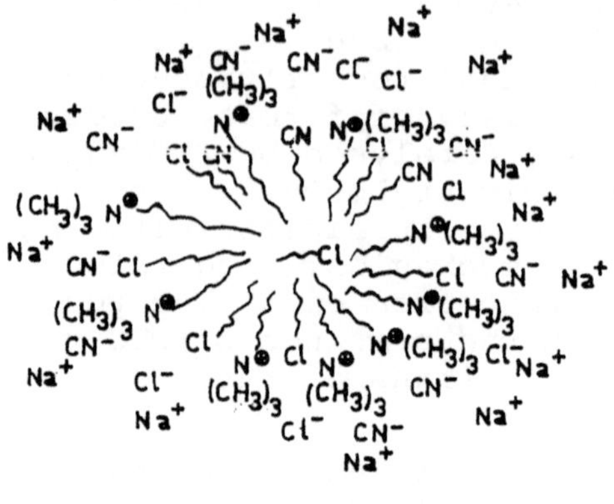

Figure 6. Hypothetical reaction schematic for quaternary ammonium surfactant-catalysed micelle reaction of alkyl chloride and aqueous sodium cyanide.

physical properties. The positively charged outer surface attracts and concentrates anions (e.g. cyanide) from the bulk aqueous solution into a counteranion layer near the surface of the micelle, strongly facilitating reaction of RCl in the micelle with CN^- just at the micelle surface. The reverse case where a central core is hydrophilic, is in general possible for reversed organic phase micelles.

The kinetics and mechanisms of micelle-catalysed reactions have been intensively studied in recent years and they are much too complicated to be discussed here in detail (80,81). Micelles may also be considered as phase transfer agents of a sort which take the organic phase reagent into the aqueous phase for reaction. Many quaternary ammonium salts containing one or two large alkyl groups, such as cetyltrimethylammonium bromide, $C_{16}H_{33}N(CH_3)_3Br$, produce micelles as well as being phase transfer agents. Indeed, there are borderline cases where a particular quaternary ammonium salt may behave as both a surfactant and a phase transfer catalyst or as either one, depending on the particular reaction conditions. Starks (6) discusses further similarities and essential differences of the two phenomena. The most important difference is that whereas the rate of phase transfer catalysed reactions are directly roportional to the catalyst concentration, the ratio of micelle-catalysed reaction is low until the critical micelle concentration of surfactant is approached, hen increases tremendously and reaches a maximum. urther increase in surfactant concentration has either o effect or decreases the rate slightly.

It is also possible to inhibit an undesirable eaction in a two-phase system. Menger and Portney (82) nd Bunton (83) have developed mathematical models on he basis of which the catalysis and inhibition of eactions can be treated quantitatively.

Recently it has been suggested that many aspects f micellar and reverse micellar catalysis are relevant in commercially important extractants involving IX63, DEHPA. Lauric acid, and Aliquat 336 (78). Osseo-sera and Keeney (78) were able to demonstrate that the otential for effective utilization of reversed micelar catalysis in liquid-liquid extraction systems deends primarily upon the ability of micelles to soluilize both extractant molecules and metal ions, and tabilize H_2O-extractant liquid exchange reactions.

Reactions in micelle system are usually difficult or synthetic applications because of the problems in andling emulsions and the need for careful regulation f concentrations. However, there are many instances e.g. emulsion polymerisation) where micelle systems re highly useful and are applied on a commercial asis.

.3.4. Dissociation Extraction. Dissociation extraction s a technique for the separation of mixtures of organic acids or bases, which depends upon differences oth in the dissociation constants and the distribuion coefficients of the components of the mixture, n a two-phase systems. It can be applied to closely elated or isomeric compounds, which are difficult if indeed possible- to separate by common methods of istillation, solvent extraction or fractional crysallization (84). Consider the case of a mixture of eak organic acids, m- and p-cresol with dissociation onstants of $9.8x10^{-11}$ and $6.70x10^{-11}$ respectively. If mixture of these compounds, dissolved in a water imiscible organic solvent, such as toluene etc., is conacted with a stoichiometrically deficient amount of queous alkali, in relation to total acids, the stroner acid, having the large dissociation constant (m-resol) will preferentially react with the alkali. Thus, here will be an enrichment of p-cresol in the organic

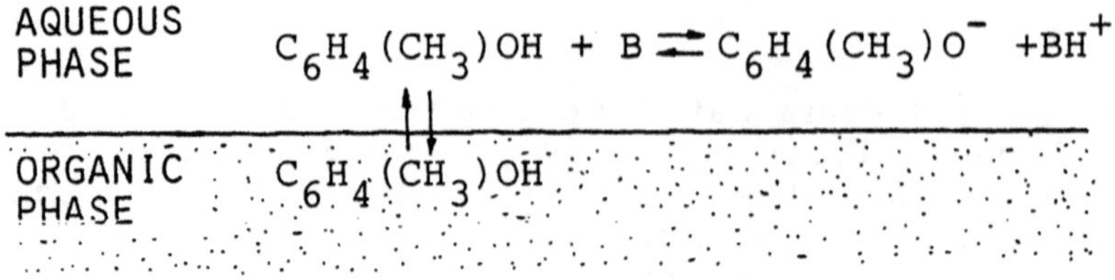

Figure 7. Dissociation extraction with a weak base.

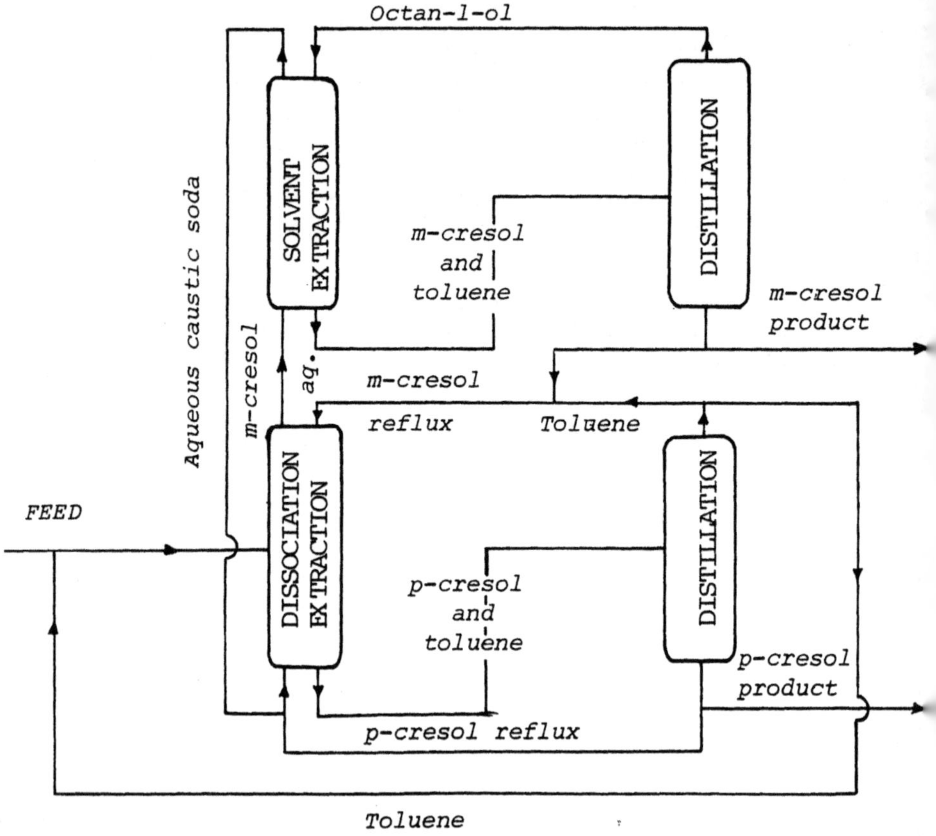

Figure 8. Flow diagram for the continuous separation of m- and p-cresols.

phase and the stronger acid will be in the aqueous phase as its dissociated salt. This process of dissociation extraction when carried out in a multistage extractor can lead to high purity components at the two ends. In the so called classical or conventional dissociation extraction process, m-cresol would be generated from its salt in the aqueous phase by treatment with strong mineral acid. Commercial applications of dissociation extraction have been considered for separation of organic acids and bases occuring in coal tar, particularly by m- and p-cresol (85), xylenols (86) and 2,6-lutidine, 3- and 4-picoline systems, separation of dichlorophenols, penicillin acids, benzoic acid derivatves, and close boiling amines, such as substituted aniline, heterocyclic amines, xylidines etc.

The dissociation extraction efficiency is measured in terms of a separation factor, α, which is defined in an analogous manner to that of relative volatility in distillation:

$$\alpha = \frac{[A]_{org}/[B]_{org}}{[A^*]_{aq}/[B^*]_{aq}} \tag{3}$$

where $[A]_{org}$ and $[B]_{org}$ are concentrations of A and B in the organic phase and $[A^*]$ and $[B^*]$ are the equilibrium concentrations of A and B in the aqueous phase resepctively; A refers to the weaker of the two acids /bases. Anwar et al. (87) have developed the theory of dissociation extraction in a somewhat similar way to that of binary distillation. Recently, Wadekar and Sharma (88) have extended the theory to multicomponent and multifunctional systems. It is possible to directly predict the value of α in such systems provided that the data on dissociation constants and distribution coefficients are available.

In the conventional dissociation extraction, the consumption of the extractant alkali or acid makes the process less attractive. To overcome this difficulty, Anwar et al. (84) have evolved a modified strategy where instead of strong acid/base, a weakly acidic/ basic extractant is used so that the reaction does not proceed to completion. Then the extract is contacted with a fresh solvent having a strong affinity for the undissociated organic component so that the reaction is reversed. The basic/acidic reagent is thereby regenerated and can be recycled (Figure 7). They have suggested both Na_3PO_4 and monomethanolamine for separation of m-cresol and p-cresol isomers. However, with weak reagents, in general, the reaction in the aqueous id incomplete, resulting in relatively low aqueous phase loadings and higher equipment and process costs. On the other hand, recently they were able to use a strong alkali, such as caustic soda, for the separation of cresol isomers leading to higher aqueous phase loadings and separation factors, in cooperation with time an organic solvent with a very high affinity for the undissociated organic component, which is sufficiently powerful to reverse the reaction despite the strong alkali. For the case of separation of cresol isomers, they used caustic soda and octan-l-ol and Figure 8 shows a flow diagram for a practical industrial separation process. No doubt similar processes could well be applied to other mixtures of acidic/basic organic compounds.

In some cases, it might also be possible to use a thermally regenrative extractant. For example, monomethylamine solution has been suggested for the separation of mixtures of p-chlorophenol/2,4-dichlorophenol By boiling the aqueous salt, monomethylamine is regenerated, and can thus be recycled.

During recent years, Sharma and coworkers (88-90) have been able to find examples of dissociation extration which have distinct features and are truly remarkable. Firstly, they have illustrated that the use of a diluent may be avoided advantageously (89). This is particularly relevant for every sparingly water soluble organic acids/bases and, here, apart from the difference an ionisation constants, the relative solubilities of species (rather than the distribution coefficients) play an important role. Typical examples are mixtures of N-ethyl-o-toluidine and NN'-diethyl-o-toluidine where higher values of the separation factors are obtained without a solvent (4,89). The same process has been applied to sparingly water soluble acidic/basic solids and such a dissociative leaching has been illustrated for separation of o- and p-chlorobenzoic acid mixtures (89). Secondly, they where able to recover and separate organic acids from dilute aqueous solutions by using modified dissociation extraction (90). For instance, they employed tri-n-octylamine, dissolved in various water immiscible solvents, as an extractant; the acid-amine complex is soluble in organic solvents such as xylene. Very promising results were obtained for mixtures such as acetic-acid/mono-chloroacetic acid, formic/oxalic acids etc.

(90). Of course, a similar strategy can be extended for the recovery and separation of aqueous solutions containing mixtures of basic compounds. The process is also interesting as it may conform to a situation where the reaction occurs in the organic phase. The work in this area so far involves only the determination od separation factors from equilibrium considerations, and there is no published information on rates of mass transfer. However, the latter is likely to be an almost trivial exercise as the reactions involved are of a relatively simple type.

6. DISCUSSION AND CONCLUSIONS

An examination of extraction with reaction processes reveals that it is an area which exploits chemistry to a greater extent than, for instance, other common separation processes. A variety of liquid-liquid reactions are encountered in practice and some illustrative examples have been presented. Further challenging examples are frequently presented in many publications; such as the special section on Journal of Separation Science, Hydrometallurgy etc. as well as the more common journals, e.g. Chemistry and Industry etc. However, the real highlights are documented mainly at the tri-annual International Solvent Extraction Conferences; ISEC's, as well as many more specialised meetings; e.g. hydrometallurgy etc.

Examination of theoretical aspects indicates that although the conventional theories developed for gas-liquid systems represents a very simplistic picture neglecting many important features of dispersion phenomena, they can still be applied for design purposes. Indeed, in a number of cases of practical relevance, a simple analytical expression can be used for calculating the process design parameter in a manner which is practically the same as those of gas-liquid systems (20). However, this should by no means suffice and the continuation of fundamental studies on dispersion phenomena and the incorporation of them with mass transfer in reacting systems is not only essential for further scientific development of this area, but also extremely desirable as they will eventually lead to more precise design methods (10,91). In particular, new measuring techniques may well yield a better understanding of interaction of fluid dynamics, interfacial phenomena and mass transfer (92).

In terms of process development, metal extraction

has probably claimed much of the limelight over the last decade. This will probably continue to be so, particularly in solvent extraction of uranium as the expansion of nuclear power generation in most industrialised countries offers inevitable. New processes, in terms of new chemical systems as well as new techniques, such as phase transfer catalysis, are likely to find interesting applications in coming years. Another relatively novel aspect which deserves full examination involves membrane processes. This includes both membrane and liquid membrane processes. Recent state of the art reviews of both processes are already available; among others, Hafez (93) discusses the use of membranes in extraction and Halwachs and Schügerl (94) review liquid membranes. The latter is also discussed in this volume by Sawistowski (1). It seems, membrane processes lend themselves readily to processes such as metal extraction, waste water treatment and pharmaceutical and medical applications. Differential reactions, as in dissociation extraction, also appear to offer scope for more selective separations and the research which has been conducted during the 1970's could well come to fruition in the form of industrial processes during the next decades.

Finally, it may be pointed out that relative increases in the cost of energy could well make liquid-liquid extraction economically attractive for some separations in the general organic field.

REFERENCES

1. Sawistowski,H. "Physical aspects of liquid-liquid extraction." (Proceesings of NATO ASI on "Mass transfer with chemical reaction in multiphase systems", Turkey, 1981).

2. Hartland,S. "Experience with liquid-liquid test systems in extraction." (Proceedings of NATO ASI on "Mass transfer with chemical reaction in multiphase systems", Turkey, 1981).

3. Sharma,M.M. "Extraction with reaction". Chapter 2a in Handbook of Extraction (To be published, Wiley and sons, New York).

4. Doraiswamy,L.K. and M.M.Sharma. Heterogeneous Reactions: Analysis, Examples and Reactor Design (Wiley Interscience, New York, 1981).

5. Laddha,G.S. and T.E.Degaleesan. Transport Phenomena in Liquid Extraction (Tata Mc Graw Hill Co., New Dehli, 1978).

6. Starks,C.M. and C.L.Liotta. Phase Transfer Catalysis: Principles and Techniques (Academic Press, New York, 1978).

7. Dehmlow,E.V. and S.S.Dehmlow. Phase Transfer Catalysis (Verlag Chemie, Weinheim, 1980).

8. Porter,K. and D.Roberts. Chem.Engng.Sci.24 (1969) 695.

9. Grosjean,P.R.L. and H.Sawistowski. Trans.Instn. Chem.Engrs. 38 (1980) 60.

10. Tavlarides,L.L. and M.Stamatoudis. "The analysis of interphase reactions and mass transfer in liquid-liquid dispersions". Advances in Chemical Engng. Vol. 11 (Academic Press, London,1981).

11. Danckwerts,P.V. Gas-Liquid-Reactions (Mc Graw Hill Co., New York, 1970).

12. Juvekar,V. and M.M.Sharma. Trans.Instn.Chem. Engrs. 55 (1972) 72.

13. Carberry,J.J. Chemical and Catalytic Reaction Engineering (Mc Graw Hill Co., New York, 1975).

14. Hanson,C. Recent Advances in Liquid-Liquid Extraction (Pergamon Press, Oxford, 1971) p. 429

15. Alper,E. "Process design aspects of gas absorbers". (Proceedings of NATO ASI on "Mass transfer with chemical reaction in multiphase systems", Turkey, 1981)

16. Van Landeghem,H. Chem.Engng.Sci. 35 (1980) 1912

17. Sharma,M.M. and P.V.Danckwerts. Brit.Chem.Engng. 5 (1970) 522.

18. Sankholkar,D.S. and M.M.Sharma. Chem.Engng.Sci. 0 (1975) 729.

19. Laddha,S.S. and M.M.Sharma. Chem.Eng.Sci. 31 1976) 843.

20. Vasudevan,T.V. and M.M.Sharma. "Some aspects of process design of liquid-liquid reactor." (Int. Symposium on "Mass Transfer with Chemical Reaction in Two-Phase Systems", ACS-Meeting, Atlanta, 1981).
21. Sarkar,S., Mumford,C.J. and C.R.Philips. Ind. Engng.Chem.Proces.Des.Dev. 19 (1980) 665.
22. Nanda,A.K. and M.M.Sharma. Chem.Engng.Sci. 22 (1967) 769.
23. Sharma,R.C. and M.M.Sharma. J.Appl.Chem.Biotechnol. 19 (1969) 162.
24. Sharma,R.C. and M.M.Sharma. Bull.Chem.Soc.Jap. 43 (1970) 43.
25. Albright,L.F. and C.Hanson (Editors). "Industrial and Laboratory Nitration". Am.Chem.Soc.Symposium Ser. 22 (1975).
26. Albright,L.F. and C.Hanson. "Loss Prevention" (CEP Technical Manual) 3 (1969) 26.
27. Cox,P.R. and A.N.Strachan. Chem.Engng.Sci. 27 (1972) 457.
28. Chapman,J.W. and A.N.Strachan. J.Chem.Soc.Chem. Commun., (1974) 293.
29. Hanson,C. and H.A.M.Ismail. J.Appl.Chem.Biotechnol. 26 (1976) 111.
30. Komasawa,I., Ionue,T. and T.Otake. J.Chem.Engng Japan. 5 (1972) 34.
31. Tiwari,R.K. and M.M.Sharma. Chem.Engng.Sci. 32 (1977) 1253.
32. Richardson,J.A. and Rase,H.F. Ind.Engng.Chem. Prod.Res.Dev. 17 (1978) 287.
33. Lee,L.M. and P.Harriot. Ind.Engng.Chem.Proc.Des Dev. 16 (1977) 282.
34. Dixon,J.K. and K.W.Saunders. Ind.Engng.Chem. 46 (1954) 652.
35. Bhave,R.R. and M.M.Sharma. J.Chem.Tech.Biotechnol. 31 (1981) 93.
36. Kothari,P.J. and M.M.Sharma. Chem.Engng.Sci. 21 (1966) 391.
37. Harnisch,H. Pure Appl.Chem. 52 (1980) 809.
38. Carr,N.L. and Y.T.Shah. Can.Jl.Chem.Engng. 57 (1979) 35.
39. Bailes,P.J., Hanson,C. and N.A.Hughes. Chem. Engng. 83 No:18 (1976) 86.
40. Fletcher,A.W. Chem.Ind. No:5 (1973) 414.
41. Lloyd,P.J. Solvent Extraction Chemistry. (Ed.D. Dyrssen, North-Holland, Amsterdam, 1967) p. 642.
42. Warner,B.F. Solvent Extraction Chemistry. (Ed.D Dyrssen, North-Holland, Amsterdam) p. 635.
43. Chalan,M.J. Chem.Ind. (1967) 1590.
44. Sawistowski,H. "Aspects of metals extraction"

(Proceedings of NATO ASI on "Mass transfer with chemical reaction in multiphase systems", Turkey, 1981).

45. Sharma,M.M. "Types of extractants and the chemistry of solvent extraction" (Unpublished work; Private communication, 1981)

46. Bautista,R.G. "Hydrometallurgy" in Advances in Chemical Eng. (Academic Press, New York, 1970).

47. Rice,N.M. Hydrometallurgy 3 (1978) 111.

48. Freiser,H. "Kinetics of solvent extraction of metal chelates" (Proceedings of Int.Solvent Extraction Conf., ISEC'80, Belgium, 1980).

49. Mühl,P. and K.Gloe. "Comparative studies on the metal extraction with different chelating extractants" (Proceedings of Int.Solvent Extraction Conf., ISEC'80, Belgium, 1980).

50. Price,R. and J.Tumilty. "An interpretation of some aspects of solvent extraction as realted to the extraction of Copper using O-Hydroxyaryl Oximes" (Proceedings of Symposium on "Hydrometallurgy", Instn. Chem.E.Symp.Series No:42, 1975).

51. Van der Zeeuw. "Selective copper extractants of the 5-alkyl-2-Hydroxyphenyl Alkyl Ketone Oxime" (Proceedings of Symposium on "Hydrometallurgy", Instn. Chem.E.Symp.Series No: 42, 1975).

52. Ritcey,G.M. CIM Transactions 74 (1973) 71.

53. Cox,M. and W.Van Bronswijk. "Chemistry of extraction of copper, cobalt and nickel with substituted 8-sulphonamideoquinilines" (Proceedings of Int.Solvent Extraction Conf., ISEC'80, Belgium, 1980).

54. Cote,G. and D.Bauer. "Kinetic aspects of the liquid-liquid extraction of germanium (IV) with the β-dodecenyl 8-hydroxyquinoline." (Proceedings of Int. Solvent Extraction Conf., ISEC'80, Belgium, 1980).

55. Guesnet,P., Sabot,J.L. and D.Bauer. "Kinetics of cobalt oxidation in solvent extraction by 8-quinolinol and KELEX 100" (Proceedings of Int.Solvent Extraction Conf., ISEC'80, Belgium, 1980)

56. B.P. 910,541 (1960); U.S.P. 2,969,275 (1961); U.S.P. 3,111,383 (1963); U.S.P. 3,479,294 (1960); U.S.P. 3,493,349 (1970) and U.S.P. 3,741,731 (1973).

57. Su,Y.F. and D.Y.Yu. "Process development of boron recovery from ascharite" (Proceedings of Int. Solvent Extraction Conf., ISEC'80, Belgium, 1980).

58. M.J. "Modern science in winning minerals and metals" (Progress in Chemistry, 1967).

59. Dietrich,B. and J.M.Lehn. Acc.Chem.Res. 11 (1978) 49.

60. McDowell,W.J.,Kinard,W.P. and R.R.Shoun. "Size-selective synergism by crown ethers in the extraction of alkali metals by di-(2-ethylhexyl) phosphoric acid"

(Proceedings of Int.Solvent Extraction Conf., ISEC'80, Belgium, 1980).

61. Marcus,Y., Asher,L.E., Hormadaly,J. and E.Pross. "Selective extraction of potassium chloride by crown ehters in substituted phenol solvents" Hydrometallurgy 1 No:. (1975) 5.

62. Ajawin,L.A., Pérez de Ortiz,S.E. and H.Sawistowski. "Kinetics of extraction of zinc by di(2-ethylhexyl) phosphoric acid in N-heptane". (Proceedings of Int. Solvent Extraction Conf., ISEC'80, Belgium, 1980).

63. R.J.Whewell, Hughes,A.M. and C.Hanson. "Aspects of the kinetics and mechanism of the extraction of copper with hydroxyoximes". (Proceedings of Int. Solvent Extraction Conf., ISEC'77, Canada, 1977).

64. Krishna,R. "Interphase mass transfer models". (Proceedings of NATO ASI on "Multiphase Chemical Reactors", Portugal, 1980).

65. Kondo,K., Takahashi,S., Tsuneyuki,T. and F.Nakashio. J.Chem.Engng.Japan 11 (1978) 193.

66. Kondo,K., Tsuneyuki,T. and F.Nakashio. "Solvent extraction kinetics of copper by benzoylacetone". (Proceedings of Solvent Extraction Conf., ISEC'80, Belgium, 1980).

67. Hanson,C., M.A.Hughes and R.J.Whewell. J.Appl. Chem.Biotechnol. 28 (1978) 435.

68. Van der Zeuw,A.J. and R.Kok. "Kinetics and mechanism of copper extraction with 5-alkyl-2-hydroxyphenyl alkyl ketoximes", (Proceedings of Int.Solvent Extraction Conf., ISEC'77, Canada, 1979).

69. Hanson,C., Hughes,M.A., Preston,J. and R.J.Whewell. J.Inorg.Nucl.Chem. 38 (1976) 2306.

70. Van der Zeuw,A.J. and R.Kok. "Ideas and practice in the design of solvent extractant reagents". (Proceedings of the Inter.Solvent Extraction Conf., ISEC'77 Canada, 1977).

71. Dalton,R.F., Hauxwell,F. and J.A.Tumilty.Chem. and Industry 6 (March) (1976) 184.

72. Gregory,D.P., Martens,R.J., Stubbs,C.E. and J.D. Wagner. J.Appl.Chem.Biotechnol. 26 (1976) 623.

73. Morgan,P.W. Condensation Polymeres: By Interfacial and Solution Methods (Interscience Publishers, New York, 1965).

74. Dehmlow,E.V. Angew.Chem. 86 (1974) 187.

75. Dehmlow,E.V. Angew.Chem. 89 (1977) 521.

76. Weber,W.P. and G.W.Gokel. Phase Transfer Catalysis in Organic Synthesis (Springer Verlag, New York, 1977).

77. Lele,S.S., Bhave,R.R. and M.M.Sharma. Chem. Engng.Sci. 36 (1981) 955.

78. Osseo-Asare,K. and M.E.Keeney. "Phase transfer and micellar catalysis in hydrometallurgical liquid-liquid extraction systems". (Proceedings of Int.Solvent Ectraction Conference, ISEC'80, Belgium, 1980).
79. Fendler,J.H. and E.J.Fendler. Catalysis in Micellar and Macromolecular Systems. (Academic Press, New York, 1975).
80. Fendler,E.J. and J.H.Fendler. Adv.Phys.Org.Chem. 8 (1970) 271.
81. Morawetz,H. Adv.Catal. 20 (1969) 341.
82. Menger,F.M. and C.E.Portnoy. J.Am.Chem.Soc. 89 (1967) 4698.
83. Bunton,C.A. Catal.Rev.Sci.Eng. 20 (1979) 1.
84. Anwar,M.M.. Pratt,M.W.T. and Snaheen,M.Y. "Developments in Dissociation Extraction", (Proceedings of Int.Solvent Extraction Conf., ISEC'80, Belgium,1980).
85. Ellis,S.R.M. and J.D.Gibbon. The Less Common Means of Separation. (Instn.Chem.Engrs., London, 1964).
86. Coleby,J. Recent Advances in Solvent Extraction (Pergamon Press, Ed.C.Hanson, London,1971).
87. Anwar,M.M., Hanson,C. and M.W.T.Pratt. Trans. Instn.Chem.Engrs. 49 (1971) 95.
88. Wadekar,V.V. and Sharma,M.M. J.Separ.Proc.Technol. 2 No:1 (1981) 1.
89. Laddha,S.S. and M.M.Sharma. J.Appl.Chem.Biotechnol. 28 (1978) 69.
90. Jagirdar,G.C. and M.M.Sharma. J.Separ.Proc. Technol. 1 No:2 (1980) 40.
91. Schügerl,K., Blaschke,H.G., Brunke,U. and R. Streicher. "Interaction of fluid dynamics, interfacial phenomena and mass transfer in extraction processes". Recent Developments in Separation Science Vol. 3 (CRC Press, Cleveland, 1977).
92. Villermaux,J. "Drop break-up and coalescence. Micromixing effects in liquid-liquid reactors". (Proceedings of NATO ASI on "Multiphase Chemical Reactors", Portugal, 1980).
93. Hafez,M.M. "Membranes in extraction. A state of art review" (Proceedings of Int.Solvent Extraction Conf., ISEC'80, Belgium 1980).
94. Halwachs,W. and K.Schügerl. Chemie-Ing.-Tech. 50 No:10 (1978) 764.

PHYSICAL ASPECTS OF LIQUID-LIQUID EXTRACTION

H. Sawistowski

Department of Chemical Engineering and Chemical Technology
Imperial College of Science and Technology, London SW7

INTRODUCTION

Liquid-liquid extraction is a separation process which relies on unequal distribution of components between two liquid phases. Mass transfer will therefore occur as a spontaneous process if the phases are not at equilibrium. The transferred components are referred to as solutes and the carrier liquids as solvents. The solvents may be practically immiscible or partially miscible. Most of the fundamental work is being conducted on the transfer of a single solute between two immiscible solvents and there is a number of such systems specially recommended for such investigations by the European Federation of Chemical Engineering (1). Partially miscible binary systems may also be used for this purpose as examples of mixtures with low interfacial tension. In practice, however, partially miscible multicomponent systems are often encountered, e.g. in the removal of aromatics from lubricating oil. This represents an area which is not particularly well researched and documented.

The process of liquid-liquid extraction is not very energy intensive. Consequently its application is on the increase although, on account of lack of proper understanding of the process, its potential is not yet fully utilized. A number of topics have therefore been selected here to highlight certain physical fundamentals of liquid-liquid extraction. Although, on purpose, no attempt will be made to deal with extraction equipment, some references to it cannot be avoided.

The best equipment for a particular separation process is that in which the hydrodynamics matches best the process in question. The hydrodynamic behaviour is a function of throughput, that is of flow rates and phase ratios, and of column design. Thus, the problems will be different in equipment with supported interfacial area, e.g. packed columns, from those encountered when the interfacial area is unsupported, as in sieve plate columns and stirred tanks. Conditions will also change if interfacial area is created by input of mechanical energy rather than by action of gravity.

The factors which are directly affected by hydrodynamic conditions are:

(a) interfacial area,
(b) mass transfer coefficients,
(c) radial and axial mixing.

Of these, only the first two will be discussed here and discussion will be focussed on the following topics:

(1) formation of dispersion as it occurs in spray columns and sieve-plate columns;
(2) behaviour and characterization of stirred dispersions;
(3) basic mass transfer phenomena including mass transfer to and from drops;
(4) mass-transfer induced interfacial convection;
(5) basic principles of the liquid membranes process.

1. FORMATION OF DISPERSIONS

Liquid-liquid dispersions are frequently encountered in a number of industrial operations such as solvent extraction, direct contact heat transfer and heterogeneous chemical reactions. They are usually formed by the application of external energy to liquid/liquid systems and, depending on their behaviour on discontinuation of energy supply, they can be divided into stable dispersions or emulsions and unstable dispersions. Only the latter, in which the phases start separating as soon as the supply of external energy is stopped, are considered here and given the generic term "dispersions".

The external force employed is either gravitational or mechanical and drops are formed either by forcing one liquid through nozzles, e.g. in a spray column, or by breaking it up in a high shear field, for instance by using an agitator in a baffled tank. The former will be used as an example of formation of a dispersion and discussed with reference to a single nozzle in a spray column.

(a) Drop Formation

It is a well-known fact that when a drop is formed at a nozzle under pseudostatic conditions, its size is determined solely by the balance of interfacial tension and gravity or buoyancy forces. This forms the basis of the drop-weight method of determination of surface and interfacial tension.

Discrete drops continue to be formed with increasing velocity of the dispersed phase up to a critical velocity u_j. The dispersion formed at each velocity is mono-disperse and the drop diameter, d_p, can be obtained from a semi-empirical equation developed by Hayworth and Treybal (2) and based on the balance of forces. The critical velocity, also called the jetting velocity, is given by (3)

$$u_j = 2\left[(1 - d_n/d_p)\ \gamma/\rho_d d_n\right]^{\frac{1}{2}} \qquad \text{(m/s)} \qquad (1)$$

where γ is the interfacial tension, d_n the nozzle diameter and ρ_d the density of the dispersed phase. Above this velocity no discrete drops will be formed at the nozzle, instead they will be produced by disintegration of jets.

A jet issuing from a nozzle will be subject to two types of hydrodynamic instabilities: symmetrical and sinuous, each characterized by a different time constant. Drops formed as a result of the fast growing symmetrical disturbance will detach along the axis of the nozzle when the amplitude of the disturbance becomes equal to the jet radius. However, drag resistance to the motion of such a drop is larger than to the motion of the jet. This results in lengthening of the jet and increase in its life time until the slow growing sinuous disturbance becomes effective. The tip of the jet begins to oscillate, drops are discharged in various directions and the jet starts to shorten. This phenomenon is referred to as "thrashing" and it begins when (4)

$$u_n = 2.83 \times 10^{-2}(\gamma/d_n\rho_c)^{\frac{1}{2}} \qquad \text{(m/s)} \qquad (2)$$

where u_n is the nozzle velocity and ρ_c the density of the continuous phase. Between jetting and thrashing the dispersion formed remains monodisperse with the drop volume equal to the volume of a liquid cylinder of diameter d_n and length λ, where λ is the dominant wave length. The latter can be obtained by solving the characteristic equation due to Tomotika (5). Beyond the thrashing velocity the dispersion becomes polydisperse and the standard deviation of drop size increases with increasing Reynolds number. This is particularly accentuated once the jet becomes turbulent.

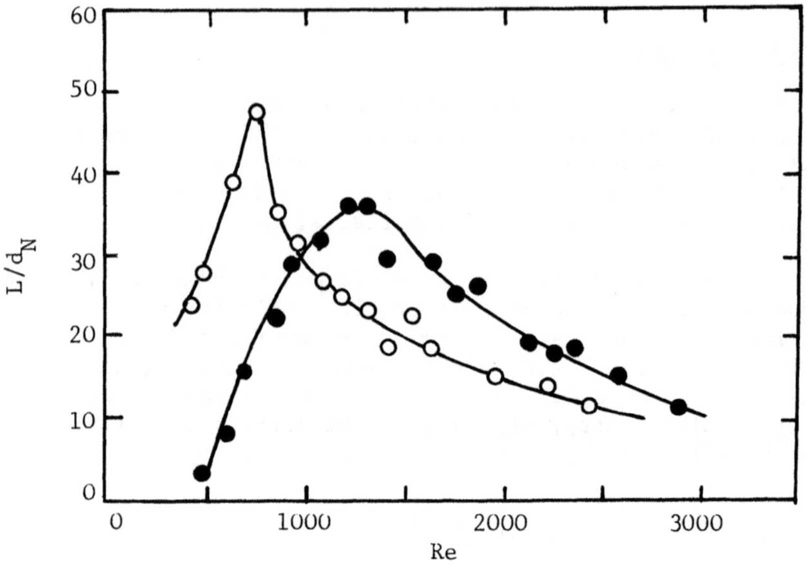

Fig. 1. Variation of dimensionless jet length with Reynolds number: • - chlorobenzene/water (γ = 36.5 mN/m), o - chlorobenzene/water with propionic acid in phase equilibrium (γ = 15 mN/m).

As seen from equations (1) and (2), a decrease in interfacial tension shifts the jetting and thrashing points towards lower nozzle velocities, i.e. lower Reynolds numbers. This has been confirmed experimentally (6) as shown in Fig. 1.

Obviously, these phenomena are affected by the presence of mass transfer but this effect will be discussed at a later stage.

(b) Drop Motion

On account of throughput requirements drops in spray and sieve-plate columns are formed by jet break-up. In fact, sieve plates will not work satisfactorily unless jet formation occurs at all holes (7). Consequently, the dispersion is polydispersed unless the interfacial tension is low when a narrow size spectrum is obtained even under jetting conditions. A polydispersed swarm of drops is difficult to describe as drops of different size move with different velocities. Such a situation is favourable to enhance coalescence as drops pass each other or get caught in the wake of another drop. The direction of mass transfer, as explained later, is here of particular importance. Jet break-up for mass transfer into the drops produces smaller droplets than in the opposite direction of transfer and mass transfer will also oppose coalescence. Hence, in this case, it is reasonable to assume that drops throughout the columns will be of formation size. The opposite will occur for mass transfer out of drops. Not only will drops produced by jet break-up be large but they will grow as the result of mass-transfer assisted coalescence until they reach a critical size. Any drop above the critical size will break up. A detailed description of such behaviour is outside the scope of this treatment which will be restricted to the behaviour of single drops.

A drop rises or falls through a liquid as the result of buoyancy or gravity forces opposed by frictional resistance to motion. The latter is characterized by a drag coefficient, C_D, the knowledge of which makes it possible to calculate drop velocity and hence the residence time of the drop in the column. Although the variation of drag coefficients with Reynolds number is well known for solid spheres, this is not the case for liquid drops, as shown by the bottom line of Fig. 2.

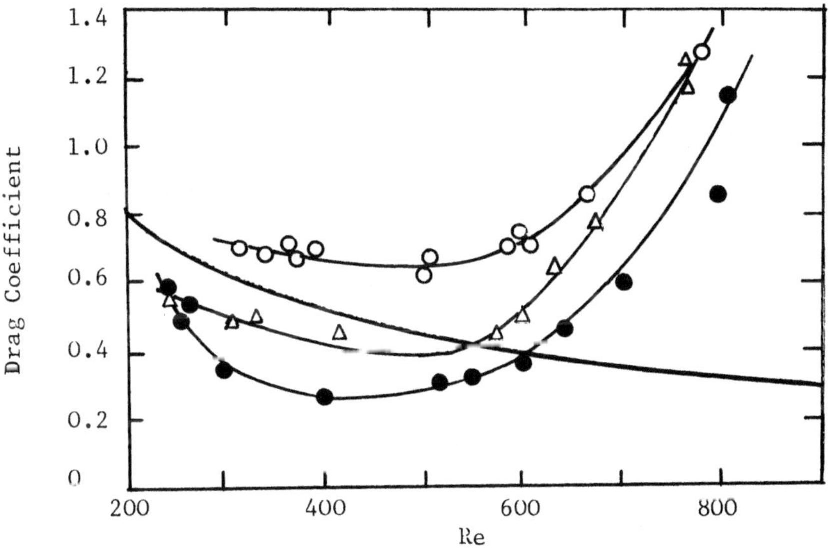

Fig. 2. Drag coefficient of nitrobenzene drops falling through water: • - pure phases, Δ - transfer of propionic acid into drops (C_C = 0.375 kmol/m^3, C_D = 0), o - transfer of acid out of drops (C_C = 0, C_D = 0.125 kmol/m^3. The curve for solid spheres is drawn in for reference purposes.

Although small drops behave like rigid spheres, this similarity of behaviour does not extend beyond a Reynolds number of around 10. For larger drops internal circulation of the Hadamard-Rybczynski type sets in which reduces the drag coefficient to a value below that of a corresponding solid sphere. However, larger drops are also subject to deformation, the extent of which depends on the Weber number (8)

$$d_h/d_v = 1.0 + 0.09/We^{0.95} \tag{3}$$

where $We = du^2\rho_c/\gamma$ (d is the diameter of equivalent sphere, subscripts h, v refer to horizontal and vertical direction respectively, u is the relative drop velocity, γ the interfacial tension and ρ_c the density of the continuous phase). Deformation becomes therefore significant at We = 1 and this will start counteracting the effect of internal circulation. According to Kintner (9) drop oscillation begins at We = 3. This overcomes the effect of internal circulation and with further increase of drop size the drag coefficient starts rising. The increase in drag coefficient becomes very pronounced since, on account of non-uniform vortex shedding, the drop ceases to move in a straight line and follows a zig-zag pattern. Finally, oscillation of drops is converted into a random change in shape and the drop disintegrates at around We = 12. It should be noted that relations $C_D = f(Re)$ are reported (10) in which the drag coefficient does not drop below the value of a corresponding solid sphere. It can be assumed that in such cases the liquids were not free of impurities.

2. BEHAVIOUR AND CHARACTERIZATION OF DISPERSIONS

(a) Drop Size

Most of the work on characterization of dispersion is restricted to the determination of the Sauter mean drop diameter, d_{32}, or the interfacial area per unit volume of dispersion, a, and of the volumetric hold-up of the dispersed phase, ϕ. These parameters are related by the equation:

$$d_{32} = 6\phi/a \tag{4}$$

The problems encountered will be discussed with reference to a fully-baffled stirred tank operating under fully turbulent conditions, i.e. $Re > 10^4$. Under these conditions the Newton number, Ne, also called the power number, is constant so that the expression for power consumption P is

$$P = \text{const}\ D^5\ N^3\ \rho_c \tag{5}$$

where D is the impeller diameter and N the stirring speed (Hz). In such a tank the dispersion can be regarded as consisting of large eddies, generated by the impeller, producing macroscopic liquid motion. However, within these macro eddies, the medium can be regarded as consisting of small eddies - energy transmitting (inertial subrange) and energy dissipating (universal or viscous range) - and be isotropically turbulent. This assumption forms the basis of application of Kolmogoroff's theory of isotropic turbulence (11) to the prediction of drop sizes and their dependence on various parameters.

The size of drops encountered in stirred dispersions is comparable to the size of energy transmitting eddies. By consideration of stresses exerted on a drop in this range of the turbulence spectrum, Hinze (12) characterized the maximum drop size by a critical Weber number

$$We_c = \rho_c \overline{u^2(d)}\ d_{max}/\gamma \tag{6}$$

Drop break-up will occur if this value is exceeded. On substitution into eqn. 6 of Batchelor's expression (13) for velocity fluctuations in the inertial subrange

$$\overline{u^2(d)} = c_3\ (\varepsilon\ d_{max})^{2/3} \tag{7}$$

where ε is the local rate of energy dissipation per unit mass of continuous phase,

$$d_{max} = c_4\ \varepsilon^{-0.4}\ (\gamma/\rho_c)^{0.6} \tag{8}$$

Assuming that under the specified conditions the local rate of energy dissipation can be replaced by the average rate P/V (power consumption per unit stirred volume) for the whole tank, equations 5 and 8 give

$$d_{max} = c_5\ N^{-1.2}\ D^{-0.8}\ \gamma^{0.6} \tag{9}$$

A similar relation was obtained by Shinnar (14) except for the use of d_{32} instead of d_{max}. Since it is generally accepted (15-17) that

$$d_{32} = c_6\ d_{max} \tag{10}$$

the functional relation proposed by Shinnar is analogous to that given by eqn. 9.

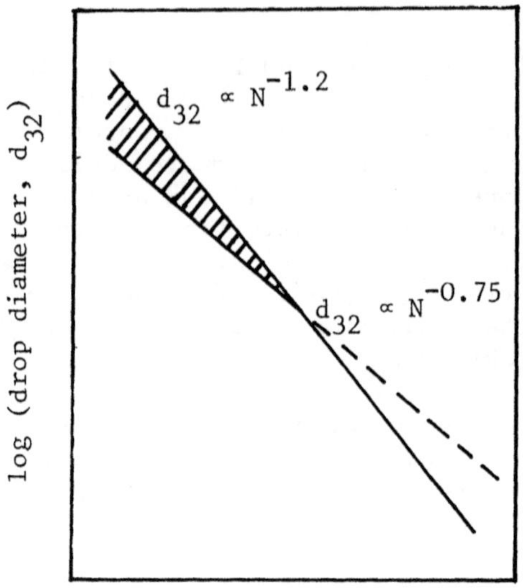

log (stirring speed, N)

Fig. 3. Limiting sizes of drop diameters as function of stirring speed.

Shinnar also considered an alternative limitation to the size of drops, i.e. their inability to coalesce. For coalescence to take place, drops must be in contact over a finite period of time. If, as a result of turbulent velocity fluctuations, the time of contact is smaller than that required for coalescence, no coalescence will take place. Again, using relations based on conditions in the inertial subrange, Shinnar obtained the following functional relation for the drop size

$$d_{32} = c_7 \, N^{-0.75} \, C^{-0.5} \, \gamma^{0.6} \tag{11}$$

Hence, Shinnar has suggested that, depending on conditions, drop size may be governed by the maximum drop size which can exist without break-up or by the inability of drops to coalesce prior to reaching the break-up size (shaded area in Fig. 3).

Experimental evidence supports this view: most of the exponents on N are between -1.0 and -1.2 but some are around -0.75. Similarly, exponents on γ range between 0.14 and 0.6 whereas effect of D is rather inconclusive with exponents showing a spread between -0.7 to -2.0. Of the many proposed correlations the best known is probably that due to Vermuelen et al (18) which, for ϕ = 0.1, is

$$d_m/D = 0.016\ We^{0.6} \tag{12}$$

or due to Thornton and Bouyatiotis (19) which for $\phi \to 0$ becomes

$$(d_{32}^{o})^3 \rho_c g/\mu_c^2 = 29.0\ (P/V\rho_c{}^2\mu_c g^4)^{-0.32}(\rho_c\gamma^3/\mu_c{}^4 g)^{0.14} \tag{13}$$

where the superscript o refers to the case $\phi \to 0$.

In general, the knowledge of the Sauter mean drop diameter is insufficient to characterize a dispersion. This requires information on the type of distribution, mean drop size and standard deviation. Data available so far indicate that distribution tends to be log normal with standard deviation decreasing with increasing stirring speed. However, any data of this type are specific to a particular point in the tank and will change with position. Hence for a total description of the dispersion it is necessary to know the variation of the distribution throughout the tank. This needs a mathematical description of the coalescence and break-up process to give not only the size but also the age distribution of drops. Such information is necessary for a proper evaluation of mass transfer in stirred tanks (20).

The discussion of drop sizes refers, strictly speaking, only to very dilute dispersions ($\phi \to 0$). An increased content of the dispersed phase produced a damping effect on the local intensity of turbulence and thus results in increased drop size. This effect is usually represented by a linear proportionality factor $(1 + b\phi)$, where b is either regarded as a constant in the relation d_m/d_m^o or according to (19)

$$b = 1.18\ (\gamma^2/\mu_c{}^2 g)\ (\mu_c{}^4 g/\Delta\rho\gamma^3)^{0.62}\ (\Delta\rho/\rho)^{0.05} \tag{14}$$

A relationship of this kind is normally applicable up to $\phi = 0.25$ although claims are made for its validity up to $\phi = 0.4$.

(b) Phase Inversion

Although increased hold-up increases the drop size, nevertheless the interfacial area increases as a result of increase in the number of drops. However, there is a limit to this increase, as eventually a critical hold-up is reached which results in phase inversion. The continuous phase suddenly becomes dispersed and the original dispersed phase becomes continuous.

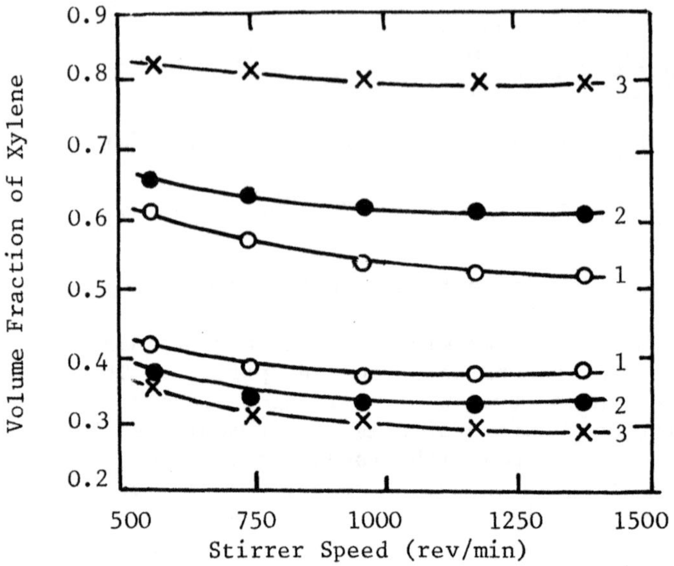

Fig. 4. Inversion characteristics for the system xylene/water. 1 - solute-free system (γ = 39.0 mN/m), 2 - acetone in phase equilibrium (γ = 36.5 mN/m), 3 - propionic acid in phase equilibrium (γ = 32.2 mN/m).

Phase inversion is best presented graphically (Fig. 4) by plotting the volume fraction at inversion of one phase against the stirring speed. Such an inversion characteristic demonstrates clearly the existence of a hysteresis effect, represented graphically by two curves defining a metastable or ambivalent region. In water/organic systems, if volume fraction of the organic phase, ϕ_o, is used as the ordinate value, the system can only exist as water dispersed/organic continuous (w/o) dispersion above the upper curve and as organic dispersed/water continuous (o/w) dispersion below the lower curve. In between the two curves either configuration is possible depending on the system's past history. The width of the hysteresis gap depends strongly on the interfacial tension. The lower the interfacial tension, the wider the gap, i.e. the greater is the resistance of the system towards phase inversion.

Little progress has been made so far to give a theoretical basis to phase inversion. Geometric considerations of Yeh et al (21) gave a ϕ_{max} of 0.74, whereas the commonly assumed criterion of minimisation of interfacial energy predicts ϕ = 0.5 at inversion. Both criteria have been shown to be incorrect by Luhning and Sawistowski (22) and Clarke and Sawistowski (23). In fact, the former have shown that inversion can be accompanied by either an increase or decrease in interfacial energy.

3. MASS TRANSFER IN LIQUID-LIQUID SYSTEMS

(a) Simple Models of Interphase Mass Transfer

In common with other mass transfer processes, solute transfer in extraction can be represented as consisting of the following steps:

(i) transfer from bulk of the raffinate phase to the interface,
(ii) crossing of the interface,
(iii) transfer from the interface to the bulk of the extract phase.

Little attention is usually given to the second step. It is generally assumed that resistance to transfer to and away from the interface is sufficiently large so that the phase boundary is in the state of thermodynamic equilibrium both with respect to composition and temperature. Such an assumption is generally valid except in cases of heavily contaminated systems when allowance has to be made for the presence of an interfacial resistance.

Another possible interfacial effect can be produced by interaction of mass transfer with interfacial tension and lead to the appearance of interfacial convection. This problem is discussed separately in section 4.

Consideration of mass transfer to and from the interface relies on the fact that, even under otherwise turbulent conditions, the rate controlling step is molecular diffusion. This forms the basis of all mass transfer theories of which the two simplest are the film theory and the penetration theory.

The film theory applies only to situations in which the phases are in fully developed turbulent motion. Since the intensity of turbulence decays towards the interface, the concentration profile is curvilinear and can be represented, as a first approximation, by two straight lines drawn tangent to the profile at the interface and in the bulk respectively. The intersection of the two lines defines the effective film thickness ℓ_D and divides the phase into the film and the bulk. Since the molar flux of the solute A is defined by

$$N_A = - (D + \varepsilon_D)(dC_A/dz) \qquad (15)$$

where D is molecular diffusivity, ε_D eddy mass diffusivity, C_A molar concentration of A and z direction normal to the interface, it follows that in the bulk $dC_A/dz = 0$. This can only be the case if $\varepsilon_D = \infty$. Hence, there is no resistance to mass transfer in the bulk. Similarly in the film a linear profile can only exist if $D + \varepsilon_D = \text{const}$, that is if $\varepsilon_D = 0$. Hence, the process is controlled by molecular diffusion through the film of effective thickness ℓ_D. Consequently, the mass transfer coefficient

$$K_L = D/\ell_D = (D/\ell_\nu)(\ell_\nu/\ell_D) \tag{16}$$

where ℓ_ν is the effective film thickness for momentum transfer. If ℓ represents a characteristic linear dimension of the system

$$Sh = K_L \ell/D = (\ell/\ell_\nu)(\ell_\nu/\ell_D) = \text{const. } Re^m\ Sc^n \tag{17}$$

Assuming that $\varepsilon \propto z^3$, it can be shown that $n = 1/3$, using certain simplifying assumptions.

The penetration theory in its simplest form represents the case of transient molecular diffusion into a semi-infinite medium. It can be applied to real situations if hydrodynamic conditions exist for which such an assumption is approximately valid. This would be the case if flow close to the interface is laminar, concentration profiles there are practically normal to the interface and time of contact of the phases is reasonably short. Under such conditions

$$K_L = 2(D/\pi\tau)^{\frac{1}{2}} \tag{18}$$

where τ is the time of contact of the phases.

Penetration theory can also be applied to turbulent conditions by assuming the turbulence spectrum to consist of large eddies, capable of surface renewal, and small eddies responsible for the presence of eddy diffusivity. The small eddies are damped when an element of liquid reaches the interface so that, during its residence time there, mass transfer occurs in accordance with the assumptions of the penetration theory. If all the eddies stay at the interface for the same interval of time we talk about penetration theory with regular surface renewal or the Higbie model. If there is random distribution of residence times with an age-independent fractional rate of surface renewal, s, the term penetration theory with random surface renewal, or the Danckwerts model, is employed. In the case of the Higbie model, the mass transfer coefficient is the same as that given by eqn (18). For the Danckwerts model it takes the form

$$K_L = (Ds)^{\frac{1}{2}} \tag{19}$$

Since s, similar to ℓ_ν in eqn (16), is an unknown hydrodynamic parameter, dependent solely on the Reynolds number, the final correlation is again empirical and has the same form as eqn (17) except that n = 0.5. Hence, the theoretical form of the penetration theory can only be applied to conditions for which eqn (18) is valid such as wetted-wall columns, flow in horizontal channels and motion of drops.

(b) Mass Transfer Coefficients

In general, mass transfer coefficients are correlated by an expression represented by eqn (17). Depending upon hydrodynamic conditions the exponent n on the Schmidt number can only vary between 1/3 and 0.5. Consequently, in the relation $K_L \propto D^r$, variation of r is limited to between 0.5 and 0.67. Correlation containing higher or lower dependence on D should therefore be viewed with suspicion.

In the case of mass transfer from drops it has been customary to express the continuous-phase mass transfer coefficient by a form of the Froessling equation

$$Sh = 2 + \text{const.}\ Re^{0.5}\ Sc^{1/3} \tag{20}$$

This equation has also been applied to swarms of drops. It should be noted that not only is its validity restricted to solid spheres but the use of the factor 2 is questionable for multiparticle systems, as it has been derived theoretically for a single particle in an infinite medium.

By comparison to solid particles, drops are not only subject to deformation but also to internal circulation and oscillation. This affects not only the values of the continuous but also of the dispersed phase mass transfer coefficients. Relevant theoretical and empirical correlations are collected in literature (24, 25). For oscillating drops the equations of Clift et al. (26) give usually a good prediction (27)

$$\text{continuous phase:} \qquad K_C = 1.2\ (f_N D_C)^{0.5} \tag{21}$$

$$\text{dispersed phase:} \qquad K_D = 1.4\ (f_N D_D)^{0.5} \tag{22}$$

$$\text{where } f_N = \{48\gamma/\pi^2 d_e^{\ 3} \rho_c (2 + 3\rho_D/\rho_c)\}^{\frac{1}{2}} \tag{23}$$

where d_e is the equivalent drop diameter.

In extraction equipment the determination of even empirical correlations for mass transfer coefficients is made difficult by the simultaneous variation in interfacial area. Hence, the coefficients obtained from experimental measurements are not only volumetric but also overall $K_o a_s$ rather than individual, i.e. $K_1 a_s$ and $K_2 a_s$, unless partially miscible binary systems are employed. The fundamentals of mass transfer are therefore often studied in equipment of constant interfacial area such as the stirred cells of Lewis (28), Austin (30-31), Prochazka (32) or Nitsch (32-35) or the drop-forming device of Goltz (36, 37). The correlation of Davies (38) seems to give best result for horizonta interfaces in stirred cells.

$$Sh_1 = B\, Sc_1^{0.5} We_1^{1} \{Re_1 + (\nu_2/\nu_1)\}^{0.5}$$

$$\times \{1 + (\rho_2 \nu_2^2/\rho_1 \nu_1^2)(Re_2/Re_1)^2\}^{0.5} \quad (24)$$

In this correlation B is a constant, ν is the kinematic viscosity, $We_1 = N_1 L(\rho_1 L\gamma_e)^{\frac{1}{2}}$, $\gamma_e = \gamma + \ell_e^2 \Delta\rho g/16$, N is the stirring speed, L the impeller diameter, g the gravitational acceleration and ℓ_e the eddy length ($0.1\ L < \ell_e < L$).

Presence of an interfacial resistance can be incorporated easily into the film theory by introduction of an additive resistance term R_s so that the expression for the overall resistance to mass transfer becomes

$$1/K_{01} = (1/K_1) + m\,(R_s + 1/K_2) \quad (25)$$

where subscripts 1 and 2 refer to the two phases, K_o is the overal mass transfer coefficient and m the distribution coefficient. Such a simple expression does not apply to the penetration theory since one of its boundary conditions is changed by the presence of an interfacial resistance.

4. MASS TRANSFER INDUCED INTERFACIAL CONVECTION

Marangoni effect is a general term for surface flow phenomena resulting from the appearance in the interface of an interfacial tension gradient. These phenomena produce traction on the adjoining sublayers and lead to two distinct and separate effects: the flux-intensive surface renewal phenomena and the area-intensive thin-film phenomena. The first phenomena predominate when the depth of the liquid is much larger than the depth of penetration of surface movement and the second are important when the two depths are comparable. In the first case the affected process parameter is the mass transfer coefficient and in the second case the interfacial area.

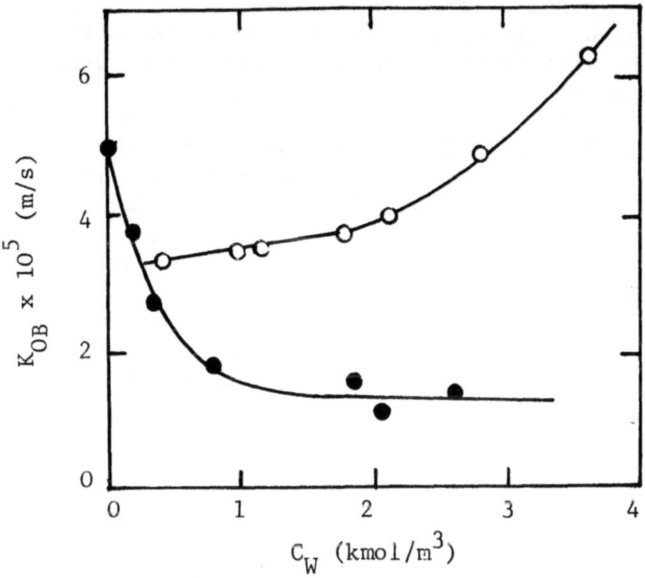

Fig. 5. Variation of overall mass transfer coefficient with concentration of propionic acid in the aqueous phase. o - transfer from water into solute-free benzene drops, • - transfer of solute from 2.40 kmol/m^3 solution in benzene into water.

The surface renewal phenomena occur as a result of Marangoni instability or forced convection. Marangoni instability results if an interfacial tension gradient produced by a perturbation is amplified by the simultaneously occurring mass transfer process. If the perturbation is within the range of creeping flow, the resulting instability is stationary, resulting in the appearance of roll cells. Otherwise the interface may become the source of interfacial turbulence. Simple mass transfer considerations lead to the so-called Sternling-Scriven instability criterion which states that for a system in which interfacial tension decreases with increase in solute concentration ($d\gamma/dC < 0$) the system will be convectionally unstable for mass transfer out of the phase of lower kinematic viscosity and higher molecular diffusivity but stable in the opposite direction of transfer. This criterion has been repeatedly confirmed experimentally for liquid-liquid systems and the relevant work is reviewed elsewhere (39, 40). The resulting phenomena are, of course, reflected in the variation of mass transfer coefficients (Fig. 5).

The forced convection effects appear if bulk turbulence is of sufficient intensity to result in surface renewal, thus bringing liquid of lower interfacial tension to the surface. In unstable systems this leads to more intense interfacial turbulence. In stable systems surface renewal tends to be suppressed but, if the intensity of turbulence is sufficiently high, the resistance to surface renewal will be overcome and an increase in mass transfer coefficients will result at high driving forces even in stable systems (Fig. 5).

A special case of Marangoni instability is its effect on the drag coefficient, that is the motion of drops. The presence of stationary instability immobilises the interface and thus increases the drag coefficient by elimination of internal circulation (Fig. 2). This effect is, however, only pronounced at a drop Reynolds number range between 200 and 600.

The thin-film phenomena are of importance if the depth of one of the phases varies locally and can decrease in places to a sufficiently small value so that depletion or accumulation of solute in that locality is significantly faster than in its surroundings. This produces a concentration gradient along the interface and hence also an interfacial tension gradient thus leading to Marangoni flow. If the Marangoni flow is directed towards the weak spot, it counteracts phase break-up. However, if the directio of the flow is reversed, break-up is accelerated. Obviously, the direction of the Marangoni flow depends on the direction of mass transfer. For a system in which $d\gamma/dC < 0$, mass transfer out of a thin spot will tend to preserve the existing interfacial area, whereas mass transfer into a thin spot will support any tendency to phase break-up and hence changes in interfacial area.

Typical examples of thin spots are the film trapped between drops (or a drop and an interface) prior to coalescence or the node formed prior to jet break-up. The former is illustrated by Fig. 6 in which two drops of anisole in water were brought into contact and acetone transferred as a solute in either direction. It can be seen that, at a sufficiently high concentration of acetone in the drops, coalescence is practically instantaneous. Similar results were obtained in a small spray column for the transfer of propionic acid between benzene drops and water (41). For each one hundred drops formed at column bottom 57 arrived at the top for transfer out of the drops, i.e. the drop size was practically doubled and interfacial area correspondingly reduced, and 99 for transfer into the drops, i.e. coalescence was practicall eliminated. Corresponding results for the effect of mass transfer on jet break-up are shown in Fig. 7 which indicates a difference of 25% in interfacial area for the two directions of transfer (6). Obviously, similar phenomena must exist in stirred tanks but they have not yet been assessed quantitatively.

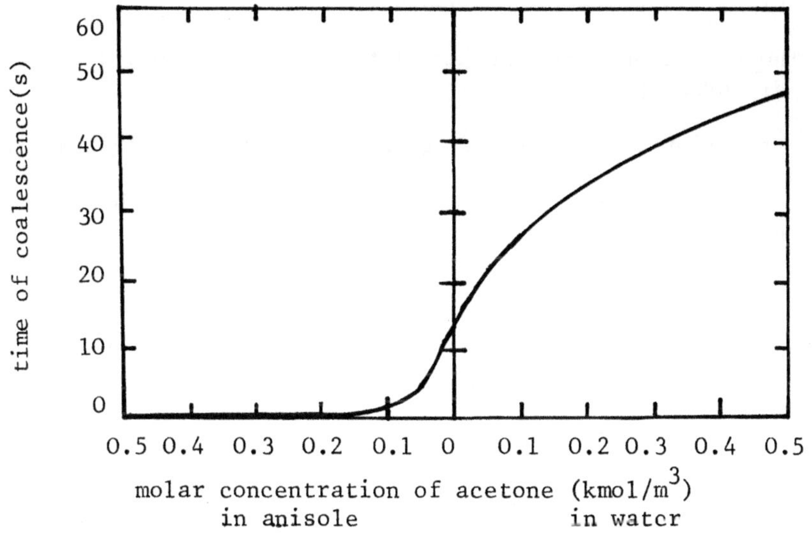

Fig. 6. Effect of direction of mass transfer on coalescence of two anisole drops in water.

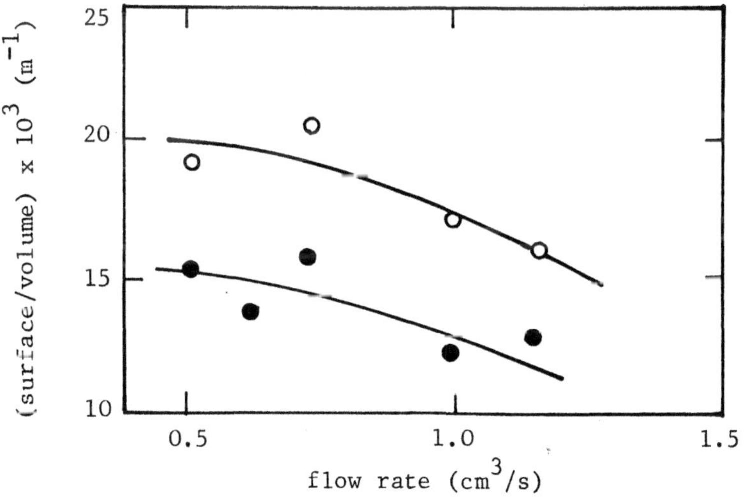

Fig. 7. Effect of direction of mass transfer of propionic acid on drop formation in break-up of a benzene jet in water (o - transfer into jet, • - transfer out of jet).

5. PRINCIPLES OF LIQUID MEMBRANE PROCESSES

The traditional solvent extraction process suffers from three major disadvantages:

(a) extraction and stripping stages are performed in separate units,
(b) equipment is bulky since large quantities of both extractive and stripping liquids are required,
(c) large quantities of stripping liquid adversely affect the running and capital cost of subsequent solvent recovery unit.

A process which could perform extraction and stripping in a single unit would, therefore, be very attractive if, in addition, it could also reduce the flow rate of the stripping phase.

Such a process is the liquid membrane process which, in its original form (42), suggested encapsulation of organic liquid droplets by an aqueous membrane. This was produced by passing organic drops through water containing a suitable membrane-stabilizing surfactant. Subsequently the droplets rose through a heavier organic phase and water-soluble solute permeated selectively through the membrane.

This arrangement suffered from several disadvantages, e.g. it could only operate with an aqueous membrane separating two organic phases. The disadvantages were overcome by developing emulsion-type membranes (43-45) in which the internal phase forms an emulsion inside the membrane phase and the whole is subsequently dispersed as globules in the external phase. The membrane phase ceases to be a membrane in its original sense but forms the encapsulating phase for a large number of embedded droplets of the internal phase. Hence the extraction process takes place at the interface between the external phase and the membrane phase (surface of globules), while the stripping process occurs at the interface between the membrane phase and the internal droplets (Fig. 8). Since one of the original aims of the process development was a reduction in volume of the stripping liquid, this liquid usually forms the internal phase. Consequently the external liquid forms the raffinate phase and the membrane liquid is the extract phase. Both oil-in-water and water-in-oil emulsions are possible but the scope of applicability of the latter is much larger.

The process (Fig. 9) consists of the following steps:

(a) selection of surfactant,
(b) production of emulsion (E),
(c) dispersion of emulsion in the continuous phase (D),

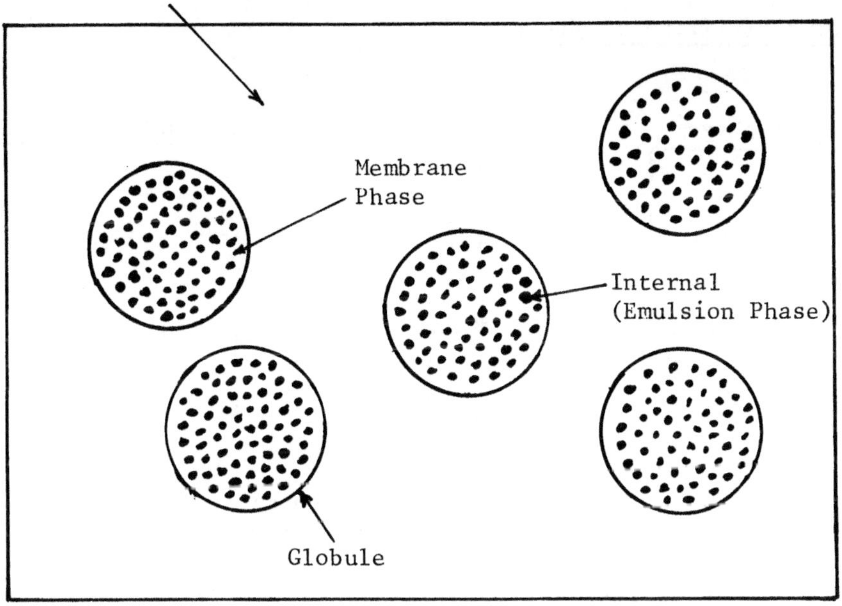

Fig. 8. Emulsion-type liquid membrane.

(d) separation of phases into raffinate (external phase) and emulsion (S),
(e) breakdown of emulsion into organic phase and extract (internal) phase (B),
(f) recycle of organic (membrane) phase,
(g) further treatment of extract phase.

For the water-in-oil emulsion the surfactant must be insoluble in water, not only on account of emulsion stability but also to minimize its losses.

The emulsion is usually produced in a stirred tank but the process can also be performed in a homogenizer. The internal droplet size varies in the range 1 to 10 μm and the internal to membrane phase ratio is around 1.

The dispersion is normally conducted in a stirred tank producing globule sizes of around 1 mm. So far mostly batch operations in stirred tanks have been investigated. Other types of extraction equipment, e.g. spray columns, could also be employed.

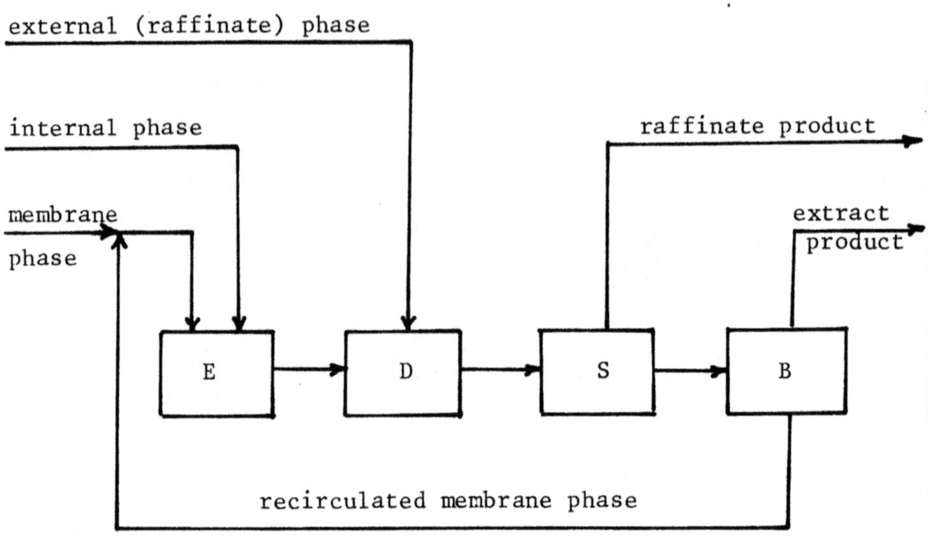

Fig. 9. Flow diagram of continuous membrane process.

If a stirred tank is used for dispersion purposes, it is followed by a gravity settler (mixer-settler process) to separate the globules from the continuous phase. In a column, phase separation occurs at column top.

The separated external phase is discharged and the emulsion broken down either electrostatically or by temperature change. The membrane phase, containing the surfactant, is recycled and the extract phase treated further for solvent recovery.

The membrane process can be kinetically or thermodynamically controlled. Thermodynamic control exists if there is a limit set by the distribution coefficient or a reaction equilibrium constant which is approached asymptotically at a slow rate. The process will be kinetically controlled if it is far away from the thermodynamic limit or if this limit is removed. The removal of ammonia from an aqueous effluent by the liquid membrane process will be used as an example to illustrate the difference between the two definitions. On account of solubility of ammonia in the membrane phase the same distribution coefficient will apply at the two surfaces of the membrane phase. The internal phase thus cannot take up more ammonia than given by the distribution coefficient at the internal surface. In turn, the maximum amount of ammonia in the membrane phase is limited by the distribution coefficient at the external phase. Consequently the process is thermodynamically limited. This limitation can be removed if excess H_2SO_4 is introduced into the internal phase. Ammonia reacts with it irreversibly so that the thermodynamic limitation is removed - the process is kinetically controlled.

The case of ammonia represents a membrane process based on physical permeation since ammonia is soluble in the membrane phase. If a solute is insoluble, a complex-forming compound (carrier agent) has to be introduced there to provide facilitated transport. Since the reaction of the solute with the carrier agent has to be reversible, reaction equilibrium will apply at both surfaces. The process will thus be thermodynamically controlled unless a reactive compound is introduced into the internal phase to remove the solute from the system by an irreversible reaction.

Most of the experimental work was conducted in stirred tanks and the results evaluated on the assumption that controlling resistance to the process is present in the globules as diffusional resistance in the membrane phase. The models presented so far represent limiting cases of globule behaviour, that is perfect mixing (46) and no mixing (46-48). In the no mixing case the droplets are assumed as being embedded in the stationary membrane phase and there is an advancing diffusional boundary, the progress of which is limited by the need for complete saturation of the internal phase droplets within the boundary. The drops are regarded as being of finite size (46) when a numerical calculation procedure is required or to be mathematical points of finite solute capacity when an analytical solution can be presented (47). An alternative model is presented by Davies (49) in which the membrane phase is considered as an extended flat membrane of effective thickness δ. On the other hand modelling by Boyadzhiev et al. (50) assumes main resistance to transfer to be located in the external phase and the model is based on diffusion accompanied by simultaneous destruction of the emulsion. However, the work of Wongswan et al. (27) conducted in a spray column indicates that, provided the process is not reaction controlled; the main resistance to transfer is located in the external phase and can be described by correlations for transfer into oscillating droplets. Obviously, the presence of large oscillating globules in the latter case, as opposed to spherical non-circulating globules in a stirred tank, may partly be responsible for the difference in interpretation. Nevertheless, there is scope for further work to elucidate the mass transfer process in emulsified membranes.

REFERENCES

1. Misek, T. (Ed.), Recommended Systems for Liquid Extraction Studies, Inst. Chem. Engrs, London (1978).
2. Hayworth, C.B. and Treybal, R.E., Ind. Engng Chem., 42, 1174 (1950).
3. Scheele, G.F. and Meister, B.J., AIChE JL, 14, 9 (1968).
4. Ranz, W.E., Can. J. Chem. Engng, 36, 1958 (1958).
5. Timotika, S., Proc. Roy. Soc., A150, 332 (1935).
6. Dzubur, I. and Sawistowski, H., Proc. ISEC '71, Society of Chemical Industry, London (1971) 379.
7. Pilhofer, T., German Chem. Engng, 2 (1979) 200.
8. Taylor, T.D. and Acrivos, A., J. Fluid Mech., (1964) 18, 466.
9. Kintner, R.C., Advances in Chemical Engineering, Vol. 4, Academic Press, New York (1963) 51.
10. Krishna, P.M., Venkateswarlu, O. and Narasimhamurthy, G.S.R., Jl Chem. Engng Data, 4 (1959) 336, 340.
11. Kolmogoroff, A.N., Dokl. Akad. Nauk SSSR, 30, 301 (1941), 31, 538 (1941), 32, 16 (1941).
12. Hinze, J.O., AIChE Jl, 1, 289 (1955).
13. Batchelor, G.K., Proc. Cambridge Phil. Soc., 47, 359 (1951).
14. Shinnar, R., J. Fluid Mech., 10, 259 (1961).
15. Chen, H.T. and Middleman, S., AIChE Jl, 13, 989 (1967).
16. Sprow, F.B., Chem. Engng Sci., 22, 435 (1967).
17. Brown, D.E. and Pitt, K., Chemca 1970, IEChemE Symp. Series, 33, 83 (1970).
18. Vermeulen, T., Williams, G.H. and Langlois, G.E., Chem. Engng Prog., 51, 85 (1955).
19. Bouyatiotis, B.A. and Thornton, J.D., IChemE Symp. Series, 26, 43 (1967).
20. Rod, V. and Misek, T., Trans. Instn Chem. Engrs, 60, 48 (1982)
21. Yeh, G.C., Haynie, C.A. and Moses, R.A., AIChE Jl, 10, 260 (1964).
22. Luhning, R.W. and Sawistowski, H., Proc. ISEC '71, Society of Chemical Industry, London 1971, 873.
23. Clarke, S.I. and Sawistowski, H., Trans. I. Chem. E., 56, 50 (1978).
24. Laddha, G.S. and Degaleesan, T.E., "Transport Phenomena in Liquid Extraction", Tata-McGraw Hill, New Delhi (1976).
25. Skelland, A.H.P., "Diffusional Mass Transfer," Wiley-Interscience New York (1974).
26. Clift, R., Grace, J.R. and Welser, M.E., "Bubbles, Drops and Particles," Academic Press, New York (1978).
27. Wongswan, S., Perez de Ortiz, E.S. and Sawistowski, H., Proc. Hydrometallurgy '81, Soc. Chem. Ind., London (1981).
28. Lewis, J.B., Chem. Engng Sci., 3 (1954) 248, 260.
29. Asutin, L.J. and Sawistowski, H., IChemE. Symp. Series 26 (1967) 3.
30. Austin, L.J. and Sawistowski, H., Chem-Ingr-Tech., 39 (1967) 224.

31. Grosjean, P.R.L. and Sawistowski, H., Trans. Instn Chem. Engrs, 58 (1980) 59.
32. Prochazka, J. and Bulicka, J., Proc. Intl Solvent Extraction Conf., Soc. Chem. Ind., London (1971) 823.
33. Nitsch, W. and Hillekamp, Chem. Ztg., 96 (1972) 254.
34. Nitsch, W. and Kahni, J.G., German Chem. Engng, 3 (1980) 86.
35. Ajawin, L.J., Perez de Ortiz, E.S. and Sawistowski, H., Proc. Intl Solvent Extraction Conf., Liege (1980), Vol. 3, 80-112.
36. Sawistowski, H. and Goltz, G.E., Trans. Instn Chem. Engrs, 4 (1963) 1974.
37. Sawistowski, H. and James, B.R., Chem-Ingr-Tech., 35 (1963) 1975.
38. Davies, J.T.,"Turbulence Phenomena," Academic Press, New York (1972).
39. Sawistowski, H.,"Interfacial Phenomena", in C. Hanson (Ed.), "Recent Advances in Liquid-Liquid Extraction", Pergamon Press, Oxford (1971).
40. Sawistowski, H., Chem. Ing. Tech., 45, 1093, 1114 (1973).
41. Sawistowski, H. and James, B.R., Chem. Ing. Tech., 35, 175 (1963).
42. Elzinga, E.R. and Li, N.N., U.S. Patent 3389078, Esso Res. & Engng Co., (1968).
43. Li, N.N., AIChE Jl, 17 (1971) 459.
44. Li, N.N. and Shrier, A.L.,"Recent Developments in Separation Science", CRC, Cleveland, 1 (1972) 163.
45. Cahn, R.P. and Li, N.N., in P. Mears (Ed.),"Membrane Separation Processes," Elsevier, Amsterdam (1976).
46. Li, N.N., Proc. of Conf. on Liquid Membrane Applications in Waste Water Treatment and Metals Recovery, UMIST (1980) 9.
47. Kopp, A.G., Marr, R.J. and Moser, F.E., IChemE. Symp. Series 54 (1978) 279.
48. Marr, R.J., Bart, H.J. and Siebenhofer, M., Proc. Intl Solvent Extraction Conf., Liege (1980) 80-144.
49. Martin, T.P. and Davies, G.A., Proc. Intl Solvent Extraction Conf., Liege (1980) 80-230.
50. Boyadzhiev, L., Sapundzhiev, T. and Bezenshek, E., Separation Science, 12 (1977) 541.

EXPERIENCE WITH LIQUID/LIQUID TEST SYSTEMS IN EXTRACTION

Stanley Hartland and Ladislav Steiner

Department of Industrial and Engineering Chemistry
Swiss Federal Institute of Technology, CH-8092 Zürich

INTRODUCTION

Liquid/liquid extraction is a unit operation with considerable industrial application. It is used as an energy-saving alternative to distillation in waste-water treatment, nuclear technology, hydrometallurgy and mineral oil processing. Extraction columns for large-scale production have diameters of several meters and volumes of the order of hundreds of cubic metres. Their design is therefore an important procedure which may save or waste considerable amounts of money. There are many different column types in practical use, ranging from the simple spray column to columns with sophisticated devices for dispersing and coalescing the liquid phases. Unfortunately the design is still done empirically, without the possibility of predicting the performance of a given column type in advance, or of selecting the optimal column construction for a given process.

To improve this situation fundamental research is necessary in order to find general design procedures and limits of operation for the process which are independent of the particular column type. This can only be achieved through cooperation between researchers in industry and universities. The European Federation of Chemical Engineers established a Working Party in Extraction for this purpose which recommended liquid systems for testing extraction columns so that the results would be comparable. Three systems were selected and their properties published (Misek, 1979). These are water-toluene-acetone, water-methyl isobutyl-ketone (MIBK)-acetic acid and water-butanol-succinic acid, representing systems with high, medium and low interfacial tensions respectively.

Table I. Column Types and Sizes

Column	Diameter	Length	Maximum Throughput	Reference
	mm	m	m^3/m^2h	
Spray	100	2	80	Horvath (1976)
Spray	75	2	80	Berger (1981)
Pulse	100	3	40	Ugarcic (1981)
Pulse	50	1	26	Toller (1981)
Packed	50	1	-	in progress
Rotary Disc	50	1	-	in progress
Kühni	150	2	20	Kumar (1981)
Kühni	150	3 stages	-	Kurt (1981)
Reciprocating Plate	75	2	90	Schläpfer(1979)
Centrifugal	100	0.5	40	Gebauer (1981)
Enhanced Coalescence	75	2	60	Ergenc (1979) von Fischer (1981)
Enhanced Coalescence	75	0.5	80	von Fischer (1981) Efthiamatou (1980) Oezdemir (1979)
Enhanced Coalescence	150	2		Berger (1981)

In our laboratory an extensive research programme has been carried out to obtain hydrodynamic and mass transfer data using these test systems in different types of column. A list of the columns used is shown in Table 1, beginning with the spray column in which we investigated the mass transfer from freely ascending drops as a function of the hydrodynamic parameters. This involved measuring the concentrations in both phases along the column, together with the hold-up of the dispersed phase, drop diameter and backmixing in both phases. For most of these measurements new experimental techniques had to be developed. The data were treated using computer-aided procedures to evaluate the true mass transfer rates. It may be assumed that such rates are independent of column type and therefore generally applicable; they are however dependent on drop diameter, interface behaviour and on local hydrodynamic conditions such as turbulence. Mathematical models were developed to enable a more realistic description of column performance using the true mass transfer rates. All three systems mentioned above were used with the same column, together with a fourth: water- o-xylene-acetone. The properties of this system are similar to those of the recommended water-toluene-acetone system.

Several, more sophisticated columns were also tested in addition to the spray column. The same parameters were measured and mass transfer rates evaluated in the same way. For some particular problem, such as the investigation of stirrer performance in agitated columns, or comparison of different plate constructions, single-purpose models were constructed on which experiments were performed. The columns were usually of pilot plant size with diameters ranging from 50-150 mm and heights between 1 and 3 m. Empirical rules exist which enable such a pilot plant to be scaled up for most types of column. Our aim is to improve the design by extracting more information from pilot-plant size data and eventually to minimize or exclude the need for extensive pilot plant operation. Design from first principles without experimental work is however, not suggested. Within the foreseeable future it will be necessary to have experimental information on the behaviour of any particular system, but the data can be obtained from laboratory apparatus and later tested on a pilot plant to confirm the results.

Using the experience gained in the investigation of conventional column types, several new extractor constructions have been developed.

EXPERIMENTAL WORK

Several experimental arrangements were used to facilitate the operation of extraction columns, two examplesbeing given in Figures 1 and 2. The former shows the arrangement for a medium-size extractor with provision for storage of small amounts of liquids, and auxiliary systems for the measurement of backmixing by the steady state method. The shared component was injected into one of the phases by a metering pump. A system of this size fits into a laboratory of normal height and enables steady state experiments to be performed in a column of 80 mm diameter for 15 to 30 minutes. Figure 2 shows a larger scale installation with four storage tanks of 400 litres capacity and a distillation column for solvent recovery. Although this arrangement needed more height it was still possible to operate extractors of up to 150 mm diameter at the steady state. Different columns can be attached to the supply system which has been used for spray, pulsed sieve plate, Kühni and EC (Enhanced Coalescence) types of column.

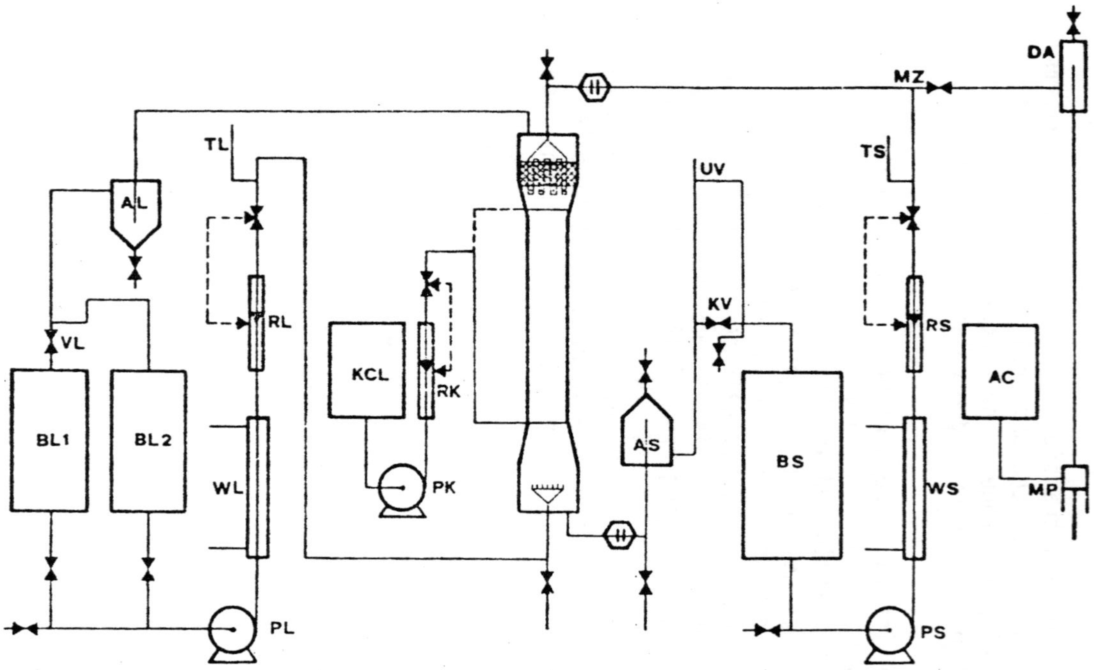

Figure 1: Flow diagram for experiments with smaller diameter columns: BL - light phase tanks, BS - heavy phase tank, WL,WS - thermostating systems, AL,AS - phase separators, AC - solute storage, MP - metering pump, DA - pressure equalizer, RS,RL,RK - flow control, KCl - tracer storage for backmixing measurement, PK,PL, PS - pumps, UV - level control.

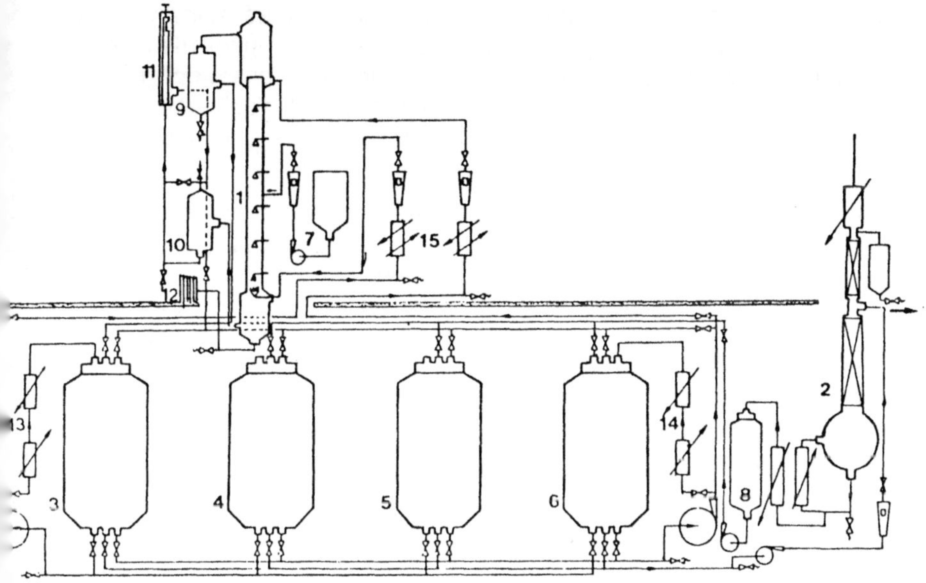

igure 2: Flow diagram of pilot plant for larger diameter columns: 1 - extraction column, 2 -distillation for solvent recovery, 3 to 6 - solvent storage, 7 - arrangement for tracer experiments, 8 - auxiliary tank (buffer between distillation and extraction), 9,10 - phase separators, 11 - level control, 12 - pulsator (when working with pulsed columns), 13,14 - thermostating, 15 - flow rate control.

The columns were specifically designed to enable the necessary measurements to be made and were built of technical glass,sometimes using parts of technical extractors. In all cases sampling ports were drilled through the glass tubes and windows provided for photographing the disperse phase. The end sections were constructed to give defined inlet and outlet conditions for both phases. Examples of the columns and measuring facilities are shown in Figures 3 and 4.

The experimental techniques for measuring the relevant parameters have been described in previous publications (Steiner et al. 1978 and 1979, Horvath 1976, Ugarcic 1981 and Berger 1981). The drop size was usually determined by evaluating photographs taken through a special window to minimize the distortion caused by the wall curvature. An example of the pictures taken, the experimental arrangement and method of evaluating the photographs can be seen in Figure 5. The drop size was evaluated by projecting the negative on a screen of a digitalizing apparatus and recording the coordinates for the lower, upper, right - hand and left-hand boundary of the drop image. The coordinates were stored in a desk-top computer which, after 100 to 500 drops had been measured, calculated the mean diameter, diameter distribution and any necessary statistical function. This procedure was reproducible as long as the drops were regular and the hold-up not too high. For more difficult cases a different technique was used in which the dispersion was sucked through a capillary tube and the length of the deformed drop measured. The capillary tubes used were of between 1 and 2 mm diameter, their ends being shaped like a funnel to avoid breaking the drops. The drop length was measured automatically with two photocells but, especially at higher agitation intensities, clusters of small drops were frequently considered to be one single large drop. The drop size was therefore measured from photographs of the dispersion as it passed through the capillary tube. Drops of any shape could be evaluated without any limitation as to hold-up or agitation inside the column. The method is shown schematically in Figure 6.

The hold-up of the dispersed phase was determined by several techniques as there is no particular method which is reliable in all cases. In spray columns the hydrostatic method gives good results but this was not applicable in agitated columns. A comparison of different possible methods and their applicability is given in Table 2.

Backmixing in the continuous phase can be determined either by steady or unsteady state trace injection. Both methods were tried and yielded comparable results. The former is based on continuous injection of a tracer near the exit of the continuous

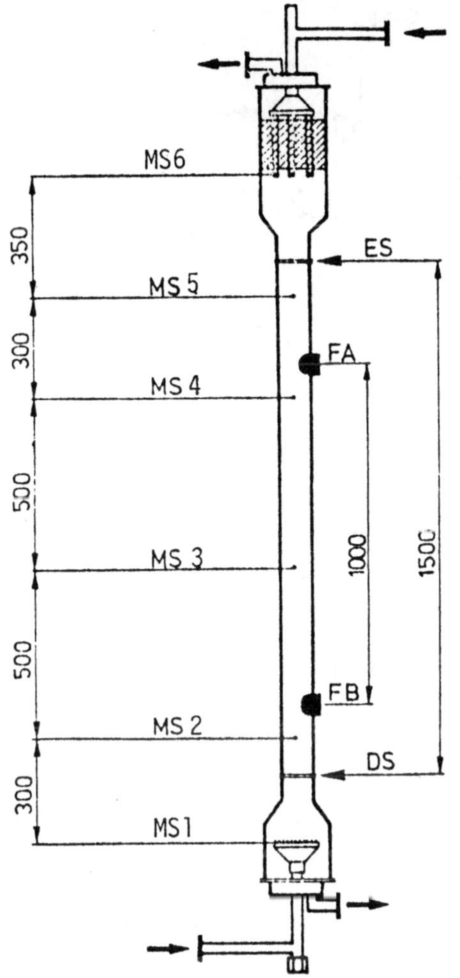

Figure 3: Example of spray column with measuring points MS 1 to 6 : Separate sampling of both phases for determination of concentration profiles, ES-DS ports for hold-up measurement by hydrostatic method, tracer injection, FA,FB windows for photographic determination of drop size.

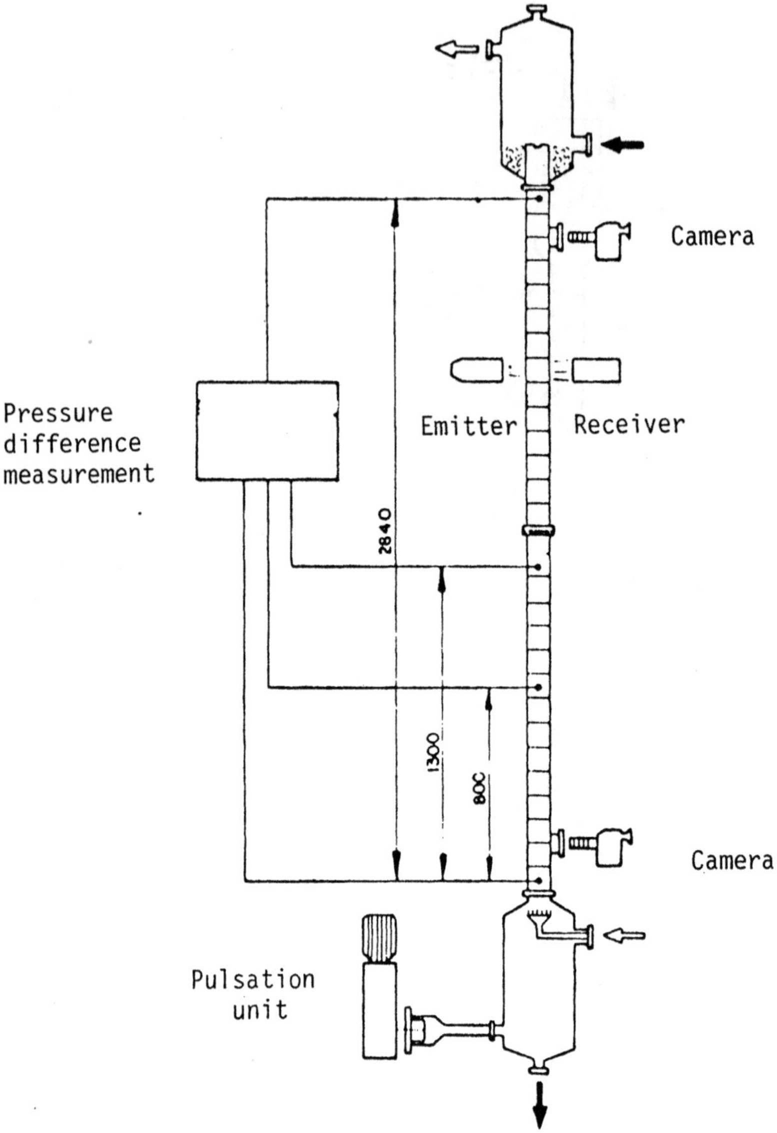

Figure 4: Example of pulsed sieve plate column
with measuring facilities.

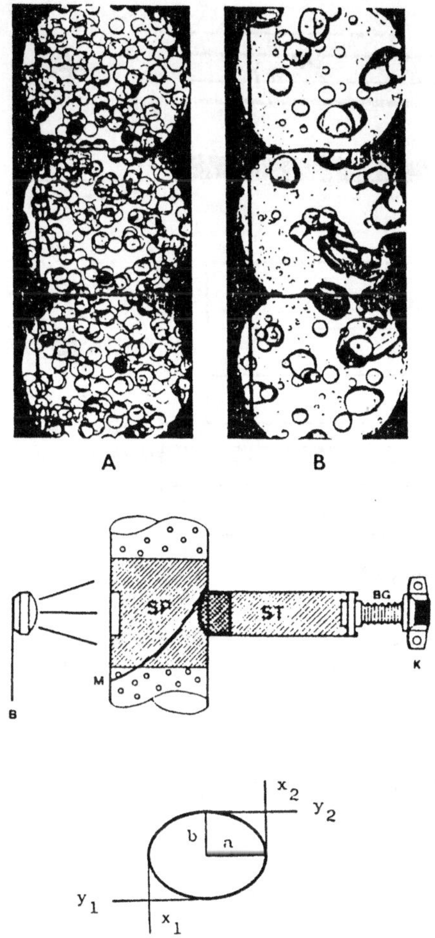

Figure 5: Photographic determination of drop size through column wall,showing non-coalescing (A) and coalescing (B) drops: arrangement for picture-taking (c) and method of evaluating drop size for non-coalescing drops (D).In C:B is a lamp, M-1mm wire used as scale, SB and SC-black screens, K-camera and BG - extension bellows.

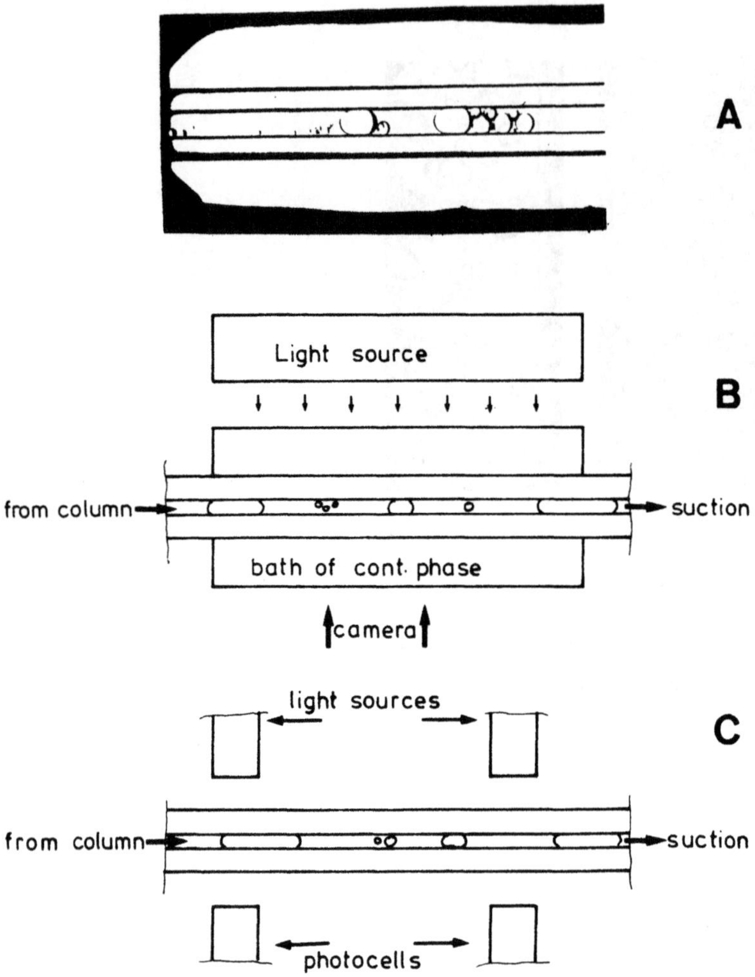

Figure 6 : Determination of drop size from length of thread in a capillary tube. A - example of photograph, B - photographic arrangement, C - principle of automatic evaluation. This last method needs well-distinguished slugs and would not work for the drops shown in A.

Table 2

Comparison of Methods for Hold-up Determination

Method	Local (L), Overall (O)	Comments
Hydrostatic	O	Good results in gravitational columns, sometimes fails in agitated columns.
γ-Ray Absorption	L	Complicated equipment,needs calibration. Applicable for small diameters only
Closing input & output , direct volumetric measurement	O	Special arrangement necessary,terminates the experiment. May be used to calibrate other methods.
Rapid Sampling	L	Applicable for columns with uniformly distributed drops (fails in spray columns). Till now no published comparison with other methods available. Favoured in large columns.Recovery after sample taking may take up to 15 minutes.
Conductivity	L	Difficult calibration; space between electrodes must be representative for entire column; little experience
Capacitance	L	As for Conductivity
Residence Time of an Impulse	O	Hold-up as "by-product" by unsteady state backmixing measurement. Fails in columns with dead spaces.

phase and measuring the tracer concentrations at several positions upstream. In the latter method a single shot of a tracer is injected at an upstream position and the variation of its concentrations with time is recorded at two positions downstream. Tracer consumption is very small and the experiment is easier to perform, but on the other hand, the evaluation of the response curves is more difficult and curve-fitting is necessary to obtain sufficiently accurate results. Several other evaluation methods were tested and compared for reliability of the results and computer time required. Unfortunately, the rather laborious method which involves fitting a numerically generated response curve to experimental data appears to be the most reliable. The arrangement for backmixing measurements is shown in Figure 7.

In the dispersed phase, only the unsteady state method is applicable. The tracer must be introduced in the form of marked drops and its concentration recorded at two positions downstream from the point where the drops were introduced. Conventional techniques proved to be too inaccurate for this purpose and special probes were therefore developed which coalesced the dispersed phase and measured the light absorption directly in the column. It was found that any attempt to withdraw drops from the column and carry out time-dependent measurements on them led to unacceptable errors.

In addition to measuring all the hydrodynamic parameters, the pure phases must be sampled at different positions along the column axis to determine the concentration profiles of the solute. Sampling the dispersed phase is not easy and special funnels were developed for this purpose, as shown in Figure 8. For normal conditions with relatively large, slowly moving drops a vertical construction (Figure 8A) was used. In this case, ascending drops are coalesced on a teflon ring in the lower part of the funnel which is completely filled by the liquid. New drops are displaced from the orifice at the top so that the liquid in the funnel is continually replaced. Pure liquid can be sampled from the device at a rate of 1 to 2 drops per second. For agitated columns with small, horizontally moving drops an alternative construction was tested, as shown in Figure 8B, in which the drops are coalesced on a teflon ring in the upstream section and the liquid escapes again from the downstream section. The sampling is done from above where a layer of continuous liquid is maintained by a baffle in the exit section. The samplers worked well and no problems were encountered when withdrawing uncontaminated phases from the column.

Different methods were used to analyse the liquid phases sampled from the column according to the system used. Acetone in toluene, xylene and water was determined from density measurements

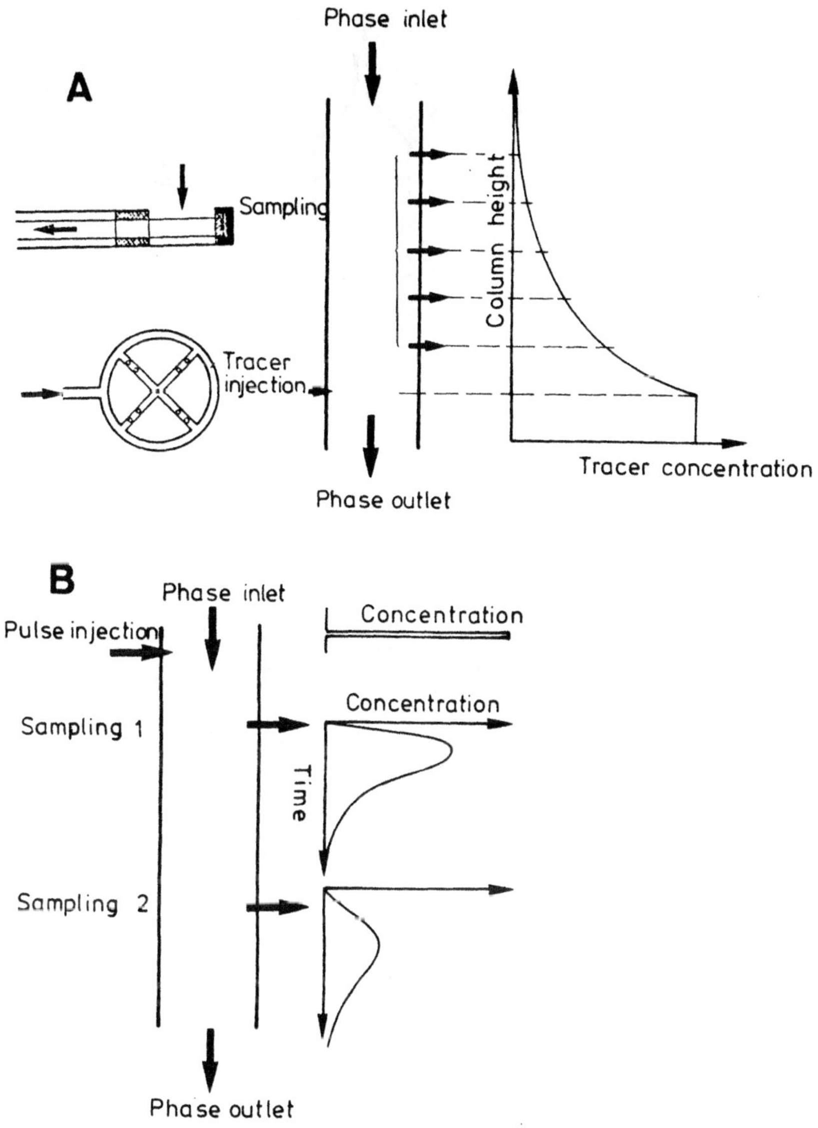

Figure 7: Determination of backmixing (eddy diffusivity) by steady state method (A), showing devices for tracer injection, sampling electrodes and arrangement of sampling points. Unsteady state arrangement is shown in B.

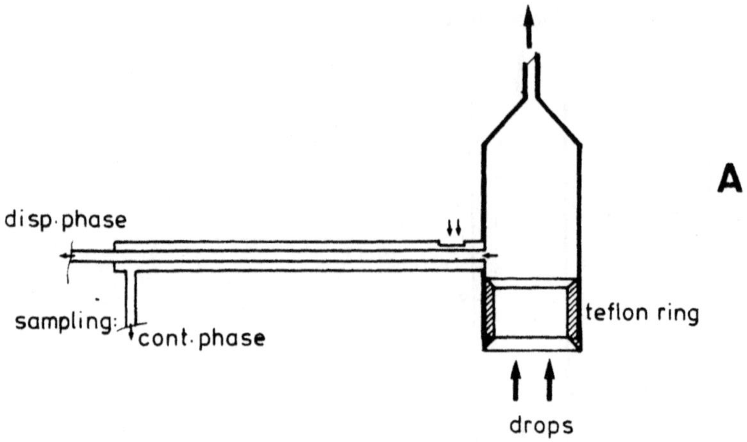

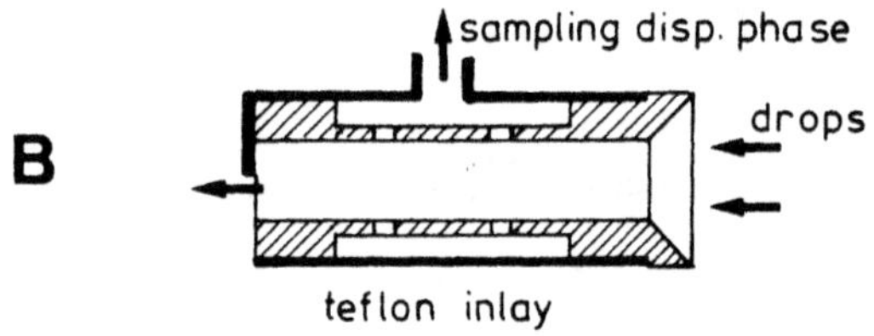

Figure 8: Sampling devices for withdrawing pure phases from a column. A - vertical device for slowly rising drops, B - alternative shape of the probe for rapidly moving drops in agitated columns.

using an automatic device manufactured by Paar of Austria. Acetic and succinic acids were titrated with sodium hydroxyde by using an automatic titration apparatus with potentiometric stop-point indication. Full analyses of partially miscible phases in systems with MIBK and butanol were carried out on a gas chromatograph.

MATHEMATICAL MODELS FOR THE PERFORMANCE OF EXTRACTION COLUMNS

The backmixing models described by Mecklenburgh and Hartland (1970) were used as a basis for simulating the column performance. Both the differential and stagewise models were tested and numerical methods of solution were developed. It was found that in most cases the stagewise model was easier to handle, and furthermore, that it could be applied for partially miscible solvents and unsteady state operation. The use of the differential model was advantageous in cases with small backmixing when too many stages are necessary if a stagewise model is used. The fundamentals of the stagewise model used are as follows :

The column is divided into a large number of discrete stages with the assumption that the liquids in each of these stages are perfectly mixed. This replaces the continuous change of concentration along the column axis by a stepwise one, but when the number of stages is large enough an approximation of this kind is acceptable, even for purely differential extractors such as spray or packed columns. It is assumed that there are additional streams in the column flowing in opposite directions to the main streams which are responsible for the backmixing, as illustrated in Figure 9. Material balances can be written for all the stages and solved numerically on a computer. At the steady state a set of algebraic equations is obtained which can be solved by some powerful procedure based on, say, the Newton-Raphson method. A linearization of the set is also possible and was discussed by Spencer et al.(1980). In the unsteady state a set of ordinary differential equations is obtained which can be solved by the Runge-Kutta method. When the equilibrium relationship is linear (which is the case with the butanol-succinic acid system), analytical solutions are available for both the differential and stagewise models(Mecklenburgh and Hartland, 1970).

Experience showed that the above models were suitable for most practical applications. For special cases new models were developed to improve the agreement with experimental measurements or make the calculations easier. A spray column has been described by a model based on the assumption that non-coalescing drops travelled in plug flow through a cascade of well-mixed stages

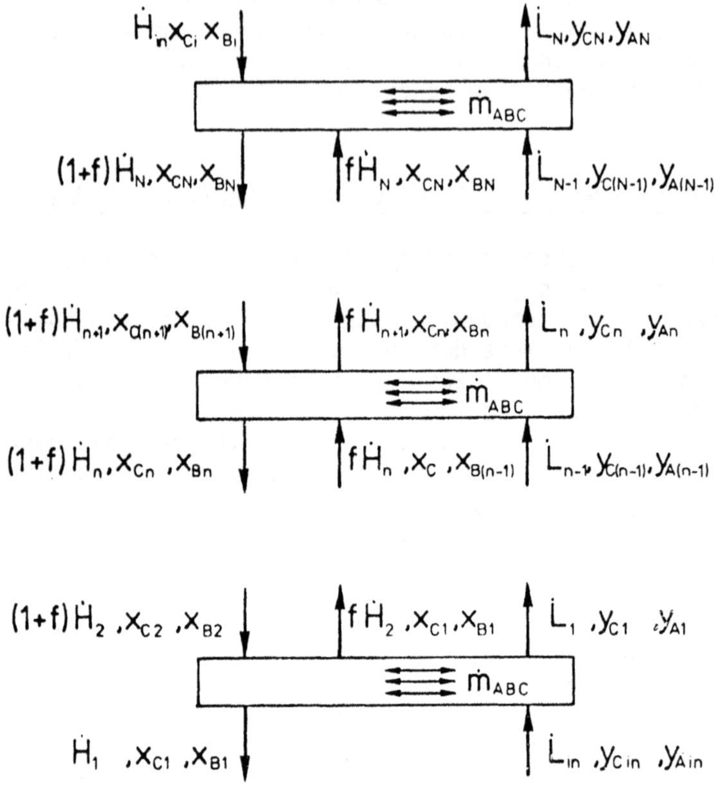

Figure 9: Derivation of backflow model for partially miscible systems. $\dot{H}$ and $\dot{L}$ are mass flow rates of the phases; A,B and C are the components, $\dot{m}$ is mass flux through the phase boundary, x and y are mass fractions in continuous and dispersed phases respectively.

filled by the continuous phase (Steiner et al. 1978, Steiner and Hartland 1980). More sophisticated models with drops moving at different velocities and with circulation in the continuous phase have also been developed.

The models were used to generate theoretical concentration profiles along the column axis which were then compared with the experimental ones. The best fitting profiles were found by varying the magnitude of the mass transfer coefficients, the resulting values being independent of the axial mixing in the column. This is not the case when the coefficients are simply determined from the end concentrations assuming plug flow; their magnitude is then dependent on the column dimensions. The variation of the simulated concentration profiles with the magnitude of the mass transfer coefficient is shown in Figure 10.

Knowing the mass transfer coefficients the performance of different extraction columns can be predicted with better accuracy than before. Using computer simulation parameter studies can be performed and the influence of different parameters investigated even before an actual column is built.

EXPERIMENTAL RESULTS

Extensive sets of experimental data were obtained on spray columns with all the systems mentioned earlier. The work was done using columns with diameters of 80 and 100 mm and lengths of 2 and 3.4 m. Examples of concentration profiles and their agreement with the best fitting simulated curves are shown in Figure 11 for an immiscible system and in Figure 12 for a system with partially miscible solvents. In the latter case three components were transferred in the column and three pairs of concentration profiles simulated.

The true mass transfer coefficients were evaluated and the Sherwood numbers correlated with the Reynolds numbers of the continuous phase, the coefficients being obtained by regression as below :

Toluene System : $Sh = 0.435\ Re^{0.92}$

MIBK System : $Sh = 1.65\ Re^{0.77}$

Butanol System : $Sh = 0.041\ Re^{1.76}$

Lumping the data for all three systems together yielded :

$$Sh = 0.0025\ Re^{0.39}\ Sc_c^{1.8}\ Sc_d^{-0.66}\ We^{0.26}$$

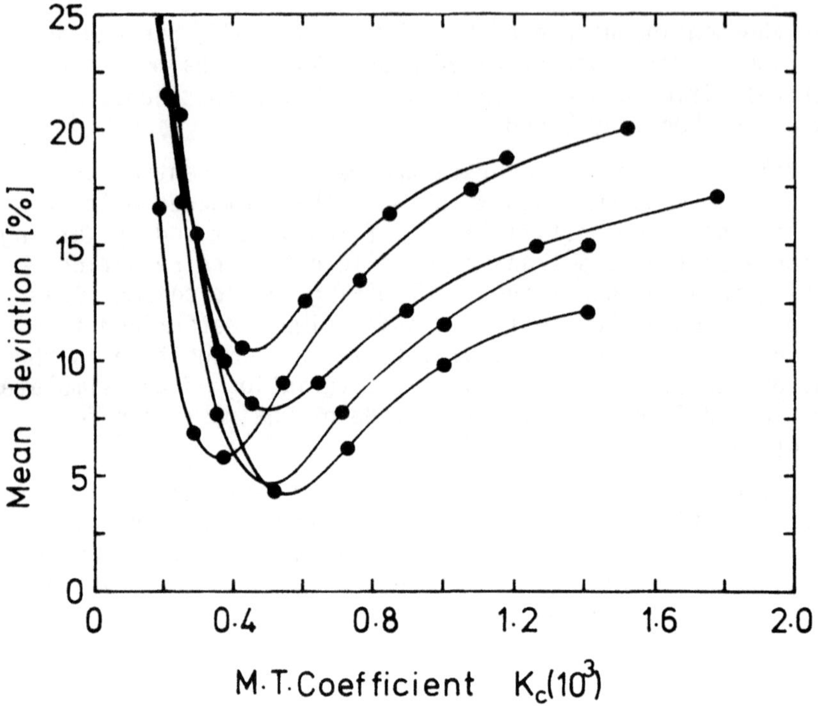

Figure 10: Effect of mass transfer coefficient on deviation of simulated concentration profiles from experimental data. The minima correspond to the best fitting curves and the corresponding coefficients are considered to be the true mass transfer coefficients for the given experimental condition. (Actual measurements with the MIBK/water/acetic acid system are shown).

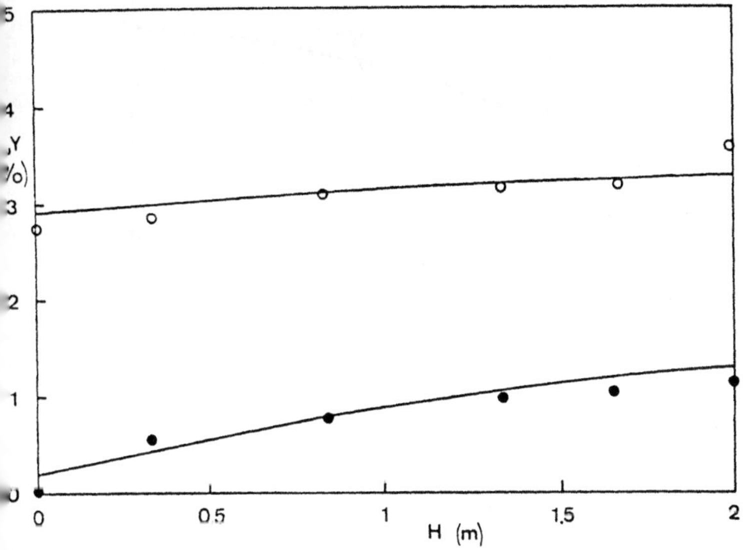

Figure 11: Comparison of simulated concentration profiles with experimental data for a system with immiscible solvents (o-xylene/ water/acetone).

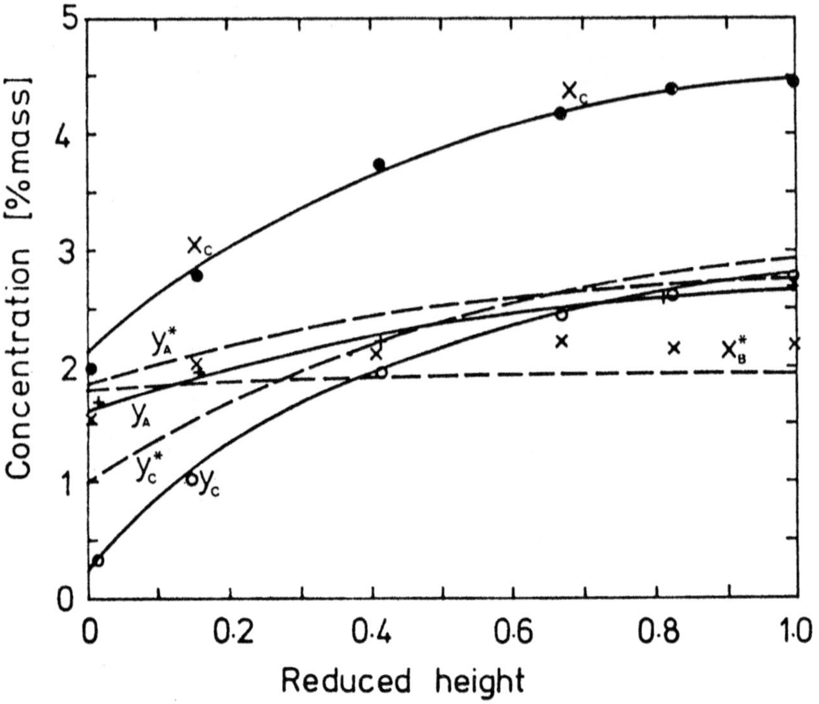

Figure 12 : Comparison of simulated concentration profiles with experimental data for a system with partially miscible solvents. x_C- solute concentration in heavy phase, y_C - solute concentration in light phase, x_B - concentration of light solvent in heavy phase, y_A - concentration of heavy solvent in light phase. Broken lines : equilibrium values.

Alternatively, the individual mass transfer coefficients can be estimated. In general, the resistance to mass transfer occurs in both phases and these partial resistances can be added together to obtain the overall value. Correlations for both sides of the phase boundaries are given in Table 3. For the inner side of the drops either perfect rigidity (Newman, 1952), laminar circulation (Kronig and Brink, 1951), or turbulent mixing (Handlos and Baron, 1957) is assumed. For the bulk phase there are different correlations based on the film or penetration theories which, however, all give rather similar results. Two formulae were selected giving the maximum and minimum mass transfer rates respectively: (Higbie, 1935 and Garner et al.,1959). Overall mass transfer coefficients were calculated using these formulae and the results compared with the values obtained experimentally. The results for solute transfer are shown in Figure 13 for the partially miscible systems and in Figure 14 for the xylene system. In all cases it seems that the overall coefficient is about 75% of that predicted by the combination of the Handlos and Baron formula for inside, and the Garner et al. formula for outside the drops.

The overall coefficient is then given by :

$$K_c = 0.75 \{d/\xi \; (1.8 \; Re^{0.5} \; Sc^{0.42} - 126) + (1 + \frac{\mu_d}{\mu_c})/0.00375m(u+v)\}^{-1}$$

the coefficients so obtained being similar to those from single drops, as can be seen in Figure 14 for the xylene system. For the MIBK system mass transfer coefficients measured on single drops (Cavers and Ewanchyna , 1975) were compared with our values for the spray column. Unexpectedly good agreement was reached, as shown in Figure 15. This figure also shows that the mass transfer coefficients may be time-dependent, especially just after the drop has been formed. In the simulation of spray columns a different mass transfer coefficient was therefore experimentally determined and used in the first stage of the mathematical model where the drops were formed. This coefficient was sometimes an order of magnitude greater than that valid for the rest of the column. This shows the importance of frequent coalescence and redistribution of the dispersed phase in the column if high mass transfer rates are to be achieved.

Other types of column are also being investigated in addition to the spray column. In a pulsed sieve plate column the same measuring techniques can be used as in a spray column. Concentration profiles and hydrodynamic parameters have been measured on columns of two sizes, namely 50 and 100 mm diameter, and the mass transfer coefficients evaluated. The results were similar to

Table 3

Equations for the Calculation of Mass Transfer Coefficients for Single Drops

a) Inside the drops :

Kronig and Brink (1951) :

$$k_d = 0.79 \frac{d}{t} + 17.66 \frac{\xi_d}{d}$$

Handlos and Baron (1957) :

$$k_d = 0.00375 (u + v)/(1 + \mu_d/\mu_c)$$

b) In the continuous phase :

Higbie (1935) :

$$Sh_c = 1.128 Re_c^{0.5} Sc_c^{0.5}$$

Garner et al. (1959) :

$$Sh_c = 1.8 Re_c^{0.5} Sc_c^{0.42} - 126$$

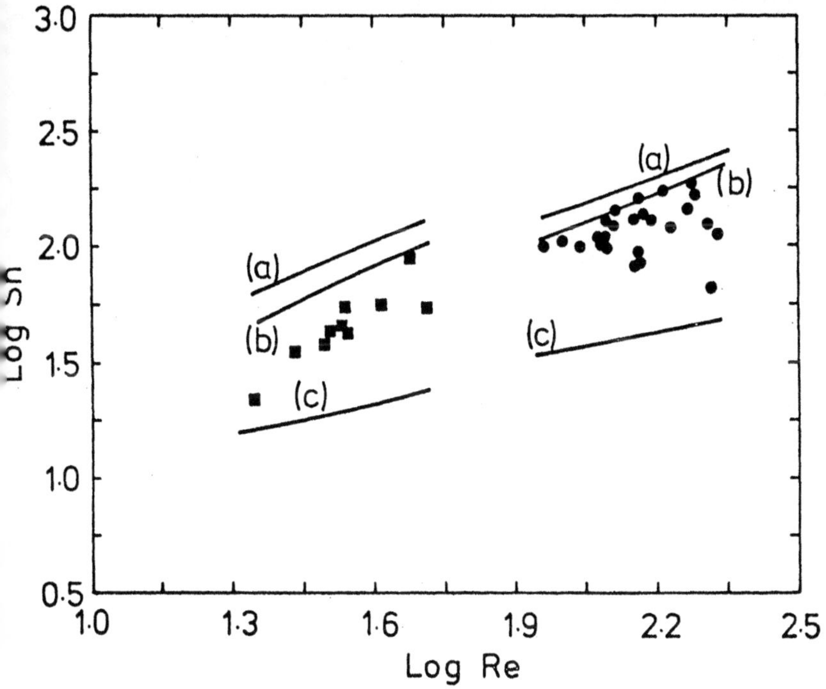

igure 13: Mass transfer coefficients for solute transfer in partially miscible systems. Left-hand side: Butanol system; right-hand side: MIBK system. Comparison of experimental values with correlations for single drops: (a) combination of Handlos and Baron with Higbie, b) Handlos and Baron with Garner et al. and c) Kroning and Brink with either Higbie or Garner.

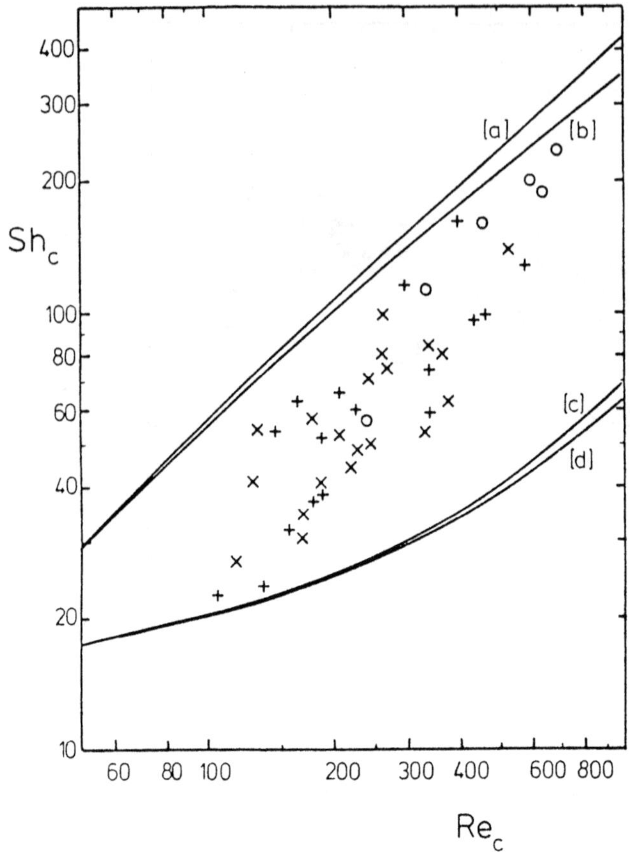

Figure 14: Mass transfer coefficients in systems with immiscible solvents. Comparison of experimental values with correlations for single drops. Systems: o-xylene/water/aceton (x) ; toluene/water/acetone (+). A few single drop measurements are also shown (o).

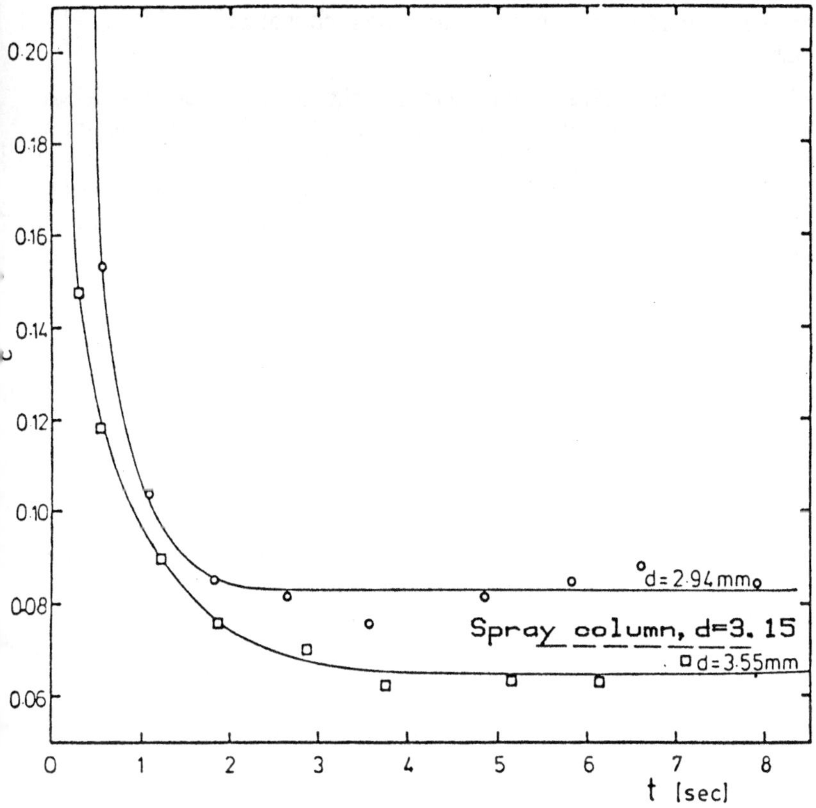

Figure 15: Time dependencies of mass transfer coefficient for single drops and comparison with coefficients obtained in spray columns. Data from Licht and Pansing (1955); spray column measurements by Cavers and Evanchyna (1975). System: MIBK/water/acetic acid.

those obtained on spray columns and the transfer rates again comparable with those measured on single drops. The same simulation programmes were used and the accuracy was comparable with that of spray columns.

Measurements with stirred columns of the Kühni type are also being carried out, placing emphasis on axial mixing and changes of hydrodynamic parameters along the column length. A considerable variation in hold-up and drop size along the column length was observed so the computer programmes were modified to take this into account.

Hydrodynamic measurements are being performed with the toluene system in a column with reciprocating plates. A new type of agitated column with coalescence-enhancing plates has been developed which combines high mass transfer rates with high throughput.

EXPERIENCE WITH THE TEST SYSTEMS

Our experiments confirm that it is impossible to have one universal system which describes all situations. Using all the systems originally recommended by the Working Party of the European Federation of Chemical Engineers gives a good picture of extractor performance provided the phase ratio remains close to 1. For more extreme ratios it is very difficult to obtain reliable mass transfer measurements as the solute concentrations must then be measured very accurately.

Doubts about the MIBK system have been recently expressed. Some investigators have reported very high transfer rates which are attributed to "interfacial instability" or dimer formation of the acetic acid in the aqueous phase. However, we suspect that it is the data evaluation rather than system properties which causes difficulties. Even very slight inaccuracies in the measurement of concentration can cause the operating line to cross the equilibrium line, particularly when only the end concentrations are measured. Furthermore, considerable difficulties were met when handling unpleasantly smelling waste waters contaminated by butanol or MIBK.

This was not the case with the toluene and xylene systems but the mass transfer rate then apparently depends on the direction of solute transfer. In this respect the butanol system was better as the rate was equal in both directions and, in addition, the equilibrium relation was linear. However, for this system the interfacial tension is very low so that general conclusions cannot be drawn from experiments carried out on this system alone. Ideally, data should be obtained on a wide variety of systems with different physical properties and different distribution coefficients. The first stage of extractor design could be done without mass transfer by using, for instance, kerosene and water, but the mass transfer tests would have to be performed with a system having similar properties to that considered. None of the systems is especially good for obtaining a general comparison of column types because of the limited range of phase ratios which may be used.

CONCLUSIONS

The performance of countercurrent extractors can be approximated by suitable mathematical models based on the stagewise backflow concept. Working with the test systems recommended by the European Federation of Chemical Engineers mass transfer rates for typical extractors were found to be similar to those measured for single drops and about 75% of the values predicted from available correlations allowing for transfer inside and outside the drops. None of the recommended systems fulfilled all the requirements so new test systems should be sought which enable columns to be operated over a wider range of phase ratios whilst retaining the present spread of interfacial tensions.

Notation

d	drop diameter (m)
K	overall mass transfer coefficient (kg/m^2 s)
m	distribution coefficient ($y^* = mx$)
u,v	actual velocities of continuous and disperse phases (m/s)
x,y	mass fractions in continuous and disperse phases respectively

$Re = \rho ud/\mu$ Reynolds number
$Sc = \mu/\rho\xi$ Schmidt number
$Sh = Kd/\xi$ Sherwood number

Greek Symbols

μ	viscosity (kg/ms)
ρ	density (kg/m^3)
ξ	diffusivity (m^2/s)

Indices

ABC	two solvents and solute respectively in partially miscible system
c	continuous phase
d	disperse phase
in	input value
n	typical stage
y*	equilibrium value.

REFERENCES

Berger, J. "Grundlagen des Stoffaustausches in einer Flüssig/Flüssig-Sprühkolonne" Dissertation ETH 6773 (1981).

Cavers, S.D., J.R. Ewanchyna, "Circulation and End Effects in Liquid Extraction Spray Columns" Can.J.Chem.Eng.,35,(1975) 113.

Ergenc, N. "Untersuchung der hydrodynamischen Vorgänge in einer gerührten Extraktionskolonne" Diplomarbeit, TCL,ETHZ 3841(1980)

Efthiamatou C. "Choice of Disperse Phase and Mass Transfer Direction in Liquid/Liquid Extraction" Internal Report,TCL,ETHZ (1980).

Fischer von, E., L. Steiner and S. Hartland, "EC Column:Highest Flexibility at Normal Loads" CHISA, Prague (1981).

Garner, F.H., A. Foord and M. Tayeban, "Mass Transfer from Circulating Liquid Drops" J.Appl.Chem., 9 (1959) 315.

Garner, F.H. and M. Tayeban, "The Importance of the Wake in Mass Transfer from both Continuous and Dispersed Phase Systems" Anal. Real Soc.,Espan. Pis.Quim., B 56, (1960) 479.

Gebauer,K., L. Steiner and S. Hartland, "A New Liquid/Liquid Centrifugal Extractor with Inner Mixers"CHISA, Prague (1981)

Handlos, A.E., T. Baron, "Mass and Heat Transfer from Drops in Liquid/Liquid Extraction" AIChE J., 3, (1957) 127.

Higbie, R. "The Rate of Absorption of a Pure Gas into Still Liquid During Short Periods of Exposure" Trans.AIChE J.31 (1935) 365.

Horvath, M. "Hydrodynamik and Stoffaustausch in einer Flüssig/Flüssig-Sprühkolonne" Dissertation ETHZ 5774 (1976).

Kronig, R., J.C. Brink, "On the Theory of Backmixing from Falling Droplets" Appl.Sci.Res., A2 (1951) 142.

Kurt, M. "Untersuchung der Rührerarbeitsweise in einer Extraktionskolonne" Diplomarbeit,TCL, ETHZ 3888 (1981)

Licht, W., W.F. Pansing, "Solute Transfer from Single Drops in Liquid/Liquid Extraction" I&EC 45, (1953) 1885.

Mecklenburgh, J.C., S. Hartland, "Theory of Backmixing" Wiley, London, (1970).

Misek, T. "Recommended Test Systems in Liquid Extraction Studies" European Federation of Chemical Engineers, Instn.of Chem.Engrs. London (1978).

Oezdemir, G. "Optimierung der koaleszenzfördernden Einbauten für Flüssig/Flüssig-Extraktionskolonnen" Diplomarbeit,TCL,ETH. Zürich, 3875 (1980).

Schläpfer, J. "Untersuchung des Verhaltens einer Karr-Kolonne" Diplomarbeit, TCL, ETH,Zürich, 3827 (1979).

Spencer, J., L. Steiner and S. Hartland, "Model-Based Analysis of Liquid/Liquid Extraction Data" AIChE J., (in Press).

Steiner, L. and S. Hartland,"Modellierung von Extraktionskolonnen unter Anwendung der Rückvermischungstheorie" Chem.-Ing.-Technik 52 (1980) 602. MS 819/80.

Steiner, L. and S. Hartland, "Agitated Liquid/Liquid Extraction" CEP 76, 60-62 (1980)

Steiner, L., M. Horvath and S. Hartland, "Mass Transfer Between Two Liquid Phases in a Spray Column at the Unsteady State" I&CE Proc.des.dev., 17, (1978) 175.

Steiner, L., M. Horvath and S. Hartland, "Determination of Actual Mass Transfer Rates in Extraction Columns" Proc. ISEC 1977, CIM, Spec.Vol. 21, (1979) 366-371

Toller, W., "Tatsächliche Stoffaustauschraten in gepulsten Siebbodenkolonnen bei der Flüssig/Flüssig-Extraktion" Diplomarbeit TCL, ETH,Zürich, 3881 (1981).

SOME ASPECTS OF METAL EXTRACTION

H. Sawistowski

Department of Chemical Engineering and Chemical Technology
Imperial College of Science and Technology, London SW7

INTRODUCTION

Metal extraction is usually referred to as hydrometallurgy and it deals with extraction of metals from acidified aqueous solutions. For this purpose the aqueous solution, that is the raffinate phase, is contacted with an organic extract phase. However, since metal ions are insoluble in organic solvents, the latter must contain a water-insoluble complexing agent which will react with the metal and extract it into the organic phase in the form of a metal complex. The reaction is reversible and the metal is re-extracted from the organic phase in the stripping process usually by an aqueous phase suitably modified to move the reaction equilibrium in the appropriate direction. This modification is usually attained by adjustment of pH of the solution. The insolubility of the complexing agent in water is very important from process considerations, since solubility would mean not only losses of the agent but its presence in the stripping solvent would introduce difficulties in subsequent processing, e.g. electrowinning. Nevertheless, the agent must contain a hydrophilic group to allow it to pick up the metal ion but this effect must be counteracted by a large hydrophobic chain to retain insolubility.

The fact that ions are insoluble in the organic phase and insolubility in water is one of the main selection criteria of complexing agents, the reaction must occur at the interface. This reasoning is also supported by experimental evidence, e.g. by Ajawin et al. (1), who demonstrated that, under conditions of chemical control, the rate of extraction is directly proportional to the interfacial area.

PHASE EQUILIBRIUM

In the presence of a reversible reaction phase equilibria are established in such a way as to satisfy the expression for the equilibrium constant of the reaction. The overall reactions may be of various form but those most commonly encountered are

$$M^{m+} + m\,\overline{HX} \rightleftharpoons \overline{MX_m} + mH^+ \quad (1)$$

for extraction by chelating agents, and

$$M^{m+} + \frac{m+n}{2}\,(\overline{HX})_2 \rightleftharpoons \overline{MX_m(HX)_n} + mH^+ \quad (2)$$

for extraction by carboxylic, sulphonic and alkylphosphoric acids, where M denotes the metal, X the anionic part of the complexing agent and bars indicate the organic diluent phase.

If m_c is the distribution coefficient, it will be defined here as

$$m_c = \overline{M}/M = C_{M1}/C_{M2} \quad (3)$$

where C_M is the concentration of the metal and subscripts 1, 2 represent the organic and aqueous phase respectively.

By introducing the concentration-based equilibrium constant K_c, eqn (1) can be represented for extraction by chelating agents as

$$K_c = C_H{}^m\, C_{M1}/C_{M2}\, C_A{}^m \quad (4)$$

where C_A is the concentration of the agent and C_H the hydrogen ion concentration in the aqueous phase.

By combining (3) and (4)

$$K_c = m_c\, C_H{}^m/C_A{}^m \quad (5)$$

and $\log m_c = \log K_c + m\,\text{pH} + m \log C_A \quad (6)$

If $C_M{}^o$ denotes the total metal concentration and $C_A{}^o$ the tota concentration of the chelating agent

$$C_{M1} = C_M{}^o - C_{M2} \quad (7)$$

and $C_A = C_A^o - m\, C_{M1}$ (8)

In general, $C_A >> C_{M1}$ so that $C_A \approx C_A^o$. Hence, since m_c can be obtained by measurement, K_c can be obtained by plotting log m_c vs pH at constant C_A^o. Eqn (2) can be treated in a similar way and the logarithmic plot gives both K_c and n. It should be noted that K_c is a function of both temperature and ionic strength of the aqueous phase. In general, K_c increases with ionic strength and the same therefore applies to the distribution coefficient.

MASS TRANSFER RATE

The mass transfer process consists of the diffusion of the reactants to the interface, a chemical reaction at that location followed by the diffusion of the products away from the interface. The rate of extraction will thus be controlled by the slowest step. Consequently, the process can be chemically controlled if the reaction is slow in comparison to the specific rate of diffusion. It can also be diffusion controlled if the rate of reaction is relatively fast. The process can also take place under condition of mixed control, when both reaction and diffusion have to be taken into account. In addition, the regime of operation may alter by changes in concentrations and other process parameters. Since the process parameters affect the diffusion regime and the chemical regime in a different way, their delineation becomes extremely important. It is particularly important to define the chemical regime, since only from studies in this regime is it possible to draw conclusions about the kinetics of the reaction and hence about its mechanism.

Most of the earlier experiments were conducted in stirred tanks in which one phase was dispersed into another. However, this type of equipment suffers from the disadvantage of difficulty in separating the effects of extraction kinetics from changes in interfacial area.

Chronologically, the next type of configuration employed was the rise or fall of a single drop. Although the interfacial area could be reasonably well estimated in this case, the hydrodynamic conditions inside and outside the drop were seldom easy to describe.

Nowadays, fundamental work on determination of characteristics of metal extraction processes is conducted almost exclusively in stirred cells, already referred to in the contribution on 'Physical Aspects of Solvent Extraction'. In such cells the two phases can be stirred independently but at a rate which produces a practically uniform concentration in the bulk of each phase while the interface remains constant and quiescent.

For each set of conditions (concentrations, temperature, ionic strength) the stirring speed is changed until the rate of extraction reaches a plateau. The determination of conditions under which a plateau is obtained delineates the chemical regime. Investigations in this regime can then follow to determine the kinetics of the reaction.

In general, it would be expected that the rate of chemical reaction per unit interfacial area, R_M, i.e. the molar flux of metal ions, N_M, would be given by an expression of the type

$$R_M = N_M = k\, C_M{}^{a} (C_A)^{b} C_H{}^{c} - k'\, C_M{}^{a'} (C_C)^{b'} C_H{}^{c'} \tag{9}$$

where k is the rate constant, the primes denote the reverse reaction and C_C is the concentration of the complex. If measurements are conducted or extrapolated to initial conditions, the reverse reaction can be neglected and the forward reaction written as

$$\log N_M = \log k + a \log C_M + b \log C_A{}^{o} - c\ \mathrm{pH} \tag{10}$$

From a proper statistical design of experiments values of a, b and c can be determined. In practice such an approach is seldom used and the constants are determined by varying one concentration at a time while keeping the other two constant. This simple procedure has proved to be quite successful. Subsequent variations in temperature and ionic strength give the effect of these parameters on the rate constant. After the rate equation for the chemical reaction has been established, experiments can be conducted in the mixed regime. Here, eqn (9) is again valid except that all concentrations assume new interfacial values so that

$$R_M = N_M = k\, C_{Mi}^{a}\ (C_{Ai}^{o})^{b}\ C_{Hi}^{c} = K_M\ (C_M - C_{Mi}) \tag{11}$$

where K_M is the mass transfer coefficient for the transfer of the metal. Similar flux equations can be set up for the other components which will be related to N_M by the stoichiometry of the reaction. C_{Mi}, C_{Ai}^{o} and C_{Hi} can thus be expressed in terms of N_M and the relevant bulk compositions. On their substitution N_M is obtained as function of C_M, $C_A{}^{o}$ and C_H in an implicit or explicit form. The flux and thus the rate of extraction can be calculated provided the values of mass transfer coefficients are available.

It is the values of mass transfer coefficients which change on studying the metal extraction process under different hydrodynamic conditions, even more so in a different equipment. As shown by Perez de Ortiz et al. (2), the assumption that the rate of chemical reaction is always the rate-controlling step can lead to wrong conclusions concerning the form of the chemical rate equations. This was demonstrated using the example of extraction of copper by hydroxyoximes. The overall extraction reaction obeys eqn (1) with m = 2, i.e.

$$Cu^{2+} + 2\ \overline{XH} \rightleftharpoons \overline{CuX_2} + 2\ H^+ \qquad (12)$$

and the rate equation reported in literature is

$$R_M = N_M = k\ C_M{}^a (C_A{}^o)^b\ C_H{}^{-c} \qquad (13)$$

Different workers employing different experimental methods proposed different values for b and c, as shown in Table 1.

Table 1. Summary of orders of reaction for copper extraction with hydroxyoximes.

Ref.	Experimental Method	c	b	a
(3)	Single drop	0	0.5	1
(4)	Quiescent interface cell	0.6	1.10	1
(5)	Stirred tank (AKUFVE)	0.9	1.01	1

By using interfacial concentrations in place of bulk concentrations Perez de Ortiz et al. obtained the rate expression in an implicit form

$$N_M = k\ C_M\ (C_A{}^o)^b\ C_H{}^{-c}\ F(N_M) \qquad (14)$$

where $F(N_M) = (1 - N_N/K_M\ C_M)(1 - 2N_M/K_A\ C_A{}^o)^b(1 + 2N_M/H_H\ C_H)^{-c}$ (15)

By assuming suitable values of k/K for each experimental method these workers were able to show that b = 1 and c = 1.

The method outlined previously has also been used by Ajawin (6) for the extraction of zinc by di(2-ethylhexyl) phosphoric acid. The relevant extraction equation is eqn (2) with m = 2 and n = 1, i.e.

$$Zn^{2+} + 1.5\ (\overline{HX})_2 \rightleftharpoons \overline{ZnX_2HX} + 2\ H^+ \qquad (16)$$

and the forward rate of chemical reaction per unit interfacial area

$$R_M = N_M = k\ C_M\ C_A^{\,o}\ C_H^{\,-1} \tag{17}$$

In such a case, for the mixed regime, an explicit equation can be obtained for the flux, as used by Wongswan et al. (7).

MECHANISM OF REACTION

The mechanisms of chemical reactions involving complex formation with metal ions are not simple and, in some cases, not yet fully understood. In addition, they may also depend on the concentration of some components (8).

A simple case will be presented here referring to the case of the use of an alkyl phosphoric acid (6, 9). Such an acid is present in the organic phase as a dimer H_2X_2. As it arrives at the interface it would be expected to ionize and dissociate:

ionization: $$\overline{H_2X_2} \rightleftharpoons \overline{HX_2^-} + H^+ \tag{18}$$

dissociation: $$\overline{H_2X_2} \rightleftharpoons 2\overline{HX} \tag{19}$$

The species $\overline{HX_2^-}$ and $\overline{HX}$ will be adsorbed at the interface in a proper orientation, that is with the hydrophilic group entering the aqueous phase. The expected interfacial reactions will therefore be

$$M^{2+} + \overline{HX_2^-} \rightleftharpoons \overline{MX_2} + H^+ \tag{20}$$

$$\overline{MX_2} + \overline{HX} \rightleftharpoons \overline{MX_2HM} \tag{21}$$

Ionization and dissociation cannot be rate controlling as they do not involve the metal. Hence they have to be assumed as instantaneous with equilibrium constants K_I and K_D respectively. If the last equation were rate controlling the exponent on hydrogen ion concentration would be -2 and the reagent 1.5, i.e. values not encountered in practice. Hence, this equation also has to be treated as instantaneous with an equilibrium constant K_B. Consequently, eqn (20) should represent the rate controlling step with a rate equation

$$R_M = k_1\ C_M \left[\overline{HX_2^-}\right] \tag{22}$$

From eqn (18)

$$\left[\overline{HX_2^-}\right] = K_I\ C_A^o\ C_H^{-1} \tag{23}$$

so that from (22) and (23)

$$R_M = k_1\ K_I\ C_M\ C_A^o\ C_H^{-1} \tag{24}$$

which in the case of extraction of zinc by DHEPA agrees with the experimentally determined values as given by eqn (17).

APPLICATION OF LIQUID MEMBRANES TO METAL EXTRACTION

It has already been stressed that the liquid membrane process allows extraction and stripping to be conducted in a single unit. In addition, its inventory of the extract (organic) phase is low and the flow rate of the stripping phase is also low resulting in saving in capital cost at the metal recovery end. In addition, the process can be made kinetically rather than thermodynamically controlled so that high metal concentrations can be attained in the stripping phase. Consequently, the process is particularly suitable for treatment of very dilute raffinates.

Extraction of metals by the liquid membrane process falls into the category of facilitated transport and any of the complexing agents can be employed as metal carrier across the membrane phase. Consider, as an example, a reaction as given by eqn (1) with the corresponding equilibrium constant given by eqn (4)

$$K_c = C_C\ C_H{}^m / C_M\ C_A{}^m \tag{25}$$

where C_C is the concentration of the complex $\overline{MX_m}$. This equilibrium constant applies at both the extracting and stripping surfaces. At the extracting surface, low concentration of $\overline{MX_m}$ will shift the reaction to the right, whereas at the stripping surface, provided a large excess of hydrogen ions is present (low pH), it will be pushed very far to the left so that the process is practically kinetically controlled throughout its whole duration.

The disadvantages of the liquid membrane process are the use of high concentration of surfactants (values range from 0.67 to 5%) which, under normal conditions, would introduce an interfacial resistance but in the presence of an interfacial reaction reduce the value of the rate constant by the physical process of surface coverage. However, this disadvantage is overcome by the highly developed surface area, particularly inside the globules. Thus, for example, a globule, 1 mm in diameter, with a 50/50 phase ratio of organic to stripping phase, will contain 5×10^8 internal phase droplets, 1 μm in diameter. This large internal surface area is another contributory factor in attainment of high concentrations of the internal phase. Other disadvantages of the membrane process are possible globule break-up, loss of internal phase droplets to the external phase and swelling of internal phase droplets by transfer of water across the membrane by the process of osmosis (large difference in pH). The first two effects are not very large and the last one is not serious provided the time of contact is not too long.

Most of the work on metal extraction has been conducted in stirred tanks and evaluated on the basis of diffusion through the membrane being the rate controlling step with no internal mixing within the globules. Since the mass transfer process in such a case is represented by transient molecular diffusion into the membrane and into the droplets, the surfaces of the globules and the internal droplets age very quickly and the rate of transfer decreases rapidly with time. It would, therefore, appear that the use of plate columns which provide periodic surface renewal and rearrangement of droplets within the globules could lead to intensification of the mass transfer process. However, the work conducted by Wongswan et al. (7) in a spray column indicates that the behaviour there is different from that assumed in stirred tanks. The globules are large, they oscillate with consequent movement of droplets inside the globules and the process is either chemical controlled or limited by the rate of diffusion in the external phase. Since it is also kinetically controlled it can be used in a semi-batch or continuous arrangement. Insufficient data are available at this stage to make a reliable comparison of stirred tanks and spray columns, except that the latter are more suitable for continuous operation.

The liquid membrane process will become accepted not only when it offers economic advantages over the traditional solvent extraction process but also if sufficient basic information is available for a reliable prediction of its performance. A cost comparison of the liquid membrane process and mixer-settler operation has been conducted for the extraction of uranium and copper (10) and the results are presented in Tables 2 and 3 (SX denotes mixer-settler operation, LM liquid membrane process).

It will be noticed that in copper recovery the savings came mainly from plant investment, whereas in uranium recovery from the reduction in organic losses. Of course, the subsequent recovery of the metal from the stripping phase is not incorporated in these estimations.

Table 2. Estimated cost of copper recovery from ore leachates.

	SX	LM
Copper recovered (ktonnes/year)	36	36
Stages	5	1
Plant investment, M $	13	8
Organic inventory, M $	2	1
Direct operating cost ¢/lb	1.8	1.7

Table 3. Estimated cost of uranium recovery from phosphoric acid.

	SX	LM
Uranium recovered (kg/year)	145	145
Stages	8	3
Plant investment, M $	38	38
Organic inventory, M $	1.6	0.2
Direct operating cost, $/kg	63	40

Apart from the work conducted by Li et al. at Exxon which is fully reported in the proceedings of a seminar on 'Liquid Membrane Applications in Waste Water Treatment and Metals Recovery' held at UMIST in May 1980, investigations on applications of liquid membranes to metals recovery are conducted at a number of other places. Thus, extraction of copper is studied at UMIST (11, 12), Graz (13-15) and Bend Research (16). The last place is also active in the use of membranes for the extraction of uranium (17), whereas work on the extraction of copper, zinc, cadmium and lead is conducted by Boyadzhiev et al. (18). Extraction of different metals has also been studied by Stelmaszek (19) and Strzelbicki (20, 21) and of zinc alone at Imperial College (7). All the accumulated data point towards membrane extraction as a promising process for the solution of separation problems, particularly in dilute solutions.

REFERENCES

1. Ajawin, L.A., Perez de Ortiz, E.S. and Sawistowski, H., Proc. Intl Solvent Extraction Conf., Liege (1980) 80-112.

2. Perez de Ortiz, E.S., Cox, M. and Flett, D.S., Proc. Intl Solvent Extraction Conf., Toronto (1977) 198.

3. Atwood, R.L., Thatcher, D.N. and Miller, J.D., Met. Trans. B. 6B (1975) 465.

4. Fleming, C.A., Nat. Inst. of Metallurgy, Report No. 1793, Johannesburg (1976).

5. Flett, D.S., Okuhara, D.N. and Spink, D.R., J. Inorg. Nucl. Chem., 35 (1973) 2471.

6. Ajawin, L.A., Ph. thesis, Univ. of London (1980).

7. Wongswan, S., Perez de Ortiz, E.S. and Sawistowski, H., Proc. Hydrometallurgy '81, Soc. Chem. Ind. (1981).
8. Kertes, A.S., "The Chemistry of Solvent Extraction," in Hanson, C. (Ed.), "Recent Advances in Liquid-Liquid Extraction," Pergamon, London (1971).
9. Ajawin, L.A., Perez de Ortiz, E.S. and Sawistowski, H., Trans. Instn. Chem. Engrs (in print).
10. Burns, W.A., in Proc. of Conf. on 'Liquid Membrane Applications in Waste Water Treatment and Metals Recovery', UMIST (1980) 72.
11. Martin, T.P. and Davies, G.A., Hydrometallurgy, 2 (1976/77) 315.
12. Martin, T.P. and Davies, G.A., Proc. Intl Solvent Extraction Conf., Liege (1980) 80-230.
13. Kopp, A., Marr, R. and Moser, F., IChemE Symp. Series, 54 (1978) 279.
14. Marr, R., Kopp, A. and Wilhelmer, J., Ber. Bunsenges. Phys. Chem., 83 (1979) 1097.
15. Mar, R. and Kopp, A., Chem-Ingr-Tech., 52 (1980) 5.
16. Baker, R.W., Tuttle, M.E., Kelly, D.J. and Lonsdale, H.K., J. Membrane Science, 2 (1977) 213.
17. Babcock, W.C., Baker, R.W., Kelly, D.J. and La Chapelle, E.D., Proc. Intl Solvent Extraction Conf., Liege (1980) 80-90.
18. Boyadzhiev, L. and Kyachoukov, J. Membrane Science, 6 (1980) 107.
19. Stelmaszek, J. and Borowska, B., Inz. Chem., 4 (1975) 869.
20. Strzelbicki, J. and Charewicz, W., J. Inorg. Nucl. Chem., 40 (1978) 1415.
21. Strzelbicki, J. and Charewicz, W., Sep. Sci. Technol., 13 (1978) 141.

PARTICIPANTS

R.C.Aiken, Department of Chemical Engineering,The University of Utah,3062 Merril Engineering Building,Salt Lake City, Utah 84112,U.S.A.

J.Akyurtlu, Chemical Engineering Department.Middle East Technical University, Ankara ,Turkey.

M.Alpbas, Chemical Engineering Dept.,Ankara University, Ankara,Turkey.

J.Andrieu,Laboratoire de cinetique et genie chimiques-404. INSA 20.avenue albert einstein,69621 villeurbanne cedex-Ølyon,France.

H.Arslan,Chemistry Department, University of Selçuk, Konya ,Turkey.

B.Beler,Department of Chemical Engineering,Bosphorous University, Bebek,Istanbul,Turkey.

P.M.M.Blauwhoff,Twente University of Technology,P.O.B.217, 7500 AE Enschede, Holland.

R.S.Carter,CIBA-GEIGY A.G.,C.H.-4002 Basle,Switzerland.

T.Çakoloz,Faculty of Food Engineering,Aegen University, Izmir,Turkey.

A.Çalimli,University of Pittsburgh,Chemical and Petroleum Engineering Department,15261 Pittsburgh,U.S.A.

A.Çetinbudaklar,DYO,Izmir,Turkey.

A.Çınar,Department of Chemical Engineering,Bosphorous University, Bebek,Istanbul,Turkey.

C.Darde,CNRS,Laboratoire des Sciences du Génie Chimique, 1.rue grandville,54042,Nancy,Cedex,France.

P.K.Demetriades,Laboratory of Unit Operations,School of Chemical Engineering,NTU,Athens 147,Greece.

S.Dinçer, Department of Chemical Engineering ,Bosphorous University,Bebek, Istanbul,Turkey.

V.Dovi,Istituto di Scienze e Tecnologie,dell'Ingegneria Chimica, University of Genova,I-16143 Genova,Italia.

P.G.Eggels,AFD Technische Scheikunde,University of Groningen, Nijenborgh 16,Groningen,Holland.

N.Eken,Chemistry Faculty,State Academy of Engineering and Architecture,Maltepe,Ankara,Turkey.

Elmaleh,Laboratoire de Génie Chimique,Université des Sciences et Techniques du Languedoc,Place Eugène Bataillon, 34060 Montpellier Cedex , France.

Eroğlu,Chemical Engineering Department,METU,Ankara.Turkey.

P.Euzen,Institut Français du Pétrole CEDI,Boite Postale 3 69390 Vernaison,France.

L.Figueiredo,Faculty of Engineering,University of Porto, 4099 Porto,Codex,Portugal.

Filippone,Collegio Vecchio,Universita Urbino,Italia.

H.Girault,Wolfson Centre for Electrochemical Science,Department of Chemistry,The University of Southampton, S09 5NH,England.

R.Gupta,Reactors and Fluid Dynamic Section,Exxon Research and Engineering Co.,P.O.Box. 101,Florham Park,NJ 07932,U.S.A.
T.Gurkaan,Chemical Engineering Dept.,METU,Ankara,Turkey.
J.Hjortkjaer,The Technical University of Denmark,Institute for Kemiindustri, DtH Building 227,DK-2800 Lyngby,Denmark.
D.K.Jain,Lehrstuhl und Institut für Chemische Verfahrenstechnic, Boblinger Strasse 72.D-7000 Stuttgart 1,F.R.Germany.
M.Jeronimo,Centro de Engenharia Quimica,Faculdade de Engenharia, Roa dos Bragas,4099 Porto,Codex ,Portugal.
S.Katnaş,Chemical Engineering Dept.,METU,Ankara ,Turkey.
F.Kayıhan,Chemical Engineering Dept., Oregon State University, Corvallis,OR 97331,U.SA.
M.A.Khidr,Mathematics Department,Centre of Science and Mathematics P.O.Box.2375,Damman,Saudi Arabia.
P.Knysh,UMIST,The University of Manchester,Chemical Engineering Dept.,PO Box 88,Manchester M 60 1QD ,England.
Ö.Kuleli,Chemical Engineering Department,Hacettepe University, Ankara,Turkey.
B.Kuryel,Chemical Engineering Department,Aegen University, İzmir,Turkey.
Ö.M.Kut,Technisch-Chemisches Laboratorium,ETH-Zentrum, CH-8092 Zurich,Switzerland.
A.Lecloux, c/o Solvay et cie ,Laboratoire Central,Rue de Ransbeek, 310-1120 Bruxelles,Belgium.
M.Lohse,Universitaet Hannover,Institut für Technische Chemie, Callinstrasse 3,D-3000 Hannover 1 ,F.R.Germany.
A.Lübbert,Universitaet Hannover,Institut für Technische Chemie, Callinstrasse 3, D-3000, Hannover 1, F.R.Germany.
M.O.Maia,Universidade do Minho.Pavilhao de Engenharia. Av.Joao XXl, 4700 Braga.Portugal.
J.Meldon,Chemical Engineering Dept.,Tufts University.Medford. Massachusetts 02155, U.SA.
L.M.Mishra,Institut für Technische Chemie der Universitaet Hannover,Callinstrasse 3,D-3000.Hannover 1,F.R.Germany.
B.I.Morsi,CNRS,Laboratoire des Sciences du Génie Chimiques, ENSIC, 1 . rue grandville , 54042, Nancy ,Cedex,France.
H.Oğuz,Chemical Engineering Dept.,Ankara University,Turkey.
A.Olcay,Department of Chemical Engineering,Faculty of Sciences, Ankara University,Ankara,Turkey.
Ö.Olgun, Chemical Engineering Dept.,Aegen University,Izmir,Turkey.
V.Oreopoulou,NTU of Athens,Laboratory of Organic Chemical Technology,42 , 28 th October Street, Athens,Greece.
Z.I.Önsan,Chemical Engineering Department,Bosphorous University, Bebek,Istanbul,Turkey.
S.Öztürk,Chemical Engineering Dept.,Ankara University,Turkey.
N.G.Papayannakos,NTU,Melissou 9-13.Pagrati,Athens,Greece.
A.Parmaliana,Via Nazionale,196 , 98050- Terme Vigliatore, Messina , Italy.
S.Peker, Chemical Engineering Dept., Aegen University.Izmir.Turkey.

Philippopoulos,NTU,Pattision 42,Athens,Greece.
N.N.C.Pinho,Departamento de Technogia Quimica,Institito Superior Technico, 1096 Lisboa Codex,Portugal.
Rizzuti,Universita di Palermo.Istituto di Ingegneria Chimica, Facolta di Ingegneria,Vialle Delle Scienze, Italia.
E.Rodrigues,Department of Chemical Engineering,University of Porto, 4099 Porto Codex,Portugal.
Rützou,Haldor Topsøe A/S,Nymøllerej 55 . P.O.Box. 213, DK-2800 Lyngby,Copenhagen ,Denmark.
M.N.Santos,Faculdade de Ciencias e Techologia, Universidade Nova de Lisboa , Quinta do Cabeco-Olivais.1899 Lisboa Codex , Portugal.
Sarıkaya, Chemistry Department,Ankara University.Ankara,Turkey.
Schumpe, Universitaet Hannover,Institut für Technische Chemie , Callinstrasse 3 , D-3000 Hannover, F.R.Germany.
Serpemen,Universitaet Hannover,Institut für Technische Chemie, Callinstrasse 3 , D-3000 Hannover 1 , F.R.Germany.
Sick,Abteilung Chemietechnik ,Lehrstuhl fur Technische Chemie B , Universitaet Dortmund, Postfach 50 05 00, D-4600 Dortmund 50 , F.R.Germany.
T. da Silva,Universidade de Coimbra,Faculdade de Ciencias e Tecnologia,Departmento de Engenharia Quimica, Portugal.
Soares,Universidade do Porto.Faculdade de Engenharia,Laboratorio de Quimica Industrial , Rua dos Bragas,4099 Porto, Codex,Portugal .
Spaninks, Badhuisweg 3 , Amsterdam-N ,Koninkljke/Shell Laboratorium , Holland.
Tiğrel, Petkim Petrochemicals Co. ,Ankara ,Turkey.
N.Upadhyay, University of Illionis at Chicago Circle,College of Engineering,Dept. of Energy Engineering ,P.O.Box.4348, Chicago.Illinois 60680,U.S.A.
Uysal,Chemical Engineering Dept., METU, Ankara ,Turkey.
S.Vaidyanathan.Department of Biophysics,State University of New York at Buffalo,114 Carry Hall,Main Campus,Buffalo New York 14214 U.S.A.
Wauters,Université de Liège,Faculté des Sciences Appliquées Chimie Industrielle.Le Rue A.Stevart,2 , B-4000 Liège , Belgique.
D.Vlaev , Bulgarian Academy of Sciences.Central Laboratory of Chemical Engineering,Geo Milev ,Bl.5, Sofia 1113,Bulgaria.
van der Wal, Koninklijke.Shell -Laboratorium,P.O.Box.3003, 1003 AA Amsterdam,Holland.
Yeniova,Chemical Engineering Dept..University of Alberta, Canada.
Yurttaş, Aegen University ,Chemical Engineering Dept.,Izmir. Turkey.

GPSR Compliance

The European Union's (EU) General Product Safety Regulation (GPSR) is a set of rules that requires consumer products to be safe and our obligations to ensure this.

If you have any concerns about our products, you can contact us on ProductSafety@springernature.com

In case Publisher is established outside the EU, the EU authorized representative is:

Springer Nature Customer Service Center GmbH
Europaplatz 3
69115 Heidelberg, Germany

Batch number: 10823951

Printed by Printforce, the Netherlands

Mass Transfer with Chemical Reaction in Multiphase Systems
Volume II: Three-Phase Systems

NATO ASI Series

Advanced Science Institutes Series

A series presenting the results of activities sponsored by the NATO Science Committee, which aims at the dissemination of advanced scientific and technological knowledge, with a view to strengthening links between scientific communities

The series is published by an international board of publishers in conjunction with NATO Scientific Affairs Division

A	**Life Sciences**	Plenum Publishing Corporation
B	**Physics**	London and New York
C	**Mathematical and Physical Sciences**	D. Reidel Publishing Company Dordrecht and Boston
D	**Behavioural and Social Sciences**	Martinus Nijhoff Publishers Boston/The Hague/Dordrecht/Lancaster
E	**Applied Sciences**	
F	**Computer and Systems Sciences**	Springer Verlag Berlin/Heidelberg/New York
G	**Ecological Sciences**	

Series E: Applied Sciences – No. 73

Mass Transfer with Chemical Reaction in Multiphase Systems

Volume II: Three-Phase Systems

edited by

Erdoğan Alper, B.Sc., Ph.D. (Cantab)

Professor of Chemical Engineering
University of Ankara, Besevler, Ankara, Turkey
Anadolu University, Eskişehir, Turkey

1983 **Martinus Nijhoff Publishers**
The Hague / Boston / Lancaster
Published in cooperation with NATO Scientific Affairs Division

Proceedings of the NATO Advanced Study Institute on
Mass Transfer with Chemical Reaction in Multiphase Systems,
Çeşme - İzmir, Turkey, August 10 - 21, 1981

Library of Congress Cataloging in Publication Data

NATO Advanced Study Institute on Mass Transfer with Chemical Reaction in Multiphase Systems (1981 : Çeşme, Turkey)
Mass transfer with chemical reaction in multiphase systems.

(NATO ASI series. Series E, Applied sciences ; no. 72-73)
"Published in cooperation with NATO Scientific Affairs Division."
"Proceedings of the NATO Advanced Study Institute on Mass Transfer with Chemical Reaction in Multiphase Systems, Çeşme--Izmir, Turkey, August 10-21, 1981"--T.p. verso.
Includes bibliographical references.
Contents: v. 1. Two-phase systems -- v. 2. Three-phase systems.
1. Mass transfer--Congresses. 2. Chemical reactions--Congresses. I. Alper, Erdoğan. II. North Atlantic Treaty Organization. Scientific Affairs Division. III. Title. IV. Series: NATO advanced science institutes series. Series E, Applied sciences ; no. 72-73.
TP156.M3N38 1981 660.2'8423 83-13285
ISBN 978-90-247-2874-9 (set)
ISBN 978-90-247-2872-5 (v. 1)
ISBN 978-90-247-2873-2 (v. 2)

ISBN 978-90-247-2872-5 (this volume)
ISBN 978-90-247-2689-9 (series)
ISBN 978-90-247-2874-9 (set)

Distributors for the United States and Canada: Kluwer Boston, Inc., 190 Old Derby Street, Hingham, MA 02043, USA

Distributors for all other countries: Kluwer Academic Publishers Group, Distribution Center, P.O. Box 322, 3300 AH Dordrecht, The Netherlands

NATO ADVANCED STUDY INSTITUTE
on
"MASS TRANSFER WITH CHEMICAL REACTION IN MULTIPHASE SYSTEMS"

DIRECTOR

E. Alper — Department of Chemical Engineering,
Faculty of Sciences,
Ankara University, Beşevler, Ankara, Turkey.

SCIENTIFIC ADVISOR

W.-D. Deckwer — Institut für Technische Chemie,
Universitaet Hannover,
D -3000 Hannover 1, F.R.Germany.

HONORARY SCIENTIFIC ADVISORS

P. V. Danckwerts — Chemical Engineering Department,
University of Cambridge,
Pembroke Street, Cambridge, England.

M. M. Sharma — Department of Chemical Technology,
University of Bombay,
Matunga Road, Bombay, India.

LECTURERS

E. Alper
Department of Chemical Engineering,
Fàculty of Sciences,
Ankara University,Beşevler,Ankara,Turkey.

G.Astarita
Istituto di principi di Ingegneria Chimica,
Piazzale Tecchio,
80125 Napoli.Italia.

J.C.Charpentier
CNRS,Laboratoire des Sciences du Genie Chimique,
1, rue grandville,
54042 Nancy Cedex,France.

W.-D.Deckwer
Institut für Technische Chemie,
Universitaet Hannover.
D -3000 Hannover 1, F.R.Germany.

A.Germain
Université de Liège,
Faculté des Sciences Appliquées,
Chimie Industrielle,
Le Rue A.Stevart,2
B -4000 Liège,Belgique.

S.Hartland
Technisch-Chemisches Laboratorium
ETH - Zentrum
CH -8092 Zurich, Switzerland.

H.Hofmann
Institut für Technische Chemie,
Egerlandstrasse 3,
d -8250 Erlangen,
F.R.Germany.

G.A.L'Homme
Université de Liège,
Laboratoire de Génie Chimique,
Institut de Chimie-Métallurgie,
2, rue A.Stevart,
B -4000 Liège,Belgique.

R.Mann
UMIST,The University of Manchester,
PO Box 88,Manchester M 60 1QD
England.

H.Sawistowski
Imperial College of Science and Technology,
Department of Chemical Engineering,
London SW7 2BY, England.

K. Schügerl
Institut für Technische Chemie
Univesitaet Hannover,
D -3000 Hannover,F.R.Germany.

Y.T. Shah
University of Pittsburgh,
School of Engineering,
Chemical and Petroleum Engineering,
1249 Benedum Hall,Pittsburgh,
P.A. 15261, U.S.A.

PREFACE

The phenomenon of "mass transfer with chemical reaction" takes place whenever one phase is brought into contact with one or more other phases not in chemical equilibrium with it. This phenomenon has industrial, biological and physiological importance. In chemical process engineering, it is encountered in both separation processes and reaction engineering. In some cases, a chemical reaction may deliberately be employed for speeding up the rate of mass transfer and/or for increasing the capacity of the solvent; in other cases the multiphase reaction system is a part of the process with the specific aim of product formation. Finally, in some cases, for instance "distillation with chemical reaction", both objectives are involved. Although the subject is clearly a chemical engineering undertaking, it requires often a good understanding of other subjects, such as chemistry and fluid mechanics etc., leading to publications in diversified areas. On the other hand, the subject has always been a major field and one of the most fruitful for chemical engineers.

It is for these reasons that the editor decided to organise a NATO Advanced Study Institute covering all aspects, with the ultimate aim of an overview of the landscape to identify features that provide orientation. After many discussions with Professors W.-D. Deckwer, P.V. Danckwerts, C. Hanson and M.M. Sharma, it was decided to limit the ASI to (1) gas-liquid, (2) liquid-liquid, and (3) gas-liquid-solid systems. Thus, the only really important area left out was fluid-solid systems, part of which was however dealt with in another NATO Advanced Study Institute on "Analysis of Fluid-Solid Catalytic Systems" under the directorship of Prof. G.F. Froment. The originally planned date for the Institute had to be postponed for one year in order to prevent a clash with another NATO Advanced Study Institute.

This two-volume book consists entirely of the proceedings of the NATO Advanced Study Institute, which was held in Cesme, İzmir, Turkey during August 10-22, 1981. It includes review lectures of the eminent scientists as presented during the Institute. Although every attempt was made by the director/editor, it was not altogether possible to realise absolute uniformity as these reviews were written in a relatively short time by authors who did not have the chance of coming together prior to the meeting. During the Institute, some short original contributions were also presented by interested participants on areas closely related to the invited reviews.

Due to the large amount of material, these Proceedings are divided into two volumes. The first volume includes the general introductory reviews involving the mathematical lay-out, physico-chemical data, reaction kinetics and transport data, gas-liquid and liquid-liquid systems, and biochemical systems. The second volume is devoted entirely to the three-phase system and its application to coal technology and Fischer-Tropsch synthesis.

Special features of this Institute reflected fully in these Proceedings, are the treatments of biological reactions, facilitated transport, reactive distillation, solvent extraction of metals and some related aspects of coal utilisation.

Here, I would very much like to compliment and thank all lecturers not only for their very clear oral and written contributions but also for wholeheartedly supporting the Institute. I feel particularly obliged to make a special acknowledgement to Prof. W. -D. Deckwer, who was involved from the very beginning to the very end, to Professors P.V. Danckwerts and M.M. Sharma who contributed immensely to the scientific organisation, and to Prof. M.M. Sharma who was also very kind in providing material prior to publication.

I would also like to thank all participants for their contributions to the Advanced Study Institute. Indeed, it was their active participation which brought a real sense of satisfaction to the director/editor.

I express, of course above all, my deepest gratitude to the Scientific Affairs Division of NATO and their officers, in particular Dr. M. di Lullo and Mr. M. Sudarskis, who not only almost entirely supported the Institute financially, but also helped a local objective of mine, i.e. promoting scientific affairs of Turkish chemical engineers. I gratefully acknowledge also the financial contributions of the Turkish Scientific and Technological Research Council and the Ankara Office of the British Council.

I would also like to thank my assistants and co-workers at various universities in Turkey for doing many tedious chores, and to thank Mrs. Bilge Göksidan for the drawings.

Last, but by no means least, my efforts in organising this ASI could not have succeeded without the patience and the understanding of my wife, Ayse, and our daughter, Gizem, who have on too many occasions been neglected during the last two years; for their co-operation and inspiration I am particularly grateful.

Ankara, Turkey

ERDOGAN ALPER

TABLE OF CONTENTS: Volume II

LECTURERS VI

PREFACE VII

G.A. L'HOMME
Introduction to Gas-Liquid-Solid Systems 1

A. GERMAIN
Industrial Applications of Three Phase
Catalytic Fixed Bed Reactors 19

H. HOFMANN
Fluiddynamics, Mass Transfer and Chemical Reaction
in Multiphase Catalytic Fixed Bed Reactors 73

M. CRINE and G.A. L'HOMME
Recent Trends in the Modelling of
Catalytic Trickle Bed Reactors 99

B.I. MORSI and J.C. CHARPENTIER
Hydrodynamics and Gas-Liquid Interfacial Parameters
with Organic and Aqueous Liquids in Catalytic and
Non Catalytic Packings in Trickle-Bed Reactors 133

İ. EROĞLU and T. DOĞU
Influence of Hydrodynamic Model Parameters on the
Estimation of Intraparticle and Interphase Transport
Rates in a Trickle Bed Reactor 161

H. HOFMANN
Reaction Engineering Problems in Slurry Reactors 171

E. ALPER and W.-D. DECKWER
Some Aspects of Gas Absorption Mechanism
in Slurry Reactors 199

O.M. KUT, G. GUT, T. BUEHLMANN and A. LUSSY
Modelling of Consecutive Hydrogenation Reactions
with Sorption and Mass Transfer Effects in a
Stirred Tank Slurry System 225

Y. SERPEMEN and W.-D. DECKWER
Influence of Nonuniform Catalyst Distribution on the Performance of the Bubble Column Slurry Reactor 239

D. ELENKOV and S.D. VLAEV
A Rotating Disc Reactor for Reaction Processes in Slurries 257

Y.T. SHAH and J. GOPAL
Slurry Reactors for Coal Technology 267

W.-D. DECKWER
Coal Liquefaction via Indirect Routes 287

Y.T. SHAH and R.S. ALBAL
Chemical Cleaning of Coal - The Oxydesulfurization Process 351

Y.T. SHAH, P.C. SINGH and A. CALIMLI
Direct Coal Liquefaction 365

List of PARTICIPANTS 397

INTRODUCTION TO GAS-LIQUID-SOLID SYSTEMS

Guy A. L'HOMME

Professor of Chemical Engineering at the Université de Liège, 4000 - LIEGE, BELGIUM

1. GAS-LIQUID-SOLID REACTIONS AND REACTORS

Three phase reacting systems are probably the most widely investigated subject in the area of chemical reaction engineering at the present time. Three phase reactors are used on a large scale in the petroleum industry for different hydroprocessing operations (hydrodesulfurization of residual oils, hydrodesulfurization and hydrocracking of gasoils) and also in the chemical industry (numerous hydrogenations and oxidations). Moreover, reactions involving gas, liquid and solid will be frequently used in the emerging fields of synthetic fuels (direct coal liquefaction, Fischer-Tropsch synthesis...) and of biotechnology.

1.1. Three phase reactions

The main advantages of three phase reactions are related to the presence of a liquid and to the operation at moderate temperature.

The presence of a liquid phase is essential to the reaction of low volatility reactants; in the other cases, it allows to save the high temperature energy related to the vaporization of the reactant and to the operation in the vapour phase. It makes possible the optimization of the selectivity by an appropriate choice of the solvent. It allows also some kind of prevention or retardation of catalyst poisoning and it provides a good help for temperature control.

The moderate temperatures associated to gas-liquid systems allow to reach a better selectivity in the valuable product, particularly when this one is heat sensitive. They are also well suited to the use of heat sensitive catalysts and (or) catalyst supports as encountered for instance in enzymatic catalysis and in homogeneous heterogenized catalysis.

There are three types of gas-liquid-solid reactions (1) :

- Reactions where gas, liquid and solid are either reactants or products. (Type I).
- Gas-liquid-solid reactions with the solid acting as a catalyst. (Type II).
- Two reacting phases with the third phase inert. (Type III).
  a. Gas-liquid reaction in packed bed - the solid imparts momentum, better transfer coefficient, and contact.
  b. Gas-solid reaction with the liquid inert - the liquid acts as a heat-transfer medium or an agent for redistributing the concentration of various reacting species at the catalyst surface.
  c. Liquid-solid reaction with inert gas - the gas provides mixing.

Examples of these different reaction types are given in *Tables 1-1, 1-2 and 1-3*. Absorption of carbon dioxide in a suspension of lime and thermal coal liquefaction are examples of Type I reactions. Type II reactions are very important in the petroleum industry : hydroprocessing reactions are characterized by reactions between hydrogen, one or more components of the oil phase (such as sulfur, nitrogen, vanadium, nickel...) and catalyst. In Type III reactions, only two of the three phases take part into the reaction, the third phase being an inert phase. For example, Fisher-Tropsch reactions are strictly gas-solid reactions : in such systems the liquid does not take an active part in the reaction but is either used as a heat transfer medium or as an agent for redistributing the reacting species at the catalyst surface.

TABLE 1-1. EXAMPLES OF GAS-LIQUID-SOLID-REACTION SYSTEMS WHERE ALL THREE PHASES ARE EITHER REACTANTS OR PRODUCTS (TYPE I). (1)

No.	Reaction system
1	Thermal coal liquefaction
2	Production of calcium acid sulfite (sulfur dioxide, water, and limestone) reacting to produce calcium bisulfite and used in the manufacture of sulfite cellulose
3	Flotation and special types of fluidized crystallization processes.
4	Production of acetylene by the reaction between water and calcium carbide - desorption of C_2H_2
5	Production of gas hydrates in desaturation processes - propane and sea water produce a solid phase
6	Melting of gas hydrate or ice crystals - reaction between gas and solid forming a liquid
7	Reaction of phosphides of Ca and Al with water (desorption of phosphine)
8	Manufacture of calcium hypophosphite by the treatment of white phosphorus with a boiling slurry of lime (desorption of phosphine, diphosphine, and hydrogen)
9	Absorption of CO_2 in a suspension of lime
10	Wet oxidation of active carbon - desorption of carbon dioxide
11	Biological and photo-oxidation of suspended organic solids in water purification

TABLE 1-2. EXAMPLES OF GAS-LIQUID-SOLID REACTION SYSTEMS WHERE THE GAS AND LIQUID ARE EITHER REACTANTS OR PRODUCTS AND THE SOLID IS A CATALYST (TYPE II). (1)

No.	Reaction system
1	Oxidation of an aqueous solution of sodium sulfite with copper ions serving as a catalyst
2	Hydrogenation of sesame seed oil with a nickel-on-silica catalyst
3	Hydrogenation of cyclohexane in an aqueous suspension of 30 μm palladium black particles
4	Hydrogenation of α-methyl styrene containing a slurry of palladium black or alumina-supported palladium catalyst
5	Hydrogenation of benzene (dilute solution of cyclohexane in benzene) by 2 percent Pt on alumina
6	Hydrogenation of ethylene by Raney nickel particles in a paraffin oil
7	Oxidation of SO_2 on wetted carbon
8	Hydrogenation of crotonaldehyde over pelleted palladium-on-alumina catalyst
9	Liquid-phase xylene isomerization on : a. H-Mordenite (zeolite) catalyst b. Silica-alumina catalyst c. Dual-function catalyst
10	Oxygen transfer in fermentation
11	Carbon dioxide absorption by an aqueous buffer in the presence of an enzyme (carbonic anhydrase)
12	Absorption of oxygen in immobilized enzyme systems
13	Absorption of oxygen in an aqueous medium containing activated carbon
14	Catalytic coal liquefaction and upgrading of coal liquids
15	Hydrogenation of acetone by Raney nickel catalysts
16	Catalytic hydrocracking of petroleum fractions
17	Hydrogenation of an aqueous solution of glucose to form sorbitol by a solid catalyst consisting of nickel on diatomaceous earth carrier
18	Production of 2-butene-1,4-diol and propargyl alcohol by reaction between acetylene and formaldehyde in aqueous solution over a copper acetyhide catalyst supported on nickel
19	Hydrodenitrogenation of a lube oil distillate
20	Hydrogenation of aromatics in a naphthenic lube oil distillate
21	Absorption of SO_2 in a suspension of magnesium oxide
22	Hydrodesulfurization of petroleum fractions
23	Hydrogenation of 1-octyne and phenylacetylene in C_1 to n-C_4 alcohols and n-C_6 to n-C_8 alkenes by palladium oxide catalysts
24	Organofunctional group hydrogenation
25	Hydrogenation of unsaturated fats using Raney nickel catalysts
26	Catalytic hydrogenation of carboxylic acid to form alcohols a. Reduction of an aqueous solution of adipic acid to produce hexane-1,6-diol b. Reduction of a reaction mixture resulting from cyclohexane oxidation to produce a mixture of hexane-1,6-diol, pentane-1,5-diol, and butane-1,4-diol
27	Conversion of the oxygen-containing products of propylene oxidation on bismuth molybdate catalyst

Table 1-2 (continued) - Examples of gas-liquid-solid-reaction systems where the gas and liquid are either reactants or products and the solid is a catalyst (Type II) (1)

No.	Reaction system
28	Hydrogenation of C_4 hydrocarbons at low temperatures (10-20°C) in the presence of a noble metal catalyst - reaction gives high yield (nearly complete hydrogenation of acetylene) and high selectivity (only a small loss of butadiene by hydrogenation), also a. Selective hydrogenation of butadiene b. Selective hydrogenation of methyl acetylene and propadiene in propylene feedstocks
29	Hydrotreating reactions
30	Denitrogenation of gas oils
31	Catalytic hydrogenation of phenylacetylene and styrene
32	Oxidation of dilute solutions (132 parts per million) of formic acid in water by a CuO.ZnO catalyst
33	Catalytic oxidation of phenol in aqueous solution over copper oxide
34	Hydrodenitrogenation of various compounds and of a catalytically cracked light furnace oil
35	Oxidation of acetic acid by copper chromite catalysts
36	Catalytic isomerization of cyclopropane
37	Reaction of phosphides of Ca and Al with water (desorption of phosphine)
38	Hydrogenation of nitro compounds in the presence of Pt or Pd catalysts (desorption of water)
39	Hydrogenation of carbonyl compounds in the presence of nickel catalyst (desorption of water)
40	Reaction between C_2H_2 and aqueous formaldehyde in the presence of copper-bismuth acetylide catalyst to give butynediol
41	Hydrogenation of aqueous butynediol to butenediol in the presence of Ni-Cu-Mn on silica-based catalyst
42	Conversion of acrylonitrile to acrylamide using copper chromite catalyst
43	Oxidation of SO_2 in water containing $MnSO_4$ as a catalyst
44	Production of acetaldehyde from oxidation of C_2H_4 in a solution of $CuCl_2$ containing $PdCl_2$ as a catalyst
45	Liquid-phase esterification of terephthalic acid with methanol
46	Hydrogenation of methyl linoleate in the presence of a palladium catalyst
47	Oxidation of sodium sulfite with cobaltous sulfate catalyst
48	Hydrogenation of allyl alcohol in the solvents water and ethanol and in the presence of Raney nickel catalyst
49	Hydrogenation of fumaric acid in the solvent ethanol and in the presence of Raney nickel catalyst
50	Hydrogenation of aniline to cyclohexylaniline by nickel catalysts
51	Hydrodesulfurization of narrow-boiling-range fractions of gas oil
52	Oxidations of sulfide ions (hydrogen sulfide) to thiosulfate ions and methyl mercaptan to dimethyl disulfide in the presence of activated carbon
53	Oxidation of aqueous solutions of sodium sulfide in the presence of activated carbon.

TABLE 1-3. EXAMPLES OF GAS-LIQUID-SOLID-REACTION SYSTEMS WHERE ONLY TWO PHASES TAKE ACTIVE PARTS IN THE REACTION. THE THIRD PHASE IS INERT. (TYPE III). (1)

No.	Reaction system
1	Polymerization of ethylene or propylene in cyclohexane
2	Catalytic hydration of olefins. Hydration of light olefins such as ethylene, propylene, and butenes (at high pressure and high water-to-olefin ratio) in the presence of catalyst
3	Hydrogenation of ethylene using a large concentration of Raney nickel catalyst suspended in solution $C_2H_4(g) + H_2(g) \rightarrow C_2H_6(g)$
4	The Fischer-Tropsch process. Reaction of carbon monoxide with hydrogen in the presence of a solid catalyst to produce a mixture of hydrocarbons, alcohols, aldehydes, ketones, and acids depending upon operating conditions and the nature of the catalyst
5	Catalytic oxidation of olefins. Production of epoxides such as ethylene oxide and higher olefin oxides by the oxidation of olefins in the presence of silver oxide on silica-gel carrier - process applicable to other organic oxidation processes
6	Catalytic hydrogenation of diolefins to form mono-olefins and saturates in the presence of a "wash oil" Diolefins $\rightarrow$ Mono-olefins $\rightarrow$ Saturates
7	Isomerization of cyclopropane
8	Hydrogenation of crotonaldehyde
9	Cleaning of sand filters in water-treating plants-gas is inert and provides stirring and mixing
10	Gas-liquid reactions in packed towers - solid is inert, e.g., a. Removal of lean H_2S from a variety of streams b. Removal of lean SO_2 from a variety of streams c. Absorption of lean SO_3 in aqueous H_2SO_4 as well as aromatic substances for sulfonation d. Absorption of nitrous gases NO_x in water and aqueous alkaline solutions e. Absorption of lean $COCl_2$ in aqueous alkaline solution f. Removal of phosphine from C_2H_2 by absorption in aqueous NaOCl or H_2SO_4.

1.2. Three phase operations and reactors

Several different types of operation and of reactor may be used in order to obtain the desired contact between the three phases. They may be grouped into two main categories, according to the state of motion of the solid particles (2) :

a. In the first one, the finely divided solid (a few tens to a few hundreds microns) is suspended in the liquid phase at a concentration of a few gr/l to a few tens gr/l, where it forms a so-called slurry. Momentum may be transferred to the solid particles in different ways :

- in the bubble-column slurry reactors, the particles are suspended by the movement of gas bubbles. In most cases the process is discontinuous with respect to the liquid phase. The operation is usually carried out in columns with high height-to-diameter ratio and it may be used for batchwise conversion of a liquid reactant, or for continuous reaction between gaseous reactants;
- in the stirred-slurry reactors, the particles are suspended by mechanical stirring as well as by the movement of gas bubbles. The operation is usually carried out in tank reactors with low height-to-diameter ratio, and it is widely used for batchwise or continuous conversion of a liquid reactant;
- in the gas-liquid fluidized reactors, the liquid flows upwards through a bed of solid particles which is fluidized by the flowing liquid, while the gaseous phase moves as discrete bubbles through the liquid fluidized bed. Relatively large particles may be employed in this operation which is suitable for the continuous processing of a liquid as well as of gaseous reactants. If the liquid flow rate is sufficiently high, the solid particles will be carried out of the reactor with the liquid, which is interesting if frequent catalyst regeneration is necessary, for instance.

b. In the second category, the solid particles (a few mm) form a fixed bed through which the fluid phases flow cocurrently, or, sometimes, countercurrently. Two main flow patterns are of interest :

- in trickle bed reactors, the liquid flows downwards over the solid surface of the particles in the form of films, rivelets and droplets. The gaseous continuous phase moves in cocurrent - and sometimes in countercurrent - flow;
- in bubble flow reactors, the gaseous phase moves upwards as discrete bubbles through the liquid continuous phase which may be in either co- or countercurrent flow.

Several practical examples using these different types of reactors are given in *Table 2*.

TABLE 2. PRACTICAL EXAMPLES OF THREE-PHASE REACTORS (1)

A. Fixed-bed reactor

1. Trickle-bed reactor

a. Catalytic hydrodesulfurization
b. Catalytic hydrocracking
c. Catalytic hydrotreating
d. Catalytic hydrogenation such as diolefin hydrogenation of various petroleum fractions, hydrogenation of lubricating oils, hydrogenation of nitro-compounds, carbonyl compounds, carboxylic acid, benzene, α-methyl styrene
e. Production of calcium acid sulfite - Jenssen tower operation
f. Synthesis of butynediol
g. Production of sorbitol
h. Oxidation of formic acid in water
i. Oxidation of sulfur dioxide in slurries of activated carbon
j. Hydrogenation of aniline to cyclohexylaniline

2. Cocurrent-upflow reactor

a. Coal liquefaction (SYNTHOIL reactor)
b. The Fischer-Tropsch process
c. Selective hydrogenation of phenylacetylene and styrene
d. Catalytic hydrodesulfurization

3. Segmented-bed reactors

a. Coal liquefaction (Gulf process)
b. Fermentation reactions

B. Gas-liquid-suspended-solid reactors

1. Slurry or fluidized-bed reactors

a. Production of calcium acid sulfite - fluidized-bed reactor
b. Catalytic hydrogenation of carboxylic acid - slurry reactor
c. The Fischer-Tropsch process - slurry reactor
d. Catalytic oxidation of olefins - slurry reactor
e. Catalytic hydration of olefins - slurry reactor
f. Polymerization of ethylene - slurry reactor
g. Cleaning of sand filters in water-treating plants - fluidized-bed reactor
h. Fluidized crystallization process
i. Coal liquefaction (H-COAL process, SRC process)-fluidized-bed reactors
j. Absorption of SO_2 in a suspension of limestone particles - slurry reactor
k. Manufacture of calcium hydrophosphite by treating white phosphorus with a boiling slurry of lime
l. Liquid-phase xylene isomerization - slurry reactor
m. Catalytic hydrogenation of α-methyl styrene
n. Catalytic oxidation of sodium sulfite

2. Agitated-slurry reactor

a. Catalytic hydrogenation of unsaturated fats and fatty oils
b. Reaction between HCl and CH_3OH in the presence of $ZnCl_2$ catalyst
c. Hydrogenation of acetone

One of the most widely-used three-phase reactors is the trickle-bed reactor which is particularly favored by the hydroprocessing industry. On the contrary, slurry systems are prefered in the chemical industry; they are used in direct coal liquefaction processes and in Fischer-Tropsch synthesis.
Tables 3A, 3B and 4 list the advantages and disadvantages of trickle-bed reactors and slurry reactors.

TABLE 3A. ADVANTAGES AND DISADVANTAGES OF TRICKLE-BED REACTORS (1)

Advantages

1. Flow is close to plug flow, allowing high conversion to be achieved in a single reactor.
2. Liquid-to-solid ratio is small, minimizing the homogeneous side reactions if possible.
3. Liquid flows as a film, thus offering very small resistance to the diffusion of gaseous reactant to the catalyst surface.
4. Flooding is not a problem. Pressure drop is lower than in cocurrent-upflow and countercurrent-flow reactors.
5. If temperature rise is significant, it may be controlled by re-cycling the liquid product or by the addition of "quenches" from the side of the reactor. The recycling of liquid would cause the reactor to behave more like a CSTR; hence, recycling will not be possible when high conversions are desired.
6. Can be operated as a partially or completely vapor-phase reactor. A trickle-bed reactor minimizes the energy costs associated with reactant vaporization.
7. Lower pressure drop will allow an essentially uniform partial pressure of reactant across the length of the reactor.
8. In the commercial reactor, uniform distribution of gas and liquid are achieved. The catalyst is uniformly and effectively wetted by the liquid.

Disadvantages

1. Poor radial mixing of heat.
2. At low liquid flow rates, flow maldistributions such as channeling, bypassing and incomplete catalyst wetting may occur. These adversely affect the reactor performance.
3. The catalyst particles cannot be very small. The intraparticle diffusion effects can be significant. The catalyst pore-mouth plugging can cause rapid deactivation.

TABLE 3B. UPFLOW VERSUS DOWNFLOW COCURRENT FIXED-BED REACTORS (1)

---

1. Larger pressure drop in an upflow reactor.
2. Better mixing in an upflow reactor. This may give better heat transfer, but larger axial mixing would give poorer conversion in an upflow reactor.
3. At low flow rates upflow behaves like a bubble column, i.e. gas as a dispersed phase, liquid as a continuous phase. In downflow trickle-bed operation, gas is a continuous phase and liquid flows as a film.
4. High pressure drop in an upflow reactor would cause significant drop in the partial pressure of the reactant across the length of the reactor.
5. Under similar flow conditions, a higher gas-liquid mass-transfer coefficient is obtained in an upflow operation than in a downflow operation.
6. High liquid holdup and liquid-to-solid ratio in an upflow reactor. High liquid holdup will offer more liquid-phase resistance to the mass transfer of the gaseous reactant to the catalyst surface. High liquid-to-solid ratio will give more importance to the role of possible homogeneous reactions.
7. At low liquid flow rates, upflow will provide better distribution of liquid and, thus, in many cases, better performance of the reactor than the downflow reactor under similar operating conditions.
8. If reaction is rapid and highly exothermic, heat transfer between liquid and solid is more effective in an upflow reactor.
9. In an upflow reactor, the catalyst must be kept in place by suitable methods, otherwise the bed will be fluidized. In a downflow reactor, the catalyst is held in place tightly by the flow. This may cause undesired cementation of the soft catalyst particles.
10. In an upflow reactor, the catalyst pores are more likely to fill completely with liquid than in a downflow reactor. The catalyst effectiveness factor is lower when the catalyst pores are completely filled with liquid compared to the case when they are only partially filled with liquid.
11. Better sweeping of the catalyst by liquid in an upflow reactor may sometimes give better aging of the catalyst. If a solid reactant is used (e.g., coal liquefaction) then an upflow would cause less solids plugging problems than the downflow operation.
12. In an upflow reactor, flooding may be a problem.

Table 4. Advantages and Disadvantages of Slurry or Fluidized-Bed Reactors (1)

Advantages

1. High heat capacity providing good temperature control.
2. Potentially high reaction rate per unit volume of reactor if the catalyst is highly active.
3. Ease of heat recovery.
4. Can be easily used as a batch (slurry) reactor or continuous-flow (fluidized-bed) reactor.
5. The catalyst can be easily removed and replaced if it decays rapidly. Steady-state operation can be achieved even in a rapidly decaying system.
6. It allows the use of very fine catalyst particles, which can give an effectiveness factor approaching unity. This is especially important if diffusion limitations cause rapid catalytic deactivation or poorer selectivity.
7. It allows three-phase gas-liquid-solid (reactant) reactions to operate in the presence of a solid catalyst without plugging of the reactor, e.g., the H-COAL process for coal liquefaction.
8. It allows more flexibility for mixing, e.g., agitated slurry reactor.

Disadvantages

1. High degree of axial mixing reduces conversion. High degree of conversion is obtained only by staging several reactors in series.
2. Catalyst separation from the product mixture by filtration may pose problems of plugging the filters. The cost of filtration may be expensive.
3. The high ratio of liquid to solid may allow homogeneous side reactions to become important, if they are possible.
4. High liquid holdup may cause the liquid-phase diffusional resistance to the gaseous reactant to be an important factor affecting the global rate of reaction.

2. APPLIED PHYSICAL CHEMISTRY AND CHEMICAL ENGINEERING PROBLEMS RAISED BY GAS-LIQUID-SOLID REACTIONS AND REACTORS

From the short review on three phase reactions and reactors presented before, it appears that such systems are of great actual and potential interest in industry. Consequently a strong research and development effort is made nowadays in this direction and three phase chemical reaction engineering is becoming an important subject in the scientific and technical litterature.

Three phase systems are of course very complicated ones, so that a perfect knowledge of all processes implied is far to be reached at present.

In order to sketch the various problems to be solved, let us analyse shortly, as an example, the three phase catalytic systems (Type II reactions) using the classical methodology of chemical reaction engineering summarized in *Table 5*.

Microkinetics concerns the chemical and physical kinetics at the scale of the particles (catalyst particles, liquid films and droplets, gas bubbles...) and their coupling in order to get the apparent reaction rate and selectivity equations. Macrokinetics refers to the transport of momentum, mass and heat at the scale of the reactor, and its goal is the establishment of a model of the reactor. Combination of this model with the apparent reaction rate and selectivity equations allows to write the equations of the reactor.

TABLE 5

CLASSICAL METHODOLOGY OF CHEMICAL REACTION ENGINEERING (3)

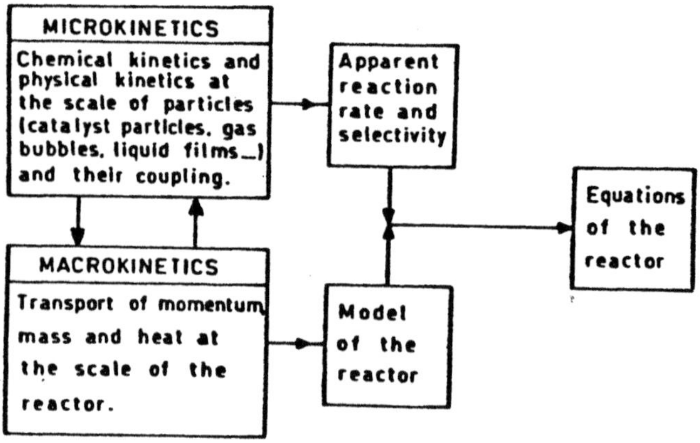

Applying first this methodology to slurry systems, let us start by considering their microkinetic processes *(Table 6)*. These processes are mass transfer of the gaseous reactant A from the gas to the liquid, through the gas-liquid interface; transport of dissolved reactant A and liquid reactant B through the liquid, in the vicinity of the solid particle; transfer of A and B at the external surface of the catalyst, through the film surrounding the solid particle and finally reaction of A and B on the catalyst surface (or sometimes diffusion and reaction of A and B in the pores of the catalyst).
The products undergo similar mass transfer steps. In principle, heat transfer steps are also to be considered, but generally, temperature gradients are negligible at the particle scale due to the small dimensions of the particle and the great heat capacity of the liquid. The coupling of these different physical and chemical kinetic processes will allow the construction of the apparent reaction rate and selectivity equations.

Macrokinetic processes for slurry systems are sketched on *Table 7*. The main points are the characteristics of the three phase dispersion (fluid holdups, interfacial areas, bubbles and catalyst particles size distributions), the state of macromixing of fluids which can be defined through the concept of residence time distribution, the state of micromixing of fluids which for the gas phase shall determine the degree of coalescence of bubbles, the heat transfer between the reactor and the environment.

TABLE 6.

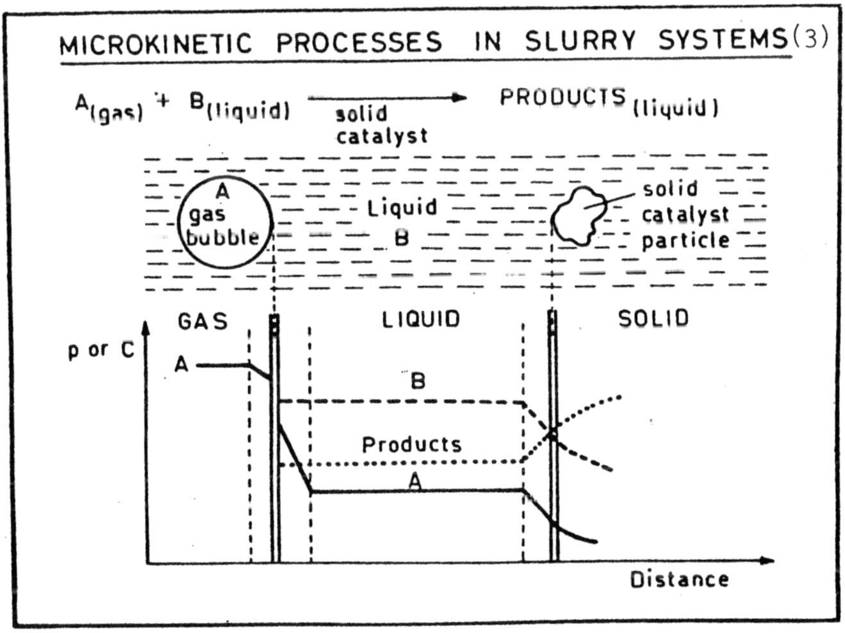

TABLE 7.

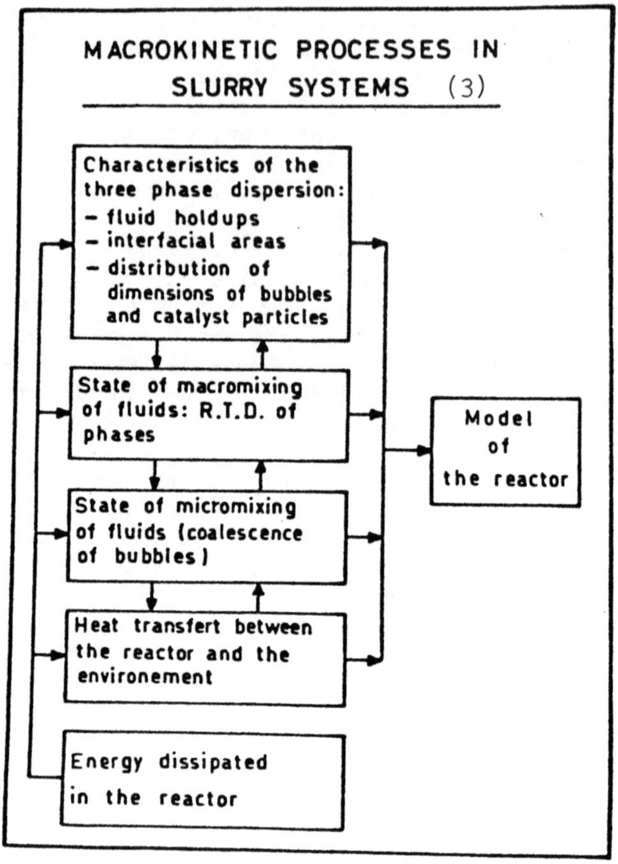

All these processes may be conveniently correlated versus the energy dissipated into the reactor.

Let us now consider the microkinetics of trickle-bed reactors : the main processes are shown on *Table 8* in the case where the liquid reactant and the liquid product are non volatile. The processes are transport of the gaseous reactant A in the vicinity of the liquid, transfer of this reactant through the gas-liquid interface into the liquid, transport of the dissolved gaseous reactant and of the liquid reactant B in the vicinity of the solid, transfer of both reactants at the external surface of the catalyst, through the liquid-solid interface, diffusion and reaction into the pores of the catalyst. The product undergoes similar mass transfer steps from the catalytic active sites to the liquid phase. In the case of a reaction with a heat effect, heat transfer steps associated to the preceeding mass transfer steps have to be considered.

TABLE 8

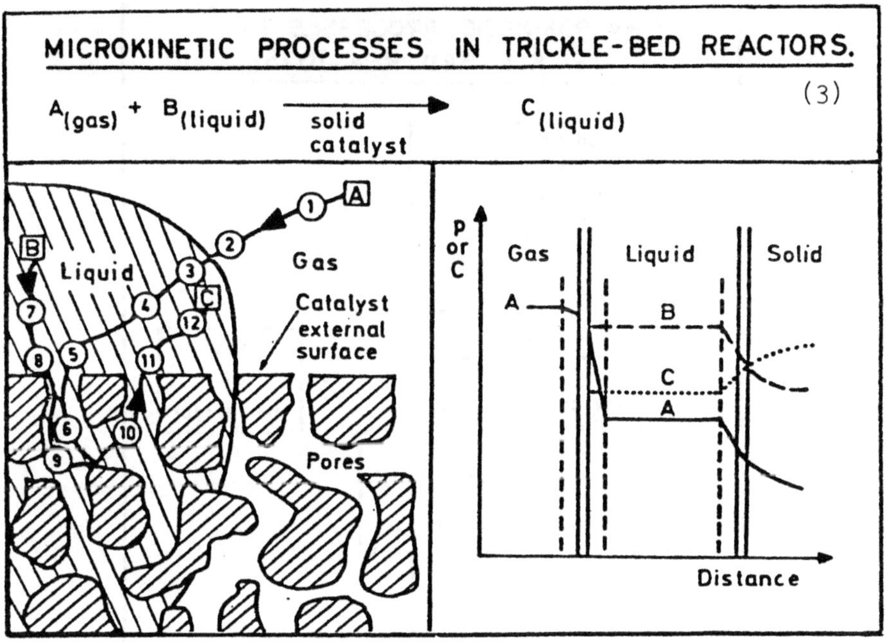

When the liquid reactant and product are non volatile, reaction can proceed only on the active sites in contact with the liquid : the wetting of the catalyst particles is therefore a very important feature of such systems. When the liquid reactant and product are relatively volatile and when the heat effect of the reaction is important, the reaction can also occur on dry fractions of the catalyst, through a classical gas phase process. As the apparent rates of reaction in both phases are generally different, it is an essential but very difficult problem to determine the contribution of the two simultaneous reaction processes : again the knowledge of the catalyst wetting appears to be the key of the problem.

As in slurry systems, the coupling of all the physical and chemical steps mentioned hereabove will allow the construction of the apparent reaction rate and selectivity equations.

Macrokinetic processes for trickle-bed reactors are summarized on *Table 9*. The main features are the hydrodynamics of fluid flows (flow regimes, pressure drops, gas-liquid-solid interfacial areas, radial distribution of fluids), the state of macromixing of fluids and the heat transfer between the reactor and the environment. All these processes may again be correlated versus the energy dissipated into the reactor, and their knowledge

TABLE 9

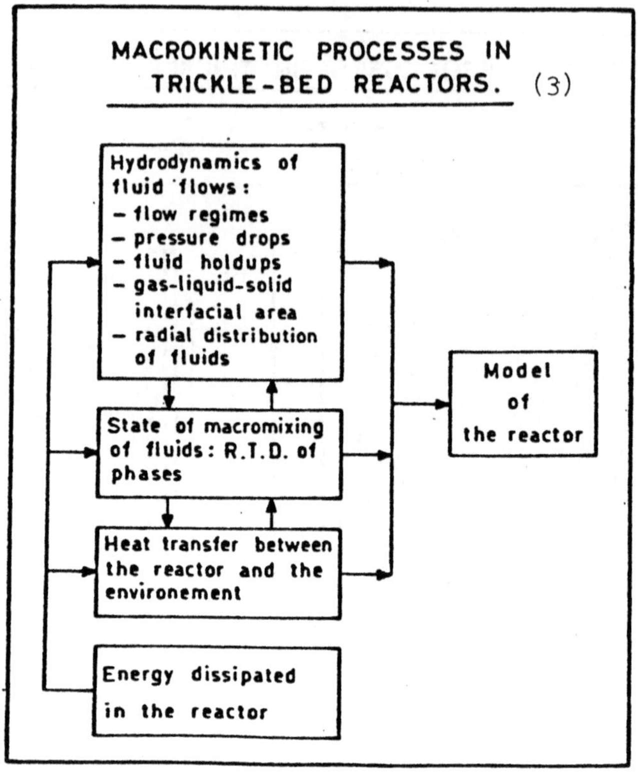

allows the establishment of the reactor model.

As it has already been stated above, three phase catalytic systems are very complicated ones, much more complicated than classical gas-solid catalyst systems, due to the presence of one more phase. Difficulties appear both at the microscopic scale with intricate interfacial phenomena and at the macroscopic scale with complex contacting patterns between the three phases.

Phenomena at the gas-liquid and at the liquid-solid interfaces are governed by properties of the gas, the liquid and the solid such as density, viscosity, surface tension, wettability... These properties are numerous to characterize a three phase system which therefore shall be very difficult to simulate by another one : let us for instance mention here the non validity for organic systems of the physical kinetics correlations determined with aqueous systems. Moreover, the interfacial phenomena and the significant physico-chemical properties to be considered are far being well known : for instance, the foaming ability of organic liquids is not determined univocally by their density, viscosity and surface tension (4).

In order to illustrate now the difficulties at the macroscopic level, let us consider an exothermal three phase reaction occuring in a trickle-bed reactor which processes a volatile liquid phase. Due to the heat of the reaction, a part of the liquid shall be vaporized and the reaction shall also occur - generally at a higher rate - on dry zones of the catalyst surface. The contacting patterns and the wetting of the catalyst shall therefore be strongly connected with the reaction process and the information obtained from "cold" hydrodynamic experiments shall not be very useful : in this case, hydrodynamic and transport parameters must be measured during chemical operation of the reactor (5).

Good chemical engineering is not only scientific *analysis* of processes, but also *synthesis* of knowledge in view of development, design and building : in this respect, three phase systems present a considerable challenge far to be taken up.

REFERENCES

1. Y.T. SHAH, *Gas-liquid-solid reactor design*, Mc. Graw Hill (1979).
2. K. ØSTERGAARD, *Gas-liquid-particle operations in Chemical Reaction Engineering*, Advances in Chemical Engineering, volume 7, 71-137 (1968).
3. G.A. L'HOMME (Edit.), *Chemical engineering of gas-liquid-solid catalyst reactions*. Proceedings of an International Symposium (Liège Belgium, March 1-3, 1978) - CEBEDOC (Liège), 1979.
4. J.C. CHARPENTIER and M. FAVIER, A.I.Ch.E. Jr., 21, 1213-1218 (1975).
5. M. CRINE and P. MARCHOT, Chem. Eng. Commun., 8, 365-371 (1981).

INDUSTRIAL APPLICATIONS OF THREE PHASE CATALYTIC FIXED BED REACTORS

Albert GERMAIN
Chargé de cours associé
Laboratoire de Chimie Industrielle
Université de Liège
4000 LIEGE BELGIUM

1. CHANGES IN CHEMICAL INDUSTRY

Reactions between a gas and a liquid catalyzed by a solid are frequently encountered in chemical processes of great economical significance. Very often, it is technically impossible to operate with just one fluid phase, gas or liquid, in the reactor. The occurence of two fluid phases mainly depends on the temperature range in which the reaction can occur. It is well known that the low temperature limit depends on kinetics and catalysis whereas the highest possible temperature is fixed either by thermodynamics either by the product and the catalyst heat sensitivity. Nevertheless, some reactants, even at rather low temperature, will never been condensed or soluble enough to eliminate the gas phase. On the other hand, other reactants will never been completely vaporized whatever the temperature. Therefore, three phase catalytic processes will always occur when the volatility of two reactants is very different. This broad class of reactions is very important in chemical industry and, as we shall see, will be more frequent in the future.

In other cases, operating with one fluid phase is possible but too expensive. As a matter of fact, the use of three phase systems is often the most economical way to realize a large number of catalytic reactions. That is the reason why increased use of two fluid phase reactors will result from economical as well as from technical reasons.

The use of new raw materials in chemical industry will have a most significant impact on its processes and the development of multiphase reactors. At the present time, chemical industy rediscovers on one hand natural raw materials of animal and vegetable origine and on the other hand, coal and other very heavy feedstocks.

The price of natural raw materials fluctuates on a small time scale but has not increased so much, on an average, as the price of oil. In addition those are renewable and more easily accepted by the market in such fields as food and cosmetics. For some oxygenated organics (fatty acids, esters and alcohols, polyols, etc...), biomass is already very competitive. No doubt that it will increase its market share as a provider of chemical industry. Almost all natural feedstocks are non-volatile heat sensitive molecules which must be processed in the liquid phase at rather low temperature. In addition they are very often over-oxygenated and will need reduction by hydrogen or other small molecules in three phase catalytic processes. Polyols from sugars, starch or cellulose are a striking example of such processes.

In western Europe's petrochemical industry, naphta has been over the last two decades, the predominant feedstock. It has been used in steam crackers to produce olefins (and aromatics), in steam reformers to produce synthesis gas [1] (i.e. methanol and ammonia) and in catalytic reformers to produce aromatics and hydrogen. Unfortunately, oil is mainly used to produce energy and the long term evolution of the fuel consumption pattern will strongly influence the raw material market of the petrochemical industry. The growth of nuclear energy for electricity generation increases significantly the availability of heavy fuels and consequently reduces the supply of lighter products. Consumption of gasoline by an increasing number of private cars leads to the same result.

In addition, very heavy fossil feedstocks , such as coal and shale, are more abundant than light petroleum and they may be available at a lower cost. Reserves are huge and distributed all over the world quite differently from petroleum. It is expected that we will use these products more and more as raw materials for the manufacture of chemicals. Of course, the refinery first will have to accommodate itself to such modifications of the market of its raw material. New hydrocracking and hydrodesulfurization plants will be needed and this is a broad field for trickle-bed reactors. But another way, perhaps the best, to produce chemicals from coal or even from oil residuals would involve gasification and production of synthesis gas. This one can supply the chemical industry with several important light gases, from which almost all organic chemicals could be manufactured. These small molecules will be brought to react with large non volatile molecules (coming perhaps from natural feedstocks) or to react together in processes, such as the Fischer Tropsch synthesis, which very often will involve three phases. It is easy to explain why the use of small molecules in industrial processes will contribute to the development of three phase systems. Small molecules

---

1. Rather recently, in Western Europe, methane from natural gas or from refineries has been preferred to feed steam reformers. Natural gas liquid has not been available to steam crackers as in the United States.

have a poor reactivity and efficient catalysts are needed. As they are scarcely soluble, reaction with a liquid or formation of a liquid product will always lead to the occurrence of two fluid phases in the reactor. In this field, the expected progress of catalysis will contribute to decrease the severity of reaction conditions and bring down the operating temperature to a level where, very often, one reaction or one product will be liquid. These low temperature, even if they are not suitable from the point of view of heat recovery, are necessary to preserve the catalyst itself which is stable and selective only at low temperature, as in the case of "heterogeneized homogeneous catalysts" and immobilized enzymes. The use of ion exchange resins as catalyst or the use of polymeric supports involves the same requirements.

In this context, it is likely that processes using fixed bed reactors will be predominent as they are well suited to continuous operation and large capacity plants which are typical of new investments.

2. THE CHOICE OF A CATALYTIC REACTOR TYPE

The simultaneous occurrence of two fluid phases in a reactor offers advantages but also disadvantages. If a choice between a gas and a gas-liquid catalytic reactor has to be made, several points must be examined. A comparison of these two classes of reactors is briefly summarized in Table 1.

The intrinsic complexity of three phase systems creates some difficulties in the scale-up and in the prediction of performances of three phase reactors. But this complexity is also often a serious advantage, as the simultaneous occurrence of three phases offers such a large number of design possibilities that almost all technical and chemical problems (heat removal, temperature control, selectivity of the catalyst, deactivation, reactants ratio etc...) can be solved by a proper choice of the equipment and of the operating conditions. For example, countercurrent flow of gas and liquid can be used to overcome thermodynamic limitations and solvent effects can be used to improve selectivity and resistance to poisoning of the catalyst.

At the present time, energy aspects are important. The reactor type and the attainable range of operating conditions have a strong influence on the specific energy consumption of a catalytic process. For exothermic reactions, it is essential to recover the heat of reaction at the highest possible temperature in order to minimize exergy dissipation.

This can explain why the gas phase ethylbenzene process seems to have superseded process where the alkylation takes places in liquid phase at low temperature (1). Of course, mechanical energy consumptions are even more important than heat losses. Nevertheless energy saving is a problem which cannot be overstated in chemical industry especially in the synthesis of fine chemicals.

Table 1

Comparison between gas and gas-liquid catalytic reactors

	Gas	Gas-liquid
Design and engineering	Simple	More elaborate
Material	Often, inexpensive material can be used	Corrosion problems can be critical
Catalyst	Possible poisoning by non volatile byproducts	Resistance to corrosion is required
Thermal	Low thermal stability and low heat capacity require internal heat exchange or low conversion	Better stability and higher heat capacity; partial vaporization is possible; better heat exchange coefficient
Reactant recycling	Often important	Stoechiometrical ratio can generally be achieved; hydrodynamics can require gas recycling
Safety	Temperature run-away and ingnition can occur. Gas mixture must lie outside the explosive range	Better stability Operation within the inflammability or explosion limits sometimes possible
Energy aspects		
Dissipated power	High pressure drop	Low pressure drop but sometimes stirring is required
Reactant preheating	Always important	Less important or unnecessary
Heat recovery	Generally at a high level but low heat transfer rate	At a lower level but high heat transfer rate; high efficiency

Table 2

Comparison between slurry and fixed bed three phase catalytic reactors

	Slurry	Fixed bed
Specific reaction rate	High for fast reactions	Rel. high for slow reactions
Catalyst	Highly active	Supported; high crushing strength, good thermal stability and long working life needed
Homogeneous side reactions	Poor selectivy	Good selectivity
Residence time distribution	Perfect mixing	Plug flow
Pressure drop	Low or medium	Low except for small particles
Temperature control	Isothermal operation	Adiabatic operation
Heat recovery	Easy	Less easy
Catalyst handling	Technical difficulties	None
Catalyst consumption	Possible losses	No loss
Maximum volume	50 m^3	300 m^3
Maximum working pressure	100 bar	High pressure possible
Process flexibility	Batch or continuous	Continuous
Investment costs	High	Low
Operating costs	High	Low
Reactor design and extrapolation	Well known	Difficult

It must be remembered that the cost of chemical intermediates is far beyond the cost of their thermodynamic energy content and that the energy needed for their manufacture is much larger than reaction heats. So, it is more important to increase selectivities than to improve heat recovery and very often low reaction temperatures will be preferred.

When a three phase system seems to be the best (or the sole) solution for a specific application, there remains the difficult task of selecting the most suitable reactor type among the numerous possibilities of contacting a gas and a liquid in the presence of a solid catalyst. Several papers have been devoted to this problem (see for example: references 2,3,4 and 5). Fundamental characteristics such as residence time distribution are as important as technological aspects such as tightness of pressure vessels. Main features on which can be based a comparison between the two broad classes of three phase reactors - slurry and fixed bed- have been collected in Tables 2 and 3. Of course, such a general comparison is very rough and each mentioned item has to be discussed for every specific case.

Table 3

Slurry and fixed bed three phase catalytic reactors. Typical properties

	Slurry	Trickle-bed	Flooded bed
Catalyst loading	0,01	0,5	0,5
Liquid hold-up	0,8	0,05-0,25	0,4
Gas hold-up	0,2	0,25-0,45	0,1
Particle diameter	0,1mm	1-5mm	1-5mm
External catalyst area	$500 m^{-1}$	$1000 m^{-1}$	$1000 m^{-1}$
Catalyst effectiveness	1	<1	<1
G/L Interfacial area	$400 m^{-1}$	$200 m^{-1}$	$200 m^{-1}$
Dissipated power	$1000 Wm^{-3}$	$100 Wm^{-3}$	$100 Wm^{-3}$

Selectivity, which is one of the most important characteristics of an industrial process, depends on several parameters : temperature control, residence time distribution, gas and liquid hold-ups, catalyst loading, catalyst type, mass transfer rates etc... If homogeneous side reactions are awkward, fixed beds give better results. But if the desired product can react further on the catalyst, small catalyst particles have to be preferred to avoid concentration gradients in the pores and slurry reactors are the best. In this last case, the poor residence time distribution in slurry reactors is a serious handicap. Of course, the chemical nature of the catalyst has the strongest influence on selectivity. As catalyst consumption is negligible in fixed bed reactors, highly expensive catalysts and rather sophisticated ones can be used at a reasonable cost. For example, Raney nickel generally prevails in hydrogenation slurry reactors whereas more selective precious metal catalysts are use in fixed beds.

Thermal control and heat recovery are easier in slurries. In fixed beds, cooling and partial recycling of the exit stream of the reactor is possible but this affects the residence time distribution. To overcome this difficulty and obtain a very high conversion level, a second reactor in series could be required. Cold injections of gas or liquid in a multibed reactor are another technique but it is thermodynamically very inefficient (Figures 1 and 2).

From the point of view of investment and operating costs, fixed bed reactors offer definite advantages. Equipment is much more simple and requires less maintenance. There are less sealing problems. Materials do not have to sustain erosion by the suspended catalyst particles. No filter nor other equipment to recover the catalyst are required. This is especially important for continuous processes and large capacity plants. We can so explain why trickle-bed reactors have been extensively used in the petroleum refineries whereas, up to now, slurry reactors have been preferred in chemical processing and synthesis. Nevertheless, we can foreseen in the near future an extension of the use of fixed bed reactors in the chemical industry, mostly due to their economical advantages. One of the main hindrance to this development is in fact the limited knowledge of their mechanisms. For this reason, the design of industrial plants and scale-up of pilot plant data are difficult.

Generally, three phase fixed bed reactors are operated with concurrent downflow of gas and liquid in the trickling regime (trickle-bed reactor). Countercurrent operation is less frequent as in this case, the possible ranges of liquid and gas flow rates are very narrow. It is used when thermodynamic equilibrium limits the extent of reaction.

The bubble flow reactor (or flooded bed reactor) does not seem to have gained wide acceptance. In this reactor liquid and gas flow rates are limited by the necessity of avoiding to carry the catalyst away. Moreover it is claimed to be unstable; reactor runaway can be initiated by hot spots which form near the gas distributor, in the case of exothermic reactions. Nevertheless flooded bed reactors have been used in a few processes where heat transfer between the cold liquid feed and the hot liquid of the reactor is realized by direct contacting and so ensures easily the initiation of the catalytic reaction (Figure 3).

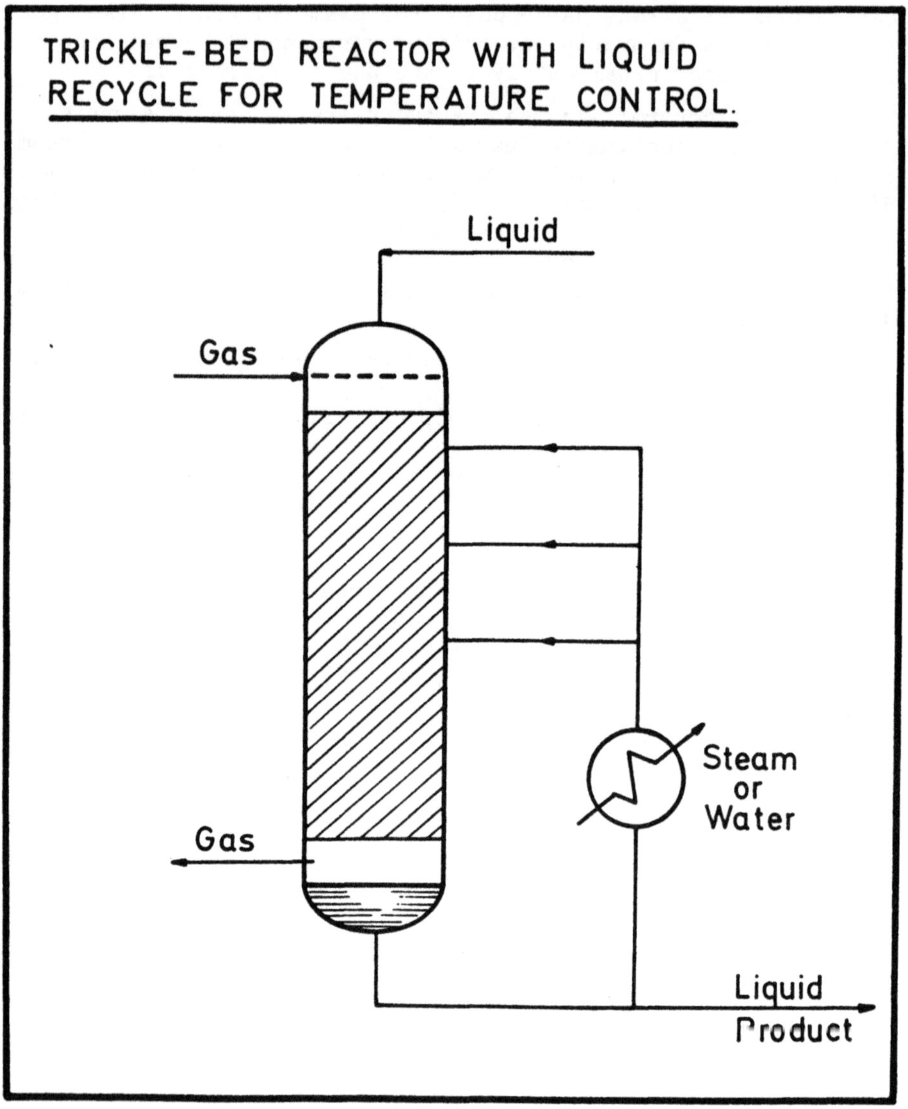
TRICKLE-BED REACTOR WITH LIQUID RECYCLE FOR TEMPERATURE CONTROL.
Liquid
Gas
Steam
or
Water
Gas
Liquid
Product

Figure 1

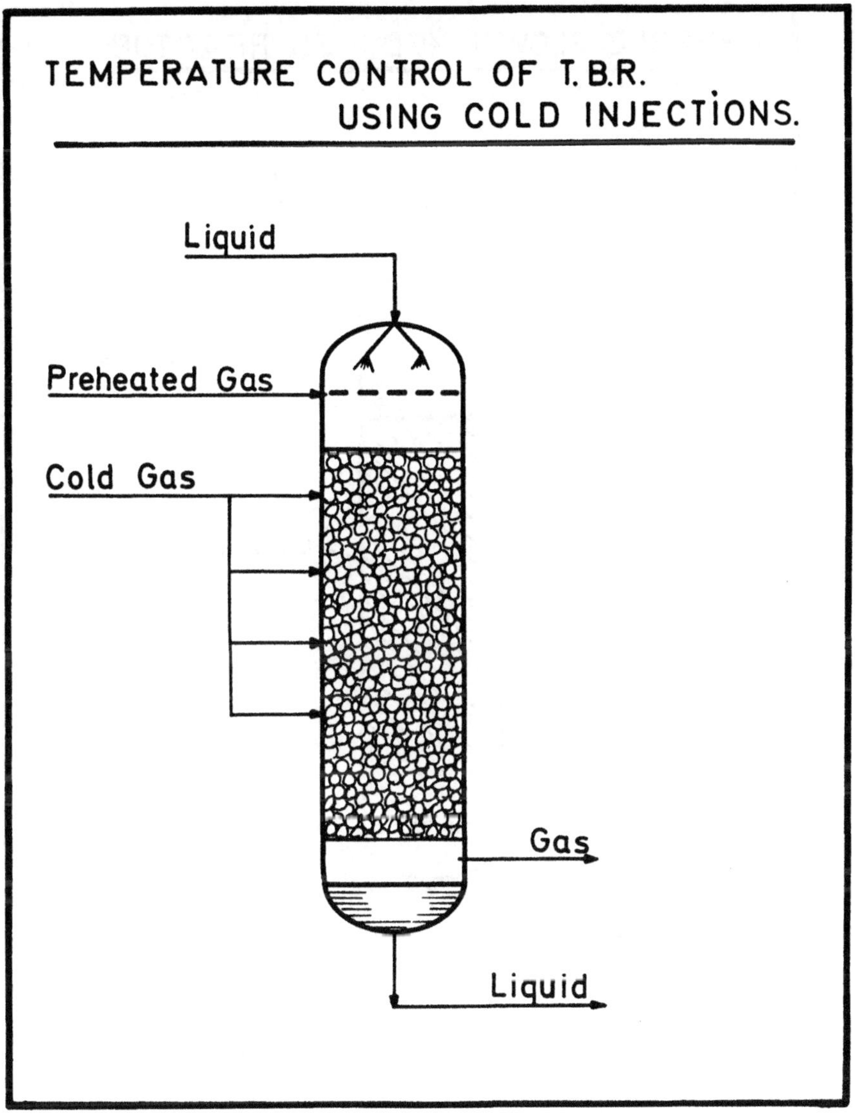
TEMPERATURE CONTROL OF T.B.R.
USING COLD INJECTIONS.
Liquid
Preheated Gas
Cold Gas
Gas
Liquid

Figure 2

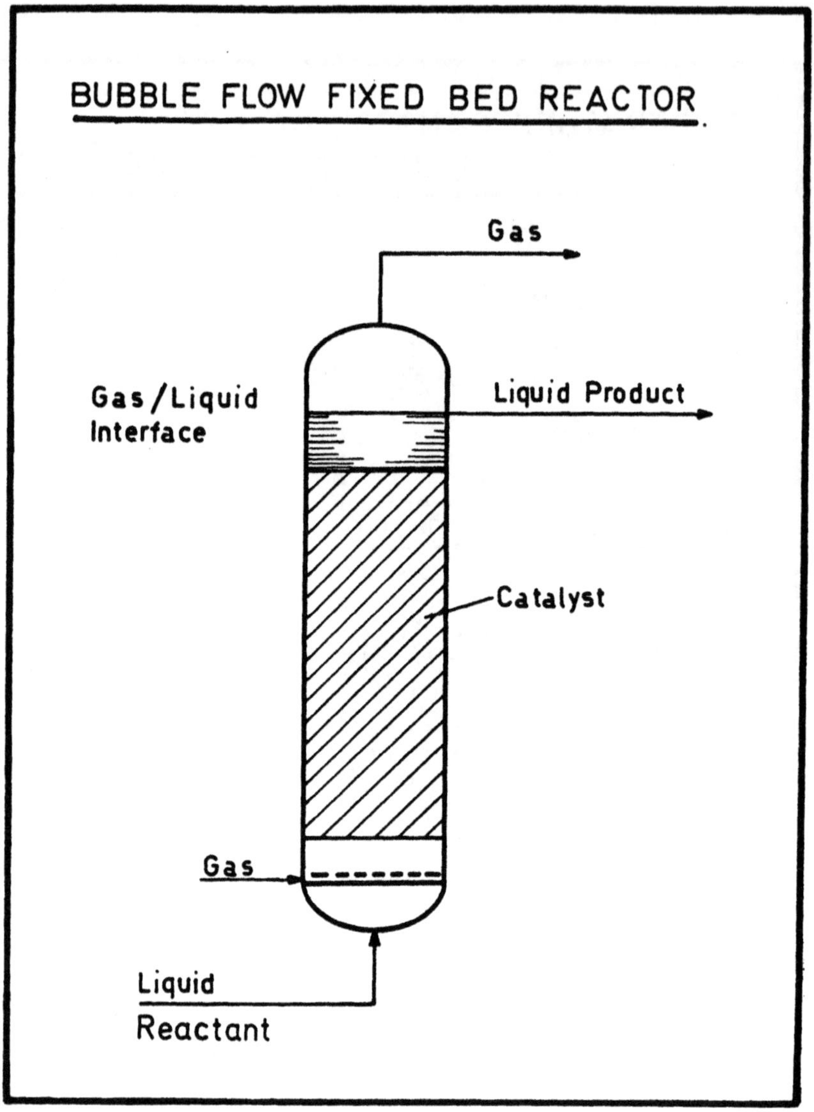
BUBBLE FLOW FIXED BED REACTOR
Gas
Gas/Liquid
Interface
Liquid Product
Catalyst
Gas
Liquid
Reactant

Figure 3

3. FIRST APPLICATIONS OF TRICKLE-BED REACTORS IN CHEMICAL INDUSTRY

The development of trickle-bed reactors in chemical industry seems to have been closely related to the first industrial processes of butadiene synthesis.

Trickle-bed reactor has been used for the first time during the twenties in the four step process for butadiene synthesis (see for example reference 6).

$$2\ C_2H_2 + 2\ H_2O \rightarrow 2\ CH_3CHO \quad (1)$$

$$2\ CH_3CHO \rightarrow CH_3CHOHCH_2CHO \quad (2)$$

$$CH_3CHOHCH_2CHO + H_2 \rightarrow CH_3CHOHCH_2CH_2OH \quad (3)$$

$$CH_3CHOHCH_2CH_2OH \rightarrow CH_2{=}CH{-}CH{=}CH_2 + 2H_2O \quad (4)$$

The third step, the hydrogenation, which needs a high pressure (300 bar) and a large residence time has been realized in such a reactor.

This process became rapidly obsolete and was superseded during the second world war, by the Reppe process, also called the three step process, in which the consumption of acetylene is considerably reduced :

$$C_2H_2 + 2\ CH_2O \rightarrow CH_2OH{-}C \equiv C - CH_2OH \qquad \Delta H = -100\ kJ/mol \quad (5)$$

$$CH_2OH - C \equiv C - CH_2OH + 2\ H_2 \rightarrow CH_2OHCH_2CH_2CH_2OH \qquad \Delta H = -245\ kJ/mol \quad (6)$$

$$CH_2OHCH_2CH_2CH_2OH \rightarrow CH_2{=}CH - CH{=}CH_2 \quad (7)$$

The reaction between acetylene and formaldehyde takes place in aqueous solution at about 100°C and 5 bar. The catalyst is copper acetylide ($Cu_2C_2\ 2\ H_2O\ 2\ C_2H_2$) supported on silica or magnesium silicate. It is prepared in situ. Bismuth is used as an additive to prevent the polymerization of acetylene. In the conditions prevailing in the reactor, acetylene can decompose spontaneously and give rise to an explosion or even to a detonation. To prevent a detonation large volume of gaseous acetylene must be avoided.[2] For this catalytic reaction, trickle-bed reactor offers obvious advantages relevant to safety and loss prevention and it is in fact the sole type of reactor to be used. In such case, the reactor is completely filled up with catalyst so that no free space remains. Design pressure of the reactor shell is 100 bar.

---

2. To prevent any detonation, all the large pipes are filled up with bundles of smaller ones whose diameter cannot exceed 12,5mm.

It can sustain an explosion without damage. Nevertheless it will not hold out in the case of a detonation. The use of a fixed bed enables to contain the dangerous catalyst 3) exclusively in the reactor which is the strongest equipment of the plant. Moreover explosive gaseous mixture can be used safely as feedstock.

The reaction between formaldehyde and acetylene is exothermic (ΔH = -100 kJ/mol) and the control of temperature is critical. To solve this problem it has been chosen to limit the temperature increase by injecting cold acetylen at different levels in the bed (7). This arrangement with the flowsheet of the Reppe's synthesis of butynediol is shown on Figure 4. Propargyl alcohol is formed according to the following equation :

$$CH_2O + C_2H_2 \longrightarrow CH \equiv C - CH_2OH \qquad (8)$$

It is recycled to the reactor where it reacts with formaldehyde and is transformed to butynediol :

$$CH \equiv C - CH_2OH + CH_2O \longrightarrow CH_2OH - C \equiv C - CH_2OH \qquad (9)$$

The second step of this butadiene synthesis, the hydrogenation of butynediol, can also use trickle-bed reactors. This will be described below.

Of course, butadiene is no more obtained commercially by this route. Nevertheless the production of butanediol by the Reppe's synthesis remains important. It is used by BASF, GAF and Dupont.

4. CLASSIFICATION OF THREE PHASE CATALYTIC PROCESSES

Reactions of a light gas with a liquid are almost always exothermic. Though very important in industry, endothermic reactions are exclusively gas phase reactions. It seems more economical to produce, in large scale endothermic processes (steam cracking, catalytic reforming and steam reforming), reactive intermediates from which, by exothermic reactions the wide variety of chemicals are synthetized. So, enthalpy of reaction cannot be used as a criterium to classify three phase industrial processes and, in the design of such plants, heat removal and temperature control are always important problems, sometimes critical ones.

It is easier to classify three phase processes according to the nature of the small molecule 4) involved in the reaction.

3. Dry copper acetylide ($Cu_2C_2H_2O$) is a very sensitive compound which can give rise to explosions.
4. The small molecule does not constitue always the gaseous phase; in hydration reactions, water is mainly present in the liquid phase.

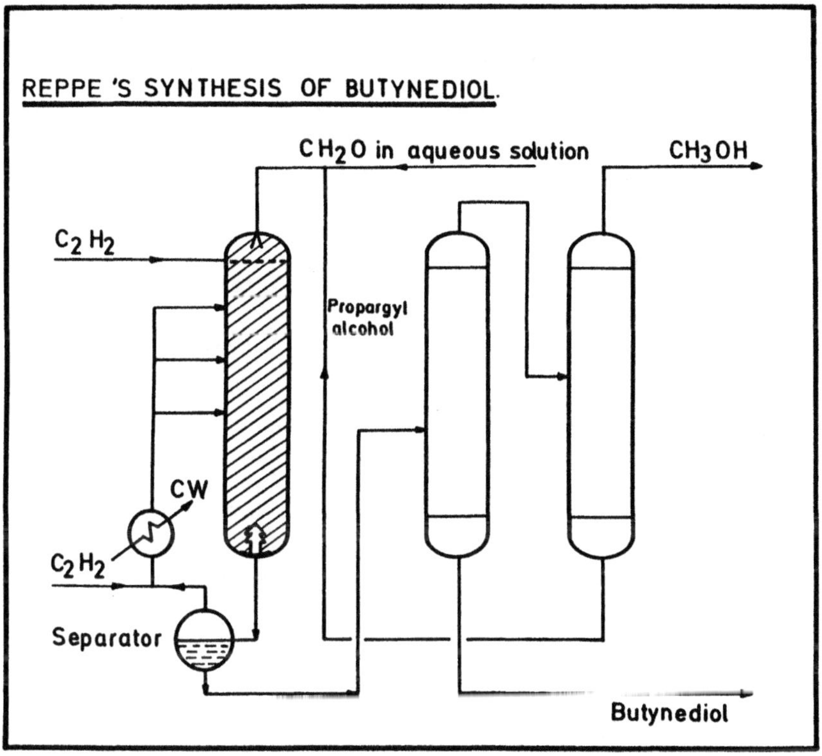
REPPE'S SYNTHESIS OF BUTYNEDIOL.
CH2O in aqueous solution
CH3OH
C2 H2
Propargyl alcohol
CW
C2 H2
Separator
Butynediol

Figure 4

We will distinguish reactions with hydrogen, with water, with ammonia and with oxygen.

We could not find any example of a heterogeneous catalytic reaction involving very reactive molecules such as chlorine or sulfur trioxide; they are non catalytic or relevant to homogeneous catalysis.

5. HYDROGENATIONS AND HYDROGENOLYSIS

Hydrogenations are by far the most important three phase catalytic reactions in chemical industry. At the present time there are strong arguments to prefer fixed beds to slurries in the following cases :

- difficult hydrogenations which require high pressure and large residence time;
- processes using expensive catalysts (platinum metals for example) in plants whose capacity is large;
- continuous processes where conversions closed to 100% are required.

Nevertheless, fixed be reactor is not always the best choice if the heat of reaction, as measured by the adiabatic temperature rise, is too large. The hydrogenation of benzene into cyclohexane is a striking example as we shall see hereafter.

We have collected in Table 4, in alphabetical order, important three phase hydrogenations which use, or could use, fixed bed reactors. The list of commercial processes is by no means complete; very often other gas-liquid processes are available and used in industry; often also pure gas phase processes exist.

Among the processes mentioned in Table 4 several are related to the synthesis of chemical intermediates and monomers for polyamide manufacture. This is shown on Figure 5 which can be useful to understand some routes to nylon 6,6. As a matter of fact, most substances involved are very polar, non volatile and heat sensitive materials which have to be processed in liquid phase. For this reason, these hydrogenations (as well as the reactions with other gases like the aminations) are always three phase ractions.

5.1. Hydrogenation of benzene to cyclohexane

The first reaction of Figure 5 is the hydrogenation of benzene to cyclohexane

$$C_6H_6 + 3\,H_2 \rightarrow C_6H_{12} \qquad (10)$$

TABLE 4 - THREE PHASE HYDROGENATIONS

Feedstock	Product	Enthalpy of reaction	Catalyst	Temperature	Pressure	Selectivity	Licensor or producer
		kJ/mol		°C	bar	%	
Acetone	Méthylisobutyl ketone	- 117	Pd on acidic exchange zeolite or on zirconium phosphate	120-160	20-50	93 95	Deutsche Texaco Tokuyama Soda
Adiponitrile	Hexamethylenediamine	- 314	Co-Cu Fe Ni or Ni-Fe or Ni-Cr	100-135 100-180 75	600-650 300-350 30	90-95 90-95 99	BASF Dupont Rhône-Poulenc Rhodiatoce
Benzoic acid	Cyclohexanecarboxylic acid		Pd on charcoal	170	10-17	c. 100	SNIA Viscosa
Benzene	Cyclohexane	- 214	Raney Ni Noble metal Pt-Li Ni	200 200-300 250	50 30	c. 100 c. 100 c. 100 c. 100	IFP Sinclair-Engelhard UOP BP
2-butyne-1,4-diol	2-butene-1,4-diol		Fe Ni,Fe Pd,Zn				
2-butyne-1,4-diol First step Second step One step process	1,4-butanediol	- 251	Supported Ni Raney Ni Supported Ni Ni-Cu-Mn on silica	80-140 80 120-140 80-160	40 20 300 200	 95	BASF GAF
γ butyrolactone	1,4-butanediol		Ni-Co-ThO_2 on silica	250	100	98	Mitsubishi Chemical
Caprolactone or whydroxycaproic acid or adipic acid	1,6-hexanediol		Raney Cu or Copper-chromite Co-Cu-Mn	250 170-240	280 150-300	C.100	Celanese BASF

TABLE 4 (continued)

Feedstock	Product	Enthalpy of reaction	Catalyst	Temperature	Pressure	Selectivity	Licensor or producer
Cyclododeca-1,3,5-triene	Cyclododecane		Ni	200	10-15		Hüls, Shell
Dinitro-toluene	Diamino-toluene	c. - 1000	Raney Ni or Pd/C	100	>50	98-99	Mobay Chem.
Dimethyl-terephtalate	Cyclohexane-1,4-dicarboxylic acid dimethylester	- 198	Supported Pd	160-180	300-400	96-98	Eastman Kodak
Cyclohexane-1,4-dicarboxylic acid dimethylester	1,4-dimethylol-cyclohexane	- 42	Copper Chromite				Eastman Kodak
Fatty esters or Fatty acids	Fatty alcohol		Copper chromite	200-300	250-300	97-100	Henkel
Furfural	Furfuryl alcohol	- 76	Copper chromite	200	50	94	Quaker Oats, Escher Wyss
Glucose	Sorbitol	- 61	Raney nickel	120-150	30-50	c. 95	Escher Wyss Roquette Frères
Maleic anhydride	Tetrahydro-furan		Ni-Re				Mitsubishi Chemical
Mesityl oxide	Methyl isobutyl ketone or carbinol		Nickel or copper	150-200	3-10		
"Oxo" aldehydes	"oxo" alcohols		Nickel	115	80		
Phenol	Cyclohexanol	- 198	Nickel	150	15		Monsanto and Allied Chemical
Phenol	cyclohexanone	- 64	Palladium	150	10		
Purification of C_3 and C_4 olefinic cuts ($C \equiv C$)	$C = C$	- 165	Pd/Al_2O_3				IFP

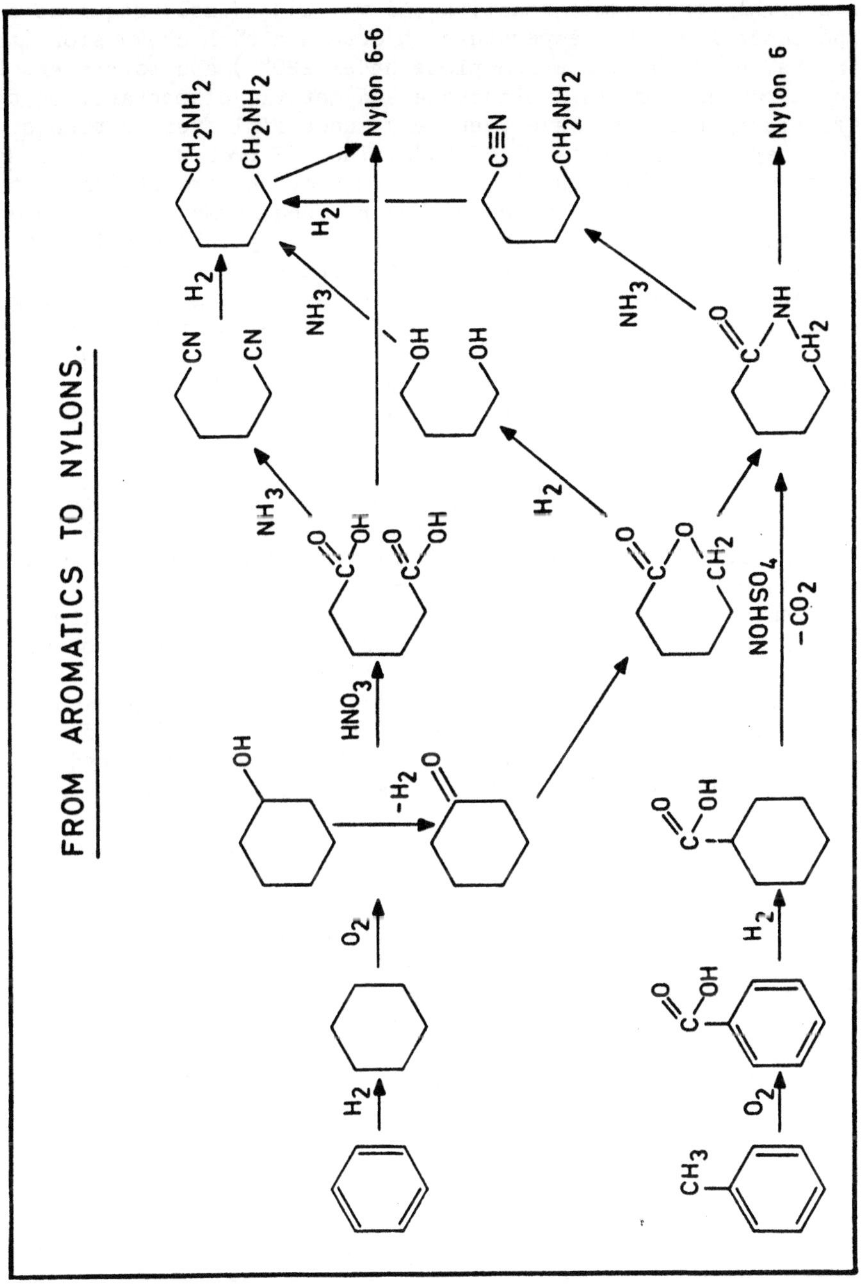
FROM AROMATICS TO NYLONS.
H2
O2
-H2
HNO3
NH3
H2
NH3
H2
Nylon 6-6
NOHSO4
-CO2
NH3
Nylon 6

Figure 5

This highly exothermic reaction (ΔH = - 214 kJ/mol) requires a good control of the temperature to obtain a high conversion (the conversion can be almost complete under 220°C) and to prevent the equilibration between cyclohexane and methylcyclopentane. In this case slurry reactors have been very successful even in plants whose capacity exceeds 100,000 t/a. The IFP process (8) uses Raney nickel catalyst in a bubble column reactor. Recirculation of the liquid in an external exchanger ensures heat removal. To avoid filtration and separation of the catalyst, cyclohexane is withdrawn form the reactor in the gaseous state; this simultaneously decreases the heat of reaction. Due to backmixing effects, high conversions are difficult to obtain at the exit of the bubble column and a finishing fixed bed reactor is needed. Of course such kind of solution to the difficult problem of filtration in large capacity plants is only possible in the case of rather volatile reaction products.

Liquid phase hydrogenation of benzene is also realized in fixed bed reactors, especially when platinum metal catalysts are used (Houdry, Hydrar U.O.P. 5) and Sinclair-Engelhard processes but also with nickel catalyst (Lummus, BP).

Generally several reactors in series with intermediate cooling are necessary to obtain the complete conversion of benzene. Simultaneously, cooled gas and liquid are recirculated at least in the first reactor. The pressure drop and the energy required in recirculating the cooled streams are higher in fixed bed than in slurry processes (8,5 kWh/t in the IFP process compared to at least 28 kWh/t for fixed bed processes). Heat recovery is also better (11).

In the Sinclair-Engelhard process (12), benzene is completely converted in one fixed bed reactor in which an internal heat exchanger is located. To our knowledge, it would be a rare example of a three phase fixed bed reactor with internal heat exchange. As far as no working temperature and pressure have been disclosed, it is impossible to verify if it is really a liquid phase hydrogenation. It is possible that the feed is purely gaseous at the reactor entry and that a liquid phase forms in the reactor as the reaction proceeds due to the hydrogen consumption and the consequent increase in hydrocarbon partial pressure. This could improve heat transfer and explain the complete conversion into cyclohexane in one reactor.

---

5. Processes such as the Hydrar process of U.O.P. are classified as liquid phase processes by Leprince et al. At the conditions prevailing in the reactor (200°C, 30 bar, hydrogen to hydrocarbon molar ratio 2/1, hydrogen diluted by inert gases) gas phase operation is more likely. U.O.P. has also developped a gas phase process based on a nickel catalyst (Hancock, 1975, p. 228).

In the Arosat process, developped by Lummus, a large part of the reaction heat is removed by the evaporation of the cyclohexane, the vapour of which is recondensed with the production of low-pressure steam. It seems that all the operations take place in a single reactor (13).

5.2. The hydrogenation of benzoic acid to cyclohexanecarboxylic acid

In 1960, S NIA Viscosa developped a process to obtain caprolactam form toluene : oxidation of toluene to benzoic acid, hydrogenation to cyclohexanecarboxylic acid and nitrosodecarboxylation to caprolactam (14).

The hydrogenation of benzoic acid is theoretically very similar to the hydrogenation of benzene. Nevertheless nickel catalysts cannot be used as they are active at a temperature at which decarboxylation occurs. For this reason, palladium is preferred.

In the first commercial plant (20,000 t/a), a cascade of stirred tank reactors are operated at 170°C between 10 and 17 bar. The catalyst (palladium supported on charcoal) is separated from the liquid product by centrifugation. Small amounts of catalyst are recovered at the bottom of the cyclohexanoe carboxylic acid distillation column. A plant of 100,000 t/a capacity is no in operation.

In this process, fixed bed reactors could be used advantageously. This would prevent any loss of catalyst and avoid the expensive equipment needed for stirring, centrifuging and filtering. Two reactors in series would be necessary to obtain a complete conversion. In the first one heat can be removed by cooling and recirculating the liquid. In the second one which could approach pistonflow behaviour, the required conversion could be achieved. Of course, in this hydrogenation hot spots in the catalyst bed would be desastrous as they would promote decarboxylation. This could explain why, at least in the 20,000 t/a plant, slurry reactors have been chosen, the more so as the operating pressure is rather low.

5.3. The hydrogenation of adiponitrile to hexamethylenediamine

α, ω-alkylenediamines are almost exclusively obtained via hydrogenation of the corresponding dinitrile. Hydrogenation of adiponitrile is the most important route to hexamethylenediamine.

$$CN-(CH_2)_4-CN + 4H_2 \rightarrow NH_2-(CH_2)_6-NH_2 \qquad (11)$$

Hydrogenation of nitrile to amine is necessarily a liquid phase process; temperatures high enough to completely vaporize the liquid feed cannot be reached due to the heat sensitivity of the product. This hydrogenation is difficult; high pressure of hydrogen and ammonia are needed. Ammonia prevents the formation of undesired secondary amines, imines and subsequent polymeric products which otherwise would be formed according to the following reactions :

$$RC \equiv N + H_2 \rightarrow RCH = NH \quad (12)$$

$$RCH = NH + H_2 \rightarrow RCH_2 - NH_2 \quad (13)$$

$$RCH_2 - NH_2 + RCH = NH + H_2 \rightarrow (RCH_2)_2 = NH + NH_3 \quad (14)$$

$$(RCH_2)_2NH + RCH = NH + H_2 \rightarrow (RCH_2)_3N + NH_3 \quad (15)$$

This hydrogenation is catalyzed by cobalt (supported or not) at temperatures above 100°C under high pressure (600-650 bar). Iron catalysts are used under slightly lower pressure (15). They are claimed to have a better mechanical resistance. Nickel catalysts are very active but their selectivity is bad unless they are used in aqueous caustic soda solution at low pressure and temperature. It is interesting to point out that fixed bed reactors are used in processes working at high pressure conditions (Co and Fe catalysts) while slurry reactors have been chosen for the low pressure processes based on nickel catalysts (16)

A flooded bed process has been patented by ICI (17) Figure 6. With this type of reactor, temperature control is claimed to be easier than with trickle-bed. Turbulence and back-mixing effects are on the other hand more important. The liquid feed, a mixture of adiponitrile and ammonia, can enter the reactor without any preheating as mixing inside the reactor is sufficient to bring them at the reaction temperature. It seems so possible to avoid any recycling of the liquid products for temperature control.

Trickle-bed reactors are used by Du Pont (18) and by BASF (19). A scheme of the BASF process is shown on Figure 7 (20). Adiponitrile and ammonia are mixed with recirculated liquid products before entering the trickle-bed. Hydrogen flows concurrently with the liquid. At the exit of the reactor, the gas and the liquid are separated successively in a hot and in a cold separator. Part of the cooled liquid is recirculated to limit the increase of temperature in the catalytic bed.

Another process to obtain hexamethylenediamine by hydrogenation has been developped by Toyo Rayon. Caprolactam and ammonia are contacted in the gas phase at 350°C to form aminocapronitrile in the presence of a phosphate catalyst. Aminocapronitrile is hydrogenated to give hexamethylenediamine in a process very simimar to that of the hydrogenation of adiponitrile.

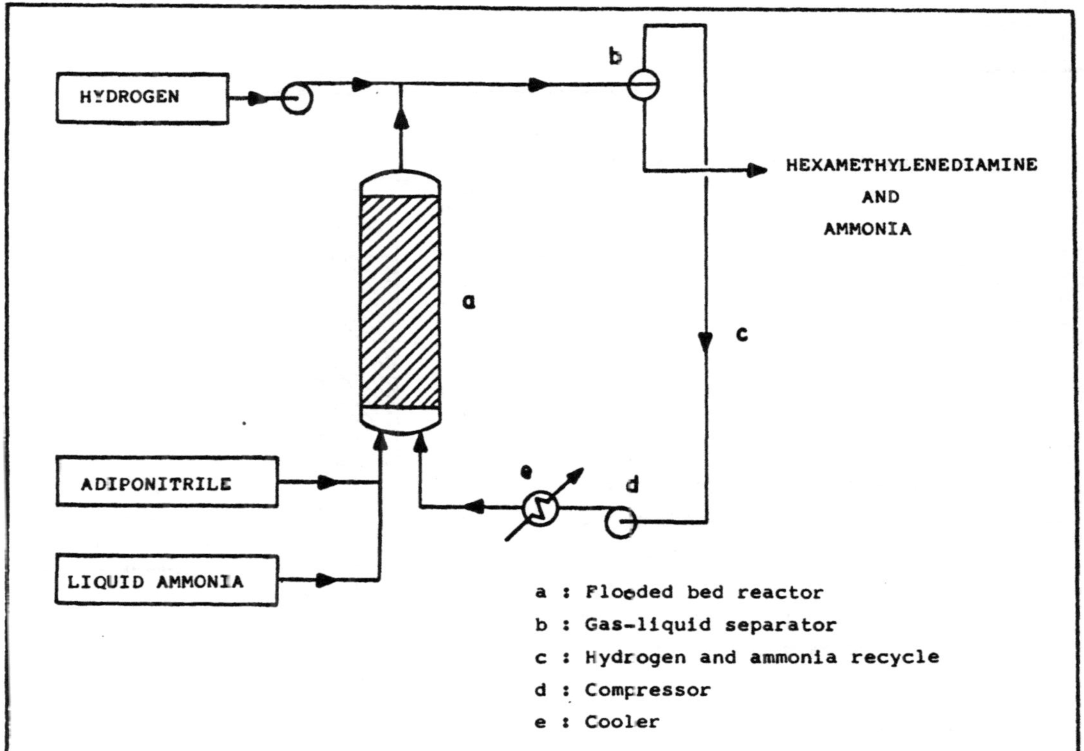

HYDROGENATION OF ADIPONITRILE IN FLOODED BED REACTOR

Figure 6

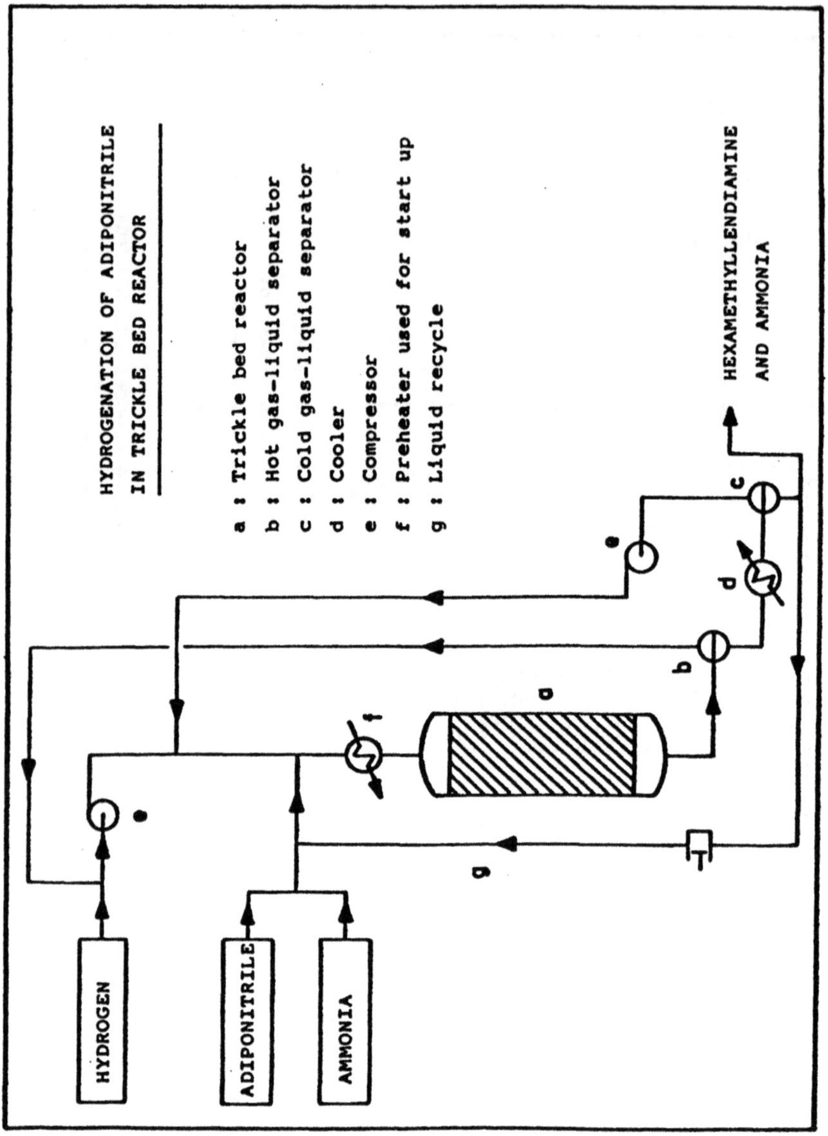
HYDROGENATION OF ADIPONITRILE
IN TRICKLE BED REACTOR
a : Trickle bed reactor
b : Hot gas-liquid separator
c : Cold gas-liquid separator
d : Cooler
e : Compressor
f : Preheater used for start up
g : Liquid recycle
HEXAMETHYLLENDIAMINE
AND AMMONIA
HYDROGEN
ADIPONITRILE
AMMONIA
a
b
c
d
e
f
g

Figure 7

5.4. The hydrogenation of caprolactone to 1,6-hexanediol

Another route to hexamethylenediamine is operated by Celanese. In this process, caprolactone is obtained by the oxidation of cyclohexanone using peracetic acid. Caprolactone is then hydrogenated to 1,6-hexanediol which is transformed to hexamethylenediamine by reaction with ammonia :

$$\underbrace{CH_2(CH_2)_4\,C}_{O}=O + 2H_2 \longrightarrow CH_2OH(CH_2)_4CH_2OH \qquad (16)$$

$$CH_2OH(CH_2)_4CH_2OH + 2NH_3 \longrightarrow CH_2NH_2(CH_2)_4CH_2NH_2 + 2H_2O \qquad (17)$$

The conditions required for the hydrogenation of lactones are always severe and the reaction is often realized in three phase systems. Copper chromite promoted by barium oxide is the preferred catalyst.

Caprolactone is continuously converted into hexanediol in a fixed bed reactor with the catalyst submerged by the liquid (flooded bed reactor). Hydrogen is charged into the bottom of the reactor in the ratio 10/1 to the caprolactone. It bubbles up through the liquid phase catalyst be mixture. The reaction conditions are 250°C and 280 bar. Backmixing, rather important in this type of reactor, and thermodynamic equilibrium impede a complete conversion. After the recovery of hydrogen in flash drums, the hydrogenated product is distilled in a series of columns where pure 1,6-hexanediol is extracted and unreacted caprolactone returned for recycle (21,22).

1,6 hexanediol can also be obtained by hydrogenation of ω-hydroxycaproic acid or adipic acid in the presence of cobalt catalysts. Cobalt promoted by copper and manganese and supported cobalt have been patented. In this case, trickle-bed reactors are used (23).

1,6-hexanediol is not only an intermediate in the synthesis of hexamethylenediamine, it finds also an outlet in the manufacture of polyurethanes.

5.5. The hydrogenation of 2-butyne-1,4-diol to 2-butene-1,4-diol and 1,4-butanediol

After these examples picked up in the main routes to some polyamide monomers, we would like to come back to the synthesis of butanediol.

Butanediol's main application is no more the synthesis of butadiene but the manufacture of polyester where it is polycondensated with terephthalic acid. It is also a starting material for tetrahydrofuran.

Several processes starting from acetylene, butadiene, maleic anhydride or propylene have been recently developped for the production of butanediol. Various possibilities are shown on Figure

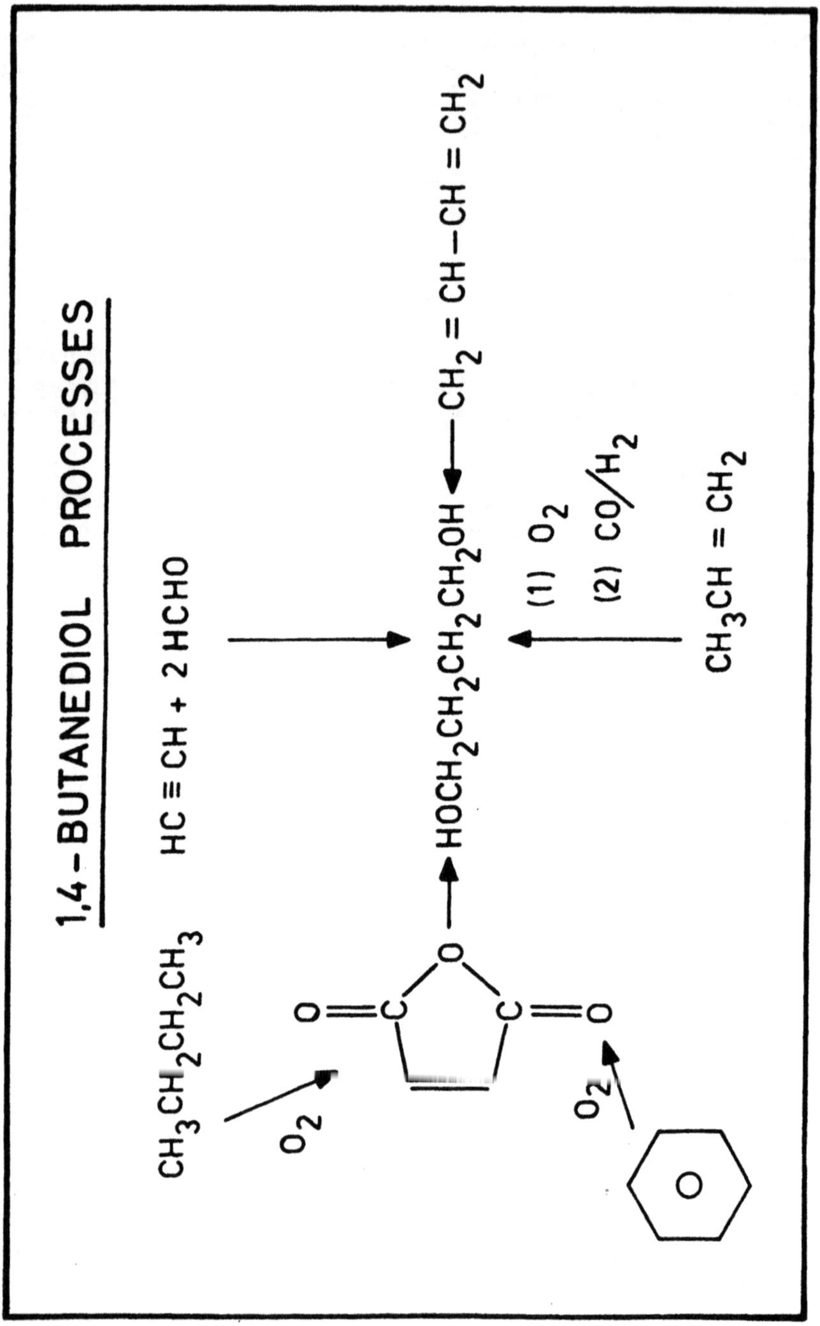
1,4 - BUTANEDIOL PROCESSES
HC ≡ CH + 2 HCHO
CH3CH2CH2CH3
O2
HOCH2CH2CH2CH2OH
CH2 = CH - CH = CH2
(1) O2
(2) CO/H2
CH3CH = CH2
O2
O
C
O

Figure 8

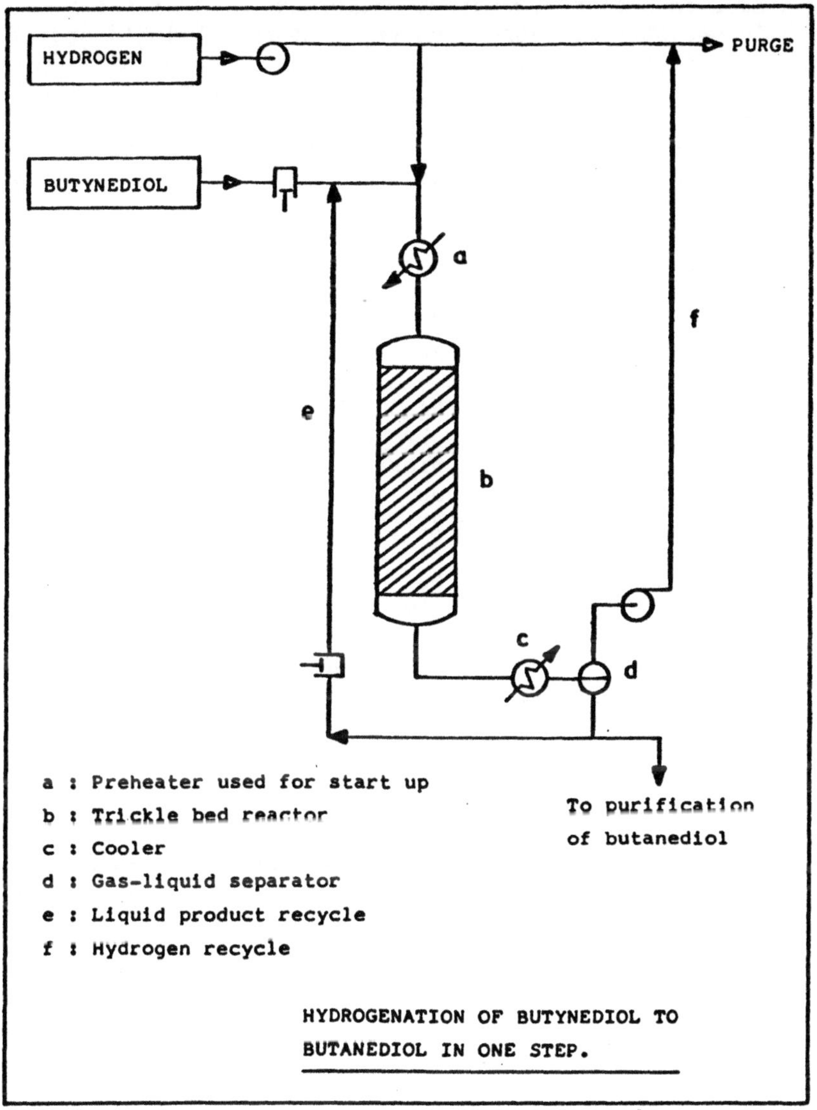
HYDROGEN
PURGE
BUTYNEDIOL
a
f
e
b
c
d
a : Preheater used for start up
b : Trickle bed reactor
c : Cooler
d : Gas-liquid separator
e : Liquid product recycle
f : Hydrogen recycle
To purification
of butanediol
HYDROGENATION OF BUTYNEDIOL TO
BUTANEDIOL IN ONE STEP.

Figure 9

8 (24). Most of them involve three phase catalytic systems.

First of all, the Reppe's synthesis of butanediol involves the hydrogenation of 2-butyne-1,4-diol to 1,4-butanediol. Using Raney nickel in a slurry reactor, butynediol is completely hydrogenated in one step at rather low temperature (70-100°C and 250-300 bar). Supported nickel catalyst with copper and chromium or manganese as promoter are also used. With this catalyst, the hydrogenation can take place in the trickle phase and though (as it is often the case with supported catalyst) higher temperature are necessary, fixed bed operation is increasingly employed (25, 26). Such a process is shown on Figure 9.

A 30-40% aqueous solution of butynediol is mixed with recirculated butanediol and charged into the top of the reactor at about 80°C. The recycle ratio controls the maximum temperature in the reactor, which cannot exceed 160°C. To improve the liquid distribution in the bed, a large hydrogen flow rate is maintained; excess hydrogen is recirculated.

This reaction can also be conducted in two steps. In the first one, butynediol is mainly hydrogenated to butenediol 6) in fixed bet at 40 bar (supported nickel) or in slurry at 20 bar (Raney nickel) (27). In the second step, the hydrogenation of butenediol to butanediol takes place at 120-140°C and 300 bar. For this, trickle-bed reactors are always employed. This is another example which shows that slurries can be competitive at medium pressure but that fixed beds are more suitable when the reaction conditions are more severe.

2-butene-1,4 diol is obtained selectively using special catalysts such as iron, nickel with iron additives or palladium partially poisoned with zinc. This reaction is also conducted in aqueous solution.

Three phase hydrogenations are also involved in other butanediol processes. Their commercial interest is less important at the present time.

Toyo Soda (28) operates plants where butanediol is obtained by chloration of butadiene to 1,4-dichloro-2-butene, hydrolysis of this compound to 2-butene-1,4-diol which is, finally, hydrogenated. This hydrogenation takes place in aqueous solution in the presence of a nickel-aluminium catalyst at 100°C and 270 bar.

Starting also from butadiene, Mitsubishi Chemical (29) has developped a process to obtain butanediol by liquid phase acetoxylation to 1,4-diacetoxy-2-butene, followed by hydrogenation and hydrolysis.

Mitsubishi Chemical produces also butanediol (as well as tetrahydrofuran) by the stepwise hydrogenation of maleic anhydride to succinic anhydride, γ butyrolactone and butanediol.

---

6. Such process cannot be used to obtain pure butenediol as the separation of butanediol, butenediol and butynediol is very difficult.

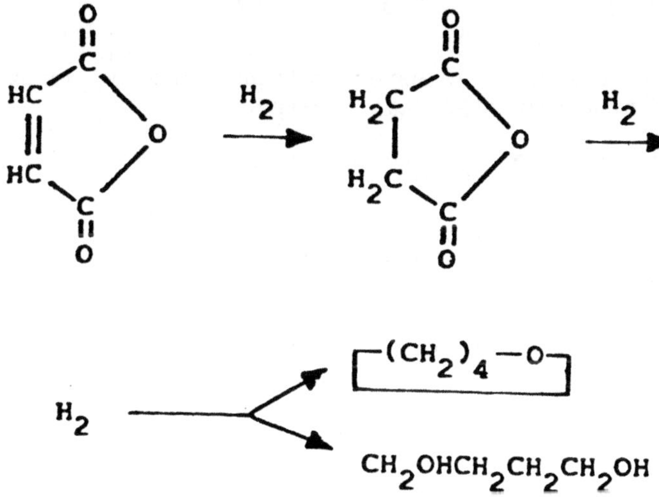

(18)(19)(20)

Tetrahydrofuran is obtained as byproduct of the hydrogenation of succinic anhydride on nickel-rhenium catalyst. The hydrogenation of γ-butyrolactone to 1,4-butanediol is conducted at 250°C and 100 bar the presence of a nickel-cobalt-thorium oxide catalyst supported on silica. Slurry reactors have been used up to now for these hydrogenations, as it is often the case when new processes are extrapolated from small pilot plant or laboratory data, the more so as the actual capacities (a few thousands tons per year)are rather small.

5.6. The hydrogenation of dimethylterephthalate to 1,4-dimethylolcyclohexane

Dimethylterephtalate is the starting material for the production of 1,4-dimethylolcyclohexane which is a component for polyesters, polyurethanes and polycarbonates (23,30).

Eastman Kodak (31) has developped a process in which dimethylterephthalate is first hydrogenated to a mixture of cis-and trans-cyclohexane-1,4-dicarboxylic acid dimethylester.

$$C_6H_4(CO_2\ CH_3)_2 + 3H_2 \rightarrow C_6H_{10}\ (CO_2\ CH_3)_2 \qquad (21)$$

In a second step, this ester is transformed by hydrogenolysis to cis-and trans-dimethylolcyclohexane.

$$C_6H_{10}(CO_2CH_3)_2 + 4H_2 \rightarrow C_6H_{10}(CH_2OH)_2 + 2CH_3OH \qquad (22)$$

This first hydrogenation, the saturation of the aromatic ring, is conducted on a supported palladium catalyst at 160-180°C and 300-400 bar. It is very similar to the hydrogenation of benzoic acid but it requires higher hydrogen pressure and it gives rise to some difficulties due to the low solubility of the feedstock in usual solvents and to its high melting point (140°C). For these reasons, fixed bed reactors, simple and reliable, have been chosen. As the heat of reaction is fairly high, two reactors in series are needed (Figure 10).In the first one, the heat evolved is removed by circulating the hydrogenated ester in which the dimethylterephthalate is fed. To obtain a good thermal stability and, consequently, a good selectivity (96-98%), the concentration of dimethylterephthalate must be low (10%) and rigorously controlled. At temperature above 200°C, several overhydrogenated products are formed such as methylcyclohexanecarboxylic acid methylester, methyl-hydroxymethylcyclohexane and dimethylcyclohexane. The recirculation is so high that the maximum temperature in the reactor is just slightly higher than the temperature at the entrance.

Without any purification, the ester is transformed into dimethylolcyclohexane in the presence of a copper chromite catalyst. Simultaneously cis-trans-isomerisation occurs, the rate of which depends on the reaction temperature and the basicity of the catalyst. The proportion of cis-and trans-isomers is important for the subsequent processing of the diol as it strongly influences the cristallinity and the mechanical properties of the polymer. That is why the choice of reaction conditions as well as the catalyst components are essential. Nevertheless, the ester hydrogenolysis is easy to operate for the reaction heat is rather small. It is realized in a single trickle-bed reactor without any recirculation.

5.7. The hydrogenation of "oxo" aldehydes to alcohols

Hydroformylation or "oxo" synthesis is an important process for the production of aldehydes. In 1977, the world installed capacity amounted to more than 4.1 millions t/a and there were projects to expand this capacity to 6.5 millions t/a.

In this process, an olefine reacts with hydrogen and carbon monoxide to give aldehydes in the presence of a homogeneous catalyst such as tetracarbonylhydridocobalt ($H\ Co(CO)_4$).

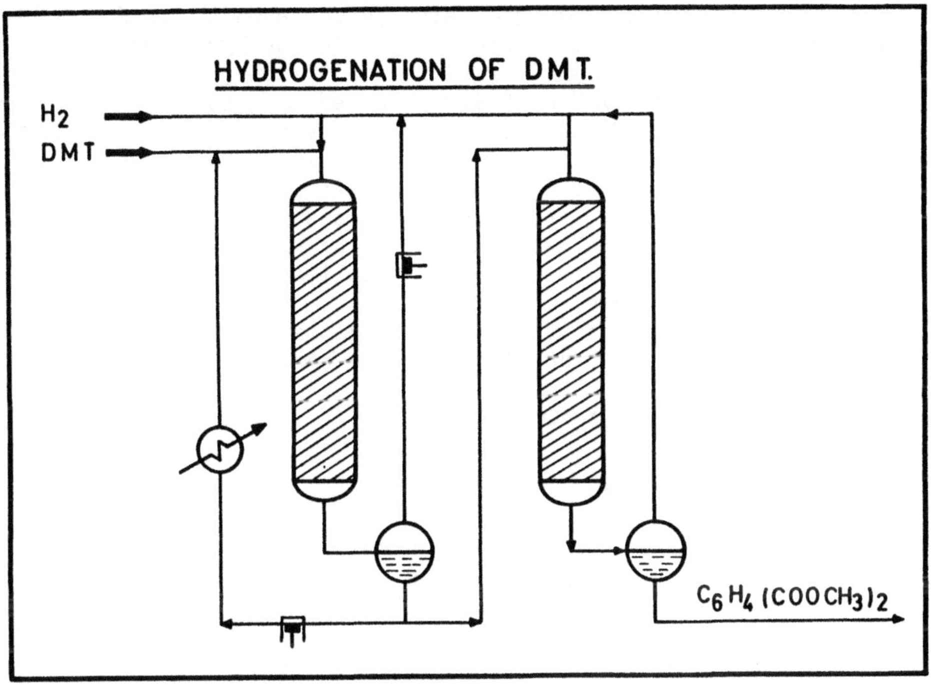

Figure 10

$$R{-}CH{=}CH_2+CO+H_2 \rightarrow R{-}CH_2{-}CH_2{-}CHO \quad (23)$$

$$R{-}CH{=}CH_2+CO+H_2 \rightarrow R{-}\underset{\displaystyle CH_3}{\underset{|}{CH}}{-}CHO \quad (24)$$

Most of the "oxo" aldehydes are hydrogenated to alcohols. This hydrogenation is sometimes favoured inside the hydroformylation reactor (32). More often, it is realized in a subsequent reactor using the purified "oxo" aldhydes or after an aldol condensations of these aldehydes. Nickel or copper are very selective catalysts. They are used in the gas phase or in the liquid phase. Purified aldehydes are treated in the liquid phase at 115°C and 80 bar in the presence of nickel. Under more rigorous conditions (200°C and 280 bar), crude "oxo" aldehydes containing small amounts of ester and acetal can be hydrogenated. This supplies further amounts of alcohol by hydrogenolysis.

The most important starting material is propylene from which n and iso-butyraldehyde and butanols can be obtained (33).

$$CH_3{-}CH{=}CH_2+H_2+CO \rightarrow CH_3CH_2CH_2CHO \quad (25)$$

$$CH_3CH_2CH_2CHO+H_2 \rightarrow CH_3CH_2CH_2CH_2OH \quad (26)$$

2-ethylhexanol is produced by aldol condensation of n-butyraldehyde with subsequent hydrogenation of the aldolisation product 2-ethylhexenal.

$$2CH_3CH_2CH_2CHO \xrightarrow{OH^-} CH_3CH_2CH_2CH{=}\underset{\displaystyle C_2H_5}{\underset{|}{C}}{-}CHO+H_2O \quad (27)$$

$$CH_3CH_2CH_2CH{=}\underset{\displaystyle C_2H_5}{\underset{|}{C}}{-}CHO+2H_2 \rightarrow CH_3CH_2CH_2CH_2\underset{\displaystyle C_2H_5}{\underset{|}{C}}HCH_2OH \quad (28)$$

The hydrogenation can be achieved in a trickle-bed reactor with copper or nickel catalysts. Sometimes, a gas phase hydrogenation at 5 bar and 100-150°C is followed by further hydrogenation in the liquid phase (34).

5.8. The hydrogenation of esters to fatty alcohols

As we mentioned previously, the use of natural raw materials opens a wide field of applications to the three phase hydrogenations.

"Natural" fatty alcohols are primary linear alcohols whose

carbon chain length ranges from six to twenty-four and higher. Alcohols of six through ten carbon atoms are mainly used as components for the manufacture of plasticizer, while those of twelve through eighteen carbon atoms are important raw materials for surfactants.

Fatty alcohols are mainly obtained by the reduction of the carboxylic function of natural triglycerides which are the essential components of animal or vegetal oils and fats 7). They are produced by transesterification of triglycerides with methanol, followed by hydrogenolysis of the esters.

$$\begin{array}{l} CH_2-O-\overset{O}{\overset{\|}{C}}-R \\ \;|\\ CH-O-\overset{O}{\overset{\|}{C}}-R' \\ \;| \\ CH_2-O-\overset{O}{\overset{\|}{C}}-R'' \end{array} + 3CH_3OH \rightarrow \begin{array}{l} R-\overset{O}{\overset{\|}{C}}-O-CH_3 \\ R'-\overset{O}{\overset{\|}{C}}-O-CH_3 \\ R''-\overset{O}{\overset{\|}{C}}-O-CH_3 \end{array} + CH_2OHCHOHCH_2OH \qquad (29)$$

$$R-\overset{O}{\overset{\|}{C}}-O-CH_3 + 2\,H_2 \rightarrow RCH_2OH + CH_3OH \qquad (30)$$

A wide wariety of processes have been developped for both steps (35,36).

The choice of one of them depends on the nature and the quality of the feedstock and also on the capacity of the plant.

Transesterification is an old process frequently described in the technical literature 8) (see for example reference 37).

---

7. Natural alcohols are also obtained commercially by saponification of natural wax esters of the higher alcohols. This production is not very important.

8. It is interesting to note that transesterification can be realized in a packed column which is operated in a way very similar to a trickle-bed, for the needed residence time is rather long. Gaseous methanol is countercurrently contacted with the liquid triglyceride using homogeneous catalysts such as sodium methylate. This rare example of a countercurrent operation with small liquid space velocity, has been preferred, because the conversion is limited by thermodynamic equilibrium.

The hydrogenation of the methyl esters is operated commercially according to three different types of process : slurry, gas phase and trickle-bed. To have a better understanding of the field of applications of tricle-beds, it is useful to describe briefly the other techniques.

In three phase processes (slurry or fixed bed) several feedstocks can be treated such as pure esters, esters containing small amounts of free acids, acids or even directly the triglycerides. Acids require more rigorous reaction conditions and give rise to higher consumptions of catalyst. When the triglycerides are directly hydrogenated, most of the glycerin is lost : it is transformed into overhydrogenated products such as propanediol.

The slurry continuous hydrogenolysis process uses a series of bubble colum reactors to provide a sufficient residence time with minimum backmixing effects and a reasonable column height. Powdered copper chromite is the usual catalyst. It is introduced at the bottom of the first reactor at 1 to 5% of the ester feed rate. At 250-300°C and 250-300 bar, depending on the feedstock, the liquid space velocity is about 1 hr^{-1}. 20 to 40 mol of hydrogen are fed per mol of ester. Conversion and selectivity can reach 97%. Such kind of plant is suitable for capacities which do not exceed 100 t per month.

Feedstocks which can be completely vaporized, can be treated in gas phase processes. At the conditions prevailing in the reactor (250°C-300 bar), several hundreds mol of hydrogen must be fed per mol of ester and, sometimes, methanol is needed to dilute the gas stream and to decrease the ester partial pressure. The energy necessary to drive the recycle compressor is therefore very important. The liquid space velocity is c. 0.35 hr^{-1}. The absence of liquid in the reactor enables to use an unsupported catalyst of high activity. This catalyst is a promoted copper oxide which is very resistant to poisoning. The consumption of catalyst is less than 0.3%. This catalyst is very pyrophoric; it must be reduced inside the reactor. As the amount of catalyst per unit volume of reactor is higher than in slurries, the selectivity is also higher; it can reach 99%. Plants of 12000 t/a capacity have been built.

The trickle-bed reactor combines the advantages of the gas phase and of the slurry processes. As there is no need to vaporize the ester, the hydrogen flow rate is significantly smaller (100 mol H_2/mol ester). Supported catalysts, which have a better mechanical resistance, must be used. Their activity is not so high as the unsupported ones. Side reactions can be catalyzed by the support which must be chosen very carefully. Silica and zeolites of optimum porosity give very good results. Supported catalysts are not pyrophoric. They can be charged safely into the reactor even after reduction. The consumption of catalyst is c.0.3% when the chlorine and sulfur content of the feedstock is less than 10 ppm.

To obtain a good selectivity, the hydrogenation is conducted at rather low temperature (200°C). For this reason, the liquid space velocity is small;it does not exceed 0,2 hr^{-1}. In these conditions, very pure fatty alcohols are obtained; selectivity is almost 100% and the conversion is at least 99.5%. Reactors have been built for plants of at least 25000 t/a capacity.

Using special catalysts (cadmium modified catalysts for example), trickle-bed reactors have also been used to obtain unsaturated fatty alcohols. The New Japan Chemical Company (38) operates a plant of 6000 t/a capacity to produce octadecenol by hydrogenation. This could decrease the consumption of sperm whale oil from which the product has been formerly obtained.

5.9. The hydrogenation and hydrogenolysis of carbohydrates

Carbohydrates are obtained from a wide variety of renewable raw materials and they are interesting feedstocks for the production of several organic chemicals (39). Hydrogenation or hydrogenolysis is the usual way to obtain polyols (40,41).

The possibility to hydrogenate carbohydrates has been discovered in 1925. (42).

At the present time, the hydrogenation of glucose to sorbitol is the most important application of this technology. World installed capacity amounts to 250000 t/a.

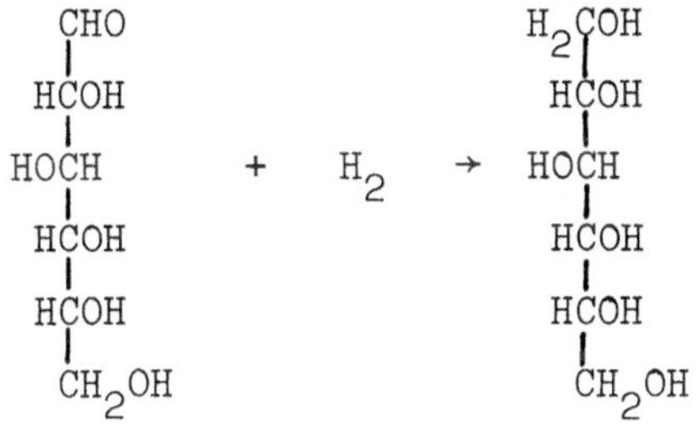

(31)

Nickel is a good catalyst for this reaction and, generally, glucose is hydrogenated in slurry reactors in the presence of Raney nickel in the range from 100 to 200°C and 20 to 200 bar, depending on the process type and the quality of the feedstock.

In small plants, batch operation is preferred. The same stirred tank reactor can be used for different hydrogenations.

Continuous processes use two or more bubble column reactors in series. About 10 mol of hydrogen are fed per mol of glucose. High superficial liquid flow rates are needed to prevent the sedimentation of the catalyst particles. This can give rise to corrosion problems in the pipes, caused by the hard Raney nickel catalyst.

High linear velocities are obtained by increasing the length-diameter ratio of the columns (at least 20/1) and by increasing the residence time. For this reason, the continuous slurry process is conducted at rather high temperature, higher than that used in the batch process where mechanical stirring is provided. The first reactor is operated at 140°C. A higher temperature is possible in the second one because sorbitol is more stable than glucose. The selectivity of the continuous process is smaller than that obtainned in a batch reactor.

Trickle-bed reactors have also been used for this hydrogenation, with supported nickel or copper catalyst. A lower reaction temperature can be applied because long contact time is easily obtained. Nevertheless, the thermal control of the reactor is difficult and in the case of local overheatings, side reactions occur, homogeneous ones as well as reactions catalyzed by the support of the catalyst (isomerisation of glucose and sorbitol). In addition, nickel catalysts are very active for the hydrogenation of carbohydrates, only when the pH of the aqueous solution is slightly basic. In basic solutions, sugars are unstable and dehydrate (caramelisation) which gives rise to the formation of coloured products. This explains why trickle-beds are not often used for hydrogenation of carbohydrates. To be able to use such reactors, we need a catalyst highly active at low temperature and, if possible, in slightly acidic medium. Ruthenium fulfils these requirements and, supported on a suitable material, it could enable the development of the hydrogenation of glucose in trickle-bed reactors. As a matter of fact, ruthenium has been frequently patented for such reactions (43).

It seems that fixed bed is the sole technology which enables the uses of ruthenium because it prevents any loss of this expensive metal. An important advantage of trickle-bed is the small liquid holdup in the reactor which considerably reduces the possibility of homogeneous side reactions. Ruthenium in trickle-bed enables, in some cases, to completely eliminate the purification step of the product leaving the reactor. This is a stricking advantage over slurries of nickel which need a centrifugation, a filtration and an adsorption of coloured materials on activated carbon.

In addition, using ruthenium on an acidic support, it is possible to achieve the hydrolysis and the hydrogenation of starch or other polysaccharides into sorbitol in one sole reactor (44).

In a similar way, xylitol, maltitol and lactitol are obtained from xylose, maltose and lactose. These sugar alcohols are used as low calory or diabetic sweetener. They are claimed to be non-cariogenic.

Under more rigorous conditions, carbohydrate hydrogenolysis enables to produce interesting smaller molecules such as glycerol or ethylene glycol (45,46). As far as the price of natural raw material has not increased so much as the price of petroleum, such processes could become economically interesting in a near future.

5.10. Miscellaneous hydrogenations

Most of the examples described hereabove have been selected in the fiels of nylon monomers, of polyalcohols and of the processing of natural feedstocks.

Numerous other hydrogenations use also three phase catalytic reactions and it is impossible to mention all of them. Nevertheless we would like not to forget the following processes which have a significant economical importance.

On Figure 11 is shown the successive reactions of the synthesis of methylisobutylketone. The use of a palladium loaded cation exchange resin or zeolite enables to realize this synthesis in one step in a trickle-bed reactor (47).

On Figure 12 are shown the reactions of the production of hydrogen peroxide. The reduction of the anthraquinone is a three phase catalytic process. Raney nickel is used in slurry reactors. If traces of nickel remain in the solution at the oxidation step, water is produced instead of hydrogen peroxide. For this reason, the use of palladium, which is a very effective catalyst, in a trickle-bed reactor could improve the yield.

Hardening of fats is also a three phase catalytic hydrogenation. On nickel or palladium, saturation of the double bonds occurs simultaneously with cis-trans-isomerization and double bond migration. As shown by the example of Figure 13, all these reactions influence the melting point of the treated fat. This very delicate hydrogenation is operated batchwise in stirred tank reactors. Trickle-bed operation is very rare.

Finally, we would like to mention the selective hydrogenation of acetylenic compounds in propylene and C_4 olefinic cuts from steam cracking units. A very efficient process, developped by IFP, uses two trickle-bed reactors in series to obtain polymer grade propylene or a feedstock from which butadiene is selectively extracted (Figure 14)(48,49).

6. HYDRATIONS

A large number of hydration reactions are catalyzed by acids Generally, homogeneous catalysis is used and mineral acids are dissolved in the reaction medium. This gives rise to some practical problems : corrosion, disposal of wasted acid etc... Such operations are expensive and cause high consumption of chemical reactants, acids and bases for neutralization. In addition, the reconcentration of the diluted acid is an energy consumming operation. Heterogeneous catalysts are sometimes preferred, at least in large capacity plants.

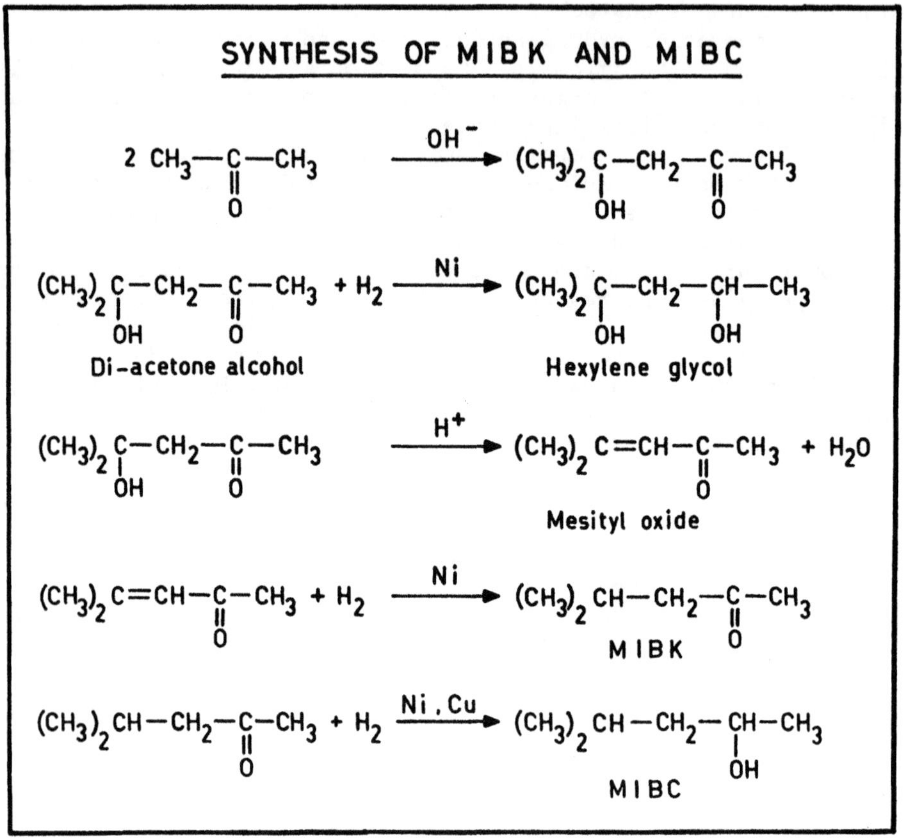
SYNTHESIS OF MIBK AND MIBC
Di-acetone alcohol
Hexylene glycol
Mesityl oxide
MIBK
MIBC

Figure 11

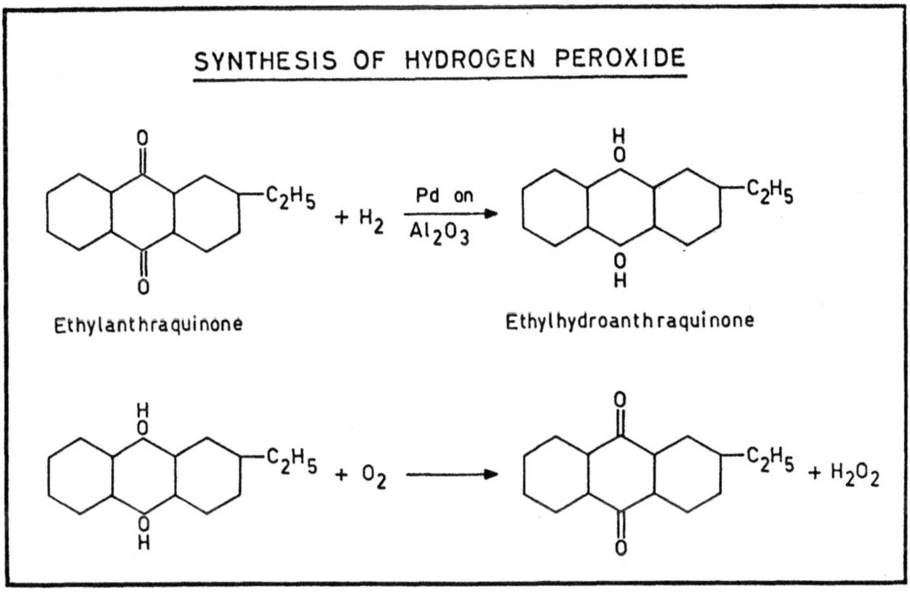
SYNTHESIS OF HYDROGEN PEROXIDE
C2H5
+ H2
Pd on
Al2O3
C2H5
Ethylanthraquinone
Ethylhydroanthraquinone
C2H5
+ O2
C2H5
+ H2O2

Figure 12

HARDENING OF FATS - HYDROGENATION OF FATTY ACIDS AND ESTERS.

	Double bond	T_m (°C) acid	T_m (°C) triglyceride
Linolenic acid	9c, 12c, 15c	−11	−24.2
↓ k_1			
Linoleic acid	9c, 12c	−5	−13.1
↓ k_2			
Oleic acid	9c	16.3	5.5
↓ k_3			
Stearic acid	—	69.6	73

$$S_1 = \frac{k_2}{k_3} \qquad S_2 = \frac{k_1}{k_2}$$

SECONDARY REACTIONS

cis-trans isomerization

Oleic acid (9c) $\xrightarrow{\text{Cat.}}$ Elaidic acid (9t)

T_m = 16.3 °C   45 °C

double bond migration ⟹ conjugated double bond

Figure 13

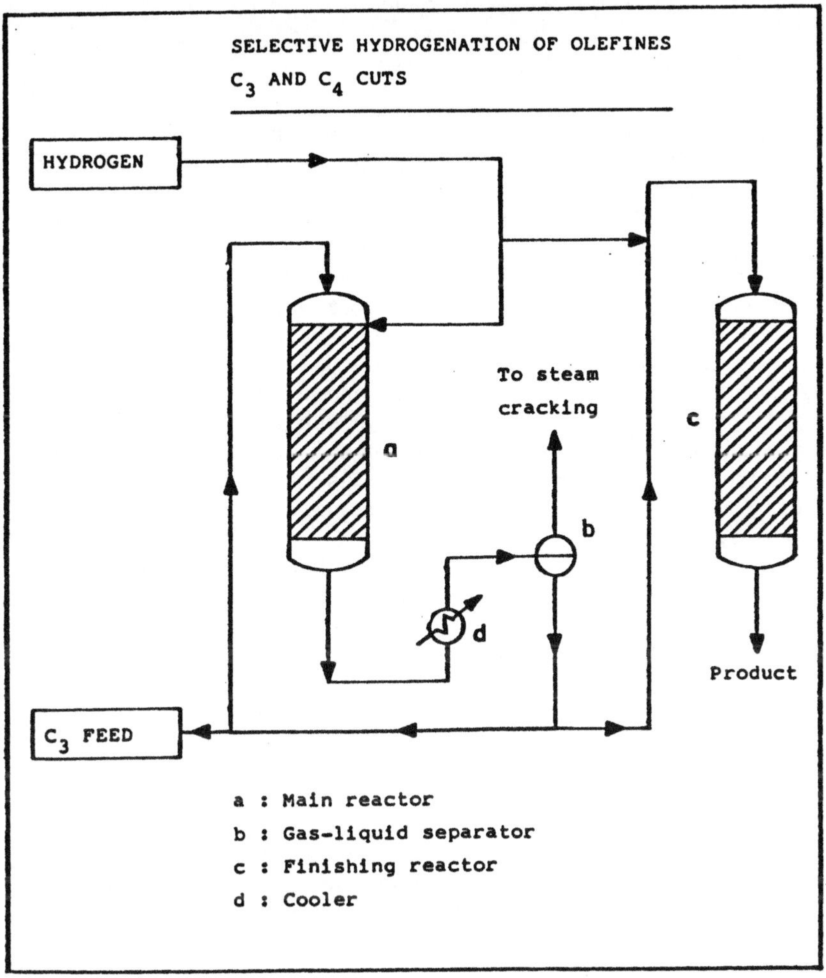
SELECTIVE HYDROGENATION OF OLEFINES
C_3 AND C_4 CUTS
HYDROGEN
To steam cracking
a
b
c
d
Product
C_3 FEED
a : Main reactor
b : Gas-liquid separator
c : Finishing reactor
d : Cooler

Figure 14

6.1. The synthesis of ethanol

The synthesis of ethanol via the hydration of ethylene is conducted on supported phosphoric acid.

$$C_2H_4 + H_2O \rightarrow C_2H_5OH \qquad (32)$$

The fixed bed reactor is operated in the gasphase at 70 bar and 280°C (50). At this temperature the conversion is limited by the thermodynamic equilibrium and a high recycle ratio is needed to ensure a sufficent conversion of the feed. With this catalyst, operation in the presence of a liquid phase is not possible because phosphoric acid would be dissolved by liquid water. But liquid water would also have a beneficial effect : it would dissolve the produced alcohol and so displace the reaction towards the ethanol side which would promote higher per pass conversion. Higher concentration of ethanol in the aqueous phase would be possible. The catalysts which can be used in the presence of an aqueous phase are very limited. Blue oxide of tungsten is effective at rather high temperature and a three phase process using this catalyst has been developped in Germany (51). More details are given by Miller (50, p.725-727).

More recently, acidic cation exchange resins have been claimed to be effective at lower temperature for the hydration of olefine. To our knowledge, they have not been used for the hydration of ethylene up to now. (52).

6.2. The hydration of propylene to isopropanol and the use of ion exchange resins

Ion exchange resins combine advantages of homogeneous catalyst (high activity at low temperature) and the technical advantages of heterogeneous catalysts. They are almost exclusively used in liquid or gas-liquid systems because their thermal stability is to low to operate at temperatures where the feed is completely vaporized. Cation exchange resins can work at c.150°C (some can sustain 200°C for limited period). Anion exchange resins are les resistant and cannot be used at temperatures higher than 60°C.

Deutsche Texaco has developped several processes in which cation exchange resins are used as catalysts : the hydrogenation of acetone methylisobutylketone, the synthesis of methylisopropylketone from methylethylketone and formaldehyde and the hydration of propylene to isopropyl alcohol (53).

In this last process (Figure 15), a mixture of liquid water and gaseous propylene is fed to the top of trickle-bed reactors. Each reactor has four catalyst beds. As the reaction is exothermal (ΔH = -51 kJ/mol), each of the beds has its own quench system. Temperature is indeed a very important parameter, on which depend the efficiency of the catalyst, its resistance to hydrolysis and

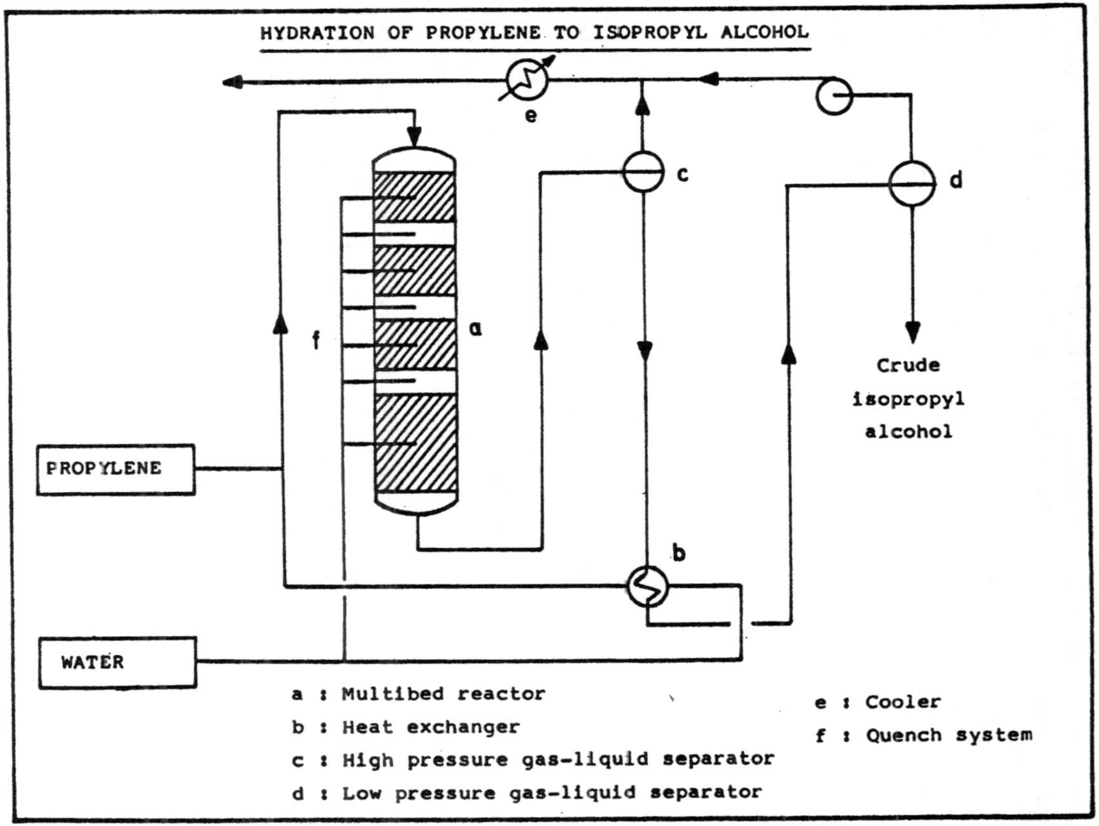
HYDRATION OF PROPYLENE TO ISOPROPYL ALCOHOL
e
c
d
a
f
Crude
isopropyl
alcohol
PROPYLENE
b
WATER
a : Multibed reactor
b : Heat exchanger
c : High pressure gas-liquid separator
d : Low pressure gas-liquid separator
e : Cooler
f : Quench system

Figure 15

the selectivity of the synthesis. Reaction temperature is regulated at 130-150°C. Optimal pressure is in the range 60-100 bar and the molar ratio of water to propylene is between 12.5 - 15.0/1. Conversion reaches 75% of the propylene feed per pass and selectivity is 94%.

It is interesting to note that the selectivity decreases sharply when distribution within the catalyst bed is poor. This is due to the formation of polymers of propylene which, in addtion, cause poisoning of the catalyst.

The use of trickle-beds enables to eliminate largely back-mixing so that high space time yield are obtained (c.1 mol/l.cat.hr)

A plant of 60000 t/a capacty is operated in Germany.

6.3. The hydration if isobutylene to isobutyl alcohol

If large amounts of isobutyl alcohol are available as by-product of the Oxirane process for propylene oxide manufacture, it can also be obtained very easily form the isobutylene contained in C_4 streams of steam cracking units. After the recovery of butadiene, the isobutylene of these streams is selectively hydrated to tertiary butyl alcohol. This process carried out in the liquid phase in the presence of a solid catalyst is certainly realized in fixed bed reactors. Very few details have been disclosed on this reaction which constitutes the first step of a new process to obtain methylmethacrylate from a spent butylene isobutylene feed (54).

A very large market for such chemical as tertiary butyl alcohol may open if lead antiknock additives are reduced in gasoline blends. Alcohols (methanol, ethanol, tertiary butyl-alcohol) and ethers (methyltertbutylether) are among these components which can be used to boost octane number. Their synthesis will involve addition of water or alcohols to olefins. This can be realized in three phase systems.

6.4. The hydration of ethylene oxide to glycol

The hydration of ethylene oxide to ethylene glycol :

$$C_2H_4O + H_2O \rightarrow CH_2OHCH_2OH \qquad (33)$$

is a difficult reaction because of the formation of polyoxyethylene glycol. Industrially it can be accomplished without catalyst (200°C and 20 bar) or with dissolved acid catalysts, such as sulphuric acid, at lower temperature and pressure.

An interesting approach involves the use of ion exchange resins in heterogeneous systems. A trickle-bed operated countercurrently has been proposed for this reaction (55).

7. AMINATIONS

Catalytic reactions involving ammonia are less numerous than hydrogenations.

By reacting ammonia and alcohols in the presence of hydrogenation catalysts, amines are formed. They are also obtained from aldehydes and ketones in nearly the same conditions, if hydrogen is fed simultaneously.

These reactions require more rigorous conditions than hydrogenations. That is the reason why, at the reaction temperature, the occurence of a liquid phase is less frequent.

Nevertheless, reactions between alcohols and ammonia are becoming more important and, rather recently, processes have been patented to produce aniline from phenol (see Figure 16) (56) and for monoethanolamine (see Figure 17) (57).

$$C_6H_5OH + NH_3 \rightarrow C_6H_5NH_2 + H_2O \qquad (34)$$

$$CH_2OHCH_2NH_2 + NH_3 \rightarrow CH_2NH_2CH_2NH_2 + H_2O \qquad (35)$$

These processes, which use catalytic fixed beds, are claimed to be operated in the gas phase, and so they are not relevant to this review 9). But, very similar processes are operated in the trickle phase. For example, fatty amines can be obtained by reacting ammonia and fatty alcohols (58).

$$ROH + NH_3 \rightarrow RNH_2 + H_2O \qquad (36)$$

It is also possible to obtain hexamethylenediamine by the amination of hexanediol. Reaction takes place at 200°C and 230 bar in the presence of a nickel catalyst.

$$CH_2OH(CH_2)_4\ CH_2OH + 2NH_3 \rightarrow CH_2NH_2(CH_2)_4CH_2NH_2 + 2H_2O \qquad (37)$$

Some by-products such as hexamehtyleneimine and 1,6-aminohexanol are formed. The yield approaches 90%. Celanese produces, by this technology, 30000 t/a of hexamethylenediamine (cf. Figure 5 and § 5.4.).

---

9. The reactor of the Leonard's process for ethylene diamine synthesis is operated at low temperature (< 300°C) and high pressure, probably more than 200 bar. It is so possible than a liquid phase occurs inside the catalytic bed.

ANILINE EX-PHENOL PROCESS

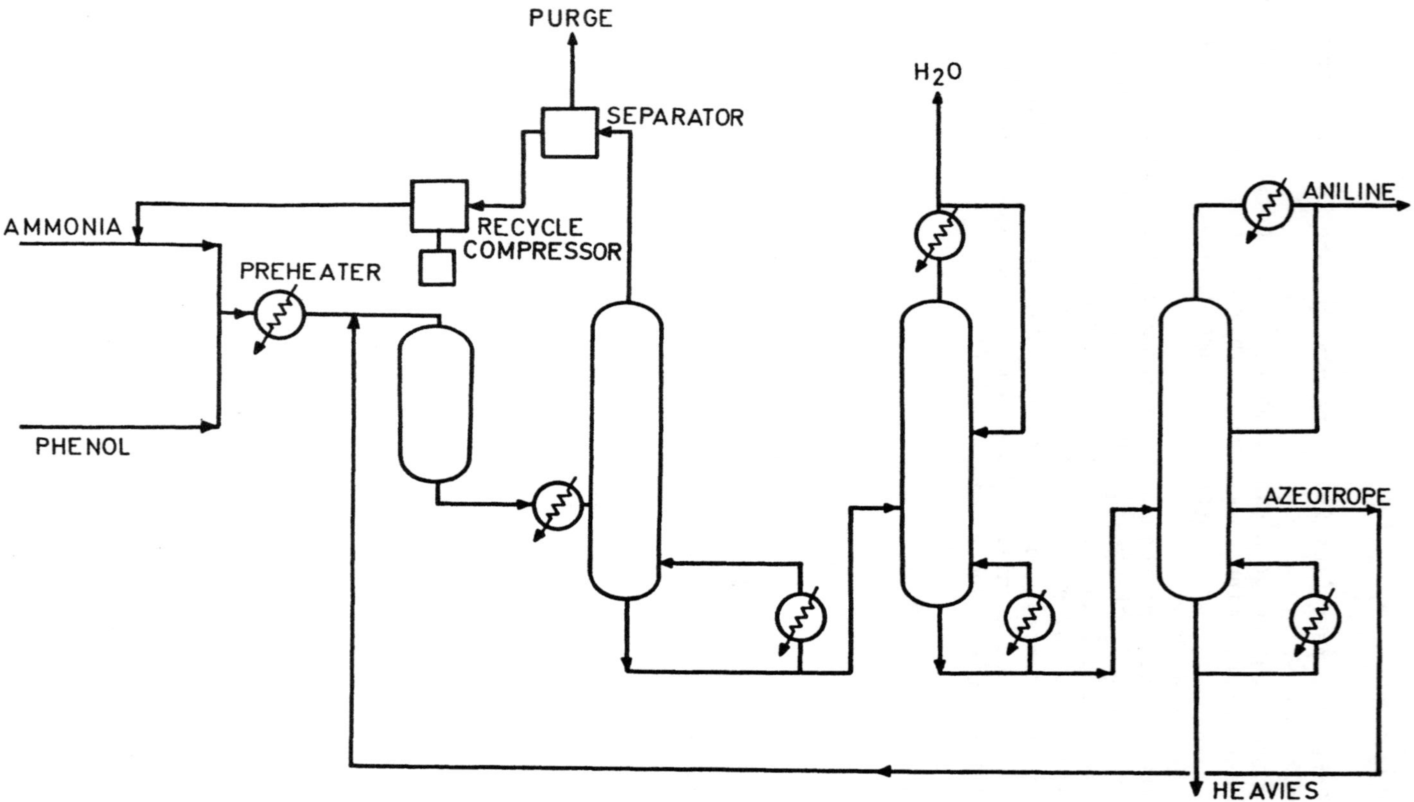

Figure 16

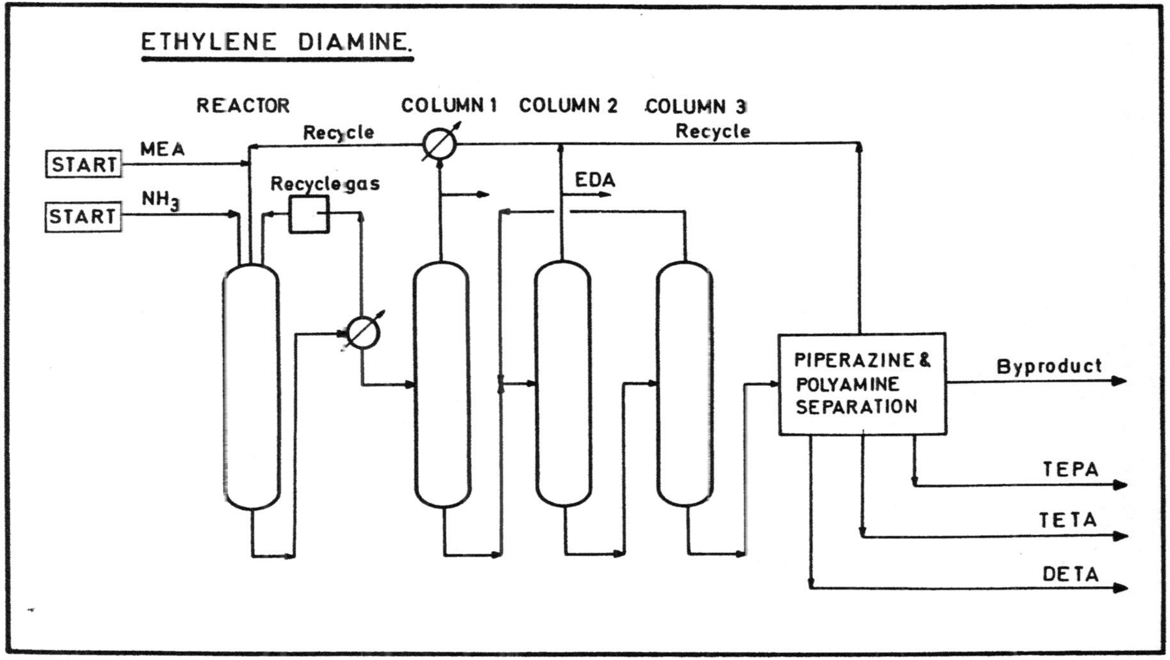
ETHYLENE DIAMINE.
REACTOR
COLUMN 1
COLUMN 2
COLUMN 3
Recycle
Recycle
START
MEA
START
NH_3
Recycle gas
EDA
PIPERAZINE & POLYAMINE SEPARATION
Byproduct
TEPA
TETA
DETA

Figure 17

8. OXIDATIONS WITH MOLECULAR OXYGEN

Three phase catalytic oxydations involving molecular oxygen are not frequent in chemical industry. In this field, most gas-liquid processes use dissolved salts as homogeneous catalyst. It is very difficult indeed to find solid catalysts which can resist to corrosion by the liquid medium in which oxidations take place.

Nevertheless a few patents can be found which claim the use of three phase reactors for the synthesis of important intermediates by oxidation. The oxidation of ethylene into ethylene oxide, for example, can be conducted in a three phase system (59).

Several laboratory studies have also been realized in gas-liquid-solid reactors. They are concerned with the oxidation of pure liquids (cyclohexane, ethylbenzene, tetralin, cumene, etc...) or of diluted aqueous solutions. In this fiels, the oxidation of wastewater in trickle-bed reactor has been proposed as an alternative to biological purification.

A review of all possible applications of trickle-bed reactors in the oxidation field, has been published by Goto et al.(60).To our knowledge, there is no commercial use of such processes.

9. HEAT TRANSFER BY A CHEMICALLY INERT LIQUID IN GAS SOLID CATALYTIC PROCESSES

We would like to end this review by mentioning a very important potential application of three phase reactors.

Three phase processes have definite advantages over gas-solid operations for temperature control and heat removal. That is the reason why the use of a chemically inert liquid in catalytic reactors can improve the thermodynamic efficiency of the reaction system.

Blum et al. (61) have proposed a three phase fluidized bed (Figure 18) for the methanation of synthesis gas into substitute natural gas.

$$CO+H_2O \rightarrow CO_2+H_2 \quad \Delta H(600K) = -39 \text{ KJ/mol} \tag{38}$$

$$CO+3H_2 \rightarrow CH_4+H_2O \quad \Delta H(600K) = -218 \text{ KJ/mol} \tag{39}$$

$$2CO+2H_2 \rightarrow CO_2+CH_4 \quad \Delta H(600K) = -257 \text{ KJ/mol} \tag{40}$$

$$4CO+8H_2 \rightarrow CO_2+3CH_4+2H_2O \quad \Delta H(600K) = -231 \text{ KJ/mol} \tag{41}$$

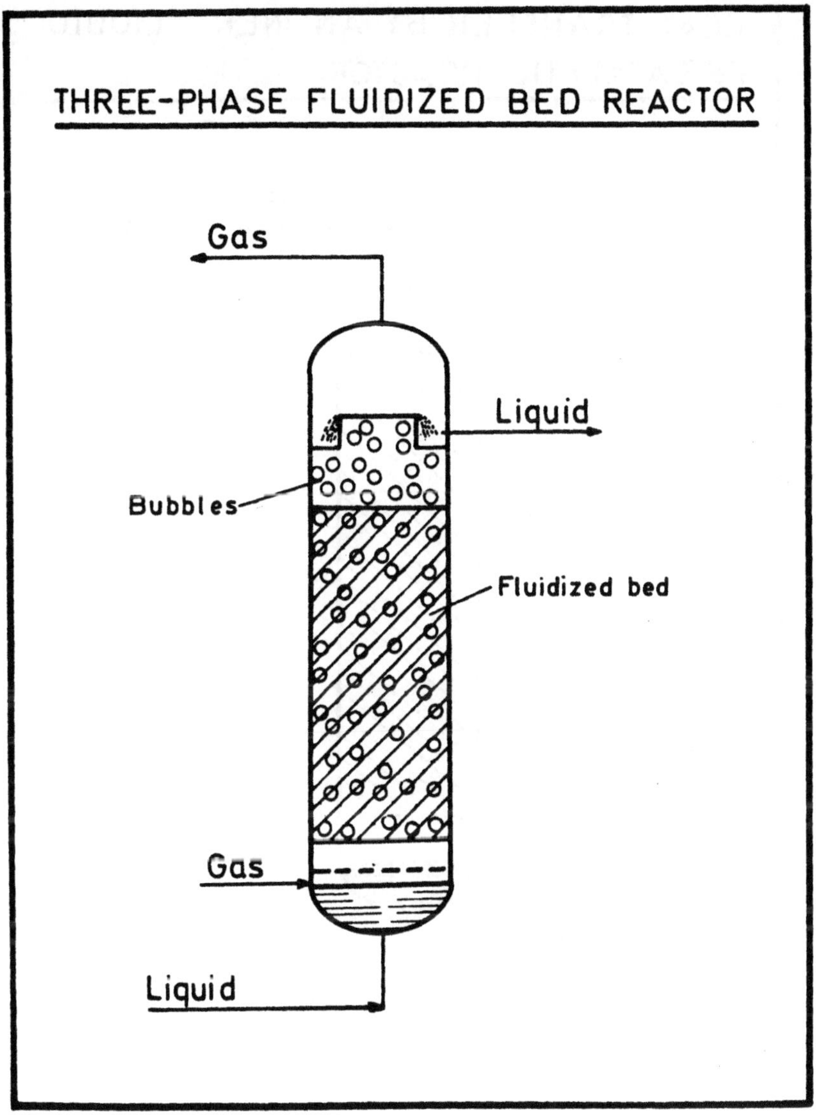
THREE-PHASE FLUIDIZED BED REACTOR
Gas
Liquid
Bubbles
Fluidized bed
Gas
Liquid

Figure 18

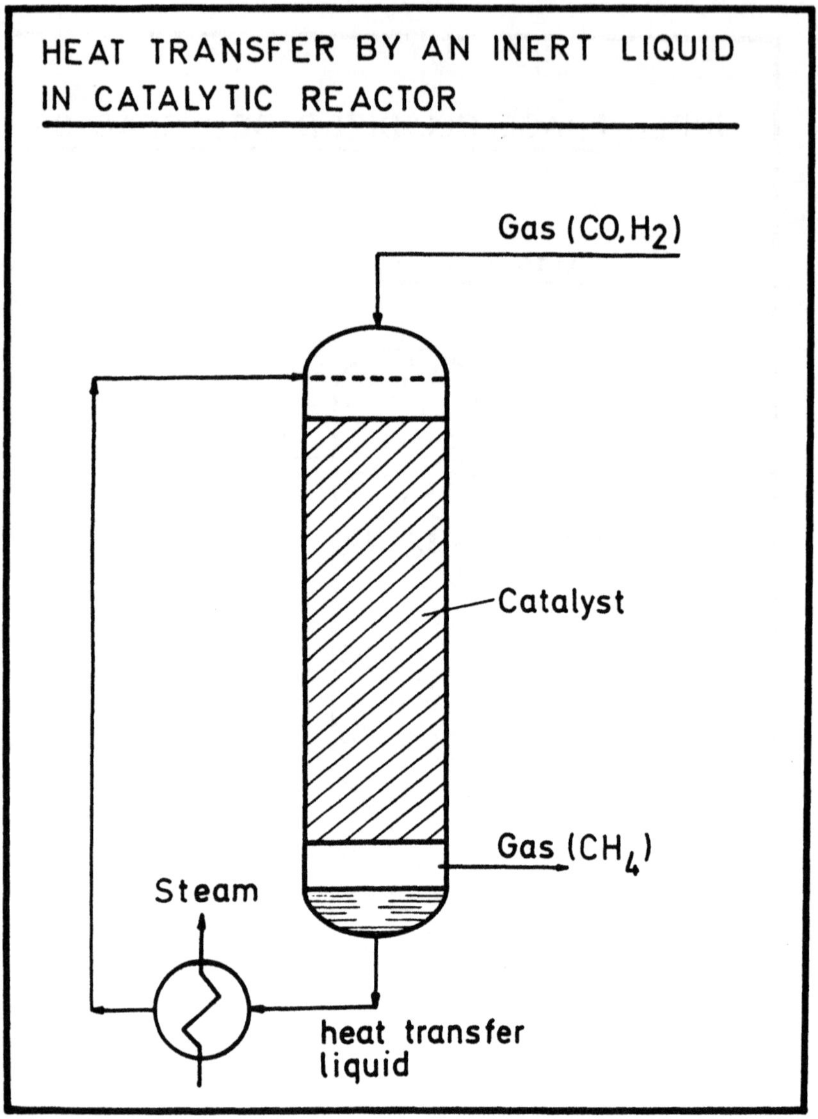
HEAT TRANSFER BY AN INERT LIQUID
IN CATALYTIC REACTOR
Gas (CO, H2)
Catalyst
Gas (CH4)
Steam
heat transfer
liquid

Figure 19

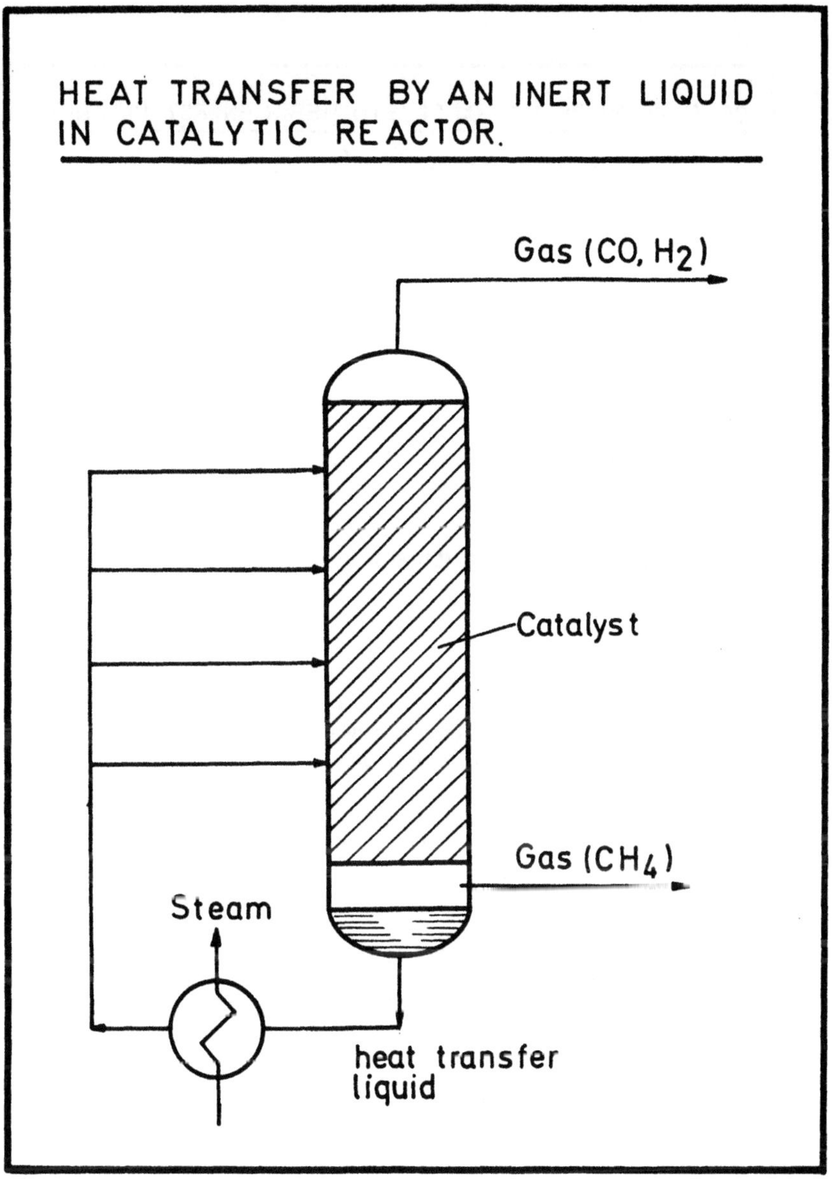
HEAT TRANSFER BY AN INERT LIQUID
IN CATALYTIC REACTOR.
Gas (CO, H2)
Catalyst
Gas (CH4)
Steam
heat transfer
liquid

Figure 20

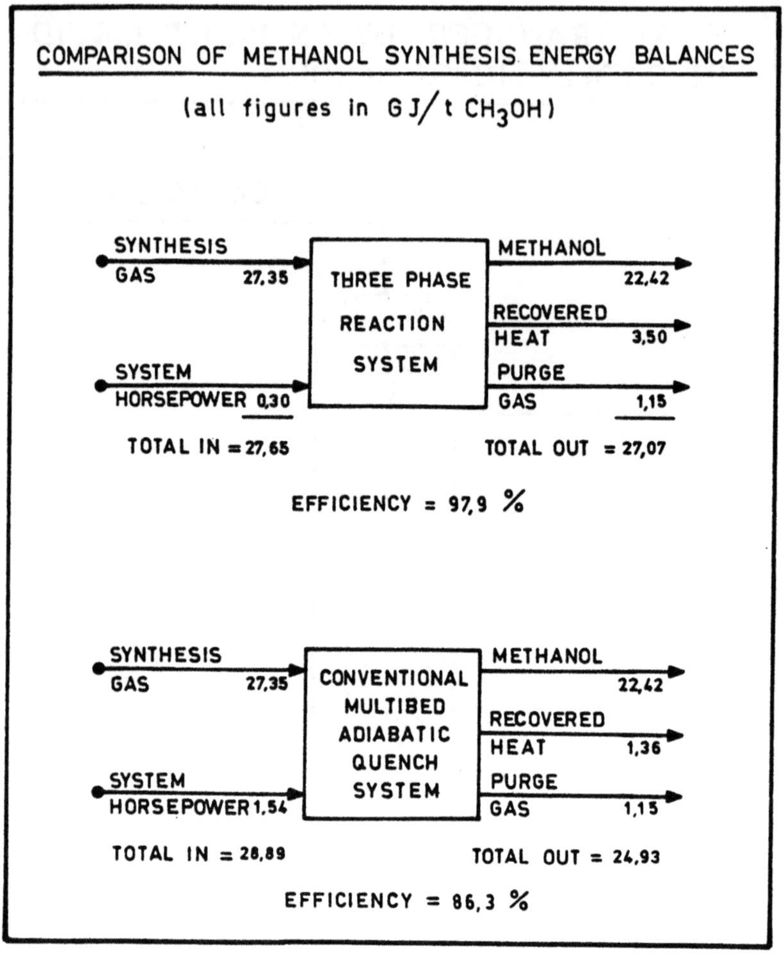
COMPARISON OF METHANOL SYNTHESIS ENERGY BALANCES
(all figures in GJ/t CH3OH)
SYNTHESIS
GAS 27,35
THREE PHASE
REACTION
SYSTEM
SYSTEM
HORSEPOWER 0,30
METHANOL
22,42
RECOVERED
HEAT 3,50
PURGE
GAS 1,15
TOTAL IN = 27,65
TOTAL OUT = 27,07
EFFICIENCY = 97,9 %
SYNTHESIS
GAS 27,35
CONVENTIONAL
MULTIBED
ADIABATIC
QUENCH
SYSTEM
SYSTEM
HORSEPOWER 1,54
METHANOL
22,42
RECOVERED
HEAT 1,36
PURGE
GAS 1,15
TOTAL IN = 28,89
TOTAL OUT = 24,93
EFFICIENCY = 86,3 %

Figure 21

The efficiency of such a reactor is high. Mixtures of carbon monoxide and hydrogen in a wide range of concentration can be used as feedstock. In addition the exothermal shift reaction can take place in the same reactor, due to the possibility of a fine control of the temperature.

Of course, it is also possible to use trickle-bed reactors for this reaction. Heat removal can be ensured by circulating the inert liquid in an external heat exchanger as shown on Figure 19. As the required liquid flow rate would be very high in this case, pressure drop could be an important problem. Another solution would involve the injection of the cooled liquid at different levels in the bed (Figure 20). This would unfortunately decrease the thermodynamic efficiency of the reaction.

A very similar process has also been proposed for the synthesis of methanol (62).

$$CO+2H_2 \rightarrow CH_3OH \qquad \Delta H(500K) = -98\ KJ/mol \qquad (42)$$

$$CO_2+3H_2 \rightarrow CH_3OH+H_2O \qquad \Delta H(500\ K) = -58\ KJ/mol \qquad (43)$$

Figure 21 compares the thermal efficiency of this process and of the conventional multibed quench reactor operated at 100 bar.

These examples show clearly that synthetic fuels can open a wide fiels for the use of three phase systems. Gas-liquid-solid reactors could so contribute to the development of "clean energy" and to the efficiency utilization of some energy sources such as coal and uranium.

10. REFERENCES

1. DWYER, F.G., LEWIS, P.J. and SCNEIDER, F.H., "Efficient , non-polluting ethylbenzene process", *Chemical Engineering*, Jan.5, p.90-91, 1976.
2. OSTERGAARD, K. "Gas-liquid-particle operations in chemical reaction engineering", *Advances in chemical Engineering*, vol.7, p.71-137, Academic Press, New York, 1968.
3. JOSCHEK, H.I, "Reaktoren für Gas-Flüssig -Fest-Reaktionen", in *Ullmans Encyklopädie der technischen Chemie*, vol.3, p.494-518, 4th Ed., Verlag Chemie, Weinheim, 1973.
4. SATTERFIELD, C.N., "Trickle-bed reactors", *AIChE Journal*, vol. 21, p. 209-228, 1975.

5. BÜSSEMEIER, B., FROHNING, C.D. and CORNILS, B., "Lower olefins via Fischer-Tropsch", *Hydrocarbon Processing*, vol.55, n°11,p. 105-112, 1976.
6. WEITZ, H.M. and HARTIG, J., "Butadien", in *Ullmanns Encyklopädie der technischen Chemie*, vol.9, pp.1-18, 4th ed., Verlag Chemie, Weinheim, 1974.
7. REPPE, W., "Chemie und Technik des Acetylen-Druck Reaktionen", Verlag Chemie, Weinheim, p.89, 1951.
8. LEPRINCE, P., CHAUVEL, A., CASTRY, J.P. and CASTEX, L., "Procédés de pétrochimie", p.165, Technip, Paris, 1971.
9. *id.*, p.164.
10. HANCOCK, E.G., "Benzene and its industrial derivatives", pp. 228-229, Ernest Benn Ltd., 1975.
11. LEPRINCE *et al.*, *op. cit.*, p.168.
12. FIELD, S. and DALSON, M.H., "Economics of making cyclohexane", *Hydrocarbon Processing*, vol.46, n°5, p.169-174, 1967.
13. HANCOCK, E.G., "Benzene and its industrial derivatives", p. 230, Ernest Benn Ltd., London, 1975.
14. WEISSERMEL, K. and ARPE, H.J., "Industrial organic chemistry. Important raw materialsand intermediates", Verlag Chemie, Weinheim, 1978.
15. DU PONT, DOS 2034380, 1970.
16. SOCIETA RHODIATOCE, DOS 2053799, 1970.
17. ICI, DOS 1543679, 1966.
18. DU PONT, US-Pat. 2225059, 1938.
19. BASF, DT-Pat. 1593767, 1967.
20. LEONHARD, K.W., "Hexamethylendiamin", in *Ullmanns Encyklopädie der technischen Chemie*, vol.12, p.665-667, 4th Ed., Verlag Chemie, Weinheim, 1976.
21. HANCOCK, E.G., "Benzene and its industrial derivatives", p. 266, Ernest Benn Ltd, London, 1975.
22. WEISSERMEL, K. and ARPE, H.J., *op.cit.*, p.215.
23. FRANZISCHKA, W., MESCH, W. and VOGES, D., "Alkohole, mehrwertige", in *Ullmanns Encyklopädie der technischen Chemie*, vol.7, p.227-235, 4th Ed., Verlag Chemie, Weinheim, 1974.
24. BROWNSTEIN, A.M. and LIST, H.L., "Which route to 1,4-butanediol", *Hydrocarbon Processing*, vol.56, n°9, p.159-162,1977.
25. WEISSERMEL, K. and ARPE, H.J., *op. cit.*, p.215.
26. FLIEGE, W., VOGES, D. and STEFFAN, G., "Butan-, Buten-und Butindiole", in *Ullmanns Encyklopädie der technischen Chemie*, vol.9, p.19-24, 4th Ed., Verlag Chemie, Weinheim, 1975.
27. GAF, US-Pat. 3449445, 1967.
28. KANETAKA, J., ASANO, T. and MASUMUNE, S., "New process for production of tetrahydrofuran", *Industrial and Engineering Chemistry*, vol. 62, p. 24-32, 1970.
29. MITSUBISHI CHEMICAL, DOS 2345160, 1974; DOS 2424539, 1974; DOS 2504637, 1975; DOS 2505749, 1975; DOS 2510088, 1975; DOS 2510089, 1975.

30. WEISSERMEL, K. and ARPE, H.J., *op.cit.*, pp. 350-351.
31. EASTMAN-KODAK, US-Pat. 3334149, 1964.
32. SHELL, DL-Pat. 1186455, 1965.
33. HANCOCK, E.G., "Propylene and its industrial derivatives",Ernest Benn Ltd, London, 1973.
34. WEISSERMEL, K. and ARPE, H.J., *op.cit.*, p. 122.
35. SCHÜTT, H., "Fettalkohole", in *Ullmanns Encyklopädie der technischen Chemie*, vol.11, p.427-445, 4th Ed., Verlag Chemie, Weinheim, 1976.
36. PETERS, R.A.,"Alcohols, higher aliphatic", in *Kirk-Othmer Encyclopedia of Chemical Technology*, vol.1, p.716-739, 3rd Ed., John Wiley, New York, 1978.
37. THOMAS, A., "Fette und Ole", in *Ullmanns Encyklopädie der technischen Chemie*, vol.11, p.455-524, 4th Ed., Verlag Chemie, Weinheim, 1976.
38. THE NEW JAPAN CHEMICAL COMPANY, GB-Pat. 1335173, 1973.
39. VLITOS, A.J., "Le sucre matière première chimique", *Informations Chimie*, n°186, p.119-123, 1979.
40. HAIDEGGER, E., "The importance of hydrogenation in the processing of carbohydrates", *Die Stärke*, vol. 29, n°12, p.430-435, 1977.
41. VON GRAEFE, G., "Zucker und Zuckeralkohole", *Die Stärke*, vol. 27, n°5, p. 160-169, 1975.
42. I.G. FARBENINDUSTRIE, DT-Pat. 544666 and 554074, 1925.
43. ENGELHARD INDUSTRIES, US-Pat, 2868857, 1959.
44. ICI, BE-Pat. 837201, 1975.
45. van LING, G. and VLUGTER, J.C., "Catalytic hydrogenolysis of saccharides", *Journal of Applied Chemistry*, vol.19, p.43-45, 1969.
46. CLARK, I.T., "Hydrogenolysis of sorbitol", *Industrial and Engineering Chemistry*, vol. 50, p.1125-1126, 1958.
47. DEUTSCHE TEXACO, DT-Pat.1260454, 1193931 and 1238453, 1962.
48. LEONARD, J. and GAILLARD, J., "Upgrading of C_3 and C_4 olefinic cuts", Institut Français du Pétrole, May 1979.
49. FRANK, J.P., LEPAGE, J.F. and de GAUDEMARIS, G., "Hydrogenate for better jet fuel", *Hydrocarbon Processing*, vol.56, n°11, p.287-289, 1977.
50. MILLER, S.A., "Ethylene and its derivatives", E.Benn, London, 1969.
51. FIAT, Final Report 968.
52. ANCILLOTI, F., "Ion exchange resin catalyzed addition of alcohols to olefins", *Journal of Catalysis*, vol.46, p.49-57,1977.
53. NEIER,W. and WOELLNER, J., "Isopropyl alcohol by direct hydration", *Chemtech*, p.95-99, Feb. 1973.
54. TOORU HASUIKE and HIDEO MATZUZAWA, "Make methyl methacrylate from spent-DD", *Hydrocarbon Processing*, vol.58, n°2, p.105-107, 1979.
55. Cie ALAIS, FROGES et CAMARGUE et AUSTERWEIL, Fr-Pat.976783, 1949.

56. GANS,M., "Which route to aniline", *Hydrocarbon Processing*, vol.55, n°11, p.145-150, 1976.
57. KOHN, P.M., "Ethylene diamine route eases pollution worries", *Chemical Engineering*, p.90-91, March 27, 1978.
58. KOCH, K., "Fettamine", in *Ullmanns Encycklopädie der technischen Chemie*, vol.11, p.447-454, 4th Ed., Verlag Chemie, Weinheim, 1976.
59. MILLER, S.A., *op.cit.*, pp.527-562.
60. GOTO, S., LEVEC, J. et SMITH, J.M., "Trickle-bed oxydation reactors", *Catalysis Review-Science and Engineering*, t.15, n°2, p.187-247, 1977.
61. BLUM, D.B., SHERWIN, M.B. and FRANK,M.E., "Liquid phase methanation of high concentration CO synthesis gas", in L. SEGLIN Ed., " Methanation of Synthesis Gas", *Advances in Chemistry Series*, n°146, American Chemical Society, Washington, 1975.
62. SHERWIN, M.B. and FRANK, M.E., "Make methanol by three phase reaction", *Hydrocarbon Processing*, vol.55, n°11, p.122-124, 1976.

FLUIDDYNAMICS, MASS TRANSFER AND CHEMICAL REACTION IN MULTIPHASE CATALYTIC FIXED BED REACTORS

Prof. Dr. H. Hofmann

University Erlangen-Nürnberg

1 INTRODUCTION

This review paper is concentrated on problems in scaling-up multiphase catalytic fixed bed reactors such as trickle-bed or packed bubble column reactors, in which two fluid phases (gas and liquid) pass concurrently through a bed of solid (usually porous) catalyst particles. These types of reactors are widely used in chemical and petrochemical industry as well as in biotechnology and waste water treatment. Typical processes are the hydrodesulphurization of petroleum fractions, the butinediol syntheses in the Reppe process for synthetic rubber, the anthrachinon/hydrochinon process for H_2O_2-production, biochemical processes with fixed enzymes or the oxidative treatment of waste water under pressure.

The design respectively scale-up of these reactors is a rather difficult problem in chemical reaction engineering, as the reactors' production capacity not only depends on the chemical kinetics and the mean residence time of the reactants but also on the complex fluiddynamic phenomena and mass transfer processes in the multiphase system. Besides the fact that there already exist monographies on this subject [1,2] and the research activity in this area has been rather extensive in recent years, many problems have not yet been satisfacturally solved and the scale-up of these reactors is still far from being a standard routine for chemical reaction engineers; there is still too much information missing.

2 SPECIAL CHARACTERISTICS OF MULTIPHASE CATALYTIC FIXED BED REACTORS

2.1 Mode of Operation and Flow Regimes

Whereas in a fixed bed reactor with a single fluid phase there exist only two modes of operation, either downflow (which is used in most cases) or upflow, and only two different flow regimes, either laminar or turbulent flow, which can be observed and characterized by a Reynolds number as the single relevant dimensionless group, the fluiddynamics in multiphase catalytic fixed bed reactors are much more complex.

Because there are two flowing phases, the possibility exists to feed them either concurrently in up- or downflow, or countercurrently. Whether a concurrent or a countercurrent flow is used depends on throughput, heat recovery and availability of driving forces for mass transfer and chemical reaction. The phenomenon of flooding, known from countercurrent flow, does not appear in concurrent flow, thus allowing for much higher flow rates.

Furthermore, not only a regime of low interaction and a regime of strong interaction of the phases exist, but also several flow patterns appear in the packed bed, which depend on shape and size of particles (and bed) as well as physical properties and flow velocities of gas and liquid. Because the characteristics of flow and transport phenomena vary according to the flow pattern, various flow charts have been proposed to show boundaries between the different regimes, which are relevant in design. For example, the flow charts for concurrent downflow and concurrent upflow, reproduced from Hirose [3] are shown in Fig. 1. They indicate at least 3 to 5 different flow regimes: bubble flow, spray flow, pulse flow and trickle flow, as well as foaming flow. Whereas the flow rate of the phases and also the phase properties of the fluids are included at least in one of these charts, packing and bed characteristics are not yet displayed.

On the other hand, there exists a significant dependence on those geometric parameters as shown in Fig. 2, where the boundary between pulsing and the other regimes is depicted for two different bed sizes. For perfect data correlation additional dimensionless groups have to be included, e.g. the product between specific surface, particle diameter and void fraction ($a_s \cdot d_p/\varepsilon_0$) and the ratio of particle to bed diameter (d_p/d_t). Those two groups have already been used successfully in other empirical correlations [4,5]. However, a better understanding of the phenomena through a more fundamental insight into the system is needed. Flow pattern e.g. should be classified - and corresponding flow charts should be prepared - on the basis of fluiddynamic considerations on the interaction between the flowing phases, such as in the work of

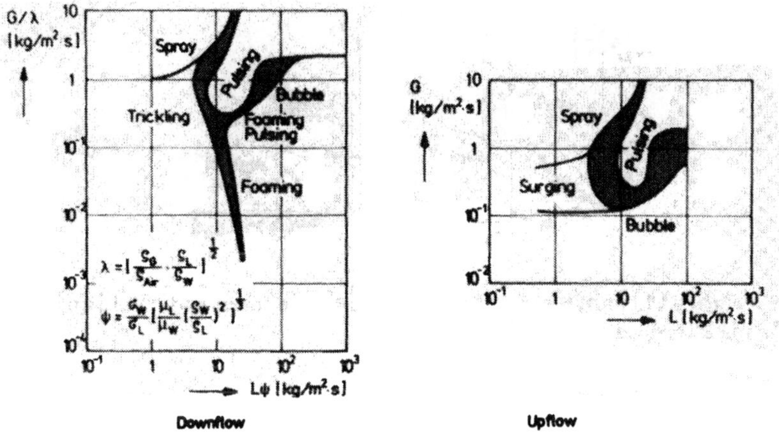

Fig. 1 Flow regimes in concurrently operated packed bed reactors

Talmor [6], Kirillov [7], Ogarkov [8] or Sicardi [9]. In the latter publication e.g. the pulsing inception in trickle bed reactors has been interpreted physically as being due to the formation of waves on the liquid surface whose amplitudes depend on flow rates etc., thus causing the occlusion of channels in the packing.

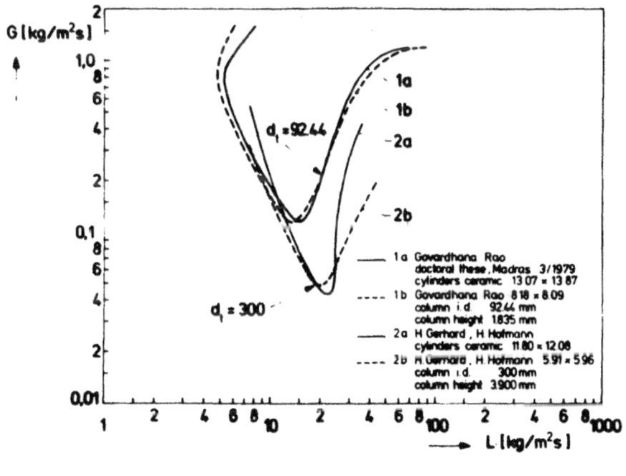

Fig. 2 Influence of bed diameter on flow regimes in trickle-bed reactors

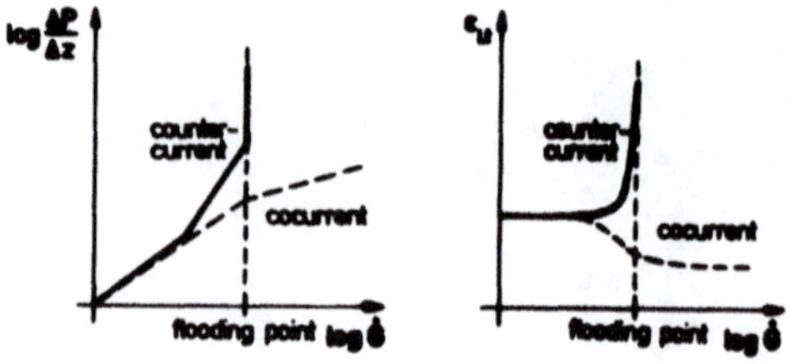

Fig. 3 Overall behavior of pressure drop and total liquid hold-up in trickle-bed reactors

2.2 Pressure drop

Pressure drop caused by the flowing phases is an important design quantity, as throughput as well as mass transfer rates depend on energy dissipation. In principle, the pressure drop (and liquid holdup) behavior of multiphase catalytic packed beds is shown in Fig. 3. As the momentum balance cannot be solved exactly, there are two empirical approaches used until now to correlate the two-phase pressure drop with the design variables, such as gas and liquid flow rate and structure of the bed.

In the first approach, the so-called Lockhart-Martinelly approach [10], the two-phase pressure drop δ_{lg} is correlated in terms of an enhancement of the pressure drop of the accompanying gas flow $\Phi_l = \sqrt{\delta_{lg}/\delta_g}$, plotted against a ratio of pressure losses $X = \sqrt{\delta_l/\delta_g}$ for single-phase flows of gas and liquid separately (Fig. 4).

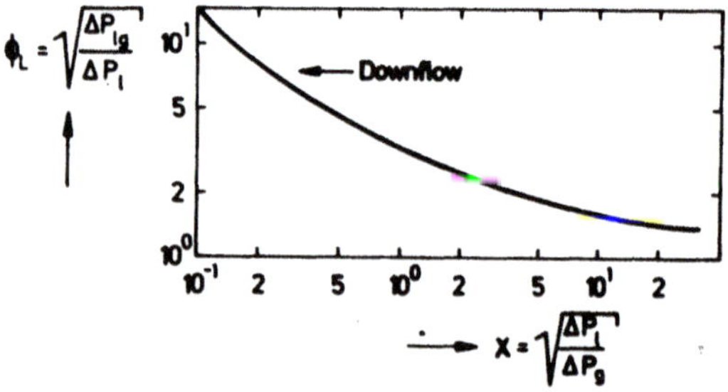

Fig. 4 Two-phase pressure drop according to Y. Sato

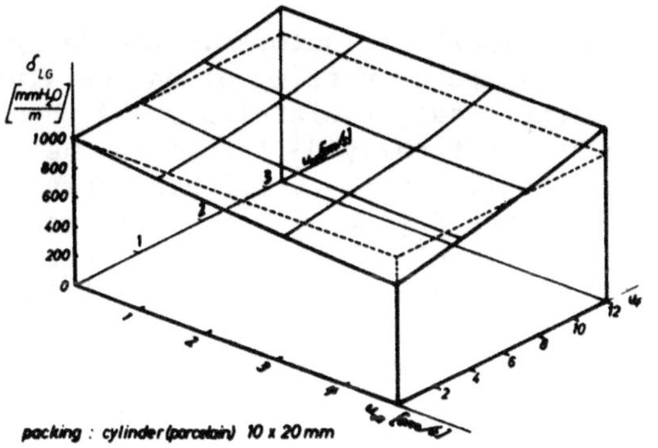

Fig. 5 Total pressure drop in bubble-flow operation of packed beds

An alternative approach [11], based on the flow resistance of gas phase, has been proposed for the regime of low interaction between gas and liquid. In this case, a two-phase friction factor, which allows for the increase of liquid holdup by increasing liquid flow rate, is defined and experimental data are correlated according to an Ergun-type equation as:

$$\delta_{lg} = k_1 \frac{[1-\varepsilon_o(1-\beta_s-\beta_D)]^2}{\varepsilon_o^{\,3}(1-\beta_s-\beta_D)^3} \mu_G u_G + k_2 \frac{1-\varepsilon_o(1-\beta_s-\beta_D)}{\varepsilon_o^{\,3}(1-\beta_s-\beta_D)^3} \varrho_G u_G^{\,2} \qquad (1)$$

with k_1 and k_2 depending on packing size and shape, obtained from δ_G from a "wetted packing" (with static holdup included). In pulsing, foaming, foaming-pulsing and spray flow such an approach is no longer valid, as pulsing phenomena increase the pressure drop, even though the liquid holdup remains approximately constant.

In case of upward bubble flow, the pressure drop at high flow rates of gas and liquid is very similar to that in downward flow (as gravity has only a minor influence), whereas at low flow rates the pressure drop is mainly due to the specific gravity of the liquid phase (Fig. 5); here again a two-phase friction factor concept seems to be advisable [12], in which the friction factor is an empirically determined, logarithmic function of a factor Z being $(Re_G^{1,167}/Re_L^{0,767})$ [13]. This again demonstrates a certain lack of fundamental understanding of fluiddynamics in such a complex system.

2.3 Degree of Wetting of the Catalyst Packing

One of the most important factors in the design of trickle bed reactors is the degree of wetting of the catalyst by the liquid film, since the catalyst efficiency and thus the degree of conversion can be drastically different in the case of incomplete wetting [14]. Still, it is astonishing that until now very little is known about the degree of wetting in catalyst packings. Incomplete wetting can result either from a faulty design of the liquid distributor at the top of the reactor [15] or from a too small liquid load and/or too high a liquid/solid interfacial tension. The minimum liquid load, necessary to wet the packing totally, depends strongly on the nature of the packing surface, the particle diameter and the surface tension of the liquid. As a rule of thumb, for complete wetting, the liquid load of a trickle-bed reactor should be in the range of 10-30 m^3/m^2, h. In case of porous catalyst particles, two types of wetting can be defined, internal wetting (pore filling) and external (or effective) wetting. Due to capillary action, pore filling is usually complete, except in case of strongly exothermic reactions, where heat of reaction causes an evaporation of the liquid [see 16]. Furthermore, external wetting in the sense of reaction engineering may be different from physical external wetting, as the catalyst particles are very likely to contact semistagnant liquid zones that contribute very little to the mass transfer and hence to the chemical reaction. Therefore, external "effective" wetting in multiphase catalytic packed bed reactors should be measured only on the base of a chemical reaction. Satterfield [17] defined, e.g., as so-called contacting effectiveness in a trickle-bed reactor the ratio of the apparent kinetic constant k_{app}, obtained in a trickle-bed reactor, to the actual kinetic constant k_v, obtained in a stirred tank reactor with complete bathing of the particles. Bondi [18] on an empirical basis found that for a number of first order reactions

$$\frac{1}{k_{app}} - \frac{1}{k_v} = \frac{A'}{\dot{L}'^{b}} , \qquad (2)$$

where $0{,}5 < b < 0{,}7$ with a median value of about 2/3. He also reports that an increase in the gas rate moderately increased conversion and suggests multiplying $\dot{L}$ in this equation by $(\varrho_G \dot{G})^{\beta}$, where $0{,}22 < \beta < 0{,}5$. A' is a constant specific to the packing.

On the other hand, Colombo [4] defined the liquid/solid contacting efficiency in terms of the ratio of the apparent diffusivity $(D_i)_{app}$ of a tracer in a porous particle, determined in a trickle-bed reactor (by assuming total wetting of the packing) and the intraparticle diffusivity D_i, determined in a liquid full reactor. This ratio was found to increase as the liquid flow rate

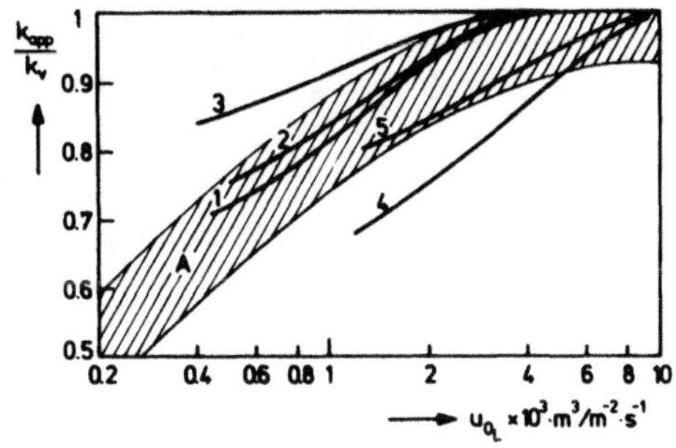

Fig. 6 Degree of wetting in trickle-bed reactors. A: range proposed by Satterfield. 1-5: experimental values for different tracers according to Colombo

increases, very similar to the increase of k_{app}/k_v considered by Satterfield (see Fig. 6). This ratio also increased with decreassing particle size. A smaller surface tension or higher viscosity of the liquid should also increase the contacting efficiency. But for the proper design of a trickle bed reactor, further experimental as well as theoretical work on this subject is needed, particularly for the case of porous packing in laboratory reactors.

2.4 Holdup of the Phases and Dispersion in the Phases

Because it has been impossible until now to solve the momentum balances for a multiphase packed bed in order to have detailed information about the velocity (and pressure) distribution in the reactor, chemical reaction engineers have defined fluiddynamic parameters relevant for the chemical conversion via tracer experiments, using appropriate models.

The mean residence time of a phase, which is most important for the degree of conversion in multiphase reactors, is directly related to one of these parameters, i.e. the phase holdup, but in addition, pressure drop and catalyst wetting are also related to holdup. In general, hold-up is expressed either as fractional bed volume ε or as fractional void volume β.

In a multiphase packed bed reactor with porous particles the holdup of a phase consists of that part which is held internally in the pores of the particles and that part which is held externally in the interstices between the packing. The external part can be either free flowing or stagnant, a subdivision which is mainly important for the dispersed phase.

Therefore, the total liquid holdup ε_{L_t} in downflow trickle operation is divided into

$$\varepsilon_{L_t} = \varepsilon_{L_{ext}} + \varepsilon_{L_{int}} = \varepsilon_{L_d} + \varepsilon_{L_s} + \varepsilon_{L_{int}} \quad , \qquad (3)$$

where ε_{L_d} is the dynamic and ε_{L_s} the stagnant holdup. The same classification is adopted in upflow bubble operation for the gas holdup, i.e.

$$\varepsilon_{G_t} = \varepsilon_{G_{ext}} + \varepsilon_{G_{int}} = \varepsilon_{G_d} + \varepsilon_{G_s} + \varepsilon_{G_{int}} \quad . \qquad (4)$$

Because the amount of the stagnant holdup depends on the degree of turbulence, i.e. on the flow rate of the dispersed phase, the only meaningful way for its determination is via tracer experiments, using appropriate mathematical models, which make allowance for stagnant zones (inside and outside of particles), such as the PE- or PDE-model (for details see the chapter on mathematical models). A residual (instead of a stagnant) and a free draining (instead of a dynamic) holdup makes little sense in reactor design.

The results depicted in Fig. 7, which concern the stagnant fraction of the dispersed phase in downflow trickle and upflow bubble operation, show the dependence on physical properties and the decrease of the stagnant holdup with increasing flow rate of the respective phase [68,25].

Using appropriate correlations for the total holdup of the dispersed phase, such as those shown in Fig. 8, which can be determined independent of any model, the dynamic holdup of the

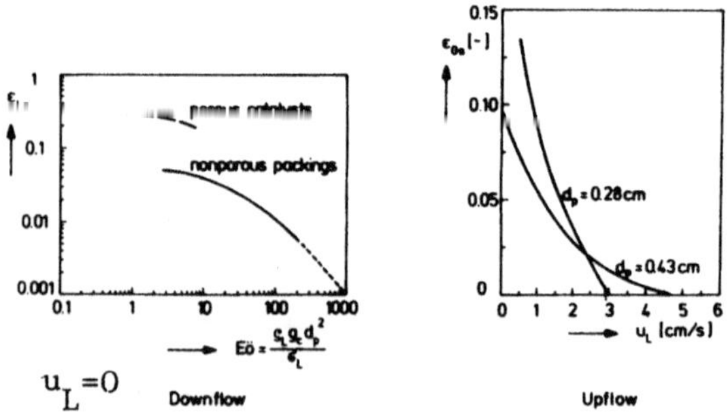

Fig. 7 Stagnant holdup of dispersed phase in concurrently operated trickle-bed and packed bubble reactors

Trickle-bed reactor (nonfoaming liquid)

$$\varepsilon_{Lt} = \frac{0.66\ \chi'^{0.81}}{1 + 0.66\ \chi'^{0.81}} \qquad 0.1 < \chi' < 80$$

$$\chi' = (\xi_L / \xi_G)^{1/2}$$

Packed bubble column reactor

$$\varepsilon_{Gt} = 21.2\,(Re_L / Re_G)^{0.2}\,(\mu_L / \mu_G)^{0.24}$$

Fig. 8 Correlations for total holdup of dispersed phase

respective phase, which finally determines the phase's mean residence time, can also be calculated (assuming that the pores of catalyst particles are completely filled with liquid). In trickle operation this dynamic liquid holdup is consequently either correlated in terms of the Lockhart-Martinelli parameter or in terms of a Reynolds number Re, a Galilei number Ga (or a modified Galilei number in which the gravity term is replaced by gravity plus pressure loss, taking into account the gas/liquid interaction at high flow rates) and a shape factor (such as $a_s d_p / \varepsilon_o$ [4]), whereas, in upflow bubble operation, the dynamic gas holdup is correlated in terms of the ratio of a gas and a liquid Reynolds number as well as a shape factor [19,20]. But here again, there are not enough experimental data available to safely correlate phase holdup for all flow regimes with phase properties and packing characteristics, as most experiments until now concern only the system air/water and non porous packings.

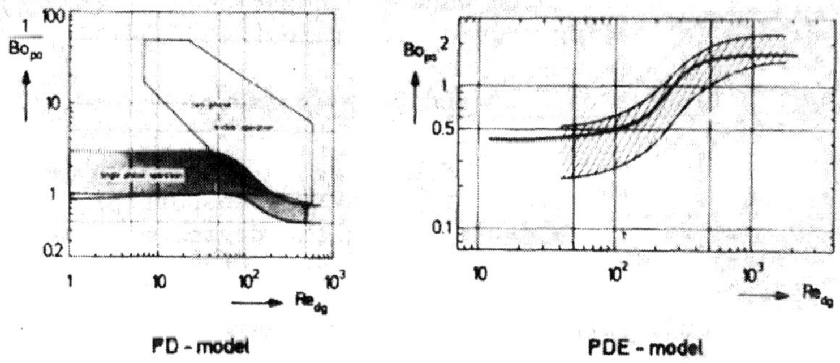

Fig. 9 Liquid Bodenstein number for trickle operation according to different models

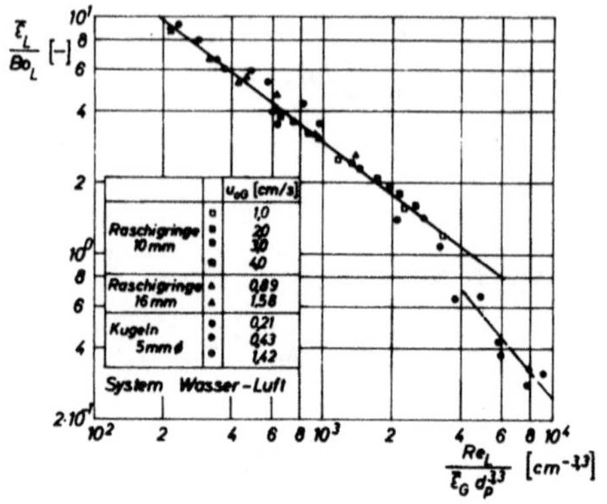

Fig. 10 Axial dispersion in the liquid phase for packed bubble reactors operated in upflow

Dispersion is also of some influence on chemical conversion, mainly in pilot units. Therefore, in the past, many efforts were made to correlate dispersion effects, characterized by a particle Peclet (or Bodenstein) number (as determined by analysing tracer experiments via a so-called PD- or PDE-model) with the operating variables of a multiphase reactor. The relation depends on the model used. E.g. in trickle operation the liquid Bodenstein number either increases in proportion to the liquid Reynolds number according to the PD-model, approaching only at high Reynolds number, a constant value equivalent to that for the single phase flow of liquid, or remains according to the PDE-model constant (and the same as in single phase flow) in each flow regime (see Fig. 9). On the other hand, in case of upflow bubble operation, the liquid Bodenstein number decreases with increasing gas flow rate [21,23] (and the decrease is accelerated when fluidization starts [22]), but increases with increasing liquid flow rate [24] (see Fig. 10).

2.5 Characteristics of Mass and Heat Transfer

In trickle bed reactor as well as in packed bubble reactor design, one is faced with mass and heat transport processes, which in many cases are decisive for the degree of conversion finally reached. The different steps encountered in the most complex type of reactions are shown in Fig. 11. Gas film resistance at the gas-liquid interface is usually neglected, since under practical operating conditions the vapor pressure of the liquid phase is very often rather low and/or the partial pressure of the gaseous reactant is rather high.

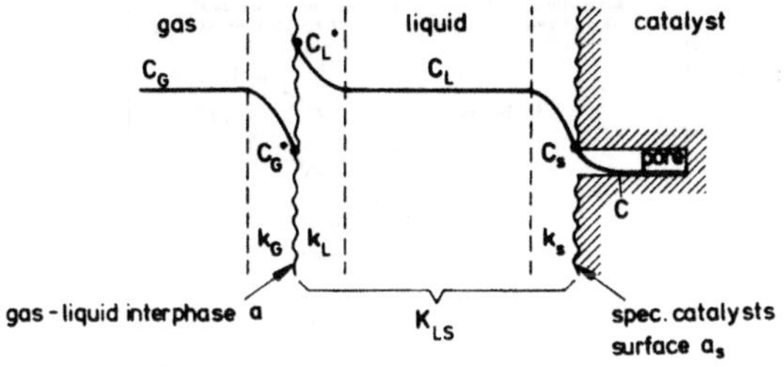

Fig. 11 Transport steps in multiphase catalytic reactions

2.5.1 The liquid side volumetric mass transfer coefficient $k_L \cdot a$ in concurrent operation is influenced by gas and liquid flow rates and increases up to about 1 [s^{-1}] at high phase loadings. Experimental investigations on this parameter have been made almost exclusively for high liquid flow rates, where an extensive momentum transfer between gas and liquid phase exists. Therefore, the concept of energy dissipation density provides the rational basis for data correlation (Fig. 12). The upflow values in packed bubble reactors, on the average, are 100 % greater than downflow values in pulsed and spray flow regimes [24], because the gravitational force leads to higher liquid holdup and pressure drop. In both types of operation, the volumetric coefficient varies at about the 0.5th power of the superficial gas velocity. In spite of the fact that data in literature show a relatively wide scatter, they all appear to be proportional to the specific gas/liquid interface area a, indicating that the liquid side mass transfer coefficient seems to have a constant value between 3-8.10^{-2}[cm/s], and energy

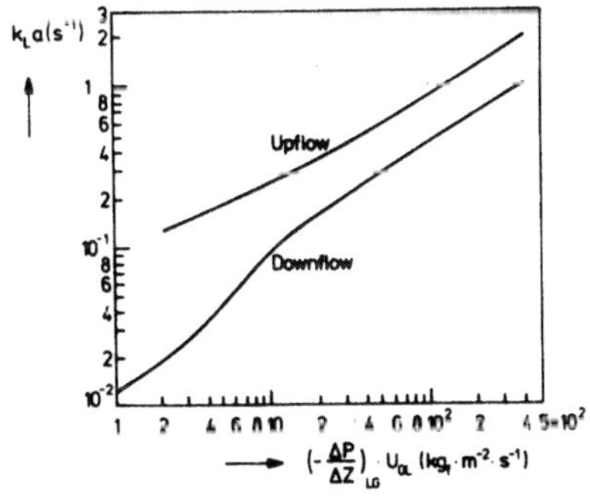

Fig. 12 k_La-values for concurrent flow in packed beds

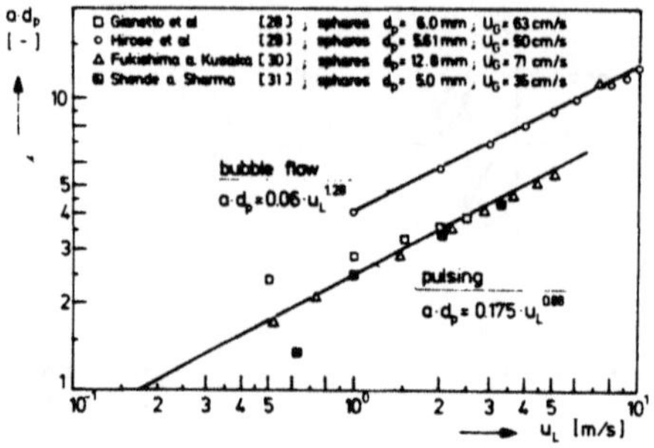

Fig. 13 Gas/liquid interfacial area for packings of spheres in pulsing and bubble flow

dissipation is mainly used for creation of new gas/liquid interface. Values of this interface differ substantially with size and type of packing and amount up to $10^3 [m^2/m^3]$ at high flow rates of gas and liquid because of the large contribution of small bubbles in pulse and bubble flow. There seems to exist a unique correlation between $(a \cdot d_p)$ and u_L for pulse and bubble flow as shown in Fig. 13. In downflow operation, the interfacial area a is proportional to the 1.2th power in trickle flow and to square root of δ_{lg} in other flow regimes. Very small particles, i.e. smaller than 2 mm, behave different with respect to all parameters discussed until now, as in this case capillary force became dominant [20,25].

2.5.2 Liquid particle mass transfer coefficients k_s are usually correlated as particle Sherwood number, depending on the 0.5-0.7th power of the liquid Reynolds number and the 1/3 power of the liquid Schmidt number and in case of bubble flow additionally on a power ratio of gas to liquid Reynolds number. Also here, the dependence of k_s on the operating variables varies (see Fig. 14) with the flow pattern [26,9]. In trickle flow and bubble flow operation the Sherwood number is comparable to single phase flow, but increases rapidly in pulsing flow. At high gas velocities surface tension of liquid becomes important and under these conditions also Weber number is a significant correlation parameter [27].

In upflow operation the liquid to particle Sherwood number is higher than in downflow operation and increases remarkably with gas flow, indicating the large contribution of the liquid turbulence caused by bubble motion. Recently attempts were made to analyse the situation also theoretically and a unified correlation has been developed for heat and mass transfer from single spheres, packed beds as well as tube wall, taking into account the diffusion of solute into a liquid film, oscillating with response to the bub-

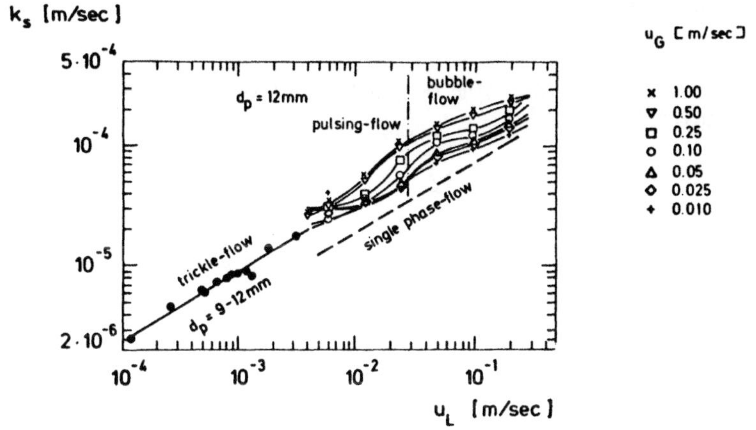

Fig. 14 Solid/liquid mass transfer as function of flow regime in trickle operation

ble passage [32].

Heat transfer performance of multiphase packed beds is characterized in terms of an effective thermal conductivity of the bed λ_{eff} and the wall heat transfer coefficient h_w. Systematic experimental studies on these parameters have been published only recently [70]. According to this paper, the effective thermal conductivity in downflow operation increases rapidly with liquid load at low liquid velocities, reaches twice the value of the single phase flow at moderate liquid loads and approaches the value of the single phase flow at high liquid flow rates [33]. The wall heat transfer coefficient is much improved in pulsing flow [34], reaches 2 - 4 times the value in single phase flow at the same liquid flow rate, but is independent of the gas flow rate in trickle flow [35]. In the upflow operation the effects of gas and liquid flow rates on the wall heat transfer coefficient are even more complicated; the coefficient has a maximum or a minimum with increasing liquid flow rate and its dependence on gas velocity is reversed at a certain liquid flow rate [36]. In addition, first correlations for the radial effective thermal conductivity were presented [37,38], showing some similarities between mass and heat transfer.

With respect to the above-given review, it certainly can be imagined that in multiphase catalytic reactions heat and mass transfer as well as phase mixing and stagnant zones have an influence on the chemical conversion in a very different way, depending on the operating variables of the reactor. But more than that, this behavior has consequences on scale-up procedures for these types of reactors, as the most appropriate scale-up procedure depends on mode of operation, flow regimes and the relative importance of mass transfer resistances compared to the chemical reaction rate.

3 SCALE-UP PROCEDURES FOR CONCURRENTLY OPERATED MULTIPHASE PACKED BED REACTORS

3.1 Principle of Similarity

The principle of similarity, i.e. the use of dimensionless groups maintained constant in the course of scale-up, has been characterized even in case of single phase catalytic reactor as limited to simple situations in chemical reactor scale-up. If the overall phenomenon is complex, as in most practical cases, more than one dimensionless group should be kept constant, sometimes leading to contradictory demands. This is certainly true in all practical cases of multiphase packed bed reactors, where also the compromise of weighing the particular effect is only of limited value. Therefore, this method in practice cannot be applied for scale-up of multiphase catalytic reactors.

3.2 Use of Mock-Ups

Mock-ups, having a scale-up factor of 1:1, are widely in use, not only in research, but also in the course of industrial scale-up of multiphase reactors, all the more so as the situation is so complex and the information available today is still incomplete, as demonstrated in the preceding chapter of this lecture. Their purpose is mainly to study special phenomena with simpler systems, such as glass walled reactors, which allow to identify flow regimes and liquid distribution visually, or water/air systems in order to lower the expenses etc. However, for a priori scale-up procedures, chemical reaction engineers prefer reactor models, although in some cases mock-ups must be used, because the a priori design is not yet safe enough.

3.3 Use of Models

The only rigorous way to scale-up multiphase catalytic reactors is the use of reactor models, or, more precisely, the use of mathematical models. Because these models in their most accurate versions are more complex than for other reactor types, this results in considerable expenditure of time and money usually not available in the design stages of a process. Therefore, these mathematical models should be simplified reasonably on the rational basis of knowledge about the system as e.g. presented in the preceding chapter.

From the standpoint of practical application, the objectives of any such simplification can be characterized through the following points:

a) the model should not be more detailed as absolutely required for the particular purpose (principle of maximum sloppyness)

b) the model should contain as few parameters as possible (principle of minimum expenditure)

In this manner, the mathematical work for the solution of model equations can be kept small, provided that reliable correlations for the parameters of the selected model really exist.

In the simplification process, the interaction between the reaction itself and the transport and/or fluiddynamic processes is no longer separated in all its elementary mechanisms, but some of these processes are lumped together into effective terms through the use of plausible "simplification rules". For these terms the effective parameters are defined by the model itself and have to be determined experimentally.

a) Continuum models

Continuum models are used very often in scale-up and design of concurrently operated multiphase reactors, because they are close to models used in heterogeneous catalytic fixed bed reactors, and therefore the numerics to solve these model equations is known to chemical engineers. They can be categorized according to the different phenomena being regarded as important for the conversion obtainable in the reactor (see Fig. 15). Entries in the different categories refer to publications in which the respective model had been used for reactor design. It is cleary to be seen that until now only the "simplest" continuum models have really been used, whereas e.g. a two-dimensional three-phase heterogeneous plug-flow dispersion model, which surely is the most accurate model has

	Pseudo-homogeneous	Heterogeneous with two phases	 with three phases
Plugflow (P)	Satterfield [17] Reiss [39] Henry [40] v. Deemter [41]		Sato [46]
Plugflow-Dispersion (PD)	Paraskos [59] Montagua [43] [60]		Goto [52]
Plugflow-Exchange (PE)	Schwartz [44]		
Plugflow-Dispersion-Exchange (PDE)			

Usually all models are written one-dimensionally, but in case of heat exchange with the wall also a two-dimensional model is recomended.

Fig. 15 Categories of continuum models for concurrently operated multiphase catalytic packed bed reactors

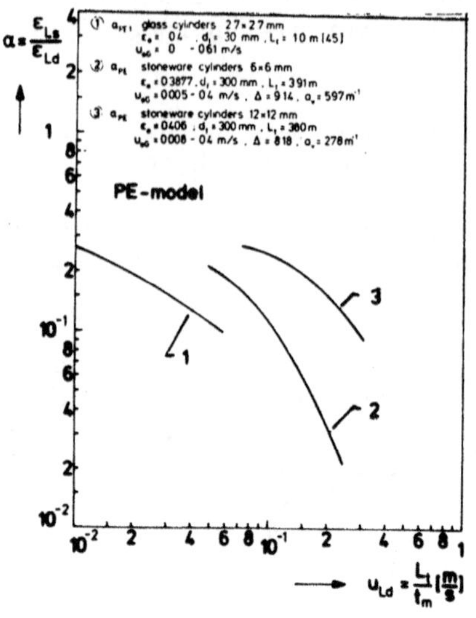

Fig. 16 Relative amount of stagnant liquid in downflow operation

never been used because of its complexity as well as because of the lack of correlations for the parameter values needed in this model.

But how to judge rationally for model simplifacations?

Let us first discuss how we can simplify reactor models with respect to fluiddynamics, i.e. with respect to stagnant parts of the dispersed phase as well as dispersion (axial and radial). Here the experimental results of mock-up studies are very interesting.

Fig. 16 shows one of the relevant results, i.e. the relative amount α of stagnant liquid in downflow operation as determined via the so called plugflow exchange (PE)-model, in which the total liquid holdup is divided into a static holdup ε_{Ls} and a dynamic holdup ε_{Ld}, where the dynamic holdup is assumed to flow through the packing in plugflow. It is interesting to see that depending on a particle number $\Delta = a_s \cdot d_p / \varepsilon_o$, but probably also depending on the (d_p/d_t)-ratio, the stagnant holdup sharply decreases to about 10 % and even less of the dynamic holdup at sufficiently high liquid velocities ($\gg 0{,}1 [m \cdot s^{-1}]$). This is, roughly speaking, the transition to pulsing operation (ignoring for the moment the influence of the gas flow rate). This leads to a first proposal for simplification:

(1) Stagnant holdup may be ignored in flow regimes with high interaction of phases (pulsing, spray, bubbling)

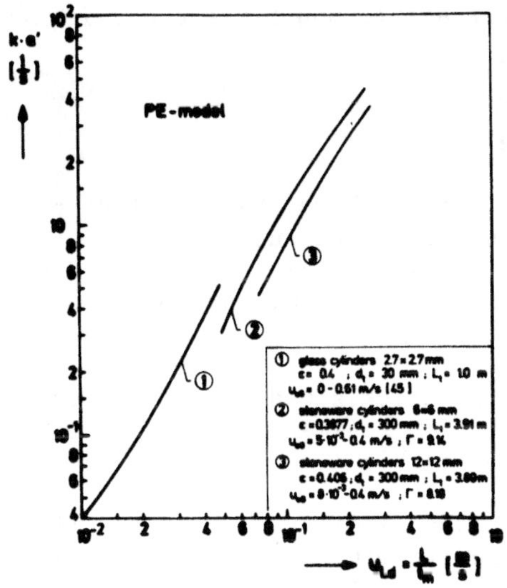

Fig. 17 ka'-values for exchange with stagnant liquid holdup in downflow operation

This proposal is supported by the ka'-values for the exchange with the stagnant part of liquid, depicted in Fig. 17, where this exchange grows by more than one order of magnitude in pulsing flow, compared to trickle flow. Besides the fact that more information is still needed to include additional packing effects (the respective values with Raschig-rings differ significantly from those of cylinders presented here), it is interesting to observe a smooth transition between the two flow regimes.

Another result, which is interesting for model simplification, concerns dispersion. In catalytic vapor phase reaction the effect of backmixing on reactor efficiency generally proves to be negligible except for cases of high conversion and short beds. For concurrent trickle-flow operation Mears [47] has developed on the basis of a perturbation solution of the one-dimensional plugflow dispersion (PD-)model, a criterion for negligible (< 5 %) influence of axial dispersion in case of first order reactions:

(2) $$\frac{L}{d_p} > \frac{DaI_1}{Bo_p(\exp[0{,}05\ DaI_1]-1)}, \tag{5}$$

which reduces for small values of the first Damköhler number (i.e. less than 90 % conversion) to

$$\frac{L}{d_p} > 20\,\frac{DaI_1}{Bo_p} = \frac{20}{Bo_p}\ln\frac{c_o}{c_e} \tag{6}$$

and which becomes for simple non-first order irreversible reactions

$$\frac{L}{d_p} > \frac{20\,n}{Bo_p} \ln \frac{c_o}{c_e} \tag{7}$$

First, this criterion shows that axial dispersion can be neglected at a high L/d_p-ratio. Because trickle-flow Bodenstein numbers for the liquid phase are only slightly influenced by the gas phase and approach the value of 0.5 at small α, one can furthermore state

(3) in concurrent trickle operation near the transition to high interaction regime, axial dispersion in the liquid phase is negligible

However, as the Bodenstein number at low liquid Reynolds numbers ($Re_{Lo} \leq 10$ as used in many pilot oparations) drops down to 0.1 or even smaller values, for freedom from significant axial dispersion in the liquid phase the L/d_p-ratio must be up to 10 times larger than in single phase flow. Furthermore, it has been shown that Mears' criterion changes significantly if mass transfer or heat effects are taken into consideration [2].

The Mears criterion can be also used for upflow operation, but only if the correct Bo_L-number is used (see e.g. Fig. 9), as no assumption is made about the flow direction. In general, axial dispersion is negligible at any operating condition for which Bo (formed with the reactor length L) is larger than a certain value [48].

There remains only one handicap, that is until now (as shown above) no Bo-Re-correlation is available which includes all packing, particle and fluid characteristics in order to define the limit of neglecting influence with high accuracy.

The final question concerning the influence of fluiddynamics on chemical conversion is the importance of radial dispersion on chemical conversion. From single phase flow it is known that radial dispersion is five times faster than axial dispersion, leading to an almost complete concentration equalization in radial direction. This is certainly true also in multiphase reactors at high phase loads [49]. But with lower phase loads, primitive feed distributors for the dispersed phase etc. radial dispersion can become important not only for heat transfer to the wall but also for mass transfer [50].Nevertheless, there is no indication in the literature of the use of a two-dimensional model for reactor design, taking consideration of the radial dispersion. On the contrary, any influence of radial dispersion on chemical conversion until now has been lumped together with axial dispersion, stagnant zones or effective wetting, perhaps because there does not exist a proved correlation which connects the radial dispersion coefficient with the operating variables of the reactor. Therefore, we conclude for the sake of simplification:

(4) radial dispersion in concurrently operated packed bed multiphase reactors is negligible at high phase loads or lumped with axial dispersion, stagnant zones or effective wetting, i.a.w. until now only one-dimensional models are in use for multiphase reactor scale-up

The degree of effective wetting, important in trickle operation, which also depends on fluiddynamics, is included correctly in the reaction rate term of the respective balance equations either by apparent rate constant or an effective pore diffusivity respectively or, more useful in reactor modeling, as a contribution to an overall efficiency η_o, which includes also the external and intraparticle mass transfer limitations [51].

Whether a pseudo-homogeneous or a heterogeneous continuum model must be used for reactor scale-up depends on the relative importance of the transport resistances. The gas phase resistance on the gas/liquid interface (as already mentioned) usually can be neglected, but this is not always the case with the other transport resistances as demonstrated experimentally by several authors [46,52].

Under these circumstances, a general stationary heterogeneous dispersion (PD-)model for an irreversible catalytic second order reaction between a gaseous and a liquid reactant in dimensionless form consists of the balance equations shown in Fig. 18. In this model the whole fluiddynamics are lumped into a single parameter, i.e. the Bodenstein number, here based on the reactor length. Even then it can be solved only numerically, but for this appropriate integration procedures and even standard routine programs are available. The model as well as its solutions can be applied to downward as well as upward concurrent operation by using the appropriate parameter values. In case of a solid catalyzed reaction, however, an additional effect of the feed concentration must be taken into account, mainly at small Damköhler numbers, since the physico-chemical equilibrium of reactants in liquid phase is not necessarily attained at the inlet.

Some simplifications are of interest, as they allow an analytical solution of the model (or at least the application of a quicker numerical routine). This is of interest for a priori selection of optimal operating conditions or for recognizing trends; e.g.:

Reaction:

$$A(gas) + B(liq.) \xrightarrow{catalyst} C(liq.)$$

Reaction rate:

$$r = k_2 \eta_0 C_{As} C_{Bs}$$

Material Balance Equations:

Reactant A:

Gas phase

$$\frac{1}{Bo_G} \frac{\partial^2 C_{AG}}{\partial Z^2} - \frac{\partial C_{AG}}{\partial Z} - \tau_G k_L a \, (HC^*_{AG} - C_{AL}) = 0$$

Liquid phase

$$\frac{1}{Bo_L} \frac{\partial^2 C_{AL}}{\partial Z^2} - \frac{\partial C_{AL}}{\partial Z} + \tau_L k_L a \, (HC^*_{AG} - C_{AL}) - \tau_L k_s a_s (C_{AL} - C_{As}) = 0$$

Solid phase

$$k_s a_s (C_{AL} - C_{As}) = k_2 \eta_0 C_{As} C_{Bs}$$

Reactant B:

Liquid phase

$$\frac{1}{Bo_L} \frac{\partial^2 C_{BL}}{\partial Z^2} - \frac{\partial C_{BL}}{\partial Z} - \tau_L k_s a_s (C_{BL} - C_{Bs}) = 0$$

Solid phase

$$k_s a_s (C_{BL} - C_{Bs}) = k_2 \eta_0 C_{As} C_{Bs}$$

with appropriate 8 boundary conditions

$C = c/c_0$; $Z = x/L$; $\tau_L = L\varepsilon_L/u_{0L}$; $\tau_G = L\varepsilon_G/u_{0G}$; $Bo_L = u_{0L}L/Da_L\varepsilon_L$; $Bo_G = u_{0G}L/Da_G\varepsilon_G$

Fig. 18 A general steady state isothermal heterogeneous dispersion (PD-)model for multiphase catalytic reactors

Pseudo first order reactions with respect to the gaseous reactant, i.e. in case of large excess of B, have been studied in detail to describe systems as oxidation of ethanol, hydrogenation of α-methyl styrene, hydrogenation of aniline etc. Limiting cases, such as plug flow of both gas and liquid phases [46] or a constant concentration in the gas phase [48], were analysed as well as the general case of finite values of dispersion coefficients in both phases [52,58].

But in literature still <u>pseudo-homogeneous models</u> are in use for scale-up in trickle-bed reactors, especially if hydrotreating processes of the petroleum industry are concerned, where the whole reaction dynamics is lumped totally either into the gas or the liquid phase.

E.g. in the so-called "pseudo-equilibrium model", developed by Sylvester [53-56], the same design procedure is used as in a single phase catalytic gas phase reaction, where the mass transfer resistance is replaced by a suitable overall term. Bulk flow and dispersion of the liquid phase are neglected and the whole transport mechanisms are lumped into the equilibrium of the reactant concentrations between gas-, liquid- and particle phase. It is an application of the same principle used successfully in fluid/fluid reactions [57]. But the necessary precondition is that the rate of reaction is slow compared to the transfer rate across the phase boundaries, so that equilibrium can really by assured. This might be justified in some of the hydrotreating processes, but certainly not in case of an aqueous liquid phase, existing in waste water treating. Earlier models used in petroleum industry have taken into account the liquid phase only and have been further simplified by neglecting any

- extraparticle mass transfer limitations in the liquid phase
- stagnant zones in the liquid phase
- homogeneous reactions and
- heat effect, i.e. isothermal operation is supposed

Furthermore, very often a first-order irreversible reaction with respect to the liquid reactant has been assumed (for a second order rate equations see [63]). Depending on the lumping of fluid-dynamics either into axial dispersion, liquid holdup or partial wetting of the catalyst; these oversimplifications result in the relations shown in Fig. 19 for the chemical conversion.

Paraskos [59] and Montagna [43,60] have tested the validity of these simplifications with hydroprocessing reactions of gas oils performed in pilot plant reactors. It resulted that log-log plots of (c/c_o) versus $1/(LHSV)$ and versus L respectively some times gave straight lines for desulphurization, demetalization and denitrogeneation reactions, but with varying slopes and dependent additionally on the nature of feed, temperature and catalyst size; i.e. an unsatisfying situation.

Physically, the effective wetting model seems to be the most appropriate one. This is supported by the fact that also hydrodesulphurization of vacuum and atmospheric residuals are better correlated by an effective catalyst wetting model than by the holdup model [60].

On the other hand, both models do not take into account axial dispersion. Those certainly can have a significant effect on reac-

Plugflow model :

$$-\ln\frac{c}{co} = (1-\varepsilon)\,\eta\,k_v(L/u_L) = (1-\varepsilon)\,\eta\,k_v\,3600\,(LHSV)^{-1}$$

Axial dispersion model : (Wehner and Wilhelm 1959)

$$-\ln\frac{c}{co} = (1-\varepsilon)\,\eta\,k_v\,3600\,(LHSV)^{-1} - \frac{1}{Bo}(1-\varepsilon)^2\,\eta^2\,k_v^2\,3600^2\,(LHSV)^{-2}$$

External holdup model : (Henry and Gilbert 1973)

$$-\ln\frac{c}{co} \propto (1-\varepsilon)\,\eta\,k_v\,(LHSV)^{-0.66}\,L^{0.33}\,d_p^{-0.66}\,v_L^{0.33}$$

Effective wetting model : (Mears 1974)

$$-\ln\frac{c}{co} \propto (1-\varepsilon)\,\eta\,k_v\,(LHSV)^{-0.68}\,L^{0.32}\,d_p^{0.18}\,v_L^{-0.05}\,(\sigma_c/\sigma_\omega)^{0.21}$$

Generalized :

$$-\ln\frac{c}{co} \propto (LHSV)^{(1-\omega)}\,L^{\omega}$$

ω depending on reaction type

Fig. 19 Pseudo-homogeneous models used in petroleum industry for scale-up (conversion as function of operating variables, 1st order reaction) tor performance in small trickle-beds, particularly if they are packed with large catalyst particles and operated at low liquid flow rates. It is interesting to note [43] that the bed length effect observed in desulphurization can be explained just as well on the basis of an axial dispersion model as on the basis of an effective wetting model. This shows therefore the insufficiency of the lumped description of the fluiddynamics of a trickle-bed reactor.

On the other hand, an experimentally obtained residence time distribution E(t), representative for fluiddynamic effects discussed in this lecture, can be easily combined with the intrinsic kinetics of a first order reaction by taking the integral

$$c/c_o = \int_0^{\infty} E(t)\, e^{-k\tau} dt \qquad (8)$$

This formula has been also used for scale-up, which is correct provided the reaction really follows the 1st order kinetics. Otherwise, also micromixing and segregation is of importance [64].

Large-scale hydroprocessing trickle-bed reactors normally operate under adiabatic conditions; therefore, heat effects caused by the reaction must also be included. Shah [61] showed that in this case the critical Bodenstein number for elimination of axial dispersion effects is a function of a heat parameter as well as a modified Damköhler number. For low Damköhler numbers smaller critical Bodenstein numbers than in isothermal reactors are sufficient to eliminate axial dispersion in adiabatic reactors, whereas the inverse is true for large Damköhler numbers.

In many hydroprocessing operations a significant evaporation of the liquid phase may also occur. The modeling of such a reactor is very complex because the catalyst has a different reaction rate on wetted and dry surface. No satisfactory scale-up method has been published until now.

b) Other models

Finally, it should be noted that more sophisticated models have been developed, either on a stagewise basis [62], similar to the Deans-Lapidus model for single phase fixed bed reactors, or on a stochastic respectively propabilistic basis [67,66]. Using data from laboratory and full-scale reactors, Schwarz and Roberts [44] have carried out parametric studies to evaluate the accuracy of the axial dispersion model. Their simulation showed that in case of first-order kinetics, dispersion in the liquid phase is frequently not of major importance. Deviations from plug flow become important only for short reactors and a high degree of conversion.

It appears at present that although crossflow and other macromixing models give a more correct description of the flow of a dispersed phase, the dispersion model predicts the conversion data satisfactorily, at least for simple reactions. In future, cell models probably will become more attractive, because they are closer to reality, making allowance for a better modeling of fluid-dynamics via the percolation theory [69].

c) Conclusions

This review on concurrently operated multiphase packed bed reactors shows that much information on the behavior of these reactor types has been accumulated in the past, but we are still far from a complete elucidation. The difficulty still exists that not enough information is available on systems different from air/water nonporous packings to safely scale-up multiphase reactors using a sophisticated mathematical model. The fact that fluid-dynamics and thermal effects may be different in laboratory units from those in technical reactors restricts the usefulness of simplified, i.e. lumped, models in reactor scale-up. On the contrary, the different mechanisms acting in multiphase catalytic reactions have to be kept separated to a certain extent, thus enabling the correct inclusion of their probably changing amount of influence during scale-up.

Because several hydroprocessing reactors operate in the pulse regime, we need experience in the application of the above-mentioned models in this range. The same is true for the modeling of nonisothermal gas/liquid catalytic reactors, where the chemical conversion is accompanied by the evolution of a considerable amount of heat, causing either a heat flux to reactor walls or the evaporation of an important part of the liquid phase.

In upflow bubble operation the consumption of the gas phase by reaction must also be considered in the model if the reactor operates under lower pressure ($<$20 bar) and if the reactor length is of technical dimensions ($L > 2$ m); additionally gas phase dispersion (radial and axial) may have an influence on conversion [65]. As this reactor type is also used in waste water treatment as well as in fermentation processes, the possible non-Newtonian behavior of the liquid phase as well as the coalescence behavior of the system must be taken into account. Finally, it should be remembered that - comparable to fluidized bed reactors - results from laboratory reactors with small column diameter and/or particle sizes smaller than 0.2 cm usually cannot be regarded as representative for technical upflow units, because capillary force as well as lare scale circulation in the liquid phase may be significantly different.

Literature

[1] P.A.Ramachandran, R.V.Chaudhari, Three-Phase Catalytic Reactors, Gordon and Breach Science Publishers 1981
[2] Shah,Y.T., Gas/Liquid/Solid Reactor Design, Mc Graw Hill Inc. 1979
[3] Hirose,T., Proc.Symp.Mult.Phase Concur. Fixed Beds, Okayama 1978, p. 103
[4] Colombo,A.J., Baldi,G., Sicardi,S., Chem.Eng.Sci.31 (1976) 1101
[5] Goto,S., Levec,J., Smith,J.M., Catal.Rev.Sci.Eng. 15 (1977) 187
[6] Talmor,E., AIChEJ 23 (1977) 868
[7] Kirillov,V.A., J.Eng.Phys. 31 (1976) 1010
[8] Ogarkov,B.L., J.Eng.Phys. 31 (1976) 1274
[9] Sicardi,S., Gerhard,H., Hofmann,H., Chem.Eng.J. 18 (1979) 173
[10] Sato,Y., Hirose,T., Ida,T., Kagaku Kogaku 38 (1974) 534
[11] Specchia,V., Baldi,G., Chem.Eng.Sci. 32 (1977) 515
[12] Heilmann,W., Hofmann,H., Proc. 4th Eur.Symp.Chem.React.Eng., Brussels 1968, p. 169
[13] Turpin,J.L., Huntington,R.L., AIChEJ 13 (1967) 1196
[14] Sylvester,N.D., Pitayagulsaru,P., Can.J.Chem.Eng. 52 (1974) 539
[15] Billet,R., Industrielle Destillation, Verlag Chemie, Weinheim 1973
[16] Sedriks,W., Kenney,C.N., Chem.Eng.Sci. 28 (1973) 559
[17] Satterfield,C.N. AIChEJ 21 (1975) 209
[18] Bondi,A., Chem.Technol. (1971) 185
[19] Ford,L.H., Ph.D. thesis Univ. London 1960
[20] Weber,H.H., Diss. TH Darmstadt 1961
[21] Hofmann,H., Chem.Eng.Sci. 14 (1961) 193
[22] Ohshima,S., Kag.Ronb. 3 (1977) 406
[23] Heilmann,W., Hofmann,H., Proc. 4th Symp.Chem.React.Eng., Amsterdam 1971, p. 169
[24] Specchia,V., Sicardi,S., Gianetto,A., AIChEJ 20 (1974) 1172
[25] Ohshima,S., J.Chem.Eng.Jap. 9 (1976) 29
[26] Hirose,T., Kagaku Kogaku Gijutsu 26 (1974) 199
[27] Specchia,V., Baldi,G., Gianetto,A., Proc. 4th Int.Symp.Chem. React.Eng., Heidelberg 1976, 390
[28] Gianetto,A., Baldi,G., Specchia,V., Ing.Chim. (Milano) 6 (1970) 125
[29] Hirose,T., Toda,M., Sato,Y., J.Chem.Eng.Jap. 7 (1974) 187
[30] Fukushima,S., Kusaka,K., J.Chem.Eng.Jap. 10 (1974) 468
[31] Shende,B.W., Sharma,H.M., Chem.Eng.Sci. 29 (1974) 1763
[32] Mochizuki,S., SCEJ. Nihama (1977) 9
[33] Hashimoto,K., Kagaku Kogaku Ronb. 2 (1976) 53
[34] Weekman jr.,V.W., AIChEJ 11 (1965) 13
[35] Muroyama,K., Kagaku Kogaku Ronb. 3 (1977) 612
[36] Kato,Y., Kagaku Kogaku Ronb. 4 (1978) 328
[37] Muroyama,K., Proc.Symp.Mult. Concur. Fixed Beds, Okayama 1978, C-1

[38] Nahamura,M., Proc.Symp.Mult. Concur. Fixed Beds, Okayama 1978, C-3
[39] Reiss,L.P., J & EC Proc.Des.Dev. 6 (1967) 486
[40] Henry,H.C., Gilbert,J.B., J & EC Proc.Des.Dev. 12 (1973) 328
[41] van Deemter,J.J., Proc. 3rd Eur.Symp.Chem.React.Eng., Amsterdam 1964, p. 215
[42] Paraskos,J.A., Shah,Y.T., Chem.Eng.Sci. 30 (1975) 1169
[43] Montagua,A.A., Shah,Y.T., J & EC Proc.Des.Dev. 14 (1975) 479
[44] Schwartz,J.G., Roberts,G.W., J & EC Proc.Des.Dev. 12 (1973) 262
[45] Sicardi,S., Baldi,G., Specchia,V., AIChEJ, to be published
[46] Sato,Y., Hirose,T., Takahashi,F., Toda,M., 1st Pac.Chem.Eng. Congr., Kyoto 1972, paper 8-3
[47] Mears,D.E., Chem.Eng.Sci. 26 (1971) 1361
[48] Goto,S., Watabe,S., Matsubara,M., Can.J.Chem.Eng. 54 (1976) 551
[49] Hochmann,J.M., Effron,E., I & EC Fund. 8 (1969) 63
[50] Sylvester,N.D., Pitayagulsarn,P., Can.J.Chem.Eng. 53 (1975) 599
[51] Dudukovic,M.P., AIChEJ 23 (1977) 940
[52] Goto,S., AIChEJ 24 (1978) 294
[53] Sylvester,N.D., Can.J.Chem.Eng. 52 (1974) 539
[54] Sylvester,N.D., AIChEJ 19 (1973) 640
[55] Sylvester,N.D., Can.J.Chem.Eng. 53 (1975) 313
[56] Sylvester,N.D., Water Res. 9 (1975) 447
[57] Hofmann,H., Proc. 1st Eur.Symp.Chem.React.Eng., Amsterdam 1957, p. 113
[58] Østergard,K., Adv.Chem.Eng. 7 (1968) 71, Adv.Chem.Ser. 26 (1971) 1361
[59] Paraskos,J.A., Frayer,J.A., Shah,Y.T., J & EC Proc.Des.Dev. 14 (1975) 315
[60] Montagua,A.A., Shah,Y.T., Paraskos,J.A., J & EC Proc.Des.Dev. 16 (1977) 152
[61] Shah,Y.T., Paraskos,J.A., Chem.Eng.Sci. 30 (1975) 1169
[62] Ramachandran,P.A., Smith,J.M., Chem.Eng.J. 17 (1979) 91
[63] Satori,H., Nishizaki,S., Int.Chem.Eng. 11 (1971) 339
[64] Ross, L.D., Chem.Eng.Progr. 61 (1965), No. 10, p. 77
[65] Deckwer,W.D., Chem.Eng.Sci. 31 (1976) 309
[66] Buffham,B.A., Gibilaro,L.G., AIChEJ 16 (1970) 218
[67] Schmalzer,D.K., Hoelscher,H.E., AIChEJ 17 (1971) 104
[68] Charpentier,J.C., Proust,C., van Swaaij,W., Le Goff,P., Chim.Ind.Gen.Chem. 99 (1968) 803
[69] Crine, Chem.Eng.Sci. 35 (1980) 51
[70] Ohshima,S., Suzuki,M., Shimada,K, Takematsu, Kuriki,Y. & Kato,J., Proc.Symp.Mult.Concur.Fixed Beds, Okayama 1978, G1-307

RECENT TRENDS IN THE MODELLING OF CATALYTIC TRICKLE-BED REACTORS

Michel CRINE
Chercheur qualifié F.N.R.S

Guy A. L'HOMME
Professeur de Génie Chimique

Département de Génie Chimique
Université de Liège
4000 LIEGE BELGIUM.

1. INTRODUCTION

A trickle-bed reactor is one in which gas and liquid flow cocurrently downward through a fixed bed of catalyst particles. In many cases, this type of reactor provides the best way of carrying out a reaction between gaseous and liquid reactants in contact with a solid catalyst or an inert packing. That is the reason why these reactors are widely used in chemical and petrochemical industries as well as in biotechnology and waste water treatment. Reviews of all these applications have been published recently by Germain et al. (1979) and Shah (1979). Industrial processes using trickle-bed reactors are also described by Germain in another section of this book.

The design and the scale-up of trickle-bed reactors are still rather difficult problems despite of the high research activity in this area for many years. As a matter of fact, an accurate modelling of these reactors should basically involve the knowledge of the fluid flow hydrodynamics as well as the various heat and mass transport resistances between the three phases. The various attempts in modelling these processes and in predicting the performance of the reactor may be classified into three categories.

1.1. Models empirically related to some hydrodynamic characteristics

In this category, the apparent reaction rate is empirically related to hydrodynamic characteristics on the basis of experimental observations. These hydrodynamic quantities are supposed to represent the quality of the contact between the liquid and the solid. Obviously, the chemical reaction is assumed to occur only in the irrigated zones of the catalytic bed in order to have a simple relationship between the apparent reaction rate and the quality of contact between liquid and solid. The models suggested by Henry et al. (1973) and by Mears (1974) belong to this category. The apparent reaction rate is assumed to be proportional to the liquid holdup in the first model and to the effective catalyst irrigation rate in the second one. These hydrodynamic quantities are estimated using empirical correlations based on experiments. Actually, both models lead to the following relation between the apparent reaction rate $\langle r_a \rangle$ and the liquid superficial velocity $\langle L \rangle$

$$\langle r_a \rangle = \alpha \, \eta_p \, r_o \, \langle L \rangle^{\beta} \qquad (1)$$

The variables under < > are bed scale averaged. $\eta_p \, r_o$ is the intrinsic reaction rate including the particle effectiveness factor. α and β are empirical factors. Henry et al. and Mears suggested a value of about 1/3 for β.
Bondi (1971) developed empirical relation

$$\langle r_a \rangle = \eta_p \, r_o \left(1 + \frac{\alpha \, \eta_p \, r_o}{\langle L \rangle^{\beta}}\right)^{-1} \qquad (2)$$

in which the second righthand term represents the ratio between the $\eta_p \, r_o$ and an effective external mass transfer resistance. The exponent β should have a value of about 2/3. Actually, when fitting Eq.1 or 2 on experimental data, the values of parameters α and β vary markedly with the operating conditions. Examples of such variations have been obtained by Paraskos et al. (1975) which used Eq.1 to correlate results of desulfurization, demetalization (Ni and V removal) and denitrogenation of petroleum atmospheric residues. The exponent β ranges between 0.468 and 0.078 as depicted in figure 1, depending on the temperature and the reaction considered. Similar results would be obtained using Eq.2. Actually, the adopted oversimplifications prevent any physical interpretation of the variations of parameters α and β.

1.2. Residence Time Distribution models

In these models, the interactions between the chemical reaction and the transport processes are described in some more details than in first category. The numerous elementary transport processes are lumped together into some effective terms, using different simplification rules. These models are generally based on the Residence Time Distribution (R.T.D.) of the fluid phases. This formulation is convenient because the R.T.D. can be determined experimentally by well established stimulus-response techniques. The resultant R.T.D. reflects bulk pheno-

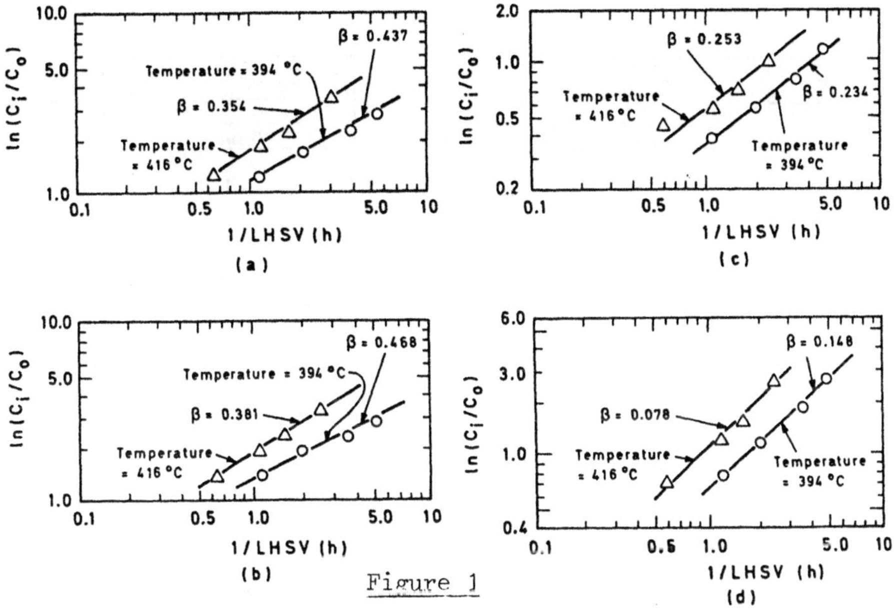

Figure 1

First-order kinetic plots for hydrotreatment reactions (after Paraskos et al., 1975).

mena such as nonuniform velocity, fluctuations due to molecular or eddy diffusivity, reactor shape and internals, backflow of fluid due to velocity differences between phases,.... The R.T.D. models may be classified into two classes :

- the differential models in which the lumped transport processes are described by differential balance equations applied to volume elements;
- the stagewise models in which the transport processes are described by a set of algebraic balance equations applied to staged regions.

The main differential R.T.D. models are schematized in figure 2:

- the piston flow model (figure 2a) in which each volume element is assumed to have an equal residence time;
- the axial dispersion model (figure 2b) in which the observed residence time dispersion is represented by a single Fick's law-type mechanism (parameter Bo_L) superimposed to the plug flow;
- the piston exchange model (figure 2c) in which mass exchange (k_s) between the flowing zone (h_d) and a stagnant one (h_s) is superimposed to the plug flow (Hochmann et al., 1969);
- the time delay models (figures 2d and 2e) which assume a plug flow region (h_d) with fluid elements randomly delayed (t_D) in stagnant zones (h_s) perfectly or not perfectly (m) mixed (Buffham et al., 1970; Oorts et al., 1974);
- the piston dispersion exchange model (figure 2f) similar to the piston exchange model (figure 2c), except that axial dis-

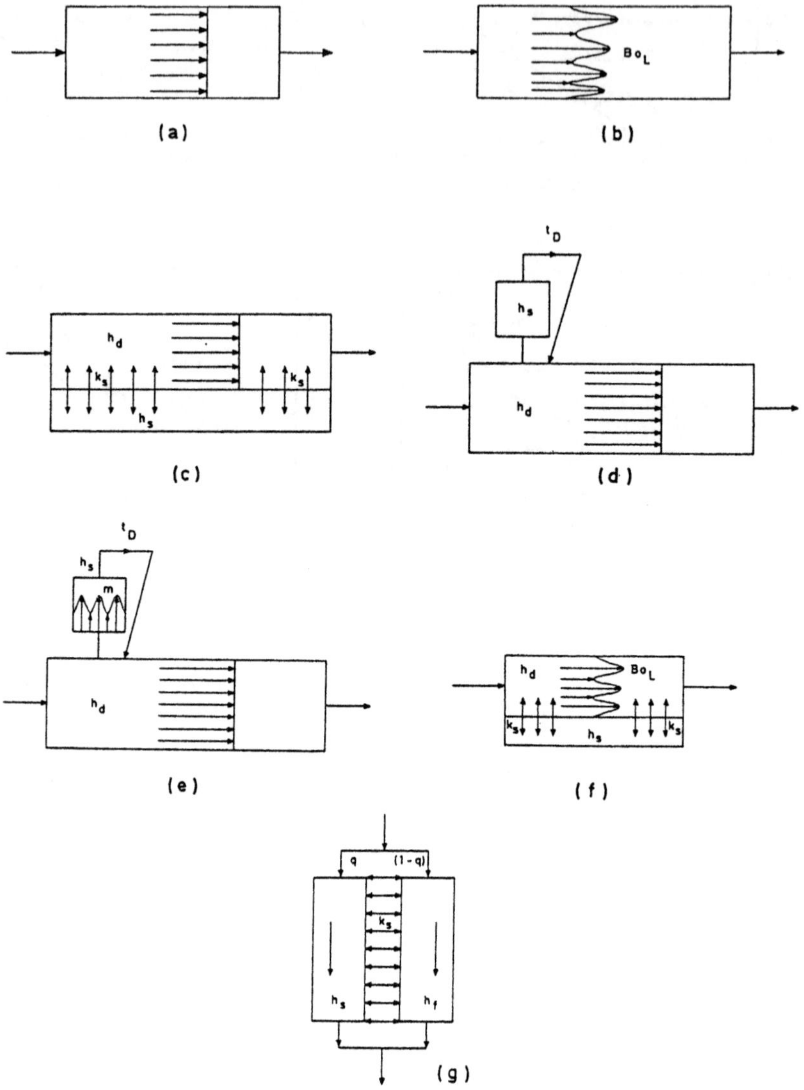

<u>Figure 2</u>

Differential R.T.D. models

persion is superimposed to the plug flow in the flowing zone (Van Swaaij et al., 1969);

- the pulsed flow model (figure 2g) in which the liquid flow is distributed (q) between two piston flow regions (h_s and h_f) with continuous mass transfer (k_s) between them (Lerou et al., 1980).

The main stagewise R.T.D. models are schematized in figure 3 :

- The cell model (figure 3a) in which the parameter is the number N of cells connected in series (Deans et al., 1980);
- The cell exchange model (figure 3b) in which mass exchange (k_s) between the flowing cells (h_d) and stagnant zones (h_s) is added (Deans, 1963);
- The parallel cell model (figure 3c) in which each stage is made of two cells (h_d, h_s) in parallel between which the liquid flow is distributed (q)(Van Swaaij et al., 1969);
- The cell exchange by-pass model (figure 3d) similar to a cell exchange model to which a by-pass (q) is added (Raghuraman et al., 1973);
- The backflow cell model (figure 3e) in which a backflow (q) is superimposed to the net flow through each cell;
- The backflow cell exchange model (figure 3f) which consists in a cell exchange model to which backflow (q) is superimposed (Popovic et al., 1976).

Combination of stagewise and differential R.T.D. models have also been proposed by Michell et al. (1972) and by Rao et al. (1976). Actually most of the models reported above are strongly related to each other, even when they apparently differ considerably (e.g. differential and stagewise models). This is primarily due to the fact that they represent, with various degrees of complexity, the lumping of similar elementary processes. Furthermore, these models are based on the same "continuum" representation of the transport processes. The hydrodynamic parameters introduced in the R.T.D. models are generally related to local gas-liquid-solid interactions and are, consequently, strongly influenced by the heterogeneous nature of gas-liquid flow.

Actually the parameters are introduced on empirical basis and have to be determined experimentally in each case. This fact is confirmed by the numerous and contradictory parameter correlations reported in the literature (see Gianetto et al. (1978)).

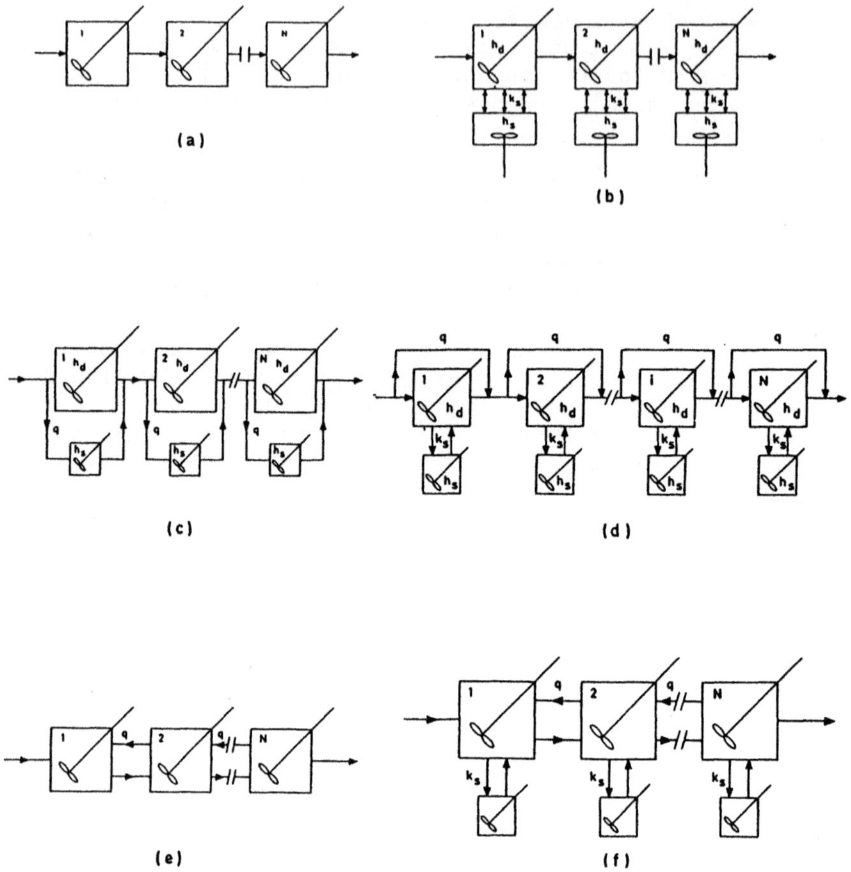

Figure 3
Stagewise R.T.D. models

1.3. Models based on a local description of the gas-liquid-solid interactions

In this category, one may classify all the attempts in describing theoretically the fluid-solid interactions. The main characteristic of these models is a local description of the fluid flow hydrodynamics. Some models for the prediction of the liquid holdup are worth noting in this category.
Davidson et al. (1959) and Buchanan (1967) treated the packing as a set of planar facets on which gas and liquid flow cocurrently. In more recent works, Reynier et al. (1971) and Clements (1976) represented the packing by a bundle of pores or capillaries. These pores join two neighbour contact points between particles and are delimited by the external surface of these particles. In all the reported models, the local hydrodynamic description is extended uniformly to the whole packing i.e., each transport cell (a facet or a pore) contains equal fractions of the overall gas and liquid flows. Generally, the agreement of these models with experiments is not very good and requires the introduction of some empirical parameters. These ones are rather sensitive to the physico-chemical properties of the system and have to be determined experimentally as for the models belonging to the two first categories.

1.4. Comments on the existing models

All the models reported above are characterized by an homogeneous representation of the fluid flows. The gas and liquid flow rates - or their local averaged values in a transport cell (third category of models) - are assumed to be independent on the position in the packing. This assumption is phenomenologically incorrect. Liquid flow maldistributions are indeed often observed in trickle-bed reactors. As a matter of fact the "homogeneous" models do not represent correctly the fluid-solid interactions they are supposed to quantify and they cannot be related to the operating conditions on sound physical basis. It is consequently logical that the most fundamental problem when using these models, is the characterization of the nature and quality of the contact between phases. For example, although there is no doubt that local liquid velocity and holdup are important in liquid-solid contacting, the fundamental relations are still not well known. Even if they were, the best way is probably not to introduce them in an "homogeneous" model but rather to develop an hydrodynamic model which would be coherent with these fundamental relations. It is not easy to predict the future developments in this field but it is sure that the new models will have to account for the random nature of the packing through which gas and liquid flow. Such a flow process is called a percolation process. We will show how the main concepts of percolation theory may lead to a phenomenological description of the fluid flow hydrodynamics and, in turn, to an accurate modelling of the transport processes in a trickle-bed reactor.

2. THE PERCOLATION PROCESS

Fluids flows through a packed bed may be analyzed at various levels each one leading to completely different observations. If one observes the bed as a whole (see figure 4), it seems reasonable to consider the medium as homogeneous, owing to the small scale at which occur the elementary transport processes. In such a case, the various heat and mass balances could be described at the bed scale by diffusion-like equations. It is indeed the approach in many models (e.g. axial and radial dispersions). However, if one looks the bed closer, e.g. at the particle scale as in the close-up of figure 4, the process representation is completely different. The liquid flow is distributed in different channels according to the local geometrical features of the packing. For example, the number of channels entering and leaving the cell represented by the close-up in figure 4, depends on these features and varies randomly form point to point. The clustering of these channels generates various flow structures at the bed scale between which the liquid velocities are randomly distributed.

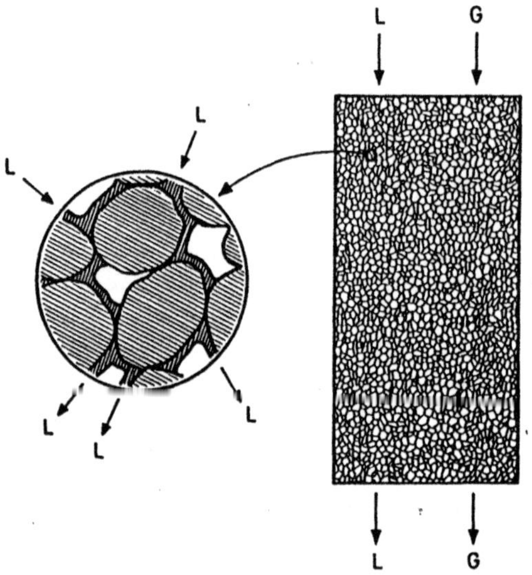

Figure 4

Global and local observation levels of a packed-bed

The flow process analyzed above may be described quantitatively using the percolation theory. The main concepts of this theory have been presented by Broadbent et al. (1957), Frisch et al. (1963) and Kirkpatrick (1973). For an exhaustive analysis of the theoretical developments concerning percolation, the reader is referred to these reviews. Actually, two aspects of a percolation problem may be considered : the static and the dynamic ones. In the first one, the morphology of the flow structures is described whereas in the second one, the flow distribution within these structures is considered.

2.1. Description of the liquid flow structure morphology

To analyze a percolation process, it is useful to represent the scattering medium (i.e. the packed bed) by a lattice as depicted in figure 5. The sites of the lattice correspond to the contact points between the particles whereas the bonds correspond to the pores connecting two neighbour contact points. The walls of these pores are delimited by the external surface of the particles. The percolation process is simulated by randomly distributing blocked and unblocked bonds within the lattice. The clustering of the unblocked bonds generates the flow structures (Crine et al., 1979).

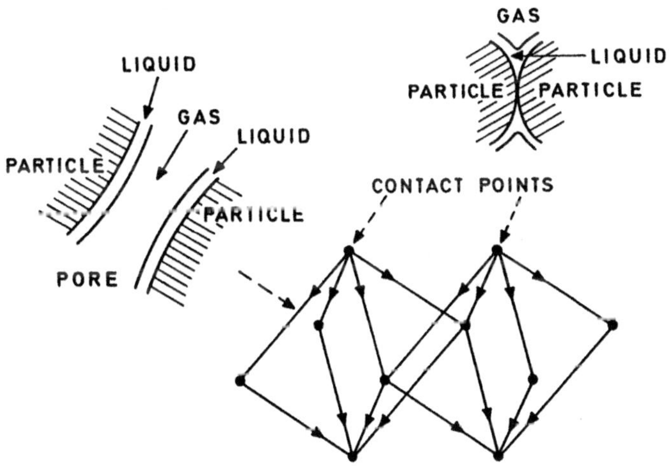

Figure 5

Lattice representation of a packed-bed

A first way in describing a percolation process consists in numerical simulations using Monte Carlo methods. These simulations are carried out by generating a very large number of random walks through the lattice, following the unblocked bonds. The clustering of the random walks represents the liquid flow structures. Further details concerning the numerical algorithm are given elsewhere (Crine et al., 1979).

This lattice representation is an interesting mathematical tool to characterize the morphology of the flow structures. For example, it allows to represent graphically the distribution of these structures in a transverse section of the packing, i.e., in a reticular plane of the lattice. In the example shown here, a centered cubic lattice has been used. In order to visualize clearly the flow structures, cubes (represented by squares in the transverse section) are centered on each bond (Scher et al., 1970). Some applications are presented in figures 6a to 6d. The black squares represent unblocked bonds belonging to the flow structures. The white zone represents the non-irrigated zone. The ratio between the area of black and white zones defines graphically the irrigation rate, i.e. the fraction of particle external surface covered by liquid. For a very small irrigation rate, one may observe isolated black squares (figure 6a). If the irrigation rate increases, a ramification of the flow structures appears. The ramification zones are represented by small clusters of black squares (see figure 6b). The threadlike structures observed in figures 6a and 6b correspond to a rivulet flow, i.e., a flow in which each trajectory is, at least partially, surrounded by non irrigated zones. Above a given threshold of the irrigation rate, the ramification zones are so wide that they form a continuous zone in the transverse section, as depicted in figures 6c and 6d. These new structures correspond to a film flow, i.e., a flow in which each trajectory is completely surrounded by irrigated zones. This transition between rivulet and film flows when increasing the irrigation rate has been actually observed in trickle-bed columns and is at the basis of a model developed some years ago by Charpentier et al. (1968).

The lattice representation allows also to define and to compute easily some quantities characterizing the morphology of the flow structures (Crine et al., 1982a).

- the fraction of bonds belonging to the flow structures;
- the fraction of sites belonging to the flow structures;
- the fraction of sites belonging to the periphery of the flow structures;
- the fraction of sites belonging to the inside of the flow structures.

Making an analogy between the sites of the lattice and the particles of the bed, one may define the following hydrodynamic quantities characterizing the local state of irrigation :

Figure 6 a, b.

Simulations of the liquid flow structure distribution

Figure 6 c, d.

Simulations of the liquid flow structure distribution

- the irrigation rate f_w, which corresponds to the fraction of bonds;
- the fraction f_p of partially or totally irrigated particles, which corresponds to the fraction of sites;
- the fraction f_{pw} of partially irrigated particles, which corresponds to the fraction of peripheral sites;
- the fraction f_{tw} of totally irrigated particles, which corresponds to the fraction of inner sites.

Some theoretical considerations allow to derive analytical correlations between these variables (see Crine et al., 1982a).

$$f_p = \frac{z\ f_w}{1 + (z-1)\ f_w^{\left(\frac{z}{z-1}\right)}} \tag{3}$$

$$f_{pw} = \frac{z^2\ f_w\ (1-f_w)}{\left[1+(z-1)\ f_w^{\left(\frac{z}{z-1}\right)}\right]\left[1+(z-1)\ (1-f_w)^{\left(\frac{z}{z-1}\right)}\right]} \tag{4}$$

and

$$f_{tw} = f_p - f_{pw} \tag{5}$$

z represents the coordination number of the lattice i.e. the number of bonds which can be used when leaving a site. For a centered cubic lattice, z equals 4. Actually, eight bonds are connected to one site in such a lattice but only four of them can be used when leaving a site because of the imposed flow direction. Eq. 3,4 and 5 have been numerically verified by the Monte Carlo methods mentionned above.

The relative importance of the three types of local irrigation is depicted in figure 7 as a function of f_w. According to this figure, in a medium range of f_w values, one may assume nearly all the particles are partially irrigated. Nevertheless, for extreme values of f_w, totally or non irrigated particles have also to be considered. This conclusion is of importance because it means that the global irrigation rate observed at the bed scale is not necessarily equal to the local irrigation rate (observed at the particle scale). The particle effectiveness factor may be very sensitive to this local irrigation rate (see e.g. Tan et al., 1980 and Goto et al., 1981). This is particularly true in extreme cases e.g. with very exothermic reactions involving highly volatile liquid reactants. In such cases, the completely non irrigated particles have a very high activity and a very low heat removal. This may explain the hot spot formation which is sometimes observed with these reactional systems.

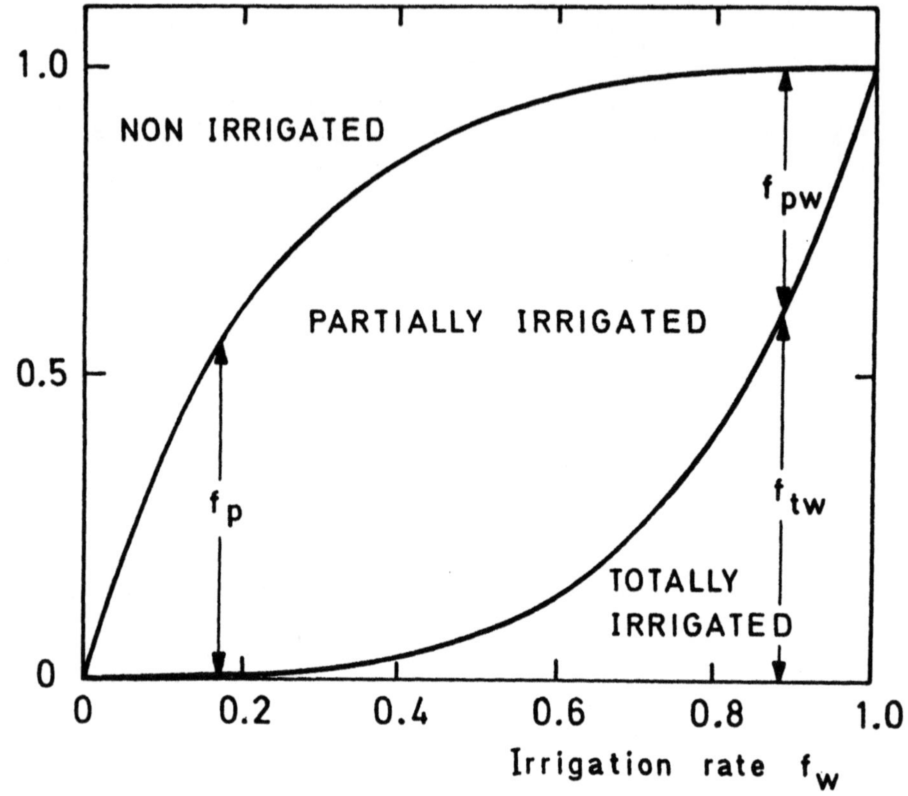

Figure 7

Repartition between non irrigated, partially and totally irrigated particles against the irrigation rate

2.2. Dynamic description of the liquid flow distribution

In the preceeding section, we have shown how the morphology of the liquid flow structures may be described using the percolation theory concepts. This description is static so that the hydrodynamic quantities f_w, f_p, f_{pw} and f_{tw} can be related between themselves but not to the liquid flow density and, consequently, to the operating conditions. As a matter of fact the lattice representation adopted above may also be used to model the liquid flow distribution at the bed scale. In this case, the local liquid flow density is represented by a density of connection through the bonds. Let us assume that this latter one takes the discrete values $1,2\ldots,\infty$ for the bonds belonging to the flow structures (irrigated zones) and 0 for the other ones (non irrigated zones). The stochastic density distribution may be computed, in a classical way, by maximizing the configurational entropy of the investigated process. Further details concerning this procedure are reported in another paper (Crine et al., 1982b).

The solution is given by

$$\alpha_i = \exp\,(a + bi) \qquad i = 0,1,2,\ldots\ldots,\infty \qquad (5)$$

where

$$\exp a = \frac{1}{\langle i\rangle+1} \qquad (6)$$

and

$$\exp b = \frac{\langle i\rangle}{\langle i\rangle+1} \qquad (7)$$

α_i represents the fraction of bonds with a density of connection equal to i. $\langle i\rangle$ is the averaged value of i for the whole lattice. The corresponding flow densities or liquid superficial velocities L_i are proportional to i, i.e.:

$$L_i = i\, L_m \qquad i = 0,1,2,\ldots\ldots\ldots,\infty \qquad (8)$$

and

$$\langle L\rangle = \langle i\rangle\, L_m \qquad (9)$$

L_m represents the minimum local liquid superficial velocity in a channel. It corresponds also to the actual velocity when the channels are completely independent. L_m is related to the dissipation of energy for the creation of an isolated channel. Actually, it characterizes the effective wettability of particles i.e., the wettability under the actual operating conditions. This wettability decreases as L_m increases.

The stochastic density distribution defined by Eq.5 has a normalized standard deviation σ given by (Crine et al., 1982b)

$$\sigma = \exp\left(-\frac{b}{2}\right) = \left(\frac{\langle i\rangle+1}{\langle i\rangle}\right)^{1/2} \qquad (10)$$

This quantity characterizes the heterogeneity of the liquid flow ; it increases with the heterogeneity. σ is very similar to the maldistribution factor defined by Bemer et al.(1978).

Actually, this latter one is given by

$$M = \frac{\sum_{i=1}^{N} \alpha_i \; |i-\langle i\rangle|}{2\langle i\rangle} , \qquad (11)$$

wherease σ may be determined numerically by

$$\sigma^2 = \frac{\sum_{i=1}^{N} \alpha_i \; |i-\langle i\rangle|^2}{\langle i\rangle^2} \qquad (12)$$

Both quantities increase when $\langle i\rangle$ or $\langle L\rangle$ decreases. The same effect is obtained when increasing L_m i.e., when decreasing the particle wettability.

These theoretical dependences are in very good agreement with many experimental observations (see e.g. Porter et al., 1968; Bemer et al., 1978). It is, for example, well known that the liquid flow maldistribution grows when decreasing the liquid flowrate.

3. ONE-DIMENSIONAL MODELLING OF TRANSPORT PROCESSES

Besides the description of gas-liquid flow maldistribution, Eq.5. may also be used to represent the effects of this distribution on the transport processes occurring at the particle scale. We have shown how fluid flows through a packed bed may be observed at various levels, leading to completely different interpretations. Actually, when modelling a transport process, it is always necessary to consider at least two observation levels : the bed scale and the particle scale. The bed scale corresponds to the whole bed or to a volume containing a large number of particles. That is the level at which we want to derive models for the investigated transport processes. However these processes are generally ruled by gas-liquid-solid interactions occurring at the particle scale. That is the reason why it is necessary to model these processes at the particle scale. We have to adopt a representation of the transport cell which is associated to each bond in the lattice defined by the percolation process (see figure 8). This cell is assumed to be exactly the same at any position within the bed. The randomness of the process is indeed accounted for by the percolation process.

The following developments will be restricted to the laminar liquid flow with weak gas-liquid interactions. However, this is not a limitation of the proposed methodology which could be easily applied to any other flow regime.

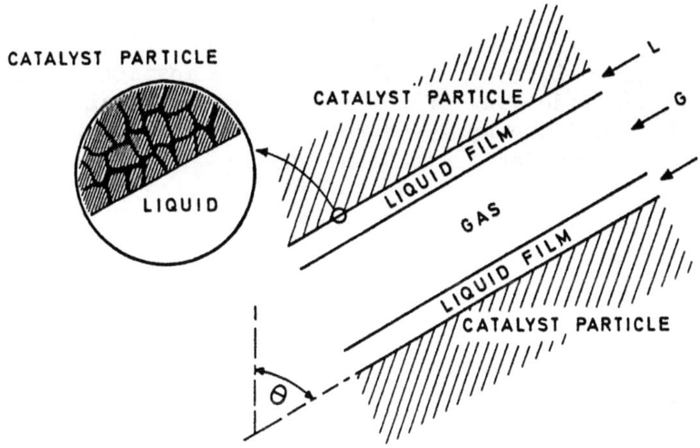

Figure 8

Gas-liquid flow modelling in a local transport cell

The local transport cell is represented by a straight pore with an inclination to the vertical characterized by angle θ (see figure 8). The curvatures of gas-liquid and liquid-solid interfaces are assumed to be negligible. The model also assumes that the laminar liquid flow is motivated only by the gravity or by a modified gravity including the pressure drop effects. In this latter case, ρ_L g is replaced by $\rho_L g + \delta_{LG}$, where δ_{LG} represents the pressure drop per unit of bed length (Specchia et al., 1977). This means that no slip and no stress occur at the gas liquid interface. The local hydrodynamic quantities may be described on the basis of this model.
The change of scale or volume averaging between the particle and bed scales is ruled by the percolation process i.e., by the velocity distribution defined by Eq.5. The averaging formula depends on the nature of the hydrodynamic quantity which has to be averaged. Applications presented hereafter will concern the bed scale averaging of extensive quantities and of the axial dispersion coefficient

3.1. Bed scale averaging of extensive quantities

If φ is an extensive transport property, its averaged value <φ> at the bed scale is given by :

$$\langle\varphi\rangle = \sum_{i=0}^{\infty} \varphi(L_i)\ \alpha_i \qquad (13)$$

where α_i is the fraction of bonds with a local liquid flow density equal to L_i (see Eq. 5 to 9). Applications will be presented for the modelling of the irrigation rate, the dynamic liquid holdup and the apparent reaction rate in the absence of external mass transfer limitations and in the case of non volatile liquid reactants (i.e. approximatively the operating conditions of the petroleum hydrotreatment).

The irrigation rate f_w - i.e. the fraction of external solid surface covered by liquid - locally equals unity when the pore is irrigated and zero in the absence of liquid.

$$f_w = 1 \qquad \text{for } L_i \neq 0 \qquad (14a)$$
$$f_w = 0 \qquad \text{for } L_i = 0 \qquad (14b)$$

The dynamic liquid holdup h_d on planar inclinated surfaces is given by (Crine et al., 1982c.).

$$h_d = \left(\frac{3}{\cos^2\theta}\right)^{1/3} a_S^{2/3} \left[\frac{\mu_L}{\rho_L(\rho_L g+\delta_{LG})}\right]^{1/3} L_i^{1/3} \qquad (15)$$

The angle θ may be estimated by assuming a random distribution of the actual pore inclinations. In this case,

$$\cos\theta = \frac{2}{\pi} \qquad (16)$$

The apparent reaction rate r_a at the level of one pore results from the exchange of mass between the liquid flow and the porous structure of the catalyst particle as depicted in the close-up of figure 8. In the absence of external mass transfer limitations, r_a equals the product of the intrinsic reaction rate r_o by the particle effectiveness factor η_P, the variables being expressed in terms of the liquid bulk concentrations :

$$r_a = \eta_P\, r_o \qquad \text{for } L_i \neq 0 \qquad (17a)$$

In the absence of liquid flow (L_i=0), r_a equals zero, because the liquid reactants are not volatile :

$$r_a = 0 \qquad \text{for } L_i = 0 \qquad (17b)$$

The local extensive quantities f_w, h_d and r_a may be averaged at the bed scale using Eq.13. The averaged value $\langle f_w \rangle$ of the irrigation rate is obtained by introducing Eq.5 , 14a and 14b into Eq.13.

$$\langle f_w \rangle = \sum_{i=1}^{\infty} \alpha_i = 1 - \alpha_o \qquad (18)$$

The value of α_o is explicited by Eq.6 and 9. Finally, one obtains :

$$\langle f_w \rangle = \frac{\langle L \rangle}{\langle L \rangle + L_m} \qquad (19)$$

$\langle f_w \rangle$ is an increasing function of $\langle L \rangle$, reaching asymptotically unity for very large values of $\langle L \rangle$. This dependence is depicted in figure 9 for different values of the parameter L_m which characterizes the wettability of the particles. Clearly, $\langle f_w \rangle$ decreases when increasing L_m. The apparent log-slope of $\langle f_w \rangle$ versus $\langle L \rangle$ equals $1-\langle f_w \rangle$. That means this slope ranges between 1 for a very small irrigation rate (linearity) and 0 for a nearly complete irrigation (asymptotic value). Eq . 19 is compared in figure 9 with empirical correlations obtained by Mills et al. (1981) (curve 1) and Colombo et al. (1976) (curves 2 and 3), by means of tracer experiments. The agreement is remarkable. The estimated values of L_m range between 0.1 and 0.3 kg/m^2s. It is worth noting that the log-slope of the empirical correlations (see figure 9) increases when the quality of irrigation decreases (from curve 1 to 3). This agrees very well with the fact that the apparent log-slope of Eq.19 equals $1-\langle f_w \rangle$.

The averaging of the dynamic liquid holdup is achieved in a similar way by introducing Eq.5 and 15 into Eq. 13.

$$\langle h_d \rangle = \left(\frac{3}{\cos^2\theta}\right)^{1/3} a_S^{\,2/3} \left[\frac{\mu_L}{\rho_L(\rho_L g+\delta_{LG})}\right]^{1/3} \sum_{i=0}^{\infty} L_i^{\,1/3} e^{(a+b\,L_i/L_m)} \qquad (20)$$

Approximating the summation in Eq.20 by an integral, leads to (Crine et al., 1982c)

$$\langle h_d \rangle = 1.74 \langle f_w \rangle^{2/3} a_S^{\,2/3} \left[\frac{\mu_L}{\rho_L(\rho_L g+\delta_{LG})}\right]^{1/3} \langle L \rangle^{1/3} \qquad (21)$$

in which θ has been replaced by the value given by Eq.16. We can put Eq.21 in dimensionless form

$$\langle h_d \rangle = 1.74 \langle f_w \rangle^{2/3} Re_L^{\,1/3} Ga_L^{*\,-1/3} (a_S d_p)^{2/3} \qquad (22)$$

This equation is reproduced in figure 10 for different values of L_m. The experimental system considered in this figure consits in water flowing through a bed of 3 mm diam. spheres with an external porosity equal to 0.35. The pressure drop is assumed to be negligible ($\delta_{LG} \ll \rho_L g$). Comparison of Eq.12 with the correlation proposed by Specchia et al. (1977), for the low gas-liquid interaction regime,

$$\langle h_d \rangle = 3.86\, \varepsilon^{0.35}\, Re_L^{\,0.545}\, Ga_L^{*\,-0.42} (a_S d_p)^{0.65} \qquad (23)$$

is shown in figure 9. The agreement between both relations is very satisfying. The apparent log-slope of Eq.22 versus Re_L equals $1-\frac{2}{3}\langle f_w \rangle$. By comparison with Eq. 23, this yields a mean irrigation rate of about 0.7 in the range 0.1 to 10 kg/m^2s, i.e., a parameter L_m of roughly 0.3 kg/m^2s. This value is in good agreement with the range found when analyzing the irrigation rate correlations (see figure 9).

The averaged value of the apparent reaction rate $\langle r_a \rangle$ is obtained readily using the same procedure as for the irrigation rate.

$$\langle r_a \rangle = \eta_p \, r_o \frac{\langle L \rangle}{\langle L \rangle + L_m} \tag{24}$$

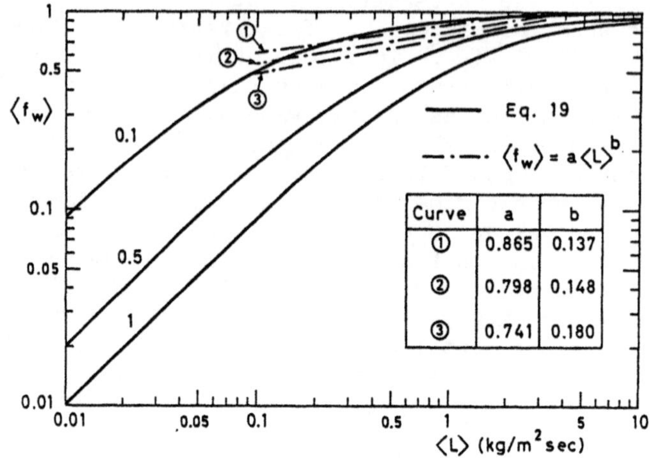

Figure 9

Bed scale irrigation rate against the liquid superficial velocity

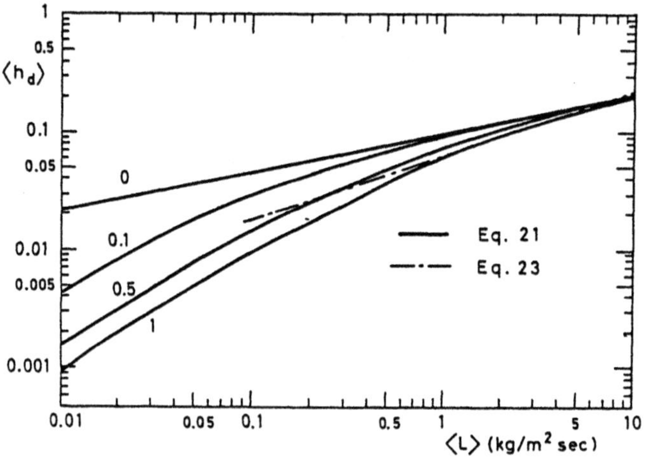

Figure 10

Bed scale dynamic liquid holdup against the liquid superficial velocity.

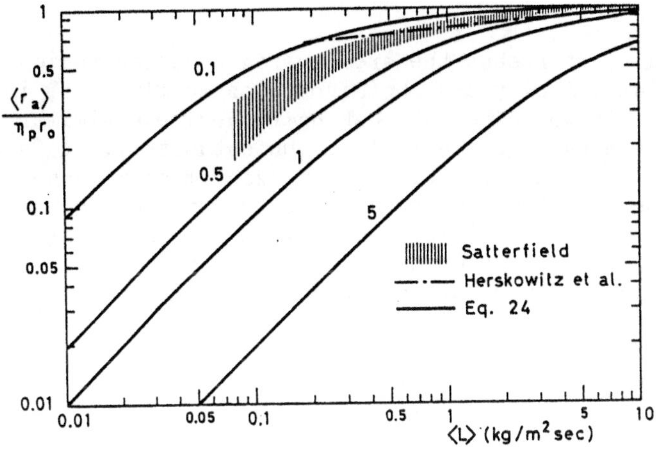

Figure 11

Bed scale apparent reaction rate against the liquid superficial velocity

This relation is represented in figure 11 for different values of L_m and compared with the experimental range of overall effectiveness proposed by Satterfield (1975). The overall effectiveness deduced by Herskowitz et al. (1979) from a study of hydrogenation of α-methylstyrene is also shown in this figure. The agreement is remarkable. The estimated values of the parameter L_m ranges between 0.1 and 0.3 kg/m^{2}s as when analyzing the irrigation rate and liquid holdup correlations. The theoretical curves depicted in figures 9 and 11 are very similar. Actually tracer and apparent reaction rate experiments are often used as two different techniques to estimate the irrigation rate. Here, comparisons are made separately in a sake of clearness.

3.2. Bed scale averaging of the axial dispersion coefficient

The axial dispersion in the liquid phase of a trickle-bed reactor results from transport processes occurring at the particle scale i.e., in each flow trajectory considered as independent, as well from the velocity distribution observed at the bed scale. If we assume that the axial dispersion at the particle scale may be represented by a diffusion-like process, the local mass balance in channel i is given by

$$\frac{\partial C_i}{\partial t} + v_i \frac{\partial C_i}{\partial Z} = D_i \frac{\partial^2 C_i}{\partial Z^2} \qquad (25)$$

C_i, v_i and D_i represent the area-averaged concentration, the intersticial velocity and the dispersion coefficient, in channel i. It would be interesting to derive a similar equation for bed scale averaged variables. Unfortunately, it is impossible to derive such equation in an exact manner because diffusion and percolation processes are ruled by fundamentally different elementary mechanisms. (see e.g. Broadbent et al., 1957). Actually, the stochastic model defined by Eq.5 , describes the liquid velocity distribution and could also be used to characterize numerically the distribution of residence times i.e., the dispersion process. Achwal et al. (1979) drew attention to a procedure using a Markov chain model which led to similar results for the velocity distribution. This model remained essentially numerical and rather cumbersome. Even if Eq.5 has a simple analytical form, its numerical application to estimate the dispersion process is also too complex for practical purposes.

An interesting alternative way could consists in fitting analytically the Residence Time Distribution model defined by Eq.25 on the velocity distribution model defined by Eq. 5. This would lead to an analytical relation between the axial dispersion coefficient and the operating conditions. This way seems to be very convenient because the R.T.D. models are generally easy to use. This methodology could also be applied to other R.T.D. models leading to correlations for the parameters involved in these models. However, the validity of these correlations is strictly limited to the assumptions at the basis of the analytical fitting. The example presented hereafter has been developed for the simulation of tracer experiments without any fluid-solid mass transfer. The derived correlations is consequently not valid when a chemical reaction occurs in the catalyst particle. It is however of interest to interpret tracer experiments in terms of hydrodynamic characteristics.

To account easily for the velocity distribution influence, a very simple model is adopted. It consists in a set of parallel channels (see figure 12). Each channel corresponds to a flow trajectory. All the independent lengths of these trajectories i.e. the lengths between two intersection points are assumed to be equal to a mean channel length L_c. Using this model, Carbonell (1981) derived an approximate expression for the area and time-averaged value of the axial dispersion coefficient $\langle D\rangle$

$$\langle D\rangle = \sum_{i=1}^{N} D_i \omega_i + \langle v\rangle L_c \sum_{i=1}^{N} \omega_i \frac{v_i}{\langle v\rangle} \left(\frac{v_i}{\langle v\rangle} - 1\right) \qquad (26)$$

This relation is obtained by fitting the zeroth, first and second moments of the approximate bed scale area-averaged solution $\langle C\rangle$ of

$$\frac{\partial <C>}{\partial t} + <v> \frac{\partial <C>}{\partial Z} = <D> \frac{\partial^2 <C>}{\partial Z^2} , \qquad (27)$$

on the exact solution given by the averaging of Eq.25.
ω_i represents the relative amount of tracer in channel i. According to Carbonell, ω_i should be equal to the fraction of total volumetric flowrate which flows throughout channel i.

$$\omega_i = \frac{\alpha_i \, v_i \, h_d}{<v> \, <h_d>} \qquad (28)$$

α_i is the fraction of channels with a velocity v_i (see Eq.5). The dynamic holdups h_d and $<h_d>$ are given by Eq.15 and 21. D_i is the dispersion coefficient in channel i. The first term of Eq.26 represents the contribution of the dispersion inside each channel weighted by the relative amount of tracer. The second term is the contribution of the velocity differences between each channel. It represents the convective effect (at the bed scale) and is independent of the dispersion process inside a channel.

Eq. 26 is general whatever the type of velocity distribution and the type of expression for D_i. v_i is related to L_i by

$$v_i = \frac{L_i}{\rho_L \, h_d} \qquad (29)$$

If we assume that the residence time in a channel between two contact points is small enough to neglect transverse molecular diffusion in the channel, D_i is given by (Crine et al., 1982c)

$$D_i = D_m + 1.307 \, v_i \, d_p \qquad (30)$$

where D_m represents the molecular diffusivity generally negligible relative to D_i.
The mean channel length L_c is assumed to be proportional to the particle dimension d_p and inversely proportional to the bed scale averaged value of the irrigation rate $<f_w>$

$$L_c = \phi \frac{d_p}{<f_w>} \qquad (31)$$

ϕ is a shape factor which accounts for the oversimplifications adopted when representing the liquid flow by a set of equal length parallel channels.
Combining Eq.5 , 15, 16, 21, 26, 28, 29, 30 and 31 leads to (see Crine et al., 1982c)

$$<D> = \frac{2 \, <v> \, d_p}{\pi} \left(2.220 + \frac{2.183}{<f_w^{\cdot}>} \right) \qquad (31)$$

We may put Eq.31 in dimensionless form

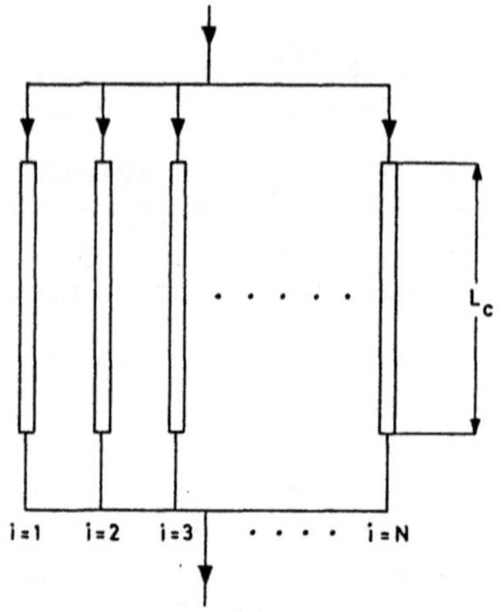

Figure 12

Liquid flow representation by a set of equal length channels

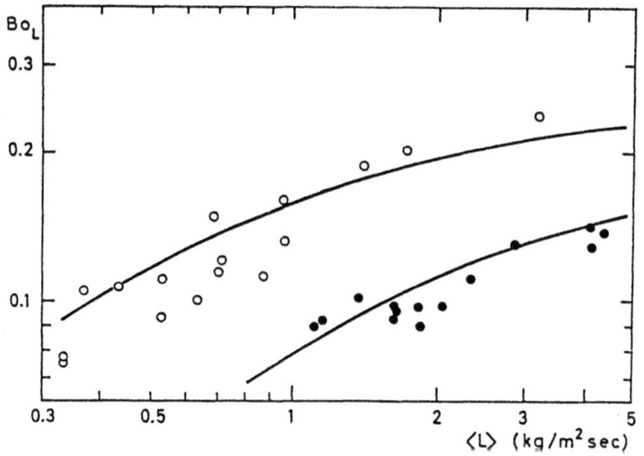

Figure 13

Bed scale Bodenstein number against the liquid superficial velocity. Experimental results after Crine et al., 1981 (○ : good wettability; • : poor wettability)

$$Bo_L = \frac{\pi}{2\ (2.220 + \frac{2.183}{<f_w>}\ \phi)} \qquad (32)$$

$<f_w>$ is given by Eq.19.
In order to assess the validity of Eq.32 as well as to estimate the order of magnitude of the parameters ϕ and L_m, we fitted the equation on two sets of data recently reported by Crine et al. (1982c). In this work, the authors used air and water flowing in a $3.8x10^{-2}$m. glass column packed with $3x10^{-3}$m. glass spheres. The fluid properties were kept constant but two different particle wettabilities were obtained by means of a surface treatment. The Bodenstein number estimates are reported in figure 13 against the liquid superficial velocity for different gas superficial velocities ranging between $2x10^{-2}$ and $5x10^{-1}$ kg/m^2s.
In view of the relatively large experimental scatter, one cannot observe any definite influence of the gas flow rate. Eq.32 has been fitted separately on the two sets of data.
The parameter estimates are reported in the following table :

Wettability	ϕ	L_m (kg/m^2s)
Good	1.7	0.7
Poor	1.7	2.9

The parameter ϕ has been kept constant for the two sets of data. This is of importance because, according to the theory, ϕ should be independent on the fluid and solid properties. Actually it must be a constant for a given packing.
The increase of L_m when decreasing the particle wettability shows clearly that this hydrodynamic parameter characterizes the particle wettability. An increase of the wettability causes the Bodenstein number to increase as indicated in figure 13. Consequently, it seems that this wettability should be accounted for when deriving correlations for the axial dispersion coefficient.

3.3. Comments on the percolation model

The description of a percolation process corresponds closely to the physical flow situation in a trickle-bed reactor. It also gives a good description of important hydrodynamic features such as the existence of non irrigated zones and the liquid velocity distribution in the irrigated ones. The few examples analyzed above indicate also that the postulated model can describe many transport processes. In fact, the list of reported applications is not limitative.

The modelling of the bed scale hydrodynamic contributions to these processes always requires the introduction of the parameter L_m. This term characterizes the interactions between the liquid and the solid. As pointed out above, it represents the local liquid velocity in an isolated rivulet under its more stable flow configuration. This configuration and consequently L_m may be dependent on the actual operating conditions i.e. the ones prevailing under reactor operation (and not the ones prevailing for isolated rivulets).

The knowledge of the influence of these operating conditions on L_m is of crucial interest to develop accurate and practical models of transport processes. Unfortunately, this influence is until now not very well known. We presented previously a first attempt in this direction, by minimizing the dissipation of energy related to the liquid flow creation (Crine (1978)). Adopting the assumptions of the laminar pore flow model (see figure 8), we obtained the following expression for L_m.

$$L_m \div \left(\frac{\mu_L \rho_L^2}{\rho_L g + \delta_{LG}}\right)^{1/5} E_S^{3/5} \qquad (33)$$

The first righthand term involves the parameters controlling the energy dissipation in the bulk of the liquid flow (laminar viscous drag). The second term (E_S) represents the surface energy dissipation due to the creation of the liquid-solid and gas-liquid interfaces. It results from a balance between the surface tensions at both interfaces and the energy of liquid-solid adhesion.

These terms are very sensitive to the local concentrations so that L_m may be strongly related to the concentration heterogeneities in the liquid film. That is the reason why it seems essential to estimate L_m under the actual operating conditions e.g. the actual concentration gradients across the liquid film, which are due to the chemical reaction. The contacting patterns and the irrigation of the catalyst particles are therefore strongly connected with the reactional system. The information obtained from cold flow experiments could be not very useful. In this case, hydrodynamics and transport processes should be measured under chemical operation of the reactor.

4. TWO-DIMENSIONAL MODELLING OF TRANSPORT PROCESSES

In large size industrial trickle-bed reactors, it is difficult and practically impossible to obtain a perfect distribution of fluid flows. In these reactors, one may consider two types of gas-liquid flow maldistribution :

- the maldistribution resulting from interactions between fluid flows and the random structure of the packing; there is no deterministic bed scale relation between the local fluid

flows and the position in the packed bed; this type of distribution may be described by the one-dimensional model described above (see Eq.19)
- the maldistribution resulting from a poor initial gas-liquid distribution (due to the distributor design); this type of distribution varies when moving from the center to the reactor wall, so that a two-dimensional model must be used.

The importance of having adequate flow distribution at the top of a trickle-bed reactor was pointed out in a classic paper by Ross (1965). By using radioactive tracers, he showed that the better performance of a pilot plant reactor than a commercial hydrotreater could be explained in terms of inadequate feed distribution over the catalytic bed. A improved design of the distributor was shown to improve both the liquid holdup and conversion.

Many studies are available dealing with radial spreading from single points or other geometries (Hoftyzer, 1964; Onda et al., 1973; Herskowitz et al., 1979). In these studies, the liquid spreading is supposed to be ruled by some diffusional mechanism, which leads to write a diffusion-like equation introducing a radial spread coefficient D_R

$$\frac{\partial \langle L \rangle}{\partial Z} = D_R \left(\frac{1}{r} \frac{\partial \langle L \rangle}{\partial r} + \frac{\partial^2 \langle L \rangle}{\partial r^2} \right) \qquad (34)$$

In the frame of the two-dimensional analysis, the variables under < > are not averaged over the whole bed but rather over volumes containing a relatively large number of particles. Actually these volumes are the smallest ones beyond which the liquid flow distribution may be considered as homogeneous. This approach was initiated by numerous workers, among them : Scott (1935), Tour et al. (1944) and Cihla et al. (1957). Later, Porter (1968) and Stanek et al. (1965) applied this homogeneous diffusional picture to the radial spreading of rivulets and replaced the liquid flow density by a rivulet density. However, Lespinasse (1962) pointed out that this diffusional picture is unable to account for the occurrence of liquid flow preferential paths stable in time (see also Bemer et al., 1978).

As a matter of fact, the radial spreading may also be represented by a percolation process. At present, there is no analytical model to describe this process but it is possible to simulate it using the lattice representation reported above. The distribution of unblocked bonds within the lattice is ruled by the type of initial liquid flow distribution. In what follows, we will simulate an initial line source distribution. For further details concerning this simulation the reader is referred to a previously published paper (Crine et al., 1980). The unblocked bonds are initially confined in a very narrow zone of the upper reticular plane of the lattice (e.g. a row of sites).

When moving down through the lattice, the accessible zone for the unblocked bond spreads progressively following a binomial law. The local probability for an unblocked bond is thus proportional to a density function P (N_x, N_z). N_x is the lateral displacement from the initial raw of sites whereas N_z is the vertical displacement. An example of flow simulation is shown in figure 14. In order to visualize the liquid flow spreading circles are centered on each unblocked site of a transverse reticular plane. The radii of these circles equal half the longest distance between neighbour sites. In that manner neighbour circles with the same flow density overlap and form a continuous zone. Three different contour lines are represented in figure 14 by the envelop curves of the circles corresponding to three different range of flow density. This figure visualizes the liquid flow spreading from the initial center line toward the outer zones of the bed. The irregular shapes of the contour lines correspond the liquid flow heterogeneities. An homogeneous spreading would indeed be represented by straight contour lines parallel to the initial center line.

It is also interesting to analyze the evolution of the liquid flow spreading when moving down through the bed and to compare the results with the diffusion model. Some typical results are depicted in figures 15 and 16.

The graphs represent the histogram of the liquid flow density along a transverse axis (in arbitrary units). The histogram is depicted together with the binomial distribution which corresponds to the diffusion model. The examples shown in figures 15 and 16 correspond to an irrigation rate of 10 percent and to vertical displacements of 28 and 140 lattice units, respectively. One may observe that the spreading obtained by using the percolation simulation is always smaller than the one predicted by the diffusion model. The gap between both results increases as the vertical displacement increases. The liquid flow seems to be trapped in a central zone of the bed.

The unaccessibility of some regions of the scattering medium is a typical property of a percolation process. For further details concerning this property and the main differences between percolation and diffusion processes, the reader is referred to the paper by Broadbent et al. (1957).

This liquid flow trapping affects the quality of distribution. Hoftyzer (1964) analyzed the dispersion from a variety of distributors e.g., a central stream, disc, ring, annular disc and an eccentric stream. This study showed that the characteristic dimensionless group for all geometries is given by

$$K = \frac{D_R \, N_z}{N_x^2} \qquad (35)$$

This relation may be used to derive a criterion for the quality of distribution. This distribution will be considered as

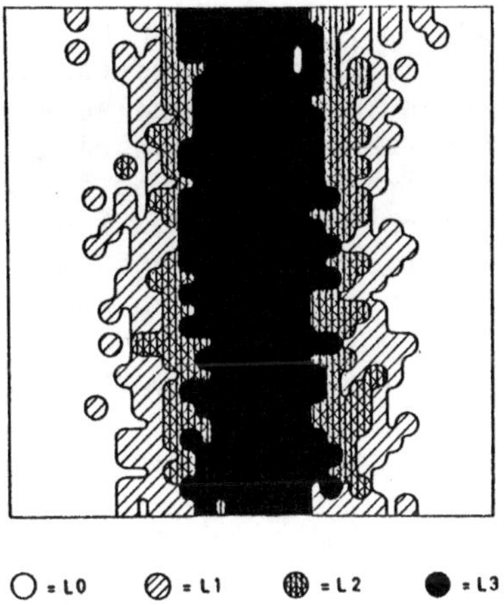

Figure 14

Simulation of the liquid flow spreading ($<f_w>_\infty$ = 0.30; L_o = 0.0; $L_1 \leqslant 2.3 <L>_\infty$; $L_2 \leqslant 3.8 <L>_\infty$; $L_3 > 3.8 <L>_\infty$)

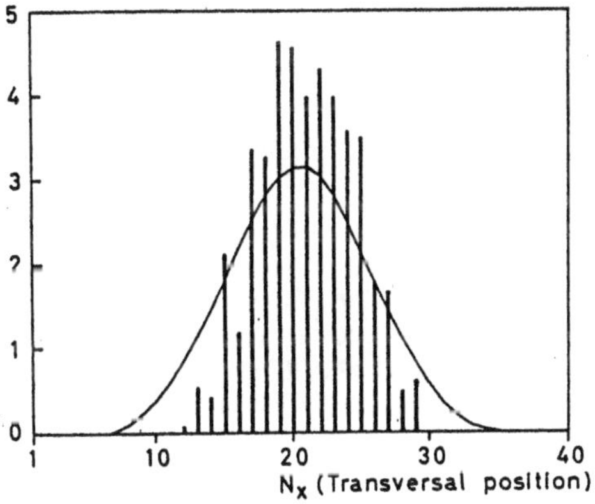

Figure 15

Lateral distribution of the liquid flow density ($<f_W>_\infty$ = 0.1; N_Z = 28)

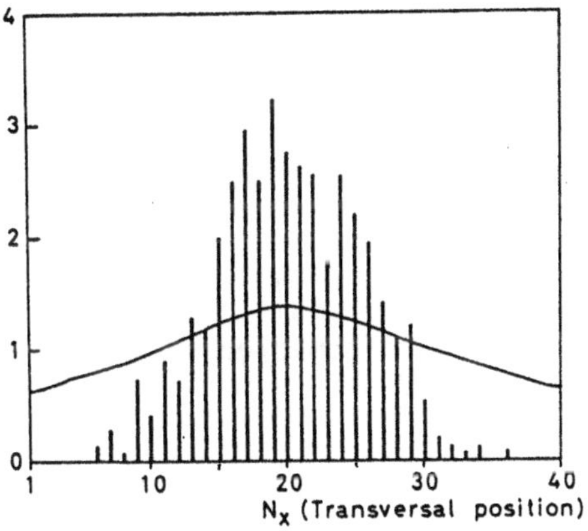

Figure 16

Lateral distribution of the liquid flow density ($<f_W>_\infty$ = 0.1; N_Z = 140) acceptable if K is beyond a given threshold. Obviously, the choice of this threshold is judgemental and arbitrary depending on what distribution is considered as acceptable. It is interesting to note that the bed length required to achieve a given flow distribution increases as the square of the distance between the distribuion points. If this length is generally insignificant in laboratory scale reactors, it may be very important in large industrial units in which the minimum distance between the distribution points is often limited by technological requirements. This remark is very important when deriving scale-up rules.

The values of K characterizing our simulations were obtained by fitting a gaussian function on the percolation simulations and the binomial function. As the simulated liquid flow distribution is far from being gaussian (e.g. figure 15 and 16), the spreading coefficient D_R estimated following this procedure is biased by a large regression error and has little physical significance. Nevertheless it allows a straighforward comparison with the diffusional model. The values of K are reproduced, in arbitrary units, in figure 17 as a function of the vertical displacement (i.e. the bed length) and for different infinite length irrigation rates $\langle f_W \rangle_\infty$ (i.e. the values reached after a complete spreading). The spreading predicted by the diffusional model increases continuously with N_Z. The values obtained with the percolation simulations are much more smaller; the smaller irrigation rates, the smaller K. For very small irrigation rates, the value of K seems to reach a constant and maximum value, indicating clearly a liquid flow trapping in the center of the bed. This accounts remarkably for various experimental results reported in the literature (Beimesch et al., 1971; Specchia et al., 1974; Crine, 1978; Van Landeghem, 1980).

As conclusion of this first approach of the two-dimensional modelling of trickle-bed reactors, it seems possible to describe the radial liquid flow spreading in terms of a percolation process. The numerical simulations presented above evidence the liquid flow trapping. This phenomenon affects the quality of the liquid flow distribution. It should be accounted for when designing a distributor, a redistributor or any other internal. A study by Jaffe (1976) shows indeed that the shadowing effect from reactor internals can cause low flow to persist for a large distance. This study was done for a hydrocracking reactor in which hot spots were observed.

The numerical simulations reported here are however to cumbersome to be applied in a practical reactor model. It would be of interest to derive an analytical expression describing the radial spreading in terms of the hydrodynamic parameter L_m (see the one-dimensional modelling section). This would allow us to relate the spreading phenomenon to the operating conditions and to the nature of the reactional system.

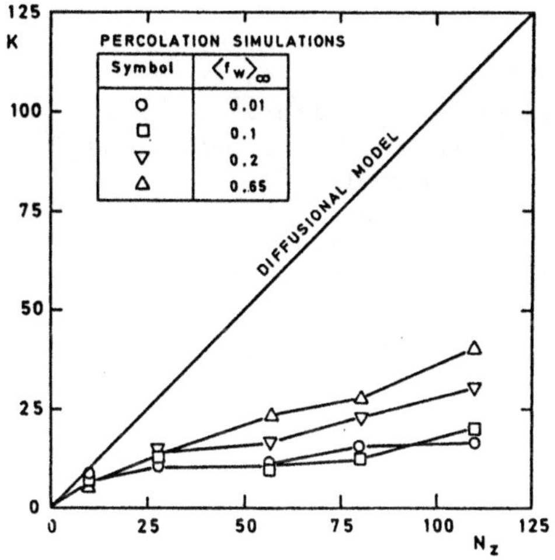

Figure 17

Distribution quality against the vertical displacement

NOMENCLATURE

a,b	variables defined in Eq.6 and 7.
a_S	specific area of the packing material per unit bed volume.
Bo_L	Bodenstein number of the liquid phase $[\frac{\langle v\rangle\ d_p}{\langle D\rangle}]$
C	area-averaged concentration.
D	area-averaged dispersion coefficient.
D_m	molecular diffusivity.
d_p	particle diameter.
D_R	radial spread coefficient of the liquid flow.
E_S	surface energy dissipation per unit liquid-solid surface.
f_w	irrigation rate.
f_p	fraction of partially or totally irrigated particles.
f_{pw}	fraction of partially irrigated particles.
f_{tw}	fraction of totally irrigated particles.
g	acceleration of gravity.
Ga_L^*	modified Galileo number of the liquid phase $[\frac{d_p^3\ \rho_L\ (\rho_L g+\delta_{LG})}{\mu_L^2}]$
h_d	dynamic liquid holdup per unit bed volume.
i	density of connection through one bond.
K	dimensionless spread factor of the liquid flow.

L	liquid mass superficial velocity.
L_c	channel length.
L_m	hydrodynamic parameter.
M	maldistribution factor defined by Eq.11.
N	number of channels.
N_x	lateral displacement.
N_z	vertical displacement.
$P(N_x,N_z)$	density function of a binomial distribution
r	radial distance from the center of the bed.
r_a	apparent reaction rate.
Re_L	Reynolds number of the liquid phase $[\frac{\langle L\rangle d_p}{\mu_L}]$
r_o	intrinsic reaction rate.
t	time
v	intersticial liquid velocity.
z	coordination number of a lattice.
Z	axial distance along the bed.

Greek letters

α,β	parameters introduced in Eq.1 and 2.
α_i	fraction of bonds with a density i.
δ_{LG}	pressure drop per unit bed length.
ε	external bed porosity.
η_P	particle effectiveness factor.
θ	mean pore inclination with vertical
μ_L	liquid viscosity.
ρ_L	liquid density.
σ	standard deviation of the density distribution.
φ	extensive transport property.
ϕ	parameter introduced in Eq. 31.
ω_i	fraction of volumetric flowrate through channel i.

Subscripts

i	variable defined in channel i.
< >	variable defined at the bed scale.
∞	value reached after a complete spreading.

REFERENCES

- Achwal, S.K. and Stepanek, J.B., *Can. J. Chem. Eng.* 57, 409 (1979).
- Beimesch, W.E. and Kessler, D.P., *AIChEJ*, 17, 1160 (1971).
- Bemer, G.G. and Zuiderweg, F.J., *Chem. Eng. Sci.*, 33, 1637 (1978).
- Bondi, A., *Chem. Technol.*, 1, 185 (1971)
- Broadbent, S.R. and Hammersley, J.M., *Proc. Camb. Phil. Soc.*, 53, 629 (1957).

- Buchanan, J.E., *Ind. Eng. Chem. Fund.*, 6, 400 (1967).
- Buffham, B.A. and Gibilaro, L.G., *Chem. Eng. J.*, 1, 31 (1970).
- Carbonell, R.G., *Chem. Eng. Sci.* 34, 1031 (1979).
- Charpentier, J.C., Prost, C., Van Swaaij, W. and Le Goff, P., *Chim. Ind. Génie Chimique*, 99, 803 (1968).
- Cihla, Z. and Schmidt, O., *Coll. Czech. Chem. Comm.*, 22, 896 (1957).
- Clements, L.D., Paper presented at *Two-Phase Flow and Heat Transfer Symposium Workshop*, Ft. Lauderdale, Florida, U.S.A., (1976).
- Colombo, A.J., Baldi, G. and Sicardi, S., *Chem. Eng. Sci.*, 31, 1101 (1976).
- Crine, M., *Doctorat en Sciences Appliquées*, Université de Liège (1978).
- Crine, M., Marchot, P. and L'Homme, G., *Comp. Chem. Eng.*, 3 515 (1979).
- Crine, M., Marchot, P. and L'Homme, G., *Chem. Eng. Comm.*, 7 377 (1980).
- Crine, M. and Marchot, P., to be published in *Entropie* (1982a).
- Crine, M. and Marchot, P., to be published in *Entropie* (1982b).
- Crine, M., Marchot, P., Asua, J.M. and L'Homme, G., to be published in *Lat. Am. J. Chem. Eng. Appl. Chem.* (1982c).
- Davidson, J.F., Cullen, E.F., Harrison, D. and Robess, D., *Trans. Instn. Chem. Eng.*, 37, 122 (1959).
- Deans, H.A. and Lapidus, L., *AIChEJ*, 6, 656 (1960).
- Deans, H.A., *Soc. Petrol. Eng. J.*, 3, 49 (1963).
- Frisch, H.L. and Hammersley, J.M., *J. Soc. Ind. Appl. Math.*, 11, 894 (1973).
- Germain, A., L'Homme, G. and Lefèbvre, A., In G. L'Homme (Ed), *Chemical Engineering of Gas-Liquid-Solid Catalyst Reactions*, Ed. Cebedoc, Liège (1979), Chap. 5, 134-171.
- Gianetto, A., Baldi, G., Specchia, V. and Sicardi, S., *AIChEJ*, 24, 1087 (1978).
- Goto, S., Lakota, A. and Levec, J., *Chem. Eng. Sci.*, 36, 157, (1981).
- Henry, H.C. and Gilbert, J.B., *Ind. Eng. Chem. Proc. Des. Dev.*, 12, 328 (1973).
- Herskowitz, M., Carbonell, R.G. and Smith, J.M., *AIChEJ*, 25, 272 (1979).
- Hochmann, J.M. and Effron, E., *Ind. Eng. Chem. Fund.*, 8, 63 (1969).
- Hoftyzer, P.J., *Trans. Instn. Chem. Eng.*, 42, 109 (1964).
- Jaffe, S.D., *Ind. Eng. Chem. Proc. Des. Dev.*, 15, 410 (1976).
- Kirkpatrick, S., *Rev. Mod. Phys.*, 45, 574 (1973).
- Lerou, J.J., Glasser, D. and Luss, D., *Ind. Eng. Chem. Fund.*, 19, 66 (1980).
- Lespinasse, B., *Rev. Inst. Français du Pétrole*, 11, 41 (1962).
- Mears, D.E., *Adv. Chem. Ser.*, 133, 218 (1974).
- Michell, R.W. and Furzer, I.A., *Trans. Instn. Chem. Eng.*, 50, 334 (1972).

- Mills P.L. and Dudukovic, M.P., Proceedings *Second World Congress on Chemical Engineering*, Montreal, Canada (1981).
- Onda, K., Takeuchi, H., Naeda, Y. and Takeuchi, N., *Chem. Eng. Sci.*, 28, 1677 (1973).
- Oorts, A.J. and Hellinckx, L.J., *Chem. Eng. J.*, 7, 147 (1974).
- Paraskos, J.A., Frayer, J.A. and Y.T. Shah, *Ind. Eng. Chem. Proc. Des. Dev.*, 14, 315 (1975).
- Popovic, M. and Deckwer, W.D., *Chem. Eng. J.*, 11, 67 (1976).
- Porter, K.E., *Trans. Instn. Chem. Eng.*, 46, 69 (1968).
- Porter, K.E. and Templeton, J., *Trans. Instn. Chem. Eng.*, 46, 86 (1968);
- Raghuraman, J. and Varma, Y.B., *Chem. Eng. Sci.*, 28, 585 (1973).
- Rao, V.G. and Varma ;Y.B., *AIChEJ.*, 22, 612 (1976).
- Reynier, J.P. and Charpentier, J.C., *Chem. Eng. Sci.*, 26, 1781 (1971).
- Ross, L.D., *Chem. Eng.Progr.*, 61, 77 (1965).
- Satterfield, C.N., *AIChEJ*, 21, 209 (1975).
- Scher, H. and Zallen, R., *J. Chem. Phys.*, 53, 3759 (1970).
- Scott, A.M., *Trans. Instn. Chem. Eng.*, 13, 211 (1935).
- Shah, Y.T., *Gas-Liquid-Solid Reactor Design*, Mc Graw-Hill, New York (1979).
- Specchia, V., Rossini, A. and Baldi, G., *Ing. Chim. Ital.*, 10, 171 (1974).
- Specchia, V. and Baldi, G., *Chem. Eng. Sci.*, 32,515 (1977).
- Stanek, V. and Kolar, V., *Coll. Czech. Chem. Comm.*, 30, 1054 (1965).
- Stanek, V. and Kolar, V., *Coll. Czech. Chem. Comm.*, 39, 2007 (1974).
- Tan, C.S. and Smith, J.M., *Chem. Eng. Sci.*, 35, 1601 (1980).
- Tour, R.S. and Lerman, F., *Trans. Am. Inst. Chem. Eng.*, 40, 79 (1944).
- Van Landeghem, H., *Chem. Eng. Sci.*, 35, 1912 (1980).
- Van Swaaij, W.P., Charpentier, J.C. and Villermaux, J., *Chem. Eng. Sci.*, 24, 1083 (1969).

HYDRODYNAMICS AND GAS-LIQUID INTERFACIAL PARAMETERS WITH ORGANIC AND AQUEOUS LIQUIDS IN CATALYTIC AND NON CATALYTIC PACKINGS IN TRICKLE-BED REACTORS

B.I. MORSI and J.C. CHARPENTIER

Laboratoire des Sciences du Génie Chimique, Centre National de la Recherche Scientifique - ENSIC, 1 rue Grandville - 54042 NANCY Cedex - France.

The term, trickle-bed means a reactor in which a liquid phase and a gas phase flow concurrently downward through a fixed bed of catalyst particles, while chemical reaction takes place. These reactors are widly used :

- *in petroleum industries*, hydrodesulfurization, hydrocracking of heavy or residual oil-stocks, hydrofinishing or hydrotreating of lubricating oils, demetallization, denitrification of gas-oils, isomerization of cyclopropane and hydrogenation of benzene and of naphthenic lube oil distillate (30, 31, 26, 6, 32).

- *in chemical processing*, syntheses of butynediol from aqueous formaldehyde and acetylene, selective hydrogenation of acetylene to remove it in the presence of butadiene in C_4 hydrocarbon streams (31), and hydrogenation of glucose to sorbitol (6).

- *in the fight against environmental air and waste water pollution* by aerobic bacterial action (6).

- *in biological and pharmaceutic industries.*

The design of these reactors requires the knowledge of both the hydrodynamic parameters (flow patterns, liquid phase holdup and two-phase pressure drop, ...) and the interfacial parameters (a, k_La, k_Ga, k_Sa).

Most of published papers regard exclusively the hydrodynamics of an air-water system and the determination of the gas-liquid interfacial area a, the gas-liquid k_La and the liquid-solid k_Sa mass transfer parameters in aqueous solutions. In this paper we present some experimental results on hydrodynamics and mass transfer parameters a and k_La, measured by the chemical technique with two organic solutions, cyclohexylamine (CHA) in toluene + 10 % by volume of isopropanol (IPA) as a solvent and diethanolamine (DEA) in ethanol as a solvent ; for three different packings, glass beads of $d_p = 1.16\times10^{-3}$ m mean diameter, spherical catalyst $Co/Mo/Al_2O_3$ of $d_p = 2.4\times10^{-3}$ m mean diameter and glass Raschig rings of 6.48×10^{-3} m nominal diameter. Some experimental gas-liquid interfacial area (a) values, for two aqueous solutions, sodium hydroxide and sodium sulfites obtained with the glass beads packing and finally some data on k_Sa obtained by electrochemical method with $d_p = 4.10^{-3}$ m mean diameter, nickel spheres packing are also presented.

EXPERIMENTAL

As regards the a and k_La, experiments have been carried out in a 0.05 m diameter column, packed to a 0.49 m height. The characteristics of the packings and the systems utilized are shown in Table (I). One may notice the small porosity (26.3 %) reported for glass beads packing of $d_p = 1.16\times10^{-3}$ m mean diameter, which is probably due both to the heterogeneous diameter distribution and to the non perfect sphericity of the particles (25). The pressure drop was measured by pressure taps located directly on top and at the bottom of the packing. The liquid phase holdup was determined by weighing method (8, 9, 19). The experimental set-up is described in Fig. (1). The liquid and the gas phases flow downward concurrently, the apparatus may operate, either in open circuit for both liquid and gas or in closed circuit for the liquid phase. The solute and inert gases are mixed and then presaturated with the solvent without reactant before the entrance in the reactor. The temperature was controlled by means of heat exchanger and was maintained at 20 °C.

For both carbamatation of CHA and DEA and carbonatation of NaOH, the reactor operated in open system with gas chromatographic analysis at reactor entrance and exit. The inlet CO_2 mole fraction varies between 2 and 4 %.

For sulfite oxidation, where the oxygen absorption flux as well as the sulfite conversion efficiency are very small, the experiments were carried out by recycling in a closed circuit a sulfite solution volume of $V_L = 0.11$ m^3 through the packed column, where the chemical reaction with the oxygen of air takes place. The sulfite concentration was measured by iodometric titration in function of time. The operating temperature was 20 °C, $C_{Bo} = 0.8$ M and pH = 8.5. The sulfite solution was catalyzed by the cobalt sulfate of a concentration = 3.56×10^{-4} $kmol/m^3$.

TABLE (I) - CHARACTERISTICS OF THE PACKINGS AND PROPERTIES OF THE LIQUID SOLUTIONS AT 20 °C
G.B. glass beads - R.R. Raschig rings - S.C. spherical catalyst - G.S. glass spheres
CHA cyclohexylamine - DEA diethanolamine - IPA isopropanol - values for k_sS at 30 °C.

COLUMN		PACKING			SYSTEM	C_{Bo}	ρ_L	$\mu_L \times 10^3$	$\sigma_L \times 10^3$	β_s	L	KEY	EXPERIMENTS
$D \times 10^2$ m	$Z \times 10^2$ m	$d_p \times 10^3$ m	ε	type		kmol/m^3	kg/m^3	Pa.s	N/m	%	kg/m^2.s		
5	49	1.16	0.263	G.B.	CO_2 + CHA + (toluene +10% IPA)	0.228	857.4	0.644	27.61	9.2	1 2 4	■ ⬓ □	Hydrodynamics and a
		4.68	0.704	R.R.		0.350	858.1	0.656	27.84	8.3	1 2 4	▲ ◭ △	
		2.40	0.385	S.C.		0.365	858.1	0.658	27.86	8.7	1 2 4	● ◒ ○	
		1.16	0.263	G.B.	CO_2+NaOH + (water)	0.200	1003.0	1.040	62.80	9.1	2 4	◇ ◆	
		1.16	0.263	G.B.	O_2+Na_2SO_3 + (water)	0.800	1079.0	1.460	66.35	10.7	1 4	▽ ▼	
		2.40	0.385	S.C.	CO_2+DEA + (ethanol)	0.045	805.5	1.530	23.19	7.2	4 6 8	▼ ▽ ▼	Hydrodynamics and k_La
4.5	154	4.00	0.380	G.S.	N_2 + NaOH	0.500	1021.0	0.933	-	-	1 2 5	× + Φ	Hydrodynamics and k_sS

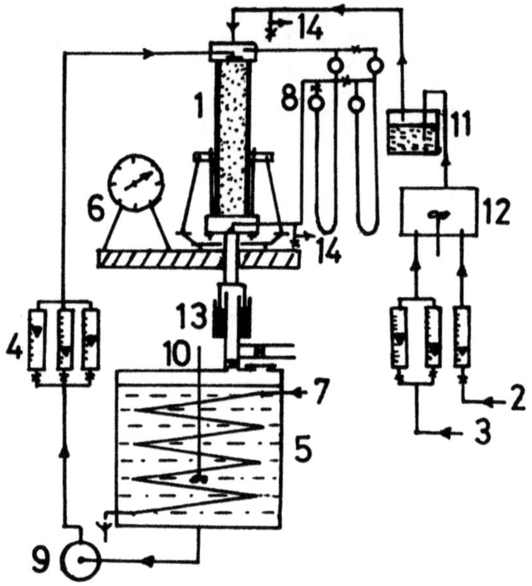

Fig. 1 : Experimental set-up

1. Packed column
2. Solute gas
3. Inert gas
4. Rotameters-liquid
5. Liquid reservoir
6. Balance
7. Heat exchanger
8. Manometer
9. Pump
10. Stirrer
11. Saturator
12. Gas mixer
13. Hydrolic joint
14. Gas sampling

KINETICS OF CHEMICAL REACTIONS

The kinetics of the carbamatation of CHA in organic solvent (toluene + 10 % by volume of IPA) have been studied by Sridharan et al. (35, 36), Belhaj (4) and Alvarez et al. (1, 2). The kinetics of the carbamatation of DEA in organic solvent (ethanol) have been studied by Alvarez (2). The kinetics of the carbonatation of NaOH in aqueous solvent (water) and oxidation of aqueous solution of Na_2SO_3 have been investigated by Laurent (17) and by us (22).

In our calculations for organic and aqueous systems we used respectively the values proposed by Alvarez (2) and by Laurent (17) ; the later's values are comparable with ours (22). These values have been obtained in our laboratory with a cylindrical wetted wall falling film contactor.

HYDRODYNAMIC EXPERIMENTAL RESULTS

It is interesting to emphasize that, our methodology consisted in simultaneous determination of both the hydrodynamic and gas-liquid-solid parameters. Most of our results were obtained in the trickling flow regime zone, i.e. weak interaction between the liquid and the gas (Fig. 2). In downward two-phase flow, at a given liquid superficial mass velocity (L), when increasing gas superficial mass velocity (G), the friction forces between the fluids increase rapidly. This result, on one hand, is an increase of pressure drop and on the other hand, is an acceleration of the liquid superficial velocity and consequently a decrease of liquid holdup. This gas-liquid interaction leads consequently to different flow patterns depending on the nature of the liquid phase. Many workers (5, 12, 15, 18, 22, 27, 36, 41) have presented different flow maps concerning downward two-phase flow through fixed beds for nonfoaming, foaming and viscous liquids. As regards the first organic solution (CHA-toluene + 10 % IPA), in Fig. (3) we present some hydrodynamic experimental data of pressure drop expressed in the head of water per meter of packed bed height, ($\Delta H/Z$) and of total liquid holdup defined as a percentage of interparticle void volume, (β) against G. This concerns two different superficial liquid mass velocities L = 1 and 4 $kg/m^2.s$, and three different packings, glass beads, spherical catalyst and Raschig rings.

Regarding the *influence of packing nature and size,* one can notice that, at the same G, the values of ($\Delta H/Z$) and (β) corresponding to Raschig rings are the smallest, while those concerning glass beads are the highest and those obtained for spherical catalyst are intermediate. This could be explained by the small void fraction reported for the glass beads, which results in relatively

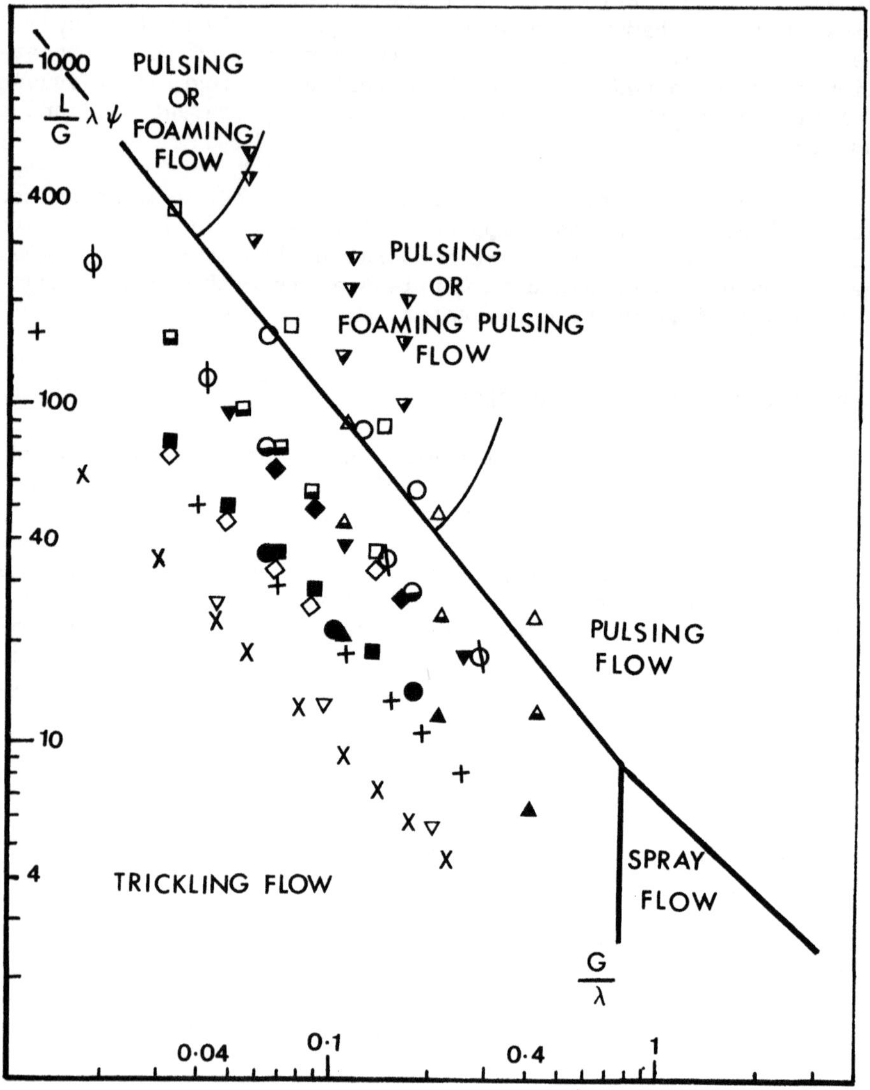

Fig. 2 : Two-phase flow patterns.
For the keys - see Table (I)

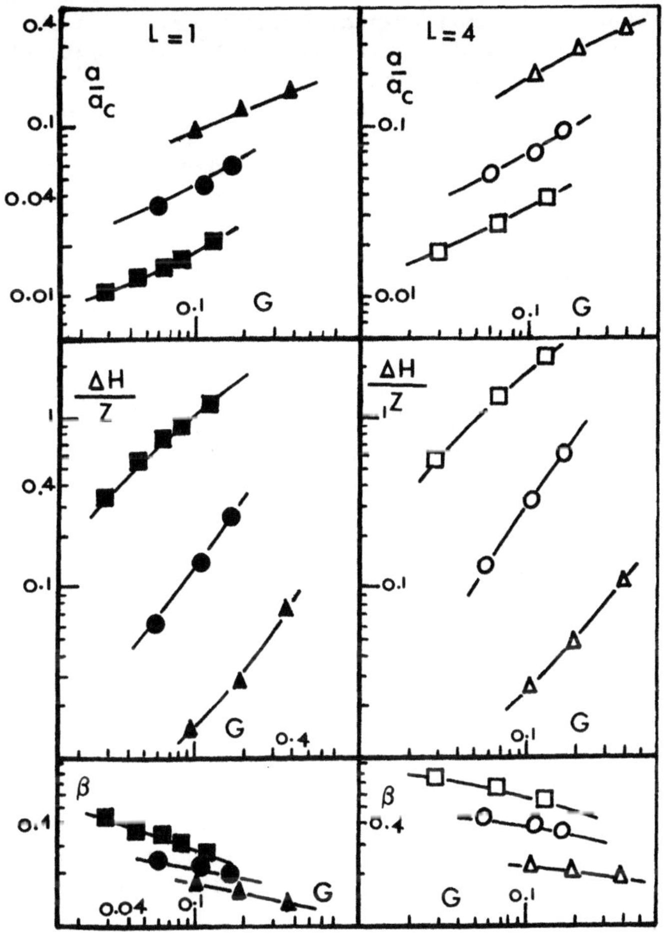

Fig. 3 : Experimental results of $\Delta H/Z$, β and a/a_c
System : CHA-toluene + 10 % IPA

Packings : $d_p = 1.16 \times 10^{-3}$, 6.48×10^{-3} and 2.4×10^{-3} m tortuous fluid paths with more contact points and consequently high capillary liquid holdup and more obstacles to the fluid flow. It is to remark also that, both (β) and ($\Delta H/Z$) increase with the augmentation of L at the same G for all three packings. Fig. (4) shows some experimental hydrodynamic data in order to investigate the *influence of the nature of the liquid phase* on the hydrodynamic parameters at L = 1, 2 and 4 kg/m^2.s, for three different solutions, organic (CHA-toluene + 10 % IPA), and aqueous (NaOH-water) and (Na_2SO_3-water). One can observe that the Na_2SO_3 solution (foaming character) shows the highest values of ($\Delta H/Z$) and (β) and the organic solution (non foaming character) gives the lowest values, while the NaOH solution (ambiguous character) is of intermediate ones. This is due to the foamability of sulfite solution. It is also to note the increase of both (β) and ($\Delta H/Z$) with the increase of L at G constant in the range of G and L experimented.

In Figure (7) we present also hydrodynamic data obtained with another organic solution (DEA-ethanol) for spherical catalyst, $d_p = 2.4\times10^{-3}$ m. We remark on one hand the systematic increase of both ($\Delta H/Z$) and β with L at G constant and on the other hand, the increase of $\Delta H/Z$ and the decrease of β with G at L constant. If we compare quantitatively ($\Delta H/Z$) and (β) values obtained for (CHA-toluene + 10 % IPA) with those for (DEA-ethanol) at L = 4 kg/m^2.s and for the same packing, $d_p = 2.4\times10^{-3}$ m, we can find that the values corresponding to the latter are much higher than those concerning the former, this is due to the foamability of the system DEA-ethanol in presence of the gas.

CORRELATION OF HYDRODYNAMIC RESULTS

Pressure drop and liquid holdup are very important parameters, indispensable for the design of trickle-bed reactor [6, 7, 13, 16, 17, 19, 23, 27, 29, 30, 32, 42]. Their values influence directly the interfacial parameters between the fluid phases and between liquid and solid phases too. Many workers have proposed different correlations for predicting the two-phase pressure drop in co-current downward flow through packed beds [6, 17, 19, 23, 29, 34, 40]. These have been recently reviewed by Perez Sosa [44]. These correlations depend essentially either on *momentum balance*, which leads to,

$$\delta_{LG} = \left(\frac{\Delta H}{Z}\right)_{LG} + \frac{\rho_L\beta+\rho_G(1-\beta)}{\rho_e} \qquad (1)$$

as a function of $\delta_G = A'\mu_G\dfrac{G}{\rho_G} + B'\dfrac{G^2}{\rho_G}$ (2)

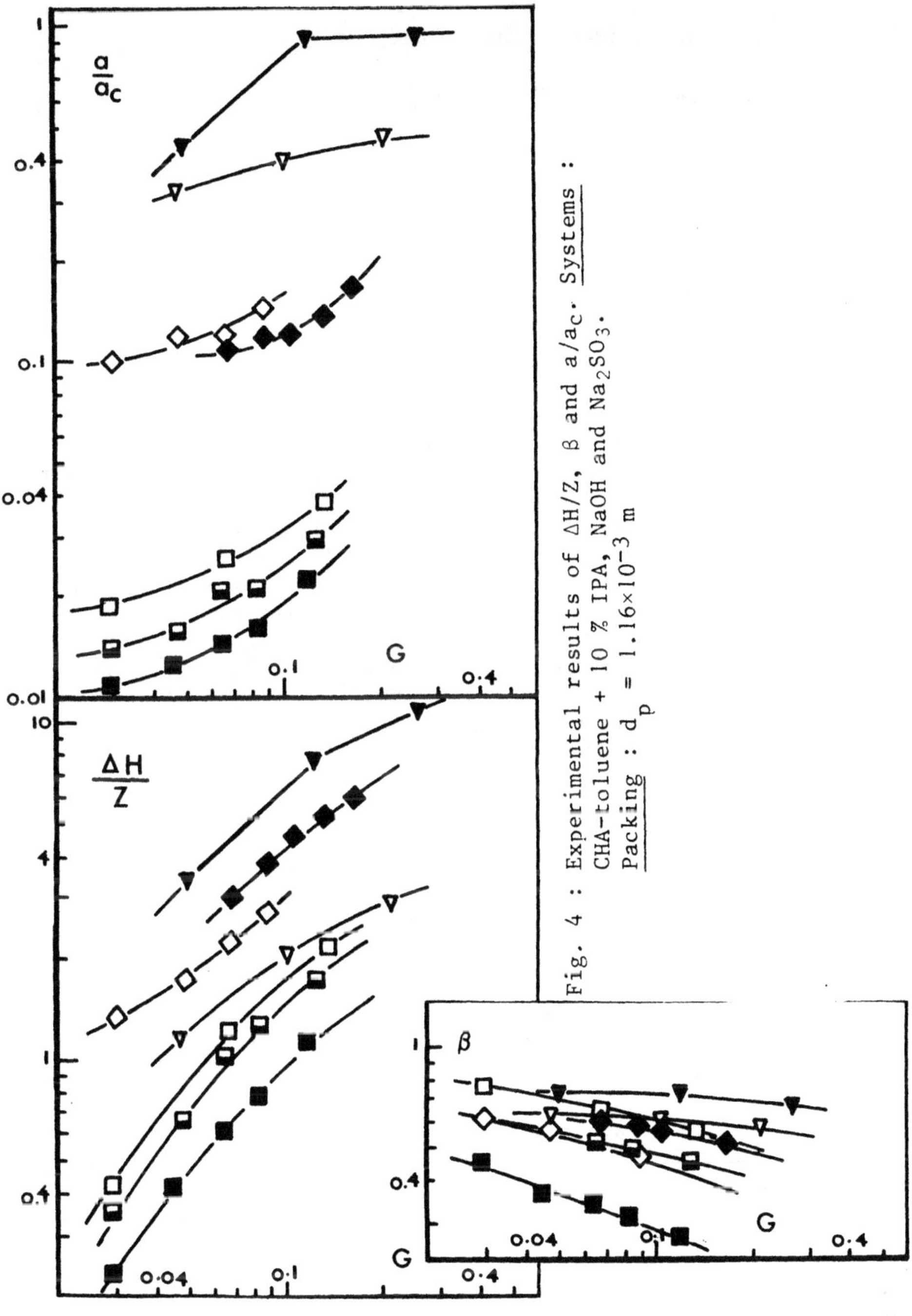

Fig. 4 : Experimental results of $\Delta H/Z$, β and a/a_c. Systems : CHA-toluene + 10 % IPA, NaOH and Na_2SO_3. Packing : $d_p = 1.16\times10^{-3}$ m and $$\delta_L = A' \mu_L \frac{L}{\rho_L} + B' \frac{L^2}{\rho_L} \qquad (3)$$

Or on *power balance*, which leads to

$$\xi_{LG} = \frac{1}{\varepsilon} \left(\frac{L}{\rho_L} + \frac{G}{\rho_G}\right) \left(\frac{\Delta H}{Z}\right)_{LG} + \frac{L+G}{\varepsilon \rho_e} \qquad (4)$$

as a function of

$$\xi_G = \frac{1}{\varepsilon} \left(\frac{G}{\rho_G}\right) \delta_G + \frac{G}{\varepsilon \rho_e} \qquad (5)$$

and $$\xi_L = \frac{L}{\varepsilon \rho_e} \qquad (6)$$

As for pressure drop, many workers (5, 12, 16, 29, 23, 34) have proposed different correlations for predicting total liquid phase holdup in two-phase concurrent downward flow. The liquid holdup depends on the nature and the flowrates of fluid phases, on the type of packing and on the eventual distribution or redistribution of the liquid phase. It is interesting to note that, both liquid holdup and two-phase pressure drop are mutually dependent, that is why many authors tried to correlate them in function of δ and ξ parameters. We present in Fig. (5) as an example, a comparison between our experimental results of two-phase pressure drop and total liquid holdup with those predicted by the correlations proposed by Midoux et al. (19)

$$\phi_L = 1 + \frac{1}{\chi} + \frac{1.14}{\chi^{0.54}} \qquad (7)$$

and $$\beta = \frac{0.66\chi^{0.81}}{1+0.66\chi^{0.81}} \qquad (8)$$

A good agreement may be observed with the values corresponding to glass beads and spherical catalyst, but total disagreement with those regarding Raschig rings and foaming liquids. This is not surprising at all, because these correlations were proposed for pellets, spherical catalytic and inert packings of small dimension, about 3×10^{-3} m, not for Raschig rings and open packings neither for foaming liquids. Moreover correlations for foaming and viscous liquids in both weak and high interaction are presented by Morsi et al. (23).

GAS-LIQUID INTERFACIAL AREA a

Experimental data on gas-liquid interfacial area in trickle bed reactors with organic solutions are scanty. Most of published results concern exclusively high ionic solutions, carbonatation of sodium hydroxide (11, 15, 20), carbonatation of carbonate-bicarbonate-arsenite (28) and oxidation of dithionite (33) or of sodium sulfite (9). Packings often used were of large sizes, the smallest used by Hirose et al. (15) were glass beads of $d_p = 2.59\times10^{-3}$ m mean diameter. Our results concerning *the carbonatation of sodium hydroxide and the carbamatation of* CHA, have been obtained by gas absorption accompanied with fast chemical reaction of pseudo m,nth order. The kinetics of this reaction are shown in Table (II).

In the case of irreversible chemical reaction of type $A+zB \rightarrow$ Product, the specific rate of absorption of solute gas A by the liquid reactant B, can be expressed by

$$\varphi = C_A^* \sqrt{\frac{2}{m+1} D_A \, k_{mn} \, (C_A^*)^{m-1} \, (C_{Bo})^n} = C_A^* \, k_L \, Ha \tag{9}$$

with the following conditions

$3 < Ha < E_i/2$

where $$Ha = \frac{1}{k_L} \sqrt{\frac{2}{m+1} D_A \, k_{mn} \, (C_A^*)^{m-1} \, (C_{Bo})^n} \tag{10}$$

and $$E_i = \sqrt{\frac{D_A}{D_B}} + \sqrt{\frac{D_B}{D_A}} \, \frac{C_{Bo}}{z C_A^*} \tag{11}$$

The knowledge of total flux of absorption of solute gas Φ allows the determination of the gas liquid interfacial area a as,

$$a = \frac{\Phi}{\varphi \Omega Z} \tag{12}$$

if we know previously both the physico-chemical parameters C_A^* ($C_A^* = p/He$), D_A and D_B and the kinetics data, k_{mn}, n and m.

In *trickle-bed reactor,* when we postulate the following hypotheses,

- the gas-side resistance ($1/k_G$) to mass transfer is negligible,
- the two fluid phases work in plug flow,
- the total pressure varies linearly with the packing height,

TABLE (II) - CHEMICAL REACTION DATA AT 20 °C

SYSTEM			C_{Bo}	$D_A \times 10^9$	$D_B \times 10^9$	$He \times 10^{-5}$	k_{mn}	m	n
solute	reactant	solvent	$kmol/m^3$	m^2/s	m^2/s	$\frac{Pa.m^3}{kmol}$	$\left(\frac{m^3}{kmol}\right)^{m+n-1}/s$		
CO_2	CHA	toluene + 10% IPA	0.228	4.08	1.80	9.60	2 900	1	2
			0.350	4.08	1.78				
			0.365	4.08	1.77				
CO_2	NaOH	water	0.200	1.48	2.00 (at 15°C)	31.93	7 850	1	1
O_2	Na_2SO_3	water	0.800	1.60	-	1581.00	$1\ 293 \times 10^7$	2	0
CO_2	DEA	ethanol	0.045	3.25	0.630	8.57	133	1	2

it is then possible to write the following conservation balance equations :

$$\phi = - \int_{Y_e}^{Y_s} \dot{N}_G \, dY = \int_o^Z C_A^* \sqrt{\frac{2}{m+1} D_A \, k_{mn} \, (C_A^*)^{m-1} (C_{Bo})^n} \, a\Omega dh \quad (13)$$

and

$$\int_{C_{Bo_e}}^{C_{Bo}} Q_L \, dC_{Bo} = \int_{Y_e}^{Y} z\dot{N}_G dY \quad (14)$$

The integration of equation (14) provides C_{Bo}, when substituting C_{Bo} and C_A^* in equation (13), the integration of the obtained differential equation enables to get an expression of the gas-liquid interfacial area (a) as a function of (Y_e) and (Y_s) of the solute gas at the entrance and the exit of the reactor respectively (22).

As regards the oxidation of Na_2SO_3, LAURENT (17) demonstrated a detailed solution, for calculating the gas-liquid interfacial area when the measurements are carried out by analysis in liquid phase. He proposed the following practical equation, which relates the evolution of the reactant concentration in time to the specific rate of gas absorption and the volume of the solution employed, as

$$\frac{-\Delta C_{Bo}}{\Delta t} = \frac{z\varphi\Omega Z}{V_L} a \quad (15)$$

This practical equation allows the determination of a.

EXPERIMENTAL GAS-LIQUID INTERFACIAL AREAS

In Figure (3) we also show some experimental values of gas-liquid interfacial area related to specific packing area (a/a_c) as a function of G for L = 1 and 4 $kg/m^2.s$. One may observe for the three packings the increase of (a/a_c) with both G and L. The (a/a_c) values obtained for the glass beads are the smallest, those for the Raschig rings are the highest while those obtained for the spherical catalyst are intermediate. It is to emphasize the deep analogy between curves representing the pressure drops and those representing the interfacial areas as a function of G at L constant. Quantitatively, there is a factor of about 2 between values of (a/a_c) for catalytic packings and those for glass beads and a factor of about 6 between values of (a/a_c) for Raschig rings and those for glass beads. This shows once again the influence of the nature and the shape of the packing on the interfacial area where a segregation between the gas and the liquid phases was provoked by the small dimension glass beads.

Figure (4) shows some experimental data obtained with different aqueous and organic systems for the glass beads packing, $d_p = 1.16\times10^{-3}$ m, in order to investigate the influence of *the nature of the liquid phase* on the gas-liquid interfacial area. One may notice that the interfacial areas increase with both liquid and gas flowrates and, at practically the same liquid flowrate, the sulfite solution always gives the highest values, the cyclohexylamine solution gives the lowest, while the sodium hydroxide solution provides the intermediate ones. Quantitatively, one may report a factor of about 5 between the experimental values, obtained with NaOH and those obtained with CHA and a factor of the same order between the values obtained with Na_2SO_3 and those with NaOH. It is interesting to lay stress on the small interfacial areas presented by the organic system, that could be explained by the presence of a continuous liquid film covering the particles and long deformed gas bubbles flowing in hydrolic channels, which leads finally to small a. Moreover this effect is amplified here by the small dimensions of the packed pores. As for Na_2SO_3 solution, one must admit that the foamability of this ionic system at even small liquid and gas flowrates and the wetting characteristics of the packing, contribute to increase the gas-liquid interfacial area.

CORRELATION OF GAS-LIQUID INTERFACIAL AREA

Our results of gas-liquid interfacial area obtained with organic solution and for glass beads and spherical catalyst, diverge completely from the correlations proposed by Gianetto et al. [11], Charpentier [6] and Fukushima and Kusaka [9] and disagree too with the mode of representation proposed by Hirose et al. [15], moreover our values concerning Raschig rings are scattered [22]. This actually is not surprising because these workers established their correlations according to data obtained with highly ionic solutions and for packings of large sizes, furthermore the higher liquid and gas rates employed. We have proposed [24] a modification of Gianetto et al's representation [11] by taking into account on one hand, the strong analogy between curves representing (a/a_c) and $(\Delta H/Z)$ in function of G at L constant shown in Figs. 3 and 4 , and on the other hand the influence of total liquid holdup (β) on gas-liquid interfacial area. This modification consisted in representing a in function of energetic term δ_{LG}, as

$$a = 7.75\times10^{5} \left(\frac{\varepsilon}{a_c}\,\delta_{LG}\right) \qquad (16)$$

for $3\times10^{-5} < \frac{\varepsilon}{a_c}\,\delta_{LG} < 7\times10^{-4}$

This correlation shown in Fig. (6) fits well our experimental data with ± 30 % accuracy.

It is also to note that the two-phase pressure drop and the liquid phase holdup necessary for the use of this correlation may be estimated from equations (7) and (8) respectively. This representation is only suitable for weak gas-liquid interaction flow in which the eventual anticoalescent characteristics of solutions do not intervene. Besides it may be noticed that the results obtained with ionic solution for glass beads do not fit with this correlation, and this occurs all the more as the anticoalescent characteristics are more pronounced (NaOH then Na_2SO_3).

LIQUID-SIDE MASS TRANSFER COEFFICIENT

Experimental data on k_La in the literature are scattered and unfortunately most of them concern exclusively aqueous solutions. Some examples, Reiss (27) investigated the absorption of O_2 into water from air at 25 °C ; Hirose et al. (15) studied the desorption of O_2 into N_2 from water at 15 °C ; Gianetto et al. (12) studied the desorption of O_2 with air from 2N NaOH solution ; Ufford and Perona (41) measured the absorption of CO_2 into water ; Sylvester and Pitayagulsarn (37) studied the absorption of pure CO_2 in water ; Goto and Smith (14) measured the desorption of O_2 in pure N_2 from distilled water saturated with pure O_2 and Turek and Lange (39) studied the hydrogenation of alpha-methylstyrene to cumene (organic solution). The packings utilized were *catalytic particles* $d_p = 2.91\times10^{-3}$ m and 0.541×10^{-3} m of CuZnO (14) and $d_p = 0.54\times10^{-3}$ m, 0.9×10^{-3} and 3×10^{-3} m of Pd/Al_2O_3 (39), *glass beads* $d_p = 2.59\times10^{-3}$ to 12.2×10^{-3} m (15) and 6×10^{-3} m (12), *cylinders*, $3.175\times10^{-3}\times3.175\times10^{-3}$ m×m (37), *Berl saddles* 6×10^{-3} m (12), 19×10^{-3} m (41) and *Raschig rings* 6×10^{-3} m (12), 6.3×10^{-3} and 12.7×10^{-3} m (41) and 7.62×10^{-3} to 40.64×10^{-3} m (27). These workers used often high liquid and gas superficial velocities up to 300 and 6 kg/m^2.s (27), 200 and 1.2 kg/m^2.s (14), 46 and 3 kg/m^2.s (12) and 31 and 0.03 kg/m^2.s (39). In this study, we present experimental data liquid phase film capacity coefficient k_La, wherein k_L is the gas-liquid mass transfer coefficient based on the individual film driving force. Our values were determined by slow chemical absorption of CO_2 into organic solution (DEA-ethanol) for spherical catalyst of $d_p = 2.4\times10^{-3}$ m mean diameter. The kinetics of this chemical reaction are shown in Table (II). This slow irreversible chemical reaction will be established if on one hand the following inequality is verified

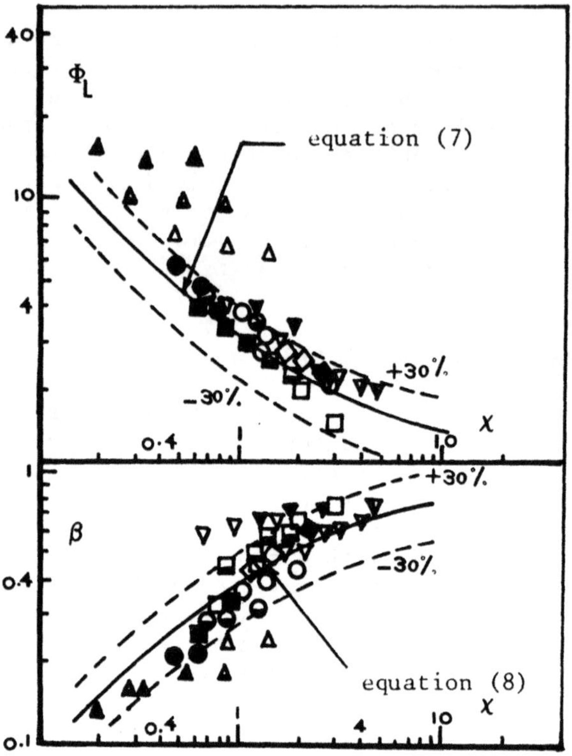

Fig. 5
Correlations of the two-phase pressure drop
and the total liquid holdup

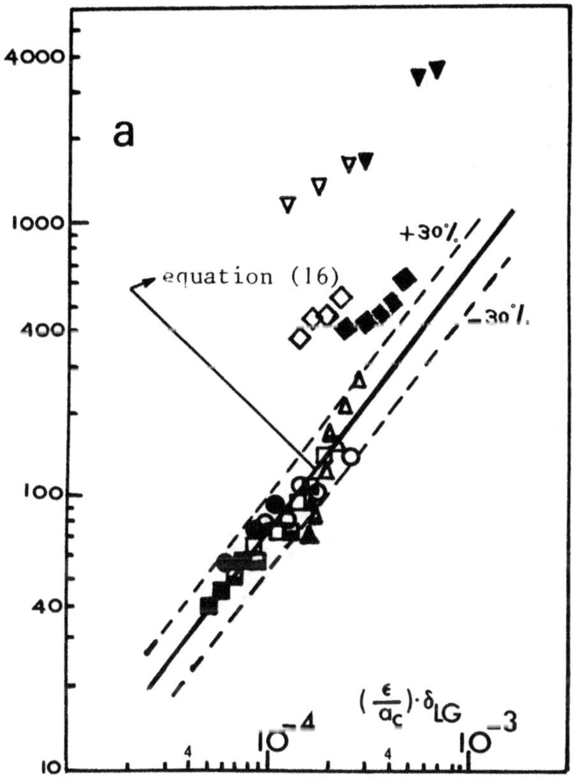

Fig. 6
Correlation of the gas liquid interfacial area

$$Ha = \frac{1}{k_L}\sqrt{D_A k_3 C_{Bo}^2} < 0.3 \tag{17}$$

and on the other hand the depletion test is achieved, i.e.

$$C_{Ao} << C_A^* \tag{18}$$

where C_{Ao} is the concentration of dissolved free solute gas in the liquid bulk. If one postulates that the chemical reaction is negligible within the liquid film, it is possible to write locally the following conservation equation,

$$k_L a(C_A^* - C_{Ao}) = k_3 \varepsilon \beta C_{Ao} C_{Bo}^2 \tag{19}$$

rearranging the previous equation, it becomes,

$$\frac{C_{Ao}}{C_A^*} = \frac{1}{1 + \dfrac{k_3 \varepsilon \beta C_{Bo}^2}{k_L a}} \tag{20}$$

In the case of $\left(\dfrac{k_L a}{k_3 \varepsilon \beta C_{Bo}^2}\right) < 0.1$, $\dfrac{C_{Ao}}{C_A^*}$ will be less than 0.1 too.

From equation (17) one obtains $C_{Bo}^2 < \dfrac{k_L^2}{10 D_A k_3}$, and finally one gets the two inequalities :

$$\frac{10 k_L a}{k_3 \varepsilon \beta} < C_{Bo}^2 < \frac{k_L^2}{10 D_A k_3} \tag{21}$$

These previous inequalities allow, on one hand to estimate an appropriate value of C_{Bo} necessary for the slow chemical reaction, if approximate values of k_L, D_A and k_3 are known, and on the other hand to verify the obtained experimental data of $k_L a$. The conservation equation of the solute in gaseous phase is,

$$\phi = -\int_{Y_e}^{Y_s} \dot{N}_G \, dY \tag{22}$$

The combination of equations (19) and (22) allows the determination of $k_L a$ in function of Y_e and Y_s.

EXPERIMENTAL RESULTS OF k_La

We present in Fig. (7) some data of k_La in function of G at different liquid rates. One can observe an increase of k_La with the increase of both liquid and gas superficial mass velocities. Our values are comprised between 0.006 and 0.09 s^{-1} in the hydrodynamic conditions employed for a packing diameter

CORRELATION OF k_La

Many workers tried to correlate k_La in function of specific energy dissipated in two-phase when neglecting the gas superficial velocity (6, 15, 27, 31, 39). There are also many empirical correlation cited (12, 14, 15, 37, 41). One may remark that the influence of the liquid viscosity has not yet been investigated. A comparison between our values and those obtained by the authors in aqueous solution is meaningless. However we should point out that our values disagree with those predicted by their correlations. This is due probably to high liquid and gas rates used when establishing their correlations. Their published values are 0.1 to 3 s^{-1} (27), 0.01 to 1 s^{-1} (12), 0.02 to 3 s^{-1} (15), 0.06 to 0.0136 s^{-1} (41) and 0.0004 to 0.002 s^{-1} (39).

LIQUID-SOLID MASS TRANSFER COEFFICIENT k_sS

Recently, Barthole et al. (3) published a complete investigation on the methods of determinating the liquid to solid mass transfer coefficient k_sS in one phase and two-phase downward flow in packed beds. Moreover they studied simultaneously, in the trickling flow zone, the hydrodynamics and k_sS parameters in a column of 0.045 m internal diameter, packed with glass spheres of 4×10^{-3} m diameter to a height of 1.54 m. Two 1.5×10^{-2} length measuring sections of spherical active nickel $d_p = 4\times10^{-3}$ m are located in the upper and lower parts of the packed column. A fine PVC grid provides an electrical insulation between the conducting packing and the anode which consists of a nickel cylinder having the same diameter as the column. The anode-cathode system and a reference electrode, which is a small platinum wire located just above the working electrode are incorporated in a classical three-electrode potentio-static circuit including a potentiostat and a current-potential curve recorder. Overall mass transfer coefficient between the flowing liquid and the packed bed of nickel spheres were deduced from the limiting current density

$$I_{lim} = nF\ k_sS\ C_s$$

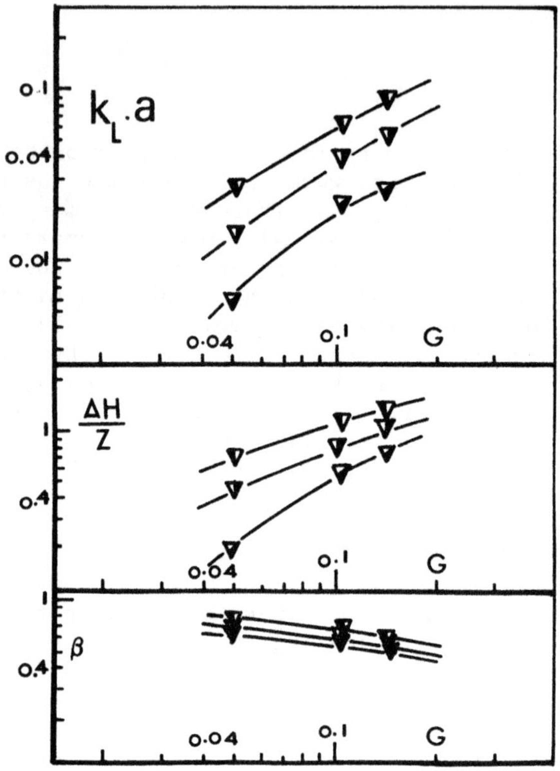

Fig. 7
Experimental results of $\Delta H/Z$, β and $k_L a$

System : DEA-ethanol

Packing : $d_p = 2.4 \times 10^{-3}$ m (catalyst)

obtained in the electrochemical reduction of potassium ferricyanide ions. nF is the Faraday number, S the active liquid-solid surface and C_s is the concentration of the reactive active species. The electrolytic solution is a mixture of 1×10^{-3} M ferricyanide ions and 5×10^{-2} M ferrocyanide ions (in order to prevent anodic polarization) in 0.5 M NaOH acting as the supporting electrolyte. The gas used was N_2 and the properties of the liquid are shown in Table (I).

The dynamic liquid holdup was measured by the classical tracer technique and an electrochemical method with probes covering a cross section of the column was employed for the determination of k_sS, S being the active liquid-solid surface.

These workers demonstrated the feasibility of the electrochemical method and emphasized the importance of the packing wettability on the k_sS measured values.

EXPERIMENTAL RESULTS of k_sS

The experimental set-up was described in ref. (3). Fig. (8) shows hydrodynamic and k_sS results in function of G at L constant. One can find also that the two-phase pressure drop increase systematically with the augmentation of G and L and both k_sS and dynamic liquid holdup increase with L at G constant while they are independent of G at L constant, this is probably due to the small superficial gas velocities experimented.

CORRELATION OF k_sS

Many experimental results and these of Barthole et al. (3) concerning k_sS were represented in dimensionless, Sherwood or modified Sherwood's number against Reynolds and Schmidt's numbers. Their values are comparable with those of Goto et al. (14) and the type of correlation proposed by Darwadkar and Sylvester (45) seems the best to correlate their data (Fig. 9)

$$Sh'\ Sc^{-1/3} = 1.637\ Re_L^{0.669} \qquad \text{for} \quad 0.2 < Re_L < 2400$$

CONCLUSION

The carbamatation of CHA in organic solvent (toluene + 10 % by volume of IPA) is usable in rather a large range of concentration 0.2 to 0.7 kmol/m^3 for the determination of gas-liquid interfacial areas (23). Glass beads of small dimension $d_p = 1.16\times10^{-3}$ m provoke, for small flowrates a segregation between the gas and

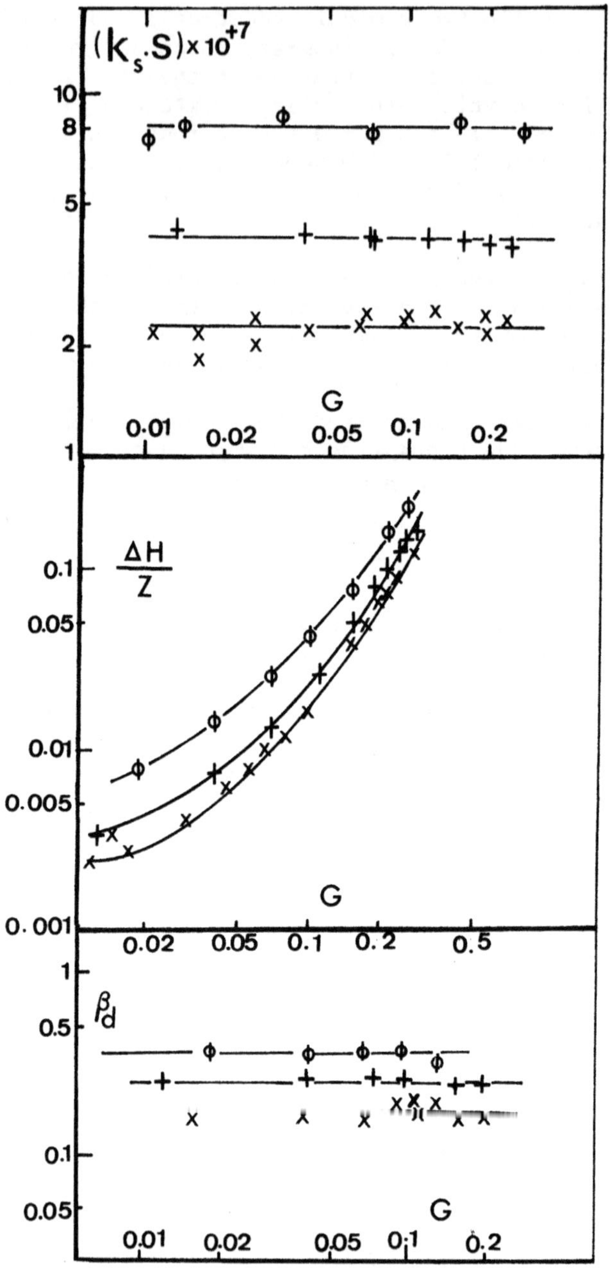

Fig. 8
Experimental results of $\Delta H/Z$, β and k_sS

System : NaOH solution

Packing : $d_p = 4\times10^{-3}$ m

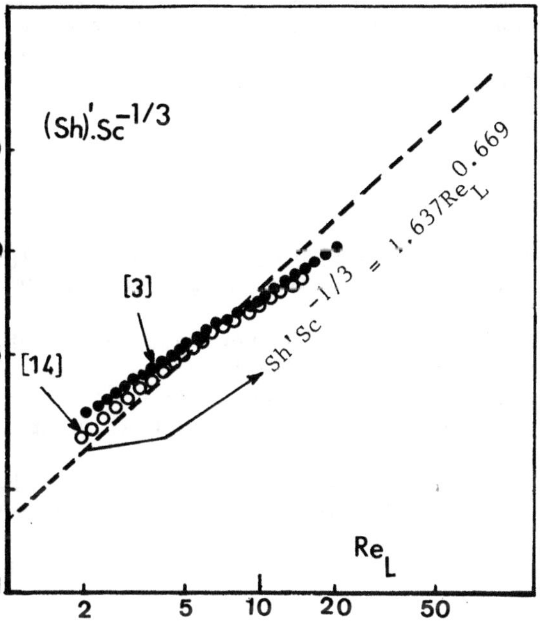

Fig. 9 : Correlation of k_sS

liquid phases which leads to small values of a. The carbamatation of DEA in organic solvent (ethanol) is a good system for the determination of k_La and the effect of liquid physico-chemical parameters on k_La in organic solutions is incomplete yet. The electrochemical technique is a good tool for the determination of k_sS values in ionic liquids and it is certainly suitable for their measurements in organic solutions.

We should emphasize that all the data which are presented here concern a small column diameter with a good initial distribution. They should be extrapolated with great care especially in the trickling flow regime for which in large size column it is often a challenge to have a uniform distribution of the liquid all over the reactor.

NOTATIONS

A'	characteristic constant of packing, (S^2/kg)
a	gas-liquid interfacial area per unit volume of the reactor (m^2/m^3)
a_c	surface area of packing per unit volume of the reactor (m^2/m^3) : $a_c = a_g(1-\varepsilon)$
a_g	specific area of the packing (m^2/m^3)
a_s	wetted active packing area per unit volume of the reactor (m^2/m^3)
B'	characteristic constant of packing $(m.s^2/kg)$
C^*_A	solubility of dissolved solute gas in the liquid $(kmol/m^3)$
C_{Ao}	concentration of free solute gas dissolved in the liquid bulk $(kmol/m^3)$
C_{Bo}, C_{Bo_e}, C_{Bo_s}	concentration of reactant, at the reactor entrance, at the reactor exit $(kmol/m^3)$
D	internal diameter of the reactor (m)
D_A, D_B	diffusion coefficient of the solute gas, of dissolved reactant in the liquid phase (m^2/s)
d_p	diameter of the particle (m)
E	enhancement factor
E_i	instantaneous enhancement factor
G	superficial gas mass velocity $(kg/m^2.s)$
Ha	HATTA number
He	HENRY constant $(Pa.m^3/kmol)$
h	axial distance from the top of the column (m)
k_{mn}	reaction rate constant with partial orders m and n, $(\|m^3/kmol\|^{m+n-1})$
k_3	reaction rate constant of third order $(m^3/kmol)^2/s$
k_G	gas-side mass transfer coefficient $(kmol/\|s.m^2.Pa\|)$
k_L	mass transfer coefficient for transfer from gas-liquid interface (m/s)

k_s	mass transfer coefficient for transfer from liquid to solid (m/s)
L	superficial liquid mass velocity ($kg/m^2.s$)
m, n	reaction partial orders with respect to solute, reactant
$\dot{N}_G$	molar inert gas rate (kmol/s)
p	partial pressure of solute gas (Pa)
Q_L	volumetric rate of liquid (m^3/s)
Re_L	Reynolds number ($L.d_p/\mu_L$)
S	active liquid-solid surface (m^2)
Sc	Schmidt's number ($\mu_L/\rho_L D_A$)
Sh	Sherwood's number ($k_s d_p/D_A$)
Sh'	Sherwood's number, modified (Sha_s/a_c)
t	time (s)
V_L	volume of the solution (m^3)
Y	gaseous molar ratio of the solute to inert
Y_e, Y_s	gaseous molar ratio of solute to inert, at the inlet, the outlet of the column
Z	packing height (m)
z	stoechiometric coefficient.

GREEK SYMBOLS

β	total liquid holdup expressed in percentage of interparticle void volume, $\beta = \beta_s+\beta_d$
β_d	dynamic liquid holdup expressed in percentage of interparticle void volume
β_s	static liquid holdup expressed in percentage of interparticle void volume
$(\Delta H/Z)_{LG}$	two-phase head loss per unit length of the packing (m.water/m)
δ_{LG}	interaction force of fluid to packing (m.water/m)
δ_G, δ_L	gas, liquid head loss by friction with packing in single-phase flow (m.water/m)
ε	void fraction in packed bed (interparticle porosity)
μ_G, μ_L	dynamic viscosity of gas, of liquid (Pa.s)
ρ_e, ρ_G, ρ_L	manometric, gas, liquid density (kg/m^3)
σ_L	liquid surface tension (N/m)
ϕ	total rate of the solute gas absorption (Kmol/s)
φ	specific rate of the solute gas absorption ($kmol/m^2.s$)
ϕ_L	two-phase correlation parameter $\lvert\phi_L = (\delta_{LG}/\delta_G)^{1/2}\rvert$
χ	parameter of correlation $\lvert\chi = (\delta_L/\delta_G)^{1/2}\rvert$
Ω	cross section area of the column (m^2)

LITERATURE CITED

(1) Alvarez-Fuster, C., Midoux, N., Laurent, A. and Charpentier, J.C., Chem. Eng. Sci., 35, 1717, (1980).

(2) Alvarez-Fuster, C., Thèse, INPL, ENSIC, Nancy, France (1980).

(3) Barthole, G.D., Storck, A., Laurent, A. and Charpentier, J.C. Entropie, 91, 38, (1980).

(4) Belhaj, S., Thèse, INPL, ENSIC, Nancy, France, (1980).

(5) Charpentier, J.C. and Favier, M., AIChE J., 21, 1213, (1975).

(6) Charpentier, J.C., Chem. Eng. J., 11, 161, (1976).

(7) Charpentier, J.C., Chem. Rea. Eng. Rev. Houston, Edited by D. Luss and V.W. Weekman, A.C.S. Symposium series 72, 223, (1978).

(8) Crine, M. and Marchot, P., Chem. Eng. Commun., 8, 365, (1981).

(9) Fukushima, J. and Kusaka, K., J. Chem. Eng. (Japan), 10, 461, (1977).

(10) Germain, A., "Etude cinétique d'une réaction de catalyse en système gaz-liquide-solide. L'hydrogénation de l'alpha-méthylstyrène par le palladium supporté", Extrait de la collection des publications de la Faculté des Sciences Appliquées de l'Université de Liège, n° 40, Liège, Belgique, (1973).

(11) Gianetto, A., Baldi, G. and Specchia, V., Ing. Chem. Ital., 8, 125, (1970).

(12) Gianetto, A., Specchia, V. and Baldi, G., AIChE J., 19, 916, (1973).

(13) Gianetto, A., Specchia, V., Baldi, G. and Sicardi, S., AIChE J., 24, 1087, (1978).

(14) Goto, S. and Smith, J.M., AIChE J., 21, 706, (1975).

(15) Hirose, J., Toda, M. and Sato, Y., J. Chem. Eng. (Japan), 7, 187, (1974).

(16) Hofmann, H., Catal. Rev. Sci. Eng., 17, 71, (1978).

(17) Larkins, P.D., White, R.R. and Jeffrey, D.W., AIChE J., 7, 231, (1961).

(18) Laurent, A., Thèse, INPL, Nancy, France, (1975).

(19) Midoux, N., Favier, M. and Charpentier, J.C., J. Chem. Eng. (Japan), 9, 350, (1976).

(20) Mitchell, M.G. and Perona, J.J., Ind. Eng. Chem. Proc. Des. Dev., 18, 316, (1979).

(21) Morsi, B.I., Midoux, N. and Charpentier, J.C., AIChE J., 24, 357, (1978).

(22) Morsi, B.I., Thèse, INPL, ENSIC, Nancy, France, (1979).

(23) Morsi, B.I., Midoux, N., Laurent, A. and Charpentier, J.C., Entropie,No:9138, (1980).

(24) Morsi, B.I., Laurent, A., Midoux, N. and Charpentier, J.C., Chem. Eng. Sci., 35, 1467, (1980).

(25) Morsi, B.I., Midoux, N., Laurent, A. and Charpentier, J.C., "Hydrodynamic and gas-liquid interfacial parameters of co-current two-phase downward flow in trickle-bed reactors", Communication presented at CHISA Congress, (1981).

[26] Paraskos, J.A., Frayer, J.A. and Shah, Y.T., Ind. Chem. Proc. Des. Dev., 14, 315, (1975).
[27] Reiss, L.P., 6, 486, (1967).
[28] Rizzuti, L., Augugliaro, V. and Locascio, G., Chem. Eng. Sci., 36, 973, (1981).
[29] Sato, Y., Hirose, F., Takahashi, Toda, M. and Hashiguchi, J. Chem. Eng. (Japan), 6, 315, (1973).
[30] Satterfield, C.N. and Way, P.F., AIChE J., 18, 305, (1972).
[31] Satterfield, C.N., AIChE J., 21, 209, (1975).
[32] Shah, Y.T., "Gas-liquid-solid reactor design", Mac Graw Hill International Book Company, (1979).
[33] Shende, D.W. and Sharma, M.M., Chem. Eng. Sci., 7, 187, (1974).
[34] Specchia, V. and Baldi, G., Chem. Eng. Sci., 23, 515, (1977).
[35] Sridharan, K., Thesis, University of Bombay, India, (1975).
[36] Sridharan, K. and Sharma, M.M., Chem. Eng. Sci., 31, 767, (1976).
[37] Sylvester, N.D. and Pitayagulsarn, P., Ind. Eng. Chem. Proc. Des. Dev., 14, 421, (1975).
[38] Talmor, E., AIChE J., 23, 868, (1977).
[39] Turek, F. and Lange, R., Chem. Eng. Sci., 36, 569, (1981).
[40] Turpin, J.L. and Huntington, R., AIChE J., 13, 1196, (1967).
[41] Ufford, R.G. and Perona, J.J., AIChE J., 19, 1223, (1973).
[42] Van Landeghem, H., Chem. Eng. Sci., 35, 1912, (1981).
[43] Weekman, V.W. and Myers, T.E., "Heat transfer and fluid flow characteristics of concurrent gas-liquid flow in packed beds", presented at the 56th annual meeting, AIChE, Texas, (1963).
[44] Perez Sosa, J.A., "Ermittlung wichtiger fluiddynamischer Grös-sen bei rieselreaktoren mit den systemen wasser-luft und cy-clohexen-kohlendioxid", Doktor-Ingenieur Thesis, Erlangen, Deutschland, (1981).
[45] Dharwadkar, A. and Sylvester, N.D., AIChE J., 23, 376, (1977).

INFLUENCE OF HYDRODYNAMIC MODEL PARAMETERS ON THE ESTIMATION OF INTRAPARTICLE AND INTERPHASE TRANSPORT RATES IN A TRICKLE BED REACTOR

İNCİ EROĞLU - TİMUR DOĞU

Middle East Technical University
Chemical Eng. Dept. Ankara-Turkey

ABSTRACT

Dynamic analysis of a trickle bed reactor is carried out with a soluble tracer. The impulse response of the tracer is given at the inlet of the column to the gas phase and the tracer concentration distributions are obtained at the effluent both from the gas phase and the liquid phase simultaneously. The overall rate process consists the rates of mass transfer between the phases, the rate of diffusion through the catalyst pores and the rate of adsorption on the solid surface. The theoretical expressions of the zero reduced and first absolute moments which are obtained for plug flow model are compared with the expressions obtained for two different liquid phase hydrodynamic models such as cross flow model and axially dispersed plug flow model. The effect of liquid phase hydrodynamic model parameters on the estimation of intraparticle and interphase transport rates by moment analysis technique are discussed.

1 INTRODUCTION

A trickle bed reactor (TBR) consists of a fixed bed of catalyst particles, where liquid and gas phases flow cocurrently downward through the bed. Although its wide application in chemical and petrochemical industry it is one of the most complicated type of reactor in its design and scale-up. Essencially, the overall rate can be controlled by one or a combination of the following processes; mass transfer between interphases, intraparticle diffusion, adsorption and surface reaction. The hydrodynamics, solid-liquid contacting efficiency and axial mixing can also affect the performance of TBR.

Laboratory studies are usually carried out with small size of TBR's where mixing in the liquid phase may be important. However it has been indicated that in pilot scale and commercial size TBR plug flow might represent the actual situation quite well |1|. Then there exists some problems in the expression of the rate parameters obtained from laboratory studies.

Dynamic analysis of TBR by sitimules response technique has been succesfully applied to determine the extent of liquid axial mixing. There are number of learning and predictive models proposed in literature |2|. Among them the ones having less number of parameters such as cross-flow model and axially dispersed plug flow ADPF model are the most adequate ones. A more realistic model profound for a TBR can be the one which includes the simultaneous effect of interphase and intraparticle transport rates, and the adequate hydrodynamic model, to minimize the relative importance of liquid mixing on these rates.

The aim of this research is to show the influence of cross-flow model and ADPF model on the estimation of mass transfer coefficient, liquid hold up and adsorption factor by dynamic analysis of a TBR.

2 THEORETICAL MODELS

The theoretical models discussed in this study are mainly based on the following assumptions;

a. All the rate processes are linear.
b. Radial dispersion in both phases and axial dispersion in gas phase is negligible.
c. Equilibrium is reached at gas-liquid interphase and Henry's law is applicable.
d. Catalyst particles are spherical and completely wetted internally.
e. There is no mass transfer between gas phase and the solid phase.
f. Reversible first order adsorption of the tracer occur on the catalyst pore surface.
g. Physical properties, flow rates and hold-ups are constant throughout the column.

When ADPF model is assumed for the liquid axial mixing in TBR, the transient mass conservation equation are for the gas phase

$$\varepsilon_G \frac{\partial C_G}{\partial t} = - U_G \frac{\partial C_G}{\partial z} - K_L a_L (HC_G - C_L), \qquad (1)$$

for the liquid phase

$$\varepsilon_L \frac{\partial C_L}{\partial t} = - U_L \frac{\partial C_L}{\partial z} + D_L \frac{\partial^2 C_L}{\partial z^2} + K_L a_L (HC_G - C_L)$$

$$- k_s a_s (C_L - C_i \Big|_{r=R}) \qquad (2)$$

for the liquid in the pores

$$\varepsilon_s \frac{\partial C_i}{\partial t} = D_e \left(\frac{\partial^2 C_i}{\partial r^2} + \frac{2}{r} \frac{\partial C_i}{\partial r} \right) - N_i \varepsilon_s \qquad (3)$$

where

$$N_i = k_a \left(C_i - \frac{C_s}{K} \right) = \frac{\partial C_s}{\partial t} . \qquad (4)$$

Initial conditions are when t=0, $C_L=C_G=C_i=C_s=0$ for all z. Boundary conditions are

$$\text{at} \quad z=0, \quad C_G = M\delta(t) \qquad (5)$$

$$\text{at} \quad z=0, \quad U_L C_L = D_L \frac{\partial C_L}{\partial z} \qquad (6)$$

$$\text{at} \quad z=Z, \quad \frac{\partial C_L}{\partial z} = 0 \qquad (7)$$

$$\text{at} \quad r=0, \quad \frac{\partial C_i}{\partial r} = 0 \qquad (8)$$

$$\text{at} \quad r=R \quad D_e \frac{\partial C_i}{\partial r} \Big|_{r=R} = k_s (C_L - C_i \Big|_{r=R}) \qquad (9)$$

Eqns. (5) and (6) are the concentration distrubances of the tracer in the gas and the liquid feed streams where impulse response is given to gas phase. Danckwerts |3| boundary conditions are applied to the liquid phase, Eqns. (6) and (7).

Eroğlu |4| has transformed Eqns. (1) to (9) into Laplace domain and has solved the theoretical expressions of zero reduced and first absolute moments of the tracer in the liquid and the gas analytically, basing on the following moment definitions applicable in moment analysis technique. The fraction of tracer absorbed in liquid phase, which can also be defined as zero reduced moment of liquid phase is

$$m^*_{OL} = \frac{U_L}{U_G M} \lim_{s \to 0} \bar{C}_L \; . \tag{10}$$

The first absolute moments of the tracer in the gas and the liquid are defined as

$$\mu_{1G} = -\lim_{s \to 0} \frac{d\bar{C}_G}{ds} \Big/ \lim_{s \to 0} \bar{C}_G \tag{11}$$

$$\mu_{1L} = -\lim_{s \to 0} \frac{d\bar{C}_L}{ds} \Big/ \lim_{s \to 0} \bar{C}_L \, . \tag{12}$$

When axial dispersion in the liquid phase is neglected, that is $N_{pe}=U_L Z/D_L$ is approaching to infinity the moment equations derived by Eroğlu |4|are in accordance with the expressions derived by Ramachandran and Smith |5| who have assumed plug flow model for the liquid phase.

In the cross-flow or piston exchange model (PE) the liquid phase is assumed to be devided into a plug flow region and stagnant region in contact with the flowing liquid with which mass can be exchanged |6|. The physical interpretation can be shown as Case I and Case II in Fig.1.

In Case I the stagnant liquid is accumulated in tops of packing however in Case II the stagnant liquid is assumed to cover the catalyst surface.

Transient equations of conservation of mass for the gas and the liquid in pores are same as Eqns. (1) and (3) both for Case I and Case II.

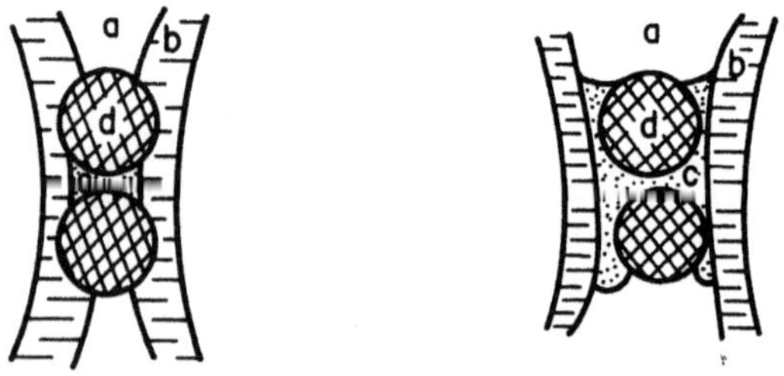

Case I

Case II

a) Un wetted region , b) Plug Flow Region , c) Stagnant region
d) Catalyst particles.

Fig. I. Physical Interpretation of cross flow model.

However the mathematical expression of the liquid phase is different. A mass balance of flowing liquid for Case I is

$$\phi\varepsilon_L \frac{\partial C_L}{\partial t} = - U_L \frac{\partial C_L}{\partial z} + K_L a_L (HC_G - C_L) - k(C_L - C_N)$$

$$- k_{sb} a_b (C_L - C_i \Big|_{r=R}) \tag{13}$$

for stagnant liquid phase

$$(1-\phi)\varepsilon_L \frac{\partial C_N}{\partial t} = k(C_L - C_N) - k_{sc} a_c (C_N - C_i \Big|_{r=R}) \tag{14}$$

Additional initial condition is when t=0, C_N=0 for all z. Although the other boundary conditions are the same Eqn. (6) is now

$$\text{at} \quad z=0, \quad C_L=0, \tag{15}$$

and Eqn. (9) is

$$\text{at} \quad r=R, \quad D_e \frac{\partial C_i}{\partial r}\Big|_{r=R} = k_{sb}(C_L - c_i \Big|_{r=R}) = k_{sc}(C_N - c_i \Big|_{r=R}). \tag{16}$$

For Case II transient equations of conservation of mass for flowing liquid is

$$\phi\varepsilon_L \frac{\partial C_L}{\partial t} = -U_L \frac{\partial C_L}{\partial z} + K_L a_L (HC_G - C_L) - k(C_L - C_N) \tag{17}$$

for stagnant liquid phase

$$(1-\phi)\varepsilon_L \frac{\partial C_N}{\partial t} = k(C_L - C_N) - k_s a_s (C_N - C_i \Big|_{r=R}) \tag{18}$$

where boundary conditions are same as Eqns. (5), (15), (7) to (9).

Solutions of C_L and C_G in Laplace domain are different for Case I and Case II. However the limits of $\bar{C}_L$ and $\bar{C}_G$ as $s \to 0$ and the limits of $d\bar{C}_L/ds$ and $d\bar{C}_G/ds$ as $s \to 0$ are same as plug flow model. Therefore consideration of cross-flow model to represent the mixing in the liquid phase does not bring any improvement on m^*_{0L}, μ_{1L} and μ_{1C} which are derived for plug flow model.

3 COMPARISON OF THE MODELS

Since m^*_{0L}, μ_{1L} and μ_{1G} derived for PE model approaches to plug flow expressions the influence of k and ϕ on these moments can not be observed. Table I shows the functional relationships between moments and model parameters. Comparison of theoretical moment expressions with experimental moments gas-liquid mass transfer coefficient, K_L, liquid hold up, ε_L, and adsorption factor, As, which is a combined parameter of adsorption equilibrium constant K and wetting efficiency,η_c, can be estimated. If K is known from independent liquid full bed experiments, then As is an important parameter to estimate wetting efficiency.

TABLE I. Functional Relationships

Models	Moments
Plug Flow Model	$m^*_{0L} = F(K_L)$
	μ_{1L} and μ_{1G} are $f(k_L, \varepsilon_L, As)$
	$\mu_2 = f(k_L, \varepsilon_L, As, k_a, k_s, D_e)$
Cross Flow Model	$m^*_{0L} = f(K_L)$
(PE)	μ_{1L} and μ_{1G} are $f(K_L, \varepsilon_L, As)$
	$\mu_2 = f(K_L, \varepsilon_L, As, D_e, k_a, k_s, \phi, k)$
Axially Dispersed	$m^*_{0L} = F(K_L, D_L)$
Plug Flow Model	μ_{1L} and μ_{1G} are $f(K_L, \varepsilon_L, As, D_L)$
(ADPF)	$\mu_2 = f(K_L, \varepsilon_L, As, D_e, k_a, k_s, D_L)$

In moment analysis technique the second moment,μ_2, is usually not considered due to its complexity and high uncertainity coming from the long tails of the residence time distribution (RTD) curves Nevertheless the functional relationships show that the transport rates such as liquid-solid mass transfer coefficient, k_s, effective diffusivity D_e, and adsorption rate, k_a, can be estimated from μ_2. It is expected that the PE model parameters ϕ and k affects the tru estimation of these parameters more successfully than ADPF model.

A fourth model to be discussed is the piston -dispersion-exchange (PDE) model which is a combination of PE model and ADPF model. For this model m^*_{0L}, μ_{1L} and μ_{1G} approach to ADPF model where only the effect of D_L is observed. For higher moments such as μ_2 other hydrodynamic model parameters are also effective. In fact this is an expected phenomena, although ADPF model fits the RTD curves at peaks well it fails to fit the tail, where PE model and PDE model are more successfully fitted.

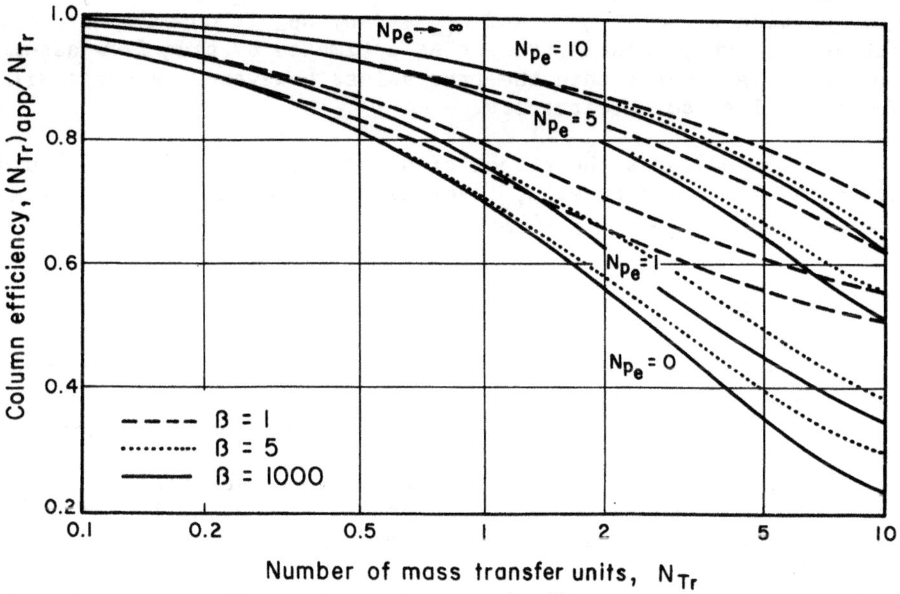

Fig. 2. Variation of column efficiency with true number of transfer units.

Ramachandran and Smith |5| have derived the theoretical expressions of m^*_{OL}, μ_{1L} and μ_{1G} also for reacting systems by considering ideal flow of the liquid phase. Unfortunately the moment expressions can not be solved analytically if axial dispersion is considered |4|. Therefore this comparison can not include reacting systems.

4 DISCUSSION

The deviation from plug flow model is most commonly observed as Peclet number, $N_{Pe}=U_LZ/D_L$, decreases, but the influence of N_{Pe} on the estimation of number of mass transfer units, $N_{Tr}=K_La_LZ/U_L$, is also related with the absorption factor, $\beta=U_LH/U_G$. The column efficiency is defined as $(N_{Tr})_{app}/N_{Tr}$, where $(N_{Tr})_{app}$ is the number of mass transfer unit estimated by plug flow model and N_{Tr} is the true number of mass transfer unit estimated by ADPF model |7|. Fig.2. shows the relation between column efficiency and N_{Tr} at different β and N_{Pe}. For very small β values column efficiency is independent of N_{Pe} and approaches to 1. For $N_{Tr}>0.02$ the effect of N_{Pe} is considerable.

The influence of N_{Pe} on μ_{1G} and μ_{1L} are shown on Fig.3. and Fig.4. respectively at different As. At very small β values μ_{1G} at $N_{Pe} \to \infty$ is equal to μ_{1G} at any N_{Pe}. The difference between them increases as β increases and N_{Pe} decreases, but if $N_{Tr}>10$, N_{Pe} has no significant effect on μ_{1G}. In contrast to μ_{1G}, the maximum deviation between μ_{1L} at $N_{Pe} \to \infty$ and μ_{1G} at any N_{Pe} is observed at small β values. For very soluble tracers As estimated by plug flow assumption will be greater than the true values however vise versa will occur for less soluble tracers.

Fig.5 indicates the effect of N_{Pe} on μ_{1G} at different liquid hold-ups. If ε_L increases μ_{1G} decreases but no significant change is observed on μ_{1L}. N_{Pe} affects μ_{1G} similar to Fig.3. at each ε_L.

5 CONCLUSIONS

The moment analysis has shown that the cross flow model parameters ϕ and k do not affect m^*_{OL}, μ_{1G} and μ_{1L}. However the axially dispersed plug flow model parameter N_{Pe} affects the estimation of the true values of N_{Tr}, ε_L and As in great deal. The difference between the true and the apparent values of these parameters depend on the system especially on the absorption factor β. If U_L/U_G is considered to be constant the most important factor is the solubility of the tracer.

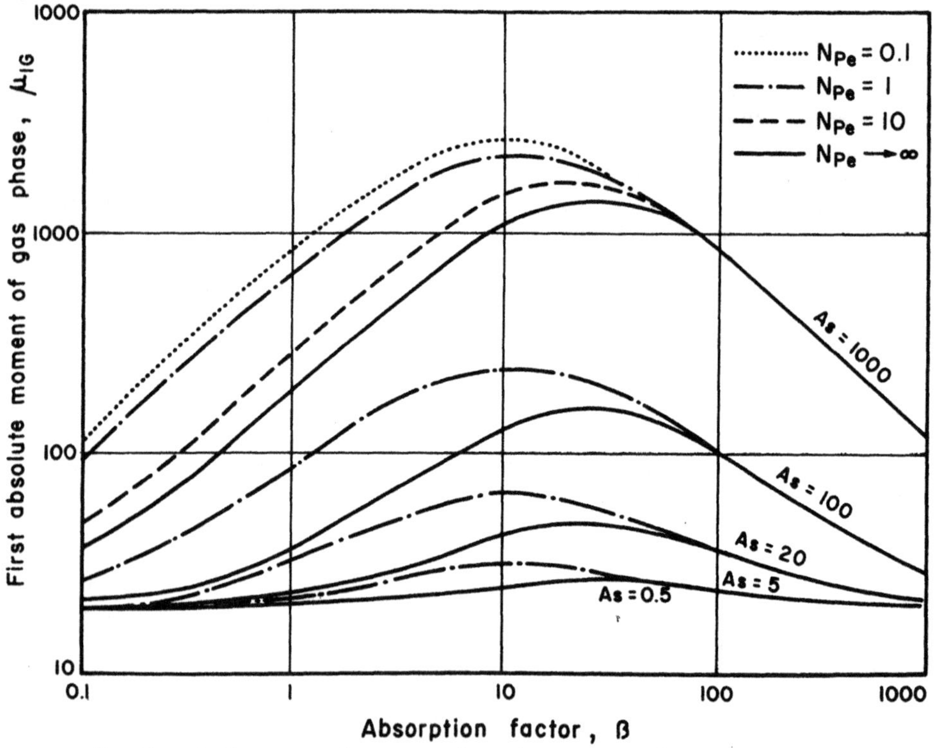

Fig. 3. Influence of absorption factor on first absolute moment of gas phase, for $N_{Tr} / \beta = 1$, ($U_L = U_G = 0.15$ cm/s, $\varepsilon_L = 0.3$, $\varepsilon_G = 0.2$, $Z = 15$ cm)

REFERENCES

1. Schwartz, J.G., and G.W. Roberts, "An Evaluation of Models for Liquid Backmixing in Trickle-Bed Reactor". Ind.Eng.Chem.Proc.Des. Dev.12 (1973) 262-270.
2. Gianetto, A., A. Baldi, V. Spechia, and S.Sicardi, "Hydrodynamics and Solid-Liquid Contacting Effectiveness in Trickle Bed Reactors". AIChE J. 24 (1978) 1087-1104.
3. Danckwerts, P.V., "Continuous Flow Systems Distribution of Residence Times". Chem.Eng. Sci., 2 (1953) 1-13.
4. Eroğlu, İ., "Dynamic Analysis of Trickle Bed Reactors". Ph D Thesis METU Ankara (1981).
5. Ramachandran, P.A., and J.M. Smith, "Dynamic Behaviour of Trickle Bed Reactors". Chem.Eng. Sci. 34 (1979) 75-91.
6. Hoogendoorn, C.J., and J. Lips, "Axial Mixing of Liquid in Gas Liquid Flow Through Packed Beds". Can. J. Chem.Eng. (1965) 125-131.
7. Sherwood, T.K., R.L. Pigford, and C.R. Wilke, "Mass Transfer" (McGraw Hill 1975).

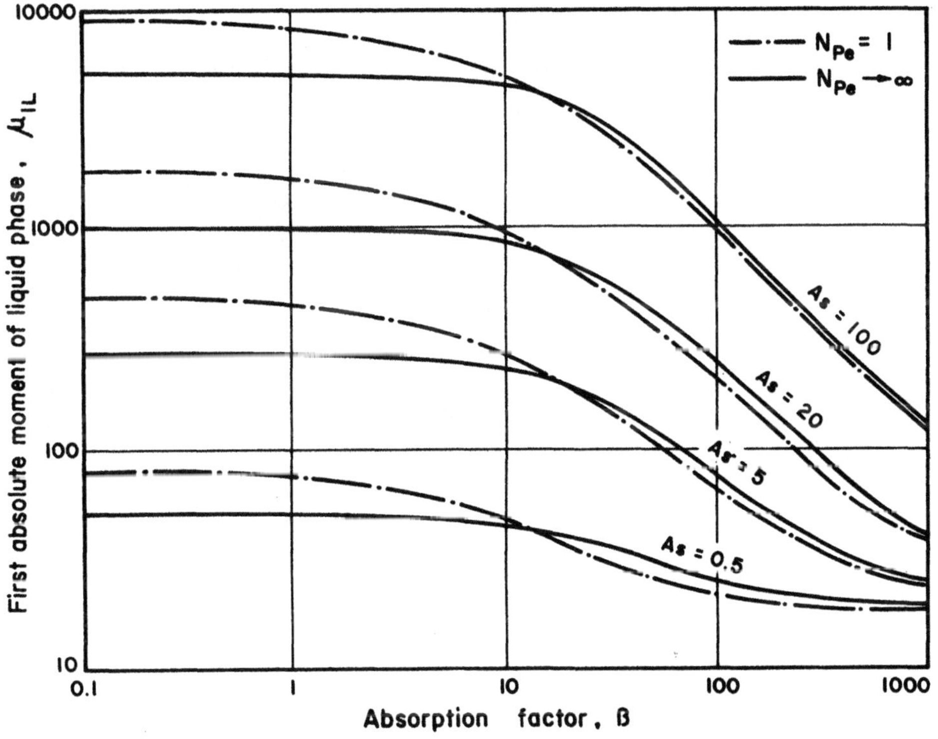

Fig. 4. Influence of absorption factor on first absolute moment of liquid phase, for $N_{Tr}/\beta = 1$ ($U_L = U_G = 0.15$ cm/s $\varepsilon_L = 0.3$, $\varepsilon_G = 0.2$, $Z = 15$ cm)

NOTATION

a effective mass transfer area per unit column volume, m^{-1}
As $= \eta_c(1-\varepsilon_L-\varepsilon_G)(\varepsilon_s+K)$ adsorption factor
$\underline{C}$ concentration of tracer, mole/m^3.
C Laplace transform of C
D_L axial dispersion coefficient, m^2/s
H Henry's law solubility constant, C_L/C_G.
K adsorption equilibrium constant
r radial distance in the catalyst particle, m.
R radius of the catalyst particle
s Laplace transform variable
t time, s.
U superficial velocity m/s.
z axial distance in the column, m.
Z height of the catalyst bed, m.

Subscripts

G refers to the gas
i refers to the liquid in the pores
L refers to the liquid
N refers to the stagnant liquid phase
s refers to the surface of catalyst pores

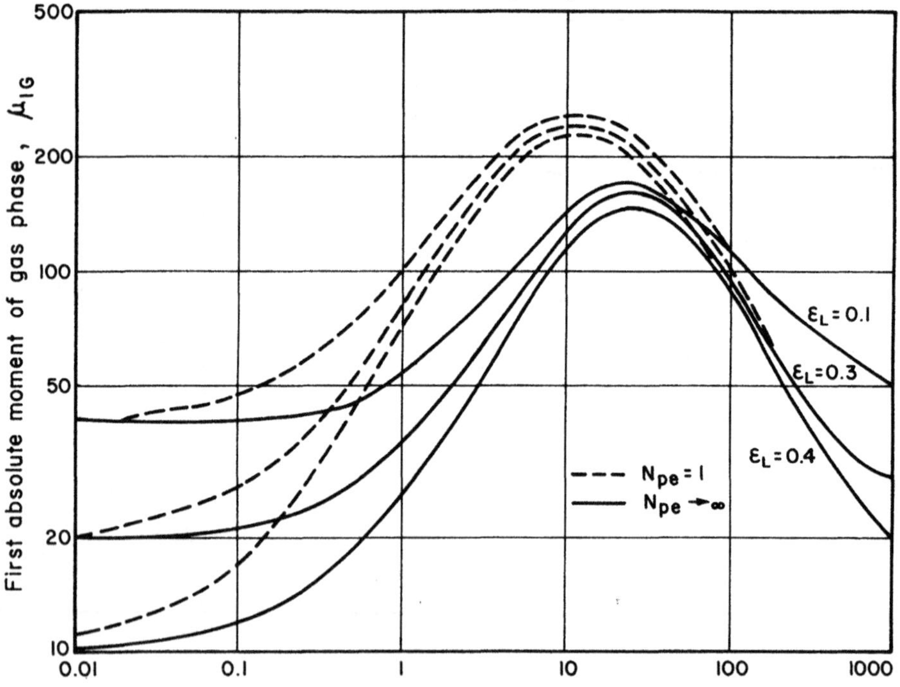

Fig. 5. Influence of absorption factor on first absolute moment of gas phase for As = 100 ($U_L = U_G = 0.15$ cm/s, $\varepsilon_L + \varepsilon_G = 0.5$, Z = 15 cm)

REACTION ENGINEERING PROBLEMS IN SLURRY REACTORS

Prof. Dr. H. Hofmann

University Erlangen-Nürnberg

1 INDUSTRIAL ASPECTS AND DIFFERENT TYPES OF SLURRY REACTORS

Three phase catalytic slurry reactors are characterized by a continuous liquid phase in which a gas phase is dispersed and a solid (catalyst) is suspended. They are commonly used for catalytic hydrogenation, oxidation, halogenation or polymerization reactions such as edible oil hydrogenation, olefin oxidation or hydroformylation etc. But also fermenters can be included into this category of multiphase reactors.

In principle, continuously (or semicontinuously) operated catalytic slurry reactors represent an alternative to catalytic fixed bed reactors, with the main difference of a much smaller size of the solid catalyst particles. From the viewpoint of chemical reaction engineering, slurry reactors seem to be the most neglected ones by researchers today, as stated by van Landeghem by occasion of his status of the art report during ISCRE 6 [1].

Slurry operations have many advantages compared to fixed bed three phase contacting, e.g.

i) a higher space time yield, because of smaller transport resistances under comparable conditions
ii) same concentration of the reactants in any volume element of the reaction mixture (i.e. lumped system). Together with a practically isothermal operation this can result in a higher selectivity of the reaction
iii) catalyst efficiency close to 100 %, because of the small particle size; uniform catalyst activity (and aging)
iv) large heat capacity of the reaction mixture and more favorable heat transfer, avoiding this way hot spots in

the reaction volume

v) possibility of continuous catalyst regeneration without any interruption of production

vi) easy to be switched from continuous into semicontinuous (or even discontinuous) operation.

In spite of these advantages, there are also some drawbacks like:

i) liquid, solid (and sometimes also gas) phase are completely back-mixed, demanding eventually for stagewise operation to get higher degree of conversion and/or selectivity in continuous operation

ii) particle attrition connected with some loss of fine catalyst particles through the liquid product stream in continuous operation, requesting additional filtration of the outlet stream

iii) large liquid to solid phase ratio, resulting eventually in a promotion of unwanted homogeneous sidereactions.

There exist four different types of three phase slurry reactors (see fig. 1)

i) the stirred tank reactor
ii) the loop reactor
iii) the bubble column
iv) the multiphase fluidized bed

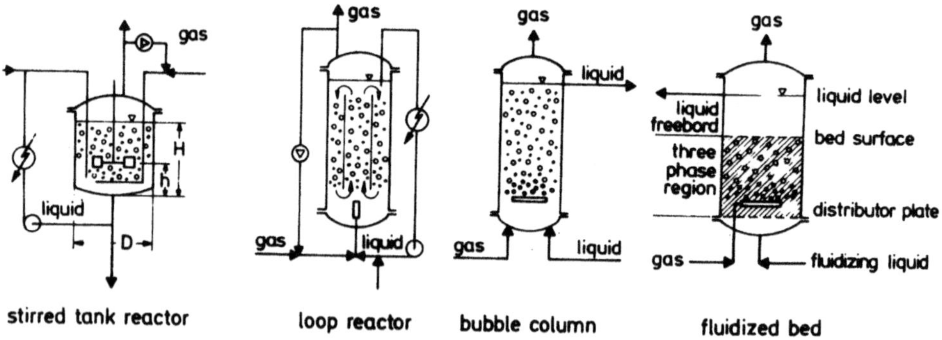

Fig. 1 Different types of threephase slurry reactors

Mechanically agitated stirred tank reactors are preferentially used in food- and bio-technology or with smaller chemical productions, where they are operated very often discontinuously. The gas phase is supplied either under pressure through a static gas feeder, by self-suction agitators or by surface aeration, sometimes assisted by an additional stirrer fitted just below the gas-liquid interface. Paddles with inclined or straight blades and turbines are the most common types of agitators. They are very effective with viscose liquids at low gas flow rates and large liquid volumes.

In loop reactors a forced and controlled circulation with flow velocities >20 m/s, is realized by a mixer device placed inside the reactor, entraining gas by suction and acting jointly with an ejector and circulation pumps for gas and liquid.

Bubble columns are very simple in their construction having the advantage of no moving parts, but do not allow a strict control of fluiddynamics in the reaction mixture. In practice they are used with coal hydrogenation and Fischer-Tropsch-synthesis and therefore reach large capacities.

Multiphase fluidzed beds are operated usually with continuous flow of gas and liquid phase, whereas the solid (catalyst) phase is either discontinuous or continuous. Here the energy needed for suspension and dispersion is provided by the flowing phases itself.

Compared to fixed bed multiphase reactors additional modes of operation are available with slurry reactors as also the solid phase can be feeded continuously and eventually recirculated; temperature control can be assured by internal cooling coils or/and external circulation through heat exchangers (for the gas as well as for the liquid phase). Furthermore also partial evaporation and external condensation with recirculation of the liquid phase can support the cooling.

Solid particle size in stirred tank reactos, loop reactors and bubble columns is usually in the range of 100 - 200 μm and solid concentration is rather low whereas in multiphase fluidized beds the solid concentration as well as solid particle size is larger. The density difference of liquid and solid usually is large, except in bioreactors, where one is faced also with agglomeration problems and nonnewtonian behavior of the suspension.

If energy needed for the suspension of solid is small, three phase slurry reactors behave very similar to twophase gas-liquid reactors and design methods developed for those types of reactors can be used as first approximation.

The following review is concerned mainly with mechanically agitated, aerated slurry tank reactors, but as information about this reactor type sometimes is limited, information about gas/liquid twophase reactors is provided, as they behave similar to threephase reactors at low solid concentrations. The paper has to be selective and additional information is available in the reviews of Chaudhari [2], Charpentier [3], Nagel [4], Nagata [5] and Wiedmann [6].

2 MECHANICALLY AGITATED, AERATED SLURRY REACTORS

The mechanically agitated, aerated slurry reactor, used in hydrogenation of unsaturated oils, production of aluminum alkyls, hydroxylamin-synthesis and fermentation processes, is the oldest type of slurry reactors. Technical dimensions of these reactors range from $< 1\ m^3$ up to several 100 m^3 reaction volume. The operating pressure usually is smaller than 100 bars in order to avoid design problems. The energy needed for solids suspension depends mainly on geometry of agitator and reactor and to a smaller amount on specific gas load. Turbines seem to be more effective as agitators than paddles. In the usual range of operation, the gas flow rate and the type of gas distributor have only a minor influence on fluiddynamics if the agitator speed is high enough. On the other hand, gas flow rate and gas distributor are important in case of smaller stirrer speeds. Gas flow rates in the range of 0.5 m/s and higher are in use and values of the gas-hold-up between 0.2 - 0.4 have been measured. For catalytic reactions typical solid concentrations are in the range of 10 - 20 g/lit. An agitator diameter of 1/3 of the reactor diameter seems to be optimal. The agitator is most often installed approximately at half the impeller (or 1/6 the reactor) diameter distant from the bottom. But it should be noted that the optimal location in twophase systems does not necessarily coincide with that in a threephase system; it is also dependent on the solid concentration and the gas supply system [5]. In case of pressure aeration, gas throughput and stirrer speed are independent process variables, whereas in surface aeration or gas suction through the stirrer shaft both are coupled. Hight to diameter ratio of the reactor is close to one, preferentially in high coalescing systems, but also H/D-ratios of 2.5 are to be found in practice; in the latter case very often several agitators are mounted on the same shaft.

2.1 Fluiddynamics in Aerated and Agitated Slurry Reactors

The most important fluiddynamic parameters of aerated stirred slurry reactors are energy dissipation and pumping efficiency of agitator and gas throughput, gas-holdup and mean bubble diameter produced (i.e. interfacial area resulting) and flooding characteristics.

2.1.1 Energy consumption and gas throughput. The energy dissipated in the aerated suspension has to be large enough

i) to suspend all solid particles and
ii) to disperse the gas-phase into small enough bubbles.

Concerning the suspension of fine solid particles, there exist two limiting states:

i) "complete" suspension, i.e. the state where just all particles have been brought into motion
ii) homogeneous suspension, i.e. the state where all the solid particles are uniformly distributed throughout the entire vessel volume.

As homogeneous suspension in nonaerated stirred vessels hardly can be achieved, even with very high stirrer speeds, mainly the minimum stirrer speed n_c, needed for "complete" suspension, is of interest for design purposes. This value, by definition, is characterized by the so called "just-suspended" criterion, i.e. the state where only a small fraction of the solids remains at the bottom of the reactor one second at maximum [7]. This state is linked with a critical Reynolds number (based on stirrer speed and diameter $Re_{nc} = n_c d/\nu_L$).

The critical stirrer speed n_c mainly depends on geometry of stirrer (d) and vessel (H/D), type of stirrer, density (ϱ_L) and viscosity (ν_L) of the liquid, solids mass ratio $\varphi_s = m_s/(m_s+m_L)$, particle diameter (dp) and other properties of the system.

If type and location (h) of the stirrer are fixed the following basic correlation may exist:

$$n_c = f\ (d,\ dp,\ \varrho_L,\ \nu_L,\ \Delta\varrho_{LS},\ \varphi_s,\ g) \tag{1}$$

On the basis of dimensional analysis one gets the dimensionless relation:

$$Re_{n_c} = f\ (Ar,\ dp/d,\ \varphi_s) \tag{2}$$

where $Ar = d^3 g \Delta\varrho_{LS}/\varrho_L \nu_L^2$ is the so called Archimedes number (also called expanded Froude number Fr^*). In most practical design cases, one is interested only in turbulent flow conditions, where $Re_{nc}(dp/d) > 10^2$.

Zlokarnik [8] has shown that in this case

$$Re_{n_c} = a \cdot Ar^{0,5}\ \varphi_s^{0,25} \tag{3}$$

or as $Ar = Re^2/Fr \cdot (\Delta\varrho_{LS}/\varrho_s)$, with $Fr = n^2 \cdot d/g$,

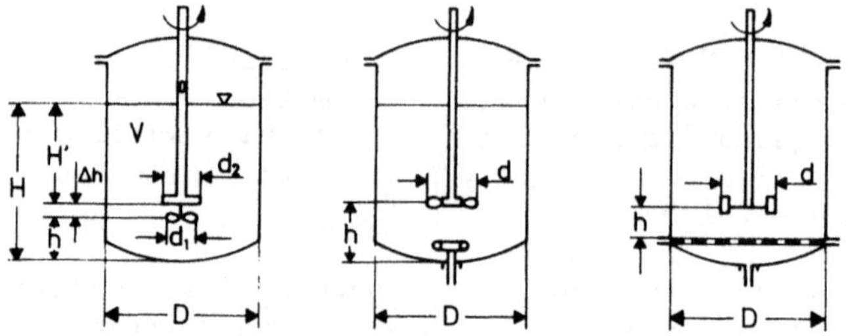

Fig. 2 Different types of dispersing/suspending devices in stirred reactors

$$Fr_c = b \cdot (\Delta\rho_{LS}/\rho_L)\ \varphi_s^{0.5} \qquad (4)$$

Therefore, with a given system (i.e. $\Delta\rho_{LS}/\rho_L$ and φ_s constant) and a fixed type and location of the stirrer, the following scale-up rule is valid

$$n_c = \sqrt{Fr_c \cdot g/d} \qquad (5)$$

Zwietering [9] has shown that for (radially acting) Rushton turbines

$$Fr_c = b\ (\Delta\rho_{LS}/\rho_L)\ (w/\rho_L)^{0.29} \qquad (6)$$

where w is mass of catalyst per unit volume of slurry and

$$b = (D/d)^{2.59}\ (dp/D)^{0.44}\ Re_n^{-0.22} \qquad (7)$$

Similar results have been published for paddles by Einenkel [10], for propellers by Pavlushchenko [11]and for paddle-ejector combination (see fig. 2) by Zlokarnik [12] . Also Zehner [13]and Baldi [14] have published similar results for paddles and Rushton turbines.

For reactor design, it is most important to know the power consumption corresponding to "complete" suspension. Power consumption is either correlated per unit volume of suspension

$(P/V)_{sp}$ in [W/m^3] or in a dimensionless way as power number (Newton number)

$$Ne = E/\varrho_L\, n^3 d^5 = f(Re_n) \tag{8}$$

in its dependence on the stirrer Reynolds number [18]. The power consumption for "complete" suspension therefore can be calculated according to

$$P_c = Ne_c \cdot n_c^{\;3} \cdot d^5 \cdot \varrho_L \tag{9}$$

From the dimensionless correlation between power number and extended Froude number [12]

$$Fr \cdot (\varrho_L/\Delta\varrho_{LS})\ \frac{n_c \cdot d}{u_{ss}} \cdot \varphi_s^{\;-1} = K(V/d^3)\ Ne^{-1} \tag{10}$$

(with V as total volume of the slurry and u_{ss} as terminal settling velocity of the solid particles) the following scale-up rules for a given suspension and constant Fr can be deducted

$$(n^2 d)_{large} = (n^2 d)_{small} \tag{11}$$

or with constant geometry of the vessel

$$n_{large} = n_{small}\ (d_{large}/d_{small})^{0.5} \tag{12}$$

i.e. the specific power consumption grows much less than proportional to size.

In aerated suspensions according to [5] , a significantly higher stirrer speed and thus power consumption per unit volume is required to establish the state of "complete" suspension. In fig. 3 it can be seen that in an aerated suspension the propeller normally requires a higher stirrer speed for suspension than the turbine. It was first Arbiter [15] who found that drastic sedimentation of suspended particles occured when a critical value of the so called gas flow number $N_Q = Q_G/n \cdot d^3$ was exeeded (here Q_G is the volumetric gas load). This critical gas flow coincided with the point, where the power drawn by the agitator decreased suddenly with a further small increase in the gas sparger rate. Thus an increase in gas rate or decrease in impeller speed have a similar sedimenting effekt, whilst decreasing gas rate or increasing stirrer speed tends to homogenize the suspension.

To achieve simultaneously suspension of solid particles and dispersion of gas, it is obviously necessary to define the state, when the gas phase is well dispersed. Nienow [16] defined this

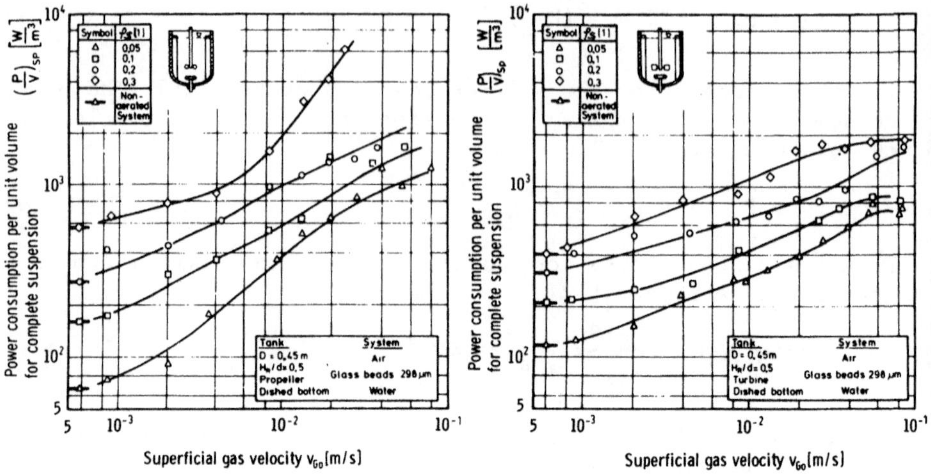

Fig. 3 Power consumption per unit volume for "complete" suspension

point as to be coincident with the minimum in the power number Ne against gas flow number N_Q relationship (see fig. 4 [17]). Accord-

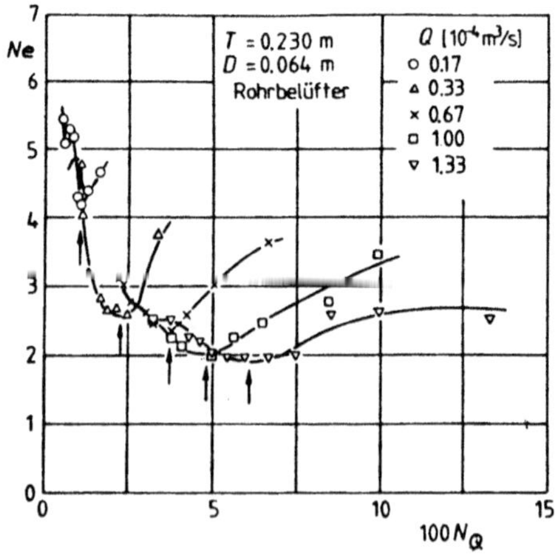

Fig. 4 Power number as function of gas flow number

ding to Chapman [18] this definition seems to be a reasonable assumption, but his study showed also that there is some critical particle density (relative to the liquid density) above which particle suspension governs the power necessary to achieve a "well-mixed" system and below which gas dispersion governs the power requirements. Therefore aeration at the critical stirrer speed for "complete" suspension of solid particles in nonaerated systems causes partial sedimentation of relatively heavy particles and aids suspension of relatively light particles. Furthermore, it seems to be possible that there may be a similar (but weaker) effect with particle size.
On the other hand, for a constant gas Reynolds number, Wiedmann [5] defines the "complete" state of suspension as to be coincident with the maximum in the Ne-Re_n-diagram (see fig. 5).

Increasing particle concentrations have a similar effect on the state of "complete" suspension in aerated and unaerated systems, i.e.

$$n_c \propto \varphi_s^{0.1} \qquad \text{with } d = D/2$$
$$n_c \propto \varphi_s^{0.14} \qquad \text{with } d = D/3 \tag{13}$$

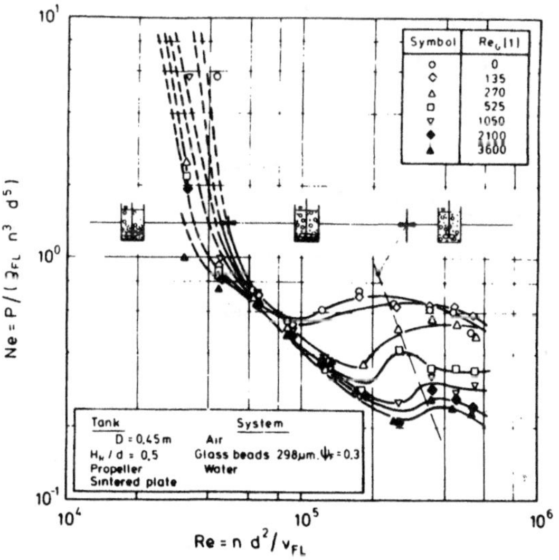

Fig. 5 Power number as function of stirring and aeration intensities in slurry reactors

But it should be noted that up to now, all these results are based exclusively on systems with air and water as gas and liquid phase respectively.

2.2.2 Energy dissipation and gas holdup. Gas holdup in aerated stirred slurry reactors is usually defined as fraction ε_G of the total volume of the aerated suspension.

As Nagel [14] has shown in solid free mechanically agitated and aerated twophase gas/liquid systems, the gas holdup depends on the power dissipartion density as well as on the gas loading according to

$$\varepsilon_G \propto \left(\frac{P}{V}\right)^q \left(\frac{Q_G}{D}\right)^p \tag{14}$$

With increasing gas load or superficial gas velocity u_{GO} [m/s] the mean gas holdup markedly increases in two phase gas/liquid systems as well as in three phase gas/liquid/solid systems as shown by Wiedmann [5]. This effect is more pronounced in case of a porous sintered plate than in the case of a sparger ring. Solid mass ratio is of little influence; higher values of φ_s cause ε_G to decrease slightly (see fig. 6 [6]); stirring intensity is of minor importance. But again, it should be pointed out that most of the data available on threephase systems concern the system air/water/glas beads and there can be expected an additional in-

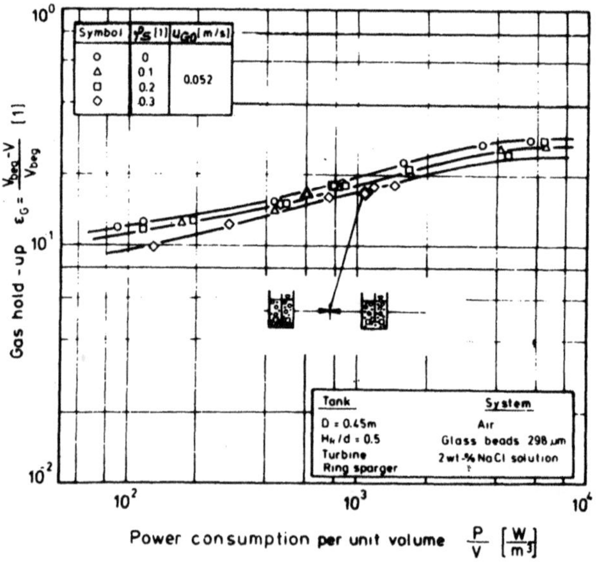

Fig. 6 Mean gas holdup as function of power consumption per unit volume and solid mass ratio

fluence from physical properties of the components like ν_L, σ_L $\Delta\rho_{LS}$ etc. as indicated e.g. by Calderbank's correlation [20] for twophase gas/liquid systems:

$$\varepsilon_G = \left(\frac{u_{Go}}{u_b}\right)^{0.5} + 0.0216 \frac{(P/V)^{0.4} \rho_L^{0.2}}{\sigma_{GL}^{0.6}} \left(\frac{u_{Go}}{u_b}\right)^{0.5} \qquad (15)$$

where σ_{GL} is the surface tension of the liquid and u_b is the terminal bubble rising velocity. As long as no better data are available Eq. (15) should be used as first approximation for threephase systems, according to the results given in fig. 6.

2.2.3 Flooding of agitators. In stirred, aerated slurries, the permissible gas load is limited by flooding. Beyond the flooding limit, the dispersing action of the mixing device becomes much less effective and the gas stream passes upwards mainly through the mixer. Gas holdup drops, mean bubble size increases and thus specific interfacial area decreases drastically as well as power requirement for the agitator. The jump in power requirement really can serve to identify the flooding point. Very little information is available in literature about gas flooding of liquid agitators. Fig. 7 about the maximum distributable superficial gas velocity is reproduced from Wiedmann [6] .

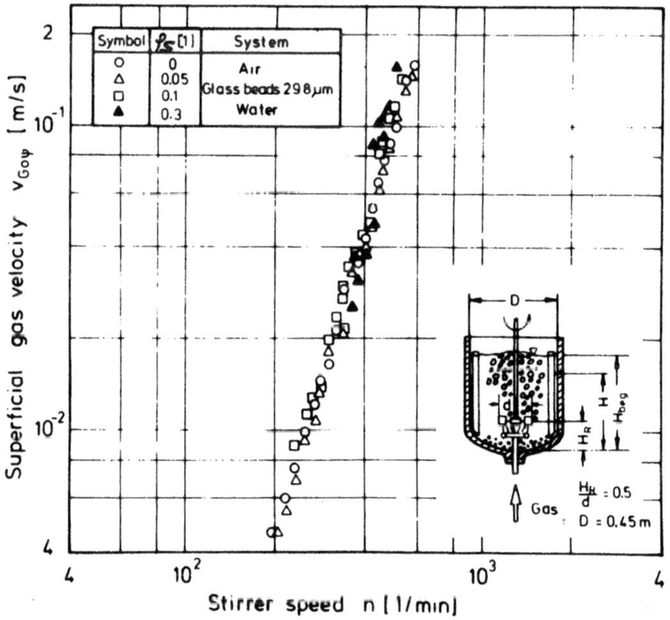

Fig. 7 Maximum distributable superficial gas velocity as function of stirrer speed and solid mass ratio

Solid mass ratio as well as liquid viscosity seems to be of little influence, but more experimental data are needed in order to develop a universal relationship.

2.2.4 Specific gas/liquid interfacial area. The specific gas/liquid interfacial area a' [m^2/m^3 of dispersion] is very important for mass transfer of reactants across this phase boundary.

In solid free systems, it depends on power dissipation density (P/V) and gas loading (Q_G/D^2) according to the general correlation

$$a' = C'\left(\frac{P}{V}\right)^{m_2}\left(\frac{Q_G}{V}\right)^{m_2+n_2-1} D^{n_2} \quad \text{with } LD^2 \sim V,\ L \sim D \tag{16a}$$

or

$$= C^*\left(\frac{P}{V}\right)^{m} \cdot \varepsilon_G^{\,n} \tag{16b}$$

as shown by Nagel [4]. E.g. Calderbank [20] correlated his experimental data as

$$a' = 1.44\,\frac{(P/V)^m \varrho_L^{\,0.2}}{\sigma_L^{\,0.6}}\,\varepsilon_G^{\,0.5} \tag{17}$$

where at lower values of (P/V) the exponent of m is dependent on the type of the gas distributor, but at (P/V) > 10 kW/m^3 is stabilizing around m=0.4. For orientation, some values of a''' [m^2/m^3 liquid] for the system O_2/aqueous salt solutions can be taken from fig. 8 [3] . a''' is increased by increases in ionic strength, ion valence number, viscosity and the presence of solids.

Nagel [4] recognizing the universal applicability of Eq.(16) has depicted the operating ranges of different types of gas/liquid reactors graphically. This diagram (reproduced in fig. 9) is very useful in selecting the appropriate reactor type for a specific reaction.

In threephase slurry reactors, the specific gas/liquid interface is decreased by an increasing solid mass ratio as demonstrated for a special system in fig. 10 [21].

But also here more experimental data are needed in order to generalize these experimental findings. As it is nearly impossible up to now to predict a priori the interfacial area, one should use for design purposes the scale-up rule that the same interfacial area requires the same total power input per unit volume of liquid, where total power input P_t is given by

$$P_t = P_m + P_s \tag{18}$$

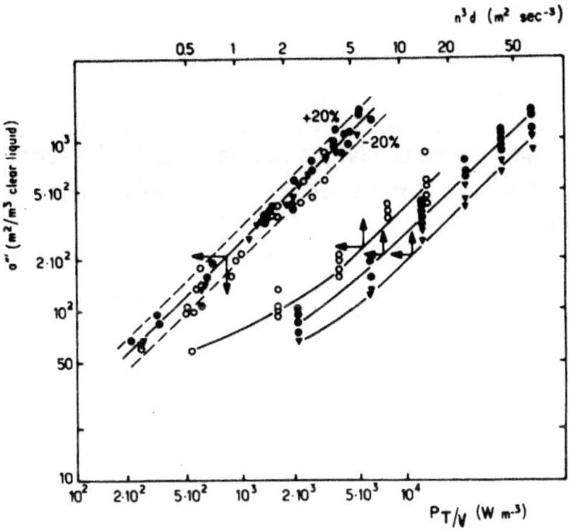

Symbol	System	μ_L (cP)	σ_L (dynes/cm)	ρ_L (gm/cm³)	D (cm)	d (cm)	u_G (cm/sec)	n (sec⁻¹)	Ref.
●	O_2–Na_2SO_3	1.5	61	1.08	19	6.4	0.64–5.50	8.33–25	L23
▼	O_2–CuCl(HCl)	1.25	44	1.15	19	6.4	0.64–4.70	8.33–25	L23
○	O_2–Na_2SO_3	1.3	61	1.09	29	12	0.28–1.62	3.33–10	L12

Fig. 8 Gas/liquid interfacial area as function of specific total power input

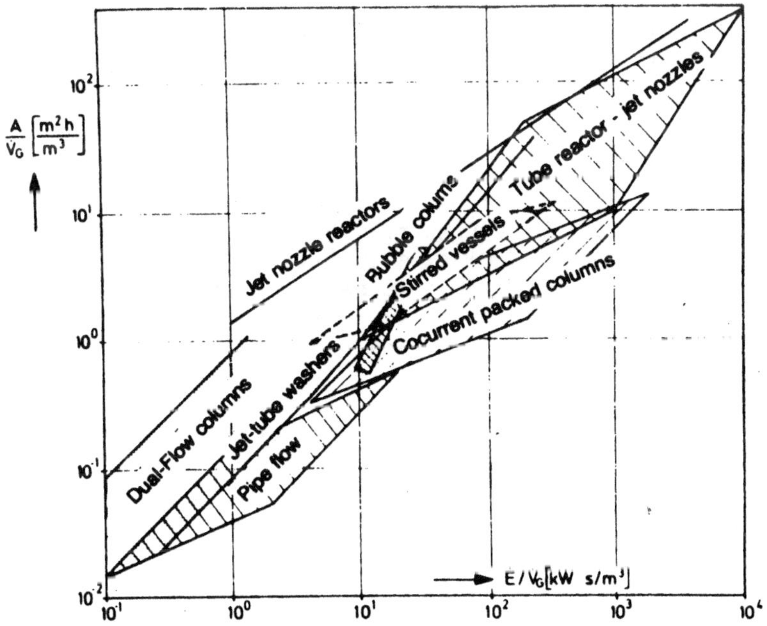

Fig. 9 Operating ranges of gas/liquid reactors

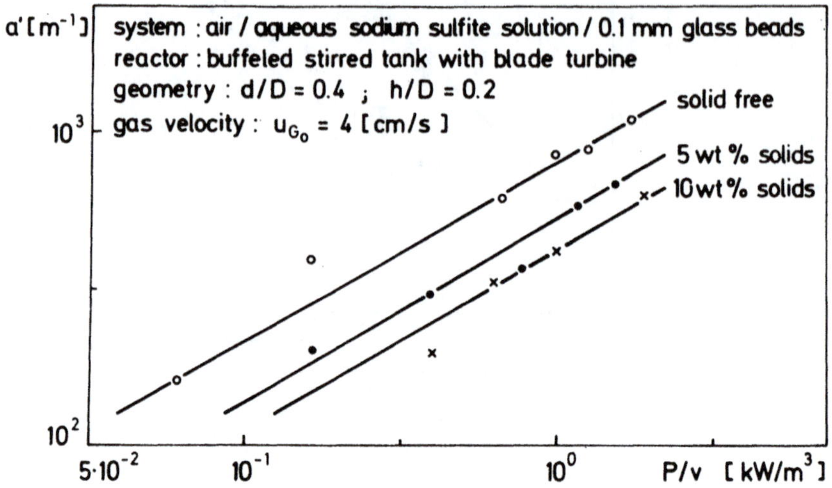

Fig. 10 Influence of solid mass ratio on specific gas/liquid interface

with P_m as mechanical agitation power [W] and P_s power expense to sparge the gas. Therefore, in case of reactor design, experiments should be performed on a small agitated and aereted reactor of the same geometry as the envisaged technical unit using the same reaction mixture as in practice.

2.2 Mass and Heat Transport Processes

Mass (and heat) transport steps sometimes can be rate limiting in agitated and aerated slurry reactors. Therefore methods have to be provided to determine the relative importance of these steps and furthermore equations have to be at hand for the estimation of the special transport parameters in their dependence on physical properties, operating conditions and reactor geometry.

2.2.1 Experimental methods for parameter estimation. In order to determine the relative importance of mass transport processes in gas/liquid/solid-reactions, it is recommended to measure the global reaction rate as function of catalyst concentration m (keeping all other operating variables constant). With the assumption of spherical catalyst particles, the specific surface of the catalyst can be calculated as

$$a_s = \frac{6m}{\rho_s\, d_p} \quad [\ m^2/m^3 \text{ suspension }] \tag{19}$$

where the catalyst load m is given in [kg/m^3 suspension] and ρ_s is the density of solid. Furthermore, if it is assumed that gas bubbles behave also as spheres, an assumption which seems to be justified with small enough bubbles, the specific gas/liquid interface can be calculated as

$$a' = \frac{6\varepsilon_G}{d_b} \quad [\,m^2/m^3 \text{ suspension}\,] \tag{20}$$

In case of a simple first order reaction with respect to the gaseous reactant (assuming that the liquid reactant is not limiting) the effective global reaction rate can be formulated as

$$\frac{1}{r_{eff}} = \frac{1}{C^*}\left[\frac{1}{K_L a'} + \frac{1}{k_s a_s} + \frac{1}{k\, a_s}\right] \tag{21}$$

with C^* the equilibrium concentration of the gaseous reactant in the suspension and k the specific first order rate constant with respect to catalyst surface [$m \cdot s^{-1}$]. In case of completely suspended, nonporous catalyst particles, Eq. (21) can be converted [using equation(19) and (20)]into

$$\frac{1}{r_{eff}} = \frac{1}{C^*}\left(\frac{d_b}{K_L 6\varepsilon_G} + \frac{\rho_s d_p}{6m}\left[\frac{1}{k_s} + \frac{1}{k}\right]\right) \tag{22}$$

Therefore if C^*/r_{eff} is plotted against 1/m a plot like fig. 11 should arise.

Slope and intersection on the ordinate usually used to determine transport parameters, naturally depend on the other operating conditions like temperatures etc.

Ruether [22] has presented another method, separating $k_s a_s$ by studying the catalytic gas/liquid reaction at low (i.e. limiting) concentrations of the liquid reactant C_L as function of the inverse of the partial pressure of the gaseous reactant. In this case C , the concentration of the gaseous reactant in the suspension is always at equilibrium with the gas phase, not only at the gas/liquid interface but also on the catalyst surface, i.e.

$$r_{eff} = k_s a_s\,(C_L - C_s) = \eta \cdot r(C_s,\, C^*) \tag{23}$$

where C_L and C_s is the concentration of the liquid reactant in the bulk of the liquid phase and on the catalyst surface. In order to arrive at a plot similar to fig. 11, r_{eff} should be linearly dependent on C_s (or C_L should be very large compared with C_s, i.e. filmtransport limitation, so that $k_s a_s$ can be determined directly

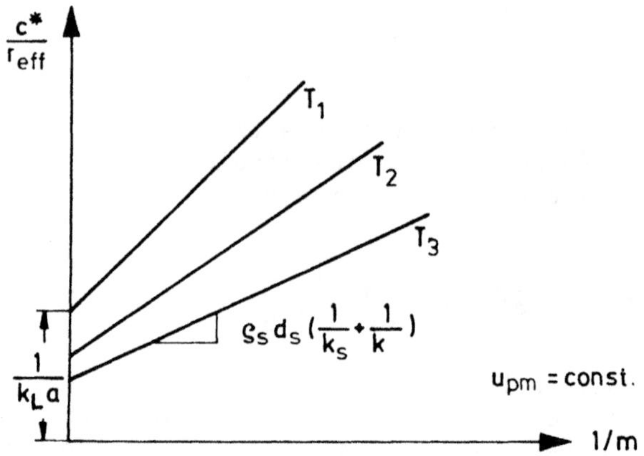

Fig. 11 Inverse effective reaction rate as function of catalyst load

as $k_s a_s = r_{eff}/C_L$). But at low substrate concentration C_L even in case of a Langmuir-Hinshelwood kinetics, one can approximate the hyperbolic rate equation by a potential rate law, leading (probably only in case of strong porediffusion limitation) to the desired pseudo-linear dependency:

$$r_{eff} = k' \cdot C^{*} \cdot C_s \tag{24}$$

with k' as apparent rate constant with respect to the catalyst load m.
By combination of Eq. (23) and (24) one gets:

$$\frac{1}{r_{eff}} = \frac{1}{C_L}\left(\frac{1}{k_s a_s} + \frac{1}{k'C^{*}}\right) = \frac{1}{C_L}\left(\frac{1}{k_s a_s} + \frac{H_1}{k'C_1}\right) \tag{25}$$

where C_1 is the gas phase concentration of the gaseous reactant and H_1 its Henry coefficient.
I.a.w., even in case of a complex rate law, there exists a certain chance to get a plot similar to fig. 12, wherefrom $k_s a_s$ can be determined.

Hofmann [23] has used just this method in order to determine the relative importance of the individual transport or reaction steps. They studied the hydrogenation of aquous glucose solution

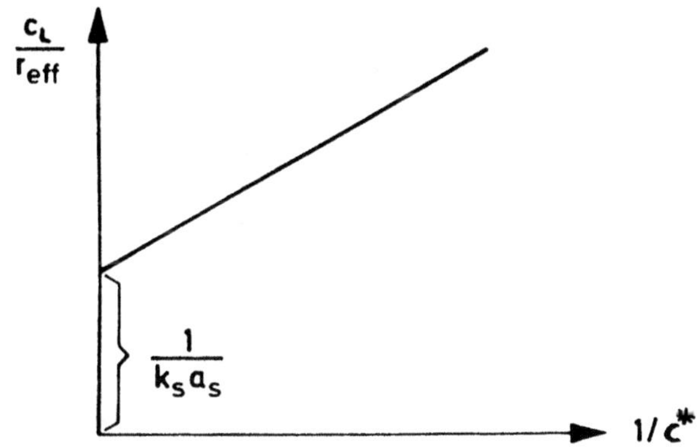

Fig. 12 Inverse effective reaction rate as function of inverse partial pressure

over Raney-nickel in the temperatur range of 80-130°C at hydrogen pressures of 10-30 bars and under variable stirrer speeds, i.e. different turbulence in the liquid phase. The effective rate of reaction could be correlated according to a first order law with respect to glucose and for a constant hydrogen pressure of 10 bars, the results given in fig. 13 have been received.

The heavy solid curves show the dependence of r_{eff}/C_L as function of temperature and stirrer speed; the nonlinear dependence on 1/T and stirrer speed respectively indicate the interaction of filmdiffusion and chemical reaction. To varify the transport limitation (for glucose) additional experiments had been performed under similar operating conditions but with a catalyst of higher activity (by adding metallic magnesium). r_{eff}/C_L -values determined in these series of experiments are connected by the thin solid lines, showing the small temperature dependence of a transport step.

Therefore, the following equations must be valid for the nonporous catalyst

$$r_{eff} = k_s \cdot \frac{6m}{\rho_s \cdot dp} (C_L - C_s) = k' \cdot C^*_{H_2} \cdot C_L \qquad (26a)$$

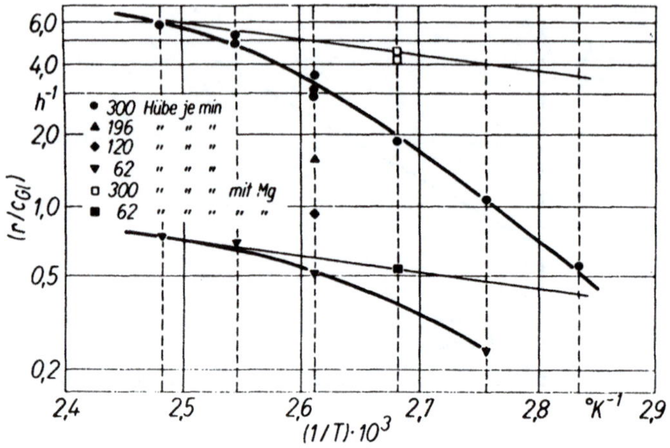

Fig. 13 Effective reaction rate for glucose hydrogenation

or

$$\frac{C_L}{r_{eff}} = \frac{\varrho_s\ d_p}{k_s \cdot 6m} + \frac{1}{k'\ C^*_{H_2}} \tag{26b}$$

By variation of the hydrogen pressure, $C^*_{H_2}$ had been varied and k_s determined the same way as shown in fig. 12.

2.2.2 Mass-transport coefficients. For the a priori design of agitated and aerated slurry reactors, correlations are needed between mass-transport coefficients at the interfaces and operating conditions.

As pointed out by Charpentier [3] almost no quantitative information is available on the gas-side mass-transport coefficient k_Ga' for mechanically agitated vessels. Fortunately this coefficient is usually quite high, leading to relatively negligible gas-phase restistance, i.a.w. $K_L \approx k_L$.

The true liquid-side mass-transfer coefficient k_L, usually measured using physical absorption, does not change significantly with agitator speed or gas load, but depends strongly on the physicochemical properties of the system. This is mainly true for gas bubbles smaller than 2,5 mm (behaving like solid spheres) whereas mass-transfer from larger bubbles seems to be enhenced by internal circulation in the bubbles, roughly porportional to d_b.

For solid free two phase systems with $d_b < 2.5$ mm, the correlation of Calderbank [24] is valid for k_L in pure liquids or solutions of nonionic solutes

$$k_L = 0.31 \cdot Sc^{-\frac{2}{3}} \cdot \left[\frac{(\rho_L-\rho_G) g \, \nu_L}{\rho_L}\right]^{\frac{1}{3}} \quad [cm/s] \tag{27}$$

This equation, in approximate accordance with experimental data from other researchers [25] , does neither contain a dependency of k_L on superficial gas velocity u_{Go} nor on specific power input (P/V). Such is not the case in electrolyte solutions, where k_L decreases with increasing specific power input corresponding to reduced bubble diameter [26]. Organic solvents behave different from aquous solutions; k_L increases with increasing diffusity of solute gas and with decreasing viscosity of the liquid [27].

In three phase systems k_L a' is changed by the solid particles. With solid volume fractions >15 % and particle diameter $d_p <$ 200 μm k_L a' does not change significantly (smaller than 10-20%) but higher solid concentrations and larger particle size drastically decrease k_L a', mainly because a' decreases [28]. But porous solids having a absorption capacity for the gas can also increase k_L a' by a factor of two and even more [29].

Concerning the liquid/solid mass transfer coefficient k_s, two different theories have been used to correlate data from solid/liquid suspensions: one uses the terminal slip velocity between particles and the liquid. The other approach is based on Kolmogoroff's theory of isotropic turbulence and uses energy dissipation rate $\varepsilon = u_{Go} g$ for correlation. Sänger [30] recommend the equation

$$Sh_s = 2 + 0.545 \, Sc^{\frac{1}{3}} \left(\frac{\varepsilon d_p^{\,4}}{\nu^3}\right)^{0.264} \tag{28}$$

for aerated suspensions and it is believed (see also [31, 32]) that this equation can be applied also on mechanically stirred and aerated suspensions if only the total power input P_t is used instead of power expense to sparge the gas, as indicated by Eq. (18) [33].

2.2.3 Heat transfer coefficients. Heat transfer rates to the reactor wall and/or cooling coil can limit the performance of agitated and aerated slurry reactors. Therefore in reactor design, the heat transfer coefficient α[kJ/m^2,h,k] for the system in its dependence on operating conditions must be known. Unfortunately no data can be found in literature concerning this parameter value. Therefore, the respective values for two phase solid free systems should be used as first approximation.

Combining Highbie's Penetration theory and Kolmogoroff's theory on isotropic turbulence, it has been shown recently [34] that experimental data on aerated liquids in stirred tank reactors can be correlated for coil as well as wall heat transfer by the equation

$$St = C\,[Re_G \cdot Fr_G \cdot Pr_G^2]^{\,m} \qquad (29)$$

with $C \approx 0.2$ and $m = -\,^1/_3$ where $St = \alpha/\varrho_L \cdot c_{pL}\, u_{Go}$ or $St = \alpha D/\varrho_L c_{pL} n d^2$ is the Stanton number and $Pr_G = nd^2 \varrho_G c_{pG}/\lambda_G$ is the Prandtel number. As long as the size of solid particles in a suspension is $d_p < 200\ \mu m$, it is believed that the same correlation can be used also for aerated suspensions in stirred tank reactors. This is supported by the finding that heat transfer in aerated suspensions (bubble columns) can be correlated very well by the same correlation with slightly different parameters ($C=0.1$, $m=-^1/_4$).

2.4 Reactor Models (this chapter follows mainly the outline given in [2])

Chemical reaction, performed in agitated and aerated suspensions, preferentially belong to the regime of absorption accompanied by slow reaction; sometimes the operating pressure is elevated. The reactions (simple reactions) are in most cases of first order with respect to the gaseous reactant and the liquid reactant is present in large excess, leading to a zero reaction order with respect to this component. The modelling of stirred slurry reactors in this case is further simplified by the practically isothermal behavior of the reactor and the assumption of a nearly homogeneous suspension of the small (mostly nonporous) catalyst particles. Therefore in a rigorous a priori design it is sufficient to use material balances only. Furthermore the liquid/solid suspension is regarded as ideally mixed, whereas for the gas phase very often plug flow is assumed.

If for the so called differential operation of a slurry reactor, i.e. the liquid phase concentrations change only slightly and the main interest is on the amount of gas reacted, then the mass balance of the gaseous reactant (A_1) in the gas phase as function of the hight of the suspension $0 < z < H$ (measured from the gas inlet) can be written as:

$$- u_{Go} \frac{dC_{G1}}{dz} = (k_L a')_1 \left[\frac{C_{G1}}{H_1} - C_{L1} \right] \tag{30}$$

with H_1 as the respective Henry coefficient and $k_L a'$ as an overall gas/liquid volumetric mass transfer coefficient. This equation can be integrated to give

$$\frac{C_{G1} - H_1 C_{L1}}{C_{gli} - H_1 C_{L1}} = \exp(-\alpha_1 z) \tag{31}$$

with

$$\alpha_1 = \frac{(k_L a')_1}{u_{Go} H_1} \tag{32}$$

Therefore the concentration of gas leaving the reactor under plug flow is

$$C_{Glo} = C_{Gli} \exp(-\alpha_1 H) + H_1 C_{L1} [1 - \exp(-\alpha_1 H)] \tag{33}$$

and the rate of gas absorbed per unit volume of slurry is

$$r_1 = \frac{H_1 Q_G}{V} [1 - \exp(-\alpha_1 H)] \left(\frac{C_{Gli}}{H_1} - C_{L1} \right) \tag{34}$$

The rate of mass transfer from the liquid to the surface of the catalyst is

$$r_1 = (k_s a_s)_1 (C_{L1} - C_{s1}) \tag{35}$$

Combining Eq. (34) and (35) the overall rate of transfer of A_1 from the gas phase to the surface of the catalyst is

$$r_1 = M_1 \left[\frac{C_{Gli}}{H_1} - C_{s1} \right] \tag{36}$$

with

$$M_1 = \left[\frac{1}{\frac{H_1 Q_G}{V} [1 - \exp(-\alpha_1 H)]} + \frac{1}{(k_s a_s)_1} \right]^{-1} \tag{37}$$

For sparingly soluble gases the first term in M_1 reduces to

$1/(k_La')_1$, indicating that varying gas phase concentrations are negligible. On the other hand, for highly soluble gases the same term reduces to $^1/(H_1Q_G/V)$ indicating that gas/liquid mass transfer resistance has no significance.

With the local apparent rate of reaction (under surface conditions) per unit volume of slurry

$$r_c = m \cdot k_n \, c_s^n \tag{38}$$

it follows

$$r_1 = M_1 \left[\frac{C_{G1i}}{H_1} - \left(\frac{r_1}{mk_n}\right)^{1/n} \right] \tag{39}$$

In case of n=1 this implicit expression for r_1 can be tranformed into an explicit on

$$r_1 = \frac{C_{G1i}}{H_1} \left[\frac{}{\frac{H_1 Q_G}{V} \left[1-\exp(-\alpha_1 H)\right]} + \frac{1}{k_s a_s} + \frac{1}{mk_1} \right] \tag{40}$$

Expressions for the overall rate of reaction in differentially operated slurry reactors in case of other reaction type can be found in [2]. In the same paper also diagrams are presented for overall effectiveness factors (including also intraparticle diffusion) and different kinetic models are given.

The effect of gas phase mixing, neglected in the development above, is likely to be important only for gases in the intermediate solubility range, i.e. $0.5 < \alpha_1 H < 10$ and when gas/liquid mass transfer is limiting. When the gas phase is completely mixed, the mass balance in the gas phase for A_1 gives

$$Q_G(C_{G1i} - C_{G1o}) = (k_La')_1 V \left[\frac{C_{G1o}}{H_1} - C_{L1} \right] \tag{41}$$

Hence the concentration of the gas leaving the reactor is

$$C_{G1o} = C_{G1i} \frac{1}{1+\alpha_1 H} + H_1 C_{L1} \left(1 - \frac{1}{1-\alpha_1 H}\right) \tag{42}$$

Compared to the plug flow situation Eq. (33) in Eq. (42) the term $\exp(-\alpha_1 H)$ is replaced by $1/(1+\alpha_1 H)$ as one might have expected. All the other equations remain the same provided the above substitution is made.

Complex reactions with consecutive or parallel steps are commonly encountered in catalytic slurry reactors, e.g. in hydration of propylene oxide, ethynylation of formaldehyde to butinediol or hydrogenation of unsatturated oils [35, 36, 37, 38, 39]. If one can assume in a simplified analysis of these reactions that the concentrations of the liquid reactants are in excess, compared to the gaseous one (A_1) the rate of reaction of the gaseous reactant (in the absence of intraparticle diffusion) can be expressed as

$$r_1 = mk_{(m_1+n_1)} C_2^{n_1} \left[C_1^* - \frac{r_1}{M_1} \right]^{m_1} + mk_{(m_2+n_2)} C_3^{n_2} \left[C_1^* - \frac{r_1}{M_1} \right]^{m_2} \tag{43}$$

where $C_1^* = C_{G1}/H_1$ is the concentration of A_1 in the liquid in equilibrium with the gas, C_2 and C_3 are the liquid concentrations of co-reactants A_2 and A_3, m is the reaction order with respect to reactant A_1,n_i are the reaction orders with respect to the liquid co-reactants and index 1 and 2 indicate the main and the side reaction respectively.
Therefore, the (differential) point selectivity with respect to the main reaction is

$$\sigma = \frac{k_{(m_1+n_1)} C_2^{n_1} \left[C_1^* - \frac{r_1}{M_1} \right]^{m_1}}{k_{(m_2+n_2)} C_3^{n_2} \left[C_1^* - \frac{r_1}{M_1} \right]^{m_2}} - 1 \tag{44}$$

Eq. (44) can be used to compute the influence of the external mass transfer parameter M_1 using r_1 calculated according Eq. (43). When reaction orders are different, mass transfer limitation of A_1 will favor the reaction of lower order (see also [35]). In case of porous particles, also intraparticle diffusion could be incorporated into the analysis by solving the relevant differential equations. Special situations can arise here, when A_1, A_2 and A_3 are present in comparable concentration inside the catalyst particle.

When, on the other hand, the conversion of liquid phase components as a function of operation time of a semi-batch slurry reactor (with a stationary liquid phase) is of interest, e.g. for a reaction

$$A_{1(g)} + \nu_2 A_{2(1)} \longrightarrow \nu_3 A_{3(1)} \tag{45}$$

with A_2 in excess, the rate of reaction per gram of catalyst can be expressed as:

$$r = k_2 C_1 C_2 \qquad (46)$$

assuming first order rate with respect to both reactants. r_1 and r_2 are related by

$$r_2 = \nu_2 r_1 \qquad (47)$$

The rate of reaction of A_2 per unit volume of slurry (in absence of significant intraparticle concentration gradients) can be written as:

$$r_2 = -\frac{dc_2}{dt} = \frac{\nu_2 C_{Gli}}{H_1} \left[\frac{1}{\frac{H_1 Q_G}{V} \quad 1-\exp(-\alpha, H)} + \frac{1}{k_s a_s} + \frac{1}{mk_2 C_2} \right]^{-1} \qquad (48a)$$

$$= \frac{\nu_2 C_{Gli}}{H_1} \left[\frac{1}{M_1} + \frac{1}{mk_2 C_2} \right]^{-1} \qquad (48b)$$

Integration of Eq. (48) for constant H_1 gives:

$$\frac{C_{2o} - C_{2f}}{M_1} + \frac{1}{mk_2} \ln \frac{C_{2o}}{C_{2f}} = \frac{\nu_2 C_{Gli}}{H_1} t \qquad (49)$$

with C_{2f} as concentration of the liquid reactant at time t. When intraparticle gradients cannot be neglected, i.e. if

$$\frac{d_p}{6} \left[\frac{k_2 \, \rho_G \, C_{2o}}{D_{1eff}} \right]^{1/2} > 0.2 \qquad (50)$$

catalyst effectiveness factor η has to be introduced into the equations by replacing the last term in Eq. (48a) through $(1/m\eta \, k_2 C_2)$. A method to estimate the time of operation to reach a given C_{2f} can be found in [40]. Equations for overall effectiveness factor η and generalized Thiele modulus in slurry reactors for different kinetic models are given in [2]. Modifications of the equations given above for the case of changing solubility due to concentration changes in the course of the reaction, i.e. $(C_{G1i}/H_1) = f(C_2, C_3$ etc.), can be found also in [2] as well as a discussion about product inhibition and temperature effect.

In continuously operated slurry reactors, where both gas and liquid phases are flowing through the reactor, also the degree of conversion of A_2 is of interest. Modelling of these reactors is based ususally on the assumption of an ideally mixed liquid phase, which is likely to be valid for any mechanically stirred reactor. Also here model equations for the conversion of A_2 are available for various kinetic schemes when intraparticle resistances are negligible [41] . These models include also possible gradients in solid concentrations as observed in cases of large H/D-ratio and/or insufficient stirrer speed.

Finally, stability problems in slurry reactors have been treated in [42]. There it has been shown that even under isothermal conditions, instabilities may exist.

To sum up, the following rules can be given for the scale-up of slurry reactors:

i) when intraparticle or liquid/solid masstransfer is limiting, no real scale-up problem exists, because the design variable is n_c and d_p respectively.

ii) when gas/liquid masstransfer is limiting k_La has to be constant during scale-up. Therefore geometrical similarity as well as constant (P/V)-ratio has to be obeyed. Both can be reached only with constant gas holdup in small and large scale; if e.g. the same u_{Go} is used, the gas residence time must be enlarged by the scale factor and the stirrer speed has to be changed by the - $^2/_3$ power of the scale factor.

Respecting those simple rules, a sufficiently safe scale-up should be possible on the base of laboratory experiments, even not all details are known about agitated and aerated slurry reactors. Additional recommandations for an optimal operation of those reactors are given in [29].

Literature

[1] van Landeghem, Chem.Eng.Sci. 35 (1980) 1912
[2] Chaudhari, R.V. and Ramachandran, P.A., AIChE J. 26 (1980) 177
[3] Charpentier, J.Ch. Adv.Chem.Eng. 11 (1981) 21
[4] Nagel, O., Hegner, B. and Kürten, H. Int.Chem.Eng. 21 (1981) 161
[5] Nagata, S. "Mixing" Wiley, New York 1975
[6] Wiedmann, J.A., Steiff, A. and Weinspack, P.M. Ger.Chem.

Eng. 3 (1980) 303
[7] Einenkel, W.D. VDI-Forschungsheft 595 (1979)
[8] Zlokarnik, M. and Judat, H. Chem.Ing.Techn. 11 (1969) 1270
[9] Zwietering, T.N., Chem.Eng.Sci. 8 (1958) 244
[10] Einenkel, W.D., VT 11 (1977) 9
[11] Pavlushchenko, Y., J.appl.Chem. USSR 30 (1957) 1235
[12] Zlokarnik, M. in Ullmann's Encyclopädie der technischen Chemie, 4th edit. vol. 2, Verlag Chemie Weinheim 1972
[13] Zehner, P., Ger.Chem.Eng. 2 (1979) 220
[14] Baldi, Q., Conti, R. and Narita, E., Chem.Eng.Sci. 33 (1978) 21
[15] Arbiter, N., Harris, C.C. and Yap, R.F., Trans. AIME 244 (1969) 134
[16] Nienow, A.W., Wisdom, D.J. and Middleton, J.C., Proc. 2nd Europ. Conf. on Mixing (BHRA Cranfield) p F1-17 (1977)
[17] Baldi, G. in Proc. NATO ASI on "Multiphase Chemical Reactors" p., Vimero 1980, Sijthoff & Noordhoff
[18] Chapman, C.M., Nienow, A.W., Middleton, J.C., Trans.I. Chem.E. 59 (1981) 134
[19] Nagel, O. and Kürten, H., Chem.Ing.Tech. 48 (1976) 513
[20] Calderbank, P.H., Trans.Instn.Chem.Eng. 37 (1958) 443
[21] Kürten, H. and Zehner, P., Ger.Chem.Eng. 2 (1979) 220
[22] Ruether, F.A. and Puri, P.S., Can.J.Chem.Eng. 51 (1973) 545
[23] Hofmann, H. and Bill, W., Chem.Ing.Techn. 31 (1959) 81
[24] Calderbank, P.H. and Moo-Young, M.B., Chem.Eng.Sci. 16 (1961) 39
[25] Deckwer, W.D., Louisi, Y., Zaidi, A. and Ralek, M., Ind. Eng.Proc.Des.Dev. 19 (1980) 699
[26] Robinson, C.W. and Wilke, C.R., AIChE J. 20 (1974)
[27] Zlokarnik, M., Adv.Biol.Eng. 8 (1978) 133
[28] Joosten, G.E.H., Schilder, J.G.M. and Janssen, J.J., Chem. Eng.Sci. 32 (1977) 563
[29] Deckwer, W.D. and Alper, E., Chem.Ing.Techn. 52 (1980) 219
[30] Sänger, P. and Deckwer, W.D. to be published in Chem.Eng. Sci. 36 (1981)
[31] Sano, Y., Yamaguchi, N. and Adachi, T., J.Chem.Eng. Japan 7 (1974) 255
[32] Sicardi, S., Conti, R., Baldi, G., Franzino, L., Chem. Ing.Techn. 53 (1981) 67
[33] Herndl, G. and Mersmann, A.B., Chem.Eng.Commun.(1981) to be published
[34] Steiff, A., Poggemann, R. and Weinspack, P.M., Chem.Ing. Techn. 52 (1980) 492
[35] Kawakani, K. and Kusunoki, K., Kogaku Ronb. 1 (1975) 559
[36] Kawakani, K., Ohgi, Y. and Kusunoki, K., J.Chem.Eng.Jap. 9 (1976) 475
[37] Kale, S.S. and Chaudhari, R.V., paper presented at 4th Nat.Symp.Catal. Bombay 1978

[38] Coenen, J.W.E., J. Amer.Oil Chem.Soc. 53 (1976) 382

[39] Cordova, W.A. and Harriott, P., Chem.Eng.Sci. 30 (1975) 1201

[40] Chaudhari, R.V. and Ramachandran, P.A., Ind.Eng.Chem.Fund. (1980)

[41] Govitarao, V.M.H., Chem.Eng.J. 9 (1975) 292

[42] Schneider, F. and Mitschka, F., Proc. 5th Europ.Symp. Chem.React.Eng. 1972, B6-1

SOME ASPECTCS OF GAS ABSORPTION MECHANISM IN SLURRY REACTORS

Erdoğan Alper*and W.-D.Deckwer**

University of Ankara,Beşevler,Ankara,Turkey*
Fachbereich Chemie,Universität Oldenburg,F.R.Germany**

1.INTRODUCTION

Slurry reactors are popular in industry where the solids either take part in the reaction or act as catalyst.Many aspects of these reactors,particularly for catalytic systems,have been discussed at length in literature(1,2).Catalytic slurry reactors are also reviewed in this proceedings by Hofmann(3).However,there are still aspects which have not been treated in the literature in sufficient detail.Firstly,until recently little attention has been paid to slurry reactors involving reactive solids.Secondly, it is often assumed that steps of diffusion of the dissolved gas from the gas-liquid interface to the bulk liquid phase;bulk liquid phase to the solid catalyst surface;and surface reaction are steps in series.This leads of course to a specific gas absorption rate which is always smaller than $k_L A^*$.While this is a representative picture in a majority of cases of industrial relevance,we can conceive situations,where the catalyst particle size may be smaller than the diffusion film(liquid film next to gas-liquid interface) thickness.We may then have steps of the transport of the dissolved gas from the gas-liquid interface and reaction on the catalyst particle in parallel,that is,while the dissolved gas diffuses it reacts on the catalyst surface.This is then in many ways analogous to normal gas-liquid reactions and may lead to the enhancement of specific gas absorption rate so that it exceeds $k_L A^*$.This point is also relevant to reactive solid systems; indeed in an earlier paper,Ramachandran and Sharma(4) had shown that the specific rate of absorption accompanied by an instantaneous reaction in a slurry containing sparingly soluble fine particle size was considerably smaller than the film thickness. Finally,there is substantial information in the literature on the combined effect of solid particles on $k_L a$.However,the information on separate effects on k_L and a is rather scanty. Indeed, recent evidence obtained by Alper et.al.(5) indicates that k_L may also be increased by the presence of highly adsorptive particles such as activated carbon.

2. SLURRY REACTORS INVOLVING REACTIVE SOLIDS

There are several industrially important gas-liquid-solid reactions such as carbonation of lime, coal liquefaction etc., where solids take part in the reaction. The common type of reactor for these reactions is normally a gas-liquid-suspended solid column with or without mechanical agitation. An extensive literature is already available on the hydrodynamic, mixing, mass transfer, and heat transfer characteristics of these reactors, and critical reviews have been published in recent years(1-3).

These three phase reactions may be classified into two distinct categories, one in which the solids are sparingly soluble in the medium and the reaction occurs between the dissolved gas and dissolved reactive species obtained from dissolution of solid particles, the other in which the particles are insoluble in medium. In each case, the product of the reaction may be soluble or insoluble in medium. Table 1 and 2 show some typical examples of these two cases.

2.1. Insoluble Solid Particles

If the solid is insoluble in the medium, a surface reaction takes place between the dissolved gas and the solid. In general, the reaction may be classified into two broad categories, one in which the solids change the particle size and the other in which the particle size is retained. In the first case, the product is soluble in the medium and hence the solid reactant shrinks. Hydrometallurgy provides a variety of examples, such as oxidative leaching, which falls in this category. For cases in other category, the product is also insoluble and stays with the particle; thus the reactant has to diffuse through the product "ash" layer. It is suggested by Joshi et.al.(6) that the oxidation of iron pyrite by dissolved oxygen in aqueous solutions falls in this category. Indeed, this forms the basis of a chemical coal cleaning method, i.e. "oxydesulfurisation"(6). Other examples are shown in Table 1.

For the corresponding cases of fluid-solid reactions, there has been considerable amount of work and it is a trivial exercise to modify them to gas-liquid-solid systems when the solids are considerably bigger than the thickness of the diffusion film at the gas-liquid interface. In these cases, the rate controlling step can be gas-liquid mass transfer, diffusion through product "ash" layer(if the reacted material is insoluble), chemical reaction or the combination of two or more steps. Here, it is assumed that all these steps occurs in series and the concentration profiles are as shown in Figure 1 and 2 for the corresponding cases of constant and diminishing solid sizes. The describing equations can be derived easily for various controlling regimes(see, for

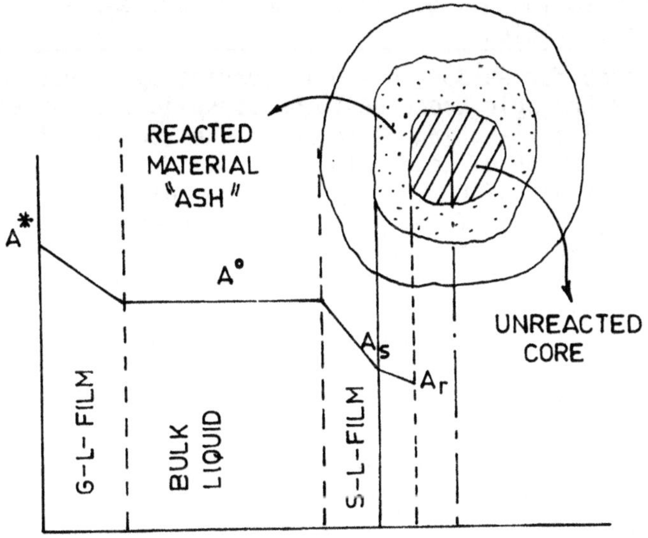

Figure 1.Concentration profile of dissolved gas A:Slurry reactor with a constant size insoluble solid reactant(Particles are larger than film thickness and the product is insoluble).

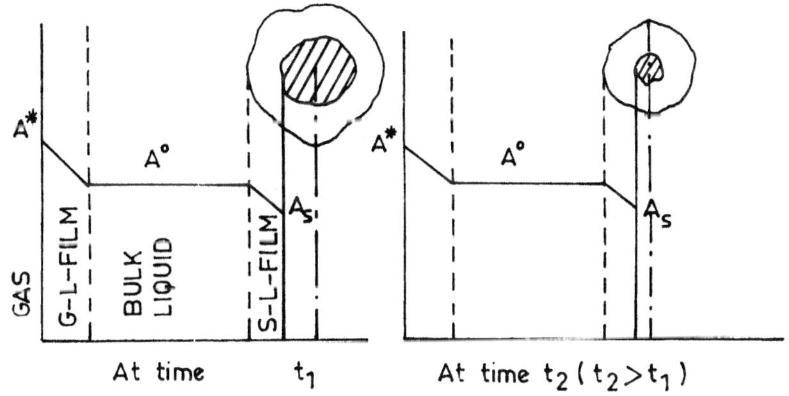

Figure 2.Concentration profile of dissolved gas A:Slurry reactor with an insoluble solid reactant(Product is soluble).

Table 1.Examples of gas-liquid-solid systems with insoluble reactant particles

- Oxydesulfurisation of coal slurries by gaseous oxygen
- Biotechnologic processes(biomass=solid substance)
- Oxidative leaching of copper,nickel,cobalt etc.
- Chlorination of wood pulp
- Absorption of CO_2 in a suspension of MgO
- Hydrogenation of polymeric substances suspended in solvents
- Chlorination of suspended polyethylene
- Manufacture of zinc hydrosulphite by reaction between SO_2 and zinc particles suspended in water
- Hydrogenation of sodium particles suspended in mineral oil to produce NaH

Table 2.Examples of gas-liquid-solid systems with sparingly soluble reactant particles

- Oxidation of calcium sulphite
- Thermal solvation of coal
- Absorption of CO_2 in aqueous suspension of lime and $Ba(OH)_2$
- Absorption of lean SO_2 in a slurry of $CaCO_3$ and a slurry of MgO
- Alkylation of naphtelene with ethylene,propylene,butylene etc. in the presence of catalysts such as BF_3-phosphoric acid
- Absorption of CO in a suspension of lime to produce $(HCOO)_2Ca$

instance,reference(6)) and the pertinent features are summarise in Table 3 and 4.

Finally,in contrast to slurries containing soluble or catal tic fine particles(see 2.2.2 and 3.2),no consideration so far ha been devoted to slurry reactions involving fine insoluble soli reactant.Indeed,in this case too,we may conceive situations,where the insoluble solid reactant size may be smaller than the liq film next to the gas-liquid interface.We may then have steps in parallel and the gas absorption rate may well be enhanced by a sufficiently fast heterogeneous reaction.

2.2.Sparingly Soluble Solid Particles

Typical examples of this case are given in Table 2.The problem here may be represented by the following scheme:

$$A(g) \rightarrow A(aq) \quad (1)$$
$$B(s) \rightarrow B(aq) \quad (2)$$
$$A(aq)+B(aq) \rightarrow \text{products} \quad (3)$$

Ramachandran and Sharma(4) were the first to discuss this problem and proposed solutions on the basis of film model(see Fi-

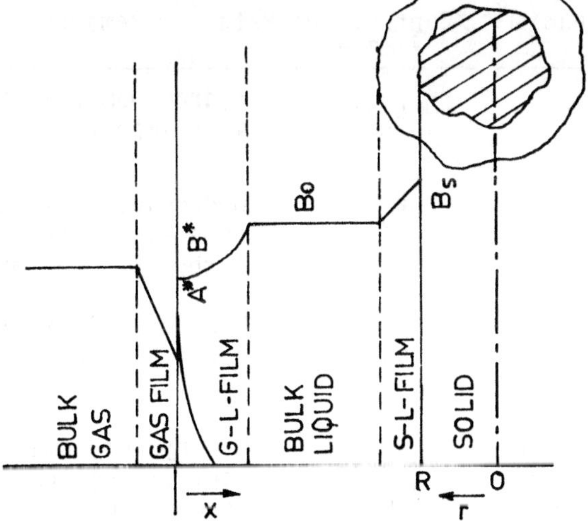

Figure 3.Concentration profiles of dissolved gas A and sparingly soluble solid B : Finite reaction rate

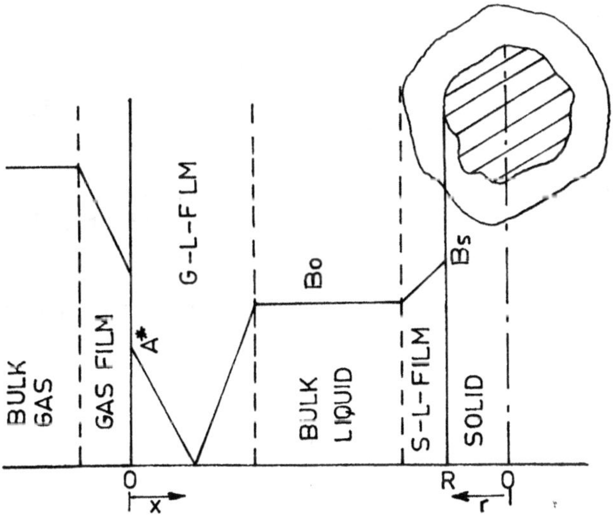

Figure 4.Concentration profiles of dissolved gas A and sparingly soluble solid B : Infinitely fast reaction rate.

Table 3. Gas-liquid-solid systems where solids are reacting but insoluble in the medium: Solids of constant size

Controlling Mechanism	Controlling Rate Term	Remarks
Gas-liquid mass transfer	$k_L a'$	Hydrodynamic factors are very important
Solid-liquid mass transfer	k_s	Hydrodynamic factors have some effect only in the turbulent regime
Diffusion through solid product "ash" layer	D_S	Hydrodynamic factors are important
Chemical reaction at the boundary of "ash" layer and unreacted shrinking core	k_R	No hydrodynamic effect, very strong temperature dependence

Table 4. Gas-liquid-solid systems where solids are reacting but insoluble in the medium, but the products are soluble: Shrinking Solids

Controlling Mechanism	Controlling Rate Term	Remarks
Gas-liquid mass transfer	$k_L a'$	Hydrodynamic factors are very important
Solid-liquid mass transfer	k_s	Hydrodynamic factors have some effect only in the turbulent regime
Chemical reaction at the surface of shrinking particles	k_R	No hydrodynamic effect, very strong temperature dependence

gures 3-6). They also pointed out that two types of problem are likely to be encountered in practice; that is, the solid particles may be either larger (Figure 3 and 4) or smaller (Figure 5) than the liquid film thickness next to gas-liquid interface.

2.2.1. Particles larger than the film thickness. It presents no problem to deal with this case as constituent steps, namely, diffusion of dissolved gas from the interface into the liquid phase and dissolution of solid particles are in series (Figure 3 and 4). Under these conditions, the extent of solid dissolution in the film next to gas-liquid interface can be neglected and the condition under which

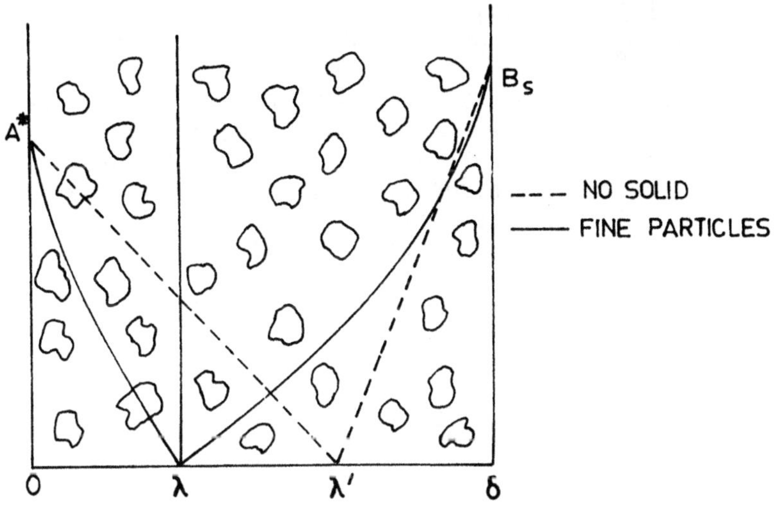

Figure 5.Concentration profiles of dissolved gas A and sparingly soluble fine particles B(Infinitely fast reaction, B_s=Saturation concentration of B).

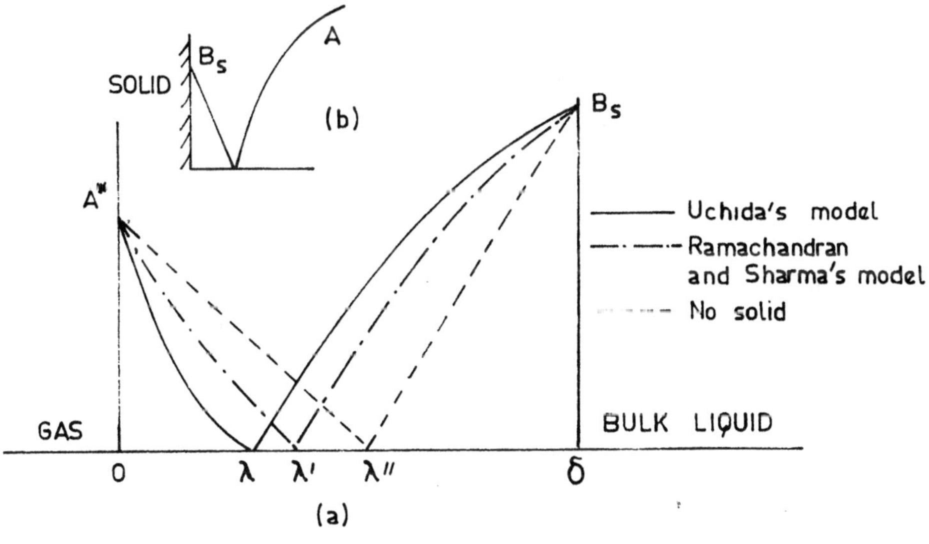

Figure 6.Concentration profiles of dissolved gas A and sparingly soluble fine particles B(Infinitely fast reaction,reaction accelerates solid dissolution)

this assumption would be valid can be shown to be the following:

$$\frac{k_s a_p D_A^2}{4k_L^2 D_B} \ll 1 \qquad (4)$$

The specific rate of mass transfer for any of the controlling regimes may then be solved easily.Depending on the particle size and loading,it may well be that the bulk liquid is saturated with respect to the dissolved solid species.However,even the case of "finite" slurry,which has a bulk concentration lower than the saturation concentration,can be solved analytically.For instance, for the case of an instantaneous reaction ,the following equation holds in the absence of any gas-side resistance:

$$R = \frac{A^* + D_B/zD_A\ B_s}{1/k_L + D_B a/D_A k_s a_p} \qquad (5)$$

2.2.2.Particles smaller than the liquid film thickness.The liquid film thickness in gas-liquid reactors may vary between 5-100 microns.Thus for a gently stirred system,the average diameter of the particles may considerably be less than the thickness of the liquid film($d_p < 0.1\ \delta$) and we can conceivably have the step of dissolution of solid particles in parallel to that of diffusion of dissolved gas from the gas-liquid interface to the bulk liquid-phase.Depending on the relative rates of diffusion and chemical reaction,the entire amount of dissolved gas may react in the film leading to considerable enhancement of gas absorption rate.Further,the concentration of dissolved reactive species in the liquid film may be uniform(in which case the simultaneous dissolution of solid particles in the film will not affect the specific rate of mass transfer),or may be zero at a location close to the gas-liquid interface.

An interesting situation arises when the reaction is instantaneous and the concentration of both species,namely the dissolved gas and the dissolved solid,is zero at a reaction plane which is very close to the gas-liquid interface(see Figure 5).In this case too,the concentration of dissolved species drops to zero from the bulk concentration which may well be the same as the saturation concentration of B_s.The problem has been analysed first by Ramachandran and Sharma(4[s]) and according to them,the relevant differential equations in terms of film model are:

$$D_A\frac{d^2A}{dx^2} - k_s a_p \frac{B_s}{z} = 0 \quad \text{for} \quad 0 < x < \lambda \qquad (6)$$

$$D_B \frac{d^2B}{dx^2} - k_s a_p (B_s - B) = 0 \quad \text{for} \quad \lambda < x < \delta \tag{7}$$

Here the second term on the left hand side of Equation 7 takes into account the simultaneous dissolution of solid particles which is only important if:

$$\frac{k_s a_p}{4k_L^2} \frac{D_A}{D_B} \gg 1 \tag{8}$$

The condition to be satisfied for the instantaneous reaction is:

$$\frac{D_A k_2 B_s}{k_L} \gg \frac{B_s}{zA^*} \left(1 + \frac{k_s a_p D_A^2}{4k_L^2 D_B} \right) \tag{9}$$

Equations 6 and 7 together with usual boundary conditions and saturated bulk phase (i.e. $x = \delta$, $B = B_s$) condition have been solved by Ramachandran and Sharma(4) to give:

$$R = \frac{D_A A^*}{\lambda} + k_s a_p B_s \frac{\lambda}{2z} \tag{10}$$

$$= B_s \sqrt{D_B k_s a_p} \coth((\delta - x)(k_s a_p / D_B)^{1/2} + k_s a_p B_s \frac{\lambda}{z} \tag{11}$$

The value of gas absorption rate per unit gas-liquid interface can then be evaluated by trial and error by assuming a value of λ ($< \delta$) to start with and proceeding until the two Equations 10 and 11 are satisfied simultaneously.

It is seen that in the absence of solid particles(that is,when $k_s a_p \rightarrow 0$),Equation 10 reduces to familiar expression for gas absorption accompanied by an instantaneous reaction.When the value of $k_s a_p$ is very large (that is for extremely fine particles),the reaction plane probably shifts to the gas-liquid interface ($\lambda \rightarrow 0$) and the specific rate of absorption becomes proportional to the square root of the amount of solids for a fixed particle size, since the surface area for dissolution ,a_p is directly proportional to the amount of solids (see Equation 11).

The above analysis does not consider the possible acceleration of solid dissolution due to the chemical reaction.This happens when the interfacial concentration of the dissolved gas is comparable or even higher than the saturation concentration of the solid particles.This case has been considered by Uchida et.al.(7), then Equation 6 becomes:

$$D_A \frac{d^2 A}{dx^2} - (k_s a_p B_s / z)(1 + (AD_A)/(B_s D_B)) = 0 \quad \text{for } 0 < x < \delta \tag{12}$$

The solution of Equation 12 gives:

$$R= mD_A\ A^*\coth(m\lambda) + \frac{mD_BD_s}{z}\left(\coth(m\lambda)- \frac{1}{\sinh(m\lambda)}\right)$$

$$=(mD_BB_s/z)\ \coth(m(\delta-\lambda) + m(D_AA^* +(2D_BD_s/z))$$

$$x(\coth(m\lambda) - 1/\sinh(m\lambda)) \quad (13)$$

where

$$m = (\ k_sa_p/D_B)^{1/2} \quad (14)$$

The solutions according to Ramachandran and Sharma(4) and Uchida et.al.(7) are both shown in Figure 6 schematically.It is seen that simultaneous acceleration in the specific rate of solid particles(that is,Uchida et.al.'s modification) results even more enhancement of gas absorption rate.

Since the earlier treatments of this problem by Ramachandran and Sharma(4) and Uchida et.al.(7),several experimental studies and verifications of predictions of enhancement factors have been reported(7,15,16);several detailed models based on film concept have also been proposed(7-12).Recently a penetration model for an instantaneous irreversible chemical reaction has also been presented,which however differs numerically only negligibly than the film model(13).The most important modification of Ramachandran and Sharma's treatment is due to Uchida et. al.(7-9) who consider that the rate of solid dissolution may be accelerated by the absorption of gas as discussed above.They have also considered the case where the concentration of solid component in the bulk liquid phase may not be maintained at the saturation solubility(that is,"finite" slurry) which occurs of course when the rate of solid dissolution is relatively slow compared with gas absorption rate(8).The case where the solid dissolution is finite was further considered by Sada et.al.(12) both theoretically and experimentally.Uchida et.al.(8) could also explain the data of Takeda et.al.(14) by their modified model.Analytical solutions presented above are for instantaneous reactions; Sada et.al.(10,15) considered the case where the reaction was finite and presented numerical solutions(An approximate solution for this case was obtained previously(4)).Sada et.al.(11) considered simultaneous absorption of two gases and presented numerical analysis and experimental data.They(16) have interpreted also their experimental results on dilute sulphur dioxide absorption into aqueous slurries of sparingly soluble fine rectant particles in terms of a "two-reaction plane" model.Sada et.al.(17, 18) considered also other interesting examples and proposed a model on the basis of above discussed theory as well as incorporating the possible solid surface reaction.In this case,the following reaction takes place:

$$A(aq) + zB(s) \longrightarrow \text{products} \quad \text{for} \quad 0 \leq x \leq \lambda \tag{15}$$

and

$$A(aq) + zB(aq) \longrightarrow \text{products} \quad \text{for} \quad \lambda \leq x \leq \delta \tag{16}$$

By this model,they(18) were able to explain the effect of solid particles for the absorption of SO_2 into aqueous slurries of $CaSO_3$.

The theoreticel values of enhancement factor which considerably exceeded the experimental values(19) obtained for the absorption of dilute SO_2 into aqueous slurries of sparingly soluble fine particles of $Mg(OH)_2$,$Ca(OH)_2$ etc.,could be explained by a reasoning(20) which was originally proposed by Alper et.al.(5) in another context.That is,the distance between the gas-liquid interface and the reaction plane(i.e. λ),decreases with decreasing SO_2 partial pressure, thus λ drops to a value which is in the same order as the average particle size.Following the ideas of Alper et.al.(5),Sada et.al.(20,21) suggested that in such cases it is not thinkable that there exists particles suspending in between the interface and the reaction plane.They proposed a model by considering an inert region ranging from the gas-liquid interface to a certain depth equivalent to particle diameter where there is no particle suspending and obtained a better agreement between the theoretical predictions and the experimental data at low SO_2 concentrations.

3.Gas Absorption Into a Slurry Containing Fine Catalyst Particles

In the majority of slurry reactors,solid particles act as catalyst. Industrial applications of such catalytic slurry reactors are numerous and examples are given in this book by L'Homme(22).Oxygen absorption into aqueous solutions in the presence of viable and aerobic microbial cells may also be considered as catalytic slurry reactors(23-26).

The usual and the most established design method of catalytic slurry reactors consists of a concept which assumes resistances in series(see Figure7).For instance,for a first order reaction this treatment given:

$$\frac{A}{Ra'} = \frac{1}{k_L a'} + \frac{1}{k_s a_p} + \frac{1}{\eta k [cat]} \tag{17}$$

there k is the catalytic reaction rate constant and η is the effectiveness factor.Thus,the global rate is dependent upon various mass transfer resistences plus a kinetic resistance and the plots of $(1/Ra')$ against $(1/[cat])$ yield various mass transfer coefficients and the rate constant,Satterfield(27),Sherwood and Farkas (28) and Deckwer and Alper(29),among others,have used this method succesfully and an authoritative review by Hofmann(3) can also be found in this book.Two main objections may,however,be raised

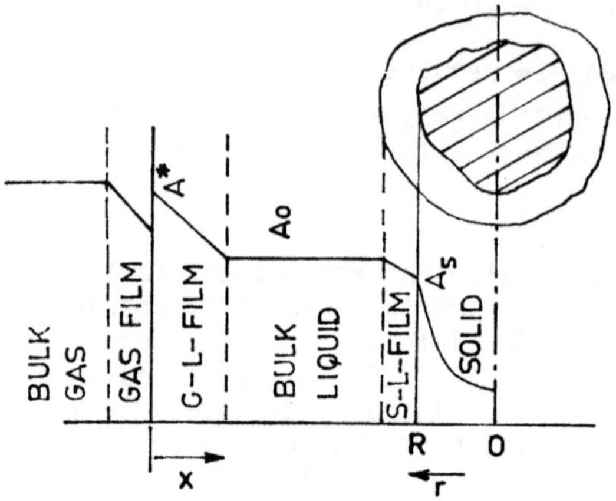

Figure 7.Concentration profile of dissolved gas A for catalytic slurry reaction(Catalyst particles are bigger than film thickness).

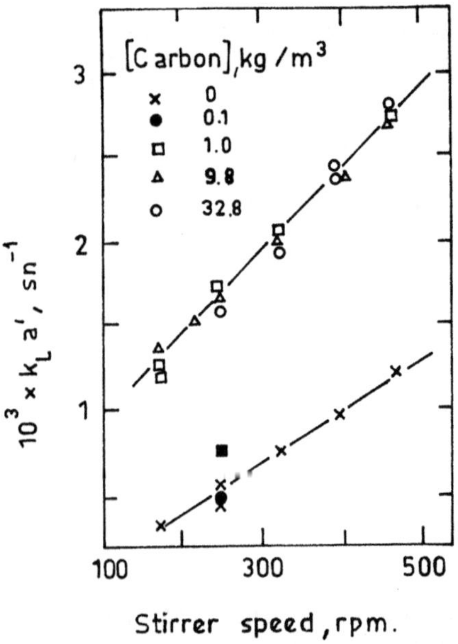

Figure 8.Effect of solid particles on k_L:Physical oxygen absorption experiments(5).

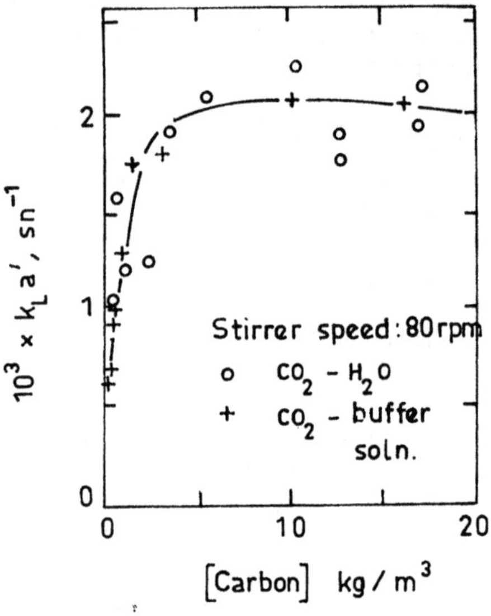

Figure 9.Effect of solid particles on k_L:Physical and chemical CO_2 absorption experiments(5).

against the above method.Firstly,the above method assumes that k_L and a′ are both independent of the catalyst particle loading. Secondly,if the particles are sufficiently small compared to liquid film thickness and for the reaction rate is sufficiently large there may be interaction of diffusion and reaction processes near the gas-liquid interface.

3.1Effect of Solid Particles on Liquid Side Mass Transfer Coefficient

Although there are substantial amount of published data in the literature on the effect of solid particles on $k_L a'$ in gas-liquid-solid systems,clear understanding of how k_L and a′ are affected is still lacking.Chandrasekaran and Sharma (28),Slesser and co-workers(31) and Joosten and co-workers(32),among others, have shown $k_L a'$ to be a function of solid loading.For instance,Joosten et.al. (32) found that the presence of suspended solid materials (of a size less than 250 µm) in stirred gas-liquid contactors hardly influences $k_L a'$,when the solid volume fraction is so low that the apparent viscosity of the slurry is not higher than four times that of the liquid.At high solid contents,which is untypical of catalytic systems,$k_L a$ may however decline sharply.Miyauchi et.al.(33) absorbed oxygen into limestone slurries with concentrations lower than 1C wt % in a stirred tank.The effect of the presence of solid particles under their experimental conditions was the slight increase in $k_L a$ values with particles size.

Sharma and Mashelkar(34) and Ganguli and van den Berg(35), among others,have shown that the particle concentration may affect a′ .Davidson et.al.(36) has considered the principal point of interest of to what extent the presence of the particles affects mass transfer on the liquid-side of the gas-liquid interface in three phase fluidised beds.After examining the bulk of the experimental evidence until 1977,they concluded that,though the particles may affect the bubble shape and velocity and in a freely bubbling system,alter the bubble size,they do not otherwise change conditions at the gas-liquid interface.In particular,they claim that surface renewal is unchanged.This conclusion is consistent with that of Calderbank and Moo-Young(37),who found previously that agitation intensity in the bulk liquid had little effect on k_L .

One of the other studies on the influence of suspended solid materials on the mass transfer coefficient is that of Alper et.al. (5).They employed various stirred cells of different designs,however,in all of the experiments gas-liquid interfacial area remained reasonably flat and could be taken as equal to the geometrical area.The values of k_L was low so that the liquid film thickness was bigger than 50 µm.In agreement with above studies, their data showed that no effect of particles on k_L values when inert particles,such as finely powdered quartz sand and oxirane acrylic beads,were employed.However,a completely different picture was obtained with highly porous particles of strong adsorbing property,such as activated carbon,which increased k_L considerably(5).

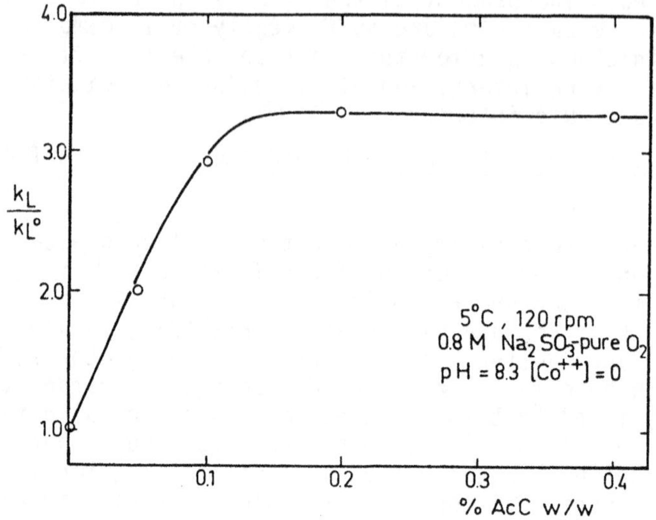

Figure 10.Oxygen absorption into sodium sulphite solutions containing finely powdered activated carbon : Effect of solid loading on (k_L/k_L^o).

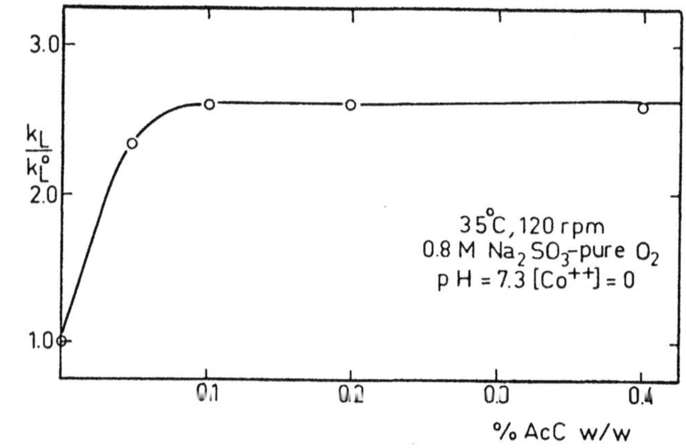

Figure 11.Oxygen absorption into sodium sulphite solutions containing finely powdered activated carbon : Effect of solid loading on (k_L/k_L^o).

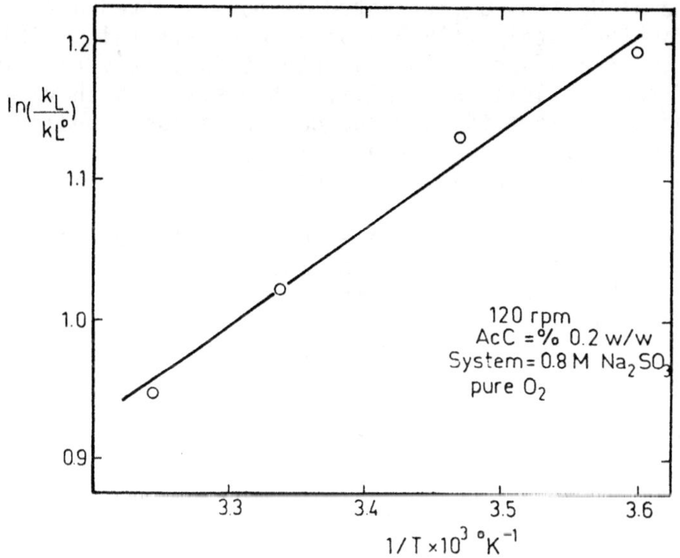

Figure 12.Effect of temperature on (k_L/k_L^o) $(k_L,k_L^o$ = Physical liquid side mass transfer coefficient with and without activated carbon particles respectively)(44).

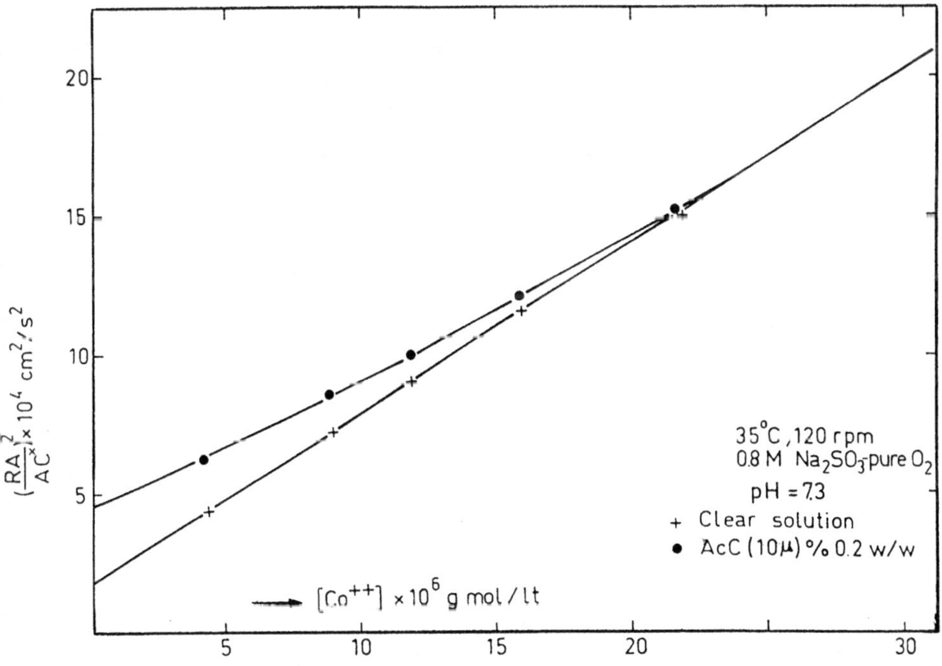

Figure 13.Absorption of oxygen into sodium sulphite solutions containing activated carbon:Effect of Co^{++} addition(44).

3.1.1. Effect of finely powdered activated carbon on the gas absorption rate of O_2 and CO_2. Activated carbon is often used either as support for the catalyst(38,39) or as a catalyst itself(30,40, 41). Due to its strong adsorbing property, it is likely that it may affect k_L if the particles are sufficiently small as they may, from time to time, move right through the concentration boundary layer and pick up adsorbate there. Wichtendahl(42) and Kars et al.(43) presented some data indicating considerable effect of activated carbon on k_L.

Alper and co-workers(5,44) have recently studied the absorption of O_2 or CO_2 into solutions containing finely powdered activated carbon in stirred cells with unbroken interfaces. They employed both non-stationary physical absorption or pseudo-stationary chemical absorption experiments. In the latter, the reaction regime was in the transition between diffusional to fast pseudo mth order regime so that the information for k_L could be deducted. For instance, for a first order reaction, the absorption rate is given by

$$Ra = a\ A^* \sqrt{Dk_1 + k_L^2} \qquad (18)$$

The chemical systems consisted of either CO_2 into Na_2CO_3 + $NaHCO_3$ buffer or O_2 into sodium sulphite solution containing no homogeneous catalyst.

Figure 8 shows the experimental results of physical absorption of oxygen; it is seen that sufficient carbon particles affect k_L considerably. There seems to be no effect of other particles which have undoubtedly little adsorbing capacity and less porous structure then the activated carbon. Figure 9 shows the results of CO_2 absorption experiments both for physical and chemical systems. By increasing carbon concentration k_L increases at first, then it remains reasonably constant. Figure 10 and 11 show the typical results obtained by absorbing pure oxygen into 0,8 M sodium sulphite at various activated carbon loadings and at two different temperatures, there k_L and k_L° are the physical liquid side mass transfer coefficients for the slurry and the clear solution respectively. The effect of temperature can be seen from Figure 12 which clearly shows that the effect of activated carbon on k_L is more significant at lower temperatures.

An interesting set of experimental results were recently obtained by Alper and co-workers(44) which showed that the effect of activated carbon decreases when the homogenous reaction between dissolved oxygen and sulphite is increased by the addition of Co^{++}. As expected there is no effect when the reaction is in the fast pseudo mth order regime so that it is not dependent on k_L. Some typical results are summarised in Figure 13 and 14.

In the literature, no similar results can be found with particles other than the activated carbon. However recently, Alper et al (44) observed that avicell cellulose of average particle diameter of 19μm has also affected k_L. The increase is, on the other hand, much less than the powdered activated carbon. Figure 14 and 15 show some

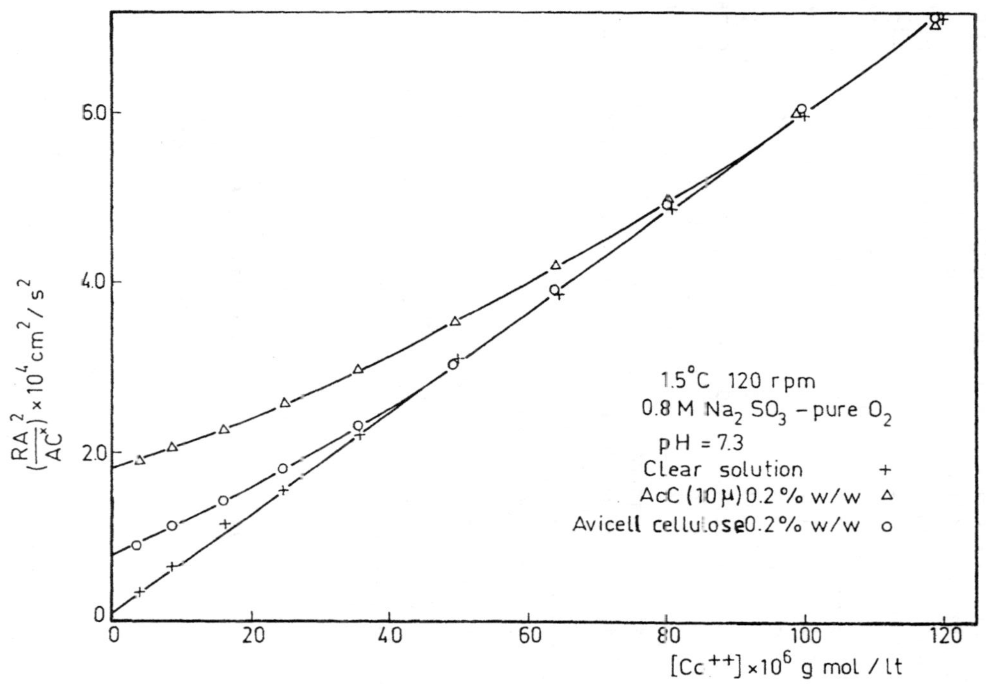

Figure 14. Oxygen absorption into sodium sulphite solutions containing either powdered activated carbon or avicell cellulose (44).

typical results.Again,by decreasing temperature,the effect on k_L increases (see Figure 15).

3.2.Gas Absorption Rate Enhancement by Fine Catalyst Particles

If the particles are sufficiently small compared to liquid film thickness and/or the reaction rate is suffuciently large,there may be interaction of diffusion and reaction processes near the gas-liquid interface,hence affect -i.e.enhance- the absorption rate. Sada et al.(45) considered this case theoretically by also considering intraparticle diffusion limitations introducing the effectiveness factor η .However,a reasonable range of film thickness δ can be taken as $10 \leq \delta \leq 100 \mu m$.In order to satisfy $\delta > dp$ (for instance,$d_p = 0.1 \delta$) the particle size should be: $1 \leq d_p \leq 10$ m. Owing to the agglomerisation of fine particles,it can also be assumed confidently that the effective particle diameters of considerably less than 1 μm are unrealistic.For then,in accordance with the experimental results of many workers(for instance,Satterfield(27),Deckwer and Alper(29) and Wictendahl et al.(46),intraparticle diffusion limitations may be neglected for particles of these sizes.Hence the effectiveness factor η will essentially be equal to unity and the treatment of Sada et al.(45) reduces to the well established diffusion model with homogeneous reactions,which is already well described by many authors(i.e.Danckwerts(47)). Figure 16 shows various possible reaction regimes in terms of film model and Figure 17 shows $(R/k_L A^*)$ against first order rate constant k_1 for a continuous reactor.The only essential difference between gas-liquid and gas-liquid-solid systems is that the condition of homogeneity,i.e. inequality $\delta \gg d_p$ cannot be maintained in an unlimited manner when the catalyst concentration increases.Alper and Deckwer(48) suggest that when the enhancement of the absorption rate is considerable,an effective film thickness (rather than the actual one) has to be considered so that the required inequality becomes : $\delta_{eff} \gg d_p$.
It is possible to calculate this effective film thickness for many cases(49);alternatively a very conservative estimate can be taken as δ/E where E is the enhancement factor.Thus,for E = 10, the effective film thickness is about $\delta eff = 0.1$ and it is obvious that the condition for pseudo-homogeneity is seriously violated.

These views were supported experimentally by Alper et al.(5). For instance,Figure 18 shows experimental enhancement factors for the oxidation of glucose solutions containing suspended Pt/C catalysts of particle sizes of μm range,i.e. $dp < 5$ μm.It is seen that the increasing catalyst concentration hence reaction rate raises the enhancement factor E to a value of approximately 1.7.However, larger catalyst concentrations than about 10 kg/m^3 did not yield any significant increase in E.This is most likely due to the fact that the condition of pseudo-homogeneity was no longer fulfilled. However,the interpretation of experimental data was further complicated by the fact that the gas absorption rate was also in-

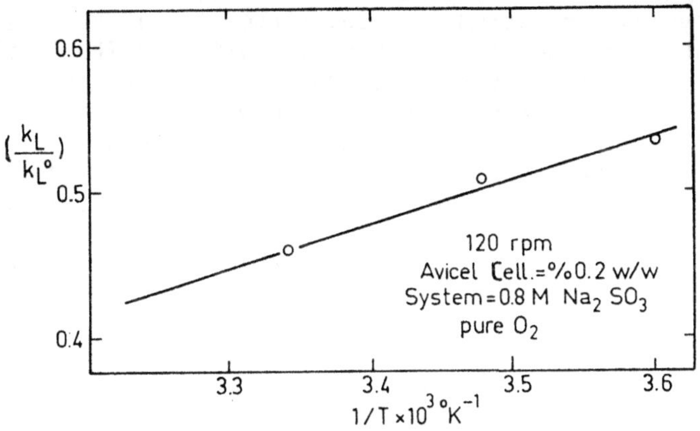

Figure 15.Oxygen absorption into slurries containing avicell cellulose : Effect of temperature on k_L/k_L^o .

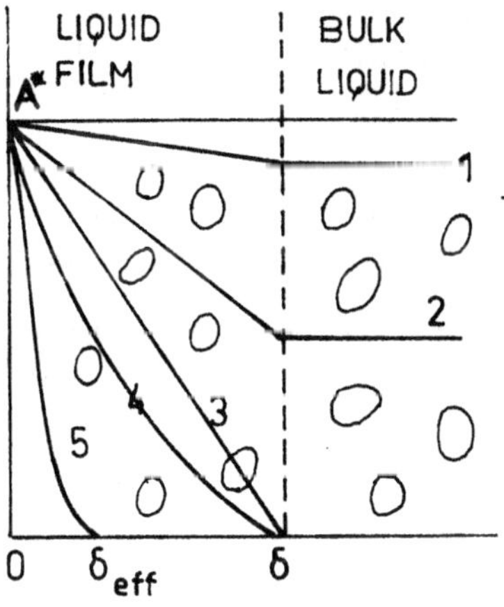

Figure 16.Concentration profile of dissolved gas in a slurry containing fine particles.

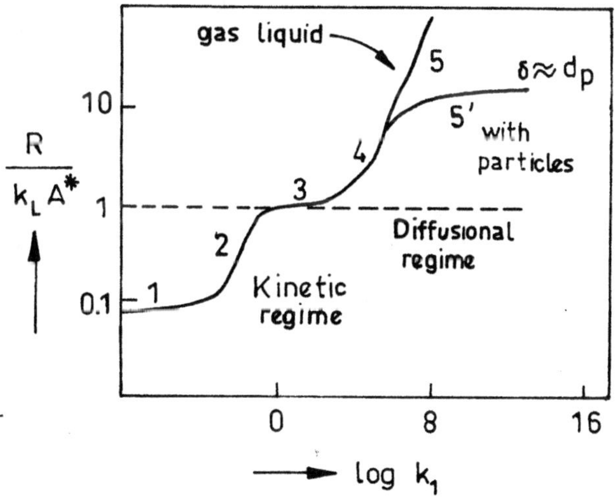

Figure 17.Regimes of absorption with chemical reaction

creased by the presence of activated carbon alone and the degree of this increase might well depend on the effective film thickness as illustrated by Alper et.al.(44)(see 3.1.1).

Alper and Deckwer(48) presented also alternative experimental evidences by another system,i.e. CO_2 carbonate/bicarbonate buffer-carbonic anhydrase immobilised on solid particles.They employed five different particle sizes,i.e. 300,200,100, <60 and approximately < 20μm.In each case,however,there were some particles of 1 - 5 μm range,but the last two contained considerable of them. Figure 19 shows the results of experiments with a solid support of dp < 20 μm which also contained considerable amount of very small particles(i.e. 1 - 3 μm in diameter).These results were from several different batches which were,however,immobilised under similar conditions.Alper and Deckwer(48) claimed that the uncertainties in the immobilization process,as well as nonuniform particle size were the main sources of considerable scatter.Despite this large scatter,Figure 19 shows that at large concentrations of particles the absorption rate reaches its limity value when $\delta_{eff} \approx d_p$.

To illustrate the effect of particle size,Figure 20 shows the results of their experiments(49) with particles of d_p < 60 μm and 100,200 and 300 μm.It is seen that with large particles (i.e.100, 200 and 300 μm) there is hardly any enhancement of gas absorption rate and the small increase may be attributed to some very small particles which were present.With particles of dp < 60 μm,there was observable enhancement,however,this was considerably less than that of dp < 20 μm particles(see Figure 20).In these experiments, the film thickness $\delta \approx 40$ μm;thus one would not expect any effect of catalyst particles when dp > δ (i.e.for 100,200 and 300 μm particles),when the particles of d_p < 20 μm were employed,the limiting value of the enhancement factor is about 4.5.Hence δ_{eff} at this value of the enhancement factor is around 9 μm and this value is indeed very near to the average diameter of the particles employed.

Recently,another interesting example was published by Pal et al.(50) who studied the absorption oxygen into sodium sulfide solutions containing fine activated carbon particles as catalyst. They could illustrate that catalyst particles,smaller than the diffusion film thickness,can enhance the specific rate of absorption substantially over that in the presence of coarser particles, even when the reaction is suffuciently fast to occur in the film in the absence of catalyst particles.They used gently stirred cells with plane interface and particles which may have an average size as low as 1,7 μm;in a number of cases,the specific rate of absorption increased by a factor of 2 to 10.

4.DISCUSSION AND CONCLUSIONS

An examination of gas absorption mechanisms of three phase slurry reactors indicates that three major points,which are not normally considered in reviews concerning this type reactors(see,for instance(1) - (3))deserve further attention.These are the slurry reactors involving reactive solids,the possible gas absorption

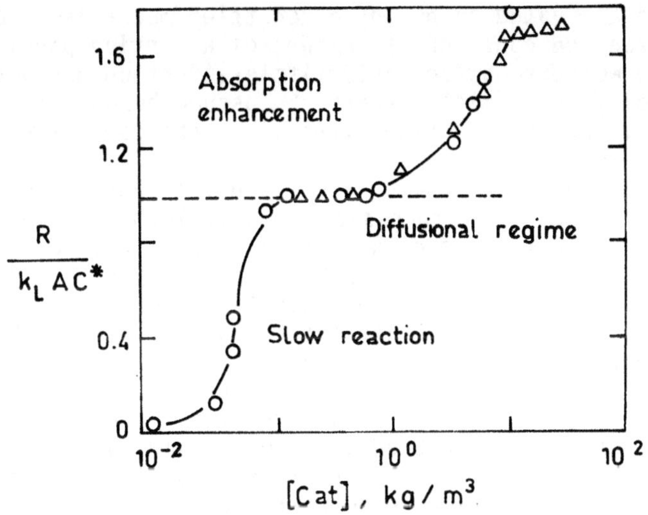

Figure 18.Absorption of oxygen into glucose solutions containing fine particles of activated carbon (5).

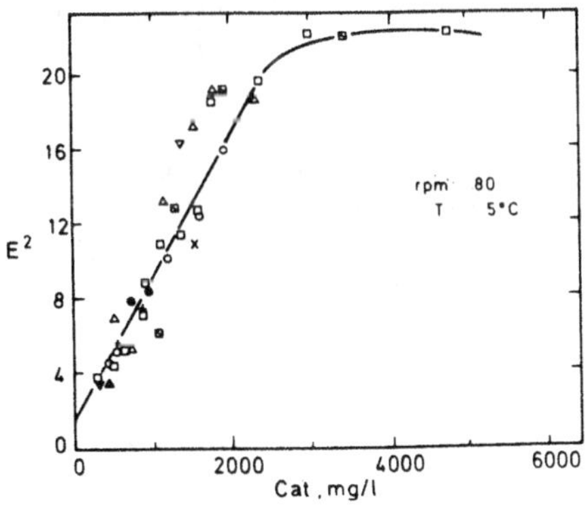

Figure 19. Plot of E^2 against catalyst concentration for particles of $d_p < 20$ μm (Points refer to different samples)(48).

rate enhancement by fine catalyst particles and the increase of physical absorption rate by fine particles of high absorbing property,such as activated carbon,by a "shuttle" mechanism.From these, the former received considerable amount of attention,particularly from Japanese workers.However,only little attention has yet paid to the case of fine particles where the pseudo-homogeneity is violated,that is,when the reaction plane is in the same order of the particle diameter.

Concerning the effect of particles on volumetric mass transfer coefficient,$k_L a$,it is necessary to establish the seperate effects on each of them.It is clear that fine particles do not seem to affect k_L except for the unusual cases reported by Sada et al. (52,53) where microstirers,made of magnetic materials spinned,at gas-liquid interface under a magnetic field.This finding has been agreed by a number of different researchers(32,33,36,51) where different chemical systems and reactors were involved.On the other hand,activated carbon(possibly also some other materials) increases k_L noticeably.This must undoubtedly be due to highly porous structure and high adsorbing property.The temperature dependence of this effect,that is,decreasing increase when the temperature is raised, supports also such a reasoning .A simple "shuttle" mechanism can therefore be postulated,i.e. a solid particle may move right into the concentration boundary,adsorb gas for sometime there and then may go back into the depth of liquid to desorb gas(see Figure 21). By this kind of a "shuttle" mechanism,increased k_L values are obtained.So far only activated carbon was investigated,however,there is evidence that some other material(for instance,avicell cellulose for oxygen absorption) may have similar effects.Such behaviours are likely to be very specific and further information - if possible also theoretical analysis - is needed.On the other hand, whatever the mechanism may be,it is important to realize that certain particles,such as activated carbon,affect k_L significantly.It may be stressed that even though carbon particles are used as catalyst support or as catalyst itself,this effect has not duly been taken into account in previous investigations.

Finally it can be conclude that for catalytic slurry reactors it is possible to observe considerable enhancement of the absorption rate if the particles are sufficiently small and/or the reaction is fast.Under these circumstances the most established design method - formulated by Eqn.(17) - does not apply and this possibility has not been considered properly until recently.Model experiments of Alper and co-workers(5,48) show that at low k_L values it is possible to observe gas absorption rate enhancement which may reach its limiting value when $d_p \approx \delta_{eff}$.However,satisfactory theoretical predictions have not so far been developed when the pseudo-homogeneity is violated.In practice,on the other hand,there are only few hydrogenation reaction where,owing to several selectivity requirements,agitation must be moderate(hence δ is large):thus with a highly active catalyst some gas absorption rate enhancement may occur.However,this enhancement will certainly not be as significant

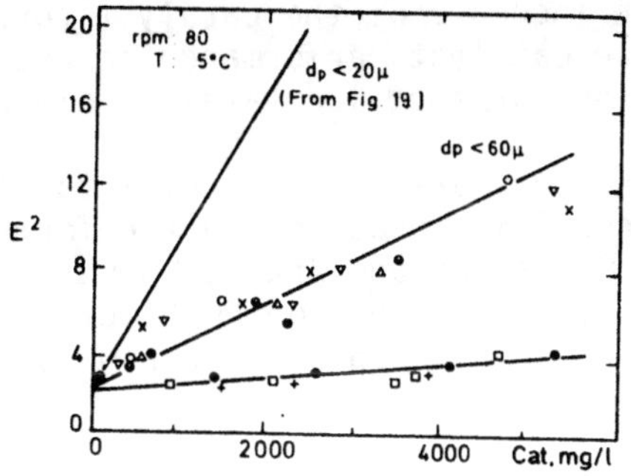

Figure 20. Plots of E^2 against catalyst concentration for different particle sizes (□ : d_p=300μm, •: d_p=200μm, + : d_p= 100μm, others: d_p< 60 m).

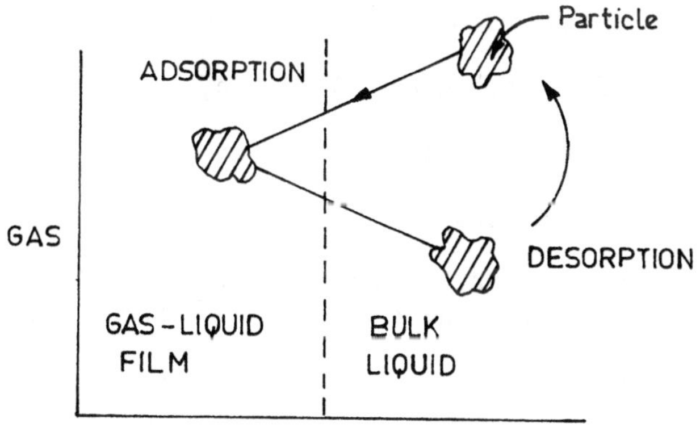

Figure 21. Schematic representation of "shuttle" mechanism .

as certain gas-liquid systems and a value of well below 10 is expected. In addition, economic reasons do not usually permit to carry out many of the catalytic slurry reactions under enhancement conditions; for instance, Figure (18) shows that a 10% increase of absorption rate requires an increase of the catalyst concentration by a factor of 100. Hence the catalyst performance decreases by a factor of 10^{-2} while reactor performance increases merely by a factor of approximately 2.

Acknowledgement: Authors are most grateful to Volkswagen Foundation for the financial assistance in the form of a "cooperation project" between Oldenburg and Ankara Universities. One of us (E.A.) is also grateful to Alexander von Humboldt Foundation for an award of research scholarship which enabled him to carry out some of the preliminary studies.

REFERENCES

1. Shah, Y.T. Gas-Liquid-Solid Reactor Design (McGraw-Hill, New York, 1979).
2. Chaudhari, R.V. and P.A. Ramachandran. AIChEJL 26 (1980) 177.
3. Hofmann, H. "Reaction Engineering Problems in Slurry Reactors" (Proceedings of NATO ASI on "Mass Transfer with Chemical Reaction in Multiphase Systems", Izmir, Turkey, 1981).
4. Ramachandran, P.A. and M.M. Sharma. Chem.Engng.Sci. 24 (1969) 1681.
5. Alper, E., B. Wichtendahl, and W.-D. Deckwer. Chem.Engng.Sci. 35 (1980) 217.
6. Joshi, J.B., J.S. Abichandani, Y.T. Shah, J.A. Ruether and H.J. Ritz. AIChEJL 27 (1981) 937
7. Uchida, S., K. Koide and M. Shindo. Chem.Engng.Sci. 30 (1975) 644.
8. Uchida, S., K. Koide and C.Y. Wen. Chem.Engng.Sci. 32 (1977) 447
9. Uchida, S. and C.Y. Wen. Chem.Engng.Sci. 32 (1977) 1277
10. Sada, E., H. Kumazawa, and M.A. Butt. Chem.Engng.Sci. 32 (1977) 1499.
11. Sada, E., H. Kumazawa, and M.A. Butt. Chem.Engng.Sci. 34 (1979) 715.
12. Sada, E., H. Kumazawa, and M.A. Butt. Chem.Engng.Sci. 34 (1979) 715.
13. Uchida, S., M. Miyauchi, and O. Ariga. Can.Jl.Chem.Engng. 59 (1981) 560.
14. Takeda, T., H. Moriguchi, S. Uchida, and K. Koide. Paper submitted to 12th Fall Meeting of the Society of Chem.Engrs., Japan, Nagoya (1976).
15. Sada, E., H. Kumazawa, and M.A. Butt. Chem.Engng.Sci. 35 (1980) 771.
16. Sada, E., H. Kumazawa, and M.A. Butt. Chem.Engng.Sci. 35 (1980) 771.

17. Sada,E., H.Kumazawa, and M.A.Butt. Chem.Engng.Sci. 35 (1980) 771.
18. Sada,E., H.Kumazawa, I.Hashizume and H.Nishimura. "Absorption of dilute SO_2 into aqueous slurries of $CaSO_3$" (Communicated to Chem.Engng.Sci. (1980)).
19. Sada,E., H.Kumazawa, Y.Sawada and I.Hashizume. Chem.Engng. Sci. 35 (1980) 945.
20. Sada,E., H.Kumazawa and I.Hashizume. Chem.Engng.Sci. 36 (1981) 639.
21. Sada,E., H.Kumazawa and I.Hashizume. "Further consideration on chemical absorption into a slurry containing fine catalyst particles or sparingly soluble fine reactant particles" (Communicated to Chem.Engng.Sci. (1981)).
22. L'Homme,G.E. "Introduction to gas-liquid-solid systems" (Proceedings of NATO ASI on "Mass Transfer with Chemical Reaction in Multiphase Systems", Izmir, Turkey, 1980)
23. Tsao,G.T. and L.L.Kempe. Biotechnol.Bioeng. 2 (1960) 347.
24. Tsao,G.T., A.Mukherjee and Y.Y.Lee. Proc.IV Intern. Fermentation Symp.1972, p. 65.
25. Bennett,G.F. and L.L.Kempe. Biotechnol.Bioeng. 6 (1964) 347.
26. Moser,A. "Prediction of enhancement of OTR in Bioreactors" (Bioconversion and Biochemical Eng.Vol.2, Symp.2., Proceedings of 2nd Symp. at New Dehli, March 1980, Editor: T.K.Ghose).
27. Satterfield,C.N. Mass Transfer in Heterogeneous Catalysis, M.I.T. Press, New York (1970).
28. Sherwood,T.K. and E.J.Farkas. Chem.Engng.Sci. 21 (1966) 573.
29. Deckwer,E.-D. and E.Alper. Chem.-Ing.Tech. 52 (1980) 219.
30. Chandrasekaran,K. and M.M.Sharma. Chem.Engng.Sci. 32 (1977) 669.
31. Slesser,C.G.M., W.T.Allen, A.R.Cumings, V.Fawlowski and J.Shields. 4. ISCRG, Brussels, Supplement to Chem.Engng.Sci. (1968) p.41.
32. Joosten,G.E.H., J.G.M.Schilder and J.J.Jansen. Chem.Engng. Sci. 32 (1977) 563.
33. Miyauchi,M., A.Iguchi, S.Uchida and K.Koide. Can.J.Chem. Engng. 59 (1981) 640.
34. Sharma,M.M. and R.A.Mashelkar. Proc.Symp.Mass Transfer with Chemical Reaction No: 28, Instn.Chem.Engrs, London (1968).
35. Ganguli,K.L. and H.J.van den Berg. Chem.Engng.Jl. 16 (1978) 193.
36. Davidson, J.F.D.Harrison, R.C.Darton and R.D.La Nauze. "The two-phase theory of fluidisation and its application to chemical reactors". Chemical Reactor Theory: A Review (Eds: L.Lapidus and N.R.Amundson, Prentice-Hall, Inc., 1977) p. 583.
37. Calderbank,P.H. and M.B.Mao-Young. Chem.Engng.Sci. 16 (1962) 39.

38. Lee,Y.Y. and G.T.Tsao. Chem.Engng.Sci. 27 (1972) 1601.
39. Fujimoto,K., O.Kuchinski and T.Kurugu. Ind.Eng.Chem.Prod. Res.Dev. 15 (1976) 259.
40. Pal,S.K., M.M.Sharma and V.A.Invekar. Chem.Engng.Sci. 37 (1982) 327.
41. Lefers,J.B., W.T.Koetsier and W.P.W. van Swaaij. Chem.Engng. Jl. 17 (1979) 201.
42. Wichtendahl,B. Diplomarbeit, Universität Hannover (1978).
43. Kars,R.L., R.J.Best, and A.A.H.Drinkenburg. Chem.Engng.Jl. 17 (1979) 201.
44. Alper,E., S.Öztürk and W.-D.Deckwer. Paper to be published (1982).
45. Sada,E., H.Kumazawa and M.A.Butt, Chem.Engng.Sci. 32 (1977) 970.
46. Wichtendahl,B., E.Alper and W.-D.Deckwer. Chemical Eng.Commun. 10 (1981) 369.
47. Danckwerts,P.V. Gas-Liquid Reactions (McGraw-Hill Co., New York, 1970).
48. Alper,E. and W.-D.Deckwer. Chem.Engng.Sci. 36 (1981) 1097.
49. Alper,E. and W.-D.Deckwer. Unpublished Work (1980).
50. Pal,S.K., M.M.Sharma and V.A.Invekar. Chem.Engng.Sci. 37 (1982) 327.
51. Sada,E., H.Kumazawa and M.A.Butt. Chem.Engng.Sci. 32 (1977) 970.
52. Sada,E., S.Katoh and M.Terashima. Biotechnol.Bioeng. 23 (1981) 1037.
53. Sada,E., S.Katoh, H.Yoshii and Y.Ban. Can.J.Chem.Engng. 57 (1979) 704.

MODELLING OF CONSECUTIVE HYDROGENATION REACTIONS WITH SORPTION AND MASS TRANSFER EFFECTS IN A STIRRED TANK SLURRY SYSTEM

O.M.Kut, G.Gut, T.Buehlmann and A.Lussy

Department of Chemical Engineering, Swiss Federal Institute of Technology (ETH), CH-8092 Zurich, Switzerland

ABSTRACT

Some examples for modelling of competitive, consecutive hydrogenation reactions in a stirred tank slurry system are presented. Based on the assumption that chemical reactions obey the Langmuir-Hinshelwood theory with adsorption of hydrogen and organic species on different active sites of the catalyst's surface, the influence of hydrogen absorption and external mass transfer of reaction components to the catalyst's surface are discussed. Starting with the chemical surface reaction, models are developed for some special cases: when surface reaction and transport of hydrogen or substrate control the kinetics, and when surface kinetics and desorption of the intermediate product control the kinetics. In each case simulated conversion curves are compared with experimental data. The effects of the reaction conditions on the selectivity to the intermediate product, as postulated by these models, are discussed in detail.

1. INTRODUCTION

In the paper presented here, we will focus our interest on irreversible bimolecular consecutive reactions, where the substrate and intermediate product are in competition for the same reactant. As an example the selective hydrogenation of o-alkylphenol on a palladium catalyst in a stirred tank slurry reactor was investigated. To properly design such a system, the chemical reaction steps and prior physical steps, such as diffusion and sorption, must be considered. The rigorous description of a multiphase reactor for a consecutive reaction system is far too complex for practical use. Therefore, some simplifying assumptions have to be introduced:

- The gas phase consists of pure hydrogen, the volatilities of the organic species are low, and the gas-side mass transfer resistance can be neglected.
- External mass transfer from gas to the liquid and from the liquid to the solid phase can be described by the use of the two-film model.
- The diffusion times are in any case much shorter than the reaction times, hence quasi steady-state concentration profiles are achieved at any time during reaction.
- The catalyst particles are so small ($< 10\ \mu m$) that intraparticle diffusion resistance can be neglected ($\eta = 1$).
- There are no temperature gradients in the system.

In a first case we will model the chemical reaction controlled regime and discuss the effects of temperature, pressure, catalyst loading and presence of adsorbable additives on the selectivity. In a second and third case we will model the reaction when mass transfer and sorption effects influence the rate and compare the selectivity changes predicted by these models with experimental data.

2. MODEL FOR THE CHEMICAL SURFACE REACTION

The partial hydrogenation of an o-alkylphenol (A) to the corresponding cyclohexanone (B) and cyclohexanole (D) can be presented by the following reaction scheme:

$$A + 2\,H_2 \xrightarrow{k_1^*} B + H_2 \xrightarrow{k_2^*} D \tag{1}$$

If all transport steps are fast and all adsorption equilibria are established, then the chemical surface reactions are rate controlling. The corresponding rate equations are given by:

$$-r_A = m_K\, k_1^*\, \theta_A\, \theta_H \tag{2}$$

$$r_D = m_K\, k_2^*\, \theta_B\, \theta_H \tag{4}$$

$$r_B = m_K\, k_1^*\, \theta_A\, \theta_H - m_K\, k_2^*\, \theta_B\, \theta_H \tag{3}$$

Based on experimental observations with different organic substrates, we can assume that the organic compounds and hydrogen are non-competitively adsorbed. Elimination of θ_A, θ_B and θ_H from the above equations give:

$$-r_A = m_K\, k_1^* \left[\frac{c_{AL}}{c_{AL} + Q_2 c_{BL} + Q_3 c_{DL}}\right]\left[\frac{(K_H/H)\, p_H}{1 + (K_H/H)\, p_H}\right] \tag{5}$$

$$r_B = r_A - r_D \tag{6}$$

$$r_D = m_K k_2^* \left[\frac{Q_2 c_{BL}}{c_{AL} + Q_2 c_{BL} + Q_3 c_{DL}}\right]\left[\frac{(K_H/H)\, p_H}{1 + (K_H/H)\, p_H}\right] \tag{7}$$

$$\text{where } Q_2 = K_B/K_A \text{ and } Q_3 = K_D/K_A \tag{8}$$

The model (eq.5-7) was successfully used to describe the hydrogenation of o-cresol [1] and o-ethylphenol [2]. It was also possible to describe the hydrogenation of sunflower seed oil with a similar set of equations [3].

2.1 Selectivity: Model Predictions and Experimental Observations

The instantaneous selectivity of the consecutive reaction for the production of the intermediate B is given by:

$$\frac{r_B}{r_A} = \frac{dc_B}{dc_A} = \frac{k_2^* K_B c_{BL}}{k_1^* K_A c_{AL}} - 1 = \frac{1}{S} \frac{c_{BL}}{c_{AL}} - 1 \tag{9}$$

Setting $dc_B/dc_A = 0$, we get the expression for the selectivity coefficient S:

$$S = (c_B/c_A)_{max} = \frac{k_1^* K_A}{k_2^* K_B} = \frac{k_1^*}{k_2^* Q_2} \tag{10}$$

The temperature dependence of S is given by the difference of the individual activation energies of the two reactions and the difference in the heats of adsorption of substrate A and intermediate B:

$$S = S'' \exp(-E_{app}/RT) \tag{11}$$

$$\text{where } E_{app} = E_{A_1} - E_{A_2} + \Delta H_A - \Delta H_B \tag{12}$$

Depending on the individual values of these terms, the apparent activation energy can be positive or negative. The regression calculation showed that the value of this apparent activation energy for the hydrogenation of o-cresol on palladium catalyst is about 4 kJ/mol. Eq.(10) predicts that the selectivity is independent of catalyst loading. A change in pressure or hydrogen gradients in the system due to mass transfer limitations will change the hydrogen coverage but not the selectivity, because of the identical dependence of the two reactions on the hydrogen coverage. Therefore, a change in hydrogen pressure or agitation intensity will only alter the measured reaction rates but not the selectivity for the production of intermediate B.

It was mentioned in the literature [5] that the selectivity of such a consecutive reaction may be improved by adding a third substance with an adsorbability between those of A and of B. This third substance can be a solvent or another additive with the required adsorption characteristics. To check this phenomena in a more quantitative manner, higher o-alkylphenols were used as additives. In this case we have a simultaneous hydrogenation as shown in the following reaction scheme:

$$A + 2\,H_2 \xrightarrow{k_1^*} B + H_2 \xrightarrow{k_2^*} D \qquad (1)$$

$$Y + 2\,H_2 \xrightarrow{k_1'^*} Z + H_2 \xrightarrow{k_2'^*} P \qquad (13)$$

The relative adsorption coefficients K_Y/K_A can be calculated if initial rates are used:

$$\frac{-r_A}{-r_Y} = \frac{dc_A}{dc_Y} = \frac{k_1^* K_A c_A}{k_1'^* K_Y c_Y} \qquad (14)$$

On integration:

$$\ln(c_A/c_{Ao}) = S' \ln(c_Y/c_{Yo}) = \frac{k_1^* K_A}{k_1'^* K_Y} \ln(c_Y/c_{Yo}) \qquad (15)$$

If the reaction rate constants are determined from individual runs, the relative adsorptivity K_Y/K_A can be calculated from eq.(15). Using adsorptivity data from the individual hydrogenation we can also calculate K_B/K_Y:

$$\frac{K_B}{K_Y} = \frac{K_B}{K_A}\,\frac{K_A}{K_Y} = Q_2\,\frac{K_A}{K_Y} \qquad (16)$$

Table 1: Relative adsorptivities.

Compound A	Compound Y	K_Y/K_A	K_Y/K_B
o-cresol	o-ethylphenol	0.89	0.5
	o-isopropylphenol	0.54	2.8
	o-tert.butylphenol	0.37	7.4

As can be seen from Table 1, where the relative adsorptivities are listed, the o-alkylphenols are "optimal" species for the postulated effect. The experimental data show, however, that substances which are competitively adsorbed on the surface, do not change the selectivity of the consecutive reaction (Fig. 1). This behaviour is easily understood if we develop the rate equations for the individual reactions assuming that the additives are also competitively adsorbed on the catalyst's surface. The denominators in these equations are complexer than in the original equations (5-7) but they are still the same for both reaction steps. Considering

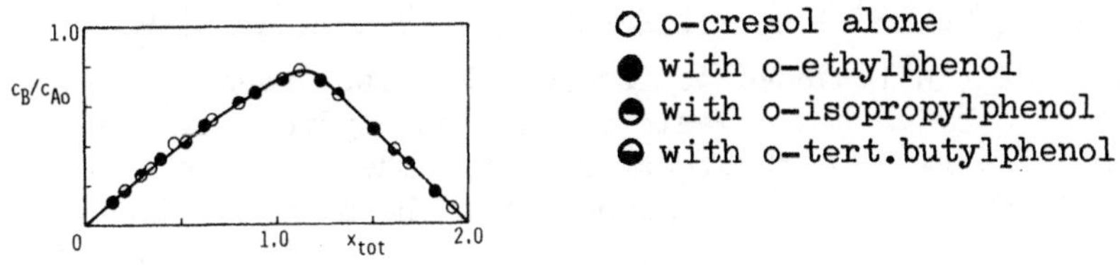

Fig. 1: Hydrogenation of o-cresol on palladium. Effect of additives on the selectivity.
$T = 120°C$; $p_H = 40$ bar; $m_K = 3$ w/w%; $N = 2400$ rpm.

the selectivity, this denominator will drop out by procedure indicated in eq.(9), and S is given by eq.(10) hence the selectivity will not change due to other adsorbed species on the catalyst (Fig. 1). The individual reaction rates will decrease because of the new inhibition terms, but these effects will be the same for both steps. Similar observations were made, when phenol and o-cresol were hydrogenated on palladium partially poisoned with thio-compounds [1, 4].

2.2 Discussion

As has been shown, the simple consecutive model, which assumes the chemical surface reaction to be rate-controlling, can be used to simulate the observed concentration curves for o-cresol and o-ethylphenol. It can also predict the effects of temperature and catalyst loading on the selectivity. This model can be expanded to the cases where other adsorbable species such as solvent or inhibitor molecules are present in the system, or if the catalyst is partially poisoned. The model equations predict that the selectivity will not change in these cases. This postulate was verified by experimental data.

In the literature [e.g. 5-7], however, there exists some examples where a parallel "shunt reaction" has to be introduced to simulate the concentration curves with a good fit:

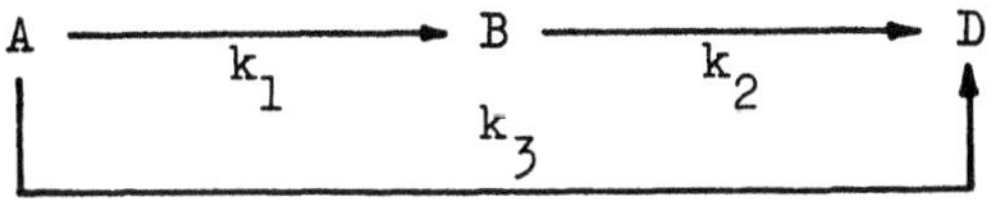

(17)

There are also several experimental observations which indicate that the model prediction, that the selectivity for intermediate production is independent of the hydrogen pressure, is not correct [8]. To get a better insight into such phenomena, let us expand the simple model to more general cases where mass transfer or sorption effects can influence the kinetics.

3. MODEL FOR EXTERNAL DIFFUSION WITH SURFACE REACTION

This model is intended to include the mass transfer steps required to bring the organic species and hydrogen to the catalyst. Normally working without any solvent, the concentrations of the organic species are much higher than the hydrogen dissolved in the liquid. Therefore, we should expect the first observable mass transfer limitation to be for the hydrogen. As it is shown in Section 2.1, such hydrogen depletion supresses both reaction rates to the same extent. Hence there should be no observable selectivity change due to hydrogen mass transfer.

If the transport of reactant A or intermediate B to or from the external catalyst surface is too slow, then there is a relative increase of the intermediate concentration c_{BK}. In this case the second step of the consecutive reaction will be accelerated and the overall selectivity will decrease. To model this situation let us make some simplifying assumptions:

- The transport steps of hydrogen are fast ($D_H > D_A$); the hydrogenation is run at constant hydrogen pressure: $c_{Hi} = c_{HL} = c_{HK}$; θ_H = constant.
- All transport parameters of organic species are equal: $k'_{AL} = k'_{BL} = k'_{DL}$.

For steady-state conditions, the following equations can be written:

$$-r_A = k_1 m_K \left(\frac{c_{AK}}{c_{AK} + Q_2 c_{BK} + Q_3 c_{DK}}\right) = k'_{AL} a_K (c_{AL} - c_{AK}) \quad (18)$$

$$r_B = r_A - r_D = k'_{AL} a_K (c_{BK} - c_{BL}) \quad (19)$$

$$r_D = k_2 m_K \left(\frac{Q_2 c_{BK}}{c_{AK} + Q_2 c_{BK} + Q_3 c_{DK}}\right) = k'_{AL} a_K (c_{DK} - c_{DL}) \quad (20)$$

Using this set of equations, it is not possible to formulate explicit expressions for the surface concentrations. Therefore, for the numerical integration, we had to use a specific program to solve the implicit algebraic equations iteratively before starting integration. The resulting simulations are represented in Fig. 2 with the Damkoehler number Da as the parameter. The simulations indicate a rapid decrease of selectivity with increasing external mass transfer limitations.

3.1 Selectivity: Model Predictions and Simulations

From simulated concentration curves the S/S_o-values can be extracted. The combination of sorption steps with chemical reaction leads to a larger decrease in selectivity at the same transport limitation level compared to a homogeneous first order reaction as it is indicated in Fig. 3. The relative adsorbabilities (eq.8) also effect the overall selectivity.

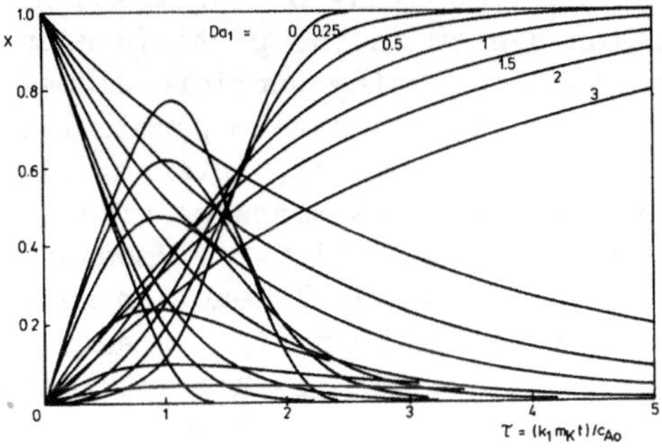

Fig. 2: Effect of liquid/solid mass transfer of substrate on the selectivity of a consecutive reaction with Langmuir-Hinshelwood kinetics (eqs. 18-20).
$S = k_1/(k_2 Q_2) = 10$; $Q_2 = 0.1$; $Q_3 = 0.01$

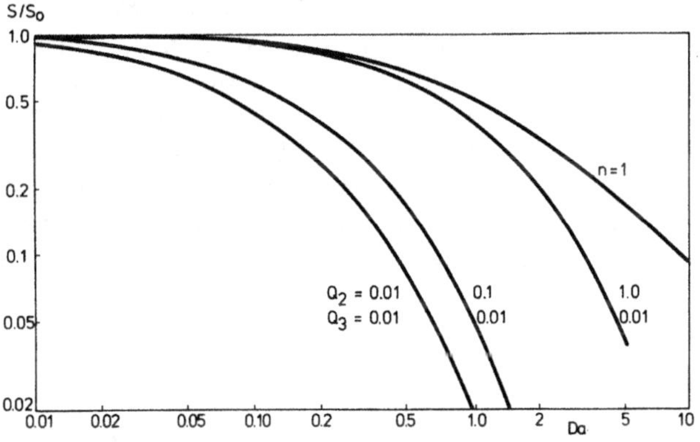

Fig. 3: Decrease of selectivity as a function of Damkoehler number for a consecutive reaction with Langmuir-Hinshelwood kinetics (eqs. 18-20).

The simulations show (Fig. 2) that with the substrate diffusion model (eqs. 18-20) it is possible to explain the presence of a "shunt" and to model the effects of external mass transfer of the organic reaction components on the selectivity. But it was not possible with this model to simulate the observed concentration curves for the hydrogenation of o-tert.butylphenol (see Fig. 4). This is easily understood if one calculates Da for the hydrogenation of o-alkylphenols on palladium catalyst (< 10 μm) using appropriate correlations. In the investigated temperature-pressure region, these Da numbers are of the magnitude 10^{-4} to 10^{-5}. Under these conditions the mass transfer of substrate will not cause a measurable decrease of the selectivity (Fig. 2 and 3).

The main parameter for the study of substrate diffusion effects for a given reaction system is the particle diameter, and therefore experiments with increasing particle diameter are desirable. Unfortunately such experiments have a particular drawback in the practice: with increasing diameter, internal diffusion effects interact with the external mass transfer effects. The use of coated catalysts can overcome this difficulty. However, to work in the required range of Da, the chemical reaction rate should not change substantially. Therefore, to maintain the same catalyst activity by weight, the velocity constant by unit surface area for a coated catalyst should be much higher for larger particle diameters. But such an intended variation of the catalyst activity is in practice almost impossible. In a trickle-bed reactor, working with coated pellets of high activity, the selectivity decrease for a consecutive reaction can be used indirectly to check whether stagnant regions exist.

4. MODEL FOR SURFACE REACTION WITH NON-EQUILIBRIUM ADSORPTION

If the desorption rate of the intermediate B from the catalyst surface is not fast enough, then the adsorbed intermediate B can react further to the final product. Therefore, more D is produced than it would be predicted by a simple consecutive reaction model. To model this system we assume that the sorption equilibria for reactant A, final product D and hydrogen are established but not for intermediate B. We assume further that all the transport steps to the catalyst surface are ocurring fast: $c_{AL} = c_{AK}$ etc.:

$$\begin{array}{ccccc} \theta_A & \xrightarrow{\quad k_1 \quad} & \theta_B & \xrightarrow{\quad k_2 \quad} & \theta_D \\ k'_A \downarrow\uparrow k_A & & k'_B \downarrow\uparrow k_B & & k'_D \downarrow\uparrow k_D \\ A & & B & & D \end{array} \qquad (21)$$

The kinetics of a consecutive reaction are given by eqs.(2-4). Under constant hydrogen pressure and steady-state conditions, we can equate the mass balance for the adsorbed species B using the reaction network above:

$$\frac{1}{m_K} \frac{d\theta_B}{dt} = 0 = k_1 \theta_A + k_B c_{BK} \theta_L - k'_B \theta_B - k_2 \theta_B \qquad (22)$$

where
$$\theta_L = 1 - \theta_A - \theta_B - \theta_D \qquad (23)$$

Combining mass balance (22) with the equilibrium relationships for components A and D, one gets:

$$-r_A = m_K\, k_1 \left[\frac{c_{AL}}{(1 + Q_1)\, c_{AL} + Q_2 c_{BL} + Q_3 c_{DL}}\right] \tag{24}$$

$$r_B = r_A - r_D \tag{25}$$

$$r_D = m_K\, k_2 \left[\frac{Q_1 c_{AL} + Q_2 c_{BL}}{(1 + Q_1)\, c_{AL} + Q_2 c_{BL} + Q_3 c_{DL}}\right] \tag{26}$$

$$\text{where: } Q_1 = \frac{k_1}{k_2 + k_B'}; \quad Q_2 = \frac{k_B}{K_A(k_2 + k_B')}; \quad Q_3 = K_D/K_A \tag{27}$$

The kinetic constants k_1 and k_2 are defined according to $k_i = k_i^* \theta_H$. Hence also Q_1 and Q_2 are pressure dependent. The curves in Fig.4 are simulated with this desorption model, the agreement with experimental points is quite satisfactory.

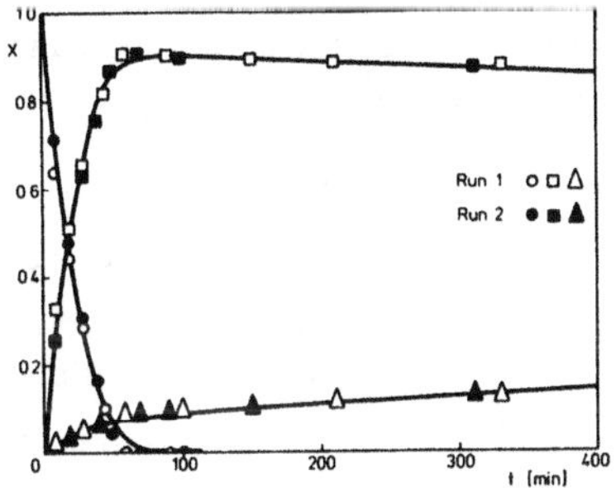

Fig. 4: Hydrogenation of o-tert.butylphenol on palladium. Curves simulated with desorption model (eqs. 24-26).
$T = 140°C$; $p_H = 40$ bar; $m_K = 3$ w/w%; N = 2400 rpm.

The simple consecutive reaction model is a special case of this desorption model. If we move towards the adsorption equilibrium for intermediate B, k_B' will become much larger than k_2, that means that Q_1 diminishes and Q_2 is now K_B/K_A as in the simple model.

4.1 Selectivity: Model Predictions, Experimental Observations and Simulations

In a similar way as shown in Section 2.1, we can evaluate S from eqs. (24-26):

$$S = \frac{k_1 - k_2 Q_1}{k_2\, Q_2} = \left[\frac{k_2\, k_B}{k_1 K_A (k_2 + k_B')}\right]^{-1} \left[1 - \frac{k_2}{k_2 + k_B'}\right] \tag{28}$$

Parameter fitting with the experimental data for the hydrogenation of different alkylphenols indicates that Q_1 has a numerical value between 10^{-1} and 10^{-3}; the values of k_1 and k_2 are generally quite similar to each other. Therefore, eq.(28) can be simplified assuming ($k_2 \ll k'_B$):

$$S = \frac{k_1 K_A}{k_2 K_B} \left(1 - \frac{k_2}{k'_B}\right) \tag{29}$$

or combining eq.(29) with (10): $S/S_o = \left(1 - \frac{k_2^* \theta_H}{k'_B}\right)$ (30)

Eq.(30) shows that the maximum selectivity is achieved when the sorption equilibrium for intermediate B is established. The catalyst loading will not affect the selectivity, which is in agreement with experimental observations. The model cannot postulate the direction of selectivity change with increasing temperature without the knowledge of the individual activation energies and the heats of adsorption. The experimental results show that S for the hydrogenation of o-tert.butylphenol on palladium increases with increasing temperature (Fig. 5).

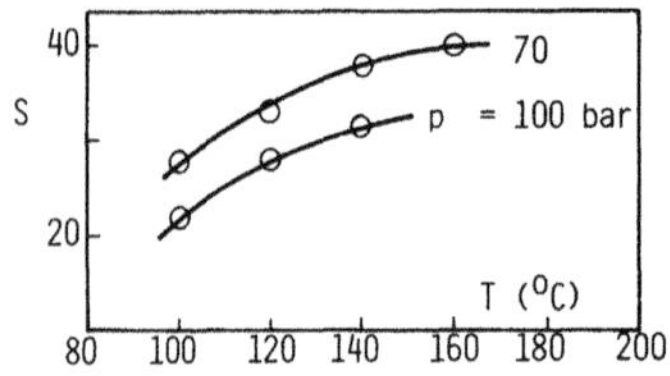

Fig. 5:

Hydrogenation of o-tert.butylphenol on palladium. Effect of temperature on S at different pressure levels.

m_K = 3 w/w%; N = 2400 rpm

Let us now consider the effects of hydrogen coverage on the selectivity. The desorption model postulates that the selectivity will decrease linearly with increasing hydrogen coverage, which is experimentally sound (Fig. 6).

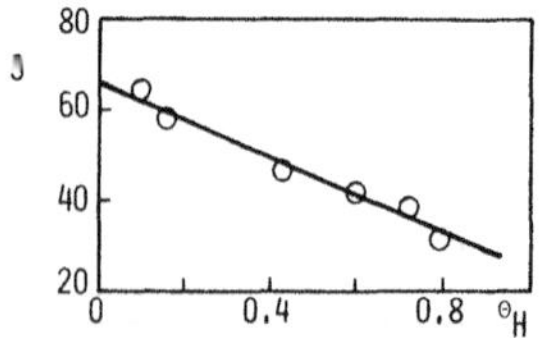

Fig. 6:

Hydrogenation of o-tert.butylphenol on palladium. S as a function of hydrogen coverage.

T = 140°C; m = 3 w/w%;
N = 2400 rpm.

The hydrogen coverage can be decreased either by working in the low pressure region or in the hydrogen diffusion regime (low agitation intensity). In Fig.7 the mole fraction of final product D is given as a function of the total conversion for the hydrogenation of o-tert.butylphenol on palladium. As can be seen from this picture, in both cases the selectivity increases significantly with decreasing hydrogen coverage, which is in agreement with the model predictions.

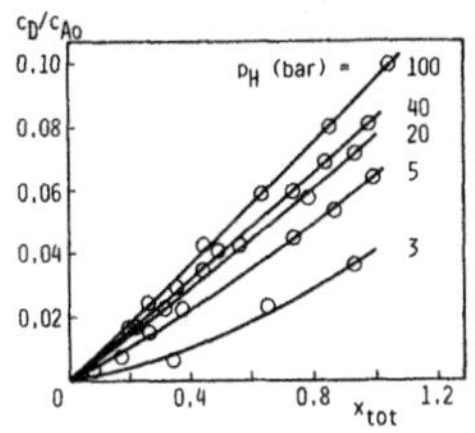

a) Effect of hydrogenation pressure on selectivity.
T = 140°C, m = 3 w/w%, N = 2400 rpm

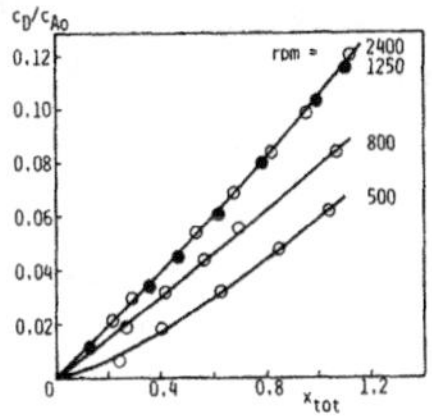

b) Effect of stirrer speed on selectivity.
T = 160°C, p_H = 40 bar

Fig. 7: Hydrogenation of o-tert.butylphenol on palladium.

5. CONCLUSIONS

The aim of the presented paper was to model a competitive, consecutive reaction in a stirred tank slurry reactor for different kinetic regimes and to check the model predictions concerning the selectivity with experimental data. It was shown that the hydrogenation of o-alkylphenols can be described by a Langmuir-Hinshelwood model assuming non-competitive adsorption of organic species and hydrogen on the active surface. Both steps of the consecutive reaction show the same kinetic behaviour especially concerning the dependence on the hydrogen coverage. In a slurry system with very fine particles, a mass transfer limitation for the organic components is not expected. As far as the adsorption equilibria are all established, a change in the hydrogen coverage does not influence the selectivity. The presence of other adsorbable molecules such as solvents or inhibitors suppress both reaction rates in the same extent and therefore they do not influence the selectivity. It was also shown that a partial poisoning of the catalyst with thio-compounds does not change the selectivity. All these phenomena can be described by the proper expansion of the chemical reaction model.

A more precise description of the concentration curves can be achieved when deviations from the adsorption equilibrium for the intermediate are not neglected. In this case the selectivity will decrease with increasing hydrogen coverage.

6. REFERENCES

1. Gut G., Meier R.U., Zwicky J.J. and Kut O.M., Chimia 29 (1975), 295.
2. Kut O.M. and Gut G., Chimia 34 (1980), 250.
3. Gut G., Kosinka J., Prabucki A. and Schuerch A., Chem.Engng Sci. 34 (1979), 1051.
4. Zwicky J.J. and Gut G., Chem.Engng Sci. 33 (1978), 1363.
5. De Boer J.H. and Van der Borg R.J.A.M., Actes de 2me Congrès Internat. de Catalyse, Paris (1960), 919.
6. Coenen J.W.E., Boerma H., Linsen B.G. and De Vries B., Proc. 3rd Int.Congress on Catalysis, Amsterdam (1965), 1378.
7. Scholfield C.R., Butterfield R.O. and Dutton H.J., J.Am.Oil Chem.Soc. 49 (1972), 586.
8. Coenen J.W.E., Chem.Ind.(London) 1978, 709.

7. NOMENCLATURE

a_K	External catalyst area (m^2/m^3)
c	Concentration ($kmol/m^3$)
c_{Hi}	Equilibrium hydrogen concentration ($kmol/m^3$)
D	Diffusivity (m^2/s)
Da	Damkoehler number
E_A	Activation energy (kJ/mol)
H	Henry's constant (m^3 bar/kmol)
ΔH	Heat of adsorption (kJ/mol)
K	Adsorption constant (m^3/kmol)
k_i	Adsorption rate constant (with one subscript)(1/s w/w% cat)
$k_i^!$	Desorption rate constant (with one subscript)($kmol/m^3$ s w/wt cat)
k	Pressure-dependent chemical rate constant ($kmol/m^3$ s w/w% cat)
k*	Pressure-independent chemical rate constant ($kmol/m^3$ s w/w% cat)
$k_{ij}^!$	Transport coefficient liquid/solid (with two subscripts)(m/s)
m_K	Amount of catalyst (w/w%)
N	Stirrer speed (rpm)
p_H	Hydrogen pressure (bar)
Q	Parameter defined by eqs.(8) or (27)
r	Reaction rate ($kmol/m^3$ s)

S Selectivity coefficient

T Temperature (K, °C)

t Time (s)

x Conversion

η Effectiveness factor

θ Fractional occupancy of active sites

Subscripts:

A Reactant
B Intermediate
D Final product
H Hydrogen
K Catalyst
L Bulk liquid
o Initial or non-disguised value
Y o-alkylphenol added

INFLUENCE OF NONUNIFORM CATALYST DISTRIBUTION ON THE PERFORMANCE OF THE BUBBLE COLUMN SLURRY REACTOR

Y. Serpemen and W.-D. Deckwer

Institut für Technische Chemie der Universität Hannover, Callinstr. 3, D-3000 Hannover 1, West Germany

INTRODUCTION

With regard to an effective use of catalyst it is necessary to realize a uniform distribution over the entire reactor. There are a number of experimental studies reported in the literature (1-5) which show that even for small particles well pronounced solid concentration profiles can be observed in the gas agitated bubble column slurry reactors (BCSR). A dispersion-sedimentation model has been proposed, which successfully describes measured data (2-4).

Although cold flow experiments indicate that nonuniform catalyst distribution may occur, it has not yet been systematically investigated how the performance of the reactor will be influenced by catalyst settling. Govindarao (6) has presented an analysis of the dynamic and steady-state behavior of BCSR with stagnant slurry phase. The calculations which were based on a three-phase dispersion model and a first order chemical surface reaction show that the catalyst distribution is mainly affected by the particle and reactor diameter and the dynamics are improved with increasing uniformity of the solid distribution. Parulekar and Shah (7) have developed a detailed model for the cocurrent BCSR which accounts for changes in phase holdups, gas and slurry velocities and for catalyst settling but neglects all mass transfer resistances in the three-phase system. Model simulations indicate that settling of catalyst particles can improve, to a certain extent, the yields.

Chaudhari and Ramachandran (8-9) give a detailed analysis of slurry reactors including all mass transfer resistances and different kinds of kinetic expressions but they do not account for catalyst settling, dispersion and conversion induced volume change in the gas phase.

OBJECTIVE

The aim of this contribution is to study the effect of the nonuniform catalyst distribution on the performance of BCSR. To this end, extensive calculations based on a rather sophisticated three-phase dispersion model were done. The effect of such parameters as gas and liquid velocity, column and particle diameter, particle density, which influence strongly the catalyst settling, is investigated.

Three reaction systems of industrial importance, i.e. Fischer-Tropsch synthesis (FTS) and the methanation of CO in batch slurries of molten wax and the continuous hydrogenation of butynediol, were used which obey first, half and second order rate laws, respectively. Kinetic expressions, rate constants and the reaction conditions are given in Table 1.

MATHEMATICAL MODEL

The proposed model is based on the following steps for the soluble component A present in the gas phase:

(1) transport of A from gas-liquid interface to the bulk liquid,
(2) transport from bulk liquid to the external catalyst surface,
(3) intraparticle diffusion in the pores of the catalyst,
(4) surface reaction to yield products.

The local rate of the reaction within the catalyst is assumed to be m-th order with respect to the concentration of the dissolved component A. In the case of a general m,n-th order reaction with a liquid phase component B the rate equation can be simplified into a pseudo-m-th order form, as usually the liquid phase component is in excess. The variation of B is then small and the concentration of B is uniform throughout the catalyst (8).

The model is further based on the following assumptions:

(1) The reactor operates under constant total pressure and isothermic conditions at steady state.
(2) Gas and liquid phase are backmixed.
(3) The nonuniform catalyst distribution will be described by the dispersion-sedimentation model.
(4) Hydrodynamic properties are spatially independent.
(5) The dependency of gas flow rate on conversion will be described by using a contraction factor, ε, following the definition given by Levenspiel (10):

$$\varepsilon = (G-G_o)/G_o$$

Making use of the foregoing assumptions material balances upon a differential volume element of the reactor yield the following equations in dimensionless form for the

gas phase

$$\frac{1}{Bo_G}\frac{d^2Y}{dz^2} - \frac{(1+\varepsilon)}{(1+\varepsilon Y)^2}\frac{dY}{dz} - St_G(Y - A_L) = 0 \qquad (1)$$

liquid phase

$$\frac{1}{Bo_L}\frac{d^2A_L}{dz^2} + f\frac{dA_L}{dz} + St_L(Y - A_L) - St_S\psi(A_L-A_S) = 0 \qquad (2)$$

$$\frac{1}{Bo_L}\frac{d^2B}{dz^2} + f\frac{dB}{dz} - Da_B\psi\eta A_S^m B^n = 0 \qquad (3)$$

and for the catalyst

$$A_L - A_S = Da_S\eta A_S^m B^n \qquad (4)$$

Intraparticle diffusional resistances are considered by means of the effectiveness factor

$$\eta = \frac{1}{\phi}\left(\coth 3\phi - \frac{1}{3\phi}\right) \qquad (5)$$

introducing a generalized form of the Thiele-modulus

(8,11)

$$\phi = \frac{Dp}{6}\left(\frac{m+1}{2}\,\frac{\rho_p k_{mn}\, A_o^{*\,m-1} B_o^n}{D_{e,A}}\, A_S^{m-1}\, B^n\right)^{1/2} \tag{6}$$

which, in the case of a nonlinear rate equation ($m \neq 1$, $n \neq 0$) depends on the concentrations and varies over the reactor length.

The nonuniform catalyst distribution due to the settling of the particles is characterized in the balance equations by the variable ψ, which represents the ratio of the local solid concentration to the reactor mean value $\bar{C}_{cat}$. The catalyst concentration profile follows from the dispersion-sedimentation model (2-4). In the case of the bubble agitated stagnant slurry (batch) one obtains

$$C_{cat} = \bar{C}_{cat}\,\frac{Bo_c\,\exp(-Bo_c\,z)}{1-\exp(-Bo_c)} \tag{7}$$

and for the cocurrent flow of gas and slurry

$$C_{cat} = C_F\,\frac{Bo_c\,\exp\,(Bo_c - Bo_L^*)(1-z)\;-\;Bo_L^*}{(Bo_c - Bo_L)} \tag{8}$$

The integration of eq. (8) over the reactor length yields for the average concentration

$$\bar{C}_{cat} = C_F\,\frac{Bo_c(\exp(Bo_c - Bo_L^*) - 1) - Bo_L^*(Bo_c - Bo_L^*)}{(Bo_c - Bo_L^*)^2} \tag{9}$$

The catalyst distribution is governed by the solid dispersion coefficient, E_c, mean settling velocity of the particles in swarm, u_{cs} and in the case of cocurrent flow by the liquid velocity, u_L, which acts against the settling.

The model equations are subject to the following boundary conditions

$$z = 0$$

$$1 = \frac{(1+\varepsilon)Y}{(1+\varepsilon Y)} - \frac{1}{Bo_G}\frac{dY}{dz} \tag{10}$$

f = 0 (batch) $$\frac{dA_L}{dz} = 0 \tag{11a}$$

f = -1 (cocurrent)

$$0 = A_L - \frac{1}{Bo_L}\frac{dA_L}{dz}$$

$$1 = B - \frac{1}{Bo_L}\frac{dB}{dz} \tag{11b}$$

$$z = 1 \qquad \frac{dY}{dz} = \frac{dA_L}{dz} = \frac{dB}{dz} = 0 \tag{12}$$

Due to the variable gas velocity and the nonlinear rate law the model equations represent a set of coupled nonlinear algebraic and differential equations of boundary value type which must be solved numerically. For this purpose the nonlinear equations are entirely linearized using the quasilinearization technique (12) and the linearized differential equations are solved using the orthogonal collocation method based on shifted Legendre polynomials (13).

MODEL PARAMETERS

For the case of FTS and CO methanation in molten wax slurry system the parameters involved in model equations, i.e. the physicochemical properties, the hydrodynamic and mass transfer parameters, can be estimated with sufficient accuracy. There exist reliable data for the physicochemical properties obtained from independent measurements and summarized by Hammer (14) and Deckwer and coworkers (15). The hydrodynamic and mass transfer parameters can be calculated from empirical correlations given by Deckwer et al. (15), which were partly established from measurements in labscale reactors under synthesis conditions and seem to be applicable for larger scale equipment (17). This data and correlations were successfully used to perform a kinetic study on the experimental data reported in the literature on the FTS (16) and to simulate the results obtained in the Rheinpreussen-Koppers demonstration plant predicting fairly well the optimal gas velocity (17,18).

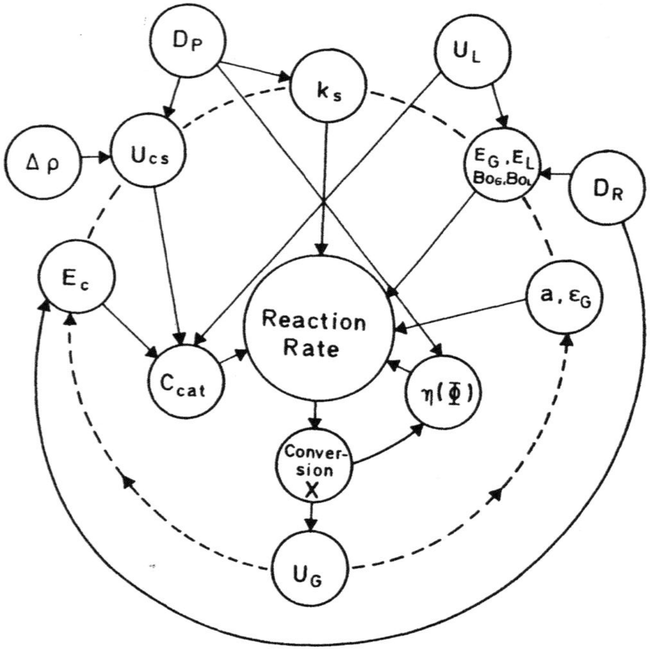

Fig. 1: Interrelations between model parameters

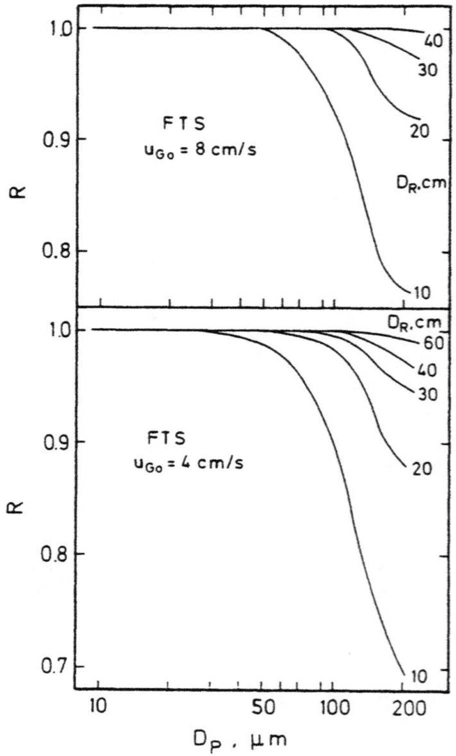

Fig. 2: Effect of catalyst distribution on reactor performance

On the basis of the correlations given in detail in reference (17,18) Fig. 1 shows the interrelations between the operating parameters (such as gas and liquid velocities, u_G, u_L, reactor and particle diameter D_R, D_P, catalyst density) and the hydrodynamic parameters (gas, liquid and solid dispersion coefficients, catalyst settling velocity, gas holdup, gas-liquid interfacial area, liquid-solid mass transfer coefficient). The influence of these parameters on the catalyst concentration, reaction rate and conversion is indicated by the arrows. As the gas velocity varies with the conversion, the hydrodynamic parameters, while depending on the gas velocity, will be influenced by the conversion as well. This causes also a modification of the catalyst concentration profile in the reactor.

RESULTS OF SIMULATIONS

While this study is mainly concerned with the influence of the catalyst concentration profile on the reactor performance, additional calculations were done under identical conditions with a uniform catalyst profile having the reactor mean value $\bar{C}_{cat}$. For the purpose of comparison the ratio R of the outlet conversions obtained with and without catalyst profile is calculated.

FT Synthesis

Fig. 2 shows for the FTS in the batch slurry of molten wax the effect of the catalyst settling on the reactor performance at two different gas velocities where the ratio R is plotted versus the particle diameter D_P with the column diameter D_R as parameter. With increasing D_P and decreasing D_R and hence decreasing solid dispersion, R drops down, which indicates that the conversion compared to a spatially uniform catalyst distribution decreases. For particle sizes less than 50 µm an influence of the catalyst profile on the conversion can be excluded altogether. For a column diameter of 30 cm and particle sizes up to 200 µm the deviation from the ideal case is only moderate and less than 5 %. The overall effect by doubling the gas velocity is less pronounced and mainly attributed to the enhanced gas-liquid mass transfer (a is higher) which leads to a higher degree of saturation in the liquid phase.

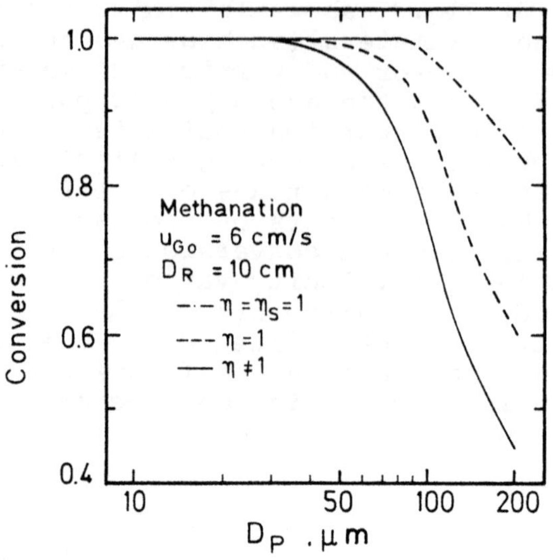

Fig. 3: Effect of mass transfer resistances on conversion

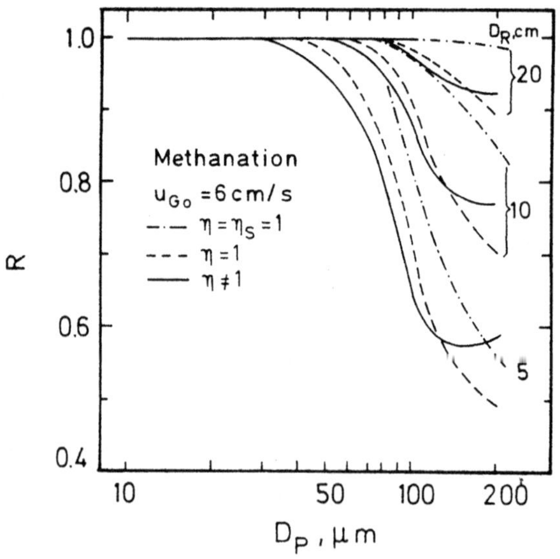

Fig. 4: Influence of nonuniform catalyst distribution on reactor performance

Methanation of CO

Detailed experimental analysis on the methanation of CO on Ni catalyst suspended in molten paraffin under similar conditions to the FTS is given by Hammer (14). The nonlinear rate equation (half order in H_2 for stoichiometric inlet mixtures of CO and H_2) is expected particularly suitable to study the combined effect of catalyst settling and intraparticle diffusional resistances on the reactor performance.

The effect of the liquid-solid mass transfer and the intraparticle diffusional resistances on the conversion are demonstrated in Fig. 3 in dependence on the catalyst particle diameter. For particle sizes larger than 70 µm the differences are remarkable.

The influence of the nonuniform solid distribution on the reactor performance is shown in Fig. 4 where the ratio R is calculated including successively the different mass transfer resistances. As in the preceding case of the FTS the reactor diameter plays a dominant role and influences mainly the solid dispersion. Great deviations from the case of uniform catalyst distribution are only expected again for larger particles (>50 um) in small size reactors. If liquid-solid ($\eta_S \neq 1$) and intraparticle mass transfer resistances are taken into account, the influence on the reactor performance is further enhanced up to the particle sizes of 120 µm. The change in the shape of R (solid curves) for larger particles are due to the sudden decrease of the reaction rate caused by the intraparticle diffusional resistances.

The effect of the particle density on the catalyst settling and its influence on the reactor performance is shown in Fig. 5 in a labscale size column, where, in addition to the computed catalyst concentration profiles, the corresponding profiles of the conversion and the intraparticle effectiveness factor are plotted. Lower particle density causes a more uniform catalyst distribution and decreases intraparticle diffusional resistances.

Hydrogenation of butynediol

The continuous hydrogenation of butynediol in a cocurrently operated bubble column reactor is chosen as an example for a second order reaction where a sparingly soluble gas phase component (H_2) reacts on the catalyst surface with a liquid phase component in excess (butyne-

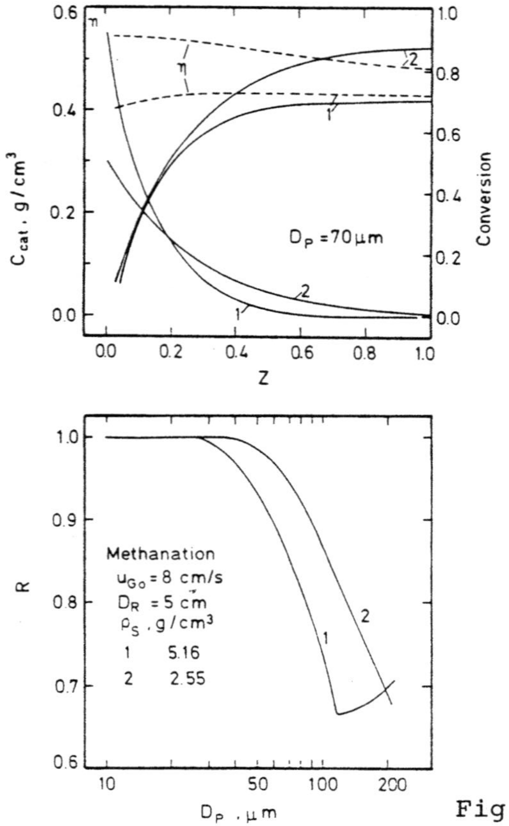

Fig. 5: Influence of the particle density on the catalyst distribution

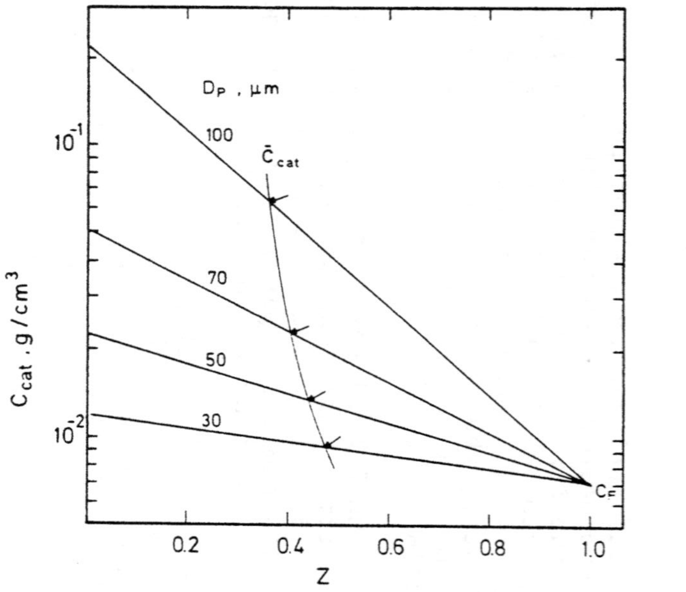

Fig. 6: Catalyst concentration profiles at cocurrent flow of gas and slurry

diol) to yield a nonvolatile product (butenediol).

In the case of cocurrent operation the catalyst settling leads to an accumulation of the solids within the reactor. This is shown in Fig. 6, where the computed catalyst concentration profiles for different particle diameters are given in a semilogarithmic plot. The catalyst concentration is larger than the feed concentration C_F, in the whole reactor. The average catalyst concentration (indicated by the arrows) increases with increasing particle diameter.

At the cocurrent upflow of the gas and the slurry the accumulation of solids is decelerated by the slurry velocity which acts against the settling. This is shown in Fig. 7 where the calculated average values of the catalyst concentrations are plotted as functions of D_P at different liquid velocities. According to the increasing solid dispersion the accumulation of the catalyst within the reactor is reduced with increasing the reactor diameter.

As has already been discussed in the two preceding cases in batch slurries, the influence of the nonuniform catalyst distribution on the reactor performance is again negligible for larger column diameters, say $D_R >$ 20 cm.

Calculated conversions of butynediol (B) and hydrogen (A) with different sizes of catalyst particles are given in Fig. 8 with D_R as a parameter. The calculations are performed with a fixed value of the volumetric mass transfer coefficient, $k_L a$, but including the liquid-solid and the intraparticle mass transfer resistances. It can be noticed that the conversions steadily decrease with increasing particle diameter, hence increasing mass transfer resistances, except for a reactor diameter of 10 cm, where they pass a minimum. This phenomenon can be explained by the fact that the catalyst accumulation in the reactor can compensate the decrease of the reaction rate caused by liquid-solid and intraparticle mass transfer resistances. If the intraparticle diffusional resistances are excluded, the overall behavior does not change but the effect is less pronounced.

The fact that the conversion of the liquid phase reactant is higher than the conversion of the gas phase reactant is in complete accordance with the previous findings of Schumpe et al. (19) which were derived from a simple model applied to the slow reaction regime

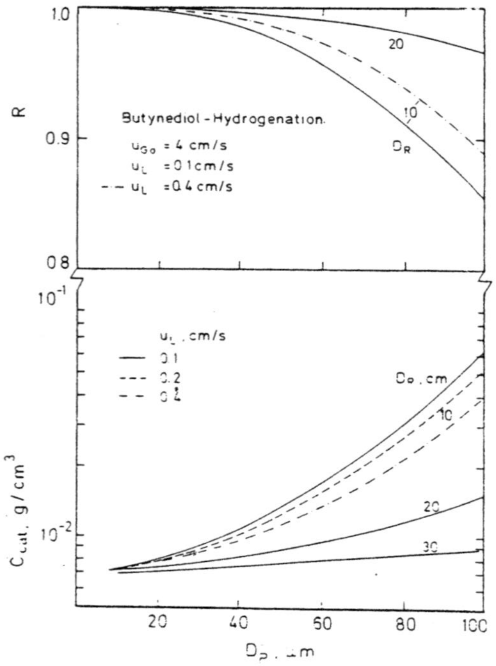

Fig. 7: Effect of catalyst settling on reactor performance and catalyst accumulation

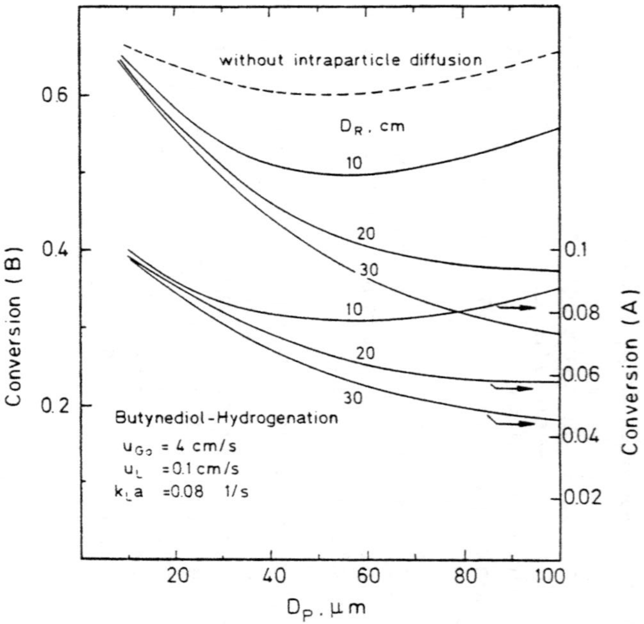

Fig. 8: Influence of catalyst settling on the conversions

of an absorption-reaction process and confirm the opposite trend of the conversions.

CONCLUSIONS

On the basis of extensive computations which were performed with a three-phase dispersion model for bubble column slurry reactors accounting for all relevant phenomena, i.e., dispersion in all phases, catalyst settling, nonlinear reaction kinetics, gas-liquid and liquid-solid mass transfer resistances, intraparticle diffusion and variable gas flow rate, the following conclusions can be stated:

(1) For reactor diameters larger or equal to D_R = 30 cm and particle sizes up to 200 um the catalyst settling can be neglected in as far as the catalyst profile does not affect the overall conversion.
(2) In small diameter BCSR the catalyst distribution can be extremely nonuniform, this in turn can reduce the conversion considerably.
(3) This effect must be considered when scaling up or scaling down BCSR.

ACKNOWLEDGMENT

The authors gratefully acknowledge the financial support from the Stiftung Volkswagenwerk.

Table 1: Reaction systems considered in simulations

	Catalyst	P g/cm^3	P atm	T C	Reaction rate	Rate constant
1) FT Synthesis	pptd. Fe	3.97	12	268	$k_1 C_{H_2,L}$	2 cm^3/gs
2) Methanation of CO	Ni-MgO	5.16	1	268	$k_{1/2} C_{H_2,L}^{0.5}$	$0.0147 \frac{cm^3}{gs} (\frac{mole}{cm^3})^{0.5}$
3) Hydrogenation of Butynediol	Pd-$CaCO_3$	1.45	14.6	35	$k_{11} C_{H_2,L} C_{Bu}$	$5.10^4 cm^6$/gs mole

NOTATION

a	gas-liquid interfacial area, cm^{-1}
$\bar{a}_s$	mean liquid-solid interfacial area, cm^{-1}
A_o^*	equilibrium concentration at reactor inlet
A_L	dimensionless liquid phase concentration, C_{AL}/A_o
A_S	dimensionless surface concentration, C_{AS}/A_o
B	dimensionless liquid phase concentration, C_B/B_o
B_o	inlet concentration of liquid phase component B
Bo_c	Bodenstein number for solid dispersion, $u_{cs}L/E_c$
Bo_G	Bodenstein number for the gas phase, $u_{Go}L/\varepsilon_G E_G$
Bo_L	Bodenstein number for the liquid phase, $u_L L/\varepsilon_L E_L$
Bo_L	Bodenstein number for the liquid phase, $u_L L/\varepsilon_L E_C$
C_F	concentration of catalyst in feed slurry
C_{AL}	liquid phase concentration of the dissolved component
C_{AS}	surface concentration of the dissolved component
C_{cat}	catalyst concentration
$\bar{C}_{cat}$	catalyst concentration, mean value
Da_B	Damköhler number, $\varepsilon_L k_{mn}\bar{C}_{cat}A_o{}^m B_o^{n-1} L/u_L(u_{Go})$
Da_S	Damköhler number, $\varepsilon_L k_{mn}C_{cat}\Lambda_o^m\ B_o^u/k_s\bar{a}_s$
$D_{e,A}$	effective diffusivity, cm^2/s
D_p	particle diameter
D_R	reactor diameter
E_G, E_L, E_C	gas, liquid and solid dispersion coefficient, cm^2/s
f	dummy variable, f=0 stagnant slurry, f=-1 co-current flow

G	molar gas flow rate at complete conversion
G_o	inlet value of molar gas flow rate
He	Henry's coefficient
k_L	liquid side mass transfer coefficient
k_s	liquid-solid mass transfer coefficient
k_{mn}	m,n-th order reaction rate constant
L	reactor length
R	ratio of conversions with and without catalyst profile, gas constant
St_G	gas phase Stanton number, $k_L a(RT/He)(L/u_{Go})$
St_L	liquid phase Stanton number, $k_L aL/u_L(u_{Go})$
St_S	solid phase Stanton number, $k_s \bar{a}_s L/u_L(u_{Go})$
T	temperature, K
u_G	variable gas velocity, $= u_{Go}(1+\varepsilon X)$
u_{Go}	inlet superficial gas velocity
u_L	liquid phase velocity
u_{CS}	terminal settling velocity of particles in swarm
X	conversion of gas phase component, $(1-Y)/(1+\varepsilon Y)$
y, y_o	actual and inlet mole fraction of the soluble component in the gas phase
Y	ratio y/y_o
z	dimensionless axial coordinate
ε	contraction factor, $(G-G_o)/G_o$
$\varepsilon_G, \varepsilon_L$	gas and liquid holdup
η	intraparticle effectiveness factor, (eq. 5)
η_s	liquid-solid mass transfer effectiveness factor
ν	stoichiometric coefficient
ρ_p	particle density
ϕ	generalized Thiele modulus, (eq. 6)
ψ	ratio of catalyst concentrations, $C_{cat}/\bar{C}_{cat}$

REFERENCES

1. Imafuku, K., T.-Y. Wang, K. Koide and H. Kubota. J. Chem. Eng. Japan 1 (1968), 153
2. Cova, D.R. Ind. Eng. Chem. Process Des. Dev. 5 (1966), 21
3. Suganuma, T. and T. Yamanishi, Kagaku Kogaku 30 (1966), 1136
4. Kato, Y., A. Nishiwaki, T. Fukuda and S. Tanaka. J. Chem. Eng. Japan 5 (1972), 112
5. Kölbel, H., M. Molzahn and H. Hammer, DECHEMA Monogr., 68 (1970) 477
6. Gowindarao, V.G.H. Chem. Eng. J. 9 (1975) 229
7. Parulekar, S.J. and Y.T. Shah. Chem. Eng. J. 20 (1980) 21
8. Chaudhari, R.V. and P.A. Ramachandran, A.I.Ch.E.J. 26 (1980) 177
9. Ramachandran, P.A. and R.V. Chaudhari. Chem. Engineer, December 1 (1980)
10. Levenspiel, O." Chemical Reaction Engineering", J. Wiley, New York 1972
11. Bischoff, K.B. A.I.Ch.E.J. 11 (1965) 351
12. Lee, E.S."Quasilinearization and Invariant Imbedding", Academic Press, New York, London, 1968
13. Villadsen, J.V. and M.L. Michelsen,"Solution of Differential Equation Models by Polynomial Approximation", Prentice Hall, Englewood Cliffs, 1978
14. Hammer, H. Habilitationsschrift, Technische Universität Berlin, 1968
15. Deckwer, W.-D., Y. Louisi, A. Zaidi and M. Ralek. Ind. Eng. Chem. Process Des. Dev., 19 (1980) 699
16. Deckwer, W.-D., Y. Serpemen, M. Ralek and B. Schmidt, Chem. Eng. Sci. 36 (1981) 765,791
17. Deckwer W.-D. "Coal Liquefaction Via Indirect Routes", (Proceedings of NATO ASI, Çeşme/Turkey, August 10-21, 1981)
18. Deckwer W.-D. Y. Serpemen, M. Ralek and B. Schmidt. Ind. Eng. Chem. Proc. Des. Dev., to be published.
19. Schumpe, A., Y. Serpemen and W.-D. Deckwer. Ger. Chem. Eng. 2 (1979) 234,267

A ROTATING DISC REACTOR FOR REACTION PROCESSES IN SLURRIES

D.Elenkov and S.D.Vlaev

Central Laboratory of Chemical Engineering,Bulgarian Academy of Sciences,Sofia 1113,Bulgaria

1 INTRODUCTION

The rotating disc contactor (RDC) [1],created originally as a liquid-liquid extractor,is notable for some attributes which back its wide application.Due to the disc impellers it exhibits lower interstage backflow rates,whereas mixing is found to be enough intensive within the individual stages. These features on the background of recent developments of slurry reaction and bioreaction engineering actualize studies on this type of equipment from the standpoint of slurry reaction process requirements.Some characteristics of a rotary-disc biocontactor were reported recently by B r a u e r [2]. In this paper we present part of our studies on the performance of a somewhat different construction of a rotating disc slurry reactor.

2 DESCRIPTION OF THE REACTOR

The reactor is shown in Fig.1.It was constructed of five one type compartments 2 with 140 mm I.D.The compartments were separated by horizontal baffles with central openings 3 .Two equal-sized discs were installed in each compartment.One of the discs 4 was centrally located while the other 5 was mounted close to the baffles at 3-5 mm over the opening.These discs carried out the mixing successfully. The diameters of the discs(size A)and those of the baffle openings(size B)are different for individual compartments of the reactor(Table 1).(Numeration of the stages coincides with direction of flow.)

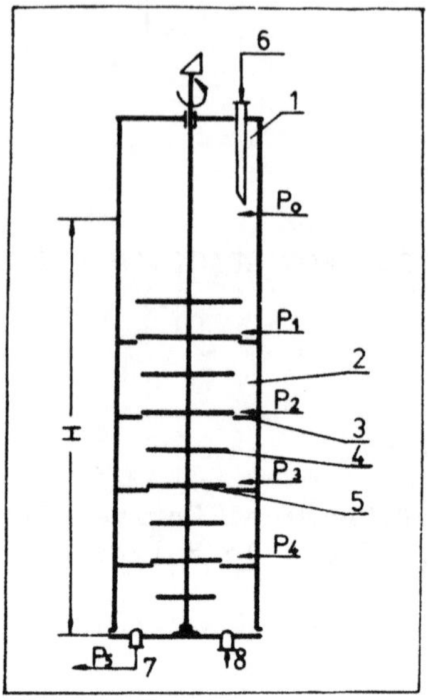

Fig.1.Schematic diagram of the reactor

Table 1

Stage No	Size A,mm	Size B,mm
1	100	93
2	91.5	82
3	80.5	71
4	68.5	60
5	59	-

By the tests suspension was introduced into the column through the opening 6 and was directed out of it through the opening 7 .Gas was introduced through the opening 8 .

Based on heterogeneous reaction process requirements studies were directed mainly on axial mixing,hold-up and interface mass transfer.

3 STUDIES ON THE REACTOR PERFORMANCE

3.1 Axial Mixing in the Reactor

These studies comprised evaluation of the axial mixing in the bed.Preliminary tests showed that axial mixing was really intensive only in the liquid phase,the solids being engaged in a motion following the liquid and moving downwards under gravity.Under these circumstances the gas phase was assumed to move in plug flow and attention has been focused on liquid mixing and solids residence time.

As the reactor was partially compartmentalized it could be well represented by the backflow cell model.In this case liquid axial mixing could be evaluated by the interstage backflow rate f_e .Following these principles backflow rates f_e were obtained by comparison of experimental RTD-curves with curves calculated from model equations (1) :

$$\frac{V_j}{V_r}\frac{dC_j}{d\Theta} = f_e(C_{j-1} + C_{j+1} - 2C_j) + (C_{j-1} - C_j) \qquad (1)$$

$(j=1,2,\ldots,N)$

where C is the tracer concentration, Θ - dimensionless time, V_r-the reactor volume and j refers to the stages.

Results on backflow rates [3,4] are shown in Fig. 2 in the form of its dependencies on linear mixing velocities nd_m

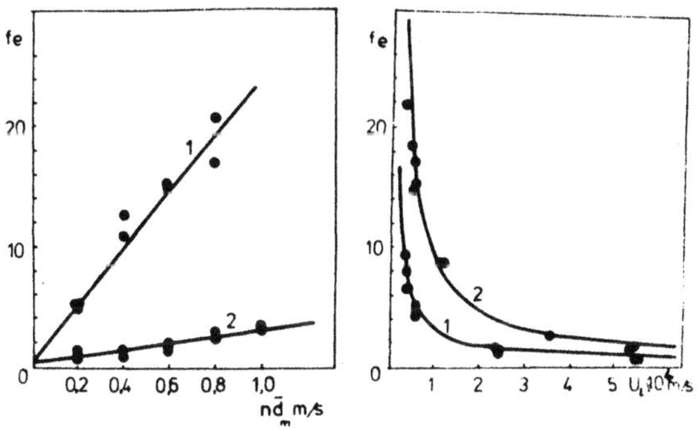

Fig.2.Correlation of the extent of liquid axial mixing

(by U_L=5.9x10^{-5}m/s(1),and U_L=5.4x10^{-4}m/s(2)) and on liquid velocities U_L (at $n=2.5\ s^{-1}$ (1) and $n=7.5\ s^{-1}$ (2)).It can be seen that these relationships do not differ substantially from their analogues of conventional RDCs.In accordance with this fact the results were correlated by a similar expression-

$$f_e - 0.552 = 1.39x10^{-3}\ nd_m/U_L \qquad (2)$$

The calculated by eqn.(2) values of f_e form the solid lines in Fig.2 .By comparison of the coefficient $k=1.39x10^{-3}$ of eqn.(2) with the coefficients $k=6x10^{-3}$ obtained for conventional RDCs [5,6] and $k=1.10^{-2}$ reported for Oldshue-Rushton columns [7,8] it can be concluded that in this case the backflow is rather low.This is one positive characteristic of the reactor under consideration.

The influence of liquid mixing and that of solids physica properties on solids residence time was studied experimentally and by means of a theoretical model [9] .The model (eq.(3)) accounted for both interstage mixing (f_j)of individual stages and gravity.

$$\frac{V_j}{V_r}\frac{d\varepsilon_{sj}}{d\Theta} = U_r'(\varepsilon_{sj-1} - \varepsilon_{sj}) + f_{j-1}\ U_r'(\varepsilon_{sj-1} - \varepsilon_{sj}) + \qquad (3)$$

$$+ f_j\ U_r'(\varepsilon_{sj+1} - \varepsilon_{sj})\ ;$$

$$U_r' = \frac{U_r + U_L}{U_L}\ .$$

The relative linear velocity of particles (U_r) was corrected for voidage by using Richardsons correlation [10]:

$$U_r = U_t(1 - \bar{\varepsilon})^{3.6}\ .$$

In these equations ε_s is the local solids hold-up and j refers to the stages; U_t is particle terminal settling velocity and $\bar{\varepsilon}$ is the average solids concentration in the bed. It is shown (Fig.3) a comparison of a model solution with the respective solids residence time distribution - the latter obtained by spectrophotometric registration of Ni-Al tracer concentration at the reactor outlet.From this and other curves generated by a computer it has been concluded that solids move essentially in backmixed flow,characterized by both liquid interstage backflow rates and solids settling velocities.

Additional study of gas influence on liquid and solids axial mixing showed a 15 % increase of interstage backflow

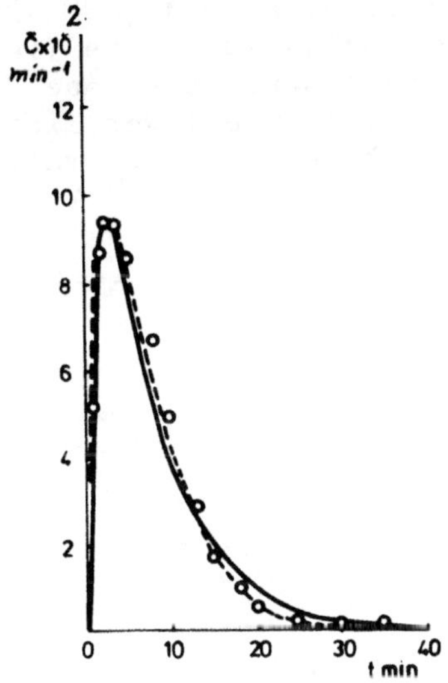

Fig.3. Measured versus predicted RTD of solids; $\bar{C} = \mathcal{E}_{s5}/(\mathcal{E}_{s5}.dt)$.

rate mainly at lower impeller revolutions and higher gas velocities ($U_G \sim 1$ cm/s).

3.2 Reactor Hold-ups

Initially it was studied the effect of various factors on the total volume of the reactor.Results showed that the main factors influencing its value were linear mixing velocity and suspension physical properties.These variables were classified in form of impeller Reynolds number of the slurry (Re_m).Its effect upon the reactor volume V_r is illustrated in Fig.4 .The experimental values in the figure are interconnected by dotted lines.They were found to be represented well by a correlation of the Darcy type:

$$V_r = V_o + V_r$$

$$V_r = K.\frac{(nd_m)^2}{2g} \qquad (4)$$

where

$$K = B.Re_m^{-m} \quad , \; B = 8.3\times10^{-3} \text{ and } m = 0.14 \; .$$

In these equations the density and viscosity of the slurry has been calculated according to [11,12] .The solution of the equations (5) are indicated in the figure with solid lines.The upper line represents the case for initial slurry concentration $\mathcal{E}_o$=1.8 % vol.;the lower line represents the case for $\mathcal{E}_o$=3.6 % vol.Both lines are valid for particles of size 10μm.

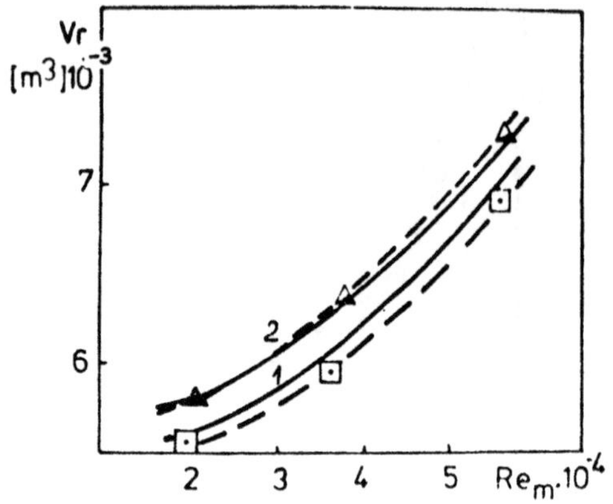

Fig.4. Effect of mixing intensity and suspension physical properties on total reactor hold-up.

Solids and gas hold-up were measured extensively by the displacement method.The main achievements of these measurements are summarized by two characteristic relationships shown in Fig. 5 .

The first relationship represents the effect of initial slurry concentration $\mathcal{E}_o$ on the average solids hold-up $\mathcal{E}_s$. Line 1 corresponds to the case for 10 micron particles , line 2 - for 27 micron and line 3 - for 58 micron particles. Different points indicate different r.p.m. of the impeller. It can be seen that for larger particles $\mathcal{E}_s$ is less dependent on $\mathcal{E}_o$ but also decreases a lot in comparison with fine fractions.Besides it doesnt depend essentially on impeller revolutions provided they are high enough to maintain the particles in suspension.Under the conditions described the actual solids hold-up in the reactor is about 1/3 of $\mathcal{E}_o$ for coarse particles and about 0.9 of $\mathcal{E}_o$ for fine ones.

The second relationship shows gas hold-up by several gas velocities U_G and mixing intensities .Lines 1 and 2 have been obtained at nd_m=0.33 m/s and $\mathcal{E}_o$ resp. 1.8 and 3.6 % vol.

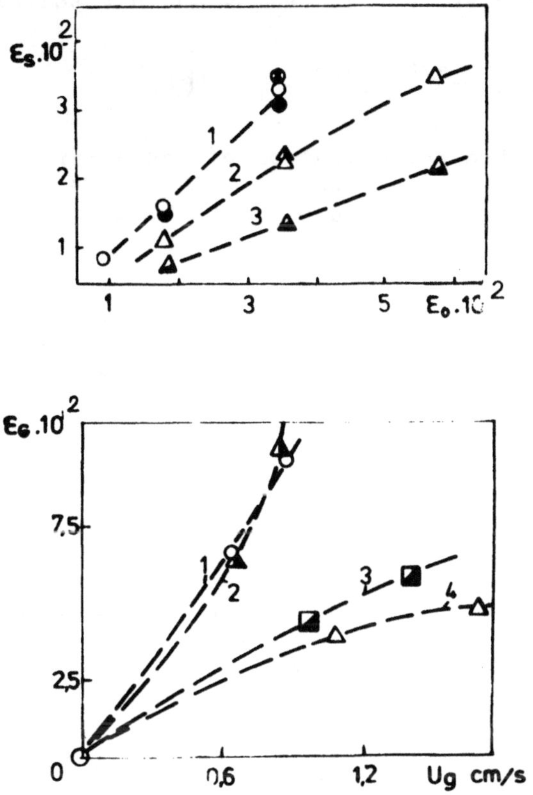

Fig. 5. Average solids and gas holdup.

Lines 3 and 4 have been obtained at ε_o=1.8 % vol. by nd_m resp.0,6 m/s and 0.94 m/s .It is clearly seen that at lower mixing intensities the gas holdup increases almost linearly with linear gas velocity approaching values(ε_G~0.1) characteristic for other agitated contactors [13] .At higher mixing intansities the gas hold-up shows mostly independent of gas velocities mainly bacause of radial stratifications.

3.3 Some Aspects of Mass Transfer in the Reactor

Three phase slurry reactors are characterized by a gas-liquid (K_La) and liquid-solid (k_s) mass transfer coefficient.These coefficients were determined for the rotating disc reactor at the appropriate operation conditions.

Experimental data for k_s of the reactor are shown in Fig. 6.These are correlated by the dimensionless equation of the form

$$Sh = 2 + k\,Re_S^{1/2}\,Sc^{1/3} \qquad [14] \qquad (5)$$

It is illustrated the deviation of the experimental points (for $nd_m = 0.8 \div 1.2$ m/s and $d_p = 0.02 \div 0.04$ cm) from that calculated according to eq.5 with k=0.6 .

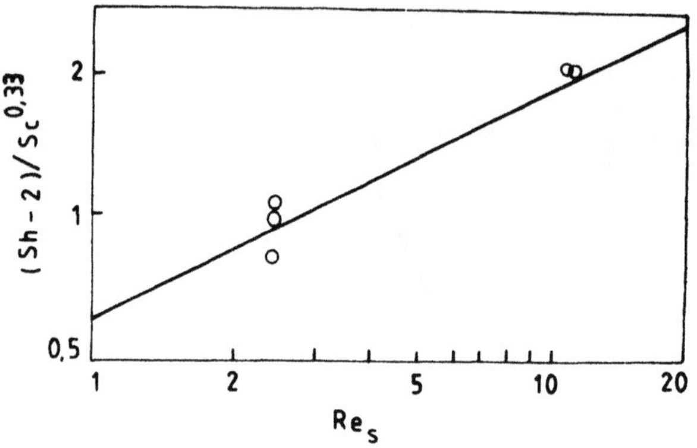

Fig.6. Slip velocity correlation for the liquid-solid mass-transfer coefficient.

The main part of the mass transfer studies was directed towards determination of the volumetric mass transfer coefficient of the gas-liquid interface (K_La).Results obtained by measurement of CO_2 absorption rates in water show that K_La changes inessentially with variation of mixing linear velocity nd_m or gas velocity U_G .A typical relationship of K_La and U_G by 0.2 m/s mixing velocity(line 1) and by nd_m=0.6 m/s (line 2) is shown in Fig.7 .It was found that the average mass transfer coefficient lies in the interval which is also characteristic of the mechanically agitated absorbers (between 0.004 - 0.007 1/s) [13] .At lower mixing velocities and higher gas velocities the coefficients obtained tend to approach higher values near to those characteristic for the bubble contactors(i.e. [15]).

The presence of solids in the reactor wasn't found to influence K_La substantially - at least for particles larger than 20-30 μm .In case of slurries containing smaller particles than these(i.e.10 μm particles of SiC), K_La was observed to increase up to 30 %, most likely because of mechanism mutations in the boundary layer [16] .

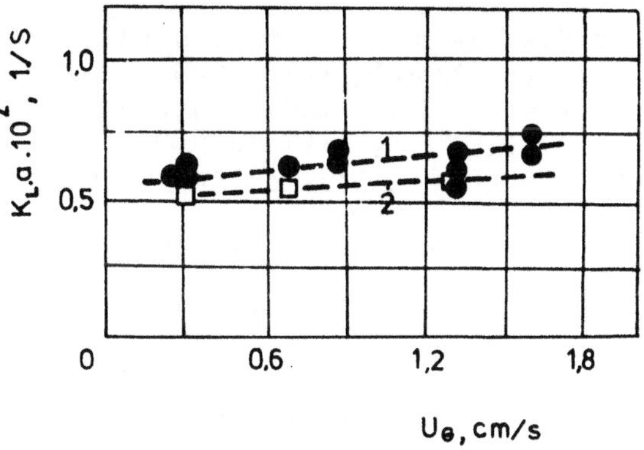

Fig. 7. The volumetric mass-transfer coefficient as a function of nominal gas velocity.

4 CONCLUSION

The performance characteristics of a modified rotating disc reactor has been studied in view of its eventual application for slurry reaction processes in practice. It has been found that the backmixing characteristic is favorable. Gas hold-up and mass transfer coefficients approach values known as characteristic for agitated bubble contactors.

REFERENCES

1. Reman, G.H. and R.B. Olney. Chem. Engng Prog. 51(1955)141-146.
2. Brauer, H. Ger. Chem. Eng. 3(1980)66.
3. Elenkov, D. and S.D. Vlaev. Commun. Dept Chem. Bulg. Acad. Sci. 10 (1977)414.
4. Elenkov, D. and S.D. Vlaev. Paper presented at CHISA´78(1978).
5. Misek, T. Coll. Czech. Chem. Comm. 40(1975)1686.
6. Stemerding, I.S. et al. Chem. Ing. Techn. 35(1963)344.
7. Miyauchi, T. et al. AIChE Jl 12(1966)509.
8. Ingham, J. in Hanson ed. Recent Advances in Liquid-Liquid Extraction. Moscow, 1974 p.162.
9. Vlaev, S.D., D. Elenkov and J. Nikov. Commun. Dept. Chem. Bulg. Acad. Sci. 12(1979)432.
10. Richardson, J.F. and W.N. Zaki. Trans. Inst. Chem. Eng. 32(1954)35
11. Brinkmann, H.J. J. Chem. Phys. 20(1952)571.
12. Roscoe, R. Brit. Appl. Phys. 3(1952)267.
13. Kramers, H. and K. Westerterp. Chemical Reactors (Moscow edtn, 1967)p.168.
14. Sherwood, Th., R. Pigford and Ch. Wilke. Mass Transfer (McGraw-Hill, 1975)p.221.
15. Hikita, H., S. Asai et al. Paper presented at CHISA´78(1978).
16. Deckwer, W.-D., E. Alper. Chem. Ing. Techn. 52(1980)219.

SLURRY REACTORS FOR COAL TECHNOLOGY

Y. T. Shah and J. Gopal

Chemical and Petroleum Engineering Department
University of Pittsburgh, Pittsburgh, PA 15261

ABSTRACT

In recent years, the use of coal as a raw material for the productions of hydrocarbons, liquid transportation fuels, chemical feedstocks and solid fuel is gaining importance. Three important processes for the achievement of this goal are: (1) direct coal liquefaction, (2) removal of sulfur from coal by oxydesulfurization and (3) indirect coal liquefaction or the Fischer-Tropsch synthesis. All of these processes employ three-phase slurry reactors. In this overview, a present state of the art for the models, scaleup, design and other operational problems associated with these processes are briefly evaluated.

1 INTRODUCTION

One of the major applications of the slurry reactor in recent years is in coal technology; in particular, coal refining and conversion processes. Three specific processes that are considered here are: (a) direct coal liquefaction (DCL) wherein coal is liquefied in the presence of a hydrogen donor solvent and hydrogen gas to produce a host of gaseous, (light hydrocarbons, water, CO, CO_2, NH_3, H_2S), liquid (of a wide boiling range components) and solid (unconverted coal, mineral matter etc.) products; (b) chemical cleaning of coal (CCC) via Pittsburgh Energy Technology Center process wherein the sulfur from coal is removed via oxidation in the presence of water; and, (c) Fischer-Tropsch slurry (FTS) process wherein carbon monoxide and hydrogen are reacted in the presence of a solvent and catalyst to produce a variety of light and heavy hydrocarbon products. This process utilizes the co-rich synthesis gas produced by second generation coal gasifiers.

Major advantages and disadvantages of a slurry reactor are discussed by Shah (1). In coal technology, slurry reactors are often preferred because of:

(a) high heat capacity due to high liquid holdup. This allows better mixing of heat which is desirable in all three exothermic processes described above. In direct coal liquefaction, good axial mixing is desirable because of the narrow temperature window for the smooth operation of the process. In Fischer-Tropsch process, the reactor temperature control appears to be easier in slurry bed operation than in fixed bed operation.

(b) high liquid holdup, which is desirable for the liquid phase reaction and which helps avoid coking during the reaction.

(c) in catalytic coal liquefaction (e.g. H-COAL Process) where four phases (gas, liquid, reactive coal and catalyst) are involved, slurry bed reactors appear to be the only logical alternative.

(d) one can change the catalyst effectiveness factor by suitably changing the particle size. The catalyst can be removed or added while the plant is in operation; in theory it is not necessary to shut down the plant for the purpose of replacing the spent catalyst. The packed bed reactor gets clogged easily and consequently the plant must be shut down for the catalyst to be regenerated.

(e) it is simple to construct and the reactor internals can be changed if so desired.

In the discussion of the application of a slurry reactor to each specific case the following topics will be considered:

(a) laboratory reactors

(b) models for the reactor

(c) scaleup and other design and operational problems.

2 LABORATORY SCALE SLURRY REACTORS

The major purpose of a laboratory scale slurry reactor used in coal technology is to evaluate the intrinsic kinetics of the process, free of extreneous mass, heat transfer and mixing effects. The most commonly used laboratory scale reactor is the continuous/batch agitated autoclave reactor. Kinetic studies for the direct coal liquefaction in such a reactor have been investigated by various workers and many modifications of the conventional agitated reactors have been proposed to overcome problems such as heating and cooling times, rapid coal-oil slurry injection, slurry sampling at short contact times, maintenance of constant pressure in the reactor, etc. A detailed analysis of different modifications of reactors has been given elsewhere (2), and it will not be repeated here. The kinetics of oxydesulfurization of coal has been studied either in batch or semi-batch agitated autoclave reactors using aqueous coal slurries and a comprehensive review on this subject is also available (3). This study has shown that one can use conventional agitated slurry reactors for the kinetic study of oxydesulfurization process. There is little information available on the use of an agitated slurry reactor for the FT synthesis. All the intrinsic kinetic rate measurements have been carried out in the vapor phase (4-9).

All three processes considered here are exothermic. Since slurry reactors on an industrial scale are operated under close to adiabatic conditions, a major scaleup problem is that of the thermal control of such reactors. Recently, Shah and Carr (10) have described a custom made agitated adiabatic slurry reactor, which can be used to evaluate the thermal behavior of the large scale reactors.

3 PILOT AND COMMERCIAL SCALE SLURRY REACTORS

Descriptions of typical slurry reactors used in the three processes considered here are available in the literature (4,5, 11,12). The operating conditions employed in these reactors are briefly outlined in Table 1. Some similarities in the operation of the three reactors can be noted. The reactors for both DCL and CCC processes are operated in a similar manner. In the FTS process, the liquid-solid (catalyst) slurry phase is usually stagnant. While the solid particle size in catalytic processes (FTS and catalytic DCL) can be large, in most practical situations, the solids are very fine and the slurry is considered to be pseudo-homogeneous. Slurry velocities for all the three processes are either small or none. In general, the gas velocities are in the same range; however, FTS uses somewhat smaller gas velocities as compared to DCL and CCC processes.

TABLE 1

COMPARISON OF OPERATING CONDITIONS IN SLURRY REACTORS USED IN COAL TECHNOLOGY

Process	Reactor	Reactor Dimensions (cm) D	Reactor Dimensions (cm) L	Solid Concen. (wt%)	Temp. T (K)	Press. P (MPa)	Slurry Space Veloc. (hr^{-1})	Gas Veloc. (cm/s)	Liquid Veloc. (cm/s)	Solids Particle Size (μm)
DCL	Coiled tube (preheater)	3-7	100-1000	33	523-673	10-15	10-20	50-100	5-21	< 70
	3 phase bubble column	30-60	200-700	33	693-738	10-17	0.5-1.0	2-4	0.08-0.4	.15-3
CCC	3 phase bubble column	2.2	183	20-26	433-503	6-8	0.5	30-120	0.06-0.5	50-1400
FTS	3 phase batch bubble column	3.8-155	70-860	up to 20	493-633	0.5-2.4	---	0.3-10	---	50

Solids play different roles in the different processes. In direct coal liquefaction, a part of the solid is dissolved in liquid (mainly in the preheater) and a part (i.e. mineral matter) may act as a catalyst for the hydrogenation reactions. In Fischer-Tropsch slurry processes, solids are catalysts. Finally, in chemical cleaning of coal, only a part of solid (i.e. sulfur) takes part in the reaction following the shrinking core diffusion/reaction mechanism. The role of solids in the design and scaleup of the reactors for the three processes is therefore different.

The fluid properties for the three processes may be considerably different. The CCC process uses a low pH solution (containing water and the acid product) whereas both DCL and FTS use hydrocarbon like liquids. Physical properties such as surface tension, viscosity, specific heat and thermal conductivity for these two cases may be quite different. Since the FTS is a very energetic process, the reactor may also contain some internals such as heat exchangers (cooling coils) etc. Some physical and thermal properties for these systems are outlined in Table 2.

One unique reactor is the preheater in the DCL process. It is a long coiled tube where the slurry is heated to the reactor temperature and as described later, for swelling coals, it may possess a peculiar thermo-hydraulic behavior. While, as described later in Table 5, available literature for various hydrodynamic, mixing, and transport parameters can be applicable to the reactors, their applicability to the preheater is questionable. More work is needed in understanding the hydrodynamic, mixing and transport characteristics of the preheater.

4 MODELS FOR THE SLURRY REACTORS

The design, scaleup and performance prediction of slurry reactors require models which must consider not only the hydrodynamic and mixing behavior of the three phases, but also the mass transfer between the phases along with the intrinsic kinetics. In the DCL and FTS processes, an axial dispersion model is applicable, with the solid phase assumed to follow sedimentation or dispersed flow model. However, in the CCC, where the solid particles take part in the reaction, dispersion model is no longer applicable. This case is evaluated with the use of the shrinking core model, wherein the solid concentration is governed by its exit age distribution.

The backmixing of the solid phase and solid concentration profile become more important when the solid loading is high (e.g. DCL process) or when the solid particles are large with significant relative slip velocity with the liquid phase. If the solid loading is low or particles are very fine, then the slurry can be assumed to be a single pseudo-homogeneous phase. In the following paragraphs, some important model features for each of the three cases are briefly outlined. A brief summary is also given in Table 3.

TABLE 2

SOME PHYSICAL AND THERMAL PROPERTIES OF SLURRIES IN COAL TECHNOLOGY

Property	DCL	CCC	FTS
Density (kg/m^3)	1270	1150	900
Viscosity (Pa·s)	(2-10)x10^{-3}	1.1x10^{-3}	4.5x10^{-3}
Surface tension (N/m)	0.025	0.070	0.024
Thermal conductivity (w/m K)	0.15	0.67	0.18
Heat capacity (J/kg K)	2700	3480	2927
Heat of reaction (kJ/mol·H_2)	50	----	290

4.1 Direct Coal Liquefaction (DCL)

The DCL process contains two steps; dissolution of coal in the preheater (accompanied by several fast reactions) and subsequent hydrogenation/hydrocracking reactions (slow reactions) in a three phase slurry reactor.

The preheater is essentially a plug flow reactor (L/D > 100) where the coal-solvent slurry and hydrogen gas are preheated to the liquefaction temperature. Extensive studies have been carried out to understand the exact nature of the processes, taking place in the preheater (2,11,13-16). Considerable work has also been performed to evaluate the dissolution process and its effect on thermal hydraulics of large-scale preheaters (13-16).

The processes occurring in the preheater have been conceptually divided into three regions. The first region is characterized by the length of the preheater used in heating the slurry with a decrease in the slurry viscosity until a local minimum in viscosity is reached. The slurry in this region behaves like a Newtonian fluid. The second region is characterized by a sharp increase in viscosity, corresponding to the region where the coal particles swell and form a gel with the solvent, exhibiting non-Newtonian behavior. The increase in viscosity and the pressure drop depends upon the shear rate applied on the slurry. The third region is characterized by disintegration and dissolution

TABLE 3

TYPES OF MODELS PROPOSED FOR SLURRY REACTORS IN COAL TECHNOLOGY

No.	Process	Model Type	Model Equation	Gas Phase Flow	Slurry Phase Behavior	Gas-Liquid Mass-Transfer	Overall Controlling Reaction Regime and Reference
1	DCL	Axial dispersion	Hydrogen mass balance	Plug flow	Partially* backmixed	Negligible	Chemical reaction (23)
		Axial dispersion	Hydrogen mass balance	Partially Backmixed	Partially backmixed	Considered	Chemical reaction (22)
		Axial dispersion	Hydrogen mass and energy balance	Plug-flow	Partially* backmixed	Absent	Chemical reaction (20)
		Axial dispersion	Hydrogen mass and energy balance	Plug-flow	Partially* backmixed	Present	Depends on the turbulence in the reactor (21)
		Axial dispersion	Mass balance for various lumped fractions	Plug-flow	Partially* backmixed	Absent	Chemical reaction (17)
2	CCC	Shrinking core model	Sulfur balance	Plug-flow	Partially* backmixed	Absent	Diffusion controlled (29)
		Shrinking core model	Sulfur balance	Plug-flow	Partially* backmixed	Absent	(a) liquid-solid mass transfer (b) ash diffusion (c) chemical reaction (30)

TABLE 3
(Concluded)

3	FTS	Plug flow/ backmixed	Hydrogen mass balance	Plug-flow	Completely backmixed	Considered	Depends on the operating temperature (32)
		Plug flow/ backmixed	Hydrogen mass balance	Plug-flow	Completely backmixed	Negligible	Chemical reaction (33)
		Dispersion	Hydrogen mass and heat balance	Partially backmixed	Partially* backmixed	Negligible	Chemical reaction (34)

*Dispersion/sedimentation model used for solid concentration profile; others assume slurry as homogeneous phase.

of the coal particles into smaller molecular weight fractions and possibly some chemical reactions which give various products found in the liquefaction process. This region is also accompanied by a decline in the slurry viscosity and the average slurry density.

The preheater models are based on plug-flow behavior of both gas and the slurry phases, as the L/D ratios in preheater are usually large. Reliable estimates of the fluid properties such as viscosity and density, pressure drop across the preheater and heat transfer coefficient are needed for an optimum design of the preheater and, these have been recently reviewed by Shah (11). Parulekar et al. (17) have proposed a kinetic model for the preheater based on certain fast reactions taking place in the preheater whereas Nunez et al. (18) evaluated the hydrogen mass balance and a heat balance on the preheater.

The three phase slurry-bubble column-reactors used in the DCL process have been modeled by using an axial dispersion model. The height to diameter ratio employed in such reactors is usually in the range of 5-20 and the axial dispersion model is applicable to both liquid and solid components in the slurry phase. Since the average coal particle size in the reactor is usually believed to be less than 5 μm (19,22), liquid and solid phases are believed to form a homogeneous slurry phase. The gas phase has been assumed to be the plug flow in the models proposed by some workers (17,20,21) while others considered the gas phase to be partially backmixed (22). Several lumped parameter kinetic models have been proposed for the DCL process and their details are given elsewhere (2,11).

An axial dispersion model for an isothermal reactor with partially backmixed liquid phase and the gas phase as plug-flow has been described by several investigators (23,24). The overall reaction rates were expressed in terms of the gas phase hydrogen concentration thereby eliminating any gas-liquid mass transfer resistance for the hydrogen transfer. The solids distribution was accounted for by the hindered-settling conditions. In the range of operating variables examined, the predictions from the above model qualitatively agreed with the performance of several pilot-scale plants. Lee et al. (22) proposed a dispersion model based on the axial-dispersed flow for both gas and liquid phases. The slurry in their case was treated as a homogeneous phase and the solid distribution was not taken into account. The model parameters were estimated from the correlations proposed for the case of no solid suspension (two-phase system). The coal dissolution, hydrogenation and hydrodesulfurization were considered as the key reactions, in the model. Also, the effect of mass transfer on the liquefaction process has been investigated. The model predictions were found to be in good agreement with the experimental facts of the Wilsonville pilot plant. Their analysis also indicated that the Wilsonville pilot plant operated in the kinetically, rather than the mass transfer controlled regime. Unlike the models proposed by other investigators (23,24) where the velocity variation was also taken into account, the above model assumed that all velocities and holdups remain unchanged throughout the reactor.

Commercial reactors are invariably operated under close to adiabatic conditions and for understanding the thermal behavior of such reactors, an energy balance along with the mass balance needs to be considered. Models have been developed (20,21) to investigate the thermal behavior of such reactors, based on the axial dispersion model. The model equations assumed that the heat generation depends on the hydrogen consumption; the hydrogenation/hydrocracking reactions are believed to be the most energetic of all the reactions taking place in the reactor. The hydrogen mass balance assumed the presence as well as the absence of the gas-liquid mass-transfer resistance. The hydrogen consumption and temperature rise predicted for the case where there is no mass transfer resistance was shown to fit the experimental data measured in the Fort Lewis SRC-II pilot plant very well (21).

4.2 Chemical Cleaning of Coal

Slurry reactors are used in the chemical cleaning process, wherein, the sulfur in coal is oxidized by air in an aqueous slurry. Unlike other two cases considered here, in this process the solids react keeping the size constant. In general, for such a type of reacting system, the resistance to the overall reaction could be the oxygen mass transfer at the gas-liquid and/or, liquid-solid interface, oxygen diffusion through the product (ash formed during the reaction) layers, chemical reaction or a combination of the above resistances. Mathematical models for such cases have been proposed by various investigators based on the shrinking core mechanism (25-29). Unlike various applications where the solids act as a catalyst in the slurry reactor, in chemical cleaning of coal, we come across a situation where the solids take part in the reaction; the usual dispersion model is not applicable and a model based on the exit age distribution of the solid particles has been developed.

Ruether (29) examined the case of oxydesulfurization for completely backmixed stirred tanks in series assuming the diffusion-controlled mechanism. The reaction in the particles was described by the shrinking core model. The results obtained on the conversion as a function of residence time were shown for various number of reactors in series. The procedure to calculate the conversion for a system having a distributed particle size has also been discussed (29).

Joshi et al. (27) investigated the kinetics of oxydesulfurization of coal assuming two alternative mechanisms, (a) continuous reaction model, assuming fine pyrite particles to be uniformly distributed in the coal particles and (b) shrinking core model where pyrite particles are assumed to be free and separate from the coal particles. The applicability of the above models were examined (25) by studying the effect of coal particle size and solid loading. It was shown that the rate controlling step in the pyrite oxidation is the intrinsic chemical reaction between the dissolved oxygen and pyrite particles. The rate constants were evaluated using the shrinking core model.

Joshi et al. (30) proposed reactor models based on the shrinking core mechanism. Since the particles take part in the reaction their role was evaluated based on the residence time distribution. For extremely fine pyrite particles, (< 100 μm), it has been shown (31) that the RTD of the solid and liquid phases can be assumed to be identical and the RTD of the solid phase is given by the diffusion-sedimentation model. Various rate controlling steps that were considered are: (1) gas-liquid mass transfer; (2) liquid-solid mass transfer; (3) ash diffusion; (4) chemical reaction; and, (5) intraparticle diffusional resistance (for particles encased in the coal matrix).

Experimental studies on the oxydesulfurization process have shown that in the practical range of operations the gas-liquid and liquid-solid mass transfer resistances are negligible (25-27). Further, it has been shown (25) that more than 85 wt% of pyrite exists in the liberated form and the intraparticle diffusional resistance can be ignored. Joshi et al. (25) also investigated the effect of particle size in the range (< 72-1410 μm) and found that the reaction is kinetically controlled.

Joshi et al. (30) developed an isothermal model for the reactor, considering the steps (3) and (4) and predicted results of conversion as a function of dimensionless residence time. It was observed that the reaction/dispersion model based on the shrinking core mechaism, with chemical reaction as the rate controlling step gave good agreement with the experimental results.

4.3 FT Synthesis

In recent years, FT synthesis has been gaining importance for the manufacture of transportation fuels. There is very little information available on the kinetics, modeling and design of a slurry reactor for the FT process. The intrinsic kinetic rate measurements have been carried out in the vapor phase. However, under certain conditions, various investigators (4-7) have suggested that the rate is independent of the CO partial pressure, and since the water content is generally low due to the water gas shift reaction, the rate is expressed only in terms of the H_2 consumption. The simplified rate expression has been used in the analysis of the FT synthesis in slurry reactors (32,33,34).

Satterfield and Huff (32) have developed a model based on the plug flow of the gas phase and completely backmixed liquid phase and the model equation is based on the summation of mass transfer and reaction resistances. These authors analyzed the data of Schlesinger et al. (35) and the pilot plant scale data of Farley and Ray (36) and concluded that the overall rate is equally influenced by mass transfer and reaction resistances at normal operating temperatures (around 503 K) and the mass transfer resistance becomes increasingly more important at higher temperatures.

Deckwer et al. (33) proposed a model for a laboratory scale slurry reactor which is essentially the same as that proposed by Satterfield and Huff (32) with the only difference that their model accounts for the contraction in the gas volume. The experimental data of various investigators were analyzed and the estimated rate constants were correlated to the iron content in the slurry which is believed to be the intrinsic catalytic component. The authors concluded that the FT synthesis is predominantly controlled by the chemical reaction provided the reactor is operated at the relevant industrial conditions and that mass transfer limitations could be important only at very low gas velocities, high catalytic concentrations and for very active catalysts.

In the design of industrial scale slurry reactors agitated by the sparging of gas, one deals with large diameter columns with length to diameter ratio usually in the range of 5 to 20. In these cases, the backmixing of all three phases may be important and a dispersion model for the reactor, considering only an overall kinetic and variable gas flow rate has also been developed by Deckwer et al. (34). The results of their computations are briefly summarized in Table 4. As mentioned above, there is a difference in the conclusions of Satterfield and Huff (32) and Deckwer et al. (33) about the mass transfer limitation. The latter authors reduced the interfacial area by 50% and the space time yield reduced by less than 4%. In addition, the values of gas holdup was reduced by 50% (with a corresponding reduction in the interfacial area by 50%) and this resulted in an increase in the liquid volume and consequently an increase in the conversion and space time yield. This led them to conclude that the FT synthesis is largely controlled by chemical reaction and significant mass transfer limitations can be expected only at very low gas velocities or if the catalyst activity and/or concentration are increased. The use of very high catalyst concentration is however not recommended as this could prove detrimental to the favorable hydrodynamic conditions (31,37). Also, the computations indicated that the space time yield runs through a maximum value depending on the gas velocity and this optimum gas velocity agreed well with that used in the Rheinpreussen-Koppers demonstration plant. Other results of the simulations were also found to be in accordance with practical experience. They have presented a diagram of space time yield and conversion as a function of the reactor diameter and length from which the design and estimation of the production capacity of a FT slurry reactor can be made.

TABLE 4

EFFECT OF OPERATING VARIABLES ON THE CONVERSION AND SPACE TIME YIELD IN FT SLURRY REACTORS

Increase in Variable	STY	X_{CO+H_2}
Pressure (0.5-3 MPa)	Increases	No influence
Column dia. (1-5 m)	---	Slight decrease
Particle dia. (25-200 μm)	No influence	Slight decrease
Sup. gas vel. (0.5-12 cm/s)	Goes through a maximum	Slight decrease

4.4 Estimation of Model Parameters

In the modeling and design of reactors, reliable estimates of the various model parameters such as holdup of the three phases, mass and heat transfer coefficients, physical and thermal properties, etc. are required. The values of these parameters depend on the prevailing flow regime, operating conditions and the type of the reactor internals used (if any) in the reactor.

There are as yet no theoretical correlations capable of predicting the viscosity, pressure drop and heat transfer coefficient in the preheater. Some empirical correlations for this purpose are available in the literature (11,13-16,38). The hydrodynamic characteristics of three phase slurry reactors have been extensively reviewed (1,39,40,41). Suitable correlations have been suggested for the estimation of model parameters encountered in the three processes discussed here (11,22,30,34). Recently, the physical and thermal properties of coal liquids have been investigated and correlations have been given, which should be useful in the estimation of the model parameters for the DCL process (42). Deckwer et al. (43) have investigated the hydrodynamic properties of FT slurry process at the relevant operating conditions and correlations for a number of model parameters have been suggested (34,43). Table 5 summarizes some of the correlations useful for the model parameter estimations for the three processes considered here.

5 SCALEUP AND OPERATIONAL PROBLEMS

Various factors should be considered during the scaleup of slurry reactors such as flow regime, backmixing in the different phases, temperature control, controlling regime of the overall reaction, etc. Details of the effects of various factors on scaleup are available in the literature (1,11,21,30,34). In this section, some of the factors which influence the scaleup of slurry reactors as applied to coal technology are briefly mentioned. Table 6 summarizes some of the important scaleup factors.

In all three processes, studies on laboratory or pilot plant scale have shown the overall reaction is generally kinetically controlled. However, under the prevailing flow regimes in the commercial reactor, the mass transfer effects may be prominent and should be considered. The thermal behavior of the DCL and FTS reactors is an important factor which needs special attention. To avoid loss in product selectivity, low product yields, and coking or repolymerization reactions, a thorough temperature control is needed either by means of quench (DCL process) or cooling coils (FTS process). The solid phase backmixing in the CCC process has to be particularly considered as the conversion is expressed in terms of the particle exit age distribution. All three processes are operated under severe conditions and proper selection of the reactor diameter and the height to diameter ratio is necessary for an economical design and scaleup. The material of construction or proper reactor lining material is important, particularly in the CCC process as sulfuric acid is produced in reaction process.

6 SUMMARY

The salient features of the slurry reactors used in three coal refining and conversion processes and the models developed for each case have been discussed. Considerable pilot plant

TABLE 5

TYPICAL CORRELATIONS USED FOR THE ESTIMATION OF MODEL AND DESIGN PARAMETERS (11,30,34,43)

Parameter	Correlation	Process
Dispersion coefficient	$E_L = 0.00108\ D^{1.4} u_G^{0.3}$ (SI units)	DCL, FTS
	$E_S = E_{SL} = 0.38\ D^{1.33}\ g(u_G - \varepsilon_g V_{b\infty})^{0.33}$ (SI units)	CCC
Heat transfer coefficient	$St = 0.1\ (Re\ Fr\ Pr^2)^{1/4}$	FTS
Gas-liquid mass transfer coefficient	$k_L Sc^{2/3} = 0.31\ (\frac{\mu_L \Delta\rho g}{\rho_L^2})^{1/3}$	DCL, FTS
Solid-liquid mass transfer coefficient	$Sh = 2 + 0.545\ Sc^{0.33}\ Re^{0.264}$	FTS
Liquid-solid Interfacial area	$a_s = \frac{6(1-\varepsilon_G)\rho_{SL}}{d_s \rho_s} C_s$ (cm^{-1})	FTS
Gas-liquid interfacial area	$a = 4.5\ u_G^{1.1}$ (cm^{-1})	FTS

TABLE 6

POSSIBLE SCALEUP PROBLEMS IN SLURRY REACTORS IN COAL TECHNOLOGY

Operating Variable	DCL	CCC	FTS
gas velocity	Not important	Not important	Important
gas-liquid mass transfer resistance	Should be low	Not important	Not important (in the normal range of operation)
backmixing			
- gas phase	Not important	Not important	Important
- liquid phase	Not important	Not important	Not important
- solid phase	Not important	Important	Not important
distributor design	Important	Important	Important
heating/cooling	Important	Not important	Important
multiple temperature steady states	Important	----	Important
- reactor startup and shutdown problems			
material of construction	Important	Critical	Not important

scale data are available for the DCL process, while for the CCC and FTS processes, data for reliable design and performance prediction of large scale slurry reactors are lacking. All the processes follow complex reaction mechanisms and the models developed so far assume overall kinetic expressions. A better understanding of the kinetics is needed for the reliable design of these reactors.

NOMENCLATURE

a	gas-liquid interfacial area
a_s	liquid-solid interfacial area
c_s	concentration of solid in the slurry
d_s	solid particle diameter
D	reactor diameter
E	dispersion coefficient
Fr	Froude number
g	acceleration due to gravity
k_L	gas-liquid mass transfer coefficient
L	reactor length
P	pressure
Pr	Prandtl number
Re	Reynolds number
Sc	Schmidt number
St	Stanton number
T	temperature
u_G	superficial gas velocity
$V_{b\infty}$	average terminal rise velocity of bubbles
X	fractional conversion
ε	fractional holdup
ρ	density

Subscripts

g	gas phase
L	liquid phase
S	solid
SL	slurry

REFERENCES

1. Shah, Y.T. "Gas-Liquid-Solid Reactor Design." (McGraw-Hill, New York, 1979).

2. Shah, Y.T., P.C. Singh and A. Calimli. "Direct Coal Liquefaction." a paper presented at NATO-ASI School, Izmir, Turkey (August 1981).

3. Shah, Y.T. and R.S. Albal. "Chemical Cleaning of Coal-The Oxydesulfurization Process." a paper presented at NATO-ASI School, Izmir, Turkey (August 1981).

4. Kolbel, H. and M. Ralek. "The Fischer-Tropsch Synthesis in the Liquid Phase." Catal. Rev. Sci. Eng. 21(2) (1980) 225-274.

5. Baird, M.J., R.R. Schehl, W.P. Haynes and J.T. Cobb, Jr. "Fischer-Tropsch Processes Investigated at the Pittsburgh Energy Technology Center Since 1944." Ind. Eng. Chem. Prod. Res. Dev. 19 (1980) 175-191.

6. Anderson, R.B. "Catalysis. Vol. 4." Ed. P. Emmett (Reinhold: New York, 1956).

7. Dry, M.E. "Advances in Fischer-Tropsch Chemistry." Ind. Eng. Chem. Prod. Res. Dev. 15 (1976) 282-286.

8. Atwood, H.E. and C.O. Bennett. "Kinetics of Fischer-Tropsch Reaction Over Iron." Ind. Eng. Chem. Process Des. Dev. 18 (1979) 163-170.

9. Zeen el Deen, A., J. Jacobs and M. Baerns. Ger. Chem. Eng. 2 (1979) 139.

10. Shah, Y.T. and N.L. Carr. "New Experimental Techniques for High Temperature High Pressure Gas-Liquid-Solid Reactors." a paper to be presented at Fifth World Congress, Montreal Canada (Oct. 1981).

11. Shah, Y.T. "Reaction Engineering in Direct Coal Liquefaction." (Addison-Wesley Publishing Co., Reading, MA, 1981).

12. Warzinski, R.P., J.A. Ruether, S. Friedman, and F.W. Steffgen. "Survey of Coals Treated by Oxydesulfurization." Proc. Symp. on Coal Cleaning to Achieve Energy and Environmental Goals." Vol. II (Hollywood, FL, Sept. 1978) 1016-1038.

13. Wen, C.Y. and K.W. Han. "Optimization Studies of Various Coal Conversion Systems: Coal Dissolution Phenomena." Report No. FE-2274-7 (1979), Univ. of W. Virginia (DOE Conatract No. EX-76-C-01-2274).

14. Weber, W.H. and A. Basu. "Coal Dissolution Studies Utilizing the Slurry Preheater at the Wilsonville SRC Pilot Plant." (Coal Liquefaction Preheater Studies, ORNL, 1979).

15. Traeger, R.K., T.C. Bickel and R.M. Curlee. "Preheater Studies in Coal Liquefaction." Annual Report No. SAND-79-0150 (Sandia Laboratories, May 1979).

16. U.S. DOE. "Coal Liquefaction Preheater Studies." Proc. of DOE Project Review Meeting (Oak Ridge, TN, March 21, 1979).

17. Parulekar, S.J., Y.T. Shah and N.L. Carr. "A Comprehensive Isothermal Model for SRC Liquefaction." a paper submitted to Ind. Eng. Chem. Proc. Des. Dev. (1981).

18. Nunez, P., A. Calimli, J. Abichandani and Y.T. Shah. "Multiple Steady States in an Adiabatic Coal Liquefaction Reactor-Role of Preheater." a paper submitted to Chem. Eng. Commun. (1981).

19. Guin, J., A. Tarrer, L. Taylor, Jr., J. Prather and S. Green, Jr. "Mechanisms of Coal Particle Dissolution." Ind. Eng. Chem. Process Des. Dev. 15 (1976) 490-494.

20. Shah, Y.T. and S.J. Parulekar. "Modeling and Simulation of the Thermal Behavior of SRC-II Reactors." Chemicals and Minerals Division Report No. 627RK017 (Gulf R&D, October 3, 1979).

21. Parulekar, S.J. and Y.T. Shah. "Steady State Thermal Behavior of an Adiabatic Three Phase Fluidized Bed Reactor-Coal Liquefaction Under Slow Hydrogen Consumption Reaction Regime." Chem. Eng. J. (1981). in press.

22. Lee, M.H., J.A. Guin and A.R. Tarrer. "A Dispersion Model for the Solvent Refined Coal Process." Ind. Eng. Chem. Process Des. Dev. 17 (1978) 127-135.

23. Parulekar, S.J. "Dynamics of Three Phase Fluidized Bed Reactors." M.S. Thesis, University of Pittsburgh, Pittsburgh, PA (1979).

24. Parulekar, S.J. and Y.T. Shah. "Steady State Behavior of Gas-Liquid-Solid Fluidized Bed Reactors." Chem. Eng. J. 20 (1980) 21-33.

25. Joshi, J.B., Y.T. Shah, J.A. Ruether and H.J. Ritz. "Particle Size Effects in Oxidation of Pyrite in Air/Water Chemical Coal Cleaning." 73rd AIChE Annual Meeting (Chicago, 1980).

26. Joshi, J.B., Y.T. Shah, R.S. Albal, H.J. Ritz and W.D. Riche. "Effect of pH on the Removal of Pyrite Sulfur from Coal by Oxydesulfurization." a paper submitted to Ind. Eng. Chem. Proc. Des. Dev. (1981).

27. Slagle, D., Y.T. Shah and J.B. Joshi. "Kinetics of Oxydesulfurization of Upper Freeport Coal." Ind. Eng. Chem. P ocess Des. Dev. 19 (1980) 294-300.

28. Chuang, K.C., M.C. Chem., R.T. Greer, R. Markuszewski, Y. Sun and T.D. Wheelock. "Pyrite Oxidation by Wet Oxidation in Alkaline Solutions." Chem. Eng. Commun. 7 (1980) 79-94.

29. Ruether, J.A. "Reaction in a Cascade of Continuous Stirred Tank Reactors of Particles Following the Shrinking Core Model." Can. J. Chem. Eng. 57 (1979) 242-245.

30. Joshi, J.B., J.G. Abichandani, Y.T. Shah, J.A. Ruether and H.J. Ritz. "Modeling of Three Phase Reactors: A Case of Oxydesulfurization of Coal." AIChE J. (in press).

31. Kato, Y., A. Nishiwaki, T. Fukuda and S. Tanaka. "The Behavior of Suspended Solid Particles and Liquid in Bubble Columns." J. Chem. Eng. Japan 5 (1972) 112-118.

32. Satterfield, C.N. and G.A. Huff. "Effects of Mass Transfer on Fischer-Tropsch Synthesis in Slurry Reactors." Chem. Eng. Sci. 35 (1980) 195-202.

33. Deckwer, W.D., Y. Serpeman, M. Ralek and B. Schmidt. "On the Relevance of Mass Transfer Limitations in the Fischer-Tropsch Slurry Process." Chem. Eng. Sci. 36 (1981) 765-771.

34. Deckwer, W.D., Y. Serpeman, M. Ralek and B. Schmidt. "Fischer Tropsch Synthesis in Slurry Phase: Analysis of Performance Data and Design Aspects." an unpublished work (1981).

35. Schlesinger, M., H. Benson, E. Murphy and H.H. Storch. "Chemicals from the Fischer-Tropsch Synthesis." Ind. Eng. Chem. 46 (1954) 1322-1326.

36. Farley, R. and D. Ray. "The Design and Operation of a Pilot Scale Plant for Hydrocarbon Synthesis in the Slurry Phase." J. Inst. Petr. 50 (1964) 27-46.

37. Joosten, G.E.H., J.G.M. Schilder and J.J. Janssen. "The Influence of Suspended Solid Material on the Gas-Liquid Mass Transfer in Stirred Gas-Liquid Contactors." Chem. Eng. Sci. 32 (1977) 563-566.

38. Thomas, M.G. and B. Granoff. "Coal Derived Product Effects on Viscosity." Fuel 57 (1978) 122-123.

39. Vasalos, I.A., E.M. Bild, D.F. Tatterson and C.C. Wallin. "H-COAL Fluid Dynamics Topical Report." Report No. FE-2588-6, Amoco Oil (U.S. DOE Contract No. EF-77-C-01-2588, May 3, 1978).

40. Shah, Y.T. and W.D. Deckwer, in "Scaleup in the Chemical Process Industries." Ed. by R. Kabel and A.J. Bisio (Wiely and Sons, New York, in press).

41. Shah, Y.T., G.J. Stiegel and M.M. Sharma. "Backmixing in Gas-Liquid Reactors." AIChE J. 24 (1978) 369-400.

42. Gray, J.A., C.J. Brady, J.R. Cunningham, J.R. Freeman and G.M. Wilson. "Selected Physical, Chemical and Thermodynamic Properties of Narrow Boiling Range Coal Liquids." a paper to be presented at the AIChE Annual Meeting, New Orleans, LA (Nov. 1981).

43. Deckwer, W.D., Y. Louisi, A. Zaidi and M. Ralek. "Hydrodynamic Properties of the Fischer-Tropsch Slurry Process." Ind. Eng. Chem. Process. Des. Dev. 19 (1980) 699-708.

COAL LIQUEFACTION VIA INDIRECT ROUTES

Wolf-Dieter Deckwer

Institut für Technische Chemie
Universität Hannover (TH)
Callinstrasse 3, D-3000 Hannover 1

1 GENERAL ASPECTS OF INDIRECT COAL LIQUEFACTION

There are two principal routes to convert coal into chemicals and transportation fuels. In direct liquefaction processes the H to C ratio is raised by hydrogenating coal under preservation of its molecular structure as much as possible. The hydrogen is required to reduce the molecular weight of the coal components and to remove heteroatoms. Indirect liquefaction means to completely gasify the coal and to convert the obtained synthesis gas in various liquid products. At the present time, it is still difficult to make a meaningful economic comparison of the various processes and to discriminate between rival routes. Recent developments as well as engineering and environmental constraints suggest that complete degradation of coal via gasification could form the basis of a coal conversion technology which might be superior to direct liquefaction. Therefore, one can suspect, at least, that a future synfuel industry will be a combination of various liquefaction methods.

1.1 Coal Gasification

Gasification determines to a large extent the cost of indirect liquefaction. Also, the thermal efficiency of gasification contributes largely to the overall thermal efficiency of the entire process. The thermal efficiency is defined as the lower heating value of the products divided by the lower heating value of the feed coal including the coal for utilities. For the gasifiers fully commercialized nowadays, i.e. Lurgi (dry ash), Koppers-Totzek and Winkler, the thermal efficiency is usually in the range of 60 % or even less. In a comprehensive study Shinnar and Kuo (1) have shown from basic scientific and engineering principles that considerable thermal efficiency advantages can be achieved if gasifiers are operated under elevated pressures and at low steam to oxygen ratios giving a synthesis of low H_2 to CO ratio. According to Shinnar and Kuo it is particularly the British Gas Corporation-Lurgi slagging gasifier which conforms closely to the basic requirement and could produce synthesis gas at a low production cost relative to other gasifiers such as the dry ash Lurgi gasifier. In addition, the Texaco gasifier and the high pressure versions of Koppers-Totzek and Winkler should be considered potential second generation gasifiers which fulfill the basic conditions established by Shinnar and Kuo. From Fig. 1, which was adapted from these authors, one can tell that the thermal efficiency of the BGC-Lurgi slagger can be as high as about 80 %.

It should also be pointed out that gasifiers operated at steam to oxygen ratios as low as possible provide the thermally most efficient and lowest cost route to the production of medium BTU gas (1). The co-production of pipeline gas in the gasification step improves the overall thermal efficiency of various liquefaction processes (Fischer-Tropsch, Methanol, Mobils MTG route) by 13 to 18 % (2).

Synthesis gases of low H_2 to CO ratio, i.e. about 0.5 as produced by second generation gasifiers have to be matched with conversion processes that can use such low ratios. Usually, processes based on synthesis gas conversion use gases of higher H_2 to CO ratios which need additional reaction of CO with steam (water -gas shift reaction). There are, however, some processes which can use syngases of a low H_2 to CO ratio requiring only little shift from the 0.5 H_2/CO ratio obtained from high efficiency gasifiers. Examples are the Fischer-Tropsch synthesis in slurry phase using gases of H_2 to CO ratios of 0.6 to 0.7 and the production of dimethylether (H_2/CO = 1) which can be processed to high octane gasoline.

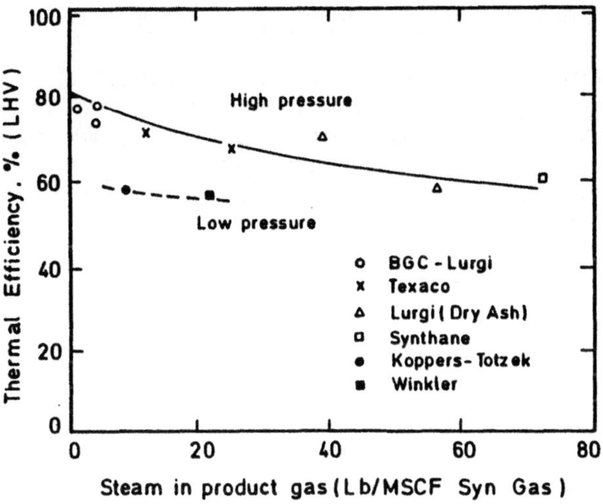

Fig. 1: Thermal efficiency as a function of excess steam in product gas (Shinnar and Kuo, 1979)

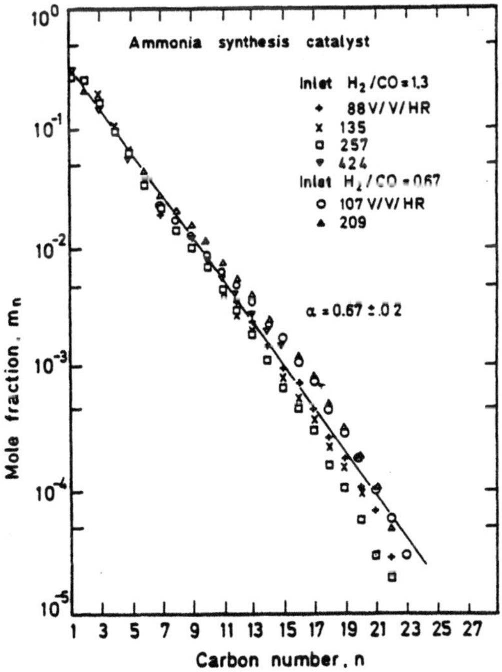

Fig. 2: Schulz-Flory plot of FT products from slurry reactor (Satterfield and Huff, 1981)

1.2 Fischer-Tropsch Synthesis (FTS)

1.2.1 General aspects of FT chemistry.

Stoichiometry: From the chemical point of view, the FT synthesis is the hydrogenation of CO which, however, does not yield one definite product but a variety of compounds. The main reactions can be summarized by the subsequent stoichiometric equation.

$$nCO + 2nH_2 \longrightarrow -(CH_2)_n- + nH_2O \qquad (1)$$

If the synthesis is carried out on iron catalysts reaction is most easily followed by conversion of water

$$CO + H_2O \longrightarrow H_2 + CO_2 \qquad (2)$$

Thus, by combining eqs. (1) and (2) the overall reaction on Fe can be written as

$$2nCO + nH_2 \longrightarrow -(CH_2)- + nCO_2 \qquad (3)$$

It should be noted that the conversion reaction, i.e. eq. (2), can reduce the hydrogen demand from $H_2/CO = 2$ in reaction (1) to $H_2/CO = 1/2$ in reaction (3). Thus, on Fe catalysts, the use of weak syngases, i.e., with a low hydrogen content, is possible. An important process parameter which characterizes the stoichiometry is the usage ratio. This presents the moles of H_2 needed to convert 1 mol CO.

Catalysts: Iron and cobalt are the principal catalysts of the FTS. Ruthenium also has catalytic activity, and manganese is claimed to be an FT catalyst as well (3). Iron as a catalyst is of utmost technical importance. It is believed that the action of the reactants on the metallic component leads to changes in composition and structure which present the active catalyst phase. An important role in catalytic activity play promoters which can be divided in structure influencing and energetic promoters. The most important energetic promoter of iron catalysts is K_2CO_3. Its exact dosage (usually less than 1 % relative to Fe) is very important as it promotes formation of longer chains, olefins and oxygenates but simultaneously leads to coke deposition. Summaries on conventional FT catalysts can be found in refs. (4,5,6).

In the second wave of FT research after the oil embargo in 1973, research activities are mainly focused on the development of new and modified catalyst systems.

Considerable success has been achieved in this area, and some aspects will be discussed later.

Products: The main products are unbranched paraffins and olefins. The molecular weight ranges from 16, i.e. methane, to above 20,000 depending on the catalyst, the operational conditions and the kind of process. The olefin content of the produced hydrocarbons may vary from 10 to 90 %. As by-products aromates and, above all, oxytenates, i.e. alcohols, ketones, acids, and esters, are formed. Under certain conditions and using nitrided fused Fe catalysts, the oxygenates content may be as high as 49 % of the C_{3+} fraction (43). It should be pointed out that, in contrast to direct liquefaction products, FT hydrocarbons are essentially nitrogen and sulfur free.

The major interest in FT synthesis is to manufacture automotive fuels, of course. However, as the straight chain α-olefin content of the overall olefin fraction may be as high as 85 % FT products are also considered starting materials for further chemical processing as, for instance, alcohols via oxo-synthesis, tensides by alkylation of aromates, and polymers from short chain olefins.

Selectivity: Two main mechanisms of the FT hydrocarbon synthesis are discussed today (5,7-12), however, the entire matter is still a subject of active research. Nevertheless, it is certain now that the FTS follows a polymerization type process. The propagation and termination of the hydrocarbon chain gives the most probable distribution of Schulz and Flory (12). If α denotes the chain growing probability defined by

$$\alpha = \frac{r_p}{r_p + r_t} \qquad (4)$$

where r_p and r_t are the rates of the propagation and termination steps then the Schulz-Flory distribution can be expressed by eq. (2)

$$x_i = i\alpha^{i-1}(1-\alpha)^2 \qquad (5)$$

This equation predicts a product distribution wich is the broader the higher the mean degree of polymerization. Indeed, this unselectivity is an intrinsic feature of the FT process if carried out on common catalysts and at high conversion levels.

Recently, Satterfield and Huff (13) determined car fully product distributions of the FTS in a slurry well mixed reactor and found striking agreement with Schulz-Flory distribution. As shown in Fig. 2 the product slat obtained under different conditions for a conventional Fe based ammonia catalyst give a chain growing probability α of about 0.67. The same value of α was reported for a Mn/Fe catalyst which was studied in fixed bed and slurry phase (14,15). Deviations from Schulz-Fbory plot are often reported in literature. But in the case of conventional catalysts such deviations are probably caused by insufficient analytical techniques and nonconsideration of oxygenates (13).

A new field of synthesis gas chemistry has been opened with the development of selective catalyst (3,16-26). Selective catalysts can produce rather narro hydrocarbon fractions. As an example, Fig. 3 presents the product distribution obtained from 1 % Ru/Al_2O_3 catalyst in fixed bed (19). There is a striking discrepancy between the experimental data (A) and the computed distribution (B) maximized for C_{10} hydrocarbons. 99 wt. % of the observed product is between C_5 and C_{25} with negligible gas and wax. Non-Schulz-Flory distributions are also obtained from synthesis gas conversion catalys comprising a CO reduction function (Fe) combined with zeolites of Y or ZSM-5 type. These bifunctional catalys are either impregnated zeolites (21,24-26) or simple physical mixtures of conventional FT catalysts and ZSM- zeolite (22,23,27). Owing to the pore sizes of the ZSM- are shape selective and this restricts the chain length to C_{11}. Though these bifunctional shape selective catalyst systems were not yet tested under industrial conditions and on a larger scale one can assume that with their application the major disadvantage of the classical FTS, i.e., the insufficient selectivity has been overcome.

Kinetics: On the basis of simplified assumptions severa investigators (7,11,28,29) proposed the followin rate expression for the vapor phase reaction on Fe catalysts

$$r = k_o \frac{p_{CO}\, p_{H_2}}{p_{CO} + A\, p_{H_2O}} \tag{6}$$

A simple derivation of this equation was given by Dry (11). Due to the water-gas shift reaction, the water concentration is usually low and, in addition, the syn-

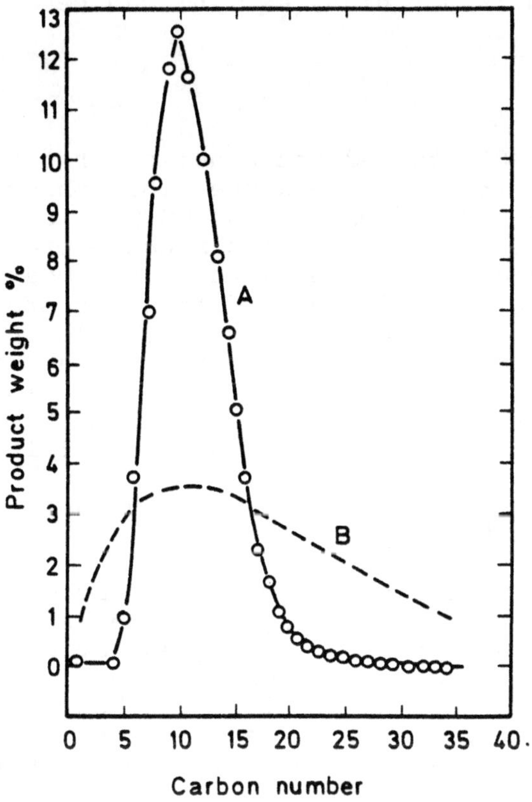

Fig. 3:
Non-Schulz-Flory distribution on Ru/Al_2O_3 (A) (Madon, 1979)

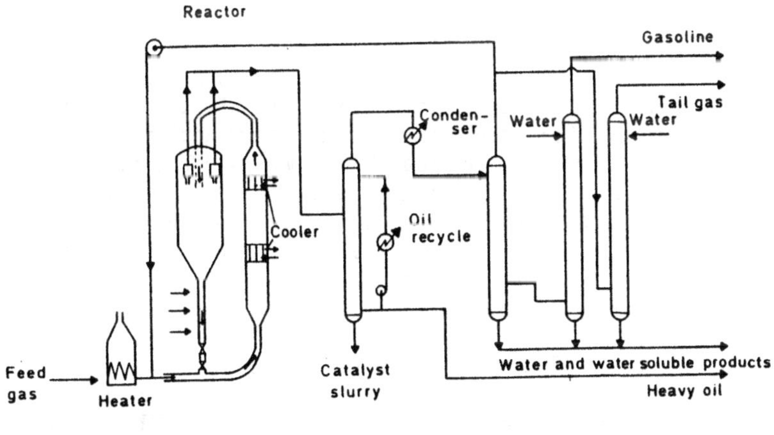

Fig. 4: Synthol process of Sasol-Kellogg

thesis gas conversion is not high, eq. (6) can be reduced to (11,28,29)

$$r = k_H p_{H_2} \tag{7}$$

Zein el Deen et al. (30) studied the kinetics of the FTS on sintered oxides of iron and manganese. They observed, too, that the rate is independent of the CO partial pressure. Bub et al. (20) developed empirical expressions for the production rate of CO_2 and C_1 to C_4 hydrocarbons on a Mn/Fe catalyst which could be used to successfully describe the conversion and selectivity in a pilot plant fixed bed reactor (2 cm ID by 80 cm length). If a catalyst like Mn/Fe gives a Schulz-Flory product distribution the hydrocarbon fraction can be calculated from the overall conversion rate and the chain growing probability α.

With regard to the FTS in slurry phase no detailed kinetic study is available as yet.

Heat generation: The hydrocarbon formation is accompanied by intense heat generation. Depending on the extent of the water-gas shift reaction(2) the heat generation amounts to (1-1.5) x 10^7 kJ per ton of hydrocarbons produced. This corresponds to about 25 % of the heat of combustion of the synthesis gas. It is therefore understood that heat removal presents the key problem of industrial FT processes.

1.2.2 Commercialized FT processes. Various FT processes have been proposed and either used commercially or tested extensively (5). They differ particularly on the type of reactor and the kind of heat removal. It is the reactor and its operational conditions which determine the product distribution and the upgrading scheme. At first, one can roughly distinguish between FT processes with stationary and mobile catalyst phase. Among the various fixed bed processes it is only the ARGE reactor (Arbeitsgemeinschaft Lurgi-Ruhrchemie) which meets the present standards of fixed bed reactor technology and offers high performance and production capacities. Reactors with mobile catalysts are fluidized beds (Hydrocarbon Research Inc.), entrained fluidized beds (Sasol - Kellogg), the three-phase slurry reactor (Rheinpreussen-Koppers), and the three-phase fluidized bed and fixed bed reactors with oil circulation (BASF, Bureau of Mines).

The Sasol I version of the fixed bed multitube reactor of ARGE contains about 40 m^3 of catalyst and produces 50 t of hydrocarbons per day. One of the main products of the fixed bed process are waxes which are collected immediately after the reactor. Detailed descriptions of the ARGE process can be found elsewhere (5,31).

The entrained fluidized bed reactor is the most successful approach to carry out the FT synthesis in large--scale operation. It was originally developed by Kellogg and has been optimized by Sasol during more than 20 years of operation.

A scheme of the Synthol process is outlined in Fig. 4. The core presents the entrained fluidized bed reactor of about 36 m height and a diameter of 2.2 m. In the reaction zone, two cooling aggregates with recirculating oil as cooling media are installed. The reaction zone is followed by a catalyst settling hopper of 5 m in diameter. Here the tail gas leaves the reactor via cyclones which separate the catalyst fines. The reactor contains about 130-140 t of a fused Fe catalyst. The catalyst recirculation rate is 8,000 t/h. To prevent formation of higher liquid hydrocarbons which would disturb a uniform operation the synthesis is carried out at temperatures of 300 to 350 ^{o}C and a pressure of 20-30 bar. The syngas enters the reactor at about 160 ^{o}C. The fresh feed gas to recycle gas ratio is about 2 to 2.5 giving a syngas composition with a H_2 to CO ratio of 6. The high ratio also suppresses formation of liquid hydrocarbon. The fresh feed flow rate amounts to 100,000 Nm^3/h giving a catalyst loading of about 700 Nm^3 syngas/m^3 cat h.

Sasol I operates 3 reactors in parallel. The mean catalyst lifetime is 42 days. On the average, 2.4 reactors are steadily under operation. Each reactor produces 60,000 tns of primary products per year. The product distribution is given in Table 1 and compared with that of the fixed bed process. The Synthol process reveals a maximum in gasoline fuel while the main products of the fixed bed process are higher hydrocarbons like diesel, heavy oils and wax.

The new Sasol II plant is intended to produce mainly gasoline and is therefore based entirely on Synthol technology. The Sasol II reactors have each a capacity of 2 1/2 times that of Sasol I. A simplified process scheme is shown in Fig. 4, and Table 2 summarizes the production figures of Sasol II (32). The productivity of Sasol II is more than twice as large as that of all plants operated in Germany in 1944 and even considerably larger than the world FT productivity at that time. How-

Table 1: Product distribution (%) of fixed and entrained bed processes (Sasol I) (32)

Components		Fixed bed	Synthol
CH_4		2.0	10.0
C_2H_4		0.1	4.0
C_2H_6		1.8	4.0
C_3H_6		2.7	12.0
C_3H_8		1.7	2.0
C_4H_8		2.8	9.0
C_4H_{10}		1.7	2.0
Petrol, C_5-C_{11}		18.0	40.0
Diesel, C_{12}-C_{18}		14.0	7.0
Heavy oil,	C_{19}-C_{23}	7.0	
	C_{24}-C_{35}	20.0	4.0
Wax	C_{35+}	25.0	
Oxygenates		3.2	6.0

Table 2: Production figures of Sasol II (32)

Coal gasified	1030 t/h
Purified synthesis gas	1.1×10^6 m^3 STP/h
Products	10^3 t/a
Motor fuels	1,400
Ethylene	180
Chemicals	175
Tar products	287
Ammonia	100
Sulphur	75

ever, it should also be pointed out that the performance of Sasol II is rather moderate compared to a conventional oil refinery and petrochemical plants.

1.2.3 Alternative routes. Though the ARGE and Synthol reactors work successfully and profitably now this must not necessarily imply that they are optimal choices under all circumstances. Other technologies may prove viable too, and,possibly, they may be better suited under different conditions to produce synthetic fuels. Already during World War II three-phase processes were considered alternatives to the fixed reactor. This finally led to the development of the Rheinpreussen-Koppers slurry process (5,33-35) and the oil circulation processes of BASF (5) and the Bureau of Mines (6, 36-38), the latter being a trickle bed reactor with upward flow of gas and liquid. In the past two decades, three-phase reactor technology has experienced a fast development, and the prospects to carry out the FTS in three-phase reactors are very promising.

In the following, only the slurry process will be described detailed as merely this route was developed to a larger scale. In 1951, a demonstration plant of the slurry process was erected by Rheinpreussen and Koppers. The process scheme is given in Fig. 5. The reactor presents essentially a bubble column which is just a pressure resistant steel cylinder of 1.5 m in diameter and a length of 8.6 m. The liquid phase is molten wax. The best is to use a hydrocarbon product fraction of the synthesis itself. In the liquid phase, a precipitated and promoted Fe catalyst is suspended by the motion of the gas bubbles. The sulfur-free preheated gas is sparged at the bottom of the column. The generated heat is removed by cooling tubes located in the slurry. The reactor temperature is simply controlled by the pressure of the saturated steam in the steam collector. The tail gas leaves the reactor at the top and is precooled in a heat exchanger which partially condenses the higher boiling products. Further cooling condenses the other liquid products which are separated and upgraded in the usual manner. The higher-boiling synthesis products remain in liquid phase and are withdrawn from the suspension by filtering off.

The slurry process offers great flexibility with respect to the product distribution which can mainly be affected by the catalyst and such important operation variables as temperature, pressure, CO/H_2 ratio of feed gas and space velocity, something that can scarcely be done in fixed and fluidized bed processes.

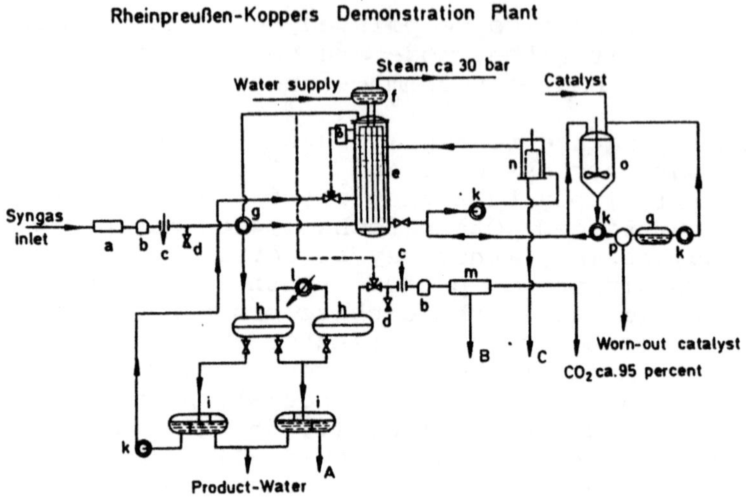

Fig. 5: Rheinpreussen-Koppers demonstration plant (5)

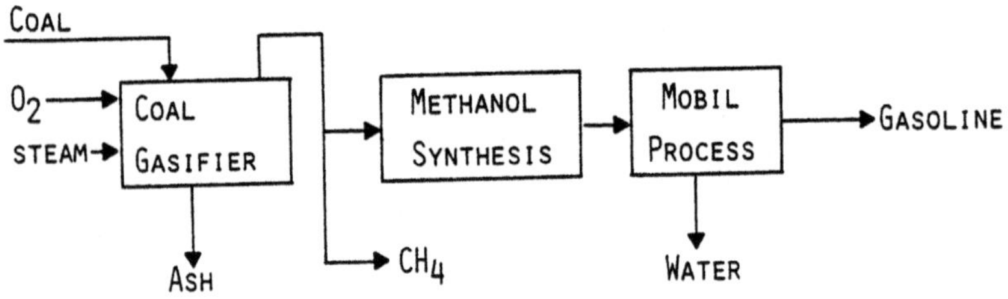

Fig. 6: Mobil's methanol to gasoline route

	Mass g/Nm^3 ($CO+H_2$)	% wt. of total product	Olefin content %
Methane + ethane	5.7	3.2	0
Ethylene	6.3	3.6	100
C_3	40.3	22.6	75-85
C_4	9.1	5.1	70-80
40 to 180° C fraction	95.5	53.6	70
180 to 220° C "	7.1	4.0	48
220 to 320° C "	10.7	6.0	37
320° C fraction	3.3	1.9	7
Total	178.0	100	

Table 2a: Product distribution from demonstration plant

A product distribution obtained under typical operational conditions is given in Table 2a. This special example leads to a maximum value of the gasoline fraction which was the main goal of the demonstration plant. Only about 4 % of the total hydrocarbons produced were in the form of methane and ethane. Other examples for maximizing the fraction of lower and higher hydrocarbons, respectively, are presented by Kölbel and Ralek (35).

The flexibility of the liquid phase process with respect to gas composition is particularly demonstrated by the fact that CO rich synthesis gases can be used which contain no hydrogen but steam (Kölbel-Engelhardt synthesis). The synthesis starts then by the formation of hydrogen from the shift reaction (2) which is followed by the hydrocarbon synthesis. The overall reaction is

$$3CO + H_2O \longrightarrow -(CH_2)- + 2CO_2$$

This synthesis carried out in slurry phase on Fe or other catalysts yields practically the same products as the FT synthesis with high conversions. It was pointed out that a combination of the synthesis step with an internal water-gas shift reaction could result in substantial energy savings as well as take advantage of gasifiers which could produce low ratio H_2 to CO synthesis gas at lower cost.

1.2.4 Comparison of FT processes. Table 3 presents a list of typical operational conditions and performance data for three main FT process modes. The table is based on data from Sasol I for the fixed and fluidized bed type reactors and from the Rheinpreussen-Koppers demonstration plant. From this table and the additional information available from the literature the advantages of the three-phase FT process can be summarized as follows:

- high single pass conversion
- high yield of C_{3+} products
- large content of transportation fuels in C_{3+} products
- high catalyst and reactor performance
- low methane formation
- high flexibility
- possibility to use synthesis gases of high CO content

Table 3: Comparison of various FT processes

	Fixed bed (ARGE) Sasol I	Entrained Fluidized bed Sasol I	Three-phase slurry reactor (Rheinpreussen-Koppers)
Temperature, °C	220-250	300-350	260-300
Pressure, bar	23-25	20-23	12 (24)
Ratio CO/H_2 (Feed)	0.5-0.8	0.36-0.42	1.5
Recycle/feed gas ratio	2.5	2-2.4	0
Catalyst loading, $Nm^3 (CO+H_2)/m^3$ cat h	500-700	700	5,000 (10,000)
$(CO+H_2)$ conversion, %	50	77-85	90
Yield of C_{3+}, $g/Nm^3 (CO+H_2)$	104	110	166
Catalyst performance tC_{3+}/t cat·day	1.35	1.85	5.3 (10.6)
Reactor performance tC_{3+}/m^3 reactor vol. day	1.25	2.1	0.93 (1.86)
Content in primary C_{1+} products			
Gasoline, % wt.	18	40	54
Diesel, % wt.	14	7	10

- simple reactor design
- no attrition and erosion problems
- easy heat removal

With the slurry process the chief technical problem of heat removal from the synthesis reactor was definitely solved. In addition, the slurry process is of great flexibility and gives high conversions. These advantages of the Rheinpreussen-Koppers process were confirmed by other investigations (39-43) particularly by studies from the Bureau of Mines (39,43). Schlesinger and coworkers demonstrated that by using a nitrided Fe catalyst about 50 % of oxygenates can be obtained from the slurry process.

The fixed bed of ARGE and the entrained bed reactors were steadily optimized and modernized to present-day technology. The three-phase FT processes were realized only in pilot and demonstration plants. They had actually no chance to prove their effectivity and superiority over a longer period of operation and were not optimized in larger-scale equipment. Although the three-phase processes were a full success from an engineering point of view they were a failure in view of the economic requirements in those days.

One should mention that the favorable results of the Rheinpreussen-Koppers demonstration plant formed the basis for an offer made to the Indian government to erect a complete liquid phase synthesis plant with an annual production of 250,000 to of hydrocarbons in 1955. Owing to the switch from coal tn petroleum, the plans for constructing such an FT plant were, however, not realized.

In the new efforts to develop a synfuel industry based on synthesis gas from coal the slurry type reactors play a dominant role. For instance, a bubble column slurry reactor was mainly applied in a recent comparative study of various catalysts carried out in Germany. In the US, several pilot plants for syngas conversion processes are under construction which make use of the advantages of the slurry technology.

1.3 Methanol Synthesis in Slurry Phase

In order to avoid some of the current limitations of the vapor phase methanol synthesis, EPRI and Chem. Systems are developing a liquid phase methanol process (44,45). In this process, the catalyst is fluidized by cocurrent upflow of synthesis gas and an inert hydrocarbon liquid. Because of its high heat capacity the liquid absorbs the liberated heat of reaction. The liquid phase without catalyst is withdrawn near the reactor top and recirculated via a heat exchanger where cooling occurs by generating high pressure steam.

The salient advantages of the slurry process over conventional multibed adiabatic quench reactors are:

- Excellent temperature control of the reactor permits to realize syngas conversion as high as 35 to 43 % as compared to a more normal figure of about 12 %. This, in turn, considerably reduces the recycle gas flow and compression energy.

- Heat can simply be recovered as high pressure steam which is the major advantage of the slurry route. An energy balance illustrates that the thermal efficiency can be raised from 86.3 % in the conventional multibed system to 97.9 % in the slurry system.

Additional advantages of the gas-liquid-catalyst system are: (1) simple reactor design, (2) small catalyst particles can be used, (3) catalyst attrition is small due to the cushioning effect of the liquid, (4) the reaction rate can be maintained constant by adding and withdrawing catalyst without shutdown.

The methanol slurry process has not yet been tested on a larger scale but a one ft. ID demonstration plant is under construction (Air Products & Chemicals Inc.)

1.4 Synthesis of Gasoline via Mobil's Route from Methanol

An entirely new route of indirect coal liquefaction has been announced by Mobil in 1976 (46). In Mobil's MTG process methanol is converted on a new type of zeolites to a C_5 to C_{11} hydrocarbon mixture which is rich in aromates. The conversion is carried out in one step at 350 to 500 ^{o}C and a pressure between 1 and 50 bar. The reaction scheme is assumed to be as follows

$$2CH_3OH \rightleftarrows CH_3OCH_3 \longrightarrow \text{lower olefins}$$

$$\downarrow$$

paraffins
cycloparaffins
aromates
(bp. $\leq$ 200 ^{o}C)

A phenomenological kinetic model of this process has been developed by Chang (47). The catalysts are zeolites of ZSM-5 type which have some unusual properties. The largest hydrocarbons which can penetrate the channel structure of ZSM-5 are in C_{10} and C_{11} range as, for instance, durol which is a tetramethyl benzene. The silica content of the ZSM-5 type zeolites is unusually high which is possibly responsible for their hydrophobic properties (48). ZSM-5 does not only convert methanol to gasoline but also other short chain oxygenates and olefins (49,50). The latter being intermediates in the MTG process. Propylene has been found to have the highest conversion rate (51).

A scheme of Mobil process starting from coal is outlined in Fig. 6 and Table 4 presents a typical product distribution (52). More than 75 % of the produced hydrocarbons are in the C_5 to C_{11} range and their content in aromates is high giving a high RON. The MTG process has been studied in fixed and fluidized bed reactors (52). Demonstration plants will be erected in New Zealand and West Germany. Process studies in bubble column slurry reactors are in preparation.

Table 4: Product distribution of MTG process (52)

	% wt.
Methane, ethane, ethylene	1.5
Propane	5.6
Isobutane, n-butane	11.9
Propylene, butylenes	4.7
C_{5+} non-aromates	49.0
Aromates	27.3
	100.0
C_{5+}	76.3

2 ANALYSIS AND STATE OF THE ART REPORT ON FT TECHNOLOGY REPORT ON FT TECHNOLOGY IN SLURRY PHASE

The performance and product slate of a multiphase process like the FTS in slurry phase is not only governed by thermodynamics and catalytic reaction rates but may also significantly be influenced by mixing, heat and interphase mass transfer. Therefore, an analysis of the FT slurry process should begin with the physical transport parameters. Owing to the particular properties of the paraffin phase used as liquid media in the FTS some important engineering parameters like holdup and interfacial area cannot be calculated from correlations established in the literature. Cold flow experiments are only of limited value and can often be considered merely as rough estimates. In general, the self-adjusting engineering parameters should be determined as closely as possible to the actual process conditions.

2.1 Hydrodynamics and Physical Transport Properties

2.1.1 Holdup, bubble size and interfacial area. Measurements on gas holdup and bubble sizes are hardly to be found for hydrocarbons at higher temperatures. Only recently were gas holdup data and bubble sizes in xylene, decalin, an n-paraffin mixture (C_{10}-C_{14}) and hard paraffin (trade name: Vestowax SH 105, mp. 105--122 °C) reported for a 9.5 cm ID bubble column with different gas spargers and for temperatures between 60 and 170 °C and atmospheric pressure (53). No new correlations are presented for the measured data but it is shown that recommended correlations of the literature fail to describe the data. Sauter diameters of the bubbles obtained by sparging the gas by means of a one--hole-nozzle are shown in Fig. 7. The data for xylene, decalin and the C_{10}-C_{14} paraffin mixture scatter around a mean value of 3 mm. In contrast, for the Vestowax the Sauter diameter decreases sharply from higher values at low gas velocities to diameters less than 1 mm if $u_G >$ 1 cm/s. This surprising decline of d_s for Vestowax is in concordance with the findings of Zaidi et al. (54) who reported a mean value of d_s = 0.7 mm for a bubble column with a porous plate sparger.

Gas holdup data for molten paraffin were determined by Deckwer et al. (56) under conditions which are relevant to the FTS. The measurements were carried out in two bubble columns of 4.1 and 10 cm ID. Both columns were equipped with a sintered sparger of about 75 μm pore width. The effect of pressure (0.4 to 1.1 MPa), temperature (143 to 285 °C), concentrations of solids (inert Al_2O_3 powder, 0 to 16 % wt.) and gas velocity (up to 3.8 cm/s) on gas holdup ε_G was studied. For the most relevant range, i.e., temperature above 250 °C, the ε_G data are presented in Fig. 8 as a function of the gas velocity. The data are independent of pressure, temperature and solids content provided T $\geq$ 250 °C and $c_s \geq$ 5.5 % wt. Empirical correlation from the literature (57-59) are not able to describe the measured holdup values for this particular system. The findings in the two columns can be well correlated by the following simple equation

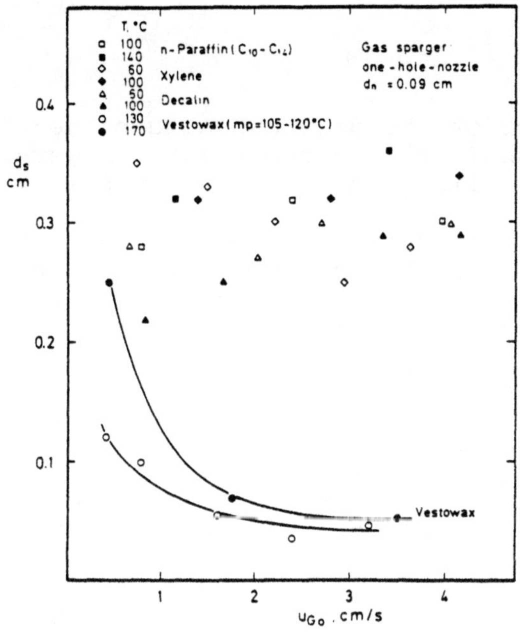

Fig. 7: Measured Sauter diameters in hydrocarbon liquids (55)

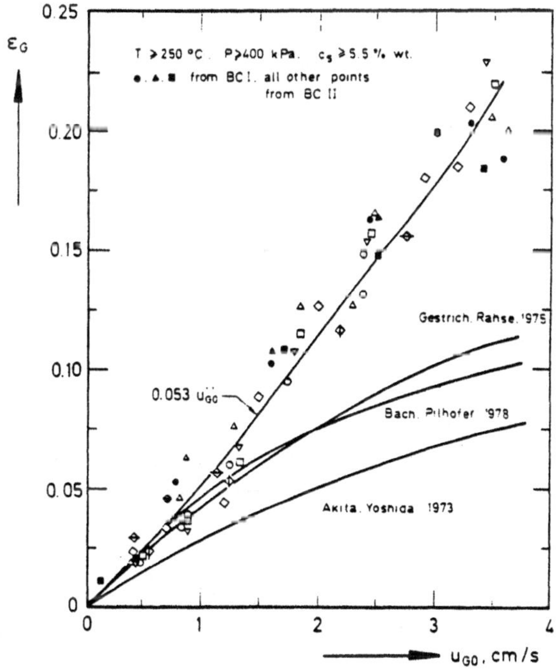

Fig. 8: Gas holdup in molten paraffin, bubble columns of 4.1 and 10 cm ID with sintered plates

$$\varepsilon_G = 0.053\ \bar{u}_G^{\,1.1} \tag{8}$$

where $\bar{u}_G$ is in cm/s. Most of the data can also be described by the ideal bubble flow concept (60). However, this does not give an explicit expression for ε_G.

Fig. 9 shows that the predictions of eq. (8) lie in between the data reported by Calderbank et al. (61) and Hammer (62) for molten wax systems at similar temperatures. The discrepancies experienced by the different authors are not reasonably explainable. It is believed that eq. (8) presents a good compromise. In addition, this relation for ε_G appears to be applicable to larger diameter columns and higher gas velocities as it predicts that holdup reported by Calderbank et al. (61) and Farley and Ray (63) for an inlet gas velocity of 7 cm/s and a 24.4 cm ID column operated under the condition of the FTS.

Knowing the Sauter mean bubble diameter and the gas holdup function, eq. (8), the specific interfacial area can be calculated from

$$a = \frac{6\,\varepsilon_G}{d_s} \tag{9}$$

As the bubble diameter does not depend on the pressure the value of $d_S = 0.07$ cm as reported by Zaidi et al. (54) can be used. Hence it follows

$$a = 4.5\ \bar{u}_G^{\,1.1} \tag{10}$$

$\bar{u}_G$ is again in cm/s and a in cm^{-1}. As compared to other gas-liquid systems, eq. (10) predicts very high interfacial areas. Eq. (10) was established from data obtained in bubble columns with sintered plate spargers. However, as shown in Fig. 10, the interfacial areas reported by Quicker and Deckwer (53,55) for a column with single orifice sparger (0.9 mm ID) are even higher if $\bar{u}_G \geq 1.5$ cm/s. Therefore, eq. (10) may be regarded as a conservative estimate for a in the molten paraffin system.

Hammer (62) measured bubble diameters in paraffin under similar conditions and found even smaller values of d_S than those shown in Fig. 7. Therefore, the interfacial areas calculated therefrom would again be larger

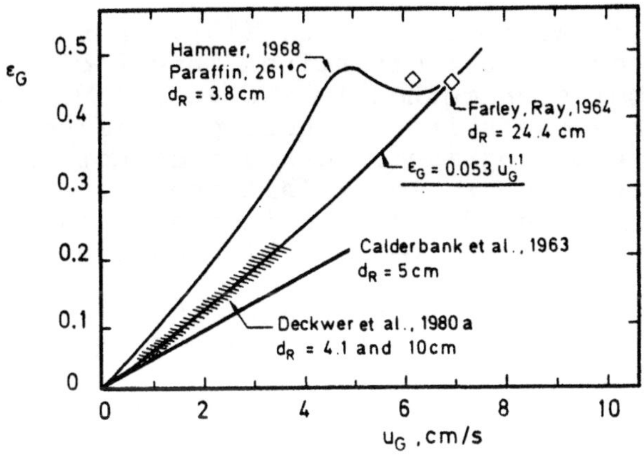

Fig. 9: Gas holdup for molten paraffin systems of various authors (56,61-63)

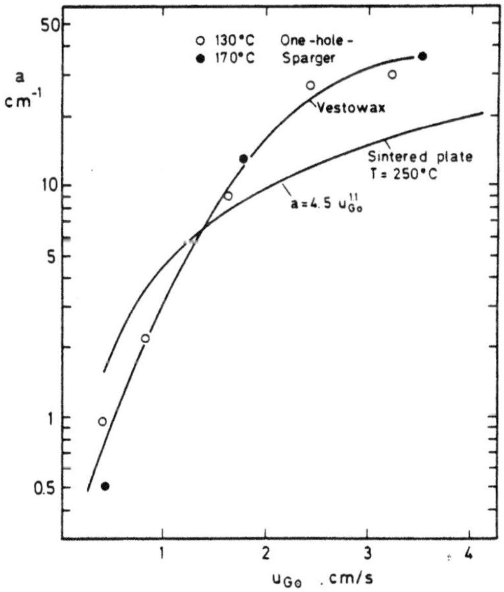

Fig. 10: Interfacial areas in molten paraffin

than those attainable from eq. (10). On the other hand, Calderbank et al.(61) determined interfacial areas by a light transmission method in wax at 265 °C. Their values of a are only one third of those calculated from eq. (10). As the photographic method does not involve such large errors which would explain the above discrepancies, it is believed that the indirect light transmission method did not give reliable results.

2.1.2 Mass transfer coefficients. Gas-liquid mass transfer in molten wax with suspended catalyst was studied by Zaidi et al. (54). The water-gas shift reaction, i.e., eq. (2), was used as a model system. The CO conversion was measured as a function of the catalyst concentration at various gas velocities, a constant pressure of 0.9 MPa and 250, 270 and 290 °C. Under the experimental conditions the following relation holds

$$\ln (1-X) = -K_o a \frac{RT}{He} \qquad (11)$$

where the reciprocal of the overall absorption coefficient $K_o a$ is given by

$$\frac{1}{K_o a} = \frac{1}{k_L a} + \frac{1}{C_{cat}} \frac{d_{cat} \rho_{cat}}{6 \rho_L \varepsilon_L} \frac{1}{k'} \qquad (12)$$

As the Henry coefficient He is known (64), $K_o a$ can be calculated from the measured CO conversion X by eq. (11). With respect to eq. (12) the plot of $1/K_o a$ vs. $1/C_{cat}$ should give a straight line for a constant gas velocity (ε_L is constant). The intercept of this line gives the reciprocal of the volumetric mass transfer coefficient $1/k_L a$. An example of such plots is shown in Fig. 11 for two gas velocities. From the volumetric mass transfer coefficients $k_L a$ which are in the reasonable range of 0.01 to 0.02 s^{-1} the liquid side mass transfer coefficient k_L can be calculated as the interfacial area a is known. The mean value of k_L is about 0.01 cm/s. The k_L value can reasonably be described by the correlations of Hughmark (65)

$$Sh = 0.0187 \left[Re^{0.484} \, Sc^{0.339} \left[\frac{d_s g^{1/3}}{D_L^{2/3}} \right]^{0.072} \right]^{1.61} \qquad (13)$$

and Calderbank and Moo-Young (66) for $d_B < 2.5$ mm

$$k_L \, Sc^{2/3} = 0.31 \left(\frac{\mu_L \Delta \rho_g}{\rho_L^2}\right)^{1/3} \tag{14}$$

The physical properties are reported in refs. (54,56, 61,62).

In general, it is also possible in a slurry reactor that a significant mass transfer resistance $1/k_s a_s$ can be observed at liquid-particle interface. However, due to the small catalyst particles which are usually applied in the FT slurry process the interfacial area a_s is large. a_s is given by

$$a_s = \frac{6(1-\varepsilon_G)}{d_{cat}\,\rho_{cat}}\,\bar{\rho}\,c_{cat} \tag{15}$$

and for typical values of c_{cat} (say 10 % wt.) and d_{cat} (say 10 μm) it follows $a_s > 100\ cm^{-1}$. Hence, $1/k_s a_s$ is small and negligible as compared to $1/k_L a$. This is valid even for the lower limit of k_s, i.e., $Sh_s = 2$. If large particles are applied and liquid-solid mass transfer may effect the overall rate the liquid-solid mass transfer coefficient k_s in bubble aerated slurries can be calculated by correlations summarized in ref. (67).

2.1.3 Mixing and solids distribution. Dispersion coefficients of gas, liquid and solid phase have not been measured under conditions prevailing in the FTS. However, the liquid phase dispersion coefficient depends only slightly on liquid phase properties. It is therefore believed that correlations based on a lot of data for low viscous media and from columns of various sizes can be applied. Such a correlation was given by Deckwer et al. (68)

$$E_L = 2.7\,\bar{u}_G^{\,0.3} d_c^{\,1.4} \tag{16}$$

where the liquid phase dispersion coefficient E_L is in cm^2/s, the mean gas velocity in cm/s and the column diameter d_c in cm. Liquid phase dispersion was recently reviewed by Shah et al. (69). These authors recommend eq. (16) and the correlation developed by Joshi and Sharma (70).

Data on gas phase dispersion are few. Mangartz and Pilhofer (71) correlated their own results under consid-

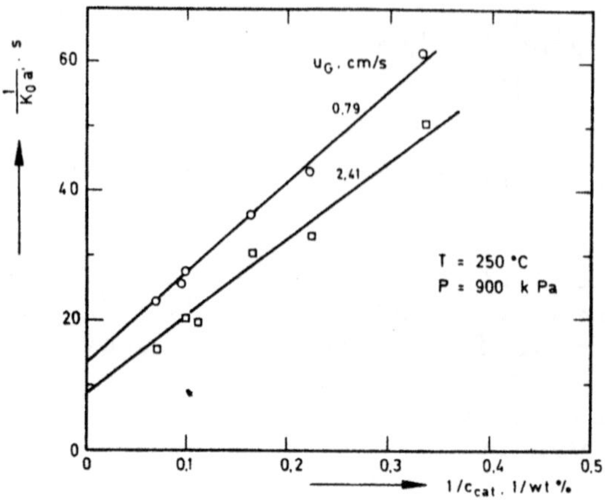

Fig. 11:
Overall resistance vs. reciprocal catalyst concentration. Water-gas shift reaction in molten paraffin

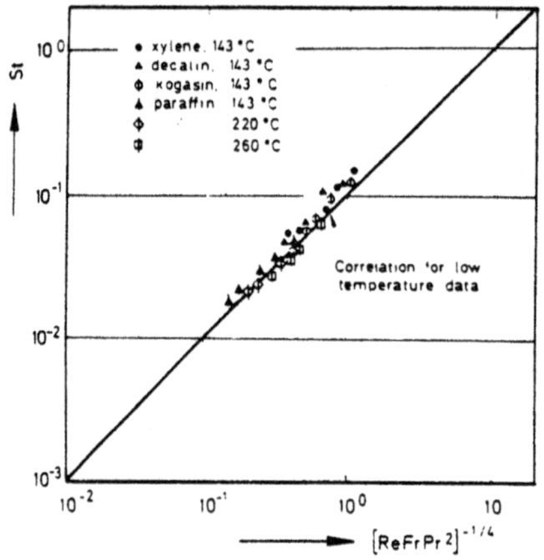

Fig. 12:
Heat transfer coefficients for hydrocarbon liquids in bubble column - check of eq. (19)

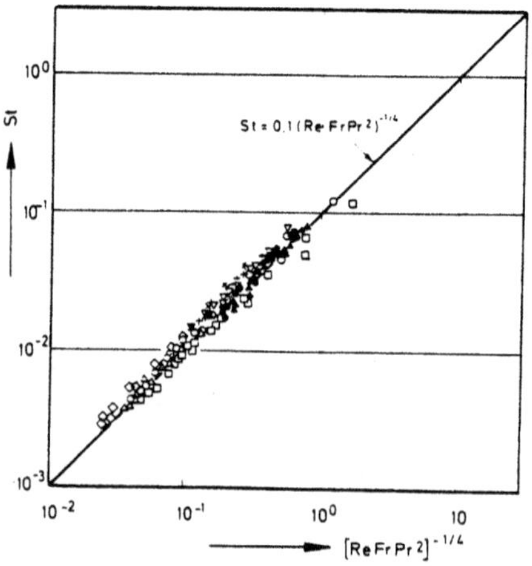

Fig. 13:
Heat transfer coefficients in slurry systems plotted as to eq. (19)

eration of literature data by

$$E_G = 5 \times 10^{-4}\ d_c^{\ 1.5}(\bar{u}_G/\varepsilon_G)^3 \qquad (17)$$

Field and Davidson (72) reported recently on gas and liquid dispersion in a bubble column of 3.2 m diameter by 18.9 m length. The authors concluded that both correlations, i.e., eq. (16) and (17) are applicable.

In general, catalyst sedimentation has to be accounted for in slurry reactors. The distribution of the catalyst along the reactor can be computed using the sedimentation-dispersion model. As to the results of Kato et al. (73), the solid dispersion coefficients do not differ much from those of the liquid phase. From the data provided by Cova (74), Imafuku et al (75), and Kato et al. (73), the solids concentration profiles can be calculated. As in the FT process the catalyst particles are usually small, according to Kölbel and Ralek (35) the diameter should be less than 50 µm, the catalyst profiles are not very pronounced, in accordance to the measurements of Cova (74).

If catalyst particles of larger size and density are used and sedimentation might be important the correlations proposed by Kato et al. (73) are recommended to calculate the solids dispersion coefficient and the settling velocity of the particle. From this data the solids concentration profile can be computed with the sedimentation-dispersion model.

2.1.4 Heat transfer. Wall-to-dispersion heat transfer coefficients in a 10 cm ID bubble column was measured by Louisi (76) for hydrocarbon liquids at temperatures between 143 and 260 °C. As pointed out by Kast (77) the heat transfer coefficients in gas-liquid dispersions are about one order of magnitude larger than in single phase liquid flow. These large heat transfer coefficients are a major advantage of bubble column reactors. Heat transfer from walls and inserted coils was analyzed theoretically and the following relation could be derived (78)

$$\frac{h}{\bar{\rho}\bar{c}_p\bar{u}_G} \propto \left(\frac{\bar{u}_G^3}{g\nu}\right)^{-1/4} \left(\frac{\overline{\mu c_p}}{\bar{\lambda}}\right)^{-1/2} \qquad (18)$$

This equation can be written in dimensionless numbers

$$St = \beta(Re\ Fr\ Pr^2)^{-1/4} \tag{19}$$

If β is taken 0.1, eq. (19) describes experimental data of various authors with striking agreement (78). Also the results reported by Louisi (76) for hydrocarbons at higher temperatures are correlated by eq. (19). A parity plot is given in Fig. 12 which illustrates good agreement.

Louisi (76) also measured heat transfer coefficients in molten paraffin at 220 and 260 °C in the presence of powdered Al_2O_3 particles. The concentration of the solid was varied from 5.5 and 16 % wt. These results of Louisi as well as other data for various slurry systems reported by Kölbel et al. (79,80) could again be well described by eq. (19) as can be discerned from Fig. 13. The plot includes data for kieselghur suspensions in water, spindle oil and machine oil, and water-sand suspensions up to particle sizes of 110 µm.

2.2 Investigations of FTS in Slurry Phase

2.2.1 Literature survey. Research work on the FTS in slurry phase already started at Rheinpreussen during World War II. After the favorable results have been published the FTS in slurry phase was studied intensively at the Bureau of Mines (6,43,98) and the UK Fuel Research Station (40,61,63). Also Indian and Japanese workers (41,42) carried out some investigations. Most of these studies confirmed in general the findings of Kölbel and coworkers (33,34) from Rheinpreussen. All these elder investigations are summarized in Table 5 and reviewed by Kölbel and Ralek (35).

Since the oil embargo in 1973 the FTS in slurry phase is studied at industrial companies (Ruhrchemie and Schering) and universities (Berlin, Hanover, Darmstadt) in Germany. [illegible]m industry only few results have been published ([illegible]). The cooperative research between the Universities [illegible] Berlin and Hanover will be discussed in more detail and Table 6 summarizes the major studies carried out in this cooperation. There is also a growing interest in the FTS in slurry phase in the USA and several comprehensive research programs are running. Until now, only few experimental results have been reported by Satterfield and Huff from MIT (13,81,82).

The primary objective of the studies carried out in Germany was to improve the selectivity with regard

Table 5: FTS in slurry phase - Summary of published studies

Authors	Catalyst	T, °C	P, bar	d_c, cm	L, cm
Kölbel, Ackermann (34) (Kölbel, Ralek (35))	Pptd. Fe Pptd. Fe	268 266	12 11	129(150) 4.7	770 352
Hall et al. (40)	NH_3 synthesis cat.	265,305	20-41	5	50-150
Schlesinger et al. (43)	Fused nitrided Fe	220-258	20.4	7.5	300
Mitra, Roy (41)	Fe pptd.on Kieselghur	240-260	6.8-10.2	5	210-300
Calderbank et al. (61) Farley, Ray (63)	Pptd. Fe	260-280	10	24.4	600
Kunugi et al. (42)	Pptd. Fe	260	10.3	5	550

Table 6: Applied catalysts in liquid phase FT synthesis in 3.8 cm ID bubble column

Catalyst	Preparation	Activation at T = 270 °C in	Composition (nonactivated)	Content in slurry % wt. (nonactivated)
(A) Red mud	-	fixed bed	48.1 % wt. Fe	20.8
(B) Mn/Fe	batch precipitation	fluidized bed	Mn/Fe=3.7:1 14.8 % wt. Fe	14.3
(C) Fe/Cu	continuous precipitation	slurry phase	Fe:Cu=19:1 55.3 % wt. Fe	15.0
(D) Mn/Fe	continuous precipitation	slurry phase	Mn/Fe = 6:1 9.2 % wt. Fe	13.8

to short chain olefins as in the mid-seventies the production of hydrocarbons could only be profitable if the FTS could be coupled with the generation of high valued olefins, i.e., ethylene and propylene. In the meantime, oil prices increased so much and still continue to rise so that even proven FT technology seems to be economical. Particular interest deserves a direct combination of the FTS with Mobil's process in slurry phase which was proposed by Poutsma (83). This concept is very promising for a diversified future synfuel industry which in addition would maintain the present structure of refineries and petrochemical plants. A brief discussion of the potential use of such a combination of FTS with Mobil's route is given in part 3 of this contribution.

2.2.2 Experimental results with a Mn/Fe catalyst. Mn/Fe catalysts were claimed to be of improved selectivity with respect to lower olefins (3). Therefore such catalysts were extensively studied in fixed bed and slurry phase and the results were compared with those from conventional Fe catalysts, see Table 5. The details of these investigations are reported elsewhere (14,84-86). In this contribution, only some findings with the Mn/Fe catalyst D in Table 6 will be presented.

The measurements with catalyst D were carried out in a 3.8 cm bubble column. The gas was sparged by a sintered plate (mean pore diameter 75 µm) and the height of expanded slurry bed varied between 60 and 90 cm. The catalyst precursor was prepared by a special procedure and 64 g of dried precipitate were suspended in 400 g molten paraffin. Activation of the catalyst was carried out in slurry phase by successive treatment with CO and H_2. The size of catalyst particles was in the range of about 10 um. The course of the conversion of synthesis gas as a function of the temperature is shown in Fig. 14. The conversion increases almost linearily with T for the range studied. Fig. 15 gives the measured conversion vs. the space velocity (volume of gas (at 0 °C and 1 atm) per catalyst volume and hour) for different CO/H_2 inlet ratios at 303 °C. High conversions were obtained for the CO/H_2 ratios of 1.35 and 1.73 whereas higher and lower values give a considerably lower overall conversion. Particularly, if the synthesis gas is rich in H_2, the attainable conversion is remarkably small.

The selectivities (yields) expressed in grams of hydrocarbons obtained per Nm^3 synthesis gas converted are plotted vs. space velocity in Fig. 16 for CO : H_2 inlet ratios larger than 1.35. It can be seen that under

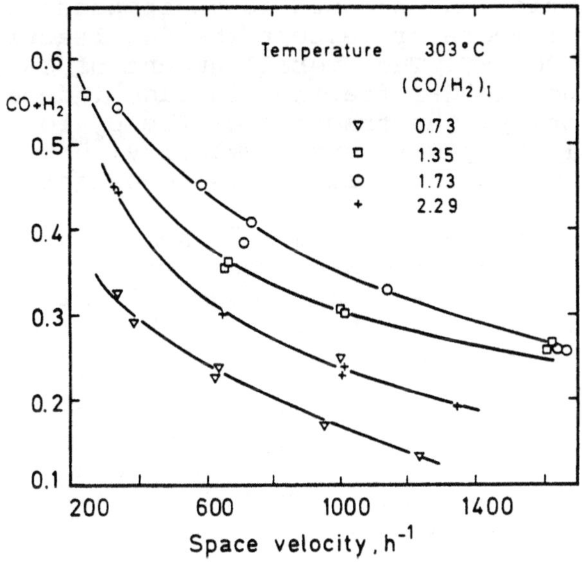

Fig. 15: Conversion vs. space velocity for different inlet gas compositions

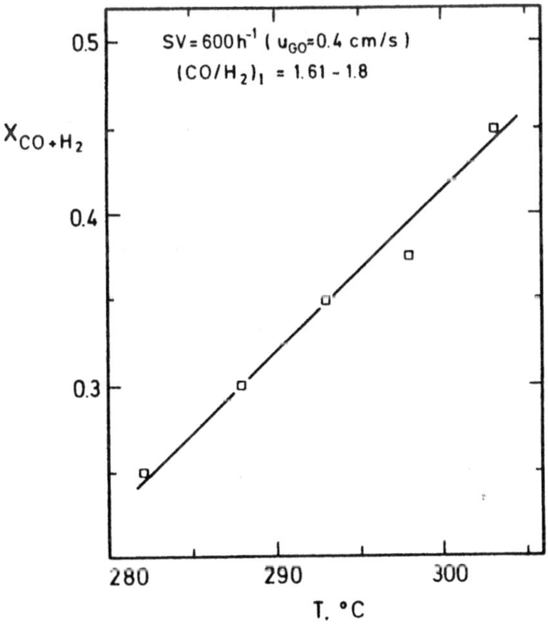

Fig. 14: Effect of temperature on conversion

this condition the product is independent of the inlet ratio and temperature except the CH_4 fraction which is higher at 303 °C. The overall amount of C_2 to C_4 hydrocarbons and the C_{5+} fraction is also independent of space velocity. The fraction of the C_2 to C_4 olefins increases slightly with space velocity, however, if SV > 600 h^{-1} the olefin yield is also constant.

At a temperature of 303 °C some measurements were also carried out with a hydrogen rich gas, i.e., (CO/H_2) = 0.73. In this case, the product distribution changes drastically as shown in Fig. 17. The CH_4 content in the product gas amounts to almost 100g/Nm^3 syngas converted and C_{5+} fraction is low. The yield of C_2 to C_4 hydrocarbons is about the same as for higher CO/H_2 inlet ratios but the olefin content is only moderate. This result clearly demonstrates that the Mn/Fe catalyst in slurry phase cannot convert H_2 rich synthesis gases to desired FT products.

An important variable of FT processes, in general, is the CO to H_2 usage ratio. In practice, the CO to H_2 feed ratio should be chosen to equal the usage ratio approximately. The investigation with the Mn/Fe catalyst indicates that the CO to H_2 usage ratio is about equal to the feed ratio if this is close to 1.5. Therefore, the Mn/Fe catalyst is exceptionally well suited to convert synthesis gases produced by second generation gasifiers.

The product slate obtained from CO to H_2 feed ratios > 1.35 and shown in Fig. 16 was examined for Schulz-Flory distribution. Fig. 18 presents a plot as to eq. (5 It can be discerned that the C number weight fractions obtained from slurry phase runs follow the Schulz-Flory distribution with striking agreement except for the C_2 and C_3 fractions which always deviate from the drawn straight line. The C_2 fraction is smaller, while the C_3 fraction is larger than the predictions of the Schulz-Flory distribution. From a least square fit of all data the chain growth probability α can be obtained. Its value was found to be α = 0.675 in this study. The deviations for the C_2 and C_3 fraction have probably to be attributed to the production of oxygenates (13). The amount of oxygenates was not determined in the investigations.

The Mn/Fe catalyst (D) was also studied in fixed bed in order to compare the two operation modes with re-

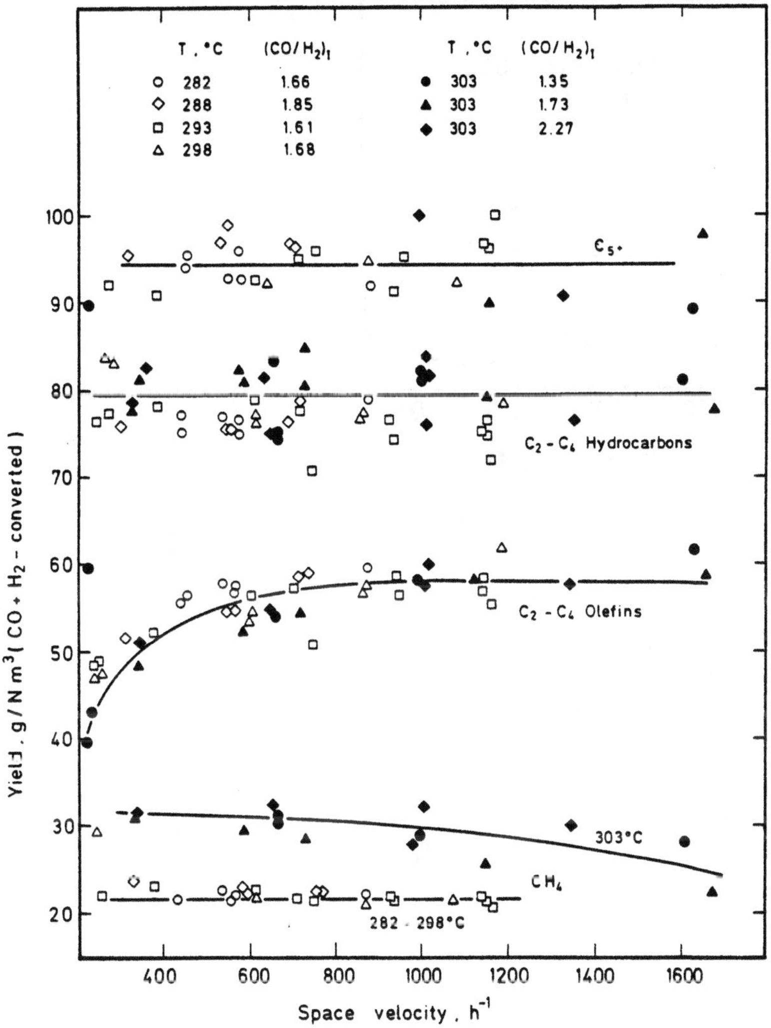

Fig. 16: Product distribution with Mn/Fe catalyst (D) in slurry phase, $(CO/H_2)_I$ 1.35

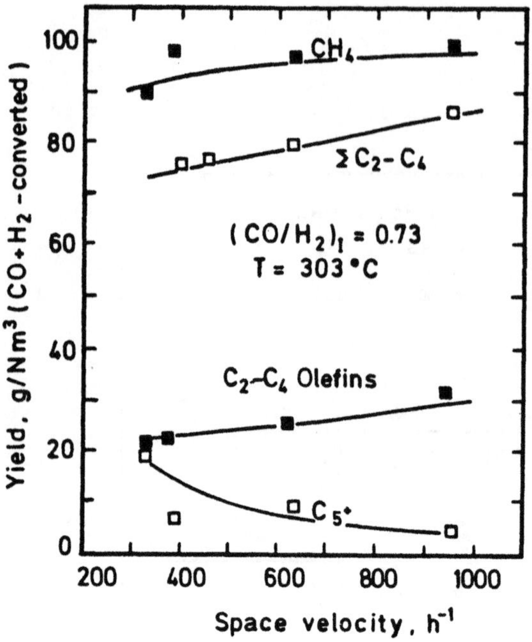

Fig. 17: Product distribution for $(CO/H_2)_I$ = 0.73 at 303°C

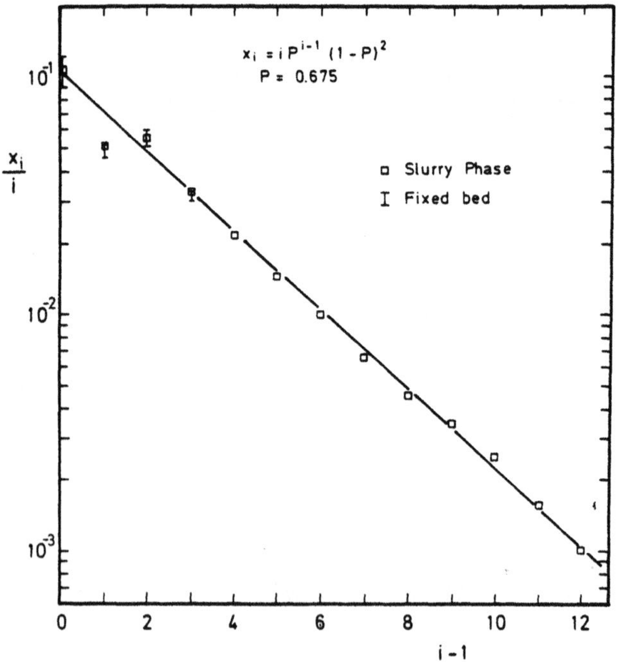

Fig. 18: Schulz-Flory plot of product distribution

gard to selectivity. Table 7 presents more detailed product distributions from slurry phase and fixed bed runs. Although the conversions attainable in the two reactors differ greatly, the product slates differ only slightly. One may therefore conclude that both operation in the fixed bed and in the slurry phase yield the same product slate for the Mn/Fe catalyst. This result also implies that the chain growth probability does not depend on the kind of operation. This is not very surprising as it is well known that in fixed bed synthesis the internal and external surface of the catalyst is covered with higher product hydrocarbons. In addition, it can be estimated on the basis of the relevant hydrodynamic properties of the slurry reactor (54) that mass transfer resistances are of minor importance. Hence, the FT synthesis in both fixed bed and slurry phase operations is mainly reaction controlled. It can therefore also be expected that the product distributions obtainable from the two operation modes are the same or similar, at least.

2.2.3 Evaluation of rate constants. From the conversion measurements the overall rate constant of the FTS in slurry phase can be evaluated on the basis of a pertinent model if the additional parameters involved in the model equations are available from independent measurements.

The reactor model is based on the following assumptions:

(i) plug flow in gas phase
(ii) stagnant liquid phase
(iii) neglect of liquid-solid mass transfer resistance
(iv) uniform distribution of catalyst
(v) the simplified rate expression eq. (7), i.e., a first order rate law, can be applied as the conditions outlined in paragraph 1.2.1 are approximately fulfilled.

Hence, the governing steady state balance equation of the gas phase is given by

$$\frac{d(u_G y)}{dx} = k_L a \frac{RT}{P} (c^{*}_{HL} - c_{HL}) \tag{20}$$

The local absorption rate in the liquid is

$$k_L a (c^{*}_{HL} - c_{HL}) = k_H \varepsilon_L c_{HL} \,. \tag{21}$$

Table 7: Product distribution of Mn/Fe catalyst in slurry phase and fixed bed

	Slurry phase		Fixed bed	
T, ^{o}C	293	298	295	295
$(CO/H_2)_I$	1.61	1.68	1.87	1.87
SV, h^{-1}	370	880	365	275
X_{CO+H_2}	0.443	0.304	0.766	0.849
Methane	23.4	22.6	25.3	24.0
Ethane	11.8	9.3	12.4	12.6
Ethylene	9.2	11.4	8.1	7.4
Propane	6.4	5.0	4.6	4.8
Propylene	26.6	27.5	30.1	27.5
Butane	7.6	5.9	6.7	6.8
Burylene-1	13.0	15.8	14.9	12.4
Butylene-2	3.5	2.3	3.4	3.0
$(C_2-C_4)_{HC}$	78.1	77.2	80.2	74.5
$(C_2-C_4)_{Ol}$	52.3	57.1	56.5	50.3
C_{5+}	91.9	94.5	95.3	105.2

in g/Nm^3 converted

By considering Henry's law

$$P_H = Py = Hec_{HL} \tag{22}$$

it follows from eqs (20) and (21)

$$-\frac{d(u_G y)}{dz} = K_A \frac{RT}{He} Ly \tag{23}$$

where K_A presents the reciprocal of the overall resistance

$$\frac{1}{K_A} = \frac{1}{k_L a} + \frac{1}{k_H \varepsilon_L} \quad . \tag{24}$$

The hydrogen conversion is given by

$$X_H = \frac{u_{Go} y_o - u_G y}{u_{Go} y_o} \tag{25}$$

and related to the overall conversion of synthesis gas by

$$X_{CO+H_2} = X_H \frac{1+U}{1+I} \tag{26}$$

where U presents the usage ratio ($\Delta N_{CO}/\Delta N_{H_2}$), and I is the inlet ratio ($N_{CO,o}/N_{H_2,o}$). Due to volume contraction the gas velocity varies with the overall conversion

$$\bar{u}_G = \bar{u}_{Go} \ (1 + \alpha \, X_{CO+H_2}) \quad . \tag{27}$$

Here α is the contraction factor defined by

$$\alpha = \frac{G(X_{H_2+CO} = 1) - G_o}{G_o} \quad . \tag{28}$$

Introducing eqs (25) and (27) into eq. (23) and integration yields

$$(1 + \alpha^*) \ln \ (1 - X_H) + \alpha^* X_H = - St_G' \quad . \tag{29}$$

In this equation α^* is given by

$$\alpha^* = \alpha \frac{1+U}{1+I} \tag{30}$$

and St_G' presents a Stanton number

$$St_G' = K_A \frac{RT}{He} \frac{L}{u_{Go}} \tag{31}$$

The present model for the FT slurry reactor differs from the treatment of Satterfield and Huff (81) by consideration of the usage ratio (stoichiometry) and a variable gas velocity due to volume contraction by reaction. The value of α depends on the selectivity, of course. α usually varies between -0.45 and -0.67. Most studies indicate that α is close to -0.5. Therefore, this value is used in the following evaluation.

With the knowledge of U and I and the measured conversion of hydrogen and synthesis gas, respectively, the Stanton number is obtained from eq. (29). From the solubility data of H_2 in molten paraffin provided by Peter and Weinert (64) the overall resistance $1/K_A$ can be computed. The liquid holdup can be estimated from eq. (8 and the specific interfacial area follows from eq. (10). The liquid side mass transfer coefficient is calculated from Calderbank and Moo-Young's correlation for small bubbles, i.e. eq. (14). The H_2 diffusivity in paraffin can be estimated from a relation given by Satterfield and Huff (81). With consideration of this relation eq. (14) reduces to

$$k_L = 0.1165 \left(\frac{\bar{\rho}}{\bar{\mu}}\right)^{1/3} \exp(-1523/T) \qquad (32)$$

Thus, the rate constant k_H specific for H_2 consumption can be obtained from eq. (24) by making use of eqs. (8) (10) and (32). For calculating ε_G and a mean gas velocity $\bar{u}_G$ is required which is calculated from the inlet gas velocity u_{Go} by

$$\bar{u}_G = u_{Go}\,(1+0.5\,\alpha X_{H_2+CO}). \qquad (33)$$

From the rate constants k_H the overall rate constants for synthesis gas conversion are obtained by

$$k_o = k_H\,(1+U). \qquad (34)$$

For an inlet ratio CO/H_2 of about 1.7 at which the majority of measurements was carried out in this study, these overall rate constants are plotted vs. the inverse temperature in Fig. 19. The least square fit of all 41 data gives an activation energy of 109 kJ/mol which is in the reasonable range (81). Though the description of the data shown in Fig. 19 is not very good, one is inclined to conclude that, in general, the first order rate law presupposed in the data analysis is obviously applicable. An Arrhenius plot of the reciprocal overall resistance K_A leads to an activation energy of 81 kJ/mol This indicates that there is some mass transfer resistance but this is moderate. Indeed, the evaluation of

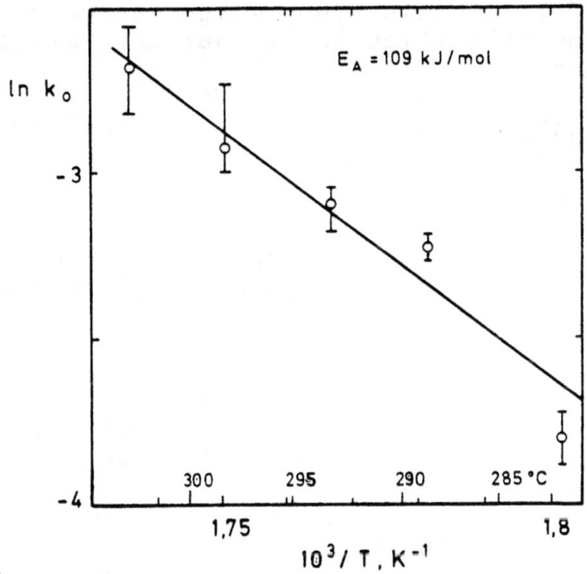

Fig. 19: Arrhenius plot, Mn/Fe catalyst (D) in slurry phase

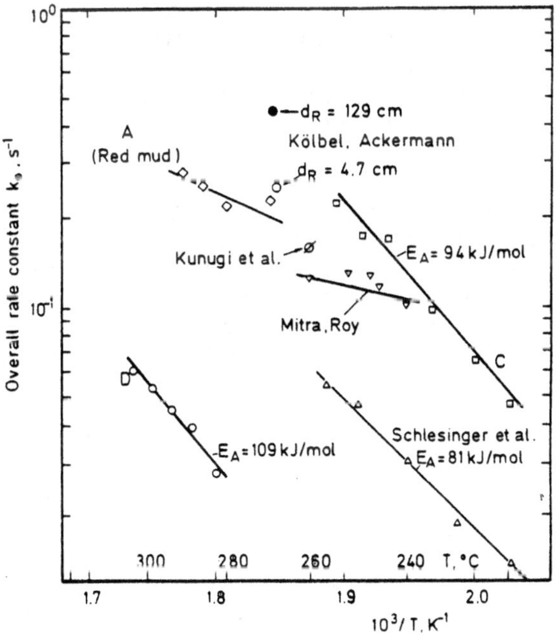

Fig. 20: Rate constants of various studies on FTS in slurry phase

the measured data of this study which was carried out at very low gas velocities shows that the relative reaction resistance

$$\beta = \frac{1/k_H \varepsilon_L}{1/K_A} \qquad (35)$$

is usually larger than 70 % except for the lowest space velocities and the highest temperature β decreases to 0.4.

The above procedure for evaluating rate constant from measured conversions was also applied to the other studies listed in Table 6. All these studies were carried out in the same 3.8 cm ID bubble column. As discussed in paragraph 2.1.1 the relations for ε_G and a can also be applied for estimations in larger scale columns. Therefore, also some of the elder studies studies given in Table 5 could be analyzed for rate constants if the required information like gas throughput could be extracted from the literature. The results of this analysis are summarized in Table 8 and Fig. 20.

Mohammed (84) used a Mn/Fe catalyst (B) which was precipitated batchwise and activated in a fluidized bed. For reasons of comparison Mohammed also did some measurements with untreated red mud activated at 270 °C in a fixed bed. The k_o value evaluated from Mohammed's results reveal a wide scatter and are not shown in Fig. 20. The red mud is very active compared to the Mn/Fe catalyst but its selectivity to C_2 to C_4 olefins is poor. Compared to the Mn/Fe catalyst (B) of Mohemmed the catalyst (B) is more active by a factor of 6 to 10. This result presents a considerable improvement and underlines the importance of catalyst preparation (14). In addition, the high activity catalyst precipitated continuously and reduced in the slurry phase is as selective with regard to C_2 to C_4 olefins as the low activity catalyst of Mohammed (84). The Fe/Cu catalyst (C) precipitated continuously is again considerably more active than the Mn/Fe catalyst (D). However, the last one reaches the same activity only at a temperature of 300 °C which the Fe/Cu catalyst already has at 220 °C. It is surprising that the Fe/Cu catalyst gives relatively large amounts of C_2 to C_4 hydrocarbons. Rough product distributions obtained with this catalyst (C) in slurry phase are shown in Fig. 21. It can be seen that the overall C_2 to C_4 fraction is larger than for the Mn/Fe

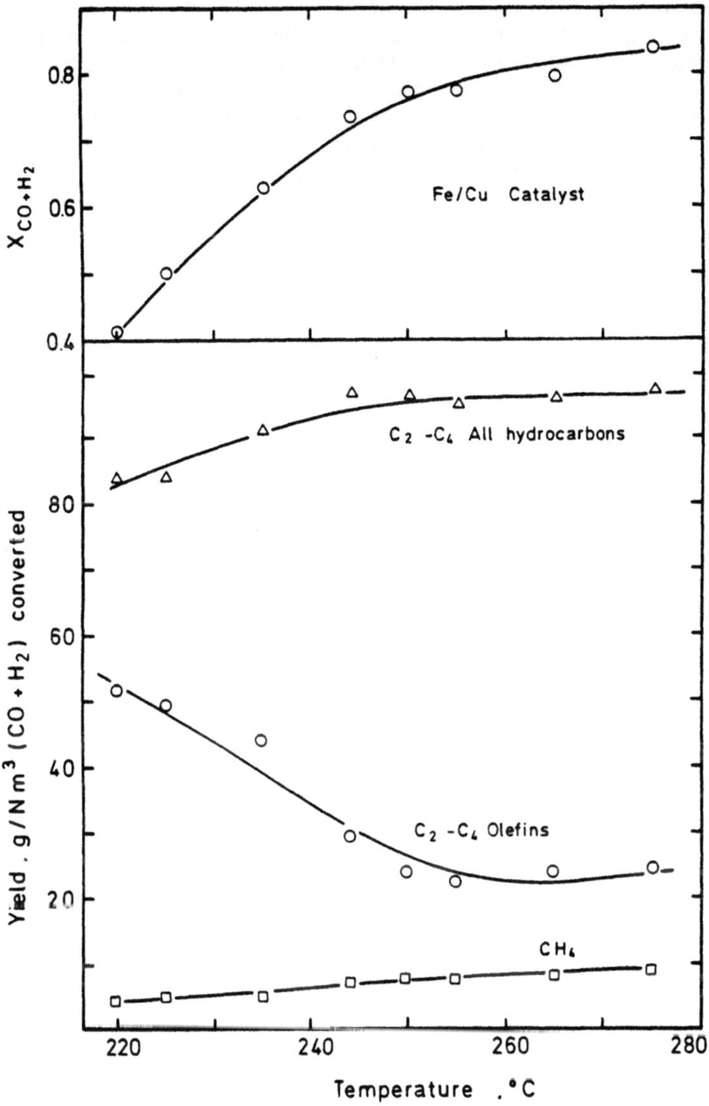

Fig. 21: Product distribution of Fe/Cu catalyst (C) as a function of T

catalyst, see Fig. 16 and at 220 °C the olefin content of the C_2 to C_4 fraction is also comparable with that obtained from the Mn/Fe catalyst. The chain growing probability calculated from the distributions shown in Fig. 21 varies between 0.63 and 0.67. Also Satterfield and Huff (13) have reported an α value of 0.67 $\pm$ 0.02 for a conventional only Fe based ammonia catalyst, see Fig. 2. It is therefore really questionable whether the claimed selectivity of the Mn/Fe catalysts with respect to short chain olefins is an actual peculiarity of this catalysts or only a result of reduced activity as compared with conventional Fe catalysts.

Schlesinger et al. (39) used a fused Fe catalyst which had a high selectivity with regard to oxygenates. The selective catalyst of Schlesinger et al. is about as active as the Mn/Fe catalyst (D) though the temperature ranges of their applicability differ by 50 °C. The Fe catalysts used by Kölbel and Ackermann (34), Kunigi et al. (42), Mitra and Roy (41), the red mud used by Mohammed and the Fe/Cu catalyst (C) give first order rate constants which vary by a factor of about 4. Table 8 lists the numerical k_o values. The data refer to different temperatures and different concentrations of the catalyst in the slurry. As the concentrations of the catylytically active component in the slurry is not known it is not possible to refer the k_o values to such a concentration. It is, however, interesting to notice that if the rate constants k_o are simply divided by the Fe content of the slurry, the k_o' values (in s^{-1} (% wt. Fe)$^{-1}$) obtained from the data reported by Kölbel and Ackermann (34) for the two slurry reactors with largely differing diameters are practically the same, and this data is also in reasonable agreement with the k_o' values evaluated for red mud and the Fe/Cu catalyst (C). It is also interesting to notice that the rate constant k_o' (referred to the overall content in the slurry) evaluated from the studies of Mitra and Roy (41) and Kunugi (42) and from the Mn/Fe catalyst (D) as well are together with the other catalysts all in the range of 0.01 to 0.05 (s % wt. Fe)$^{-1}$. This rather small range is surprising in view of the different preparation methods and the different temperatures applied. But, it supports the view that in all of the above catalysts Fe presents the main catalytically active component.

Table 8: Rate constants for overall synthesis gas conversion

Author	% wt. Fe (in slurry)	T ^{o}C	k_o s^{-1}	k_o' $(s\%\ wt.\ Fe)^{-1}$
Kölbel, Ackermann				
d_R=129 cm	23.3	268	0.4420	0.0190
d_R=4.7 cm	11.0	266	0.2450	0.0223
Schlesinger et al.	14.0	250	0.0560	0.0040
Mitra, Roy	10.1	250	0.1017	0.0100
Kunigi et al.	3.2	260	0.1470	0.0459
Mohammed				
(A) red mud	10.0	280	0.2179	0.0218
(B) Mn/Fe	2.12	300	0.0068	0.0032
This study				
(C) Fe/Cu	8.3	250	0.2306	0.0278
(D) Mn/Fe	1.26	303	0.0690	0.0544

2.4 Modeling the FTS in Slurry Phase

The main difficulties and uncertainties of simulating the behavior of multiphase reactors are concerned with the estimation of parameters involved in the model equations, i.e., the physico-chemical properties, the kinetic data, the hydrodynamic and mass transfer parameters. For the case of the FTS in slurry, the situation is, however, not so bad and it is believed that the above parameters can be estimated with sufficient accuracy. The physico-chemical properties have been measured independently and are summarized in refs. (56,62,85). As discussed in part 2.1 the hydrodynamic and mass transfer properties can be calculated from correlations which were at least partly established from measurements under synthesis conditions. There is also enough experimental information that these correlations may be applicable for larger scale equipment. In the foregoing section 2.3 it was shown that first order rate constants for syngas conversion could be evaluated which seem to be reasonable and consistent. The kinetic law applied is very simplified indeed and, for instance, cannot account for changes of k_H with variations of the inlet gas composition. Nevertheless it is thought that the available information should be sufficient to develop a more sophisticated reactor model for the FTS in slurry phase which is open for further improvements, of course.

2.4.1 Model development. A model of a FTS slurry reactor should be based on the following phenomena and assumptions:

(1) Axial dispersion in the gas and the liquid phase
(2) The catalyst must not be uniformly distributed over the entire suspension volume which is considered by introducing the sedimentation-dispersion model (73-75).
(3) The rate limiting step is first order in H_2 and zero order in CO as proposed by Dry (11) and Atwood and Bennett (29) and applied by Satterfield and Huff (81 and in the previous analysis of FTS labscale reactors (85,87).
(4) Absorption enhancement at the gas-liquid interface due to chemical reaction on the surface of fine catalyst particles can be neglected (88,89).
(5) The unknown stoichiometry of the FT synthesis will be considered by introducing the usage ratio $U = \Delta N_{CO}/\Delta N_{H_2}$.
(6) The total pressure within the reactor is constant, i.e., the influence of hydrostatic head on gas expansion is neglected.
(7) The variability of the molar gas flow rate will be accounted for by applying a contraction factor α which is defined by eq. (28). Experimental data suggest that α is about -0.5, and this value will be used in the computations throughout.
(8) The hydrodynamic properties, i.e., gas holdup, interfacial area, heat and mass transfer coefficients and dispersion coefficients are assumed to be spatially independent.
(9) Owing to the low heat capacity of the gas phase compared to the slurry phase a heat balance on the suspension (liquid plus solid) will only be considered.
(10) Preliminary calculations have indicated that the temperature profiles within the slurry phase are flat. Therefore, the variation of the physico-chemical properties (i.e., density, viscosity, diffusivity, and solubility) with temperature along the reactor is not considered. The physico-chemical properties are calculated for the mean reactor temperature. The temperature influence on the gas flow rate within the reactor is also neglected.
(11) As the catalyst particles are usually small, pore diffusion limitations are neglected. In addition, no temperature difference between the catalyst and the liquid is assumed.

As the rate of synthesis gas conversion depends only on the hydrogen concentration the balance equations of the gas and liquid phase need only be formulated for this key component. The derivation of model equations is given elsewhere (85). The final result is given in dimensionless form by the following equations

Gas phase H_2 balance

$$\frac{1}{Bo_G}\frac{d^2\bar{y}}{dz^2} - \frac{1+\alpha^*}{(1+\alpha^*\bar{y})^2}\frac{d\bar{y}}{dz} - St_G\theta(\bar{y}-\bar{x}) = 0 \qquad (35)$$

Liquid phase H_2 balance

$$\frac{1}{Bo_L}\frac{d^2\bar{x}}{dz^2} + St_L(\bar{y}-\bar{x}) - Da\,\eta_S(\theta)C_{cat}(z)d^{-\gamma/\theta}\bar{x} = 0 \qquad (36)$$

Heat balance

$$\frac{1}{Pe}\frac{d^2\theta}{dz^2} - St_H(\theta-1) + BeDa\,\eta_S(\theta)C_{cat}(z)e^{-\gamma/\theta}\bar{x} = 0 \qquad (37)$$

The balance equations are subject to boundary conditions.

$$z = 0 \qquad 1 = \frac{(1+\alpha^*)\bar{y}}{1+\alpha^*\bar{y}} - \frac{1}{Bo_G}\frac{d\bar{y}}{dz} \qquad (38)$$

$$\frac{d\bar{x}}{dz} = \frac{d\theta}{dz} = 0 \qquad (39)$$

$$z = 1 \qquad \frac{d\bar{y}}{dz} = \frac{d\bar{x}}{dz} = \frac{d\theta}{dz} = 0 \qquad (40)$$

2.4.2 Computational method and estimation of parameters. The system of three differential equations which presents the design model is nonlinear and subject to boundary conditions. For solving numerically the model equations the method of orthogonal collocation was used (90, 91). As collocation functions the so-called shifted Legendre polymonials were applied. As a rule the collocation was done for 5 inner points. The lumped equations were solved by means of the Newton-Raphson iteration method.

The program developed calculates the profiles of the following quantities: hydrogen concentration in gas and liquid phase, temperature, gas velocity, catalyst concentration, conversion and relative reaction resistance. In addition, some other quantities, as for instance, space-time-yield STY in Nm^3 synthesis gas converted per hour and m^3 reactor volume were computed. Operational conditions, rate parameters, and geometrical sizes, which are read by the program, are listed in Table 9. The relations for calculating the physico-chemical properties and hydrodynamic parameters which were applied in the computation program are not discussed in detail here. They and the correlations for estimating all the parameters involved are summarized in ref. (85).

Table 9: Quantities required in computations

Geometry	
Reactor diameter, d_R, cm	(150)
Reactor length, L, cm	(800)
Particle diameter, d_p, cm	(0.005)
Operational conditions	
Inlet gas velocity, u_{Go}, cm/s	(1-12)
Inlet hydrogen mole fraction, y_o	(0.4)
CO/H_2 inlet ratio, I	(1.5)
Pressure, P, MPa	(1.2)
Wall (cooling) temperature, T_w, °C	(258)
Specific heat exchange area, a_H, cm^{-1}	(0.1)
Reaction parameters	
CO/H_2 usage ratio, U	(1.5)
Contraction factor	(-0.5)
Concentration of catalyst in slurry, %wt.	(20)
Concentration of Fe in catalyst, % wt. Fe	(10-20)
Preexponential factor, k'_{o}, (s % wt. Fe in slurry)$^{-1}$ (referred to synthesis gas conversion)	(1.12×10^5)
Activation energy, E_A, kJ/mol	(70)

2.4.3 Results of simulations. Comprehensive computations have been carried out to detect the behavior of the FT slurry bubble column reactors (15, 85). Owing to space limitations only few results will be presented here which are related to reactor scaleup.

Effect of pressure: For isothermic conditions the operation pressure does not influence conversion, of course, and the space-time-yield increases linearly with P. If the reactor operates under nonisothermic conditons the temperature rises and this causes the conversion to increase as is shown in Fig. 22. The space-time-yield is higher than under isothermic conditions but increases yet almost linearly with pressure. The rise in temperature from 258 to 288 °C under nonisothermic conditions and a pressure of 3 MPa brings a gain in overall conversion from 0.78 to 0.96. The linear increase of the space-time-yield with synthesis pressure is in full agreement with the findings of Hall et al. (40) and Benson et al. (38) and is mainly a result of the first-order rate expression, of course. It should be pointed out that the linear dependency of STY on P facilitates design calculations considerably.

Effect of column diameter: The column diameter was varied between 1 and 5 m. The results with respect to conversion are shown in Fig. 23. A slight decrease in conversion is found when increasing d_R. As the dispersion coefficients are mainly affected by d_R the conversion drop has to be attributed to enlarged dispersion, particularly, that of the gas phase. The overall effect is, however, only moderate.

Due to high dispersion in the liquid phase the temperature profile is usually flat. The largest temperature difference was found to be 5 °C for the 1 m diameter reactor which reduces for larger diameter columns.

Influence of gas velocity: The FTS in slurry phase takes place in the absorption-with-slow-reaction regime as any significant absorption enhancement cannot be expected (88,89). In this case, the space-time-yields obtainable in two-phase bubble columns run through a maximum value as a function of the gas velocity (92). Fig. 24 presents some computed results for the three--phase FT process and various catalyst concentrations. Here also the space-time-yield goes via a maximum value in dependence on the inlet gas velocity. The optimum gas velocity, i.e., that at which the STY is highest, increases with the reaction rate, i.e., catalyst content in the slurry. Of course, under nonisothermic conditions when the temperature is allowed to increase slightly the optimum gas velocity rises slightly as well. It is interesting to note that for 20 % wt. Fe in slurry which corresponds to the conditions

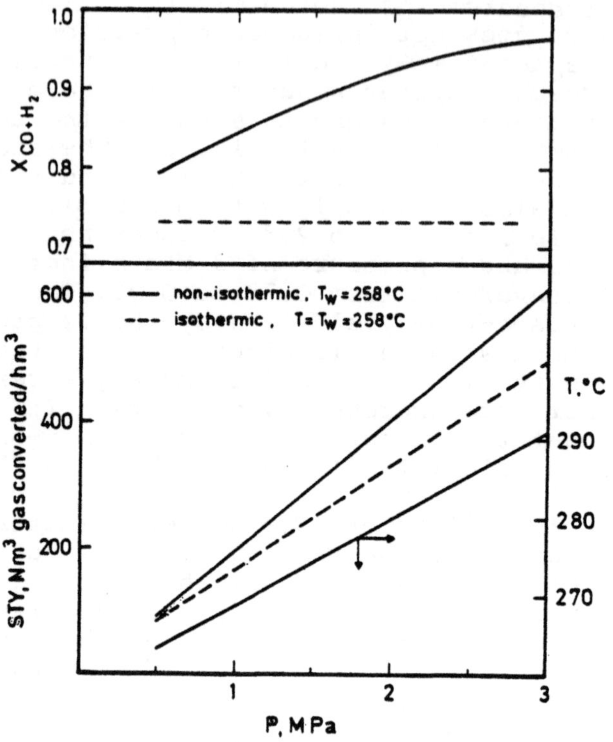

Fig. 22: Effect of pressure on space-time-yield and conversion

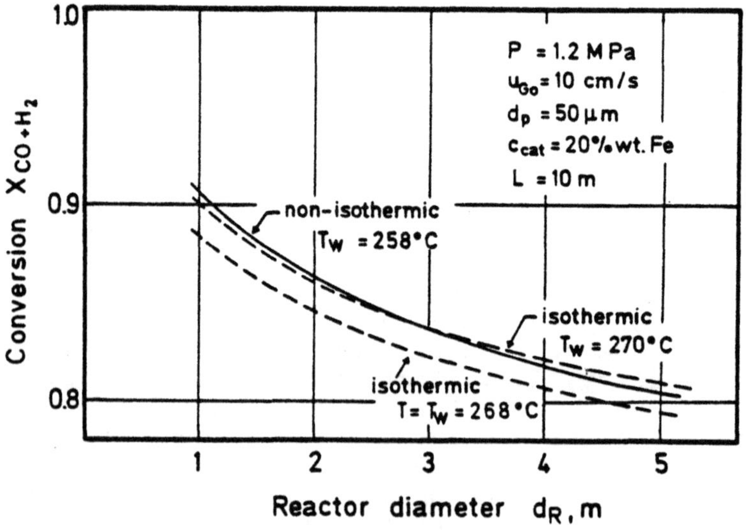

Fig. 23: Influence of reactor diameter on conversion

applied in the Rheinpreussen-Koppers demonstration plant the optimum inlet gas velocity is between 9 and 11 cm/s which is in complete accordance with the gas velocity applied usually in that plant (35) and recommended by Kölbel (93). The model also simulates qualitatively correct the observation that increase of gas throughput decreases only very moderately the conversion (93). For instance, if the inlet gas velocity is increased from 4 to 8 cm/s the conversion drops only from 0.95 to 0.91. This is partly due to the fact that under nonisothermic conditions the temperature simultaneously rises from 265 to 269.5 °C.

Effect of gas-liquid mass transfer: Different opinions on the importance of mass transfer limitations have been uttered by Satterfield and Huff (81,82,94) and Zaidi et al. (54) and Deckwer et al. (87). Satterfield and Huff (81) assumed a bubble diameter of about 2 mm and concluded that the FTS in slurry phase may be significantly limited by gas-liquid mass transfer. New experimental results do, however, confirm that in the molten paraffin system bubble diameters are less than 1 mm (53-55). With this low value and the high gas hold-up observed in FT liquid phase, i.e., eq. (8), high interfacial areas are obtained. Therefore significant mass transfer can be excluded and for the catalyst systems studied until now the FT process in slurry phase is mainly reaction controlled (54,87). In addition, it follows from the reactor model used by Satterfield and Huff (81) that the relative mass transfer resistance is given by (95)

$$1 - \beta = \frac{1/k_L a}{1/K_A} = \frac{X_{H_2}}{1 - e^{-\alpha_H L}} \tag{41}$$

which certainly is an unreasonable result.

Numerical simulations on the basis of the above model also indicate that mass transfer resistances only slightly effect reactor performance. The full-drawn curve in Fig. 25 presents space-time-yields calculated with ε_G and a from eqs. (8) and (10). If, now, the values of a predicted from eq. (10) are reduced to one half only, little effect on the STY can be observed, see Fig. 25. Actually, the reduction of a by 50 % reduces the maximum attainable STY by less than 4 %. In both cases, i.e., with the full values of a and the reduced ones, the overall process is predominantly reaction con-

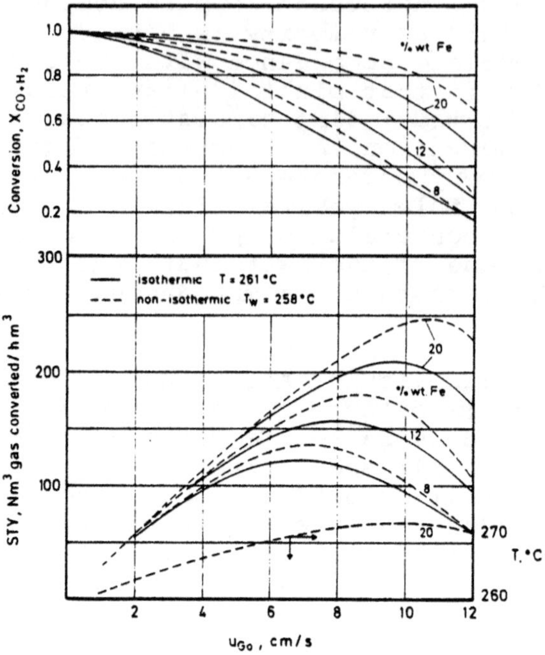

Fig. 24: Conversion and space-time-yield as a function of inlet gas velocity

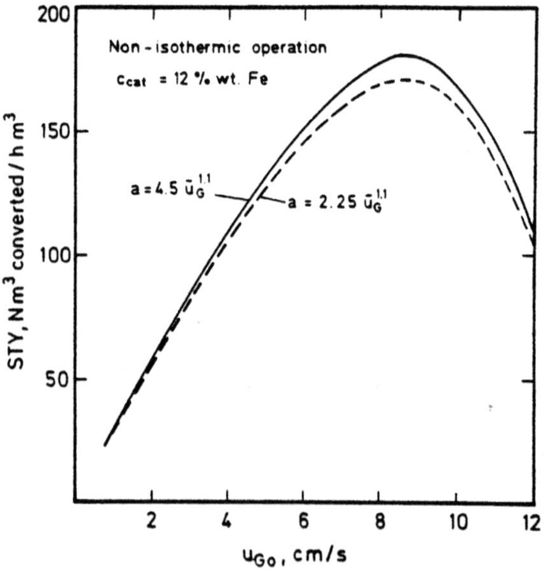

Fig. 25: Effect of reduced interfacial area on STY

trolled. For instance, reduction of a decreases the relative reaction resistance (average over reactor length) from 0.95 to 0.91 if u_{GO} = 10 cm/s. However, if the reactor does not operate under optimum conditions with regard to high space-time-yield the relative reaction resistance may be smaller, hence the overall process is more and more controlled by mass transfer limitations. For instance, if the inlet gas velocity is u_{GO} = 3 cm/s corresponding to a mean gas velocity of only 1.6 cm/s in the reactor as the conversion is 0.94, then the reaction resistance is 0.7 if a from eq. (10) is used. If a is reduced to half the value from eq. (10) the reaction resistance decreases to 0.55, i.e., the overall process is now controlled by mass transfer to an extent of 45 %. These examples illustrate that it is not possible to say in general, that a certain slurry process is mass transfer or reaction rate controlled without stating the operational conditions, i.e., above all the gas velocity. The computations indicate, at least, that under favorable operational conditions, which lead to high space-time-yields, the FT slurry process is mainly reaction controlled for the conventional Fe catalysts.

2.4.4 Design considerations on a large scale FT slurry plant. The presented model predicts reasonably and, at least, qualitatively in a correct manner the behavior of FT slurry reactors operated under industrial conditions (Rheinpreussen-Koppers demonstration plant). Also other observations in lab-scale apparatuses are in agreement with model simulations. Therefore, the model is thought to offer the possibility to estimate production capacities rather reliably.

Fig. 26 presents overall synthesis gas conversions and space-time-yields as a function of the reactor length for 3 diameters. The calculations are again based on the data set of Table 9. To be conservative, an inlet gas velocity of 9 cm/s was used which guarantees operation in the vicinity of maximum achievable space-time-yields. The information given in this figure can easily be used to calculate production heights. For instance, if the desired conversion level is fixed to about 0.9 in order to dispense with gas recycling a 3-m diameter reactor would require a height of 12 m and the achievable space-time-yield (at 1.2 MPa operational pressure) would be 160 Nm^3 synthesis gas converted per hour and m^3 reactor volume. Hence, the entire reactor would hourly convert 13,570 Nm^3 which corresponds with an annual production of 24,600 t of hydrocarbons (C_{1+}).

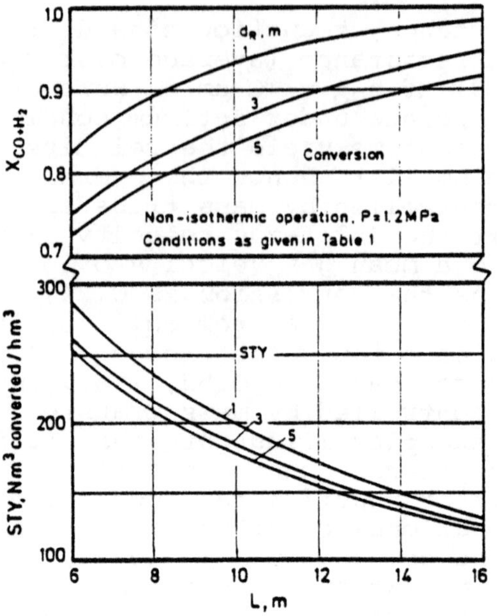

Fig. 26: Conversion and space-time yield as a function of reactor length

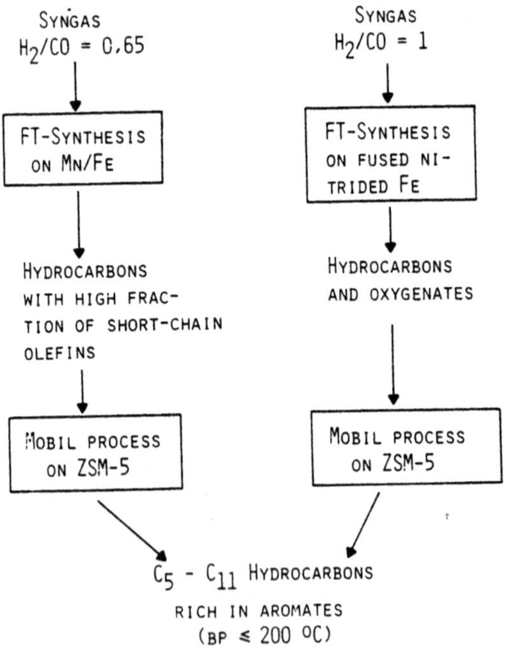

Fig. 27: Combination of FTS and Mobil's process

If the pressure can be increased threefold, for example, which appears to be particularly timely if synthesis gas is manufactured by high-pressure gasifiers the production height could be increased to 75,000 t/a as STY increases almost linearly with pressure, see Fig. 22, which is in full agreement with the results of Hall et al. (40) and Benson et al. (38). It should be emphasized once again that the calculations refer to the catalytic rate constant evaluated from the overall conversion of the Rheinpreussen-Koppers plant. If the rate constant evaluated from Kunugi's study would be used the result would be even more promising, i.e., higher conversion and larger space-time-yields would be obtainable.

Provided a reactor 5 m in diameter can uniformly be supplied with gas by introducing pertinent spargers such a reactor operated at 3.6 MPa and a conversion of 0.9 would require a length of 14 m corresponding with a STY of 3x136 = 408 $Nm^3/h\ m^3$. This gives an annual production of about 200,000 t of hydrocarbons. It is interesting to calculate the number of reactors of this size required to achieve the production height of Sasol II. The new Sasol plant is intended to produce 1.4 million tons of motor fuels per year in 8 synthol reactors (entrained fluidized beds) of an enormous size. Typical product distributions of the slurry reactor reported by Kölbel and Ralek (35) show about 65 % wt. motor fuels (gasoline and diesel). Therefore, annual production of $1.4x10^6$ t of motor fuels would require only about 11 slurry reactors of 5 m in diameter and 14 m in height. One can expect that operation of such a battery of slurry reactors is considerably easier and more economical than production with synthol reactors. In addition, it should be pointed out that in contrast to Sasol the slurry reactor processes favorably weak synthesis gas, i.e., gases of high CO content, which is also advantageous from an economic point of view (83,96).

2.5 Summary and Recommendation for Further Work

In this chapter it was shown that the major engineering parameters which might affect the performance of a FT slurry reactor can be estimated from rather reliable correlations. There are, however, some controversial results in the literature which concern gas holdup and interfacial area (bubble diameter). Additional studies would be valuable for further clarification of this point. However, one can state, at least, that gas holdup and interfacial area are surprisingly large in the molten paraffin slurry system. It is thought that eqs. (8) and (10) can be used giving conservative estimates of ε_G and a. Mixing and heat transfer coefficients can be calculated from well established correlations with sufficient accuracy.

Recent studies on the FTS in slurry phase using various catalysts (Fe/Cu, Mn/Fe, Fe based ammonia synthesis catalysts) indicate that the product slate follows the most probable distribution of Schulz-Flory, the chain growing probability being in the range 0.63 to 0.7. Hence, the C_2 to C_4 fraction is about 80 to 100 g per Nm^3 syngas converted corresponding to 40 to 50 % wt. of product. The C_2 to C_4 olefin fraction amounts to a maximum of about 60 g per Nm^3 converted. Compared to the processes operated by Sasol and the classical K promoted Fe precipitation catalyst this is a considerable increase in selectivity. Further reduction of the chain growing probability would possibly increase the C_2 to C_4 fraction but simultaneously the CH_4 fraction would also rise which is generally undesirable. The reaction studies confirm that the slurry is exceptionally suited to process syngases of low hydrogen content, i. e., $CO/H_2 \approx 1.5$.

From the conversion measurements and the known engineering parameters kinetic constants for syngas conversion could be evaluated on the basis of a very simplified kinetic law and if the CO to H_2 feed ratio is in the range of 1.5. If referred to the Fe content the rate constants lie approximately in the same range. This could be an indication that Fe is the principal catalytically active component, also for the case of Mn/Fe catalyst.

A design model has been developed which can be used to simulate the performance of larger scale FT slurry reactors. Predictions of this model are in accordance with practical experience as far as this is reported in the literature.

The results on the FTS in slurry phase presented here show where additional investigations are required. First of all, a thorough kinetic study of the synthesis in slurry phase with different types of catalysts is needed. Such a kinetic study should also account for the adsorption equilibria of reactants and products on the suspended catalyst. Such a kinetic analysis could probably explain some particular features of the slurry operation. For instance, it is not clearly understood yet why only the slurry reactor uses favorably CO rich syngas while such gases cannot be operated in fixed and fluidized beds. High mixing and excellent heat transfer in the slurry reactor might explain in part this peculiarity but it is believed that the major reason should be due to different kinetics and/or adsorption equilibria.

In recent years, catalysts have been developed which give a non-Schulz-Flory product distribution in fixed beds. Such catalysts should be tested in slurry phase operation.

For an economic evaluation of the FT slurry process it is timely to carry out studies on catalyst life-time and deactivation. This should be accompanied by the development of strategies how to minimize the detrimental effects of catalyst deactivation. Another important point concerns the check of competitive regeneration techniques of spent catalysts.

The FT studies in slurry phase reported so far are not optimized with regard to today's technological requirements and the design of an industrial FT plant given in 2.4.4 referred closely to the operational conditions of the Rheinpreussen-Koppers demonstration plant. One can certainly assume that the performance, i.e., particularly the space-time yield can considerably be improved. From the point of view of reactor performance alone a slurry reactor gives highest space-time yields if it is operated close to the diffusional regime of mass transfer theory. The entire process is then predominantly mass transfer controlled. To achieve this kind of control the reaction rate in the suspension phase must be high. Therefore the development of highly active catalysts is an important objective but the comparative analysis of various catalysts has shown that rate constants of 0.05 $(s\ \%\ wt.\ Fe)^{-1}$ seem to be an upper limit. On the other hand the reaction rate can be raised by increasing the catalyst concentration. This possibility has not yet been systematically studied but looks very promising as cold flow experiments have shown that slurry reactors can be operated with a solid concentration up to 20 % vol. (corresponding to 40 to 50 % wt.) without significant loss in gas holdup and interfacial area. Therefore, FT reaction studies with higher catalyst concentrations in the slurry are urgently needed.

Studies carried out in the fifties and model simulations show that slurry reactor performance is almost proportional to operation pressure. This result is especially relevant in view of high pressure gasifiers and should be confirmed by new experimental studies.

3 COMBINATION OF FTS AND MOBIL PROCESS

Owing to the high contribution of syngas costs to the economy of all indirect liquefaction routes a process is desired which could directly use purified syngas from second-generation gasifiers, i.e., gasifiers which operate cost effectively at high pressure and low amounts of steam but produce a syngas of low H_2 hydrogen content (CO to H_2 ratio of about 1.5 to 1.7). The Fe catalysts used in the FTS have not only the ability to hydrogenate CO but also sufficient water-gas shift activity. However only the process conditions of the FTS in slurry phase permit the use of syngases of the above composition. This has not only been proven for the classical K promoted precipitated Fe catalyst but also by recent studies in slurry phase with nonpromoted Mn/Fe and Fe/Cu catalysts. Though the product slates of these catalysts follow closely Schulz-Flory distribution the selectivity with regard to lower olefins is high as compared to classical K promoted catalysts.

It has been pointed out (83,96,97) that the direct conversion of syngases of low H_2 to CO ratio does not only save the investment of the shift reactor but reduce additionally operation costs due to savings in steam demand. Hence, the overall thermal efficiency can be improved.

It is therefore obvious to combine the ability of the FTS slurry reactor to use syngases of low H_2 to CO ratio with the high selectivity of the Mobil route in order to produce a C_5 to C_{11} hydrocarbon mixture of a high RON. In particular, advantageous results can be expected if the FTS is carried out on catalysts which possess improved selectivity either to short chain olefins or to lower oxygenates. Mn/Fe and nonpromoted Fe/Cu catalysts can be used in case of short-chain olefins, while nitrided fused Fe catalysts have proven to yield high selectivities to oxygenates (39,98).

Both short-chain olefins and oxygenates can be converted on zeolites of ZSM-5 type to C_5 to C_{11} hydrocarbons with a high content of aromates at temperatures between 350 to 400 ^{o}C. Fig. 27 presents a scheme of the proposed combination of the FT slurry process with Mobil's process. The advantage of such a combination is that the conversion of syngas from low to high H_2 content and the methanol synthesis is substituted by only one process, i.e., the FTS in slurry phase. One can expect that the two-step process (FT-Mobil) from syngas of low H_2 content to high quality gasoline is more profitable than the original MTG route of Mobil. In Table 10 hydrocarbon fractions are given which were obtained with a Mn/Fe catalyst in slurry phase. The olefin content in the C_2 to C_4 fraction is about 75 %. If these olefins are converted on ZSM-5 to C_5 to C_{11} hydrocarbons, the C_5 to C_{11} fraction increases to about 70 % wt. and is in the same range as that attainable in the MTG route.

Table 10: Product slate (% wt.) FTS in slurry phase on Mn/Fe catalyst (A) and after conversion on ZSM-5 (75 % wt. of the C_2 to C_4 fraction are taken as olefins)

	A	B
C_1	10.56	10.56
ΣC_2-C_4	38.36	9.59
ΣC_5-C_{11}	41.66	70.43
C_{12+}	9.42	9.42

An advanced process scheme would place both catalysts in one single slurry reactor. Studies in fixed beds with a layer of Mn/Fe catalyst followed by a layer of ZSM-5 zeolite and with physical mixtures of these two catalysts as well have given promising results (22, 27). About 120 g of C_5 to C_{11} hydrocarbons per Nm^3 syngas converted (CO to H_2 feed ratio 1.43) could be obtained. 83 % wt. of this fraction are aromates and branched hydrocarbons (27). In the one-stage process each catalyst has to be operated somehow removed from its optimum temperature range. To find out the optimum temperature range for such a bifunctional catalytic system is a challenging task for reaction engineers.

Another unsolved problem concerns the regeneration of two component catalysts because each catalytic function surely requires its own regeneration conditions.

The development of a combined FT-Mobil process is still at the very beginning and involves a lot of speculation. Much additional work is required to prove the economic viability of such a combination either in a one-stage or a two-stage process. The envisioned route from coal to hydrocarbons has the primary advantage of high C_5 to C_{11} selectivity. The produced hydrocarbons can be used either as high quality motor fuels or as a raw material source for the chemical industry. It should be pointed out that the FT-Mobil route would require only little effort for environmental protection and no or little modification of the present structure of petrochemical industry.

Notation

a	gas-liquid interfacial area
a_H	specific heat transfer area
a_S	liquid-solid specific interfacial area
Be	dimensionless group, $(-\Delta H_R/\bar{\rho}\bar{C}_p)(Py_o/HeT_w)$
Bo_G	gas phase Bodenstein number, $u_{Go}L/E_G \varepsilon_G$
Bo_L	liquid phase Bodenstein number, $u_{Go}L/E_L \varepsilon_L$
C_H, C_{HL}	hydrogen concentration in liquid phase
C_H^*, C_{HL}^*	equilibrium concentration of H_2 in liquid phase
C_{cat}	catalyst concentration in suspension
$\bar{C}_p$	heat capacity of suspension
Da	Damköhler number, $k_f \varepsilon_L/u_{Go}$
d_C	reactor diameter
d_{cat}	catalyst particle diameter
D_L	diffusion coefficient in liquid phase
d_S	Sauter mean bubble diameter
E_G	gas phase dispersion coefficient
E_L	liquid phase dispersion coefficient
Fr	Froude number, $\bar{u}_G^2/d_C g$
g	gravitation constant
G	volumetric gas flow rate
He	Henry coefficient

h	heat transfer coefficient (to the cooling wall)
$(-\Delta H_R)$	reaction enthalpy
I	CO/H_2 inlet molar ratio
k, k'	rate constant
K_A	overall absorption-reaction parameter, eq. (24)
k_f	frequency factor for hydrogen consumption rate referred to wt. % Fe in slurry, (1/s wt. % Fe)
k_H	rate constant for hydrogen consumption
k_o	rate constant for synthesis gas consumption
k_o'	rate constant for synthesis gas consumption referred to wt. % Fe in slurry
$K_o a$	overall absorption reaction parameter, eq. (12)
k_L	liquid side mass transfer coefficient
k_S	liquid-solid mass transfer coefficient
L	reactor length
N	mole flow rate
p	partial pressure
P	total pressure
Pe	Pèclet number for heat, $u_{Go} \bar{\rho} \bar{c}_p L/(\varepsilon_L \lambda_{ax})$
Pr	Prandtl number, $\bar{u}\, \bar{c}_p/\bar{\lambda}$
R	universal gas constant
r	reaction rate
r_p	rate of chain propagation step
r_t	rate of chain termination step
Re	Reynolds number of particles or bubbles
Re	Reynold number, $\bar{u}_G d_c/\nu$, in eq. (19)
Sc	Schmidt number
Sh	Sherwood number, $k_L d_s/D_L$
St	Stanton number for heat transfer, $h/(\bar{\rho}\bar{c}_p \bar{u}_G)$
St_H	Stanton number for heat transfer, $h\, a_H L/(u_{Go}\bar{\rho}\bar{c}_p)$
St_G	gas phase Stanton number, $(K_L a)_H (L/u_{Go})(RT_W/He)$
St_G'	Stanton number, $K_A (RT/He)(L/u_{Go})$
St_L	liquid phase Stanton number, $(k_L a)_H (L/u_{Go})$
SV	space velocity, volume of gas per catalyst volume and hour

STY — space-time yield, Nm^3 synthesis gas converted per m^3 reactor volume and hour

T — temperature

T_w — cooling wall temperature

u_G — linear gas velocity

$\bar{u}_G$ — mean linear velocity or dimensionless gas velocity u_G/u_{Go}

u_{Go} — inlet linear gas velocity

U — usage ratio, $\Delta N_{CO}/\Delta N_{H_2}$

x — axial coordinate

$\bar{x}$ — dimensionless hydrogen concentration in liquid phase, $C_{HL}/(Py_o/He)$

X_H — hydrogen conversion

X_{CO+H_2} — synthesis gas conversion

y — hydrogen mole fraction in gas phase

$\bar{y}$ — dimensionless hydrogen concentration, y/y_o

y_o — inlet hydrogen mole fraction

z — dimensionless axial coordinate

<u>Greek symbols</u>

α — contraction factor, eq. (28) or chain growing probability

α^* — modified contraction factor, eq. (30)

$\bar{\beta}$ — relative reaction resistance (average value), eq. (35)

γ — Arrhenius number, E_A/RT_w

ε_G — gas holdup

ε_L — liquid holdup

η_S — liquid-solid mass transfer effectiveness factor, $(1 + kC_{cat}\varepsilon_L/k_s a_s)^{-1}$

$\bar{\lambda}$ — heat conductivity of suspension

λ_L — heat conductivity of liquid

λ_{cat} — heat conductivity of catalyst

λ_{ax} — effective heat conductivity of suspension

$\bar{\mu}$	viscosity of suspension, g/cm s
μ_L	viscosity of liquid, g/cm s
$\bar{\rho}$	density of suspension, g/cm^3
ρ_L	density of liquid, g/cm^3
ρ_{cat}	density of catalyst, g/cm^3
θ	dimensionless temperature, T/T_w
ν_L	kinematic viscosity of liquid, cm^2/s

References

1. Shinnar, R. and J.C.W. Kuo, DOE Report No. FE-2766-13, 1978.

2. Eisenlohr, K.H. and H. Gaensslen, Fuel Processing Techn. 4 (1981) 43

3. Kölbel, H. and K.D. Tillmetz, Belg. Patent 837.628 (1976)

4. Shah, Y.T. and A.J. Perotta, Ind. Eng. Chem. Proc. Des. Dev. 15 (1976) 123

5. Frohning, C.D., H. Kölbel, M. Ralek, W. Rottig, F. Schnur and H. Schulz in "Chemierohstoffe aus Kohle", ed. by J. Falbe, G. Thieme Verlag, Stuttgart, 1977

6. Baird, M.J., R.R. Schehl, W.P. Haynes and J.T. Caleb, Ind. Eng. Chem. Proc. Des. Dev. 19 (1980) 175

7. Anderson, R.B., in "Catalysis" Vol. IV, ed. by P.H. Emmett, Reinhold Publ. Corp., New York, 1956

8. Anderson, R.B., L.J.E. Hofer and H.H. Storch, Chem.-Ing.-Techn. 30 (1958) 560

9. Pichler, H. and H. Schulz, Chem.-Ing.-Tech. 42 (1970) 1162

10. Schulz, H. and A. Zein el Deen, Fuel Processing Tech. 1 (1977) 31, 45

11. Dry, M.E., Ind. Eng. Chem. Prod. Res. Dev. 15 (1976) 282

12. Henrici-Olivé, G. and S. Olivé, Ang. Chem. Int. Ed. Engl. 15 (1976) 136

13. Satterfield, C.N., and G.A. Huff, to be published 1981

14. Lehmann, H.-J., B. Schmidt, M. Ralek and W.-D. Deckwer, paper (No. 103d) presented at AIChE 73rd Annual Meeting, Chicago, Nov. 1980

15. Deckwer, W.-D., Y. Serpemen, M. Ralek and B. Schmidt, submitted to Ind. Eng. Chem.Process Des. Dev., 1981

16. Büssemeier, B., C.D. Frohning and B. Cornils, Hydrocarbon Process. 11 (1976) 105
17. Kitzelmann, D., W. Vielstich and T. Dittrich, Chem.-Ing.-Techn. 19 (1977) 463
18. Bhasin, M.M., W.J. Bartley, P.C. Elegen and T. P. Wilson, J. Catal. 54 (1978) 120
19. Madon, R.J., J. Catal. 57 (1979) 183
20. Bub, G., M. Baerns, B. Büssemeier and C. Frohning, Chem. Eng. Sci. 35 (1980) 348
21. Nijs, H.H., P.A. Jacobs and J.B. Uytterhoeven, J.C.S. Chem. Comm. 1979, 1095
22. Caesar, P.D., J.A. Brennan, W.E. Garwood and J. Ciric, J. Catal. 56 (1979) 274
23. Chang, C.D., W.H. Lang and A.J. Silvestri, ibid. 56 (1979) 268
24. Jacobs, P.A., Proc. Int. Symp. on Catalysis by Zeolites, Ecully (Lyon), France, 1980, p. 293
25. Ballivet-Tkatchenko, D., G. Coudurier and H. Mozzanega, ibid., p. 309
26. Rao, V.U.S. and R.J. Gormley, Hydrocarbon Process Nov. 1980, p. 139
27. Müller, K. Diplomarbeit, TU Berlin, 1981
28. Dry, M.E., T. Shingles and L.J. Boshoff, J. Catal. 25 (1972) 99
29. Atwood, H.E. and C.O. Bennett, Ind. Eng. Chem. Process Des. Dev. 18 (1979) 163
30. Zein el Deen, A., J. Jacobs and M. Baerns, Ger. Chem. Eng. 2 (1979) 139
31. Deckwer, W.-D., Oil Gas J., Nov. 10, 1980, 198; Proc. 7th Ann. Int. Conf. on Coal Gasification and Liquefaction, Pittsburgh, August 1980
32. Dry, M.E., Energy Spectrum, Oct 1977, 298
33. Kölbel, H., P. Ackermann and F. Engelhardt, Erdöl und Kohle 9 (1956) 153, 225, 303
34. Kölbel, H. and P. Ackermann, Chem.-Ing.-Tech. 28 (1956) 381
35. Kölbel, H. and M. Ralek, Cat. Rev.-Sci. Eng. 21 (1980) 54
36. Crowell, J.H., H.E. Benson, J.H. Field and H.H. Storch, Ind. Eng. Chem. 42, 2376 (1950)
37. Kastens, M.L., L.L. Hirst and R.G. Dressler, Ind. Eng. Chem. 44, 450 (1952)
38. Benson, H.E., J.H. Field, D. Bienstock and H.H. Storch, Ind. Eng. Chem., 46, 2278 (1954)
39. Schlesinger, M.D., H.E. Benson, F.M. Murphy and H.H. Storch, Ind. Eng. Chem. 46 (1954) 1322
40. Hall, C.C., D. Gall and S.L. Smith. J. Inst. Petroleum, 38 (1952) 845
41. Mitra, A.K. and A.N. Roy. Indian Chemical Engineer 1963, 127

42. Kunugi, T., T. Sakai and N. Negishi. Sekiyu Gakkai Shi., 11 (1968) 636

43. Schlesinger, M.D., J.H. Crowell, M. Leva and H. H. Storch, Ind. Eng. Che., 43 (1951) 1474

44. Sherwin, M.B. and M.E. Franck, Hydrocarbon Processing Now. 1976, 122

45. Sherwin, M.B. and D. Blum, EPRI Report AF-693, Project 317-2, May 1978

46. Meisel, S.L., J.O. McCullough, C.H. Lechthaler and P.B. Weisz, Chemtech. February 1976, 86

47. Chang, C.D., Chem. Eng. Sci. 35 (1980) 619

48. Olson, D.H., W.O. Haag and R.M. Lago, J. Catal. 61 (1980) 390

49. Chang, C.D. and A. Silvestri, J. Catal. 47 (1977) 249

50. Derouane, E.G., J.B. Nagy, P. Dejaifve, J.H.C. van Hoof, B.P. Spekman, J.C. Vendrine and C. Naccache, J. Catal. 53 (1978) 40

51. Dejaifve, P., J.C. Vendrine, V. Bolis and E.G. Derouane, J. Catal. 63 (1980) 331

52. Chang, C.D., J.C.W. Kuo, W.H. Lang, S.M. Jacob, J.J. Wise and A.J. Silvestri, Ind. Eng. Chem. Proc. Des. Dev. 17 (1978) 255

53. Quicker, G. and W.-D. Deckwer, Chem.-Ing.-Tech., to be published 1981

54. Zaidi, A., Y. Louisi, M. Ralek and W.-D. Deckwer, Ger. Chem. Eng. 2 (1979) 94

55. Quicker, G. and W.-D. Deckwer, Chem. Eng. Sci., to be published

56. Deckwer, W.-D., Y. Louisi, A. Zaidi and M. Ralek, Ind. Eng. Chem. Process Des. Dev. 19 (1980) 699

57. Akita, K. and F. Yoshida, Ind. Eng. Chem. Proc. Des. Dev. 12 (1973) 76

58. Gestrich, W. and W. Rähse, Chem.-Ing.-Tech. 47 (1975) 8

59. Bach, H.F. and T. Pilhofer, Ger. Chem. Eng. 1 (1978) 270

60. Deckwer, W.-D., Y. Louisi, A. Zaidi and M. Ralek, paper (No. 76a) presented at AIChE 72nd Annual Meeting, San Francisco, Nov. 25-28, 1979

61. Calderbank, P.H., F. Evans, R. Farley, G. Jepson and A. Poll, "Catalysis in Practice", Instn. Chem. Engrs., London, 1963, 66

62. Hammer, H., "Zur Reaktionstechnik in Blasensäulen-Reaktoren mit suspendiertem Kontakt", Habilitationsschrift, Technische Universität Berlin, 1968

63. Farley, R. and D.J. Ray, J. Inst. Petr. 50 (1964) 27

64. Peter, S. and M. Weinert, Z. Phys. Chem. N.F. 5 (1955) 114

65. Hughmark, G.A., Ind. Eng. Chem. Proc. Des. Dev. 6 (1967) 218
66. Calderbank, P.H. and M. Moo-Young, Chem. Eng. Sci. 16 (1961) 39
67. Sänger, P. and W.-D. Deckwer, Chem. Eng. J., to be published, 1981
68. Deckwer, W.-D., R. Burckhart and G. Zoll, Chem. Eng. Sci. 29 (1974) 2177
69. Shah, Y.T., G.J. Stiegel and M.M. Sharma, AIChE J. 24 (1978) 369
70. Joshi, J.B. and M.M. Sharma, Trans. Instn. Chem. Engrs. 57 (1979) 244
71. Mangartz, K.-H. and T. Pilhofer, Verfahrenstechnik (Mainz) 14 (1980) 40
72. Field, R.W. and J.F. Davidson, Trans. Instn. Chem. Engrs. 58 (1980) 228
73. Kato, Y., A. Nishiwaki, T. Kago, T. Fukuda and S. Tanaka, Int. Chem. Eng. 13 (1973) 82
74. Cova, D.R., Ind. Eng. Chem. Process Des. Dev. 5 (1966) 21
75. Imafuku, K., T.-Y. Wang, K. Koide and H. Kubota, J. Chem. Eng. Japan, 1 (1968) 153
76. Louisi, Y., Dr. Ing. thesis, Technische Universität Berlin, 1979
77. Kast, W., Int. J. Heat Mass Transfer 5 (1962) 329
78. Deckwer, W.-D., Chem. Eng. Sci. 35 (1980) 1341
79. Kölbel, H., E. Borchers and K. Müller, Chem.-Ing.-Tech. 30 (1958) 729
80. Kölbel, H., E. Borchers and J. Martins, ibid. 32 (1960) 84
81. Satterfield, C.N. and G.A. Huff, Chem. Eng. Sci. 35 (1980) 195
82. Satterfield, C.N., J.P. Longwell and G.A. Huff, paper (No. 102a) presented at AIChE 73rd Annual Meeting Nov. 1980, Chicago
83. Poutsma, M.L., Oak Ridge Natl. Lab, Report No. 5635 (1980)
84. Mohammed, M.S., Dr.-Ing. thesis, TU Berlin, 1977
85. Deckwer, W.-D., Y. Serpemen, M. Ralek and B. Schmidt, Ind. Eng. Chem. Process Des. Dev., submitted 1981
86. Lehmann, H.-J., Dr.-Ing. thesis, TU Berlin 1981
87. Deckwer, W.-D., Y. Serpemen, M. Ralek and B. Schmidt, Chem. Eng. Sci. 36 (1981) 765, 791
88. Alper, E., B. Wichtendahl and W.-D. Deckwer, Chem. Eng. Sci. 35 (1980) 217
89. Alper, E. and W.-D. Deckwer, Chem. Eng. Sci. 36 (1981) 1097
90. Finlayson, B.A., The Method of Weighted Residuals and Variational Principles, Acad. Press, New York, 1972

91. Villadsen, J.V. and M.L. Michelsen, Solution of Differential Equation Models by Polynomial Approximation, Prentice Hall, Englewood Cliffs, 1978

92. Schumpe, A., Y. Serpemen and W.-D. Deckwer, Ger. Chem. Eng. 2 (1979) 234, 267

93. Kölbel, H., private communication

94. Satterfield, C.N. and G.A. Huff, Chem. Eng. Sci. 36 (1981) 790

95. Bukur, D., Mobil Research and Development Corporation, Paulsboro, N.J., to be published

96. Fischer, R.H. and R.E. Hildebrand, paper presented at Methanol Symp. 13th Middle Atlantic Regional Meeting ACS, March 23, 1979

97. R.E. Hildebrand and L.M. Joseph, ibid.

98. R.B. Anderson, Cat. Rev.-Sci. Eng. 21 (1980) 4

CHEMICAL CLEANING OF COAL - THE OXYDESULFURIZATION PROCESS

Y. T. Shah and R. S. Albal

Department of Chemical and Petroleum Engineering
University of Pittsburgh, Pittsburgh, PA 15261

ABSTRACT

One of the major uses of coal is to burn it directly in power plants; our objective is to burn coal in an environmentally acceptable manner. This requires the removal of sulfur so that EPA standards for the omission of sulfur oxides in power plants are met. This paper briefly reviews the present state of the art for the chemical removal of sulfur from coal via an oxidation process. A brief summary of the existing sulfur removal processes and their economics along with the chemistry and kinetics of inorganic and organic sulfur removal from coal and the reactor design considerations are outlined.

1. INTRODUCTION

The worsening energy situation, amplified by the increasing prices of oil, has forced the world to search for alternate sources of energy. In this country, coal is the largest single energy source with the distinct advantage of long term availability. Unfortunately, it is also a resource with high sulfur and mineral matter content.

There are some environmental and technological problems which need to be solved for the increasing use of coal as an energy source. The major problem facing coal burning power facilities is the production of noxious sulfur oxides (SO_x's). Current Environmental Protection Agency (EPA) regulations restrict SO_2 emissions for a new facility to 1.2 lbs per million BTU equivalent of coal combusted (1).

The sulfur in coal is found in three forms - pyritic, organic and sulfates. Pyritic sulfur, classified as compounds with the formula FeS_2, represents the bulk of the sulfur in many types of coal. Organic sulfur is a broad classification containing any sulfur which is chemically bound to the actual organic coal matrix. Sulfates constitute less than a few percent of the total sulfur in most of the coals.

There are several potential methods for clean coal combustion, including physical coal cleaning, flue gas desulfurization by scrubbing, fluidized bed combustion, conversion of coal into clean gaseous or liquid fuels, etc. However, none are capable of handling special circumstances and none can be applied to coals with diverse physical and chemical properties. Physical cleaning processes such as separation by density difference, magnetic separation etc. are capable, at best, of removing 30-70 percent of pyritic sulfur in the coal with no removal of organic sulfur. Hence, although these techniques are relatively inexpensive and simple to operate, they are not very effective for large organic sulfur coals. Flue gas desulfurization by scrubbing is impractical for small plants, and uneconomical and energy intensive for plants which operate intermittently. Finally, coal conversion to clean gaseous or liquid fuels is very expensive.

Chemical cleaning methods, i.e. removal of sulfur by means of chemical reagents, provide a good alternative approach, which may be an efficient and a cost effective way of solving the air pollution problem. In general, chemically cleaned coal is of uniform quality, low in sulfur and low in mineral matter. It may be supplied as a powder or slurry, or in the form of uniformly sized pellets or granules which are better for good transportation and storage. Chemically cleaned coal will be especially well-suited for small sized plants requiring a highly reliable method for pollution control, and coal-oil mixtures which require fine sized coal of uniform quality.

There are various chemical cleaning processes currently under development. They differ in chemical reagents used, operating conditions, amount of pyritic and organic sulfur removed and overall economics. A brief summary of the operating conditions, reagents used, capability of sulfur removal etc. for some of these processes is given in Table 1 and further details can be found in the list of references. Some other processes besides those outlined in Table 1 are the TRW or Meyers Process (22-33), the ARCO Promoted Oxydesulfurization Process (22-24), the JPL Chlorinolysis Process (34-37), the G.E. Microwave Treatment Process (38-40), the Maynex Process (41-47), and the IGT Hydrodesulfurization Process (48,49).

TABLE 1

SUMMARY OF VARIOUS PROCESSES AVAILABLE FOR THE CHEMICAL CLEANING OF COAL

Process	Reagents Used and Operating Conditions	Capability of Sulfur Removal	Particular Features	References
Battelle Hydrothermal Process	Dry crushed coal is treated with a caustic solution containing 10 wt% NaOH and 2.3 wt% $Ca(OH)_2$ for 15 to 30 min at 2.41 to 17.24 MPa and 250 to 350°C.	90% of pyritic and over 50% of organic sulfur along with the trace materials like arsenic, lead, vanadium etc.	Product recovery circuit requires large number of filtration and washing stages, material of construction is a problem because of hot caustic solution.	2-6
KVB Process	Dry pulverized coal at 100°C is treated with a gas stream containing NO_2 and O_2 under pressure for 30 to 60 min. Then it is washed with hot water and caustic solution.	Over 90% of pyritic and up to 40% of organic sulfur.	Process must be operated under carefully chosen operating conditions so as to avoid nitration of coal.	7-8

TABLE 1
(Continued)

Ledgemont Process	Ground coal is leached with a hot acidic solution containing dissolved oxygen under pressure for reaction time of 15 min to 2 hr at 6.87 MPa and 130 to 230°C.	Over 90% of pyritic and up to 20% of organic sulfur.		9–14
PETC Oxydesulfurization Process	Crushed coal is leached under acidic conditions at pressure of 1.52 to 10.34 MPa and temp. of 150–230°C using air/oxygen mixture for about 1 hr.	Over 90% of pyritic and 40% of organic sulfur.	Coking and swelling properties of the coal are also destroyed. Because of severe operating conditions, leaching rates are faster.	15–17
Ames Wet Oxidation Process	Pulverized coal is leached with alkaline solution of Na_2CO_3 containing dissolved oxygen or air under pressure at temp of about 150°C. Oxygen partial pressure 1.34 MPa. Residence time 1 hr.	Over 90% of pyritic and 20% of organic sulfur.		18–21

2. ECONOMICS OF CHEMICAL COAL CLEANING

Recently, Contos et al. (50) and McCandless and Contos (51) have made a comprehensive assessment of the economics of the major chemical coal cleaning processes. Since pilot plant data were not available for most processes, the costs were based on preliminary conceptual designs. Taking bituminous coal from the Pittsburgh seam as a representative coal (which contained 1.93 wt% total sulfur, about 66% of which was pyritic sulfur), capital and operating costs for a feed rate of 7,200 metric ton/day plant were evaluated based on the first quarter of 1977 prices. A brief summary of their study is shown in Table 2.

It can be seen that the plant capital cost is high for all the processes shown and the processing cost is generally about equal to the cost of raw coal. However, since all the processes are still in the development stage and the optimum operating conditions are unknown, the above cost estimates may change considerably in the future.

One other independent economic survey has been carried out by Oder et al. (1), and they found that the most promising of these processes are Ledgemont and PETC processes.

3. CHEMISTRY

The chemical reactions for sulfur removal are different for different processes depending on the chemical reagent involved. For example, the principal overall reactions for Ames wet oxidation process involving pyrite are (52)

TABLE 2

COST DATA FOR SELECTED PROCESSES (50,51)

	Processes			
	Ledgemont	PETC	Meyers	Battelle
Product sulfur content, wt%	0.83	0.65	0.83	0.65
Energy recovered, %	94	94	94	88
Plant capital cost, million $	114	167	109	169
Processing cost, $/metric ton	21	26	17	33
Processing cost, $/metric ton[a]	52	57	48	62

[a]Includes cost of coal at $27.6/metric ton

$$FeS_2 + 3.75O_2 + 2H_2O = 0.5Fe_2O_3 + 2H_2SO_4 \quad (1)$$

$$H_2SO_4 + 2Na_2CO_3 = Na_2SO_4 + 2NaHCO_3 \quad (2)$$

The oxydesulfurization process includes several reactions such as pyritic and organic sulfur oxidation, oxidation of carbon, and hydroperoxide formation at benzylic positions. A brief discussion of these reactions is given below.

The primary steps in the pyritic sulfur oxidation are (53)

$$FeS_2 + 2O_2 \rightarrow FeSO_4 + S \quad (3)$$

$$2FeS_2 + 7O_2 + 2H_2O \rightarrow 2Fe^{2+} + 4SO_4^{2-} + 4H^+ \quad (4)$$

$$2FeS_2 + 7.5O_2 + H_2O \rightarrow 2Fe^{3+} + 4SO_4^{2-} + 2H^+ \quad (5)$$

$$FeS_2 + 3.75O_2 + 2H_2O \rightarrow 0.5Fe_2O_3 + 4H^+ + 2SO_4^{2-} \quad (6)$$

$$2S + 3O_2 + 2H_2O \rightarrow 4H^+ + 2SO_4^{2-} \quad (7)$$

Vracar and Vucurovic (54-56) have found that the rate of reaction 7 is very slow and generally reactions 3 and 7 do not occur. Friedman and Warzinski (15) have confirmed the above observations. The mechanism of the aqueous oxidation of C is complex. Most of the oxygen which reacts with the organic matrix remains form hydroperoxides and their decomposition products.

Since "organic sulfur" is a broad classification for any sulfur atom which is bonded to the organic matrix of the coal, it is difficult to outline the possible reaction mechanisms for the organic sulfur reactions. Compounds which fall under the heading of organic sulfur include mercaptans, sulfides, disulfides, thiophenes, thiols, and thiopyrones (26). Due to the nonhomogeneity of coals, it is possible that any given sample may contain one or more of the above compounds. Friedman et al. (57) have provided some insight into the chemistry of organic sulfur removal. They found that although the treatment of coal with compressed air and steam resulted in a 25% decrease in its organic sulfur content, identical treatment of dibenzothiophene (an organic sulfur compound) produced no reactions.

4. KINETICS OF DESULFURIZATION REACTIONS

4.1 Pyritic Sulfur Removal

For pyritic sulfur oxidation, various dependence of rate on partial pressure of oxygen and pyritic sulfur concentration are reported in the literature. Within the temperature range of 100-130°C and oxygen partial pressures up to 0.4 MPa, McKay and Halpern (53) reported the rate of pyritic sulfur oxidation to be first order in oxygen partial pressure and zero order with respect to pyritic sulfur. A kinetic study of the pyrite oxidation at Kennecott Copper Corporation (58) indicated a square root relationship between oxygen partial pressure and the rate of pyrite oxidation with an activation energy of 58.6 kJ/mole for the rate constant.

Slagle et al. (59) carried out a detailed kinetic study for the oxydesulfurization reactions for PETC process using Upper Freeport coal. The experimental data were taken in the temperature range of 150-210°C, total pressure range of 3.44-6.88 MPa, oxygen partial pressure range of 0.69-3.44 MPa and for batch times up to 2400 seconds. All data were taken at a stirrer speed of 1000 rpm. For pyritic sulfur oxidation, they found a second order dependence on pyritic sulfur concentration and an activation energy of 46.5 kJ/mole (11.2 kcal/mole) for the rate constant. Same order of magnitude values of the activation energy are found by McKay and Halpern (53), Vracer and Vucurovic (54) and Sareen et al. (58). The extent of sulfur removal was found to be independent of total pressure at constant oxygen partial pressure. Also, the data fitted the shrinking core model where the overall rate was controlled by the diffusion through ash layer.

4.1.1 Particle size effect. Joshi et al. (63) investigated the effect of coal particle size on pyritic sulfur removal by PETC oxydesulfurization process. Coal particles in the size range <72, 72-99, 150-208, 208-430 and 700-1410 microns were used and the experimental data were taken in the temperature range of 120-230°C, oxygen partial pressures of 0.32 to 1.36 MPa, total pressures of 1.68-6.7 MPa and batch times up to 5400 seconds. The results showed that the rate of pyritic sulfur removal decreased with an increase in coal particle size. A stepwise procedure indicated the rate controlling step to be the intrinsic chemical reaction between the dissolved oxygen and pyrite. The shrinking core model (65) was successfully applied for the calculation of a rate constant which was found to be practically independent of the coal particle size and oxygen partial pressure. Experiments at various temperatures indicated an activation energy of 35 kJ/mole. The fractional conversion was found to be independent of the solid loading.

Chuang et al. (52) found similar results for their Ames wet oxidation process. For desulfurization of pyrite with dissolved oxygen in sodium carbonate solutions between 120-180°C, they observed that the conversion achieved in a given time increased as the particle size was reduced or as either temperature or oxygen partial pressure was raised.

4.1.2 Effect of pH. In another work, Joshi et al. (64) investigated the effect of acidic, alkaline as well as nautral pH conditions on the removal of pyritic sulfur in oxydesulfurization. The pH was varied by adding sulfuric acid or sodium carbonate to a coal-water slurry. The effect of neutral pH ($\simeq$ 7) was determined in a buffer solution of sodium dihydrogen phosphate and disodium hydrogen phosphate in the temperature range from 130° to 190°C. The oxygen partial pressure and the reaction time were varied from 0.32-1.36 MPa and 0-3600 seconds, respectively. The results indicated that at neutral pH, the rate of pyritic oxidation was found to be the slowest, with the overall reaction being controlled by the surface reaction between the pyrite and dissolved oxygen. The activation energy under neutral conditions was found to be 33 kJ/mole.

Under acidic conditions, pyrite oxidation was enhanced by a factor of two to three over that at neutral pH. The order of reaction was found to be 0.7 with respect to oxygen and the rate controlling step was found to be the surface reaction. Under alkaline conditions, the rate controlling step was probably the combined effect of the diffusion of oxygen through the product ash layer and the surface chemical reaction. For the Ames wet oxidation process also, which operates under alkaline conditions, the rate controlling step was found to be the diffusion of dissolved oxygen through the hematite shell (52). A good agreement between the effective diffusivity based on the shell diffusion process and the molecular diffusivity for dissolved oxygen in water was observed.

4.2 Organic Sulfur Removal

For high organic sulfur coals, to meet emission standards, the severity of the operating conditions should be increased. However, this is often accompanied by a greater heating-value loss. Warzinski et al. (17) observed a dependency on coal particle size, suggesting a surface effect. They observed a higher organic sulfur removal for larger sizes of coal. In the absence of such an effect, the smaller particles would be expected to exhibit a higher organic sulfur removal since more sulfur would be exposed to the oxidant.

Based on the work of Attar (60), who has reported the distribution of sulfur functional groups in the Lower Freeport coal, Joshi and Shah (61) have investigated the overall kinetics of organic sulfur removal for the Lower Freeport coal in the temperature range of 130-190°C, oxygen partial pressure of 0.32-1.36 MPa and batch times up to 3600 seconds. The effect of pH of the medium on organic sulfur removal was also studied.

Depending upon the temperature, oxygen partial pressure, particle size and the pH of the medium, organic sulfur removal reactions showed three distinct regions. Under acidic conditions and at relatively low oxygen partial pressure (<0.68 MPa) and temperature (<150°C) the removal of organic sulfur was practically negligible for relatively large particle sizes (>150 microns). For the case of smaller particle sizes (<100 microns), initially the organic sulfur content of the coal was found to increase and this was followed by a decrease. Finally, in the third region of relatively high oxygen partial pressures (>0.68 MPa) and temperatures (>150°C) the removal of organic sulfur was found to follow first order kinetics with respect to removable concentration of organic sulfur. Under neutral and alkaline conditions, the first two regions were found to be absent. In general, the rate of organic sulfur removal was found to increase with increase in pH of the medium. In all the regions the reaction was first order in organic sulfur, a result different from the zero order mechanism found by Slagle et al. (59) for the Upper Freeport coal. The data of Slagle et al. (59) were, however, quite scattered. The activation energy for the zero order rate constant was 78.7 kJ/mole. Joshi and Shah (61) found the extent of the organic sulfur removal to be independent of the oxygen partial pressure.

4.3 Carbon Oxidation and Heating Value Loss

Because of high temperatures and the presence of oxygen in the oxydesulfurization process, carbon oxidation is inevitable. Excessive carbon and heating value loss can make the process economically unattractive.

Thring and Essenhigh (62) reviewed the kinetics and chemistry of solid carbon oxidation. They pointed out that gasification of carbon goes to completion in the presence of oxygen at temperatures above 300°C; and at high pressures and temperatures, the reaction is zero-order with respect to carbon. Carbon oxidation in the Ledgemont process was examined by Sareen et al. (58). At temperatures between 100-130°C and 2.0 MPa pressure, they observed a first order relation between the partial pressure of oxygen and the rate of CO_2 formation; but they found zero order kinetics with respect to carbon. Joshi et al. (63) observed a zero order dependence with respect to oxygen partial pressure and carbon concentration for coal oxidation loss. Slagle et al. (59) found that the kinetic mechanism of the carbon reaction can be divided into two regions. Initially the rate of oxidation was observed to be very high; and at each temperature, a sharp change in the rate of oxidation was observed. In both regions, the order with respect to carbon was found to be zero, but a strong temperature dependence was observed with activation energies of 85.5 and 380 kJ/mole in the two regions respectively.

5. REACTOR MODELING, DESIGN AND SCALE-UP CONSIDERATIONS

Oxydesulfurization essentially involves a three phase (gas-liquid-solid) reactor in which the solids take part in the reaction. The rate controlling step can be gas-liquid mass transfer, liquid-solid mass transfer, diffusion through product ash layer, chemical reaction, or the combination of two or more steps. Under the assumption of constant particle size, the rate determining step can be determined on the basis of a 'Shrinking Core Model,' details of which are described by Levenspiel (65). Joshi et al. (66) modelled the reactor based on exit age distribution. They collected their data in a continuous bubble column slurry reactor as well as a semi-batch agitated reactor and presented performance charts for all the controlling mechanism in the form of average conversion of the solid phase with the parameters; the average residence time, the solid phase Peclet number and the time required for complete conversion of a single particle. They observed that the oxidation of pyritic sulfur followed the shrinking core mechanism in which the rate controlling step was the chemical reaction. The performance charts generated by Joshi et al. (66) can be effectively utilized for the design and scaleup of such reactors knowing the process parameters such as treating capacity, slurry concentration, temperature and pressure of operation, oxygen partial pressure and amount of sulfur removal desired. More details are given by Shah and Gopal (67).

6. SUMMARY

At present, pyritic sulfur from coal can be easily removed by oxidation. The removal of organic sulfur generally requires severe operating conditions which also cause undesirable carbon loss. Future work must consider the removal of organic sulfur by the use of homogeneous catalysts under mild conditions. The processing of a low pH slurry could be an important factor in the commercialization of an oxydesulfurization process.

REFERENCES

1. Oder, R.R., L. Kulpaditharom, A.K. Lee and E.L. Ekholm. "Technical and Cost Comparisons for Chemical Coal Cleaning Processes." Mining Congress Journal 63 (1977) 42.

2. Stambaugh, E.P., J.F. Miller, S.S. Tam, S.P. Chauhan, H.F. Feldman, H.E. Carlton, J.F. Foster, H. Nack and J.H. Oxley. "Hydrothermal Process Produces Clean Fuel." Hydro. Processing 54(7) (July 1975) 115-116.

3. Stambaugh, E.P. "Coal Desulfurization, Chemical and Physical Methods." ACS Symp. Series 64 (Am. Chem. Soc., Washington, D.C., 1977) 198-205.

4. Stambaugh, E.P., H.N. Conkle, J.F. Miller, E.J. Mezey and B.C. Kim. Proceedings: Symp. on Coal Cleaning to Achieve Energy and Environmental Goals, Vol. 2 (Sept. 1978) 991-1015. EPA-600/7-79-098b (April 1979).

5. Stambaugh, E.P., J.F. Miller, H.N. Conkle, E.J. Mezey and R.K. Smith. A draft report for EPA prepared by Battelle, Columbus, OH (1980).

6. Stambaugh, E.P. presented at "Second Conference on Air Quality Management in the Electrical Power Industry." Austin, TX (Jan. 22-25, 1980).

7. Diaz, A.F. and E.D. Guth, U.S. Pat. 3,909,211 (Sept. 30, 1975).

8. Guth, E.D. Proceedings: Symp. on Coal Cleaning to Achieve Energy and Environmental Goals, Vol. 2 (Sept. 1978) 1141-1164. EPA-600/7-79-098b (April 1979).

9. Agarwal, J.C., R.A. Giberti, P.F. Irminger, L.J. Petrovic and S.S. Sareen. Mining Congress Journal 61(3) (March 1975) 40-43.

10. Sareen, S.S., R.A. Giberti, P.F. Irminger and L.J. Petrovic. presented at Meeting of Am. Inst. Chem. Eng., Boston, MA (Sept. 7-10, 1975).

11. Agarwal, J.C., R.A. Giberti and L.J. Petrovic. U.S. Pat. 3,960,513 (June 1, 1976).

12. Sareen, S.S. "Coal Desulfurization, Chemical and Physical Methods." ACS Symp. Series 64 (Am. Chem. Soc., Washington, D.C., 1977) 173-181.

13. Giberti, R.A., R.S. Opalanko and J.R. Sinek. Proceedings: Symp. on Coal Cleaning to Achieve Energy and Environmental Goals, Vol. 2 (Sept. 1978) 1064-1095. EPA-600/7-79-098b (April 1979).

14. DeVaux, G.R. presented at "Second Conference on Air Quality Management in the Electrical Power Industry," Austin, TX (Jan. 22-25, 1980).

15. Friedman, S. and R.P. Warzinski. "Chemical Cleaning of Coal." Engineering for Power 99(3) (July 1977) 361-364.

16. Friedman, S., R.B. LaCount and R.P. Warzinski. "Coal Desulfurization, Chemical and Physical Methods." ACS Symp. Series 64, (Am. Chem. Soc., Washington, D.C., 1977) 164-172.

17. Warzinski, R.P., J.A. Ruether, S. Friedman and F.W. Steffgen. Proceedings: Symp. on Coal Cleaning to Achieve Energy and Environmental Goals, Vol. 2 (Sept. 1978) 1016-1038. EPA-600/7-79-098b (April 1979).

18. Tai, C.Y., G.V. Graves and T.D. Wheelock. "Coal Desulfurization, Chemical and Physical Methods." ACS Symp. Series 64 (Am. Chem. Soc., Washington, D.C., 1977) 182-197.

19. Markuszewski, R., K.-C. Chuang and T.D. Wheelock. Proceedings: Symp. on Coal Cleaning to Achieve Energy and Environmental Goals, Vol. 2 (Sept. 1978) 1039-1063. EPA-600/7-79-098b (April 1979).

20. Chuang, K.-C., M.-C. Chen, R.T. Greer, R. Markuszewski, Yu Sun and T.D. Wheelock. presented at Meeting of Am. Inst. of Chem. Eng., San Francisco, CA (Nov. 25-29, 1979).

21. Wheelock, T.D. and R. Markuszewski. Fossil Energy Annual Report, Oct. 1, 1978-Sept. 30, 1979, IS-4714, Iowa State University, Ames, IA (Jan. 1980).

22. Burk, E.H., Jr., J.S. Yoo and J.A. Karch. U.S. Pat. 4,097,244 (June 27, 1978).

23. Burk, E.H., Jr., J.S. Yoo and J.A. Karch. U.S. Pat. 4,158,548 (June 19, 1979).

24. Beckberger, L.H., E.H. Burk, Jr., M.P. Grosboll and J.S. Yoo. EPRI EM-1044 Project 833-1 Final Report, Atlantic Richfield Co., Harvey, IL (April 1979).

25. Meyers, R.A, "Desulfurize Coal Chemically." Hydrocarbon Processing 54(6) (June 1975) 93-95.

26. Meyers, R.A. "Coal Desulfurization." (Marcel Dekker, Inc., New York, 1977).

27. Hamersma, J.W., M.L. Kraft and R.A. Meyers. "Coal Desulfurization, Chemical and Physical Methods." ACS Symp. Series 64 (Am. Chem. Soc., Washington, D.C., 1977) 143-152.

28. Van Nice, L.J., M.J. Santy, E.P. Koutsoukos, R.A. Orsini and R.A. Meyers. "Coal Desulfurization, Chemical and Physical Methods." ACS Symp. Series 64 (Am. Chem. Soc., Washington, D.D., 1977) 153-163.

29. Santy, M.J. and L.J. Van Nice. Proceedings: Symp. on Coal Cleaning to Achieve Energy and Environmental Goals, Vol. 2 (Sept. 1978) 960-990. EPA-600/7-79-098b (April 1979).

30. Meyers, R.A., J.J. Santy, W.D. Hart, L.C. McClanathan and R.A. Orsini. EPA-600/7-39-013a (Jan. 1979).

31. Meyers, R.A., E.P. Koutsoukos, M.J. Santy and R. Orsini. EPA-600/7-79-012 (Jan. 1979).

32. Meyers, R.A. "System Optimizes Coal Desulfurization." Hydro. Processing 58(6) (June 1979) 123-166.

33. Hart, W.D., L.C. McClanathan, R.A. Meyers and D.M. Wever. EPA-600/7-79-240 (Nov. 1979).

34. Hsu, G.C., J.J. Kalvinskas, P.S. Ganguli and G.R. Gavalas. "Coal Desulfurization, Chemical and Physical Methods." ACS Symp. Series 64 (Am. Chem. Soc., Washington, D.C., 1977) 206-217.

35. Hsu, G.C., G.R. Gavalas, P.S. Ganguli and S.H. Kalfayan, U.S. Pat. 4,081,250 (March 28, 1978).

36. Kalvinskas, J.J. and G.C. Hsu. Proceedings: Symp. on Coal Cleaning to Achieve Energy and Environmental Goals, Vol. 2 (Sept. 1978) 1096-1140. EPA-600/7-79-098b (April 1979).

37. Kalvinskas, J.J., K. Grohmann and N. Rohatgi. presented at "Coal Age Conference," Louisville, KY (Oct. 23-25, 1979).

38. Zavitsanos, P.D., J.A. Golden, K.W. Bleiler and W.K. Kinkhead. EPA-600/7-78-089 (June 1978).

39. Zavitsanos, P.D. and K.W. Bleiler. U.S. Pat. 4,076,607 (Feb. 28, 1978).

40. Zavitsanos, P.D., K.W. Bleiler and J.A. Golden. U.S. Pat. 4,152,120 (May 1, 1979).

41. Kindig, J.K. and R.L. Turner, U.S. Pat. 3,938,966 (Feb. 17, 1976).

42. Kindig, J.K. and R.L. Turner, presented at "Society of Mining Engineers of AIME," Denver, CO (Sept. 1-3, 1976).

43. Porter, C.R. and D.N. Goens, Mining Engineering 31 (Feb. 1979) 175-180.

44. Kindig, J.K. and D.N. Goens. Proceedings: Symp. on Coal Cleaning to Achieve Energy and Environmental Goals, Vol. 2 (Sept. 1978) 1165-1195. EPA-600/7-79-098b (April 1979).

45. Porter, C.R. presented at "Symposium on Coal Preparation and Utilization," Louisiville, KY (Oct. 23-25, 1979).

46. Kindig, J.K. and R.L. Turner. U.S. Pat. 4,119, 410 (Oct. 10, 1978).

47. Kindig, J.K. and R.L. Turner. U.S. Pat. 4,120,665 (Oct. 17, 1978).

48. Fleming, D.K., R.D. Smith and Rosario Y. Aquino. "Coal Desulfurization, Chemical and Physical Methods." ACS Symp. Series 64 (Am. Chem. Soc., Washington, D.C., 1977) 267-279.

49. Fleming, D.K. and R.D. Smith. EPA-600/7-79-016 (Jan. 1979).

50. Contos, G.Y., I.F. Frankel and L.C. McCandless. EPA-600/7-78-173a (Aug. 1978).

51. McCandless, L.C. and G.Y. Contos. Proceedings: Symp. on Coal Cleaning to Achieve Energy and Environmental Goals, Vol. 2 (Sept. 1978) 934-959. EPA-600/7-79-098b (April 1979).

52. Chuang, K.C., M.C. Chen, R.T. Greer, R. Markuszewski, T. Sun and T.D. Wheelock. "Pyrite Oxidation by Wet Oxidation in Alkaline Solutions." Chemical Engineering Communications 7 (1980) 79-94.

53. McKay, D.R. and J. Halpern. Trans. Metall. Soc. AIME 212 (1958) 301.

54. Vracar, R. and D. Vucurovic. "Oxidation of Pyrites by Gaseous Oxygen from an Aqueous Suspension at Elevated Temperatures in an Autoclave (I)." Rudarstvo I Metalursija (1970).

55. Vracar, R. and D. Vucurovic. "Oxidation of Pyrites by Gaseous Oxygen from an Aqueous Suspension at Elevated Temperatures in an Autoclave (II)." Ruderstvo I Metalursija (1971).

56. Vracer, R. and D. Vucurovic. "Oxidation of Pyrites by Gaseous Oxygen from an Aqueous Suspension at Elevated Temperatures in an Autoclave (III)." Redarstvo I Metalursija (1972).

57. Friedman, S., R.B. Lacount and R.P. Warzinski. Proceedings of the National Meeting of the Division of Fuel Chemistry, New Orleans (March 1977).

58. Sareen, S.S., R.A. Gilberti, P.F. Irringer and L.J. Petrovic. AIChE Symposium Series 73 (1977) 183-189.

59. Slagle, D.J., Y.T. Shah and J.B. Joshi. "Kinetics of Oxydesulfurization of Upper Freeport Coal." Ind. Engg. Chem. Process Des. Dev. 19 (1980) 294-300.

60. Attar, A. "Thermkinetic Analysis of Three Coal Samples." Final Report submitted by Coal Gas (August 1979).

61. Joshi, J.B. and Y.T. Shah. "Kinetics of Organic Sulfur Removal from Coal by Oxydesulfurization." Fuel (in press).

62. Thring, M.W. and R.H. Essenhigh. "Thermodynamics and Kinetics of Combustion of Solid Fuels." H.H. Lowry, editor (John Wiley and Sons, Inc., New York, 1963).

63. Joshi, J.B., Y.T. Shah, J.A. Ruether and H.J. Ritz. "Particle Size Effects in Oxidation of Pyrite in Air/Water Chemical Coal Cleaning." AIChE Symposium Series on Coal Technology (1980) (in press).

64. Joshi, J.B., Y.T. Shah, R.S. Albal, H.J. Ritz and W.D. Richey. "Effect of pH on the Removal of Pyritic Sulfur from Coal by Oxydesulfurization." a paper submitted to I&EC Process Design and Dev. (1981).

65. Levenspiel, O. "Chemical Reaction Engineering," Second Edition (John Wiley and Sons, Inc., New York, 1972).

66. Joshi, J.B., J.S. Abichandani, Y.T. Shah, J. Ruether and H. Ritz. "Modeling of Three Phase Reactors: A Case of Oxydesulfurization of Coal." AIChE J. (in press).

67. Shah, Y.T. and J. Gopal. "Slurry Reactors for Coal Technology." a paper presented at NATO-ASI School, Izmir, Turkey (August 1981).

DIRECT COAL LIQUEFACTION

Y. T. Shah, P. C. Singh and A. Calimli

Department of Chemical and Petroleum Engineering
University of Pittsburgh, Pittsburgh, PA 15261

ABSTRACT

This paper presents a brief state of the art review of direct coal liquefaction. The review includes important pilot scale processes available for the liquefaction and a brief description of the structure of coal and the chemistry, mechanism and available lumped kinetic models for the liquefaction process. It also includes some discussions on the role of catalysts during coal liquefaction and on the use of model compounds for the understanding of coal liquefaction kinetics. Reactor design aspects are covered in a separate paper and will not be repeated here.

1 INTRODUCTION

The production of liquid fuels from coal can be divided into three broad categories: pyrolysis, indirect liquefaction and direct liquefaction. In pyrolysis, coal is heated in the absence of air to a temperature such that it gives off liquids and gases leaving a large amount of char. In indirect liquefaction, coal is gasified and the resulting gases are catalytically converted to liquid fuels.

This paper deals with the subject of solvent extraction and hydrogenation of coal (i.e. direct liquefaction). The main purpose of liquefaction is to produce clean fuel (both liquid and solid), eliminating the mineral matter and heteroatoms from the parent coal. Most of the processes aim towards high liquid yields. In all processes, crushed coal, mixed with the process solvent is contacted with hydrogen gas under pressure. The reaction products, after cooling, are separated from remaining solids. Different processes use different separation schemes. A large number of coal liquefaction processes are currently being developed. Some of the important ones are briefly described below.

2 PROCESSES

2.1 The SRC-I Process

In this process, pulverized coal, dissolved in a process-derived solvent, is reacted at high temperature and pressure in the presence of hydrogen. In the dissolution step, the coal molecules are fragmented, freeing organic sulfur and light hydrocarbons which are evolved as gases. The undissolved solid residue is then separated from the liquid stream which is distilled to recover process solvent and to produce an additional side stream of light liquid fuel products. The remaining heavier liquid is solvent refined coal (SRC) which, if cooled to ambient temperature, becomes a solid. The undissolved solid residue, supplemented with additional feed coal, is sent to a gasifier to produce the hydrogen required by the process. The clean residue from the gasifier is expected to be environmentally acceptable (1-3).

2.2 The SRC-II Process

In this process, the pulverized coal, dissolved in a recycle slurry containing process solvent, SRC, and undissolved solid residue is reacted at high temperature and pressure in the presence of hydrogen. In the dissolution step, the coal molecules are severely hydrocracked to gaseous and liquid fuels. A major portion of the sulfur and some nitrogen and oxygen are converted via hydrogenation to hydrogen sulfide, ammonia and water. The cooled reaction products are physically separated to recover fuel gases, liquid fuel products, and a product slurry containing solvent, SRC and undissolved solid residue. Product slurry, after removal of recycle slurry for the coal dissolution step, is distilled to recover additional liquid fuel products from the residue slurry. The residue slurry is sent to a gasifier to produce the hydrogen required by the process and additional quantities of fuel gas are also produced. The clean residue from the gasifier is expected to be environmentally acceptable (2-4).

2.3 The TSL (Two Stage Liquefaction) Process

In contrast to the SRC-II Process, where coal dissolution and coal and solvent hydrocracking are achieved non-selectively in a single unit under largely non-optimal thermal conditions, the TSL process minimizes unnecessary degradation of solvent and lighter boiling range material through staged processing of segregated streams, thereby decreasing the hydrogen consumption. In this process, pulverized coal is thermally upgraded to a low ash and low sulfur content SRC fuel product using conventional SRC-I technology. The resulting hot and fluid SRC, after separation of by-products, and gaseous and liquid fuels, is charged to an LC-Finer for conversion to the specified grade liquid and solid fuel products. In the LC-Finer, the hot SRC dissolved in an internally produced solvent, is catalytically hydrocracked in an expanded bed of catalyst in the presence of hydrogen at elevated temperature and pressure. In the hydrocracking step, the SRC is selectively cracked to gases and distillate fuels. A major portion of sulfur and some nitrogen and oxygen are converted via hydrogenation to hydrogen sulfide, ammonia and water. Following conversion, the cooled reaction products are physically separated to recover fuel gas, liquid fuel products, LC-Finer solvent and unconverted SRC (5,6).

2.4 The EXXON Donor Solvent (EDS) Process

This process is designed to maximize liquid products. The feed coal is crushed, dried and mixed with hydrogenated recycle solvent (i.e. donor solvent) and fed to the liquefaction reactor along with gaseous hydrogen. The reactor is upward plug flow type operating at 723 K and approximately 10-13.6 MPa total pressure. The reactor effluents are separated by a series of distillation steps into gaseous, liquid and solid products. The recycle solvent is hydrogenated in a fixed bed catalytic reactor employing "off-the-shelf" hydrotreating catalysts.

The heavy bottoms from vacuum distillation may be sent to a FLEXICOKING unit along with air and steam to produce additional distilled liquid products and a low quality fuel gas for process furnaces. Light hydrocarbon gases coming from the distillation unit are steam reformed to produce hydrogen. The total liquid yield is thus a blend of streams from liquefaction and flexicoking.

2.5 The H-COAL Process

The "H-COAL" process was developed by Hydrocarbon Research Inc. to convert all types of coal to high octane gasoline, petrochemicals, LPG, low sulfur distillate fuels and low sulfur heavy boiler fuel oil.

Hydrogen and a slurry of coal and recycle oil are introduced to a plenum chamber at the bottom of the ebullated bed reactor operating at 10-20 MPa and 700-755 K. They pass up through a

TABLE 1

TYPICAL COMPARISON OF DIFFERENT LIQUEFACTION PROCESSES

	SRC-I	SRC-II	EDS	H-COAL
Operating Conditions				
1. Coal (type)	Western Kentucky 9/14	Western Kentucky 9/14	Ill. No. 6, Wyodak	Any Type
2. Reactor pressure (MPa)	10.3	13.34	10.3	12.1-12.6
3. Reactor temperature (K)	724	734	722	726
4. Hydrogen consumption (wt% maf coal)	2.4	4.8	4.3	3.8-5.25
Typical Product Yield (wt% maf coal)				
1. C_1-C_4 hydrocarbon gas	3.7	18.4	7.3-9.3	8.6-11.8
2. Light oil	5.1	14.2	---	16.9-23.6
3. Middle & heavy oil	8.0	28.2	33.3	18-23
4. Ash	9.6	--	---	10.9-11.67
5. Unreacted coal	5.4	6.6	---	6.3-7.5

distributor tray into the ebullated bed of cobalt molybdate catalyst. Since there is a sharp interface at the top of the catalyst bed, the catalyst level is detected by sending a beam of gamma rays through the reactor. The catalyst replacement system used here continuously removes the carbon deposited on the catalyst particles, avoiding a build up of pressure across the bed, a problem otherwise serious in other conventional units.

Table 1 presents a comparison of different liquefaction processes with respect to their operating conditions and product distribution.

3 ON THE STRUCTURE OF COAL

Coal structure has been studied using techniques like pyrolysis (7-10), alkylphenol determination (11-17), liquefaction and oxidation (18-21). After a considerable effort in this area, the compositions of some typical bituminous and subbituminous coals have been approximated as (19).

Bituminous: $C_{50.8}H_{43}O_{4.8}S_{0.8}N_{0.6}$

Subbituminous: $C_{48.4}H_{42.4}O_{8.6}S_{0.07}N_{0.5}$

Along with carbon and hydrogen, oxygen is the most abundant heteroatom found in coal in functional groups such as phenols, carboxylic acids, ethers etc. Sulfur is present as thiophenes, sulfides, disulfides and thiols while nitrogen is present as pyridines, quinolines, carbazoles and pyrroles.

3.1 On the Mechanism of Coal Liquefaction

Coal liquefaction has been assumed to be occurring in three steps: dissolution, hydrogen transfer and hydrogenation. Whitehurst et al. (21) have given a conceptual picture of coal dissolution in which the weak bonds (activation energy $\leq$ 210 kJ/mole) are broken at low temperatures ($\leq$523 K) and extracts of up to 40-50% of bituminous coals are obtained. As the temperature is raised to about 673 K, the formation of free radicals takes place. If hydrogen is available at this stage from the organic matrix or from the solvent (donor), these radicals will combine with the hydrogen forming stable species with molecular weights varying in the range of 300 to 1000. However, if there is insufficient hydrogen, the radicals will recombine forming high molecular weight compounds and coke. Han and Wen (22) presented a three step mechanism. Farcasiu et al. (23) have also presented similar explanations giving the name 'asphaltols' for preasphaltenes.

Oele et al. (24) described the extraction of Dutch bituminous coals at temperatures from 473 K to 673 K. They have shown that the extractive disintegration can be compared with a thermal decomposition requiring activation energies of 80 to 160 kJ/mol. According to them, the extraction process is greatly governed by the following factors.

1. Extraction Rate and Agent: Initially, the extraction process proceeds rapidly and becomes slow after a few hours (25). An extracting agent is effective if at 473 K the liquid is capable of dissolving 20 to 40% of a bituminous coal. Effective extracting agents are pyridine, picolines, aliphatic amines, ethylenediamine, phenol, cresol, o-phenylphenol, acetophenone, furfural etc. while benzene, trichloroethylene etc. are less effective.

2. Particle Size of Coal: In less effective solvents, coal of 1 μm particle size yields thirty times as much extract as coal of coarser particle size. With less effective solvents, the retarding action on the diffusion path through the already extracted part of coal particles increases to such a high value that further penetration of the solvent becomes very difficult.

3. Temperature of Extraction: Since it is generally assumed that coal constituents behave like a gel which is held together by secondary valancey forces, this gel shows only a limited degree of swelling at low temperature. Although there are insufficient data to verify a clear cut demarkation line, extraction with phenolic solvents becomes appreciable at temperatures above 473 K.

Morita et al. (26) pointed out that the extraction rate parameters calculated by many workers under different hydrogen pressures on the basis of isothermal and isobaric conditions may be erroneous. In batch experiments, the hydrogen absorption rate showed curious behavior. They carried out experiments at temperatures up to 713 K, residence time from 120 secs to 7200 secs and initial hydrogen pressures from 5 MPa to 11 MPa and observed that the absorption of hydrogen was initiated at about 573 K. When the temperature reached 713 K, the hydrogen pressure began to decrease at constant rate. Also, there was a tendency for the increased extraction rate and lower coke formation with the increased initial hydrogen pressure. For coals rich in oxygen (approximately 30%), hydrogen pressure was found to have little effect on the extraction rate.

4 LUMPED KINETIC MODELS FOR COAL LIQUEFACTION

Kinetic modeling of the coal liquefaction is complex because the liquefaction process depends on many variables viz., temperature and pressure of the reactor, nature and amount of solvent, presence of mineral matters and/or externally added catalysts and also the nature and rank of coals. The process is further complicated as samples of coal from the same seam differ in their response to various operating conditions.

Coal is a nonhomogeneous material and its liquefaction produces a very large number of products. A completely detailed kinetic analysis involving all chemical species is therefore impossible. All the studies reported in the literature evaluate kinetic models using different types of lumped reacting species. A number of different types of reaction paths are evaluated. Unfortunately, the kinetic parameters evaluated for a certain coal under particular conditions vary significantly from that of a different coal under identical conditions. It is therefore obvious that the models developed in this manner do not possess global applicability and their use is limited. Several kinetic models have been reported in the literature and these have been reviewed by Lee (27) and Shah (28). Here we only summarize them in some order of their intricate details as shown in Table 2. For the details the reader is advised to refer to the original references.

5 CATALYSIS IN COAL LIQUEFACTION

The study regarding catalysis in coal liquefaction can be broadly divided into two groups: (1) effects of mineral matter present in the coal slurry itself and (2) effects of externally added catalysts.

In either mode, the catalyst (or mineral matter) serves to: (a) improve the liquid yield, including enhancement in the hydrogenation and hydrocracking rates and (b) improve heteroatom removal.

Inherent coal minerals are readily available and inexpensive catalysts for liquefaction, hydrogenation/hydrocracking and heteroatom removal reactions. In recent years, experimental work has been carried out to determine: (a) liquefaction behavior of various coals with different mineral matter contents, (b) liquefaction behavior by adding various mineral matter in or to a particular coal or by reducing the mineral matter contents of a coal by some physical means and (c) liquefaction behavior in the presence of a variety of externally added catalysts. Some of these studies are briefly described below.

Given et al. (44) studied the liquefaction behavior of a number of vitrinite rich coals in batch autoclaves at 658-698 K and 8.6 MPa hydrogen pressure. In one set of experiments,

TABLE 2

SUMMARY OF LUMPED KINETIC MODELS FOR COAL LIQUEFACTION

Kinetic Scheme*	Coal	Solvent	Catalyst	References
$C \rightarrow A \rightarrow O$	Pittsburgh Anthraxylon	None	SnS, NH_4Cl	Weller et al. (29)
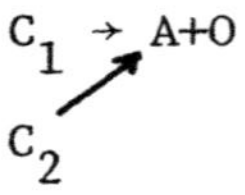	Spitsbergen	None	Ca-Cu-Cr	Falkum and Glenn (30)
	Pittsburgh	Tetralin	None	Curran et al. (33)
	Pittsburgh extract	None	$ZnO/ZnCl_2$	Struck et al. (34)
$C \rightarrow A+O$	Wyoming	None	None	Pelipetz et. al. (31)
	Utah	Tetralin	None	Hill et al. (32)
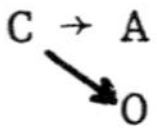	Bituminous	Tetralin	None	Liebenberg and Potgieter (35)
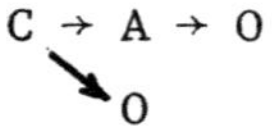	Japan	Decrystallized anthracene oil	Red mud and sulfur	Yoshida et al. (36)
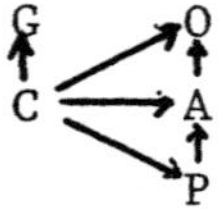	Belle Ayr	Hydrogenated phenanthrene	None	Cronauer et al. (37)

TABLE 2
(Continued)

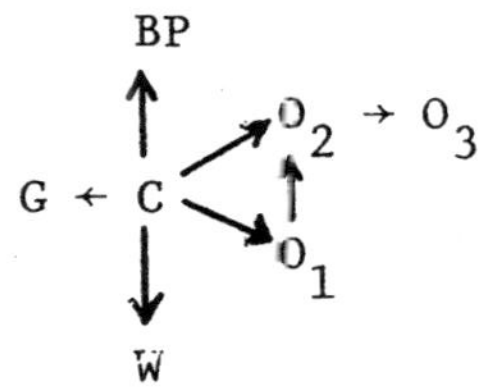	Subbituminous	Recycle solvent and hydrogenated anthracene oil	$CoMo/Al_2O_3$	Shah et al. (38)
C → P → A → O 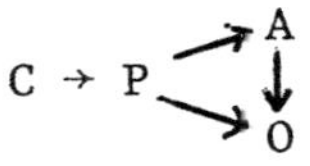 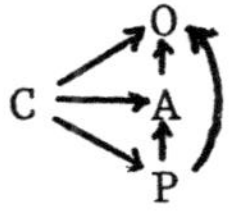 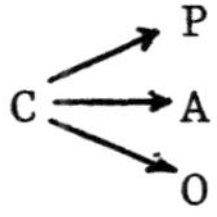	Kentucky	Tetralin	None	Shalabi et al. (39)
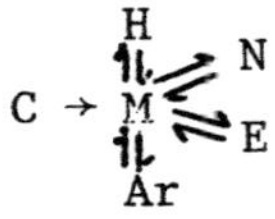	Illinois No. 6	Tetralin	None	Govindon and Silla (40)

TABLE 2
(Continued)

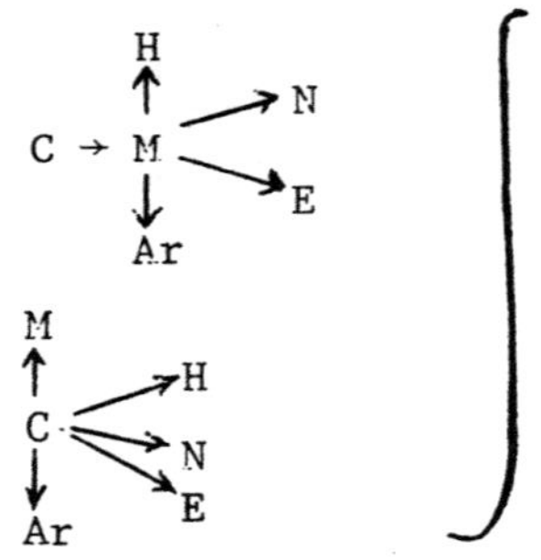				
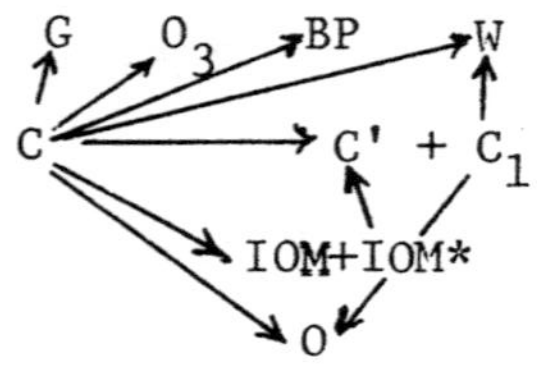	Kentucky No. 11	Recycle solvent	None	Parulekar et al. (41)
C + O → C' 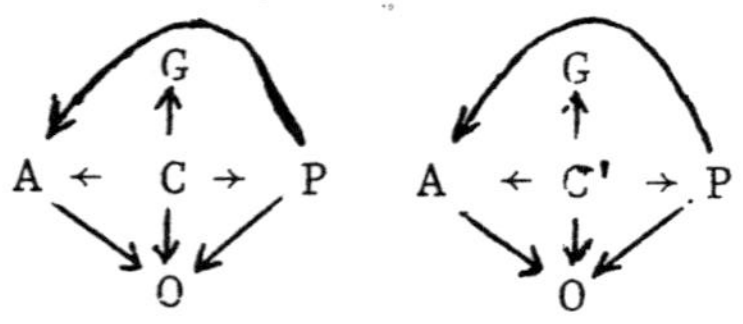	Belle Ayr Burning Star	Anthracene oil hydrogenated anthracene oil hydrogenated phenanthrene	None	Abichandani et al. (42)

TABLE 2
(Continued)

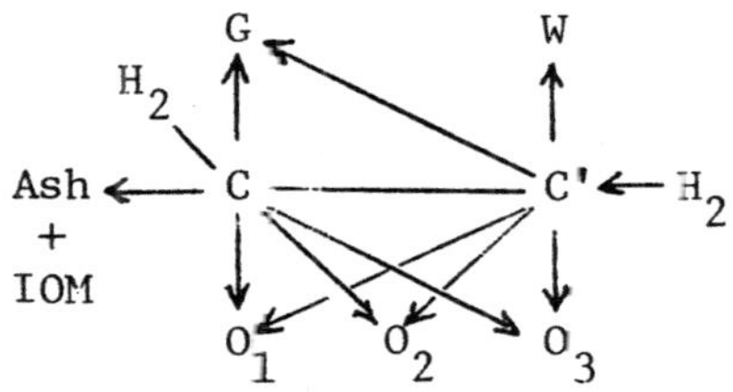	Powhatan	SRC-recycle solvent	None	Singh et al. (43)
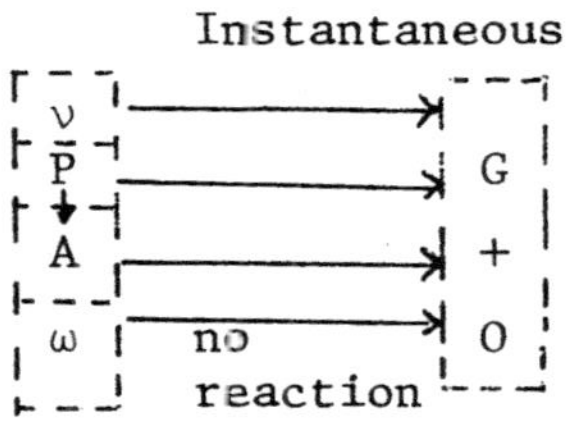	Illinois No. 6	---	---	Brunson (107)
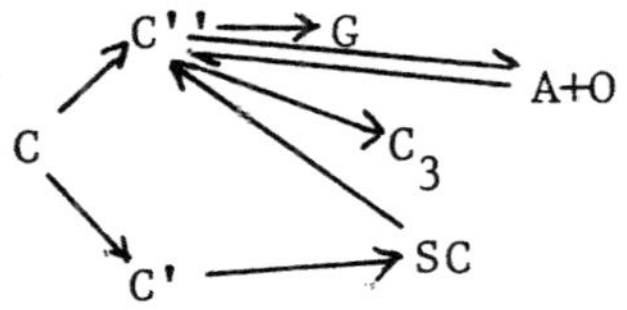	---	---	---	Shinn et al. (108) and Shinn & Vermeulen (109)
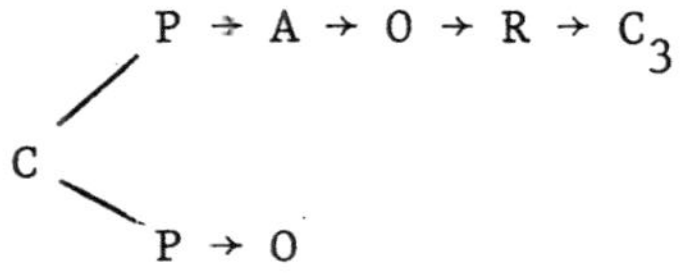	Japanese Australian & Indonesian	---	H_2MoO_4	Morita et al. (26)

TABLE 2
(Concluded)

*Legends

A - asphaltenes (benzene solubles but pentane insolubles)

Ar - aromatics

BP - by-products

C - coal, moisture and ash free

C_1, C_2 - two reactive parts of coal

C_3 - coke or char

C' - SRC

C'' - highly activated coal

C_1' - active SRC

E - ethers

G - gases (H_2O, CO, CO_2, H_2S, NH_3, light hydrocarbons)

H - hydroxyls

IOM - insoluble organic matter

IOM* - active insoluble organic matter

M - multifunctionals

N - nitrogens

O - oils (pentane soluble)

O_1 - oil (heavy distillate)

O_2 - oil (middle distillate)

O_3 - oil (light distillate)

P - preasphaltenes (pyridine solubles but benzene insolubles)

R - resin

SC - soluble coal

W - water

ν - volatile portion of coal

ω - unreactive portion of coal

impregnated ammonium molybdate was used as catalyst with no added liquid as vehicle while in a second set, a proprietary catalyst was used with anthracene oil serving as vehicle. Their data indicated lower yields of oil from the lignites and subbituminous coals than from coals of higher rank. However, yields fell off again at the upper end of bituminous rank (> 90% C).

Mukherjee and Chowdhury (45) presented plots showing the catalytic effects of iron and titanium on the conversion of Assam (India) coal to oils. Their data indicated an increase in conversion with mineral matter content corresponding to an ash content of 27% and then the conversion was found to drop. This finding is corroborated by Granoff et al. (46) who also found no effect of mineral matter corresponding to ash contents beyond 20%. The reason for this drop in conversion is supposed to be due to the excessive increase in inertinites. Iron as a reduced sulfide is supposed to be active for the catalysis of the liquefaction reaction. Titanium was added in the form of a hydroxide. They found that the total iron acts as a catalyst, a finding in contrast to that of Tarrer et al. (47) who concluded that only pyritic iron acts as a catalyst for the liquefaction. It is also interesting to note that according to Given et al. (44) the organic complexes of titanium poison the catalysts for liquefaction. Kawa et al. (48) concluded that the tin catalysts were the best for conversion of coal to oil and iron catalysts were only moderately active for the same purpose.

Guin et al. (49) studied the hydrogenation of creosote oil at 683 K and 6.8 MPa initial hydrogen pressure using different catalysts. The catalytic activity was defined in terms of hydrogen consumption. They found a $CoMo/Al_2O_3$ catalyst to be the most effective while calcite, quartz, dolomite and kaolin had no effect at all. The hydrogen consumption for the demineralized coal was lower than that for untreated coal. An unexplained phenomenon observed by Guin et al. (49) was that slurrying coal with water prior to hydrogenation decreased its rate of liquefaction.

Guin et al. (50) also examined hydrogen transfer activity of tetralin under liquefaction conditions. Tetralin donates hydrogen to coal derived free radicals producing naphthalene and hydrogenated free radicals as

$$\text{tetralin} + \text{free radicals (F.R.)} \rightarrow \text{naphthalene} + \text{H}\cdot\text{(F.R.)}$$

The dehydrogenated solvent can be regenerated in the presence of a catalyst and gaseous hydrogen as

$$\text{naphthalene} + 2H_2 \xrightarrow{\text{catalyst}} \text{tetralin}$$

Using a mixture of Kentucky No. 9/14 and Illinois No. 6 coals and light recycle oil from the Wilsonville, Alabama SRC pilot plant, in place of tetralin, they also showed that the presence of mineral matter decreased benzene insolubles by about 15% and pyridine insolubles by about 24%.

Shah (28) evaluated the use of catalysts NiMoTi on Al_2O_3 and $AlPO_4 \cdot Al_2O_3$ and NiCoMo on $AlPO_4 \cdot Al_2O_3$. The study showed that Al_2O_3 support was better than $AlPO_4 \cdot Al_2O_3$ support for both catalysts. The study also showed that NiCoMo was more reactive as such than when deposited on Torvex and the catalysts supported on magnesium aluminate had low coke deposition.

Catalyst aging is one of the most important problems in catalytic coal liquefaction. The catalyst is coked very readily because of the highly aromatic nature of coal liquids. Coking causes a significant decline in the catalytic activity for hydrogenation/hydrocracking, heteroatom removal and the production of liquid fuel. Unlike the catalysts used in petroleum cracking, the regeneration of coal liquefaction catalysts is very difficult. Utmost care is, therefore, needed to protect the liquefaction catalysts from possible poisoning and coking.

Hildebrand and Tsai (51) studied the aging of a $NiTiMo/Al_2O_3$ catalyst in a Gulf-patented catalytic coal liquefaction reactor (CCL) using Big Horn coal and anthracene oil/vacuum tower overhead as solvent at 661-678 K and 28 MPa total pressure (corresponding to about 26.5 MPa hydrogen partial pressure). Their results showed that while coal conversion to pyridine soluble materials remained almost constant, the hydrocracking declined by more than 50% in one month's period. Similarly, in the syncrude mode of the H-COAL process with $CoMo/Al_2O_3$ catalyst and Illinois No. 6 and Wyodak coals required a catalyst replacement rate of 0.5-0.55 Kg/metric ton of coal. Such high catalyst consumption indicates that further research is needed to improve the catalyst age and reduce its consumption.

Table 3 presents a brief summary of some of the catalytic coal liquefaction studies.

6. HETEROATOM REMOVAL

While the organic coal matrix contains methylene bridges of the 9,10-dehydroanthracene type (54), the compounds of nitrogen, oxygen and sulfur are also constituents of coal and their presence affect the nature of the coal and the liquid fuel produced from it.

TABLE 3

CATALYSIS IN COAL LIQUEFACTION

Type of Coal	Solvent	Catalyst	Operating Conditions	References
1. Pittsburgh seam (hVab) and Indiana No. 5	Coal tar	Co,Mo,Ni,Sn,Fe	673 K, 13 MPa, 1800 sec	Kawa et al. (48)
2. Vitrinite rich coals from Appalachian Province, Interior Province, North Great Plains Province, Rocky Mountain Province, Pacific Province & Gulf Province	(1) none (2) anthracene oil	(1) impregnated ammonium molybdate (2) proprietary catalyst	658–698 K 8.6 MPa 1.08×10^4 secs	Given et al. (44)
3. North Assam (India)	None	Representative coal mineral matters	673 K, 10 MPa 1.08×10^4 secs	Mukherjee & Chowdhury (45)
4. Kentucky No. 9/14	Creosote oil	Different coal minerals	683 K, 6.8 MPa 900–7200 secs 1000 rpm stirrer speed	Tarrer et al. (47)
5. Kentucky No. 9/14	Creosote oil	$CoMo/Al_2O_3$ catalyst and almost all the coal minerals	683 K, 6.8–17 MPa 900–7200 secs	Guin et al. (49)

TABLE 3
(Concluded)

6. Mixture of Kentucky No. 9/14 and Illinois No. 6	Tetralin and light recycle oil	$CoMo/Al_2O_3$ and representative coal minerals	673 K, 13.6 MPa 7200 secs	Guin et al. (50)
7. Brown coal	Tetralin	Alumina, silica gel, red haematite, heavy magnesium carbonate, calcium carbonate, anhyd. Na_2CO_3	704 K, 9.66 MPa 3600 secs	Jackson et al. (52)
8. Illinois No. 6 and Wyodak	Recycle oil	$CoMo/Al_2O_3$	727 K, 20 MPa	U.S. DOE Report (53)
9. Wyodak	Anthracene oil	(1) NiMoTi on Al_2O_3 and $AlPO_4 \cdot Al_2O_3$ (2) NiCoMo on $AlPO_4 \cdot Al_2O_3$ and Torvex (3) NiW on Al_2O_3	683 K, 24.13 MPa	Shah (28)

Sulfur is an objectionable element due to its harmful environmental and catalytic poisoning effects. It is generally believed that mercaptan, sulfide, disulfide and thiophene are the major organic sulfur containing functional groups and minerals like pyrite and marcasite are mostly responsible for inorganic sulfur in coal. A small amount of inorganic sulfur is also present as sulfate minerals like melanterite ($FeSO_4 \cdot 7H_2O$) and jarasite ($(Na,K)Fe_3(SO_4)_2(OH)_6$) as well as gypsum.

Nitrogen compounds in coal liquids can cause storage and processing problems. During storage, they cause polymerization and hence, gum formation and deposition. Similarly, oxygen compounds will enhance the polymerization tendency of coal liquids. It is thus very important to remove S, N, and O compounds to the maximum possible extent to render coal liquids fit for end uses without causing environmental or storage problems.

An extensive study of heteroatom removal has been carried out at the Gulf Research and Development Company. This and a few other studies have been summarized by Shah and Cronauer (55) and Shah and Krishnamurthy (56) and they will not be repeated here. Instead, in the following pages, a few other reported works on hydrodesulfurization (HDS), hydrodeoxygenation (HDO), and hydrodenitrogenation (HDN) of real systems and model compounds, that may be relevant to coal liquefaction, are briefly discussed.

Guin et al. (49) studied the HDS of creosote oil and Kentucky No. 9/14 coal mixture at 683 K and 6.8 MPa and 17 MPa initial hydrogen pressures in the presence of $CoMo/Al_2O_3$ catalyst and other mineral matter. The results indicated that $CoMo/Al_2O_3$ was the best catalyst (removing almost all the sulfur) and ankerite was the worst.

Pyrite has been found to be a relatively poor catalyst for HDS. The reason for this may be due to the fact that pyrite is reduced rapidly during hydrogenation to the sulfide form and some reverse reaction by H_2S generated may occur. The H_2S formed may react with organic compounds forming more sulfur. Iron, on the other hand, acts as a sulfur scavenger, suppressing the reverse reaction and thus reducing the overall sulfur content.

Jackson et al. (52) extensively studied the catalytic effect of additives such as alumina, red haematite, magnesium carbonate, silica gel, calcium carbonate and anhydrous sodium carbonate on HDS. Surprisingly, except for haematite none of the additives increased the conversion of sulfur compounds. Haematite increased the HDS rate by 20%, almost on a par with the CoMo catalyst.

Betrolacini et al. (57) studied the liquefaction of Illinois No. 6 coal in trimethylnaphthalene at 13.79 MPa and 700 K and concluded that desulfurization increased with the addition of MoO_3 and CoO. After an addition of about 10% of MoO_3, the desulfurization attained a constant value and it dropped suddenly from its peak value after addition of 2% of CoO. They also found that the desulfurization increased steeply with the surface area of the catalyst; but the conversion dropped after a surface area of about 80 m^2/gm was attained. No explanation is presently available for this peculiar trend.

Garg et al. (58) studied the effects of haematite on HDS of Western Kentucky No. 9/14 coal at 658-693 K for 15-20 min at hydrogen partial pressures varying from 7.0 to 20.8 MPa. They found that after 15 minutes of reaction time in the presence of haematite, the same amount of sulfur was removed as is removed in 120 mins without haematite. Haematite was found to be very active during short reaction times; however, it was not very effective at the large reaction times. They also concluded that the desulfurization rate was independent of catalyst particle size but it depended upon the surface area.

Significant hydrodenitrogenation (HDN) does not take place during liquefaction unless externally added good denitrogenation catalysts are used. The denitrogenation is usually achieved by the separate refining of coal liquids. Both denitrogenation and deoxygenation increase with increase in temperature, hydrogen partial pressure and a decrease in feed coal concentration. Hydrodeoxygenation (HDO) also depends upon the nature and rank of coal. Both HDN and HDO studies have been extensively reviewed by Shah (28) and Shah and Cronauer (55). Hildebrand and Tsai (51) also showed that HDS, HDO and HDN all decline rapidly with catalyst aging, the effect being most pronounced for HDN. A brief summary of the reported heteroatom removal studies with real systems is given in Table 4.

7.1 Model Compound Studies

7.1.1 Hydrodesulfurization. Several model compound studies have been undertaken to understand the mechanism of sulfur removal during coal liquefaction. Hydrodesulfurization of thiophene has been reported by Amberg and his co-workers (86-91), Schuit and Gates (92), Lipsch and Schuit (67) and Shah and Cronauer (55). The last authors indicated that the hydrodesulfurization of thiophene proceeds through butadiene and not through a hydrogenation-hydrogenolysis sequence. Further, they claimed that the reactivity of the thiophene ring is decreased by an addition of a benzene ring, as in benzothiophene, resulting in a hydrogenation-hydrogenolysis route for the sulfur removal.

The HDS of thiophenes and their derivatives have been investigated by Schuit and Gates (92), Guin et al. (50), Houalla et al. (93), and Givens and Venuto (94). They concluded that for these reactions $CoMo/Al_2O_3$ was a better catalyst than Fe, pyrite, SRC residue, SRC ash and reduced pyrites. Besides thiophenes and their derivatives, Schuit and Gates (92) examined hydrodesulfurization of phenyl sulfide and Cronauer et al. (95) investigated the reaction of dibenzyl sulfide in solvents like tetralin or mesitylene. Shah and Cronauer (55) studied reactions between a variety of sulfur compounds and cyclohexane. Their study indicates the order of reactivity of different sulfur compounds to be disulfide (aliphatic or aromatic) > diarylsulfide > aliphatic sulfide > thiophene. Further studies in the area is warranted with other donor solvents like tetralin and hydrophenanthrene.

Burow et al. (96) recently explored the utility of liquid SO_2 for the removal of organic sulfur from Eastern bituminous coals. Liquid SO_2 is supposed to be an excellent solvent for aromatic heterocyclic and alkyl sulfides derived from coal. They have considered the mild Lewis acid characteristics of SO_2 and presented the following scheme for reaction

$$-\ddot{\underset{..}{S}}: + SO_2 \rightarrow -\underset{..}{S} : SO_2$$

Products from this reaction are usually highly colored and highly soluble in liquid SO_2.

7.1.2 Hydrodenitrogenation. Heterocyclic compounds containing nitrogen in coal liquids are either basic (pyridines, quinolines and acridines) or non-basic (pyrroles, indoles and carbazoles). Attempts have been made to study these model compounds to highlight the mechanism involved in hydrodenitrogenation process. The important reported studies are those of Sonnemans et al. (84,97), Goudriaan et al. (77), Satterfield et al. (79,82,83), McIlvried (75) and Cox and Berg (81) for the denitrogenation of pyridine and its derivatives, Doelman and Vlugter (72), Madkour et al. (74), Larson (98), Shih et al. (78), and Satterfield et al. (79,82,83) for the denitrogenation of quinolines, Hartung et al. (99) for the indole denitrogenation, Flinn et al. (73) for the hydrogenation of aniline, n-butylamine, indole and quinoline, and Aboul-Gheit and Abdou (76) for the denitrification of pyridine, quinoline, aniline, pyrrole and indole. In many cases, the overall nitrogen removal reaction was found to be of first order with respect to the nitrogen containing species. Some of these studies are briefly described below.

Gourdiaan et al. (77) studied $CoMo/Al_2O_3$ catalyst for pyridine hydrodenitrogenation at about 8 MPa pressure and 523-673 K temperature. They concluded that the conversion was 25-45% higher on the presulfided catalyst than on the oxide catalyst and the hydrogen sulfide pressure was found to have little effect on conversion. Satterfield and Cocheto (83) studied NiMo/Al_2O_3 catalyst for pyridine hydrodenitrogenation at 1.1 MPa pressure and 673 K temperature. Their conclusion was that NiMo/Al_2O_3 catalyst has greater activity for hydrogenation-dehydrogenation than CoMo/Al_2O_3 but the latter appears to have greater hydrogenolysis activity than NiMo/Al_2O_3 at the temperatures below 573 K. Satterfield et al. (79,82) also studied intermediate reactions in the HDN of quinoline at pressures of 3.4 MPa and 6.8 MPa and at temperatures ranging from 503 K to 693 K. The catalyst was American Cyanamid Aero HDS-3A NiMo/Al_2O_3 extrudates (3.1 wt% NiO, 15.0 wt% MoO_3). Its performance was compared with those of CoMo/Al_2O_3 used by Doelman and Vlugter (72) and NiMo/Al_2O_3 used by Shih et al. (78). They concluded that CoMo catalyst was less active for the first step of hydrogenation of quinoline to pytetrahydroquinoline than the NiMo catalyst. Similarly for the HDN of pyridine, the first step of hydrogenation to piperidine was more rapid on a NiMo/Al_2O_3 catalyst than on a CoMo/Al_2O_3 catalyst. Madkour et al. (74) found that the presence of HCl accelerated HDN on a CoMo/Al_2O_3 catalyst suggesting the possibility of a catalyst with stronger acid sites to be more active for the overall rate of HDN.

Zawadski et al. (100) studied the denitrogenation of acridine whereas Stern (101) studied the hydrodenitrogenation of pyrroles, indole and carbazole using commercially available CoMo/Al_2O_3 catalyst containing 3% CoO, 15% MoO_3 on alumina containing 5% SiO_2; NiMo/Al_2O_3 containing 3.8% NiO, 16.8% MoO_3 and some novel catalysts such as: Re/Al_2O_3 containing 5% Re as Re_2S_7 on alumina and CoRe/Al_2O_3 containing 0.79% CO, 5% Re as $Co(ReO_4)_2$ on alumina. Conversion over each of the catalysts decreased as the five membered ring of pyrrole was increasingly substituted. The Re catalysts, which were somewhat more reactive for the conversion of pyrrole than CoMo and NiMo catalysts, were less reactive for indole conversion and had the same activity as the commercial catalysts for the conversion of carbazole.

7.1.4 Hydrodeoxygenation. Davies and Lawson (102) demonstrated the presence of oxygen compounds such as

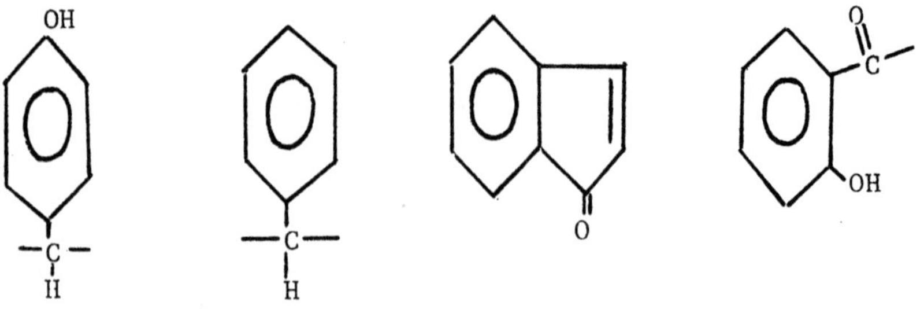

and even more complex compounds in coal liquids. Almost half of the oxygen in coal liquids is present as ethers and all the carboxylic acids are probably esters in the original coal. The largest unknown and indeterminate parameter of solid coals is oxygen incorporated in water of hydration which is erroneously assumed to be "organic" in nature.

The severity requirement of a liquefaction process can very well be known before hand due to the fact that removal of an (OH) group requires two hydrogen atoms whereas removal of (-C=O) and (C-O-C) groups require 4 atoms. Some of the relevant model compound studies are briefly described below.

Cronauer et al. (95) presented a scheme for deoxygenation of dibenzyl ether and concluded overall reaction to be a second order. They also gave a mechanism for thermal dehydration of α-tetralone and Eisenbraum et al. (103) determined that this reaction would normally take place above 673 K without a catalyst but in the presence of an alkali or noble metal catalyst it may proceed at lower temperatures.

Roberti (68,69) and Polozov (70) studied the catalytic activity of commercial catalysts: CoS, MoS_2 for the hydrodeoxygenation (HDO) of phenol to cyclohexane. Their conclusion was that the reaction followed a path via cyclohexanol while Moldavskii and Livshits (104) found the direct dehydration rate to dominate at least at low pressures. Hall and Cawley (71) studied the HDO of dibenzofuran on a MoS_2 catalyst and presented two different possible schemes. Benjamin et al. (105) have presented a summary of reactions of oxygen compounds (phenols and ethers) in tetralin at 673 K for 18 hour reaction time. A brief summary of some of the reported model compound studies is given in Table 4.

8 REACTOR DESIGN

Reactor design considerations for coal liquefaction are discussed in another paper presented at this NATO School (106) and hence they will not be repeated here.

9 SUMMARY

It is obvious that in spite of vast efforts being put forth on the development of various processes, the basic understanding needs further work. The areas of most importance are (a) analytical chemistry for the product distribution, (b) hydrogen transfer mechanism and (c) sophisticated lumped kinetic model. Future work needed for the reactor design is discussed by Shah and Gopal (106).

TABLE 4

SUMMARY OF HETEROATOM REMOVAL STUDIES

Part A - Study Related to Real Systems

Feedstock	Pressure (MPa)	Temperature (K)	Catalyst	References
Illinois No. 6 coal/ creosote solvent	6.8	678	FeS, Montmorillonite Fe_2S_3, pyrite, ZnS SRC-residue	Granoff et al. (46)
Illinois No. 6 coal/ trimethylnaphthalene solvent	13.79	700	MoO_3, CoO	Betrolacini et al. (57)
Western Kentucky No. 9/14 coal	7-20.8	658-693	Fe_2O_3	Garg et al. (58)
Petroleum fractions ranging from naphthas to residues	2-41	478-700	Ni-W/Al_2O_3	Cited in (55)
Raw anthracene oil, COED filtered oil, synthoil liquid	3.4-10	589-700	CoMo/Al_2O_3 (presulfided)	Ahmed and Crynes (59)
Raw anthracene oil	3.4-13.6	589-700	CoMo/Al_2O_3 (presulfided)	Wan and Crynes (60)
SRC liquids	19	705	NiMo/Al_2O_3 (presulfided)	Kang and Gendler (61)
Athabasca bitumen	13.9	713	CoMo/Al_2O_3	Hardin et al. (62)

TABLE 4
(Continued)

Low temperature tar	10	573-773	WS_2	Qader et al. (63)
Heavy gas oil	13.9	673-723	$NiMo/Al_2O_3$ (presulfided)	Ternan and Whaley (64)
Athabasca bitumen	7.0-13.9	713-743	$CoMo/Al_2O_3$ (presulfided)	Aarts et al. (65)
Athabasca bitumen distillates	13.9	593-693	CoO, NiO and MoO_3, on several Al_2O_3 supports (presulfided)	Furimsky et al. (66)
		Part B - Model Compound Studies		
Benzothiophene, thiophene, phenyl sulfide, dibenzothiophene	13.6	683	$CoMo/Al_2O_3$, 10% iron, pyrite, SRC residue, SRC ash	Guin et al. (50)
Thiophene	atmospheric pressure	773	$CoOMoO_3/Al_2O_3$	Lipsch and Schuit (67)
Phenol	----	---	CoS, MoS_2	Roberti and Polzov (68-70)
Dibenzofuran	----	---	MoS_2	Hall and Cawley (71)
Quinoline isoquinoline	8.1	423-673	$CoMo/Al_2O_3$	Doelman and Vlugter (72)
Aniline, indole n-butylamine and quinoline	6.8	873	$Ni\text{-}W/Al_2O_3$	Flinn et al. (73)

TABLE 4
(Continued)

Quinoline	8.1	473-723	$CoMo/Al_2O_3$	Madkour et al. (74)
Pyridine, piperidine and a hexylamine in mixed xylenes	5.1-6.8	583	$CoNiMo/Al_2O_3$ (presulfided)	McIlvried (75)
Pyridine, quinoline, aniline, pyrrole and indole in high purity paraffin oil	20	623-673	$CoMo/Al_2O_3$	Aboul-Gheit & Abdou (76)
Pyridine in p-xylene	8	523-673	$CoMo/Al_2O_3$ (presulfided, H_2S)	Goudriaan et al. (77)
Quinoline, acridine in a highly paraffinic white oil	3.4-13.6	615-640	NiOMo, CoMo, $Ni\text{-}W/Al_2O_3$ (unsulfided and presulfided)	Shih et al. (78)
Quinoline	3.4-6.8	503-693	$NiMo/Al_2O_3$ (presulfided)	Satterfield et al. (79)
5,6 benzoquinoline 7,8 benzoquinoline	19.5	473-653	$Ni\text{-}W/Al_2O_3$ (presulfided)	Shabtai et al. (80)
Twenty-nine heterocyclic nitrogen compounds	1.7	643		Cox and Berg (81)
Thiophene, pyridine	0.4-1.1	373-773	CoMo, NiMo, Ni-W/ Al_2O_3 and Ni-W/ $SiO_2\text{-}Al_2O_3$	Satterfield et al. (82)

TABLE 4
(Concluded)

Pyridine, piperidine	1.1	473-698	NiMo, CoMo/Al_2O_3	Satterfield and Cocchetto (83)
Pyridine	15.75-75.31	523-648	Mo, CoMo/Al_2O_3	Sonnemans et al. (84)
Mixture of fused ring thiophenes furans, quinolines indole and alkylphenols	2-10	573-723	CoMo (presulfided)	Rollmann (85)

REFERENCES

1. Thorogood, R.M. and C.L. Yeh. "Coal Reactor Design for the SRC-I Process." The 72nd Annual AIChE Meeting, San Francisco, California (November 28, 1979).

2. Flask, A.P. and J.A. Pryor. "SRC Solids Boiler Fuel and Building Block." The 6th Energy Technology Conference, Washington, D.C. (February 27, 1979).

3. Morris, S.M., W.L. Hollands and A.S. Gupta. "The SRC-I Project." The 86th National AIChE Meeting, Houston, Texas (April 3, 1979).

4. Schmid, B.K. and D.M. Jackson. "Recycle SRC Processing for Liquid and Solid Fuels." The 4th International Conference on Coal Gasification, Liquefaction and Conversion to Electricity, Pittsburgh, Pennsylvania (August 2-4, 1977).

5. Tao, J.C., C.L. Yeh, S.M. Morris, M.B. Dell, K.A. Schowalter and K.E. Hastings. "Solid SRC-A Building Block." The 6th International Conference on Coal Gasification, Liquefaction and Conversion to Electricity, Pittsburgh, Pennsylvania (August 2, 1979).

6. Gupta, A.S. and F.L. Moore, "SRC-I Coal Refinery: Implication for the Coal Industry." Coal Exposition V. Conference, Louisville, Kentucky (October 23-25, 1979).

7. Francis, W. "Coal." (Edward Arnold Publishers Ltd., 1961).

8. Lowry, H.H. ed. "The Chemistry of Coal Utilization," Vol. I&II (John Wiley & Sons, Inc., New York, 1945).

9. Lowry, H.H. ed. "The Chemistry of Coal Utilization," Supplementary Vol. (Wiley, New York, 1963).

10. Van Krevelen, D.W. "Coal." (Elsevier, Amsterdam, 1961) 393.

11. Heredy, L.A. and M.B. Neuworth. "Low Temperature Depolymerization of Bituminous Coal." Fuel 41 (1962) 221.

12. Heredy, L.A., A.E. Kostyo and M.B. Neuworth. "Chemical Identification of Methylene Bridges in Bituminous Coal." Fuel 43 (1964) 414.

13. Heredy, L.A., A.E. Kostyo and M.B. Neuworth. "Studies on the Structure of Coals of Different Rank." Fuel 44 (1965) 125.

14. Heredy, L.A. "The Chemistry of Acid-Catalyzed Coal Depolymerization." ACS Div. Fuel Chemistry Preprints 24(1) (1979) 142.

15. Hooper, R.J. and D.G. Evans. "Deduction of the Structure of Brown Coal by Reaction with Phenol." ACS Div. Fuel Chemistry Preprints 24(1) (1979) 131.

16. Larsen, J.W. and E.W. Kuemmerle. "Alkylation and Depolymerization Reactions of Coal." Fuel 55 (1976) 162.

17. Larsen, J.W. and P. Chaudhury. "Molecular Weight of Depolymerized Bruceton Coal." J. Org. Chem. 44 (1979) 2856.

18. Farcasiu, M., T.O. Mitchell and D.D. Whitehurst. "Asphaltols-Keys to Coal Liquefaction." Chemtech. (November, 1977) 680.

19. Farcasiu, M. "Short Time Reaction Products of Coal Liquefaction and Their Relevance to the Structure of Coal." ACS Div. Fuel Chem. Preprints 24(1) (1979) 121.

20. Whitehurst, D.D. "A Premier on the Chemistry and Constitution of Coal." Paper presented at the ACS Industrial and Engineering Chemistry Division Meeting, Chicago (Aug. 29-Sept. 1, 1977).

21. Whitehurst, D.D., M. Farcasiu, T.O. Mitchell and J.S. Dickert, Jr. "The Nature and Origin of Asphaltenes in Processed Coals." EPRI Report AF-480, II Annual Report Under Project RP-410 (July, 1977).

22. Han, K.W. and C.Y. Wen. "Initial Stage (Short Residence Time) Coal Dissolution." Fuel 58 (1979) 779.

23. Farcasiu, M., T.O. Mitchell and D.D. Whitehurst. "On the Chemical Nature of the Benzene Insoluble Components of Solvent Refined Coals." ACS Fuel Chem. Div. 21(7) (1976) 11.

24. Oele, A.P., H.I. Waterman, M.L. Goldkoop and D.W. Van Krevelen. "Extractive Disintegration of Bituminous Coals." Fuel 20 (1951) 169.

25. Peters, K. Chem. Fabr. 7 (1934) 21.

26. Morita, M., S. Sato and T. Hashimoto. ACS Div. Fuel Chem. Preprints 24(2) (1976) 270.

27. Lee, E.S. "Coal Liquefaction." Chapter 5 in "Coal-Conversion Technology." ed. C.Y. Wen and E.S. Lee (Addison Wesley Publishing Co. Inc., Reading, MA, 1979).

28. Shah, Y.T. "Reaction Engineering in Direct Coal Liquefaction." (Addison Wesley Publishing Co., Reading, MA, 1981).

29. Weller, S., M.G. Pelipetz and S. Friedman. "Conversion of Anthraxylon." IEC 43(7) (1951) 1575.

30. Falkum, E. and R.A. Glenn. Fuel 31 (1952) 133.

31. Pelipetz, M.G., J.R. Salmon, J. Bayer and E.L. Clark. "Uncatalyzed Coal Hydrogenolysis." IEC 47(10) (1955) 2101.

32. Hill, G.R., H. Hariri, R.I. Redd and L.L. Anderson. Adv. Chem. Ser. No. 55 (1966) 427.

33. Curran, G.P., R.T. Struck and E. Gorin. "Mechanism of the Hydrogen Transfer Process to Coal and Coal Extract." IEC Proc. Des. Dev. 6 (1967) 166.

34. Struck, R.T., W.E. Clark, P.J. Dudt, W.A. Rosenhoover, C.W. Zielke and E. Gorin. "Kinetics of Hydrocracking of Coal Extract with Molten Zinc Chloride Catalysts in Batch and Continuous Systems." IEC Proc. Des. Dev. 8 (1969) 546.

35. Liebenberg, B.J. and H.G.J. Potgieter. "The Uncatalyzed Hydrogenation of Coal." Fuel 52 (1973) 130.

36. Yoshida, R., Y. Maekawa, T. Ishii and G. Tayeka. "Mechanism of High-Pressure Hydrogenolysis of Hokkaido Coals (Japan), 1. Simulation of Product Distributions." Fuel 55(10) (1976) 337.

37. Cronauer, D.C., Y.T. Shah and R.G. Ruberto. "Kinetics of Thermal Liquefaction of Belle Ayr Subbituminous Coal." IEC Proc. Des. Dev. 17 (1978) 281.

38. Shah, Y.T., D.C. Cronauer, H.G. McIlvried and J.A. Paraskos. "Kinetics of Catalytic Liquefaction of Big Horn Coal in a Segmented Bed Reactor." IEC Proc. Des. Dev. 17 (1978) 288.

39. Shalabi, M.A., R.M. Baldwin, R.L. Bain, J.H. Gary and J.O. Golden. "Noncatalytic Coal Liquefaction in a Donor Solvent. Rate of Formation of Oil, Asphaltenes, and Preasphaltenes." IEC Proc. Des. Dev. 18 (1979) 474.

40. Govindan, Mohan and H. Silla. "Kinetics of Donor-Solvent Liquefaction of Bituminous Coals in Nonisothermal Experiments." IEC Proc. Des. Dev. 20 (1981) 349.

41. Parulekar, S.J., Y.T. Shah and N.L. Carr. "A Comprehensive Isothermal Model for SRC Liquefaction." paper submitted to IEC Proc. Des. Dev. (1981).

42. Abichandani, J., Y.T. Shah, D.C. Cronauer and R.G. Ruberto. "Kinetics of Thermal Liquefaction of Coal." paper submitted to Fuel (1981).

43. Singh, C.P.P., Y.T. Shah, N.L. Carr and M.E. Prudich. "Liquefaction of Coal by SRC-II Process - Part 1: A New Model for Coal Liquefaction." a paper submitted to The Can. J. of Chem. Eng. (1981).

44. Given, P.H., D.C. Cronauer, W. Spackman, H.L. Lovell, A. Davis and B. Biswas. "Dependence of Coal Liquefaction Behavior on Coal Characteristics. 1. Vitrinite-Rich Samples." Fuel 54 (1975) 34.

45. Mukherjee, D.K. and P.B. Chowdhury. "Catalytic Effect of Mineral Matter Constituents in a North Assam Coal on Hydrogenation." Fuel 55 (1976) 4.

46. Granoff, B., M.G. Thomas, P.M. Baca and G.T. Noltes. "Effects of Mineral Matter on the Hydroliquefaction of Coal." ACS Div. Fuel Chem. Preprints 23(1) (March 12-17, 1978) 23.

47. Tarrer, A.R., J.A. Guin, W.S. Pitts, J.P. Henley, J.W. Prather and G.A. Styles. "Effect of Coal Minerals on Reaction Rates During Coal Liquefaction." ACS Div. of Fuel Chem. Preprints 21(5) (1976) 59.

48. Kawa, W., S. Friedman, W.R.K. Wu, L.V. Frank and P.M. Yavorsky. "Evaluation of Catalysts for Hydrodesulfurization and Liquefaction of Coal." ACS, Division of Fuel Chemistry, Preprints 19(1) (1980) 192.

49. Guin, J.A., A.R. Tarrer, J.W. Parther, D.R. Johnson and J.M. Lee. "Effects of Coal Minerals on the Hydrogenation, Desulfurization and Solvent Extraction of Coal." IEC Proc. Des. Dev. 17(2) (1978) 118.

50. Guin, J.A., A.R. Tarrer, J.M. Lee, L. Lo and C.W. Curtis. "Further Studies of Catalytic Activity of Coal Minerals in Coal Liquefaction. 1. Verification of Catalytic Activity of Mineral Matter by Model Compound Studies." IEC Proc. Des. Dev. 18(3) (1979) 371.

51. Hildebrand, R.E. and R.C. Tsai. "CCL Processing of Big Horn Coal in P98 Using a Predissolver before the Reactor." (Gulf R&D, Report No. 624RH008, September 30, 1977).

52. Jackson, W.R., F.P. Larkins, M. Marshall, D. Rash and N. White. "Hydrogenation of Brown Coal. 1. The Effects of Additional Quantities of the Inorganic Constituents." Fuel 58 (1979) 281.

53. "Coal Conversion." 1978 Technical Report (U.S. Department of Energy, Sept., 1979).

54. Meyers, R.A. "Coal Desulfurization." (Marcel Dekker Inc., New York, 1977).

55. Shah, Y.T. and D.C. Cronauer. Catalysis Reviews Science and Engineering 20(2) (1979) 209.

56. Shah, Y.T. and S. Krishnamurthy. 3rd Chapter, Shah, Y.T. "Reaction Engineering in Direct Coal Liquefaction." (Addison Wesley Pub. Co., Reading, MA, 1981).

57. Betrolacini, R.J., L.C. Gutberlet, D.K. Kim and K.K. Robinson. ACS Div. Fuel Chem. 23(1) Anaheim, CA (March 12-17, 1978) 1.

58. Garg, D., A.R. Tarrer, J.A. Guin, C.W. Curtis and J.H. Clinton. "Selective Action of Haematite in Coal Desulfurization." IEC Proc. Des. Dev. 19 (1980) 572.

59. Ahmed, M.M. and B.L. Crynes. "Coal Liquids Hydrotreatment-Catalyst Pore Size Effects." Preprints of ACS Symposium on Effect of Pore Size on Catalytic Behavior. Floride 23(4) (1978) 1376.

60. Wan, K.T. and B.L. Crynes. "Catalyst Tailoring for Coal-Derived Liquids." cited in Shah, Y.T. and D.C. Cronauer. Oxygen, Nitrogen and Sulfur Removal Reactions in Donor Solvent Coal Liquefaction. Catalysis Reviews Science and Eng. 20(2) (1979) 209.

61. Kang, C.C. and J. Gendler. ACS Div. Fuel Chem. 23(4) (1978) 1412.

62. Hardin, A.H., R.H. Packwood and M. Ternan. "Effect of Medium Pore Diameters in $CoMo/Al_2O_3$ Catalysts on the Conversion of Athabasca Bitumen." ACS Div. Fuel Chem. Preprints 23(1) (1978) 1450.

63. Qader, S.A., W.H. Wiser and G.R. Hill. "Kinetics of Hydro-removal of Sulfur, Oxygen and Nitrogen from a Low Temperature Tar." IEC Proc. Des. Dev. 7(3) (1968) 390.

64. Ternan, M. and M.J. Whalley. "Presulfiding of Catalysts for Hydrodesulfurization and Hydrodenitrogenation in a Bottom Feed Liquid Phase Reaction." Canad. J. Chem. Engg. 54 (Dec. 1976) 642.

65. Arts, J.J.B., M. Ternan and B.I. Parsons. "Catalytic Desulfurization of Athabasca Bitumen." Fuel 57 (1978) 473.

66. Furimsky, E., R. Ranganathan and B.O. Parsons. "Catalytic Hydrodenitrogenation of Basic and Non-basic Nitrogen Compounds in Athabasca Bitumen Distillates." Fuel 57 (1978) 427.

67. Lipsch, J.M.J.G. and G.C.A. Schuit. "The $CoO-MoO_3-Al_2O_3$ Catalyst." Catalysis 15 (1969) 179.

68. Roberti, G. Ann. Chim. Applicata 21 (1931) 217.

69. Roberti, G. Ann. Chim. Applicata 22 (1932) 3.

70. Polozov, V.F. Khim. Tverd. Topliva 6 (1932) 78.

71. Hall, C.C. and C.M. Cawley. J. Soc. Chem. Ind. 58 (1939) 7.

72. Doelman, J. and J.C. Vlugter, Proc. Sixth World Petr. Cong. Section III. The Hague, Netherlands (1973) 247.

73. Flinn, R.A., O.A. Larsen and H. Beuther. "How Easy is Hydrodenitrogenation." Hydrocarbon Processing and Petroleum Refinery 42(9) (1963) 129.

74. Madkour, M.M., B.H. Mahmoud, I.K. Abdou, J.C. Vlugter, Indian Chem. Soc. 46 (1969) 720.

75. McIlvried, H.G. "Kinetics of the Hydrodenitrogenation of Pyridine." Paper presented at Houston Meeting of the ACS Division of Petr. Chem. (Feb. 1970).

76. Aboul-Gheit, A.K. and I.K. Abdou. J. Inst. Petr. 59(568) (1973) 188.

77. Goudriaan, F., H. Gierman and J.C. Vlugter. J. Inst. Petr. 59(565) (1973) 40.

78. Shih, S., E. Reiff, R. Zawadski and J.R. Katzer. "Effect of Catalyst Composition on Quinoline and Acridine Hydrodenitrogenation." ACS Div. Fuel Chem. 23(1) (March, 1978) 99.

79. Satterfield, C.N., M. Modell and J.R. Mayer. "Interaction between Catalytic Hydrodesulfurization of Thiophene and Hydrodenitrogenation of Pyridine." AIChE J. 21(6) (1975) 1100.

80. Shabtai, J., L. Veluswamy and A.G. Oblad. "Steric Effects in the Hydrogenation-Hydrodenitrogenation of Isomeric Benzoquinolines Catalyzed by Sulfided Ni-W/Al_2O_3." ACS Div. Fuel Chem. 23(4) Florida (Sept., 1978) 114.

81. Cox, K.E. and L. Berg. Chem. Eng. Prog. 58 (1962) 12.

82. Satterfield, C.N., M. Modell, R.A. Hites and C.J. Dederck. "Intermediate Reactions in the Catalytic Hydrodenitrogenation of Quinoline." IEC Proc. Des. Dev. 17(2) (1978) 141.

83. Satterfield, C.N. and J.F. Cocchetto. "Pyridine Hydrodenitrogenation: An Equilibrium Limitation on the Formation of Piperidine Intermediate." AIChE J. 21(6) (1975) 1107.

84. Sonnemans, J., G.H. Van Den Berg and P. Mars. "The Mechanism of Pyridine Hydrogenolysis on Molybdenum-Containing Catalysts. II. Hydrogenation of Pyridine to Piperidine." J. Catalysis 31 (1973) 220.

85. Rollman, L.D. "Catalytic Hydrogenation of Model Nitrogen, Sulfur and Oxygen Compounds." J. Catalysis 46 (1977) 243.

86. Owens, P.J. and C.H. Amberg. Adv. Chem. Ser. 33 (1961) 182.

87. Owens, P.J. and C.H. Amberg. "Hydrodesulfurization of Thiophene. II. Reactions over a Chromia Catalyst." Can. J. Chem. 40 (1962a) 941.

88. Owens, P.J. and C.H. Amberg. "Hydrodesulfurization of Thiophene. III. Adsorption of Reactants and Products on Chromia." Can. J. Chem. 40 (1962b) 947.

89. Desikan, P. and C.H. Amberg. "Catalytic Hydrodesulfurization of Thiophene. IV. The Methylthiophenes." Can. J. Chem. 41 (1963) 1966.

90. Desikan, P. and C.H. Amberg. "Catalytic Hydrodesulfurization of Thiophene. V. The Hydrothiophenes, Selective Poisoning and Acidity of Catalyst Surface." Can. J. Chem. 42 (1964) 843.

91. Kolboe, S. and C.H. Amberg. "Catalytic Hydrodesulfurization of Thiophene. VI. Comparisons over Molybdenum Disulfide, Cobalt Molybdate, and Chromia Catalysts." Can. J. Chem. 44 (1966) 2623.

92. Schuit, G.C.A. and B.C. Gates. "Chemistry and Engineering of Catalytic Hydrodesulfurization." AIChE J. 19 (1973) 417.

93. Houalla, M., N.K. Nag, A.V. Sapre, D.H. Broderick and B.C. Gates. "Hydrodesulfurization of Dibenzothiophene Catalyzed by Sulfided $CoO-MoO_3-\gamma Al_2O_3$: The Reaction Network." AIChE J. 24(6) (1978) 1015.

94. Given, E.N. and P.B. Venuto, ACS Div. Petrol. Chem. Preprints 15(4) (1970) A183.

95. Cronauer, D.C., D.M. Jewell, Y.T. Shah and R.J. Modi, "Investigation of Mechanisms of Hydrogen Transfer in Coal Hydrogenation." Phase I final report under DOE Contract No. E(49-18)-2305 (February, 1978).

96. Burow, D.F. and B.M. Glavincevski. "Removal of Organic Sulfur from Coal: The Use of Liquid Sulfur Dioxide." ACS Division of Fuel Chemistry, Preprints 25 (1980) 153.

97. Sonnemans, J., F. Goudriaan and P. Mars. Paper 76, Fifth International Congress on Catalysis, Palm Beach, Florida (1972).

98. Larson, O.A. ACS Division of Pet. Chem. Prepr. 12(4) (1967) B-123.

99. Hartung, G.K., D.M. Jewell, D.A. Larson and R.A. Flinn, General Papers, ACS Division of Fuel Chemistry, Preprints (1960) 27.

100. Zawadski, R., S.S. Shih, J.R. Katzer and H. Kwart. "Kinetics of Acridine Hydrodenitrogenation." a paper submitted to J. Catal.

101. Stern, E.W. "Reaction Networks in Catalytic Hydrodenitrogenation." J. Catalysis 57 (1979) 390.

102. Davies, L. and G.J. Lawson. Fuel 46 London (1967) 95.

103. Eisenbraum, E.J. "The Reaction of 1-Tetralones with Potassium Hydroxide-Sodium Hydroxide. The Reaction of 1-Tetralones with Palladium/Carbon." J. Org. Chem. 35 (1970) 1260 and ibid. 36 (1971) 686.

104. Moldavskii, B.L. and S.E. Livshits. Sh. Obshch. Khim. 3 (1933) 603.

105. Benjamin, B.M., V.F. Raaen, P.H. Maupin, L.L. Brown and C.D. Collins. "Thermal Cleavage of Chemical Bonds in Selected Coal-Related Structures." Fuel 57 (1978) 269.

106. Shah, Y.T. and J. Gopal. "Slurry Reactors for Coal Technology." a paper presented at NATO School on Mass Transfer with Chemical Reaction, Cesme/Izmir, Turkey (Aug., 1981).

107. Brunson, R.J. "Kinetics of Donor-Vehicle Coal Liquefaction in a Flow Reactor." Fuel 58 (1979) 203.

108. Shinn, J.H., F. Hershkowitz, R.R. Holten, T. Vermeulen and E.A. Greus, paper presented at the Annual Meeting of the AIChE, Miami, FL (Nov., 1977).

109. Shinn, J.H. and T. Vermeulen. ACS Div. Fuel Chem. Preprints 24(2) (1976) 80.

PARTICIPANTS

R.C.Aiken, Department of Chemical Engineering,The University of Utah,3062 Merril Engineering Building,Salt Lake City, Utah 84112,U.S.A.

J.Akyurtlu, Chemical Engineering Department.Middle East Technical University, Ankara ,Turkey.

M.Alpbas, Chemical Engineering Dept.,Ankara University, Ankara,Turkey.

J.Andrieu,Laboratoire de cinetique et genie chimiques-404, INSA 20.avenue albert einstein,69621 villeurbanne cedex-Ølyon,France.

H.Arslan,Chemistry Department, University of Selçuk, Konya ,Turkey.

B.Beler,Department of Chemical Engineering,Bosphorous University, Bebek,Istanbul,Turkey.

P.M.M.Blauwhoff,Twente University of Technology,P.O.B.217, 7500 AE Enschede, Holland.

R.S.Carter,CIBA-GEIGY A.G.,C.H.-4002 Basle,Switzerland.

T.Çakoloz,Faculty of Food Engineering,Aegen University, Izmir,Turkey.

A.Çalimli,University of Pittsburgh,Chemical and Petroleum Engineering Department,15261 Pittsburgh,U.S.A.

A.Çetinbudaklar,DYO,Izmir,Turkey.

A.Çınar,Department of Chemical Engineering,Bosphorous University, Bebek,Istanbul,Turkey.

T.Darde,CNRS,Laboratoire des Sciences du Génie Chimique, 1.rue grandville,54042,Nancy,Cedex,France.

P.K.Demetriades,Laboratory of Unit Operations,School of Chemical Engineering,NTU,Athens 147,Greece.

S.Dinçer, Department of Chemical Engineering ,Bosphorous University,Bebek, Istanbul,Turkey.

V.Dovi,Istituto di Scienze e Tecnologie,dell'Ingegneria Chimica, University of Genova,I-16143 Genova,Italia.

P.G.Eggels,AFD Technische Scheikunde,University of Groningen, Nijenborgh 16,Groningen,Holland.

N.Eken,Chemistry Faculty,State Academy of Engineering and Architecture,Maltepe,Ankara,Turkey.

S.Elmaleh,Laboratoire de Génie Chimique,Université des Sciences et Techniques du Languedoc,Place Eugène Bataillon, 34o60 Montpellier Cedex , France.

I.Eroğlu,Chemical Engineering Department,METU,Ankara.Turkey.

J.P.Euzen,Institut Français du Pétrole CEDI,Boite Postale 3 69390 Vernaison,France.

J.L.Figueiredo,Faculty of Engineering,University of Porto, 4099 Porto,Codex,Portugal.

P.Filippone,Collegio Vecchio,Universita Urbino,Italia.

H.H.Girault,Wolfson Centre for Electrochemical Science,Department of Chemistry,The University of Southampton, SO9 5NH,England.

R.Gupta,Reactors and Fluid Dynamic Section,Exxon Research and Engineering Co.,P.O.Box. 101,Florham Park,NJ 07932,U.S.A.
T.Gurkaan,Chemical Engineering Dept.,METU,Ankara,Turkey.
J.Hjortkjaer,The Technical University of Denmark,Institutet for Kemiindustri, DtH Building 227,DK-2800 Lyngby,Denmark.
D.K.Jain,Lehrstuhl und Institut für Chemische Verfahrenstechnic, Boblinger Strasse 72.D-7000 Stuttgart 1,F.R.Germany.
M.Jeronimo,Centro de Engenharia Quimica,Faculdade de Engenharia, Roa dos Bragas,4099 Porto,Codex ,Portugal.
S.Katnaş,Chemical Engineering Dept.,METU,Ankara ,Turkey.
F.Kayıhan,Chemical Engineering Dept., Oregon State University, Corvallis,OR 97331,U.SA.
M.A.Khidr,Mathematics Department,Centre of Science and Mathematics P.O.Box.2375,Damman,Saudi Arabia.
P.Knysh,UMIST,The University of Manchester,Chemical Engineering Dept.,PO Box 88,Manchester M 60 1QD ,England.
Ö.Kuleli,Chemical Engineering Department,Hacettepe University, Ankara,Turkey.
B.Kuryel,Chemical Engineering Department,Aegen University, İzmir,Turkey.
Ö.M.Kut,Technisch-Chemisches Laboratorium,ETH-Zentrum, CH-8092 Zurich,Switzerland.
A.Lecloux, c/o Solvay et cie ,Laboratoire Central,Rue de Ransbeek, 310-1120 Bruxelles,Belgium.
M.Lohse,Universitaet Hannover,Institut für Technische Chemie, Callinstrasse 3,D-3000 Hannover 1 ,F.R.Germany.
A.Lübbert,Universitaet Hannover,Institut für Technische Chemie, Callinstrasse 3, D-3000, Hannover 1, F.R.Germany.
M.O.Maia,Universidade do Minho.Pavilhao de Engenharia. Av.Joao XX1, 4700 Braga.Portugal.
J.Meldon,Chemical Engineering Dept.,Tufts University.Medford. Massachusetts 02155, U.SA.
L.M.Mishra,Institut für Technische Chemie der Universitaet Hannover,Callinstrasse 3,D-3000.Hannover 1,F.R.Germany.
B.I.Morsi,CNRS,Laboratoire des Sciences du Génie Chimiques, ENSIC, 1 . rue grandville , 54042, Nancy ,Cedex,France.
H.Oğuz,Chemical Engineering Dept.,Ankara University,Turkey.
A.Olcay,Department of Chemical Engineering,Faculty of Sciences, Ankara University,Ankara,Turkey.
Ö.Olgun, Chemical Engineering Dept.,Aegen University,Izmir,Turkey.
V.Oreopoulou,NTU of Athens,Laboratory of Organic Chemical Technology,42 , 28 th October Street, Athens,Greece.
Z.I.Önsan,Chemical Engineering Department,Bosphorous University, Bebek,Istanbul,Turkey.
S.Öztürk,Chemical Engineering Dept.,Ankara University,Turkey.
N.G.Papayannakos,NTU,Melissou 9-13.Pagrati,Athens,Greece.
A.Parmaliana,Via Nazionale,196 , 98050- Terme Vigliatore, Messina , Italy.
S.Peker, Chemical Engineering Dept., Aegen University.Izmir.Turkey.

C.Philippopoulos,NTU,Pattision 42,Athens,Greece.
M.N.N.C.Pinho,Departamento de Technogia Quimica,Institito Superior Tecnico, 1096 Lisboa Codex,Portugal.
L.Rizzuti,Universita di Palermo.Istituto di Ingegneria Chimica, Facolta di Ingegneria,Vialle Delle Scienze, Italia.
A.E.Rodrigues,Department of Chemical Engineering,University of Porto, 4099 Porto Codex,Portugal.
C.Rützou,Haldor Topsøe A/S,Nymøllerej 55 . P.O.Box. 213, DK-2800 Lyngby,Copenhagen ,Denmark.
A.M.N.Santos,Faculdade de Ciencias e Techologia, Universidade Nova de Lisboa , Quinta do Cabeco-Olivais.1899 Lisboa Codex , Portugal.
Y.Sarıkaya, Chemistry Department,Ankara University.Ankara,Turkey.
A.Schumpe, Universitaet Hannover,Institut für Technische Chemie , Callinstrasse 3 , D-3000 Hannover, F.R.Germany.
Y.Serpemen,Universitaet Hannover,Institut für Technische Chemie, Callinstrasse 3 , D-3000 Hannover 1 , F.R.Germany.
R.Sick,Abteilung Chemietechnik ,Lehrstuhl fur Technische Chemie B , Universitaet Dortmund, Postfach 50 05 00, D-4600 Dortmund 50 , F.R.Germany.
A.T. da Silva,Universidade de Coimbra,Faculdade de Ciencias e Tecnologia,Departmento de Engenharia Quimica, Portugal.
M.Soares,Universidade do Porto.Faculdade de Engenharia,Laboratorio de Quimica Industrial , Rua doe Bragas,4099 Porto, Codex,Portugal .
J.Spaninks, Badhuisweg 3 , Amsterdam-N ,Koninkljke/Shell Laboratorium , Holland.
A.Tiğrel, Petkim Petrochemicals Co. ,Ankara ,Turkey.
S.N.Upadhyay, University of Illionis at Chicago Circle,College of Engineering,Dept. of Energy Engineering ,P.O.Box.4348, Chicago.Illinois 60680,U.S.A.
Z.Uysal,Chemical Engineering Dept., METU, Ankara ,Turkey.
V.S.Vaidyanathan.Department of Biophysics,State University of New York at Buffalo,114 Carry Hall,Main Campus,Buffalo New York 14214 U.S.A.
M.Wauters,Université de Liège,Faculté des Sciences Appliquées Chimie Industrielle.Le Rue A.Stevart,2 , B-4000 Liège , Belgique.
S.D.Vlaev , Bulgarian Academy of Sciences.Central Laboratory of Chemical Engineering,Geo Milev ,Bl.5, Sofia 1113,Bulgaria.
H.van der Wal, Koninklijke.Shell -Laboratorium,P.O.Box.3003, 1003 AA Amsterdam,Holland.
H.Yeniova,Chemical Engineering Dept..University of Alberta, Canada.
L.Yurttaş, Aegen University ,Chemical Engineering Dept.,Izmir. Turkey.

GPSR Compliance

The European Union's (EU) General Product Safety Regulation (GPSR) is a set of rules that requires consumer products to be safe and our obligations to ensure this.

If you have any concerns about our products, you can contact us on ProductSafety@springernature.com

In case Publisher is established outside the EU, the EU authorized representative is:

Springer Nature Customer Service Center GmbH
Europaplatz 3
69115 Heidelberg, Germany

Batch number: 10823957

Printed by Printforce, the Netherlands